Applied Linear Statistical Models

Regression, Analysis of Variance, and Experimental Designs

Applied Linear Statistical Models

Regression, Analysis of Variance, and Experimental Designs

THIRD EDITION

John Neter
University of Georgia

William Wasserman
Syracuse University

Michael H. Kutner
Emory University

IRWIN

Homewood, IL 60430
Boston, MA 02116

To
Dorothy, Ron, David,
    Mona, Cathy, Jordan, Chloe
Cathy, Christopher, Timothy,
    Randall, Erin, Fiona
Nancy, Michelle, Allison

© RICHARD D. IRWIN, INC., 1974, 1985, and 1990

Sponsoring editor: Richard T. Hercher, Jr.
Project editor: Ethel Shiell
Production manager: Bette K. Ittersagen
Designer: Robyn Basquin
Compositor: Interactive Composition Corporation
Typeface: 10½/12 Times Roman
Printer: R. R. Donnelley & Sons Company

Library of Congress Cataloging-in-Publication Data

Neter, John.
    Applied linear statistical models: regression, analysis of
variance, and experimental designs / John Neter, William Wasserman,
Michael H. Kutner. — 3rd ed.
        p.   cm.
    ISBN 0-256-08338-X
    1. Regression analysis.   2. Analysis of variance.   3. Experimental
design.   4. Linear models (Statistics)   I. Wasserman, William.
II. Kutner, Michael H.   III. Title.
QA278.2.N47   1990
519.5′36—dc20   89–49309
                                                              CIP

Printed in the United States of America

2 3 4 5 6 7 8 9 0 DO 7 6 5 4 3 2 1 0

Preface

Linear statistical models for regression, analysis of variance, and experimental designs are widely used today in business administration, economics, and the social, health, and biological sciences. Successful applications of these models require a sound understanding of both the underlying theory and the practical problems that are encountered in using the models in real-life situations. While *Applied Linear Statistical Models,* Third Edition, is basically an applied book, it seeks to blend theory and applications effectively, avoiding the extremes of presenting theory in isolation and of giving elements of applications without the needed understanding of the theoretical foundations.

The third edition differs from the second in a number of important respects.

1. We have added a new chapter on repeated measures designs because they are of great importance in the behavioral and life sciences. The new Chapter 28 introduces the reader to repeated measures designs and takes up single-factor repeated measures designs, two-factor designs with repeated measures on one or both factors, and split-plot designs.

In addition, Chapter 12 on regression model building has been largely recast and greatly expanded. We develop the model-building process in detail in this chapter to integrate the many elements of this process considered in earlier chapters. We also include a much expanded treatment of the validation of regression models.

2. We have greatly expanded the discussion of regression and analysis of variance diagnostics throughout the text. In the area of regression analysis, we now take up the DFBETAS, DFFITS, and PRESS measures among the diagnostic measures considered. We have also added a discussion of partial regression plots and take up the Box-Cox transformation as a remedial measure.

We have also increased the emphasis on diagnostics for the analysis of variance and experimental designs. We provide many more diagnostic plots, and we have added a discussion of normal probability plots of estimated factor main effects.

3. We have reorganized and expanded a number of topics. In the area of regression analysis, the discussion of weighted least squares is now unified and taken up in conjunction with multiple regression. The discussion of standardized regression models has been reorganized, and the developments of extra sums of squares and multicollinearity have been strengthened by extensive reorganization. We have expanded Chapter 13 on autocorrelation by also taking up the Hildreth-Lu procedure for estimating the autocorrelation parameter and by adding a section on prediction intervals for forecasting a new observation. We have also included a brief discussion of response surface methodology in Chapter 9 on polynomial regression.

In the area of analysis of variance and experimental designs, we have greatly expanded the explanation of ANOVA models, especially for random and mixed effects models for randomized block designs, nested designs, repeated measures designs, and latin square designs. In particular, we have emphasized the correspondence between an ANOVA model and the correlation structure of the observations.

In addition, we have consolidated the discussion of power and planning of sample sizes in view of the close relations between these two topics.

We have also expanded the discussion of multifactor ANOVA when the treatment means are of unequal importance.

4. We have strengthened the integration of experimental designs and observational studies throughout the book, beginning with the discussion of obtaining data for regression analysis in Chapter 2.

5. Throughout the text, we have made extensive revisions in the exposition on the basis of classroom experience to improve the clarity of the presentation.

The first 13 chapters of the third edition of *Applied Linear Statistical Models* have also been published as a separate book under the title *Applied Linear Regression Models,* Second Edition. This latter book includes three additional chapters on correlation analysis (Chapter 14), nonlinear regression (Chapter 15), and regression techniques when the dependent variable is binary (Chapter 16).

A key feature of *Applied Linear Statistical Models,* Third Edition, is its unified approach to the application of linear statistical models in regression, analysis of variance, and experimental designs. Instead of treating these areas in isolated fashion, we seek to show the interrelationships between them. Use of a common notation for regression, on the one hand, and analysis of variance and experimental designs, on the other, facilitates a unified view. The notion of a general linear statistical model, which arises naturally in the context of regression models, is carried over to analysis of variance and experimental design models to bring out their relation to regression models. This unified approach also has the advantage of simplified presentation.

We have included in this book not only the more conventional topics in regression, analysis of variance, and basic experimental designs, but also topics that are frequently slighted, though important in practice. Thus, we devote a full chapter (Chapter 10) to independent indicator variables. Another chapter (Chapter 12) takes up the model-building process for regression, including computer-assisted selection procedures for identifying "good" subsets of independent variables for thorough analysis before a final selection is made of the regression model, and validation of the chosen regression model. The use of residual analysis and other diagnostics for examining the aptness of a statistical model is a recurring theme throughout this book. So is the use of remedial measures that may be helpful when the model is not appropriate. In the analysis of the results of a study, we emphasize the use of estimation procedures rather than significance tests, because estimation is often more meaningful in practice. Also, since practical problems seldom are concerned with a single estimate, we stress the use of simultaneous estimation procedures.

Theoretical ideas are presented to the degree needed for good understanding in making sound applications. Proofs are given in those instances where we feel they serve to demonstrate an important method of approach. Emphasis is placed on a thorough understanding of the models, particularly the meaning of the model parameters, since such understanding is basic to proper applications. A wide variety of case examples is presented to illustrate the use of the theoretical principles, to show the great diversity of applications of linear statistical models, and to demonstrate how analyses are carried out for different problems.

We use "Notes" and "Comments" sections in each chapter to present additional discussion and matters related to the mainstream of development. In this way, the basic ideas in a chapter are presented concisely and without distraction.

Applications of linear statistical models frequently require extensive computations. We take the position that a computer is available in most applied work. Further, almost every computer user has access to program packages for regression analysis and analysis of variance of different types. Hence, we explain the basic mathematical steps in fitting a linear statistical model but do not dwell on computational details. This approach permits us to avoid many complex formulas and enables us to focus on basic principles. We make extensive use in this text of computer capabilities for performing computations, and we illustrate a variety of computer printouts and explain how they are used for analysis.

A selection of problems is provided at the end of each chapter (except Chapter 1). Here the reader can reinforce his or her understanding of the methodology and use the concepts learned to analyze data. We have been careful to supply data-analysis problems that typify genuine applications. In most problems the calculations are best handled on a calculator or computer.

We assume that the reader of *Applied Linear Statistical Models,* Third Edition, has had an introductory course in statistical inference, covering the material outlined in Chapter 1. Should some gaps in the reader's background exist, he or she can read the relevant portions of an introductory text, or the instructor of the class may use supplemental materials for covering the missing segments. Chapter 1 is primarily intended as a reference chapter of basic statistical results for continuing use as the reader progresses through the book.

Calculus is not required for reading *Applied Linear Statistical Models,* Third Edition. We sometimes use calculus to demonstrate how some important results are obtained, but these demonstrations are confined to supplementary comments or notes and can be omitted without any loss of continuity. Readers who do know calculus will find these comments and notes in natural sequence so that the benefits of the mathematical developments are obtained in their immediate context. Some basic elements of matrix algebra are needed for linear models in general and for multiple regression in particular. Chapter 6 introduces these elements of matrix algebra in the context of simple regression for easy learning.

Applied Linear Statistical Models, Third Edition, is intended for use in undergraduate or graduate courses in linear statistical models and in second courses in applied statistics. The extent to which material presented in this text is used in a particular course depends upon the amount of time available and the objectives of the course. Some possible courses include:

1. A two-quarter or two-semester course in regression, analysis of variance, and basic experimental designs might be based on the following chapters:

    Regression: 2, 3, 4, 5 (Sections 5.1–5.3), 6, 7, 8, 10 (Sections 10.1–10.4), 11 (Sections 11.1–11.6), 12.

    Analysis of variance: 14, 15, 16, 18, 19, 23.

    Experimental designs: 24, 25, 26, 29.

2. A one-quarter or one-semester course in regression analysis might be based on the following chapters: 2, 3, 4, 5 (Sections 5.1–5.3), 6, 7, 8, 9, 10 (Sections 10.1–10.4), 11 (selected topics), 12, 13.

3. A one-quarter or one-semester course in analysis of variance might be based on the following chapters: 14, 15, 16, 17 (selected topics), 18, 19, 20, 21 (selected topics), 22, 23.

4. A one-quarter or one-semester course in regression and analysis of variance might be based on the following chapters:
   Regression: 2, 3, 4, 5 (Sections 5.1–5.3), 6, 7, 8, 10 (Sections 10.1–10.4).
   Analysis of variance: 14, 15, 16, 18, 19.

5. A one-quarter or one-semester course in basic experimental designs might be based on the following chapters: 24, 25, 26, 27, 28, 29.

As time permits, the instructor could cover additional topics in the text.

This book can also be used for self-study by persons engaged in the fields of business administration, economics, and the social, health, and biological sciences who desire to obtain competence in the application of linear statistical models.

Instructors can obtain a solutions manual from the publisher, Irwin. Included in the solutions manual is a diskette that contains the data for all problems, exercises, and projects, and for the data sets in Appendix B.

A book such as this cannot be written without substantial assistance from others. We are indebted to the many contributors who have developed the theory and practice discussed in this book. We also would like to acknowledge appreciation to our students, who helped us in a variety of ways to fashion the method of presentation contained herein. We are grateful to the many users of *Applied Linear Statistical Models* and *Applied Linear Regression Models* who have provided us with comments and suggestions based on their teaching with these texts. We are also indebted to Professors James E. Holstein, University of Missouri, and David L. Sherry, University of West Florida, for their review of *Applied Linear Statistical Models,* First Edition, to Professors Samuel Kotz, University of Maryland, Ralph P. Russo, University of Iowa, and Peter F. Thall, The George Washington University, for their review of *Applied Linear Regression Models,* and to Professors John S. Y. Chiu, University of Washington, James A. Calvin, University of Iowa, and Michael F. Driscoll, Arizona State University, for their review of *Applied Linear Statistical Models,* Second Edition. These reviews provided many important suggestions, for which we are most grateful.

George Cotsonis, Margarette S. Kolczak, and Alvin H. Rampey assisted us diligently in the checking of the manuscript, in preparing computer-generated plots, and in other ways. Almost all of the typing was done by Jane Disney and Sandra June-Hatfield, who ably handled the preparation of a difficult manuscript. We are most grateful to all of these persons for their help and assistance.

Finally, our families bore patiently the pressures caused by our commitment to complete this revision. We are appreciative of their understanding.

John Neter
William Wasserman
Michael H. Kutner

Contents

Chapter 1

Some Basic Results in Probability and Statistics

This chapter contains some basic results in probability and statistics. It is intended as a reference chapter to which you may refer as you read this book. Sometimes, specific references to results in this chapter are made in the text. At other times, you may wish to refer on your own to particular results in this chapter as you feel the need.

You may prefer to scan the results on probability and statistical inference in this chapter before reading Chapter 2, or you may proceed directly to the next chapter.

1.1 SUMMATION AND PRODUCT OPERATORS

Summation Operator

The summation operator Σ is defined as follows:

$$(1.1) \qquad \sum_{i=1}^{n} Y_i = Y_1 + Y_2 + \cdots + Y_n$$

Some important properties of this operator are:

$$(1.2a) \qquad \sum_{i=1}^{n} k = nk \qquad \text{where } k \text{ is a constant}$$

$$(1.2b) \qquad \sum_{i=1}^{n} (Y_i + Z_i) = \sum_{i=1}^{n} Y_i + \sum_{i=1}^{n} Z_i$$

$$(1.2c) \qquad \sum_{i=1}^{n} (a + cY_i) = na + c \sum_{i=1}^{n} Y_i \qquad \text{where } a \text{ and } c \text{ are constants}$$

The double summation operator $\Sigma\Sigma$ is defined as follows:

$$(1.3) \quad \sum_{i=1}^{n} \sum_{j=1}^{m} Y_{ij} = \sum_{i=1}^{n} (Y_{i1} + \cdots + Y_{im})$$

$$= Y_{11} + \cdots + Y_{1m} + Y_{21} + \cdots + Y_{2m} + \cdots + Y_{nm}$$

An important property of the double summation operator is:

$$(1.4) \qquad \sum_{i=1}^{n} \sum_{j=1}^{m} Y_{ij} = \sum_{j=1}^{m} \sum_{i=1}^{n} Y_{ij}$$

Product Operator

The product operator Π is defined as follows:

$$(1.5) \qquad \prod_{i=1}^{n} Y_i = Y_1 \cdot Y_2 \cdot Y_3 \cdot \cdot \cdot Y_n$$

1.2 PROBABILITY

Addition Theorem

Let A_i and A_j be two events defined on a sample space. Then:

$$(1.6) \qquad P(A_i \cup A_j) = P(A_i) + P(A_j) - P(A_i \cap A_j)$$

where $P(A_i \cup A_j)$ denotes the probability of either A_i or A_j or both occurring; $P(A_i)$ and $P(A_j)$ denote, respectively, the probability of A_i and the probability of A_j; and $P(A_i \cap A_j)$ denotes the probability of both A_i and A_j occurring.

Multiplication Theorem

Let $P(A_i | A_j)$ denote the conditional probability of A_i occurring, given that A_j has occurred. This conditional probability is defined as follows:

$$(1.7) \qquad P(A_i | A_j) = \frac{P(A_i \cap A_j)}{P(A_j)} \qquad P(A_j) \neq 0$$

The multiplication theorem states:

$$(1.8) \qquad P(A_i \cap A_j) = P(A_i)P(A_j | A_i)$$

$$= P(A_j)P(A_i | A_j)$$

Complementary Events

The complementary event of A_i is denoted by $\overline{A}_i$. The following results for complementary events are useful:

(1.9)
$$P(\overline{A}_i) = 1 - P(A_i)$$

(1.10)
$$P(\overline{A_i \cup A_j}) = P(\overline{A}_i \cap \overline{A}_j)$$

1.3 RANDOM VARIABLES

Throughout this section, except as noted, we assume that the random variable Y assumes a finite number of outcomes.

Expected Value

Let the random variable Y assume the outcomes $Y_1, \ldots, Y_k$ with probabilities given by the probability function:

(1.11)
$$f(Y_s) = P(Y = Y_s) \qquad s = 1, \ldots, k$$

The expected value of Y, denoted by $E\{Y\}$, is defined by:

(1.12)
$$E\{Y\} = \sum_{s=1}^{k} Y_s f(Y_s)$$

$E\{\ \}$ is called the *expectation operator*.

An important property of the expectation operator is:

(1.13) $E\{a + cY\} = a + cE\{Y\}$ where a and c are constants

Special cases of this are:

(1.13a)
$$E\{a\} = a$$

(1.13b)
$$E\{cY\} = cE\{Y\}$$

(1.13c)
$$E\{a + Y\} = a + E\{Y\}$$

Note

If the random variable Y is continuous, with density function $f(Y)$, $E\{Y\}$ is defined as follows:

(1.14)
$$E\{Y\} = \int_{-\infty}^{\infty} Y f(Y)\, dY$$

Variance

The variance of the random variable Y is denoted by $\sigma^2\{Y\}$ and is defined as follows:

(1.15) $$\sigma^2\{Y\} = E\{(Y - E\{Y\})^2\}$$

An equivalent expression is:

(1.15a) $$\sigma^2\{Y\} = E\{Y^2\} - (E\{Y\})^2$$

$\sigma^2\{\ \}$ is called the *variance operator*.

The variance of a linear function of Y is frequently encountered. We denote the variance of $a + cY$ by $\sigma^2\{a + cY\}$ and have:

(1.16) $$\sigma^2\{a + cY\} = c^2\sigma^2\{Y\} \qquad \text{where } a \text{ and } c \text{ are constants}$$

Special cases of this result are:

(1.16a) $$\sigma^2\{a + Y\} = \sigma^2\{Y\}$$

(1.16b) $$\sigma^2\{cY\} = c^2\sigma^2\{Y\}$$

Note

If Y is continuous, $\sigma^2\{Y\}$ is defined as follows:

(1.17) $$\sigma^2\{Y\} = \int_{-\infty}^{\infty} (Y - E\{Y\})^2 f(Y)\, dY$$

Joint, Marginal, and Conditional Probability Distributions

Let the joint probability function for the two random variables Y and Z be denoted by $g(Y, Z)$:

(1.18) $\quad g(Y_s, Z_t) = P(Y = Y_s \cap Z = Z_t) \qquad s = 1, \ldots, k; t = 1, \ldots, m$

The marginal probability function of Y, denoted by $f(Y)$, is:

(1.19a) $$f(Y_s) = \sum_{t=1}^{m} g(Y_s, Z_t) \qquad s = 1, \ldots, k$$

and the marginal probability function of Z, denoted by $h(Z)$, is:

(1.19b) $$h(Z_t) = \sum_{s=1}^{k} g(Y_s, Z_t) \qquad t = 1, \ldots, m$$

The conditional probability function of Y, given $Z = Z_t$, is:

(1.20a) $$f(Y_s \mid Z_t) = \frac{g(Y_s, Z_t)}{h(Z_t)} \qquad h(Z_t) \neq 0; s = 1, \ldots, k$$

and the conditional probability function of Z, given $Y = Y_s$, is:

(1.20b) $$h(Z_t \mid Y_s) = \frac{g(Y_s, Z_t)}{f(Y_s)} \qquad f(Y_s) \neq 0; t = 1, \ldots, m$$

Covariance

The covariance of Y and Z is denoted by $\sigma\{Y, Z\}$ and is defined by:

(1.21) $$\sigma\{Y, Z\} = E\{(Y - E\{Y\})(Z - E\{Z\})\}$$

An equivalent expression is:

(1.21a) $$\sigma\{Y, Z\} = E\{YZ\} - (E\{Y\})(E\{Z\})$$

$\sigma\{\ ,\ \}$ is called the *covariance operator*.

The covariance of $a_1 + c_1 Y$ and $a_2 + c_2 Z$ is denoted by $\sigma\{a_1 + c_1 Y, a_2 + c_2 Z\}$, and we have:

(1.22) $\sigma\{a_1 + c_1 Y, a_2 + c_2 Z\} = c_1 c_2 \sigma\{Y, Z\}$ where a_1, a_2, c_1, c_2 are constants

Special cases of this are:

(1.22a) $$\sigma\{c_1 Y, c_2 Z\} = c_1 c_2 \sigma\{Y, Z\}$$

(1.22b) $$\sigma\{a_1 + Y, a_2 + Z\} = \sigma\{Y, Z\}$$

By definition, we have:

(1.23) $$\sigma\{Y, Y\} = \sigma^2\{Y\}$$

where $\sigma^2\{Y\}$ is the variance of Y.

Independent Random Variables

(1.24)   Random variables Y and Z are independent if and only if:

$$g(Y_s, Z_t) = f(Y_s)h(Z_t) \qquad s = 1, \ldots, k; t = 1, \ldots, m$$

If Y and Z are independent random variables:

(1.25) $\sigma\{Y, Z\} = 0$ when Y, Z are independent

(In the special case where Y and Z are jointly normally distributed, $\sigma\{Y, Z\} = 0$ implies that Y and Z are independent.)

Functions of Random Variables

Let $Y_1, \ldots, Y_n$ be n random variables. Consider the function $\Sigma a_i Y_i$, where the a_i are constants. We then have:

(1.26a) $$E\left\{\sum_{i=1}^{n} a_i Y_i\right\} = \sum_{i=1}^{n} a_i E\{Y_i\} \qquad \text{where the } a_i \text{ are constants}$$

(1.26b) $$\sigma^2\left\{\sum_{i=1}^{n} a_i Y_i\right\} = \sum_{i=1}^{n}\sum_{j=1}^{n} a_i a_j \sigma\{Y_i, Y_j\} \qquad \text{where the } a_i \text{ are constants}$$

Specifically, we have for $n = 2$:

(1.27a) $E\{a_1 Y_1 + a_2 Y_2\} = a_1 E\{Y_1\} + a_2 E\{Y_2\}$

(1.27b) $\sigma^2\{a_1 Y_1 + a_2 Y_2\} = a_1^2 \sigma^2\{Y_1\} + a_2^2 \sigma^2\{Y_2\} + 2a_1 a_2 \sigma\{Y_1, Y_2\}$

If the random variables Y_i are independent, we have:

(1.28) $\sigma^2\left\{ \sum_{i=1}^{n} a_i Y_i \right\} = \sum_{i=1}^{n} a_i^2 \sigma^2\{Y_i\}$ when the Y_i are independent

Special cases of this are:

(1.28a) $\sigma^2\{Y_1 + Y_2\} = \sigma^2\{Y_1\} + \sigma^2\{Y_2\}$ when Y_1, Y_2 are independent

(1.28b) $\sigma^2\{Y_1 - Y_2\} = \sigma^2\{Y_1\} + \sigma^2\{Y_2\}$ when Y_1, Y_2 are independent

When the Y_i are independent random variables, the covariance of two linear functions $\Sigma a_i Y_i$ and $\Sigma c_i Y_i$ is:

(1.29) $\sigma\left\{ \sum_{i=1}^{n} a_i Y_i, \sum_{i=1}^{n} c_i Y_i \right\} = \sum_{i=1}^{n} a_i c_i \sigma^2\{Y_i\}$ when the Y_i are independent

Central Limit Theorem

(1.30)   If $Y_1, \ldots, Y_n$ are independent random observations from a population with probability function $f(Y)$ for which $\sigma^2\{Y\}$ is finite, the sample mean $\bar{Y}$:

$$\bar{Y} = \frac{\sum_{i=1}^{n} Y_i}{n}$$

is approximately normally distributed when the sample size n is reasonably large, with mean $E\{Y\}$ and variance $\sigma^2\{Y\}/n$.

1.4 NORMAL PROBABILITY DISTRIBUTION AND RELATED DISTRIBUTIONS

Normal Probability Distribution

The density function for a normal random variable Y is:

(1.31) $f(Y) = \dfrac{1}{\sqrt{2\pi}\sigma} \exp\left[-\dfrac{1}{2}\left(\dfrac{Y - \mu}{\sigma} \right)^2 \right]$ $-\infty < Y < +\infty$

where μ and σ are the two parameters of the normal distribution and $\exp(a)$ denotes e^a.

The mean and variance of a normal random variable Y are:

(1.32a) $E\{Y\} = \mu$

(1.32b) $\sigma^2\{Y\} = \sigma^2$

Function of Normal Random Variable.   A linear function of a normal random variable Y has the following property:

(1.33)   If Y is a normal random variable, the transformed variable $Y' = a + cY$ (a and c are constants) is normally distributed, with mean $a + cE\{Y\}$ and variance $c^2\sigma^2\{Y\}$.

Standard Normal Variable.   The standard normal variable z:

(1.34) $$z = \frac{Y - \mu}{\sigma}$$ where Y is a normal random variable

is normally distributed, with mean 0 and variance 1. We denote this as follows:

(1.35) z is $N(0, 1)$

                                  ↗   ↖

                            Mean Variance

Table A.1 in Appendix A contains the cumulative probabilities A for percentiles $z(A)$ where:

(1.36) $P\{z \leq z(A)\} = A$

For instance, when $z(A) = 2.00$, $A = .9772$. Because the normal distribution is symmetrical about 0, when $z(A) = -2.00$, $A = 1 - .9772 = .0228$.

Function of Independent Normal Random Variables.   Let $Y_1, \ldots, Y_n$ be independent normal random variables. We then have:

(1.37)   When $Y_1, \ldots, Y_n$ are independent normal random variables, the linear combination $a_1 Y_1 + a_2 Y_2 + \cdots + a_n Y_n$ is normally distributed, with mean $\Sigma a_i E\{Y_i\}$ and variance $\Sigma a_i^2 \sigma^2\{Y_i\}$.

χ^2 Distribution

Let $z_1, \ldots, z_\nu$ be ν independent standard normal variables. We then define:

(1.38) $\chi^2(\nu) = z_1^2 + z_2^2 + \cdots + z_\nu^2$ where the z_i are independent

The χ^2 distribution has one parameter, ν, which is called the *degrees of freedom* (*df*). The mean of the χ^2 distribution with ν degrees of freedom is:

(1.39) $E\{\chi^2(\nu)\} = \nu$

Table A.3 in Appendix A contains percentiles of various χ^2 distributions. We define $\chi^2(A; \nu)$ as follows:

(1.40) $P\{\chi^2(\nu) \leq \chi^2(A; \nu)\} = A$

Suppose $\nu = 5$. The 90th percentile of the χ^2 distribution with 5 degrees of freedom is $\chi^2(.90; 5) = 9.24$.

t Distribution

Let z and $\chi^2(\nu)$ be independent random variables (standard normal and χ^2, respectively). We then define:

(1.41) $$t(\nu) = \frac{z}{\left[\dfrac{\chi^2(\nu)}{\nu}\right]^{1/2}}$$ where z and $\chi^2(\nu)$ are independent

The t distribution has one parameter, the *degrees of freedom* ν. The mean of the t distribution with ν degrees of freedom is:

(1.42) $$E\{t(\nu)\} = 0$$

Table A.2 in Appendix A contains percentiles of various t distributions. We define $t(A; \nu)$ as follows:

(1.43) $$P\{t(\nu) \leq t(A; \nu)\} = A$$

Suppose $\nu = 10$. The 90th percentile of the t distribution with 10 degrees of freedom is $t(.90; 10) = 1.372$. Because the t distribution is symmetrical about 0, we have $t(.10; 10) = -1.372$.

F Distribution

Let $\chi^2(\nu_1)$ and $\chi^2(\nu_2)$ be two independent χ^2 random variables. We then define:

(1.44) $$F(\nu_1, \nu_2) = \frac{\chi^2(\nu_1)}{\nu_1} \div \frac{\chi^2(\nu_2)}{\nu_2}$$ where $\chi^2(\nu_1)$ and $\chi^2(\nu_2)$ are independent

Numerator   Denominator
df   *df*

The F distribution has two parameters, the *numerator degrees of freedom* and the *denominator degrees of freedom*, here ν_1 and ν_2, respectively.

Table A.4 in Appendix A contains percentiles of various F distributions. We define $F(A; \nu_1, \nu_2)$ as follows:

(1.45) $$P\{F(\nu_1, \nu_2) \leq F(A; \nu_1, \nu_2)\} = A$$

Suppose $\nu_1 = 2$, $\nu_2 = 3$. The 90th percentile of the F distribution with 2 and 3 degrees of freedom, respectively, in the numerator and denominator is $F(.90; 2, 3) = 5.46$.

Percentiles below 50 percent can be obtained by utilizing the relation:

(1.46) $$F(A; \nu_1, \nu_2) = \frac{1}{F(1 - A; \nu_2, \nu_1)}$$

Thus, $F(.10; 3, 2) = 1/F(.90; 2, 3) = 1/5.46 = .183$.

The following relation exists between the t and F random variables:

(1.47a) $$[t(\nu)]^2 = F(1, \nu)$$

and the percentiles of the t and F distibutions are related as follows:

(1.47b) $$[t(.5 + A/2; \nu)]^2 = F(A; 1, \nu)$$

Note

Throughout this text, we consider $z(A)$, $\chi^2(A; \nu)$, $t(A; \nu)$, and $F(A; \nu_1, \nu_2)$ as $A(100)$ percentiles. Equivalently, they can be considered as A fractiles.

1.5 STATISTICAL ESTIMATION

Properties of Estimators

(1.48)   An estimator $\hat{\theta}$ of the parameter θ is *unbiased* if:

$$E\{\hat{\theta}\} = \theta$$

(1.49)   An estimator $\hat{\theta}$ is a *consistent estimator* of θ if:

$$\lim_{n \to \infty} P(|\hat{\theta} - \theta| \geq \varepsilon) = 0 \qquad \text{for any } \varepsilon > 0$$

(1.50)   An estimator $\hat{\theta}$ is a *sufficient estimator* of θ if the conditional joint probability function of the sample observations, given $\hat{\theta}$, does not depend on the parameter θ.

(1.51)   An estimator $\hat{\theta}$ is a *minimum variance estimator* of θ if for any other estimator $\hat{\theta}*$:

$$\sigma^2\{\hat{\theta}\} \leq \sigma^2\{\hat{\theta}*\} \qquad \text{for all } \hat{\theta}*$$

Maximum Likelihood Estimators

The method of maximum likelihood is a general method of finding estimators. Suppose we are sampling a population whose probability function $f(Y; \theta)$ involves one parameter, θ. Given independent observations $Y_1, \ldots, Y_n$, the joint probability function of the sample observations is:

(1.52a) $$g(Y_1, \ldots, Y_n) = \prod_{i=1}^{n} f(Y_i; \theta)$$

When this joint probability function is viewed as a function of θ, with the observations given, it is called the *likelihood function $L(\theta)$*.

(1.52b) $$L(\theta) = \prod_{i=1}^{n} f(Y_i; \theta)$$

Maximizing $L(\theta)$ with respect to θ yields the maximum likelihood estimator of θ.

Under quite general conditions, maximum likelihood estimators are consistent and sufficient.

Least Squares Estimators

The method of least squares is another general method of finding estimators. The sample observations are assumed to be of the form (for the case of a single parameter θ):

$$(1.53) \qquad Y_i = f_i(\theta) + \varepsilon_i \qquad i = 1, \ldots, n$$

where $f_i(\theta)$ is a known function of the parameter θ and the ε_i are random variables, usually assumed to have expectation $E\{\varepsilon_i\} = 0$.

With the method of least squares, for the given sample observations, the sum of squares:

$$(1.54) \qquad Q = \sum_{i=1}^{n} [Y_i - f_i(\theta)]^2$$

is considered as a function of θ. The least squares estimator of θ is obtained by minimizing Q with respect to θ. In many instances, least squares estimators are unbiased and consistent.

1.6 INFERENCES ABOUT POPULATION MEAN—NORMAL POPULATION

We have a random sample of n observations $Y_1, \ldots, Y_n$ from a normal population with mean μ and standard deviation σ. The sample mean and sample standard deviation are:

$$(1.55a) \qquad \bar{Y} = \frac{\sum_i Y_i}{n}$$

$$(1.55b) \qquad s = \left[\frac{\sum_i (Y_i - \bar{Y})^2}{n-1} \right]^{1/2} = \left[\frac{\sum_i Y_i^2 - \dfrac{\left(\sum_i Y_i\right)^2}{n}}{n-1} \right]^{1/2}$$

and the estimated standard deviation of the sampling distribution of $\bar{Y}$, denoted by $s\{\bar{Y}\}$, is:

$$(1.55c) \qquad s\{\bar{Y}\} = \frac{s}{\sqrt{n}}$$

We then have:

$$(1.56) \qquad \frac{\bar{Y} - \mu}{s\{\bar{Y}\}}$$ is distributed as t with $n - 1$ degrees of freedom when the random sample is from a normal population.

Interval Estimation

The confidence limits for μ with a confidence coefficient of $1 - \alpha$ are obtained by means of (1.56):

(1.57) $$\bar{Y} \pm t(1 - \alpha/2; n - 1)s\{\bar{Y}\}$$

Example 1. Obtain a 95 percent confidence interval for μ when:

$$n = 10 \qquad \bar{Y} = 20 \qquad s = 4$$

We require:

$$s\{\bar{Y}\} = \frac{4}{\sqrt{10}} = 1.265 \qquad t(.975; 9) = 2.262$$

so that the confidence limits are $20 \pm 2.262(1.265)$. Hence, the 95 percent confidence interval for μ is:

$$17.1 \leq \mu \leq 22.9$$

Tests

One-sided and two-sided tests concerning the population mean μ are constructed by means of (1.56), based on the test statistic:

(1.58) $$t^* = \frac{\bar{Y} - \mu_0}{s\{\bar{Y}\}}$$

Table 1.1 contains the decision rules for each of three possible cases, with the risk of making a Type I error controlled at α.

TABLE 1.1   Decision Rules for Tests Concerning Mean μ of Normal Population

Alternatives	Decision Rule
	(a)
$H_0: \mu = \mu_0$	If $\lvert t^* \rvert \leq t(1 - \alpha/2; n - 1)$, conclude H_0
$H_a: \mu \neq \mu_0$	If $\lvert t^* \rvert > t(1 - \alpha/2; n - 1)$, conclude H_a
	where:
	$t^* = \dfrac{\bar{Y} - \mu_0}{s\{\bar{Y}\}}$
	(b)
$H_0: \mu \geq \mu_0$	If $t^* \geq t(\alpha; n - 1)$, conclude H_0
$H_a: \mu < \mu_0$	If $t^* < t(\alpha; n - 1)$, conclude H_a
	(c)
$H_0: \mu \leq \mu_0$	If $t^* \leq t(1 - \alpha; n - 1)$, conclude H_0
$H_a: \mu > \mu_0$	If $t^* > t(1 - \alpha; n - 1)$, conclude H_a

Example 2. Choose between the alternatives:

$$H_0: \mu \leq 20$$

$$H_a: \mu > 20$$

when α is to be controlled at .05 and:

$$n = 15 \qquad \bar{Y} = 24 \qquad s = 6$$

We require:

$$s\{\bar{Y}\} = \frac{6}{\sqrt{15}} = 1.549$$

$$t(.95; 14) = 1.761$$

so that the decision rule is:

If $t^* \leq 1.761$, conclude H_0

If $t^* > 1.761$, conclude H_a

Since $t^* = (24 - 20)/1.549 = 2.58 > 1.761$, we conclude H_a.

Example 3. Choose between the alternatives:

$$H_0: \mu = 10$$

$$H_a: \mu \neq 10$$

when α is to be controlled at .02 and:

$$n = 25 \qquad \bar{Y} = 5.7 \qquad s = 8$$

We require:

$$s\{\bar{Y}\} = \frac{8}{\sqrt{25}} = 1.6$$

$$t(.99; 24) = 2.492$$

so that the decision rule is:

If $|t^*| \leq 2.492$, conclude H_0

If $|t^*| > 2.492$, conclude H_a

where the symbol $|\ |$ stands for the absolute value. Since $|t^*| = |(5.7 - 10)/1.6|$ $= |-2.69| = 2.69 > 2.492$, we conclude H_a.

P-Value for Sample Outcome.   The P-value for a sample outcome is the probability that the sample outcome could have been more extreme than the observed one when $\mu = \mu_0$. Large P-values support H_0 while small P-values support H_a. A test can be carried out by comparing the P-value with the specified α risk. If the P-value equals or is greater than the specified α, H_0 is concluded. If the P-value is less than α, H_a is concluded.

Example 4. In Example 2, $t^* = 2.58$. The P-value for this sample outcome is the probability $P\{t(14) > 2.58\}$. From Table A.2, we find $t(.985; 14) = 2.415$ and $t(.990; 14) = 2.624$. Hence, the P-value is between .010 and .015. In fact, it can be shown to be .011. Thus, for $\alpha = .05$, H_a is concluded.

Example 5. In Example 3, $t^* = -2.69$. We find from Table A.2 that the P-value $P\{t(24) < -2.69\}$ is between .005 and .0075. In fact, it can be shown to be .0064. Because the test is two-sided and the t distribution is symmetrical, the two-sided P-value is twice the one-sided value, or $2(.0064) = .013$. Hence, for $\alpha = .02$, we conclude H_a.

Relation between Tests and Confidence Intervals. There is a direct relation between tests and confidence intervals. For example, the two-sided confidence limits (1.57) can be used for testing:

$$H_0: \mu = \mu_0$$

$$H_a: \mu \neq \mu_0$$

If μ_0 is contained within the $1 - \alpha$ confidence interval, then the two-sided decision rule in Table 1.1a, with level of significance α, will lead to conclusion H_0, and vice versa. If μ_0 is not contained within the confidence interval, the decision rule will lead to H_a, and vice versa.

There are similar correspondences between one-sided confidence intervals and one-sided decision rules.

1.7 COMPARISONS OF TWO POPULATION MEANS— NORMAL POPULATIONS

Independent Samples

There are two normal populations, with means μ_1 and μ_2, respectively, and with the same standard deviation σ. The means μ_1 and μ_2 are to be compared on the basis of independent samples for each of the two populations:

$$\text{Sample 1: } Y_1, \ldots, Y_{n_1}$$

$$\text{Sample 2: } Z_1, \ldots, Z_{n_2}$$

Estimators of the two population means are the sample means:

(1.59a)
$$\bar{Y} = \frac{\sum_i Y_i}{n_1}$$

(1.59b)
$$\bar{Z} = \frac{\sum_i Z_i}{n_2}$$

and an estimator of $\mu_1 - \mu_2$ is $\bar{Y} - \bar{Z}$.

An estimator of the common variance σ^2 is:

$$(1.60) \qquad s^2 = \frac{\sum_i (Y_i - \bar{Y})^2 + \sum_i (Z_i - \bar{Z})^2}{n_i + n_2 - 2}$$

and an estimator of $\sigma^2\{\bar{Y} - \bar{Z}\}$, the variance of the sampling distribution of $\bar{Y} - \bar{Z}$, is:

$$(1.61) \qquad s^2\{\bar{Y} - \bar{Z}\} = s^2\left(\frac{1}{n_1} + \frac{1}{n_2}\right)$$

We have:

$(1.62) \quad \dfrac{(\bar{Y} - \bar{Z}) - (\mu_1 - \mu_2)}{s\{\bar{Y} - \bar{Z}\}}$ is distributed as t with $n_1 + n_2 - 2$ degrees of freedom when the two independent samples come from normal populations with the same standard deviation.

Interval Estimation.   The confidence limits for $\mu_1 - \mu_2$ with confidence coefficient $1 - \alpha$ are obtained by means of (1.62):

$$(1.63) \qquad (\bar{Y} - \bar{Z}) \pm t(1 - \alpha/2; n_1 + n_2 - 2)s\{\bar{Y} - \bar{Z}\}$$

Example 6.   Obtain a 95 percent confidence interval for $\mu_1 - \mu_2$ when:

$$n_1 = 10 \qquad \bar{Y} = 14 \qquad \sum(Y_i - \bar{Y})^2 = 105$$
$$n_2 = 20 \qquad \bar{Z} = 8 \qquad \sum(Z_i - \bar{Z})^2 = 224$$

We require:

$$s^2 = \frac{105 + 224}{10 + 20 - 2} = 11.75$$

$$s^2\{\bar{Y} - \bar{Z}\} = 11.75\left(\frac{1}{10} + \frac{1}{20}\right) = 1.7625$$

$$s\{\bar{Y} - \bar{Z}\} = 1.328$$

$$t(.975; 28) = 2.048$$

$$3.3 = (14 - 8) - 2.048(1.328) \leq \mu_1 - \mu_2 \leq (14 - 8) + 2.048(1.328) = 8.7$$

Tests.   One-sided and two-sided tests concerning $\mu_1 - \mu_2$ are constructed by means of (1.62). Table 1.2 contains the decision rules for each of three possible cases, based on the test statistic:

$$(1.64) \qquad t^* = \frac{\bar{Y} - \bar{Z}}{s\{\bar{Y} - \bar{Z}\}}$$

with the risk of making a Type I error controlled at α.

TABLE 1.2 Decision Rules for Tests Concerning Means μ_1 and μ_2 of Two Normal Populations ($\sigma_1 = \sigma_2 = \sigma$)—Independent Samples

Alternatives	Decision Rule
	(a)
$H_0 : \mu_1 = \mu_2$	If $\lvert t^* \rvert \leq t(1 - \alpha/2; n_1 + n_2 - 2)$, conclude H_0
$H_a : \mu_1 \neq \mu_2$	If $\lvert t^* \rvert > t(1 - \alpha/2; n_1 + n_2 - 2)$, conclude H_a
	where:
	$$t^* = \frac{\bar{Y} - \bar{Z}}{s\{\bar{Y} - \bar{Z}\}}$$
	(b)
$H_0 : \mu_1 \geq \mu_2$	If $t^* \geq t(\alpha; n_1 + n_2 - 2)$, conclude H_0
$H_a : \mu_1 < \mu_2$	If $t^* < t(\alpha; n_1 + n_2 - 2)$, conclude H_a
	(c)
$H_0 : \mu_1 \leq \mu_2$	If $t^* \leq t(1 - \alpha; n_1 + n_2 - 2)$, conclude H_0
$H_a : \mu_1 > \mu_2$	If $t^* > t(1 - \alpha; n_1 + n_2 - 2)$, conclude H_a

Example 7. Choose between the alternatives:

$$H_0 : \mu_1 = \mu_2$$

$$H_a : \mu_1 \neq \mu_2$$

when α is to be controlled at .10 and the data are those of Example 6. We require $t(.95; 28) = 1.701$, so that the decision rule is:

$$\text{If } \lvert t^* \rvert \leq 1.701, \text{ conclude } H_0$$

$$\text{If } \lvert t^* \rvert > 1.701, \text{ conclude } H_a$$

Since $\lvert t^* \rvert = \lvert (14 - 8)/1.328 \rvert = \lvert 4.52 \rvert = 4.52 > 1.701$, we conclude H_a.

The one-sided P-value here is the probability $P\{t(28) > 4.52\}$. We see from Table A.2 that this P-value is less than .0005. In fact, it can be shown to be .00005. Hence, the two-sided P-value is .0001. For $\alpha = .10$, the appropriate conclusion therefore is H_a.

Paired Observations

When the observations in the two samples are paired (e.g., attitude scores Y_i and Z_i for the ith sample employee before and after a year's experience on the job), we use the differences:

$$(1.65) \qquad W_i = Y_i - Z_i \qquad i = 1, \ldots, n$$

in the fashion of a sample from a single population. Thus, when the W_i can be treated as observations from a normal population, we have:

(1.66) $\dfrac{\overline{W} - (\mu_1 - \mu_2)}{s\{\overline{W}\}}$ is distributed as t with $n - 1$ degrees of freedom when the differences W_i can be considered to be observations from a normal population and:

$$\overline{W} = \dfrac{\sum\limits_i W_i}{n}$$

$$s^2\{\overline{W}\} = \left(\dfrac{\sum\limits_i (W_i - \overline{W})^2}{n - 1}\right) \div n$$

1.8 INFERENCES ABOUT POPULATION VARIANCE— NORMAL POPULATION

When sampling from a normal population, the following holds for the sample variance s^2, where s is defined in (1.55b):

(1.67) $\dfrac{(n - 1)s^2}{\sigma^2}$ is distributed as χ^2 with $n - 1$ degrees of freedom when the random sample is from a normal population.

Interval Estimation

The lower confidence limit L and the upper confidence limit U in a confidence interval for the population variance σ^2 with confidence coefficient $1 - \alpha$ are obtained by means of (1.67):

(1.68) $L = \dfrac{(n - 1)s^2}{\chi^2(1 - \alpha/2; n - 1)}$ $\qquad$ $U = \dfrac{(n - 1)s^2}{\chi^2(\alpha/2; n - 1)}$

Example 8. Obtain a 98 percent confidence interval for σ^2, using the data of Example 1 ($n = 10$, $s = 4$).

We require:

$$s^2 = 16 \qquad \chi^2(.01; 9) = 2.09 \qquad \chi^2(.99; 9) = 21.67$$

$$6.6 = \dfrac{9(16)}{21.67} \le \sigma^2 \le \dfrac{9(16)}{2.09} = 68.9$$

Tests

One-sided and two-sided tests concerning the population variance σ^2 are constructed by means of (1.67). Table 1.3 contains the decision rule for each of three possible cases, with the risk of making a Type I error controlled at α.

TABLE 1.3   Decision Rules for Tests Concerning Variance σ^2 of Normal Population

Alternatives	*Decision Rule*
	(a)
$H_0: \sigma^2 = \sigma_0^2$	If $\chi^2(\alpha/2; n - 1) \leq \dfrac{(n - 1)s^2}{\sigma_0^2} \leq \chi^2(1 - \alpha/2; n - 1),$
$H_a: \sigma^2 \neq \sigma_0^2$	conclude H_0
	Otherwise conclude H_a
	(b)
$H_0: \sigma^2 \geq \sigma_0^2$	If $\dfrac{(n - 1)s^2}{\sigma_0^2} \geq \chi^2(\alpha; n - 1)$, conclude H_0
$H_a: \sigma^2 < \sigma_0^2$	If $\dfrac{(n - 1)s^2}{\sigma_0^2} < \chi^2(\alpha; n - 1)$, conclude H_a
	(c)
$H_0: \sigma^2 \leq \sigma_0^2$	If $\dfrac{(n - 1)s^2}{\sigma_0^2} \leq \chi^2(1 - \alpha; n - 1)$, conclude H_0
$H_a: \sigma^2 > \sigma_0^2$	If $\dfrac{(n - 1)s^2}{\sigma_0^2} > \chi^2(1 - \alpha; n - 1)$, conclude H_a

1.9 COMPARISONS OF TWO POPULATION VARIANCES— NORMAL POPULATIONS

Independent samples are selected from two normal populations, with means and variances μ_1 and σ_1^2 and μ_2 and σ_2^2, respectively. Using the notation of Section 1.7, the two sample variances are:

$$(1.69a) \qquad s_1^2 = \frac{\sum_i (Y_i - \bar{Y})^2}{n_1 - 1}$$

$$(1.69b) \qquad s_2^2 = \frac{\sum_i (Z_i - \bar{Z})^2}{n_2 - 1}$$

We have:

$(1.70) \qquad \dfrac{s_1^2}{\sigma_1^2} \div \dfrac{s_2^2}{\sigma_2^2}$ is distributed as $F(n_1 - 1, n_2 - 1)$ when the two independent samples come from normal populations.

Interval Estimation

The lower and upper confidence limits L and U for σ_1^2/σ_2^2 with confidence coefficient $1 - \alpha$ are obtained by means of (1.70):

$$(1.71) \qquad L = \frac{s_1^2}{s_2^2}\left[\frac{1}{F(1 - \alpha/2; n_1 - 1, n_2 - 1)}\right]$$

$$U = \frac{s_1^2}{s_2^2}\left[\frac{1}{F(\alpha/2; n_1 - 1, n_2 - 1)}\right]$$

Example 9. Obtain a 90 percent confidence interval for σ_1^2/σ_2^2 when the data are:

$$n_1 = 16 \qquad n_2 = 21$$
$$s_1^2 = 54.2 \qquad s_2^2 = 17.8$$

We require:

$$F(.05; 15, 20) = 1/F(.95; 20, 15) = 1/2.33 = .429$$
$$F(.95; 15, 20) = 2.20$$
$$1.4 = \frac{54.2}{17.8}\left(\frac{1}{2.20}\right) \le \frac{\sigma_1^2}{\sigma_2^2} \le \frac{54.2}{17.8}\left(\frac{1}{.429}\right) = 7.1$$

Tests

One-sided and two-sided tests concerning σ_1^2/σ_2^2 are constructed by means of (1.70). Table 1.4 contains the decision rules for each of three possible cases, with the risk of making a Type I error controlled at α.

TABLE 1.4 Decision Rules for Tests Concerning Variances σ_1^2 and σ_2^2 of Two Normal Populations—Independent Samples

Alternatives	Decision Rule
	(a)
$H_0: \sigma_1^2 = \sigma_2^2$	If $F(\alpha/2; n_1 - 1, n_2 - 1) \le \dfrac{s_1^2}{s_2^2}$
$H_a: \sigma_1^2 \neq \sigma_2^2$	$\le F(1 - \alpha/2; n_1 - 1, n_2 - 1)$, conclude H_0 Otherwise conclude H_a
	(b)
$H_0: \sigma_1^2 \ge \sigma_2^2$	If $\dfrac{s_1^2}{s_2^2} \ge F(\alpha; n_1 - 1, n_2 - 1)$, conclude H_0
$H_a: \sigma_1^2 < \sigma_2^2$	If $\dfrac{s_1^2}{s_2^2} < F(\alpha; n_1 - 1, n_2 - 1)$, conclude H_a
	(c)
$H_0: \sigma_1^2 \le \sigma_2^2$	If $\dfrac{s_1^2}{s_2^2} \le F(1 - \alpha; n_1 - 1, n_2 - 1)$, conclude H_0
$H_a: \sigma_1^2 > \sigma_2^2$	If $\dfrac{s_1^2}{s_2^2} > F(1 - \alpha; n_1 - 1, n_2 - 1)$, conclude H_a

Example 10.   Choose between the alternatives:

$$H_0: \sigma_1^2 = \sigma_2^2$$

$$H_a: \sigma_1^2 \neq \sigma_2^2$$

when α is to be controlled at .02 and the data are those of Example 9.
We require:

$$F(.01; 15, 20) = 1/F(.99; 20, 15) = 1/3.37 = .297$$

$$F(.99; 15, 20) = 3.09$$

so that the decision rule is:

$$\text{If } .297 \leq \frac{s_1^2}{s_2^2} \leq 3.09, \text{ conclude } H_0$$

Otherwise conclude H_a

Since $s_1^2/s_2^2 = 54.2/17.8 = 3.04$, we conclude H_0.

Part I

Simple Linear Regression

Chapter 2

Linear Regression with One Independent Variable

Regression analysis is a statistical tool that utilizes the relation between two or more quantitative variables so that one variable can be predicted from the other, or others. For example, if one knows the relation between advertising expenditures and sales, one can predict sales by regression analysis once the level of advertising expenditures has been set.

In Part I of this book, we take up regression analysis when a single predictor variable is used for predicting the variable of interest. In this chapter specifically, we consider the basic ideas of regression analysis and discuss the estimation of the parameters of the regression model.

2.1 RELATIONS BETWEEN VARIABLES

The concept of a relation between two variables, such as between family income and family expenditures for housing, is a familiar one. We distinguish between a *functional relation* and a *statistical relation,* and consider each of these in turn.

Functional Relation between Two Variables

A functional relation between two variables is expressed by a mathematical formula. If X is the *independent variable* and Y the *dependent variable,* a functional relation is of the form:

$$Y = f(X)$$

Given a particular value of X, the function f indicates the corresponding value of Y.

FIGURE 2.1   Example of Functional Relation

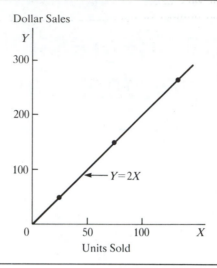

Example.   Consider the relation between dollar sales (Y) of a product sold at a fixed price and number of units sold (X). If the selling price is \$2 per unit, the relation is expressed by the equation:

$$Y = 2X$$

This functional relation is shown in Figure 2.1. Number of units sold and dollar sales during three recent periods (while the unit price remained constant at \$2) were as follows:

Period	Number of Units Sold	Dollar Sales
1	75	\$150
2	25	50
3	130	260

These observations are plotted also in Figure 2.1. Note that all fall directly on the line of functional relationship. This is characteristic of all functional relations.

Statistical Relation between Two Variables

A statistical relation, unlike a functional relation, is not a perfect one. In general, the observations for a statistical relation do not fall directly on the curve of relationship.

Example 1.   A spare part is manufactured by the Westwood Company once a month in lots that vary in size as demand fluctuates. Table 2.1, page 40, contains data on lot size and number of man-hours of labor for 10 recent production runs performed under similar production conditions. These data are plotted in Figure 2.2a. Man-hours are taken as the *dependent* or *response variable Y*, and lot size as the *independent* or *predictor variable X*. The plotting is done as before. For instance, the first production run results are plotted as $X = 30$, $Y = 73$.

Figure 2.2a clearly suggests that there is a relation between lot size and number of man-hours, in the sense that the larger the lot size, the greater tends to be the number of man-hours. However, the relation is not a perfect one. There is a scattering of points, suggesting that some of the variation in man-hours is not accounted for by lot size. For instance, two production runs (1 and 8) consisted of 30 parts, yet they required somewhat different numbers of man-hours. Because of the scattering of points in a statistical relation, Figure 2.2a is called a *scatter diagram* or *scatter plot*. In statistical terminology, each point in the scatter diagram represents a *trial* or a *case*.

In Figure 2.2b, we have plotted a line of relationship that describes the statistical relation between man-hours and lot size. It indicates the general tendency by which man-hours vary with changes in lot size. Note that most of the points do not fall directly on the line of statistical relationship. This scattering of points around the line represents variation in man-hours that is not associated with the lot size and that is

FIGURE 2.2   Statistical Relation between Lot Size and Number of Man-Hours—
Westwood Company Example

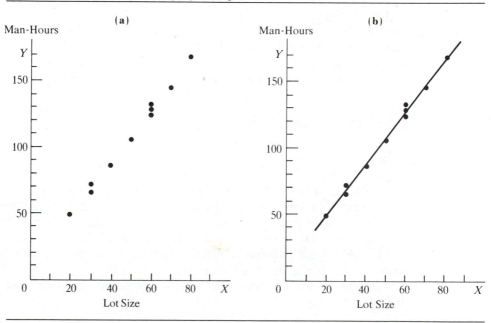

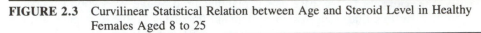

FIGURE 2.3   Curvilinear Statistical Relation between Age and Steroid Level in Healthy Females Aged 8 to 25

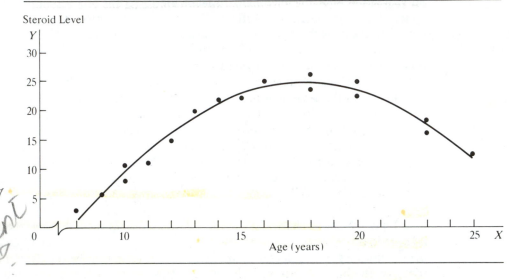

usually considered to be of a random nature. Statistical relations can be highly useful, even though they do not have the exactitude of a functional relation.

Example 2.   Figure 2.3 presents data on age and level of a steroid in plasma for 17 healthy females between 8 and 25 years old. The data strongly suggest that the statistical relationship is *curvilinear* (not linear). The curve of relationship has also been drawn in Figure 2.3. It implies that as age becomes increasingly higher, steroid level increases up to a point and then begins to decline. Note again the scattering of points around the curve of statistical relationship, typical of all statistical relations.

2.2 REGRESSION MODELS AND THEIR USES

Historical Origins

Regression analysis was first developed by Sir Francis Galton in the latter part of the 19th century. Galton had studied the relation between heights of fathers and sons and noted that the heights of sons of both tall and short fathers appeared to "revert" or "regress" to the mean of the group. He considered this tendency to be a regression to "mediocrity." Galton developed a mathematical description of this regression tendency, the precursor of today's regression models.

The term *regression* persists to this day to describe statistical relations between variables.

Basic Concepts

A regression model is a formal means of expressing the two essential ingredients of a statistical relation:

1. A tendency of the dependent variable Y to vary with the independent variable in a systematic fashion.
2. A scattering of points around the curve of statistical relationship.

These two characteristics are embodied in a regression model by postulating that:

1. There is a probability distribution of Y for each level of X.
2. The means of these probability distributions vary in some systematic fashion with X.

Example. Consider again the Westwood Company lot size example. The number of man-hours Y is treated in a regression model as a random variable. For each lot size, there is postulated a probability distribution of Y. Figure 2.4 shows such a probability distribution for $X = 30$, which is the lot size for the first production run in Table 2.1. The actual number of man-hours Y (73 in our example in Table 2.1) is then viewed as a random selection from this probability distribution.

FIGURE 2.4 Pictorial Representation of Linear Regression Model

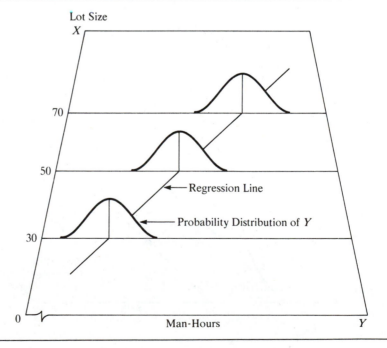

Figure 2.4 also shows probability distributions of Y for lot sizes $X = 50$ and $X = 70$. Note that the means of the probability distributions have a systematic relation to the level of X. This systematic relationship is called the *regression function of Y on X*. The graph of the regression function is called the *regression curve*. Note that in Figure 2.4 the regression function is linear. This would imply for our example that the expected (mean) number of man-hours varies linearly with lot size.

There is, of course, no a priori reason why man-hours need be linearly related to lot size. Figure 2.5 shows another possible regression model for our example. Here the regression function is curvilinear, with a shape reflecting economies of scale with larger lot sizes. Figure 2.5 differs in orientation from Figure 2.4 in that the X and Y axes are plotted conventionally in Figure 2.5. While this makes it not quite as easy to view the probability distributions, the orientation of Figure 2.5 shows the regression curve in the perspective to be utilized from here on.

Regression models may differ in the form of the regression function, as in Figures 2.4 and 2.5, in the shape of the probability distributions of Y, and in other ways. Whatever the variation, the concept of a probability distribution of Y for given X is the formal counterpart to the empirical scatter in a statistical relation. Similarly, the regression curve, which describes the relation between the means of the probability distributions and X, is the counterpart to the general tendency of Y to vary with X systematically in a statistical relation.

Note

The expressions "independent variable" or "predictor variable" for X and "dependent variable" or "response variable" for Y in a regression model simply are conventional labels. There is no implication that Y causally depends on X in a given case. No matter how strong the

FIGURE 2.5   Pictorial Representation of Curvilinear Regression Model

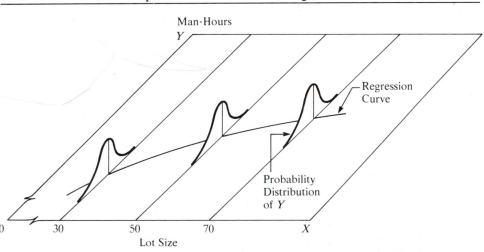

statistical relation, no cause-and-effect pattern is necessarily implied by the regression model. In some applications, an independent variable actually is dependent causally on the response variable, as when we estimate temperature (the response) from the height of mercury (the independent variable) in a thermometer.

Regression Models with More than One Independent Variable.   Regression models may contain more than one independent variable.

1. In an application of regression analysis pertaining to 67 branch offices of a consumer finance chain, the response variable was direct operating cost for the year just ended. There were four independent variables—average size of loan outstanding during the year, average number of loans outstanding, total number of new loan applications processed, and office salary scale index.
2. In a tractor purchase study, the response variable was volume (in horsepowers) of tractor purchases in a sales territory of a farm equipment firm. There were nine independent variables, including average age of tractors on farms in the territory, number of farms in the territory, and a quantity index of crop production in the territory.
3. In a medical study of short children, the response variable was the peak plasma growth hormone level. There were 14 independent variables, including age, gender, height, weight, and 10 skinfold measurements.

The features represented in Figures 2.4 and 2.5 must be extended into further dimensions when there is more than one independent variable. With two independent variables X_1 and X_2, for instance, a probability distribution of Y for each (X_1, X_2) combination is assumed by the regression model. The systematic relation between the means of these probability distributions and the independent variables X_1 and X_2 is then given by a regression surface.

Construction of Regression Models

Selection of Independent Variables.   Since reality must be reduced to manageable proportions whenever we construct models, only a limited number of independent or predictor variables can—or should—be included in a regression model for any situation of interest. A central problem therefore is that of choosing, for a regression model, a set of independent variables that is "good" in some sense for the purposes of the analysis. A major consideration in making this choice is the extent to which a chosen variable contributes to reducing the remaining variation in Y after allowance is made for the contributions of other independent variables that have tentatively been included in the regression model. Other considerations include the importance of the variable as a causal agent in the process under analysis; the degree to which observations on the variable can be obtained more accurately, or quickly, or economically than on competing variables; and the degree to which the variable can be controlled. In Chapter 12, we shall discuss procedures and problems in choosing the independent variables to be included in a regression model.

Functional Form of Regression Relation. The choice of the functional form of the regression relation is tied to the choice of the independent variables. Sometimes, relevant theory may indicate the appropriate functional form. Learning theory, for instance, may indicate that the regression function relating unit production cost to the number of previous times the item has been produced should have a specified shape with particular asymptotic properties.

More frequently, however, the functional form of the regression relation is not known in advance and must be decided upon once the data have been collected and analyzed. Thus, linear or quadratic regression functions are often used as satisfactory first approximations to regression functions of unknown nature. Indeed, these simple types of regression functions may be used even when theory provides the relevant functional form, notably when the known form is highly complex but can be reasonably approximated by a linear or quadratic regression function. Figure 2.6a illustrates a case where a complex regression function may be reasonably approximated by a linear regression function. Figure 2.6b provides an example where two linear regression functions may be used "piecewise" to approximate a complex regression function.

Scope of Model. In formulating a regression model, we usually need to restrict the coverage of the model to some interval or region of values of the independent variable or variables. The scope is determined either by the design of the investigation or by the range of data at hand. For instance, a company studying the effect of price on sales volume investigated six price levels, ranging from $4.95 to $6.95. Here, the scope of the model would be limited to price levels ranging from near $5 to near $7. The shape of the regression function would be in serious doubt substantially outside this range because the investigation provided no evidence as to the nature of the statistical relation below $4.95 or above $6.95.

FIGURE 2.6 Uses of Linear Regression Functions to Approximate Complex Regression Functions

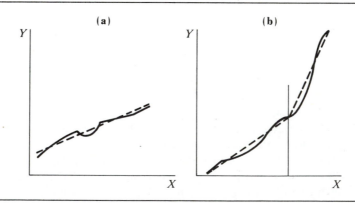

Uses of Regression Analysis

Regression analysis serves three major purposes: (1) description, (2) control, and (3) prediction. These purposes are illustrated by the three examples cited earlier. The tractor purchase study served a descriptive purpose. In the study of branch office operating costs, the purpose was administrative control; by developing a usable statistical relation between costs and independent variables in the system, management was able to set cost standards for each branch office in the company chain. In the medical study of short children, the purpose was prediction. Clinicians were able to use the statistical relation to predict growth hormone deficiencies in short children using simple measurements of the children.

The several purposes of regression analysis frequently overlap in practice. The Westwood Company lot size example is a case in point. Knowledge of the relation between lot size and man-hours in past production runs enables management to predict the man-hours for the next production run of given lot size, for purposes of cost estimation and production scheduling. After the run is completed, management can compare the actual man-hours against the predicted hours for purposes of administrative control.

2.3 SIMPLE LINEAR REGRESSION MODEL WITH DISTRIBUTION OF ERROR TERMS UNSPECIFIED

Formal Statement of Model

In Part I of this book, we consider a basic regression model where there is only one independent variable and the regression function is linear. The model can be stated as follows:

$$(2.1) \qquad Y_i = \beta_0 + \beta_1 X_i + \varepsilon_i$$

where:

Y_i is the value of the response variable in the ith trial
β_0 and β_1 are parameters
X_i is a known constant, namely, the value of the independent variable in the ith trial
ε_i is a random error term with mean $E\{\varepsilon_i\} = 0$ and variance $\sigma^2\{\varepsilon_i\} = \sigma^2$; ε_i and ε_j are uncorrelated so that the covariance $\sigma\{\varepsilon_i, \varepsilon_j\} = 0$ for all $i, j; i \neq j$
$i = 1, \ldots, n$

Regression model (2.1) is said to be *simple*, *linear in the parameters*, and *linear in the independent variable*. It is "simple" in that there is only one independent variable, "linear in the parameters" because no parameter appears as an exponent or is multiplied or divided by another parameter, and "linear in the independent variable" because this variable appears only in the first power. A model which is linear in the parameters and the independent variable is also called a *first-order model*.

Important Features of Model

1. The observed value of Y in the ith trial is the sum of two components: (1) the constant term $\beta_0 + \beta_1 X_i$ and (2) the random term ε_i. Hence, Y_i is a random variable.

2. Since $E\{\varepsilon_i\} = 0$, it follows from (1.13c) that:

$$E\{Y_i\} = E\{\beta_0 + \beta_1 X_i + \varepsilon_i\} = \beta_0 + \beta_1 X_i + E\{\varepsilon_i\} = \beta_0 + \beta_1 X_i$$

Note that $\beta_0 + \beta_1 X_i$ plays the role of the constant a in theorem (1.13c).

Thus, the response Y_i, when the level of X in the ith trial is X_i, comes from a probability distribution whose mean is:

(2.2) $$E\{Y_i\} = \beta_0 + \beta_1 X_i$$

We therefore know that the regression function for model (2.1) is:

(2.3) $$E\{Y\} = \beta_0 + \beta_1 X$$

since the regression function relates the means of the probability distributions of Y for given X to the level of X.

3. The observed value of Y in the ith trial exceeds or falls short of the value of the regression function by the error term amount ε_i.

4. The error terms ε_i are assumed to have constant variance σ^2. It therefore follows that the responses Y_i have the same constant variance:

(2.4) $$\sigma^2\{Y_i\} = \sigma^2$$

since, using theorem (1.16a), we have:

$$\sigma^2\{\beta_0 + \beta_1 X_i + \varepsilon_i\} = \sigma^2\{\varepsilon_i\} = \sigma^2$$

Thus, regression model (2.1) assumes that the probability distributions of Y have the same variance σ^2, regardless of the level of the independent variable X.

5. The error terms are assumed to be uncorrelated. Hence, the outcome in any one trial has no effect on the error term for any other trial—as to whether it is positive or negative, small or large. Since the error terms ε_i and ε_j are uncorrelated, so are the responses Y_i and Y_j.

6. In summary, regression model (2.1) implies that the response variable observations Y_i come from probability distributions whose means are $E\{Y_i\} = \beta_0 + \beta_1 X_i$ and whose variances are σ^2, the same for all levels of X. Further, any two observations Y_i and Y_j are uncorrelated.

Example

Suppose that regression model (2.1) is applicable for the Westwood Company lot size example and is as follows:

$$Y_i = 9.5 + 2.1X_i + \varepsilon_i$$

Figure 2.7 contains a presentation of the regression function:

$$E\{Y\} = 9.5 + 2.1X$$

FIGURE 2.7   Illustration of Simple Linear Regression Model (2.1)

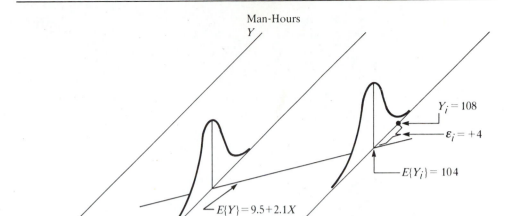

Suppose that in the ith trial, a lot of $X_i = 45$ units is produced and the actual number of man-hours is $Y_i = 108$. In that case, the error term value is $\varepsilon_i = +4$, for we have:

$$E\{Y_i\} = 9.5 + 2.1(45) = 104$$

and:

$$Y_i = 108 = 104 + 4$$

Figure 2.7 displays the probability distribution of Y when $X = 45$, and indicates from where in this distribution the observation $Y_i = 108$ came. Note again that the error term ε_i is simply the deviation of Y_i from its mean value $E\{Y_i\}$.

Figure 2.7 also shows the probability distribution of Y when $X = 25$. Note that this distribution exhibits the same variability as the probability distribution when $X = 45$, in conformance with the requirements of regression model (2.1).

Meaning of Regression Parameters

The parameters β_0 and β_1 in regression model (2.1) are called *regression coefficients*. β_1 is the slope of the regression line. It indicates the change in the mean of the probability distribution of Y per unit increase in X. The parameter β_0 is the Y intercept of the regression line. If the scope of the model includes $X = 0$, β_0 gives the mean of the probability distribution of Y at $X = 0$. When the scope of the model

---

FIGURE 2.8   Meaning of Parameters of Simple Linear Regression Model

---

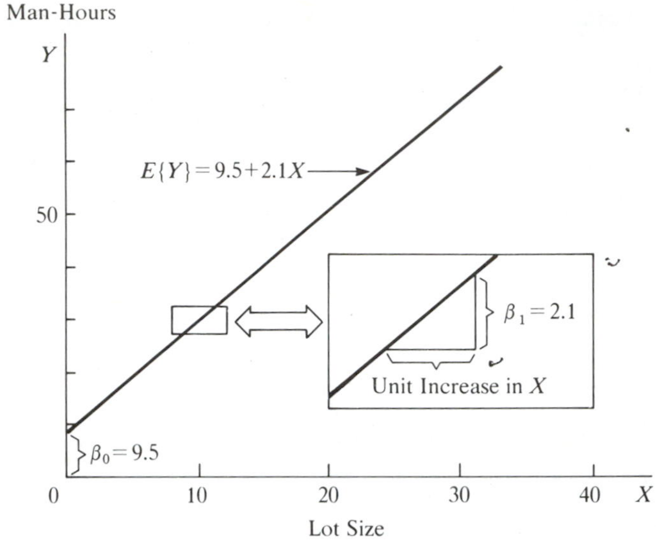

does not cover $X = 0$, β_0 does not have any particular meaning as a separate term in the regression model.

Example.   Figure 2.8 shows the regression function:

$$E\{Y\} = 9.5 + 2.1X$$

for the previous Westwood Company lot size example. The slope $\beta_1 = 2.1$ indicates that an increase of one unit in lot size leads to an increase in the mean of the probability distribution of Y of 2.1 man-hours.

The intercept $\beta_0 = 9.5$ indicates the value of the regression function at $X = 0$. However, since the linear regression model was formulated to apply to lot sizes ranging from 20 to 80 units, β_0 does not have any intrinsic meaning of its own here. In particular, it does not necessarily indicate the average setup time for the process (the average man-hours before actual output begins). A model with a curvilinear regression function and some different value of β_0 than that in the linear model might well be required if the scope of the model were to extend to lot sizes down to zero.

Alternative Versions of Regression Model

Sometimes it is convenient to write the simple linear regression model (2.1) in somewhat different, though equivalent, forms. Let X_0 be a *dummy variable* identically equal to one. Then, we can write (2.1) as follows:

(2.5) $$Y_i = \beta_0 X_0 + \beta_1 X_i + \varepsilon_i \qquad \text{where } X_0 \equiv 1$$

An alternative modification is to use for the independent variable the deviation $X_i - \bar{X}$ rather than X_i. To leave model (2.1) unchanged, we need to write:

$$Y_i = \beta_0 + \beta_1(X_i - \bar{X}) + \beta_1\bar{X} + \varepsilon_i$$
$$= (\beta_0 + \beta_1\bar{X}) + \beta_1(X_i - \bar{X}) + \varepsilon_i$$
$$= \beta_0^* + \beta_1(X_i - \bar{X}) + \varepsilon_i$$

Thus, this alternative model version is:

$$(2.6) \qquad Y_i = \beta_0^* + \beta_1(X_i - \bar{X}) + \varepsilon_i$$

where:

$$(2.6a) \qquad \beta_0^* = \beta_0 + \beta_1\bar{X}$$

We shall use models (2.1), (2.5), and (2.6) interchangeably as convenience dictates.

2.4 DATA FOR REGRESSION ANALYSIS

Ordinarily, we do not know the values of the regression parameters β_0 and β_1 in regression model (2.1), and we need to estimate them from relevant data. Indeed, as we noted earlier, we frequently do not have adequate a priori knowledge of the appropriate independent variables and of the functional form of the regression relation (e.g., linear or curvilinear), and we need to rely on the characteristics of the data for developing a suitable regression model.

Data for regression analysis may be obtained by nonexperimental or experimental means. We shall consider each of these in turn.

Observational Data

Observational data are nonexperimental data. Such data are obtained without controlling the independent variable(s) of interest. For example, company officials wished to study the relation between age of employee (X) and number of days of illness last year (Y). They used data obtained from personnel records for the regression analysis. Such data are observational data since the independent variable is not controlled.

The Westwood Company in our earlier lot size example relied on historical data. These data are also observational data because the lot sizes were determined in accordance with the demand for the product rather than controlled under experimental conditions.

Regression analyses are frequently based on observational data, since often it is not feasible to conduct controlled experimentation. In the company personnel example, for instance, it would not be possible to control age by assigning ages to persons.

A major limitation of observational data is that they often do not provide adequate information about cause-and-effect relationships. For example, a positive relation between age of employee and number of days of illness in the company personnel example may not imply that number of days of illness is the direct result of age. It might be that younger employees of the company primarily work outdoors while older employees usually work indoors, and that work location is more directly responsible for the number of days of illness.

Whenever regression analyses undertaken for purposes of description are based on observational data, one should investigate whether independent variables other than the independent variable(s) considered in the regression model might more directly explain cause-and-effect relationships.

Experimental Data

Frequently, it is possible to conduct a controlled experiment to provide data from which the regression parameters can be estimated. Consider, for instance, an insurance company that wishes to study the relation between productivity of its analysts in processing claims and length of training. Eight analysts are to be used in the study. Three of them selected at random are to be trained for two weeks, two for three weeks, and three for five weeks. The productivity of the analysts during the next 10 weeks is then to be observed. The data so obtained will be experimental data because control is exercised over the independent variable, length of training.

When control over the independent variable(s) is exercised through random assignments, as in the productivity study example, the resulting experimental data provide much stronger information about cause-and-effect relationships than do observational data. The reason is that randomization tends to balance out the effects of any other variables that might affect the dependent variable, such as the effect of aptitude of the employee on productivity.

In the terminology of experimental design, the length of training assigned to an analyst in the productivity study example is called a *treatment*. The analysts to be included in the study are called the *experimental units*. Control over the independent variable(s) then consists of assigning a treatment to each of the experimental units by means of randomization.

Completely Randomized Design

The most basic type of statistical design for making randomized assignments of treatments to experimental units (or vice versa) is the *completely randomized design*. With this design, the assignments are made completely at random. This complete randomization provides that every experimental unit has an equal chance to receive any one of the treatments or, equivalently, that all combinations of experimental units assigned to the different treatments are equally likely.

A completely randomized design is particularly useful when the experimental units are quite homogeneous. This design is very flexible; it accommodates any number of treatments and permits different sample sizes for different treatments. Its chief disadvantage is that, when the experimental units are heterogeneous, this design is not as efficient as some other statistical designs.

Use of Table of Random Digits or Uniform Random Number Generator for Randomization. Randomization for a completely randomized design requires that a series of experimental units (or treatments) be placed in a random order. To illustrate this, consider again the productivity study example. Here three treatments (T_1—two weeks training, T_2—three weeks training, T_3—five weeks training) are considered and eight analysts are to be included in the study, with $n_1 = 3$ to be assigned treatment T_1, $n_2 = 2$ to be assigned treatment T_2, and $n_3 = 3$ to be assigned treatment T_3. The situation thus is:

Treatments: T_1, T_2, T_3

Sample sizes: $n_1 = 3, n_2 = 2, n_3 = 3$

The eight treatment assignments to the analysts are now listed (the order is arbitrary):

$$T_1 \quad T_1 \quad T_1 \quad T_2 \quad T_2 \quad T_3 \quad T_3 \quad T_3$$

To randomize the treatments to the experimental units, we number the analysts from 1 to 8. Next, we need to obtain a random arrangement or permutation of the digits 1, 2, . . . , 8. We do this by utilizing either a table of random digits or a uniform random number generator. First, we list the eight digits in sequence:

$$1 \quad 2 \quad 3 \quad 4 \quad 5 \quad 6 \quad 7 \quad 8$$

Next, we generate eight random numbers between 000 and 999 (three-digit numbers are used here to make the probability of a tie small) and enter them over the digits 1 to 8. The random numbers generated happen to be as follows:

737	325	349	084	279	362	107	263
1	2	3	4	5	6	7	8

Now we rearrange the pairs of numbers in ascending sequence for the random numbers:

084	107	263	279	325	349	362	737
4	7	8	5	2	3	6	1

Thus, we obtain the following randomized assignment of the eight analysts to the three treatments:

Analyst:	4	7	8	5	2	3	6	1
Treatment:	T_1	T_1	T_1	T_2	T_2	T_3	T_3	T_3

Analysts 4, 7, and 8 therefore are to receive treatment T_1 (two weeks training), and so on.

If a tie should be found between two random numbers obtained, a selection or generation of another random digit can be used to break the tie.

2.5 OVERVIEW OF REGRESSION ANALYSIS

The regression analyses to be considered in this and following chapters can be utilized for either observational data or experimental data from a completely randomized design. (Regression analysis can also utilize data from other types of experimental designs, but the regression models presented here will need to be modified.) Whether the data are observational or experimental, it is essential that the conditions of the regression model be appropriate for the data at hand.

We begin our discussion of regression analysis by considering inferences about the regression parameters for the simple linear regression model (2.1). For the rare occasion where prior knowledge or theory alone enables us to determine the appropriate regression model, inferences based on the regression model are the first step in the regression analysis. In the usual situation, however, where we do not have adequate knowledge to specify the appropriate regression model in advance, the first step is an exploratory study of the data, as shown in the flowchart in Figure 2.9. Based on this initial exploratory analysis, one or more preliminary regression models are developed. These regression models are then examined for their appropriateness for the data at hand and revised, or new models are developed, until the investigator is satisfied with the suitability of a particular regression model. Only then are inferences made on the basis of this regression model, such as inferences about the regression parameters of the model or predictions of new observations.

For pedagogic reasons, we begin with inferences based on the regression model that is finally considered to be appropriate before we take up how a suitable regression model is developed. One must have an understanding of regression models and how they can be utilized before the issues involved in the development of an appropriate regression model can be fully understood.

2.6 ESTIMATION OF REGRESSION FUNCTION

Example

We shall use the Westwood Company example to illustrate the estimation of a simple linear regression function. The observational data for 10 recent production runs on size of production run (X) and number of man-hours required for the run are presented in Table 2.1 (p. 40). We shall denote the (X, Y) observations for the first trial as (X_1, Y_1), for the second trial as (X_2, Y_2), and in general for the ith trial as (X_i, Y_i), where $i = 1, \ldots, n$. For the data in Table 2.1, $X_1 = 30$, $Y_1 = 73$, and so on, and $n = 10$.

Method of Least Squares

To find "good" estimators of the regression parameters β_0 and β_1, we shall employ the method of least squares. For each sample obervation (X_i, Y_i), the method of least squares considers the deviation of Y_i from its expected value:

(2.7)
$$Y_i - (\beta_0 + \beta_1 X_i)$$

FIGURE 2.9   Typical Strategy for Regression Analysis

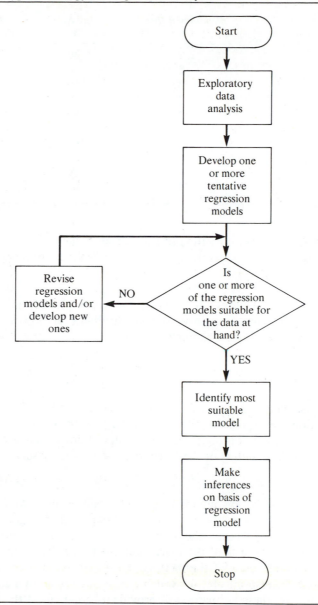

In particular, the method of least squares requires that we consider the sum of the n squared deviations. This criterion is denoted by Q:

(2.8)
$$Q = \sum_{i=1}^{n} (Y_i - \beta_0 - \beta_1 X_i)^2$$

TABLE 2.1   Data on Lot Size and Number of Man-Hours—Westwood Company Example

Production Run i	Lot Size X_i	Man-Hours Y_i
1	30	73
2	20	50
3	60	128
4	80	170
5	40	87
6	50	108
7	60	135
8	30	69
9	70	148
10	60	132

According to the method of least squares, the estimators of β_0 and β_1 are those values b_0 and b_1, respectively, that minimize the criterion Q for the given sample observations (X_i, Y_i).

Example.   Figure 2.10a repeats the scatter plot of the sample data of Table 2.1 for the Westwood Company example. In Figure 2.10b is plotted a fitted regression line using the arbitrary estimates:

$$b_0 = 30$$

$$b_1 = 0$$

Also shown in Figure 2.10b are the deviations $Y_i - 30 - (0)X_i$. Note that each deviation corresponds to the vertical distance between Y_i and the fitted regression line. Clearly, the fit is poor. Hence, the deviations are large and so are the squared deviations. The sum Q of the squared deviations is (observations in ascending order):

$$Q = (50 - 30)^2 + (69 - 30)^2 + \cdots + (170 - 30)^2 = 77,660$$

Figure 2.10c contains the deviations $Y_i - b_0 - b_1 X_i$ for the estimates $b_0 = 15$, $b_1 = 1.5$. Here, the fit is better (though still not good), the deviations are much smaller, and hence the sum of the squared deviations is reduced to $Q = 4,910$. Thus, a better fit of the regression line to the data corresponds to a smaller sum Q.

The objective of the method of least squares is to find estimates b_0 and b_1 for β_0 and β_1, respectively, for which Q is a minimum. In a certain sense, to be discussed shortly, these estimates will provide a "good" fit of the linear regression function.

Least Squares Estimators.   The estimators b_0 and b_1 that satisfy the least squares criterion can be found in two basic ways. First, numerical search procedures can be used that evaluate in a systematic fashion the least squares criterion Q for different estimates b_0 and b_1 until the ones that minimize Q are found. The second approach is to find analytically the values of b_0 and b_1 that minimize Q. The analytical approach

FIGURE 2.10 Deviations from Different Fitted Regression Lines—Westwood Company Example

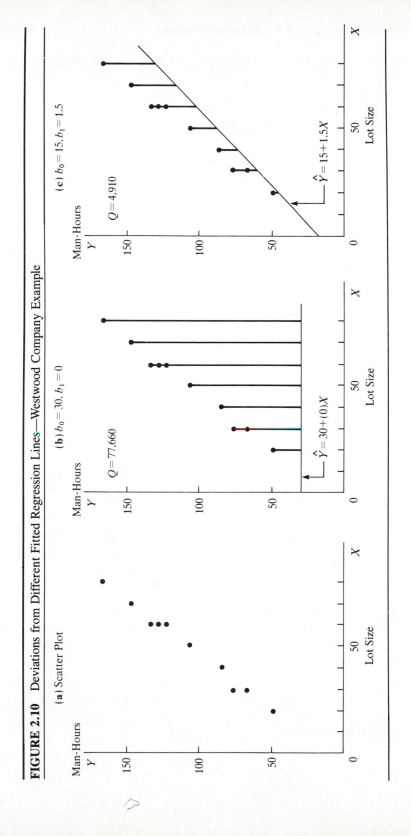

(a) Scatter Plot

(b) $b_0 = 30, b_1 = 0$

(c) $b_0 = 15, b_1 = 1.5$

$\hat{Y} = 30 + (0)X$

$Q = 77,660$

$\hat{Y} = 15 + 1.5X$

$Q = 4,910$

is feasible when the regression model is not complex mathematically, as here. It can be shown that the values b_0 and b_1 that minimize Q for any particular set of sample data are given by the following simultaneous equations:

$$(2.9a) \qquad \sum Y_i = nb_0 + b_1 \sum X_i$$

$$(2.9b) \qquad \sum X_i Y_i = b_0 \sum X_i + b_1 \sum X_i^2$$

The equations (2.9a) and (2.9b) are called *normal equations;* b_0 and b_1 are called *point estimators* of β_0 and β_1, respectively.

The quantities ΣY_i, ΣX_i, and so on in (2.9) are calculated from the sample observations (X_i, Y_i). The equations then can be solved simultaneously for b_0 and b_1. Alternatively, b_0 and b_1 can be obtained directly as follows:

$$(2.10a) \qquad b_1 = \frac{\Sigma X_i Y_i - \dfrac{\Sigma X_i \Sigma Y_i}{n}}{\Sigma X_i^2 - \dfrac{(\Sigma X_i)^2}{n}} = \frac{\Sigma (X_i - \bar{X})(Y_i - \bar{Y})}{\Sigma (X_i - \bar{X})^2}$$

$$(2.10b) \qquad b_0 = \frac{1}{n}\left(\sum Y_i - b_1 \sum X_i\right) = \bar{Y} - b_1 \bar{X}$$

where $\bar{X}$ and $\bar{Y}$ are the means of the X_i and the Y_i observations, respectively.

Note

The normal equations (2.9) can be derived by calculus. For given sample observations (X_i, Y_i), the quantity Q in (2.8) is a function of β_0 and β_1. The values of β_0 and β_1 that minimize Q can be derived by differentiating (2.8) with respect to β_0 and β_1. We obtain:

$$\frac{\partial Q}{\partial \beta_0} = -2 \sum (Y_i - \beta_0 - \beta_1 X_i)$$

$$\frac{\partial Q}{\partial \beta_1} = -2 \sum X_i(Y_i - \beta_0 - \beta_1 X_i)$$

We then set these partial derivatives equal to zero, using b_0 and b_1 to denote the particular values of β_0 and β_1, respectively, that minimize Q:

$$-2 \sum (Y_i - b_0 - b_1 X_i) = 0$$

$$-2 \sum X_i(Y_i - b_0 - b_1 X_i) = 0$$

Simplifying, we obtain:

$$\sum_{i=1}^{n} (Y_i - b_0 - b_1 X_i) = 0$$

$$\sum_{i=1}^{n} X_i(Y_i - b_0 - b_1 X_i) = 0$$

Expanding, we have:

$$\sum Y_i - nb_0 - b_1 \sum X_i = 0$$

$$\sum X_i Y_i - b_0 \sum X_i - b_1 \sum X_i^2 = 0$$

from which the normal equations (2.9) are obtained by rearranging terms.

A test of the second partial derivatives will show that a minimum is obtained with the least squares estimators b_0 and b_1.

Properties of Least Squares Estimators.   An important theorem, called the *Gauss-Markov theorem*, states:

(2.11)   Under the conditions of regression model (2.1), the least squares estimators b_0 and b_1 in (2.10) are unbiased and have minimum variance among all unbiased linear estimators.

This theorem, which is proven in the next chapter, states first that both b_0 and b_1 are unbiased estimators. Hence:

$$E\{b_0\} = \beta_0$$

$$E\{b_1\} = \beta_1$$

so that neither estimator tends to overestimate or underestimate systematically.

Second, the theorem states that the sampling distributions of b_0 and b_1 have smaller variability than those of any other estimators belonging to a particular class of estimators. Thus, the least squares estimators are more precise than any of these other estimators. The class of estimators for which the least squares estimators are "best" consists of all unbiased estimators that are linear functions of the observations $Y_1, \ldots, Y_n$. The estimators b_0 and b_1 are such linear functions of the Y_i. Consider, for instance, b_1. We have from (2.10a):

$$b_1 = \frac{\Sigma(X_i - \bar{X})(Y_i - \bar{Y})}{\Sigma(X_i - \bar{X})^2}$$

It will be shown in Chapter 3 that this expression is equal to:

$$b_1 = \frac{\Sigma(X_i - \bar{X})Y_i}{\Sigma(X_i - \bar{X})^2} = \sum k_i Y_i$$

where:

$$k_i = \frac{X_i - \bar{X}}{\Sigma(X_i - \bar{X})^2}$$

Since the k_i are known constants (because the X_i are known constants), b_1 is a linear combination of the Y_i and hence is a linear estimator.

In the same fashion, it can be shown that b_0 is a linear estimator.

Among all linear estimators that are unbiased then, b_0 and b_1 have the smallest variability in repeated samples in which the X levels remain unchanged.

Example.   To illustrate the calculation of the least squares estimators b_0 and b_1, we will use the Westwood Company example. The sample data are given in Table 2.1 and plotted in Figure 2.10a. Table 2.2 gives the basic results required to calculate b_0 and b_1. We have: $\Sigma Y_i = 1,100$, $\Sigma X_i = 500$, $\Sigma X_i Y_i = 61,800$, $\Sigma X_i^2 = 28,400$, $n = 10$. Using (2.10) we obtain:

$$b_1 = \frac{\Sigma X_i Y_i - \dfrac{\Sigma X_i \Sigma Y_i}{n}}{\Sigma X_i^2 - \dfrac{(\Sigma X_i)^2}{n}} = \frac{61,800 - \dfrac{500(1,100)}{10}}{28,400 - \dfrac{(500)^2}{10}} = 2.0$$

$$b_0 = \frac{1}{n}\left(\sum Y_i - b_1 \sum X_i\right) = \frac{1}{10}[1,100 - 2.0(500)] = 10.0$$

Thus, we estimate that the mean number of man-hours increases by 2.0 hours for each unit increase in lot size.

Point Estimation of Mean Response

Estimated Regression Function.   Given sample estimators b_0 and b_1 of the parameters in the regression function (2.3):

$$E\{Y\} = \beta_0 + \beta_1 X$$

we estimate the regression function as follows:

(2.12) $$\hat{Y} = b_0 + b_1 X$$

TABLE 2.2   Basic Calculations to Obtain b_0 and b_1—Westwood Company Example

i	Y_i	X_i	$X_i Y_i$	X_i^2	(for later use) Y_i^2
1	73	30	2,190	900	5,329
2	50	20	1,000	400	2,500
3	128	60	7,680	3,600	16,384
4	170	80	13,600	6,400	28,900
5	87	40	3,480	1,600	7,569
6	108	50	5,400	2,500	11,664
7	135	60	8,100	3,600	18,225
8	69	30	2,070	900	4,761
9	148	70	10,360	4,900	21,904
10	132	60	7,920	3,600	17,424
Total	1,100	500	61,800	28,400	134,660

where $\hat{Y}$ (read Y hat) is the value of the estimated regression function at the level X of the independent variable.

We will call a *value* of the response variable a *response* and will call $E\{Y\}$ the *mean response*. Thus, the mean response is the mean of the probability distribution of Y corresponding to the level X of the independent variable. $\hat{Y}$ then is a point estimator of the mean response when the level of the independent variable is X. It can be shown as an extension of the Gauss-Markov theorem (2.11) that $\hat{Y}$ is an unbiased estimator of $E\{Y\}$, with minimum variance in the class of unbiased linear estimators.

For the cases in the study, we will call $\hat{Y}_i$:

$$(2.13) \qquad \hat{Y}_i = b_0 + b_1 X_i \qquad i = 1, \ldots, n$$

the *fitted value* for the ith case. Thus, the fitted value $\hat{Y}_i$ is to be viewed in distinction to the *observed value* Y_i.

Example. For the Westwood Company example, we found that the least squares estimates of the regression coefficients were:

$$b_0 = 10.0 \qquad b_1 = 2.0$$

Hence, the estimated regression function is:

$$\hat{Y} = 10.0 + 2.0X$$

If we are interested in the mean number of man-hours when the lot size is $X = 55$, our point estimate is:

$$\hat{Y} = 10.0 + 2.0(55) = 120$$

Thus, we estimate that the mean number of man-hours for production runs of $X = 55$ units is 120. We interpret this to mean that if many runs of size 55 are produced under the conditions of the 10 runs in the sample, the mean labor time for these many runs is about 120 hours. Of course, the labor time for any one run of size 55 is likely to fall above or below the mean response because of inherent variability in the system, as represented by the error term in the model.

Figure 2.11 contains a computer plot of the estimated regression function $\hat{Y} = 10.0 + 2.0X$, as well as the original data. Note the improved fit of the least squares regression line over the arbitrary lines in Figure 2.10. Indeed, we will show shortly that the criterion Q for the least squares regression line now is only $Q = 60$, a much smaller value than the values of Q for the arbitrary fitted lines in Figure 2.10.

Fitted values for the sample data are obtained by substituting the X values in the sample into the estimated regression function. For example, for our sample data, $X_1 = 30$. Hence, the fitted value for the first case is:

$$\hat{Y}_1 = 10.0 + 2.0(30) = 70$$

This compares with the observed man-hours of $Y_1 = 73$. Table 2.3 (p. 47) contains all the observed and fitted values for the Westwood Company data in columns 2 and 3, respectively.

FIGURE 2.11   Observations and Least Squares Regression Line for Westwood Company Example: $b_0 = 10.0$, $b_1 = 2.0$

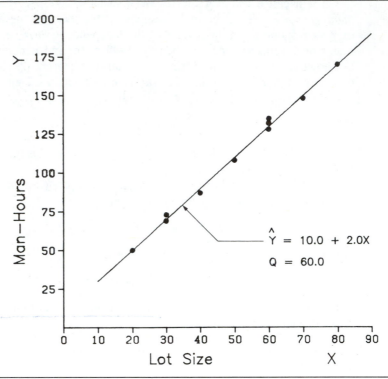

Alternative Model (2.6).   If the alternative regression model (2.6):

$$Y_i = \beta_0^* + \beta_1(X_i - \bar{X}) + \varepsilon_i$$

is to be utilized, the least squares estimator b_1 of β_1 is the same as before. The least squares estimator of $\beta_0^* = \beta_0 + \beta_1\bar{X}$ is, using (2.10b):

(2.14) $$b_0^* = b_0 + b_1\bar{X} = (\bar{Y} - b_1\bar{X}) + b_1\bar{X} = \bar{Y}$$

Hence, the estimated regression equation for alternative model (2.6) is:

(2.15) $$\hat{Y} = \bar{Y} + b_1(X - \bar{X})$$

In the Westwood Company example, $\bar{Y} = 1{,}100/10 = 110$ and $\bar{X} = 500/10 = 50$ (Table 2.2). Hence, the estimated regression function in alternative form is:

$$\hat{Y} = 110.0 + 2.0(X - 50)$$

For our sample data, $X_1 = 30$; hence, we estimate the mean response to be:

$$\hat{Y}_1 = 110.0 + 2.0(30 - 50) = 70$$

which, of course, is identical to our earlier result.

TABLE 2.3   Fitted Values, Residuals, and Squared Residuals—Westwood Company Example

	(1)	(2)	(3)	(4)	(5)
			Estimated		
Production	Lot	Man-	Mean		Squared
Run	Size	Hours	Response	Residual	Residual
i	X_i	Y_i	$\hat{Y}_i$	$Y_i - \hat{Y}_i = e_i$	$(Y_i - \hat{Y}_i)^2 = e_i^2$
1	30	73	70	+3	9
2	20	50	50	0	0
3	60	128	130	−2	4
4	80	170	170	0	0
5	40	87	90	−3	9
6	50	108	110	−2	4
7	60	135	130	+5	25
8	30	69	70	−1	1
9	70	148	150	−2	4
10	60	132	130	+2	4
Total	500	1,100	1,100	0	60

Residuals

The ith *residual* is the difference between the observed value Y_i and the corresponding fitted value $\hat{Y}_i$. Denoting this residual by e_i, we can write:

(2.16)
$$e_i = Y_i - \hat{Y}_i = Y_i - b_0 - b_1 X_i$$

Figure 2.12 shows the 10 residuals for the Westwood Company example. The magnitudes of the residuals are shown by the vertical lines between an observed Y-value and the fitted value on the estimated regression line. The residuals are calculated in Table 2.3, column 4.

We need to distinguish between the model error term value $\varepsilon_i = Y_i - E\{Y_i\}$ and the residual $e_i = Y_i - \hat{Y}_i$. The former involves the vertical deviation of Y_i from the unknown true regression line and hence is unknown. On the other hand, the residual is the vertical deviation of Y_i from the fitted value $\hat{Y}_i$ on the estimated regression line, and it is known.

Residuals are highly useful for studying whether a given regression model is appropriate for the data at hand. We shall discuss this use in Chapter 4.

Properties of Fitted Regression Line

The regression line fitted by the method of least squares has a number of properties worth noting:

1. The sum of the residuals is zero:

(2.17)
$$\sum_{i=1}^{n} e_i = 0$$

FIGURE 2.12   Least Squares Regression Line and Residuals—Westwood Company Example (observed values and residuals not plotted to scale)

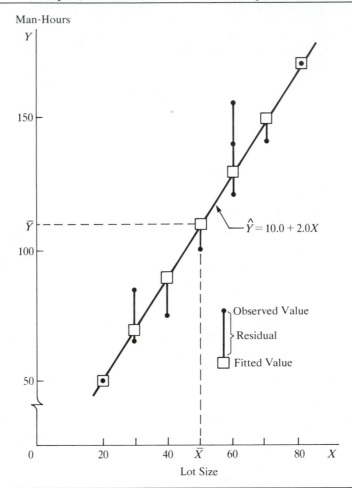

Table 2.3, column 4, illustrates this property for the Westwood Company example. Rounding errors may, of course, be present in any particular case, resulting in a sum of the residuals that does not equal zero exactly.

2. The sum of the squared residuals, Σe_i^2, is a minimum. This was the requirement to be satisfied in deriving the least squares estimators of the regression parameters.

3. The sum of the observed values Y_i equals the sum of the fitted values $\hat{Y}_i$:

(2.18)
$$\sum_{i=1}^{n} Y_i = \sum_{i=1}^{n} \hat{Y}_i$$

This property is illustrated in Table 2.3, columns 2 and 3, for the Westwood Company example. It follows that the mean of the $\hat{Y}_i$ is the same as the mean of the Y_i, namely, $\bar{Y}$.

4. The sum of the weighted residuals is zero when the residual in the ith trial is weighted by the level of the independent variable in the ith trial:

$$(2.19) \qquad \sum_{i=1}^{n} X_i e_i = 0$$

5. The sum of the weighted residuals is zero when the residual in the ith trial is weighted by the fitted value of the response variable for the ith trial:

$$(2.20) \qquad \sum_{i=1}^{n} \hat{Y}_i e_i = 0$$

6. The regression line always goes through the point $(\bar{X}, \bar{Y})$. Figure 2.12 demonstrates this property for the Westwood Company example.

Comments

1. The six properties of the residuals follow directly from the least squares normal equations (2.9). For example, property 1 in (2.17) is proven as follows:

$$\sum e_i = \sum (Y_i - b_0 - b_1 X_i) = \sum Y_i - nb_0 - b_1 \sum X_i$$

$$= 0 \qquad \text{by the first normal equation (2.9a)}$$

Property 6, that the regression line always goes through the point $(\bar{X}, \bar{Y})$, can be demonstrated easily from the alternative form (2.15) of the estimated regression line. When $X = \bar{X}$, we have:

$$\hat{Y} = \bar{Y} + b_1(X - \bar{X}) = \bar{Y} + b_1(\bar{X} - \bar{X}) = \bar{Y}$$

2. The properties of the least squares residuals noted here apply to regression model (2.1). They do not apply to all regression models, as we will see in Chapter 5.

2.7 ESTIMATION OF ERROR TERMS VARIANCE σ^2

The variance σ^2 of the error terms ε_i in the regression model (2.1) needs to be estimated for a variety of purposes. Frequently, we would like to obtain an indication of the variability of the probability distributions of Y. Further, as we shall see in the next chapter, a variety of inferences concerning the regression function and the prediction of Y require an estimate of σ^2.

Point Estimator of σ^2

Single Population.   To lay the basis for developing an estimator of σ^2 for the regression model (2.1), let us consider for a moment the simpler problem of sampling from a single population. In obtaining the sample variance s^2, we begin by considering the deviation of an observation Y_i from the estimated mean $\bar{Y}$, squaring it, and then summing all such squared deviations:

$$\sum_{i=1}^{n} (Y_i - \bar{Y})^2$$

Such a sum is called a *sum of squares*. The sum of squares is then divided by the degrees of freedom associated with it. This number is $n - 1$ here, because one degree of freedom is lost by using the estimate $\bar{Y}$ instead of the population mean μ. The resulting estimator is the usual sample variance:

$$s^2 = \frac{\sum_{i=1}^{n} (Y_i - \bar{Y})^2}{n - 1}$$

which is an unbiased estimator of the variance σ^2 of an infinite population. The sample variance is often called a *mean square*, because a sum of squares has been divided by the appropriate number of degrees of freedom.

Regression Model.   The logic of developing an estimator of σ^2 for the regression model is the same as when sampling a single population. Recall in this connection from (2.4) that the variance of each observation Y_i is σ^2, the same as that of each error term ε_i. We again need to calculate a sum of squared deviations, but must recognize that the Y_i come from different probability distributions with different means, depending upon the level X_i. Thus, the deviation of an observation Y_i must be calculated around its own estimated mean $\hat{Y}_i$. Hence, the deviations are the residuals:

$$Y_i - \hat{Y}_i = e_i$$

and the appropriate sum of squares, denoted by *SSE*, is:

$$(2.21) \qquad SSE = \sum_{i=1}^{n} (Y_i - \hat{Y}_i)^2 = \sum_{i=1}^{n} (Y_i - b_0 - b_1 X_i)^2 = \sum_{i=1}^{n} e_i^2$$

where *SSE* stands for *error sum of squares* or *residual sum of squares*.

The sum of squares *SSE* has $n - 2$ degrees of freedom associated with it. Two degrees of freedom are lost because both β_0 and β_1 had to be estimated in obtaining the estimated means $\hat{Y}_i$. Hence, the appropriate mean square, denoted by *MSE*, is:

$$(2.22) \qquad MSE = \frac{SSE}{n - 2} = \frac{\Sigma(Y_i - \hat{Y}_i)^2}{n - 2}$$

$$= \frac{\Sigma(Y_i - b_0 - b_1 X_i)^2}{n - 2} = \frac{\Sigma e_i^2}{n - 2}$$

where *MSE* stands for *error mean square* or *residual mean square*.

It can be shown that *MSE* is an unbiased estimator of σ^2 for the regression model (2.1):

$$(2.23) \qquad E\{MSE\} = \sigma^2$$

An estimator of the standard deviation σ is simply the positive square root of *MSE*.

Alternative Computational Formulas

There are a number of alternative computational formulas for *SSE*. Three of these formulas are as follows:

(2.24a) $$SSE = \sum Y_i^2 - b_0 \sum Y_i - b_1 \sum X_i Y_i$$

(2.24b) $$SSE = \sum (Y_i - \bar{Y})^2 - \frac{[\sum(X_i - \bar{X})(Y_i - \bar{Y})]^2}{\sum(X_i - \bar{X})^2}$$

(2.24c) $$SSE = \left[\sum Y_i^2 - \frac{(\sum Y_i)^2}{n} \right] - \frac{\left(\sum X_i Y_i - \frac{\sum X_i \sum Y_i}{n} \right)^2}{\sum X_i^2 - \frac{(\sum X_i)^2}{n}}$$

Comments

1. Formula (2.24a) is useful if b_0 and b_1 have already been calculated. Otherwise, (2.24b) and (2.24c) are more direct.

2. In (2.24a), the estimates b_0 and b_1 should be carried to a large number of digits in order to yield reliable results for *SSE*.

3. None of the three alternative formulas explicitly provides the residuals e_i. As noted earlier, the residuals are useful for studying the appropriateness of the model.

Example

Returning to the Westwood Company example, we will calculate *SSE* by (2.21). The residuals were obtained earlier in Table 2.3, column 4. This table also shows the squared residuals in column 5. From these results, we obtain:

$$SSE = 60$$

Since $10 - 2 = 8$ degrees of freedom are associated with *SSE*, we find:

$$MSE = \frac{60}{8} = 7.5$$

Finally, a point estimate of σ, the standard deviation of the probability distribution of Y for any X, is $\sqrt{7.5} = 2.74$ man-hours.

Consider again the case where the lot size is $X = 55$ units. We estimated earlier that the probability distribution of Y for this lot size has a mean of 120 man-hours. Now, we have the additional information that the standard deviation of this distribution is estimated to be 2.74 man-hours.

If we wished to use, say, (2.24a) for calculating *SSE*, we would need $\sum Y_i^2$. This sum is calculated in Table 2.2. We then obtain, using the results in Table 2.2 and the estimates $b_0 = 10.0$, $b_1 = 2.0$:

$$SSE = \sum Y_i^2 - b_0 \sum Y_i - b_1 \sum X_i Y_i$$

$$= 134{,}660 - 10.0(1{,}100) - 2.0(61{,}800) = 60$$

which is, of course, the same result (except sometimes for rounding errors) as obtained earlier.

2.8 NORMAL ERROR REGRESSION MODEL

No matter what may be the functional form of the distribution of ε_i (and hence of Y_i), the least squares method provides unbiased point estimators of β_0 and β_1 that have minimum variance among all unbiased linear estimators. To set up interval estimates and make tests, however, we do need to make an assumption about the functional form of the distribution of the ε_i. The standard assumption is that the error terms are normally distributed, and we will adopt it here. A normal error term greatly simplifies the theory of regression analysis and is justifiable in many real world situations where regression analysis is applied.

Model

The normal error regression model is as follows:

$$(2.25) \qquad\qquad Y_i = \beta_0 + \beta_1 X_i + \varepsilon_i$$

where:

Y_i is the observed response in the ith trial
X_i is a known constant, the level of the independent variable in the ith trial
β_0 and β_1 are parameters
ε_i are independent $N(0, \sigma^2)$
$i = 1, \ldots, n$

Comments

1. The symbol $N(0, \sigma^2)$ stands for "normally distributed, with mean 0 and variance σ^2."

2. The normal error model (2.25) is the same as regression model (2.1) with unspecified error distribution, except that model (2.25) assumes that the errors ε_i are normally distributed.

3. Because regression model (2.25) assumes that the errors are normally distributed, the assumption of uncorrelatedness of the ε_i in regression model (2.1) becomes one of independence in the normal error model.

4. Regression model (2.25) implies that the Y_i are independent normal random variables, with mean $E\{Y_i\} = \beta_0 + \beta_1 X_i$ and variance σ^2. Figure 2.4 pictures this normal error model. Each of the probability distributions of Y there is normally distributed, with constant variability, and the regression function is linear.

5. A major reason why the normality assumption for the error terms is justifiable in many situations is that the error terms frequently represent the effects of many factors omitted explicitly from the model, that do affect the response to some extent and that vary at random without reference to the independent variable X. For instance, in the Westwood Company example, the effects of such factors as time lapse since the last production run, particular machines used, season of the year, and personnel employed, could vary more or less at random from run to run, independent of lot size. Also, there might be random measurement errors in recording Y. Insofar as these random effects have a degree of mutual independence, the composite error term ε_i representing all these factors would tend to comply with the central limit theorem and the error term distribution would approach normality as the number of factor effects becomes large.

A second reason why the normality assumption for the error terms is frequently justifiable is that the estimation and testing procedures to be discussed in the next chapter are based on the t distribution, which is not sensitive to moderate departures from normality. Thus, unless the departures from normality are serious, particularly with respect to skewness, the actual confidence coefficients and risks of errors will be close to the levels for exact normality.

Estimation of Parameters by Method of Maximum Likelihood

When the functional form of the probability distribution of the error terms is specified, estimators of the parameters β_0, β_1, and σ^2 can be obtained by the *method of maximum likelihood*. This method utilizes the joint probability distribution of the sample observations. When this joint probability distribution is viewed as a function of the parameters, given the particular sample observations, it is called the *likelihood function*. The likelihood function for the normal error regression model (2.25), given the sample observations $Y_1, \ldots, Y_n$, is:

$$(2.26) \qquad L(\beta_0, \beta_1, \sigma^2) = \prod_{i=1}^{n} \frac{1}{(2\pi\sigma^2)^{1/2}} \exp\left[-\frac{1}{2\sigma^2}(Y_i - \beta_0 - \beta_1 X_i)^2\right]$$

$$= \frac{1}{(2\pi\sigma^2)^{n/2}} \exp\left[-\frac{1}{2\sigma^2} \sum_{i=1}^{n} (Y_i - \beta_0 - \beta_1 X_i)^2\right]$$

The values of β_0, β_1, and σ^2 that maximize this likelihood function are the maximum likelihood estimators. These are:

	Parameter	Maximum Likelihood Estimator
	β_0	b_0 same as (2.10b)
(2.27)	β_1	b_1 same as (2.10a)
	σ^2	$\hat{\sigma}^2 = \dfrac{\Sigma(Y_i - \hat{Y}_i)^2}{n}$

Thus, the maximum likelihood estimators of β_0 and β_1 are the same estimators as provided by the method of least squares. The maximum likelihood estimator $\hat{\sigma}^2$ is biased, and ordinarily the unbiased estimator *MSE* as given in (2.22) is used. Note that the unbiased estimator *MSE* differs but slightly from the maximum likelihood estimator $\hat{\sigma}^2$, especially if n is not small:

$$(2.28) \qquad\qquad MSE = \frac{n}{n-2}\hat{\sigma}^2$$

Comments

1. Since the maximum likelihood estimators b_0 and b_1 are the same as the least squares estimators, they have the properties of all least squares estimators:

a. They are unbiased.
b. They have minimum variance among all unbiased linear estimators.

In addition, the maximum likelihood estimators b_0 and b_1 for the normal error regression model (2.25) have other desirable properties:

c. They are consistent, as defined in (1.49).
d. They are sufficient, as defined in (1.50).
e. They are minimum variance unbiased; i.e., they have minimum variance in the class of all unbiased estimators (linear or otherwise).

Thus, for the normal error model, the estimators b_0 and b_1 have many desirable properties.

2. We find the values of β_0, β_1, and σ^2 that maximize the likelihood function L in (2.26) by taking partial derivatives of L with respect to β_0, β_1, and σ^2, equating each of the partials to zero, and solving the system of equations thus obtained. We can work with $\log_e L$, rather than L, because both L and $\log_e L$ are maximized for the same values of β_0, β_1, and σ^2.

$$(2.29) \qquad \log_e L = -\frac{n}{2}\log_e 2\pi - \frac{n}{2}\log_e \sigma^2 - \frac{1}{2\sigma^2}\sum (Y_i - \beta_0 - \beta_1 X_i)^2$$

Partial differentiation of this logarithmic likelihood is much easier; it yields:

$$\frac{\partial(\log_e L)}{\partial\beta_0} = \frac{1}{\sigma^2}\sum (Y_i - \beta_0 - \beta_1 X_i)$$

$$\frac{\partial(\log_e L)}{\partial\beta_1} = \frac{1}{\sigma^2}\sum X_i(Y_i - \beta_0 - \beta_1 X_i)$$

$$\frac{\partial(\log_e L)}{\partial\sigma^2} = -\frac{n}{2\sigma^2} + \frac{1}{2\sigma^4}\sum (Y_i - \beta_0 - \beta_1 X_i)^2$$

We now set these partial derivatives equal to zero, replacing β_0, β_1, and σ^2 by the estimators b_0, b_1, and $\hat{\sigma}^2$. We obtain after some simplification:

$$(2.30a) \qquad\qquad\qquad \sum (Y_i - b_0 - b_1 X_i) = 0$$

$$(2.30b) \qquad\qquad\qquad \sum X_i(Y_i - b_0 - b_1 X_i) = 0$$

$$(2.30c) \qquad\qquad\qquad \frac{\sum(Y_i - b_0 - b_1 X_i)^2}{n} = \hat{\sigma}^2$$

Formulas (2.30a) and (2.30b) are identical to the earlier least squares normal equations (2.9), and (2.30c) is the biased estimator of σ^2 given earlier in (2.27).

PROBLEMS

2.1. Refer to the sales volume example on page 24. Suppose that the number of units sold is measured accurately but clerical errors are frequently made in determining the dollar sales. Would the relation between the number of units sold and dollar sales still be a functional one? Discuss.

2.2. The members of a health spa pay annual membership dues of $300 plus a charge of $2 for each visit to the spa. Let Y denote the total dollar cost for the year for a member and X the number of visits by the member during the year. Express the relation between X and Y mathematically. Is it a functional or a statistical relation?

2.3. Experience with a certain type of plastic indicates that a relation exists between the hardness (measured in Brinell units) of items molded from the plastic (Y) and the elapsed time since termination of the molding process (X). It is proposed to study this relation by means of regression analysis. A participant in the discussion objects, pointing out that the hardening of the plastic "is the result of a natural chemical process that doesn't leave anything to chance, so the relation must be mathematical and regression analysis is not appropriate." Evaluate this objection.

2.4. In Table 2.1, the lot size X is the same in production runs 1 and 8 but the man-hours Y differ. What feature of regression model (2.1) is illustrated by this?

2.5. When asked to state the simple linear regression model, a student wrote it as follows: $E\{Y_i\} = \beta_0 + \beta_1 X_i + \varepsilon_i$. Do you agree?

2.6. Consider the normal error regression model (2.25). Suppose that the parameter values are $\beta_0 = 200$, $\beta_1 = 5.0$, and $\sigma = 4$.
   a. Plot this normal error regression model in the fashion of Figure 2.7. Show the distributions of Y for $X = 10$, 20, and 40.
   b. Explain the meaning of the parameters β_0 and β_1. Assume that the scope of the model includes $X = 0$.

2.7. In a simulation exercise, regression model (2.1) applies with $\beta_0 = 100$, $\beta_1 = 20$, and $\sigma^2 = 25$. An observation on Y will be made for $X = 5$.
   a. Can you state the exact probability that Y will fall between 195 and 205? Explain.
   b. If the normal error regression model (2.25) is applicable, can you now state the exact probability that Y will fall between 195 and 205? If so, state it.

2.8. In Figure 2.7, suppose another Y observation is obtained at $X = 45$. Would $E\{Y\}$ for this new observation still be 104? Would the Y value for this new case again be 108?

2.9. A student in accounting enthusiastically declared: "Regression is a very powerful tool. We can isolate fixed and variable costs by fitting a linear regression model, even when we have no data for small lots." Discuss.

2.10. An analyst in a large corporation studied the relation between current annual salary (Y) and age (X) for the 46 computer programmers presently employed in the company. She concluded that the relation is curvilinear, reaching a maximum at 47 years. Does this imply that the salary for a programmer increases until age 47 and then decreases? Explain.

2.11. The regression function relating production output by an employee after taking a training program (Y) to the production output before the training program (X) is $E\{Y\} = 20 + .95X$, where X ranges from 40 to 100. An observer concludes that the training program does not raise production output on the average because β_1 is not greater than 1.0. Comment.

2.12. Refer to Problem 2.3. The hardnesses after four different elapsed times since termination of the molding process (treatments) are to be studied. Sixteen batches (experimental units) are available for the study. Each treatment is to be assigned to four experimental units selected at random. Use a table of random digits or a random number generator to make an appropriate randomization of assignments.

2.13. The effects of five dose levels are to be studied in a completely randomized design, and 20 experimental units are available. Each dose level is to be assigned to four experimental units selected at random. Use a table of random digits or a random number generator to make an appropriate randomization of assignments.

2.14. Evaluate the following statement: "For the least squares method to be fully valid, it is required that the distribution of Y is normal."

2.15. A person states that b_0 and b_1 in the fitted regression function (2.12) can be estimated by the method of least squares. Comment.

2.16. According to (2.17), $\Sigma e_i = 0$ when regression model (2.1) is fitted to a set of n cases by the method of least squares. Is it also true that $\Sigma \varepsilon_i = 0$? Comment.

2.17. Grade point average.   The director of admissions of a small college administered a newly designed entrance test to 20 students selected at random from the new freshman class in a study to determine whether a student's grade point average (GPA) at the end of the freshman year (Y) can be predicted from the entrance test score (X). The results of the study follow. Assume that the first-order regression model (2.1) is appropriate.

i:	1	2	3	4	5	6	7	8	9	10
X_i:	5.5	4.8	4.7	3.9	4.5	6.2	6.0	5.2	4.7	4.3
Y_i:	3.1	2.3	3.0	1.9	2.5	3.7	3.4	2.6	2.8	1.6

i:	11	12	13	14	15	16	17	18	19	20
X_i:	4.9	5.4	5.0	6.3	4.6	4.3	5.0	5.9	4.1	4.7
Y_i:	2.0	2.9	2.3	3.2	1.8	1.4	2.0	3.8	2.2	1.5

Summary calculational results are: $\Sigma X_i = 100.0$, $\Sigma Y_i = 50.0$, $\Sigma X_i^2 = 509.12$, $\Sigma Y_i^2 = 134.84$, $\Sigma X_i Y_i = 257.66$.

a. Obtain the least squares estimates of β_0 and β_1, and state the estimated regression function.

b. Plot the estimated regression function and the data. Does the estimated regression function appear to fit the data well?

c. Obtain a point estimate of the mean freshman GPA for students with entrance test score $X = 5.0$.

d. What is the point estimate of the change in the mean response when the entrance test score increases by one point?

2.18. Calculator maintenance.   The Tri-City Office Equipment Corporation sells an imported desk calculator on a franchise basis and performs preventive maintenance and repair service on this calculator. The data below have been collected from 18 recent calls on users to perform routine preventive maintenance service; for each call, X is the number of machines serviced and Y is the total number of minutes spent by the service person. Assume that the first-order regression model (2.1) is appropriate.

i:	1	2	3	4	5	6	7	8	9
X_i:	7	6	5	1	5	4	7	3	4
Y_i:	97	86	78	10	75	62	101	39	53

i:	10	11	12	13	14	15	16	17	18
X_i:	2	8	5	2	5	7	1	4	5
Y_i:	33	118	65	25	71	105	17	49	68

Summary calculational results are: $\Sigma Y_i = 1{,}152$, $\Sigma X_i = 81$, $\Sigma(Y_i - \bar{Y})^2 = 16{,}504$, $\Sigma(X_i - \bar{X})^2 = 74.5$, $\Sigma(X_i - \bar{X})(Y_i - \bar{Y}) = 1{,}098$.

a. Obtain the estimated regression function.

b. Plot the estimated regression function and the data. How well does the estimated regression function fit the data?

c. Interpret b_0 in your estimated regression function. Does b_0 provide any relevant information here? Explain.

d. Obtain a point estimate of the mean service time when $X = 5$ machines are serviced.

2.19. **Airfreight breakage.** A substance used in biological and medical research is shipped by airfreight to users in cartons of 1,000 ampules. The data below, involving 10 shipments, were collected on the number of times the carton was transferred from one aircraft to another over the shipment route (X) and the number of ampules found to be broken upon arrival (Y). Assume that the first-order regression model (2.1) is appropriate.

i:	1	2	3	4	5	6	7	8	9	10
X_i:	1	0	2	0	3	1	0	1	2	0
Y_i:	16	9	17	12	22	13	8	15	19	11

a. Obtain the estimated regression function. Plot the estimated regression function and the data. Does a linear regression function appear to give a good fit here?

b. Obtain a point estimate of the expected number of broken ampules when $X = 1$ transfer is made.

c. Estimate the increase in the expected number of ampules broken when there are 2 transfers as compared to 1 transfer.

d. Verify that your fitted regression line goes through the point $(\bar{X}, \bar{Y})$.

2.20. **Plastic hardness.** Refer to Problems 2.3 and 2.12. Sixteen batches of the plastic were made, and from each batch one test item was molded. Each test item was randomly assigned to one of the four predetermined time levels, and the hardness was measured after the assigned elapsed time. The results are shown below: X is the elapsed time in hours, and Y is hardness in Brinell units. Assume that the first-order regression model (2.1) is appropriate.

i:	1	2	3	4	5	6	7	8
X_i:	16	16	16	16	24	24	24	24
Y_i:	199	205	196	200	218	220	215	223

i:	9	10	11	12	13	14	15	16
X_i:	32	32	32	32	40	40	40	40
Y_i:	237	234	235	230	250	248	253	246

a. Obtain the estimated regression function. Plot the estimated regression function and the data. Does a linear regression function appear to give a good fit here?

b. Obtain a point estimate of the mean hardness when $X = 40$ hours.

c. Obtain a point estimate of the change in mean hardness when X increases by one hour.

2.21. Refer to **Grade point average** Problem 2.17.

   a. Obtain the residuals e_i. Do they sum to zero in accord with (2.17)?

   b. Estimate σ^2 and σ. In what units is σ expressed?

2.22. Refer to **Calculator maintenance** Problem 2.18.

   a. Obtain the residuals e_i and the sum of the squared residuals Σe_i^2. What is the relation between the sum of the squared residuals here and the quantity Q in (2.8)?

   b. Obtain point estimates of σ^2 and σ. In what units is σ expressed?

2.23. Refer to **Airfreight breakage** Problem 2.19.

   a. Obtain the residual for the first case. What is its relation to ε_1?

   b. Compute Σe_i^2 and MSE. What is estimated by MSE?

2.24. Refer to **Plastic hardness** Problem 2.20.

   a. Obtain the residuals e_i. Do they sum to zero in accord with (2.17)?

   b. Estimate σ^2 and σ. In what units is σ expressed?

2.25. **Muscle mass.**   A person's muscle mass is expected to decrease with age. To explore this relationship in women, a nutritionist randomly selected four women from each 10-year age group, beginning with age 40 and ending with age 79. The results follow; X is age, and Y is a measure of muscle mass. Assume that the first-order regression model (2.1) is appropriate.

i:	1	2	3	4	5	6	7	8
X_i:	71	64	43	67	56	73	68	56
Y_i:	82	91	100	68	87	73	78	80

i:	9	10	11	12	13	14	15	16
X_i:	76	65	45	58	45	53	49	78
Y_i:	65	84	116	76	97	100	105	77

a. Obtain the estimated regression function. Plot the estimated regression function and the data. Does a linear regression function appear to give a good fit here? Does your plot support the anticipation that muscle mass decreases with age?

b. Obtain the following: (1) a point estimate of the difference in the mean muscle mass for women differing in age by one year, (2) a point estimate of the mean muscle mass for women aged $X = 60$ years, (3) the value of the residual for the eighth case, (4) a point estimate of σ^2.

2.26. **Robbery rate.**   A criminologist studying the relationship between population density and robbery rate in medium-sized U.S. cities collected the following data for a random sample of 16 cities; X is the population density of the city (number of people per unit area), and Y is the robbery rate last year (number of robberies per 100,000 people). Assume that the first-order regression model (2.1) is appropriate.

i:	1	2	3	4	5	6	7	8
X_i:	59	49	75	54	78	56	60	82
Y_i:	209	180	195	192	215	197	208	189

i:	9	10	11	12	13	14	15	16
X_i:	69	83	88	94	47	65	89	70
Y_i:	213	201	214	212	205	186	200	204

a. Obtain the estimated regression function. Plot the estimated regression function and the data. Does the linear regression function appear to give a good fit here? Discuss.

b. Obtain point estimates of the following: (1) the difference in the mean robbery rate in cities that differ by one unit in population density, (2) the mean robbery rate last year in cities with population density $X = 60$, (3) ε_{10}, (4) σ^2.

EXERCISES

2.27. Refer to regression model (2.1). Assume that $X = 0$ is within the scope of the model. What is the implication for the regression function if $\beta_0 = 0$ so that the model is $Y_i = \beta_1 X_i + \varepsilon_i$? How would the regression function plot on a graph?

2.28. Refer to regression model (2.1). What is the implication for the regression function if $\beta_1 = 0$ so that the model is $Y_i = \beta_0 + \varepsilon_i$? How would the regression function plot on a graph?

2.29. Refer to **Plastic hardness** Problem 2.20. Suppose one test item was molded from a single batch of plastic and the hardness of this one item was measured at 16 different points in time. Would the error term in the regression model for this case still reflect the same effects as for the experiment initially described? Would you expect the error terms for the different points in time to be uncorrelated? Discuss.

2.30. Derive the expression for b_1 in (2.10a) from the normal equations in (2.9).

2.31. (Calculus needed.) Refer to the regression model $Y_i = \beta_0 + \varepsilon_i$ in Exercise 2.28. Derive the least squares estimator of β_0 for this model.

2.32. Prove that the least squares estimator of β_0 obtained in Exercise 2.31 is unbiased.

2.33. Prove the result in (2.18)—that the sum of the Y observations is the same as the sum of the fitted values.

2.34. Prove the result in (2.20)—that the sum of the residuals weighted by the fitted values is zero.

2.35. Prove the result for *SSE* in (2.24a).

2.36. Refer to Table 2.1 for the Westwood Company example. When asked to present a point estimate of the expected man-hours for runs of 60 pieces, a person gave the estimate 131.7 because this is the mean number of man-hours in the three runs of size 60 in the study. A critic states that this person's approach "throws away" most of the data in the study because cases with lot sizes other than 60 are ignored. Comment.

2.37. In **Airfreight breakage** Problem 2.19, the least squares estimates are $b_0 = 10.20$ and $b_1 = 4.00$, and $\Sigma e_i^2 = 17.60$. Evaluate the least squares criterion Q in (2.8) for the estimates: (1) $b_0 = 9$, $b_1 = 3$; (2) $b_0 = 11$, $b_1 = 5$. Is the criterion Q larger for these estimates than for the least squares estimates?

2.38. Two observations on Y were obtained at each of three X levels—namely, $X = 5$, $X = 10$, and $X = 15$.

    a. Show that the least squares regression line fitted to the *three* points $(5, \bar{Y}_1)$, $(10, \bar{Y}_2)$, and $(15, \bar{Y}_3)$, where $\bar{Y}_1$, $\bar{Y}_2$, and $\bar{Y}_3$ denote the means of the Y observations at the three X levels, is identical to the least squares regression line fitted to the original six cases.

    b. In this study, could the error term variance σ^2 be estimated without fitting a regression line? Explain.

2.39. In the Westwood Company example, the Y observations at $X = 20$ and $X = 80$ fall directly on the fitted regression line (Table 2.3 and Figure 2.2b). If these two cases were deleted, would the least squares regression line fitted to the remaining eight cases be changed? [*Hint:* What is the contribution of the two cases to the least squares criterion Q in (2.8)?]

2.40. (Calculus needed.) Refer to the regression model $Y_i = \beta_1 X_i + \varepsilon_i$, $i = 1, \ldots, n$, in Exercise 2.27.

    a. Find the least squares estimator of β_1.

    b. Assume that the error terms ε_i are independent $N(0, \sigma^2)$ and that σ^2 is known. State the likelihood function for the n sample observations on Y and obtain the maximum likelihood estimator of β_1. Is it the same as the least squares estimator?

    c. Show that the maximum likelihood estimator of β_1 is unbiased.

2.41. Shown below are the number of galleys for a manuscript (X) and the dollar cost of correcting typographical errors (Y) in a random sample of recent orders handled by a firm specializing in technical manuscripts. Assume that the regression model $Y_i = \beta_1 X_i + \varepsilon_i$ is appropriate, with normally distributed independent error terms whose variance is $\sigma^2 = 16$.

i:	1	2	3	4	5	6
X_i:	7	12	4	14	25	30
Y_i:	128	213	75	250	446	540

    a. State the likelihood function for the six Y observations, for $\sigma^2 = 16$.

    b. Evaluate the likelihood function for $\beta_1 = 17$, 18, and 19. For which of these β_1 values is the likelihood function largest?

    c. The maximum likelihood estimator is $b_1 = \Sigma X_i Y_i / \Sigma X_i^2$. Find the maximum likelihood estimate. Are your results in part (b) consistent with this estimate?

PROJECTS

2.42. Refer to the **SMSA** data set in Appendix B.2. The number of active physicians in an SMSA (Y) is expected to be related to total population, land area, and total personal income. Assume that the first-order regression model (2.1) is appropriate for each of the three independent variables.

    a. Regress the number of active physicians in turn on each of the three independent variables. State the estimated regression functions.

    b. Plot the three estimated regression functions and data on separate graphs. Does a linear regression relation appear to provide a good fit for each of the three independent variables?

    c. Calculate *MSE* for each of the three independent variables. Which independent variable leads to the smallest variability around the fitted regression line?

2.43. Refer to the **SMSA** data set in Appendix B.2.

    a. For each geographic region, regress total serious crimes in an SMSA (Y) against total population (X). Assume that the first-order regression model (2.1) is appropriate for each region. State the estimated regression functions.

    b. Are the estimated regression functions similar for the four regions? Discuss.

    c. Calculate *MSE* for each region. Is the variability around the fitted regression line approximately the same for the four regions? Discuss.

✓ **2.44.** Refer to the **SENIC** data set in Appendix B.1. The average length of stay in a hospital (Y) is anticipated to be related to infection risk, available facilities and services, and routine chest X-ray ratio. Assume that the first-order regression model (2.1) is appropriate for each of the three independent variables.

    a. Regress average length of stay on each of the three independent variables. State the estimated regression functions.

    b. Plot the three estimated regression functions and data on separate graphs. Does a linear relation appear to provide a good fit for each of the three independent variables?

    c. Calculate *MSE* for each of the three independent variables. Which independent variable leads to the smallest variability around the fitted regression line?

2.45. Refer to the **SENIC** data set in Appendix B.1.

    a. For each geographic region, regress average length of stay in hospital (Y) against infection risk (X). Assume that the first-order regression model (2.1) is appropriate for each region. State the estimated regression functions.

    b. Are the estimated regression functions similar for the four regions? Discuss.

    c. Calculate *MSE* for each region. Is the variability around the fitted regression line approximately the same for the four regions? Discuss.

Chapter 3

Inferences in Regression Analysis

In this chapter, we first take up inferences concerning the regression parameters β_0 and β_1, considering both interval estimation of these parameters and tests about them. We then discuss interval estimation of the mean $E\{Y\}$ of the probability distribution of Y, for given X, and prediction intervals for a new observation Y, given X. Finally, we take up the analysis of variance approach to regression analysis, the general linear test approach, and descriptive measures of association.

Throughout this chapter, and in the remainder of Part 1 unless otherwise stated, we assume that the normal error regression model (2.25) is applicable. This model is:

$$(3.1) \qquad Y_i = \beta_0 + \beta_1 X_i + \varepsilon_i$$

where:

β_0 and β_1 are parameters
X_i are known constants
ε_i are independent $N(0, \sigma^2)$

3.1 INFERENCES CONCERNING β_1

Frequently, we are interested in drawing inferences about β_1, the slope of the regression line in model (3.1). For instance, a market research analyst studying the relation between sales (Y) and advertising expenditures (X) may wish to obtain an interval estimate of β_1 because it will provide information as to how many additional sales dollars, on the average, are generated by an additional dollar of advertising expenditure.

At times, tests concerning β_1 are of interest, particularly one of the form:

$$H_0: \beta_1 = 0$$

$$H_a: \beta_1 \neq 0$$

FIGURE 3.1   Regression Model (3.1) when $\beta_1 = 0$

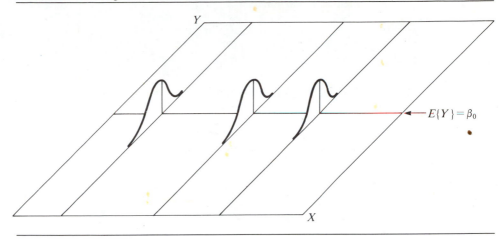

The reason for interest in testing whether or not $\beta_1 = 0$ is that $\beta_1 = 0$ indicates that there is no linear association between Y and X. Figure 3.1 illustrates the case when $\beta_1 = 0$ for the normal error regression model (3.1). Note that the regression line is horizontal and that the means of the probability distributions of Y are therefore all equal, namely:

$$E\{Y\} = \beta_0 + (0)X = \beta_0$$

Since regression model (3.1) assumes normality of the probability distributions of Y with constant variance, and since the means are equal when $\beta_1 = 0$, it follows that the probability distributions of Y are identical when $\beta_1 = 0$. This is shown in Figure 3.1. Thus, $\beta_1 = 0$ for regression model (3.1) implies not only that there is no linear association between Y and X but also that there is no relation of any type between Y and X, since the probability distributions of Y are then identical at all levels of X.

Before discussing inferences concerning β_1 further, we need to consider the sampling distribution of b_1, the point estimator of β_1.

Sampling Distribution of b_1

The point estimator b_1 was given in (2.10a) as follows:

(3.2)
$$b_1 = \frac{\Sigma(X_i - \bar{X})(Y_i - \bar{Y})}{\Sigma(X_i - \bar{X})^2}$$

The sampling distribution of b_1 refers to the different values of b_1 that would be obtained with repeated sampling when the levels of the independent variable X are held constant from sample to sample.

(3.3)   For regression model (3.1), the sampling distribution of b_1 is normal, with mean and variance:

(3.3a) $$E\{b_1\} = \beta_1$$

(3.3b) $$\sigma^2\{b_1\} = \frac{\sigma^2}{\Sigma(X_i - \bar{X})^2}$$

To show this, we need to recognize that b_1 is a linear combination of the observations Y_i.

b_1 as Linear Combination of the Y_i.   It can be shown that b_1, as defined in (3.2), can be expressed as follows:

(3.4) $$b_1 = \sum k_i Y_i$$

where:

(3.4a) $$k_i = \frac{X_i - \bar{X}}{\Sigma(X_i - \bar{X})^2}$$

Observe that the k_i are fixed quantities since the X_i are fixed. Hence, b_1 is a linear combination of the Y_i where the coefficients are solely a function of the fixed X_i.

The coefficients k_i have a number of interesting properties that will be used later:

(3.5) $$\sum k_i = 0$$

(3.6) $$\sum k_i X_i = 1$$

(3.7) $$\sum k_i^2 = \frac{1}{\Sigma(X_i - \bar{X})^2}$$

Comments

1. To show that b_1 is a linear combination of the k_i, we first prove:

(3.8) $$\sum (X_i - \bar{X})(Y_i - \bar{Y}) = \sum (X_i - \bar{X})Y_i$$

This follows since:

$$\sum (X_i - \bar{X})(Y_i - \bar{Y}) = \sum (X_i - \bar{X})Y_i - \sum (X_i - \bar{X})\bar{Y}$$

But $\Sigma(X_i - \bar{X})\bar{Y} = \bar{Y}\Sigma(X_i - \bar{X}) = 0$ since $\Sigma(X_i - \bar{X}) = 0$. Hence, (3.8) holds.

We now express b_1, using (3.8) and (3.4a):

$$b_1 = \frac{\Sigma(X_i - \bar{X})(Y_i - \bar{Y})}{\Sigma(X_i - \bar{X})^2} = \frac{\Sigma(X_i - \bar{X})Y_i}{\Sigma(X_i - \bar{X})^2} = \sum k_i Y_i$$

2. The proofs of the properties of the k_i are direct. For example, property (3.5) follows because:

$$\sum k_i = \sum \left[\frac{X_i - \bar{X}}{\Sigma(X_i - \bar{X})^2} \right] = \frac{\Sigma(X_i - \bar{X})}{\Sigma(X_i - \bar{X})^2} = \frac{0}{\Sigma(X_i - \bar{X})^2} = 0$$

Similarly, property (3.7) follows because:

$$\sum k_i^2 = \sum \left[\frac{(X_i - \bar{X})}{\Sigma(X_i - \bar{X})^2} \right]^2 = \frac{1}{[\Sigma(X_i - \bar{X})^2]^2} \sum (X_i - \bar{X})^2$$

$$= \frac{1}{\Sigma(X_i - \bar{X})^2}$$

Normality. We return now to the sampling distribution of b_1 for the normal error regression model (3.1). The normality of the sampling distribution of b_1 follows at once from the fact that b_1 is a linear combination of the Y_i. The Y_i are independently, normally distributed according to model (3.1), and theorem (1.37) states that a linear combination of independent normal random variables is normally distributed.

Mean. The unbiasedness of the point estimator b_1, stated earlier in the Gauss-Markov theorem (2.11), is easy to show:

$$E\{b_1\} = E\left\{ \sum k_i Y_i \right\} = \sum k_i E\{Y_i\} = \sum k_i(\beta_0 + \beta_1 X_i)$$

$$= \beta_0 \sum k_i + \beta_1 \sum k_i X_i$$

Hence, by (3.5) and (3.6) $E\{b_1\} = \beta_1$.

Variance. The variance of b_1 can be derived readily. We need only remember that the Y_i are independent random variables, each with variance σ^2, and that the k_i are constants. Hence, we obtain by (1.28):

$$\sigma^2\{b_1\} = \sigma^2\{\Sigma k_i Y_i\} = \sum k_i^2 \sigma^2\{Y_i\}$$

$$= \sum k_i^2 \sigma^2 = \sigma^2 \sum k_i^2$$

$$= \sigma^2 \frac{1}{\Sigma(X_i - \bar{X})^2}$$

The last step follows from (3.7).

Estimated Variance. We can estimate the variance of the sampling distribution of b_1:

$$\sigma^2\{b_1\} = \frac{\sigma^2}{\Sigma(X_i - \bar{X})^2}$$

by replacing the parameter σ^2 with the unbiased estimator of σ^2, namely, *MSE:*

(3.9)
$$s^2\{b_1\} = \frac{MSE}{\Sigma(X_i - \bar{X})^2} = \frac{MSE}{\Sigma X_i^2 - \dfrac{(\Sigma X_i)^2}{n}}$$

The point estimator $s^2\{b_1\}$ is an unbiased estimator of $\sigma^2\{b_1\}$. Taking the square root, we obtain $s\{b_1\}$, the point estimator of $\sigma\{b_1\}$.

Note

We stated in theorem (2.11) that b_1 has minimum variance among all unbiased linear estimators of the form:

$$\hat{\beta}_1 = \sum c_i Y_i$$

where the c_i are arbitrary constants. We shall now prove this. Since $\hat{\beta}_1$ must be unbiased, the following must hold:

$$E\{\hat{\beta}_1\} = E\left\{ \sum c_i Y_i \right\} = \sum c_i E\{Y_i\} = \beta_1$$

Now $E\{Y_i\} = \beta_0 + \beta_1 X_i$ by (2.2) so that the above condition becomes:

$$E\{\hat{\beta}_1\} = \sum c_i(\beta_0 + \beta_1 X_i) = \beta_0 \sum c_i + \beta_1 \sum c_i X_i = \beta_1$$

For the unbiasedness condition to hold, the c_i must follow the restrictions:

$$\sum c_i = 0 \qquad \sum c_i X_i = 1$$

Now the variance of $\hat{\beta}_1$ is by (1.28):

$$\sigma^2\{\hat{\beta}_1\} = \sum c_i^2 \sigma^2\{Y_i\} = \sigma^2 \sum c_i^2$$

Let us define $c_i = k_i + d_i$, where the k_i are the least squares constants in (3.4a) and the d_i are arbitrary constants. We can then write:

$$\sigma^2\{\hat{\beta}_1\} = \sigma^2 \sum c_i^2 = \sigma^2 \sum (k_i + d_i)^2 = \sigma^2\left(\sum k_i^2 + \sum d_i^2 + 2 \sum k_i d_i \right)$$

We know that $\sigma^2 \Sigma k_i^2 = \sigma^2\{b_1\}$ from our proof above. Further, $\Sigma k_i d_i = 0$ because of the restrictions on the k_i and c_i above:

$$\sum k_i d_i = \sum k_i(c_i - k_i)$$

$$= \sum c_i k_i - \sum k_i^2$$

$$= \sum c_i \left[\frac{X_i - \bar{X}}{\Sigma(X_i - \bar{X})^2} \right] - \frac{1}{\Sigma(X_i - \bar{X})^2}$$

$$= \frac{\Sigma c_i X_i - \bar{X} \Sigma c_i}{\Sigma(X_i - \bar{X})^2} - \frac{1}{\Sigma(X_i - \bar{X})^2} = 0$$

Hence, we have:

$$\sigma^2\{\hat{\beta}_1\} = \sigma^2\{b_1\} + \sigma^2 \Sigma d_i^2$$

Note that the smallest value of Σd_i^2 is zero. Hence, the variance of $\hat{\beta}_1$ is at a minimum when $\Sigma d_i^2 = 0$. But this can only occur if all $d_i = 0$, which implies $c_i \equiv k_i$. Thus, the least squares estimator b_1 has minimum variance among all unbiased linear estimators.

Sampling Distribution of $(b_1 - \beta_1)/s\{b_1\}$

Since b_1 is normally distributed, we know that the standardized statistic $(b_1 - \beta_1)/\sigma\{b_1\}$ is a standard normal variable. Ordinarily, of course, we need to estimate $\sigma\{b_1\}$ by $s\{b_1\}$, and hence are interested in the distribution of the standardized statistic $(b_1 - \beta_1)/s\{b_1\}$. An important theorem in statistics states:

$$(3.10) \qquad \frac{b_1 - \beta_1}{s\{b_1\}} \text{ is distributed as } t(n - 2) \text{ for regression model (3.1)}$$

Intuitively, this result should not be unexpected. We know that if the observations Y_i come from the same normal population, $(\bar{Y} - \mu)/s\{\bar{Y}\}$ follows the t distribution with $n - 1$ degrees of freedom. The estimator b_1, like $\bar{Y}$, is a linear combination of the observations Y_i. The reason for the difference in the degrees of freedom is that two parameters (β_0 and β_1) need to be estimated for the regression model; hence, two degrees of freedom are lost here.

Note

We can show that $(b_1 - \beta_1)/s\{b_1\}$ is distributed as t with $n - 2$ degrees of freedom by relying on the following theorem:

$$(3.11) \qquad \text{For regression model (3.1), } SSE/\sigma^2 \text{ is distributed as } \chi^2 \text{ with } n - 2 \text{ degrees of freedom, and is independent of } b_0 \text{ and } b_1.$$

First, let us rewrite $(b_1 - \beta_1)/s\{b_1\}$ as follows:

$$\frac{b_1 - \beta_1}{\sigma\{b_1\}} \div \frac{s\{b_1\}}{\sigma\{b_1\}}$$

The numerator is a standard normal variable z. The nature of the denominator can be seen by first considering:

$$\frac{s^2\{b_1\}}{\sigma^2\{b_1\}} = \frac{\dfrac{MSE}{\Sigma(X_i - \bar{X})^2}}{\dfrac{\sigma^2}{\Sigma(X_i - \bar{X})^2}} = \frac{MSE}{\sigma^2} = \frac{\dfrac{SSE}{n - 2}}{\sigma^2}$$

$$= \frac{SSE}{\sigma^2(n-2)} \sim \frac{\chi^2(n-2)}{n-2}$$

where the symbol $\sim$ stands for "is distributed as." The last step follows from (3.11). Hence, we have:

$$\frac{b_1 - \beta_1}{s\{b_1\}} \sim \frac{z}{\sqrt{\dfrac{\chi^2(n-2)}{n-2}}}$$

But by theorem (3.11), z and χ^2 are independent, since z is a function of b_1 and b_1 is independent of $SSE/\sigma^2 \sim \chi^2$. Hence, by definition (1.41), it follows that:

$$\frac{b_1 - \beta_1}{s\{b_1\}} \sim t(n-2)$$

This result places us in a position to readily make inferences concerning β_1.

Confidence Interval for β_1

Since $(b_1 - \beta_1)/s\{b_1\}$ follows a t distribution, we can make the following probability statement:

(3.12) $P\{t(\alpha/2; n-2) \leq (b_1 - \beta_1)/s\{b_1\} \leq t(1 - \alpha/2; n-2)\} = 1 - \alpha$

Here, $t(\alpha/2; n-2)$ denotes the $(\alpha/2)100$ percentile of the t distribution with $n-2$ degrees of freedom. Because of the symmetry of the t distribution, it follows that:

(3.13) $t(\alpha/2; n-2) = -t(1 - \alpha/2; n-2)$

Rearranging the inequalities in (3.12) and using (3.13), we obtain:

(3.14) $P\{b_1 - t(1 - \alpha/2; n-2)s\{b_1\}$
                $\leq \beta_1 \leq b_1 + t(1 - \alpha/2; n-2)s\{b_1\}\} = 1 - \alpha$

Since (3.14) holds for all possible values of β_1, the $1 - \alpha$ confidence limits for β_1 are:

(3.15) $b_1 \pm t(1 - \alpha/2; n-2)s\{b_1\}$

Example.   Let us return to the Westwood Company lot size example of Chapter 2. Management wishes an estimate of β_1 with a 95 percent confidence coefficient. We summarize in Table 3.1 the needed results obtained earlier. First, we need to obtain $s\{b_1\}$:

$$s^2\{b_1\} = \frac{MSE}{\Sigma(X_i - \bar{X})^2} = \frac{7.5}{3,400} = .002206$$

and:

$$s\{b_1\} = .04697$$

TABLE 3.1   Results for Westwood Company Example Obtained in Chapter 2

$$n = 10 \qquad\qquad \bar{X} = 50$$
$$b_0 = 10.0 \qquad\qquad b_1 = 2.0$$
$$\hat{Y} = 10.0 + 2.0\,X \qquad\qquad SSE = 60$$
$$\Sigma X_i^2 = 28{,}400 \qquad\qquad MSE = 7.5$$

$$\Sigma X_i^2 - \frac{(\Sigma X_i)^2}{n} = \Sigma(X_i - \bar{X})^2 = 3{,}400$$

$$\Sigma X_i Y_i - \frac{\Sigma X_i \Sigma Y_i}{n} = \Sigma(X_i - \bar{X})(Y_i - \bar{Y}) = 6{,}800$$

$$\Sigma Y_i^2 - \frac{(\Sigma Y_i)^2}{n} = \Sigma(Y_i - \bar{Y})^2 = 13{,}660$$

For a 95 percent confidence coefficient, we require $t(.975; 8)$. From Table A.2 in Appendix A, we find $t(.975; 8) = 2.306$. The 95 percent confidence interval, by (3.15), then is:

$$2.0 - 2.306(.04697) \le \beta_1 \le 2.0 + 2.306(.04697)$$

$$1.89 \le \beta_1 \le 2.11$$

Thus, with confidence coefficient .95, we estimate that the mean number of man-hours increases by somewhere between 1.89 and 2.11 for each increase of one part in the lot size.

Note

In Chapter 2, we noted that the scope of a regression model is restricted ordinarily to some interval of values of the independent variable. This is particularly important to keep in mind when using estimates of the slope β_1. In our lot size example, a linear regression model appeared appropriate for lot sizes between 20 and 80, the range of the independent variable in the recent past. It may not be reasonable to use the estimate of the slope to infer the effect of lot size on number of man-hours far outside this range since the regression relation may not be linear there.

Tests Concerning β_1

Since $(b_1 - \beta_1)/s\{b_1\}$ is distributed as t with $n - 2$ degrees of freedom, tests concerning β_1 can be set up in ordinary fashion using the t distribution.

Example 1: Two-Sided Test.   A cost analyst in the Westwood Company is interested in testing whether or not there is a linear association between man-hours and lot size, using regression model (3.1). The two alternatives then are:

(3.16)
$$H_0: \beta_1 = 0$$
$$H_a: \beta_1 \ne 0$$

If the analyst wishes to control the risk of a Type I error at .05, he could indeed conclude H_a at once by referring to the 95 percent confidence interval for β_1 constructed earlier, since the interval does not include 0.

An explicit test of the alternatives (3.16) is based on the test statistic:

$$(3.17) \qquad t^* = \frac{b_1}{s\{b_1\}}$$

The decision rule with this test statistic when controlling the level of significance at α is:

$$(3.17a) \qquad \begin{array}{l} \text{If } |t^*| \le t(1 - \alpha/2; n - 2), \text{ conclude } H_0 \\[6pt] \text{If } |t^*| > t(1 - \alpha/2; n - 2), \text{ conclude } H_a \end{array}$$

For the Westwood Company example, where $\alpha = .05$, $b_1 = 2.0$, and $s\{b_1\} = .04697$, we require $t(.975; 8) = 2.306$. Thus, the decision rule for testing alternatives (3.16) is:

$$\text{If } |t^*| \le 2.306, \text{ conclude } H_0$$

$$\text{If } |t^*| > 2.306, \text{ conclude } H_a$$

Since $|t^*| = |2.0/.04697| = 42.58 > 2.306$, we conclude H_a, that $\beta_1 \ne 0$ or that there is a linear association between man-hours and lot size.

The P-value for the sample outcome is obtained by finding the probability $P\{t(8) > t^* = 42.58\}$. We see from Table A.2 that this probability is less than .0005. Indeed, it can be shown to be almost 0, to be denoted by 0+. Thus, the two-sided P-value is $2(0+) = 0+$. Since the two-sided P-value is less than the specified level of significance $\alpha = .05$, we could conclude H_a directly.

Example 2: One-Sided Test.   If the analyst had wished to test whether or not β_1 is positive, controlling the level of significance at $\alpha = .05$, the alternatives would have been:

$$H_0: \beta_1 \le 0$$

$$H_a: \beta_1 > 0$$

and the decision rule based on test statistic (3.17) would have been:

$$\text{If } t^* \le t(1 - \alpha; n - 2), \text{ conclude } H_0$$

$$\text{If } t^* > t(1 - \alpha; n - 2), \text{ conclude } H_a$$

For $\alpha = .05$, we require $t(.95; 8) = 1.860$. Since $t^* = 42.58 > 1.860$, we would conclude H_a, that β_1 is positive.

This same conclusion could be reached directly from the one-sided P-value, which was noted in Example 1 to be 0+. Since this P-value is less than .05, we would conclude H_a.

Comments

1. Many computer packages and scientific publications commonly report the P-value together with the value of the test statistic. In this way, one can conduct a test at any desired level of significance α by comparing the P-value with the specified level α. Users of computer packages need to be careful to ascertain whether one-sided or two-sided P-values are furnished.

2. Occasionally, it is desired to test whether or not β_1 equals some specified nonzero value β_{10}, which may be a historical norm, the value for a comparable process, or an engineering specification. For such a test, the appropriate test statistic is:

$$(3.18) \qquad t^* = \frac{b_1 - \beta_{10}}{s\{b_1\}}$$

The decision rule to be used for the alternatives:

$$H_0: \beta_1 = \beta_{10}$$

$$H_a: \beta_1 \neq \beta_{10}$$

is still (3.17a), but it now is based on t^* defined in (3.18).

Note that test statistic (3.18) simplifies to test statistic (3.17) when the test involves $H_0: \beta_1 = \beta_{10} = 0$.

3.2 INFERENCES CONCERNING β_0

As noted in Chapter 2, there are only infrequent occasions when we wish to make inferences concerning β_0, the intercept of the regression line. These occur when the scope of the model includes $X = 0$.

Sampling Distribution of b_0

The point estimator b_0 was given in (2.10b) as follows:

$$(3.19) \qquad b_0 = \bar{Y} - b_1 \bar{X}$$

The sampling distribution of b_0 refers to the different values of b_0 that would be obtained with repeated sampling when the levels of the independent variable X are held constant from sample to sample.

(3.20)   For regression model (3.1), the sampling distribution of b_0 is normal, with mean and variance:

$$(3.20a) \qquad E\{b_0\} = \beta_0$$

$$(3.20b) \qquad \sigma^2\{b_0\} = \sigma^2 \frac{\Sigma X_i^2}{n \Sigma (X_i - \bar{X})^2} = \sigma^2 \left[\frac{1}{n} + \frac{\bar{X}^2}{\Sigma (X_i - \bar{X})^2} \right]$$

The normality of the sampling distribution of b_0 follows because b_0, like b_1, is a linear combination of the observations Y_i. The results for the mean and variance of the sampling distribution of b_0 can be obtained in similar fashion as those for b_1.

An estimator of $\sigma^2\{b_0\}$ is obtained by replacing σ^2 by its point estimator *MSE*:

$$(3.21) \qquad s^2\{b_0\} = MSE\frac{\Sigma X_i^2}{n\Sigma(X_i - \bar{X})^2} = MSE\left[\frac{1}{n} + \frac{\bar{X}^2}{\Sigma(X_i - \bar{X})^2}\right]$$

The square root, $s\{b_0\}$, is an estimator of $\sigma\{b_0\}$.

Sampling Distribution of $(b_0 - \beta_0)/s\{b_0\}$

Analogous to theorem (3.10) for b_1, there is a theorem for b_0 that states:

$$(3.22) \qquad \frac{b_0 - \beta_0}{s\{b_0\}} \text{ is distributed as } t(n-2) \text{ for regression model (3.1)}$$

Hence, confidence intervals for β_0 and tests concerning β_0 can be set up in ordinary fashion, using the t distribution.

Confidence Interval for β_0

The $1 - \alpha$ confidence limits for β_0 are obtained in the same manner as those for β_1 derived earlier. They are:

$$(3.23) \qquad\qquad b_0 \pm t(1 - \alpha/2; n - 2)s\{b_0\}$$

Example.   As noted earlier, the scope of the model for the Westwood Company example does not extend to lot sizes of $X = 0$. Hence, the regression parameter β_0 may not have intrinsic meaning here. If, nevertheless, a 90 percent confidence interval for β_0 were desired, we would proceed by finding $t(.95; 8)$ and $s\{b_0\}$. From Table A.2, we find $t(.95; 8) = 1.860$. Using the earlier results summarized in Table 3.1, we obtain by (3.21):

$$s^2\{b_0\} = MSE\frac{\Sigma X_i^2}{n\Sigma(X_i - \bar{X})^2} = 7.5\left[\frac{28,400}{10(3,400)}\right] = 6.26471$$

or:

$$s\{b_0\} = 2.50294$$

Hence, the 90 percent confidence interval for β_0 is:

$$10.0 - 1.860(2.50294) \leq \beta_0 \leq 10.0 + 1.860(2.50294)$$

$$5.34 \leq \beta_0 \leq 14.66$$

We caution again that this confidence interval does not necessarily provide meaningful information. For instance, it does not necessarily provide information about the "setup" costs of producing a lot of parts (costs incurred in setting up the production process, no matter what is the lot size), since we are not certain whether a linear regression model is appropriate when the scope of the model is extended to $X = 0$.

3.3 SOME CONSIDERATIONS ON MAKING INFERENCES CONCERNING β_0 AND β_1

Effect of Departures from Normality

If the probability distributions of Y are not exactly normal but do not depart seriously, the sampling distributions of b_0 and b_1 will be approximately normal, and the use of the t distribution will provide approximately the specified confidence coefficient or level of significance. Even if the distributions of Y are far from normal, the estimators b_0 and b_1 generally have the property of *asymptotic normality*—their distributions approach normality under very general conditions as the sample size increases. Thus, with sufficiently large samples, the confidence intervals and decision rules given earlier still apply even if the probability distributions of Y depart far from normality. For large samples, the t value is, of course, replaced by the z value for the standard normal distribution.

Interpretation of Confidence Coefficient and Risks of Errors

Since regression model (3.1) assumes that the X_i are known constants, the confidence coefficient and risks of errors are interpreted with respect to taking repeated samples in which the X observations are kept at the same levels as in the observed sample. For instance, we constructed a confidence interval for β_1 with confidence coefficient .95 in the Westwood Company example. This coefficient is interpreted to mean that if many independent samples are taken where the levels of X (the lot sizes) in the first sample are repeated in these other samples and a 95 percent confidence interval is constructed for each sample, 95 percent of the intervals will contain the true value of β_1.

Spacing of the X Levels

Inspection of formulas (3.3b) and (3.20b) for the variances of b_1 and b_0, respectively, indicates that for given n and σ^2, these variances are affected by the spacing of the X levels in the observed data. For example, the greater is the spread in the X levels, the larger is the quantity $\Sigma(X_i - \bar{X})^2$ and the smaller is the variance of b_1. We will discuss in Section 5.7 how the X observations should be spaced in experiments where spacing can be controlled.

Power of Tests

The power of tests on β_0 and β_1 can be obtained from Table A.5 in Appendix A, which contains charts of the power function of the t test. Consider, for example, the general decision problem:

(3.24)

$$H_0: \beta_1 = \beta_{10}$$

$$H_a: \beta_1 \neq \beta_{10}$$

for which the general test statistic (3.18) is employed:

(3.24a)
$$t^* = \frac{b_1 - \beta_{10}}{s\{b_1\}}$$

and the decision rule for level of significance α is:

(3.24b)
$$\text{If } |t^*| \leq t(1 - \alpha/2; n - 2), \text{ conclude } H_0$$
$$\text{If } |t^*| > t(1 - \alpha/2; n - 2), \text{ conclude } H_a$$

The power of this test is the probability that the decision rule will lead to conclusion H_a when H_a in fact holds. Specifically, the power is given by:

(3.25)
$$\text{Power} = P\{|t^*| > t(1 - \alpha/2; n - 2)|\delta\}$$

where δ is a measure of *noncentrality*—i.e., how far the true value of β_1 is from β_{10}:

(3.26)
$$\delta = \frac{|\beta_1 - \beta_{10}|}{\sigma\{b_1\}}$$

Table A.5 presents the power of the two-sided t test (in percent) for $\alpha = .01$ and $\alpha = .05$, for various degrees of freedom df. To illustrate the use of this table, let us return to the Westwood Company example where we tested:

$$H_0: \beta_1 = \beta_{10} = 0$$

$$H_a: \beta_1 \neq \beta_{10} = 0$$

Suppose we wish to know the power of the test when $\beta_1 = .25$. To ascertain this, we need to know σ^2, the variance of the error terms. Assume that $\sigma^2 = 10.0$ so that $\sigma^2\{b_1\}$ for our example would be:

$$\sigma^2\{b_1\} = \frac{\sigma^2}{\Sigma(X_i - \bar{X})^2} = \frac{10.0}{3,400} = .002941$$

or $\sigma\{b_1\} = .05423$. Then $\delta = |.25 - 0| \div .05423 = 4.6$. We enter the graph for $\alpha = .05$ (the level of significance used in the test) and approximate visually the curve for eight degrees of freedom. Reading the ordinate at $\delta = 4.6$, we obtain 97 percent approximately. Thus, if $\beta_1 = .25$, the probability would be about .97 that we would be led to conclude H_a ($\beta_1 \neq 0$). In other words, if $\beta_1 = .25$, we would be almost certain to conclude that there is a linear relation between man-hours and lot size.

The power of tests concerning β_0 can be obtained from Table A.5 in completely analogous fashion. For one-sided tests, Table A.5 should be entered so that one half the level of significance shown there is the level of significance of the one-sided test.

3.4 INTERVAL ESTIMATION OF $E\{Y_h\}$

In regression analysis, one of the major goals usually is to estimate the mean for one or more probability distributions of Y. Consider, for example, a study of the relation between level of piecework pay (X) and worker productivity (Y). The mean productivity at high and medium levels of piecework pay may be of particular interest for purposes of analyzing the benefits obtained from an increase in the pay. As another example, the Westwood Company may be interested in the mean response (mean number of man-hours) for lot sizes of $X = 40$ parts, $X = 55$ parts, and $X = 70$ parts for purposes of choosing appropriate lot sizes for production.

Let X_h denote the level of X for which we wish to estimate the mean response. X_h may be a value which occurred in the sample, or it may be some other value of the independent variable within the scope of the model. The mean response when $X = X_h$ is denoted by $E\{Y_h\}$. Formula (2.12) gives us the point estimator $\hat{Y}_h$ of $E\{Y_h\}$:

$$(3.27) \qquad \hat{Y}_h = b_0 + b_1 X_h$$

We consider now the sampling distribution of $\hat{Y}_h$.

Sampling Distribution of $\hat{Y}_h$

The sampling distribution of $\hat{Y}_h$, like the earlier sampling distributions discussed, refers to the different values of $\hat{Y}_h$ which would be obtained if repeated samples were selected, each holding the levels of the independent variable X constant, and calculating $\hat{Y}_h$ for each sample.

(3.28) For regression model (3.1), the sampling distribution of $\hat{Y}_h$ is normal, with mean and variance:

$$(3.28a) \qquad E\{\hat{Y}_h\} = E\{Y_h\}$$

$$(3.28b) \qquad \sigma^2\{\hat{Y}_h\} = \sigma^2 \left[\frac{1}{n} + \frac{(X_h - \bar{X})^2}{\Sigma(X_i - \bar{X})^2} \right]$$

Normality. The normality of the sampling distribution of $\hat{Y}_h$ follows directly from the fact that $\hat{Y}_h$ is a linear combination of the observations Y_i.

Mean. To prove that $\hat{Y}_h$ is an unbiased estimator of $E\{Y_h\}$, we proceed as follows:

$$E\{\hat{Y}_h\} = E\{b_0 + b_1 X_h\} = E\{b_0\} + X_h E\{b_1\}$$

$$= \beta_0 + \beta_1 X_h$$

by (3.3a) and (3.20a).

FIGURE 3.2   Effect on $\hat{Y}_h$ of Variation in b_1 from Sample to Sample in Two Samples with Same Means $\bar{Y}$ and $\bar{X}$

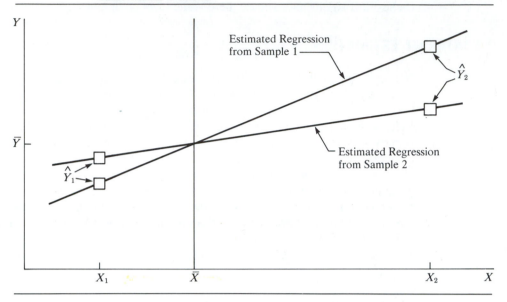

Variance.   Note that the variability of the sampling distribution of $\hat{Y}_h$ is affected by how far X_h is from $\bar{X}$, through the term $(X_h - \bar{X})^2$. The further is X_h from $\bar{X}$, the greater is the quantity $(X_h - \bar{X})^2$ and the larger is the variance of $\hat{Y}_h$. An intuitive explanation of this effect is found in Figure 3.2. Shown there are two sample regression lines, based on two samples for the same set of X values. The two regression lines are assumed to go through the same $(\bar{X}, \bar{Y})$ point to isolate the effect of interest, namely, the effect of variation in the estimated slope b_1 from sample to sample. Note that at X_1, near $\bar{X}$, the fitted values $\hat{Y}_1$ for the two sample regression lines are close to each other. At X_2, which is far from $\bar{X}$, the situation is different. Here, the fitted values $\hat{Y}_2$ differ substantially. Thus, variation in the slope b_1 from sample to sample has a much more pronounced effect on $\hat{Y}_h$ for X levels far from the mean $\bar{X}$ than for X levels near $\bar{X}$. Hence, the variation in the $\hat{Y}_h$ values from sample to sample will be greater when X_h is far from the mean than when X_h is near the mean.

When *MSE* is substituted for σ^2 in (3.28b), we obtain $s^2\{\hat{Y}_h\}$, the estimated variance of $\hat{Y}_h$:

(3.29)
$$s^2\{\hat{Y}_h\} = MSE\left[\frac{1}{n} + \frac{(X_h - \bar{X})^2}{\Sigma(X_i - \bar{X})^2}\right]$$

The estimated standard deviation of $\hat{Y}_h$ then is $s\{\hat{Y}_h\}$, the square root of $s^2\{\hat{Y}_h\}$.

Note

To derive $\sigma^2\{\hat{Y}_h\}$, we first show that b_1 and $\bar{Y}$ are uncorrelated and, hence, for regression model (3.1), independent:

(3.30)
$$\sigma\{\bar{Y}, b_1\} = 0$$

where $\sigma\{\bar{Y}, b_1\}$ denotes the covariance between $\bar{Y}$ and b_1. We begin with the definitions:

$$\bar{Y} = \sum \left(\frac{1}{n}\right) Y_i$$

$$b_1 = \sum k_i Y_i$$

where k_i is as defined in (3.4a). We now use theorem (1.29), with $a_i = 1/n$ and $c_i = k_i$; remember that the Y_i are independent random variables:

$$\sigma\{\bar{Y}, b_1\} = \sum \left(\frac{1}{n}\right) k_i \sigma^2\{Y_i\} = \frac{\sigma^2}{n} \sum k_i$$

But we know from (3.5) that $\Sigma k_i = 0$. Hence, the covariance is 0.

Now we are ready to find the variance of $\hat{Y}_h$. We shall use the estimator in the alternative form (2.15):

$$\sigma^2\{\hat{Y}_h\} = \sigma^2\{\bar{Y} + b_1(X_h - \bar{X})\}$$

Since $\bar{Y}$ and b_1 are independent and X_h and $\bar{X}$ are constants, we obtain:

$$\sigma^2\{\hat{Y}_h\} = \sigma^2\{\bar{Y}\} + (X_h - \bar{X})^2 \sigma^2\{b_1\}$$

Now $\sigma^2\{b_1\}$ is given in (3.3b), and:

$$\sigma^2\{\bar{Y}\} = \frac{\sigma^2\{Y_i\}}{n} = \frac{\sigma^2}{n}$$

Hence:

$$\sigma^2\{\hat{Y}_h\} = \frac{\sigma^2}{n} + (X_h - \bar{X})^2 \frac{\sigma^2}{\Sigma(X_i - \bar{X})^2}$$

which, upon a slight rearrangement of terms, yields (3.28b).

Sampling Distribution of $(\hat{Y}_h - E\{Y_h\})/s\{\hat{Y}_h\}$

Since we have encountered the t distribution in each type of inference for regression model (3.1) up to this point, it should not be surprising that:

(3.31) $\dfrac{\hat{Y}_h - E\{Y_h\}}{s\{\hat{Y}_h\}}$ is distributed as $t(n - 2)$ for regression model (3.1)

Hence, all inferences concerning $E\{Y_h\}$ are carried out in the usual fashion with the t distribution. We illustrate the construction of confidence intervals, since in practice these are more frequently used than tests.

Confidence Interval for $E\{Y_h\}$

A confidence interval for $E\{Y_h\}$ is constructed in the standard fashion, making use of the t distribution as indicated by theorem (3.31). The $1 - \alpha$ confidence limits are:

$$(3.32) \qquad \hat{Y}_h \pm t(1 - \alpha/2; n - 2)s\{\hat{Y}_h\}$$

Example 1.   Returning to the Westwood Company lot size example, let us find a 90 percent confidence interval for $E\{Y_h\}$ when the lot size is $X_h = 55$ parts. Using the earlier results in Table 3.1, we find the point estimate $\hat{Y}_h$:

$$\hat{Y}_h = 10.0 + 2.0(55) = 120$$

Next, we need to find the estimated standard deviation $s\{\hat{Y}_h\}$. We obtain, using (3.29):

$$s^2\{\hat{Y}_h\} = 7.5\left[\frac{1}{10} + \frac{(55 - 50)^2}{3,400}\right] = .80515$$

so that:

$$s\{\hat{Y}_h\} = .89730$$

For a 90 percent confidence coefficient, we require $t(.95; 8) = 1.860$. Hence, our confidence interval with confidence coefficient .90 is by (3.32):

$$120 - 1.860(.89730) \leq E\{Y_h\} \leq 120 + 1.860(.89730)$$

$$118.3 \leq E\{Y_h\} \leq 121.7$$

We conclude with confidence coefficient .90 that the mean number of man-hours required when lots of 55 parts are produced is somewhere between 118.3 and 121.7.

Example 2.   Suppose the Westwood Company wishes to estimate $E\{Y_h\}$ when $X_h = 80$ parts with a 90 percent confidence interval. We require:

$$\hat{Y}_h = 10.0 + 2.0(80) = 170$$

$$s^2\{\hat{Y}_h\} = 7.5\left[\frac{1}{10} + \frac{(80 - 50)^2}{3,400}\right] = 2.73529$$

$$s\{\hat{Y}_h\} = 1.65387$$

$$t(.95; 8) = 1.860$$

Hence, the 90 percent confidence interval is:

$$170 - 1.860(1.65387) \leq E\{Y_h\} \leq 170 + 1.860(1.65387)$$

$$166.9 \leq E\{Y_h\} \leq 173.1$$

Note that this confidence interval is somewhat wider than that for example 1, since the X_h level here ($X_h = 80$) is substantially farther from the mean $\bar{X} = 50$ than the X_h level for example 1 ($X_h = 55$).

Comments

1. Since the X_i are known constants in regression model (3.1), the interpretation of confidence intervals and risks of errors in inferences on the mean response is in terms of taking repeated samples in which the X observations are at the same levels as in the sample actually taken. We noted this same point earlier in connection with inferences on β_0 and β_1.

2. We see from formula (3.28b) that for given sample results, the variance of $\hat{Y}_h$ is smallest when $X_h = \bar{X}$. Thus, in an experiment to estimate the mean response at a particular level X_h of the independent variable, the precision of the estimate will be greatest if (everything else remaining equal) the observations on X are spaced so that $\bar{X} = X_h$.

3. When the sample size is large, the t value in the confidence limits (3.32) may be replaced by the standard normal z value, since the t distribution approaches the standard normal distribution with increasing sample size.

4. The usual relationship between confidence intervals and tests applies in inferences concerning the mean response. Thus, the two-sided confidence limits (3.32) can be utilized for two-sided tests concerning the mean response at X_h. Alternatively, a regular decision rule can be set up.

5. The confidence limits (3.32) for a mean response $E\{Y_h\}$ are not sensitive to moderate departures from the assumption that the error terms are normally distributed. Indeed, the limits are not sensitive to substantial departures from normality if the sample size is large. This robustness in estimating the mean response is related to the robustness of the confidence limits for β_0 and β_1, noted earlier.

6. Confidence limits (3.32) apply when a single mean response is to be estimated from the sample. We discuss in Chapter 5 how to proceed when several mean responses are to be estimated from the same sample.

3.5 PREDICTION OF NEW OBSERVATION

We consider now the prediction of a new observation Y corresponding to a given level X of the independent variable. In the Westwood Company illustration, for instance, the next lot to be produced consists of 55 parts and management wishes to predict the number of man-hours for this particular lot. As another example, an economist has estimated the regression relation between company sales and number of persons 16 or more years old, based on data for the past 10 years. Given a reliable demographic projection of the number of persons 16 or more years old for next year, the economist wishes to predict next year's company sales.

The new observation on Y is viewed as the result of a new trial, independent of the trials on which the regression analysis is based. We shall denote the level of X for the new trial as X_h and the new observation on Y as $Y_{h(new)}$. Of course, we assume that the underlying regression model applicable for the basic sample data continues to be appropriate for the new observation.

The distinction between estimation of the mean response $E\{Y_h\}$, discussed in the preceding section, and prediction of a new response $Y_{h(new)}$, discussed now, is basic. In the former case, we estimate the *mean* of the distribution of Y. In the present case, we predict an *individual outcome* drawn from the distribution of Y. Of course, the great majority of individual outcomes deviate from the mean response, and this must be allowed for in the procedure for predicting $Y_{h(new)}$.

Prediction Interval when Parameters Known

To illustrate the nature of a *prediction interval* for a new observation $Y_{h(new)}$ in as simple a fashion as possible, we shall first assume that all regression parameters are known. Later we shall drop this assumption and make appropriate modifications.

Suppose that the Westwood Company plans to produce a lot of $X_h = 40$ parts in a few weeks and that the relevant parameters of the regression model are known to be:

$$\beta_0 = 9.5 \qquad \beta_1 = 2.1$$
$$E\{Y\} = 9.5 + 2.1X$$
$$\sigma^2 = 10.0$$

Thus, for $X_h = 40$ parts, we have:

$$E\{Y_h\} = 9.5 + 2.1(40) = 93.5$$

Figure 3.3 shows the probability distribution of Y for $X_h = 40$ parts. Its mean is $E\{Y_h\} = 93.5$, and its standard deviation is $\sigma = \sqrt{10.0} = 3.162$. Further, the distribution is normal in accord with regression model (3.1).

Suppose we were to predict that the number of man-hours for the next lot of $X_h = 40$ parts will be between:

$$E\{Y_h\} \pm 3\sigma$$
$$93.5 \pm 3(3.162)$$

FIGURE 3.3   Prediction of $Y_{h(new)}$ when Parameters Known

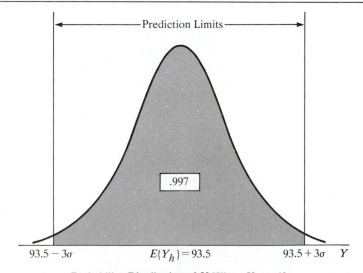

Probability Distribution of Y When $X_h = 40$

so that the prediction interval would be:

$$84.0 \leq Y_{h(new)} \leq 103.0$$

Since 99.7 percent of the area in a normal probability distribution falls within three standard deviations from the mean, the probability is .997 that this prediction interval will give a correct prediction for the next production run of 40 parts.

The basic idea of a prediction interval is thus to choose a range in the distribution of Y wherein most of the observations will fall, and to declare that the next observation will fall in this range. The usefulness of the prediction interval depends, as always, on the width of the interval and the needs for precision by the user.

In general, when the regression parameters are known, the $1 - \alpha$ prediction limits for $Y_{h(new)}$ are:

$$(3.33) \qquad E\{Y_h\} \pm z(1 - \alpha/2)\sigma$$

In centering the limits around $E\{Y_h\}$, we obtain the narrowest interval consistent with the specified probability of a correct prediction.

Prediction Interval for $Y_{h(new)}$ when Parameters Unknown

When the regression parameters are unknown, they must be estimated. The mean of the distribution of Y is estimated by $\hat{Y}_h$, as usual, and the variance of the distribution of Y is estimated by MSE. We cannot, however, simply use the prediction limits (3.33) with the parameters replaced by the corresponding point estimators. The reason is illustrated intuitively in Figure 3.4. Shown there are two probability distributions of Y, corresponding to the upper and lower limits of a confidence interval for $E\{Y_h\}$. In other words, the distribution of Y could be located as far left as the one shown, as far right as the other one shown, or anywhere in between. Since we do not know the mean $E\{Y_h\}$ and only estimate it by a confidence interval, we cannot be certain of the location of the distribution of Y.

Figure 3.4 also shows the prediction limits for each of the two probability distributions of Y presented there. Since we cannot be certain of the location of the distribution of Y, prediction limits for $Y_{h(new)}$ clearly must take account of two elements, as shown in Figure 3.4:

1. Variation in possible location of the distribution of Y.
2. Variation within the probability distribution of Y.

Prediction limits for a new observation Y at a given level X_h are obtained by means of the following theorem:

$$(3.34) \qquad \frac{\hat{Y}_h - Y}{s\{Y_{h(new)}\}} \text{ is distributed as } t(n - 2) \text{ for regression model (3.1)}$$

Note that the standardized statistic (3.34) uses the point estimator $\hat{Y}_h$ in the numerator rather than the true mean $E\{Y_h\}$ because the true mean is unknown and cannot be used in making a prediction. The estimated standard deviation $s\{Y_{h(new)}\}$ in the denominator of the standardized statistic will be defined shortly.

FIGURE 3.4   Prediction of $Y_{h(\text{new})}$ when Parameters Unknown

---

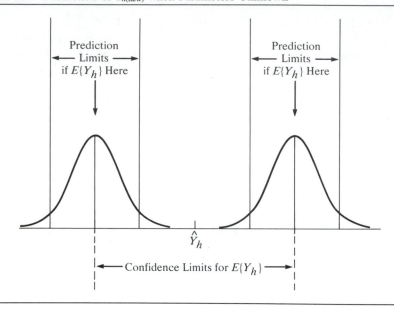

Prediction
← Limits →
if $E\{Y_h\}$ Here

Prediction
← Limits →
if $E\{Y_h\}$ Here

$\hat{Y}_h$

←——Confidence Limits for $E\{Y_h\}$——→

From theorem (3.34), it follows in the usual fashion that the $1 - \alpha$ prediction limits for a new observation are [for instance, compare (3.34) to (3.10) and relate $\hat{Y}_h$ to b_1 and Y to β_1]:

(3.35)
$$\hat{Y}_h \pm t(1 - \alpha/2; n - 2)s\{Y_{h(\text{new})}\}$$

The variance of the numerator of the standardized statistic (3.34) is readily obtained, utilizing the independence of the new observation Y and the original sample observations on which $\hat{Y}_h$ is based. We shall denote the variance of the numerator by $\sigma^2\{Y_{h(\text{new})}\}$, and we obtain by (1.28b):

(3.36)
$$\sigma^2\{Y_{h(\text{new})}\} = \sigma^2\{\hat{Y}_h - Y\} = \sigma^2\{\hat{Y}_h\} + \sigma^2\{Y\} = \sigma^2\{\hat{Y}_h\} + \sigma^2$$

Note that $\sigma^2\{Y_{h(\text{new})}\}$ has two components:

1. The variance of the sampling distribution of $\hat{Y}_h$.
2. The variance of the distribution of Y at $X = X_h$.

An unbiased estimator of $\sigma^2\{Y_{h(\text{new})}\}$ is:

(3.37)
$$s^2\{Y_{h(\text{new})}\} = s^2\{\hat{Y}_h\} + MSE$$

which can be expressed, using (3.29), as follows:

(3.37a)
$$s^2\{Y_{h(\text{new})}\} = MSE\left[1 + \frac{1}{n} + \frac{(X_h - \bar{X})^2}{\Sigma(X_i - \bar{X})^2}\right]$$

Example.   The Westwood Company wishes to predict the number of man-hours required in the forthcoming production run of size 55 with a 90 percent prediction interval; the parameter values are unknown. We require $t(.95; 8) = 1.860$. From earlier work, we have:

$$\hat{Y}_h = b_0 + b_1 X_h$$

$$\hat{Y}_h = 120 \qquad s^2\{\hat{Y}_h\} = .80515$$

$$\frac{E(Y_i - \hat{Y}_i)^2}{n-2} \qquad \leftarrow MSE = 7.5 \qquad 3.29$$

Using (3.37), we obtain:

$$\leftarrow s^2\{Y_{h(new)}\} = .80515 + 7.5 = 8.30515$$

so that:

$$s\{Y_{h(new)}\} = 2.88187$$

Hence, the 90 percent prediction interval for $Y_{h(new)}$ is by (3.35):

$$120 - 1.860(2.88187) \le Y_{h(new)} \le 120 + 1.860(2.88187)$$

$$114.6 \le Y_{h(new)} \le 125.4$$

With confidence coefficient .90, we predict that the number of man-hours for the next production run of 55 parts will be somewhere between 114.6 and 125.4.

Comments

1. The 90 percent prediction interval for $Y_{h(new)}$ just obtained is wider than the 90 percent confidence interval for $E\{Y_h\}$ obtained in Example 1 on page 78. The reason is that when predicting a new observation, we encounter both the variability in $\hat{Y}_h$ from sample to sample as well as the variation within the probability distribution of Y.

2. Formula (3.37a) indicates that the prediction interval is wider the further X_h is from $\bar{X}$. The reason for this is that the estimate of the mean $\hat{Y}_h$, as noted earlier, is less precise as X_h is located further away from $\bar{X}$.

3. The prediction limits (3.35), unlike the confidence limits (3.32) for a mean response $E\{Y_h\}$, are sensitive to departures from normality of the error terms distribution. In Chapter 4, we discuss diagnostic procedures for examining the nature of the probability distribution of the error terms, and we describe remedial measures if the departure from normality is serious.

4. The confidence coefficient for the prediction limits (3.35) refers to the taking of repeated samples based on the same set of X values, and calculating prediction limits for $Y_{h(new)}$ for each sample.

5. Prediction limits lend themselves to statistical control uses. In the Westwood Company example, suppose that the new production run of 55 parts, for which the prediction limits were 114.6 and 125.4 hours, actually required 135 hours. Management here would have an indication that a change in the production process may have occurred, and may wish to initiate a search for the assignable cause.

6. When the sample size is large, the last two terms inside the brackets in (3.37a) are small compared to 1, the first term in the brackets. Also, of course, the t distribution is then approximately normal. Hence, approximate $1 - \alpha$ prediction limits for $Y_{h(new)}$ when n is large are:

(3.38)
$$\hat{Y}_h \pm z(1 - \alpha/2)\sqrt{MSE}$$

7. Prediction limits (3.35) apply for a single prediction based on the sample data. Next, we discuss how to predict the mean of several new observations at a given X_h; and in Chapter 5, we take up how to make several predictions at different X_h levels.

8. Prediction intervals resemble confidence intervals. However, they differ conceptually. A confidence interval represents an inference on a parameter, and is an interval which is intended to cover the value of the parameter. A prediction interval, on the other hand, is a statement about the value to be taken by a random variable.

Prediction of Mean of *m* New Observations for Given X_h

Occasionally, one would like to predict the mean of m new observations on Y for a given level of the independent variable. Suppose the Westwood Company has been asked to bid on a contract that calls for $m = 3$ independent production runs of $X_h = 55$ parts during the next few months. Management would like to predict the mean man-hours per run for these three runs, and then convert this into a prediction of the total man-hours required to fill the contract.

We shall denote the mean value of Y to be predicted as $\bar{Y}_{h(new)}$. It can be shown that the appropriate $1 - \alpha$ prediction limits are:

$$(3.39) \qquad \hat{Y}_h \pm t(1 - \alpha/2; n - 2)s\{\bar{Y}_{h(new)}\}$$

where:

$$(3.39a) \qquad s^2\{\bar{Y}_{h(new)}\} = s^2\{\hat{Y}_h\} + \frac{MSE}{m}$$

or equivalently:

$$(3.39b) \qquad s^2\{\bar{Y}_{h(new)}\} = MSE\left[\frac{1}{m} + \frac{1}{n} + \frac{(X_h - \bar{X})^2}{\Sigma(X_i - \bar{X})^2}\right]$$

Note from (3.39a) that the variance $s^2\{\bar{Y}_{h(new)}\}$ has two components:

1. The variance of the sampling distribution of $\hat{Y}_h$.
2. The variance of the mean of m observations from the probability distribution of Y at $X = X_h$.

Example.   In the Westwood Company example, let us find the 90 percent prediction interval for the mean number of man-hours $\bar{Y}_{h(new)}$ in three new production runs, each for $X_h = 55$ parts. From previous work, we have:

$$\hat{Y}_h = 120 \qquad s^2\{\hat{Y}_h\} = .80515$$
$$MSE = 7.5 \qquad t(.95; 8) = 1.860$$

Hence, we obtain:

$$s^2\{\bar{Y}_{h(new)}\} = .80515 + \frac{7.5}{3} = 3.30515$$

or:

$$s\{\bar{Y}_{h(new)}\} = 1.81801$$

The prediction interval for the mean man-hours per run then is:

$$120 - 1.860(1.81801) \leq \bar{Y}_{h(\text{new})} \leq 120 + 1.860(1.81801)$$

$$116.6 \leq \bar{Y}_{h(\text{new})} \leq 123.4$$

Note that these prediction limits are somewhat narrower than those for predicting the man-hours for a single lot of 55 parts because they involve a prediction of the mean man-hours for three lots.

We obtain the prediction interval for the total number of man-hours in the three production runs by multiplying the prediction limits for $\bar{Y}_{h(\text{new})}$ by three:

$$349.8 = 3(116.6) \leq \text{Total man-hours} \leq 3(123.4) = 370.2$$

Thus, it can be predicted with 90 percent confidence that between 350 and 370 man-hours will be needed to fill the contract for three lots of 55 parts each.

3.6 CONSIDERATIONS IN APPLYING REGRESSION ANALYSIS

We have now discussed the major uses of regression analysis—to make inferences about the regression parameters, to estimate the mean response for a given X, and to predict a new observation Y for a given X. It remains to make a few cautionary remarks about implementing applications of regression analysis.

1. Frequently, regression analysis is used to make inferences for the future. For instance, the Westwood Company may wish to estimate expected man-hours for given lot sizes for purposes of planning future production. In applications of this type, it is important to remember that the validity of the regression application depends upon whether basic causal conditions in the period ahead will be similar to those in existence during the period upon which the regression analysis is based. This caution applies whether mean responses are to be estimated, new observations predicted, or regression parameters estimated.

2. In predicting new observations on Y, the independent variable X itself often has to be predicted. For instance, we mentioned earlier the prediction of company sales for next year from a demographic projection of the number of persons 16 years of age or older next year. A prediction of company sales under these circumstances is a conditional prediction, dependent upon the correctness of the population projection. It is easy to forget the conditional nature of this type of prediction. (below we must not)

3. Another caution deals with inferences pertaining to levels of the independent variable that fall outside the range of observations. Unfortunately, this situation frequently occurs in practice. A company that predicts its sales from a regression relation of company sales to disposable personal income will often find the level of disposable personal income of interest (e.g., for the year ahead) to fall beyond the range of past data. If the X level does not fall far beyond this range, one may have reasonable confidence in the application of the regression analysis. On the other hand, if the X level falls far beyond the range of past data, extreme caution should be exercised since one cannot be sure that the regression function which fits the past data is appropriate over the wider range of the independent variable.

4. A statistical test that leads to the conclusion that $\beta_1 \neq 0$ does not establish a cause-and-effect relation between the independent and dependent variables. For instance, with nonexperimental data, both the X and Y variables may be simultaneously influenced by other variables not included in the regression model. Thus, data on grade school children's vocabulary (X) and writing speed (Y) may show a clear linear association, but this could be largely the result of a child's age, amount of education, and similar factors that affect both X and Y. On the other hand, the existence of a regression relation in controlled experiments is often good evidence of a cause-and-effect relation.

5. Finally, we should note again that special problems arise when one wishes to estimate the mean response or predict a new observation for a number of different levels of the independent variable, as is frequently the case. The confidence coefficients for the limits (3.32) for estimating a mean response and for the prediction limits (3.35) for a new observation apply for a single level of X for a given sample. In Chapter 5, we discuss how to make multiple inferences from a given sample.

3.7 CASE WHEN *X* IS RANDOM

The normal error regression model (3.1), which has been used throughout this chapter and will continue to be used, assumes that the X values are known constants. As a consequence of this, the confidence coefficients and risks of errors refer to repeated sampling when the X values are kept the same from sample to sample.

Frequently, it may not be appropriate to consider the X values as known constants. For instance, consider regressing daily bathing suit sales by a department store on mean daily temperature. Surely, the department store cannot control daily temperatures, so it would not be meaningful to think of repeated sampling where the temperature levels are the same from sample to sample.

In this type of situation, it may be preferable to consider both Y and X as random variables. Does this mean then that all of our earlier results are not applicable here? Not at all. It can be shown that all results on estimation, testing, and prediction obtained for regression model (3.1) still apply if the following conditions hold:

(3.40)   1. The conditional distributions of the Y_i, given X_i, are normal and independent, with conditional means $\beta_0 + \beta_1 X_i$ and conditional variance σ^2.

   2. The X_i are independent random variables, whose probability distribution $g(X_i)$ does not involve the parameters β_0, β_1, σ^2.

These conditions require only that regression model (3.1) is appropriate for each *conditional* distribution of Y_i, and that the probability distribution of X_i does not involve the regression parameters. If these conditions are met, all earlier results on estimation, testing, and prediction still hold even though the X_i are now random variables. The major modification occurs in the interpretation of confidence coefficients and specified risks of error. When X is random, these refer to repeated sampling of pairs of (X_i, Y_i) values where the X_i values as well as the Y_i values change from sample to sample. Thus, in our bathing suit sales illustration, a confidence coefficient would refer to the proportion of correct interval estimates if repeated samples of n days' sales and temperatures were obtained and the confidence interval calculated for each sample. Another modification occurs in the power of tests, which is different when X is a random variable.

3.8 ANALYSIS OF VARIANCE APPROACH TO REGRESSION ANALYSIS

We now have developed the basic regression model and demonstrated its major uses. At this point, we will view the relationship in a regression model from the perspective of analysis of variance. This new perspective will not enable us to do anything new with the basic regression model, but the analysis of variance approach will come into its own when we take up more complex regression models and additional types of linear statistical models.

Partitioning of Total Sum of Squares

Basic Notions. The analysis of variance approach is based on the partitioning of sums of squares and degrees of freedom associated with the response variable Y. To explain the motivation of this approach, consider again the Westwood Company lot size example. Figure 3.5a shows the man-hours required for the 10 production runs presented earlier in Table 2.1. There is variation in the number of man-hours, as in all statistical data. Indeed, if all observations Y_i were identically equal, in which case $Y_i \equiv \bar{Y}$, there would be no statistical problems. The variation of the Y_i is conventionally measured in terms of the deviations:

$$(3.41) \qquad\qquad Y_i - \bar{Y}$$

These deviations are shown in Figure 3.5a, and one is labeled explicitly. The measure of total variation, denoted by *SSTO*, is the sum of the squared deviations (3.41):

$$(3.42) \qquad\qquad SSTO = \sum (Y_i - \bar{Y})^2$$

Here *SSTO* stands for *total sum of squares*. If *SSTO* = 0, all observations are the same. The greater is *SSTO*, the greater is the variation among the Y observations.

When we utilize the regression approach, the variation reflecting the uncertainty in the data is that of the Y observations around the fitted regression line:

$$(3.43) \qquad\qquad Y_i - \hat{Y}_i$$

These deviations are shown in Figure 3.5b. The measure of variation in the data with the regression model is the sum of the squared deviations (3.43), which is the familiar *SSE* of (2.21):

$$(3.44) \qquad\qquad SSE = \sum (Y_i - \hat{Y}_i)^2$$

FIGURE 3.5   Partitioning of Total Deviations $Y_i - \bar{Y}$ (Y values not plotted to scale)

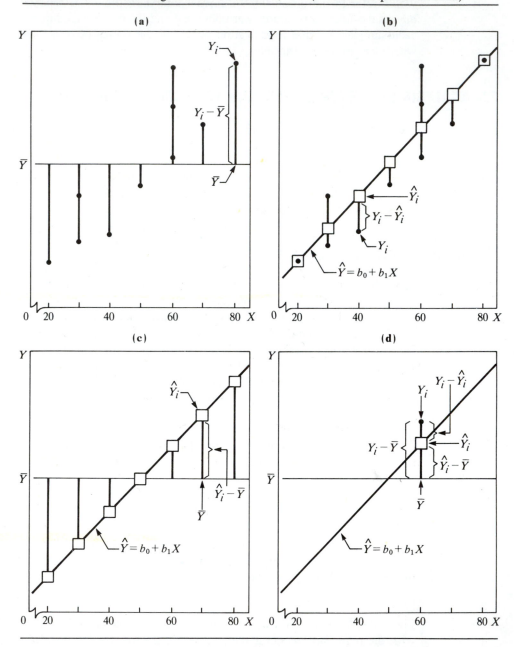

Again, *SSE* denotes *error sum of squares*. If $SSE = 0$, all observations fall on the fitted regression line. The larger is *SSE*, the greater is the variation of the Y observations around the fitted regression line.

For the Westwood Company example, we know from earlier work (Table 3.1) that:

$$SSTO = 13,660$$

$$SSE = 60$$

What accounts for the substantial difference between these two sums of squares? The difference, as we shall show shortly, is another sum of squares:

$$(3.45) \qquad SSR = \sum (\hat{Y}_i - \bar{Y})^2$$

where *SSR* stands for *regression sum of squares*. Note that *SSR* is a sum of squared deviations, the deviations being:

$$(3.46) \qquad \hat{Y}_i - \bar{Y}$$

These deviations are shown in Figure 3.5c. Each deviation is simply the difference between the fitted value on the regression line and the mean of the fitted values $\bar{Y}$. [Recall from (2.18) that the mean of the fitted values $\hat{Y}_i$ is $\bar{Y}$.] If the regression line is horizontal so that $\hat{Y}_i - \bar{Y} \equiv 0$, $SSR = 0$. Otherwise, *SSR* is positive.

SSR may be considered a measure of the variability of the Y_i that is associated with the regression line. The larger is *SSR* in relation to *SSTO*, the greater is the effect of the regression relation in accounting for the total variation in the Y observations.

For the Westwood Company lot size example, we have:

$$SSR = SSTO - SSE = 13,660 - 60 = 13,600$$

which indicates that most of the total variability in man-hours is accounted for by the relation between lot size and man-hours.

Formal Development of Partitioning.   Consider the total deviation $Y_i - \bar{Y}$, the basic quantity measuring the total variation of the observations Y_i. We can decompose this deviation as follows:

$$(3.47) \qquad \underbrace{Y_i - \bar{Y}}_{\substack{\text{Total} \\ \text{deviation}}} = \underbrace{\hat{Y}_i - \bar{Y}}_{\substack{\text{Deviation} \\ \text{of fitted} \\ \text{regression} \\ \text{value} \\ \text{around mean}}} + \underbrace{Y_i - \hat{Y}_i}_{\substack{\text{Deviation} \\ \text{around} \\ \text{fitted} \\ \text{regression} \\ \text{line}}}$$

Thus, the total deviation $Y_i - \bar{Y}$ can be viewed as the sum of two components:

1. The deviation of the fitted value $\hat{Y}_i$ around the mean $\bar{Y}$.
2. The deviation of Y_i around the fitted regression line.

Figure 3.5d shows this decomposition for one of the observations.

It is a remarkable property that the sums of these squared deviations have the same relationship:

$$(3.48) \qquad \sum (Y_i - \bar{Y})^2 = \sum (\hat{Y}_i - \bar{Y})^2 + \sum (Y_i - \hat{Y}_i)^2$$

or, using the notation in (3.42), (3.44), and (3.45):

$$(3.48a) \qquad SSTO = SSR + SSE$$

To prove this basic result in the analysis of variance, we proceed as follows:

$$\sum (Y_i - \bar{Y})^2 = \sum [(\hat{Y}_i - \bar{Y}) + (Y_i - \hat{Y}_i)]^2$$

$$= \sum [(\hat{Y}_i - \bar{Y})^2 + (Y_i - \hat{Y}_i)^2 + 2(\hat{Y}_i - \bar{Y})(Y_i - \hat{Y}_i)]$$

$$= \sum (\hat{Y}_i - \bar{Y})^2 + \sum (Y_i - \hat{Y}_i)^2 + 2 \sum (\hat{Y}_i - \bar{Y})(Y_i - \hat{Y}_i)$$

The last term on the right equals zero, as we can see by expanding it:

$$2 \sum (\hat{Y}_i - \bar{Y})(Y_i - \hat{Y}_i) = 2 \sum \hat{Y}_i(Y_i - \hat{Y}_i) - 2\bar{Y} \sum (Y_i - \hat{Y}_i)$$

The first summation on the right equals zero by (2.20), and the second equals zero by (2.17). Hence, (3.48) follows.

Computational Formulas.   The definitional formulas for *SSTO, SSR,* and *SSE* presented above are often not convenient for hand computation. Useful computational formulas for *SSTO* and *SSR,* which are algebraically equivalent to the definitional formulas, are:

$$(3.49) \qquad SSTO = \sum Y_i^2 - \frac{(\sum Y_i)^2}{n} = \sum Y_i^2 - n\bar{Y}^2$$

$$(3.50a) \qquad SSR = b_1 \left(\sum X_i Y_i - \frac{\sum X_i \sum Y_i}{n} \right)$$

$$= b_1 \left[\sum (X_i - \bar{X})(Y_i - \bar{Y}) \right]$$

or:

$$(3.50b) \qquad SSR = b_1^2 \sum (X_i - \bar{X})^2$$

Computational formulas for *SSE* were given earlier in (2.24).

Using the results for the Westwood Company example summarized in Table 3.1, we obtain for *SSR* by (3.50a):

$$SSR = 2.0(6,800) = 13,600$$

This, of course, is the same result obtained previously by taking the difference $SSTO - SSE$, except sometimes for a slight difference due to rounding effects.

Breakdown of Degrees of Freedom

Corresponding to the partitioning of the total sum of squares $SSTO$, there is a partitioning of the associated degrees of freedom (abbreviated df). We have $n - 1$ degrees of freedom associated with $SSTO$. One degree of freedom is lost because the deviations $Y_i - \bar{Y}$ are not independent in that they must sum to zero. Equivalently, one degree of freedom is lost because the sample mean $\bar{Y}$ is used to estimate the population mean.

SSE, as noted earlier, has $n - 2$ degrees of freedom associated with it. Two degrees of freedom are lost because the two parameters β_0 and β_1 were estimated in obtaining the fitted values $\hat{Y}_i$.

SSR has one degree of freedom associated with it. There are two parameters in the regression equation, but the deviations $\hat{Y}_i - \bar{Y}$ are not independent because they must sum to zero; hence, one degree of freedom is lost from the possible degrees of freedom.

Note that the degrees of freedom are additive:

$$n - 1 = 1 + (n - 2)$$

For the Westwood Company example, these degrees of freedom are:

$$9 = 1 + 8$$

Mean Squares

A sum of squares divided by its associated degrees of freedom is called a *mean square* (abbreviated *MS*). For instance, an ordinary sample variance is a mean square since a sum of squares, $\Sigma(Y_i - \bar{Y})^2$, is divided by its associated degrees of freedom, $n - 1$. We are interested here in the *regression mean square*, denoted by *MSR*:

(3.51)
$$MSR = \frac{SSR}{1} = SSR$$

and in the *error mean square, MSE*, defined earlier in (2.22):

(3.52)
$$MSE = \frac{SSE}{n - 2}$$

For the Westwood Company example, we have $SSR = 13{,}600$ and $SSE = 60$. Hence:

$$MSR = \frac{13{,}600}{1} = 13{,}600$$

Also, we obtained earlier:

$$MSE = \frac{60}{8} = 7.5$$

Note

The two mean squares MSR and MSE do not add to $SSTO \div (n - 1) = 13,660 \div 9 = 1,518$. Thus, mean squares are not additive.

Analysis of Variance Table

Basic Table. The breakdowns of the total sum of squares and associated degrees of freedom are displayed in the form of an analysis of variance table (ANOVA table) in Table 3.2. Mean squares of interest also are shown. In addition, there is a column of expected mean squares which will be utilized below. The ANOVA table for the Westwood Company example is shown in Table 3.3.

Modified Table. Sometimes, an ANOVA table showing one additional element of decomposition is utilized. Recall that by (3.49):

$$SSTO = \sum (Y_i - \bar{Y})^2 = \sum Y_i^2 - n\bar{Y}^2$$

In the modified ANOVA table, the *total uncorrected sum of squares*, denoted by *SSTOU*, is defined as:

(3.53) $$SSTOU = \sum Y_i^2$$

and the *correction for the mean sum of squares*, denoted by SS(correction for mean), is defined as:

(3.54) $$SS(\text{correction for mean}) = n\bar{Y}^2$$

TABLE 3.2 ANOVA Table for Simple Linear Regression

Source of Variation	SS	df	MS	E{MS}
Regression	$SSR = \sum (\hat{Y}_i - \bar{Y})^2$	1	$MSR = \dfrac{SSR}{1}$	$\sigma^2 + \beta_1^2 \sum (X_i - \bar{X})^2$
Error	$SSE = \sum (Y_i - \hat{Y}_i)^2$	$n - 2$	$MSE = \dfrac{SSE}{n - 2}$	σ^2
Total	$SSTO = \sum (Y_i - \bar{Y})^2$	$n - 1$		

TABLE 3.3   ANOVA Table for Westwood Company Example

Source of Variation	SS	df	MS
Regression	13,600	1	13,600
Error	60	8	7.5
Total	13,660	9	

Table 3.4 shows this modified ANOVA table. The general format is presented in part (a) and the Westwood Company results in part (b). Both types of ANOVA tables are widely used. Ordinarily, we shall utilize the basic type of table.

TABLE 3.4   Modified ANOVA Table for Simple Linear Regression and Results for Westwood Company Example

(a) General

Source of Variation	SS	df	MS
Regression	$SSR = \sum (\hat{Y}_i - \bar{Y})^2$	1	$MSR = \dfrac{SSR}{1}$
Error	$SSE = \sum (Y_i - \hat{Y}_i)^2$	$n - 2$	$MSE = \dfrac{SSE}{n - 2}$
Total	$SSTO = \sum (Y_i - \bar{Y})^2$	$n - 1$	
Correction for mean	SS (correction for mean) $= n\bar{Y}^2$	1	
Total, uncorrected	$SSTOU = \sum Y_i^2$	n	

(b) Westwood Company Example

Source of Variation	SS	df	MS
Regression	13,600	1	13,600
Error	60	8	7.5
Total	13,660	9	
Correction for mean	121,000	1	
Total, uncorrected	134,660	10	

Expected Mean Squares

In order to be able to make inferences based on the analysis of variance approach, we need to know the expected value of each of the mean squares. The expected value of a mean square tells us what is being estimated by the mean square. Statistical theory provides the following results:

$$(3.55) \qquad E\{MSE\} = \sigma^2$$

$$(3.56) \qquad E\{MSR\} = \sigma^2 + \beta_1^2 \sum (X_i - \bar{X})^2$$

The expected mean squares in (3.55) and (3.56) are shown in the analysis of variance table in Table 3.2. Note that result (3.55) is in accord with our earlier statement that MSE is an unbiased estimator of σ^2.

Two important implications of the expected mean squares in (3.55) and (3.56) are the following:

1. The expectation of MSE is σ^2 whether or not X and Y are linearly related, i.e., whether or not $\beta_1 = 0$.

2. The expectation of MSR is also σ^2 when $\beta_1 = 0$. On the other hand, when $\beta_1 \neq 0$, $E\{MSR\}$ is greater than σ^2 since the term $\beta_1^2 \Sigma (X_i - \bar{X})^2$ in (3.56) then must be positive. Thus, for testing whether or not $\beta_1 = 0$, a comparison of MSR and MSE suggests itself. If MSR and MSE are of the same order of magnitude, this would suggest that $\beta_1 = 0$. On the other hand, if MSR is substantially greater than MSE, this would suggest that $\beta_1 \neq 0$. This indeed is the basic idea underlying the analysis of variance test to be discussed next.

Note

The derivation of (3.55) follows from theorem (3.11), which states that $SSE/\sigma^2 \sim \chi^2(n - 2)$ for regression model (3.1). Hence, it follows from property (1.39) of the chi-square distribution that:

$$E\left\{\frac{SSE}{\sigma^2}\right\} = n - 2$$

or that:

$$E\left\{\frac{SSE}{n - 2}\right\} = E\{MSE\} = \sigma^2$$

To find the expected value of MSR, we begin with (3.50b):

$$SSR = b_1^2 \sum (X_i - \bar{X})^2$$

Now by (1.15a), we have:

$$(3.57) \qquad \sigma^2\{b_1\} = E\{b_1^2\} - (E\{b_1\})^2$$

We know from (3.3a) that $E\{b_1\} = \beta_1$, and from (3.3b) that:

$$\sigma^2\{b_1\} = \frac{\sigma^2}{\Sigma(X_i - \bar{X})^2}$$

Hence, substituting into (3.57), we obtain:

$$E\{b_1^2\} = \frac{\sigma^2}{\Sigma(X_i - \bar{X})^2} + \beta_1^2$$

It now follows that:

$$E\{SSR\} = E\{b_1^2\} \sum (X_i - \bar{X})^2 = \sigma^2 + \beta_1^2 \sum (X_i - \bar{X})^2$$

Finally, $E\{MSR\}$ is:

$$E\{MSR\} = E\left\{\frac{SSR}{1}\right\} = \sigma^2 + \beta_1^2 \sum (X_i - \bar{X})^2$$

F Test of $\beta_1 = 0$ versus $\beta_1 \neq 0$

The general analysis of variance approach provides us with a battery of highly useful tests for regression models (and other linear statistical models). For the simple linear regression case considered here, the analysis of variance provides us with a test for:

(3.58)
$$H_0: \beta_1 = 0$$
$$H_a: \beta_1 \neq 0$$

Test Statistic.   The test statistic for the analysis of variance approach is denoted by F^*. As just mentioned, it compares MSR and MSE in the following fashion:

(3.59)
$$F^* = \frac{MSR}{MSE}$$

The earlier motivation, based on the expected mean squares in Table 3.2, suggests that large values of F^* support H_a and values of F^* near 1 support H_0. In other words, the appropriate test is an upper-tail one.

Distribution of F^*.   In order to be able to construct a statistical decision rule and examine its properties, we need to know the sampling distribution of F^*. We begin by considering the sampling distribution of F^* when H_0 ($\beta_1 = 0$) holds. *Cochran's theorem* will be most helpful in this connection. For our purposes, this theorem can be put as follows:

(3.60)   If all n observations Y_i come from the same normal distribution with mean μ and variance σ^2, and $SSTO$ is decomposed into k sums of squares SS_r, each with degrees of freedom df_r, then the SS_r/σ^2 terms are independent χ^2 variables with df_r degrees of freedom if:

$$\sum_{r=1}^{k} df_r = n - 1$$

Note from Table 3.2 that we have decomposed $SSTO$ into the two sums of squares SSR and SSE, and that their degrees of freedom are additive. Hence:

If $\beta_1 = 0$ so that all Y_i have the same mean $\mu = \beta_0$ and the same variance σ^2, SSE/σ^2 and SSR/σ^2 are independent χ^2 variables.

Now consider the test statistic F^*, which we can write as follows:

$$F^* = \frac{\dfrac{SSR}{\sigma^2}}{1} \div \frac{\dfrac{SSE}{\sigma^2}}{n-2} = \frac{MSR}{MSE}$$

But by Cochran's theorem, we have when H_0 holds:

$$F^* \sim \frac{\chi^2(1)}{1} \div \frac{\chi^2(n-2)}{n-2}$$

where the χ^2 variables are independent. Thus, when H_0 holds, F^* is the ratio of two independent χ^2 variables, each divided by its degrees of freedom. But this is the definition of an F random variable in (1.44).

We have thus established that if H_0 holds, F^* follows the F distribution, specifically the $F(1, n-2)$ distribution.

When H_a holds, it can be shown that F^* follows the noncentral F distribution, a complex distribution that we need not consider further at this time.

Note

Even if $\beta_1 \neq 0$, SSR and SSE are independent and $SSE/\sigma^2 \sim \chi^2$. But that both SSR/σ^2 and SSE/σ^2 are χ^2 random variables requires $\beta_1 = 0$.

Construction of Decision Rule.   Since the test is upper-tailed and F^* is distributed as $F(1, n-2)$ when H_0 holds, the decision rule is as follows when the risk of a Type I error is to be controlled at α:

(3.61)
$$\text{If } F^* \leq F(1 - \alpha; 1, n-2), \text{ conclude } H_0$$
$$\text{If } F^* > F(1 - \alpha; 1, n-2), \text{ conclude } H_a$$

where $F(1 - \alpha; 1, n-2)$ is the $(1 - \alpha)100$ percentile of the appropriate F distribution.

Example.   Using the Westwood Company lot size example again, let us repeat the earlier test on β_1. This time we will use the F test. The alternative conclusions are:

$$H_0: \beta_1 = 0$$

$$H_a: \beta_1 \neq 0$$

As before, let $\alpha = .05$. Since $n = 10$, we require $F(.95; 1, 8)$. We find from Table A.4 in Appendix A that $F(.95; 1, 8) = 5.32$. The decision rule is:

$$\text{If } F^* \leq 5.32, \text{ conclude } H_0$$

$$\text{If } F^* > 5.32, \text{ conclude } H_a$$

We have from Table 3.3 that $MSR = 13,600$ and $MSE = 7.5$. Hence, F^* is:

$$F^* = \frac{13,600}{7.5} = 1,813$$

Since $F^* = 1,813 > 5.32$, we conclude H_a, that $\beta_1 \neq 0$, or that there is a linear association between man-hours and lot size. This is the same result as when the t test was employed, as it must be according to our discussion below.

The P-value for the test statistic is the probability $P\{F(1, 8) > F^* = 1,813\}$. From Table A.4 it can be seen that this P-value is less than .001 since $F(.999; 1, 8) = 25.4$.

Equivalence of F Test and t Test. For a given α level, the F test of $\beta_1 = 0$ versus $\beta_1 \neq 0$ is equivalent algebraically to the two-tailed t test. To see this, recall from (3.50b) that:

$$SSR = b_1^2 \sum (X_i - \bar{X})^2$$

Thus, we can write:

$$F^* = \frac{SSR \div 1}{SSE \div (n - 2)} = \frac{b_1^2 \Sigma (X_i - \bar{X})^2}{MSE}$$

But since $s^2\{b_1\} = MSE/\Sigma(X_i - \bar{X})^2$, we obtain:

$$(3.62) \qquad F^* = \frac{b_1^2}{s^2\{b_1\}} = \left(\frac{b_1}{s\{b_1\}}\right)^2$$

Now, we know from earlier discussion that the t^* statistic for testing whether or not $\beta_1 = 0$ is by (3.17):

$$t^* = \frac{b_1}{s\{b_1\}}$$

In squaring, we obtain the expression for F^* in (3.62). Thus:

$$(t^*)^2 = \left(\frac{b_1}{s\{b_1\}}\right)^2 = F^*$$

In our illustrative problem, we just calculated that $F^* = 1,813$. From earlier work, we have: $b_1 = 2.0$, $s\{b_1\} = .04697$. Thus:

$$(t^*)^2 = \left(\frac{2.0}{.04697}\right)^2 = 1,813$$

Corresponding to the relation between t^* and F^*, we have the following relation between the required percentiles of the t and F distributions in the tests: $[t(1 - \alpha/2; n - 2)]^2 = F(1 - \alpha; 1, n - 2)$. In our tests on β_1, these percentiles were: $[t(.975; 8)]^2 = (2.306)^2 = 5.32 = F(.95; 1, 8)$. Remember that the t test is two-tailed while the F test is one-tailed.

Thus, at a given α level, we can use either the t test or the F test for testing $\beta_1 = 0$ versus $\beta_1 \neq 0$. Whenever one test leads to H_0, so will the other, and correspondingly for H_a. The t test, however, is more flexible since it can be used for one-sided alternatives involving $\beta_1 (\leq \geq)0$ versus $\beta_1 (> <)0$, while the F test cannot.

3.9 GENERAL LINEAR TEST APPROACH

The analysis of variance test of $\beta_1 = 0$ versus $\beta_1 \neq 0$ is an example of a general test for a linear statistical model. We shall explain this general test approach in terms of the simple linear regression model. We do so at this time because of the generality of the approach and the wide use we shall make of it, and because of the simplicity of understanding the approach in terms of our present problem.

The general linear test approach involves three basic steps, which we will now describe in turn.

Full Model

We begin with the model that is considered to be appropriate for the data, which in this context is called the *full* or *unrestricted model*. For the simple linear regression case, the full model is:

$$(3.63) \qquad\qquad Y_i = \overset{b_0}{\beta_0} + \overset{b_1}{\beta_1} X_i + \varepsilon_i \qquad\qquad \text{Full model}$$

We fit this full model by the method of least squares and obtain the error sum of squares. The error sum of squares is the sum of the squared deviations of each observation Y_i around its estimated expected value. In this context, we shall denote this sum of squares by $SSE(F)$ to indicate that it is the error sum of squares for the full model. Here, we have:

$$(3.64) \qquad\qquad SSE(F) = \sum [Y_i - (b_0 + b_1 X_i)]^2$$

$$= \sum (Y_i - \hat{Y}_i)^2 = SSE$$

Thus, for the full model (3.63), the error sum of squares is simply SSE, which measures the variability of the Y_i observations around the fitted regression line.

Reduced Model

Next, we consider H_0. In this instance, we have:

$$(3.65) \qquad\qquad \begin{array}{l} H_0: \beta_1 = 0 \\ H_a: \beta_1 \neq 0 \end{array}$$

The model when H_0 holds is called the *reduced* or *restricted model*. When $\beta_1 = 0$, model (3.63) reduces to:

(3.66) $$Y_i = \beta_0 + \varepsilon_i$$ Reduced model

We fit this reduced model by the method of least squares and obtain the error sum of squares for this reduced model, denoted by $SSE(R)$. When we fit the particular reduced model (3.66), it can be shown that the least squares estimator of β_0 is $\bar{Y}$. Hence, the estimated expected value for each observation is $b_0 = \bar{Y}$, and the error sum of squares for this reduced model is:

(3.67) $$SSE(R) = \sum (Y_i - b_0)^2 = \sum (Y_i - \bar{Y})^2 = SSTO$$

Test Statistic

The logic now is to compare the two error sums of squares $SSE(F)$ and $SSE(R)$. It can be shown that $SSE(F)$ never is greater than $SSE(R)$:

(3.68) $$SSE(F) \leq SSE(R)$$

The reason is that the more parameters there are in the model, the better one can fit the data and the smaller are the deviations around the fitted regression function. If $SSE(F)$ is not much less than $SSE(R)$, using the full model does not account for much more of the variability of the Y_i than does the reduced model, in which case the data suggest that H_0 holds. To put this another way, if $SSE(F)$ is close to $SSE(R)$, the variation of the observations around the fitted regression function for the full model is almost as great as the variation around the fitted regression function for the reduced model, in which case the added parameters in the full model really do not help to reduce the variation in the Y_i. Thus, a small difference $SSE(R) - SSE(F)$ suggests that H_0 holds. On the other hand, a large difference suggests that H_a holds because the additional parameters in the model do help to reduce substantially the variation of the observations Y_i around the fitted regression function.

The actual test statistic used is a function of $SSE(R) - SSE(F)$, namely:

(3.69) $$F^* = \frac{SSE(R) - SSE(F)}{df_R - df_F} \div \frac{SSE(F)}{df_F}$$

which follows the F distribution when H_0 holds. The degrees of freedom df_R and df_F are those associated with the reduced and full model error sums of squares, respectively. Large values of F^* lead to H_a because a large difference $SSE(R) - SSE(F)$ suggests that H_a holds. The decision rule therefore is:

(3.70)
$$\text{If } F^* \leq F(1 - \alpha; df_R - df_F, df_F), \text{ conclude } H_0$$
$$\text{If } F^* > F(1 - \alpha; df_R - df_F, df_F), \text{ conclude } H_a$$

For testing whether or not $\beta_1 = 0$, we therefore have:

$$SSE(R) = SSTO \qquad SSE(F) = SSE$$
$$df_R = n - 1 \qquad df_F = n - 2$$

so that we obtain when substituting into (3.69):

$$F^* = \frac{SSTO - SSE}{(n-1) - (n-2)} \div \frac{SSE}{n-2} = \frac{SSR}{1} \div \frac{SSE}{n-2} = \frac{MSR}{MSE}$$

which is identical to the analysis of variance test statistic (3.59).

Summary

The general linear test approach can be used for highly complex tests of linear statistical models, as well as for simpler tests. The basic steps again are:

1. Fit the full model and obtain the error sum of squares $SSE(F)$.
2. Fit the reduced model under H_0 and obtain the error sum of squares $SSE(R)$.
3. Use the test statistic (3.69) and decision rule (3.70).

3.10 DESCRIPTIVE MEASURES OF ASSOCIATION BETWEEN *X* AND *Y* IN REGRESSION MODEL

We have discussed the major uses of regression analysis—estimation of parameters and means and prediction of new observations—without mentioning the "degree of linear association" between X and Y, or similar terms. The reason is that the usefulness of estimates or predictions depends upon the width of the interval and the user's needs for precision, which vary from one application to another. Hence, no single descriptive measure of the "degree of linear association" can capture the esssential information as to whether a given regression relation is useful in any particular application.

Nevertheless, there are times when the degree of linear association is of interest in its own right. We shall now briefly discuss two descriptive measures that are frequently used in practice to describe the degree of linear association between X and Y.

Coefficient of Determination

We saw earlier that $SSTO$ measures the variation in the observations Y_i, or the uncertainty in predicting Y, when no account of the independent variable X is taken. Thus, $SSTO$ is a measure of the uncertainty in predicting Y when X is not considered. Similarly, SSE measures the variation in the Y_i when a regression model utilizing the independent variable X is employed. A natural measure of the effect of X in reducing the variation in Y, i.e., the uncertainty in predicting Y, is therefore:

$$(3.71) \qquad r^2 = \frac{SSTO - SSE}{SSTO} = \frac{SSR}{SSTO} = 1 - \frac{SSE}{SSTO}$$

The measure r^2 is called the *coefficient of determination*. Since $0 \le SSE \le SSTO$, it follows that:

(3.72) $$0 \leq r^2 \leq 1$$

We may interpret r^2 as the proportionate reduction of total variation associated with the use of the independent variable X. Thus, the larger is r^2, the more is the total variation of Y reduced by introducing the independent variable X. The limiting values of r^2 occur as follows:

1. If all observations fall on the fitted regression line, $SSE = 0$ and $r^2 = 1$. This case is shown in Figure 3.6a. Here, the independent variable X accounts for all variation in the observations Y_i.
2. If the slope of the fitted regression line is $b_1 = 0$ so that $\hat{Y}_i \equiv \bar{Y}$, $SSE = SSTO$ and $r^2 = 0$. This case is shown in Figure 3.6b. Here, there is no linear association between X and Y in the sample data, and the independent variable X is of no help in reducing the variation in the observations Y_i with linear regression.

In practice, r^2 is not likely to be 0 or 1, but rather somewhere in between these limits. The closer it is to 1, the greater is said to be the degree of linear association between X and Y.

Coefficient of Correlation

The square root of r^2:

(3.73) $$r = \pm \sqrt{r^2}$$

is called the *coefficient of correlation*. A plus or minus sign is attached to this measure according to whether the slope of the fitted regression line is positive or nega-

---

FIGURE 3.6   Scatter Plots when $r^2 = 0$ and $r^2 = 1$

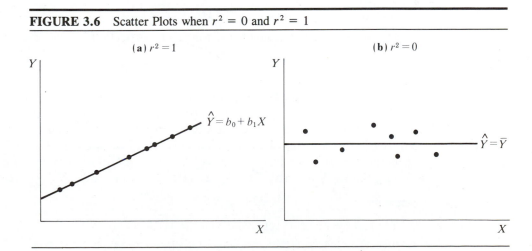

(a) $r^2 = 1$ (b) $r^2 = 0$

$\hat{Y} = b_0 + b_1 X$

$\hat{Y} = \bar{Y}$

tive. Thus, the range of r is:

(3.74)
$$-1 \leq r \leq 1$$

Whereas r^2 indicates the proportional reduction in the variability of Y attained by the use of information about X, the square root, r, does not have such a clear-cut operational interpretation. Nevertheless, there is a tendency to use r instead of r^2 in much applied work.

It is worth noting that since for any r^2 other than 0 or 1, $r^2 < |r|$, r may give the impression of a "closer" relationship between X and Y than does the corresponding r^2. For instance, $r^2 = .10$ indicates that the total variation in Y is reduced by only 10 percent when X is introduced, yet $|r| = .32$ may give an impression of greater linear association between X and Y.

Example

For the Westwood Company example, we obtained $SSTO = 13,660$ and $SSE = 60$. Hence:

$$r^2 = \frac{13,660 - 60}{13,660} = .996$$

Thus, the variation in man-hours is reduced by 99.6 percent when lot size is considered.

The correlation coefficient here is:

$$r = +\sqrt{.996} = +.998$$

The plus sign is affixed since b_1 is positive.

Computational Formula for r

A direct computational formula for r, which automatically furnishes the proper sign, is:

(3.75)
$$r = \frac{\Sigma(X_i - \bar{X})(Y_i - \bar{Y})}{[\Sigma(X_i - \bar{X})^2 \Sigma(Y_i - \bar{Y})^2]^{1/2}}$$

$$= \frac{\Sigma X_i Y_i - \dfrac{\Sigma X_i \Sigma Y_i}{n}}{\left[\left(\Sigma X_i^2 - \dfrac{(\Sigma X_i)^2}{n}\right)\left(\Sigma Y_i^2 - \dfrac{(\Sigma Y_i)^2}{n}\right)\right]^{1/2}}$$

Comments

1. The following relation between b_1 and r is worth noting:

(3.76)
$$b_1 = \left[\frac{\Sigma(Y_i - \bar{Y})^2}{\Sigma(X_i - \bar{X})^2}\right]^{1/2} r = \left(\frac{s_Y}{s_X}\right)r$$

where $s_Y = [\Sigma(Y_i - \bar{Y})^2/(n - 1)]^{1/2}$ and $s_X = [\Sigma(X_i - \bar{X})^2/(n - 1)]^{1/2}$ are the sample standard deviations for the Y and X observations, respectively. Note that $b_1 = 0$ when $r = 0$, and vice versa. Thus, $r = 0$ implies a horizontal fitted regression line, and vice versa.

2. The value taken by r^2 in a given sample tends to be affected by the spacing of the X observations. This is implied in (3.71). SSE is not affected systematically by the spacing of the X_i since for regression model (3.1), $\sigma^2\{Y_i\} = \sigma^2$ at all X levels. However, the wider the spacing is of the X_i in the sample when $b_1 \neq 0$, the greater will tend to be the spread of the observed Y_i around $\bar{Y}$ and hence the greater will be $SSTO$. Consequently, the wider the X_i are spaced, the higher will tend to be r^2.

3. The regression sum of squares SSR is often called the "explained variation" in Y. The residual sum of squares SSE is then called the "unexplained variation," and the total sum of squares the "total variation." The coefficient r^2 then is interpreted in terms of the proportion of the total variation in Y which has been "explained" by X. Unfortunately, this terminology frequently is taken literally, hence misunderstood. Remember that in a regression model there is no implication that Y necessarily depends on X in a causal or explanatory sense.

4. A value of r or r^2 relatively close to 1 sometimes is taken as an indication that sufficiently precise inferences on Y can be made from knowledge of X. As mentioned earlier, the usefulness of the regression relation depends upon the width of the confidence or prediction interval and the particular needs for precision, which vary from one application to another. Hence, no single measure is an adequate indicator of the usefulness of the regression relation.

5. Regression models do not contain any parameter to be estimated by r or r^2. These coefficients simply are descriptive measures of the degree of linear association between X and Y in the sample observations which may, or may not, be useful in any one instance.

3.11 COMPUTER INPUTS AND OUTPUTS

Regression calculations used to be tedious chores, especially when the number of observations was large and when there were several independent variables. Today computers can be used quite easily to perform regression calculations, with one of many available packaged programs. Also, a number of calculators contain regression routines.

The inputting of data will vary from program to program. With some, the X and Y observations are entered as separate sets. In other cases, the inputting of the data is in the form: X_1, Y_1, X_2, Y_2, etc.

The computer output will also vary from one program package to another. Figure 3.7 illustrates a typical output format when the simple linear regression model is fitted to the Westwood Company data in Table 2.1 by the SPSS^X computer program (Ref. 3.1). The 10 observations on lot size and man-hours are printed at the top. This enables one to verify that the observations were entered into the computer accurately. Next, r, r^2 and $\sqrt{MSE}$ are given, followed by the estimated regression coefficients and the estimated standard deviation of b_1, as well as the test statistic F^*. Finally, descriptive statistics for the X and Y variables are provided, followed by the analysis of variance table.

We have annotated the output in Figure 3.7 in terms of the notation used in this book. All results in Figure 3.7 agree with our earlier calculations, except for the number of digits shown.

FIGURE 3.7 Segment of Computer Output for Regression Run on Westwood Company Data (SPSS^X, Ref. 3.1)

X_i Y_i

VARIABLES

1 SIZE	2 HOURS
30.0000	73.0000
20.0000	50.0000
60.0000	128.0000
80.0000	170.0000
40.0000	87.0000
50.0000	108.0000
60.0000	135.0000
30.0000	69.0000
70.0000	148.0000
60.0000	132.0000

DEPENDENT VARIABLE..   HOURS

VARIABLE(S) ENTERED ON STEP NUMBER 1..   SIZE

MULTIPLE R	0.99780 ← r
R SQUARE	0.99561 ← r^2

STANDARD ERROR   2.73861 ← $\sqrt{MSE}$

------------------ VARIABLES IN THE EQUATION ------------------

VARIABLE	B	STD ERROR B	F
SIZE	2.000000 ← b_1	$s\{b_1\}$ → 0.04697	1813.333 ← F^*
(CONSTANT)	10.00000 ← b_0		

VARIABLE	MEAN	STANDARD DEV	CASES
SIZE	$\bar{X}$ → 50.0000	s_X → 19.4365	10
HOURS	$\bar{Y}$ → 110.0000	s_Y → 38.9587	10 ← n

ANALYSIS OF VARIANCE	DF	SUM OF SQUARES	MEAN SQUARE
REGRESSION	1.	SSR → 13600.00000	MSR → 13600.00000
RESIDUAL ← Error	8.	SSE → 60.00000	MSE → 7.50000

The computer package output illustrated in Figure 3.7 does not provide $s\{b_0\}$, the estimated standard deviation of b_0. However, this estimate can be easily calculated from the data given in the computer output. Note in this connection that the denominator term $\Sigma(X_i - \bar{X})^2$ in (3.21) is equal to $(n-1)s_X^2$, and that s_X is given in the computer output.

Computer printouts for regression analysis programs differ substantially in format from one program to another. In addition, differences in the computed results may occur because different program packages do not control roundoff errors equally well. Before using a computer program the first time, it is a good idea to check it on a set of data for which the exact results are known.

CITED REFERENCE

3.1. *SPSS^X User's Guide*. 2nd ed. Chicago: SPSS, Inc., 1986.

PROBLEMS

3.1. A student, working on a summer internship in the economic research office of a large corporation, studied the relation between sales of a product (Y, in million dollars) and population (X, in million persons) in the firm's 50 marketing districts. The normal error regression model (3.1) was employed. The student first wished to test whether or not a linear association between Y and X existed. The student accessed an interactive simple linear regression program and obtained the following information on the regression coefficients:

Parameter	Estimated Value	95 Percent Confidence Limits	
Intercept	7.43119	−1.18518	16.0476
Slope	.755048	.452886	1.05721

  a. The student concluded from these results that there is a linear association between Y and X. Is the conclusion warranted? What is the implied level of significance?

  b. Someone questioned the negative lower confidence limit for the intercept, pointing out that dollar sales cannot be negative even if the population in a district is zero. Discuss.

3.2. In a test of the alternatives H_0: $\beta_1 \leq 0$ versus H_a: $\beta_1 > 0$, an analyst concluded H_0. Does this conclusion imply that there is no linear association between X and Y? Explain.

3.3. A member of a student team playing an interactive marketing game received the following computer output when studying the relation between advertising expenditures (X) and sales (Y) for one of the team's products:

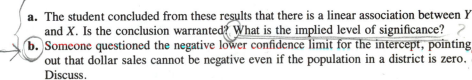

Estimated regression equation: $\hat{Y} = 350.7 - .18X$
Two-sided P-value for estimated slope: .91

The student stated: "The message I get here is that the more we spend on advertising this product, the fewer units we sell!" Comment.

3.4. Refer to **Grade point average** Problem 2.17. Some additional results are: $b_0 = -1.700$, $s\{b_0\} = .7267$, $b_1 = .8399$, $s\{b_1\} = .144$, $MSE = .1892$.

  a. Obtain a 99 percent confidence interval for β_1. Interpret your confidence interval. Does it include zero? Why might the director of admissions be interested in whether the confidence interval includes zero?

  b. Test, using the test statistic t^*, whether or not a linear association exists between student's entrance test score (X) and GPA at the end of the freshman year (Y). Use a level of significance of .01. State the alternatives, decision rule, and conclusion.

  c. What is the P-value of your test in part (b)? How does it support the conclusion reached in part (b)?

3.5. Refer to **Calculator maintenance** Problem 2.18. Some additional results are: $b_0 = -2.3221$, $s\{b_0\} = 2.564$, $b_1 = 14.738$, $s\{b_1\} = .519$, $MSE = 20.086$.

   a. Estimate the change in the mean service time when the number of machines serviced increases by one. Use a 90 percent confidence interval. Interpret your confidence interval.

   b. Conduct a t test to determine whether or not there is a linear association between X and Y here; control the α risk at .10. State the alternatives, decision rule, and conclusion. What is the P-value of your test?

   c. Are your results in parts (a) and (b) consistent? Explain.

   d. The manufacturer has suggested that the mean required time should not increase by more than 14 minutes for each additional machine that is serviced on a service call. Conduct a test to decide whether this standard is being satisfied by Tri-City. Control the risk of a Type I error at .05. State the alternatives, decision rule, and conclusion. What is the P-value of the test?

   e. Does b_0 give any relevant information here about the "start-up" time on calls—i.e., about the time required before service work is begun on the machines at a customer location?

3.6. Refer to **Airfreight breakage** Problem 2.19.

   a. Estimate β_1 with a 95 percent confidence interval. Interpret your interval estimate.

   b. Conduct a t test to decide whether or not there is a linear association between number of times a carton is transferred (X) and number of broken ampules (Y). Use a level of significance of .05. State the alternatives, decision rule, and conclusion. What is the P-value of the test?

   c. β_0 represents here the mean number of ampules broken when no transfers of the shipment are made—i.e., when $X = 0$. Obtain a 95 percent confidence interval for β_0 and interpret it.

   d. A consultant has suggested, based on previous experience, that the mean number of broken ampules should not exceed 9 when no transfers are made. Conduct an appropriate test using $\alpha = .025$. State the alternatives, decision rule, and conclusion. What is the P-value of the test?

   e. Obtain the power of your test in part (b) if actually $\beta_1 = 2.0$. Assume $\sigma\{b_1\} = .50$. Also obtain the power of your test in part (d) if actually $\beta_0 = 11$. Assume $\sigma\{b_0\} = .75$.

3.7 Refer to **Plastic hardness** Problem 2.20.

   a. Estimate the change in the mean hardness when the elapsed time increases by one hour. Use a 99 percent confidence interval. Interpret your interval estimate.

   b. The plastic manufacturer has stated that the mean hardness should increase by 2 Brinell units per hour. Conduct a two-sided test to decide whether this standard is being satisfied; use $\alpha = .01$. State the alternatives, decision rule, and conclusion. What is the P-value of the test?

   c. Obtain the power of your test in part (b) if the standard actually is being exceeded by .3 Brinell unit per hour. Assume $\sigma\{b_1\} = .1$.

3.8. Refer to Figure 3.7 for the Westwood Company example. A consultant has advised that an increase of one unit in lot size should require an increase of 1.8 in the expected number of man-hours for the given production item.

   a. Conduct a test to decide whether or not the increase in the expected number of man-hours in the Westwood Company equals this standard. Use $\alpha = .05$. State the alternatives, decision rule, and conclusion.

    b. Obtain the power of your test in part (a) if the consultant's standard actually is being exceeded by .1 hour. Assume $\sigma\{b_1\} = .05$.

    c. Why is $F^* = 1813.333$, given in the printout, not relevant for the test in part (a)?

3.9. Refer to Figure 3.7. A student, noting that $s\{b_1\}$ is furnished in the printout, asks why $s\{\hat{Y}_h\}$ is not also given. Discuss.

3.10. For each of the following questions, explain whether a confidence interval for a mean response or a prediction interval for a new observation is appropriate.

    a. What will be the humidity level in this greenhouse tomorrow when we set the temperature level at 31°C?

    b. How much do families whose disposable income is $23,500 spend, on the average, for meals away from home?

    c. How many kilowatt-hours of electricity will be consumed next month by commercial and industrial users in the Twin Cities service area, given that the index of business acitivity for the area remains at its present level?

3.11. A person asks if there is a difference between the "mean response at $X = X_h$" and the "mean of m new observations at $X = X_h$." Reply.

3.12. Can $\sigma^2\{Y_{h(new)}\}$ in (3.36) be brought increasingly close to 0 as n becomes large? Is this also the case for $\sigma^2\{\hat{Y}_h\}$ in (3.28b)? What is the implication of this difference?

3.13. Refer to **Grade point average** Problems 2.17 and 3.4.

    a. Obtain a 95 percent interval estimate of the mean freshman GPA for students whose entrance test score is 4.7. Interpret your confidence interval.

    b. Mary Jones obtained a score of 4.7 on the entrance test. Predict her freshman GPA using a 95 percent prediction interval. Interpret your prediction interval.

    c. Is the prediction interval in part (b) wider than the confidence interval in part (a)? Should it be?

3.14. Refer to **Calculator maintenance** Problems 2.18 and 3.5.

    a. Obtain a 90 percent confidence interval for the mean service time on calls in which six machines are serviced. Interpret your confidence interval.

    b. Obtain a 90 percent prediction interval for the service time on the next call in which six machines are serviced. Is your prediction interval wider than the corresponding confidence interval in part (a)? Should it be?

    c. Suppose that management wishes to estimate the expected service time *per machine* on calls in which six machines are serviced. Obtain an appropriate confidence interval by converting the interval obtained in part (a). Interpret the converted confidence interval.

3.15. Refer to **Airfreight breakage** Problem 2.19.

    a. Because of changes in airline routes, shipments may have to be transferred more frequently than in the past. Estimate the mean breakage for the following numbers of transfers: $X = 2, 4$. Use separate 99 percent confidence intervals. Interpret your results.

    b. The next shipment will entail two transfers. Obtain a 99 percent prediction interval for the number of broken ampules for this shipment. Interpret your prediction interval.

    c. In the next several days, three independent shipments will be made, each entailing two transfers. Obtain a 99 percent prediction interval for the mean number of ampules broken in the three shipments. Convert this interval into a 99 percent prediction interval for the total number of ampules broken in the three shipments.

3.16. Refer to **Plastic hardness** Problem 2.20.

    a. Obtain a 98 percent confidence interval for the mean hardness of molded items with an elapsed time of 30 hours. Interpret your confidence interval.

    b. Obtain a 98 percent prediction interval for the hardness of a newly molded test item with an elapsed time of 30 hours.

    c. Obtain a 98 percent prediction interval for the mean hardness of 10 newly molded test items, each with an elapsed time of 30 hours.

    d. Is the prediction interval in part (c) narrower than the one in part (b)? Should it be?

3.17. An analyst fitted normal error regression model (3.1) and conducted an F test of $\beta_1 = 0$ versus $\beta_1 \neq 0$. The P-value of the test was .033, and the analyst concluded H_a: $\beta_1 \neq 0$. Was the α level used by the analyst greater than or smaller than .033? If the α level had been .01, what would have been the appropriate conclusion?

3.18. For conducting statistical tests concerning the parameter β_1, why is the t test more versatile than the F test?

3.19. When testing whether or not $\beta_1 = 0$, why is the F test a one-sided test even though H_a includes both $\beta_1 < 0$ and $\beta_1 > 0$? [*Hint:* Refer to (3.56).]

3.20. A student asks whether r^2 is a point estimator of any parameter in the normal error regression model (3.1). Respond.

3.21. A value of r^2 near 1 is sometimes interpreted to imply that the relation between Y and X is sufficiently close so that suitably precise predictions of Y can be made from knowledge of X. Is this implication a necessary consequence of the definition of r^2?

3.22. Using the normal error regression model (3.1) in an engineering safety experiment, a researcher found for the first 10 cases that r^2 was zero. Is it possible that for the complete set of 30 cases r^2 will not be zero? Could r^2 not be zero for the first 10 cases, yet equal zero for all 30 cases? Explain.

3.23. Refer to **Grade point average** Problems 2.17 and 3.4. Some additional calculational results are: $SSE = 3.406$, $SSR = 6.434$.

    a. Set up the ANOVA table.

    b. What is estimated by MSR in your ANOVA table? By MSE? Under what condition do MSR and MSE estimate the same quantity?

    c. Conduct an F test of whether or not $\beta_1 = 0$. Control the α risk at .01. State the alternatives, decision rule, and conclusion.

    d. What is the absolute magnitude of the reduction in the variation of Y when X is introduced into the regression model? What is the relative reduction? What is the name of the latter measure?

    e. Obtain r and attach the appropriate sign.

    f. Which measure, r^2 or r, has the more clear-cut operational interpretation? Explain.

3.24. Refer to **Calculator maintenance** Problems 2.18 and 3.5. Some additional calculational results are: $SSE = 321.396$, $SSR = 16{,}182.604$.

    a. Set up the basic ANOVA table in the format of Table 3.2. Which elements of your table are additive? Also set up the ANOVA table in the format of Table 3.4a. How do the two tables differ?

    b. Conduct an F test to determine whether or not there is a linear association between time spent and number of machines serviced; use $\alpha = .10$. State the alternatives, decision rule, and conclusion.

    c. By how much, relatively, is the total variation in number of minutes spent on a call reduced when the number of machines serviced is introduced into the analysis? Is this a relatively small or large reduction? What is the name of this measure?

  d. Calculate r and attach the appropriate sign.

  e. Which measure, r or r^2, has the more clear-cut operational interpretation?

3.25. Refer to **Airfreight breakage** Problem 2.19.

  a. Set up the ANOVA table. Which elements are additive?

  b. Conduct an F test to decide whether or not there is a linear association between the number of times a carton is transferred and the number of broken ampules; control the α risk at .05. State the alternatives, decision rule, and conclusion.

  c. Obtain the t^* statistic for the test in part (b) and demonstrate numerically its equivalence to the F^* statistic obtained in part (b).

  d. Calculate r^2 and r. What proportion of the variation in Y is accounted for by introducing X into the regression model?

3.26. Refer to **Plastic hardness** Problem 2.20.

  a. Set up the ANOVA table.

  b. Test by means of an F test whether or not there is a linear association between the hardness of the plastic and the elapsed time. Use $\alpha = .01$. State the alternatives, decision rule, and conclusion.

  c. Plot the deviations $Y_i - \hat{Y}_i$ against X_i on a graph. Plot the deviations $\hat{Y}_i - \bar{Y}$ against X_i on another graph. From your two graphs, does SSE or SSR appear to be the larger component of $SSTO$?

  d. Calculate r^2 and r.

3.27. Refer to **Muscle mass** Problem 2.25.

  a. Conduct a test to decide whether or not there is a negative linear association between amount of muscle mass and age. Control the risk of Type I error at .05. State the alternatives, decision rule, and conclusion. What is the P-value of the test?

  b. The two-sided P-value for the test whether $\beta_0 = 0$ is 0+. Can it now be concluded that b_0 provides relevant information on the amount of muscle mass at birth for a female child?

  c. Estimate with a 95 percent confidence interval the difference in expected muscle mass for women whose ages differ by one year. Why is it not necessary to know the specific ages to make this estimate?

3.28. Refer to **Muscle mass** Problem 2.25.

  a. Obtain a 95 percent confidence interval for the mean muscle mass for women of age 60. Interpret your confidence interval.

  b. Obtain a 95 percent prediction interval for the muscle mass of a woman whose age is 60. Is the prediction interval relatively precise?

3.29. Refer to **Muscle mass** Problem 2.25.

  a. Plot the deviations $Y_i - \hat{Y}_i$ against X_i on one graph. Plot the deviations $\hat{Y}_i - \bar{Y}$ against X_i on another graph. From your graphs, does SSE or SSR appear to be the larger component of $SSTO$?

  b. Set up the ANOVA table.

  c. Test whether or not $\beta_1 = 0$ using an F test with $\alpha = .10$. State the alternatives, decision rule, and conclusion.

  d. What proportion of the total variation in muscle mass remains "unexplained" when age is introduced into the analysis? Is this proportion relatively small or large?

  e. Obtain r^2 and r.

3.30. Refer to **Robbery rate** Problem 2.26.

   a. Test whether or not there is a linear association between robbery rate and population density using a t test with $\alpha = .01$. State the alternatives, decision rule, and conclusion. What is the P-value of the test?

   b. Test whether or not $\beta_0 = 0$; control the risk of Type I error at .01. State the alternatives, decision rule, and conclusion. Why might there be interest in testing whether or not $\beta_0 = 0$?

   c. Estimate β_1 with a 99 percent confidence interval. Interpret your interval estimate.

3.31. Refer to **Robbery rate** Problem 2.26.

   a. Set up the ANOVA table.

   b. Carry out the test in Problem 3.30a by means of the F test. Show the numerical equivalence of the two test statistics and decision rules. Is the P-value for the F test the same as that for the t test?

   c. By how much is the total variation in robbery rate reduced when population density is introduced into the analysis? Is this a relatively large or small reduction?

   d. Obtain r.

3.32. Refer to **Robbery rate** Problems 2.26 and 3.30. Suppose that the test in Problem 3.30a is to be carried out by means of a general linear test.

   a. State the full and reduced models.

   b. Obtain: (1) $SSE(F)$, (2) $SSE(R)$, (3) df_F, (4) df_R, (5) test statistic F^* for the general linear test, (6) decision rule.

   c. Are the test statistic F^* and the decision rule for the general linear test numerically equivalent to those in Problem 3.30a?

3.33. In empirically developing a cost function from observed data on a complex chemical experiment, an analyst employed the normal error regression model (3.1). β_0 was interpreted here as the cost of setting up the experiment. The analyst hypothesized that this cost should be \$7.5 thousand and wished to test the hypothesis by means of a general linear test.

   a. Indicate the alternative conclusions for the test.

   b. Specify the full and reduced models.

   c. Without additional information, can you tell what the quantity $df_R - df_F$ in test statistic (3.69) will equal in the analyst's test? Explain.

3.34. Refer to **Grade point average** Problem 2.17.

   a. Would it be more reasonable to consider the X_i as known constants or as random variables here? Explain.

   b. If the X_i were considered to be random variables, would this have any effect on prediction intervals for new applicants? Explain.

3.35. Refer to **Calculator maintenance** Problems 2.18 and 3.5. How would the meaning of the confidence coefficient in Problem 3.5a change if the independent variable were considered a random variable and the conditions in (3.40) were applicable?

EXERCISES

3.36. Derive the property in (3.6) for the k_i.

3.37. Show that b_0 as defined in (3.19) is an unbiased estimator of β_0.

3.38. Derive the expression in (3.20b) for the variance of b_0, making use of theorem (3.30). Also explain how variance (3.20b) is a special case of variance (3.28b).

3.39. (Calculus needed.)
   a. Obtain the likelihood function for the sample observations $Y_1, \ldots, Y_n$ given $X_1, \ldots, X_n$, if the conditions in (3.40) apply.
   b. Obtain the maximum likelihood estimators of β_0, β_1, and σ^2. Are the estimators of β_0 and β_1 the same as those in (2.27) when the X_i are fixed?

3.40. Suppose that the normal error regression model (3.1) is applicable except that the error variance is not constant; rather the variance is larger, the larger is X. Does $\beta_1 = 0$ still imply that there is no linear association between X and Y? That there is no association between X and Y? Explain.

3.41. Derive the expression for SSR in (3.50b).

3.42. In a small-scale regression study, five observations on Y were obtained corresponding to $X = 1, 4, 10, 11$, and 14. Assume that $\sigma = .6$, $\beta_0 = 5$, and $\beta_1 = 3$.
   a. What are the expected values of MSR and MSE here?
   b. For purposes of determining whether or not a regression relation exists, would it have been better or worse to have made the five observations at $X = 6, 7, 8, 9$, and 10? Why? Would the same answer apply if the principal purpose were to estimate the mean response for $X = 8$? Discuss.

3.43. The normal error regression model (3.1) is assumed to be applicable.
   a. When testing $H_0: \beta_1 = 5$ versus $H_a: \beta_1 \neq 5$ by means of a general linear test, what is the reduced model? What are the degrees of freedom df_R?
   b. When testing $H_0: \beta_0 = 2, \beta_1 = 5$ versus H_a: not both $\beta_0 = 2$ and $\beta_1 = 5$ by means of a general linear test, what is the reduced model? What are the degrees of freedom df_R?

3.44. Derive (3.75) from (3.71), using the result in Exercise 3.41.

PROJECTS

3.45. Refer to the **SMSA** data set and Project 2.42. Using r^2 as the criterion, which independent variable accounts for the largest reduction in the variability in the number of active physicians?

3.46. Refer to the **SMSA** data set and Project 2.43. Obtain a separate interval estimate of β_1 for each region. Use a 90 percent confidence coefficient in each case. Do the regression lines for the different regions appear to have similar slopes?

3.47. Refer to the **SENIC** data set and Project 2.44. Using r^2 as the criterion, which independent variable accounts for the largest reduction in the variability of the average length of stay?

3.48. Refer to the **SENIC** data set and Project 2.45. Obtain a separate interval estimate of β_1 for each region. Use a 95 percent confidence coefficient in each case. Do the regression lines for the different regions appear to have similar slopes?

3.49. Five observations on Y are to be taken when $X = 4, 8, 12, 16$, and 20, respectively. The true regression function is $E\{Y\} = 20 + 4X$, and the ε_i are independent $N(0, 25)$.

a. Generate five normal random numbers, with mean 0 and variance 25. Consider these random numbers as the error terms for the five Y observations at $X = 4$, 8, 12, 16, and 20, and calculate Y_1, Y_2, Y_3, Y_4, and Y_5. Obtain the least squares estimates b_0 and b_1 when fitting a straight line to the five cases. Also calculate $\hat{Y}_h$ when $X_h = 10$.

b. Repeat part (a) 200 times, generating new random numbers each time.

c. Make a frequency distribution of the 200 estimates b_1. Calculate the mean and standard deviation of the 200 estimates b_1. Are the results consistent with theoretical expectations?

d. For each of the 200 replications, calculate a 95 percent confidence interval for $E\{Y_h\}$ when $X_h = 10$. What proportion of the 200 confidence intervals include $E\{Y_h\}$? Is this result consistent with theoretical expectations?

Chapter 4

Diagnostics and Remedial Measures—I

When a regression model, such as the simple linear regression model (3.1), is selected for an application, one can usually not be certain in advance that the model is appropriate for that application. Any one, or several, of the features of the model, such as linearity of the regression function or normality of the error terms, may not be appropriate for the particular data at hand. Hence, it is important to examine the aptness of the model for the data before further analysis based on that model is undertaken. In this chapter, we discuss some simple graphic methods for studying the aptness of a model, as well as some formal statistical tests for doing so. We conclude with a consideration of some techniques whereby the simple linear regression model (3.1) can be made appropriate when the data do not accord with the conditions of the model.

While the discussion in this chapter is in terms of the aptness of the simple linear regression model (3.1), the basic principles apply to all statistical models discussed in this book. In later chapters, additional material concerning the aptness of a model and remedial measures will be presented.

4.1 DIAGNOSTICS FOR INDEPENDENT VARIABLE

We begin by considering some graphic diagnostics for the independent variable. Figure 4.1a contains a simple *box plot* for the lot sizes in the Westwood Company example in Table 2.1. The box plot in Figure 4.1a shows the minimum and maximum lot sizes, the first and third quartiles, and the median lot size. We see from Figure 4.1a that there are no lot sizes that are far outlying. We also see, because the median is near the upper end of the box, that the central portion of the distribution of lot sizes is skewed. Additional information about the pattern of lot sizes is provided by the lengths of the "whiskers" from the quartiles to the extremes. Here again, the whiskers suggest some skewness in the right tail.

FIGURE 4.1 Diagnostic Plots for Independent Variable—Westwood Company Example

(a) Box Plot

(b) Time Plot

(c) Stem-and-Leaf Plot

(d) Dot Plot

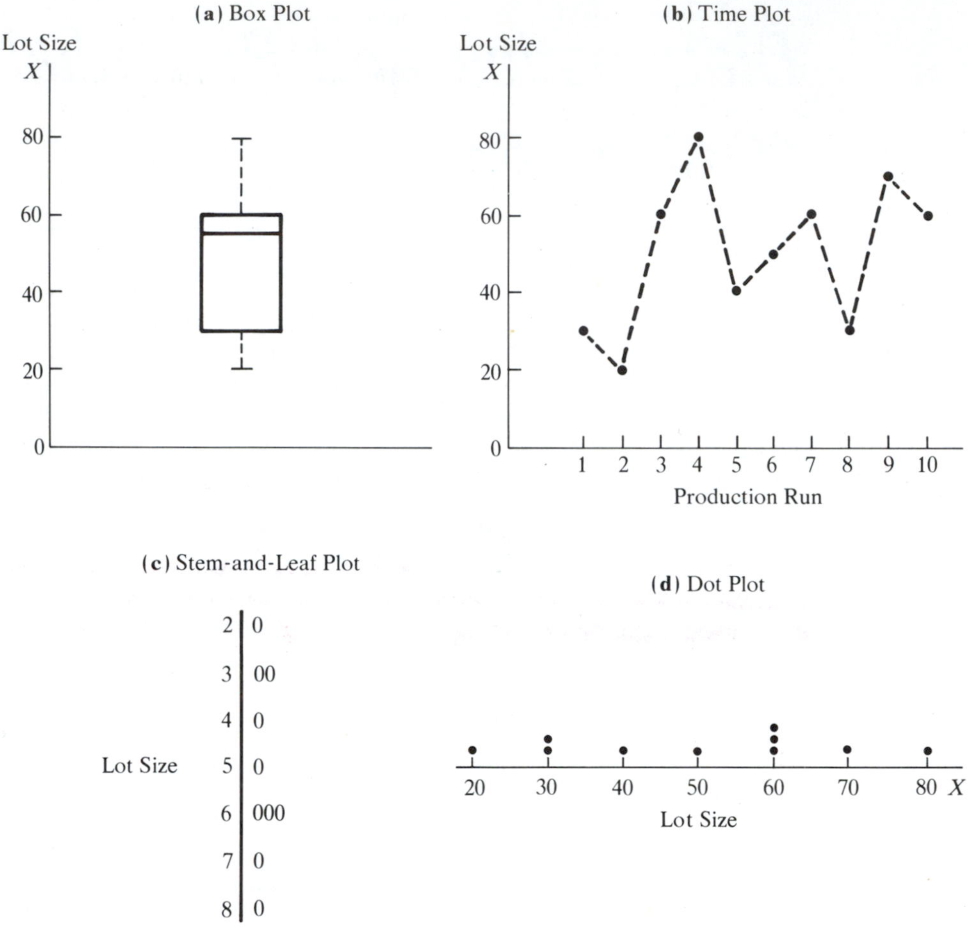

A second useful diagnostic for the independent variable is a *time plot*. Figure 4.1b contains a time plot of lot sizes for the Westwood Company example. Lot size is here plotted against production run (i.e., against time). The points in the plot are connected to show more effectively the time sequence. Time plots should be utilized whenever data are obtained in a time sequence. The time plot in Figure 4.1b contains no special pattern. If, say, the plot had shown that smaller lot sizes had been utilized early on and larger lot sizes later on, this information could be very helpful for subsequent diagnostic studies of the aptness of the fitted regression model.

Figures 4.1c and 4.1d contain two other diagnostic plots that present information similar to the box plot in Figure 4.1a. The *stem-and-leaf plot* in Figure 4.1c provides information similar to a frequency histogram. By displaying the last digits, this plot also indicates here that all lot sizes in the Westwood Company example were multiples of 10.

The *dot plot* in Figure 4.1d is helpful when there are only a few observations in the data set or when there are only a limited number of outcomes in the data.

Both the stem-and-leaf plot and the dot plot in Figure 4.1 indicate, like the box plot, that the lot sizes in the Westwood Company example are not entirely symmetrically distributed. The two plots also show that several runs were made for lot sizes 30 and 60.

4.2 RESIDUALS

Ordinarily, direct diagnostic plots for the dependent variable Y are not too useful in regression analysis because the values of the observations on the dependent variable are a function of the level of the independent variable. Instead, diagnostics for the dependent variable are usually carried out indirectly through an examination of the residuals.

The residual e_i, as defined in (2.16), is the difference between the observed value and the fitted value:

$$(4.1) \qquad e_i = Y_i - \hat{Y}_i$$

As such, the residual may be regarded as the observed error, in distinction to the unknown true error ε_i in the regression model:

$$(4.2) \qquad \varepsilon_i = Y_i - E\{Y_i\}$$

For regression model (3.1), the ε_i are assumed to be independent normal random variables, with mean 0 and constant variance σ^2. If the model is appropriate for the data at hand, the observed residuals e_i should then reflect the properties assumed for the ε_i. This is the basic idea underlying *residual analysis,* a highly useful means of examining the aptness of a model.

Properties of Residuals

Mean.   The mean of the n residuals e_i for the simple linear regression model (3.1) is, by (2.17):

$$(4.3) \qquad \bar{e} = \frac{\Sigma e_i}{n} = 0$$

where $\bar{e}$ denotes the mean of the residuals. Thus, since $\bar{e}$ is always 0, it provides no information as to whether the true errors ε_i have expected value $E\{\varepsilon_i\} = 0$.

Variance.   The variance of the n residuals e_i is defined as follows for regression model (3.1):

$$(4.4) \qquad \frac{\Sigma(e_i - \bar{e})^2}{n - 2} = \frac{\Sigma e_i^2}{n - 2} = \frac{SSE}{n - 2} = MSE$$

If the model is appropriate, MSE is, as noted earlier, an unbiased estimator of the variance of the error terms σ^2.

Nonindependence.   The residuals e_i are not independent random variables because they involve the fitted values $\hat{Y}_i$ which are based on the sample estimates b_0 and b_1. Thus, the residuals for regression model (3.1) are associated with only $n - 2$ degrees of freedom. As a result, we know from (2.17) that the sum of the e_i must be 0 and from (2.19) that the products $X_i e_i$ must sum to 0.

When the sample size is large in comparison to the number of parameters in the regression model, the dependency effect among the residuals e_i is relatively unimportant and can be ignored for most purposes.

Standardized Residuals

Standardized residuals are used at times in residual analysis. Since the standard deviation of the error terms ε_i is σ, which is estimated by $\sqrt{MSE}$, we shall define here the standardized residual as follows:

$$(4.5) \qquad \frac{e_i - \bar{e}}{\sqrt{MSE}} = \frac{e_i}{\sqrt{MSE}}$$

As we shall see, standardized residuals can be particularly helpful in identifying outlying observations.

There are still other measures based on residuals, besides the standardized residuals, that are helpful in studying the aptness of the regression model. We shall take these up in Chapter 11.

Departures from Model to Be Studied by Residuals

We shall consider the use of residuals for examining six important types of departures from the simple linear regression model (3.1) with normal errors:

1. The regression function is not linear.
2. The error terms do not have constant variance.
3. The error terms are not independent.
4. The model fits all but one or a few outlier observations.
5. The error terms are not normally distributed.
6. One or several important independent variables have been omitted from the model.

4.3 DIAGNOSTICS FOR RESIDUALS

We take up now some informal diagnostic plots of residuals to provide information on whether any of the six types of departures from the simple linear regression model (3.1) just mentioned are present. The following plots of residuals (or standardized residuals) will be utilized here for this purpose:

1. Plot of residuals against independent variable
2. Plot of residuals against fitted values
3. Plot of residuals against time
4. Plot of residuals against omitted independent variable
5. Box plot of residuals
6. Normal probability plot of residuals

Figure 4.2 contains, for the Westwood Company example, plots of the residuals in Table 2.3 against the independent variable and against time, a box plot, and a nor-

FIGURE 4.2   Diagnostic Residual Plots—Westwood Company Example

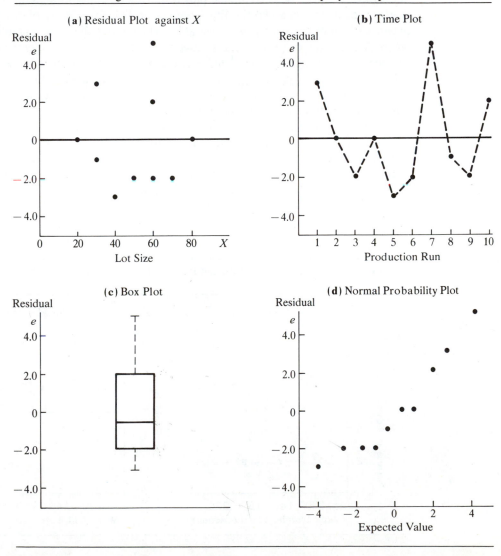

mal probability plot. All of these plots, as we shall see, support the aptness of regression model (3.1) for the lot size data.

We turn now to consider how residual analysis can be helpful in studying each of the six departures from regression model (3.1).

Nonlinearity of Regression Function

Whether a linear regression function is appropriate for the data being analyzed can be studied from a *residual plot against the independent variable* or from a *residual plot against the fitted values,* and also from a *scatter plot*. The latter, however, is not always as effective as a residual plot. Figure 4.3a contains a computer-generated scatter plot of the data and the fitted regression line for a study of the relation between amount of transit information and bus ridership in eight comparable test cities, where X is the number of bus transit maps distributed free to residents of the city at the beginning of the test period and Y is the increase during the test period in average daily bus ridership during nonpeak hours. The original data and fitted values are given in Table 4.1, columns 1, 2, and 3. The graph suggests strongly that a linear regression function is not appropriate.

Figure 4.3b presents for this same example a computer plot of the residuals e, shown in Table 4.1, column 4, plotted against the independent variable X. The lack of fit of the linear regression function is also strongly suggested by the residual plot against X in Figure 4.3b, since the residuals depart from 0 in a systematic fashion. Note that they are negative for smaller X values, positive for medium-size X values, and negative again for large X values.

FIGURE 4.3   Scatter Plot and Residual Plot Illustrating Nonlinear Regression Function—Transit Example

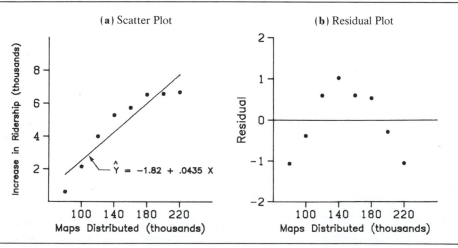

TABLE 4.1 Number of Maps Distributed and Increase in Ridership—Transit Example

City i	(1) Increase in Ridership (thousands) Y_i	(2) Maps Distributed (thousands) X_i	(3) Fitted Value $\hat{Y}_i$	(4) Residual $Y_i - \hat{Y}_i = e_i$	(5) Standardized Residual $e_i/\sqrt{MSE}$
1	.60	80	1.66	−1.06	−1.22
2	6.70	220	7.75	−1.05	−1.21
3	5.30	140	4.27	1.03	1.18
4	4.00	120	3.40	.60	.69
5	6.55	180	6.01	.54	.62
6	2.15	100	2.53	−.38	−.44
7	6.60	200	6.88	−.28	−.32
8	5.75	160	5.14	.61	.70

$$\hat{Y} = -1.82 + .0435X$$
$$MSE = .756$$

In this case, both Figures 4.3a and 4.3b are effective means of examining the appropriateness of the linearity of the regression function. In general, however, the residual plot has some important advantages over the scatter plot. First, the residual plot can easily be used for examining other facets of the aptness of the model. Second, there are occasions when the scaling of the scatter plot places the Y_i observations close to the fitted values $\hat{Y}_i$, for instance, when there is a steep slope. It then becomes more difficult to study the appropriateness of a linear regression function from the scatter plot. A residual plot, on the other hand, can clearly show any systematic pattern in the deviations around the fitted regression line under these conditions.

Figure 4.4a shows a prototype situation of the residual plot against X when the linear model is appropriate. The residuals should tend to fall within a horizontal band centered around 0, displaying no systematic tendencies to be positive and negative. This is the case in Figure 4.2a for the Westwood Company example.

Figure 4.4b shows a prototype situation of a departure from the linear regression model indicating the need for a curvilinear regression function. Here the residuals tend to vary in a systematic fashion between being positive and negative. This is the case in Figure 4.3b for the transit example. A different type of departure from linearity would, of course, lead to a picture different from the prototype pattern in Figure 4.4b.

Note

A plot of residuals against the fitted values $\hat{Y}$ provides equivalent information as the plot of residuals against X for the simple linear regression model. The reason is that the fitted values $\hat{Y}_i$ for the simple linear regression model are a linear function of the values X_i for the independent variable; thus, only the X scale values, not the basic pattern of the plotted points, are af-

FIGURE 4.4   Prototype Residual Plots

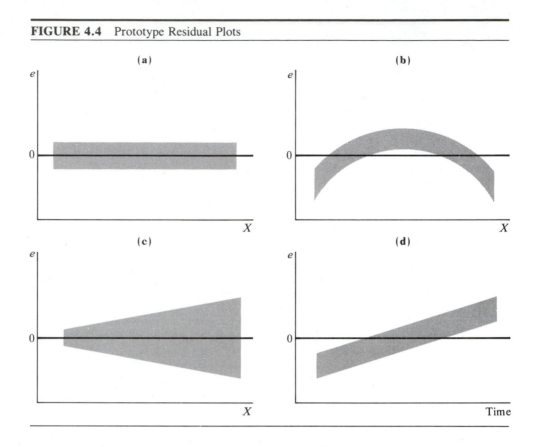

fected. For curvilinear regression and multiple regression, separate plots of the residuals against the fitted values and against the independent variable(s) are often helpful.

Nonconstancy of Error Variance

Plots of the residuals against the independent variable or against the fitted values are not only helpful to study whether a linear regression function is appropriate but also to examine whether the variance of the error terms is constant. Figure 4.5 shows a residual plot against the fitted values $\hat{Y}$ for an application involving the regression of the diastolic blood pressure of female children (Y) against their age (X). The plot was generated by the BMDP package (Ref. 4.1). Note that the horizontal axis is labeled PREDICTD, which stands for "predicted," an alternative term often used for "fitted" value. The numerical values shown in the graph indicate the number of residuals falling on or near a point. We have added the flared lines to highlight the tendency that the larger the fitted value $\hat{Y}$ is, the more spread out the residuals are. Since the relation between blood pressure and age is positive, this suggests that the error variance is larger for older children than for younger ones.

FIGURE 4.5 Residual Plot Illustrating Nonconstant Error Variance—Blood Pressure
Example (BMDP2R, Ref. 4.1)

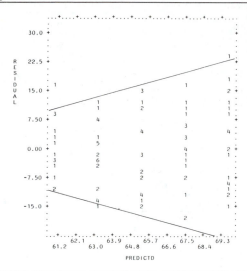

The prototype plot in Figure 4.4a exemplifies a residual plot when the error term
variance is constant. The residual plot in Figure 4.2a for the Westwood Company example is of this type, suggesting that the error terms have constant variance here.

Figure 4.4c shows a prototype picture of a residual plot when the error variance
increases with X. In many business, social science, and biological science applications, departures from constancy of the error variance tend to be of the trapezoidal
type shown in Figure 4.4c, as in the blood pressure example in Figure 4.5. One can
also encounter error variances decreasing with increasing levels of the independent
variable or varying in some other fashion.

Presence of Outliers

Outliers are extreme observations. Residual outliers can be identified from *residual
plots against X or $\hat{Y}$* (preferably based on standardized residuals), as well as from *box
plots, stem-and-leaf plots,* and *dot plots.* In a standardized residual plot, outliers are
points that lie far beyond the scatter of the remaining residuals, perhaps four or more
standard deviations from zero. The residual plot in Figure 4.6 presents standardized
residuals and contains one outlier, which is circled. Note that this residual represents
an observation almost six standard deviations from the fitted value.

Outliers can create great difficulty. When we encounter one, our first suspicion is
that the observation resulted from a mistake or other extraneous effect, and hence
should be discarded. A major reason for discarding it is that under the least squares
method, a fitted line may be pulled disproportionately toward an outlying observation
because the sum of the *squared* deviations is minimized. This could cause a mislead-

FIGURE 4.6 Residual Plot with Outlier

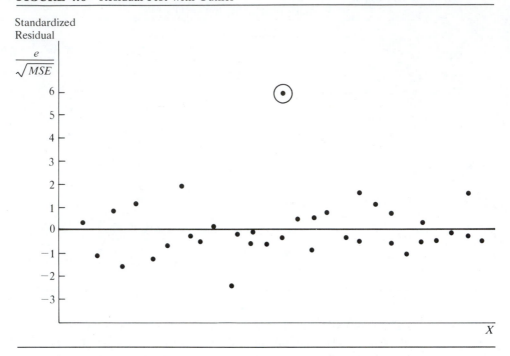

ing fit if indeed the outlier observation resulted from a mistake or other extraneous cause. On the other hand, outliers may convey significant information, as when an outlier occurs because of an interaction with another independent variable omitted from the model. A safe rule frequently suggested is to discard an outlier only if there is direct evidence that it represents an error in recording, a miscalculation, a malfunctioning of equipment, or a similar type of circumstance.

Note

When a linear regression model is fitted to a data set with a small number of observations and an outlier is present, the fitted regression may be so distorted by the outlier that the residual plot suggests a lack of fit of the linear regression model in addition to flagging the outlier. Figure 4.7 illustrates this situation. The scatter plot in Figure 4.7a presents a case where all observations except the outlier fall around a straight-line statistical relationship. When a linear regression function is fitted to these data, the outlier causes such a shift in the fitted regression line as to lead to a systematic pattern of deviations from the fitted line for the other observations, as evidenced by the residual plot in Figure 4.7b.

Nonindependence of Error Terms

Whenever data are obtained in a time sequence, it is a good idea to prepare a *time plot of the residuals*. The purpose of plotting the residuals against time is to see if there is any correlation between the error terms over time. Figure 4.8 contains a

FIGURE 4.7   Distorting Effect on Residuals Caused by an Outlier when Remaining Data
Follow Linear Regression

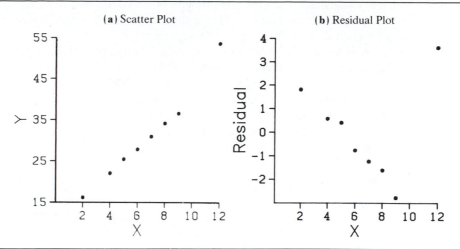

(a) Scatter Plot   (b) Residual Plot

time plot of the residuals in an experiment to study the relation between the diameter
of a weld (X) and the shear strength of the weld (Y). An evident correlation between
the error terms stands out. Negative residuals are associated mainly with the early
trials, and positive residuals with the later trials. Apparently, some effect connected
with time was present, such as learning by the welder or a gradual change in the
welding equipment, so the shear strength tended to be greater in the later welds because of this effect.

FIGURE 4.8   Residual Time Plot Illustrating Nonindependence of Error Terms—Welding
Example

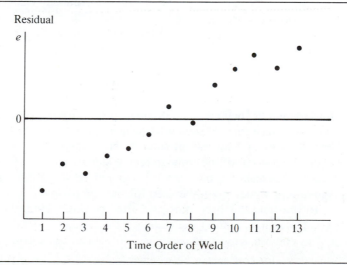

A prototype residual plot showing a time-related effect is presented in Figure 4.4d, which portrays a linear time-related effect, as in the welding example. It is sometimes useful to view the problem of nonindependence of the error terms as one in which an important variable (in this case, time) has been omitted from the model. We shall discuss this type of problem shortly.

When the error terms are independent, we expect the residuals to fluctuate in a more or less random pattern around the base line 0, such as the scattering shown in Figure 4.2b for the Westwood Company example. Lack of randomness can take the form of too much alternation of points around the zero line, or too little alternation. In practice, there is little concern with the former case because it does not arise frequently. Too little alternation, in contrast, frequently occurs, as in the welding example in Figure 4.8.

Note

When the residuals are plotted against X, as in Figure 4.3b, the scatter may not appear to be random. For this plot, however, the basic problem is probably not lack of independence of the error terms but rather a poorly fitting regression function. This, indeed, is the situation portrayed in the scatter plot in Figure 4.3a.

Nonnormality of Error Terms

As we noted earlier, small departures from normality do not create any serious problems. Major departures, on the other hand, should be of concern. The normality of the error terms can be studied informally by examining the residuals in a variety of graphic ways.

Distribution Plots.   A *box plot* is helpful for obtaining summary information about the symmetry of the residuals and about possible outliers. Figure 4.2c contains the box plot for the residuals in the Westwood Company example. No serious departures from normality are suggested by this plot. One can also construct a *histogram,* a *dot plot,* or a *stem-and-leaf plot* of the residuals to see if gross departures from normality are shown by such a plot. However, the number of cases in the regression study must be reasonably large for any one of these plots to convey reliable information about the shape of the distribution of the error terms.

Comparison of Frequencies.   Another possibility is to compare actual frequencies of the residuals against expected frequencies under normality. For example, if the number of cases in a regression study is fairly large, one can determine whether, say, about 68 percent of the standardized residuals $e_i/\sqrt{MSE}$ fall between -1 and 1, or about 90 percent fall between -1.645 and 1.645. If the sample size is small, corresponding t values would be used for the comparison.

To illustrate this procedure, we shall consider again the transit example in Table 4.1. Column 5 contains the standardized residuals, obtained by dividing each residual in column 4 by $\sqrt{MSE} = \sqrt{.756}$. Since n is small, we shall use for the comparison the t distribution with $n - 2 = 8 - 2 = 6$ degrees of freedom. The 95th percentile of this distribution, from Table A.2, is 1.943. Hence, under normality we would expect about 90 percent of the standardized residuals to fall between -1.943 and 1.943. Here, all eight standardized residuals fall within these limits. Similarly, under normality, we would expect about 60 percent of the standardized residuals to fall between $-.906$ and .906. The actual percentage here is 62.5 percent. Thus, the actual frequencies here are reasonably consistent with those expected under normality.

Normal Probability Plot.   Still another possibility is to prepare a *normal probability plot of the residuals*. Here each residual is plotted against its expected value when the distribution is normal. A plot that is nearly linear suggests agreement with normality, whereas a plot that departs substantially from linearity suggests that the error distribution is not normal.

Table 4.2, column 1, contains the residuals, in ascending order, for the Westwood Company example (from Table 2.3). To find the expected values of the ordered residuals under normality, we utilize the facts that (1) the expected value of the error terms for regression model (3.1) is zero, and (2) the standard deviation of the error terms is estimated by $\sqrt{MSE}$. Statistical theory has shown that for a normal random variable with mean 0 and estimated standard deviation $\sqrt{MSE}$, a good approximation of the expected value of the ith smallest observation in a random sample of n is:

$$(4.6) \qquad \sqrt{MSE}\left[z\left(\frac{i - .375}{n + .25}\right)\right]$$

where $z(A)$ as usual denotes the $(A)100$ percentile of the standard normal distribution.

TABLE 4.2   Residuals and Expected Values under Normality for Westwood Company Example

Ascending Order i	(1) Ordered Residual e_i	(2) Expected Value under Normality
1	-3.0	-4.24
2	-2.0	-2.74
3	-2.0	-1.79
4	-2.0	-1.02
5	-1.0	$-.33$
6	0.0	.33
7	0.0	1.02
8	2.0	1.79
9	3.0	2.74
10	5.0	4.24

Using this approximation, let us calculate the expected values of the ordered residuals under normality for the Westwood Company example. We found earlier (Table 3.3) that $MSE = 7.5$. For the smallest residual, we have $i = 1$. Hence, $(i - .375)/(n + .25) = (1 - .375)/(10 + .25) = .061$, and the expected value of the smallest residual under normality is:

$$\sqrt{7.5}[z(.061)] = \sqrt{7.5}(-1.55) = -4.24$$

Similarly, the expected value of the second smallest residual under normality is obtained by finding, for $i = 2$, $(i - .375)/(n + .25) = (2 - .375)/(10 + .25) = .159$ so that:

$$\sqrt{7.5}[z(.159)] = \sqrt{7.5}(-1.00) = -2.74$$

Because of the symmetry of a normal probability distribution, the expected values of the largest and second largest residuals are 4.24 and 2.74, respectively.

Table 4.2, column 2, contains all 10 expected values under the assumption of normality. Figure 4.2d presents a plot of the residuals against their expected values under normality. Note that the points in Figure 4.2d fall reasonably close to a straight line, suggesting that the distribution of the error terms does not depart substantially from a normal distribution. The steps in the plot in Figure 4.2d are due to the rounded nature of the data in the Westwood Company example.

Many computer packages will prepare normal probability plots at the option of the user. Some of these plots utilize standardized residuals, but this does not affect the basic nature of the plot.

In addition to visually assessing the approximate linearity of the points plotted in a normal probability plot, one can also calculate the coefficient of correlation (3.73) relating the residuals e_i to their expected values under normality. A high value of the correlation coefficient is indicative of normality. Table 4.3 contains critical values (percentiles) for various sample sizes for the distribution of the coefficient of correlation between the ordered residuals and their expected values under normality when the error terms are normally distributed. For a given α level, if the observed coefficient of correlation is at least as large as the tabled value, one can conclude that the error terms are reasonably normally distributed. For the Westwood Company example in Table 4.2, the coefficient of correlation is .955. Controlling the α risk at .05, we see from Table 4.3 that the entry for $n = 10$ is .918. Since the observed coefficient exceeds this level, we have support for our earlier conclusion that the distribution of the error terms does not depart substantially from a normal distribution.

Figure 4.9 contains a normal probability plot of the residuals in a regression study where the error terms follow a highly skewed distribution. This plot was generated by the Minitab statistical package (Ref. 4.2). Note the substantial departure from linearity of the points in Figure 4.9. The coefficient of correlation between the 14 ordered residuals and their expected values under normality is only .88, also suggesting a departure from a normal distribution since the critical value for $\alpha = .05$ and $n = 14$ according to Table 4.3 is greater than .92.

TABLE 4.3   Critical Values for Coefficient of Correlation between Ordered Residuals and Expected Values under Normality when Distribution of Error Terms Is Normal

	Level of Significance α		
n	.10	.05	.01
5	.903	.880	.826
10	.934	.918	.879
15	.951	.939	.910
20	.960	.951	.926
25	.966	.959	.939
30	.971	.964	.947
40	.977	.972	.959
50	.981	.977	.966
75	.987	.984	.976
100	.989	.987	.982

Source: Reprinted, with permission, from S. W. Looney and T. R. Gulledge, Jr., "Use of the Correlation Coefficient with Normal Probability Plots," *The American Statistician* 39 (1985), pp. 75–79.

FIGURE 4.9   Example of Normal Probability Plot when Distribution of Error Terms Is Highly Skewed (Minitab, Ref. 4.2)

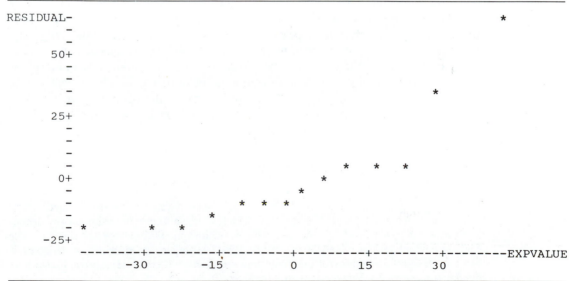

Note

The analysis for model departures with respect to normality is, in many respects, more difficult than that for other types of departures. In the first place, random variation can be particularly mischievous when one studies the nature of a probability distribution unless the sample size is quite large. Even worse, other types of departures can and do affect the distribution of the residuals. For instance, residuals may appear to be not normally distributed because an inappropriate regression function is used or because the error variance is not constant. Hence, it is usually a good strategy to investigate these other types of departures first, before concerning oneself with the normality of the error terms.

Omission of Important Independent Variables

Residuals should also be plotted against variables omitted from the model that might have important effects on the response, data being available. The time variable cited earlier in the welding application is an example. The purpose of this additional analysis is to determine whether there are any other key independent variables that could provide important additional descriptive and predictive power to the model.

As another example, in a study to predict output by piece rate workers in an assembling operation, the relation between output (Y) and age (X) of worker was studied for a sample of employees. The plot of the residuals against X is shown in Figure 4.10a, and indicates no ground for suspecting the appropriateness of the linearity of the regression function or the constancy of the error variance.

Machines produced by two companies (A and B) are used in the assembling operation. Residual plots against X by type of machine were undertaken and are shown in Figures 4.10b and 4.10c. Note that the residuals for machines made by Company A tend to be positive, while those for machines made by Company B tend to be negative. Thus, type of machine appears to have a definite effect on productivity, and output predictions may turn out to be far superior when this independent variable is added to the model. While this example dealt with a classification variable (type of machine), the residual analysis for an additional quantitative variable is completely analogous. One would simply plot the residuals against the additional independent variable and see whether or not the residuals tend to vary systematically with the level of the additional independent variable.

Note

We do not say that the original model is "wrong" when it can be improved materially by adding one or more independent variables. Only a few of the factors operating on any dependent variable Y in real-world situations can be included explicitly in a regression model. The chief purpose of residual analysis in identifying other important independent variables is therefore to test the adequacy of the model and see whether it could be improved materially by adding one or a few independent variables.

FIGURE 4.10   Residual Plots for Possible Omission of Important Independent Variable—
Productivity Example

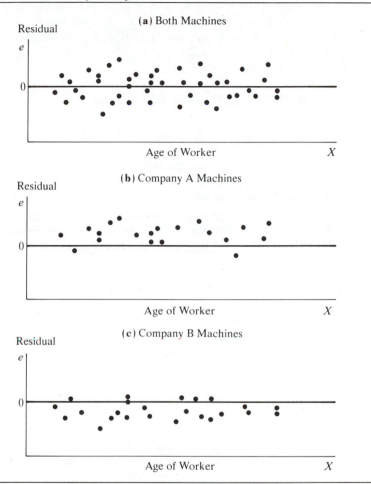

(a) Both Machines

Residual

e

0

Age of Worker X

(b) Company A Machines

Residual

e

0

Age of Worker X

(c) Company B Machines

Residual

e

0

Age of Worker X

Some Final Comments

1. We discussed the model departures one at a time. In actuality, several types of departures may occur together. For instance, a linear regression function may be a poor fit and the variance of the error terms may not be constant. In these cases, the prototype patterns of Figure 4.4 can still be useful, but they would need to be combined into composite patterns.

2. While graphic analysis of residuals is only an informal method of analysis, in many cases it suffices for examining the aptness of a model.

3. The basic approach to residual analysis applies not only to simple linear regression but also to more complex regression and other types of statistical models.

4. Most of the routine work in residual analysis can be handled on computers. Almost all regression programs supply the fitted values and corresponding residuals, and routines are generally available whereby the various types of residual plots can be obtained.

4.4 OVERVIEW OF TESTS INVOLVING RESIDUALS

Graphic analysis of residuals is inherently subjective. Nevertheless, subjective analysis of a variety of interrelated residual plots will frequently reveal difficulties in the model more clearly than particular formal tests. There are occasions, however, when one wishes to put specific questions to a test. We now review some of the relevant tests briefly, and take up one new type of test.

Most statistical tests require independent observations. As we have seen, however, the residuals are dependent. Fortunately, the dependency becomes quite small for large samples, so that one can usually then ignore it.

Tests for Randomness

A runs test is frequently used to test for lack of randomness in the residuals arranged in time order. Another test, specifically designed for lack of randomness in least squares residuals, is the Durbin-Watson test. This test is discussed in Chapter 13.

Tests for Constancy of Variance

When a residual plot gives the impression that the variance may be increasing or decreasing in a systematic manner related to X or $E\{Y\}$, a simple test is to fit separate regression functions to each half of the observations arranged by level of X, calculate error mean squares for each, and test for equality of the error variances by an F test. Another simple test is by means of rank correlation between the absolute value of the residual and the value of the independent variable.

Tests for Outliers

A simple test for an outlier observation involves fitting a new regression line to the other $n - 1$ observations. The suspect observation, which was not used in fitting the new line, can now be regarded as a new observation. One can calculate the probability that in n observations, a deviation from the fitted line as great as that of the outlier will be obtained by chance. If this probability is sufficiently small, the outlier can be rejected as not having come from the same population as the other $n - 1$ observations. Otherwise, the outlier is retained.

Many other tests to aid in evaluating outliers have been developed. These are discussed in specialized references such as Reference 4.3 and in statistical journals.

Tests for Normality

Goodness of fit tests can be used for examining the normality of the error terms. For instance, the chi-square test or the Kolmogorov-Smirnov test and its modification, the Lilliefors test, can be employed for testing the normality of the error terms by analyzing the residuals.

Note

The runs test, rank correlation, and goodness of fit tests are commonly used statistical procedures and are discussed in many basic statistics texts.

4.5 *F* TEST FOR LACK OF FIT

We now take up a formal test for determining whether a specified regression function adequately fits the data. We illustrate this test for ascertaining whether a linear regression function is a good fit for the data.

Assumptions

The lack of fit test assumes that the observations Y for given X are (1) independent and (2) normally distributed, and that (3) the distributions of Y have the same variance σ^2.

The lack of fit test also requires repeat observations at one or more X levels. In nonexperimental data, these may occur fortuitously, as when in a productivity study relating workers' output and age, several workers of the same age happen to be included in the study. In an experiment, one can assure by design that there are repeat observations. For instance, in an experiment on the effect of size of salesperson bonus on sales, three salespersons can be offered a particular size of bonus, for each of six bonus sizes, and their sales then observed.

Repeated trials for the same level of the independent variable, of the type described, are called *replications*. The resulting observations are called *replicates*.

Example

In an experiment involving 12 similar but scattered suburban branch offices of a commercial bank, holders of checking accounts at the offices were offered gifts for setting up savings accounts. The initial deposit in the new savings account had to be for a specified minimum amount to qualify for the gift. The value of the gift was di-

TABLE 4.4   Data for Bank Example

Branch i	Size of Minimum Deposit (dollars) X_i	Number of New Accounts Y_i	Branch i	Size of Minimum Deposit (dollars) X_i	Number of New Accounts Y_i
1	125	160	7	75	42
2	100	112	8	175	124
3	200	124	9	125	150
4	75	28	10	200	104
5	150	152	11	100	136
6	175	156			

rectly proportional to the specified minimum deposit. Various levels of minimum deposit and related gift values were used in the experiment in order to ascertain the relation between the specified minimum deposit and gift value on the one hand and number of accounts opened at the office on the other. Altogether, six levels of minimum deposit and proportional gift value were used, with two of the branch offices assigned at random to each level. One branch office had a fire during the period and was dropped from the study. Table 4.4 contains the results, where X is the amount of minimum deposit and Y is the number of new savings accounts that were opened and qualified for the gift during the test period.

A linear regression function was fitted in the usual fashion; it is (calculations not shown):

$$\hat{Y} = 50.72251 + .48670X$$

The analysis of variance table also was obtained and is shown in Table 4.5. A scatter plot, together with the fitted regression line, is shown in Figure 4.11. The indications are strong that a linear regression function is inappropriate. To test this formally, we shall use the general linear test approach described in Section 3.9.

TABLE 4.5   ANOVA Table for Bank Example

Source of Variation	SS	df	MS
Regression	$SSR = 5,141.3$	1	$MSR = 5,141.3$
Error	$SSE = 14,741.6$	9	$MSE = 1,638.0$
Total	$SSTO = 19,882.9$	10	

FIGURE 4.11   Scatter Plot and Fitted Regression Line—Bank Example

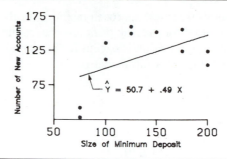

Notation

First, we need to modify our notation to recognize the existence of replications at some levels of X. Table 4.6 presents the same data as Table 4.4, but in an arrangement that recognizes the replicates. We shall denote the different X levels in the study, whether or not replicated observations are present, as $X_1, \ldots, X_c$. For the bank example, $c = 6$ since there are six minimum deposit size levels in the study, for five of which there are two observations and for one there is a single observation. We shall let $X_1 = 75$ (the smallest minimum deposit level), $X_2 = 100, \ldots,$ $X_6 = 200$. Further, we shall denote the number of replicates for the jth level of X as n_j; for our example, $n_1 = n_2 = n_3 = n_5 = n_6 = 2$ and $n_4 = 1$. Thus, the total number of observations n is given by:

$$(4.7) \qquad\qquad n = \sum_{j=1}^{c} n_j$$

We shall denote the observed value of the dependent variable for the ith replicate for the jth level of X by Y_{ij}, where $i = 1, \ldots, n_j, j = 1, \ldots, c$. For the bank example (Table 4.6), $Y_{11} = 28$, $Y_{21} = 42$, $Y_{12} = 112$, and so on. Finally, we shall denote the mean of the Y observations at the level $X = X_j$ by $\bar{Y}_j$. Thus, $\bar{Y}_1 = (28 + 42)/2 = 35$ and $\bar{Y}_4 = 152/1 = 152$.

TABLE 4.6   Data for Bank Example, Arranged by Replicate Number and Minimum Deposit

	Size of Minimum Deposit (dollars)					
Replicate	$j = 1$ $X_1 = 75$	$j = 2$ $X_2 = 100$	$j = 3$ $X_3 = 125$	$j = 4$ $X_4 = 150$	$j = 5$ $X_5 = 175$	$j = 6$ $X_6 = 200$
$i = 1$	28	112	160	152	156	124
$i = 2$	42	136	150		124	104
Mean $\bar{Y}_j$	35	124	155	152	140	114

Full Model

The general linear test approach begins with the specification of the full model. The full model used for the lack of fit test makes the same assumptions as the simple linear regression model (3.1) except for assuming a linear regression relation, the subject of the test. This full model is:

(4.8) $Y_{ij} = \mu_j + \varepsilon_{ij}$ Full model

where:

  μ_j are parameters $j = 1, \ldots, c$
  ε_{ij} are independent $N(0, \sigma^2)$

Since the error terms have expectation zero, it follows that:

(4.9) $E\{Y_{ij}\} = \mu_j$

Thus, the parameter μ_j $(j = 1, \ldots, c)$ is the mean response when $X = X_j$.

The full model (4.8) is like the regression model (3.1) in stating that each response Y is made up of two components—the mean response when $X = X_j$ and a random error term. The difference between the two models is that in the full model (4.8) there are no restrictions on the means μ_j, whereas in the regression model (3.1) the mean responses are linearly related to X (i.e., $E\{Y\} = \beta_0 + \beta_1 X$).

To fit the full model to the data, we require the least squares estimators for the parameters μ_j. It can be shown that the least squares estimators of μ_j are simply the sample means $\bar{Y}_j$:

(4.10) $\hat{\mu}_j = \bar{Y}_j$

Thus, the estimated expected value for observation Y_{ij} is $\bar{Y}_j$, and the error sum of squares for the full model therefore is:

(4.11) $SSE(F) = \sum_j \sum_i (Y_{ij} - \bar{Y}_j)^2 = SSPE$

In the context of the test for lack of fit, the full model error sum of squares (4.11) is called the *pure error sum of squares* and is denoted by *SSPE*.

Note that *SSPE* is made up of the sums of squared deviations at each X level. At level $X = X_j$, this sum of squared deviations is:

(4.12) $\sum_i (Y_{ij} - \bar{Y}_j)^2$

These sums of squares are then added over all of the X levels $(j = 1, \ldots, c)$. For the bank example, we have:

$$SSPE = (28 - 35)^2 + (42 - 35)^2 + (112 - 124)^2 + (136 - 124)^2$$
$$+ (160 - 155)^2 + (150 - 155)^2 + (152 - 152)^2$$
$$+ (156 - 140)^2 + (124 - 140)^2 + (124 - 114)^2$$
$$+ (104 - 114)^2 = 1{,}148$$

Note that any X level where there is no replication makes no contribution to *SSPE* because $\bar{Y}_j = Y_{1j}$ then. Thus, $(152 - 152)^2 = 0$ for $j = 4$ in the bank example.

The degrees of freedom associated with *SSPE* can be obtained by recognizing that the sum of squared deviations (4.12) at a given level of X is like an ordinary total sum of squares based on n observations, which has $n - 1$ degrees of freedom associated with it. Here, there are n_j observations when $X = X_j$, hence the degrees of freedom are $n_j - 1$. Just as *SSPE* is the sum of the sums of squares (4.12), so the number of degrees of freedom associated with *SSPE* is the sum of the component degrees of freedom:

$$(4.13) \qquad df_F = \sum_j (n_j - 1) = \sum_j n_j - c = n - c$$

For the bank example, we have $df_F = 11 - 6 = 5$. Note that any X level with no replication makes no contribution to df_F because $n_j - 1 = 1 - 1 = 0$ then, just as such an X level makes no contribution to *SSPE*.

Reduced Model

The general linear test approach next requires consideration of the reduced model under H_0. For testing the aptness of a linear regression relation, the alternatives are:

$$(4.14) \qquad \begin{aligned} H_0 &: E\{Y\} = \beta_0 + \beta_1 X \\ H_a &: E\{Y\} \neq \beta_0 + \beta_1 X \end{aligned}$$

Thus, H_0 postulates that μ_j in the full model (4.8) is linearly related to X_j:

$$\mu_j = \beta_0 + \beta_1 X_j$$

The reduced model therefore is:

$$(4.15) \qquad Y_{ij} = \beta_0 + \beta_1 X_j + \varepsilon_{ij} \qquad\qquad \text{Reduced model}$$

Note that the reduced model is the ordinary simple linear regression model (3.1), with the subscripts modified to recognize the existence of replications. We know that the estimated expected value for observation Y_{ij} with regression model (3.1) is the fitted value $\hat{Y}_{ij}$:

$$(4.16) \qquad \hat{Y}_{ij} = b_0 + b_1 X_j$$

Hence, the error sum of squares for the reduced model is the usual error sum of squares *SSE*:

$$(4.17) \qquad SSE(R) = \sum\sum [Y_{ij} - (b_0 + b_1 X_j)]^2$$

$$= \sum\sum (Y_{ij} - \hat{Y}_{ij})^2 = SSE$$

We also know that the degrees of freedom associated with $SSE(R)$ are:

$$df_R = n - 2$$

For the bank example, we have from Table 4.5:

$$SSE(R) = SSE = 14{,}741.6$$

$$df_R = 9$$

Test Statistic

The general linear test statistic (3.69):

$$F^* = \frac{SSE(R) - SSE(F)}{df_R - df_F} \div \frac{SSE(F)}{df_F}$$

here becomes:

(4.18)
$$F^* = \frac{SSE - SSPE}{(n-2) - (n-c)} \div \frac{SSPE}{n-c}$$

The difference between the two error sums of squares is called the *lack of fit sum of squares* here, and is denoted by $SSLF$:

(4.19)
$$SSLF = SSE - SSPE$$

We can then express the test statistic as follows:

(4.20)
$$F^* = \frac{SSLF}{c-2} \div \frac{SSPE}{n-c}$$

$$= \frac{MSLF}{MSPE}$$

where $MSLF$ denotes the *lack of fit mean square* and $MSPE$ denotes the *pure error mean square*.

We know that large values of F^* lead to conclusion H_a in the general linear test. Decision rule (3.70) here becomes:

(4.21)

If $F^* \le F(1 - \alpha; c - 2, n - c)$, conclude H_0

If $F^* > F(1 - \alpha; c - 2, n - c)$, conclude H_a

For the bank example, the test statistic can be constructed easily from our earlier results:

$$SSPE = 1{,}148.0 \qquad\qquad n - c = 11 - 6 = 5$$

$$SSE = 14{,}741.6$$

$$SSLF = 14{,}741.6 - 1{,}148.0 = 13{,}593.6 \qquad c - 2 = 6 - 2 = 4$$

$$F* = \frac{13,593.6}{4} \div \frac{1,148.0}{5}$$

$$= \frac{3,398.4}{229.6} = 14.80$$

If the level of significance is to be $\alpha = .01$, we require $F(.99; 4, 5) = 11.4$. Since $F* = 14.80 > 11.4$, we conclude H_a, that the regression function is not linear. This, of course, accords with our visual impression from Figure 4.11. To report the *P*-value for the test statistic, we note that $F* = 14.80$ lies between $F(.99; 4, 5) = 11.4$ and $F(.995; 4, 5) = 15.6$ and thus the *P*-value must be between .005 and .01. The exact *P*-value can be shown to be .006.

ANOVA Table

The definition of the lack of fit sum of squares *SSLF* in (4.19) indicates that we have, in fact, decomposed the error sum of squares into two components:

(4.22) $$SSE = SSPE + SSLF$$

This decomposition follows from the identity:

(4.23) $$\underbrace{Y_{ij} - \hat{Y}_{ij}}_{\substack{\text{Error} \\ \text{deviation}}} = \underbrace{Y_{ij} - \bar{Y}_j}_{\substack{\text{Pure error} \\ \text{deviation}}} + \underbrace{\bar{Y}_j - \hat{Y}_{ij}}_{\substack{\text{Lack of fit} \\ \text{deviation}}}$$

This identity shows that the error deviations in *SSE* are made up of a pure error component and a lack of fit component. Figure 4.12 illustrates this partitioning for the case $Y_{13} = 160$, $X_3 = 125$ in the bank example.

When (4.23) is squared and summed over all observations, we obtain (4.22) since the cross-product sum equals zero:

(4.24) $$\sum\sum (Y_{ij} - \hat{Y}_{ij})^2 = \sum\sum (Y_{ij} - \bar{Y}_j)^2 + \sum\sum (\bar{Y}_j - \hat{Y}_{ij})^2$$

$$SSE \qquad = \qquad SSPE \qquad + \qquad SSLF$$

Note from (4.24) that we can define the lack of fit sum of squares directly as follows:

(4.25) $$SSLF = \sum\sum (\bar{Y}_j - \hat{Y}_{ij})^2$$

Formula (4.25) indicates clearly why *SSLF* measures lack of fit. If the linear regression function is appropriate, then the means $\bar{Y}_j$ will be near the fitted values $\hat{Y}_{ij}$ calculated from the estimated linear regression function and *SSLF* will be small. On the other hand, if the linear regression function is not appropriate, the means $\bar{Y}_j$ will not be near the fitted values calculated from the estimated linear regression function, as in Figure 4.11 for the bank example, and *SSLF* will be large.

FIGURE 4.12   Illustration of Decomposition of Error Deviation $Y_{ij} - \hat{Y}_{ij}$—Bank Example

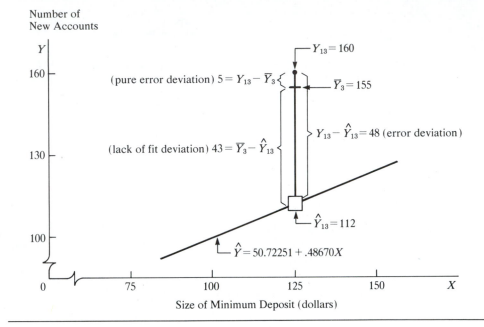

Formula (4.25) also indicates why $c - 2$ degrees of freedom are associated with *SSLF*. There are c means $\bar{Y}_j$ in the sum of squares, and two degrees of freedom are lost in estimating the parameters β_0 and β_1 of the linear regression function to obtain the fitted values $\hat{Y}_{ij}$.

An ANOVA table can be constructed for the decomposition of *SSE*. Table 4.7a contains the general ANOVA table, including the decomposition of *SSE* just explained and the mean squares of interest, and Table 4.7b contains the ANOVA decomposition for the bank example.

Comments

1. As was shown by the bank example, not all levels of X need have repeat observations for the F test for lack of fit to be applicable. Repeat observations at only one or some levels of X are sufficient.

2. It can be shown that the mean squares *MSPE* and *MSLF* have the following expectations:

(4.26)
$$E\{MSPE\} = \sigma^2$$

(4.27)
$$E\{MSLF\} = \sigma^2 + \frac{\Sigma n_j[\mu_j - (\beta_0 + \beta_1 X_j)]^2}{c - 2}$$

The reason for the term "pure error" is that *MSPE* is always an unbiased estimator of the error term variance σ^2, no matter what is the true regression function. The expected value of

TABLE 4.7   ANOVA Table for Testing Lack of Fit of Simple Linear Regression Function and Bank Example Table

(a) General

Source of Variation	SS	df	MS
Regression	SSR	1	MSR
Error	SSE	$n - 2$	MSE
Lack of fit	SSLF	$c - 2$	MSLF
Pure error	SSPE	$n - c$	MSPE
Total	SSTO	$n - 1$	

(b) Bank Example

Source of Variation	SS	df	MS
Regression	$SSR = 5{,}141.3$	1	$MSR = 5{,}141.3$
Error	$SSE = 14{,}741.6$	9	$MSE = 1{,}638.0$
Lack of fit	$SSLF = 13{,}593.6$	4	$MSLF = 3{,}398.4$
Pure error	$SSPE = 1{,}148.0$	5	$MSPE = 229.6$
Total	$SSTO = 19{,}882.9$	10	

MSLF also is σ^2 if the regression function is linear, because $\mu_j = \beta_0 + \beta_1 X_j$ then and the second term in (4.27) becomes zero. On the other hand, if the regression function is not linear, $\mu_j \neq \beta_0 + \beta_1 X_j$, so $E\{MSLF\}$ will be greater than σ^2. Hence, a value of F^* near 1 accords with a linear regression function; large values of F^* indicate that the regression function is not linear.

3. Suppose that prior to any analysis of the aptness of the model, we had wished to test whether or not $\beta_1 = 0$ for the bank example (Table 4.5). The test statistic (3.59) would be:

$$F^* = \frac{MSR}{MSE} = \frac{5{,}141.3}{1{,}638.0} = 3.14$$

For $\alpha = .10$, $F(.90; 1, 9) = 3.36$, and we would conclude H_0, that $\beta_1 = 0$ or that there is no *linear association* between minimum deposit size (and value of gift) and number of new accounts. A conclusion that there is no *relation* between these variables would be improper, however. Such an inference requires that regression model (3.1) be appropriate. Here it is not, as we have seen, because the regression function is not linear. There exists indeed a (curvilinear) relation between minimum deposit size and number of new accounts, and testing whether or not $\beta_1 = 0$ under these circumstances has entirely different implications. This illustrates the importance of always examining the aptness of a model before further inferences are drawn.

4. The general linear test approach just explained can be used to test the aptness of other regression functions, not just the simple linear one in (4.14). Only the degrees of freedom for *SSLF* will need be modified. In general, $c - p$ degrees of freedom are associated with *SSLF*, where p is the number of parameters in the regression function. For the test of a simple linear regression function, $p = 2$ because there are two parameters, β_0 and β_1, in the regression function.

5. The alternative H_a in (4.14) includes all regression functions other than a linear one. For instance, it includes a quadratic regression function or a logarithmic one. If H_a is concluded, a study of residuals can be helpful in identifying an appropriate function.

6. If we conclude that the employed model in H_0 is appropriate, the usual practice is to use the error mean square MSE as an estimator of σ^2 in preference to the pure error mean square $MSPE$, since the former contains more degrees of freedom.

7. Observations at the same level of X are genuine repeats only if they involve independent trials with respect to the error term. Suppose that in a regression analysis of the relation between hardness (Y) and amount of carbon (X) in specimens of an alloy, the error term in the model covers, among other things, random errors in the measurement of hardness by the analyst and effects of uncontrolled production factors which vary at random from specimen to specimen and affect hardness. If the analyst takes two readings on the hardness of a specimen, this will not provide a genuine replication because the effects of random variation in the production factors are fixed in any given specimen. For genuine replications, different specimens with the same carbon content (X) would have to be measured by the analyst so that *all* the effects covered in the error term could vary at random from one repeated observation to the next.

8. Clearly, repeat observations are most valuable whenever we are not certain of the nature of the regression function. If at all possible, provision should be made for some replications. If it is not possible to have replications, an approximate test for lack of fit can sometimes be conducted. For the approximate test, it is necessary to have some cases at adjacent X levels for which the mean responses are quite close to each other. Such adjacent cases are then grouped together and treated as pseudoreplicates for purposes of conducting an approximate test for lack of fit.

4.6 OVERVIEW OF REMEDIAL MEASURES

If the simple linear regression model (3.1) is not appropriate for a data set, there are two basic choices:

1. Abandon regression model (3.1) and search for a more appropriate model.
2. Use some transformation on the data so that regression model (3.1) is appropriate for the transformed data.

Each approach has advantages and disadvantages. The first approach may entail a more complex model that could yield better insights, but may also lead into difficulties in estimating the parameters. Successful use of transformations, on the other hand, leads to relatively simple methods of estimation and may involve fewer parameters than a complex model, an advantage when the sample size is small. Yet transformations may obscure the fundamental interconnections between the variables, though at other times may illuminate them.

We shall consider the use of transformations in this chapter and the use of more complex models in later chapters. First, we provide a brief overview of remedial measures.

Nonlinearity of Regression Function

If the regression function is not linear, a direct approach is to modify regression model (3.1) with respect to the nature of the regression function. For instance, a quadratic regression function might be used:

$$E\{Y\} = \beta_0 + \beta_1 X + \beta_2 X^2$$

or an exponential regression function:

$$E\{Y\} = \beta_0 \beta_1^X$$

In Chapter 9, we discuss models where the regression function is a polynomial.

The transformation approach uses a transformation to linearize, at least approximately, a nonlinear regression function. We discuss the use of transformations to linearize regression functions in the next section.

Nonconstancy of Error Variance

If the error variance is not constant but varies in a systematic fashion, a direct approach is to modify the model to allow for this and use the method of *weighted least squares* to obtain the estimators of the parameters. We discuss the use of weighted least squares for this purpose in Chapter 11.

Transformations can also be effective in stabilizing the variance. Some of these are discussed in the next section.

Nonindependence of Error Terms

If the error terms are correlated, a direct remedial measure is to work with a model that calls for correlated error terms. We discuss such a model in Chapter 13. A simple remedial transformation that is often helpful is to work with first differences, a topic also discussed in Chapter 13.

Nonnormality of Error Terms

Lack of normality and nonconstant error variances frequently go hand in hand. Fortunately, it is often the case that the same transformation that helps stabilize the variance is also helpful in normalizing the error terms. It is therefore desirable that the transformation for stabilizing the error variance be utilized first, and then the residuals studied to see if serious departures from normality are still present. We discuss transformations to achieve normality in the next section.

Omission of Important Independent Variables

When residual analysis indicates that an important independent variable has been omitted from the model, the solution is to modify the model. In Chapter 7 and later chapters, we discuss multiple regression analysis in which two or more independent variables are utilized.

4.7 TRANSFORMATIONS

We now consider in more detail the use of transformations of one or both of the original variables before carrying out the regression analysis. Simple transformations of either the dependent variable Y or the independent variable X, or of both, are often sufficient to make the simple linear regression model appropriate for the transformed data.

Transformations for Nonlinear Relation Only

We first consider transformations for linearizing a nonlinear regression relation when the distribution of the error terms is reasonably close to a normal distribution and the error terms have approximately constant variance. In this situation, transformations on X should be attempted. The reason why transformations on Y may not be desirable here is that a transformation on Y, such as $Y' = \sqrt{Y}$, may materially change the shape of the distribution of the error terms from the normal distribution and may also lead to substantially differing error term variances.

Figure 4.13 contains some prototype nonlinear regression relations with constant error term variance, and presents some simple transformations on X that may be helpful to linearize the regression relationship without affecting the distributions of Y. Alternative transformations may be tried. Scatter plots and residual plots based on each transformation should then be prepared and analyzed to decide which transformation is most effective.

Example.   In columns 1 and 2 of Table 4.8 are presented experimental data for 10 sales trainees on number of days of training received (X) and performance score (Y) in a battery of simulated sales situations. These observations are shown as a scatter plot in Figure 4.14a. Clearly the regression relation appears to be curvilinear, so the simple linear regression model (3.1) does not seem to be appropriate. Since the variability at the different X levels appears to be fairly constant, we shall consider a transformation on X. Based on the prototype plot in Figure 4.13a, we shall consider initially the square root transformation $X' = \sqrt{X}$. The transformed values are shown in column 3 of Table 4.8.

In Figure 4.14b, the same data are plotted with the independent variable transformed to $X' = \sqrt{X}$. Note that the scatter plot now shows a reasonably linear rela-

FIGURE 4.13 Prototype Nonlinear Regression Patterns with Constant Error Variance and Simple Transformations of X

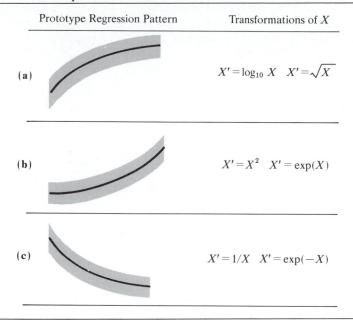

Prototype Regression Pattern	Transformations of X
(a)	$X' = \log_{10} X \quad X' = \sqrt{X}$
(b)	$X' = X^2 \quad X' = \exp(X)$
(c)	$X' = 1/X \quad X' = \exp(-X)$

TABLE 4.8 Regression Calculations with Square Root Transformation of X—Sales Training Example

Sales Trainee i	(1) Days of Training X_i	(2) Performance Score Y_i	(3) $X_i' = \sqrt{X_i}$	(4) $X_i' Y_i$	(5) $(X_i')^2$
1	.5	46	.70711	32.527	.5
2	.5	51	.70711	36.062	.5
3	1.0	71	1.00000	71.000	1.0
4	1.0	75	1.00000	75.000	1.0
5	1.5	92	1.22474	112.677	1.5
6	1.5	99	1.22474	121.250	1.5
7	2.0	105	1.41421	148.492	2.0
8	2.0	112	1.41421	158.392	2.0
9	2.5	121	1.58114	191.318	2.5
10	2.5	125	1.58114	197.643	2.5
Total	15.0	897	11.85440	1,144.361	15.0

FIGURE 4.14   Scatter Plots and Residual Plots—Sales Training Example

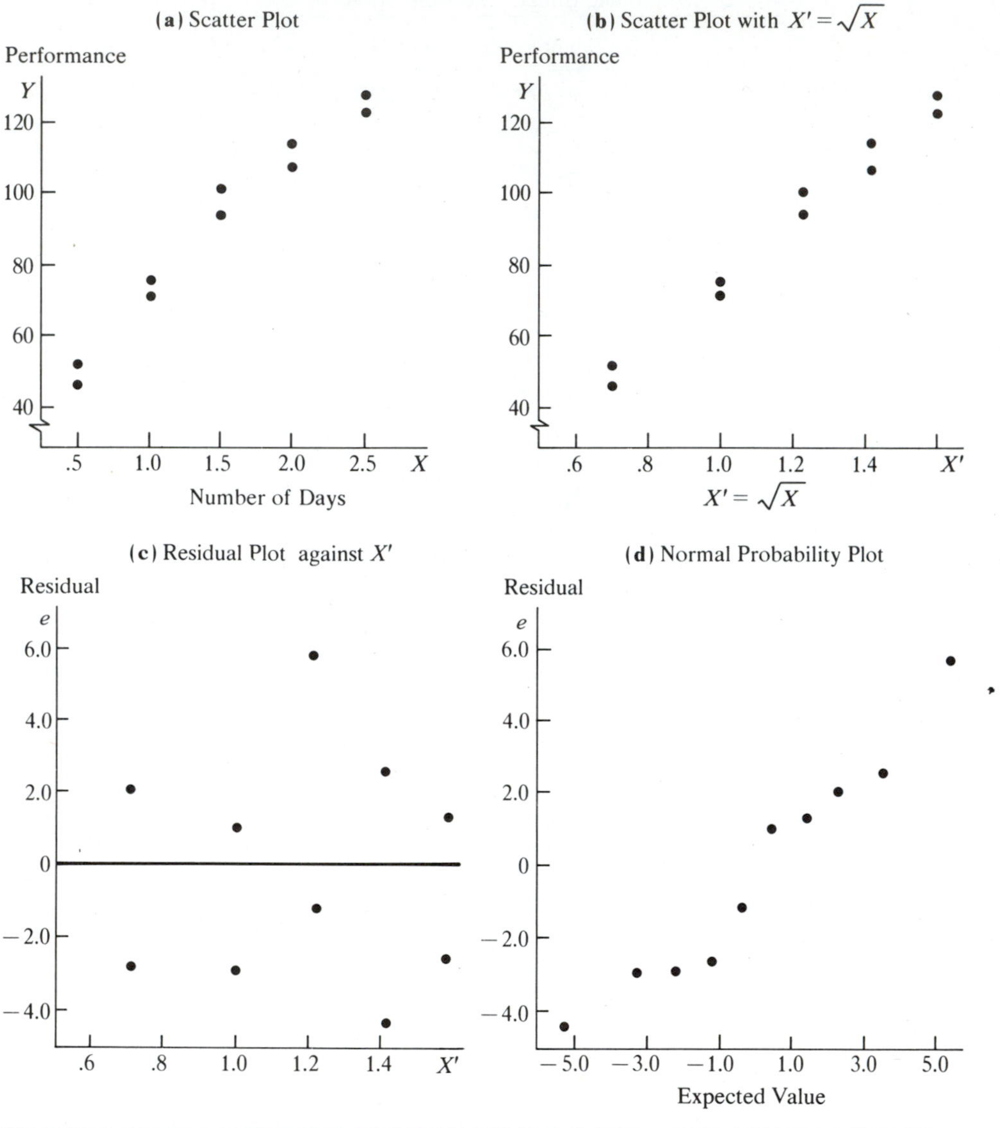

(a) Scatter Plot

(b) Scatter Plot with $X' = \sqrt{X}$

(c) Residual Plot against X'

(d) Normal Probability Plot

tion. The variability of the scatter at the different X levels is the same as before, since we did not make a transformation on Y.

To examine further whether the simple linear regression model (3.1) is appropriate now, we fit it to the transformed X data. The regression calculations with the transformed X data are carried out in the usual fashion. Table 4.8 contains the necessary least squares calculations. Since X' now plays the role of X in all earlier formulas, we obtain:

$$b_1 = \frac{\sum X_i' Y_i - \dfrac{\sum X_i' \sum Y_i}{n}}{\sum (X_i')^2 - \dfrac{(\sum X_i')^2}{n}} = \frac{1,144.361 - \dfrac{11.85440(897)}{10}}{15.0 - \dfrac{(11.85440)^2}{10}}$$

$$= 85.5259$$

$$b_0 = \frac{1}{n}\left(\sum Y_i - b_1 \sum X_i'\right) = \frac{1}{10}[897 - 85.5259(11.85440)]$$

$$= -11.6858$$

The fitted regression function therefore is:

$$\hat{Y} = -11.69 + 85.53X'$$

Figure 4.14c contains a plot of the residuals against X'. There is no evidence of lack of fit or of strongly unequal error variances. Figure 4.14d contains a normal probability plot of the residuals. No strong indications of substantial departures from normality are indicated by this plot. This conclusion is supported by the high coefficient of correlation between the ordered residuals and their expected values under normality, .972 (see Table 4.3). Thus, the simple linear regression model (3.1) appears to be appropriate here for the transformed data.

The fitted regression function in the original units of X can easily be obtained, if desired:

$$\hat{Y} = -11.69 + 85.53\sqrt{X}$$

Note

At times, it may be helpful to introduce a constant into the transformation. For example, if some of the X data are near zero and the reciprocal transformation is desired, we can shift the origin by using the transformation $X' = 1/(X + k)$, where k is an appropriately chosen constant.

Transformations for Nonnormality and Unequal Error Variances

Unequal error variances and nonnormality of the error terms frequently appear together. To remedy this departure from the simple linear regression model (3.1), we need a transformation on Y, since the shapes and spreads of the distributions of Y need to be changed. Such a transformation on Y may also at the same time help to linearize a curvilinear regression relation. At other times, a simultaneous transformation on X may also be needed to obtain or maintain a linear regression relation.

Frequently, the nonnormality and unequal variances departures from regression model (3.1) take the form of increasing skewness and increasing variability of the distributions of the error terms as the mean response $E\{Y\}$ increases. For example, in a regression of yearly household expenditures for vacations (Y) on household in-

FIGURE 4.15   Prototype Regression Patterns with Unequal Error Variances and Simple Transformations
of Y

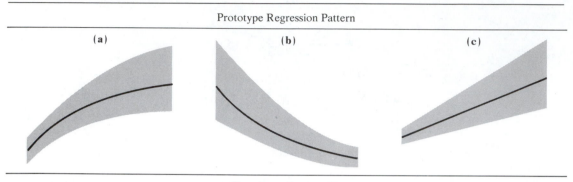

Prototype Regression Pattern

(a)   (b)   (c)

Transformations on Y

$$Y' = \sqrt{Y}$$

$$Y' = \log_{10} Y$$

$$Y' = 1/Y$$

Note: A simultaneous transformation on X may also be helpful or necessary.

come (X), there will tend to be more variation and positive skewness (i.e., some
very high yearly expenditures) for high-income households than for low-income
households, who tend to consistently spend much less. Figure 4.15 contains some
prototype regression relations where the skewness and the error variance increase
with the mean response $E\{Y\}$. This figure also presents for these cases some simple
transformations on Y that may be helpful. Several alternative transformations on Y
may be tried, as well as some simultaneous transformations on X. Scatter plots and
residual plots should be prepared to determine which transformation is most effective.

Example.   In columns 1 and 2 of Table 4.9 are presented data on age (X) and
plasma level of a polyamine (Y) for 25 healthy children. These data are plotted in
Figure 4.16a as a scatter plot. Note the distinct curvilinear regression relationship, as
well as the greater variability for younger children than for older ones.

Based on the prototype regression pattern in Figure 4.15b, we shall first try the
logarithmic transformation $Y' = \log_{10} Y$. The transformed Y values are shown in
column 3 of Table 4.9. Figure 4.16b contains the scatter plot with this transformation. Note that the transformation not only has led to a reasonably linear regression
relation, but the variability at the different levels of X also has become reasonably
constant.

To further examine the reasonableness of the transformation $Y' = \log_{10} Y$, we

TABLE 4.9 Data for Plasma Levels Example and Logarithmic Transformation of Y

Child i	(1) Age X_i	(2) Plasma Level Y_i	(3) $Y_i' = \log_{10} Y_i$
1	0 (newborn)	13.44	1.1284
2	0 (newborn)	12.84	1.1086
3	0 (newborn)	11.91	1.0759
4	0 (newborn)	20.09	1.3030
5	0 (newborn)	15.60	1.1931
6	1.0	10.11	1.0048
7	1.0	11.38	1.0561
8	1.0	10.28	1.0120
9	1.0	8.96	.9523
10	1.0	8.59	.9340
11	2.0	9.83	.9926
12	2.0	9.00	.9542
13	2.0	8.65	.9370
14	2.0	7.85	.8949
15	2.0	8.88	.9484
16	3.0	7.94	.8998
17	3.0	6.01	.7789
18	3.0	5.14	.7110
19	3.0	6.90	.8388
20	3.0	6.77	.8306
21	4.0	4.86	.6866
22	4.0	5.10	.7076
23	4.0	5.67	.7536
24	4.0	5.75	.7597
25	4.0	6.23	.7945

fitted the simple linear regression model (3.1) to the transformed Y data and obtained:

$$\hat{Y}' = 1.135 - .1023X$$

The regression calculations are not shown since they are the same as always, except that Y is replaced by Y' in formula (2.10). A plot of the residuals against X is shown in Figure 4.16c, and a normal probability plot of the residuals is shown in Figure 4.16d. The coefficient of correlation between the ordered residuals and their expected values under normality is .981. All of this evidence supports the appropriateness of regression model (3.1) for the transformed Y data.

If it is desired to express the estimated regression function in the original units of Y, we simply take the antilog of $\hat{Y}'$ and obtain:

$$\hat{Y} = \text{antilog}_{10}(1.135 - .1023X)$$

FIGURE 4.16 Scatter Plots and Residual Plots—Plasma Levels Example

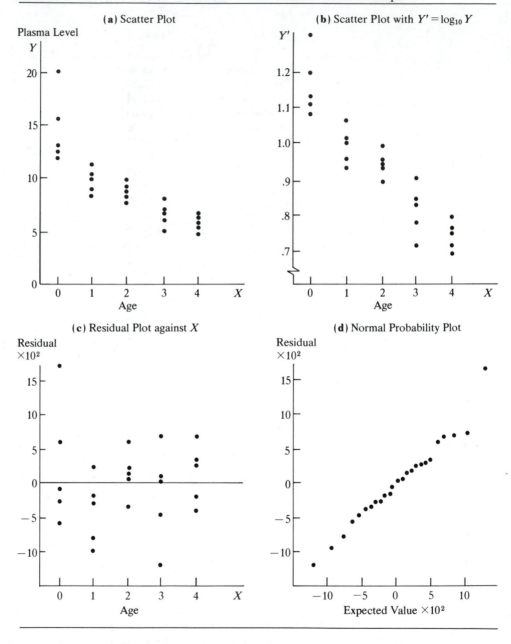

(a) Scatter Plot

(b) Scatter Plot with $Y' = \log_{10} Y$

(c) Residual Plot against X

(d) Normal Probability Plot

Note

At times, it may be desirable to introduce a constant into a transformation of Y, such as when Y may be negative. For instance, the logarithmic transformation to shift the origin in Y and make all Y observations positive would be $Y' = \log_{10}(Y + k)$, where k is an appropriately chosen constant.

Box–Cox Transformations

Box and Cox (Ref. 4.4) have developed a procedure for choosing a transformation from the family of power transformations on Y. This procedure is useful for correcting skewness of the distributions of error terms, unequal error variances, and non-linearity of the regression function. The family of power transformations is of the form:

$$(4.28) \qquad\qquad Y' = Y^\lambda$$

where λ is a parameter to be determined from the data. Note that this family encompasses the following simple transformations:

$$\lambda = 2 \qquad Y' = Y^2$$
$$\lambda = .5 \qquad Y' = \sqrt{Y}$$
$$\lambda = 0 \qquad Y' = \log_e Y \qquad\qquad \text{(by definition)}$$
$$\lambda = -.5 \qquad Y' = \frac{1}{\sqrt{Y}}$$
$$\lambda = -1.0 \qquad Y' = \frac{1}{Y}$$

The criterion for determining the appropriate parameter λ of the transformation of Y in the Box-Cox approach is to find the value of λ that minimizes the error sum of squares SSE for a linear regression based on that transformation. Computer programs are available for finding the appropriate value of λ. Alternatively, one can select a number of values of λ, make the corresponding transformation for each, fit the linear regression function to the transformed Y data, and calculate SSE for each fit. That value of λ is then chosen that minimizes SSE.

If desired, a finer search can be conducted in the neighborhood of the λ value that minimizes SSE. However, the Box-Cox procedure ordinarily is only used to provide a guide for selecting a transformation, so overly precise results are not needed. In any case, scatter and residual plots should be utilized to examine the appropriateness of the transformation identified by the Box-Cox procedure.

Since the power of the transformation affects the magnitude of the error sum of squares SSE, one needs either to adjust SSE for this or to use a standardized variable that leaves SSE unaffected by the value of λ. Usually, the latter approach is used, and the following standardized variable is employed:

$$(4.29) \qquad W = \begin{array}{ll} K_1(Y^\lambda - 1) & \lambda \neq 0 \\ \\ K_2(\log_e Y) & \lambda = 0 \end{array}$$

where:

$$(4.29a) \qquad K_2 = \left(\prod_{i=1}^{n} Y_i \right)^{\frac{1}{n}}$$

$$(4.29b) \qquad K_1 = \frac{1}{\lambda K_2^{\lambda - 1}}$$

Note that K_2 is the geometric mean of the Y observations.

Example.   Table 4.10 contains the Box-Cox results for the plasma levels example. Selected values of λ, ranging from -1.0 to 1.0, were chosen, the transformations were made, and the linear regressions were fitted. For instance, for $\lambda = .5$, the transformation $W = K_1(\sqrt{Y} - 1)$ was made, and the linear regression of W on X was fitted. For this fitted linear regression, we have $SSE = 48.4$.

Note from Table 4.10 that the Box-Cox procedure identifies a power value near $\lambda = -.50$ as being appropriate. However, SSE as a function of λ is fairly stable in the range from near 0 to -1.0, so the earlier choice of the logarithmic transformation $Y' = \log_{10} Y$ for the plasma levels example is not unreasonable according to the Box-Cox approach. One reason the logarithmic transformation was chosen here is because of the ease of interpreting it. The use of logarithms to base 10 rather than natural logarithms does not, of course, affect the appropriateness of the logarithmic transformation.

TABLE 4.10   Box-Cox Results for Plasma Levels Example

λ	SSE
1.0	78.0
.9	70.4
.7	57.8
.5	48.4
.3	41.4
.1	36.4
0	34.5
$-.1$	33.1
$-.3$	31.2
$-.4$	30.7
$-.5$	30.6
$-.6$	30.7
$-.7$	31.1
$-.9$	32.7
-1.0	33.9

Comments

1. At times, theoretical or a priori considerations can be utilized to help in choosing an appropriate transformation. For example, when the shape of the scatter in a study of the relation between price of a commodity (X) and quantity demanded (Y) is that in Figure 4.15b, economists may prefer a logarithmic transformation on both Y and X because the slope of the regression line for the transformed variables then measures the price elasticity of demand. The slope is then commonly interpreted as showing the percent change in quantity demanded per 1 percent change in price, where it is understood that the changes are in opposite directions.

Similarly, scientists may prefer logarithmic transformations of both Y and X when studying the relation between radioactive decay (Y) of a substance and time (X) for a curvilinear relation of the type illustrated in Figure 4.15b because the slope of the regression line for the transformed variables then measures the decay rate.

2. After a transformation has been tentatively selected, residual plots and other analyses described earlier need to be employed to ascertain that the simple linear regression model (3.1) is appropriate for the transformed data.

3. When transformed models are employed, the estimators b_0 and b_1 obtained by least squares have the least squares properties with respect to the transformed observations, not the original ones.

CITED REFERENCES

4.1. Dixon, W. J., chief editor. *BMDP Statistical Software Manual,* vols. 1 and 2. Berkeley, Calif.: University of California Press, 1988.

4.2. *MINITAB Reference Manual, Release 7.* State College, Pa.: Minitab, Inc., 1989.

4.3. Barnett, V., and T. Lewis. *Outliers in Statistical Data.* 2nd ed. New York: John Wiley & Sons, 1984.

4.4. Box, G. E. P., and D. R. Cox. "An Analysis of Transformations." *Journal of the Royal Statistical Society B* 26 (1964), pp. 211–43.

PROBLEMS

4.1. Distinguish between: (1) residual and standardized residual, (2) $E\{\varepsilon_i\} = 0$ and $\bar{e} = 0$, (3) error term and residual.

4.2. Prepare a prototype residual plot for each of the following cases: (1) error variance decreases with X; (2) true regression function is $\cup$ shaped, but a linear regression function is fitted.

4.3. Refer to **Grade point average** Problem 2.17. The fitted values and residuals are:

i:	1	2	3	4	5	6	7	8	9	10
$\hat{Y}_i$:	2.92	2.33	2.25	1.58	2.08	3.51	3.34	2.67	2.25	1.91
e_i:	.18	−.03	.75	.32	.42	.19	.06	−.07	.55	−.31

i:	11	12	13	14	15	16	17	18	19	20
$\hat{Y}_i$:	2.42	2.84	2.50	3.59	2.16	1.91	2.50	3.26	1.74	2.25
e_i:	−.42	.06	−.20	−.39	−.36	−.51	−.50	.54	.46	−.75

a. Prepare a box plot for the entrance test scores X_i. Are there any noteworthy features in this plot?

b. Prepare a dot plot of the residuals. What information does this plot provide?

c. Plot the residuals e_i against the fitted values $\hat{Y}_i$. What departures from regression model (3.1) can be studied from this plot? What are your findings?

d. Prepare a normal probability plot of the residuals. Also obtain the coefficient of correlation between the ordered residuals and their expected values under normality. Test the reasonableness of the normality assumption here using Table 4.3 and $\alpha = .05$. What do you conclude?

e. Information is given below for each student on two variables not included in the model, namely, intelligence test score (X_2) and high school average (X_3). Plot the residuals against X_2 and X_3 on separate graphs to ascertain whether the model can be improved by including either of these variables. What do you conclude?

i:	1	2	3	4	5	6	7	8	9	10
X_2:	105	113	118	107	110	125	115	121	117	111
X_3:	2.9	2.8	3.1	2.4	3.0	2.4	3.5	3.1	3.1	2.9

i:	11	12	13	14	15	16	17	18	19	20
X_2:	123	114	120	132	122	110	119	109	116	108
X_3:	3.2	3.3	3.4	2.6	3.0	2.8	3.3	3.4	2.6	2.7

4.4. Refer to **Calculator maintenance** Problem 2.18. The fitted values and residuals are:

i:	1	2	3	4	5	6	7	8	9
$\hat{Y}_i$:	100.8	86.1	71.4	12.4	71.4	56.6	100.8	41.9	56.6
e_i:	−3.8	−.1	6.6	−2.4	3.6	5.4	.2	−2.9	−3.6

i:	10	11	12	13	14	15	16	17	18
$\hat{Y}_i$:	27.2	115.6	71.4	27.2	71.4	100.8	12.4	56.6	71.4
e_i:	5.8	2.4	−6.4	−2.2	−.4	4.2	4.6	−7.6	−3.4

a. Prepare a dot plot for the number of machines serviced X_i. What information is provided by this plot? Are there any outlying cases with respect to this variable?

b. The cases are given in time order. Prepare a time plot for the number of machines serviced. What does your plot show?

c. Prepare a stem-and-leaf plot of the residuals. Are there any noteworthy features in this plot?

d. Prepare residual plots of e_i versus $\hat{Y}_i$ and e_i versus X_i on separate graphs. Do these plots provide the same information? What departures from regression model (3.1) can be studied from these plots? State your findings.

e. Prepare a normal probability plot of the residuals. Also obtain the coefficient of correlation between the ordered residuals and their expected values under normality. Does the normality assumption appear to be tenable here? Use Table 4.3 and $\alpha = .10$.

f. Prepare a time plot of the residuals to ascertain whether the error terms are correlated over time. What is your conclusion?

g. Information is given below on two variables not included in the regression model, namely, mean operational age of machines serviced on the call (X_2, in months) and years of experience of the service person making the call (X_3). Plot the residuals against X_2 and X_3 on separate graphs to ascertain whether the model can be improved by including either or both of these variables. What do you conclude?

i:	1	2	3	4	5	6	7	8	9
X_2:	12	21	38	16	25	32	18	14	12
X_3:	3	6	2	2	3	4	5	2	3

i:	10	11	12	13	14	15	16	17	18
X_2:	35	20	8	15	17	28	29	9	14
X_3:	6	5	3	5	6	3	5	3	6

4.5. Refer to **Airfreight breakage** Problem 2.19.

a. Prepare a dot plot for the number of transfers X_i. Does the distribution of number of transfers appear to be asymmetrical?

b. The cases are given in time order. Prepare a time plot for the number of transfers. Is any systematic pattern evident in your plot? Discuss.

c. Obtain the residuals e_i and prepare a stem-and-leaf plot of the residuals. What information is provided by your plot?

d. Plot the residuals e_i against X_i to ascertain whether any departures from regression model (3.1) are evident. What is your conclusion?

e. Prepare a normal probability plot of the residuals. Also obtain the coefficient of correlation between the ordered residuals and their expected values under normality to ascertain whether the normality assumption is reasonable here. Use Table 4.3 and a level of significance of .01. What do you conclude?

f. Prepare a time plot of the residuals. What information is provided by your plot?

4.6. Refer to **Plastic hardness** Problem 2.20.

a. Obtain the residuals e_i and prepare a box plot of the residuals. What information is provided by your plot?

b. Plot the residuals e_i against the fitted values $\hat{Y}_i$ to ascertain whether any departures from regression model (3.1) are evident. State your findings.

c. Prepare a normal probability plot of the residuals. Also obtain the coefficient of correlation between the ordered residuals and their expected values under normality. Does the normality assumption appear to be reasonable here? Use Table 4.3 and $\alpha = .05$.

d. Obtain the standardized residuals. Compare the actual frequencies against the expected frequencies under normality, using the 25th, 50th, and 75th percentiles of the relevant t distribution. Is the information provided by these comparisons consistent with the findings from the normal probability plot in part (c)?

4.7. Refer to **Muscle mass** Problem 2.25.

   a. Prepare a stem-and-leaf plot for the ages X_i. Is this plot consistent with the random selection of women from each 10-year age group? Explain.

   b. Obtain the residuals e_i and prepare a frequency dot plot of the residuals. What does your plot show?

   c. Plot the residuals e_i against $\hat{Y}_i$ and also against X_i on separate graphs to ascertain whether any departures from regression model (3.1) are evident. Do the two plots provide the same information? State your conclusions.

   d. Prepare a normal probability plot of the residuals. Also obtain the coefficient of correlation between the ordered residuals and their expected values under normality to ascertain whether the normality assumption is tenable here. Use Table 4.3 and a level of significance of .10. What do you conclude?

   e. Obtain the standardized residuals. Compare the actual frequencies against the expected frequencies under normality, using the 25th, 75th, and 90th percentiles of the relevant t distribution. Is the information provided by these comparisons consistent with the findings from the normal probability plot in part (d)?

4.8. Refer to **Robbery rate** Problem 2.26.

   a. Prepare a stem-and-leaf plot for the city population densities X_i. What information does your plot provide?

   b. Obtain the residuals e_i and prepare a box plot of the residuals. Does the distribution of the residuals appear to be symmetrical?

   c. Make a residual plot of e_i versus $\hat{Y}_i$. What does the plot show?

   d. Prepare a normal probability plot of the residuals. Also obtain the coefficient of correlation between the ordered residuals and their expected values under normality. Test the reasonableness of the normality assumption using Table 4.3 and $\alpha = .05$. What do you conclude?

   e. Obtain the standardized residuals. Compare the actual frequencies against the expected frequencies under normality, using the 50th, 75th, and 95th percentiles of the relevant t distribution. Is the information provided by these comparisons consistent with the findings from the normal probability plot in part (d)?

4.9. Electricity consumption. An economist studying the relation between household electricity consumption (Y) and number of rooms in the home (X) employed linear regression model (3.1) and obtained the following residuals:

i:	1	2	3	4	5	6	7	8	9	10
X_i:	2	3	4	5	6	7	8	9	10	11
e_i:	3.2	2.9	−1.7	−2.0	−2.3	−1.2	−.9	.8	.7	.5

Plot the residuals e_i against X_i. What problem appears to be present here? Might a transformation alleviate this problem?

4.10. Per capita earnings. A sociologist employed linear regression model (3.1) to relate per capita earnings (Y) to average number of years of schooling (X) for 12 cities. The fitted values $\hat{Y}_i$ and the standardized residuals $e_i/\sqrt{MSE}$ follow.

i:	1	2	3	4	5	6	7	8	9	10	11	12
$\hat{Y}_i$:	9.9	9.3	10.2	9.6	10.2	12.4	14.3	9.6	9.2	15.6	11.2	13.1
$e_i/\sqrt{MSE}$:	−1.12	.81	−.76	.43	.65	−.17	1.62	1.79	−.53	−3.78	.74	.32

a. Plot the standardized residuals against the fitted values. What does the plot suggest?

b. How many standardized residuals are outside ± 1 standard deviation? Approximately how many would you expect to see if the model is apt?

4.11. Drug concentration. A pharmacologist employed linear regression model (3.1) to study the relation between the concentration of a drug in plasma (Y) and the log-dose of the drug (X). The residuals and log-dose levels follow.

i:	1	2	3	4	5	6	7	8	9
X_i:	-1	0	1	-1	0	1	-1	0	1
e_i:	.5	2.1	-3.4	.3	-1.7	4.2	$-.6$	2.6	-4.0

Plot the residuals e_i against X_i. What conclusions do you draw from the plot?

4.12. A student states that she doesn't understand why the sum of squares defined in (4.11) is called a pure error sum of squares "since the formula looks like one for an ordinary sum of squares." Explain.

4.13. Refer to **Calculator maintenance** Problem 2.18. Some additional calculational results are: $SSR = 16,182.6$, $SSE = 321.4$.

a. In an F test for lack of fit of a linear regression function, what are the alternative conclusions?

b. Perform the test indicated in part (a). Control the risk of Type I error at .05. State the decision rule and conclusion.

c. Does your test in part (b) detect other departures from regression model (3.1), such as lack of constant variance or lack of normality in the error terms? Could the results of the test be affected by such departures? Discuss.

4.14. Refer to **Plastic hardness** Problem 2.20.

a. Perform an F test to determine whether or not there is lack of fit of a linear regression function. Use a level of significance of .01. State the alternatives, decision rule, and conclusion.

b. Is there any advantage of having an equal number of replications at each of the X levels? Is there any disadvantage?

c. Does the test in part (a) indicate what regression function is appropriate when it leads to the conclusion that the regression function is not linear? How would you proceed?

4.15. Solution concentration. A chemist studied the concentration of a solution (Y) over time (X). Fifteen identical solutions were prepared. The 15 solutions were randomly divided into five sets of three, and the five sets were measured, respectively, after 1, 3, 5, 7, and 9 hours. The results follow.

i:	1	2	3	4	5	6	7	8	9	10	11	12	13	14	15
X_i:	9	9	9	7	7	7	5	5	5	3	3	3	1	1	1
Y_i:	.07	.09	.08	.16	.17	.21	.49	.58	.53	1.22	1.15	1.07	2.84	2.57	3.10

a. Fit a linear regression function.

b. Perform an F test to determine whether or not there is lack of fit of a linear regression function; use $\alpha = .025$. State the alternatives, decision rule, and conclusion.

c. Does the test in part (b) indicate what regression function is appropriate when it leads to the conclusion that lack of fit of a linear regression function exists? Explain.

4.16. Refer to **Solution concentration** Problem 4.15.

a. Prepare a scatter plot of the data. What transformation of Y might you try based on the prototype patterns in Figure 4.15 to achieve constant variance and linearity?

b. Use the Box-Cox procedure and standardization (4.29) to find an appropriate power transformation. Evaluate SSE for $\lambda = -.2, -.1, 0, .1, .2$. What transformation of Y is suggested?

c. Use the transformation $Y' = \log_{10} Y$ and obtain the estimated linear regression function for the transformed data.

d. Plot the estimated regression line and the transformed data. Does the regression line appear to be a good fit to the transformed data?

e. Obtain the residuals and plot them against the fitted values. Also prepare a normal probability plot. What do your plots show?

f. Express the estimated regression function in the original units.

4.17. Sales growth. A marketing researcher studied annual sales of a product that had been introduced 10 years ago. The data were as follows, where X is the year (coded) and Y is sales in thousands of units:

i:	1	2	3	4	5	6	7	8	9	10
X_i:	0	1	2	3	4	5	6	7	8	9
Y_i:	98	135	162	178	221	232	283	300	374	395

a. Prepare a scatter plot of the data. Does a linear relation appear adequate here?

b. Use the Box-Cox procedure and standardization (4.29) to find an appropriate power transformation of Y. Evaluate SSE for $\lambda = .3, .4, .5, .6, .7$. What transformation of Y is suggested?

c. Use the transformation $Y' = \sqrt{Y}$ and obtain the estimated linear regression function for the transformed data.

d. Plot the estimated regression line and the transformed data. Does the regression line appear to be a good fit to the transformed data?

e. Obtain the residuals and plot them against the fitted values. Also prepare a normal probability plot. What do your plots show?

f. Express the estimated regression function in the original units.

4.18. Response errors. In a small-scale study of response errors when expenditures incurred during the most recent hunting trip are recalled, the following data on response errors (Y) and number of months since last hunting trip (X) were obtained:

i:	1	2	3	4	5	6
X_i:	12	5	1	8	1	3
Y_i:	−94	−68	−41	−78	−36	−55

i:	7	8	9	10	11	12
X_i:	15	9	7	2	10	4
Y_i:	−106	−90	−78	−49	−85	−70

a. Prepare a scatter plot of the data. Does a linear relation appear adequate here? Would a transformation on X or Y be more appropriate here? Why?

b. Use the transformation $X' = \sqrt{X}$ and obtain the estimated linear regression function for the transformed data.

c. Plot the estimated regression line and the transformed data. Does the regression line appear to be a good fit to the transformed data?

d. Obtain the residuals and plot them against the fitted values. Also prepare a normal probability plot. What do your plots show?

e. Express the estimated regression function in the original units.

EXERCISES

4.19. A student fitted a linear regression function for a class assignment. Some results follow.

i:	1	2	3	4	5
Y_i:	35	17	42	28	53
$\hat{Y}_i$:	42	29	32	32	40
e_i:	−7	−12	10	−4	13

The student plotted the residuals e_i against Y_i and found a positive relation. When he plotted the residuals against the fitted values $\hat{Y}_i$, he found no relation. Why is there this difference, and which is the more meaningful plot?

4.20. If the error terms in a regression model are independent $N(0, \sigma^2)$, what can be said about the error terms after transformation $X' = 1/X$ is used? Is the situation the same after transformation $Y' = 1/Y$ is used?

4.21. Derive the result in (4.24).

4.22. Using theorems (1.67), (1.38), and (1.39), show that $E\{MSPE\} = \sigma^2$ for the normal error regression model (3.1).

4.23. A linear regression model with the intercept $\beta_0 = 0$ is under consideration. Data have been obtained which contain replications. State the full and reduced models for testing the appropriateness of the regression function under consideration. What are the degrees of freedom associated with the full and reduced models if $n = 20$ and $c = 10$?

PROJECTS

4.24. Blood pressure. The following data were obtained in a study of the relation between diastolic blood pressure (Y) and age (X) for boys 5 to 13 years old.

i:	1	2	3	4	5	6	7	8
X_i:	5	8	11	7	13	12	12	6
Y_i:	63	67	74	64	75	69	90	60

    a. Assuming the normal error regression model (3.1) is appropriate, obtain the estimated regression function and plot the residuals e_i against X_i. What does your residual plot show?

    b. Omit case 7 from the data and obtain the estimated regression function based on the remaining seven cases. Compare this estimated regression function to that obtained in part (a). What can you conclude about the effect of case 7?

    c. Using your fitted regression function in part (b), obtain a 99 percent prediction interval for a new Y observation at $X = 12$. Does observation Y_7 fall outside this prediction interval? What is the significance of this?

4.25. Refer to the **SMSA** data set and Project 2.42. For each of the three fitted regression models, obtain the residuals and prepare a residual plot against X and a normal probability plot. Summarize your conclusions. Is linear regression model (3.1) more apt in one case than in the others?

4.26. Refer to the **SMSA** data set and Project 2.43. For each geographic region, obtain the residuals and prepare a residual plot against X and a normal probability plot. Do the four regions appear to have similar error variances? What other conclusions do you draw from your plots?

4.27. Refer to the **SENIC** data set and Project 2.44.

    a. For each of the three fitted regression models, obtain the residuals and prepare a residual plot against X and a normal probability plot. Summarize your conclusions. Is linear regression model (3.1) more apt in one case than in the others?

    b. Obtain the fitted regression function for the relation between length of stay and infection risk after deleting cases 47 ($X_{47} = 6.5$, $Y_{47} = 19.56$) and 112 ($X_{112} = 5.9$, $Y_{112} = 17.94$). From this fitted regression function obtain separate 95 percent prediction intervals for new Y observations at $X = 6.5$ and $X = 5.9$, respectively. Do observations Y_{47} and Y_{112} fall outside these prediction intervals? Discuss the significance of this.

4.28. Refer to the **SENIC** data set and Project 2.45. For each geographic region, obtain the residuals and prepare a residual plot against X and a normal probability plot. Do the four regions appear to have similar error variances? What other conclusions do you draw from your plots?

Chapter 5

Simultaneous Inferences and Other Topics in Regression Analysis

In this chapter, we take up a variety of topics in simple linear regression analysis. Several of the topics pertain to the problem of how to make simultaneous inferences from the same set of sample observations.

5.1 JOINT ESTIMATION OF β_0 AND β_1

Need for Joint Estimation

A market research analyst conducted a study of the relation between level of advertising (X) and sales (Y), in which there was no advertising ($X = 0$) for some cases while for other cases the level of advertising was varied. The scatter plot suggested a linear relationship in the range of the advertising expenditures levels studied. The analyst wished to draw inferences about both the intercept β_0 and the slope β_1. The analyst could use the methods of Chapter 3 to construct separate 95 percent confidence intervals for β_0 and β_1. The difficulty is that these would not provide 95 percent confidence that the conclusions for *both* β_0 and β_1 are correct. If the inferences were independent, the probability of both being correct would be $(.95)^2$, or only .9025. The inferences are not, however, independent, coming as they do from the same set of sample data, which makes the determination of the probability of both inferences being correct much more difficult.

Analysis of data frequently requires a series of estimates (or tests) where the analyst would like to have an assurance about the correctness of the entire set of estimates (or tests). We shall call the set of estimates (or tests) of interest the *family* of estimates (or tests). In our illustration, the family consists of two estimates, for β_0

and β_1, respectively. We then distinguish between a statement confidence coefficient and a family confidence coefficient. A *statement confidence coefficient* is the familiar type of confidence coefficient discussed earlier, which indicates the proportion of correct estimates that are obtained when repeated samples are selected and the specified confidence interval is calculated for each sample. A *family confidence coefficient*, on the other hand, indicates the proportion of families of estimates that are entirely correct when repeated samples are selected and the specified confidence intervals for the entire family are calculated for each sample. Thus, a family confidence coefficient corresponds to the probability, in advance of sampling, that the entire family of statements will be correct.

To illustrate the meaning of a family confidence coefficient further, let us return to the joint estimation of β_0 and β_1. A family confidence coefficient of, say, .95 would indicate for this situation that if repeated samples are selected and interval estimates for both β_0 and β_1 are calculated for each sample by specified procedures, 95 percent of the samples would lead to a family of estimates where *both* confidence intervals are correct. For 5 percent of the samples, either one or both of the interval estimates would be incorrect.

Bonferroni Joint Confidence Intervals

Clearly, a procedure that provides a family confidence coefficient is often highly desirable since it permits the analyst to weave the separate results together into an integrated set of conclusions, with an assurance that the entire set of estimates is correct. The Bonferroni method of developing joint confidence intervals with a specified family confidence coefficient is a very simple one: each statement confidence coefficient is adjusted to be higher than $1 - \alpha$ so that the family confidence coefficient is at least $1 - \alpha$. The method is a general one that can be applied in many cases, as we shall see, not just for the joint estimation of β_0 and β_1. Here, we explain the Bonferroni method as it applies for estimating β_0 and β_1 jointly.

We start with ordinary confidence limits for β_0 and β_1 with statement confidence coefficients $1 - \alpha$ each. These limits are:

$$b_0 \pm t(1 - \alpha/2; n - 2)s\{b_0\}$$
$$b_1 \pm t(1 - \alpha/2; n - 2)s\{b_1\}$$

We then ask what is the probability that both sets of limits are correct. Let A_1 denote the event that the first confidence interval does not cover β_0 and A_2 denote the event that the second confidence interval does not cover β_1. We know:

$$P(A_1) = \alpha \qquad P(A_2) = \alpha$$

Probability theorem (1.6) states:

$$P(A_1 \cup A_2) = P(A_1) + P(A_2) - P(A_1 \cap A_2)$$

and hence:

(5.1) $$1 - P(A_1 \cup A_2) = 1 - P(A_1) - P(A_2) + P(A_1 \cap A_2)$$

Now by probability theorems (1.9) and (1.10), we have:

$$1 - P(A_1 \cup A_2) = P(\overline{A_1 \cup A_2}) = P(\bar{A}_1 \cap \bar{A}_2)$$

$P(\bar{A}_1 \cap \bar{A}_2)$ is the probability that both confidence intervals are correct. We thus have from (5.1):

(5.2) $$P(\bar{A}_1 \cap \bar{A}_2) = 1 - P(A_1) - P(A_2) + P(A_1 \cap A_2)$$

Since $P(A_1 \cap A_2) \geq 0$, we obtain from (5.2) the *Bonferroni inequality:*

(5.3) $$P(\bar{A}_1 \cap \bar{A}_2) \geq 1 - P(A_1) - P(A_2)$$

which for our situation is:

(5.3a) $$P(\bar{A}_1 \cap \bar{A}_2) \geq 1 - \alpha - \alpha = 1 - 2\alpha$$

Thus, if β_0 and β_1 are separately estimated with, say, 95 percent confidence intervals, the Bonferroni inequality guarantees us a family confidence coefficient of at least 90 percent that both intervals based on the same sample are correct.

We can easily use the Bonferroni inequality (5.3a) to obtain a family confidence coefficient of at least $1 - \alpha$ for estimating β_0 and β_1. We do this by estimating β_0 and β_1 separately with statement confidence coefficients of $1 - \alpha/2$ each. This yields the Bonferroni bound $1 - \alpha/2 - \alpha/2 = 1 - \alpha$. Thus, the $1 - \alpha$ family confidence limits for β_0 and β_1, often called a *confidence set,* are by the Bonferroni procedure:

(5.4)
$$b_0 \pm Bs\{b_0\}$$
$$b_1 \pm Bs\{b_1\}$$

where:

(5.4a) $$B = t(1 - \alpha/4; n - 2)$$

Note that a statement confidence coefficient of $1 - \alpha/2$ requires the $(1 - \alpha/4)100$ percentile of the t distribution for a two-sided confidence interval.

Example

For the Westwood Company lot size application, 90 percent family confidence intervals for β_0 and β_1 require $B = t(1 - .10/4; 8) = t(.975; 8) = 2.306$. We have from Chapter 3:

$$b_0 = 10.0 \qquad s\{b_0\} = 2.50294$$
$$b_1 = 2.0 \qquad s\{b_1\} = .04697$$

Hence, the respective confidence limits for β_0 and β_1 are $10.0 \pm 2.306(2.50294)$ and $2.0 \pm 2.306(.04697)$, and the joint confidence intervals are:

$$4.2282 \leq \beta_0 \leq 15.7718$$
$$1.8917 \leq \beta_1 \leq 2.1083$$

Thus, we conclude that β_0 is between 4.23 and 15.77 *and* β_1 is between 1.89 and 2.11. The family confidence coefficient is at least .90 that the procedure leads to correct pairs of interval estimates.

Comments

1. We reiterate that the Bonferroni $1 - \alpha$ family confidence coefficient is actually a lower bound on the true (but often unknown) family confidence coefficient. To the extent that incorrect interval estimates of β_0 and β_1 tend to pair up in the family, the families of statements will tend to be correct more than $(1 - \alpha)100$ percent of the time. Because of this conservative nature of the Bonferroni procedure, family confidence coefficients are frequently specified at lower levels (e.g., 90 percent) than when a single estimate is made.

2. The Bonferroni inequality (5.3a) can easily be extended to g simultaneous confidence intervals with family confidence coefficient $1 - \alpha$:

$$(5.5) \qquad\qquad P\left(\bigcap_{i=1}^{g} \overline{A}_i \right) \geq 1 - g\alpha$$

Thus, if g interval estimates are desired with a family confidence coefficient $1 - \alpha$, constructing each interval estimate with statement confidence coefficient $1 - \alpha/g$ will suffice.

3. For a given family confidence coefficient, the larger the number of confidence intervals in the family, the greater becomes the multiple B, which may make some or all of the confidence intervals too wide to be helpful. The Bonferroni technique is ordinarily most useful when the number of simultaneous estimates is not too large.

4. It is not necessary with the Bonferroni procedure that the confidence intervals have the same statement confidence coefficient. Different statement confidence coefficients can be used, depending on the importance of each estimate. For instance, in our earlier illustration β_0 might be estimated with a 92 percent confidence interval and β_1 with a 98 percent confidence interval. The family confidence coefficient by (5.3) will still be at least 90 percent.

5. The joint confidence intervals can be used directly for testing. To illustrate this use, suppose an industrial engineer working for the Westwood Company theorized that the regression function should have an intercept of 13.0 and a slope of 2.50. While 13.0 falls in the confidence interval for β_0, 2.50 does not fall in the confidence interval for β_1. Thus, the engineer's theoretical expectations are not correct, at the $\alpha = .10$ family level of significance.

6. The estimators b_0 and b_1 are usually correlated, but the Bonferroni simultaneous confidence limits in (5.4) only recognize this correlation by means of the bound on the family confidence coefficient. It can be shown that the covariance between b_0 and b_1 is:

$$(5.6) \qquad\qquad \sigma\{b_0, b_1\} = -\overline{X}\sigma^2\{b_1\}$$

Note that if $\overline{X}$ is positive, b_0 and b_1 are negatively correlated, implying that if the estimate b_1 is too high, the estimate b_0 is likely to be too low, and vice versa.

In the Westwood Company example, $\overline{X} = 50$; hence the covariance is negative. This implies that the estimators b_0 and b_1 here tend to err in opposite directions. We expect this intuitively. Since the observed points (X_i, Y_i) fall in the first quadrant (see Figure 2.2a), we anticipate that if the slope of the fitted regression line is too steep (b_1 overestimates β_1), the intercept is most likely to be too low (b_0 underestimates β_0), and vice versa.

When the independent variable is $X_i - \overline{X}$, as in the alternative model (2.6), b_0^* and b_1 are uncorrelated according to (5.6) because the mean of the $X_i - \overline{X}$ observations is zero.

5.2 SIMULTANEOUS ESTIMATION OF MEAN RESPONSES

Often one would like to estimate the mean responses at a number of X levels from the same sample data. The Westwood Company, for instance, may wish to estimate the mean number of man-hours for lots of 30, 55, and 80 parts. We already know how to do this for any one level of X with given statement confidence coefficient. Now we shall discuss two approaches for simultaneous estimation of mean responses with a family confidence coefficient, so that there is a known assurance of all estimates of mean responses being correct. The two approaches are the Working-Hotelling approach and the Bonferroni approach.

The reason for concern with a family confidence coefficient is that separate interval estimates of $E\{Y_h\}$ at various levels X_h need not all be correct or all be incorrect, even though they are all based on the same sample data and the same fitted regression line. The combination of sampling errors in b_0 and b_1 may be such that the interval estimates of $E\{Y_h\}$ will be correct over some range of X levels and incorrect elsewhere.

Working-Hotelling Approach

The Working-Hotelling approach is applicable when the family of estimates consists of all possible X levels. Ordinarily, of course, we wish to estimate the mean response only for a limited number of X levels. As a result, the family confidence coefficient with the Working-Hotelling approach when only a limited number of mean responses are estimated will actually be larger than indicated.

The simultaneous confidence limits for g mean responses $E\{Y_h\}$ with the Working-Hotelling approach are of the form:

$$(5.7) \qquad \hat{Y}_h \pm Ws\{\hat{Y}_h\}$$

where:

$$(5.7a) \qquad W^2 = 2F(1 - \alpha; 2, n - 2)$$

and $\hat{Y}_h$ and $s\{\hat{Y}_h\}$ are defined in (3.27) and (3.29), respectively.

Example. For the Westwood Company lot size example, we require a family of estimates of the mean number of man-hours at the following levels of lot size: 30, 55, 80. The family confidence coefficient is to be .90. In Chapter 3 we obtained $\hat{Y}_h$ and $s\{\hat{Y}_h\}$ for $X_h = 55$. In similar fashion, we can obtain the needed results for the other lot sizes. We summarize them here, without showing the calculations:

X_h	$\hat{Y}_h$	$s\{\hat{Y}_h\}$
30	70.0	1.27764
55	120.0	.89730
80	170.0	1.65387

For a family confidence coefficient of .90, we require $F(.90; 2, 8) = 3.11$. Hence:

$$W^2 = 2(3.11) = 6.22 \qquad W = 2.494$$

We can now obtain the confidence intervals for the mean number of man-hours at $X_h = 30, 55,$ and 80:

$$66.8 = 70.0 - 2.494(1.27764) \leq E\{Y_h\} \leq 70.0 + 2.494(1.27764) = 73.2$$

$$117.8 = 120.0 - 2.494(.89730) \leq E\{Y_h\} \leq 120.0 + 2.494(.89730) = 122.2$$

$$165.9 = 170.0 - 2.494(1.65387) \leq E\{Y_h\} \leq 170.0 + 2.494(1.65387) = 174.1$$

With family confidence coefficient .90, we conclude that the mean number of man-hours for lots of 30 parts is between 66.8 and 73.2, for lots of 55 parts is between 117.8 and 122.2, and for lots of 80 parts is between 165.9 and 174.1. The family confidence coefficient .90 provides assurance that the procedure leads to all correct estimates in the family of estimates.

Bonferroni Approach

The Bonferroni approach, discussed earlier for simultaneous estimation of β_0 and β_1, is a completely general approach. To construct a family of confidence intervals for mean responses at different X levels, we calculate the usual confidence limits for a single mean response $E\{Y_h\}$, given in (3.32), and adjust the statement confidence coefficient to yield the specified family confidence coefficient.

When $E\{Y_h\}$ is to be estimated for g levels X_h, with family confidence coefficient $1 - \alpha$, the Bonferroni confidence limits are:

(5.8) $$\hat{Y}_h \pm Bs\{\hat{Y}_h\}$$

where:

(5.8a) $$B = t(1 - \alpha/2g; n - 2)$$

g is the number of confidence intervals in the family

Example.   The estimates of the mean number of man-hours for lot sizes of 30, 55, and 80 parts with family confidence coefficient .90 by the Bonferroni approach require the same data as the Working-Hotelling approach presented above. In addition, we require $B = t[1 - .10/2(3); 8] = t(.983; 8)$. By linear interpolation in Table A.2, we obtain $t(.983; 8) = 2.56$ (see the following Comment 4).

We thus obtain the confidence intervals, with a 90 percent family confidence coefficient, for the mean number of man-hours for lot sizes $X_h = 30, 55,$ and 80:

$$66.7 = 70.0 - 2.56(1.27764) \leq E\{Y_h\} \leq 70.0 + 2.56(1.27764) = 73.3$$

$$117.7 = 120.0 - 2.56(.89730) \leq E\{Y_h\} \leq 120.0 + 2.56(.89730) = 122.3$$

$$165.8 = 170.0 - 2.56(1.65387) \leq E\{Y_h\} \leq 170.0 + 2.56(1.65387) = 174.2$$

Comments

1. In this instance the Working-Hotelling confidence limits are slightly tighter than the Bonferroni limits. In other cases where the number of statements is small, the Bonferroni limits may be tighter. For larger families, the Working-Hotelling confidence limits will always be the tighter, since W in (5.7a) stays the same for any number of statements in the family whereas B in (5.8a) becomes larger as the number of statements increases. In practice, once the family confidence coefficient has been decided upon, one can calculate the W and B multiples to determine which procedure leads to tighter confidence limits.

2. Both the Working-Hotelling and Bonferroni approaches to multiple estimation of mean responses provide lower bounds to the actual family confidence coefficient.

3. Sometimes it is not known in advance for which levels of the independent variable to estimate the mean response. Instead, this is determined as the analysis proceeds. In such cases, it is better to use the Working-Hotelling approach because the family for this approach encompasses all possible levels of X.

4. To obtain an untabled percentile of the t distribution, linear interpolation in Table A.2 ordinarily will give a reasonably close approximation as long as the degrees of freedom are not minimal. In our illustration of the Bonferroni method, we required $t(.983; 8)$. From Table A.2, we know that:

$$t(.980; 8) = 2.449 \qquad t(.985; 8) = 2.634$$

Linear interpolation therefore gives:

$$t(.983; 8) = 2.449 + \left(\frac{.983 - .980}{.985 - .980}\right)(2.634 - 2.449) = 2.56$$

Many computer packages provide percentiles of the t distribution and other distributions, thus obviating the need for interpolating in tables.

5.3 SIMULTANEOUS PREDICTION INTERVALS FOR NEW OBSERVATIONS

Now we consider the simultaneous predictions of g new observations on Y in g independent trials at g different levels of X. To illustrate this type of application, let us suppose the Westwood Company plans to produce the next three lots in sizes of 30, 55, and 80 parts, and wishes to predict the man-hours for each of these lots with a family confidence coefficient of .95.

Two procedures will be considered here, the Scheffé procedure and the Bonferroni procedure. Both utilize the same type of limits as for predicting a single observation, given in (3.35), and only the multiple of the estimated standard deviation is changed. The Scheffé procedure uses the F distribution, while the Bonferroni procedure uses the t distribution. The simultaneous prediction limits for g predictions with the Scheffé procedure with family confidence coefficient $1 - \alpha$ are:

(5.9)
$$\hat{Y}_h \pm Ss\{Y_{h(\text{new})}\}$$

where:

(5.9a)
$$S^2 = gF(1 - \alpha; g, n - 2)$$

With the Bonferroni procedure, the $1 - \alpha$ simultaneous prediction limits are:

$$(5.10) \qquad \hat{Y}_h \pm Bs\{Y_{h(\text{new})}\}$$

where:

$$(5.10a) \qquad B = t(1 - \alpha/2g; n - 2)$$

We can evaluate the S and B multiples to see which procedure provides tighter prediction limits. For our example, we have:

$$S^2 = 3F(.95; 3, 8) = 3(4.07) = 12.21 \qquad S = 3.49$$

$$B = t[1 - .05/2(3); 8] = t(.992; 8) = 3.04$$

so that the Bonferroni method will be used here. From earlier results, we obtain (calculations not shown):

X_h	$\hat{Y}_h$	$s\{Y_{h(\text{new})}\}$	$Bs\{Y_{h(\text{new})}\}$
30	70.0	3.02198	9.18682
55	120.0	2.88187	8.76088
80	170.0	3.19926	9.72575

The simultaneous prediction limits for the next three lots, with family confidence coefficient .95, when $X_h = 30$, 55, and 80 then are:

$$60.8 = 70.0 - 9.18682 \leq Y_{h(\text{new})} \leq 70.0 + 9.18682 = 79.2$$

$$111.2 = 120.0 - 8.76088 \leq Y_{h(\text{new})} \leq 120.0 + 8.76088 = 128.8$$

$$160.3 = 170.0 - 9.72575 \leq Y_{h(\text{new})} \leq 170.0 + 9.72575 = 179.7$$

With family confidence coefficient at least .95, we can predict that the man-hours for the next three production runs all will be within the above limits.

Comments

1. Simultaneous prediction intervals for g new observations on Y at g different levels of X with a $1 - \alpha$ family confidence coefficient are wider than the corresponding single prediction intervals of (3.35). When the number of simultaneous predictions is not large, however, the difference in the width is only moderate. For instance, a single 95 percent prediction interval for the Westwood Company example would have utilized the t multiple $t(.975; 8) = 2.306$, which is only moderately smaller than the multiple $B = 3.04$ for three simultaneous predictions.

2. Note that both B and S become larger as g increases. This contrasts with simultaneous estimation of mean responses where B becomes larger but not W. When g is large, both the B and S multiples may become so large that the prediction intervals will be too wide to be useful. Other simultaneous estimation techniques could then be considered, as discussed in Reference 5.1.

5.4 REGRESSION THROUGH THE ORIGIN

Sometimes the regression line is known to go through the origin at $(0, 0)$. This occurs, for instance, when X is units of output and Y is variable cost, so Y is zero by definition when X is zero. Another example is where X is the number of brands of cigarettes stocked in a supermarket in an experiment (including some supermarkets with no brands stocked) and Y is the volume of cigarette sales in the supermarket.

The normal error model for these cases is the same as regression model (3.1) except $\beta_0 = 0$:

$$(5.11) \qquad Y_i = \beta_1 X_i + \varepsilon_i$$

where:

> β_1 is a parameter
> X_i are known constants
> ε_i are independent $N(0, \sigma^2)$

The regression function for model (5.11) is:

$$(5.12) \qquad E\{Y\} = \beta_1 X$$

The least squares estimator of β_1 is obtained by minimizing:

$$(5.13) \qquad Q = \sum (Y_i - \beta_1 X_i)^2$$

with respect to β_1. The resulting normal equation is:

$$(5.14) \qquad \sum X_i(Y_i - b_1 X_i) = 0$$

leading to the point estimator:

$$(5.15) \qquad b_1 = \frac{\sum X_i Y_i}{\sum X_i^2}$$

b_1 as given in (5.15) is also the maximum likelihood estimator.

An unbiased estimator of $E\{Y\}$ is:

$$(5.16) \qquad \hat{Y} = b_1 X$$

The residuals are defined, as usual, as the difference between the observed and fitted values:

$$(5.17) \qquad e_i = Y_i - \hat{Y}_i = Y_i - b_1 X_i$$

An unbiased estimator of σ^2 is:

$$(5.18) \qquad MSE = \frac{\sum(Y_i - \hat{Y}_i)^2}{n - 1} = \frac{\sum(Y_i - b_1 X_i)^2}{n - 1} = \frac{\sum e_i^2}{n - 1}$$

The reason for the denominator $n - 1$ is that only one degree of freedom is lost in estimating the single parameter of the regression function (5.12).

TABLE 5.1   Confidence Limits for Regression through Origin

Estimate of—	Estimated Variance		Confidence Limits
β_1	$s^2\{b_1\} = \dfrac{MSE}{\Sigma X_i^2}$	(5.19)	$b_1 \pm ts\{b_1\}$
$E\{Y_h\}$	$s^2\{\hat{Y}_h\} = \dfrac{X_h^2 MSE}{\Sigma X_i^2}$	(5.20)	$\hat{Y}_h \pm ts\{\hat{Y}_h\}$
$PI \leftarrow Y_{h(new)}$	$s^2\{Y_{h(new)}\} = MSE\left(1 + \dfrac{X_h^2}{\Sigma X_i^2}\right)$	(5.21)	$\hat{Y}_h \pm ts\{Y_{h(new)}\}$

where:
$$t = t(1 - \alpha/2; n - 1)$$

Confidence limits for β_1, $E\{Y_h\}$, and a new observation $Y_{h(new)}$ are shown in Table 5.1. Note that the t multiple has $n - 1$ degrees of freedom here, the degrees of freedom associated with MSE. The results in Table 5.1 are derived in analogous fashion to the earlier results for regression model (3.1). Whereas for model (3.1) with an intercept, we encounter terms $(X_i - \bar{X})^2$ or $(X_h - \bar{X})^2$, here we find X_i^2 and X_h^2 because of the regression through the origin.

Example

The Charles Plumbing Supplies Company operates 12 warehouses. In an attempt to tighten procedures for planning and control, a consultant studied the relation between number of work units performed (X) and total variable labor cost (Y) in the warehouses during a test period. The data are given in Table 5.2, columns 1 and 2, and the observations are shown as a scatter plot in Figure 5.1.

Model (5.11) for regression through the origin was employed since Y involves variable costs only and the other conditions of the model appeared to be satisfied as well. From Table 5.2, columns 3 and 4, we have $\Sigma X_i Y_i = 894,714$ and $\Sigma X_i^2 = 190,963$. Hence:

$$b_1 = \frac{\Sigma X_i Y_i}{\Sigma X_i^2} = \frac{894,714}{190,963} = 4.68527$$

and the estimated regression function is:

$$\hat{Y} = 4.68527X$$

In Table 5.2, the fitted values are shown in column 5 and the residuals in column 6. The fitted regression line is plotted in Figure 5.1.

To illustrate inferences for regression through the origin, suppose an interval estimate of β_1 is desired with a 95 percent confidence coefficient. By squaring the residuals in Table 5.2, column 6, and then summing them, we obtain (calculations not shown):

TABLE 5.2   Regression through Origin—Warehouse Example

Warehouse i	(1) Work Units Performed X_i	(2) Variable Labor Cost (dollars) Y_i	(3) $X_i Y_i$	(4) X_i^2	(5) $\hat{Y}_i$	(6) e_i
1	20	114	2,280	400	93.71	20.29
2	196	921	180,516	38,416	918.31	2.69
3	115	560	64,400	13,225	538.81	21.19
4	50	245	12,250	2,500	234.26	10.74
5	122	575	70,150	14,884	571.60	3.40
6	100	475	47,500	10,000	468.53	6.47
7	33	138	4,554	1,089	154.61	−16.61
8	154	727	111,958	23,716	721.53	5.47
9	80	375	30,000	6,400	374.82	.18
10	147	670	98,490	21,609	688.74	−18.74
11	182	828	150,696	33,124	852.72	−24.72
12	160	762	121,920	25,600	749.64	12.36
Total	1,359	6,390	894,714	190,963	6,367.28	22.72

$$MSE = \frac{\Sigma e_i^2}{n - 1} = \frac{2,457.6}{11} = 223.42$$

From Table 5.2, column 4, we have $\Sigma X_i^2 = 190,963$. Hence:

$$s^2\{b_1\} = \frac{MSE}{\Sigma X_i^2} = \frac{223.42}{190,963} = .0011700 \qquad s\{b_1\} = .034205$$

For a 95 percent confidence coefficient, we require $t(.975; 11) = 2.201$. The

FIGURE 5.1   Scatter Plot and Fitted Regression through Origin—Warehouse Example

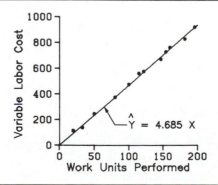

confidence limits, by (5.19) in Table 5.1, are $b_1 \pm ts\{b_1\}$ or $4.68527 \pm 2.201(.034205)$. The 95 percent confidence interval for β_1 therefore is:

$$4.61 \leq \beta_1 \leq 4.76$$

Thus, with 95 percent confidence, it is estimated that the mean of the distribution of total variable labor costs increases by somewhere between \$4.61 and \$4.76 for each additional work unit performed.

Comments

1. In linear regression through the origin, there is no least squares property of the form $\Sigma e_i = 0$. Consequently, the residuals usually will not sum to zero, as is illustrated in Table 5.2, column 6 for the warehouse example. The only property of the residuals comes from the normal equation (5.14), namely, $\Sigma X_i e_i = 0$.

2. In interval estimation of $E\{Y_h\}$ or prediction of $Y_{h(\text{new})}$, note that the intervals (5.20) and (5.21) in Table 5.1 widen, the further X_h is from the origin. The reason is that the value of the true regression function is known precisely at the origin, so the effect of the sampling error in the slope b_1 becomes increasingly important the farther X_h is from the origin.

3. Since only one regression parameter, β_1, must be estimated for the regression function (5.12), simultaneous estimation methods are not required to make a family of statements about several mean responses. For a given confidence coefficient $1 - \alpha$, formula (5.20) can be used repetitively with the given sample results for different levels of X to generate a family of statements for which the family confidence coefficient is still $1 - \alpha$.

4. Like any other statistical model, regression model (5.11) should be evaluated for aptness. Even when it is known that the regression function must go through the origin, the function might not be linear or the variance of the error terms might not be constant. Often one cannot be sure in advance that the regression line goes through the origin, and it is then safe practice to use the intercept regression model (3.1). If the regression line does go through the origin, b_0 will differ from 0 only by a small sampling error, and unless the sample size is very small, use of regression model (3.1) has no disadvantages of any consequence. If the regression line does not go through the origin, use of regression model (3.1) will avoid potentially serious difficulties resulting from forcing the regression line through the origin when this is not appropriate.

5.5 EFFECT OF MEASUREMENT ERRORS

In our discussion of regression models up to this point, we have not explicitly considered the presence of measurement errors in either X or Y. We now examine briefly the effect of measurement errors.

Measurement Errors in Y

If random measurement errors are present in the dependent variable Y, no new problems are created when these errors are uncorrelated and not biased (positive and negative measurement errors tend to cancel out). Consider, for example, a study of the relation between the time required to complete a task (Y) and the complexity of the task (X). The time to complete the task may not be measured accurately because the person operating the stopwatch may not do so at the precise instants called for. As long as such measurement errors are of a random nature, uncorrelated, and not biased, these measurement errors can simply be absorbed in the model error term ε. The model error term reflects the composite effects of a large number of factors not considered in the model, one of which simply would be random errors due to inaccuracy in the process of measuring Y.

Measurement Errors in *X*

Unfortunately, a different situation holds if the independent variable X is known only with measurement error. Frequently, to be sure, X is known without measurement error, as when the independent variable is price of a product, number of variables in an optimization problem, or wage rate for a class of employees. At other times, however, measurement errors may enter the value observed for the independent variable, for instance, when it is pressure, temperature, production line speed, or person's age.

We shall use the latter illustration in our development of the nature of the problem. Suppose we are regressing employees' piecework earnings on age. Let X_i denote the true age of the ith employee and X_i^* the age given by the employee on his or her employment record. Needless to say, the two are not always the same. We define the measurement error δ_i as follows:

$$(5.22) \qquad \delta_i = X_i^* - X_i$$

The regression model we would like to study is:

$$(5.23) \qquad Y_i = \beta_0 + \beta_1 X_i + \varepsilon_i$$

Since, however, we only observe X_i^*, model (5.23) becomes:

$$(5.24) \qquad Y_i = \beta_0 + \beta_1(X_i^* - \delta_i) + \varepsilon_i$$

where we make use of (5.22) in replacing X_i. We can rewrite (5.24) as follows:

$$(5.25) \qquad Y_i = \beta_0 + \beta_1 X_i^* + (\varepsilon_i - \beta_1 \delta_i)$$

Model (5.25) may appear like an ordinary regression model, with independent variable X^* and error term $\varepsilon - \beta_1 \delta$, but it is not. The independent variable observation X_i^* is a random variable, which, as we shall see, is correlated with the error term $\varepsilon_i - \beta_1 \delta_i$. Theorem (3.40) for the case of random independent variables requires that the error term be independent of the independent variable. Hence, the standard regression results are not applicable for model (5.25).

Intuitively, we know that $\varepsilon_i - \beta_1 \delta_i$ is not independent of X_i^* since (5.22) constrains $X_i^* - \delta_i$ to equal X_i. To determine the dependence formally, let us assume:

$$(5.26a) \qquad E\{\delta_i\} = 0$$

(5.26b) $$E\{\varepsilon_i\} = 0$$

(5.26c) $$E\{\delta_i \varepsilon_i\} = 0$$

Note that (5.26a) implies that $E\{X_i^*\} = E\{X_i + \delta_i\} = X_i$, and that (5.26c) assumes the measurement error δ_i is not correlated with the model error ε_i because by (1.21a) we have $\sigma\{\delta_i, \varepsilon_i\} = E\{\delta_i \varepsilon_i\}$ since $E\{\delta_i\} = E\{\varepsilon_i\} = 0$ by (5.26a) and (5.26b). We now wish to find the covariance:

$$\sigma\{X_i^*, \varepsilon_i - \beta_1 \delta_i\} = E\{[X_i^* - E\{X_i^*\}][(\varepsilon_i - \beta_1 \delta_i) - E\{\varepsilon_i - \beta_1 \delta_i\}]\}$$

$$= E\{(X_i^* - X_i)(\varepsilon_i - \beta_1 \delta_i)\}$$

$$= E\{\delta_i(\varepsilon_i - \beta_1 \delta_i)\}$$

$$= E\{\delta_i \varepsilon_i - \beta_1 \delta_i^2\}$$

Now $E\{\delta_i \varepsilon_i\} = 0$ by (5.26c), and $E\{\delta_i^2\} = \sigma^2\{\delta_i\}$ by (1.15a) because $E\{\delta_i\} = 0$ by (5.26a). We therefore obtain:

(5.27) $$\sigma\{X_i^*, \varepsilon_i - \beta_1 \delta_i\} = -\beta_1 \sigma^2\{\delta_i\}$$

This covariance is not zero if there is a linear regression relation between X and Y.

If standard least squares procedures are applied to model (5.25), the estimators b_0 and b_1 are biased and also lack the property of consistency. Great difficulties are encountered in developing unbiased estimators when there are measurement errors in X. One approach is to impose severe conditions on the problem—for example, to make fairly strong assumptions about the properties of the distributions of δ_i, the covariance of δ_i and ε_j, and so on. Another approach is to use additional variables that are known to be related to the true value of X but not with the errors of measurement δ. Such variables are called *instrumental variables* because they are used as an instrument in studying the relation between X and Y. Instrumental variables make it possible to obtain consistent estimators of the regression parameters.

Discussions of possible approaches and further references will be found in specialized texts such as Reference 5.2.

Note

It may be asked what is the distinction between the case when X is a random variable, considered in Chapter 3, and the case when X is subject to random measurement errors, and why are there special problems with the latter. When X is a random variable, it is not under the control of the analyst and will vary at random from trial to trial, as when X is the number of persons entering a store in a day. If this random variable X is not subject to measurement errors, however, it can be accurately ascertained for a given trial. Thus, if there are no measurement errors in counting the number of persons entering a store in a day, the analyst has accurate information to study the relation between number of persons entering the store and sales, even though the levels of number of persons entering the store that actually occur cannot be controlled. If, on the other hand, measurement errors are present in the number of persons entering the store, a distorted picture of the relation between number of persons and sales occurs because the sales observations will frequently be matched against an incorrect number of persons. This distorting effect of measurement errors is present whether X is fixed or random.

Berkson Model

There is one situation where measurement errors in X are no problem. This case was first noted by Berkson (Ref. 5.3). Frequently in an experiment, the independent variable is set at a target value. For instance, in an experiment on the effect of temperature on typist productivity, the temperature may be set at target levels of 68°F, 70°F, and so on, according to the temperature control on the thermostat. The observed temperature X_i^* is fixed here, while the actual temperature X_i is a random variable since the thermostat may not be completely accurate. Similar situations exist when water pressure is set according to a gauge, or employees of specified ages according to their employment records are selected for a study.

In all of these cases, the observation X_i^* is a fixed quantity, while the unobserved true value X_i is a random variable. The measurement error is, as before:

$$(5.28) \qquad \qquad \delta_i = X_i^* - X_i$$

Here, however, there is no constraint on the relation between X_i^* and δ_i, since X_i^* is a fixed quantity. Again, we assume that $E\{\delta_i\} = 0$.

Model (5.25), which we obtained when replacing X_i by $X_i^* - \delta_i$, is still applicable for the Berkson case:

$$(5.29) \qquad \qquad Y_i = \beta_0 + \beta_1 X_i^* + (\varepsilon_i - \beta_1 \delta_i)$$

The expected value of the error term, $E\{\varepsilon_i - \beta_1 \delta_i\}$, is zero as before, since $E\{\varepsilon_i\} = 0$ and $E\{\delta_i\} = 0$. Further, $\varepsilon_i - \beta_1 \delta_i$ is now uncorrelated with X_i^*, since X_i^* is a constant for the Berkson case. Hence, the following conditions of an ordinary regression model are met:

1. The error terms have expectation zero.
2. The independent variable is a constant, and hence the error terms are not correlated with it.

Thus, least squares procedures can be applied for the Berkson case without modification, and the estimators b_0 and b_1 will be unbiased. If we can make the standard normality and constant variance assumptions for the errors $\varepsilon_i - \beta_1 \delta_i$, the usual tests and interval estimates can be utilized.

5.6 INVERSE PREDICTIONS

At times, a regression model of Y on X is used to make a prediction of the value of X which gave rise to a new observation Y. This is known as an *inverse prediction*. We illustrate inverse predictions by two examples:

1. A trade association analyst has regressed the selling price of a product (Y) on its cost (X) for the 15 member firms of the association. The selling price $Y_{h(\text{new})}$ for another firm not belonging to the trade association is known, and it is desired to estimate the cost $X_{h(\text{new})}$ for this firm.

2. A regression analysis of the decrease in cholesterol level (Y) against dosage of a new drug (X) has been conducted, based on observations for 50 patients. A physician is treating a new patient for whom the cholesterol level should decrease by $Y_{h(new)}$. It is desired to estimate the appropriate dosage level $X_{h(new)}$ to be administered to bring about the desired cholesterol level decrease $Y_{h(new)}$.

In inverse predictions, regression model (3.1) is assumed as before:

$$(5.30) \qquad\qquad Y_i = \beta_0 + \beta_1 X_i + \varepsilon_i$$

The estimated regression function based on n observations is obtained as usual:

$$(5.31) \qquad\qquad \hat{Y} = b_0 + b_1 X$$

A new observation $Y_{h(new)}$ becomes available, and it is desired to estimate the level $X_{h(new)}$ that gave rise to this new observation. A natural point estimator is obtained by solving (5.31) for X, given $Y_{h(new)}$:

$$(5.32) \qquad\qquad \hat{X}_{h(new)} = \frac{Y_{h(new)} - b_0}{b_1} \qquad b_1 \neq 0$$

where $\hat{X}_{h(new)}$ denotes the point estimator of the new level $X_{h(new)}$. Figure 5.2 contains a representation of this point estimator for an example to be discussed shortly. $\hat{X}_{h(new)}$ is, indeed, the maximum likelihood estimator of $X_{h(new)}$ for regression model (3.1).

FIGURE 5.2   Scatter Plot and Fitted Regression Line—Calibration Example

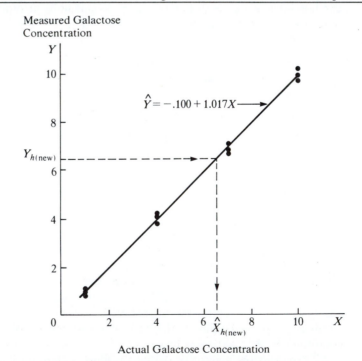

It can be shown that approximate $1 - \alpha$ confidence limits for $X_{h(new)}$ are:

(5.33)
$$\hat{X}_{h(new)} \pm t(1 - \alpha/2; n - 2)s\{\hat{X}_{h(new)}\}$$

where:

(5.33a)
$$s^2\{\hat{X}_{h(new)}\} = \frac{MSE}{b_1^2}\left[1 + \frac{1}{n} + \frac{(\hat{X}_{h(new)} - \bar{X})^2}{\Sigma(X_i - \bar{X})^2}\right]$$

Example

A medical researcher in an experiment employed a new method for measuring low concentration of galactose (sugar) in the blood (Y) on 12 samples containing known concentrations (X). Altogether, four concentration levels were used in the experiment. Linear regression model (3.1) was fitted with the following results:

$$n = 12 \qquad b_0 = -.100 \qquad b_1 = 1.017 \qquad MSE = .0272$$

$$s\{b_1\} = .0142 \qquad \bar{X} = 5.500 \qquad \bar{Y} = 5.492 \qquad \Sigma(X_i - \bar{X})^2 = 135$$

$$\hat{Y} = -.100 + 1.017X$$

The data and the estimated regression line are plotted in Figure 5.2.

The researcher first wished to make sure that there is a linear association between the two variables. A test of $H_0: \beta_1 = 0$ versus $H_a: \beta_1 \neq 0$ utilizing test statistic $t^* = b_1/s\{b_1\} = 1.017/.0142 = 71.6$ was conducted for $\alpha = .05$. Since $t(.975; 10) = 2.228$ and $|t^*| = 71.6 > 2.228$, it was concluded that $\beta_1 \neq 0$, or that a linear association exists between the measured concentration and the actual concentration.

The researcher now wishes to use the regression relation for a new patient for whom the new measurement procedure yielded $Y_{h(new)} = 6.52$. It is desired to estimate the actual concentration $X_{h(new)}$ for this patient by means of a 95 percent confidence interval.

Using (5.32) and (5.33a), we obtain:

$$\hat{X}_{h(new)} = \frac{6.52 - (-.100)}{1.017} = 6.509$$

$$s^2\{\hat{X}_{h(new)}\} = \frac{.0272}{(1.017)^2}\left[1 + \frac{1}{12} + \frac{(6.509 - 5.500)^2}{135}\right] = .0287$$

so that $s\{\hat{X}_{h(new)}\} = .1694$. We require $t(.975; 10) = 2.228$, and using (5.33), we obtain the confidence limits $6.509 \pm 2.228(.1694)$. Hence, the 95 percent confidence interval is:

$$6.13 \leq X_{h(new)} \leq 6.89$$

Thus, it can be concluded with 95 percent confidence that the actual galactose concentration is between 6.13 and 6.89. This is approximately a ±6 percent error, which was considered reasonable by the researcher.

Comments

1. The inverse prediction problem is also known as a *calibration problem* since it is applicable when inexpensive, quick, and approximate measurements (Y) are related to precise, often expensive, and time-consuming measurements (X) based on n observations. The resulting regression model is then used to estimate for a new approximate measurement $Y_{h(new)}$ what is the precise measurement $X_{h(new)}$. We illustrated this use in the calibration example.

2. The approximate confidence interval (5.33) is appropriate if the quantity:

$$(5.34) \qquad \frac{[t(1 - \alpha/2; n - 2)]^2 MSE}{b_1^2 \Sigma(X_i - \bar{X})^2}$$

is small, say less than .1. For the calibration example, this quantity is:

$$\frac{(2.228)^2(.0272)}{(1.017)^2(135)} = .00097$$

so that the approximate confidence interval is appropriate here.

3. Simultaneous prediction intervals based on g different new observed measurements $Y_{h(new)}$, with a $1 - \alpha$ family confidence coefficient, are easily obtained using either the Bonferroni or the Scheffé procedures discussed in Section 5.3. The value of $t(1 - \alpha/2; n - 2)$ in (5.33) is replaced by either $B = t(1 - \alpha/2g; n - 2)$ or $S = [gF(1 - \alpha; g, n - 2)]^{1/2}$.

5.7 CHOICE OF X LEVELS

When regression data are obtained by experiment, the levels of X at which observations on Y are to be taken are under the control of the experimenter. Among other things, the experimenter will have to consider:

1. How many levels of X should be investigated?
2. What shall the two extreme levels be?
3. How shall the other levels of X, if any, be spaced?
4. How many observations should be taken at each level of X?

There is no single answer to these questions, since different purposes of the regression analysis lead to different answers. The possible objectives in regression analysis are varied, as we have noted earlier. The main objective may be to estimate the slope of the regression line, or in some cases to estimate the intercept. In many cases, the main objective is to predict one or more new observations or to estimate one or more mean responses. When the regression function is curvilinear, the main objective may be to locate the maximum or minimum mean response. At still other times, the main purpose is to determine the nature of the regression function.

To illustrate how the purpose affects the design, consider the variances of b_0, b_1, $\hat{Y}_h$, and $Y_{h(new)}$, which were developed earlier for regression model (3.1):

$$(5.35) \qquad \sigma^2\{b_0\} = \sigma^2 \frac{\Sigma X_i^2}{n\Sigma(X_i - \bar{X})^2} = \sigma^2\left[\frac{1}{n} + \frac{\bar{X}^2}{\Sigma(X_i - \bar{X})^2}\right]$$

$$(5.36) \qquad \sigma^2\{b_1\} = \frac{\sigma^2}{\Sigma(X_i - \bar{X})^2}$$

$$(5.37) \qquad \sigma^2\{\hat{Y}_h\} = \sigma^2\left[\frac{1}{n} + \frac{(X_h - \bar{X})^2}{\Sigma(X_i - \bar{X})^2}\right]$$

$$(5.38) \qquad \sigma^2\{Y_{h(\text{new})}\} = \sigma^2\left[1 + \frac{1}{n} + \frac{(X_h - \bar{X})^2}{\Sigma(X_i - \bar{X})^2}\right]$$

The variance of the slope is minimized if $\Sigma(X_i - \bar{X})^2$ is maximized. This is accomplished by using two levels of X, at the two extremes for the scope of the model, and placing half of the observations at each of the two levels. Of course, if one were not sure of the linearity of the regression function, one would be hesitant to use only two levels since they would provide no information about possible departures from linearity.

If the main purpose is to estimate β_0, the number and placement of levels does not matter as long as $\bar{X} = 0$. On the other hand, to estimate the mean response or predict a new observation at X_h, it is best to use X levels so that $\bar{X} = X_h$. If a number of mean responses are to be estimated or a number of new observations are to be predicted, it would be best to spread out the X levels such that $\bar{X}$ is in the center of the X_h levels of interest.

Although the number and spacing of X levels depends very much on the major purpose of the regression analysis, some general advice can be given, at least to be used as a point of departure. D. R. Cox suggests as follows:

Use two levels when the object is primarily to examine whether or not . . . (the independent variable) . . . has an effect and in which direction that effect is. Use three levels whenever a description of the response curve by its slope and curvature is likely to be adequate; this should cover most cases. Use four levels if further examination of the shape of the response curve is important. Use more than four levels when it is required to estimate the detailed shape of the response curve, or when the curve is expected to rise to an asymptotic value, or in general to show features not adequately described by slope and curvature. Except in these last cases it is generally satisfactory to use equally spaced levels with equal numbers of observations per level [Ref. 5.4].

CITED REFERENCES

5.1. Miller, R. G., Jr. *Simultaneous Statistical Inference.* 2nd ed. New York: Springer-Verlag, 1981, pp. 114–16.

5.2. Fuller, W. A. *Measurement Error Models.* New York: John Wiley & Sons, 1987.

5.3. Berkson, J. "Are There Two Regressions?" *Journal of the American Statistical Association* 45 (1950), pp. 164–80.

5.4. Cox, D. R. *Planning of Experiments.* New York: John Wiley & Sons, 1958, pp. 141–42.

PROBLEMS

5.1. When joint confidence intervals for β_0 and β_1 are developed by the Bonferroni method with a family confidence coefficient of 90 percent, does this imply that 10 percent of the time the confidence intervals for β_0 will be incorrect? That 5 percent of the time the confidence interval for β_0 will be incorrect and 5 percent of the time that for β_1 will be incorrect? Discuss.

5.2. Refer to Problem 3.1. Suppose the student combines the two confidence intervals into a confidence set. What can you say about the family confidence coefficient for this set?

5.3. Refer to **Calculator maintenance** Problems 2.18 and 3.5.
   a. Will b_0 and b_1 tend to err in the same direction or in opposite directions here? Explain.
   b. Obtain Bonferroni joint confidence intervals for β_0 and β_1 using a 95 percent family confidence coefficient.
   c. A consultant has suggested that β_0 should be 0 and β_1 should equal 14.0. Do your joint confidence intervals in part (b) support this view?

5.4. Refer to **Airfreight breakage** Problem 2.19.
   a. Will b_0 and b_1 tend to err in the same direction or in opposite directions here? Explain.
   b. Obtain Bonferroni joint confidence intervals for β_0 and β_1 using a 99 percent family confidence coefficient. Interpret your confidence intervals.

5.5. Refer to **Plastic hardness** Problem 2.20.
   a. Obtain Bonferroni joint confidence intervals for β_0 and β_1 using a 90 percent family confidence coefficient. Interpret your confidence intervals.
   b. Are b_0 and b_1 positively or negatively correlated here? Is this reflected in your joint confidence intervals in part (a)?
   c. What is the meaning of the family confidence coefficient in part (a)?

5.6. Refer to **Muscle mass** Problem 2.25.
   a. Obtain Bonferroni joint confidence intervals for β_0 and β_1 using a 99 percent family confidence coefficient. Interpret your confidence intervals.
   b. Will b_0 and b_1 tend to err in the same direction or in opposite directions here? Explain.
   c. A researcher has suggested that β_0 should equal approximately 160 and that β_1 should be between -1.9 and -1.5. Do the joint confidence intervals in part (a) support this expectation?

5.7. Refer to **Calculator maintenance** Problems 2.18 and 3.5.
   a. Estimate the expected number of minutes spent when there are 3, 5, and 7 machines to be serviced, respectively. Use interval estimates with a 90 percent family confidence coefficient based on the Working-Hotelling approach.
   b. Two service calls for preventive maintenance are scheduled in which the numbers of machines to be serviced are 4 and 7, respectively. A family of prediction intervals for the times to be spent on these calls is desired with a 90 percent family confidence coefficient. Which procedure, Scheffé or Bonferroni, will provide tighter prediction limits here?
   c. Obtain the family of prediction limits required in part (b) using the more efficient procedure.

5.8. Refer to **Airfreight breakage** Problem 2.19.
   a. It is desired to obtain interval estimates of the mean number of broken ampules when there are 0, 1, and 2 transfers for the shipment using a 95 percent family confidence coefficient. Obtain the desired confidence intervals using the Working-Hotelling approach.
   b. Are the confidence intervals obtained in part (a) more efficient than Bonferroni intervals here? Explain.

   c. The next three shipments will make 0, 1, and 2 transfers, respectively. Obtain prediction limits for the number of broken ampules for each of these three shipments using the Scheffé procedure and a 95 percent family confidence coefficient.

   d. Would the Bonferroni procedure have been more efficient in developing the prediction intervals in part (c)? Explain.

5.9. Refer to **Plastic hardness** Problem 2.20.

   a. Management wishes to obtain interval estimates of the mean hardness when the elapsed time is 20, 30, and 40 hours, respectively. Calculate the desired confidence intervals using the Bonferroni procedure and a 90 percent family confidence coefficient. What is the meaning of the family confidence coefficient here?

   b. Is the Bonferroni procedure employed in part (a) the most efficient one that could be employed here? Explain.

   c. The next two test items will be measured after 30 and 40 hours of elapsed time, respectively. Predict the hardness for each of these two items using the most efficient procedure and a 90 percent family confidence coefficient.

5.10. Refer to **Muscle mass** Problem 2.25.

   a. The nutritionist is particularly interested in the mean muscle mass for women aged 45, 55, and 65. Obtain joint confidence intervals for the means of interest using the Working-Hotelling procedure and a 95 percent family confidence coefficient.

   b. Is the Working-Hotelling approach the most efficient one to be employed in part (a)? Explain.

   c. Three additional women aged 48, 59, and 74 have contacted the nutritionist. Predict the muscle mass for each of these three women using the Bonferroni approach and a 95 percent family confidence coefficient.

   d. Subsequently, the nutritionist wishes to predict the muscle mass for a fourth woman aged 64, with a family confidence coefficient of 95 percent for the four predictions. Will the three prediction intervals in part (c) have to be recalculated? Would this also be true if the Scheffé method had been used in constructing the prediction intervals?

5.11. A behavioral scientist stated recently: "I am never sure whether the regression line goes through the origin. Hence, I will not use such a model." Comment.

5.12. **Typographical errors.** Shown below are the number of galleys for a manuscript (X) and the total dollar cost of correcting typographical errors (Y) in a random sample of recent orders handled by a firm specializing in technical manuscripts. Since Y involves variable costs only, an analyst wished to determine whether the regression through the origin model (5.11) is apt for studying the relation between the two variables.

i:	1	2	3	4	5	6	7	8	9	10	11	12
X_i:	7	12	10	10	14	25	30	25	18	10	4	6
Y_i:	128	213	191	178	250	446	540	457	324	177	75	107

   a. Fit regression model (5.11) and state the estimated regression function.

   b. Plot the estimated regression function and the data. Does a linear regression function through the origin appear to provide a good fit here? Comment.

   c. In estimating costs of handling prospective orders, management has used a standard of $17.50 per galley for the cost of correcting typographical errors. Test whether or not this standard should be revised; use $\alpha = .02$. State the alternatives, decision rule, and conclusion.

      d. Obtain a prediction interval for the correction cost on a forthcoming job involving 10 galleys. Use a confidence coefficient of 98 percent.

5.13. Refer to **Typographical errors** Problem 5.12.

      a. Obtain the residuals e_i. Do they sum to zero? Plot the residuals against the fitted values $\hat{Y}_i$. What conclusions can be drawn from your plot?

      b. Conduct a formal test for lack of fit of linear regression through the origin; use $\alpha = .01$. State the alternatives, decision rule, and conclusion. What is the P-value of the test?

5.14. Refer to **Grade point average** Problem 2.17. Assume that linear regression through the origin model (5.11) is appropriate.

      a. Fit regression model (5.11) and state the estimated regression function.

      b. Estimate β_1 with a 95 percent confidence interval. Interpret your interval estimate.

      c. Estimate the mean freshman GPA for students whose entrance test score is 5.7. Use a 95 percent confidence interval.

5.15. Refer to **Grade point average** Problem 5.14.

      a. Plot the fitted regression line and the data. Does the linear regression function through the origin appear to be a good fit here?

      b. Obtain the residuals e_i. Do they sum to zero? Plot the residuals against the fitted values $\hat{Y}_i$. What conclusions can be drawn from your plot?

      c. Conduct a formal test for lack of fit of linear regression through the origin; use $\alpha = .005$. State the alternatives, decision rule, and conclusion. What is the P-value of the test?

5.16. Refer to **Calculator maintenance** Problem 2.18. Assume that linear regression through the origin model (5.11) is appropriate.

      a. Obtain the estimated regression function.

      b. Estimate β_1 with a 90 percent confidence interval. Interpret your interval estimate.

      c. Predict the service time on a new call in which six machines are to be serviced. Use a 90 percent prediction interval.

5.17. Refer to **Calculator maintenance** Problem 5.16.

      a. Plot the fitted regression line and the data. Does the linear regression function through the origin appear to be a good fit here?

      b. Obtain the residuals e_i. Do they sum to zero? Plot the residuals against the fitted values $\hat{Y}_i$. What conclusions can be drawn from your plot?

      c. Conduct a formal test for lack of fit of linear regression through the origin; use $\alpha = .01$. State the alternatives, decision rule, and conclusion. What is the P-value of the test?

5.18. Refer to **Plastic hardness** Problem 2.20. Suppose that errors arise in X because the laboratory technician is instructed to measure the hardness of the ith specimen (Y_i) at a prerecorded elapsed time (X_i), but the timing is imperfect so the true elapsed time varies at random from the prerecorded elapsed time. Will ordinary least squares estimates be biased here? Discuss.

5.19. Refer to **Grade point average** Problems 2.17 and 3.4. A new student earned a grade point average of 3.4 in the freshman year.

      a. Obtain a 90 percent confidence interval for the student's entrance test score. Interpret your confidence interval.

      b. Is criterion (5.34) met as to the appropriateness of the approximate confidence interval?

5.20. Refer to **Plastic hardness** Problem 2.20. The measurement of a new test item showed 238 Brinell units of hardness.

    a. Obtain a 99 percent confidence interval for the elapsed time before the hardness was measured. Interpret your confidence interval.

    b. Is criterion (5.34) met as to the appropriateness of the approximate confidence interval?

EXERCISES

5.21. If the independent variable is so coded that $\bar{X} = 0$ and the normal error regression model (3.1) applies, are b_0 and b_1 independent? Are the joint confidence intervals for β_0 and β_1 then independent?

5.22. Derive an extension of the Bonferroni inequality (5.3a) for the case of three statements, each with statement confidence coefficient $1 - \alpha$.

5.23. Show that for the fitted least squares regression line through the origin (5.16), $\Sigma X_i e_i = 0$.

5.24. Show that $\hat{Y}$ as defined in (5.16) for linear regression through the origin is an unbiased estimator of $E\{Y\}$.

5.25. Derive the formula for $s^2\{\hat{Y}_h\}$ given in Table 5.1 for linear regression through the origin.

PROJECTS

5.26. Refer to the **SMSA** data set and Project 2.42. Consider the regression relation of number of active physicians to total population.

    a. Obtain Bonferroni joint confidence intervals for β_0 and β_1 using a 95 percent family confidence coefficient.

    b. An investigator has suggested that β_0 should be -400 and β_1 should be 2.25. Do the joint confidence intervals in part (a) support this view? Discuss.

    c. It is desired to estimate the expected number of active physicians for SMSAs with total population of $X = 500$, 1,000, 5,000 thousands with family confidence coefficient .90. Which procedure, the Working-Hotelling or the Bonferroni, is more efficient here?

    d. Obtain the family of interval estimates required in part (c) using the more efficient procedure. Interpret your confidence intervals.

5.27. Refer to the **SENIC** data set and Project 2.44. Consider the regression relation of average length of stay to infection risk.

    a. Obtain Bonferroni joint confidence intervals for β_0 and β_1 using a 90 percent family confidence coefficient.

    b. A researcher suggested that β_0 should be approximately 7 and β_1 should be approximately 1. Do the joint intervals in part (a) support this expectation? Discuss.

    c. It is desired to estimate the expected hospital stay for persons with infection risks $X = 2, 3, 4, 5$ with family confidence coefficient .95. Which procedure, the Working-Hotelling or the Bonferroni, is more efficient here?

    d. Obtain the family of interval estimates required in part (c) using the more efficient procedure. Interpret your confidence intervals.

Chapter 6

Matrix Approach to Simple Linear Regression Analysis

Matrix algebra is widely used for mathematical and statistical analysis. The matrix approach is practically a necessity in multiple regression analysis, since it permits extensive systems of equations and large arrays of data to be denoted compactly and operated upon efficiently.

In this chapter, we first take up a brief introduction to matrix algebra. (A fuller treatment of matrix algebra may be found in specialized texts such as Reference 6.1.) Then we apply matrix methods to the simple linear regression model discussed in previous chapters. While matrix algebra is not really required for simple linear regression, the application of matrix methods to this case will provide a useful transition to multiple regression, which will be taken up in Part II.

Readers who are familiar with matrix algebra may wish to scan the introductory parts of this chapter and focus upon the parts dealing with the use of matrix methods in regression analysis.

6.1 MATRICES

Definition of Matrix

A matrix is a rectangular array of elements arranged in rows and columns. An example of a matrix is:

$$
\begin{array}{cc}
 & \text{Column} \quad \text{Column} \\
 & 1 \qquad\quad 2 \\
\begin{array}{c} \text{Row 1} \\ \text{Row 2} \\ \text{Row 3} \end{array} &
\begin{bmatrix}
16{,}000 & 23 \\
33{,}000 & 47 \\
21{,}000 & 35
\end{bmatrix}
\end{array}
$$

The *elements* of this particular matrix are numbers representing income (column 1) and age (column 2) of three persons. The elements are arranged by row (person) and column (characteristic of person). Thus, the element in the first row and first column (16,000) represents the income of the first person. The element in the first row and second column (23) represents the age of the first person. The *dimension* of the matrix is 3 × 2, i.e., 3 rows by 2 columns. If we wanted to present income and age for 1,000 persons in a matrix with the same format as the one earlier, we would require a 1,000 × 2 matrix.

Other examples of matrices are:

$$\begin{bmatrix} 1 & 0 \\ 5 & 10 \end{bmatrix} \qquad \begin{bmatrix} 4 & 7 & 12 & 16 \\ 3 & 15 & 9 & 8 \end{bmatrix}$$

These two matrices have dimensions of 2 × 2 and 2 × 4, respectively. Note that we always specify the number of rows first and then the number of columns in giving the dimension of a matrix.

As in ordinary algebra, we may use symbols to identify the elements of a matrix:

$$\begin{array}{ccc} j=1 & j=2 & j=3 \\ \begin{array}{c} i=1 \\ i=2 \end{array} \begin{bmatrix} a_{11} & a_{12} & a_{13} \\ a_{21} & a_{22} & a_{23} \end{bmatrix} \end{array}$$

Note that the first subscript identifies the row number and the second the column number. We shall use the general notation a_{ij} for the element in the ith row and the jth column. In our above example, $i = 1, 2$ and $j = 1, 2, 3$.

A matrix may be denoted by a symbol such as **A**, **X**, or **Z**. The symbol is in **bold-face** to identify that it refers to a matrix. Thus, we might define for the above matrix:

$$\mathbf{A} = \begin{bmatrix} a_{11} & a_{12} & a_{13} \\ a_{21} & a_{22} & a_{23} \end{bmatrix}$$

Reference to the matrix **A** then implies reference to the 2 × 3 array just given.

Another notation for the matrix **A** just given is:

$$\mathbf{A} = [a_{ij}] \qquad i = 1, 2; j = 1, 2, 3$$

which avoids the need for writing out all elements of the matrix by stating only the general element. This notation can only be used, of course, when the elements of a matrix are symbols.

To summarize, a matrix with r rows and c columns will be represented either in full:

$$(6.1) \qquad \mathbf{A} = \begin{bmatrix} a_{11} & a_{12} & \cdots & a_{1j} & \cdots & a_{1c} \\ a_{21} & a_{22} & \cdots & a_{2j} & \cdots & a_{2c} \\ \vdots & \vdots & & \vdots & & \vdots \\ a_{i1} & a_{i2} & \cdots & a_{ij} & \cdots & a_{ic} \\ \vdots & \vdots & & \vdots & & \vdots \\ a_{r1} & a_{r2} & \cdots & a_{rj} & \cdots & a_{rc} \end{bmatrix}$$

or in the abbreviated form:

(6.2) $\mathbf{A} = [a_{ij}]$ $i = 1, \ldots, r; j = 1, \ldots, c$

or simply by a boldface symbol, such as $\mathbf{A}$.

Comments

1. Do not think of a matrix as a number. It is a set of elements arranged in an array. Only when the matrix has dimension 1×1 is there a single number in the matrix, in which case one *can* think of it interchangeably as either a matrix or a number.

2. The following is *not* a matrix:

$$\begin{bmatrix} & 14 & \\ & 8 & \\ 10 & & 15 \\ 9 & & 16 \end{bmatrix}$$

since the numbers are not arranged in columns and rows.

Square Matrix

A matrix is said to be square if the number of rows equals the number of columns. Two examples are:

$$\begin{bmatrix} 4 & 7 \\ 3 & 9 \end{bmatrix} \qquad \begin{bmatrix} a_{11} & a_{12} & a_{13} \\ a_{21} & a_{22} & a_{23} \\ a_{31} & a_{32} & a_{33} \end{bmatrix}$$

Vector

A matrix containing only one column is called a *column vector* or simply a *vector*. Two examples are:

$$\mathbf{A} = \begin{bmatrix} 4 \\ 7 \\ 10 \end{bmatrix} \qquad \mathbf{C} = \begin{bmatrix} c_1 \\ c_2 \\ c_3 \\ c_4 \\ c_5 \end{bmatrix}$$

The vector $\mathbf{A}$ is a 3×1 matrix, and the vector $\mathbf{C}$ is a 5×1 matrix.

A matrix containing only one row is called a *row vector*. Two examples are:

$$\mathbf{B}' = [15 \quad 25 \quad 50] \qquad \mathbf{F}' = [f_1 \quad f_2]$$

We use the prime symbol for row vectors for reasons to be seen shortly. Note that the row vector $\mathbf{B}'$ is a 1×3 matrix and the row vector $\mathbf{F}'$ is a 1×2 matrix.

A single subscript suffices to identify the elements of a vector.

Transpose

The transpose of a matrix **A** is another matrix, denoted by **A′**, that is obtained by interchanging corresponding columns and rows of the matrix **A**.

For example, if:

$$\mathbf{A}_{3\times 2} = \begin{bmatrix} 2 & 5 \\ 7 & 10 \\ 3 & 4 \end{bmatrix}$$

then the transpose **A′** is:

$$\mathbf{A}'_{2\times 3} = \begin{bmatrix} 2 & 7 & 3 \\ 5 & 10 & 4 \end{bmatrix}$$

Note that the first column of **A** is the first row of **A′**, and similarly the second column of **A** is the second row of **A′**. Correspondingly, the first row of **A** has become the first column of **A′**, and so on. Note that the dimension of **A**, indicated under the symbol **A**, becomes reversed for the dimension of **A′**.

As another example, consider:

$$\mathbf{C}_{3\times 1} = \begin{bmatrix} 4 \\ 7 \\ 10 \end{bmatrix} \qquad \mathbf{C}'_{1\times 3} = [4 \quad 7 \quad 10]$$

Thus, the transpose of a column vector is a row vector, and vice versa. This is the reason why we used the symbol **B′** earlier to identify a row vector, since it may be thought of as the transpose of a column vector **B**.

In general, we have:

(6.3)

$$\mathbf{A}_{r\times c} = \begin{bmatrix} a_{11} & \cdots & a_{1c} \\ \vdots & & \vdots \\ a_{r1} & \cdots & a_{rc} \end{bmatrix} = [a_{ij}] \qquad i = 1, \ldots, r; j = 1, \ldots, c$$

Row index   Column index

$$\mathbf{A}'_{c\times r} = \begin{bmatrix} a_{11} & \cdots & a_{r1} \\ \vdots & & \vdots \\ a_{1c} & \cdots & a_{rc} \end{bmatrix} = [a_{ji}] \qquad j = 1, \ldots, c; i = 1, \ldots, r$$

Row index   Column index

Thus, the element in the ith row and the jth column in **A** is found in the jth row and ith column in **A′**.

Equality of Matrices

Two matrices **A** and **B** are said to be equal if they have the same dimension and if all corresponding elements are equal. Conversely, if two matrices are equal, their corresponding elements are equal. For example, if:

$$\mathbf{A}_{3 \times 1} = \begin{bmatrix} a_1 \\ a_2 \\ a_3 \end{bmatrix} \qquad \mathbf{B}_{3 \times 1} = \begin{bmatrix} 4 \\ 7 \\ 3 \end{bmatrix}$$

then $\mathbf{A} = \mathbf{B}$ implies:

$$a_1 = 4$$
$$a_2 = 7$$
$$a_3 = 3$$

Similarly, if:

$$\mathbf{A}_{3 \times 2} = \begin{bmatrix} a_{11} & a_{12} \\ a_{21} & a_{22} \\ a_{31} & a_{32} \end{bmatrix} \qquad \mathbf{B}_{3 \times 2} = \begin{bmatrix} 17 & 2 \\ 14 & 5 \\ 13 & 9 \end{bmatrix}$$

then $\mathbf{A} = \mathbf{B}$ implies:

$$a_{11} = 17 \qquad a_{12} = 2$$
$$a_{21} = 14 \qquad a_{22} = 5$$
$$a_{31} = 13 \qquad a_{32} = 9$$

Regression Examples

In regression analysis, one basic matrix is the vector $\mathbf{Y}$, consisting of the n observations on the dependent variable:

$$(6.4) \qquad \mathbf{Y}_{n \times 1} = \begin{bmatrix} Y_1 \\ Y_2 \\ \cdot \\ \cdot \\ \cdot \\ Y_n \end{bmatrix}$$

Note that the transpose $\mathbf{Y}'$ is the row vector:

$$(6.5) \qquad \mathbf{Y}'_{1 \times n} = [Y_1 \quad Y_2 \quad \cdots \quad Y_n]$$

Another basic matrix in regression analysis is the $\mathbf{X}$ matrix, which is defined as follows for simple linear regression analysis:

$$(6.6) \qquad \mathbf{X}_{n \times 2} = \begin{bmatrix} 1 & X_1 \\ 1 & X_2 \\ \cdot & \cdot \\ \cdot & \cdot \\ \cdot & \cdot \\ 1 & X_n \end{bmatrix}$$

The matrix $\mathbf{X}$ consists of a column of 1s and a column containing the n values of the independent variable X. Note that the transpose of $\mathbf{X}$ is:

$$(6.7) \qquad \mathbf{X}'_{2 \times n} = \begin{bmatrix} 1 & 1 & \cdots & 1 \\ X_1 & X_2 & \cdots & X_n \end{bmatrix}$$

For the Westwood Company lot size example, the $\mathbf{Y}$ and $\mathbf{X}$ matrices are (Table 2.1):

$$\mathbf{Y}_{10 \times 1} = \begin{bmatrix} 73 \\ 50 \\ . \\ . \\ . \\ 132 \end{bmatrix} \qquad \mathbf{X}_{10 \times 2} = \begin{bmatrix} 1 & 30 \\ 1 & 20 \\ . & . \\ . & . \\ . & . \\ 1 & 60 \end{bmatrix}$$

6.2 MATRIX ADDITION AND SUBTRACTION

Adding or subtracting two matrices requires that they have the same dimension. The sum, or difference, of two matrices is another matrix whose elements each consist of the sum, or difference, of the corresponding elements of the two matrices. Suppose:

$$\mathbf{A}_{3 \times 2} = \begin{bmatrix} 1 & 4 \\ 2 & 5 \\ 3 & 6 \end{bmatrix} \qquad \mathbf{B}_{3 \times 2} = \begin{bmatrix} 1 & 2 \\ 2 & 3 \\ 3 & 4 \end{bmatrix}$$

then:

$$\mathbf{A} + \mathbf{B}_{3 \times 2} = \begin{bmatrix} 1+1 & 4+2 \\ 2+2 & 5+3 \\ 3+3 & 6+4 \end{bmatrix} = \begin{bmatrix} 2 & 6 \\ 4 & 8 \\ 6 & 10 \end{bmatrix}$$

Similarly:

$$\mathbf{A} - \mathbf{B}_{3 \times 2} = \begin{bmatrix} 1-1 & 4-2 \\ 2-2 & 5-3 \\ 3-3 & 6-4 \end{bmatrix} = \begin{bmatrix} 0 & 2 \\ 0 & 2 \\ 0 & 2 \end{bmatrix}$$

In general, if:

$$\mathbf{A}_{r \times c} = [a_{ij}] \qquad \mathbf{B}_{r \times c} = [b_{ij}] \qquad i = 1, \ldots, r; j = 1, \ldots, c$$

then:

(6.8) $$\mathbf{A} + \mathbf{B}_{r \times c} = [a_{ij} + b_{ij}] \quad \text{and} \quad \mathbf{A} - \mathbf{B}_{r \times c} = [a_{ij} - b_{ij}]$$

Formula (6.8) generalizes in an obvious way to addition and subtraction of more than two matrices. Note also that $\mathbf{A} + \mathbf{B} = \mathbf{B} + \mathbf{A}$, as in ordinary algebra.

Regression Example

The regression model:

$$Y_i = E\{Y_i\} + \varepsilon_i \qquad i = 1, \ldots, n$$

can be written compactly in matrix notation. First, let us define the vector of mean responses:

$$(6.9) \qquad \underset{n\times 1}{\mathbf{E\{Y\}}} = \begin{bmatrix} E\{Y_1\} \\ E\{Y_2\} \\ \cdot \\ \cdot \\ \cdot \\ E\{Y_n\} \end{bmatrix}$$

and the vector of the error terms:

$$(6.10) \qquad \underset{n\times 1}{\boldsymbol{\varepsilon}} = \begin{bmatrix} \varepsilon_1 \\ \varepsilon_2 \\ \cdot \\ \cdot \\ \cdot \\ \varepsilon_n \end{bmatrix}$$

Recalling the definition of the **Y** observation vector (6.4), we can write the regression model as follows:

$$\underset{n\times 1}{\mathbf{Y}} = \underset{n\times 1}{\mathbf{E\{Y\}}} + \underset{n\times 1}{\boldsymbol{\varepsilon}}$$

because:

$$\begin{bmatrix} Y_1 \\ Y_2 \\ \cdot \\ \cdot \\ \cdot \\ Y_n \end{bmatrix} = \begin{bmatrix} E\{Y_1\} \\ E\{Y_2\} \\ \cdot \\ \cdot \\ \cdot \\ E\{Y_n\} \end{bmatrix} + \begin{bmatrix} \varepsilon_1 \\ \varepsilon_2 \\ \cdot \\ \cdot \\ \cdot \\ \varepsilon_n \end{bmatrix} = \begin{bmatrix} E\{Y_1\} + \varepsilon_1 \\ E\{Y_2\} + \varepsilon_2 \\ \cdot \\ \cdot \\ \cdot \\ E\{Y_n\} + \varepsilon_n \end{bmatrix}$$

Thus, the observations vector **Y** equals the sum of two vectors, a vector containing the expected values and another containing the error terms.

6.3 MATRIX MULTIPLICATION

Multiplication of a Matrix by a Scalar

A *scalar* is an ordinary number or a symbol representing a number. In multiplication of a matrix by a scalar, every element of the matrix is multiplied by the scalar. For example, suppose the matrix **A** is given by:

$$\mathbf{A} = \begin{bmatrix} 2 & 7 \\ 9 & 3 \end{bmatrix}$$

Then 4**A**, where 4 is the scalar, equals:

$$4\mathbf{A} = 4\begin{bmatrix} 2 & 7 \\ 9 & 3 \end{bmatrix} = \begin{bmatrix} 8 & 28 \\ 36 & 12 \end{bmatrix}$$

Similarly, $\lambda \mathbf{A}$ equals:

$$\lambda \mathbf{A} = \lambda \begin{bmatrix} 2 & 7 \\ 9 & 3 \end{bmatrix} = \begin{bmatrix} 2\lambda & 7\lambda \\ 9\lambda & 3\lambda \end{bmatrix}$$

where λ denotes a scalar.

If every element of a matrix has a common factor, this factor can be taken outside the matrix and treated as a scalar. For example:

$$\begin{bmatrix} 9 & 27 \\ 15 & 18 \end{bmatrix} = 3 \begin{bmatrix} 3 & 9 \\ 5 & 6 \end{bmatrix}$$

Similarly:

$$\begin{bmatrix} \dfrac{5}{\lambda} & \dfrac{2}{\lambda} \\[2mm] \dfrac{3}{\lambda} & \dfrac{8}{\lambda} \end{bmatrix} = \frac{1}{\lambda} \begin{bmatrix} 5 & 2 \\ 3 & 8 \end{bmatrix}$$

In general, if $\mathbf{A} = [a_{ij}]$ and λ is a scalar, we have:

(6.11) $$\lambda \mathbf{A} = \mathbf{A}\lambda = [\lambda a_{ij}]$$

Multiplication of a Matrix by a Matrix

Multiplication of a matrix by a matrix may appear somewhat complicated at first, but a little practice will make it into a routine operation.

Consider the two matrices:

$$\mathbf{A}_{2\times2} = \begin{bmatrix} 2 & 5 \\ 4 & 1 \end{bmatrix} \qquad \mathbf{B}_{2\times2} = \begin{bmatrix} 4 & 6 \\ 5 & 8 \end{bmatrix}$$

The product **AB** will be a 2×2 matrix whose elements are obtained by finding the cross products of rows of **A** with columns of **B** and summing the cross products. For instance, to find the element in the first row and the first column of the product **AB**, we work with the first row of **A** and the first column of **B**, as follows:

We take the cross products and sum:

$$2(4) + 5(5) = 33$$

The number 33 is the element in the first row and first column of the matrix **AB**.

To find the element in the first row and second column of **AB**, we work with the first row of **A** and the second column of **B**:

$$
\begin{array}{ccc}
\mathbf{A} & \mathbf{B} & \mathbf{AB} \\
\text{Row 1} \begin{bmatrix} \boxed{2} & \boxed{5} \\ 4 & 1 \end{bmatrix} \begin{bmatrix} 4 & \boxed{6} \\ 5 & \boxed{8} \end{bmatrix} & & \text{Row 1} \begin{bmatrix} 33 & 52 \\ & \end{bmatrix}
\end{array}
$$

Column 1   Column 2   Column 1   Column 2

The sum of the cross products is:

$$2(6) + 5(8) = 52$$

Continuing this process, we find the product **AB** to be:

$$
\underset{2\times2}{\mathbf{AB}} = \begin{bmatrix} 2 & 5 \\ 4 & 1 \end{bmatrix} \begin{bmatrix} 4 & 6 \\ 5 & 8 \end{bmatrix} = \begin{bmatrix} 33 & 52 \\ 21 & 32 \end{bmatrix}
$$

Let us consider another example:

$$
\underset{2\times3}{\mathbf{A}} = \begin{bmatrix} 1 & 3 & 4 \\ 0 & 5 & 8 \end{bmatrix} \qquad \underset{3\times1}{\mathbf{B}} = \begin{bmatrix} 3 \\ 5 \\ 2 \end{bmatrix}
$$

$$
\underset{2\times1}{\mathbf{AB}} = \begin{bmatrix} 1 & 3 & 4 \\ 0 & 5 & 8 \end{bmatrix} \begin{bmatrix} 3 \\ 5 \\ 2 \end{bmatrix} = \begin{bmatrix} 26 \\ 41 \end{bmatrix}
$$

When obtaining the product **AB**, we say that **A** is *postmultiplied* by **B** or **B** is *premultiplied* by **A**. The reason for this precise terminology is that multiplication rules for ordinary algebra do not apply to matrix algebra. In ordinary algebra, $xy = yx$. In matrix algebra, $\mathbf{AB} \neq \mathbf{BA}$ usually. In fact, even though the product **AB** may be defined, the product **BA** may not be defined at all.

In general, the product **AB** is only defined when the number of columns in **A** equals the number of rows in **B** so that there will be corresponding terms in the cross products. Thus, in our previous two examples, we had:

$$
\begin{array}{cc}
\overset{\text{Equal}}{\underset{2\times2}{\mathbf{A}}\quad\underset{2\times2}{\mathbf{B}}} = \underset{2\times2}{\mathbf{AB}} & \overset{\text{Equal}}{\underset{2\times3}{\mathbf{A}}\quad\underset{3\times1}{\mathbf{B}}} = \underset{2\times1}{\mathbf{AB}} \\
\text{Dimension of product} & \text{Dimension of product}
\end{array}
$$

Note that the dimension of the product **AB** is given by the number of rows in **A** and the number of columns in **B**. Note also that in the second case the product **BA** would not be defined since the number of columns in **B** is not equal to the number of rows in **A**:

$$
\overset{\text{Unequal}}{\underset{3\times1}{\mathbf{B}}\quad\underset{2\times3}{\mathbf{A}}}
$$

Here is another example of matrix multiplication:

$$\mathbf{AB} = \begin{bmatrix} a_{11} & a_{12} & a_{13} \\ a_{21} & a_{22} & a_{23} \end{bmatrix} \begin{bmatrix} b_{11} & b_{12} \\ b_{21} & b_{22} \\ b_{31} & b_{32} \end{bmatrix}$$

$$= \begin{bmatrix} a_{11}b_{11} + a_{12}b_{21} + a_{13}b_{31} & a_{11}b_{12} + a_{12}b_{22} + a_{13}b_{32} \\ a_{21}b_{11} + a_{22}b_{21} + a_{23}b_{31} & a_{21}b_{12} + a_{22}b_{22} + a_{23}b_{32} \end{bmatrix}$$

In general, if $\mathbf{A}$ has dimension $r \times c$ and $\mathbf{B}$ has dimension $c \times s$, the product $\mathbf{AB}$ is a matrix of dimension $r \times s$ whose element in the ith row and jth column is:

$$\sum_{k=1}^{c} a_{ik}b_{kj}$$

so that:

$$(6.12) \qquad \mathbf{AB}_{r \times s} = \left[\sum_{k=1}^{c} a_{ik}b_{kj} \right] \qquad i = 1, \ldots, r; j = 1, \ldots, s$$

Thus, in the foregoing example, the element in the first row and second column of the product $\mathbf{AB}$ is:

$$\sum_{k=1}^{3} a_{1k}b_{k2} = a_{11}b_{12} + a_{12}b_{22} + a_{13}b_{32}$$

as indeed we found by taking the cross products of the elements in the first row of $\mathbf{A}$ and second column of $\mathbf{B}$ and summing.

Additional Examples.

1.
$$\begin{bmatrix} 4 & 2 \\ 5 & 8 \end{bmatrix} \begin{bmatrix} a_1 \\ a_2 \end{bmatrix} = \begin{bmatrix} 4a_1 + 2a_2 \\ 5a_1 + 8a_2 \end{bmatrix}$$

2.
$$[2 \quad 3 \quad 5] \begin{bmatrix} 2 \\ 3 \\ 5 \end{bmatrix} = [2^2 + 3^2 + 5^2] = [38]$$

Here, the product is a 1×1 matrix, which is equivalent to a scalar. Thus, the matrix product here equals the number 38.

3.
$$\begin{bmatrix} 1 & X_1 \\ 1 & X_2 \\ 1 & X_3 \end{bmatrix} \begin{bmatrix} \beta_0 \\ \beta_1 \end{bmatrix} = \begin{bmatrix} \beta_0 + \beta_1 X_1 \\ \beta_0 + \beta_1 X_2 \\ \beta_0 + \beta_1 X_3 \end{bmatrix}$$

Regression Examples. A product frequently needed is $\mathbf{Y'Y}$, where $\mathbf{Y}$ is the vector of observations on the dependent variable as defined in (6.4):

$$(6.13) \quad \mathbf{Y'Y}_{1 \times 1} = [Y_1 \quad Y_2 \quad \cdots \quad Y_n] \begin{bmatrix} Y_1 \\ Y_2 \\ \vdots \\ Y_n \end{bmatrix} = [Y_1^2 + Y_2^2 + \cdots + Y_n^2] = [\Sigma Y_i^2]$$

Note that $\mathbf{Y'Y}$ is a 1×1 matrix, or a scalar. We thus have a compact way of writing a sum of squared terms: $\mathbf{Y'Y} = \Sigma Y_i^2$.

We also will need $\mathbf{X'X}$, which is a 2×2 matrix:

$$(6.14) \qquad \underset{2\times2}{\mathbf{X'X}} = \begin{bmatrix} 1 & 1 & \cdots & 1 \\ X_1 & X_2 & \cdots & X_n \end{bmatrix} \begin{bmatrix} 1 & X_1 \\ 1 & X_2 \\ \vdots & \vdots \\ 1 & X_n \end{bmatrix} = \begin{bmatrix} n & \Sigma X_i \\ \Sigma X_i & \Sigma X_i^2 \end{bmatrix}$$

and $\mathbf{X'Y}$, which is a 2×1 matrix:

$$(6.15) \qquad \underset{2\times1}{\mathbf{X'Y}} = \begin{bmatrix} 1 & 1 & \cdots & 1 \\ X_1 & X_2 & \cdots & X_n \end{bmatrix} \begin{bmatrix} Y_1 \\ Y_2 \\ \vdots \\ Y_n \end{bmatrix} = \begin{bmatrix} \Sigma Y_i \\ \Sigma X_i Y_i \end{bmatrix}$$

6.4 SPECIAL TYPES OF MATRICES

Certain special types of matrices arise regularly in regression analysis. We shall consider the most important of these.

Symmetric Matrix

If $\mathbf{A} = \mathbf{A'}$, $\mathbf{A}$ is said to be symmetric. Thus, $\mathbf{A}$ below is symmetric:

$$\underset{3\times3}{\mathbf{A}} = \begin{bmatrix} 1 & 4 & 6 \\ 4 & 2 & 5 \\ 6 & 5 & 3 \end{bmatrix} \qquad \underset{3\times3}{\mathbf{A'}} = \begin{bmatrix} 1 & 4 & 6 \\ 4 & 2 & 5 \\ 6 & 5 & 3 \end{bmatrix}$$

Clearly, a symmetric matrix necessarily is square. Symmetric matrices arise typically in regression analysis when we premultiply a matrix, say, $\mathbf{X}$, by its transpose, $\mathbf{X'}$. The resulting matrix, $\mathbf{X'X}$, is symmetric, as can readily be seen from (6.14).

Diagonal Matrix

A diagonal matrix is a square matrix whose off-diagonal elements are all zeros, such as:

$$\underset{3\times3}{\mathbf{A}} = \begin{bmatrix} a_1 & 0 & 0 \\ 0 & a_2 & 0 \\ 0 & 0 & a_3 \end{bmatrix} \qquad \underset{4\times4}{\mathbf{B}} = \begin{bmatrix} 4 & 0 & 0 & 0 \\ 0 & 1 & 0 & 0 \\ 0 & 0 & 10 & 0 \\ 0 & 0 & 0 & 5 \end{bmatrix}$$

We will often not show all zeros for a diagonal matrix, presenting it in the form:

$$
\mathbf{A}_{3\times3} = \begin{bmatrix} a_1 & & 0 \\ & a_2 & \\ 0 & & a_3 \end{bmatrix} \qquad \mathbf{B}_{4\times4} = \begin{bmatrix} 4 & & & \\ & 1 & & 0 \\ & & 10 & \\ 0 & & & 5 \end{bmatrix}
$$

Two important types of diagonal matrices are the identity matrix and the scalar matrix.

Identity Matrix. The identity matrix or unit matrix is denoted by $\mathbf{I}$. It is a diagonal matrix whose elements on the main diagonal are all 1s. Premultiplying or postmultiplying any $r \times r$ matrix $\mathbf{A}$ by the $r \times r$ identity matrix $\mathbf{I}$ leaves $\mathbf{A}$ unchanged. For example:

$$
\mathbf{IA} = \begin{bmatrix} 1 & 0 & 0 \\ 0 & 1 & 0 \\ 0 & 0 & 1 \end{bmatrix} \begin{bmatrix} a_{11} & a_{12} & a_{13} \\ a_{21} & a_{22} & a_{23} \\ a_{31} & a_{32} & a_{33} \end{bmatrix} = \begin{bmatrix} a_{11} & a_{12} & a_{13} \\ a_{21} & a_{22} & a_{23} \\ a_{31} & a_{32} & a_{33} \end{bmatrix}
$$

Similarly, we have:

$$
\mathbf{AI} = \begin{bmatrix} a_{11} & a_{12} & a_{13} \\ a_{21} & a_{22} & a_{23} \\ a_{31} & a_{32} & a_{33} \end{bmatrix} \begin{bmatrix} 1 & 0 & 0 \\ 0 & 1 & 0 \\ 0 & 0 & 1 \end{bmatrix} = \begin{bmatrix} a_{11} & a_{12} & a_{13} \\ a_{21} & a_{22} & a_{23} \\ a_{31} & a_{32} & a_{33} \end{bmatrix}
$$

Note that the identity matrix $\mathbf{I}$ therefore corresponds to the number 1 in ordinary algebra, since we have there that $1 \cdot x = x \cdot 1 = x$.

In general, we have for any $r \times r$ matrix $\mathbf{A}$:

(6.16) $$\mathbf{AI} = \mathbf{IA} = \mathbf{A}$$

Thus, the identity matrix can be inserted or dropped from a matrix expression whenever it is convenient to do so.

Scalar Matrix. A scalar matrix is a diagonal matrix whose main-diagonal elements are the same. Two examples of scalar matrices are:

$$
\begin{bmatrix} 2 & 0 \\ 0 & 2 \end{bmatrix} \qquad \begin{bmatrix} \lambda & 0 & 0 \\ 0 & \lambda & 0 \\ 0 & 0 & \lambda \end{bmatrix}
$$

A scalar matrix can be expressed as $\lambda \mathbf{I}$, where λ is the scalar. For instance:

$$
\begin{bmatrix} 2 & 0 \\ 0 & 2 \end{bmatrix} = 2 \begin{bmatrix} 1 & 0 \\ 0 & 1 \end{bmatrix} = 2\mathbf{I}
$$

$$
\begin{bmatrix} \lambda & 0 & 0 \\ 0 & \lambda & 0 \\ 0 & 0 & \lambda \end{bmatrix} = \lambda \begin{bmatrix} 1 & 0 & 0 \\ 0 & 1 & 0 \\ 0 & 0 & 1 \end{bmatrix} = \lambda \mathbf{I}
$$

Multiplying an $r \times r$ matrix $\mathbf{A}$ by the $r \times r$ scalar matrix $\lambda \mathbf{I}$ is equivalent to multiplying $\mathbf{A}$ by the scalar λ.

Vector and Matrix with All Elements Unity

A column vector with all elements 1 will be denoted by **1**:

(6.17)
$$\mathbf{1}_{r \times 1} = \begin{bmatrix} 1 \\ 1 \\ \cdot \\ \cdot \\ \cdot \\ 1 \end{bmatrix}$$

and a square matrix with all elements 1 will be denoted by **J**:

(6.18)
$$\mathbf{J}_{r \times r} = \begin{bmatrix} 1 & \cdots & 1 \\ 1 & & 1 \\ \cdot & & \cdot \\ \cdot & & \cdot \\ \cdot & & \cdot \\ 1 & \cdots & 1 \end{bmatrix}$$

For instance, we have:

$$\mathbf{1}_{3 \times 1} = \begin{bmatrix} 1 \\ 1 \\ 1 \end{bmatrix} \qquad \mathbf{J}_{3 \times 3} = \begin{bmatrix} 1 & 1 & 1 \\ 1 & 1 & 1 \\ 1 & 1 & 1 \end{bmatrix}$$

Note that for an $n \times 1$ vector **1** we obtain:

$$\mathbf{1'1}_{1 \times 1} = \begin{bmatrix} 1 & \cdots & 1 \end{bmatrix} \begin{bmatrix} 1 \\ \cdot \\ \cdot \\ \cdot \\ 1 \end{bmatrix} = [n] = n$$

and:

$$\mathbf{11'}_{n \times n} = \begin{bmatrix} 1 \\ \cdot \\ \cdot \\ \cdot \\ 1 \end{bmatrix} \begin{bmatrix} 1 & \cdots & 1 \end{bmatrix} = \begin{bmatrix} 1 & \cdots & 1 \\ \cdot & & \cdot \\ \cdot & & \cdot \\ \cdot & & \cdot \\ 1 & \cdots & 1 \end{bmatrix} = \mathbf{J}_{n \times n}$$

Zero Vector

A zero vector is a vector containing only zeros. The zero column vector will be denoted by **0**:

(6.19)
$$\mathbf{0}_{r \times 1} = \begin{bmatrix} 0 \\ 0 \\ \cdot \\ \cdot \\ \cdot \\ 0 \end{bmatrix}$$

For example, we have:

$$\underset{3\times 1}{\mathbf{0}} = \begin{bmatrix} 0 \\ 0 \\ 0 \end{bmatrix}$$

6.5 LINEAR DEPENDENCE AND RANK OF MATRIX

Linear Dependence

Consider the following matrix:

$$\mathbf{A} = \begin{bmatrix} 1 & 2 & 5 & 1 \\ 2 & 2 & 10 & 6 \\ 3 & 4 & 15 & 1 \end{bmatrix}$$

Let us think now of the columns of this matrix as vectors. Thus, we view $\mathbf{A}$ as being made up of four column vectors. It happens here that the columns are interrelated in a special manner. Note that the third column vector is a multiple of the first column vector:

$$\begin{bmatrix} 5 \\ 10 \\ 15 \end{bmatrix} = 5 \begin{bmatrix} 1 \\ 2 \\ 3 \end{bmatrix}$$

We say that the columns of $\mathbf{A}$ are linearly dependent. They contain redundant information, so to speak, since one column can be obtained as a linear combination of the others.

We define a set of column vectors to be linearly dependent if one vector can be expressed as a linear combination of the others. If no vector in the set can be so expressed, we define the set of vectors to be linearly independent. A more general, though equivalent, definition for the c column vectors $\mathbf{C}_1, \ldots, \mathbf{C}_c$ in an $r \times c$ matrix is:

(6.20)   When c scalars $\lambda_1, \ldots, \lambda_c$, not all zero, can be found such that:

$$\lambda_1 \mathbf{C}_1 + \lambda_2 \mathbf{C}_2 + \cdots + \lambda_c \mathbf{C}_c = \mathbf{0}$$

where $\mathbf{0}$ denotes the zero column vector, the c column vectors are *linearly dependent*. If the only set of scalars for which the equality holds is $\lambda_1 = 0, \ldots, \lambda_c = 0$, the set of c column vectors is *linearly independent*.

To illustrate for our example, $\lambda_1 = 5$, $\lambda_2 = 0$, $\lambda_3 = -1$, $\lambda_4 = 0$ leads to:

$$5 \begin{bmatrix} 1 \\ 2 \\ 3 \end{bmatrix} + 0 \begin{bmatrix} 2 \\ 2 \\ 4 \end{bmatrix} - 1 \begin{bmatrix} 5 \\ 10 \\ 15 \end{bmatrix} + 0 \begin{bmatrix} 1 \\ 6 \\ 1 \end{bmatrix} = \begin{bmatrix} 0 \\ 0 \\ 0 \end{bmatrix}$$

Hence, the column vectors are linearly dependent. Note that some of the $\lambda_j = 0$ here. It is only required for linear dependence that not all λ_j are zero.

Rank of a Matrix

The rank of a matrix is defined to be the maximum number of linearly independent columns in the matrix. We know that the rank of $\mathbf{A}$ in our earlier example cannot be 4, since the four columns are linearly dependent. We can, however, find 3 columns (1, 2, and 4) which are linearly independent. There are no scalars λ_1, λ_2, λ_4 such that $\lambda_1 \mathbf{C}_1 + \lambda_2 \mathbf{C}_2 + \lambda_4 \mathbf{C}_4 = \mathbf{0}$ other than $\lambda_1 = \lambda_2 = \lambda_4 = 0$. Thus, the rank of $\mathbf{A}$ in our example is 3.

The rank of a matrix is unique and can equivalently be defined as the maximum number of linearly independent rows. It follows that the rank of an $r \times c$ matrix cannot exceed $\min(r, c)$, the minimum of the two values r and c.

6.6 INVERSE OF A MATRIX

In ordinary algebra, the inverse of a number is its reciprocal. Thus, the inverse of 6 is $\frac{1}{6}$. A number multiplied by its inverse always equals 1:

$$6 \cdot \tfrac{1}{6} = 1$$

$$x \cdot \frac{1}{x} = x \cdot x^{-1} = x^{-1} \cdot x = 1$$

In matrix algebra, the inverse of a matrix $\mathbf{A}$ is another matrix, denoted by $\mathbf{A}^{-1}$, such that:

(6.21) $$\mathbf{A}^{-1}\mathbf{A} = \mathbf{A}\mathbf{A}^{-1} = \mathbf{I}$$

where $\mathbf{I}$ is the identity matrix. Thus, again, the identity matrix $\mathbf{I}$ plays the same role as the number 1 in ordinary algebra. An inverse of a matrix is defined only for square matrices. Even so, many square matrices do not have an inverse. If a square matrix does have an inverse, the inverse is unique.

Examples

1. The inverse of the matrix:

$$\mathop{\mathbf{A}}_{2\times 2} = \begin{bmatrix} 2 & 4 \\ 3 & 1 \end{bmatrix}$$

is:

$$\mathop{\mathbf{A}^{-1}}_{2\times 2} = \begin{bmatrix} -.1 & .4 \\ .3 & -.2 \end{bmatrix}$$

since:

$$\mathbf{A}^{-1}\mathbf{A} = \begin{bmatrix} -.1 & .4 \\ .3 & -.2 \end{bmatrix} \begin{bmatrix} 2 & 4 \\ 3 & 1 \end{bmatrix} = \begin{bmatrix} 1 & 0 \\ 0 & 1 \end{bmatrix}$$

or:

$$\mathbf{A}\mathbf{A}^{-1} = \begin{bmatrix} 2 & 4 \\ 3 & 1 \end{bmatrix} \begin{bmatrix} -.1 & .4 \\ .3 & -.2 \end{bmatrix} = \begin{bmatrix} 1 & 0 \\ 0 & 1 \end{bmatrix}$$

2. The inverse of the matrix:

$$\mathop{\mathbf{A}}_{3\times3} = \begin{bmatrix} 3 & 0 & 0 \\ 0 & 4 & 0 \\ 0 & 0 & 2 \end{bmatrix}$$

is:

$$\mathop{\mathbf{A}^{-1}}_{3\times3} = \begin{bmatrix} \frac{1}{3} & 0 & 0 \\ 0 & \frac{1}{4} & 0 \\ 0 & 0 & \frac{1}{2} \end{bmatrix}$$

since:

$$\mathbf{A}^{-1}\mathbf{A} = \begin{bmatrix} \frac{1}{3} & 0 & 0 \\ 0 & \frac{1}{4} & 0 \\ 0 & 0 & \frac{1}{2} \end{bmatrix} \begin{bmatrix} 3 & 0 & 0 \\ 0 & 4 & 0 \\ 0 & 0 & 2 \end{bmatrix} = \begin{bmatrix} 1 & 0 & 0 \\ 0 & 1 & 0 \\ 0 & 0 & 1 \end{bmatrix}$$

Note that the inverse of a diagonal matrix is a diagonal matrix consisting simply of the reciprocals of the elements on the diagonal.

Finding the Inverse

Up to this point, the inverse of a matrix **A** has been given, and we have only checked to make sure it is the inverse by seeing whether or not $\mathbf{A}^{-1}\mathbf{A} = \mathbf{I}$. But how does one find the inverse, and when does it exist?

An inverse of a square $r \times r$ matrix exists if the rank of the matrix is r. Such a matrix is said to be *nonsingular*. An $r \times r$ matrix with rank less than r is said to be *singular*, and does not have an inverse.

Finding the inverse of a matrix can often require a tremendous amount of computing. We shall take the approach in this book that the inverse of a 2 × 2 matrix and a 3 × 3 matrix can be calculated by hand. For any larger matrix, one ordinarily uses a computer or a programmable calculator to find the inverse, unless the matrix is of a special form such as a diagonal matrix. It can be shown that the inverses for 2 × 2 and 3 × 3 matrices are as follows:

1. If:

$$\mathop{\mathbf{A}}_{2\times2} = \begin{bmatrix} a & b \\ c & d \end{bmatrix}$$

then:

(6.22)

$$\underset{2\times2}{\mathbf{A}^{-1}} = \begin{bmatrix} a & b \\ c & d \end{bmatrix}^{-1} = \begin{bmatrix} \dfrac{d}{D} & \dfrac{-b}{D} \\ \dfrac{-c}{D} & \dfrac{a}{D} \end{bmatrix}$$

where:

$$D = ad - bc$$

D is called the *determinant* of the matrix $\mathbf{A}$. If $\mathbf{A}$ were singular, its determinant would equal zero and no inverse of $\mathbf{A}$ would exist.

2. If:

$$\underset{3\times3}{\mathbf{B}} = \begin{bmatrix} a & b & c \\ d & e & f \\ g & h & k \end{bmatrix}$$

then:

(6.23)

$$\underset{3\times3}{\mathbf{B}^{-1}} = \begin{bmatrix} a & b & c \\ d & e & f \\ g & h & k \end{bmatrix}^{-1} = \begin{bmatrix} A & B & C \\ D & E & F \\ G & H & K \end{bmatrix}$$

where:

$$A = (ek - fh)/Z \qquad B = -(bk - ch)/Z \qquad C = (bf - ce)/Z$$
$$D = -(dk - fg)/Z \qquad E = (ak - cg)/Z \qquad F = -(af - cd)/Z$$
$$G = (dh - eg)/Z \qquad H = -(ah - bg)/Z \qquad K = (ae - bd)/Z$$

and:

$$Z = a(ek - fh) - b(dk - fg) + c(dh - eg)$$

Z is called the determinant of the matrix $\mathbf{B}$.

Let us use (6.22) to find the inverse of:

$$\mathbf{A} = \begin{bmatrix} 2 & 4 \\ 3 & 1 \end{bmatrix}$$

We have:

$$a = 2 \qquad b = 4$$
$$c = 3 \qquad d = 1$$
$$D = ad - bc = 2(1) - 4(3) = -10$$

Hence:

$$\mathbf{A}^{-1} = \begin{bmatrix} \dfrac{1}{-10} & \dfrac{-4}{-10} \\ \dfrac{-3}{-10} & \dfrac{2}{-10} \end{bmatrix} = \begin{bmatrix} -.1 & .4 \\ .3 & -.2 \end{bmatrix}$$

as was given in an earlier example.

When an inverse $\mathbf{A}^{-1}$ has been obtained, either by hand calculations or from a computer run, it is usually wise to compute $\mathbf{A}^{-1}\mathbf{A}$ to check whether the product equals the identity matrix, allowing for minor rounding departures from 0 and 1.

Regression Example

The principal inverse matrix encountered in regression analysis is the inverse of the matrix $\mathbf{X}'\mathbf{X}$ in (6.14):

$$\mathbf{X}'\mathbf{X}_{2\times2} = \begin{bmatrix} n & \Sigma X_i \\ \Sigma X_i & \Sigma X_i^2 \end{bmatrix}$$

Using rule (6.22), we have:

$$a = n \qquad b = \sum X_i$$
$$c = \sum X_i \qquad d = \sum X_i^2$$

so that:

$$D = n \sum X_i^2 - \left(\sum X_i\right)\left(\sum X_i\right) = n\left[\sum X_i^2 - \frac{(\Sigma X_i)^2}{n}\right] = n \sum (X_i - \bar{X})^2$$

Hence:

(6.24)
$$(\mathbf{X}'\mathbf{X})^{-1}_{2\times2} = \begin{bmatrix} \dfrac{\Sigma X_i^2}{n\Sigma(X_i - \bar{X})^2} & \dfrac{-\Sigma X_i}{n\Sigma(X_i - \bar{X})^2} \\[3mm] \dfrac{-\Sigma X_i}{n\Sigma(X_i - \bar{X})^2} & \dfrac{n}{n\Sigma(X_i - \bar{X})^2} \end{bmatrix}$$

Since $\Sigma X_i = n\bar{X}$, we can simplify (6.24):

(6.25)
$$(\mathbf{X}'\mathbf{X})^{-1}_{2\times2} = \begin{bmatrix} \dfrac{\Sigma X_i^2}{n\Sigma(X_i - \bar{X})^2} & \dfrac{-\bar{X}}{\Sigma(X_i - \bar{X})^2} \\[3mm] \dfrac{-\bar{X}}{\Sigma(X_i - \bar{X})^2} & \dfrac{1}{\Sigma(X_i - \bar{X})^2} \end{bmatrix}$$

Uses of Inverse Matrix

In ordinary algebra, we solve an equation of the type:

$$5y = 20$$

by multiplying both sides of the equation by the inverse of 5, namely:

$$\tfrac{1}{5}(5y) = \tfrac{1}{5}(20)$$

We obtain:

$$y = \tfrac{1}{5}(20) = 4$$

In matrix algebra, if we have an equation:

$$\mathbf{AY} = \mathbf{C}$$

we correspondingly premultiply both sides by $\mathbf{A}^{-1}$, assuming $\mathbf{A}$ has an inverse:

$$\mathbf{A}^{-1}\mathbf{AY} = \mathbf{A}^{-1}\mathbf{C}$$

Since $\mathbf{A}^{-1}\mathbf{AY} = \mathbf{IY} = \mathbf{Y}$, we obtain:

$$\mathbf{Y} = \mathbf{A}^{-1}\mathbf{C}$$

To illustrate this use, suppose we have two simultaneous equations:

$$2y_1 + 4y_2 = 20$$

$$3y_1 + \ \ y_2 = 10$$

which can be written as follows in matrix notation:

$$\begin{bmatrix} 2 & 4 \\ 3 & 1 \end{bmatrix} \begin{bmatrix} y_1 \\ y_2 \end{bmatrix} = \begin{bmatrix} 20 \\ 10 \end{bmatrix}$$

The solution of these equations then is:

$$\begin{bmatrix} y_1 \\ y_2 \end{bmatrix} = \begin{bmatrix} 2 & 4 \\ 3 & 1 \end{bmatrix}^{-1} \begin{bmatrix} 20 \\ 10 \end{bmatrix}$$

Earlier we found the required inverse, so we obtain:

$$\begin{bmatrix} y_1 \\ y_2 \end{bmatrix} = \begin{bmatrix} -.1 & .4 \\ .3 & -.2 \end{bmatrix} \begin{bmatrix} 20 \\ 10 \end{bmatrix} = \begin{bmatrix} 2 \\ 4 \end{bmatrix}$$

Hence, $y_1 = 2$ and $y_2 = 4$ satisfy these two equations.

6.7 SOME BASIC THEOREMS FOR MATRICES

We list here, without proof, some basic theorems for matrices which we will utilize in later work.

(6.26)	$\mathbf{A} + \mathbf{B} = \mathbf{B} + \mathbf{A}$
(6.27)	$(\mathbf{A} + \mathbf{B}) + \mathbf{C} = \mathbf{A} + (\mathbf{B} + \mathbf{C})$
(6.28)	$(\mathbf{AB})\mathbf{C} = \mathbf{A}(\mathbf{BC})$
(6.29)	$\mathbf{C}(\mathbf{A} + \mathbf{B}) = \mathbf{CA} + \mathbf{CB}$
(6.30)	$\lambda(\mathbf{A} + \mathbf{B}) = \lambda\mathbf{A} + \lambda\mathbf{B}$
(6.31)	$(\mathbf{A}')' = \mathbf{A}$
(6.32)	$(\mathbf{A} + \mathbf{B})' = \mathbf{A}' + \mathbf{B}'$
(6.33)	$(\mathbf{AB})' = \mathbf{B}'\mathbf{A}'$
(6.34)	$(\mathbf{ABC})' = \mathbf{C}'\mathbf{B}'\mathbf{A}'$

(6.35) $$(\mathbf{AB})^{-1} = \mathbf{B}^{-1}\mathbf{A}^{-1}$$

(6.36) $$(\mathbf{ABC})^{-1} = \mathbf{C}^{-1}\mathbf{B}^{-1}\mathbf{A}^{-1}$$

(6.37) $$(\mathbf{A}^{-1})^{-1} = \mathbf{A}$$

(6.38) $$(\mathbf{A}')^{-1} = (\mathbf{A}^{-1})'$$

6.8 RANDOM VECTORS AND MATRICES

A random vector or a random matrix contains elements that are random variables. Thus, the observation vector $\mathbf{Y}$ in (6.4) is a random vector since the Y_i elements are random variables.

Expectation of Random Vector or Matrix

Suppose we have $n = 3$ observations and are concerned with the observation vector:

$$\mathbf{Y}_{3 \times 1} = \begin{bmatrix} Y_1 \\ Y_2 \\ Y_3 \end{bmatrix}$$

The expected value of $\mathbf{Y}$ is a vector, denoted by $\mathbf{E}\{\mathbf{Y}\}$, that is defined as follows:

$$\mathbf{E}\{\mathbf{Y}\}_{3 \times 1} = \begin{bmatrix} E\{Y_1\} \\ E\{Y_2\} \\ E\{Y_3\} \end{bmatrix}$$

Thus, the expected value of a random vector is a vector whose elements are the expected values of the random variables that are the elements of the random vector. Similarly, the expectation of a random matrix is a matrix whose elements are the expected values of the corresponding random variables in the original matrix. We encountered a vector of expected values earlier in (6.9).

In general, for a random vector $\mathbf{Y}$ the expectation is:

(6.39) $$\mathbf{E}\{\mathbf{Y}\}_{n \times 1} = [E\{Y_i\}] \qquad i = 1, \ldots, n$$

and for a random matrix $\mathbf{Y}$ with dimension $n \times p$, the expectation is:

(6.40) $$\mathbf{E}\{\mathbf{Y}\}_{n \times p} = [E\{Y_{ij}\}] \qquad i = 1, \ldots, n; j = 1, \ldots, p$$

Regression Example. Suppose the number of cases in a regression application is $n = 3$. The three error terms ε_1, ε_2, ε_3 each have expectation zero. For the error vector:

$$\boldsymbol{\varepsilon}_{3 \times 1} = \begin{bmatrix} \varepsilon_1 \\ \varepsilon_2 \\ \varepsilon_3 \end{bmatrix}$$

we have:

$$\mathbf{E\{\epsilon\}} = \underset{3\times1}{\mathbf{0}}$$
$$\underset{3\times1}{}$$

since:

$$\mathbf{E\{\epsilon\}} = \begin{bmatrix} E\{\varepsilon_1\} \\ E\{\varepsilon_2\} \\ E\{\varepsilon_3\} \end{bmatrix} = \begin{bmatrix} 0 \\ 0 \\ 0 \end{bmatrix} = \mathbf{0}$$

Variance-Covariance Matrix of a Random Vector

Consider again the random vector $\mathbf{Y}$ consisting of three observations Y_1, Y_2, Y_3. Each random variable has a variance, $\sigma^2\{Y_i\}$, and any two random variables have a covariance, $\sigma\{Y_i, Y_j\}$. We can assemble these in a matrix called the *variance-covariance matrix of* $\mathbf{Y}$, denoted by $\boldsymbol{\sigma}^2\{\mathbf{Y}\}$:

(6.41)
$$\boldsymbol{\sigma}^2\{\mathbf{Y}\} = \begin{bmatrix} \sigma^2\{Y_1\} & \sigma\{Y_1, Y_2\} & \sigma\{Y_1, Y_3\} \\ \sigma\{Y_2, Y_1\} & \sigma^2\{Y_2\} & \sigma\{Y_2, Y_3\} \\ \sigma\{Y_3, Y_1\} & \sigma\{Y_3, Y_2\} & \sigma^2\{Y_3\} \end{bmatrix}$$

Note that the variances are on the main diagonal and the covariance $\sigma\{Y_i, Y_j\}$ is found in the ith row and jth column of the matrix. Thus, $\sigma\{Y_2, Y_1\}$ is found in the second row, first column, and $\sigma\{Y_1, Y_2\}$ is found in the first row, second column. Remember, of course, that $\sigma\{Y_2, Y_1\} = \sigma\{Y_1, Y_2\}$. Since in general $\sigma\{Y_i, Y_j\} = \sigma\{Y_j, Y_i\}$ for $i \neq j$, $\boldsymbol{\sigma}^2\{\mathbf{Y}\}$ is a symmetric matrix.

It follows readily that:

(6.42)
$$\boldsymbol{\sigma}^2\{\mathbf{Y}\} = \mathbf{E}\{[\mathbf{Y} - \mathbf{E}\{\mathbf{Y}\}] [\mathbf{Y} - \mathbf{E}\{\mathbf{Y}\}]'\}$$

For our illustration, we have:

$$\boldsymbol{\sigma}^2\{\mathbf{Y}\} = \mathbf{E}\left\{ \begin{bmatrix} Y_1 - E\{Y_1\} \\ Y_2 - E\{Y_2\} \\ Y_3 - E\{Y_3\} \end{bmatrix} [Y_1 - E\{Y_1\} \quad Y_2 - E\{Y_2\} \quad Y_3 - E\{Y_3\}] \right\}$$

Multiplying the two matrices and then taking expectations, we obtain:

Location in Product	Term	Expected Value
Row 1, column 1	$(Y_1 - E\{Y_1\})^2$	$\sigma^2\{Y_1\}$
Row 1, column 2	$(Y_1 - E\{Y_1\})(Y_2 - E\{Y_2\})$	$\sigma\{Y_1, Y_2\}$
Row 1, column 3	$(Y_1 - E\{Y_1\})(Y_3 - E\{Y_3\})$	$\sigma\{Y_1, Y_3\}$
Row 2, column 1	$(Y_2 - E\{Y_2\})(Y_1 - E\{Y_1\})$	$\sigma\{Y_2, Y_1\}$
etc.	etc.	etc.

This, of course, leads to the variance-covariance matrix in (6.41). Remember the definitions of variance and covariance in (1.15) and (1.21), respectively, when taking expectations.

To generalize, the variance-covariance matrix for an $n \times 1$ random vector $\mathbf{Y}$ is:

$$(6.43) \qquad \underset{n \times n}{\boldsymbol{\sigma}^2\{\mathbf{Y}\}} = \begin{bmatrix} \sigma^2\{Y_1\} & \sigma\{Y_1, Y_2\} & \cdots & \sigma\{Y_1, Y_n\} \\ \sigma\{Y_2, Y_1\} & \sigma^2\{Y_2\} & \cdots & \sigma\{Y_2, Y_n\} \\ \vdots & \vdots & & \vdots \\ \sigma\{Y_n, Y_1\} & \sigma\{Y_n, Y_2\} & \cdots & \sigma^2\{Y_n\} \end{bmatrix}$$

Note again that $\boldsymbol{\sigma}^2\{\mathbf{Y}\}$ is a symmetric matrix.

Regression Example. Let us return to the example based on $n = 3$ cases. Suppose that the three error terms have constant variance, $\sigma^2\{\varepsilon_i\} = \sigma^2$, and are uncorrelated so that $\sigma\{\varepsilon_i, \varepsilon_j\} = 0$ for $i \neq j$. We can then write the variance-covariance matrix for the random vector $\boldsymbol{\varepsilon}$ of the previous example as follows:

$$\underset{3 \times 3}{\boldsymbol{\sigma}^2\{\boldsymbol{\varepsilon}\}} = \underset{3 \times 3}{\sigma^2 \mathbf{I}}$$

since:

$$\sigma^2 \mathbf{I} = \sigma^2 \begin{bmatrix} 1 & 0 & 0 \\ 0 & 1 & 0 \\ 0 & 0 & 1 \end{bmatrix} = \begin{bmatrix} \sigma^2 & 0 & 0 \\ 0 & \sigma^2 & 0 \\ 0 & 0 & \sigma^2 \end{bmatrix}$$

Note that all variances are σ^2 and all covariances are zero.

Some Basic Theorems

Frequently, we shall encounter a random vector $\mathbf{W}$ which is obtained by premultiplying the random vector $\mathbf{Y}$ by a constant matrix $\mathbf{A}$ (a matrix whose elements are fixed):

$$(6.44) \qquad\qquad \mathbf{W} = \mathbf{AY}$$

Some basic theorems for this case are:

$$(6.45) \qquad\qquad E\{\mathbf{A}\} = \mathbf{A}$$

$$(6.46) \qquad\qquad E\{\mathbf{W}\} = E\{\mathbf{AY}\} = \mathbf{A}E\{\mathbf{Y}\}$$

$$(6.47) \qquad\qquad \boldsymbol{\sigma}^2\{\mathbf{W}\} = \boldsymbol{\sigma}^2\{\mathbf{AY}\} = \mathbf{A}\boldsymbol{\sigma}^2\{\mathbf{Y}\}\mathbf{A}'$$

where $\boldsymbol{\sigma}^2\{\mathbf{Y}\}$ is the variance-covariance matrix of $\mathbf{Y}$.

Example. As a simple illustration of the use of these theorems, consider:

$$\underset{\substack{\mathbf{W} \\ 2 \times 1}}{\begin{bmatrix} W_1 \\ W_2 \end{bmatrix}} = \underset{\substack{\mathbf{A} \\ 2 \times 2}}{\begin{bmatrix} 1 & -1 \\ 1 & 1 \end{bmatrix}} \underset{\substack{\mathbf{Y} \\ 2 \times 1}}{\begin{bmatrix} Y_1 \\ Y_2 \end{bmatrix}} = \begin{bmatrix} Y_1 - Y_2 \\ Y_1 + Y_2 \end{bmatrix}$$

We then have by (6.46):

$$\mathbf{E}\{\mathbf{W}\}_{2\times 1} = \begin{bmatrix} 1 & -1 \\ 1 & 1 \end{bmatrix} \begin{bmatrix} E\{Y_1\} \\ E\{Y_2\} \end{bmatrix} = \begin{bmatrix} E\{Y_1\} - E\{Y_2\} \\ E\{Y_1\} + E\{Y_2\} \end{bmatrix}$$

and by (6.47):

$$\boldsymbol{\sigma}^2\{\mathbf{W}\}_{2\times 2} = \begin{bmatrix} 1 & -1 \\ 1 & 1 \end{bmatrix} \begin{bmatrix} \sigma^2\{Y_1\} & \sigma\{Y_1, Y_2\} \\ \sigma\{Y_2, Y_1\} & \sigma^2\{Y_2\} \end{bmatrix} \begin{bmatrix} 1 & 1 \\ -1 & 1 \end{bmatrix}$$

$$= \begin{bmatrix} \sigma^2\{Y_1\} + \sigma^2\{Y_2\} - 2\sigma\{Y_1, Y_2\} & \sigma^2\{Y_1\} - \sigma^2\{Y_2\} \\ \sigma^2\{Y_1\} - \sigma^2\{Y_2\} & \sigma^2\{Y_1\} + \sigma^2\{Y_2\} + 2\sigma\{Y_1, Y_2\} \end{bmatrix}$$

Thus:

$$\sigma^2\{W_1\} = \sigma^2\{Y_1 - Y_2\} = \sigma^2\{Y_1\} + \sigma^2\{Y_2\} - 2\sigma\{Y_1, Y_2\}$$

$$\sigma^2\{W_2\} = \sigma^2\{Y_1 + Y_2\} = \sigma^2\{Y_1\} + \sigma^2\{Y_2\} + 2\sigma\{Y_1, Y_2\}$$

$$\sigma\{W_1, W_2\} = \sigma\{Y_1 - Y_2, Y_1 + Y_2\} = \sigma^2\{Y_1\} - \sigma^2\{Y_2\}$$

6.9 SIMPLE LINEAR REGRESSION MODEL IN MATRIX TERMS

We are now ready to develop simple linear regression in matrix terms. Remember again that we will not present any new results, but shall only state in matrix terms the results obtained earlier. We shall begin with the regression model (3.1):

(6.48) $$Y_i = \beta_0 + \beta_1 X_i + \varepsilon_i \qquad i = 1, \ldots, n$$

This implies:

$$Y_1 = \beta_0 + \beta_1 X_1 + \varepsilon_1$$

(6.48a) $$Y_2 = \beta_0 + \beta_1 X_2 + \varepsilon_2$$

$$\vdots$$

$$Y_n = \beta_0 + \beta_1 X_n + \varepsilon_n$$

We defined earlier the observation vector $\mathbf{Y}$ in (6.4), the $\mathbf{X}$ matrix in (6.6), and the $\boldsymbol{\varepsilon}$ vector in (6.10). Let us repeat these definitions and define the $\boldsymbol{\beta}$ vector of the regression coefficients:

(6.49) $$\mathbf{Y}_{n\times 1} = \begin{bmatrix} Y_1 \\ Y_2 \\ \vdots \\ Y_n \end{bmatrix} \quad \mathbf{X}_{n\times 2} = \begin{bmatrix} 1 & X_1 \\ 1 & X_2 \\ \vdots & \vdots \\ 1 & X_n \end{bmatrix} \quad \boldsymbol{\beta}_{2\times 1} = \begin{bmatrix} \beta_0 \\ \beta_1 \end{bmatrix} \quad \boldsymbol{\varepsilon}_{n\times 1} = \begin{bmatrix} \varepsilon_1 \\ \varepsilon_2 \\ \vdots \\ \varepsilon_n \end{bmatrix}$$

Now we can write (6.48a) in matrix terms compactly as follows:

(6.50) $$\mathbf{Y}_{n\times 1} = \mathbf{X}_{n\times 2} \, \boldsymbol{\beta}_{2\times 1} + \boldsymbol{\varepsilon}_{n\times 1}$$

since:

$$
\begin{bmatrix} Y_1 \\ Y_2 \\ \cdot \\ \cdot \\ Y_n \end{bmatrix} = \begin{bmatrix} 1 & X_1 \\ 1 & X_2 \\ \cdot & \cdot \\ \cdot & \cdot \\ 1 & X_n \end{bmatrix} \begin{bmatrix} \beta_0 \\ \beta_1 \end{bmatrix} + \begin{bmatrix} \varepsilon_1 \\ \varepsilon_2 \\ \cdot \\ \cdot \\ \varepsilon_n \end{bmatrix}
$$

$$
= \begin{bmatrix} \beta_0 + \beta_1 X_1 \\ \beta_0 + \beta_1 X_2 \\ \cdot \\ \cdot \\ \beta_0 + \beta_1 X_n \end{bmatrix} + \begin{bmatrix} \varepsilon_1 \\ \varepsilon_2 \\ \cdot \\ \cdot \\ \varepsilon_n \end{bmatrix} = \begin{bmatrix} \beta_0 + \beta_1 X_1 + \varepsilon_1 \\ \beta_0 + \beta_1 X_2 + \varepsilon_2 \\ \cdot \\ \cdot \\ \beta_0 + \beta_1 X_n + \varepsilon_n \end{bmatrix}
$$

Note that $\mathbf{X\beta}$ is the vector of the expected values of the Y_i observations since $E\{Y_i\} = \beta_0 + \beta_1 X_i$; hence:

(6.51) $$E\{\mathbf{Y}\} = \mathbf{X\beta}$$
$$\underset{n\times 1}{} \quad \underset{n\times 1}{}$$

where $E\{\mathbf{Y}\}$ is defined in (6.9).

The column of 1s in the $\mathbf{X}$ matrix may be viewed as consisting of the dummy variable $X_0 \equiv 1$ in the alternative regression model (2.5):

$$
Y_i = \beta_0 X_0 + \beta_1 X_i + \varepsilon_i \qquad \text{where } X_0 \equiv 1
$$

Thus, the $\mathbf{X}$ matrix may be considered to contain a column vector of the dummy variable X_0 and another column vector consisting of the independent variable observations X_i.

With respect to the error terms, regression model (3.1) assumes that $E\{\varepsilon_i\} = 0$, $\sigma^2\{\varepsilon_i\} = \sigma^2$, and that ε_i are independent normal random variables. The condition $E\{\varepsilon_i\} = 0$ in matrix terms is:

(6.52) $$E\{\mathbf{\varepsilon}\} = \mathbf{0}$$
$$\underset{n\times 1}{} \quad \underset{n\times 1}{}$$

since (6.52) states:

$$
E\{\mathbf{\varepsilon}\} = \begin{bmatrix} E\{\varepsilon_1\} \\ E\{\varepsilon_2\} \\ \cdot \\ \cdot \\ E\{\varepsilon_n\} \end{bmatrix} = \begin{bmatrix} 0 \\ 0 \\ \cdot \\ \cdot \\ 0 \end{bmatrix}
$$

The condition that the error terms have constant variance σ^2 and that all covariances $\sigma\{\varepsilon_i, \varepsilon_j\}$ for $i \neq j$ are zero (since the ε_i are independent) is expressed in matrix terms through the variance-covariance matrix:

(6.53) $$\mathbf{\sigma}^2\{\mathbf{\varepsilon}\} = \sigma^2\mathbf{I}$$
$$\underset{n\times n}{} \quad \underset{n\times n}{}$$

since (6.53) states:

$$\sigma^2\{\boldsymbol{\varepsilon}\} = \sigma^2 \begin{bmatrix} 1 & 0 & 0 & \cdots & 0 \\ 0 & 1 & 0 & \cdots & 0 \\ \cdot & \cdot & \cdot & & \cdot \\ \cdot & \cdot & \cdot & & \cdot \\ \cdot & \cdot & \cdot & & \cdot \\ 0 & 0 & 0 & \cdots & 1 \end{bmatrix} = \begin{bmatrix} \sigma^2 & 0 & 0 & \cdots & 0 \\ 0 & \sigma^2 & 0 & \cdots & 0 \\ \cdot & \cdot & \cdot & & \cdot \\ \cdot & \cdot & \cdot & & \cdot \\ \cdot & \cdot & \cdot & & \cdot \\ 0 & 0 & 0 & \cdots & \sigma^2 \end{bmatrix}$$

Thus, the normal error regression model (3.1) in matrix terms is:

(6.54)
$$\mathbf{Y} = \mathbf{X\beta} + \boldsymbol{\varepsilon}$$

where:

$\boldsymbol{\varepsilon}$ is a vector of independent normal random variables with $\mathbf{E}\{\boldsymbol{\varepsilon}\} = \mathbf{0}$ and $\sigma^2\{\boldsymbol{\varepsilon}\} = \sigma^2\mathbf{I}$

6.10 LEAST SQUARES ESTIMATION OF REGRESSION PARAMETERS

Normal Equations

The normal equations (2.9):

$$nb_0 + b_1 \sum X_i = \sum Y_i$$

(6.55)

$$b_0 \sum X_i + b_1 \sum X_i^2 = \sum X_i Y_i$$

in matrix terms are:

(6.56)
$$\mathbf{X'X}\,\mathbf{b} = \mathbf{X'Y}$$
$${\scriptstyle 2\times2\ 2\times1}{\scriptstyle 2\times1}$$

where $\mathbf{b}$ is the vector of the least squares regression coefficients:

(6.56a)
$$\mathbf{b}_{\substack{2\times1}} = \begin{bmatrix} b_0 \\ b_1 \end{bmatrix}$$

To see this, recall that we obtained $\mathbf{X'X}$ in (6.14) and $\mathbf{X'Y}$ in (6.15). Equation (6.56) thus states:

$$\begin{bmatrix} n & \Sigma X_i \\ \Sigma X_i & \Sigma X_i^2 \end{bmatrix} \begin{bmatrix} b_0 \\ b_1 \end{bmatrix} = \begin{bmatrix} \Sigma Y_i \\ \Sigma X_i Y_i \end{bmatrix}$$

or:

$$\begin{bmatrix} nb_0 + b_1 \Sigma X_i \\ b_0 \Sigma X_i + b_1 \Sigma X_i^2 \end{bmatrix} = \begin{bmatrix} \Sigma Y_i \\ \Sigma X_i Y_i \end{bmatrix}$$

These are precisely the normal equations in (6.55).

Estimated Regression Coefficients

To obtain the estimated regression coefficients from the normal equations:

$$\mathbf{X'Xb} = \mathbf{X'Y}$$

by matrix methods, we premultiply both sides by the inverse of $\mathbf{X'X}$ (we assume this exists):

$$(\mathbf{X'X})^{-1}\mathbf{X'Xb} = (\mathbf{X'X})^{-1}\mathbf{X'Y}$$

so that we find, since $(\mathbf{X'X})^{-1}\mathbf{X'X} = \mathbf{I}$ and $\mathbf{Ib} = \mathbf{b}$:

(6.57)
$$\underset{2\times1}{\mathbf{b}} = \underset{2\times2}{(\mathbf{X'X})^{-1}}\underset{2\times1}{\mathbf{X'Y}}$$

The estimators b_0 and b_1 in $\mathbf{b}$ are the same as those given earlier in (2.10a) and (2.10b). We shall demonstrate this by an example.

Example. Let us find the estimated regression coefficients for the Westwood Company lot size example by matrix methods. From earlier work, we have (Table 2.2):

$$n = 10 \qquad \sum Y_i = 1{,}100 \qquad \sum X_i = 500 \qquad \sum X_i^2 = 28{,}400$$

$$\sum X_i Y_i = 61{,}800$$

Let us now use (6.24) to evaluate $(\mathbf{X'X})^{-1}$. We have:

$$n \sum (X_i - \bar{X})^2 = n\left[\sum X_i^2 - \frac{(\sum X_i)^2}{n}\right] = 10\left[28{,}400 - \frac{(500)^2}{10}\right] = 34{,}000$$

Therefore:

$$(\mathbf{X'X})^{-1} = \begin{bmatrix} \dfrac{\sum X_i^2}{n\sum(X_i - \bar{X})^2} & \dfrac{-\sum X_i}{n\sum(X_i - \bar{X})^2} \\ \dfrac{-\sum X_i}{n\sum(X_i - \bar{X})^2} & \dfrac{n}{n\sum(X_i - \bar{X})^2} \end{bmatrix} = \begin{bmatrix} \dfrac{28{,}400}{34{,}000} & \dfrac{-500}{34{,}000} \\ \dfrac{-500}{34{,}000} & \dfrac{10}{34{,}000} \end{bmatrix}$$

$$= \begin{bmatrix} .83529412 & -.01470588 \\ -.01470588 & .00029412 \end{bmatrix}$$

We also wish to make use of (6.15) to evaluate $\mathbf{X'Y}$:

$$\mathbf{X'Y} = \begin{bmatrix} \sum Y_i \\ \sum X_i Y_i \end{bmatrix} = \begin{bmatrix} 1{,}100 \\ 61{,}800 \end{bmatrix}$$

Hence, by (6.57):

$$\mathbf{b} = \begin{bmatrix} b_0 \\ b_1 \end{bmatrix} = (\mathbf{X'X})^{-1}\mathbf{X'Y} = \begin{bmatrix} .83529412 & -.01470588 \\ -.01470588 & .00029412 \end{bmatrix}\begin{bmatrix} 1{,}100 \\ 61{,}800 \end{bmatrix}$$

$$= \begin{bmatrix} 10.0 \\ 2.0 \end{bmatrix}$$

or $b_0 = 10.0$ and $b_1 = 2.0$. This agrees with the results in Chapter 2. Any difference would have been due to rounding errors.

To reduce the effect of rounding errors when obtaining the vector **b** by hand calculations, it is often desirable to move the constant in the denominator of the elements of $(\mathbf{X'X})^{-1}$ outside the matrix, and do the division as the last step. For our example, this would lead to:

$$(\mathbf{X'X})^{-1} = \frac{1}{n\Sigma(X_i - \bar{X})^2} \begin{bmatrix} \Sigma X_i^2 & -\Sigma X_i \\ -\Sigma X_i & n \end{bmatrix}$$

$$= \frac{1}{34,000} \begin{bmatrix} 28,400 & -500 \\ -500 & 10 \end{bmatrix}$$

$$\mathbf{b} = \frac{1}{34,000} \begin{bmatrix} 28,400 & -500 \\ -500 & 10 \end{bmatrix} \begin{bmatrix} 1,100 \\ 61,800 \end{bmatrix}$$

$$= \frac{1}{34,000} \begin{bmatrix} 340,000 \\ 68,000 \end{bmatrix} = \begin{bmatrix} 10.0 \\ 2.0 \end{bmatrix}$$

In this instance, the two methods of calculation lead to identical results. Often, however, postponing division by $n\Sigma(X_i - \bar{X})^2$ until the end yields more accurate results.

Comments

1. To derive the normal equations by the method of least squares, we minimize the quantity:

$$Q = \Sigma [Y_i - (\beta_0 + \beta_1 X_i)]^2$$

In matrix notation:

(6.58) $$Q = (\mathbf{Y} - \mathbf{X\beta})'(\mathbf{Y} - \mathbf{X\beta})$$

Expanding out, we obtain

$$Q = \mathbf{Y'Y} - \mathbf{\beta'X'Y} - \mathbf{Y'X\beta} + \mathbf{\beta'X'X\beta}$$

since $(\mathbf{X\beta})' = \mathbf{\beta'X'}$ by (6.33). Note now that $\mathbf{Y'X\beta}$ is 1×1, hence is equal to its transpose, which according to (6.34) is $\mathbf{\beta'X'Y}$. Thus, we find:

(6.59) $$Q = \mathbf{Y'Y} - 2\mathbf{\beta'X'Y} + \mathbf{\beta'X'X\beta}$$

To find the value of $\mathbf{\beta}$ that minimizes Q, we differentiate with respect to β_0 and β_1. Let:

(6.60) $$\frac{\partial}{\partial \mathbf{\beta}}(Q) = \begin{bmatrix} \dfrac{\partial Q}{\partial \beta_0} \\[2mm] \dfrac{\partial Q}{\partial \beta_1} \end{bmatrix}$$

Then it follows that:

(6.61) $$\frac{\partial}{\partial \mathbf{\beta}}(Q) = -2\mathbf{X'Y} + 2\mathbf{X'X\beta}$$

Equating to zero, dividing by 2, and substituting $\mathbf{b}$ for $\boldsymbol{\beta}$ gives the matrix form of the least squares normal equations:

$$\mathbf{X'Xb} = \mathbf{X'Y}$$

2. A comparison of the normal equations and $\mathbf{X'X}$ shows that whenever the columns of $\mathbf{X'X}$ are linearly dependent, the normal equations will be linearly dependent also. No unique solutions can then be obtained for b_0 and b_1. Fortunately, in most regression applications, the columns of $\mathbf{X'X}$ are linearly independent, leading to unique solutions for b_0 and b_1.

6.11 FITTED VALUES AND RESIDUALS

Fitted Values

Let the vector of the fitted values $\hat{Y}_i$ be denoted by $\hat{\mathbf{Y}}$:

(6.62)
$$\underset{n \times 1}{\hat{\mathbf{Y}}} = \begin{bmatrix} \hat{Y}_1 \\ \hat{Y}_2 \\ \cdot \\ \cdot \\ \cdot \\ \hat{Y}_n \end{bmatrix}$$

In matrix notation, we then have:

(6.63)
$$\underset{n \times 1}{\hat{\mathbf{Y}}} = \underset{n \times 2}{\mathbf{X}} \ \underset{2 \times 1}{\mathbf{b}}$$

because:

$$\begin{bmatrix} \hat{Y}_1 \\ \hat{Y}_2 \\ \cdot \\ \cdot \\ \hat{Y}_n \end{bmatrix} = \begin{bmatrix} 1 & X_1 \\ 1 & X_2 \\ \cdot & \cdot \\ \cdot & \cdot \\ 1 & X_n \end{bmatrix} \begin{bmatrix} b_0 \\ b_1 \end{bmatrix} = \begin{bmatrix} b_0 + b_1 X_1 \\ b_0 + b_1 X_2 \\ \cdot \\ \cdot \\ b_0 + b_1 X_n \end{bmatrix}$$

Hat Matrix.   We can express the matrix result for $\hat{\mathbf{Y}}$ in (6.63) as follows by using the expression for $\mathbf{b}$ in (6.57):

$$\hat{\mathbf{Y}} = \mathbf{X(X'X)}^{-1}\mathbf{X'Y}$$

or, equivalently:

(6.64)
$$\underset{n \times 1}{\hat{\mathbf{Y}}} = \underset{n \times n}{\mathbf{H}} \ \underset{n \times 1}{\mathbf{Y}}$$

where:

(6.64a)
$$\underset{n \times n}{\mathbf{H}} = \mathbf{X(X'X)}^{-1}\mathbf{X'}$$

We thus see from (6.64) that the fitted values $\hat{Y}_i$ can be expressed as linear combinations of the dependent variable observations Y_i, with the coefficients being elements of the matrix $\mathbf{H}$. The $\mathbf{H}$ matrix involves only the observations on the independent variable X, as is evident from (6.64a).

The square $n \times n$ matrix $\mathbf{H}$ is called the *hat matrix*. It plays an important role in regression analysis, as we shall see in Chapter 11 when we consider whether regression results are unduly influenced by one or a few observations. The matrix $\mathbf{H}$ is symmetric and has the special property (called idempotency):

$$(6.65) \qquad\qquad\qquad \mathbf{HH} = \mathbf{H}$$

In general, a matrix $\mathbf{M}$ is said to be *idempotent* if $\mathbf{MM} = \mathbf{M}$.

Residuals

Let the vector of the residuals $e_i = Y_i - \hat{Y}_i$ be denoted by $\mathbf{e}$:

$$(6.66) \qquad\qquad \underset{n \times 1}{\mathbf{e}} = \begin{bmatrix} e_1 \\ e_2 \\ \cdot \\ \cdot \\ \cdot \\ e_n \end{bmatrix}$$

In matrix notation, we then have:

$$(6.67) \qquad\qquad \underset{n \times 1}{\mathbf{e}} = \underset{n \times 1}{\mathbf{Y}} - \underset{n \times 1}{\hat{\mathbf{Y}}} = \underset{n \times 1}{\mathbf{Y}} - \underset{n \times 1}{\mathbf{Xb}}$$

The residuals e_i, like the fitted values $\hat{Y}_i$, also can be expressed as linear combinations of the dependent variable observations Y_i, using the result in (6.64) for $\hat{\mathbf{Y}}$:

$$\mathbf{e} = \mathbf{Y} - \hat{\mathbf{Y}} = \mathbf{Y} - \mathbf{HY} = (\mathbf{I} - \mathbf{H})\mathbf{Y}$$

We thus have the important result:

$$(6.68) \qquad\qquad \underset{n \times 1}{\mathbf{e}} = \underset{n \times n}{(\mathbf{I} - \mathbf{H})} \underset{n \times 1}{\mathbf{Y}}$$

where $\mathbf{H}$ is the hat matrix defined in (6.64a).

The matrix $\mathbf{I} - \mathbf{H}$, like the matrix $\mathbf{H}$, is symmetric and idempotent.

It can be shown that the variance-covariance matrix of the vector of residuals $\mathbf{e}$ also involves the matrix $\mathbf{I} - \mathbf{H}$:

$$(6.69) \qquad\qquad \underset{n \times n}{\sigma^2\{\mathbf{e}\}} = \sigma^2(\mathbf{I} - \mathbf{H})$$

and is estimated by:

$$(6.70) \qquad\qquad \underset{n \times n}{s^2\{\mathbf{e}\}} = MSE(\mathbf{I} - \mathbf{H})$$

Note

The variance-covariance matrix of $\mathbf{e}$ can be derived by means of (6.47). Since $\mathbf{e} = (\mathbf{I} - \mathbf{H})\mathbf{Y}$, we obtain:

$$\sigma^2\{\mathbf{e}\} = (\mathbf{I} - \mathbf{H})\sigma^2\{\mathbf{Y}\}(\mathbf{I} - \mathbf{H})'$$

Now $\sigma^2\{\mathbf{Y}\} = \sigma^2\{\boldsymbol{\varepsilon}\} = \sigma^2\mathbf{I}$ for the normal error model according to (6.53). Also, $(\mathbf{I} - \mathbf{H})' = \mathbf{I} - \mathbf{H}$ because of the symmetry of the matrix. Hence:

$$\sigma^2\{e\} = \sigma^2(I - H)I(I - H)$$

$$= \sigma^2(I - H)(I - H)$$

In view of the fact that the matrix $I - H$ is idempotent, we know that $(I - H)(I - H) = I - H$ and we obtain formula (6.69):

$$\sigma^2\{e\} = \sigma^2(I - H)$$

6.12 ANALYSIS OF VARIANCE RESULTS

Sums of Squares

To see how the sums of squares are expressed in matrix notation, we begin with *SSTO*. We know from (3.49) that:

(6.71)
$$SSTO = \sum Y_i^2 - n\bar{Y}^2 = \sum Y_i^2 - \frac{(\sum Y_i)^2}{n}$$

We also know from (6.13) that:

$$Y'Y = \sum Y_i^2$$

The subtraction term $n\bar{Y}^2 = (\sum Y_i)^2/n$ in matrix form uses J, the matrix of 1s defined in (6.18), as follows:

(6.72)
$$\frac{(\sum Y_i)^2}{n} = \left(\frac{1}{n}\right)Y'JY$$

For instance, if $n = 2$, we have:

$$\left(\frac{1}{2}\right)[Y_1 \quad Y_2]\begin{bmatrix} 1 & 1 \\ 1 & 1 \end{bmatrix}\begin{bmatrix} Y_1 \\ Y_2 \end{bmatrix} = \frac{(Y_1 + Y_2)(Y_1 + Y_2)}{2}$$

Hence, it follows that:

(6.73a)
$$SSTO = Y'Y - \left(\frac{1}{n}\right)Y'JY$$

Just as $\sum Y_i^2$ is represented by $Y'Y$ in matrix terms, so $SSE = \sum e_i^2 = \sum(Y_i - \hat{Y}_i)^2$ can be represented as follows:

(6.73b)
$$SSE = e'e = (Y - Xb)'(Y - Xb)$$

which can be shown to equal:

(6.73c)
$$SSE = Y'Y - b'X'Y$$

Finally, it can be shown that:

(6.73d)
$$SSR = b'X'Y - \left(\frac{1}{n}\right)Y'JY$$

Example. Let us find *SSE* for the Westwood Company lot size example by matrix methods, using (6.73c). We know from earlier results:

$$\mathbf{Y'Y} = \sum Y_i^2 = 134{,}660$$

We also know from earlier that:

$$\mathbf{b} = \begin{bmatrix} 10.0 \\ 2.0 \end{bmatrix} \qquad \mathbf{X'Y} = \begin{bmatrix} 1{,}100 \\ 61{,}800 \end{bmatrix}$$

Hence:

$$\mathbf{b'X'Y} = \begin{bmatrix} 10.0 & 2.0 \end{bmatrix} \begin{bmatrix} 1{,}100 \\ 61{,}800 \end{bmatrix} = 134{,}600$$

and:

$$SSE = \mathbf{Y'Y} - \mathbf{b'X'Y} = 134{,}660 - 134{,}600 = 60$$

which is the same result as that obtained in Chapter 2. Any difference would have been due to rounding errors.

Similarly, we can find *SSR* using (6.73d):

$$SSR = \mathbf{b'X'Y} - \left(\frac{1}{n}\right)\mathbf{Y'JY}$$

$$= 134{,}600 - \frac{1}{10}(1{,}100)^2 = 13{,}600$$

since $\Sigma Y_i = 1{,}100$ for the Westwood Company example.

Note

To illustrate the derivation of the sums of squares expressions in matrix notation, consider *SSE*:

$$SSE = \mathbf{e'e} = (\mathbf{Y} - \mathbf{Xb})'(\mathbf{Y} - \mathbf{Xb}) = \mathbf{Y'Y} - 2\mathbf{b'X'Y} + \mathbf{b'X'Xb}$$

In substituting for the right-most **b** we obtain by (6.57):

$$SSE = \mathbf{Y'Y} - 2\mathbf{b'X'Y} + \mathbf{b'X'X(X'X)^{-1}X'Y}$$

$$= \mathbf{Y'Y} - 2\mathbf{b'X'Y} + \mathbf{b'IX'Y}$$

In dropping **I** and subtracting, we obtain the result in (6.73c):

$$SSE = \mathbf{Y'Y} - \mathbf{b'X'Y}$$

Sums of Squares as Quadratic Forms

The ANOVA sums of squares can be shown to be *quadratic forms*. An example of a quadratic form of the observations Y_i when $n = 2$ is:

(6.74) $$5Y_1^2 + 6Y_1 Y_2 + 4Y_2^2$$

Note that this expression is a second-degree polynomial containing terms involving the squares of the observations and the cross product. We can express (6.74) in matrix terms as follows:

(6.74a)
$$[Y_1 \quad Y_2] \begin{bmatrix} 5 & 3 \\ 3 & 4 \end{bmatrix} \begin{bmatrix} Y_1 \\ Y_2 \end{bmatrix} = \mathbf{Y'AY}$$

where $\mathbf{A}$ is a symmetric matrix.

In general, a quadratic form is defined as:

(6.75)
$$\underset{1 \times 1}{\mathbf{Y'AY}} = \sum_{i=1}^{n} \sum_{j=1}^{n} a_{ij} Y_i Y_j \qquad \text{where } a_{ij} = a_{ji}$$

$\mathbf{A}$ is a symmetric $n \times n$ matrix and is called the *matrix of the quadratic form.*

The ANOVA sums of squares *SSTO*, *SSE*, and *SSR* are all quadratic forms. To see this, we need to express the matrix forms for these sums of squares in (6.73) more compactly. We do this by reexpressing $\mathbf{b'X'}$. From (6.33) and (6.63), we know that:

$$\mathbf{b'X'} = (\mathbf{Xb})' = \hat{\mathbf{Y}}'$$

We now use the result in (6.64) to obtain:

$$\mathbf{b'X'} = (\mathbf{HY})'$$

Since $\mathbf{H}$ is a symmetric matrix so that $\mathbf{H'} = \mathbf{H}$, we finally obtain, using (6.33):

(6.76)
$$\mathbf{b'X'} = \mathbf{Y'H}$$

This result enables us to express the sums of squares in (6.73) as follows:

(6.77a)
$$SSTO = \mathbf{Y'} \left[\mathbf{I} - \left(\frac{1}{n} \right) \mathbf{J} \right] \mathbf{Y}$$

(6.77b)
$$SSE = \mathbf{Y'(I - H)Y}$$

(6.77c)
$$SSR = \mathbf{Y'} \left[\mathbf{H} - \left(\frac{1}{n} \right) \mathbf{J} \right] \mathbf{Y}$$

Each of these sums of squares can now be seen to be of the form $\mathbf{Y'AY}$. It can be shown that the three $\mathbf{A}$ matrices:

(6.78a)
$$\mathbf{I} - \left(\frac{1}{n} \right) \mathbf{J}$$

(6.78b)
$$\mathbf{I} - \mathbf{H}$$

(6.78c)
$$\mathbf{H} - \left(\frac{1}{n} \right) \mathbf{J}$$

are symmetric. Hence, *SSTO*, *SSE*, and *SSR* are quadratic forms, with the matrices of the quadratic forms given in (6.78). Quadratic forms play an important role in statistics because all sums of squares in the analysis of variance for linear statistical models can be expressed as quadratic forms.

6.13 INFERENCES IN REGRESSION ANALYSIS

As we saw in earlier chapters, all interval estimates are of the following form: point estimator plus and minus a certain number of estimated standard deviations of the point estimator. Similarly, all tests require the point estimator and the estimated standard deviation of the point estimator or, in the case of analysis of variance tests, various sums of squares. Matrix algebra is of principal help in inference making when obtaining the estimated standard deviations and sums of squares. We have already given the matrix equivalents of the sums of squares for the analysis of variance. Hence, we focus here chiefly on the matrix expressions for the estimated variances of point estimators of interest.

Regression Coefficients

The variance-covariance matrix of **b**:

$$(6.79) \qquad \underset{2\times2}{\sigma^2\{\mathbf{b}\}} = \begin{bmatrix} \sigma^2\{b_0\} & \sigma\{b_0, b_1\} \\ \sigma\{b_1, b_0\} & \sigma^2\{b_1\} \end{bmatrix}$$

is:

$$(6.80) \qquad \underset{2\times2}{\sigma^2\{\mathbf{b}\}} = \sigma^2(\mathbf{X'X})^{-1}$$

or, using (6.25):

$$(6.80a) \qquad \underset{2\times2}{\sigma^2\{\mathbf{b}\}} = \begin{bmatrix} \dfrac{\sigma^2\Sigma X_i^2}{n\Sigma(X_i - \bar{X})^2} & \dfrac{-\bar{X}\sigma^2}{\Sigma(X_i - \bar{X})^2} \\ \dfrac{-\bar{X}\sigma^2}{\Sigma(X_i - \bar{X})^2} & \dfrac{\sigma^2}{\Sigma(X_i - \bar{X})^2} \end{bmatrix}$$

When *MSE* is substituted for σ^2 in (6.80a) we have:

$$(6.81) \qquad \underset{2\times2}{s^2\{\mathbf{b}\}} = MSE(\mathbf{X'X})^{-1} = \begin{bmatrix} \dfrac{MSE\Sigma X_i^2}{n\Sigma(X_i - \bar{X})^2} & \dfrac{-\bar{X}MSE}{\Sigma(X_i - \bar{X})^2} \\ \dfrac{-\bar{X}MSE}{\Sigma(X_i - \bar{X})^2} & \dfrac{MSE}{\Sigma(X_i - \bar{X})^2} \end{bmatrix}$$

where $s^2\{\mathbf{b}\}$ is the estimated variance-covariance matrix of **b**. In (6.80a), you will recognize the variances of b_0 in (3.20b) and b_1 in (3.3b) and the covariance of b_0 and b_1 in (5.6). Likewise, the estimated variances in (6.81) are familiar from earlier chapters.

Mean Response

To estimate the mean response at X_h, let us define the vector:

$$(6.82) \qquad \underset{2\times1}{\mathbf{X}_h} = \begin{bmatrix} 1 \\ X_h \end{bmatrix} \quad \text{or} \quad \underset{1\times2}{\mathbf{X}_h'} = [1 \quad X_h]$$

The fitted value in matrix notation then is:

(6.83)
$$\hat{Y}_h = \mathbf{X}'_h \mathbf{b}$$

since:

$$\mathbf{X}'_h \mathbf{b} = \begin{bmatrix} 1 & X_h \end{bmatrix} \begin{bmatrix} b_0 \\ b_1 \end{bmatrix} = [b_0 + b_1 X_h] = [\hat{Y}_h] = \hat{Y}_h$$

Note that $\mathbf{X}'_h\mathbf{b}$ is a 1×1 matrix; hence, we can write the final result as a scalar.

The variance of $\hat{Y}_h$, given earlier in (3.28b), is in matrix notation:

(6.84)
$$\sigma^2\{\hat{Y}_h\} = \sigma^2 \mathbf{X}'_h(\mathbf{X}'\mathbf{X})^{-1}\mathbf{X}_h = \mathbf{X}'_h \sigma^2\{\mathbf{b}\}\mathbf{X}_h$$

where $\sigma^2\{\mathbf{b}\}$ is the variance-covariance matrix of the regression coefficients in (6.80). Note, therefore, that $\sigma^2\{\hat{Y}_h\}$ is a function of the variances $\sigma^2\{b_0\}$ and $\sigma^2\{b_1\}$ and of the covariance $\sigma\{b_0, b_1\}$.

The estimated variance of $\hat{Y}_h$, given earlier in (3.29), is in matrix notation:

(6.85)
$$s^2\{\hat{Y}_h\} = MSE(\mathbf{X}'_h(\mathbf{X}'\mathbf{X})^{-1}\mathbf{X}_h) = \mathbf{X}'_h s^2\{\mathbf{b}\}\mathbf{X}_h$$

where $s^2\{\mathbf{b}\}$ is the estimated variance-covariance matrix of the regression coefficients in (6.81).

Prediction of New Observation

The estimated variance $s^2\{Y_{h(\text{new})}\}$, given earlier in (3.37), is in matrix notation:

(6.86)
$$s^2\{Y_{h(\text{new})}\} = MSE + s^2\{\hat{Y}_h\} = MSE + \mathbf{X}'_h s^2\{\mathbf{b}\}\mathbf{X}_h$$
$$= MSE(1 + \mathbf{X}'_h(\mathbf{X}'\mathbf{X})^{-1}\mathbf{X}_h)$$

Examples

1. We wish to find $s^2\{b_0\}$ and $s^2\{b_1\}$ for the Westwood Company lot size example by matrix methods. We found earlier that $MSE = 7.5$ and:

$$(\mathbf{X}'\mathbf{X})^{-1} = \begin{bmatrix} .83529412 & -.01470588 \\ -.01470588 & .00029412 \end{bmatrix}$$

Hence, by (6.81):

$$s^2\{\mathbf{b}\} = MSE(\mathbf{X}'\mathbf{X})^{-1} = 7.5 \begin{bmatrix} .83529412 & -.01470588 \\ -.01470588 & .00029412 \end{bmatrix}$$
$$= \begin{bmatrix} 6.264706 & -.1102941 \\ -.1102941 & .0022059 \end{bmatrix}$$

Thus, $s^2\{b_0\} = 6.26471$ and $s^2\{b_1\} = .002206$. These are the same as the results obtained in Chapter 3.

Note how simple it is to find the estimated variances of the regression coefficients as soon as $(\mathbf{X}'\mathbf{X})^{-1}$ has been obtained. This inverse is needed in the first place to find the regression coefficients, so that practically no extra work is required to obtain their estimated variances.

2. We wish to find $s^2\{\hat{Y}_h\}$ for the Westwood Company example when $X_h = 55$. We define:

$$\mathbf{X}'_h = [1 \quad 55]$$

and obtain by (6.85):

$$s^2\{\hat{Y}_h\} = \mathbf{X}'_h s^2\{\mathbf{b}\}\mathbf{X}_h$$

$$= [1 \quad 55] \begin{bmatrix} 6.264706 & -.1102941 \\ -.1102941 & .0022059 \end{bmatrix} \begin{bmatrix} 1 \\ 55 \end{bmatrix} = .80520$$

This is the same result as that obtained in Chapter 3, except for a minor difference due to rounding.

Comments

1. To illustrate a derivation in matrix terms, let us find the variance-covariance matrix of **b**. Recall that:

$$\mathbf{b} = (\mathbf{X}'\mathbf{X})^{-1}\mathbf{X}'\mathbf{Y} = \mathbf{A}\mathbf{Y}$$

where **A** is a constant matrix:

$$\mathbf{A} = (\mathbf{X}'\mathbf{X})^{-1}\mathbf{X}'$$

Hence by (6.47), we have:

$$\sigma^2\{\mathbf{b}\} = \mathbf{A}\sigma^2\{\mathbf{Y}\}\mathbf{A}'$$

Now $\sigma^2\{\mathbf{Y}\} = \sigma^2\mathbf{I}$. Further, it follows from (6.33) and the fact that $(\mathbf{X}'\mathbf{X})^{-1}$ is symmetric that:

$$\mathbf{A}' = \mathbf{X}(\mathbf{X}'\mathbf{X})^{-1}$$

We find therefore:

$$\sigma^2\{\mathbf{b}\} = (\mathbf{X}'\mathbf{X})^{-1}\mathbf{X}'\sigma^2\mathbf{I}\mathbf{X}(\mathbf{X}'\mathbf{X})^{-1}$$

$$= \sigma^2(\mathbf{X}'\mathbf{X})^{-1}\mathbf{X}'\mathbf{X}(\mathbf{X}'\mathbf{X})^{-1}$$

$$= \sigma^2(\mathbf{X}'\mathbf{X})^{-1}\mathbf{I}$$

$$= \sigma^2(\mathbf{X}'\mathbf{X})^{-1}$$

2. Since $\hat{Y}_h = \mathbf{X}'_h\mathbf{b}$, it follows at once from (6.47) that:

$$\sigma^2\{\hat{Y}_h\} = \mathbf{X}'_h\sigma^2\{\mathbf{b}\}\mathbf{X}_h$$

Hence:

$$\sigma^2\{\hat{Y}_h\} = [1 \quad X_h] \begin{bmatrix} \sigma^2\{b_0\} & \sigma\{b_0, b_1\} \\ \sigma\{b_1, b_0\} & \sigma^2\{b_1\} \end{bmatrix} \begin{bmatrix} 1 \\ X_h \end{bmatrix}$$

or:

(6.87) $$\sigma^2\{\hat{Y}_h\} = \sigma^2\{b_0\} + 2X_h\sigma\{b_0, b_1\} + X_h^2\sigma^2\{b_1\}$$

Using the results from (6.80a), we obtain:

$$\sigma^2\{\hat{Y}_h\} = \frac{\sigma^2\Sigma X_i^2}{n\Sigma(X_i - \bar{X})^2} + \frac{2X_h(-\bar{X})\sigma^2}{\Sigma(X_i - \bar{X})^2} + \frac{X_h^2\sigma^2}{\Sigma(X_i - \bar{X})^2}$$

which reduces to the familiar expression:

(6.88) $$\sigma^2\{\hat{Y}_h\} = \sigma^2\left[\frac{1}{n} + \frac{(X_h - \bar{X})^2}{\Sigma(X_i - \bar{X})^2}\right]$$

Thus, we see explicitly that the variance expression in (6.88) contains contributions from $\sigma^2\{b_0\}$, $\sigma^2\{b_1\}$, and $\sigma\{b_0, b_1\}$, which it must according to theorem (1.27b) since $\hat{Y}_h$ is a linear combination of b_0 and b_1:

$$\hat{Y}_h = b_0 + b_1 X_h$$

3. We do not show the results in matrix terms for other types of inferences, such as simultaneous prediction of several new observations on Y at different X_h levels, since these are based on results we have developed.

CITED REFERENCE

6.1. Graybill, F. A. *Matrices with Applications in Statistics.* 2nd ed. Belmont, Calif.: Wadsworth, 1983.

PROBLEMS

6.1. For the matrices below, obtain: (1) $\mathbf{A} + \mathbf{B}$, (2) $\mathbf{A} - \mathbf{B}$, (3) $\mathbf{AC}$, (4) $\mathbf{AB}'$, (5) $\mathbf{B}'\mathbf{A}$.

$$\mathbf{A} = \begin{bmatrix} 1 & 4 \\ 2 & 6 \\ 3 & 8 \end{bmatrix} \quad \mathbf{B} = \begin{bmatrix} 1 & 3 \\ 1 & 4 \\ 2 & 5 \end{bmatrix} \quad \mathbf{C} = \begin{bmatrix} 3 & 8 & 1 \\ 5 & 4 & 0 \end{bmatrix}$$

State the dimension of each resulting matrix.

6.2. For the matrices below, obtain: (1) $\mathbf{A} + \mathbf{C}$, (2) $\mathbf{A} - \mathbf{C}$, (3) $\mathbf{B}'\mathbf{A}$, (4) $\mathbf{AC}'$, (5) $\mathbf{C}'\mathbf{A}$.

$$\mathbf{A} = \begin{bmatrix} 2 & 1 \\ 3 & 5 \\ 5 & 7 \\ 4 & 8 \end{bmatrix} \quad \mathbf{B} = \begin{bmatrix} 6 \\ 9 \\ 3 \\ 1 \end{bmatrix} \quad \mathbf{C} = \begin{bmatrix} 3 & 8 \\ 8 & 6 \\ 5 & 1 \\ 2 & 4 \end{bmatrix}$$

State the dimension of each resulting matrix.

6.3. Show how the following expressions are written in terms of matrices: (1) $Y_i - \hat{Y}_i = e_i$, (2) $\Sigma X_i e_i = 0$. Assume $i = 1, \ldots, 4$.

6.4. **Flavor deterioration.** The results shown below were obtained in a small-scale experiment to study the relation between °F of storage temperature (X) and number of weeks before flavor deterioration of a food product begins to occur (Y).

i:	1	2	3	4	5
X_i:	8	4	0	−4	−8
Y_i:	7.8	9.0	10.2	11.0	11.7

Assume that the first-order regression model (3.1) is applicable. Using matrix methods, find: (1) $\mathbf{Y'Y}$, (2) $\mathbf{X'X}$, (3) $\mathbf{X'Y}$.

6.5. Consumer finance. The data below show for a consumer finance company operating in six cities, the number of competing loan companies operating in the city (X) and the number per thousand of the company's loans made in that city that are currently delinquent (Y):

i:	1	2	3	4	5	6
X_i:	4	1	2	3	3	4
Y_i:	16	5	10	15	13	22

Assume that the first-order regression model (3.1) is applicable. Using matrix methods, find: (1) $\mathbf{Y'Y}$, (2) $\mathbf{X'X}$, (3) $\mathbf{X'Y}$.

6.6. Refer to **Airfreight breakage** Problem 2.19. Using matrix methods, find: (1) $\mathbf{Y'Y}$, (2) $\mathbf{X'X}$, (3) $\mathbf{X'Y}$.

6.7. Refer to **Plastic hardness** Problem 2.20. Using matrix methods, find: (1) $\mathbf{Y'Y}$, (2) $\mathbf{X'X}$, (3) $\mathbf{X'Y}$.

6.8. Let $\mathbf{B}$ be defined as follows:

$$\mathbf{B} = \begin{bmatrix} 1 & 5 & 0 \\ 1 & 0 & 5 \\ 1 & 0 & 5 \end{bmatrix}$$

   a. Are the column vectors of $\mathbf{B}$ linearly dependent?
   b. What is the rank of $\mathbf{B}$?
   c. What must be the determinant of $\mathbf{B}$?

6.9. Let $\mathbf{A}$ be defined as follows:

$$\mathbf{A} = \begin{bmatrix} 0 & 1 & 8 \\ 0 & 3 & 1 \\ 0 & 5 & 5 \end{bmatrix}$$

   a. Are the column vectors of $\mathbf{A}$ linearly dependent?
   b. Restate definition (6.20) in terms of row vectors. Are the row vectors of $\mathbf{A}$ linearly dependent?
   c. What is the rank of $\mathbf{A}$?
   d. Calculate the determinant of $\mathbf{A}$.

6.10. Find the inverse of each of the following matrices:

$$\mathbf{A} = \begin{bmatrix} 2 & 4 \\ 3 & 1 \end{bmatrix} \quad \mathbf{B} = \begin{bmatrix} 4 & 3 & 2 \\ 6 & 5 & 10 \\ 10 & 1 & 6 \end{bmatrix}$$

Check in each case that the resulting matrix is indeed the inverse.

6.11. Find the inverse of the following matrix:

$$A = \begin{bmatrix} 5 & 1 & 3 \\ 4 & 0 & 5 \\ 1 & 9 & 6 \end{bmatrix}$$

Check that the resulting matrix is indeed the inverse.

6.12. Refer to **Flavor deterioration** Problem 6.4. Find $(X'X)^{-1}$.

6.13. Refer to **Consumer finance** Problem 6.5. Find $(X'X)^{-1}$.

6.14. Consider the simultaneous equations:

$$4y_1 + 7y_2 = 25$$
$$2y_1 + 3y_2 = 12$$

a. Write these equations in matrix notation.
b. Using matrix methods, find the solutions for y_1 and y_2.

6.15. Consider the simultaneous equations:

$$5y_1 + 2y_2 = 8$$
$$23y_1 + 7y_2 = 28$$

a. Write these equations in matrix notation.
b. Using matrix methods, find the solutions for y_1 and y_2.

6.16. Consider the estimated linear regression function in the form of (2.15). Write expressions in this form for the fitted values $\hat{Y}_i$ in matrix terms for $i = 1, \ldots, 5$.

6.17. Consider the following functions of the random variables Y_1, Y_2, and Y_3:

$$W_1 = Y_1 + Y_2 + Y_3$$
$$W_2 = Y_1 - Y_2$$
$$W_3 = Y_1 - Y_2 - Y_3$$

a. State the above in matrix notation.
b. Find the expectation of the random vector **W**.
c. Find the variance-covariance matrix of **W**.

6.18. Consider the following functions of the random variables Y_1, Y_2, Y_3, and Y_4:

$$W_1 = \frac{1}{4}(Y_1 + Y_2 + Y_3 + Y_4)$$

$$W_2 = \frac{1}{2}(Y_1 + Y_2) - \frac{1}{2}(Y_3 + Y_4)$$

a. State the above in matrix notation.
b. Find the expectation of the random vector **W**.
c. Find the variance-covariance matrix of **W**.

6.19. Find the matrix **A** of the quadratic form:

$$3Y_1^2 + 10Y_1Y_2 + 17Y_2^2$$

6.20. Find the matrix **A** of the quadratic form:

$$7Y_1^2 - 8Y_1 Y_2 + 8Y_2^2$$

6.21. For the matrix:

$$\mathbf{A} = \begin{bmatrix} 5 & 2 \\ 2 & 1 \end{bmatrix}$$

find the quadratic form of the observations Y_1 and Y_2.

6.22. For the matrix:

$$\mathbf{A} = \begin{bmatrix} 1 & 0 & 4 \\ 0 & 3 & 0 \\ 4 & 0 & 9 \end{bmatrix}$$

find the quadratic form of the observations Y_1, Y_2, and Y_3.

6.23. Refer to **Flavor deterioration** Problems 6.4 and 6.12.
  a. Using matrix methods, obtain the following: (1) vector of estimated regression coefficients, (2) vector of residuals, (3) *SSR*, (4) *SSE*, (5) estimated variance-covariance matrix of **b**, (6) point estimate of $E\{Y_h\}$ when $X_h = -6$, (7) estimated variance of $\hat{Y}_h$ when $X_h = -6$.
  b. What simplifications arose from the spacing of the X levels in the experiment?
  c. Find the hat matrix **H**.
  d. Find $s^2\{e\}$.

6.24. Refer to **Consumer finance** Problems 6.5 and 6.13.
  a. Using matrix methods, obtain the following: (1) vector of estimated regression coefficients, (2) vector of residuals, (3) *SSR*, (4) *SSE*, (5) estimated variance-covariance matrix of **b**, (6) point estimate of $E\{Y_h\}$ when $X_h = 4$, (7) estimated variance of $Y_{h(new)}$ when $X_h = 4$.
  b. From your estimated variance-covariance matrix in part (a5), obtain the following: (1) $s\{b_0, b_1\}$; (2) $s^2\{b_0\}$; (3) $s\{b_1\}$.
  c. Find the hat matrix **H**.
  d. Find $s^2\{e\}$.

6.25. Refer to **Airfreight breakage** Problems 2.19 and 6.6.
  a. Using matrix methods, obtain the following: (1) $(\mathbf{X}'\mathbf{X})^{-1}$, (2) **b**, (3) **e**, (4) **H**, (5) *SSE*, (6) $s^2\{\mathbf{b}\}$, (7) $\hat{Y}_h$ when $X_h = 2$, (8) $s^2\{\hat{Y}_h\}$ when $X_h = 2$.
  b. From part (a6), obtain the following: (1) $s^2\{b_1\}$; (2) $s\{b_0, b_1\}$; (3) $s\{b_0\}$.
  c. Find the matrix of the quadratic form for *SSR*.

6.26. Refer to **Plastic hardness** Problems 2.20 and 6.7.
  a. Using matrix methods, obtain the following: (1) $(\mathbf{X}'\mathbf{X})^{-1}$, (2) **b**, (3) $\hat{\mathbf{Y}}$, (4) **H**, (5) *SSE*, (6) $s^2\{\mathbf{b}\}$, (7) $s^2\{Y_{h(new)}\}$ when $X_h = 30$.
  b. From part (a6), obtain the following: (1) $s^2\{b_0\}$; (2) $s\{b_0, b_1\}$; (3) $s\{b_1\}$.
  c. Obtain the matrix of the quadratic form for *SSE*.

EXERCISES

6.27. Refer to regression through the origin model (5.11). Set up the expectation vector for $\boldsymbol{\varepsilon}$. Assume that $i = 1, \ldots, 4$.

6.28. Consider model (5.11) for regression through the origin and the estimator b_1 given in (5.15). Obtain (5.15) by utilizing (6.57) with $\mathbf{X}$ suitably defined.

6.29. Consider the least squares estimator $\mathbf{b}$ given in (6.57). Using matrix methods, show that $\mathbf{b}$ is an unbiased estimator.

6.30. Show that $\hat{Y}_h$ in (6.83) can be expressed in matrix terms as $\mathbf{b}'\mathbf{X}_h$.

6.31. Obtain an expression for the variance-covariance matrix of the fitted values $\hat{Y}_i$, $i = 1, \ldots, n$, in terms of the hat matrix.

Part II

General Linear Regression

Chapter 7

Multiple Regression—I

Multiple regression analysis is one of the most widely used of all statistical tools. In this chapter, we first discuss a variety of multiple regression models. Then we present the basic statistical results for multiple regression in matrix form. Since the matrix expressions for multiple regression are the same as for simple linear regression, we state the results without much discussion. We then give an example, illustrating a variety of inferences and residual analyses in multiple regression analysis.

7.1 MULTIPLE REGRESSION MODELS

Need for Several Independent Variables

When we first introduced regression analysis in Chapter 2, we spoke of regression models containing a number of independent variables. We mentioned a regression model where the dependent variable was direct operating cost for a branch office of a consumer finance chain, and four independent variables were considered, including average number of loans outstanding at the branch and total number of new loan applications processed by the branch. We also mentioned a tractor purchase study where the response variable was volume of tractor purchases in a sales territory, and the nine independent variables included number of farms in the territory and quantity of crop production in the territory. In addition, we mentioned a study of short children where the response variable was the peak plasma growth hormone level, and the 14 independent variables included gender, age, and various body measurements. In all these examples, a single independent variable in the model would have provided an inadequate description since a number of key independent variables affect the response variable in important and distinctive ways. Furthermore, in situations of this type, one will frequently find that predictions of the response variable based on a model containing only a single independent variable are too imprecise to be useful. A more complex model, containing additional independent variables, typically is more helpful in providing sufficiently precise predictions of the response variable.

In each of the examples mentioned, the analysis is based on observational data because some or all of the independent variables are not susceptible to direct control. Multiple regression analysis is also highly useful in experimental situations where the experimenter can control the independent variables. An experimenter typically will wish to investigate a number of independent variables simultaneously because almost always more than one key independent variable influences the response. For example, in a study of productivity of work crews, the experimenter may wish to control both the size of the crew and the level of bonus pay. Similarly, in a study of responsiveness to a drug, the experimenter may wish to control both the dose of the drug and the method of administration.

The multiple regression models which we shall now describe can be utilized for either observational data or for experimental data from a completely randomized design.

First-Order Model with Two Independent Variables

When there are two independent variables X_1 and X_2, the regression model:

$$(7.1) \qquad Y_i = \beta_0 + \beta_1 X_{i1} + \beta_2 X_{i2} + \varepsilon_i$$

is called a first-order model with two independent variables. A first-order model, it will be recalled from Chapter 2, is linear in the parameters and linear in the independent variables. Y_i denotes as usual the response in the ith trial, and X_{i1} and X_{i2} are the values of the two independent variables in the ith trial. The parameters of the model are β_0, β_1, and β_2, and the error term is ε_i.

Assuming that $E\{\varepsilon_i\} = 0$, the regression function for model (7.1) is:

$$(7.2) \qquad E\{Y\} = \beta_0 + \beta_1 X_1 + \beta_2 X_2$$

Analogous to simple linear regression, where the regression function $E\{Y\} = \beta_0 + \beta_1 X$ is a line, the regression function (7.2) is a plane. Figure 7.1 contains a representation of a portion of the response plane:

$$(7.3) \qquad E\{Y\} = 20.0 + .95X_1 - .50X_2$$

Note that a point on the response plane (7.3) corresponds to the mean response $E\{Y\}$ at the given combination of levels of X_1 and X_2.

Figure 7.1 also shows a series of observations Y_i corresponding to given levels (X_{i1}, X_{i2}) of the two independent variables. Note that each vertical rule in Figure 7.1 represents the difference between Y_i and the mean $E\{Y_i\}$ of the probability distribution for (X_{i1}, X_{i2}) on the response plane. Hence, the vertical distance from Y_i to the response plane represents the error term $\varepsilon_i = Y_i - E\{Y_i\}$.

Frequently the regression function in multiple regression is called a *regression surface* or a *response surface*. In Figure 7.1, the response surface is just a plane, but in other cases the response surface may be complex in nature.

Meaning of Regression Coefficients.   Let us now consider the meaning of the regression coefficients in the multiple regression function (7.3). The parameter $\beta_0 = 20.0$

FIGURE 7.1   Example of Response Surface—a Response Plane with Observations
Scattered about It

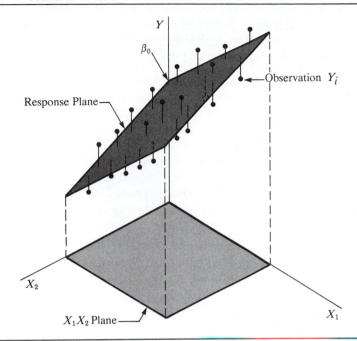

is the Y intercept of the regression plane. If the scope of the model includes $X_1 = 0$, $X_2 = 0$, $\beta_0 = 20.0$ gives the mean response at $X_1 = 0$, $X_2 = 0$. Otherwise, β_0 does not have any particular meaning as a separate term in the regression model.

The parameter β_1 indicates the change in the mean response per unit increase in X_1 when X_2 is held constant. Likewise, β_2 indicates the change in the mean response per unit increase in X_2 when X_1 is held constant. To see this for our example, suppose X_2 is held at the level $X_2 = 20$. The regression function (7.3) now is:

$$(7.4) \qquad E\{Y\} = 20.0 + .95X_1 - .50(20) = (20.0 - 10.0) + .95X_1$$

$$= 10.0 + .95X_1$$

Note that for $X_2 = 20$, the response function is a straight line with slope .95. The same is true for any other value of X_2; only the intercept of the response function differs. Hence, $\beta_1 = .95$ indicates that the mean response increases by .95 with a unit increase in X_1 when X_2 is constant, no matter what the level of X_2. More loosely speaking, we state that β_1 indicates the change in $E\{Y\}$ with a unit increase in X_1 when X_2 is held constant.

Similarly, $\beta_2 = -.50$ in regression function (7.3) indicates that the mean response decreases by .50 with a unit increase in X_2 when X_1 is held constant.

When the effect of X_1 on the mean response does not depend on the level of X_2, and correspondingly the effect of X_2 does not depend on the level of X_1, the two independent variables are said to have *additive effects* or *not to interact*. Thus, the first-order regression model (7.1) is designed for independent variables whose effects on the mean response are additive or do not interact.

The parameters β_1 and β_2 are frequently called *partial regression coefficients* because they reflect the partial effect of one independent variable when the other independent variable is included in the model and is held constant.

Example.   Suppose that the response surface in (7.3) pertains to urban full-service stations of a major oil company and shows the effects of variety and adequacy of services (X_1) and average time taken to reach a car (X_2) on the ratio of actual gallonage of gasoline sold to potential gallonage (Y), where X_1 is expressed as an index with $100 =$ average, X_2 is in seconds, and Y is stated as a percent. Increasing the index of adequacy of services by one point while holding average time to reach a car constant leads to an increase of .95 percent point in the expected ratio of actual to potential gallonage. If the index of adequacy of services is held constant and the average time to reach a car is increased by one second, the expected ratio of actual to potential gallonage decreases by .50 percent point.

Comments

1. A regression model for which the response surface is a plane can be used either in its own right when it is appropriate, or as an approximation to a more complex response surface. Many complex response surfaces can be approximated well by a plane for limited ranges of X_1 and X_2.

2. We can readily establish the meaning of β_1 and β_2 by calculus, taking partial derivatives of the response surface (7.2) with respect to X_1 and X_2 in turn:

$$\frac{\partial E\{Y\}}{\partial X_1} = \beta_1 \qquad \frac{\partial E\{Y\}}{\partial X_2} = \beta_2$$

The partial derivatives measure the rate of change in $E\{Y\}$ with respect to one independent variable when the other is held constant.

First-Order Model with More than Two Independent Variables

We consider now the case where there are $p - 1$ independent variables $X_1, \ldots, X_{p-1}$. The regression model:

$$(7.5) \qquad Y_i = \beta_0 + \beta_1 X_{i1} + \beta_2 X_{i2} + \cdots + \beta_{p-1} X_{i,p-1} + \varepsilon_i$$

is called a first-order model with $p - 1$ independent variables. It can also be written:

$$(7.5a) \qquad Y_i = \beta_0 + \sum_{k=1}^{p-1} \beta_k X_{ik} + \varepsilon_i$$

or, if we let $X_{i0} \equiv 1$, it can be written as:

$$(7.5b) \qquad Y_i = \sum_{k=0}^{p-1} \beta_k X_{ik} + \varepsilon_i \qquad \text{where } X_{i0} \equiv 1$$

Assuming that $E\{\varepsilon_i\} = 0$, the response function for regression model (7.5) is:

$$(7.6) \qquad E\{Y\} = \beta_0 + \beta_1 X_1 + \beta_2 X_2 + \cdots + \beta_{p-1} X_{p-1}$$

This response function is a *hyperplane,* which is a plane in more than two dimensions. It is no longer possible to picture this response surface, as we were able to do in Figure 7.1 for the case of two indepenent variables. Nevertheless, the meaning of the parameters is analogous to the two independent variables case. The parameter β_k indicates the change in the mean response $E\{Y\}$ with a unit increase in the independent variable X_k, when all other independent variables in the regression model are held constant. Note again that the effect of any independent variable on the mean response is the same for regression model (7.5) no matter what are the levels at which the other indepenent variables are held. Hence, the first-order regression model (7.5) is designed for independent variables whose effects on the mean response are additive and therefore do not interact.

Note

If $p - 1 = 1$, regression model (7.5) reduces to:

$$Y_i = \beta_0 + \beta_1 X_{i1} + \varepsilon_i$$

which is the simple linear regression model considered in earlier chapters.

General Linear Regression Model

In general, the variables $X_1, \ldots, X_{p-1}$ in a regression model do not have to represent different independent variables, as we shall shortly see. We therefore define the general linear regression model, with normal error terms, simply in terms of X variables:

$$(7.7) \qquad Y_i = \beta_0 + \beta_1 X_{i1} + \beta_2 X_{i2} + \cdots + \beta_{p-1} X_{i,p-1} + \varepsilon_i$$

where:

$\beta_0, \beta_1, \ldots, \beta_{p-1}$ are parameters
$X_{i1}, \ldots, X_{i,p-1}$ are known constants
ε_i are independent $N(0, \sigma^2)$
$i = 1, \ldots, n$

If we let $X_{i0} \equiv 1$, regression model (7.7) can be written as follows:

$$(7.7a) \qquad Y_i = \beta_0 X_{i0} + \beta_1 X_{i1} + \beta_2 X_{i2} + \cdots + \beta_{p-1} X_{i,p-1} + \varepsilon_i$$

$$\text{where } X_{i0} \equiv 1$$

or:

$$(7.7b) \qquad Y_i = \sum_{k=0}^{p-1} \beta_k X_{ik} + \varepsilon_i \qquad \text{where } X_{i0} \equiv 1$$

The response function for regression model (7.7) is, since $E\{\varepsilon_i\} = 0$:

(7.8)
$$E\{Y\} = \beta_0 + \beta_1 X_1 + \beta_2 X_2 + \cdots + \beta_{p-1} X_{p-1}$$

Thus, the general linear regression model with normal error terms implies that the observations Y_i are independent normal variables, with mean $E\{Y_i\}$ as given by (7.8) and with constant variance σ^2.

This general linear model encompasses a vast variety of situations. We shall consider a few of these now.

$p - 1$ Independent Variables.   When $X_1, \ldots, X_{p-1}$ represent $p - 1$ different independent variables, the general linear regression model (7.7) is, as we have seen, a first-order model in which there are no interaction effects between the independent variables.

Polynomial Regression.   Consider the curvilinear regression model with one independent variable:

(7.9)
$$Y_i = \beta_0 + \beta_1 X_i + \beta_2 X_i^2 + \varepsilon_i$$

If we let $X_{i1} = X_i$ and $X_{i2} = X_i^2$, we can write (7.9) as follows:

$$Y_i = \beta_0 + \beta_1 X_{i1} + \beta_2 X_{i2} + \varepsilon_i$$

so that model (7.9) is a particular case of the general linear regression model (7.7). While (7.9) illustrates a curvilinear regression model where the response function is quadratic, models with higher degree polynomial response functions are also particular cases of the general linear regression model.

Transformed Variables.   Consider the model:

(7.10)
$$\log Y_i = \beta_0 + \beta_1 X_{i1} + \beta_2 X_{i2} + \beta_3 X_{i3} + \varepsilon_i$$

Here, the response surface is complex, yet model (7.10) can be treated as a general linear regression model. If we let $Y_i' = \log Y_i$, we can write regression model (7.10) as follows:

$$Y_i' = \beta_0 + \beta_1 X_{i1} + \beta_2 X_{i2} + \beta_3 X_{i3} + \varepsilon_i$$

which is in the form of the general linear regression model (7.7). The dependent variable just happens to be the logarithm of Y.

Many models can be transformed into general linear regression models. Thus, the model:

(7.11)
$$Y_i = \frac{1}{\beta_0 + \beta_1 X_{i1} + \beta_2 X_{i2} + \varepsilon_i}$$

can be transformed to a general linear regression model by letting $Y_i' = 1/Y_i$. We then have:

$$Y_i' = \beta_0 + \beta_1 X_{i1} + \beta_2 X_{i2} + \varepsilon_i$$

Interaction Effects. Consider the regression model in two independent variables X_1 and X_2:

$$(7.12) \qquad Y_i = \beta_0 + \beta_1 X_{i1} + \beta_2 X_{i2} + \beta_3 X_{i1} X_{i2} + \varepsilon_i$$

The meaning of β_1 and β_2 here is not the same as that given earlier because of the cross-product term $\beta_3 X_{i1} X_{i2}$. It can be shown that the change in the mean response with a unit increase in X_1 when X_2 is held constant is:

$$(7.13) \qquad \beta_1 + \beta_3 X_2$$

Similarly, the change in the mean response with a unit change in X_2 when X_1 is held constant is:

$$(7.14) \qquad \beta_2 + \beta_3 X_1$$

Hence, in regression model (7.12) both the effect of X_1 for given level of X_2 and the effect of X_2 for given level of X_1 depend on the level of the other independent variable.

In Figure 7.2, we illustrate the effect of the cross-product term in regression model (7.12). In Figure 7.2a, we consider a response function without a cross-product term:

$$E\{Y\} = 10 + 2X_1 + 5X_2$$

and show there the response function $E\{Y\}$ when $X_2 = 1$ and when $X_2 = 3$. Note that the two response functions are parallel—that is, the mean response increases by the same amount $\beta_1 = 2$ with a unit increase of X_1 whether $X_2 = 1$ or $X_2 = 3$.

FIGURE 7.2   Effect of Cross-Product Term in Response Function with Two
Independent Variables

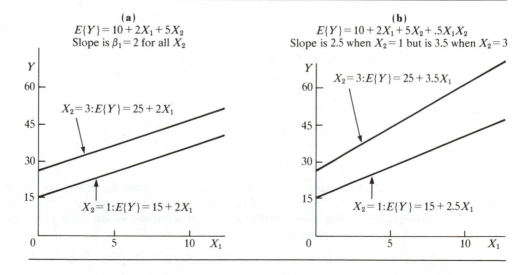

In Figure 7.2b, we consider the same response function but with the cross-product term $.5X_1X_2$ added:

$$E\{Y\} = 10 + 2X_1 + 5X_2 + .5X_1X_2$$

and show the response function $E\{Y\}$ when $X_2 = 1$ and when $X_2 = 3$. Note that the slopes of the response functions when plotted against X_1 now differ for $X_2 = 1$ and $X_2 = 3$. The slope of the response function when $X_2 = 1$ is by (7.13):

$$\beta_1 + \beta_3X_2 = 2 + .5(1) = 2.5$$

and when $X_2 = 3$, the slope is:

$$\beta_1 + \beta_3X_2 = 2 + .5(3) = 3.5$$

Hence, β_1 in regression model (7.12) containing a cross-product term no longer indicates the change in the mean response for a unit increase in X_1 for any given X_2 level. That effect in this model depends on the level of X_2. Regression model (7.12) with the cross-product term is therefore designed for independent variables whose effects on the dependent variable *interact*. The cross-product term $\beta_3X_{i1}X_{i2}$ is called an *interaction term*. While the mean response in regression model (7.12) when X_2 is constant is still a linear function of X_1, now both the intercept and the slope of the response function change as the level at which X_2 is held constant is varied. The same holds when the mean response is regarded as a function of X_2, with X_1 constant.

Despite these complexities of regression model (7.12), it can still be regarded as a general linear regression model. Let $X_{i3} = X_{i1}X_{i2}$. We can then write (7.12) as follows:

$$Y_i = \beta_0 + \beta_1X_{i1} + \beta_2X_{i2} + \beta_3X_{i3} + \varepsilon_i$$

which is in the form of the general linear regression model (7.7).

Note

To derive (7.13) and (7.14), we differentiate:

$$E\{Y\} = \beta_0 + \beta_1X_1 + \beta_2X_2 + \beta_3X_1X_2$$

with respect to X_1 and X_2, respectively:

$$\frac{\partial E\{Y\}}{\partial X_1} = \beta_1 + \beta_3X_2 \qquad \frac{\partial E\{Y\}}{\partial X_2} = \beta_2 + \beta_3X_1$$

Combination of Cases.   A regression model may combine a number of the elements we have just noted and still can be treated as a general linear regression model. Consider a regression model with two independent variables, each in quadratic form, with an interaction term:

$$(7.15) \qquad Y_i = \beta_0 + \beta_1X_{i1} + \beta_2X_{i1}^2 + \beta_3X_{i2} + \beta_4X_{i2}^2 + \beta_5X_{i1}X_{i2} + \varepsilon_i$$

FIGURE 7.3   Additional Examples of Response Functions

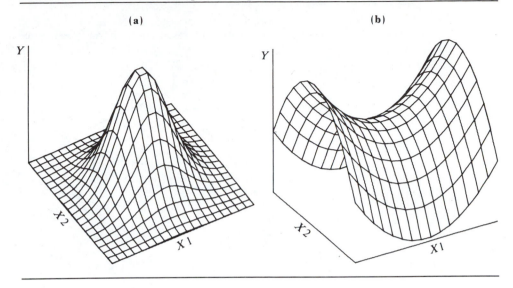

Let us define:

$$Z_{i1} = X_{i1} \quad Z_{i2} = X_{i1}^2 \quad Z_{i3} = X_{i2} \quad Z_{i4} = X_{i2}^2 \quad Z_{i5} = X_{i1}X_{i2}$$

We can then write regression model (7.15) as follows:

$$Y_i = \beta_0 + \beta_1 Z_{i1} + \beta_2 Z_{i2} + \beta_3 Z_{i3} + \beta_4 Z_{i4} + \beta_5 Z_{i5} + \varepsilon_i$$

which is in the form of the general linear regression model (7.7).

Comments

1. It should be clear from the various examples that the general linear regression model (7.7) is not restricted to linear response surfaces. The term *linear model* refers to the fact that (7.7) is linear in the parameters, not to the shape of the response surface.

2. Figure 7.3 illustrates some complex response surfaces, when there are two independent variables, that can be represented by the general linear regression model (7.7).

Interactions and Nature of Response Surface

We introduced the concept of interacting independent variables earlier, and now shall illustrate further how the response surface differs when two independent variables do not interact and when they do interact.

Figure 7.4a contains a representation of a response surface in which the two independent variables (mean season temperature, amount of rainfall) do not interact on the dependent variable (corn yield). The absence of interactions can be seen by con-

---

FIGURE 7.4 Response Surfaces for Additive and Interacting Independent Variables

---

(**a**) Independent Variables Do Not Interact
Yield of corn as function of season rainfall and mean temperature

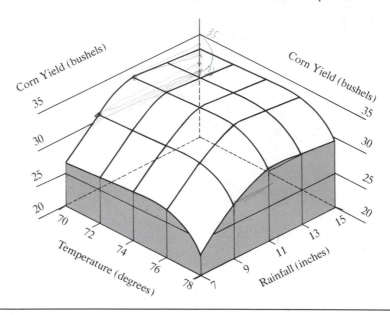

---

sidering the corn yield curves, for given mean season temperatures, as a function of rainfall. These curves all have the same shape and differ only by a constant. Thus, each ordinate of the corn yield curve when the mean temperature is 70° is a constant number of units higher than the corresponding ordinate for the corn yield curve when the mean temperature is 78°.

Equivalently, one can note the absence of interactions by considering the corn yield curves, for given amounts of rainfall, as a function of temperature. Again, these curves are the same in shape and differ only by a constant.

Absence of interactions therefore implies that the mean response $E\{Y\}$ can be expressed in the form:

$$(7.16) \qquad E\{Y\} = f_1(X_1) + f_2(X_2)$$

where f_1 and f_2 can be any functions, not necessarily simple ones.

Figure 7.4b illustrates a case where the two independent variables (age, percent of normal weight) interact on the dependent variable (mortality ratio). Here, the shape of the mortality ratio curve as a function of percent of normal weight varies for different ages. For men 22 years old, both underweight and overweight persons have higher mortality rates than normal (normal = 100) for that age. On the other hand, for men 52 years old, the mortality rate is above normal for that age for overweight

FIGURE 7.4 (*concluded*)

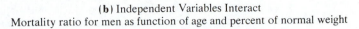

(**b**) Independent Variables Interact
Mortality ratio for men as function of age and percent of normal weight

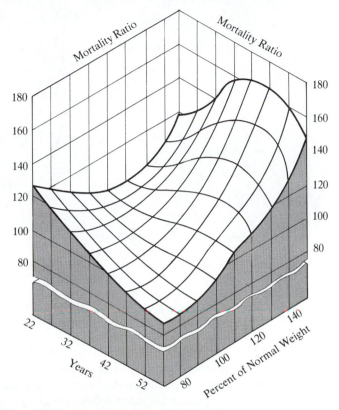

Source: Reprinted, with permission, from M. Ezekiel and K. A. Fox, *Methods of Correlation and Regression Analysis,* 3rd ed. (New York: John Wiley & Sons, 1959), pp. 349–50.

persons but not for underweight persons. Similarly, the mortality ratio curves as a function of age vary in shape for different weights.

We can illustrate the difference in the shape of the response function when the two independent variables do and do not interact in yet another way, namely, by representing the response surface by means of a contour diagram. Such a diagram shows, for a number of different response levels, the various combinations of the two independent variables that yield the same level of response. Figure 7.5a shows a contour diagram for the response surface portrayed in Figure 7.1:

$$E\{Y\} = 20.0 + .95X_1 - .50X_2$$

Note that the independent variables do not interact in this response function and that

FIGURE 7.5   Response Contour Diagrams

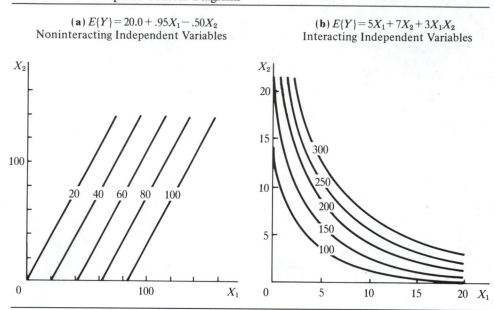

(a) $E\{Y\} = 20.0 + .95X_1 - .50X_2$
Noninteracting Independent Variables

(b) $E\{Y\} = 5X_1 + 7X_2 + 3X_1X_2$
Interacting Independent Variables

the contour lines are parallel. Figure 7.5b shows a contour diagram for the response function:

$$E\{Y\} = 5X_1 + 7X_2 + 3X_1X_2$$

where the two independent variables interact and the contour curves are not parallel.

In general, additive or noninteracting independent variables lead to parallel contour curves while interacting independent variables lead to nonparallel contour curves.

7.2 GENERAL LINEAR REGRESSION MODEL IN MATRIX TERMS

We shall now present the principal results for the general linear regression model (7.7) in matrix terms. This model, as we have noted, encompasses a wide variety of particular cases. The results to be presented are applicable to all of these.

It is a remarkable property of matrix algebra that the results for the general linear regression model (7.7) appear exactly the same in matrix notation as those for the simple linear regression model (6.54). Only the degrees of freedom and other constants related to the number of independent variables and the dimensions of some matrices will be different. Hence, we shall be able to present the results very concisely.

The matrix notation, to be sure, may hide enormous computational complexities. The inverse of a 10×10 matrix **A** requires tremendous amounts of computation, yet is simply represented as $\mathbf{A}^{-1}$. Our reason for emphasizing matrix algebra is that it indicates the essential conceptual steps in the solution. The actual computations will in all but the very simplest cases be done by programmable calculator or computer. Hence, it does not matter for us whether $(\mathbf{X}'\mathbf{X})^{-1}$ represents finding the inverse of a 2×2 or a 10×10 matrix. The important point is to know what the inverse of the matrix represents.

To express the general linear regression model (7.7):

$$Y_i = \beta_0 + \beta_1 X_{i1} + \beta_2 X_{i2} + \cdots + \beta_{p-1} X_{i,p-1} + \varepsilon_i$$

in matrix terms, we need to define the following matrices:

(7.17)

(7.17a)

$$\underset{n\times 1}{\mathbf{Y}} = \begin{bmatrix} Y_1 \\ Y_2 \\ \vdots \\ Y_n \end{bmatrix}$$

(7.17b)

$$\underset{n\times p}{\mathbf{X}} = \begin{bmatrix} 1 & X_{11} & X_{12} & \cdots & X_{1,p-1} \\ 1 & X_{21} & X_{22} & \cdots & X_{2,p-1} \\ \vdots & \vdots & \vdots & & \vdots \\ 1 & X_{n1} & X_{n2} & \cdots & X_{n,p-1} \end{bmatrix}$$

(7.17c)

$$\underset{p\times 1}{\boldsymbol{\beta}} = \begin{bmatrix} \beta_0 \\ \beta_1 \\ \vdots \\ \beta_{p-1} \end{bmatrix}$$

(7.17d)

$$\underset{n\times 1}{\boldsymbol{\varepsilon}} = \begin{bmatrix} \varepsilon_1 \\ \varepsilon_2 \\ \vdots \\ \varepsilon_n \end{bmatrix}$$

Note that the $\mathbf{Y}$ and $\boldsymbol{\varepsilon}$ vectors are the same as for simple linear regression. The $\boldsymbol{\beta}$ vector contains additional regression parameters, and the $\mathbf{X}$ matrix contains a column of 1s as well as a column of the n values for each of the $p - 1$ X variables in the regression model. The row subscript for each element X_{ik} in the $\mathbf{X}$ matrix identifies the trial or case, and the column subscript identifies the X variable.

In matrix terms, the general linear regression model (7.7) is:

(7.18)

$$\underset{n\times 1}{\mathbf{Y}} = \underset{n\times p}{\mathbf{X}} \underset{p\times 1}{\boldsymbol{\beta}} + \underset{n\times 1}{\boldsymbol{\varepsilon}}$$

where:

$\mathbf{Y}$ is a vector of responses
$\boldsymbol{\beta}$ is a vector of parameters
$\mathbf{X}$ is a matrix of constants
$\boldsymbol{\varepsilon}$ is a vector of independent normal random variables with expectation
    $\mathbf{E}\{\boldsymbol{\varepsilon}\} = \mathbf{0}$ and variance-covariance matrix $\sigma^2\{\boldsymbol{\varepsilon}\} = \sigma^2 \mathbf{I}$

Consequently, the random vector $\mathbf{Y}$ has expectation:

(7.18a)

$$\underset{n\times 1}{\mathbf{E}\{\mathbf{Y}\}} = \mathbf{X}\boldsymbol{\beta}$$

and the variance-covariance matrix of $\mathbf{Y}$ is:

$$(7.18b) \qquad\qquad \underset{n\times n}{\sigma^2\{\mathbf{Y}\}} = \sigma^2\mathbf{I}$$

7.3 LEAST SQUARES ESTIMATORS

Let us denote the vector of estimated regression coefficients $b_0, b_1, \ldots, b_{p-1}$ as $\mathbf{b}$:

$$(7.19) \qquad\qquad \underset{p\times 1}{\mathbf{b}} = \begin{bmatrix} b_0 \\ b_1 \\ b_2 \\ \vdots \\ b_{p-1} \end{bmatrix}$$

The least squares normal equations for the general linear regression model (7.18) are:

$$(7.20) \qquad\qquad \mathbf{X}'\mathbf{X}\mathbf{b} = \mathbf{X}'\mathbf{Y}$$

and the least squares estimators are:

$$(7.21) \qquad\qquad \underset{p\times 1}{\mathbf{b}} = \underset{p\times p}{(\mathbf{X}'\mathbf{X})^{-1}}\underset{p\times 1}{\mathbf{X}'\mathbf{Y}}$$

For regression model (7.18), these least squares estimators are also maximum likelihood estimators and have all the properties mentioned in Chapter 2: they are unbiased, minimum variance unbiased, consistent, and sufficient.

7.4 FITTED VALUES AND RESIDUALS

Let the vector of the fitted values $\hat{Y}_i$ be denoted by $\hat{\mathbf{Y}}$ and the vector of the residual terms $e_i = Y_i - \hat{Y}_i$ be denoted by $\mathbf{e}$:

$$(7.22) \qquad (7.22a) \quad \underset{n\times 1}{\hat{\mathbf{Y}}} = \begin{bmatrix} \hat{Y}_1 \\ \hat{Y}_2 \\ \vdots \\ \hat{Y}_n \end{bmatrix} \qquad (7.22b) \quad \underset{n\times 1}{\mathbf{e}} = \begin{bmatrix} e_1 \\ e_2 \\ \vdots \\ e_n \end{bmatrix}$$

The fitted values are represented by:

$$(7.23) \qquad\qquad \underset{n\times 1}{\hat{\mathbf{Y}}} = \mathbf{X}\mathbf{b}$$

and the residual terms by:

$$(7.24) \qquad\qquad \underset{n\times 1}{\mathbf{e}} = \mathbf{Y} - \hat{\mathbf{Y}} = \mathbf{Y} - \mathbf{X}\mathbf{b}$$

The vector of the fitted values $\hat{\mathbf{Y}}$ can be expressed in terms of the hat matrix $\mathbf{H}$ as follows:

$$(7.25) \qquad \underset{n \times 1}{\hat{\mathbf{Y}}} = \mathbf{HY}$$

where:

$$(7.25a) \qquad \underset{n \times n}{\mathbf{H}} = \mathbf{X}(\mathbf{X'X})^{-1}\mathbf{X'}$$

Similarly, the vector of residuals can be expressed as follows:

$$(7.26) \qquad \underset{n \times 1}{\mathbf{e}} = (\mathbf{I} - \mathbf{H})\mathbf{Y}$$

The variance-covariance matrix of the residuals is:

$$(7.27) \qquad \underset{n \times n}{\sigma^2\{\mathbf{e}\}} = \sigma^2(\mathbf{I} - \mathbf{H})$$

which is estimated by:

$$(7.28) \qquad \underset{n \times n}{s^2\{\mathbf{e}\}} = MSE\,(\mathbf{I} - \mathbf{H})$$

7.5 ANALYSIS OF VARIANCE RESULTS

Sums of Squares and Mean Squares

The sums of squares for the analysis of variance in matrix terms are:

$$(7.29) \qquad SSTO = \mathbf{Y'Y} - \left(\frac{1}{n}\right)\mathbf{Y'JY} = \mathbf{Y'}\left[\mathbf{I} - \left(\frac{1}{n}\right)\mathbf{J}\right]\mathbf{Y}$$

$$(7.30) \qquad SSE = \mathbf{e'e} = (\mathbf{Y} - \mathbf{Xb})'(\mathbf{Y} - \mathbf{Xb}) = \mathbf{Y'Y} - \mathbf{b'X'Y}$$
$$= \mathbf{Y'}(\mathbf{I} - \mathbf{H})\mathbf{Y}$$

$$(7.31) \qquad SSR = \mathbf{b'X'Y} - \left(\frac{1}{n}\right)\mathbf{Y'JY} = \mathbf{Y'}\left[\mathbf{H} - \left(\frac{1}{n}\right)\mathbf{J}\right]\mathbf{Y}$$

where $\mathbf{J}$ is an $n \times n$ matrix of 1s defined in (6.18) and $\mathbf{H}$ is the hat matrix defined in (7.25a).

SSTO, as usual, has $n - 1$ degrees of freedom associated with it. *SSE* has $n - p$ degrees of freedom associated with it since p parameters need to be estimated in the regression function for model (7.18). Finally, *SSR* has $p - 1$ degrees of freedom associated with it, representing the number of X variables $X_1, \ldots, X_{p-1}$.

Table 7.1 shows these analysis of variance results, as well as the mean squares *MSR* and *MSE*:

$$(7.32) \qquad MSR = \frac{SSR}{p - 1}$$

TABLE 7.1   ANOVA Table for General Linear Regression Model (7.18)

Source of Variation	SS	df	MS
Regression	$SSR = \mathbf{b'X'Y} - \left(\dfrac{1}{n}\right)\mathbf{Y'JY}$	$p - 1$	$MSR = \dfrac{SSR}{p-1}$
Error	$SSE = \mathbf{Y'Y} - \mathbf{b'X'Y}$	$n - p$	$MSE = \dfrac{SSE}{n-p}$
Total	$SSTO = \mathbf{Y'Y} - \left(\dfrac{1}{n}\right)\mathbf{Y'JY}$	$n - 1$	

$$(7.33) \qquad MSE = \frac{SSE}{n-p}$$

The expectation of MSE is σ^2, as for simple linear regression. The expectation of MSR is σ^2 plus a quantity that is nonnegative. For instance, when $p - 1 = 2$, we have:

$$E\{MSR\} = \sigma^2 + \left[\beta_1^2 \sum (X_{i1} - \bar{X}_1)^2 + \beta_2^2 \sum (X_{i2} - \bar{X}_2)^2 \right.$$

$$\left. + 2\beta_1\beta_2 \sum (X_{i1} - \bar{X}_1)(X_{i2} - \bar{X}_2) \right] \Big/ 2$$

Note that if both β_1 and β_2 equal zero, $E\{MSR\} = \sigma^2$. Otherwise $E\{MSR\} > \sigma^2$.

F Test for Regression Relation

To test whether there is a regression relation between the dependent variable Y and the set of X variables $X_1, \ldots, X_{p-1}$, i.e., to choose between the alternatives:

$$(7.34a) \qquad \begin{aligned} &H_0: \beta_1 = \beta_2 = \cdots = \beta_{p-1} = 0 \\ &H_a: \text{not all } \beta_k \ (k = 1, \ldots, p - 1) \text{ equal zero} \end{aligned}$$

we use the test statistic:

$$(7.34b) \qquad F^* = \frac{MSR}{MSE}$$

The decision rule to control the Type I error at α is:

$$(7.34c) \qquad \begin{aligned} &\text{If } F^* \leq F(1 - \alpha; p - 1, n - p), \text{ conclude } H_0 \\ &\text{If } F^* > F(1 - \alpha; p - 1, n - p), \text{ conclude } H_a \end{aligned}$$

The existence of a regression relation by itself does not of course assure that useful predictions can be made by using it.

Note that when $p - 1 = 1$, this test reduces to the F test in (3.61) for testing in simple linear regression whether or not $\beta_1 = 0$.

Coefficient of Multiple Determination

The coefficient of multiple determination, denoted by R^2, is defined as follows:

$$(7.35) \qquad R^2 = \frac{SSR}{SSTO} = 1 - \frac{SSE}{SSTO}$$

It measures the proportionate reduction of total variation in Y associated with the use of the set of X variables $X_1, \ldots, X_{p-1}$. The coefficient of multiple determination R^2 reduces to the coefficient of determination r^2 in (3.71) for simple linear regression when $p - 1 = 1$, i.e., when one independent variable is in regression model (7.18). Just as for r^2, we have:

$$(7.36) \qquad 0 \le R^2 \le 1$$

R^2 assumes the value 0 when all $b_k = 0$ $(k = 1, \ldots, p - 1)$. R^2 takes on the value 1 when all Y observations fall directly on the fitted response surface, i.e., when $Y_i = \hat{Y}_i$ for all i.

Comments

1. To distinguish between the coefficients of determination for simple and multiple regression, we shall from now on call r^2 the *coefficient of simple determination*.

2. It can be shown that the coefficient of multiple determination R^2 can be viewed as a coefficient of simple determination r^2 between the responses Y_i and the fitted values $\hat{Y}_i$.

3. A large R^2 does not necessarily imply that the fitted model is a useful one. For instance, observations may have been taken at only a few levels of the independent variables. Despite a high R^2 in this case, the fitted model may not be useful because most predictions would require extrapolations outside the region of observations. Again, even though R^2 is large, *MSE* may still be too large for inferences to be useful when high precision is required.

4. Adding more independent variables to the model can only increase R^2 and never reduce it, because *SSE* can never become larger with more indepenent variables and *SSTO* is always the same for a given set of responses. Since R^2 often can be made large by including a large number of independent variables, it is sometimes suggested that a modified measure be used that adjusts for the number of independent variables in the model. The *adjusted coefficient of multiple determination*, denoted by R_a^2, adjusts R^2 by dividing each sum of squares by its associated degrees of freedom; thus:

$$(7.37) \qquad R_a^2 = 1 - \frac{\dfrac{SSE}{n - p}}{\dfrac{SSTO}{n - 1}} = 1 - \left(\frac{n - 1}{n - p}\right)\frac{SSE}{SSTO}$$

This adjusted coefficient of multiple determination may actually become smaller when another independent variable is introduced into the model, because the decrease in *SSE* may be more than offset by the loss of a degree of freedom in the denominator $n - p$.

Coefficient of Multiple Correlation

The coefficient of multiple correlation R is the positive square root of R^2:

$$(7.38) \qquad\qquad R = \sqrt{R^2}$$

It equals in absolute value the correlation coefficient r in (3.73) for simple correlation when $p - 1 = 1$, i.e., when there is one independent variable in regression model (7.18).

Note

From now on, we shall call r the *coefficient of simple correlation* to distinguish it from the coefficient of multiple correlation.

7.6 INFERENCES ABOUT REGRESSION PARAMETERS

The least squares estimators in **b** are unbiased:

$$(7.39) \qquad\qquad \mathbf{E\{b\}} = \boldsymbol{\beta}$$

The variance-covariance matrix $\boldsymbol{\sigma}^2\{\mathbf{b}\}$:

$$(7.40) \qquad \underset{p\times p}{\boldsymbol{\sigma}^2\{\mathbf{b}\}} = \begin{bmatrix} \sigma^2\{b_0\} & \sigma\{b_0, b_1\} & \cdots & \sigma\{b_0, b_{p-1}\} \\ \sigma\{b_1, b_0\} & \sigma^2\{b_1\} & \cdots & \sigma\{b_1, b_{p-1}\} \\ \vdots & \vdots & & \vdots \\ \sigma\{b_{p-1}, b_0\} & \sigma\{b_{p-1}, b_1\} & \cdots & \sigma^2\{b_{p-1}\} \end{bmatrix}$$

is given by:

$$(7.41) \qquad\qquad \underset{p\times p}{\boldsymbol{\sigma}^2\{\mathbf{b}\}} = \sigma^2(\mathbf{X'X})^{-1}$$

The estimated variance-covariance matrix $\mathbf{s}^2\{\mathbf{b}\}$:

$$(7.42) \qquad \underset{p\times p}{\mathbf{s}^2\{\mathbf{b}\}} = \begin{bmatrix} s^2\{b_0\} & s\{b_0, b_1\} & \cdots & s\{b_0, b_{p-1}\} \\ s\{b_1, b_0\} & s^2\{b_1\} & \cdots & s\{b_1, b_{p-1}\} \\ \vdots & \vdots & & \vdots \\ s\{b_{p-1}, b_0\} & s\{b_{p-1}, b_1\} & \cdots & s^2\{b_{p-1}\} \end{bmatrix}$$

is given by:

$$(7.43) \qquad\qquad \underset{p\times p}{\mathbf{s}^2\{\mathbf{b}\}} = MSE\,(\mathbf{X'X})^{-1}$$

From $s^2\{\mathbf{b}\}$, one can obtain $s^2\{b_0\}$, $s^2\{b_1\}$ or whatever other variance is needed, or any needed covariances.

Interval Estimation of β_k

For the normal error regression model (7.18), we have:

$$(7.44) \qquad \frac{b_k - \beta_k}{s\{b_k\}} \sim t(n - p) \qquad k = 0, 1, \ldots, p - 1$$

Hence, the confidence limits for β_k with $1 - \alpha$ confidence coefficient are:

$$(7.45) \qquad b_k \pm t(1 - \alpha/2; n - p)s\{b_k\}$$

Tests for β_k

Tests for β_k are set up in the usual fashion. To test:

$$(7.46a) \qquad H_0: \beta_k = 0$$
$$H_a: \beta_k \neq 0$$

we may use the test statistic:

$$(7.46b) \qquad t^* = \frac{b_k}{s\{b_k\}}$$

and the decision rule:

$$(7.46c) \qquad \text{If } |t^*| \leq t(1 - \alpha/2; n - p), \text{ conclude } H_0$$
$$\text{Otherwise conclude } H_a$$

The power of the t test can be obtained as explained in Chapter 3, with the degrees of freedom modified to $n - p$.

As with simple linear regression, the test whether or not $\beta_k = 0$ in multiple regression models can also be conducted by means of an F test. We discuss this test in Chapter 8.

Joint Inferences

The Bonferroni joint confidence intervals can be used to estimate several regression coefficients simultaneously. If g parameters are to be estimated jointly (where $g \leq p$), the confidence limits with family confidence coefficient $1 - \alpha$ are:

$$(7.47) \qquad b_k \pm Bs\{b_k\}$$

where:

(7.47a) $B = t(1 - \alpha/2g; n - p)$

In Chapter 8, we discuss tests concerning subsets of the regression parameters.

7.7 INFERENCES ABOUT MEAN RESPONSE

Interval Estimation of $E\{Y_h\}$

For given values of $X_1, \ldots, X_{p-1}$, denoted by $X_{h1}, \ldots, X_{h,p-1}$, the mean response is denoted by $E\{Y_h\}$. We define the vector $\mathbf{X}_h$:

$$(7.48) \qquad \underset{p \times 1}{\mathbf{X}_h} = \begin{bmatrix} 1 \\ X_{h1} \\ X_{h2} \\ \cdot \\ \cdot \\ \cdot \\ X_{h,p-1} \end{bmatrix}$$

so that the mean response to be estimated is:

$$(7.49) \qquad E\{Y_h\} = \mathbf{X}'_h \boldsymbol{\beta}$$

The estimated mean response corresponding to $\mathbf{X}_h$, denoted by $\hat{Y}_h$, is:

$$(7.50) \qquad \hat{Y}_h = \mathbf{X}'_h \mathbf{b}$$

This estimator is unbiased:

$$(7.51) \qquad E\{\hat{Y}_h\} = \mathbf{X}'_h \boldsymbol{\beta} = E\{Y_h\}$$

and its variance is:

$$(7.52) \qquad \sigma^2\{\hat{Y}_h\} = \sigma^2 \mathbf{X}'_h(\mathbf{X}'\mathbf{X})^{-1}\mathbf{X}_h = \mathbf{X}'_h \sigma^2\{\mathbf{b}\}\mathbf{X}_h$$

Note that the variance $\sigma^2\{\hat{Y}_h\}$ is a function of the variances $\sigma^2\{b_k\}$ of the regression coefficients and of the covariances $\sigma\{b_k, b_{k'}\}$ between pairs of regression coefficients, just as in simple linear regression. The estimated variance $s^2\{\hat{Y}_h\}$ is given by:

$$(7.53) \qquad s^2\{\hat{Y}_h\} = MSE(\mathbf{X}'_h(\mathbf{X}'\mathbf{X})^{-1}\mathbf{X}_h) = \mathbf{X}'_h s^2\{\mathbf{b}\}\mathbf{X}_h$$

The $1 - \alpha$ confidence limits for $E\{Y_h\}$ are:

$$(7.54) \qquad \hat{Y}_h \pm t(1 - \alpha/2; n - p)s\{\hat{Y}_h\}$$

Simultaneous Confidence Intervals for Several Mean Responses

When it is desired to estimate a number of mean responses $E\{Y_h\}$ corresponding to different $\mathbf{X}_h$ vectors, one can employ two basic approaches to control the family confidence coefficient at $1 - \alpha$:

1. Use Working-Hotelling type confidence bounds for the several $\mathbf{X}_h$ vectors of interest:

(7.55)
$$\hat{Y}_h \pm Ws\{\hat{Y}_h\}$$

where:

(7.55a)
$$W^2 = pF(1 - \alpha; p, n - p)$$

2. Use Bonferroni simultaneous confidence intervals. When g interval estimates are to be made, the Bonferroni confidence limits are:

(7.56)
$$\hat{Y}_h \pm Bs\{\hat{Y}_h\}$$

where:

(7.56a)
$$B = t(1 - \alpha/2g; n - p)$$

For any particular application, one should compare W and B to see which procedure will lead to narrower confidence intervals. If the $\mathbf{X}_h$ levels are not specified in advance but are determined as the analysis proceeds, it is better to use the Working-Hotelling type limits (7.55).

F Test for Lack of Fit

To test whether the multiple regression response function:

$$E\{Y\} = \beta_0 + \beta_1 X_1 + \cdots + \beta_{p-1} X_{p-1}$$

is an appropriate response surface for a data set requires repeat observations, as for simple linear regression analysis. Repeat observations in multiple regression are replicate observations on Y corresponding to levels of each of the X variables that are constant from trial to trial. Thus, with two independent variables repeat observations require that X_1 and X_2 each remain at given levels from trial to trial.

The procedures described in Chapter 4 for the F test for lack of fit are applicable to multiple regression. Once the ANOVA table, shown in Table 7.1, has been obtained, SSE is decomposed into pure error and lack of fit components. The pure error sum of squares $SSPE$ is obtained by first calculating for each replicate group the sum of squared deviations of the Y observations around the group mean, where a replicate group has the same values for each of the X variables. Suppose there are c replicate groups with distinct sets of levels for the X variables, and let the mean of the Y observations for the jth group be denoted by $\bar{Y}_j$. Then the sum of squares for the jth group is given by (4.12), and the pure error sum of squares is the sum of these sums of squares, as given by (4.11). The lack of fit sum of squares $SSLF$ equals the difference $SSE - SSPE$, as indicated by (4.19).

The number of degrees of freedom associated with $SSPE$ is $n - c$, and the number of degrees of freedom associated with $SSLF$ is $(n - p) - (n - c) = c - p$.

The F test is conducted as described in Chapter 4, but with the degrees of freedom modified to those just stated. Thus, for testing the alternatives:

(7.57a)
$$H_0: E\{Y\} = \beta_0 + \beta_1 X_1 + \cdots + \beta_{p-1} X_{p-1}$$
$$H_a: E\{Y\} \neq \beta_0 + \beta_1 X_1 + \cdots + \beta_{p-1} X_{p-1}$$

the appropriate test statistic is:

(7.57b)
$$F^* = \frac{SSLF}{c-p} \div \frac{SSPE}{n-c} = \frac{MSLF}{MSPE}$$

where *SSLF* and *SSPE* are given by (4.19) and (4.11), respectively, and the appropriate decision rule is:

(7.57c)
$$\text{If } F^* \leq F(1-\alpha; c-p, n-c), \text{ conclude } H_0$$
$$\text{If } F^* > F(1-\alpha; c-p, n-c), \text{ conclude } H_a$$

7.8 PREDICTIONS OF NEW OBSERVATIONS

Prediction of New Observation $Y_{h(\text{new})}$

The prediction limits with $1 - \alpha$ confidence coefficient for a new observation $Y_{h(\text{new})}$ corresponding to $\mathbf{X}_h$, the specified values of the X variables, are:

(7.58)
$$\hat{Y}_h \pm t(1-\alpha/2; n-p) s\{Y_{h(\text{new})}\}$$

where:

(7.58a)
$$s^2\{Y_{h(\text{new})}\} = MSE + s^2\{\hat{Y}_h\} = MSE + \mathbf{X}_h' s^2\{\mathbf{b}\} \mathbf{X}_h$$
$$= MSE(1 + \mathbf{X}_h'(\mathbf{X}'\mathbf{X})^{-1} \mathbf{X}_h)$$

Prediction of Mean of *m* New Observations at X_h

When m new observations are to be selected at $\mathbf{X}_h$ and their mean $\bar{Y}_{h(\text{new})}$ is to be predicted, the $1 - \alpha$ prediction limits are:

(7.59)
$$\hat{Y}_h \pm t(1-\alpha/2; n-p) s\{\bar{Y}_{h(\text{new})}\}$$

where:

(7.59a)
$$s^2\{\bar{Y}_{h(\text{new})}\} = \frac{MSE}{m} + s^2\{\hat{Y}_h\} = \frac{MSE}{m} + \mathbf{X}_h' s^2\{\mathbf{b}\} \mathbf{X}_h$$
$$= MSE\left(\frac{1}{m} + \mathbf{X}_h'(\mathbf{X}'\mathbf{X})^{-1} \mathbf{X}_h\right)$$

Predictions of *g* New Observations

Simultaneous Scheffé prediction limits for g new observations at g different levels of $\mathbf{X}_h$ with family confidence coefficient $1 - \alpha$ are given by:

$$(7.60) \qquad\qquad \hat{Y}_h \pm Ss\{Y_{h(\text{new})}\}$$

where:

$$(7.60a) \qquad\qquad S^2 = gF(1 - \alpha; g, n - p)$$

and $s^2\{Y_{h(\text{new})}\}$ is given by (7.58a).

Alternatively, Bonferroni simultaneous prediction limits can be used. For g predictions with a $1 - \alpha$ family confidence coefficient, they are:

$$(7.61) \qquad\qquad \hat{Y}_h \pm Bs\{Y_{h(\text{new})}\}$$

where:

$$(7.61a) \qquad\qquad B = t(1 - \alpha/2g; n - p)$$

A comparison of S and B in advance of any particular use will indicate which procedure will lead to narrower prediction intervals.

7.9 RESIDUAL PLOTS, OTHER DIAGNOSTICS, AND REMEDIAL MEASURES

The diagnostic methods discussed in Chapter 4 for simple linear regression are also useful for multiple regression. Thus, box plots, time plots, stem-and-leaf plots, and dot plots for each of the independent variables can provide helpful, preliminary information about these variables. Similarly, scatter plots of the dependent variable against each of the independent variables can aid in determining the nature and strength of the relationship between the independent variable and the dependent variable and in identifying gaps in the data points as well as outlying data points. Scatter plots of each independent variable against each of the other independent variables are helpful for studying relationships between the independent variables and for finding gaps and detecting outliers.

A plot of the residuals against the fitted values is useful for assessing the appropriateness of the regression function and the constancy of the variance of the error terms, as well as for providing information about outliers. Similarly, a plot of the residuals against time can provide diagnostic information about possible correlations between the error terms. Box plots and normal probability plots of the residuals are useful for examining whether the error terms are reasonably normally distributed.

In addition, residuals should be plotted against each of the independent variables. Each of these plots can provide further information about the adequacy of the regression function with respect to that independent variable (e.g., whether a curvature effect is required for that variable) and about possible variation in the magnitude of the error variance in relation to that independent variable.

Finally, residuals should be plotted against important independent variables omitted from the model to see if the omitted variables have important additional effects on the dependent variable not yet recognized in the regression model. Also, residuals should be plotted against interaction terms not included in the regression model, such as $X_1 X_2$, $X_1 X_3$, and $X_2 X_3$, to see whether some or all of these interaction terms are required in the model.

The remedial measures described in Chapter 4 are also applicable for multiple regression. If a more complex model is required to recognize curvature or interaction effects, the multiple regression model can be expanded to include these effects. For example, X_2^2 might be added as a variable to take into account a curvature effect of X_2, or $X_1 X_3$ might be added as a variable to recognize an interaction effect between X_1 and X_3 on the dependent variable. Alternatively, transformations on the dependent and/or the independent variables can be made, following the principles discussed in Chapter 4, to remedy any model deficiencies. Transformations on the dependent variable Y may be helpful when the distributions of the error terms are quite skewed and the variance of the error terms is not constant. Transformations of some of the independent variables may be helpful when the effects of these independent variables are curvilinear. In addition, transformations on Y and/or the independent variables may be helpful in eliminating or substantially reducing interaction effects.

As with simple linear regression, the usefulness of transformations needs to be examined by means of residual plots and other diagnostic tools to determine whether the multiple regression model for the transformed data is appropriate.

Note

In Chapter 4, we described the Box-Cox approach for determining an appropriate power transformation on Y for simple regression models. This approach is also applicable to multiple regression models. In addition, Box and Tidwell (Ref. 7.1) have developed an iterative approach for ascertaining appropriate power transformations for each of the independent variables in a multiple regression model.

7.10 AN EXAMPLE—MULTIPLE REGRESSION WITH TWO INDEPENDENT VARIABLES

In this section, we shall develop a multiple regression application with two independent variables. We shall illustrate a number of different diagnostic procedures and several types of inferences that might be made for this application.

Setting

The Zarthan Company sells a special skin cream through fashion stores exclusively. It operates in 15 marketing districts and is interested in predicting district sales. Table 7.2 contains data on sales by district, as well as district data on target popula-

TABLE 7.2 Basic Data—Zarthan Company Example

District i	Sales (gross of jars; 1 gross = 12 dozen) Y_i	Target Population (thousands of persons) X_{i1}	Per Capita Discretionary Income (dollars) X_{i2}
1	162	274	2,450
2	120	180	3,254
3	223	375	3,802
4	131	205	2,838
5	67	86	2,347
6	169	265	3,782
7	81	98	3,008
8	192	330	2,450
9	116	195	2,137
10	55	53	2,560
11	252	430	4,020
12	232	372	4,427
13	144	236	2,660
14	103	157	2,088
15	212	370	2,605

tion and per capita discretionary income. Sales are to be treated as the dependent variable Y, and target population and per capita discretionary income as independent variables X_1 and X_2, respectively, in an exploration of the feasibility of predicting district sales from target population and per capita discretionary income. The first-order regression model:

$$(7.62) \qquad Y_i = \beta_0 + \beta_1 X_{i1} + \beta_2 X_{i2} + \varepsilon_i$$

with normal error terms is expected to be appropriate.

Basic Calculations

The **Y** and **X** matrices for the Zarthan Company illustration are shown in Table 7.3. We shall require:

1.

$$\mathbf{X'X} = \begin{bmatrix} 1 & 1 & \cdots & 1 \\ 274 & 180 & \cdots & 370 \\ 2{,}450 & 3{,}254 & \cdots & 2{,}605 \end{bmatrix} \begin{bmatrix} 1 & 274 & 2{,}450 \\ 1 & 180 & 3{,}254 \\ \vdots & \vdots & \vdots \\ 1 & 370 & 2{,}605 \end{bmatrix}$$

TABLE 7.3 Y and X Matrices—Zarthan Company Example

$$
Y = \begin{bmatrix} 162 \\ 120 \\ 223 \\ 131 \\ 67 \\ 169 \\ 81 \\ 192 \\ 116 \\ 55 \\ 252 \\ 232 \\ 144 \\ 103 \\ 212 \end{bmatrix}
\qquad
X = \begin{bmatrix} 1 & 274 & 2,450 \\ 1 & 180 & 3,254 \\ 1 & 375 & 3,802 \\ 1 & 205 & 2,838 \\ 1 & 86 & 2,347 \\ 1 & 265 & 3,782 \\ 1 & 98 & 3,008 \\ 1 & 330 & 2,450 \\ 1 & 195 & 2,137 \\ 1 & 53 & 2,560 \\ 1 & 430 & 4,020 \\ 1 & 372 & 4,427 \\ 1 & 236 & 2,660 \\ 1 & 157 & 2,088 \\ 1 & 370 & 2,605 \end{bmatrix}
$$

which yields:

(7.63)
$$
X'X = \begin{bmatrix} 15 & 3,626 & 44,428 \\ 3,626 & 1,067,614 & 11,419,181 \\ 44,428 & 11,419,181 & 139,063,428 \end{bmatrix}
$$

2.

$$
X'Y = \begin{bmatrix} 1 & 1 & \cdots & 1 \\ 274 & 180 & \cdots & 370 \\ 2,450 & 3,254 & \cdots & 2,605 \end{bmatrix} \begin{bmatrix} 162 \\ 120 \\ \vdots \\ 212 \end{bmatrix}
$$

which yields:

(7.64)
$$
X'Y = \begin{bmatrix} 2,259 \\ 647,107 \\ 7,096,619 \end{bmatrix}
$$

3.

$$
(X'X)^{-1} = \begin{bmatrix} 15 & 3,626 & 44,428 \\ 3,626 & 1,067,614 & 11,419,181 \\ 44,428 & 11,419,181 & 139,063,428 \end{bmatrix}^{-1}
$$

Using (6.23), we define:

$$a = 15 \qquad b = 3,626 \qquad c = 44,428$$

$$d = 3,626 \qquad e = 1,067,614 \qquad f = 11,419,181$$

$$g = 44,428 \qquad h = 11,419,181 \qquad k = 139,063,428$$

so that:

$$Z = 14,497,044,060,000$$

$$A = 1.246348416$$

$$B = .0002129664176$$

and so on. We obtain:

(7.65) $(\mathbf{X'X})^{-1} =$

$$\begin{bmatrix} 1.2463484 & 2.1296642E-4 & -4.1567125E-4 \\ 2.1296642E-4 & 7.7329030E-6 & -7.0302518E-7 \\ -4.1567125E-4 & -7.0302518E-7 & 1.9771851E-7 \end{bmatrix}$$

Note that some of the results in the $(\mathbf{X'X})^{-1}$ matrix are given in the E format, where, say, $E-4$ stands for $10^{-4} = 1/10^4$. Thus, $2.1296642E-4$ stands for .00021296642.

Algebraic Equivalents. Note that $\mathbf{X'X}$ for the first-order regression model (7.62) with two independent variables is:

$$\mathbf{X'X} = \begin{bmatrix} 1 & 1 & \cdots & 1 \\ X_{11} & X_{21} & \cdots & X_{n1} \\ X_{12} & X_{22} & \cdots & X_{n2} \end{bmatrix} \begin{bmatrix} 1 & X_{11} & X_{12} \\ 1 & X_{21} & X_{22} \\ \vdots & \vdots & \vdots \\ 1 & X_{n1} & X_{n2} \end{bmatrix}$$

or:

(7.66) $$\mathbf{X'X} = \begin{bmatrix} n & \Sigma X_{i1} & \Sigma X_{i2} \\ \Sigma X_{i1} & \Sigma X_{i1}^2 & \Sigma X_{i1}X_{i2} \\ \Sigma X_{i2} & \Sigma X_{i2}X_{i1} & \Sigma X_{i2}^2 \end{bmatrix}$$

Thus, for the Zarthan Company example:

$$n = 15$$

$$\sum X_{i1} = 274 + 180 + \cdots = 3,626$$

$$\sum X_{i1}X_{i2} = 274(2,450) + 180(3,254) + \cdots = 11,419,181$$

etc.

These elements are found in (7.63).

Also note that $\mathbf{X'Y}$ for the first-order regression model (7.62) with two independent variables is:

(7.67) $$\mathbf{X'Y} = \begin{bmatrix} 1 & 1 & \cdots & 1 \\ X_{11} & X_{21} & \cdots & X_{n1} \\ X_{12} & X_{22} & \cdots & X_{n2} \end{bmatrix} \begin{bmatrix} Y_1 \\ Y_2 \\ \vdots \\ Y_n \end{bmatrix} = \begin{bmatrix} \Sigma Y_i \\ \Sigma X_{i1}Y_i \\ \Sigma X_{i2}Y_i \end{bmatrix}$$

For the Zarthan Company example, we have:

$$\sum Y_i = 162 + 120 + \cdots = 2{,}259$$

$$\sum X_{i1} Y_i = 274(162) + 180(120) + \cdots = 647{,}107$$

$$\sum X_{i2} Y_i = 2{,}450(162) + 3{,}254(120) + \cdots = 7{,}096{,}619$$

These are the elements found in (7.64).

Estimated Regression Function

The least squares estimates **b** are readily obtained by (7.21), given our basic calculations in (7.64) and (7.65):

$$\mathbf{b} = (\mathbf{X}'\mathbf{X})^{-1}\mathbf{X}'\mathbf{Y}$$

$$= \begin{bmatrix} 1.2463484 & 2.1296642E-4 & -4.1567125E-4 \\ 2.1296642E-4 & 7.7329030E-6 & -7.0302518E-7 \\ -4.1567125E-4 & -7.0302518E-7 & 1.9771851E-7 \end{bmatrix}$$

$$\times \begin{bmatrix} 2{,}259 \\ 647{,}107 \\ 7{,}096{,}619 \end{bmatrix}$$

$$= \begin{bmatrix} 3.4526127900 \\ .4960049761 \\ .009199080867 \end{bmatrix}$$

Thus:

$$\begin{bmatrix} b_0 \\ b_1 \\ b_2 \end{bmatrix} = \begin{bmatrix} 3.4526127900 \\ .4960049761 \\ .009199080867 \end{bmatrix}$$

and the estimated regression function is:

$$\hat{Y} = 3.453 + .496X_1 + .00920X_2$$

This estimated regression function indicates that mean jar sales are expected to increase by .496 gross when the target population increases by 1 thousand, holding per capita discretionary income constant, and that mean jar sales are expected to increase by .0092 gross when per capita discretionary income increases by one dollar, holding population constant.

Algebraic Version of Normal Equations.   The normal equations in algebraic form for the case of two independent variables can be obtained readily from (7.66) and

(7.67). We have:

$$(\mathbf{X'X})\mathbf{b} = \mathbf{X'Y}$$

$$\begin{bmatrix} n & \Sigma X_{i1} & \Sigma X_{i2} \\ \Sigma X_{i1} & \Sigma X_{i1}^2 & \Sigma X_{i1} X_{i2} \\ \Sigma X_{i2} & \Sigma X_{i2} X_{i1} & \Sigma X_{i2}^2 \end{bmatrix} \begin{bmatrix} b_0 \\ b_1 \\ b_2 \end{bmatrix} = \begin{bmatrix} \Sigma Y_i \\ \Sigma X_{i1} Y_i \\ \Sigma X_{i2} Y_i \end{bmatrix}$$

from which we obtain the normal equations:

$$\sum Y_i = nb_0 \qquad\quad + b_1 \sum X_{i1} \quad + b_2 \sum X_{i2}$$

(7.68) $$\sum X_{i1} Y_i = b_0 \sum X_{i1} + b_1 \sum X_{i1}^2 \quad + b_2 \sum X_{i1} X_{i2}$$

$$\sum X_{i2} Y_i = b_0 \sum X_{i2} + b_1 \sum X_{i1} X_{i2} + b_2 \sum X_{i2}^2$$

Fitted Values and Residuals

To examine the aptness of regression model (7.62) for the data at hand, we require the fitted values $\hat{Y}_i$ and the residuals $e_i = Y_i - \hat{Y}_i$. We obtain by (7.23):

$$\hat{\mathbf{Y}} = \mathbf{Xb}$$

$$\begin{bmatrix} \hat{Y}_1 \\ \hat{Y}_2 \\ \vdots \\ \hat{Y}_{15} \end{bmatrix} = \begin{bmatrix} 1 & 274 & 2{,}450 \\ 1 & 180 & 3{,}254 \\ \vdots & \vdots & \vdots \\ 1 & 370 & 2{,}605 \end{bmatrix} \begin{bmatrix} 3.4526127900 \\ .4960049761 \\ .009199080867 \end{bmatrix} = \begin{bmatrix} 161.896 \\ 122.667 \\ \vdots \\ 210.938 \end{bmatrix}$$

Further, by (7.24) we find:

$$\mathbf{e} = \mathbf{Y} - \hat{\mathbf{Y}}$$

$$\begin{bmatrix} e_1 \\ e_2 \\ \vdots \\ e_{15} \end{bmatrix} = \begin{bmatrix} 162 \\ 120 \\ \vdots \\ 212 \end{bmatrix} - \begin{bmatrix} 161.896 \\ 122.667 \\ \vdots \\ 210.938 \end{bmatrix} = \begin{bmatrix} .104 \\ -2.667 \\ \vdots \\ 1.062 \end{bmatrix}$$

Analysis of Aptness of Model

We begin our analysis of the appropriateness of regression model (7.62) for the Zarthan Company data by considering the plot of the residuals e against the fitted values $\hat{Y}$ in Figure 7.6a. This plot does not suggest any systematic deviations from the response plane, nor that the variance of the error terms varies with the level of $\hat{Y}$. Plots of the residuals e against X_1 and X_2 in Figures 7.6b and 7.6c, respectively,

FIGURE 7.6   Diagnostic Residual Plots—Zarthan Company Example

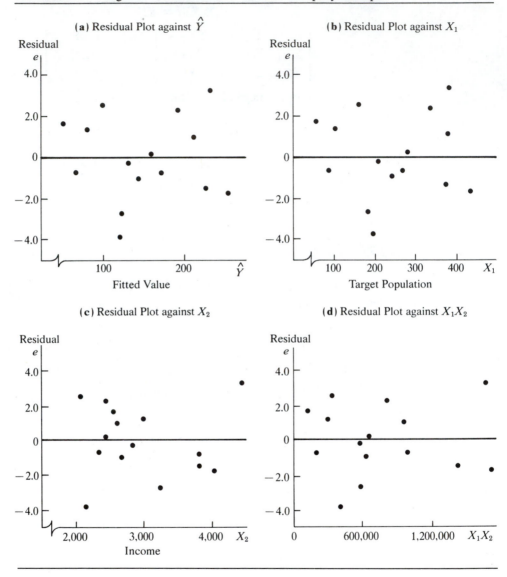

are entirely consistent with the conclusions of good fit by the response function and constant variance of the error terms.

In multiple regression applications, there is frequently the possibility of interaction effects being present. To examine this for the Zarthan Company example, we plotted the residuals e against the interaction term $X_1 X_2$ in Figure 7.6d. A systematic pattern

FIGURE 7.7   Normal Probability Plot—Zarthan Company Example

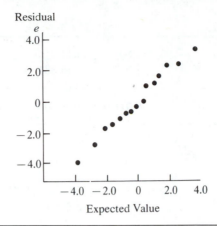

of this plot would suggest that an interaction effect may be present, so a response function of the type:

$$E\{Y\} = \beta_0 + \beta_1 X_1 + \beta_2 X_2 + \beta_3 X_1 X_2$$

might be more appropriate. Figure 7.6d does not exhibit any systematic pattern; hence no interaction effects reflected by the model term $\beta_3 X_1 X_2$ appear to be present.

Finally, Figure 7.7 contains a normal probability plot of the residuals. The pattern is reasonably linear, consistent with a normal distribution of the error terms. The coefficient of correlation between the ordered residuals and their expected values under normality is .993. This high value (see Table 4.3) helps to confirm the reasonableness of the conclusion that the error terms are fairly normally distributed.

Since the Zarthan Company data are cross-sectional and do not involve a time sequence, time plots are not relevant here. Thus, all of the diagnostics support the use of regression model (7.62) for the Zarthan Company data.

Analysis of Variance

To test whether sales are related to population and per capita discretionary income, we construct the ANOVA table in Table 7.4. The basic quantities needed are:

$$\mathbf{Y'Y} = \begin{bmatrix} 162 & 120 & \cdots & 212 \end{bmatrix} \begin{bmatrix} 162 \\ 120 \\ \cdot \\ \cdot \\ 212 \end{bmatrix}$$

TABLE 7.4   ANOVA Table—Zarthan Company Example

Source of Variation	SS	df	MS
Regression	$SSR = 53,844.716$	2	$MSR = 26,922.358$
Error	$SSE = 56.884$	12	$MSE = 4.740$
Total	$SSTO = 53,901.600$	14	

$$= (162)^2 + (120)^2 + \cdots + (212)^2$$

$$= 394,107.000$$

$$\left(\frac{1}{n}\right)\mathbf{Y'JY} = \frac{1}{15}[162 \quad 120 \quad \cdots \quad 212]\begin{bmatrix} 1 & 1 & \cdots & 1 \\ 1 & 1 & \cdots & 1 \\ \vdots & \vdots & & \vdots \\ 1 & 1 & \cdots & 1 \end{bmatrix}\begin{bmatrix} 162 \\ 120 \\ \vdots \\ 212 \end{bmatrix}$$

$$= \frac{(2,259)^2}{15} = 340,205.400$$

Thus:

$$SSTO = \mathbf{Y'Y} - \left(\frac{1}{n}\right)\mathbf{Y'JY} = 394,107.000 - 340,205.400 = 53,901.600$$

and using our result in (7.64):

$$SSE = \mathbf{Y'Y} - \mathbf{b'X'Y}$$

$$= 394,107.000 - [3.4526127900 \quad .4960049761 \quad .009199080867]$$

$$\times \begin{bmatrix} 2,259 \\ 647,107 \\ 7,096,619 \end{bmatrix}$$

$$= 394,107.000 - 394,050.116 = 56.884$$

Finally, we obtain by subtraction:

$$SSR = SSTO - SSE = 53,901.600 - 56.884 = 53,844.716$$

The degrees of freedom and mean squares are entered in Table 7.4. Note that three regression parameters had to be estimated; hence, $15 - 3 = 12$ degrees of freedom are associated with SSE. Also, the number of degrees of freedom associated with SSR is 2—the number of X variables in the model.

Test of Regression Relation. To test whether sales are related to population and per capita discretionary income:

$$H_0: \beta_1 = 0 \text{ and } \beta_2 = 0$$

$$H_a: \text{not both } \beta_1 \text{ and } \beta_2 \text{ equal zero}$$

we use test statistic (7.34b):

$$F^* = \frac{MSR}{MSE} = \frac{26{,}922.358}{4.740} = 5{,}680$$

Assuming α is to be .05, we require $F(.95; 2, 12) = 3.89$. Since $F^* = 5{,}680 > 3.89$, we conclude H_a, that sales are related to population and per capita discretionary income. The P-value for this test is less than .001 since we note from Table A.4 that $F(.999; 2, 12) = 13.0$.

Whether the regression relation is useful for making predictions of sales or estimates of mean sales still remains to be seen.

Coefficient of Multiple Determination. For our example, we have by (7.35):

$$R^2 = \frac{SSR}{SSTO} = \frac{53{,}844.716}{53{,}901.600} = .9989$$

Thus, when the two independent variables population and per capita discretionary income are considered, the variation in sales is reduced by 99.9 percent.

Algebraic Expression for *SSE*. The error sum of squares for the case of two independent variables in algebraic terms is:

$$SSE = \mathbf{Y'Y} - \mathbf{b'X'Y} = \sum Y_i^2 - [b_0 \quad b_1 \quad b_2] \begin{bmatrix} \Sigma Y_i \\ \Sigma X_{i1} Y_i \\ \Sigma X_{i2} Y_i \end{bmatrix}$$

or:

$$(7.69) \qquad SSE = \sum Y_i^2 - b_0 \sum Y_i - b_1 \sum X_{i1} Y_i - b_2 \sum X_{i2} Y_i$$

Note how this expression is a straightforward extension of (2.24a) for the case of one independent variable.

Estimation of Regression Parameters

The Zarthan Company is not interested in the parameter β_0 since it falls far outside the scope of the model. It is desired to estimate β_1 and β_2 jointly with family confidence coefficient .90. We shall use the simultaneous Bonferroni confidence limits in (7.47).

First, we need the estimated variance-covariance matrix $s^2\{\mathbf{b}\}$:

$$s^2\{\mathbf{b}\} = MSE(\mathbf{X}'\mathbf{X})^{-1}$$

MSE is given in Table 7.4, and $(\mathbf{X}'\mathbf{X})^{-1}$ was obtained in (7.65). Hence:

(7.70)

$$s^2\{\mathbf{b}\} = 4.7403 \begin{bmatrix} 1.2463484 & 2.1296642\mathrm{E} - 4 & -4.1567125\mathrm{E} - 4 \\ 2.1296642\mathrm{E} - 4 & 7.7329030\mathrm{E} - 6 & -7.0302518\mathrm{E} - 7 \\ -4.1567125\mathrm{E} - 4 & -7.0302518\mathrm{E} - 7 & 1.9771851\mathrm{E} - 7 \end{bmatrix}$$

$$= \begin{bmatrix} 5.9081 & 1.0095\mathrm{E} - 3 & -1.9704\mathrm{E} - 3 \\ 1.0095\mathrm{E} - 3 & 3.6656\mathrm{E} - 5 & -3.3326\mathrm{E} - 6 \\ -1.9704\mathrm{E} - 3 & -3.3326\mathrm{E} - 6 & 9.3725\mathrm{E} - 7 \end{bmatrix}$$

The two estimated variances we require are:

$$s^2\{b_1\} = .000036656 \quad \text{or} \quad s\{b_1\} = .006054$$
$$s^2\{b_2\} = .00000093725 \quad \text{or} \quad s\{b_2\} = .0009681$$

Next, we require for $g = 2$ simultaneous estimates:

$$B = t[1 - .10/2(2); 12] = t(.975; 12) = 2.179$$

Thus the two simultaneous confidence limits are $.4960 \pm 2.179(.006054)$ and $.009199 \pm 2.179(.0009681)$, which yield the confidence intervals:

$$.483 \leq \beta_1 \leq .509$$
$$.0071 \leq \beta_2 \leq .0113$$

With family confidence coefficient .90, we conclude that β_1 falls between .483 and .509 and that β_2 falls between .0071 and .0113.

Note that the simultaneous confidence intervals suggest that both β_1 and β_2 are positive, which is in accord with theoretical expectations that sales should increase with higher target population and higher per capita discretionary income, the other variable being held constant.

Estimation of Mean Response

The Zarthan Company would like to estimate expected (mean) sales in a district with target population $X_{h1} = 220$ thousand persons and per capita discretionary income $X_{h2} = \$2,500$. We define:

$$\mathbf{X}_h = \begin{bmatrix} 1 \\ 220 \\ 2,500 \end{bmatrix}$$

The point estimate of mean sales is by (7.50):

$$\hat{Y}_h = \mathbf{X}'_h\mathbf{b} = [1 \quad 220 \quad 2{,}500] \begin{bmatrix} 3.4526 \\ .4960 \\ .009199 \end{bmatrix} = 135.57$$

The estimated variance by (7.53) and using the results in (7.70) is:

$$s^2\{\hat{Y}_h\} = \mathbf{X}'_h s^2\{\mathbf{b}\}\mathbf{X}_h$$

$$= [1 \quad 220 \quad 2{,}500]$$

$$\times \begin{bmatrix} 5.9081 & 1.0095E-3 & -1.9704E-3 \\ 1.0095E-3 & 3.6656E-5 & -3.3326E-6 \\ -1.9704E-3 & -3.3326E-6 & 9.3725E-7 \end{bmatrix} \begin{bmatrix} 1 \\ 220 \\ 2{,}500 \end{bmatrix}$$

$$= .46638$$

or:

$$s\{\hat{Y}_h\} = .68292$$

Assume that the confidence coefficient for the interval estimate of $E\{Y_h\}$ is to be .95. We then need $t(.975; 12) = 2.179$, and obtain by (7.54) the confidence limits $135.57 \pm 2.179(.68292)$. The confidence interval for $E\{Y_h\}$ therefore is:

$$134.1 \le E\{Y_h\} \le 137.1$$

Thus, with confidence coefficient .95, we estimate that mean jar sales in a district with target population of 220 thousand and per capita discretionary income of \$2,500 are somewhere between 134.1 and 137.1 gross.

Algebraic Version of Estimated Variance $s^2\{\hat{Y}_h\}$. Since by (7.53):

$$s^2\{\hat{Y}_h\} = \mathbf{X}'_h s^2\{\mathbf{b}\}\mathbf{X}_h$$

it follows for the case of two independent variables:

$$(7.71) \quad s^2\{\hat{Y}_h\} = s^2\{b_0\} + X_{h1}^2 s^2\{b_1\} + X_{h2}^2 s^2\{b_2\} + 2X_{h1}s\{b_0, b_1\}$$
$$+ 2X_{h2}s\{b_0, b_2\} + 2X_{h1}X_{h2}s\{b_1, b_2\}$$

When we substitute in (7.71), utilizing the estimated variances and covariances from (7.70), we obtain the same result as before, namely, $s^2\{\hat{Y}_h\} = .46638$.

Prediction Limits for New Observations

The Zarthan Company would like to predict sales in two districts with the following characteristics:

	District A	District B
X_{h1}	220	375
X_{h2}	2,500	3,500

To determine which simultaneous prediction intervals are best here, we shall find S as given in (7.60a) and B as given in (7.61a) for $g = 2$, assuming the family confidence coefficient is to be .90:

$$S^2 = 2F(.90; 2, 12) = 2(2.81) = 5.62 \qquad S = 2.37$$

and:

$$B = t[1 - .10/2(2); 12] = t(.975; 12) = 2.179$$

Hence, the Bonferroni limits are more efficient here.

For district A, we shall use the results we found when estimating mean sales, since the levels of the independent variables are the same here. We have from earlier:

$$\hat{Y}_h = 135.57 \qquad s^2\{\hat{Y}_h\} = .46638 \qquad MSE = 4.7403$$

Hence, by (7.58a):

$$s^2\{Y_{h(\text{new})}\} = MSE + s^2\{\hat{Y}_h\} = 4.7403 + .46638 = 5.20668$$

or:

$$s\{Y_{h(\text{new})}\} = 2.28182$$

In similar fashion, we obtain for district B (calculations not shown):

$$\hat{Y}_h = 221.65 \qquad s\{Y_{h(\text{new})}\} = 2.34536$$

We found before that the Bonferroni multiple is $B = 2.179$. Hence, by (7.61) the simultaneous Bonferroni prediction limits with family confidence coefficient .90 are $135.57 \pm 2.179(2.28182)$ and $221.65 \pm 2.179(2.34536)$, leading to the simultaneous prediction intervals:

$$\text{District A:} \quad 130.6 \leq Y_{h(\text{new})} \leq 140.5$$

$$\text{District B:} \quad 216.5 \leq Y_{h(\text{new})} \leq 226.8$$

With family confidence coefficient .90, we predict that sales in the two districts will be within the indicated limits. The Zarthan Company considers these prediction limits sufficiently precise, and hence useful.

Computer Printout

Figure 7.8 contains an illustrative computer printout for the Zarthan Company example, obtained by using the GLM (general linear model) program of the SAS (Statistical Analysis System) computer package (Ref. 7.2). Regression analysis printouts differ in format from one computer program to another, as may be seen by comparing the output in Figure 7.8 with other outputs presented in earlier chapters. However, the basic information presented in the different outputs is essentially the same for the major statistical regression packages.

FIGURE 7.8 Computer Printout for Zarthan Company Example (SAS, Ref. 7.2)

THE X'X MATRIX

	INTERCEPT	TARGTP	INCOME
INTERCEPT ← X_0	15.00	3626.00	44428.00
TARGTP ← X_1	3626.00	1067614.00	11419181.00
INCOME ← X_2	44428.00	11419181.00	139063428.00

X'X INVERSE MATRIX

	INTERCEPT	TARGTP	INCOME
INTERCEPT	1.24634842	0.00021297	-0.00041567
TARGTP	0.00021297	0.00000773	-0.00000070
INCOME	-0.00041567	-0.00000070	0.00000020

PARAMETER	ESTIMATE	T FOR H0: PARAMETER=0	PR > |T|	STD ERROR OF ESTIMATE
INTERCEPT	3.45261279	1.42	0.1809	2.43065049
TARGTP	0.49600498	81.92	0.0001	0.00605444
INCOME	0.00919908	9.50	0.0001	0.00096811

b_k $t_k^* = b_k/s\{b_k\}$ Two-sided P-value $s\{b_k\}$

SOURCE	DF	SUM OF SQUARES	MEAN SQUARE	F VALUE
MODEL	2	SSR → 53844.71643444	MSR → 26922.35821722	F^* → 5679.47
ERROR	12	SSE → 56.88356556	MSE → 4.74029713	
CORRECTED TOTAL	14	53901.60000000		

SSTO

One-sided P-value

PR > F	R-SQUARE
→ 0.0001	0.998945 ← R^2

STD DEV

2.17722234 ← $\sqrt{MSE}$

OBSERVATION	Y_i OBSERVED VALUE	$\hat{Y}_i$ PREDICTED VALUE	e_i RESIDUAL
1	162.00000000	161.89572437	0.10427563
2	120.00000000	122.66731763	-2.66731763
3	223.00000000	224.42938429	-1.42938429
4	131.00000000	131.24062439	-0.24062439
5	67.00000000	67.69928353	-0.69928353
6	169.00000000	169.68485530	-0.68485530
7	81.00000000	79.73193570	1.26806430
8	192.00000000	189.67200303	2.32799697
9	116.00000000	119.83201895	-3.83201895
10	55.00000000	53.29052354	1.70947646
11	252.00000000	253.71505760	-1.71505760
12	232.00000000	228.69079490	3.30920510
13	144.00000000	144.97934226	-0.97934226
14	103.00000000	100.53307489	2.46692511
15	212.00000000	210.93805961	1.06194039

We have annotated the output in Figure 7.8 to tie in with the notation of this book. The first two blocks of information contain intermediate regression analysis results in matrix form, specifically the $\mathbf{X'X}$ and $(\mathbf{X'X})^{-1}$ matrices. The label "intercept" in these matrices refers to $X_{i0} \equiv 1$ in the alternative regression model (7.7b).

The next block presents information about the estimated regression coefficients b_k. Shown, in turn, are the estimates b_k, the test statistics $t_k^* = b_k/s\{b_k\}$ for testing whether or not $\beta_k = 0$, the two-sided P-values for the test statistics, and the estimated standard deviations $s\{b_k\}$.

The fourth block contains ANOVA information: the ANOVA table, the F^* value for the test of whether or not a regression relation exists, the P-value for this test, $\sqrt{MSE}$, and R^2.

The final block shows the observed values Y_i, the fitted values $\hat{Y}_i$, and the residuals e_i.

Because of rounding, some results in Figure 7.8 do not coincide precisely with the corresponding results given earlier. In this connection, it should be noted that different computer regression packages may lead to somewhat different results because final results are rounded to different extents, and even more importantly because rounding errors are not handled equally well by all packages. Particularly when there are a number of independent variables, some of which are highly correlated, rounding errors can be a serious source of difficulty. It is a wise policy to investigate a computer regression package before using it, for instance, by comparing its output for a test problem against results known to be accurate.

Caution about Hidden Extrapolations

Before concluding this illustration of multiple regression analysis, we should caution again about making estimates or predictions outside the scope of the model. The danger, of course, is that the model may not be appropriate when extended outside the region of the observations. In multiple regression, it is particularly easy to lose track of this region since the levels of $X_1, \ldots, X_{p-1}$ *jointly* define the region. Thus, one cannot merely look at the ranges of each independent variable. Consider Figure 7.9, where the shaded region is the region of observations for a multiple regression application with two independent variables. The circled dot is within the ranges of the independent variables X_1 and X_2 individually, yet is well outside the joint region of observations.

CITED REFERENCES

7.1. Box, G. E. P., and P. W. Tidwell. "Transformations of the Independent Variables," *Technometrics* 4 (1962), pp. 531–50.

7.2. *SAS User's Guide: Statistics*. Version 6 edition. Cary, N.C.: SAS Institute, 1987.

FIGURE 7.9   Region of Observations on X_1 and X_2 Jointly, Compared with Ranges of X_1 and X_2 Individually

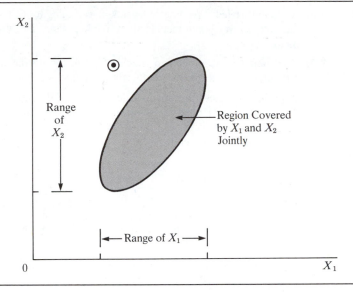

PROBLEMS

7.1. Refer to Figure 7.4a. By how much approximately does mean yield increase when rainfall increases from 9 to 11 inches and temperature is held constant? Could you have answered this question if rainfall and temperature interact in their effects on crop yield?

7.2. Consider the response function $E\{Y\} = 25 + 3X_1 + 4X_2 + 1.5X_1X_2$.
   a. Plot the response function against X_1 when $X_2 = 3$ and when $X_2 = 6$. How is the interaction effect of X_1 and X_2 on Y apparent from this graph?
   b. Sketch a set of contour curves for the response surface. How is the interaction effect of X_1 and X_2 on Y apparent from this graph?

7.3. Consider the response function $E\{Y\} = 14 + 7X_1 - 5X_2$.
   a. Plot the response function against X_2 when $X_1 = 1$ and when $X_1 = 4$. How does the graph indicate that the effects of X_1 and X_2 on Y are additive?
   b. Sketch a set of contour curves for the response surface. How does the graph indicate that the effects of X_1 and X_2 on Y are additive?

7.4. Set up the **X** matrix and **β** vector for each of the following regression models (assume $i = 1, \ldots, 4$):

   a. $Y_i = \beta_0 + \beta_1 X_{i1} + \beta_2 X_{i1} X_{i2} + \varepsilon_i$

   b. $\log Y_i = \beta_0 + \beta_1 X_{i1} + \beta_2 X_{i2} + \varepsilon_i$

7.5. Set up the **X** matrix and **β** vector for each of the following regression models (assume $i = 1, \ldots, 5$):

a. $Y_i = \beta_1 X_{i1} + \beta_2 X_{i2} + \beta_3 X_{i1}^2 + \varepsilon_i$

b. $\sqrt{Y_i} = \beta_0 + \beta_1 X_{i1} + \beta_2 \log_{10} X_{i2} + \varepsilon_i$

7.6. A student stated: "Adding independent variables to a regression model can never reduce R^2, so we should include all available independent variables in the model." Comment.

7.7. Why is it not meaningful to attach a sign to the coefficient of multiple correlation R, although we do so for the coefficient of simple correlation r?

7.8. **Brand preference.** In a small-scale experimental study of the relation between degree of brand liking (Y) and moisture content (X_1) and sweetness (X_2) of the product, the following results were obtained from the experiment based on a completely randomized design (data are coded):

i:	1	2	3	4	5	6	7	8	9	10	11	12	13	14	15	16
X_{i1}:	4	4	4	4	6	6	6	6	8	8	8	8	10	10	10	10
X_{i2}:	2	4	2	4	2	4	2	4	2	4	2	4	2	4	2	4
Y_i:	64	73	61	76	72	80	71	83	83	89	86	93	88	95	94	100

a. Fit regression model (7.1) to the data. State the estimated regression function. How is b_1 interpreted here?

b. Obtain the residuals and prepare a box plot of the residuals. What information does this plot provide?

c. Plot the residuals against $\hat{Y}$, X_1, X_2, and $X_1 X_2$ on separate graphs. Also prepare a normal probability plot. Analyze the plots and summarize your findings.

d. Conduct a formal test for lack of fit of the first-order regression function; use $\alpha = .01$. State the alternatives, decision rule, and conclusion.

7.9. Refer to **Brand preference** Problem 7.8. Assume that regression model (7.1) with independent normal error terms is appropriate.

a. Test whether there is a regression relation using a level of significance of .01. State the alternatives, decision rule, and conclusion. What does your test imply about β_1 and β_2?

b. What is the P-value of the test in part (a)?

c. Estimate β_1 and β_2 jointly by the Bonferroni procedure using a 99 percent family confidence coefficient. Interpret your results.

7.10. Refer to **Brand preference** Problem 7.8.

a. Calculate the coefficient of multiple determination R^2. How is it interpreted here?

b. Calculate the coefficient of simple determination r^2 between Y_i and $\hat{Y}_i$. Does it equal R^2?

7.11. Refer to **Brand preference** Problem 7.8. Assume that regression model (7.1) with independent normal error terms is appropriate.

a. Obtain an interval estimate of $E\{Y_h\}$ when $X_{h1} = 5$ and $X_{h2} = 4$. Use a 99 percent confidence coefficient. Interpret your interval estimate.

b. Obtain a prediction interval for a new observation $Y_{h(new)}$ when $X_{h1} = 5$ and $X_{h2} = 4$. Use a 99 percent confidence coefficient.

7.12. **Chemical shipment.** The data to follow, taken on 20 incoming shipments of chemicals in drums arriving at a warehouse, show number of drums in shipment (X_1), total weight of shipment (X_2, in hundred pounds), and number of minutes required to handle shipment (Y).

i:	1	2	3	4	5	6	7	8	9	10
X_{i1}:	7	18	5	14	11	5	23	9	16	5
X_{i2}:	5.11	16.72	3.20	7.03	10.98	4.04	22.07	7.03	10.62	4.76
Y_i:	58	152	41	93	101	38	203	78	117	44

i:	11	12	13	14	15	16	17	18	19	20
X_{i1}:	17	12	6	12	8	15	17	21	6	11
X_{i2}:	11.02	9.51	3.79	6.45	4.60	13.86	13.03	15.21	3.64	9.57
Y_i:	121	112	50	82	48	127	140	155	39	90

a. Prepare separate stem-and-leaf plots for the numbers of drums in the shipments X_{i1} and the weights of the shipments X_{i2}. Are there any outlying cases present? Are there any gaps in the data?

b. The cases are given in the order that the incoming shipments arrived. Prepare a time plot for each of the independent variables. What do the plots show?

c. Fit regression model (7.1) to the data. State the estimated regression function. How are b_1 and b_2 interpreted here?

d. Obtain the residuals and prepare a box plot of the residuals. What information does this plot provide?

e. Plot the residuals against $\hat{Y}$, X_1, X_2, and $X_1 X_2$ on separate graphs. Also prepare a normal probability plot. Analyze the plots and summarize your findings.

f. Prepare a time plot of the residuals. Is there any indication that the error terms are correlated? Discuss.

7.13. Refer to **Chemical shipment** Problem 7.12. Assume that regression model (7.1) with independent normal error terms is appropriate.

a. Test whether there is a regression relation, using a level of significance of .05. State the alternatives, decision rule, and conclusion. What does your test result imply about β_1 and β_2? What is the P-value of the test?

b. Estimate β_1 and β_2 jointly by the Bonferroni procedure using a 95 percent family confidence coefficient. Interpret your results.

c. Calculate the coefficient of multiple determination R^2. How is this measure interpreted here?

7.14. Refer to **Chemical shipment** Problem 7.12. Assume that regression model (7.1) with independent normal error terms is appropriate.

a. Management desires simultaneous interval estimates of the mean handling times for five typical shipments specified to be as follows:

	1	2	3	4	5
X_1:	5	6	10	14	20
X_2:	3.20	4.80	7.00	10.00	18.00

Obtain the family of estimates using a 95 percent family confidence coefficient. Employ the Working-Hotelling type bounds or the Bonferroni procedure, whichever is more efficient.

b. For the cases in Problem 7.12, would you consider a shipment of 20 drums with a weight of 5 hundred pounds to be within the scope of the model? What about a shipment of 20 drums with a weight of 19 hundred pounds? Support your answers by preparing a relevant plot.

7.15. Refer to **Chemical shipment** Problem 7.12. Assume that regression model (7.1) with independent normal error terms is appropriate. Four separate shipments with the following characteristics will arrive in the next day or two:

	1	2	3	4
X_1:	9	12	15	18
X_2:	7.20	9.00	12.50	16.50

Management desires predictions of the handling times for these shipments so that the actual handling times can be compared with the predicted times to determine whether any are "out of line." Develop the needed predictions using the most efficient approach and a family confidence coefficient of 95 percent.

7.16. Refer to **Chemical shipment** Problem 7.12. Assume that regression model (7.1) with independent normal error terms is appropriate. Three new shipments are to be received, each with $X_{h1} = 7$ and $X_{h2} = 6$.

a. Obtain a 95 percent prediction interval for the mean handling time for these shipments.

b. Convert the interval obtained in part (a) into a 95 percent prediction interval for the total handling time for the three shipments.

7.17. **Patient satisfaction.** A hospital administrator wished to study the relation between patient satisfaction (Y) and patient's age (X_1, in years), severity of illness (X_2, an index), and anxiety level (X_3, an index). She randomly selected 23 patients and collected the data presented below, where larger values of Y, X_2, and X_3 are, respectively, associated with more satisfaction, increased severity of illness, and more anxiety.

i:	1	2	3	4	5	6	7	8	9	10	11	12
X_{i1}:	50	36	40	41	28	49	42	45	52	29	29	43
X_{i2}:	51	46	48	44	43	54	50	48	62	50	48	53
X_{i3}:	2.3	2.3	2.2	1.8	1.8	2.9	2.2	2.4	2.9	2.1	2.4	2.4
Y_i:	48	57	66	70	89	36	46	54	26	77	89	67

i:	13	14	15	16	17	18	19	20	21	22	23
X_{i1}:	38	34	53	36	33	29	33	55	29	44	43
X_{i2}:	55	51	54	49	56	46	49	51	52	58	50
X_{i3}:	2.2	2.3	2.2	2.0	2.5	1.9	2.1	2.4	2.3	2.9	2.3
Y_i:	47	51	57	66	79	88	60	49	77	52	60

a. Prepare a stem-and-leaf plot for each of the independent variables. Are any noteworthy features revealed by these plots?

b. Fit regression model (7.5) for three independent variables to the data and state the estimated regression function. How is b_2 interpreted here?

    c. Obtain the residuals and prepare a box plot of the residuals. Do there appear to be any outliers?

    d. Plot the residuals against $\hat{Y}$, each of the independent variables, and each two-factor interaction on separate graphs. Also prepare a normal probability plot. Analyze your plots and summarize your findings.

    e. Can you conduct a formal test for lack of fit here?

7.18. Refer to **Patient satisfaction** Problem 7.17. Assume that regression model (7.5) for three independent variables with independent normal error terms is appropriate.

    a. Test whether there is a regression relation; use a .10 level of significance. State the alternatives, decision rule, and conclusion. What does your test imply about β_1, β_2, and β_3? What is the P-value of the test?

    b. Obtain joint interval estimates of β_1, β_2, and β_3 using a 90 percent family confidence coefficient. Interpret your results.

    c. Calculate the coefficient of multiple correlation. What does it indicate here?

7.19. Refer to **Patient satisfaction** Problem 7.17. Assume that regression model (7.5) for three independent variables with independent normal error terms is appropriate.

    a. Obtain an interval estimate of the mean satisfaction when $X_{h1} = 35$, $X_{h2} = 45$, and $X_{h3} = 2.2$. Use a 90 percent confidence coefficient. Interpret your confidence interval.

    b. Obtain a prediction interval for a new patient's satisfaction when $X_{h1} = 35$, $X_{h2} = 45$, and $X_{h3} = 2.2$. Use a 90 percent confidence coefficient. Interpret your prediction interval.

7.20. Mathematicians' salaries. A researcher in a scientific foundation wished to evaluate the relation between intermediate and senior level annual salaries of research mathematicians (Y, in thousand dollars) and an index of publication quality (X_1), number of years of experience (X_2), and an index of success in obtaining grant support (X_3). The data for a sample of 24 intermediate and senior level research mathematicians follow.

i:	1	2	3	4	5	6	7	8	9	10	11	12
X_{i1}:	3.5	5.3	5.1	5.8	4.2	6.0	6.8	5.5	3.1	7.2	4.5	4.9
X_{i2}:	9	20	18	33	31	13	25	30	5	47	25	11
X_{i3}:	6.1	6.4	7.4	6.7	7.5	5.9	6.0	4.0	5.8	8.3	5.0	6.4
Y_i:	33.2	40.3	38.7	46.8	41.4	37.5	39.0	40.7	30.1	52.9	38.2	31.8

i:	13	14	15	16	17	18	19	20	21	22	23	24
X_{i1}:	8.0	6.5	6.6	3.7	6.2	7.0	4.0	4.5	5.9	5.6	4.8	3.9
X_{i2}:	23	35	39	21	7	40	35	23	33	27	34	15
X_{i3}:	7.6	7.0	5.0	4.4	5.5	7.0	6.0	3.5	4.9	4.3	8.0	5.0
Y_i:	43.3	44.1	42.8	33.6	34.2	48.0	38.0	35.9	40.4	36.8	45.2	35.1

    a. Prepare a stem-and-leaf plot for each of the independent variables. What information do these plots provide?

    b. Fit regression model (7.5) for three independent variables to the data. State the estimated regression function.

    c. Obtain the residuals and prepare a box plot of the residuals. Does the distribution appear to be fairly symmetrical?

    d. Plot the residuals against $\hat{Y}$, each of the independent variables, and each two-factor interaction on separate graphs. Also prepare a normal probability plot. Analyze your plots and summarize your findings.

e. Can you conduct a formal test for lack of fit here?

7.21. Refer to **Mathematicians' salaries** Problem 7.20. Assume that regression model (7.5) for three independent variables with independent normal error terms is appropriate.

a. Test whether there is a regression relation; use $\alpha = .05$. State the alternatives, decision rule, and conclusion. What does your test imply about β_1, β_2, and β_3? What is the P-value of the test?

b. Estimate β_1, β_2, and β_3 jointly by the Bonferroni procedure using a 95 percent family confidence coefficient. Interpret your results.

c. Calculate R^2 and interpret this measure.

7.22. Refer to **Mathematicians' salaries** Problem 7.20. Assume that regression model (7.5) for three independent variables with independent normal error terms is appropriate. The researcher wishes to obtain simultaneous interval estimates of the mean salary levels for four typical research mathematicians specified as follows:

	1	2	3	4
X_1:	5.0	6.0	4.0	7.0
X_2:	20	30	10	50
X_3:	5.0	6.0	4.0	7.0

Obtain the family of estimates using a 95 percent family confidence coefficient. Employ the most efficient procedure.

7.23. Refer to **Mathematicians' salaries** Problem 7.20. Assume that regression model (7.5) for three independent variables with independent normal error terms is appropriate. Three research mathematicians with the following characteristics did not provide any salary information in the study.

	1	2	3
X_1:	5.4	6.2	6.4
X_2:	17	12	21
X_3:	6.0	5.8	6.1

Develop separate prediction intervals for the annual salaries of these mathematicians using a 95 percent statement confidence coefficient in each case. Can the salaries of these three mathematicians be predicted fairly precisely?

EXERCISES

7.24. For each of the following regression models, indicate whether it is a general linear regression model. If it is not, state whether it can be expressed in the form of (7.7) by a suitable transformation:

a. $Y_i = \beta_0 + \beta_1 X_{i1} + \beta_2 \log_{10} X_{i2} + \beta_3 X_{i1}^2 + \varepsilon_i$

b. $Y_i = \varepsilon_i \exp(\beta_0 + \beta_1 X_{i1} + \beta_2 X_{i2}^2)$

c. $Y_i = \beta_0 + \log_{10}(\beta_1 X_{i1}) + \beta_2 X_{i2} + \varepsilon_i$

d. $Y_i = \beta_0 \exp(\beta_1 X_{i1}) + \varepsilon_i$

e. $Y_i = [1 + \exp(\beta_0 + \beta_1 X_{i1} + \varepsilon_i)]^{-1}$

7.25. (Calculus needed.) Consider the multiple regression model:

$$Y_i = \beta_1 X_{i1} + \beta_2 X_{i2} + \varepsilon_i \qquad i = 1, \ldots, n$$

where the ε_i are uncorrelated, with $E\{\varepsilon_i\} = 0$ and $\sigma^2\{\varepsilon_i\} = \sigma^2$.
a. Derive the least squares estimators of β_1 and β_2.
b. Assuming that the ε_i are independent normal random variables, state the likelihood function and obtain the maximum likelihood estimators of β_1 and β_2. Are these the same as the least squares estimators?

7.26. (Calculus needed.) Consider the multiple regression model:

$$Y_i = \beta_0 + \beta_1 X_{i1} + \beta_2 X_{i1}^2 + \beta_3 X_{i2} + \varepsilon_i \qquad i = 1, \ldots, n$$

where the ε_i are independent $N(0, \sigma^2)$. Derive the least squares normal equations. Will these yield the same estimators of the regression coefficients as the maximum likelihood estimators?

7.27. An analyst wanted to fit the regression model $Y_i = \beta_0 + \beta_1 X_{i1} + \beta_2 X_{i2} + \beta_3 X_{i3} + \varepsilon_i$, $i = 1, \ldots, n$, by the method of least squares when it is known that $\beta_2 = 4$. How can the analyst obtain the desired fit using a multiple regression computer program?

7.28. For regression model (7.1), show that the coefficient of simple determination r^2 between Y_i and $\hat{Y}_i$ equals the coefficient of multiple determination R^2.

7.29. In a small-scale regression study, the following data were obtained:

i:	1	2	3	4	5	6
X_{i1}:	7	4	16	3	21	8
X_{i2}:	33	41	7	49	5	31
Y_i:	42	33	75	28	91	55

Assume that regression model (7.1) with independent normal error terms is appropriate. Using matrix methods, obtain (a) **b**; (b) **e**; (c) **H**; (d) *SSR*; (e) $s^2\{\mathbf{b}\}$; (f) $\hat{Y}_h$ when $X_{h1} = 10$, $X_{h2} = 30$; (g) $s^2\{\hat{Y}_h\}$ when $X_{h1} = 10$, $X_{h2} = 30$.

PROJECTS

7.30. Refer to the **SMSA** data set. You have been asked to evaluate two alternative models for predicting the number of active physicians (Y) in an SMSA. Proposed model I includes as independent variables total population (X_1), land area (X_2), and total personal income (X_3). Proposed model II includes as independent variables population density (X_1, total population divided by land area), percent of population in central cities (X_2), and total personal income (X_3).
a. Prepare a stem-and-leaf plot for each of the independent variables. What noteworthy information is provided by your plots?
b. For each of the two proposed models, fit the first-order regression model (7.5) with three independent variables.
c. Calculate R^2 for each model. Is one model clearly preferable in terms of this measure?
d. For each model, obtain the residuals and plot them against $\hat{Y}$, each of the three independent variables, and each of the two-factor interactions. Also prepare a normal

probability plot for each of the two fitted models. Analyze your plots and state your findings. Is one model clearly preferable in terms of aptness?

7.31. Refer to the **SMSA** data set.

a. For each geographic region, regress the number of serious crimes in an SMSA (Y) against population density (X_1, total population divided by land area), total personal income (X_2), and percent high school graduates (X_3). Use the first-order regression model (7.5) with three independent variables. State the estimated regression functions.

b. Are the estimated regression functions similar for the four regions? Discuss.

c. Calculate MSE and R^2 for each region. Are these measures similar for the four regions? Discuss.

d. Obtain the residuals for each fitted model and prepare a box plot of the residuals for each fitted model. Analyze your plots and state your findings.

7.32. Refer to the **SENIC** data set. Two models have been proposed for predicting the average length of patient stay in a hospital (Y). Model I utilizes as independent variables age (X_1), infection risk (X_2), and available facilities and services (X_3). Model II uses as independent variables number of beds (X_1), infection risk (X_2), and available facilities and services (X_3).

a. Prepare a stem-and-leaf plot for each of the independent variables. What information do these plots provide?

b. For each of the two proposed models, fit the first-order regression model (7.5) with three independent variables.

c. Calculate R^2 for each model. Is one model clearly preferable in terms of this measure?

d. For each model, obtain the residuals and plot them against $\hat{Y}$, each of the three independent variables, and each of the two-factor interactions. Also prepare a normal probability plot of the residuals for each of the two fitted models. Analyze your plots and state your findings. Is one model clearly preferable in terms of aptness?

7.33. Refer to the **SENIC** data set.

a. For each geographic region, regress infection risk (Y) against the independent variables age (X_1), routine culturing ratio (X_2), average daily census (X_3), and available facilities and services (X_4). Use the first-order regression model (7.5) with four independent variables. State the estimated regression functions.

b. Are the estimated regression functions similar for the four regions? Discuss.

c. Calculate MSE and R^2 for each region. Are these measures similar for the four regions? Discuss.

d. Obtain the residuals for each fitted model and prepare a box plot of the residuals for each fitted model. Analyze the plots and state your findings.

Chapter 8

Multiple Regression—II

In this chapter, we take up several topics that are unique to multiple regression. First, we consider extra sums of squares, which are useful for conducting a variety of tests about the regression coefficients. Then we take up a standardized version of the multiple regression model and introduce multicollinearity, a condition where the independent variables are highly correlated. Finally, we present the general linear test in matrix formulation.

8.1 EXTRA SUMS OF SQUARES

Basic Ideas

An extra sum of squares measures the marginal reduction in the error sum of squares when one or several independent variables are added to the regression model, given that other independent variables are already in the model. Equivalently, one can view an extra sum of squares as measuring the marginal increase in the regression sum of squares when one or several independent variables are added to the regression model.

We shall first utilize an example to illustrate these ideas, and then present definitions of extra sums of squares and discuss a variety of uses of extra sums of squares in tests about regression coefficients.

Example. Table 8.1a contains data for a study of the relation of amount of body fat (Y) to several possible explanatory, independent variables, based on a sample of 20 healthy females 25–34 years old. The possible independent variables are triceps skinfold thickness (X_1), thigh circumference (X_2), and midarm circumference (X_3).

Table 8.2 contains regression results when body fat (Y) is regressed (1) on triceps skinfold thickness (X_1) alone, (2) on thigh circumference (X_2) alone, (3) on X_1 and X_2 only, and (4) on all three independent variables. To keep track of the regression model which is fitted, we shall modify our notation slightly. The regression sum of

TABLE 8.1   Data and Correlation Matrix of the X Variables for Body Fat Example

(a) Data

Subject i	Triceps Skinfold Thickness X_{i1}	Thigh Circumference X_{i2}	Midarm Circumference X_{i3}	Body Fat Y_i
1	19.5	43.1	29.1	11.9
2	24.7	49.8	28.2	22.8
3	30.7	51.9	37.0	18.7
4	29.8	54.3	31.1	20.1
5	19.1	42.2	30.9	12.9
6	25.6	53.9	23.7	21.7
7	31.4	58.5	27.6	27.1
8	27.9	52.1	30.6	25.4
9	22.1	49.9	23.2	21.3
10	25.5	53.5	24.8	19.3
11	31.1	56.6	30.0	25.4
12	30.4	56.7	28.3	27.2
13	18.7	46.5	23.0	11.7
14	19.7	44.2	28.6	17.8
15	14.6	42.7	21.3	12.8
16	29.5	54.4	30.1	23.9
17	27.7	55.3	25.7	22.6
18	30.2	58.6	24.6	25.4
19	22.7	48.2	27.1	14.8
20	25.2	51.0	27.5	21.1

(b) Correlation Matrix of X Variables

$$\mathbf{r}_{XX} = \begin{bmatrix} 1.0 & .92 & .46 \\ .92 & 1.0 & .08 \\ .46 & .08 & 1.0 \end{bmatrix}$$

TABLE 8.2   Regression Results for Several Fitted Models for Body Fat Example

(a) Regression of Y on X_1

$$\hat{Y} = -1.496 + .8572X_1$$

Source of Variation	SS	df	MS
Regression	352.27	1	352.27
Error	143.12	18	7.95
Total	495.39	19	

Variable	Estimated Regression Coefficient	Estimated Standard Deviation	t^*
X_1	$b_1 = .8572$	$s\{b_1\} = .1288$	6.66

TABLE 8.2 (*concluded*)

(b) Regression of Y on X_2
$$\hat{Y} = -23.634 + .8565X_2$$

Source of Variation	SS	df	MS
Regression	381.97	1	381.97
Error	113.42	18	6.30
Total	495.39	19	

Variable	Estimated Regression Coefficient	Estimated Standard Deviation	t^*
X_2	$b_2 = .8565$	$s\{b_2\} = .1100$	7.79

(c) Regression of Y on X_1 and X_2
$$\hat{Y} = -19.174 + .2224X_1 + .6594X_2$$

Source of Variation	SS	df	MS
Regression	385.44	2	192.72
Error	109.95	17	6.47
Total	495.39	19	

Variable	Estimated Regression Coefficient	Estimated Standard Deviation	t^*
X_1	$b_1 = .2224$	$s\{b_1\} = .3034$	.73
X_2	$b_2 = .6594$	$s\{b_2\} = .2912$	2.26

(d) Regression of Y on X_1, X_2, and X_3
$$\hat{Y} = 117.08 + 4.334X_1 - 2.857X_2 - 2.186X_3$$

Source of Variation	SS	df	MS
Regression	396.98	3	132.33
Error	98.41	16	6.15
Total	495.39	19	

Variable	Estimated Regression Coefficient	Estimated Standard Deviation	t^*
X_1	$b_1 = 4.334$	$s\{b_1\} = 3.016$	1.44
X_2	$b_2 = -2.857$	$s\{b_2\} = 2.582$	-1.11
X_3	$b_3 = -2.186$	$s\{b_3\} = 1.596$	-1.37

squares when X_1 only is in the model is, according to Table 8.2a, 352.27. This sum of squares will be denoted by $SSR(X_1)$. The error sum of squares for this model will be denoted by $SSE(X_1)$; according to Table 8.2a it is $SSE(X_1) = 143.12$.

Similarly, Table 8.2c indicates that when X_1 and X_2 are in the regression model, the regression sum of squares is $SSR(X_1, X_2) = 385.44$ and the error sum of squares is $SSE(X_1, X_2) = 109.95$.

Notice that the error sum of squares when X_1 and X_2 are in the model, $SSE(X_1, X_2) = 109.95$, is smaller than when the model contains only X_1, $SSE(X_1) = 143.12$. The difference is called an *extra sum of squares* and will be denoted by $SSR(X_2 \mid X_1)$:

$$SSR(X_2 \mid X_1) = SSE(X_1) - SSE(X_1, X_2)$$

$$= 143.12 - 109.95 = 33.17$$

This reduction in the error sum of squares is the result of adding X_2 to the regression model when X_1 is already included in the model. Thus, the extra sum of squares $SSR(X_2 \mid X_1)$ measures the marginal effect of adding X_2 to the regression model when X_1 is already in the model. The notation $SSR(X_2 \mid X_1)$ reflects this additional or extra reduction in the error sum of squares associated with X_2, given that X_1 is already included in the model.

The extra sum of squares $SSR(X_2 \mid X_1)$ equivalently measures the marginal increase in the regression sum of squares:

$$SSR(X_2 \mid X_1) = SSR(X_1, X_2) - SSR(X_1)$$

$$= 385.44 - 352.27 = 33.17$$

The reason for the equivalence of the marginal reduction in the error sum of squares and the marginal increase in the regression sum of squares is the basic analysis of variance identity (3.48a):

$$SSTO = SSR + SSE$$

Since $SSTO$ measures the variability of the Y_i observations and hence does not depend on the regression model fitted, any reduction in SSE implies an identical increase in SSR.

We can consider other extra sums of squares, such as the marginal effect of adding X_3 to the regression model when X_1 and X_2 are already in the model. We find from Tables 8.2c and 8.2d that:

$$SSR(X_3 \mid X_1, X_2) = SSE(X_1, X_2) - SSE(X_1, X_2, X_3)$$

$$= 109.95 - 98.41 = 11.54$$

or, equivalently:

$$SSR(X_3 \mid X_1, X_2) = SSR(X_1, X_2, X_3) - SSR(X_1, X_2)$$

$$= 396.98 - 385.44 = 11.54$$

We can even consider the marginal effect of adding several variables, such as adding both X_2 and X_3 to the regression model already containing X_1 (see Tables 8.2a and 8.2d):

$$SSR(X_2, X_3 \mid X_1) = SSE(X_1) - SSE(X_1, X_2, X_3)$$
$$= 143.12 - 98.41 = 44.71$$

or, equivalently:

$$SSR(X_2, X_3 \mid X_1) = SSR(X_1, X_2, X_3) - SSR(X_1)$$
$$= 396.98 - 352.27 = 44.71$$

Definitions

We assemble now our earlier definitions of extra sums of squares and provide some additional ones. As we noted earlier, an extra sum of squares always involves the difference between the error sum of squares for the regression model containing the X variable(s) already in the model and the error sum of squares for the regression model containing both the original X variable(s) and the new X variable(s). Equivalently, an extra sum of squares involves the difference between the two corresponding regression sums of squares.

Thus, we have:

(8.1a) $$\qquad SSR(X_1 \mid X_2) = SSE(X_2) - SSE(X_1, X_2)$$

or, equivalently:

(8.1b) $$\qquad SSR(X_1 \mid X_2) = SSR(X_1, X_2) - SSR(X_2)$$

If X_2 is the extra variable, we have:

(8.2a) $$\qquad SSR(X_2 \mid X_1) = SSE(X_1) - SSE(X_1, X_2)$$

or, equivalently:

(8.2b) $$\qquad SSR(X_2 \mid X_1) = SSR(X_1, X_2) - SSR(X_1)$$

Extensions for three or more variables are straightforward. For example, we have:

(8.3a) $$\qquad SSR(X_3 \mid X_1, X_2) = SSE(X_1, X_2) - SSE(X_1, X_2, X_3)$$

(8.3b) $$\qquad SSR(X_3 \mid X_1, X_2) = SSR(X_1, X_2, X_3) - SSR(X_1, X_2)$$

and:

(8.4a) $$\qquad SSR(X_2, X_3 \mid X_1) = SSE(X_1) - SSE(X_1, X_2, X_3)$$

(8.4b) $$\qquad SSR(X_2, X_3 \mid X_1) = SSR(X_1, X_2, X_3) - SSR(X_1)$$

Decomposition of *SSR* into Extra Sums of Squares

In multiple regression, unlike simple regression, we can obtain a variety of decompositions of the regression sum of squares *SSR* into extra sums of squares. Let us consider the case of two X variables. We begin with the identity (3.48a) for variable X_1:

(8.5)
$$SSTO = SSR(X_1) + SSE(X_1)$$

where the notation now shows explicitly that X_1 is the X variable in the model. Replacing $SSE(X_1)$ by its equivalent in (8.2a), we obtain:

(8.6)
$$SSTO = SSR(X_1) + SSR(X_2 \mid X_1) + SSE(X_1, X_2)$$

We have the same identity for multiple regression with two X variables as in (8.5) for a single X variable, namely:

(8.7)
$$SSTO = SSR(X_1, X_2) + SSE(X_1, X_2)$$

Solving (8.7) for $SSE(X_1, X_2)$ and using this expression in (8.6) leads to:

(8.8)
$$SSR(X_1, X_2) = SSR(X_1) + SSR(X_2 \mid X_1)$$

Thus, we have decomposed the regression sum of squares $SSR(X_1, X_2)$ into two marginal components: (1) $SSR(X_1)$, measuring the contribution by including X_1 alone in the model, and (2) $SSR(X_2 \mid X_1)$, measuring the additional contribution when X_2 is included, given that X_1 is already in the model.

Of course, the order of the X variables is arbitrary. Here, we can also obtain the decomposition:

(8.9)
$$SSR(X_1, X_2) = SSR(X_2) + SSR(X_1 \mid X_2)$$

We show in Figure 8.1 schematic representations of the two decompositions of $SSR(X_1, X_2)$ for the body fat example. The total bar on the left represents $SSTO$ and presents the decomposition (8.9). The unshaded component of this bar is $SSR(X_2)$, and the combined shaded area represents $SSE(X_2)$. The latter area in turn is the combination of the extra sum of squares $SSR(X_1 \mid X_2)$ and the error sum of squares $SSE(X_1, X_2)$ when both X_1 and X_2 are included in the model. Similarly, the bar on the right in Figure 8.1 shows the decomposition (8.8). Note in both cases how the extra sum of squares can be viewed either as a reduction in the error sum of squares or as an increase in the regression sum of squares when the second variable is added to the regression model.

When the regression model contains three X variables, a variety of decompositions of $SSR(X_1, X_2, X_3)$ can be obtained. We illustrate three of these:

(8.10a) $\quad SSR(X_1, X_2, X_3) = SSR(X_1) + SSR(X_2 \mid X_1) + SSR(X_3 \mid X_1, X_2)$

(8.10b) $\quad SSR(X_1, X_2, X_3) = SSR(X_2) + SSR(X_3 \mid X_2) + SSR(X_1 \mid X_2, X_3)$

(8.10c) $\quad SSR(X_1, X_2, X_3) = SSR(X_1) + SSR(X_2, X_3 \mid X_1)$

It is obvious that the number of possible decompositions becomes vast as the number of X variables in the regression model increases.

FIGURE 8.1   Schematic Representation of Extra Sums of Squares—Body Fat Example

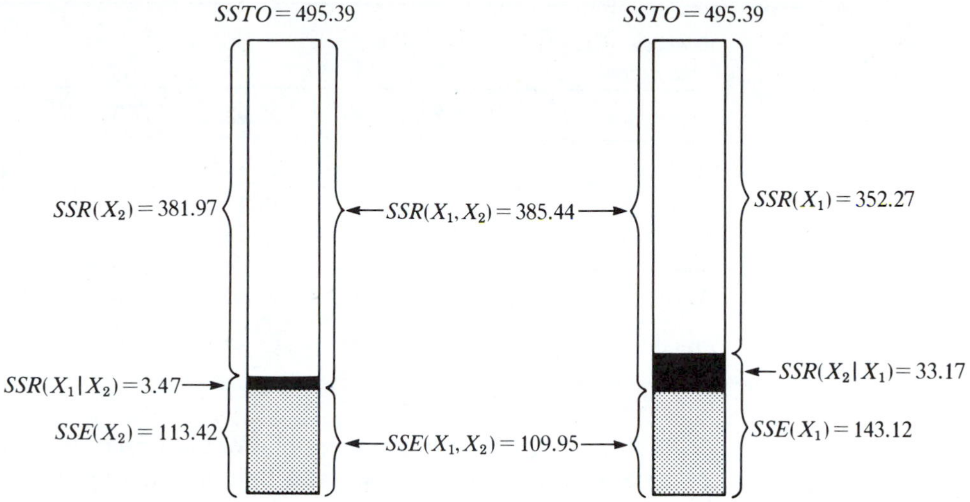

ANOVA Table Containing Decomposition of *SSR*

ANOVA tables can be constructed containing decompositions of the regression sum of squares into extra sums of squares. Table 8.3 contains the ANOVA table for one possible decomposition for the case of three independent variables, and Table 8.4 contains this same decomposition for the body fat example.

Note that each extra sum of squares involving a single extra variable has associated with it one degree of freedom. Extra sums of squares involving two extra variables,

TABLE 8.3   Example of ANOVA Table with Decomposition of *SSR* for Three Independent Variables

Source of Variation	SS	df	MS
Regression	$SSR(X_1, X_2, X_3)$	3	$MSR(X_1, X_2, X_3)$
X_1	$SSR(X_1)$	1	$MSR(X_1)$
$X_2 \mid X_1$	$SSR(X_2 \mid X_1)$	1	$MSR(X_2 \mid X_1)$
$X_3 \mid X_1, X_2$	$SSR(X_3 \mid X_1, X_2)$	1	$MSR(X_3 \mid X_1, X_2)$
Error	$SSE(X_1, X_2, X_3)$	$n - 4$	$MSE(X_1, X_2, X_3)$
Total	$SSTO$	$n - 1$	

TABLE 8.4   ANOVA Table with Decomposition of SSR—Body Fat Example with Three Independent Variables

Source of Variation	SS	df	MS
Regression	$SSR(X_1, X_2, X_3) = 396.98$	3	132.33
X_1	$SSR(X_1) = 352.27$	1	352.27
$X_2 \mid X_1$	$SSR(X_2 \mid X_1) = 33.17$	1	33.17
$X_3 \mid X_1, X_2$	$SSR(X_3 \mid X_1, X_2) = 11.54$	1	11.54
Error	$SSE(X_1, X_2, X_3) = 98.41$	16	6.15
Total	$SSTO = 495.39$	19	

such as $SSR(X_2, X_3 \mid X_1)$, have two degrees of freedom associated with them. This follows because we can express such an extra sum of squares as a sum of two extra sums of squares, each associated with one degree of freedom. For example, by definition of the extra sums of squares, we have:

$$(8.11) \qquad SSR(X_2, X_3 \mid X_1) = SSR(X_2 \mid X_1) + SSR(X_3 \mid X_1, X_2)$$

A number of computer regression packages provide decompositions of SSR into single-degree-of-freedom extra sums of squares, usually in the order in which the X variables are entered into the model. Thus, if the independent variables are entered in the order X_1, X_2, X_3, the extra sums of squares given in the output are:

$$SSR(X_1)$$

$$SSR(X_2 \mid X_1)$$

$$SSR(X_3 \mid X_1, X_2)$$

If an extra sum of squares involving several extra X variables is desired, it can be obtained by summing appropriate single-degree-of-freedom extra sums of squares. For instance, to obtain $SSR(X_2, X_3 \mid X_1)$ in our earlier illustration, we would utilize (8.11) and simply add $SSR(X_2 \mid X_1)$ and $SSR(X_3 \mid X_1, X_2)$.

If the extra sum of squares $SSR(X_1, X_3 \mid X_2)$ were desired with a computer package that provides single-degree-of-freedom extra sums of squares in the order in which the X variables are entered, the independent variables would need to be entered in the order X_2, X_1, X_3 or X_2, X_3, X_1. The first ordering would give:

$$SSR(X_2)$$

$$SSR(X_1 \mid X_2)$$

$$SSR(X_3 \mid X_1, X_2)$$

The sum of the last two extra sums of squares will yield $SSR(X_1, X_3 \mid X_2)$.

The reason why extra sums of squares are of interest is that they occur in a variety of tests about regression coefficients where the question of concern is whether certain X variables can be dropped from the regression model. We turn next to this use of extra sums of squares.

Use of Extra Sum of Squares in Test Whether a Single $\beta_k = 0$

When we wish to test whether the term $\beta_k X_k$ can be dropped from a multiple regression model, we are interested in the alternatives:

$$H_0: \beta_k = 0$$

$$H_a: \beta_k \neq 0$$

We already know that test statistic (7.46b):

$$t^* = \frac{b_k}{s\{b_k\}}$$

is appropriate for this test.

Equivalently, we can use the general linear test approach described in Section 3.9. We shall now show that this approach involves an extra sum of squares. Let us consider the first-order regression model with three independent variables:

$$(8.12) \qquad Y_i = \beta_0 + \beta_1 X_{i1} + \beta_2 X_{i2} + \beta_3 X_{i3} + \varepsilon_i \qquad \text{Full model}$$

To test the alternatives:

$$(8.13) \qquad \begin{aligned} H_0: \beta_3 &= 0 \\ H_a: \beta_3 &\neq 0 \end{aligned}$$

we fit the full model and obtain the error sum of squares $SSE(F)$. We shall now explicitly show the variables in the full model, as follows:

$$SSE(F) = SSE(X_1, X_2, X_3)$$

The degrees of freedom associated with $SSE(F)$ are $df_F = n - 4$; remember that there are four parameters in the regression function for the full model (8.12).

The reduced model when H_0 in (8.13) holds is:

$$(8.14) \qquad Y_i = \beta_0 + \beta_1 X_{i1} + \beta_2 X_{i2} + \varepsilon_i \qquad \text{Reduced model}$$

We next fit this reduced model and obtain:

$$SSE(R) = SSE(X_1, X_2)$$

There are $df_R = n - 3$ degrees of freedom associated with the reduced model.

The general linear test statistic (3.69):

$$F^* = \frac{SSE(R) - SSE(F)}{df_R - df_F} \div \frac{SSE(F)}{df_F}$$

here becomes:

$$F^* = \frac{SSE(X_1, X_2) - SSE(X_1, X_2, X_3)}{(n - 3) - (n - 4)} \div \frac{SSE(X_1, X_2, X_3)}{n - 4}$$

Note that the difference between the two error sums of squares in the numerator term is the extra sum of squares (8.3a):

$$SSE(X_1, X_2) - SSE(X_1, X_2, X_3) = SSR(X_3 \mid X_1, X_2)$$

Hence the general linear test statistic here is:

$$(8.15) \qquad F^* = \frac{SSR(X_3 \mid X_1, X_2)}{1} \div \frac{SSE(X_1, X_2, X_3)}{n - 4}$$

$$= \frac{MSR(X_3 \mid X_1, X_2)}{MSE(X_1, X_2, X_3)}$$

We thus see that the test whether or not $\beta_3 = 0$ is a marginal test, given that X_1 and X_2 are already in the model. We also note that the extra sum of squares $SSR(X_3 \mid X_1, X_2)$ has one degree of freedom associated with it, just as was pointed out earlier.

Test statistic (8.15) shows that we do not need to fit both the full model and the reduced model to use the general linear test approach here. A single computer run can provide a fit of the full model and the appropriate extra sum of squares.

Example.   In the body fat example with all three independent variables in Table 8.1, we wish to test whether midarm circumference (X_3) can be dropped from the model. The test alternatives are those of (8.13). Table 8.4 contains the ANOVA results from a computer fit of the full regression model (8.12), including the extra sums of squares when the independent variables are entered in the order X_1, X_2, X_3. Hence test statistic (8.15) here is:

$$F^* = \frac{SSR(X_3 \mid X_1, X_2)}{1} \div \frac{SSE(X_1, X_2, X_3)}{n - 4}$$

$$= \frac{11.54}{1} \div \frac{98.41}{16} = 1.88$$

For $\alpha = .01$, we require $F(.99; 1, 16) = 8.53$. Since $F^* = 1.88 \le 8.53$, we conclude H_0, that X_3 can be dropped from the regression model that already contains X_1 and X_2.

Note from Table 8.2d that the t^* test statistic here is:

$$t^* = \frac{b_3}{s\{b_3\}} = \frac{-2.186}{1.596} = -1.37$$

Since $(t^*)^2 = (-1.37)^2 = 1.88 = F^*$, we see that the two test statistics are equivalent, just as for simple linear regression.

Note

The F^* test statistic (8.15) to test whether or not $\beta_3 = 0$ is called a *partial F test* statistic to distinguish it from the F^* statistic in (7.34b) for testing whether *all* $\beta_k = 0$, i.e., whether or not there is a regression relation between Y and the set of independent variables. The latter test is called the *overall F test*.

Use of Extra Sum of Squares in Test Whether Several $\beta_k = 0$

Frequently, in multiple regression there is interest in whether several terms in the regression model can be dropped. For example, one may wish to know whether both $\beta_2 X_2$ and $\beta_3 X_3$ can be dropped from the full model (8.12). The alternatives here are:

(8.16)
$$H_0: \beta_2 = \beta_3 = 0$$
$$H_a: \text{not both } \beta_2 \text{ and } \beta_3 \text{ equal zero}$$

With the general linear test approach, the reduced model under H_0 is:

(8.17)
$$Y_i = \beta_0 + \beta_1 X_{i1} + \varepsilon_i \qquad \text{Reduced model}$$

and the error sum of squares for the reduced model is:

$$SSE(R) = SSE(X_1)$$

This error sum of squares has $df_R = n - 2$ degrees of freedom associated with it. Test statistic (3.69) thus becomes here:

$$F^* = \frac{SSE(X_1) - SSE(X_1, X_2, X_3)}{(n - 2) - (n - 4)} \div \frac{SSE(X_1, X_2, X_3)}{n - 4}$$

Again the difference between the two error sums of squares in the numerator term is an extra sum of squares, namely:

$$SSE(X_1) - SSE(X_1, X_2, X_3) = SSR(X_2, X_3 \mid X_1)$$

Hence, the test statistic becomes:

(8.18)
$$F^* = \frac{SSR(X_2, X_3 \mid X_1)}{2} \div \frac{SSE(X_1, X_2, X_3)}{n - 4}$$
$$= \frac{MSR(X_2, X_3 \mid X_1)}{MSE(X_1, X_2, X_3)}$$

Note that $SSR(X_2, X_3 \mid X_1)$ has two degrees of freedom associated with it, as we pointed out earlier.

Example. We wish to test in the body fat example with three independent variables whether both thigh circumference (X_2) and midarm circumference (X_3) can be dropped from the full regression model (8.12). The alternatives are those in (8.16).

The appropriate extra sum of squares can be obtained from Table 8.4, using (8.11):

$$SSR(X_2, X_3 \mid X_1) = SSR(X_2 \mid X_1) + SSR(X_3 \mid X_1, X_2)$$

$$= 33.17 + 11.54 = 44.71$$

Test statistic (8.18) therefore is:

$$F^* = \frac{SSR(X_2, X_3 \mid X_1)}{2} \div MSE(X_1, X_2, X_3)$$

$$= \frac{44.71}{2} \div 6.15 = 3.63$$

For $\alpha = .01$, we require $F(.99; 2, 16) = 6.23$. Since $F^* = 3.63 \leq 6.23$, we conclude H_0, that X_2 and X_3 can be dropped from the regression model that already contains X_1.

Note

For testing whether a single β_k equals zero, two equivalent test statistics are available: the t^* test statistic (7.46b) and the F^* general linear test statistic (3.69). When testing whether several β_k equal zero, only the general linear test statistic F^* is available.

8.2 TESTING HYPOTHESES CONCERNING REGRESSION COEFFICIENTS IN MULTIPLE REGRESSION

We have already discussed how to conduct several types of tests concerning the regression coefficients in a multiple regression model. For completeness, we summarize these tests here and then take up some additional types of tests.

Test Whether All $\beta_k = 0$

This is the *overall F test* (7.34) of whether or not there is a regression relation between the dependent variable Y and the set of independent variables. The alternatives are:

(8.19)
$$H_0: \beta_1 = \beta_2 = \cdots = \beta_{p-1} = 0$$
$$H_a: \text{not all } \beta_k \ (k = 1, \ldots, p - 1) \text{ equal zero}$$

and the test statistic is:

(8.20)
$$F^* = \frac{SSR(X_1, \ldots, X_{p-1})}{p - 1} \div \frac{SSE(X_1, \ldots, X_{p-1})}{n - p}$$

$$= \frac{MSR}{MSE}$$

If H_0 holds, $F^* \sim F(p - 1, n - p)$. Large values of F^* lead to conclusion H_a.

Test Whether a Single $\beta_k = 0$

This is a *partial F test* of whether a particular regression coefficient β_k equals zero. The alternatives are:

(8.21)
$$H_0: \beta_k = 0$$
$$H_a: \beta_k \neq 0$$

and the test statistic is:

(8.22)
$$F^* = \frac{SSR(X_k \mid X_1, \ldots, X_{k-1}, X_{k+1}, \ldots, X_{p-1})}{1} \div \frac{SSE(X_1, \ldots, X_{p-1})}{n - p}$$
$$= \frac{MSR(X_k \mid X_1, \ldots, X_{k-1}, X_{k+1}, \ldots, X_{p-1})}{MSE}$$

If H_0 holds, $F^* \sim F(1, n - p)$. Large values of F^* lead to conclusion H_a. Computer packages which provide extra sums of squares permit use of this test without having to fit the reduced model.

An equivalent test statistic, as we have seen, is (7.46b):

(8.23)
$$t^* = \frac{b_k}{s\{b_k\}}$$

If H_0 holds, $t^* \sim t(n - p)$. Large values of $|t^*|$ lead to conclusion H_a.

Since the two tests are equivalent, the choice is usually made in terms of available information provided by the computer package output.

Test Whether Some $\beta_k = 0$

This is another *partial F test*. Here, the alternatives are:

(8.24)
$$H_0: \beta_q = \beta_{q+1} = \cdots = \beta_{p-1} = 0$$
$$H_a: \text{not all of the } \beta_k \text{ in } H_0 \text{ equal zero}$$

where for convenience, we arrange the model so that the last $p - q$ coefficients are the ones to be tested. The test statistic is:

(8.25)
$$F^* = \frac{SSR(X_q, \ldots, X_{p-1} \mid X_1, \ldots, X_{q-1})}{p - q} \div \frac{SSE(X_1, \ldots, X_{p-1})}{n - p}$$
$$= \frac{MSR(X_q, \ldots, X_{p-1} \mid X_1, \ldots, X_{q-1})}{MSE}$$

If H_0 holds, $F^* \sim F(p - q, n - p)$. Large values of F^* lead to conclusion H_a.

Note that test statistic (8.25) actually encompasses the two earlier cases. If $q = 1$, the test is whether all regression coefficients equal zero. If $q = p - 1$, the test is whether a single regression coefficient equals zero. Also note that test statistic (8.25) can be calculated without having to fit the reduced model if the computer program provides the needed extra sums of squares:

$$(8.26) \quad SSR(X_q, \ldots, X_{p-1} \mid X_1, \ldots, X_{q-1}) = SSR(X_q \mid X_1, \ldots, X_{q-1})$$
$$+ \cdots + SSR(X_{p-1} \mid X_1, \ldots, X_{p-2})$$

Other Tests

When tests about regression coefficients are desired that do not involve testing whether one or several β_k equal zero, extra sums of squares cannot be used and the general linear test approach requires separate fittings of the full and reduced models. For instance, for the full model containing three X variables:

$$(8.27) \qquad Y_i = \beta_0 + \beta_1 X_{i1} + \beta_2 X_{i2} + \beta_3 X_{i3} + \varepsilon_i \qquad \text{Full model}$$

we might wish to test:

$$(8.28) \qquad \begin{aligned} H_0&: \beta_1 = \beta_2 \\ H_a&: \beta_1 \neq \beta_2 \end{aligned}$$

The procedure would be to fit the full model (8.27), and then the reduced model:

$$(8.29) \qquad Y_i = \beta_0 + \beta_c(X_{i1} + X_{i2}) + \beta_3 X_{i3} + \varepsilon_i \qquad \text{Reduced model}$$

where β_c denotes the common coefficient for β_1 and β_2 under H_0 and $X_{i1} + X_{i2}$ is the corresponding new X variable. We then use the general F^* test statistic (3.69) with 1 and $n - 4$ degrees of freedom.

Another example where extra sums of squares cannot be used is in the following test for regression model (8.27):

$$(8.30) \qquad \begin{aligned} H_0&: \beta_1 = 3, \ \beta_3 = 5 \\ H_a&: \text{not both equalities in } H_0 \text{ hold} \end{aligned}$$

Here, the reduced model would be:

$$(8.31) \qquad Y_i - 3X_{i1} - 5X_{i3} = \beta_0 + \beta_2 X_{i2} + \varepsilon_i \qquad \text{Reduced model}$$

Note the new dependent variable $Y - 3X_1 - 5X_3$ in the reduced model, since $\beta_1 X_1$ and $\beta_3 X_3$ are known constants under H_0. We then use the general linear test statistic F^* in (3.69) with 2 and $n - 4$ degrees of freedom.

8.3 COEFFICIENTS OF PARTIAL DETERMINATION

Extra sums of squares are not only useful for tests on the regression coefficients of a multiple regression model but are also encountered in descriptive measures of relationship called coefficients of partial determination. A coefficient of multiple determination R^2, it will be recalled, measures the proportionate reduction in the variation of Y achieved by the introduction of the entire set of X variables considered in the model. A *coefficient of partial determination,* in contrast, measures the marginal contribution of one X variable, when all others are already included in the model.

Two Independent Variables

Let us consider a first-order multiple regression model with two independent variables, as given in (7.1):

$$Y_i = \beta_0 + \beta_1 X_{i1} + \beta_2 X_{i2} + \varepsilon_i$$

$SSE(X_2)$ measures the variation in Y when X_2 is included in the model. $SSE(X_1, X_2)$ measures the variation in Y when both X_1 and X_2 are included in the model. Hence, the relative marginal reduction in the variation in Y associated with X_1 when X_2 is already in the model is:

$$\frac{SSE(X_2) - SSE(X_1, X_2)}{SSE(X_2)} = \frac{SSR(X_1 \mid X_2)}{SSE(X_2)}$$

This measure is the coefficient of partial determination between Y and X_1, given that X_2 is in the model. We denote this measure by $r_{Y1.2}^2$:

$$(8.32) \qquad r_{Y1.2}^2 = \frac{SSE(X_2) - SSE(X_1, X_2)}{SSE(X_2)} = \frac{SSR(X_1 \mid X_2)}{SSE(X_2)}$$

$r_{Y1.2}^2$ thus measures the proportionate reduction in the variation of Y remaining after X_2 is included in the model that is gained by also including X_1 in the model.

The coefficient of partial determination between Y and X_2, given that X_1 is in the model, is defined:

$$(8.33) \qquad r_{Y2.1}^2 = \frac{SSR(X_2 \mid X_1)}{SSE(X_1)}$$

General Case

The generalization of coefficients of partial determination to three or more independent variables in the model is immediate. For instance:

$$(8.34) \qquad r_{Y1.23}^2 = \frac{SSR(X_1 \mid X_2, X_3)}{SSE(X_2, X_3)}$$

$$(8.35) \qquad r_{Y2.13}^2 = \frac{SSR(X_2 \mid X_1, X_3)}{SSE(X_1, X_3)}$$

$$(8.36) \qquad r_{Y3.12}^2 = \frac{SSR(X_3 \mid X_1, X_2)}{SSE(X_1, X_2)}$$

$$(8.37) \qquad r_{Y4.123}^2 = \frac{SSR(X_4 \mid X_1, X_2, X_3)}{SSE(X_1, X_2, X_3)}$$

Note that in the subscripts to r^2, the entries to the left of the dot show in turn the variable taken as the response and the variable being added. The entries to the right of the dot show the X variables already in the model.

Example

For the body fat example, we can obtain a variety of coefficients of partial determination. Here are three (Tables 8.2 and 8.4):

$$r^2_{Y2.1} = \frac{SSR(X_2 \mid X_1)}{SSE(X_1)} = \frac{33.17}{143.12} = .232$$

$$r^2_{Y3.12} = \frac{SSR(X_3 \mid X_1, X_2)}{SSE(X_1, X_2)} = \frac{11.54}{109.95} = .105$$

$$r^2_{Y1.2} = \frac{SSR(X_1 \mid X_2)}{SSE(X_2)} = \frac{3.47}{113.42} = .031$$

Thus, when X_2 is added to the regression model containing X_1 here, the error sum of squares $SSE(X_1)$ is reduced by 23.2 percent. The error sum of squares for the model containing both X_1 and X_2 is only reduced by another 10.5 percent when X_3 is added to the model. Finally, if the regression model already contains X_2, adding X_1 only reduces $SSE(X_2)$ by 3.1 percent.

Comments

1. The coefficients of partial determination can take on values between 0 and 1, as the definitions readily indicate.

2. A coefficient of partial determination can be interpreted as a coefficient of simple determination. Consider a multiple regression model with two independent variables. Suppose we regress Y on X_2 and obtain the residuals:

$$Y_i - \hat{Y}_i(X_2)$$

where $\hat{Y}_i(X_2)$ denotes the fitted values of Y when X_2 is in the model.

Suppose we further regress X_1 on X_2 and obtain the residuals:

$$X_{i1} - \hat{X}_{i1}(X_2)$$

where $\hat{X}_{i1}(X_2)$ denotes the fitted values of X_1 in the regression of X_1 on X_2. The coefficient of simple determination r^2 between these two sets of residuals equals the coefficient of partial determination $r^2_{Y1.2}$. Thus, this coefficient measures the relation between Y and X_1 when both of these variables have been adjusted for their linear relationships to X_2.

Coefficients of Partial Correlation

The square root of a coefficient of partial determination is called a *coefficient of partial correlation*. It is given the same sign as that of the corresponding regression coefficient in the fitted regression function. Coefficients of partial correlation are frequently used in practice, although they do not have as clear a meaning as coefficients of partial determination.

For the body fat example, we have:

$$r_{Y2.1} = \sqrt{.232} = .482$$

$$r_{Y3.12} = -\sqrt{.105} = -.324$$

$$r_{Y1.2} = \sqrt{.031} = .176$$

Note that the coefficients $r_{Y2.1}$ and $r_{Y1.2}$ are positive because we see from Table 8.2c that $b_2 = .6594$ and $b_1 = .2224$. Similarly, $r_{Y3.12}$ is negative because we see from Table 8.2d that $b_3 = -2.186$.

Partial correlation coefficients are frequently used in computer routines for finding the best independent variable to be selected next for inclusion in the regression model. We shall discuss this use in Chapter 12.

Note

Coefficients of partial determination can be expressed in terms of simple or other partial correlation coefficients. For example:

$$(8.38) \qquad r_{Y2.1}^2 = \frac{(r_{Y2} - r_{12}r_{Y1})^2}{(1 - r_{12}^2)(1 - r_{Y1}^2)}$$

$$(8.39) \qquad r_{Y2.13}^2 = \frac{(r_{Y2.3} - r_{12.3}r_{Y1.3})^2}{(1 - r_{12.3}^2)(1 - r_{Y1.3}^2)}$$

where r_{Y1} denotes the coefficient of simple correlation between Y and X_1, r_{12} denotes the coefficient of simple correlation between X_1 and X_2, and so on. Extensions are straightforward.

8.4 STANDARDIZED MULTIPLE REGRESSION MODEL

A standardized form of the general multiple regression model (7.7) is employed to control roundoff errors in the least squares calculations and to permit comparisons of the estimated regression coefficients in common units.

Roundoff Errors in Least Squares Calculations

Least squares results can be sensitive to rounding of data in intermediate stages of calculations. When the number of X variables is small—say, three or less—roundoff effects can be controlled by carrying a sufficient number of digits in intermediate calculations. Indeed, most computer regression programs use double precision arithmetic (e.g., use of 16 digits instead of 8 digits) in all computations to control roundoff effects. Still, with a large number of X variables, serious roundoff effects can arise despite the use of many digits in intermediate calculations.

Roundoff errors tend to enter into least squares calculations primarily when the inverse of $\mathbf{X'X}$ is taken. Of course, any errors in $(\mathbf{X'X})^{-1}$ may be magnified when calculating **b** and other subsequent statistics. The danger of serious roundoff errors in $(\mathbf{X'X})^{-1}$ is particularly great when (1) $\mathbf{X'X}$ has a determinant that is close to zero and/or (2) the elements of $\mathbf{X'X}$ differ substantially in order of magnitude. The first condition arises when some or all of the independent variables are highly intercorrelated. We shall discuss this situation in Section 8.5.

The second condition arises when the variables have substantially different magnitudes so that the entries in the $\mathbf{X'X}$ matrix cover a wide range, say, from 15 to 49,000,000. A solution for this condition is to transform the variables and thereby reparameterize the regression model.

The transformation that we shall take up is called the *correlation transformation*. It makes all entries in the $\mathbf{X'X}$ matrix for the transformed variables fall between -1 and $+1$ inclusive, so that the calculation of the inverse matrix becomes much less subject to roundoff errors due to dissimilar orders of magnitudes than with the original variables. Many computer regression packages automatically use this transformation to obtain the basic regression results and then retransform to the original variables.

Lack of Comparability in Regression Coefficients

A second difficulty with the nonstandardized multiple regression model (7.7) is that ordinarily regression coefficients cannot be compared because of differences in the units involved. We cite two examples.

1. When considering the fitted response function:

$$\hat{Y} = 200 + 20,000X_1 + .2X_2$$

one may be tempted to conclude that X_1 is the only important independent variable, and that X_2 has little effect on the dependent variable Y. A little reflection should make one wary of this conclusion. The reason is that we do not know the units involved. Suppose the units are:

Y in dollars
X_1 in thousand dollars
X_2 in cents

In that event, the effect on the mean response of a \$1,000 increase in X_1 when X_2 is constant would be exactly the same as the effect of a \$1,000 increase in X_2 when X_1 is constant, despite the difference in the regression coefficients.

2. In the Zarthan Company example of Table 7.2, we cannot make any comparison between b_1 and b_2 because b_1 is in units of one gross of jars per thousand persons while b_2 is in units of one gross of jars per dollar of per capita discretionary income.

Correlation Transformation

Use of the correlation transformation helps with controlling roundoff errors and makes the units of the regression coefficients comparable. We shall first describe the correlation transformation and then the resulting standardized regression model.

The correlation transformation is a simple modification of the usual standardization of a variable. Standardizing a variable, as in (1.34), involves taking the difference between each observation and the mean of all observations and then expressing the differences in units of the standard deviation of the observations. Thus, the usual standardizations of the dependent variable Y and the independent variables $X_1, \ldots, X_{p-1}$ are as follows:

(8.40a)
$$\frac{Y_i - \bar{Y}}{s_Y}$$

(8.40b)
$$\frac{X_{ik} - \bar{X}_k}{s_k} \qquad (k = 1, \ldots, p-1)$$

where $\bar{Y}$ and $\bar{X}_k$ are the respective means of Y and X_k, and s_Y and s_k are the respective standard deviations defined as follows:

(8.40c)
$$s_Y = \sqrt{\frac{\sum_i (Y_i - \bar{Y})^2}{n-1}}$$

(8.40d)
$$s_k = \sqrt{\frac{\sum_i (X_{ik} - \bar{X}_k)^2}{n-1}} \qquad (k = 1, \ldots, p-1)$$

The correlation transformation uses the following function of the standardized variables in (8.40):

(8.41a)
$$Y_i' = \frac{1}{\sqrt{n-1}} \left(\frac{Y_i - \bar{Y}}{s_Y} \right)$$

(8.41b)
$$X_{ik}' = \frac{1}{\sqrt{n-1}} \left(\frac{X_{ik} - \bar{X}_k}{s_k} \right) \qquad (k = 1, \ldots, p-1)$$

Standardized Regression Model

The regression model with the transformed variables Y' and X_k' as defined by the correlation transformation in (8.41) is called a *standardized regression model* and is as follows:

(8.42)
$$Y_i' = \beta_1' X_{i1}' + \cdots + \beta_{p-1}' X_{i,p-1}' + \varepsilon_i'$$

The reason why there is no intercept parameter in the standardized regression model (8.42) is that the least squares calculations always would lead to an estimated intercept term of zero if an intercept parameter were present in the model.

It is easy to show that the new parameters $\beta_1', \ldots, \beta_{p-1}'$ and the original parameters $\beta_0, \beta_1, \ldots, \beta_{p-1}$ in the ordinary multiple regression model (7.7) are related as follows:

(8.43a)
$$\beta_k = \left(\frac{s_Y}{s_k} \right) \beta_k' \qquad (k = 1, \ldots, p-1)$$

(8.43b)
$$\beta_0 = \bar{Y} - \beta_1 \bar{X}_1 - \cdots - \beta_{p-1} \bar{X}_{p-1}$$

Thus, the new regression coefficients β_k' and the original regression coefficients β_k $(k = 1, \ldots, p - 1)$ are related by simple scaling factors involving ratios of standard deviations.

X′X Matrix for Transformed Variables

In order to be able to study the special nature of the $\mathbf{X}'\mathbf{X}$ matrix and the least squares normal equations when the variables have been transformed by the correlation transformation, we shall need to define two matrices involving coefficients of simple correlation. The first matrix, denoted by $\mathbf{r}_{XX}$, is called the *correlation matrix of the X variables*. It has as its elements the coefficients of simple correlation between all pairs of the X variables. This matrix is defined as follows:

$$(8.44) \qquad \mathbf{r}_{XX} \atop {\scriptstyle (p-1)\times(p-1)} = \begin{bmatrix} 1 & r_{12} & \cdots & r_{1,p-1} \\ r_{21} & 1 & \cdots & r_{2,p-1} \\ \vdots & \vdots & & \vdots \\ r_{p-1,1} & r_{p-1,2} & \cdots & 1 \end{bmatrix}$$

Here, r_{12} denotes the coefficient of simple correlation between X_1 and X_2, r_{13} denotes the coefficient of simple correlation between X_1 and X_3, and so on. Note that the main diagonal consists of 1s because the coefficient of simple correlation between a variable and itself is 1. The correlation matrix $\mathbf{r}_{XX}$ is symmetric; remember that $r_{kk'} = r_{k'k}$. Because of the symmetry of this matrix, computer printouts frequently will omit the lower or upper triangular block of elements.

The second matrix, to be denoted by $\mathbf{r}_{YX}$, is a vector containing the coefficients of simple correlation between the dependent variable Y and each of the X variables, denoted by r_{Y1}, r_{Y2}, etc.:

$$(8.45) \qquad \mathbf{r}_{YX} \atop {\scriptstyle (p-1)\times 1} = \begin{bmatrix} r_{Y1} \\ r_{Y2} \\ \vdots \\ r_{Y,p-1} \end{bmatrix}$$

Now we are ready to consider the $\mathbf{X}'\mathbf{X}$ matrix for the transformed variables in the standardized regression model (8.42). The $\mathbf{X}$ matrix here is:

$$(8.46) \qquad \mathbf{X} \atop {\scriptstyle n\times(p-1)} = \begin{bmatrix} X_{11}' & \cdots & X_{1,p-1}' \\ X_{21}' & \cdots & X_{2,p-1}' \\ \vdots & & \vdots \\ X_{n1}' & \cdots & X_{n,p-1}' \end{bmatrix}$$

Remember that the standardized regression model (8.42) does not contain an intercept term; hence, there is no column of 1s in the $\mathbf{X}$ matrix. It can be shown that the $\mathbf{X}'\mathbf{X}$ matrix for the transformed variables then is simply the correlation matrix of the

X variables defined in (8.44):

(8.47)
$$\underset{(p-1)\times(p-1)}{\mathbf{X'X}} = \mathbf{r}_{XX}$$

Since the $\mathbf{X'X}$ matrix for the transformed variables consists of coefficients of correlation between the X variables, all of its elements are between -1 and 1 and thus are of the same order of magnitude. As we pointed out earlier, this can be of great help in controlling roundoff errors when inverting the $\mathbf{X'X}$ matrix.

Note

We illustrate that the $\mathbf{X'X}$ matrix for the transformed variables is the correlation matrix of the X variables by considering two entries in the matrix:

1. In the upper left corner of $\mathbf{X'X}$ we have:

$$\sum (X'_{i1})^2 = \sum \left(\frac{X_{i1} - \bar{X}_1}{\sqrt{n-1}\, s_1} \right)^2 = \frac{\Sigma(X_{i1} - \bar{X}_1)^2}{n-1} \div s_1^2 = 1$$

2. In the first row, second column of $\mathbf{X'X}$, we have:

$$\sum X'_{i1} X'_{i2} = \sum \left(\frac{X_{i1} - \bar{X}_1}{\sqrt{n-1}\, s_1} \right)\left(\frac{X_{i2} - \bar{X}_2}{\sqrt{n-1}\, s_2} \right)$$

$$= \frac{1}{n-1} \frac{\Sigma(X_{i1} - \bar{X}_1)(X_{i2} - \bar{X}_2)}{s_1 s_2}$$

$$= \frac{\Sigma(X_{i1} - \bar{X}_1)(X_{i2} - \bar{X}_2)}{[\Sigma(X_{i1} - \bar{X}_1)^2 \Sigma(X_{i2} - \bar{X}_2)^2]^{1/2}}$$

But this equals r_{12}, the coefficient of correlation between X_1 and X_2 by (3.75).

Estimated Standardized Regression Coefficients

The least squares normal equations (7.20) for the ordinary multiple regression model:

$$\mathbf{X'Xb} = \mathbf{X'Y}$$

and the least squares estimators (7.21):

$$\mathbf{b} = (\mathbf{X'X})^{-1}\mathbf{X'Y}$$

can be expressed more simply for the transformed variables. It is easy to show that for the transformed variables, $\mathbf{X'Y}$ becomes:

(8.48)
$$\underset{(p-1)\times 1}{\mathbf{X'Y}} = \mathbf{r}_{YX}$$

where $\mathbf{r}_{YX}$ is defined in (8.45) as the vector of the coefficients of simple correlation between Y and each X variable. It now follows from (8.47) and (8.48) that the least squares normal equations and estimators of the regression coefficients of the standardized regression model (8.42) are as follows:

(8.49a)
$$\mathbf{r}_{XX}\mathbf{b} = \mathbf{r}_{YX}$$

(8.49b)
$$\mathbf{b} = \mathbf{r}_{XX}^{-1}\mathbf{r}_{YX}$$

where:

(8.49c)
$$\underset{(p-1)\times 1}{\mathbf{b}} = \begin{bmatrix} b_1' \\ b_2' \\ \cdot \\ \cdot \\ \cdot \\ b_{p-1}' \end{bmatrix}$$

The regression coefficients b_1', . . . , b_{p-1}' are often called *standardized regression coefficients*.

The return to the estimated regression coefficients for regression model (7.7) in the original variables is accomplished by employing the relations:

(8.50a)
$$b_k = \left(\frac{s_Y}{s_k}\right) b_k' \qquad (k = 1, \ldots, p-1)$$

(8.50b)
$$b_0 = \bar{Y} - b_1\bar{X}_1 - \cdots - b_{p-1}\bar{X}_{p-1}$$

Note

When the number of X variables in the regression model is $p - 1 = 2$, we can readily see the algebraic form of the standardized regression coefficients. We have:

(8.51a)
$$\mathbf{r}_{XX} = \begin{bmatrix} 1 & r_{12} \\ r_{12} & 1 \end{bmatrix}$$

(8.51b)
$$\mathbf{r}_{YX} = \begin{bmatrix} r_{Y1} \\ r_{Y2} \end{bmatrix}$$

(8.51c)
$$\mathbf{r}_{XX}^{-1} = \frac{1}{1 - r_{12}^2}\begin{bmatrix} 1 & -r_{12} \\ -r_{12} & 1 \end{bmatrix}$$

Hence, by (8.49b) we obtain:

(8.52)
$$\mathbf{b} = \frac{1}{1 - r_{12}^2}\begin{bmatrix} 1 & -r_{12} \\ -r_{12} & 1 \end{bmatrix}\begin{bmatrix} r_{Y1} \\ r_{Y2} \end{bmatrix} = \frac{1}{1 - r_{12}^2}\begin{bmatrix} r_{Y1} - r_{12}r_{Y2} \\ r_{Y2} - r_{12}r_{Y1} \end{bmatrix}$$

Thus:

(8.52a)
$$b_1' = \frac{r_{Y1} - r_{12}r_{Y2}}{1 - r_{12}^2}$$

(8.52b)
$$b_2' = \frac{r_{Y2} - r_{12}r_{Y1}}{1 - r_{12}^2}$$

TABLE 8.5   Correlation Transformation for the Zarthan Company Data

(a) Original Data

District i	Sales Y_i	Target Population X_{i1}	Per Capita Discretionary Income X_{i2}
1	162	274	2,450
2	120	180	3,254
⋮	⋮	⋮	⋮
15	212	370	2,605

$$\bar{Y} = 150.60 \qquad \bar{X}_1 = 241.73 \qquad \bar{X}_2 = 2,961.9$$
$$s_Y = 62.049 \qquad s_1 = 116.83 \qquad s_2 = 730.64$$

(b) Transformed Data

i	Y_i'	X_{i1}'	X_{i2}'
1	.04910	.07381	−.18724
2	−.13180	−.14122	.10686
⋮	⋮	⋮	⋮
15	.26447	.29342	−.13054

(c) Fitted Standardized Model
$$\hat{Y}' = .9339X_1' + .1083X_2'$$

Example

Table 8.5a repeats (partially) the original data for the Zarthan Company example (Table 7.2), and Table 8.5b contains the data transformed according to the correlation transformation (8.41). To illustrate the calculation of the transformed data, we have, using the means and standard deviations in Table 8.5a:

$$Y_1' = \frac{1}{\sqrt{n-1}}\left(\frac{Y_1 - \bar{Y}}{s_Y}\right) \qquad\qquad X_{11}' = \frac{1}{\sqrt{n-1}}\left(\frac{X_{11} - \bar{X}_1}{s_1}\right)$$

$$= \frac{1}{\sqrt{15-1}}\left(\frac{162 - 150.60}{62.049}\right) \qquad\qquad = \frac{1}{\sqrt{15-1}}\left(\frac{274 - 241.73}{116.83}\right)$$

$$= .04910 \qquad\qquad\qquad\qquad\qquad\quad = .07381$$

$$X'_{12} = \frac{1}{\sqrt{n-1}}\left(\frac{X_{12} - \bar{X}_2}{s_2}\right) = \frac{1}{\sqrt{15-1}}\left(\frac{2{,}450 - 2{,}961.9}{730.64}\right)$$

$$= -.18724$$

When fitting the standardized regression model (8.42) to the transformed data by means of a computer regression package, we obtain the fitted model in Table 8.5c:

$$\hat{Y}' = .9339X'_1 + .1083X'_2$$

Thus, an increase of one standard deviation of X_1 (target population) when X_2 is fixed leads to a much larger increase in expected sales (in units of standard deviations of Y) than does an increase of one standard deviation of X_2 (per capita discretionary income) when X_1 is fixed.

To shift from the standardized regression coefficients b'_1 and b'_2 back to the regression coefficients for the model with the original variables, we employ (8.50). Using the data in Table 8.5, we obtain:

$$b_1 = \left(\frac{s_Y}{s_1}\right)b'_1 = \frac{62.049}{116.83}(.9339) = .496$$

$$b_2 = \left(\frac{s_Y}{s_2}\right)b'_2 = \frac{62.049}{730.64}(.1083) = .00920$$

$$b_0 = \bar{Y} - b_1\bar{X}_1 - b_2\bar{X}_2$$

$$= 150.60 - .496(241.73) - .00920(2{,}961.9) = 3.45$$

The estimated regression function for the multiple regression model in the original variables therefore is:

$$\hat{Y} = 3.45 + .496X_1 + .00920X_2$$

This is the same fitted regression function we obtained in Chapter 7. Here, b_1 and b_2 cannot be compared directly because X_1 is in units of thousands of persons and X_2 is in units of dollars.

Sometimes the standardized regression coefficients $b'_1 = .9339$ and $b'_2 = .1083$ are interpreted as showing that target population (X_1) has a much greater impact on sales than per capita discretionary income (X_2) because b'_1 is much larger than b'_2. However, as we will see in the next section, one must be cautious about interpreting regression coefficients, whether standardized or not. The reason is that when the independent variables are correlated among themselves, as here, the regression coefficients are affected by the other independent variables in the model. For the Zarthan Company data, the correlation between X_1 and X_2 is $r_{12} = .569$.

Not only does the presence of correlations among the independent variables affect the magnitude of the standardized regression coefficients but so do the spacings of the observations on the independent variables. Sometimes these spacings may be quite arbitrary.

Hence, it is ordinarily not wise to interpret the magnitudes of standardized regression coefficients as reflecting the comparative importance of the independent variables.

Comments

1. Some computer packages present both the regression coefficients b_k for the model in the original variables as well as the standardized coefficients b'_k. The latter are sometimes labeled *beta coefficients* in printouts.

2. Some computer printouts show the magnitude of the determinant of the correlation matrix of the X variables. A near-zero value for this determinant implies both a high degree of linear association among the X variables and a high potential for roundoff errors. For the case of two X variables, this determinant is seen from (8.51c) to be $1 - r_{12}^2$, which approaches 0 as r_{12}^2 approaches 1.

3. When the correlation matrix of the X variables is augmented by a row and column for Y, it is called the *correlation matrix*. A correlation matrix shows the coefficients of correlation for all pairs of dependent and X variables. This information is useful in a variety of tasks—for instance, in selecting the final independent variables to be included in the model. Many computer programs display the correlation matrix in the printout.

For the case of two X variables, the correlation matrix is as follows:

$$\begin{bmatrix} 1 & r_{Y1} & r_{Y2} \\ r_{Y1} & 1 & r_{12} \\ r_{Y2} & r_{12} & 1 \end{bmatrix}$$

Since the correlation matrix is symmetric, the lower (or upper) triangular block of elements is frequently omitted in computer printouts.

4. It is possible to use the correlation transformation with a computer package that does not permit regression through the origin, because the intercept coefficient b'_0 will always be zero for data so transformed. The other regression coefficients will also be correct.

5. Use of the standardized variables in (8.40a and b) without the correlation transformation modification in (8.41) will lead to the same standardized regression coefficients as those in (8.49b) for the correlation-transformed variables. However, the elements of the $\mathbf{X'X}$ matrix will not then be bounded between -1 and 1.

8.5 MULTICOLLINEARITY AND ITS EFFECTS

In multiple regression analysis, one is often concerned with the nature and significance of the relations between the independent variables and the dependent variable. Questions that are frequently asked include:

1. What is the relative importance of the effects of the different independent variables?
2. What is the magnitude of the effect of a given independent variable on the dependent variable?
3. Can any independent variable be dropped from the model because it has little or no effect on the dependent variable?
4. Should any independent variables not yet included in the model be considered for possible inclusion?

If the independent variables included in the model are (1) uncorrelated among themselves and (2) uncorrelated with any other independent variables that are related to the dependent variable but omitted from the model, relatively simple answers can be given to these questions. Unfortunately, in many nonexperimental situations in business, economics, and the social and biological sciences, the independent variables tend to be correlated among themselves and with other variables that are related to the dependent variable but are not included in the model. For example, in a regression of family food expenditures on the independent variables family income, family savings, and age of head of household, the independent variables will be correlated among themselves. Further, the independent variables will also be correlated with other socioeconomic variables not included in the model that do affect family food expenditures, such as family size.

When the independent variables are correlated among themselves, *intercorrelation* or *multicollinearity* among them is said to exist. (Sometimes the latter term is reserved for those instances when the correlation among independent variables is very high.) We shall explore now a variety of interrelated problems that are created by multicollinearity among the independent variables. First, however, we examine the situation when the independent variables are not correlated.

Effects when Independent Variables Are Uncorrelated

Table 8.6 contains data for a small-scale experiment on the effect of work crew size (X_1) and level of bonus pay (X_2) on crew productivity score (Y). It is easy to show that X_1 and X_2 are uncorrelated here, i.e., $r_{12}^2 = 0$, where r_{12}^2 denotes the coefficient of simple determination between X_1 and X_2. Table 8.7a contains the fitted regression function and analysis of variance table when both X_1 and X_2 are included in the model. Table 8.7b contains the same information when only X_1 is included in the model, and Table 8.7c contains this information when only X_2 is in the model.

An important feature to note in Table 8.7 is that the regression coefficient for X_1, $b_1 = 5.375$, is the same whether only X_1 is included in the model or both independent variables are included. The same holds for $b_2 = 9.250$. This is a result of the two independent variables being uncorrelated.

TABLE 8.6   Data for Work Crew Productivity Example with Uncorrelated Independent Variables

Trial i	Crew Size X_{i1}	Bonus Pay X_{i2}	Crew Productivity Score Y_i
1	4	$2	42
2	4	2	39
3	4	3	48
4	4	3	51
5	6	2	49
6	6	2	53
7	6	3	61
8	6	3	60

TABLE 8.7   ANOVA Tables for Work Crew Productivity Example with Uncorrelated Independent Variables

(a) Regression of Y on X_1 and X_2
$$\hat{Y} = .375 + 5.375X_1 + 9.250X_2$$

Source of Variation	SS	df	MS
Regression	402.250	2	201.125
Error	17.625	5	3.525
Total	419.875	7	

(b) Regression of Y on X_1
$$\hat{Y} = 23.500 + 5.375X_1$$

Source of Variation	SS	df	MS
Regression	231.125	1	231.125
Error	188.750	6	31.458
Total	419.875	7	

(c) Regression of Y on X_2
$$\hat{Y} = 27.250 + 9.250X_2$$

Source of Variation	SS	df	MS
Regression	171.125	1	171.125
Error	248.750	6	41.458
Total	419.875	7	

Thus, when the independent variables are uncorrelated, the effects ascribed to them by a first-order regression model are the same no matter which other independent variables are included in the model. This is a strong argument for controlled experiments whenever possible, since experimental control permits making the independent variables uncorrelated.

Another important feature of Table 8.7 is related to the error sums of squares. Note from Table 8.7 that the extra sum of squares $SSR(X_1 \mid X_2)$ equals the regression sum of squares $SSR(X_1)$ when only X_1 is in the regression model:

$$SSR(X_1 \mid X_2) = SSE(X_2) - SSE(X_1, X_2)$$
$$= 248.750 - 17.625 = 231.125$$
$$SSR(X_1) = 231.125$$

Similarly, the extra sum of squares $SSR(X_2 \mid X_1)$ equals $SSR(X_2)$, the regression sum of squares when only X_2 is in the regression model:

$$SSR(X_2 \mid X_1) = SSE(X_1) - SSE(X_1, X_2)$$

$$= 188.750 - 17.625 = 171.125$$

$$SSR(X_2) = 171.125$$

In general, when two or more independent variables are uncorrelated, the marginal contribution of one independent variable in reducing the error sum of squares when the other independent variables are in the model is exactly the same as when this independent variable is in the model alone.

Note

To show that the regression coefficient of X_1 is unchanged when X_2 is added to the regression model in the case where X_1 and X_2 are uncorrelated, consider the algebraic expression for b_1 in the multiple regression model with two independent variables:

$$(8.53) \qquad b_1 = \frac{\dfrac{\Sigma(X_{i1} - \bar{X}_1)(Y_i - \bar{Y})}{\Sigma(X_{i1} - \bar{X}_1)^2} - \left[\dfrac{\Sigma(Y_i - \bar{Y})^2}{\Sigma(X_{i1} - \bar{X}_1)^2}\right]^{1/2} r_{Y2}r_{12}}{1 - r_{12}^2}$$

where, as before, r_{Y2} denotes the coefficient of simple correlation between Y and X_2, and r_{12} denotes the coefficient of simple correlation between X_1 and X_2.

If X_1 and X_2 are uncorrelated, $r_{12} = 0$, and (8.53) reduces to:

$$(8.53a) \qquad b_1 = \frac{\Sigma(X_{i1} - \bar{X}_1)(Y_i - \bar{Y})}{\Sigma(X_{i1} - \bar{X}_1)^2}$$

But (8.53a) is the estimator of the slope for the simple linear regression of Y on X_1, per (2.10a).

Hence, if X_1 and X_2 are uncorrelated, adding X_2 to the regression model does not change the regression coefficient for X_1; correspondingly, adding X_1 to the regression model does not change the regression coefficient for X_2.

Nature of Problem when Independent Variables Are Perfectly Correlated

To see the essential nature of the problem of multicollinearity, we shall employ a simple example where the two independent variables are perfectly correlated. The data in Table 8.8 refer to four sample observations on a dependent variable and two independent variables. Mr. A was asked to fit the multiple regression function:

$$(8.54) \qquad E\{Y\} = \beta_0 + \beta_1 X_1 + \beta_2 X_2$$

He returned in a short time with the fitted response function:

$$(8.55) \qquad \hat{Y} = -87 + X_1 + 18X_2$$

He was proud because the response function fits the data perfectly. The fitted values are shown in Table 8.8.

TABLE 8.8 Example of Perfectly Correlated Independent Variables

| Case | | | | Fitted Values for Response Function | |
				(8.55)	(8.56)
i	X_{i1}	X_{i2}	Y_i		
1	2	6	23	23	23
2	8	9	83	83	83
3	6	8	63	63	63
4	10	10	103	103	103

Response Functions:
$$(8.55) \quad \hat{Y} = -87 + X_1 + 18X_2$$
$$(8.56) \quad \hat{Y} = -7 + 9X_1 + 2X_2$$

It so happened that Ms. B also was asked to fit the response function (8.54) to the same data, and she proudly obtained:

$$(8.56) \qquad\qquad \hat{Y} = -7 + 9X_1 + 2X_2$$

Her response function also fits the data perfectly, as shown in Table 8.8.

Indeed, it can be shown that infinitely many response functions will fit the data in Table 8.8 perfectly. The reason is that the independent variables X_1 and X_2 are perfectly related, according to the relation:

$$(8.57) \qquad\qquad X_2 = 5 + .5X_1$$

Note carefully that fitted response functions (8.55) and (8.56) are entirely different response surfaces. The regression coefficients are different, and the fitted values will differ when X_1 and X_2 do not follow relation (8.57). For example, the fitted value for response function (8.55) when $X_1 = 5$ and $X_2 = 5$ is:

$$\hat{Y} = -87 + 5 + 18(5) = 8$$

while the fitted value for response function (8.56) is:

$$\hat{Y} = -7 + 9(5) + 2(5) = 48$$

Thus, when X_1 and X_2 are perfectly related and, as in our example, the data do not contain any random error component, many different response functions will lead to the same perfectly fitted values for the observations and to the same fitted values for any other (X_1, X_2) combinations following the relation between X_1 and X_2. Yet these response functions are not the same and will lead to different fitted values for (X_1, X_2) combinations that do not follow the relation between X_1 and X_2.

Two key implications of this example are:

1. The perfect relation between X_1 and X_2 did not inhibit our ability to obtain a good fit to the data.
2. Since many different response functions provide the same good fit, one cannot interpret any one set of regression coefficients as reflecting the effects of the dif-

ferent independent variables. Thus, in response function (8.55), $b_1 = 1$ and $b_2 = 18$ do not imply that X_2 is the key independent variable and X_1 plays little role, because response function (8.56) provides an equally good fit and its regression coefficients have opposite comparative magnitudes.

Effects of Multicollinearity

In actual practice, we seldom find independent variables that are perfectly related or data that do not contain some random error component. Nevertheless, the implications just noted for our idealized example still have relevance.

1. The fact that some or all independent variables are correlated among themselves does not, in general, inhibit our ability to obtain a good fit nor does it tend to affect inferences about mean responses or predictions of new observations, provided these inferences are made within the region of observations. (Figure 7.9 on p. 263 provides an illustration of the concept of the region of observations for the case of two independent variables.)

2. The counterpart in real life to the many different regression functions providing equally good fits to the data in our idealized example is that the estimated regression coefficients tend to have large sampling variability when the independent variables are highly correlated. Thus, the estimated regression coefficients tend to vary widely from one sample to the next when the independent variables are highly correlated. As a result, only imprecise information may be available about the individual true regression coefficients. Indeed, each of the estimated regression coefficients individually may be statistically not significant even though a definite statistical relation exists between the dependent variable and the set of independent variables.

3. The common interpretation of regression coefficients as measuring the change in the expected value of the dependent variable when the corresponding independent variable is increased by one unit while all other independent variables are held constant is not fully applicable when multicollinearity exists. While it may be conceptually possible to vary one independent variable and hold the others constant, it may not be possible in practice to do so for independent variables that are highly correlated. For example, in a regression model for predicting crop yield from amount of rainfall and hours of sunshine, the relation between the two independent variables makes it unrealistic to consider varying one while holding the other constant. Therefore, the simple interpretation of the regression coefficients as measuring marginal effects is often unwarranted with highly correlated independent variables.

We shall illustrate these effects of multicollinearity by returning to the body fat example. The basic data were given in Table 8.1, and the regression results for different fitted models were presented in Table 8.2. Independent variables X_1 and X_2 are highly correlated, as may be seen from the correlation matrix of the X variables in Table 8.1b. The coefficient of simple correlation is $r_{12} = .92$. X_3, on the other hand, is not so highly correlated to X_1 and X_2 individually, the correlation coefficients being $r_{13} = .46$ and $r_{23} = .08$. (But X_3 is highly correlated with X_1 and

X_2 together; the coefficient of multiple determination when X_3 is regressed on X_1 and X_2 is .998.)

Effects on Regression Coefficients. Note from Table 8.2 that the regression coefficient for X_1, triceps skinfold thickness, varies markedly depending upon which other variables are included in the model:

Variables in Model	b_1	b_2
X_1	.8572	—
X_2	—	.8565
X_1, X_2	.2224	.6594
X_1, X_2, X_3	4.334	−2.857

The story is the same for the regression coefficient for X_2. Indeed, the regression coefficient b_2 even changes sign when X_3 is added to the model that includes X_1 and X_2.

The important conclusion we must draw is: When independent variables are correlated, the regression coefficient of any independent variable depends on which other independent variables are included in the model and which ones are left out. Thus, a regression coefficient does not reflect any inherent effect of the particular independent variable on the dependent variable but only a marginal or partial effect, given whatever other correlated independent variables are included in the model.

Note

The fact that intercorrelated independent variables that are omitted from the regression model can influence the regression coefficients in the regression model is illustrated by an analyst who was perplexed about the sign of a regression coefficient in the fitted regression model. He had found in a regression of territory company sales on territory population size, per capita income, and some other independent variables that the confidence interval for the regression coefficient for population size indicated it is negative. The analyst should have considered some of the omitted independent variables in search of an explanation. A consultant noted that the analyst did not include the major competitor's market penetration in the model. Since the competitor was most active and effective in territories with large populations, and thereby kept company sales down in these territories, the result of the omission of this independent variable from the model was a negative coefficient for the population size variable.

Effects on Extra Sums of Squares. Like a regression coefficient, the marginal contribution of an independent variable in reducing the error sum of squares varies, depending on which other variables are already in the regression model, when the independent variables are correlated. For example, Table 8.2 provides the following extra sums of squares for X_1:

$$SSR(X_1) = 352.27$$

$$SSR(X_1 \mid X_2) = 3.47$$

The reason why $SSR(X_1 \mid X_2)$ is so small compared with $SSR(X_1)$ is that X_1 and X_2 are highly correlated. Thus, when X_2 is already in the regression model, the marginal contribution of X_1 in reducing the error sum of squares is comparatively small because X_2 contains much of the same information as X_1.

The same story is found in Table 8.2 for X_2. Here $SSR(X_2 \mid X_1) = 33.17$, which is much smaller than $SSR(X_2) = 381.97$. The important conclusion is this: When independent variables are correlated, there is no unique sum of squares that can be ascribed to an independent variable as reflecting its effect in reducing the total variation in Y. The reduction in the total variation ascribed to an independent variable must be viewed in the context of the other independent variables included in the model, whenever the independent variables are correlated.

Note

Multicollinearity also affects the coefficients of partial determination through its effects on the extra sums of squares. Note from Table 8.2 for the body fat example, for instance, that X_1 is highly correlated with Y:

$$r_{Y1}^2 = \frac{SSR(X_1)}{SSTO} = \frac{352.27}{495.39} = .71$$

However, the coefficient of partial determination between Y and X_1, when X_2 is already in the regression model, is much smaller:

$$r_{Y1.2}^2 = \frac{SSR(X_1 \mid X_2)}{SSE(X_2)} = \frac{3.47}{113.42} = .03$$

The reason for the small coefficient of partial determination here is, as we have seen, that X_1 and X_2 are highly correlated. Hence, X_1 provides only relatively limited additional information beyond that furnished by X_2.

Effects on $s\{b_k\}$.   Note from Table 8.2 how much more imprecise the estimated regression coefficients b_1 and b_2 become as more independent variables are added to the regression model:

Variables in Model	$s\{b_1\}$	$s\{b_2\}$
X_1	.1288	—
X_2	—	.1100
X_1, X_2	.3034	.2912
X_1, X_2, X_3	3.016	2.582

Again, the high degree of multicollinearity among the independent variables is responsible for the inflated variability of the estimated regression coefficients.

Effects on Fitted Values and Predictions.   Notice in Table 8.2 that the high multicollinearity among the independent variables does not prevent the mean square error, measuring the variability of the error terms, from being steadily reduced as additional variables are added to the regression model:

Variables in Model	MSE
X_1	7.95
X_1, X_2	6.47
X_1, X_2, X_3	6.15

Furthermore, the precision of fitted values within the range of the observations on the independent variables is not eroded with the addition of correlated independent variables into the regression model. Consider the estimation of mean body fat when the only independent variable included in the model is triceps skinfold thickness (X_1) for $X_{h1} = 25.0$. The fitted value and its estimated standard deviation are (calculations not shown):

$$\hat{Y}_h = 19.93 \qquad s\{\hat{Y}_h\} = .632$$

When the highly correlated independent variable thigh circumference (X_2) is also included in the model, the estimated mean body fat, together with its estimated standard deviation, are as follows for $X_{h1} = 25.0$ and $X_{h2} = 50.0$:

$$\hat{Y}_h = 19.36 \qquad s\{\hat{Y}_h\} = .624$$

Thus, the precision of the estimated mean response is equally good as before, despite the addition of the second independent variable which is highly correlated with the first one. This stability in the precision of the estimated mean response occurred despite the fact that the estimated standard deviation of b_1 became substantially larger when X_2 was added to the model (Table 8.2). The essential reason for the stability is that the covariance between b_1 and b_2 is negative, and plays a strong counteracting influence to the increase in $s^2\{b_1\}$ in determining the value of $s^2\{\hat{Y}_h\}$ as given in (7.71).

When all three independent variables are included in the model, the estimated mean body fat, together with its estimated standard deviation, are as follows for $X_{h1} = 25.0$, $X_{h2} = 50.0$, and $X_{h3} = 29.0$:

$$\hat{Y}_h = 19.19 \qquad s\{\hat{Y}_h\} = .621$$

Thus, the addition of the third independent variable, which is highly correlated with the first two independent variables together, also does not materially affect the precision of the estimated mean response.

Effects on Simultaneous Tests of β_k.   A not infrequent abuse in the analysis of multiple regression models is to examine the t^* statistic in (7.46b):

$$t^* = \frac{b_k}{s\{b_k\}}$$

for each regression coefficient in turn to decide whether $\beta_k = 0$ for $k = 1, \ldots,$ $p - 1$. Even if a simultaneous inference procedure is used, and often it is not, problems still exist when the independent variables are highly correlated.

Suppose we wish to test whether $\beta_1 = 0$ and $\beta_2 = 0$ in the body fat example regression model with two independent variables of Table 8.2c. Controlling the family level of significance at .05, we require with the Bonferroni method that each of the two t tests be conducted with level of significance .025. Hence, we need $t(.9875;$ $17) = 2.46$. Since both t^* statistics in Table 8.2c have absolute values that do not exceed 2.46, we would conclude from the two separate tests that $\beta_1 = 0$ and $\beta_2 = 0$. Yet the proper F test for H_0: $\beta_1 = \beta_2 = 0$ would conclude H_a, that not both coefficients equal zero. This can be seen from Table 8.2c, where we find $F^* = MSR/MSE = 192.72/6.47 = 29.8$, which far exceeds $F(.95; 2, 17) = 3.59$.

The reason for this apparently paradoxical result is that the t^* test is a marginal test, as we have seen in (8.15) from the perspective of the general linear test approach. Thus, when X_1 and X_2 are highly correlated, $SSR(X_1 \mid X_2)$ is small, indicating that X_1 does not provide much more information than does X_2, which already is in the model; hence, we are led to the conclusion $\beta_1 = 0$. Similarly, we are led to conclude $\beta_2 = 0$ here because $SSR(X_2 \mid X_1)$ is small, indicating that X_2 does not contain much more additional information when X_1 is already in the model. But the two tests of the marginal effects of X_1 and X_2 together are not equivalent to testing whether there is a regression relation between Y and the independent variables X_1 and X_2. The reason is that the reduced model for each of the separate tests contains the other independent variable, whereas the reduced model for testing whether both $\beta_1 = 0$ and $\beta_2 = 0$ would contain neither independent variable. The proper F test shows that there is a definite regression relation here between Y and X_1 and X_2.

The same paradox would be encountered in Table 8.2d for the regression model with three independent variables if three simultaneous tests on the regression coefficients were conducted at the family level of significance of .05.

Note

We have just seen that it is possible that a set of independent variables is related to the dependent variable, yet all of the individual tests on the regression coefficients will lead to the conclusion that they equal zero because of the multicollinearity among the independent variables. This apparently paradoxical result is also possible under special circumstances when there is no multicollinearity among the independent variables. The special circumstances are not likely to be found in practice, however.

Diagnostics for Multicollinearity and Remedial Measures

As we have seen, multicollinearity among the independent variables can have important consequences for interpreting and using a fitted regression model. The diagnostic tool considered here for identifying multicollinearity—namely, the pairwise coefficients of simple correlation between the independent variables—is often helpful. There are times, however, when serious multicollinearity exists without being disclosed by the pairwise correlation coefficients. In Chapter 11, we present a more powerful tool for identifying the existence of serious multicollinearity, and we will also consider there a number of remedial measures for lessening the effects of multicollinearity.

Comments

1. It was noted in Section 8.4 that a near-zero determinant of $\mathbf{X'X}$ is a potential source of serious roundoff errors in least squares results. Severe multicollinearity has the effect of making this determinant come close to zero. Thus, under severe multicollinearity, the regression coefficients may be subject to large roundoff errors as well as large sampling variances. Hence, it is particularly advisable to employ the correlation transformation (8.41) in fitting the regression model when multicollinearity is present.

2. Just as high intercorrelations between the independent variables tend to make the estimated regression coefficients imprecise (i.e., erratic from sample to sample), so do the coefficients of partial correlation between the dependent variable and each of the independent variables tend to become erratic from sample to sample when the independent variables are highly correlated.

3. The effect of intercorrelations between the independent variables on the standard deviations of the estimated regression coefficients can be seen readily when the variables in the model are transformed by means of the correlation transformation (8.41). Consider the first-order model with two independent variables:

$$(8.58) \qquad Y_i = \beta_0 + \beta_1 X_{i1} + \beta_2 X_{i2} + \varepsilon_i$$

This model in the variables transformed by (8.41) becomes:

$$(8.59) \qquad Y_i' = \beta_1' X_{i1}' + \beta_2' X_{i2}' + \varepsilon_i'$$

The $(\mathbf{X'X})^{-1}$ matrix for this standardized model is given by (8.51c):

$$(8.60) \qquad (\mathbf{X'X})^{-1} = \mathbf{r}_{XX}^{-1} = \frac{1}{1 - r_{12}^2} \begin{bmatrix} 1 & -r_{12} \\ -r_{12} & 1 \end{bmatrix}$$

Hence, the variance-covariance matrix of the estimated regression coefficients is by (7.41):

$$(8.61) \qquad \sigma^2\{\mathbf{b}\} = (\sigma')^2 \mathbf{r}_{XX}^{-1} = (\sigma')^2 \frac{1}{1 - r_{12}^2} \begin{bmatrix} 1 & -r_{12} \\ -r_{12} & 1 \end{bmatrix}$$

where $(\sigma')^2$ is the error term variance for the standardized model (8.59).

Thus, the estimated regression coefficients b_1' and b_2' have the same variance here:

$$(8.62) \qquad \sigma^2\{b_1'\} = \sigma^2\{b_2'\} = \frac{(\sigma')^2}{1 - r_{12}^2}$$

which becomes larger as the correlation between X_1 and X_2 increases. Indeed, as X_1 and X_2 approach perfect correlation (i.e., as r_{12}^2 approaches 1), the variances of b_1' and b_2' become larger without limit.

8.6 MATRIX FORMULATION OF GENERAL LINEAR TEST

The procedures for tests about regression coefficients summarized in Section 8.2 are generally adequate. We can use extra sums of squares whenever we wish to test whether some regression coefficients are equal to zero; otherwise we can fit the full and reduced models in conducting the general linear test about the regression coefficients.

Occasionally, however, it is necessary to carry out the general linear test in matrix form, such as for certain types of tests in the analysis of variance. We now explain how the general linear test statistic (3.69):

$$F^* = \frac{SSE(R) - SSE(F)}{df_R - df_F} \div \frac{SSE(F)}{df_F}$$

can be represented in matrix form.

Full Model

The full regression model with $p - 1$ predictor variables is given by (7.18):

$$(8.63) \qquad \mathbf{Y} = \mathbf{X\beta} + \mathbf{\varepsilon}$$

The least squares estimators for the full model will now be denoted by $\mathbf{b}_F$ and are, as before, given by (7.21):

$$(8.64) \qquad \mathbf{b}_F = (\mathbf{X'X})^{-1}\mathbf{X'Y}$$

Also, the error sum of squares is given by (7.30):

$$(8.65) \qquad SSE(F) = (\mathbf{Y} - \mathbf{Xb}_F)'(\mathbf{Y} - \mathbf{Xb}_F) = \mathbf{Y'Y} - \mathbf{b}_F'\mathbf{X'Y}$$

which has associated with it $df_F = n - p$ degrees of freedom.

Statement of Hypothesis H_0

A linear test hypothesis H_0 is represented in matrix form as follows:

$$(8.66) \qquad H_0: \underset{s \times p}{\mathbf{C}} \ \underset{p \times 1}{\mathbf{\beta}} = \underset{s \times 1}{\mathbf{h}}$$

where $\mathbf{C}$ is a specified $s \times p$ matrix of rank s and $\mathbf{h}$ is a specified $s \times 1$ vector.

Example 1. The regression model contains two X variables, and we wish to test $H_0: \beta_1 = 0$. Then:

$$\underset{1 \times 3}{\mathbf{C}} = [0 \quad 1 \quad 0] \qquad \underset{1 \times 1}{\mathbf{h}} = [0]$$

and we have:

$$H_0: \mathbf{C}\boldsymbol{\beta} = \begin{bmatrix} 0 & 1 & 0 \end{bmatrix} \begin{bmatrix} \beta_0 \\ \beta_1 \\ \beta_2 \end{bmatrix} = [0]$$

or $H_0: \beta_1 = 0$.

Example 2.   The regression model contains two X variables, and we wish to test H_0: $\beta_1 = \beta_2 = 0$. Then:

$$\underset{2\times3}{\mathbf{C}} = \begin{bmatrix} 0 & 1 & 0 \\ 0 & 0 & 1 \end{bmatrix} \qquad \underset{2\times1}{\mathbf{h}} = \begin{bmatrix} 0 \\ 0 \end{bmatrix}$$

and we have:

$$H_0: \mathbf{C}\boldsymbol{\beta} = \begin{bmatrix} 0 & 1 & 0 \\ 0 & 0 & 1 \end{bmatrix} \begin{bmatrix} \beta_0 \\ \beta_1 \\ \beta_2 \end{bmatrix} = \begin{bmatrix} 0 \\ 0 \end{bmatrix}$$

or $H_0: \beta_1 = 0, \beta_2 = 0$.

Example 3.   The regression model contains three X variables, and we wish to test $H_0: \beta_1 = \beta_2$. Then:

$$\underset{1\times4}{\mathbf{C}} = \begin{bmatrix} 0 & 1 & -1 & 0 \end{bmatrix} \qquad \underset{1\times1}{\mathbf{h}} = [0]$$

and we have:

$$H_0: \mathbf{C}\boldsymbol{\beta} = \begin{bmatrix} 0 & 1 & -1 & 0 \end{bmatrix} \begin{bmatrix} \beta_0 \\ \beta_1 \\ \beta_2 \\ \beta_3 \end{bmatrix} = [0]$$

or $H_0: \beta_1 - \beta_2 = 0$.

Reduced Model

The reduced model is:

(8.67) $\mathbf{Y} = \mathbf{X}\boldsymbol{\beta} + \boldsymbol{\varepsilon}$ where $\mathbf{C}\boldsymbol{\beta} = \mathbf{h}$

It can be shown that the least squares estimators under the reduced model, to be denoted by $\mathbf{b}_R$, are:

(8.68) $\mathbf{b}_R = \mathbf{b}_F - (\mathbf{X}'\mathbf{X})^{-1}\mathbf{C}'(\mathbf{C}(\mathbf{X}'\mathbf{X})^{-1}\mathbf{C}')^{-1}(\mathbf{C}\mathbf{b}_F - \mathbf{h})$

and the error sum of squares is:

$$(8.69) \qquad\qquad SSE(R) = (\mathbf{Y} - \mathbf{Xb}_R)'(\mathbf{Y} - \mathbf{Xb}_R)$$

which has associated with it $df_R = n - (p - s)$ degrees of freedom.

Test Statistic

It can be shown that the difference $SSE(R) - SSE(F)$ can be expressed as follows:

$$(8.70) \qquad SSE(R) - SSE(F) = (\mathbf{Cb}_F - \mathbf{h})'(\mathbf{C}(\mathbf{X'X})^{-1}\mathbf{C'})^{-1}(\mathbf{Cb}_F - \mathbf{h})$$

which has associated with it $df_R - df_F = (n - p + s) - (n - p) = s$ degrees of freedom.

Hence, the test statistic is:

$$(8.71) \qquad\qquad F^* = \frac{SSE(R) - SSE(F)}{s} \div \frac{SSE(F)}{n - p}$$

where $SSE(R) - SSE(F)$ is given by (8.70) and $SSE(F)$ is given by (8.65).

To confirm that the degrees of freedom associated with $SSE(R) - SSE(F)$ are s, consider the earlier three examples.

1. In Example 1, $s = 1$. This is consistent with the numerator degrees of freedom in test statistic (8.22).
2. In Example 2, $s = 2$. This is consistent with the numerator degrees of freedom in test statistic (8.20).
3. In Example 3, $s = 1$. This is consistent with the numerator degrees of freedom in the test statistic for the example on page 284.

Note

The least squares estimators $\mathbf{b}_R$ under the reduced model, given in (8.68), can be derived by minimizing the least squares criterion $Q = (\mathbf{Y} - \mathbf{X\beta})'(\mathbf{Y} - \mathbf{X\beta})$ subject to the constraint $\mathbf{C\beta} - \mathbf{h} = \mathbf{0}$, using Lagrangian multipliers.

PROBLEMS

8.1. State the number of degrees of freedom that are associated with each of the following extra sums of squares: (1) $SSR(X_1 \mid X_2)$; (2) $SSR(X_2 \mid X_1, X_3)$; (3) $SSR(X_1, X_2 \mid X_3, X_4)$; (4) $SSR(X_1, X_2, X_3 \mid X_4, X_5)$.

8.2. Explain in what sense the regression sum of squares $SSR(X_1)$ is an extra sum of squares.

8.3. Refer to **Brand preference** Problem 7.8.
   a. Obtain the analysis of variance table that decomposes the regression sum of squares into extra sums of squares associated with X_1 and with X_2, given X_1.
   b. Test whether X_2 can be dropped from the regression model given that X_1 is retained. Use the F^* test statistic and level of significance .01. State the alternatives, decision rule, and conclusion. What is the P-value of the test?

8.4. Refer to **Chemical shipment** Problem 7.12.

    a. Obtain the analysis of variance table that decomposes the regression sum of squares into extra sums of squares associated with X_2 and with X_1, given X_2.

    b. Test whether X_1 can be dropped from the regression model given that X_2 is retained. Use the F^* test statistic and $\alpha = .05$. State the alternatives, decision rule, and conclusion. What is the P-value of the test?

    c. Does $SSR(X_1) + SSR(X_2 \mid X_1)$ equal $SSR(X_2) + SSR(X_1 \mid X_2)$ here? Must this always be the case?

8.5. Refer to **Patient satisfaction** Problem 7.17.

    a. Obtain the analysis of variance table that decomposes the regression sum of squares into extra sums of squares associated with X_2; with X_1, given X_2; and with X_3, given X_2 and X_1.

    b. Test whether X_3 can be dropped from the regression model given that X_1 and X_2 are retained. Use the F^* test statistic and level of significance .025. State the alternatives, decision rule, and conclusion. What is the P-value of the test?

8.6. Refer to **Patient satisfaction** Problem 7.17. Test whether both X_2 and X_3 can be dropped from the regression model given that X_1 is retained. Use $\alpha = .025$. State the alternatives, decision rule, and conclusion. What is the P-value of the test?

8.7. Refer to **Mathematicians' salaries** Problem 7.20.

    a. Obtain the analysis of variance table that decomposes the regression sum of squares into extra sums of squares associated with X_2; with X_3, given X_2; and with X_1, given X_2 and X_3.

    b. Test whether X_1 can be dropped from the regression model given that X_2 and X_3 are retained. Use the F^* test statistic and level of significance .01. State the alternatives, decision rule, and conclusion. What is the P-value of the test?

8.8. Refer to **Mathematicians' salaries** Problems 7.20 and 8.7. Test whether both X_1 and X_3 can be dropped from the regression model given that X_2 is retained; use $\alpha = .01$. State the alternatives, decision rule, and conclusion. What is the P-value of the test?

8.9. Refer to **Patient satisfaction** Problem 7.17. Test whether $\beta_1 = -1.0$ and $\beta_2 = 0$; use $\alpha = .025$. State the alternatives, full and reduced models, decision rule, and conclusion.

8.10. Refer to **Mathematicians' salaries** Problem 7.20. Test whether $\beta_1 = \beta_3$; use $\alpha = .01$. State the alternatives, full and reduced models, decision rule, and conclusion.

8.11. Refer to the work crew productivity example on page 296.

    a. Calculate r_{Y1}^2, r_{Y2}^2, r_{12}^2, $r_{Y1.2}^2$, $r_{Y2.1}^2$, and R^2. Explain what each coefficient measures and interpret your results.

    b. Are any of the results obtained in part (a) special because the two independent variables are uncorrelated?

8.12. Refer to **Brand preference** Problem 7.8. Calculate r_{Y1}^2, r_{Y2}^2, r_{12}^2, $r_{Y1.2}^2$, $r_{Y2.1}^2$, and R^2. Explain what each coefficient measures and interpret your results.

8.13. Refer to **Chemical shipment** Problem 7.12. Calculate r_{Y1}^2, r_{Y2}^2, r_{12}^2, $r_{Y1.2}^2$, $r_{Y2.1}^2$, and R^2. Explain what each coefficient measures and interpret your results.

8.14. Refer to **Patient satisfaction** Problem 7.17.

    a. Calculate r_{Y1}^2, $r_{Y1.2}^2$, and $r_{Y1.23}^2$. How is the degree of linear association between Y and X_1 affected as other independent variables are already included in the model?

b. Make a similar analysis as in part (a) for the degree of linear association between Y and X_2. Are your findings similar to those in part (a) for Y and X_1?

8.15. Refer to **Mathematicians' salaries** Problem 7.20. Calculate r_{Y1}^2, r_{Y3}^2, r_{23}^2, $r_{Y1.2}^2$, $r_{Y1.23}^2$, and R^2. Explain what each coefficient measures and interpret your results. How is the degree of linear association between Y and X_1 affected as other independent variables are already included in the model?

8.16. Refer to **Brand preference** Problem 7.8.
  a. Transform the variables by means of the correlation transformation (8.41) and fit the standardized regression model (8.42).
  b. Interpret the standardized regression coefficient b_1'.
  c. Transform the estimated standardized regression coefficients by means of (8.50) back to the ones for the fitted regression model in the original variables. Verify that they are the same as the ones obtained in Problem 7.8a.

8.17. Refer to **Chemical shipment** Problem 7.12.
  a. Transform the variables by means of the correlation transformation (8.41) and fit the standardized regression model (8.42).
  b. Calculate the coefficient of determination between the two independent variables. Is it meaningful here to consider the standardized regression coefficients to reflect the effect of one independent variable when the other is held constant?
  c. Transform the estimated standardized regression coefficients by means of (8.50) back to the ones for the fitted regression model in the original variables. Verify that they are the same as the ones obtained in Problem 7.12c.

8.18. Refer to **Patient satisfaction** Problem 7.17.
  a. Transform the variables by means of the correlation transformation (8.41) and fit the standardized regression model (8.42).
  b. Calculate the coefficients of determination between all pairs of independent variables. Do these indicate that it is meaningful here to consider the standardized regression coefficients as indicating the effect of one independent variable when the others are held constant?
  c. Transform the estimated standardized regression coefficients by means of (8.50) back to the ones for the fitted regression model in the original variables. Verify that they are the same as the ones obtained in Problem 7.17b.

8.19. Refer to **Mathematicians' salaries** Problem 7.20.
  a. Transform the variables by means of the correlation transformation (8.41) and fit the standardized regression model (8.42).
  b. Interpret the standardized regression coefficient b_2'.
  c. Transform the estimated standardized regression coefficients by means of (8.50) back to the ones for the fitted regression model in the original variables. Verify that they are the same as the ones obtained in Problem 7.20b.

8.20. A speaker stated in a workshop on applied regression analysis: "In business and the social sciences, some degree of multicollinearity in survey data is practically inevitable." Does this statement apply equally to experimental data?

8.21. Refer to the Zarthan Company example on page 248. The company's sales manager has suggested that the predictive ability of the model could be greatly improved if promotional expenditures were added to the model, since these expenditures are known to have a substantial impact on sales. The company allocates its total promotional budget proportionately to the target population in the districts. Thus, a district containing 4.7

percent of the total target population receives 4.7 percent of the total promotional budget. Evaluate the sales manager's suggestion.

8.22. Refer to the example of perfectly correlated independent variables in Table 8.8.
   a. Develop another response function, like response functions (8.55) and (8.56), that fits the data perfectly.
   b. What is the intersection of the infinitely many response surfaces that fit the data perfectly?

8.23. The progress report of a research analyst to the supervisor stated: "All the estimated regression coefficients in our model with three independent variables to predict sales are statistically significant. Our new preliminary model with seven independent variables, which includes the three variables of our smaller model, is less satisfactory because only two of the seven regression coefficients are statistically significant. Yet in some initial trials the expanded model is giving more precise sales predictions than the smaller model. The reasons for this anomaly are now being investigated." Comment.

8.24. Two authors wrote as follows: "Our research utilized a multiple regression model. Two of the independent variables important in our theory turned out to be highly correlated in our data set. This made it difficult to assess the individual effects of each of these variables separately. We retained both variables in our model, however, because the high coefficient of multiple determination makes this difficulty unimportant." Comment.

8.25. Refer to **Brand preference** Problem 7.8.
   a. Fit the first-order simple linear regression model (3.1) for relating brand liking (Y) to moisture content (X_1). State the fitted regression function.
   b. Compare the estimated regression coefficient for moisture content obtained in part (a) with the corresponding coefficient obtained in Problem 7.8a. What do you find?
   c. Does $SSR(X_1)$ equal $SSR(X_1 \mid X_2)$ here? If not, is the difference substantial?
   d. Obtain the correlation matrix of the X variables. What bearing does this have on your findings in parts (b) and (c)?

8.26. Refer to **Chemical shipment** Problem 7.12.
   a. Fit the first-order simple linear regression model (3.1) for relating number of minutes required to handle shipment (Y) to total weight of shipment (X_2). State the fitted regression function.
   b. Compare the estimated regression coefficient for total weight of shipment obtained in part (a) with the corresponding coefficient obtained in Problem 7.12c. What do you find?
   c. Does $SSR(X_2)$ equal $SSR(X_2 \mid X_1)$ here? If not, is the difference substantial?
   d. Obtain the correlation matrix of the X variables. What bearing does this have on your findings in parts (b) and (c)?

8.27. Refer to **Patient satisfaction** Problem 7.17.
   a. Fit the first-order linear regression model (7.1) for relating patient satisfaction (Y) to patient's age (X_1) and severity of illness (X_2). State the fitted regression function.
   b. Compare the estimated regression coefficients for patient's age and severity of illness obtained in part (a) with the corresponding coefficients obtained in Problem 7.17b. What do you find?
   c. Does $SSR(X_1)$ equal $SSR(X_1 \mid X_3)$ here? Does $SSR(X_2)$ equal $SSR(X_2 \mid X_3)$?
   d. Obtain the correlation matrix of the X variables. What bearing does it have on your findings in parts (b) and (c)?

8.28. Refer to **Mathematicians' salaries** Problem 7.20.

  a. Fit the first-order linear regression model (7.1) for relating annual salaries (Y) to publication quality (X_1) and experience (X_2). State the fitted regression function.

  b. Compare the estimated regression coefficients for publication quality and experience with the corresponding coefficients obtained in Problem 7.20b. What do you find?

  c. Does $SSR(X_2)$ equal $SSR(X_2 \mid X_3)$ here? Does $SSR(X_1)$ equal $SSR(X_1 \mid X_3)$?

  d. Obtain the correlation matrix of the X variables. What bearing does this have on your findings in parts (b) and (c)?

EXERCISES

8.29. a. Define each of the following extra sums of squares: (1) $SSR(X_5 \mid X_1)$; (2) $SSR(X_3, X_4 \mid X_1)$; (3) $SSR(X_4 \mid X_1, X_2, X_3)$.

  b. For a multiple regression model with five X variables, what is the relevant extra sum of squares for testing whether or not $\beta_5 = 0$? Whether or not $\beta_2 = \beta_4 = 0$?

8.30. Show that:

  a. $SSR(X_1, X_2, X_3, X_4) = SSR(X_1) + SSR(X_2, X_3 \mid X_1) + SSR(X_4 \mid X_1, X_2, X_3)$

  b. $SSR(X_1, X_2, X_3, X_4) = SSR(X_2, X_3) + SSR(X_1 \mid X_2, X_3) + SSR(X_4 \mid X_1, X_2, X_3)$

8.31. Refer to **Brand preference** Problem 7.8.

  a. Regress Y on X_2 using the simple linear regression model (3.1) and obtain the residuals.

  b. Regress X_1 on X_2 using the simple linear regression model (3.1) and obtain the residuals.

  c. Calculate the coefficient of simple correlation between the two sets of residuals and show that it equals $r_{Y1.2}$.

8.32. An undergraduate working for a campus apparel shop serving student customers studied the relation between monthly allowance received by customer (X_1), number of years customer is in college (X_2), and dollar sales to customer to date (Y). The predictive model considered was:

$$Y_i = \beta_0 + \beta_1 X_{i1} + \beta_2 X_{i2} + \beta_3 X_{i1}^2 + \varepsilon_i$$

State the reduced models for testing whether or not: (1) $\beta_1 = \beta_3 = 0$, (2) $\beta_0 = 0$, (3) $\beta_3 = 5$, (4) $\beta_0 = 10$, (5) $\beta_1 = \beta_2$.

8.33. The following regression model is being considered in a water resources study:

$$Y_i = \beta_0 + \beta_1 X_{i1} + \beta_2 X_{i2} + \beta_3 X_{i1} X_{i2} + \beta_4 \sqrt{X_{i3}} + \varepsilon_i$$

State the reduced models for testing whether or not: (1) $\beta_3 = \beta_4 = 0$, (2) $\beta_3 = 0$, (3) $\beta_1 = \beta_2 = 5$, (4) $\beta_4 = 7$.

8.34. Show the equivalence of the expressions in (8.33) and (8.38) for $r_{Y2.1}^2$.

8.35. Refer to the work crew productivity example data in Table 8.6.

  a. For the variables transformed according to (8.41), obtain: (1) $\mathbf{X'X}$, (2) $\mathbf{X'Y}$, (3) $\mathbf{b}$, (4) $s^2\{\mathbf{b}\}$.

b. Show that the standardized regression coefficients obtained in part (a3) are related to the regression coefficients for the regression model in the original variables according to (8.50).

8.36. Derive the relations between the β_k and β'_k in (8.43a) for $p - 1 = 2$.

8.37. Derive the expression for $\mathbf{X'Y}$ in (8.48) for the standardized regression model (8.42) for $p - 1 = 2$.

8.38. Refer to Exercise 8.32. For each of the cases, state the hypothesis H_0 using the matrix formulation (8.66).

8.39. Refer to Exercise 8.33. For each of the cases, state the hypothesis H_0 using the matrix formulation (8.66).

8.40. (Calculus needed.) Derive the least squares estimator under the reduced model in (8.68), where $\mathbf{C}\boldsymbol{\beta} = \mathbf{h}$. [*Hint:* The Lagrangian function is:

$$L = (\mathbf{Y} - \mathbf{X}\boldsymbol{\beta})'(\mathbf{Y} - \mathbf{X}\boldsymbol{\beta}) + \boldsymbol{\lambda}'(\mathbf{C}\boldsymbol{\beta} - \mathbf{h}), \quad \text{where } \boldsymbol{\lambda}' = (\lambda_1, \dots, \lambda_s).]$$

8.41. Derive (8.70). [*Hint:* Show that $SSE(R) - SSE(F) = (\mathbf{b}_F - \mathbf{b}_R)'\mathbf{X'X}(\mathbf{b}_F - \mathbf{b}_R)$ and obtain an expression for $\mathbf{b}_F - \mathbf{b}_R$ from (8.68).]

PROJECTS

8.42. Refer to the **SMSA** data set. For predicting the number of active physicians (Y) in an SMSA, it has been decided to include total population (X_1) and total personal income (X_2) as independent variables. The question now is whether an additional independent variable would be helpful in the model, and if so, which variable would be most helpful. Assume that a first-order multiple regression model is appropriate.

a. For each of the following variables, calculate the coefficient of partial determination given that X_1 and X_2 are included in the model: land area (X_3), percent of population 65 or older (X_4), number of hospital beds (X_5), and total serious crimes (X_6).

b. On the basis of the results in part (a), which of the four additional independent variables is best? Is the extra sum of squares associated with this variable larger than those for the other three variables?

c. Using the F^* test statistic, test whether or not the variable determined to be best in part (b) is helpful in the regression model when X_1 and X_2 are included in the model; use $\alpha = .01$. State the alternatives, decision rule, and conclusion. Would the F^* test statistics for the other three potential independent variables be as large as the one here? Discuss.

8.43. Refer to the **SENIC** data set. For predicting the average length of stay of patients in a hospital (Y), it has been decided to include age (X_1) and infection risk (X_2) as independent variables. The question now is whether an additional independent variable would be helpful in the model, and if so, which variable would be most helpful. Assume that a first-order multiple regression model is appropriate.

a. For each of the following variables, calculate the coefficient of partial determination given that X_1 and X_2 are included in the model: routine culturing ratio (X_3), average daily census (X_4), number of nurses (X_5), and available facilities and services (X_6).

b. On the basis of the results in part (a), which of the four additional independent variables is best? Is the extra sum of squares associated with this variable larger than those for the other three variables?

c. Using the F^* test statistic, test whether or not the variable determined to be best in part (b) is helpful in the regression model when X_1 and X_2 are included in the model; use $\alpha = .05$. State the alternatives, decision rule, and conclusion. Would the F^* test statistics for the other three potential independent variables be as large as the one here? Discuss.

8.44. Refer to Exercise 7.29. It is desired to test whether or not $\beta_1 = \beta_2$. Using matrix methods, obtain $SSE(R) - SSE(F)$ according to (8.70).

Chapter 9

Polynomial Regression

In this chapter, we consider one important type of curvilinear response model, namely, the polynomial regression model. This is the most frequently used curvilinear response model in practice, because of its ease in handling as a special case of the general linear regression model (7.18). First, we discuss some commonly used polynomial regression models. Then we present two cases to illustrate some of the major problems encountered with polynomial regression. We conclude this chapter with a brief discussion of response surface methodology.

9.1 POLYNOMIAL REGRESSION MODELS

Polynomial regression models can contain one, two, or more than two independent variables. Further, each independent variable can be present in various powers. We illustrate now some major possibilities.

One Independent Variable—Second Order

The regression model:

$$(9.1) \qquad Y_i = \beta_0 + \beta_1 x_i + \beta_2 x_i^2 + \varepsilon_i$$

where:

$$x_i = X_i - \overline{X}$$

is called a *second-order model with one independent variable* because the single independent variable appears to the first and second powers. Note that the independent variable is expressed as a deviation around its mean $\overline{X}$, and that the ith observation deviation is denoted by x_i. The reason for using deviations around the mean in polynomial regression models is that X, X^2, and higher-power terms often will be highly correlated. This, as we noted in Chapter 8, can cause serious computational difficulties when the $\mathbf{X'X}$ matrix is inverted for estimating the regression coefficients. Expressing the independent variable as a deviation from its mean reduces the multicollinearity substantially, as we shall illustrate in a following example, and tends to avoid computational difficulties.

The regression coefficients in polynomial regression are frequently written in a slightly different fashion, to reflect the pattern of the exponents:

(9.1a)
$$Y_i = \beta_0 + \beta_1 x_i + \beta_{11} x_i^2 + \varepsilon_i$$

We shall employ this latter notation in this chapter.

The response function for regression model (9.1a) is:

(9.2)
$$E\{Y\} = \beta_0 + \beta_1 x + \beta_{11} x^2$$

which is a parabola and is frequently called a *quadratic response function*. Figure 9.1 contains two examples of second-order polynomial response functions.

The regression coefficient β_0 represents the mean response of Y when $x = 0$, i.e., when $X = \bar{X}$. The regression coefficient β_1 is often called the *linear effect coefficient* while β_{11} is called the *quadratic effect coefficient*.

Uses of Second-Order Model.   The second-order polynomial response function (9.2) has two basic types of uses:

1. When the true response function is indeed a second-degree polynomial, containing additive linear and quadratic effect components.

FIGURE 9.1   Examples of Second-Order Polynomial Response Functions

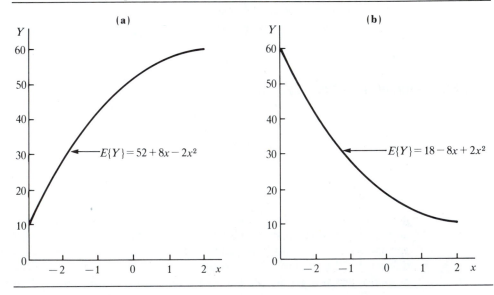

FIGURE 9.2   Extrapolation of Second-Order Polynomial Response Function in Figure 9.1a

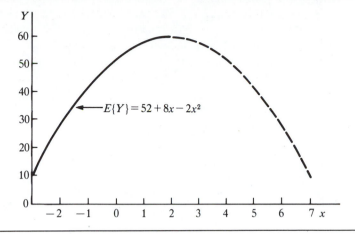

2. When the true response function is unknown (or complex) but a second-order polynomial is a good approximation to the true function.

The second type of use is the more common one, but it entails a special danger, that of extrapolation. Consider again Figure 9.1a. This response function may fit the data at hand very well. If, however, information about $E\{Y\}$ is sought for a larger value of x, extrapolation of this response function leads to the result shown in Figure 9.2, namely, a turning down of the response function, which may not be in accord with reality. Polynomial regressions of all types, especially those of higher order, share this danger of extrapolation. They may provide good fits for the data at hand, but may turn in unexpected directions when extrapolated beyond the range of the data.

One Independent Variable—Third Order

The regression model:

(9.3) $$Y_i = \beta_0 + \beta_1 x_i + \beta_{11} x_i^2 + \beta_{111} x_i^3 + \varepsilon_i$$

where:

$$x_i = X_i - \bar{X}$$

is a *third-order model with one independent variable*. The response function for regression model (9.3) is:

(9.4) $$E\{Y\} = \beta_0 + \beta_1 x + \beta_{11} x^2 + \beta_{111} x^3$$

Figure 9.3 contains two examples of third-order polynomial response functions.

FIGURE 9.3 Examples of Third-Order Polynomial Response Functions

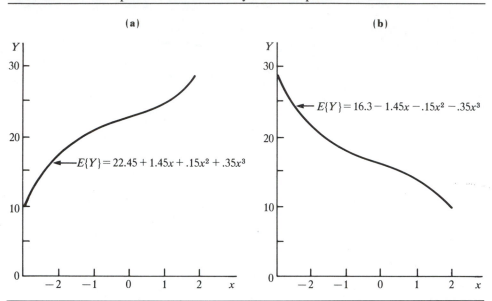

(a)

$E\{Y\} = 22.45 + 1.45x + .15x^2 + .35x^3$

(b)

$E\{Y\} = 16.3 - 1.45x - .15x^2 - .35x^3$

One Independent Variable—Higher Orders

Polynomial models with the independent variable present in higher powers than the third are not often employed. The interpretation of the coefficients becomes difficult for such models, and they may be highly erratic for interpolations and even small extrapolations. It must be recognized in this connection that a polynomial model of sufficiently high order can always be found to fit the data perfectly. For instance, the fitted polynomial regression function for one independent variable of order $n - 1$ will pass through all n observed Y values. One needs to be wary therefore of using high-order polynomials for the sole purpose of obtaining a good fit. Such regression functions may not show clearly the basic elements of the regression relation between X and Y and may lead to erratic interpolations and extrapolations.

Two Independent Variables—Second Order

The regression model:

$$(9.5) \qquad Y_i = \beta_0 + \beta_1 x_{i1} + \beta_2 x_{i2} + \beta_{11} x_{i1}^2 + \beta_{22} x_{i2}^2 + \beta_{12} x_{i1} x_{i2} + \varepsilon_i$$

where:

$$x_{i1} = X_{i1} - \bar{X}_1$$
$$x_{i2} = X_{i2} - \bar{X}_2$$

is a *second-order model with two independent variables*. The response function is:

$$(9.6) \qquad E\{Y\} = \beta_0 + \beta_1 x_1 + \beta_2 x_2 + \beta_{11} x_1^2 + \beta_{22} x_2^2 + \beta_{12} x_1 x_2$$

which is the equation of a conic section. Note that regression model (9.5) contains separate linear and quadratic components for each of the two independent variables and a cross-product term. The latter represents the interaction effects between x_1 and x_2, as we noted in Chapter 7. The coefficient β_{12} is often called the *interaction effect coefficient*.

The second-order response function for two independent variables in (9.6) represents the two basic types of surfaces illustrated in Figure 7.3. Stationary and rising ridges constitute limiting cases of these two basic types of response surfaces.

Often, it is easiest to portray the second-order response function (9.6) in terms of contour curves. Figure 9.4 contains a representation in terms of contour curves of the response function:

$$(9.7) \qquad E\{Y\} = 1{,}740 - 4x_1^2 - 3x_2^2 - 3x_1 x_2$$

Note that this response surface has a maximum at $x_1 = 0$ and $x_2 = 0$.

FIGURE 9.4 Example of a Quadratic Response Surface in Terms of Contour Curves: $E\{Y\} = 1{,}740 - 4x_1^2 - 3x_2^2 - 3x_1 x_2$

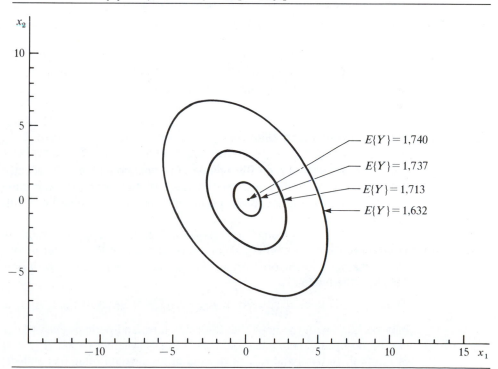

Polynomial models in two (or more) independent variables are well adapted to situations where the response function is unknown and a suitable model is to be developed empirically.

Note

The cross-product term $\beta_{12}x_1x_2$ in (9.6) is considered to be a second-order term, the same as $\beta_{11}x_1^2$ or $\beta_{22}x_2^2$. The reason can be seen readily by writing the latter terms as $\beta_{11}x_1x_1$ and $\beta_{22}x_2x_2$, respectively.

Three Independent Variables—Second Order

The *second-order regression model with three independent variables* is:

$$(9.8) \qquad Y_i = \beta_0 + \beta_1 x_{i1} + \beta_2 x_{i2} + \beta_3 x_{i3} + \beta_{11}x_{i1}^2 + \beta_{22}x_{i2}^2 + \beta_{33}x_{i3}^2$$
$$+ \beta_{12}x_{i1}x_{i2} + \beta_{13}x_{i1}x_{i3} + \beta_{23}x_{i2}x_{i3} + \varepsilon_i$$

where:

$$x_{i1} = X_{i1} - \bar{X}_1$$
$$x_{i2} = X_{i2} - \bar{X}_2$$
$$x_{i3} = X_{i3} - \bar{X}_3$$

The response function for this regression model is:

$$(9.9) \qquad E\{Y\} = \beta_0 + \beta_1 x_1 + \beta_2 x_2 + \beta_3 x_3 + \beta_{11}x_1^2 + \beta_{22}x_2^2 + \beta_{33}x_3^2$$
$$+ \beta_{12}x_1x_2 + \beta_{13}x_1x_3 + \beta_{23}x_2x_3$$

The coefficients β_{12}, β_{13}, and β_{23} are interaction effects coefficients for interactions between pairs of independent variables.

Uses of Polynomial Regression Models

Fitting of polynomial regression models presents no new problems since, as we have seen in Chapter 7, they are special cases of the general linear regression model (7.18). Hence, all earlier results on fitting apply, as do the earlier results on making inferences.

When using a polynomial regression model as an approximation to the true regression function, one will often fit a second-order or third-order model and then explore whether a lower-order model is adequate. For instance, with one independent variable, the model:

$$Y_i = \beta_0 + \beta_1 x_i + \beta_{11}x_i^2 + \beta_{111}x_i^3 + \varepsilon_i$$

may be fitted with the hope that the cubic term and perhaps even the quadratic term can be dropped. Thus, one would wish to test whether or not $\beta_{111} = 0$, or whether or not both $\beta_{11} = 0$ and $\beta_{111} = 0$. Similar tests would often be conducted with polynomial regression models for two or more independent variables.

9.2 EXAMPLE 1—ONE INDEPENDENT VARIABLE

We illustrate now some of the major types of analyses usually conducted with polynomial regression models with one independent variable.

Setting

A staff analyst for a cafeteria chain wished to investigate the relation between the number of self-service coffee dispensers in a cafeteria line and sales of coffee. Fourteen cafeterias that are similar in terms of volume of business, type of clientele, and location were chosen for the experiment. The number of self-service dispensers placed in the test cafeterias varied from zero (coffee is dispensed here by a line attendant) to six and was assigned randomly to each cafeteria.

Table 9.1 contains the results of the experimental study. Sales are measured in hundreds of gallons of coffee sold.

Fitting of Model

The analyst believes that the relation between sales and number of self-service dispensers is quadratic in the range of observations; sales should increase as the number of dispensers is greater, but if the space is cluttered with dispensers, this increase becomes retarded. Hence, she would like to fit the quadratic regression model:

$$(9.10) \qquad Y_i = \beta_0 + \beta_1 x_i + \beta_{11} x_i^2 + \varepsilon_i$$

TABLE 9.1 Data for Cafeteria Coffee Sales Example

Cafeteria i	Number of Dispensers X_i	Coffee Sales (hundred gallons) Y_i
1	0	508.1
2	0	498.4
3	1	568.2
4	1	577.3
5	2	651.7
6	2	657.0
7	3	713.4
8	3	697.5
9	4	755.3
10	4	758.9
11	5	787.6
12	5	792.1
13	6	841.4
14	6	831.8

where:

$$x_i = X_i - \overline{X}$$

She further anticipates that the error terms ε_i will be fairly normally distributed with constant variance.

The $\mathbf{Y}$ and $\mathbf{X}$ matrices for this application are given in Table 9.2. Note that the $\mathbf{X}$ matrix contains a column of 1s, a column of the independent variable observations x (expressed as deviations around their mean $\overline{X} = 3$), and a column of the x^2 values.

Note also in Table 9.2 that both small and large values of x are associated with large values of x^2. As a result, and because of the symmetry of the deviations, the coefficient of simple correlation between x and x^2 in Table 9.2 is $r = 0$. If the polynomial regression model had used the original X variable:

$$X: \quad 0 \quad 0 \quad 1 \quad 1 \quad \ldots \quad 6 \quad 6$$

$$X^2: \quad 0 \quad 0 \quad 1 \quad 1 \quad \ldots \quad 36 \quad 36$$

small values of X would have been associated only with small values of X^2, and large values of X only with large values of X^2. In that case, the coefficient of simple correlation between X and X^2 would have been $r = .961$. This illustrates that use of deviations from the mean for the independent variable leads to substantially lower correlations between the X variables in polynomial regression models. In this case, indeed, x and x^2 are uncorrelated because of the symmetry of the x deviations.

From this point on, the calculations are routine. We could do the matrix calculations manually, as illustrated in Chapter 7, or use a computer multiple regression program. Since no new problems are encountered, we simply present the basic computer output in Table 9.3, including needed extra sums of squares and the $\mathbf{s}^2\{\mathbf{b}\}$ matrix.

TABLE 9.2   Data Matrices for Cafeteria Coffee Sales Example

$$
\mathbf{Y} =
\begin{bmatrix}
508.1 \\
498.4 \\
568.2 \\
577.3 \\
651.7 \\
657.0 \\
713.4 \\
697.5 \\
755.3 \\
758.9 \\
787.6 \\
792.1 \\
841.4 \\
831.8
\end{bmatrix}
\qquad
\mathbf{X} =
\begin{bmatrix}
 & x & x^2 \\
1 & -3 & 9 \\
1 & -3 & 9 \\
1 & -2 & 4 \\
1 & -2 & 4 \\
1 & -1 & 1 \\
1 & -1 & 1 \\
1 & 0 & 0 \\
1 & 0 & 0 \\
1 & 1 & 1 \\
1 & 1 & 1 \\
1 & 2 & 4 \\
1 & 2 & 4 \\
1 & 3 & 9 \\
1 & 3 & 9
\end{bmatrix}
$$

TABLE 9.3   Regression Results for Cafeteria Coffee Sales Example

(a) Regression Coefficients

Regression Coefficient	Estimated Regression Coefficient	Estimated Standard Deviation	t^*
β_0	705.474	3.208	219.91
β_1	54.893	1.050	52.28
β_{11}	−4.249	.606	−7.01

(b) Analysis of Variance

Source of Variation	SS	df	MS
Regression	171,773	2	85,887
x	168,741	1	168,741
$x^2 \mid x$	3,033	1	3,033
Error	679	11	61.7
Total	172,453	13	

(c) $s^2\{b\}$ Matrix

$$\begin{bmatrix} 10.2912 & 0 & -1.4702 \\ 0 & 1.1026 & 0 \\ -1.4702 & 0 & .3675 \end{bmatrix}$$

The fitted regression function is:

$$(9.11) \qquad \hat{Y} = 705.47 + 54.89x - 4.25x^2$$

This response function is plotted in Figure 9.5, together with the original data. We show the horizontal scale expressed at the bottom in the deviation units x and at the top in the original units X.

Algebraic Version of Normal Equations.   The algebraic version of the least squares normal equations:

$$\mathbf{X'Xb = X'Y}$$

for the second-order polynomial regression model (9.10) can be readily obtained from (7.68) by replacing X_{i1} by x_i and X_{i2} by x_i^2. Since $\Sigma x_i = 0$, this yields the normal equations:

FIGURE 9.5   Fitted Second-Order Polynomial Regression–Cafeteria Coffee Sales Example

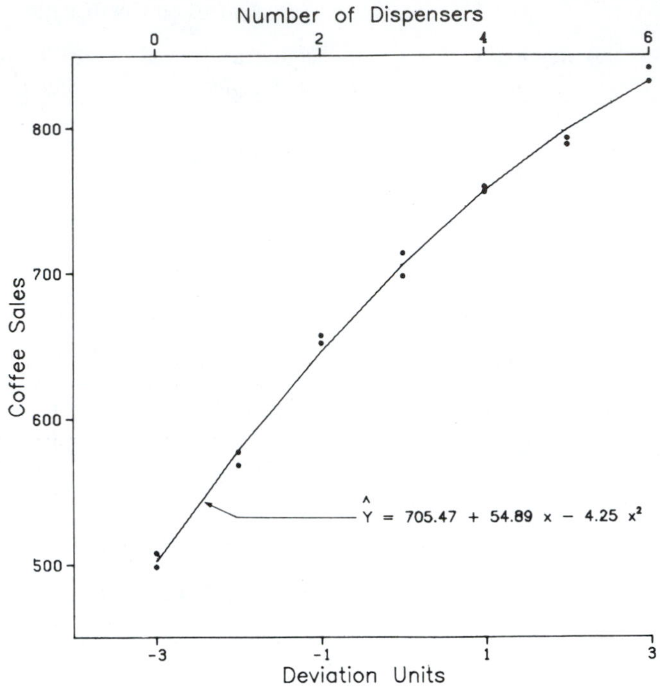

$$\sum Y_i = nb_0 + b_{11} \sum x_i^2$$

(9.12)
$$\sum x_i Y_i = b_1 \sum x_i^2 + b_{11} \sum x_i^3$$

$$\sum x_i^2 Y_i = b_0 \sum x_i^2 + b_1 \sum x_i^3 + b_{11} \sum x_i^4$$

Analysis of Aptness of Model

Residual Analysis.   To study the aptness of regression model (9.10) for her data, the analyst plotted the residuals e_i against the fitted values, as shown in Figure 9.6a, and also against the independent variable x_i expressed in deviation units, as shown in Figure 9.6b. We do not present the calculations of the e_i, as these are routine.

There are no systematic departures from 0 evident in the residuals as either $\hat{Y}$ or x increases, suggesting that the quadratic response function is a good fit. The scatter plot in Figure 9.5 makes this point also. Further, there is no tendency in Figures 9.6a and 9.6b for the spread of the residuals to vary systematically, so it appears that the constant error variance assumption is reasonable. A normal probability plot, not shown here, did not provide any strong evidence that the distribution of the error terms is far from normal.

FIGURE 9.6   Residual Plots for Cafeteria Coffee Sales Example

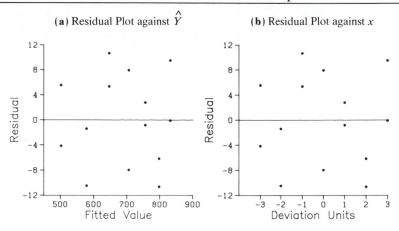

Based on this study of the aptness of the model, the analyst was willing to conclude that the normal error regression model (9.10) with constant error variance is appropriate here.

Test for Quadratic Response Function.   Since there are two repeat observations for each level of x, the analyst could have used the formal test (7.57) of the aptness of the regression function, the alternatives here being:

(9.13)
$$H_0: E\{Y\} = \beta_0 + \beta_1 x + \beta_{11} x^2$$
$$H_a: E\{Y\} \neq \beta_0 + \beta_1 x + \beta_{11} x^2$$

The basic ANOVA results were presented earlier in Table 9.3b. The pure error sum of squares (4.11) is obtained as follows from the data in Table 9.2:

$$SSPE = (508.1 - 503.25)^2 + (498.4 - 503.25)^2 + (568.2 - 572.75)^2$$
$$+ \cdots + (831.8 - 836.6)^2 = 292$$

Note that $\bar{Y}_1 = 503.25$ for $x = -3$, $\bar{Y}_2 = 572.75$ for $x = -2$, and so on. There are $c = 7$ distinct levels of x here and thus $n - c = 14 - 7 = 7$ degrees of freedom are associated with $SSPE$. Hence, we have:

$$MSPE = \frac{SSPE}{7} = \frac{292}{7} = 41.7$$

Now we are in a position to obtain the lack of fit sum of squares by (4.19):

$$SSLF = SSE - SSPE = 679 - 292 = 387$$

There are $c - p = 7 - 3 = 4$ degrees of freedom associated with $SSLF$. (Remember that $p = 3$ parameters had to be estimated for the fitted regression function.) Hence, we have:

$$MSLF = \frac{SSLF}{4} = \frac{387}{4} = 96.8$$

Thus, test statistic (7.57b) here is:

$$F^* = \frac{MSLF}{MSPE} = \frac{96.8}{41.7} = 2.32$$

Assuming the level of significance is to be .05, we require $F(.95; 4, 7) = 4.12$. Since $F^* = 2.32 \leq 4.12$, we conclude H_0, that the quadratic response function is appropriate.

Test Whether β_{11} Equals Zero

t Test.   The analyst next studied whether the quadratic term could be dropped from the model. She therefore wished to test:

(9.14)
$$H_0: \beta_{11} = 0$$
$$H_a: \beta_{11} \neq 0$$

H_0 implies that there is no quadratic effect in the response function.

Table 9.3a indicates that:

$$t^* = \frac{b_{11}}{s\{b_{11}\}} = \frac{-4.249}{.606} = -7.012$$

For a level of significance of .05, we require $t(.975; 11) = 2.201$. The decision rule is:

$$\text{If } |t^*| \leq 2.201, \text{ conclude } H_0$$
$$\text{If } |t^*| > 2.201, \text{ conclude } H_a$$

Since $|t^*| = 7.012 > 2.201$, we conclude H_a, that a quadratic effect does exist, so that the quadratic term should be retained in the model.

Partial F Test.   The analyst could also have used the partial F test to choose the appropriate conclusion in (9.14). Indeed, she had specified the order of entering the variables x and x^2 into the computer regression fit so that she would obtain the extra sums of squares $SSR(x)$ and $SSR(x^2 \mid x)$ in the output. Utilizing the partial F test statistic (8.22) and the results in Table 9.3b, we obtain:

$$F^* = \frac{MSR(x^2 \mid x)}{MSE} = \frac{3{,}033}{61.7} = 49.2$$

For a 5 percent level of significance, we need $F(.95; 1, 11) = 4.84$. Since $F^* = 49.2 > 4.84$, we are led to conclude H_a, as by the t test.

Note

One may observe here the relation discussed in the previous chapter between the t and partial F tests as to whether a regression coefficient equals zero. We have for the two test statistics:

$$(t^*)^2 = (-7.012)^2 = 49.2 = F^*$$

Estimation of Regression Coefficients

The analyst next wished to obtain confidence bounds for the two regression coefficients β_1 and β_{11} with family confidence coefficient .90 by means of the Bonferroni method.

Here $g = 2$ statements are desired; hence, by (7.47a), we have:

$$B = t[1 - .10/2(2); 11] = t(.975; 11) = 2.201$$

From Table 9.3a, we find:

$$b_1 = 54.893 \qquad s\{b_1\} = 1.050$$

$$b_{11} = -4.249 \qquad s\{b_{11}\} = .606$$

The Bonferroni confidence limits by (7.47) therefore are $54.893 \pm 2.201(1.050)$ and $-4.249 \pm 2.201(.606)$, yielding the confidence intervals:

$$52.58 \leq \beta_1 \leq 57.20$$

$$-5.58 \leq \beta_{11} \leq -2.92$$

The analyst was satisfied with the precision of these two statements, feeling that the intervals are narrow enough to give her reliable simultaneous information about the comparative magnitudes of the linear and quadratic effects.

Coefficient of Multiple Determination

For a descriptive measure of the degree of relation between coffee sales and number of dispensing machines, the analyst calculated the coefficient of multiple determination in (7.35), using the data in Table 9.3b:

$$R^2 = \frac{SSR}{SSTO} = \frac{171,773}{172,453} = .996$$

This measure shows that the variation in coffee sales is reduced by 99.6 percent when the quadratic relation to the number of dispensing machines is utilized.

Note that the coefficient of multiple determination R^2 is the relevant measure here, not the coefficient of simple determination r^2, since model (9.10) is a multiple regression model even though it contains only one independent variable. Sometimes in curvilinear regression, the coefficient of multiple correlation R is called the *correlation index*.

Estimation of Mean Response

The analyst was particularly interested in the mean response for $X_h = 3$ dispensing machines. She wished to estimate this mean response with a 98 percent confidence coefficient. The proper interval estimate is given by (7.54). For our example, where $x_h = X_h - \bar{X} = 3 - 3 = 0$, we have:

$$\mathbf{X}_h = \begin{bmatrix} 1 \\ x_h \\ x_h^2 \end{bmatrix} = \begin{bmatrix} 1 \\ 0 \\ 0 \end{bmatrix}$$

The estimated mean response $\hat{Y}_h$ corresponding to $\mathbf{X}_h$ is by (7.50):

$$\hat{Y}_h = \mathbf{X}_h'\mathbf{b} = \begin{bmatrix} 1 & 0 & 0 \end{bmatrix} \begin{bmatrix} 705.474 \\ 54.893 \\ -4.249 \end{bmatrix} = 705.474$$

Next, using the results in Table 9.3c for $s^2\{\mathbf{b}\}$, we obtain when substituting into (7.53):

$$s^2\{\hat{Y}_h\} = \mathbf{X}_h's^2\{\mathbf{b}\}\mathbf{X}_h$$

$$= \begin{bmatrix} 1 & 0 & 0 \end{bmatrix} \begin{bmatrix} 10.2912 & 0 & -1.4702 \\ 0 & 1.1026 & 0 \\ -1.4702 & 0 & .3675 \end{bmatrix} \begin{bmatrix} 1 \\ 0 \\ 0 \end{bmatrix}$$

$$= 10.2912$$

or $s\{\hat{Y}_h\} = 3.208$. We require $t(.99; 11) = 2.718$. Hence, the confidence limits are $705.474 \pm 2.718(3.208)$, and we obtain the confidence interval:

$$696.8 \le E\{Y_h\} \le 714.2$$

With confidence coefficient .98, the analyst can conclude that the mean coffee sales when three dispensing machines are used are somewhere between 696.8 and 714.2 hundred gallons.

Regression Function in Terms of *X*

The analyst wishes for reporting purposes to express the fitted regression function (9.11) in terms of X rather than in terms of deviations $x = X - \bar{X}$. The equivalent fitted regression function in the original X variable is:

(9.15) $$\hat{Y} = b_0' + b_1'X + b_{11}'X^2$$

where the coefficients are obtained as follows:

(9.15a) $$b_0' = b_0 - b_1\bar{X} + b_{11}\bar{X}^2$$

(9.15b) $$b_1' = b_1 - 2b_{11}\bar{X}$$

(9.15c) $$b_{11}' = b_{11}$$

For our example, where $\bar{X} = 3$, we obtain:

$$b_0' = 705.474 - 54.893(3) + (-4.249)(3)^2 = 502.554$$

$$b_1' = 54.893 - 2(-4.249)(3) = 80.387$$

$$b_{11}' = -4.249$$

so that the fitted regression function in terms of X is:

$$\hat{Y} = 502.554 + 80.387X - 4.249X^2$$

The fitted values and residuals for the regression function in terms of X are exactly the same as for the regression function in terms of the deviations x. The reason, as we noted earlier, for utilizing model (9.10), which is expressed in terms of deviations x, is to avoid potential calculational difficulties due to multicollinearity between X and X^2, inherent in polynomial regression.

Note

The estimated standard deviations of the regression coefficients in Table 9.3a do not apply to the regression coefficients in terms of X in (9.15). If the estimated standard deviations for the regression coefficients in terms of X are desired, they may be obtained from $s^2\{\mathbf{b}\}$ in Table 9.3c by using theorem (6.47), where the transformation matrix $\mathbf{A}$ is developed from (9.15a, b, c).

9.3 EXAMPLE 2—TWO INDEPENDENT VARIABLES

We shall discuss now another example of polynomial regression, this one involving two independent variables. Rather than carrying this example through all of the various analytical stages as we did the first example, we shall focus here primarily on the analysis of interaction effects and quadratic effects.

Setting

A researcher studied the effects of the charge rate and temperature on the life of a new type of power cell. An experiment was conducted where the charge rate (X_1) was controlled at three levels (.6, 1.0, and 1.4 amperes) and the ambient temperature (X_2) was controlled at three levels (10, 20, 30 °C). Factors pertaining to the discharge of the power cell were held at fixed levels. The life of the power cell (Y) was measured in terms of the number of discharge-charge cycles that a power cell underwent before it failed. The data obtained in the study are contained in Table 9.4, columns 1, 2, and 5.

The researcher was not sure about the nature of the response function in the range of the factors studied. Hence, he decided to fit the second-order polynomial regression model (9.5):

$$(9.16) \qquad Y_i = \beta_0 + \beta_1 x_{i1} + \beta_2 x_{i2} + \beta_{11} x_{i1}^2 + \beta_{22} x_{i2}^2 + \beta_{12} x_{i1} x_{i2} + \varepsilon_i$$

TABLE 9.4 Data for Power Cells Example

Cell	(1) Charge Rate	(2) Temperature	(3)	(4) Coded Values	(5) Number of Cycles
i	X_{i1}	X_{i2}	x_{i1}	x_{i2}	Y_i
1	.6	10	-1	-1	150
2	1.0	10	0	-1	86
3	1.4	10	1	-1	49
4	.6	20	-1	0	288
5	1.0	20	0	0	157
6	1.0	20	0	0	131
7	1.0	20	0	0	184
8	1.4	20	1	0	109
9	.6	30	-1	1	279
10	1.0	30	0	1	235
11	1.4	30	1	1	224
	$\bar{X}_1 = 1.0$	$\bar{X}_2 = 20$			

Setting adapted from: S. M. Sidik, H. F. Leibecki, and J. M. Bozek, *Cycles Till Failure of Silver-Zinc Cells with Competing Failure Modes—Preliminary Data Analysis,* NASA Technical Memorandum 81556, 1980.

for which the response function is:

$$(9.16a) \qquad E\{Y\} = \beta_0 + \beta_1 x_1 + \beta_2 x_2 + \beta_{11} x_1^2 + \beta_{22} x_2^2 + \beta_{12} x_1 x_2$$

Because of the balanced nature of the X_1 and X_2 levels studied, the researcher defined the deviations of X_1 and X_2 from their respective means in convenient units, as follows:

$$x_{i1} = \frac{X_{i1} - \bar{X}_1}{.4} = \frac{X_{i1} - 1.0}{.4}$$

(9.16b)

$$x_{i2} = \frac{X_{i2} - \bar{X}_2}{10} = \frac{X_{i2} - 20}{10}$$

where the denominator for each variable represents the difference between adjacent levels. These deviation variables are shown in columns 3 and 4 of Table 9.4. Note that the deviations defined in (9.16b) lead to a simple coding of the X_1 and X_2 levels in terms of -1, 0, and 1.

The researcher was particularly interested in whether interaction effects and curvature effects are required in the model for the range of the X variables considered.

Development of Model

Table 9.5a contains the basic regression results for the fit of model (9.16). The researcher first investigated the appropriateness of this regression model for the data at hand. Plots of the residuals against $\hat{Y}$, x_1, and x_2 are shown in Figure 9.7, as is also

TABLE 9.5 Regression Results for Second-Order Polynomial Model (9.16)—Power Cells Example

(a) Regression Coefficients

Regression Coefficient	Estimated Regression Coefficient	Estimated Standard Deviation	t^*
β_0	162.84	16.61	9.81
β_1	-55.83	13.22	-4.22
β_2	75.50	13.22	5.71
β_{11}	27.39	20.34	1.35
β_{22}	-10.61	20.34	$-.52$
β_{12}	11.50	16.19	.71

(b) Analysis of Variance

Source of Variation	SS	df	MS
Regression	55,366	5	11,073
x_1	18,704	1	18,704
$x_2 \mid x_1$	34,201	1	34,201
$x_1^2 \mid x_1, x_2$	1,646	1	1,646
$x_2^2 \mid x_1, x_2, x_1^2$	285	1	285
$x_1 x_2 \mid x_1, x_2, x_1^2, x_2^2$	529	1	529
Error	5,240	5	1,048
Total	60,606	10	

a normal probability plot. None of these plots suggest any gross inadequacies of regression model (9.16). The coefficient of correlation between the ordered residuals and their expected values under normality is .974, which supports the assumption of normality of the error terms (see Table 4.3).

Another indication of the adequacy of regression model (9.16) can be obtained by the formal test in (7.57) of the goodness of fit of the regression function (9.16a), since there are three replications at $x_1 = 0$, $x_2 = 0$. The pure error sum of squares (4.11) is simple here, because there is only one combination of levels at which replications occur:

$$SSPE = (157 - 157.33)^2 + (131 - 157.33)^2 + (184 - 157.33)^2$$
$$= 1,404.67$$

Since there are $c = 9$ distinct combinations of levels of the X variables here, $n - c = 11 - 9 = 2$ degrees of freedom are associated with $SSPE$. Further, $SSE = 5,240$ according to Table 9.5b; hence the lack of fit sum of squares (4.19) is:

$$SSLF = SSE - SSPE = 5,240 - 1,404.67 = 3,835.33$$

FIGURE 9.7   Residual Plots for Power Cells Example

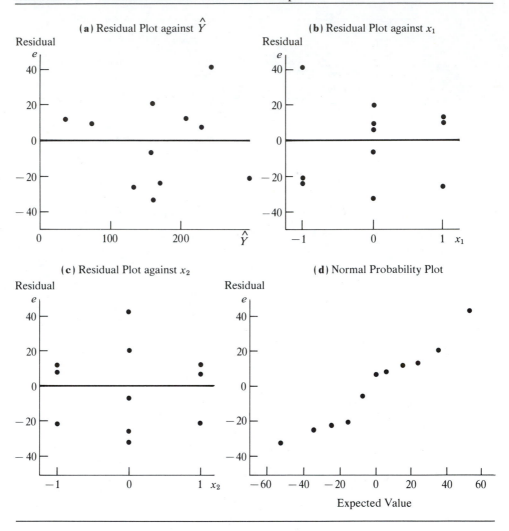

with which $c - p = 9 - 6 = 3$ degrees of freedom are associated. Hence, the test statistic (7.57b) for testing the adequacy of the regression function (9.16a) is:

$$F^* = \frac{SSLF}{c - p} \div \frac{SSPE}{n - c} = \frac{3,835.33}{3} \div \frac{1,404.67}{2} = 1.82$$

For $\alpha = .05$, we require $F(.95; 3, 2) = 19.2$. Since $F^* = 1.82 \leq 19.2$, we conclude according to decision rule (7.57c) that the second-order polynomial regression function (9.16a) is adequate.

The researcher now turned to consider whether a first-order model would be adequate. The test alternatives are:

$$H_0: \beta_{11} = \beta_{22} = \beta_{12} = 0$$

$$H_a: \text{not all } \beta\text{s in } H_0 \text{ equal zero}$$

The partial F test statistic (8.25) here is:

$$F^* = \frac{SSR(x_1^2, x_2^2, x_1x_2 \mid x_1, x_2)}{3} \div MSE$$

In anticipation of this test, the researcher entered the X variables in the computer regression program in the order $x_1, x_2, x_1^2, x_2^2, x_1x_2$. The analysis of variance table, including the extra sums of squares, is shown in Table 9.5b. The required extra sum of squares can therefore be obtained as follows:

$$SSR(x_1^2, x_2^2, x_1x_2 \mid x_1, x_2) = SSR(x_1^2 \mid x_1, x_2) + SSR(x_2^2 \mid x_1, x_2, x_1^2)$$

$$+ SSR(x_1x_2 \mid x_1, x_2, x_1^2, x_2^2)$$

$$= 1{,}646 + 285 + 529 = 2{,}460$$

The test statistic is:

$$F^* = \frac{2{,}460}{3} \div 1{,}048 = .78$$

For level of significance $\alpha = .05$, we require $F(.95; 3, 5) = 5.41$. Since $F^* = .78 \leq 5.41$, we conclude H_0, that no curvature and interaction effects exist, so a first-order model is adequate for the range of the charge rates and temperatures considered.

Hence, the researcher decided to employ the first-order model:

$$(9.17) \qquad Y_i = \beta_0 + \beta_1 x_{i1} + \beta_2 x_{i2} + \varepsilon_i$$

and obtained the fitted response function:

$$(9.18) \qquad \hat{Y} = 172.00 - 55.83x_1 + 75.50x_2$$

Note that the regression coefficients b_1 and b_2 are the same as in Table 9.5a for the fitted second-order model. This is a result of the choices of the X_1 and X_2 levels studied. We shall comment on this further shortly.

The fitted regression function (9.18) can be transformed back to the original variables by utilizing (9.16b). We obtain:

$$(9.18a) \qquad \hat{Y} = 160.58 - 139.58X_1 + 7.55X_2$$

Figure 9.8 contains a three-dimensional computer-generated plot of the fitted response plane. The researcher used this fitted response surface for investigating the effects of charge rate and temperature on the life of this new type of power cell.

FIGURE 9.8   Three-Dimensional Computer-Generated Plot of Fitted Response Plane (9.18a)—Power Cells Example

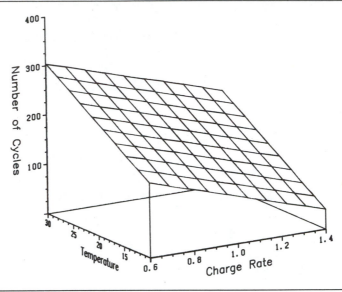

9.4 RESPONSE SURFACE METHODOLOGY

The use of polynomial response functions to approximate complex response surfaces is common in many experimental situations. The name *response surface methodology* has been given to the statistical methodology concerned with the design of studies to estimate response surfaces and with the actual estimation of response surfaces and the interpretation of the results.

Response surface methodology is employed for two principal purposes: (1) to provide a description of the response pattern in the region of the X observations studied, and (2) to assist in finding the region where the optimal response occurs (i.e., where the response surface is at a maximum or a minimum).

Methods for evaluating the aptness of a fitted response surface have already been discussed, as have tests for deciding whether interaction and curvature effects are required in the model. We shall now briefly discuss two new aspects of response surface methodology: (1) the design of response surface studies, and (2) the search for optimal response conditions.

Design of Response Surface Studies

A large variety of experimental designs have been developed for estimating response surfaces efficiently. The design for the power cells example is presented in Figure 9.9a. Note that every charge rate level was considered with every temperature level.

FIGURE 9.9   Full and Fractional Factorial Designs—Power Cells Example (numbers in parentheses are estimated standard derivations $s\{\hat{Y}_h\}$)

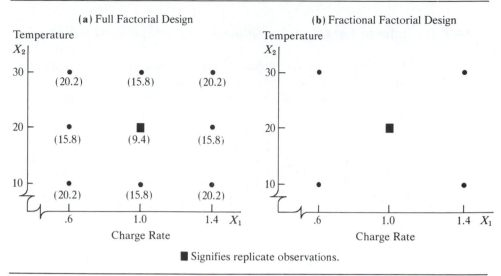

(a) Full Factorial Design

(b) Fractional Factorial Design

■ Signifies replicate observations.

Such a study design is called a *full factorial design*. Extra replications were made at the center point (1.0, 20) to provide a measure of pure error and to assist in evaluating the aptness of the model fitted.

When experimental runs are expensive, a *fractional factorial design* may be used. Here, not every level of one independent variable is studied with each level of every other independent variable. Figure 9.9b shows a fractional factorial design that might have been used for the power cells example. Here, selected (X_1, X_2) combinations are not studied. The design in Figure 9.9b provides information about the linear effects of X_1 and X_2 as well as some information about curvature and interaction effects. The fractional factorial design in Figure 9.9b can also be considered to be a full factorial design with each factor at two levels, augmented by replications at the center point of the design.

Both of the designs in Figure 9.9 are *rotatable designs*. Such designs have the property that the estimated standard deviation of the fitted value, $s\{\hat{Y}_h\}$, is the same for any given distance from the center of the design, regardless of direction. In Figure 9.9a, the estimated standard deviations of the fitted values at the experimental points are given in parentheses. Note that the points (1.0, 10), (.6, 20), (1.4, 20), and (1.0, 30) all have $s\{\hat{Y}_h\} = 15.8$ because they are all equally distant from the center point. Similarly, all four corner points have $s\{\hat{Y}_h\} = 20.2$ because they are equally distant from the center point. As usual, $s\{\hat{Y}_h\}$ increases, the further the experimental point is from the center.

The property of equal precision at any given distance from the center of rotatable designs is desirable because it is not usually known in advance which direction from the center point will be of later interest. A rotatable design provides assurance that the precision of the fitted values is not affected by the direction, only by the distance from the center point.

Search for Optimal Response Conditions—One Independent Variable

Response surfaces are frequently fitted for purposes of finding optimal response conditions. In the power cells example, for instance, management may wish to know which combination of charge rate and temperature maximizes the expected life of the power cells. Generally, a sequence of experiments is required to find the optimal response conditions, as we shall now illustrate—first for the case of one independent variable and then when there are two independent variables.

When only one independent variable is involved and the estimated response function is quadratic:

$$(9.19) \qquad \hat{Y} = b_0 + b_1 x + b_{11} x^2$$

where:

$$x = X - \overline{X}$$

the maximum (minimum) occurs at the level x_m:

$$(9.20) \qquad x_m = -\frac{b_1}{2b_{11}}$$

In terms of the original variable X, the maximum (minimum) occurs at the level X_m:

$$(9.20a) \qquad X_m = \overline{X} - \frac{b_1}{2b_{11}}$$

The estimated mean response at X_m is:

$$(9.21) \qquad \hat{Y}_m = b_0 - \frac{b_1^2}{4b_{11}}$$

$\hat{Y}_m$ is a maximum if b_{11} is negative, and a minimum if b_{11} is positive.

Example.   For the earlier cafeteria coffee sales example, the fitted regression curve was:

$$\hat{Y} = 705.47 + 54.89x - 4.25x^2$$

If the quadratic regression function is appropriate for larger x values than those in the study, we could estimate that maximum mean coffee sales occur at:

$$X_m = 3 - \frac{54.89}{2(-4.25)} = 9$$

and the estimated mean response there is:

$$\hat{Y}_m = 705.47 + 54.89(9) - 4.25(9)^2 = 855$$

Since the largest number of dispensers in the study was only $X = 6$, it would be highly desirable to extend the study by investigating coffee sales with up to, say, 12 dispensers. This information could then be used to examine the appropriateness of a quadratic response function in the larger range of the independent variable and to confirm the location of the optimal response condition. A study extension based on $X = 8, 10, 12$ dispensers might be sufficient for these purposes.

Comments

1. To derive (9.20), we differentiate $\hat{Y}$ in (9.19) with respect to x, and set this derivative equal to 0:

$$\frac{d\hat{Y}}{dx} = \frac{d}{dx}(b_0 + b_1 x + b_{11} x^2) = b_1 + 2b_{11} x = 0$$

and obtain:

$$x_m = -\frac{b_1}{2b_{11}}$$

Substituting this value into the fitted response function (9.19), we find:

$$\hat{Y}_m = b_0 + b_1 \left(\frac{-b_1}{2b_{11}}\right) + b_{11}\left(\frac{-b_1}{2b_{11}}\right)^2$$

$$= b_0 - \frac{b_1^2}{4b_{11}}$$

2. For large samples, the approximate estimated variance of X_m in (9.20a) is:

(9.22)
$$s^2\{X_m\} = \frac{b_1^2}{4b_{11}^2}\left[\frac{s^2\{b_1\}}{b_1^2} + \frac{s^2\{b_{11}\}}{b_{11}^2} - \frac{2s\{b_1, b_{11}\}}{b_1 b_{11}}\right]$$

This approximate estimated variance can be used to construct a confidence interval for the true X level at which the maximum (minimum) occurs. Approximate confidence intervals for $E\{Y_m\}$ can also be obtained. These are discussed in Reference 9.1.

Search for Optimal Response Conditions—Two Independent Variables

When the researcher does not know in advance the region near where the optimal response exists for two independent variables, the strategy usually employed is sequential. First, a simple experimental design is run in a restricted space of the (X_1, X_2) variables. On the basis of this initial experiment, information is obtained about the direction where the optimal response is located. This direction is called the *path of steepest ascent (descent)*. Some additional experimental runs are then made to identify more clearly where the optimal response is located, and then an additional experimental design is run to identify the optimal region more precisely. This search process is iterated until the optimal region is identified with sufficient precision.

We shall illustrate these basic ideas with the power cells example.

FIGURE 9.10   Response Contours and Line of Steepest Ascent for Fitted Response
Function (9.18)—Power Cells Example

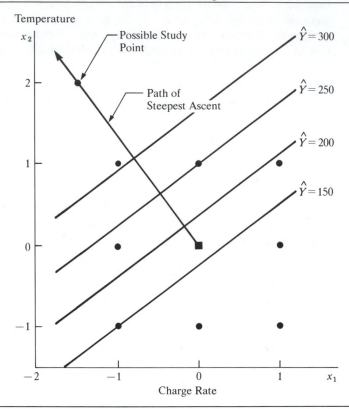

Example.   Figure 9.10 contains the design points of the initial study of the effects of
the charge rate and temperature on the life of the power cells. It also contains some
contour lines for the fitted response plane (9.18), which was determined to be an appropriate fit in the range of the X variables considered.

To search for the conditions maximizing the expected life of the power cells, we
need to study the responses along the path of steepest ascent. This path is perpendic-
ular to the contour lines. Using fitted regression model (9.18) in coded units, for every $b_1 = -55.83$ units in the negative direction along the x_1 axis, the path increases
by $b_2 = 75.50$ units along the x_2 axis. Thus, its slope is $b_2/b_1 = 75.50/(-55.83) =$
-1.35 units of x_2 for every unit increase of x_1. The path of steepest ascent with
slope -1.35, starting at the center point of the design $(x_1 = 0, x_2 = 0)$, is shown in
Figure 9.10. One convenient next study point might be at $x_1 = -1.5$. The corresponding value of x_2 on the path of steepest ascent would be $x_2 = -1.5(-1.35) =$
2.025. This possible study point is identified in Figure 9.10. The actual values of
charge rate and temperature at this study point can be obtained by means of (9.16b);
they are $X_1 = .4$, $X_2 = 40.25$.

After exploration along the line of steepest ascent, another experimental design should be conducted in the region that the additional study points have identified as being nearer the optimal conditions. It may well be that a second-order model may then be required to describe adequately the response function in that region. This process would be repeated as necessary until the region of optimal response conditions has been precisely identified.

Note
Response surface methodology is discussed in specialized texts such as Reference 9.2.

9.5 SOME FURTHER COMMENTS ON POLYNOMIAL REGRESSION

1. The use of polynomial models in X is not without drawbacks. Such models can be more expensive in degrees of freedom than alternative nonlinear models or linear models with transformed variables. Another potential drawback is that multicollinearity is unavoidable. Indeed, if the levels of X are restricted to a narrow range, the degree of multicollinearity in the columns of the $\mathbf{X}$ matrix can be quite high, especially for higher-degree polynomials. It is for this reason that all polynomial regression models in this chapter are formulated in terms of deviations $x_i = X_i - \bar{X}$.

2. An alternative to using variables expressed in deviations from the mean in polynomial regression is to use *orthogonal polynomials*. Orthogonal polynomials are uncorrelated. Some computer packages use orthogonal polynomials in their polynomial regression routines and present the final fitted results in terms of both the orthogonal polynomials and the original polynomials. Orthogonal polynomials are discussed in specialized texts such as Reference 9.3.

3. Sometimes a quadratic response function is fitted for the purpose of establishing the linearity of the response function when repeat observations are not available for directly testing the linearity of the response function. Fitting the quadratic model:

$$(9.23) \qquad Y_i = \beta_0 + \beta_1 x_i + \beta_{11} x_i^2 + \varepsilon_i$$

and testing whether $\beta_{11} = 0$ does not, however, necessarily establish that a linear response function is appropriate. Figure 9.11 provides an example. If sample data were obtained for the response function in Figure 9.11, model (9.23) fitted, and a test on β_{11} made, it likely would lead to the conclusion that $\beta_{11} = 0$. Yet a linear response function clearly might not be appropriate. Examination of residuals would disclose this lack of fit, and should always accompany formal testing of polynomial regression coefficients.

4. When a polynomial regression model with one independent variable is employed, one ordinarily fits a polynomial to the highest power expected to be appropriate, and the decomposition of *SSR* into extra sum of squares components proceeds as follows:

FIGURE 9.11   Example of Curvilinear Response Function

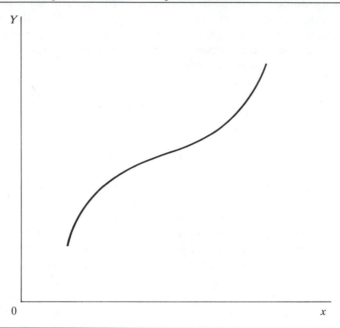

$$SSR(x)$$

$$SSR(x^2 \mid x)$$

$$SSR(x^3 \mid x, x^2)$$

etc.

The reason for this approach is that generally one is most interested in whether higher-order terms can be dropped from the model. Thus, when a cubic model is fitted because it is expected that a third-order model will be sufficient and one wishes to test whether $\beta_{111} = 0$, the appropriate extra sum of squares is $SSR(x^3 \mid x, x^2)$. If, instead, one wishes to test whether a linear term is adequate so that $\beta_{11} = \beta_{111} = 0$, the appropriate extra sum of squares is $SSR(x^2, x^3 \mid x) = SSR(x^2 \mid x) + SSR(x^3 \mid x, x^2)$. Ordinarily, one would not fit a third-order model and test first whether a lower-order coefficient is zero, say, whether $\beta_{11} = 0$. The reason is that it is usually desired to employ as simple a regression model as possible, which in the case of polynomial regression means a lower-order model.

CITED REFERENCES

9.1. Williams, E. J. *Regression Analysis*. New York: John Wiley & Sons, 1959.

9.2. Box, G. E. P., and N. R. Draper. *Empirical Model-Building and Response Surfaces*. New York: John Wiley & Sons, 1987.

9.3. Draper, N. R., and H. Smith. *Applied Regression Analysis*. 2nd ed. New York: John Wiley & Sons, 1981.

PROBLEMS

9.1. A speaker stated: "In developing third-order or higher-order polynomial regression models in social science and managerial applications, inferences on the βs usually take the form of direct tests. There is relatively little interest in estimating the βs to assess effects of the individual polynomial terms." Why might this be so?

9.2. Plot several contour curves for the quadratic response surface $E\{Y\} = 140 + 4x_1^2 - 2x_2^2 + 5x_1x_2$.

9.3. A junior investment analyst used a polynomial regression model of relatively high order in a research seminar on municipal bonds. She obtained an R^2 of .991 in the regression of net interest yield of bond (Y) on industrial diversity index of municipality (X) for seven bond issues. A classmate, unimpressed, said: "You overfitted. Your curve follows the random effects in the data."

   a. Comment on the criticism.

   b. Might R_a^2 defined in (7.37) be more appropriate than R^2 as a descriptive measure here?

9.4. A student in a class demonstration of how to fit a second-order polynomial model in one independent variable entered the X variables in the form X, X^2. He was disturbed when the computer program would not include X^2 in the regression model and regressed Y on X only. The output contained the message: X-SQUARE IS REDUNDANT VARIABLE. X'X IS NEAR-SINGULAR WHEN X-SQUARE IS INCLUDED. Explain the situation. What should the student have done?

9.5. **Mileage study.** The effectiveness of a new experimental overdrive gear in reducing gasoline consumption was studied in 12 trials with a light truck equipped with this gear. In the data that follow, X_i denotes the constant speed (in miles per hour) on the test track in the ith trial and Y_i denotes miles per gallon obtained.

i:	1	2	3	4	5	6	7	8	9	10	11	12
X_i:	35	35	40	40	45	45	50	50	55	55	60	60
Y_i:	22	20	28	31	37	38	41	39	34	37	27	30

The second-order regression model (9.1a) with independent normal error terms is expected to be appropriate.

   a. Fit regression model (9.1a). Plot the fitted regression function and the data. Does the quadratic regression function appear to be a good fit here? Find R^2.

   b. Test whether or not there is a regression relation. Control the risk of a Type I error at .05. State the alternatives, decision rule, and conclusion.

   c. Estimate the mean miles per gallon for test runs at a speed of 48 miles per hour; use a 95 percent confidence interval. Intepret your interval.

   d. Predict the miles per gallon in the next test run at 48 miles per hour; use a 95 percent prediction interval. Interpret your interval.

   e. Test whether the quadratic term can be dropped from the regression model; use $\alpha = .05$. State the alternatives, decision rule, and conclusion.

f. Express the fitted regression function obtained in part (a) in terms of the original variable X.

g. Calculate the coefficient of simple correlation between X and X^2 and between x and x^2. Is the use of a deviation variable helpful here? Explain.

9.6. Refer to **Mileage study** Problem 9.5.

a. Obtain the residuals and plot them against $\hat{Y}$ and against x on separate graphs. Also prepare a normal probability plot. Interpret your plots.

b. Test formally for lack of fit of the quadratic regression function; use $\alpha = .05$. State the alternatives, decision rule, and conclusion. What assumptions did you implicitly make in this test?

c. Fit the third-order model (9.3) and test whether or not $\beta_{111} = 0$; use $\alpha = .05$. State the alternatives, decision rule, and conclusion. Is your conclusion consistent with your finding in part (b)?

9.7. Piecework operation. An operations analyst in a multinational electronics firm studied factors affecting production in a piecework operation where earnings are based on the number of pieces produced. Two employees each were selected from various age groups and data on their productivity last year were obtained (X is age of employee, in years; Y is employee's productivity, coded):

i:	1	2	3	4	5	6	7	8	9
X_i:	20	20	25	25	30	30	35	35	40
Y_i:	97	93	99	105	109	106	109	111	100

i:	10	11	12	13	14	15	16	17	18
X_i:	40	45	45	50	50	55	55	60	60
Y_i:	105	97	101	105	103	105	109	112	110

The analyst recognized that the relation between age and productivity is complex, in part because earnings targets (which he could not measure) shift in complex ways with age. However, he believed that for purposes of estimating mean responses, the response function can be approximated suitably by a polynomial of third order and that the error terms are independent and approximately normally distributed.

a. Fit regression model (9.3). Plot the fitted regression function and the data. Does the cubic regression function appear to be a good fit here? Find R^2.

b. Test whether or not there is a regression relation; use a level of significance of .01. State the alternatives, decision rule, and conclusion. What is the P-value of the test?

c. Obtain joint interval estimates for the mean productivity of employees aged 53, 58, and 62, respectively. Use the most efficient simultaneous estimation procedure and a 99 percent family confidence coefficient. Interpret your intervals.

d. Predict the productivity of an employee aged 53 using a 99 percent prediction interval. Interpret your interval.

e. Express the fitted regression function obtained in part (a) in terms of the original variable X.

f. Calculate the coefficient of simple correlation between X and X^3 and between x and x^3. Is the use of a deviation variable helpful here? Explain.

9.8. Refer to **Piecework operation** Problem 9.7.

a. Test whether both the quadratic and cubic terms can be dropped from the regression model; use $\alpha = .01$. State the alternatives, decision rule, and conclusion.

b. Test whether the cubic term alone can be dropped from the regression model; use $\alpha = .01$. State the alternatives, decision rule, and conclusion.

9.9. Refer to **Piecework operation** Problem 9.7.
  a. Obtain the residuals and plot them against the fitted values and against x on separate graphs. Also prepare a normal probability plot. What do your plots show?
  b. Test formally for lack of fit. Control the risk of a Type I error at .01. State the alternatives, decision rule, and conclusion. What assumptions did you implicitly make in this test?

9.10. Sales forecasting. The Wheaton Company introduced a new product in 1980. Annual sales of this product (Y, in thousand units) follow; the time period (X) is coded, with $X = 1$ for 1980.

i:	1	2	3	4	5	6	7	8	9
X_i:	1	2	3	4	5	6	7	8	9
Y_i:	3.49	3.78	4.05	4.41	4.73	5.12	5.56	5.99	6.44

Assume that the second-order polynomial regression model (9.1a) with independent normal error terms is appropriate.
  a. Fit regression model (9.1a). Plot the fitted regression function and the data. Does the quadratic regression function appear to be a good fit here? What is R^2? Do you believe that the quadratic regression function is appropriate for making projections to year 2000? Discuss.
  b. Obtain simultaneous Bonferroni confidence intervals for β_1 and β_{11} with a 90 percent family confidence coefficient. Interpret your intervals.
  c. Predict sales of the product for 1990 using a 90 percent prediction interval. Interpret your interval.
  d. Express the fitted regression function obtained in part (a) in the original X units.
  e. Calculate the coefficient of simple correlation between X and X^2 and between x and x^2. Is the use of a deviation variable helpful here? Explain.

9.11. Refer to **Sales forecasting** Problem 9.10.
  a. Test whether the quadratic term can be dropped from the regression model. Control the risk of a Type I error at .10. State the alternatives, decision rule, and conclusion. What is the P-value of the test?
  b. Obtain the residuals. Plot the residuals against the fitted values and against time on separate graphs. What do these plots show?

9.12. Crop yield. An agronomist studied the effects of moisture (X_1, in inches) and temperature (X_2, in °C) on the yield of a new hybrid tomato (Y). The experimental data follow.

i:	1	2	3	4	5	6	7	8	9	10	11	12	13
X_{i1}:	6	6	6	6	6	8	8	8	8	8	10	10	10
X_{i2}:	20	21	22	23	24	20	21	22	23	24	20	21	22
Y_i:	49.2	48.1	48.0	49.6	47.0	51.5	51.7	50.4	51.2	48.4	51.1	51.5	50.3

i:	14	15	16	17	18	19	20	21	22	23	24	25
X_{i1}:	10	10	12	12	12	12	12	14	14	14	14	14
X_{i2}:	23	24	20	21	22	23	24	20	21	22	23	24
Y_i:	48.9	48.7	48.6	47.0	48.0	46.4	46.2	43.2	42.6	42.1	43.9	40.5

The agronomist expects that the second-order polynomial regression model (9.5) with independent normal error terms is appropriate here.

a. Fit regression model (9.5). Plot the Y observations against the fitted values. Does the response function provide a good fit?

b. Calculate R^2. What information does this measure provide?

c. Test whether or not there is a regression relation; use $\alpha = .05$. State the alternatives, decision rule, and conclusion. What is the P-value of the test?

d. Estimate the mean yield when $X_1 = 7$ and $X_2 = 22$; use a 95 percent confidence interval. Interpret your interval.

e. Express the fitted response function obtained in part (a) in the original X variables.

9.13. Refer to **Crop yield** Problem 9.12.

a. Test whether or not the interaction term can be dropped from the regression model. Control the α risk at .005. State the alternatives, decision rule, and conclusion.

b. Assuming that the interaction term has been dropped from the regression model, test whether or not the quadratic effect term for temperature can be dropped from the model; control the α risk at .005. State the alternatives, decision rule, and conclusion. What is the combined α risk for both the test here and the one in part (a)?

c. Fit the second-order polynomial regression model omitting the interaction term and the quadratic effect term for temperature. Obtain the residuals and plot them against the fitted values, against x_1, and against x_2 on separate graphs. What do your plots show?

9.14. Computerized game. Students comprising firm A in a computerized marketing game have approached you for assistance in analyzing the relation between promotional expenditures (X) and demand for their firm's product (Y) in the firm's home territory. They believe that the following characteristics hold in this relation: (1) demand in the home territory is affected primarily by promotional expenditures, (2) the relation is either quadratic or linear within the range of X levels of interest to the firm. The team has provided the data shown below for the 14 periods covered in the game to date (X in thousand dollars, Y in thousand units), and has stated that these data span the X levels of interest.

i:	1	2	3	4	5	6	7
X_i:	17	15	25	10	18	15	20
Y_i:	56.15	54.50	55.27	52.54	56.23	55.97	55.55

i:	8	9	10	11	12	13	14
X_i:	25	17	13	20	23	25	16
Y_i:	54.32	55.14	54.28	55.78	55.65	54.96	55.06

Assume that the second-order regression model (9.1a) with independent normal error terms applies.

a. Fit this model and test whether a regression relation exists. Use a level of significance of .01. State the alternatives, decision rule, and conclusion.

b. Test whether the quadratic term can be dropped from the regression model. Use a level of significance of .01. State the alternatives, decision rule, and conclusion.

c. Obtain the residuals and plot them against $\hat{Y}$ and against x on separate graphs. Also obtain a normal probability plot. What do your plots show?

d. Conduct a formal test for lack of fit using a level of significance of .01. State the alternatives, decision rule, and conclusion. Does your conclusion imply that the model cannot be improved further? Discuss.

9.15. Refer to **Computerized game** Problem 9.14. Someone who is familiar with this computerized marketing game enters the discussion. She states that in the system of equations on which the game is based, a quadratic relation does hold between promotional expenditures and mean demand in the firm's home territory. She believes that another significant variable related to expected demand in the home territory is the ratio of the firm's selling price to the average competitive selling price; however she does not recall whether this price ratio has both linear and quadratic effects. She also does not recall whether price ratio and promotional expenditures interact in affecting demand. The firm's price ratios for the 14 periods are as follows:

i:	1	2	3	4	5	6	7
Ratio:	.931	.976	1.045	.939	1.010	1.059	1.000

i:	8	9	10	11	12	13	14
Ratio:	.950	.995	1.011	1.008	.947	1.000	1.017

a. Fit the second-order polynomial regression model (9.5) with promotional expenditures (X_1) and price ratio (X_2) as independent variables. How much has R^2 increased by adding the price ratio as an independent variable?

b. Test whether the price ratio variable should be retained in the regression model. Control the risk of Type I error at .05. State the alternatives, decision rule, and conclusion.

c. Assuming that the price ratio variable is to be retained in the regression model, test whether the interaction term is needed in the model; use $\alpha = .01$. State the alternatives, decision rule, and conclusion. What is the P-value of the test?

d. The team has decided to adopt regression model (9.5) without interaction effects. Fit this model and obtain the residuals. Plot the residuals against $\hat{Y}$ and against the time order of the observations on separate graphs. Also, prepare a normal probability plot. Interpret these plots and state your findings.

9.16. Refer to **Mileage study** Problem 9.5.
   a. At what speed is the estimated quadratic response function a maximum? What is the estimated mean mileage at this speed?
   b. Does the maximum of the response function occur within the scope of the model?

9.17. Refer to **Sales forecasting** Problem 9.10.
   a. In what year does the minimum of the estimated quadratic response function occur? What are the estimated mean sales for this year?
   b. Does the minimum of the response function occur within the scope of the model?

9.18. Biochemical yield. A laboratory investigator wanted to find the settings of pressure (X_1) and temperature (X_2) that produce the maximum biochemical process yield (Y). An initial factorial design with four added center points at $X_1 = 70$, $X_2 = 135$ was utilized. The data follow:

i:	1	2	3	4	5	6	7	8
X_{i1}:	60	60	80	80	70	70	70	70
X_{i2}:	130	140	130	140	135	135	135	135
Y_i:	56.2	60.4	62.5	66.7	60.2	61.8	61.3	58.9

The investigator was fairly certain that the first-order regression model (7.1) would be appropriate.

a. Obtain the fitted first-order response function, using the coded X variables $x_{i1} = (X_{i1} - 70)/10$ and $x_{i2} = (X_{i2} - 135)/5$.

b. Determine the path of steepest ascent from the center point of the design. Plot several contour lines and the path of steepest ascent.

c. One convenient next study point might be $x_1 = 2$. Determine the corresponding value of x_2 on the path of steepest ascent and find the actual values of pressure and temperature at this study point.

d. Calculate the estimated variances of the fitted values at $x_1 = -1$, $x_2 = 1$ and at $x_1 = 1$, $x_2 = -1$. Are the variances consistent with those for a rotatable design? Explain.

9.19. Refer to **Crop yield** Problem 9.12. Consider the first 15 cases to constitute the results of an exploratory investigation of the response surface. The agronomist wishes to fit the first-order regression model (7.1) to determine the path of steepest ascent.

a. Obtain the fitted first-order response function, using the coded variables $x_{i1} = (X_{i1} - 8)/2$ and $x_{i2} = (X_{i2} - 22)/1$.

b. Determine the path of steepest ascent from the center point of the design. Plot several contour lines and the path of steepest ascent.

c. Where might be the next useful study point? Explain.

d. Calculate the estimated variances of the fitted values at $x_1 = -1$, $x_2 = -2$ and at $x_1 = 1$, $x_2 = 2$. Are the variances consistent with those for a rotatable design? Explain.

EXERCISES

9.20. Consider the second-order regression model with one independent variable (9.1a) and the following two sets of X values:

$$\text{Set 1:}\quad 1.0 \quad 1.5 \quad 1.1 \quad 1.3 \quad 1.9 \quad .8 \quad 1.2 \quad 1.4$$
$$\text{Set 2:}\quad 12 \quad 1 \quad 123 \quad 17 \quad 415 \quad 71 \quad 283 \quad 38$$

For each set, calculate the coefficient of correlation between X and X^2, then between x and x^2. Also calculate the coefficients of correlation between X and X^3 and between x and x^3. What generalizations are suggested by your results?

9.21. (Calculus needed.) Refer to the second-order response function (9.2). Explain precisely the meaning of the linear effect coefficient β_1 and the quadratic effect coefficient β_{11}.

9.22. a. Derive the expressions for b'_0, b'_1, and b'_{11} in (9.15).

b. Using theorem (6.47), obtain the variance-covariance matrix for the regression coefficients pertaining to the original X variable in terms of the variance-covariance matrix for the regression coefficients pertaining to the transformed x variable.

9.23. How are the normal equations (9.12) simplified if the X values are equally spaced, such as the time series representation $X_1 = 1$, $X_2 = 2$, . . . , $X_n = n$?

PROJECTS

9.24. Refer to the **SMSA** data set. It is desired to fit the second-order regression model (9.1a) for relating number of active physicians (Y) against total population (X).
   a. Fit the second-order regression model. Plot the residuals against the fitted values. How well does the second-order model appear to fit the data?
   b. Obtain R^2 for the second-order regression model. Also obtain the coefficient of simple determination r^2 for the first-order regression model. Has the addition of the quadratic term in the regression model substantially increased the coefficient of determination?
   c. Test whether the quadratic term can be dropped from the regression model; use $\alpha = .05$. State the alternatives, decision rule, and conclusion.
   d. Omit case 1 (New York City) from the data set. Fit the second-order regression model (9.1a) based on the remaining 140 SMSAs. Repeat the test in part (c). Has the omission of the outlying case affected your conclusion about whether the quadratic term can be dropped from the model?

9.25. Refer to the **SMSA** data set. A regression model relating serious crime rate (Y, total serious crimes divided by total population) to population density (X_1, total population divided by land area) and percent of population in central cities (X_3) is to be constructed.
   a. Fit the second-order regression model (9.5). Plot the residuals against the fitted values. How well does the second-order model appear to fit the data? What is R^2?
   b. Test whether or not all quadratic and interaction terms can be dropped from the regression model; use $\alpha = .01$. State the alternatives, decision rule, and conclusion.
   c. Instead of using the independent variable population density, total population (X_1) and land area (X_2) are to be employed as separate independent variables, in addition to percent of population in central cities (X_3). The regression model should contain linear and quadratic terms for total population, and linear terms only for land area and percent of population in central cities. (No interaction terms are to be included in this model.) Fit this regression model and obtain R^2. Is this coefficient of multiple determination substantially different from the one for the regression model in part (a)?

9.26. Refer to the **SENIC** data set. The second-order regression model (9.1a) is to be fitted for relating number of nurses (Y) to available facilities and services (X).
   a. Fit the second-order regression model. Plot the residuals against the fitted values. How well does the second-order model appear to fit the data?
   b. Obtain R^2 for the second-order regression model. Also obtain the coefficient of simple determination r^2 for the first-order regression model. Has the addition of the quadratic term in the regression model substantially increased the coefficient of determination?
   c. Test whether the quadratic term can be dropped from the regression model; use $\alpha = .10$. State the alternatives, decision rule, and conclusion.

9.27. Refer to **Sales forecasting** Problem 9.10. Instead of using a polynomial regression model here, it has been suggested that a transformation of variables might yield an equally good fit and would be more desirable since forecasting requires extrapolation.

a. Fit a linear regression model for relating $Y' = \sqrt{Y}$ against X. Plot the fitted regression function and the transformed data. How effective does the use of the transformed variable appear to be here?

b. Obtain the fitted values and transform them to the original variable Y. Calculate the residuals in the original variable and plot these residuals against X. Interpret your plot.

c. Square the residuals in the original variable obtained in part (b), sum, and obtain MSE. Compare this with MSE for the quadratic regression model in Problem 9.10a. How does the variability around the fitted regression function, as measured by MSE, compare for the two approaches?

d. Repeat parts (a) through (c) using the transformation $Y' = \log_{10} Y$. Is either the square-root or the logarithmic transformation clearly preferable here?

Chapter 10

Qualitative Independent Variables

Throughout the previous chapters on regression analysis, we have utilized quantitative variables in the regression models considered. Quantitative variables take on values on a well-defined scale; examples are income, age, temperature, and amount of liquid assets.

Many variables of interest in business, economics, and the social and biological sciences, however, are not quantitative but are qualitative. Examples of qualitative variables are gender (male, female), purchase status (purchase, no purchase), and disability status (not disabled, partly disabled, fully disabled).

Qualitative variables can also be used in multiple regression models. In this chapter, we take up the case where some or all of the independent variables are qualitative.

10.1 ONE QUALITATIVE INDEPENDENT VARIABLE

An economist wished to relate the speed with which a particular insurance innovation is adopted (Y) to the size of the insurance firm (X_1) and the type of firm. The dependent variable is measured by the number of months elapsed between the time the first firm adopted the innovation and the time the given firm adopted the innovation. The first independent variable, size of firm, is quantitative, and is measured by the amount of total assets of the firm. The second independent variable, type of firm, is qualitative and is composed of two classes—stock companies and mutual companies. In order that such a qualitative variable can be used in a regression model, quantitative indicators for the classes of the qualitative variable must be found.

Indicator Variables

There are many ways of quantitatively identifying the classes of a qualitative variable. We shall use indicator variables that take on the values 0 and 1. These indicator variables are easy to use and are widely employed, but they are by no means the only way to quantify a qualitative variable.

For the insurance innovation example, where the qualitative variable has two classes, we might define two indicator variables X_2 and X_3 as follows:

(10.1)

$$X_2 = \begin{array}{ll} 1 & \text{if stock company} \\ 0 & \text{otherwise} \end{array}$$

$$X_3 = \begin{array}{ll} 1 & \text{if mutual company} \\ 0 & \text{otherwise} \end{array}$$

Assuming that a first-order model is to be employed, it would be:

(10.2) $\qquad Y_i = \beta_0 X_{i0} + \beta_1 X_{i1} + \beta_2 X_{i2} + \beta_3 X_{i3} + \varepsilon_i \qquad$ where $X_{i0} \equiv 1$

This intuitive approach of setting up an indicator variable for each class of the qualitative variable unfortunately leads to computational difficulties. To see why, suppose we have $n = 4$ observations, the first two being stock firms for which $X_2 = 1$ and $X_3 = 0$, and the second two being mutual firms for which $X_2 = 0$ and $X_3 = 1$. The $\mathbf{X}$ matrix would then be:

$$
\mathbf{X} = \begin{array}{cccc} X_0 & X_1 & X_2 & X_3 \end{array}
\begin{bmatrix}
1 & X_{11} & 1 & 0 \\
1 & X_{21} & 1 & 0 \\
1 & X_{31} & 0 & 1 \\
1 & X_{41} & 0 & 1
\end{bmatrix}
$$

Note that the X_0 column is equal to the sum of the X_2 and X_3 columns, so that the columns are linearly dependent according to definition (6.20). This has a serious effect on the $\mathbf{X'X}$ matrix:

$$
\mathbf{X'X} =
\begin{bmatrix}
1 & 1 & 1 & 1 \\
X_{11} & X_{21} & X_{31} & X_{41} \\
1 & 1 & 0 & 0 \\
0 & 0 & 1 & 1
\end{bmatrix}
\begin{bmatrix}
1 & X_{11} & 1 & 0 \\
1 & X_{21} & 1 & 0 \\
1 & X_{31} & 0 & 1 \\
1 & X_{41} & 0 & 1
\end{bmatrix}
$$

$$
= \begin{bmatrix}
4 & \sum_{i=1}^{4} X_{i1} & 2 & 2 \\
\sum_{i=1}^{4} X_{i1} & \sum_{i=1}^{4} X_{i1}^2 & \sum_{i=1}^{2} X_{i1} & \sum_{i=3}^{4} X_{i1} \\
2 & \sum_{i=1}^{2} X_{i1} & 2 & 0 \\
2 & \sum_{i=3}^{4} X_{i1} & 0 & 2
\end{bmatrix}
$$

It is quickly apparent that the first column of the $\mathbf{X'X}$ matrix equals the sum of the last two columns, so that the columns are linearly dependent. Hence, the $\mathbf{X'X}$ matrix does not have an inverse, and no unique estimators of the regression coefficients can be found.

A simple way out of this difficulty is to drop one of the indicator variables. In our example, for instance, we might drop X_3. While dropping one indicator variable is not the only way out of the difficulty, it leads to simple interpretations of the parameters. In general, therefore, we shall follow the principle:

(10.3)   A qualitative variable with c classes will be represented by $c - 1$ indicator variables, each taking on the values 0 and 1.

Note

Indicator variables are frequently also called *dummy variables* or *binary variables*. The latter term has reference to the binary number system containing only 0 and 1.

Interpretation of Regression Coefficients

Returning to the insurance innovation example, suppose that we drop the indicator variable X_3 from regression model (10.2) so that the model becomes:

(10.4) $$Y_i = \beta_0 + \beta_1 X_{i1} + \beta_2 X_{i2} + \varepsilon_i$$

where:

$X_{i1} = $ Size of firm

$$X_{i2} = \begin{matrix} 1 & \text{if stock company} \\ 0 & \text{otherwise} \end{matrix}$$

The response function for this regression model is:

(10.5) $$E\{Y\} = \beta_0 + \beta_1 X_1 + \beta_2 X_2$$

To understand the meaning of the regression coefficients in this model, consider first the case of a mutual firm. For such a firm, $X_2 = 0$ and we have:

(10.5a) $$E\{Y\} = \beta_0 + \beta_1 X_1 + \beta_2(0) = \beta_0 + \beta_1 X_1 \qquad \text{Mutual firms}$$

Thus, the response function for mutual firms is a straight line, with Y intercept β_0 and slope β_1. This response function is shown in Figure 10.1.

For a stock firm, $X_2 = 1$ and the response function (10.5) is:

(10.5b) $$E\{Y\} = \beta_0 + \beta_1 X_1 + \beta_2(1) = (\beta_0 + \beta_2) + \beta_1 X_1 \qquad \text{Stock firms}$$

This also is a straight line, with the same slope β_1 but with Y intercept $\beta_0 + \beta_2$. This response function is also shown in Figure 10.1.

The meaning of the regression coefficients in the response function (10.5) is now clear. With reference to the insurance innovation example, the mean time elapsed before the innovation is adopted, $E\{Y\}$, is a linear function of size of firm (X_1), with

FIGURE 10.1   Illustration of Meaning of Regression Coefficients for Regression Model
(10.4) with Indicator Variable X_2—Insurance Innovation Example

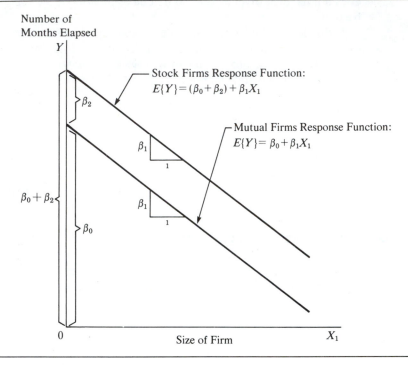

the same slope β_1 for both types of firms. β_2 indicates how much higher (lower) the
response function for stock firms is than the one for mutual firms, for any given size
of firm. Thus, β_2 measures the differential effect of type of firm. In general, β_2
shows how much higher (lower) the mean response line is for the class coded 1 than
the line for the class coded 0, for any given level of X_1.

Example

In the insurance innovation example, the economist studied 10 mutual firms and 10
stock firms. The data are shown in Table 10.1. The **Y** and **X** data matrices are
shown in Table 10.2. Note that $X_2 = 1$ for each stock firm and $X_2 = 0$ for each mutual firm.

Given the **Y** and **X** matrices in Table 10.2, fitting the regression model (10.4) is
straightforward. Table 10.3 (p. 355) presents the key results from a computer run.
The fitted response function is:

$$\hat{Y} = 33.87407 - .10174X_1 + 8.05547X_2$$

Figure 10.2 (p. 354) contains the fitted response function for each type of firm, together with the actual observations.

TABLE 10.1 Data for Insurance Innovation Study

Firm i	Number of Months Elapsed Y_i	Size of Firm (million dollars) X_{i1}	Type of Firm
1	17	151	Mutual
2	26	92	Mutual
3	21	175	Mutual
4	30	31	Mutual
5	22	104	Mutual
6	0	277	Mutual
7	12	210	Mutual
8	19	120	Mutual
9	4	290	Mutual
10	16	238	Mutual
11	28	164	Stock
12	15	272	Stock
13	11	295	Stock
14	38	68	Stock
15	31	85	Stock
16	21	224	Stock
17	20	166	Stock
18	13	305	Stock
19	30	124	Stock
20	14	246	Stock

TABLE 10.2 Data Matrices for Insurance Innovation Example

$$
Y = \begin{bmatrix} 17 \\ 26 \\ 21 \\ 30 \\ 22 \\ 0 \\ 12 \\ 19 \\ 4 \\ 16 \\ 28 \\ 15 \\ 11 \\ 38 \\ 31 \\ 21 \\ 20 \\ 13 \\ 30 \\ 14 \end{bmatrix}
\qquad
X = \begin{bmatrix}
X_0 & X_1 & X_2 \\
1 & 151 & 0 \\
1 & 92 & 0 \\
1 & 175 & 0 \\
1 & 31 & 0 \\
1 & 104 & 0 \\
1 & 277 & 0 \\
1 & 210 & 0 \\
1 & 120 & 0 \\
1 & 290 & 0 \\
1 & 238 & 0 \\
1 & 164 & 1 \\
1 & 272 & 1 \\
1 & 295 & 1 \\
1 & 68 & 1 \\
1 & 85 & 1 \\
1 & 224 & 1 \\
1 & 166 & 1 \\
1 & 305 & 1 \\
1 & 124 & 1 \\
1 & 246 & 1 \end{bmatrix}
$$

FIGURE 10.2 Fitted Regression Functions for Regression Model (10.4)—Insurance Innovation Example

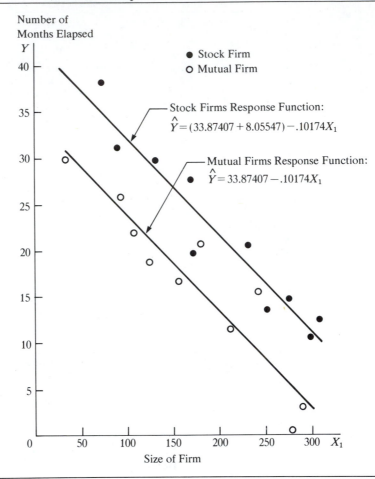

Number of Months Elapsed

Stock Firms Response Function:
$$\hat{Y} = (33.87407 + 8.05547) - .10174X_1$$

Mutual Firms Response Function:
$$\hat{Y} = 33.87407 - .10174X_1$$

Size of Firm

The economist was most interested in the effect of type of firm (X_2) on the elapsed time for the innovation to be adopted. He therefore desired to obtain a 95 percent confidence interval for β_2. We require $t(.975; 17) = 2.110$ and obtain from the data in Table 10.3 the confidence limits $8.05547 \pm 2.110(1.45911)$. The confidence interval for β_2 therefore is:

$$4.98 \leq \beta_2 \leq 11.13$$

Thus, with 95 percent confidence, we conclude that stock companies tend to adopt the innovation somewhere between 5 and 11 months later, on the average, than mutual companies, for any given size of firm.

TABLE 10.3   Regression Results for Regression Model (10.4) Fit—Insurance Innovation Example

(a) Regression Coefficients

Regression Coefficient	Estimated Regression Coefficient	Estimated Standard Deviation	t^*
β_0	33.87407	1.81386	18.68
β_1	$-.10174$	.00889	-11.44
β_2	8.05547	1.45911	5.52

(b) Analysis of Variance

Source of Variation	SS	df	MS
Regression	1,504.41	2	752.20
Error	176.39	17	10.38
Total	1,680.80	19	

A formal test of:

$$H_0: \beta_2 = 0$$

$$H_a: \beta_2 \neq 0$$

with level of significance .05 would lead to H_a, that type of firm has an effect, since the 95 percent confidence interval for β_2 does not include zero.

The economist also carried out other analyses, some of which will be described shortly.

Note

The reader may wonder why we did not simply fit separate regressions for stock firms and mutual firms in our example, and instead adopted the approach of fitting one regression with an indicator variable. There are two reasons for this. Since the model assumed equal slopes and the same constant error term variance for each type of firm, the common slope β_1 can best be estimated by pooling the two types of firms. Also, other inferences, such as ones pertaining to β_0 and β_2, can be made more precisely by working with one regression model containing an indicator variable since more degrees of freedom will then be associated with *MSE*.

10.2 MODEL CONTAINING INTERACTION EFFECTS

In the insurance innovation example, the economist actually did not begin his analysis with regression model (10.4) since he expected interaction effects between size and type of firm. Even though one of the independent variables in the regression model is qualitative, interaction effects can be introduced into the model in the usual manner, by including cross-product terms. A first-order regression model with an interaction term for the insurance innovation example is:

(10.6) $$Y_i = \beta_0 + \beta_1 X_{i1} + \beta_2 X_{i2} + \beta_3 X_{i1} X_{i2} + \varepsilon_i$$

where:

$$X_{i1} = \text{Size of firm}$$

$$X_{i2} = \begin{matrix} 1 & \text{if stock company} \\ 0 & \text{otherwise} \end{matrix}$$

The response function for this regression model is:

(10.7) $$E\{Y\} = \beta_0 + \beta_1 X_1 + \beta_2 X_2 + \beta_3 X_1 X_2$$

Meaning of Regression Coefficients

The meaning of the regression coefficients in response function (10.7) can best be understood by examining the nature of this function for each type of firm. For a mutual firm, $X_2 = 0$ and hence $X_1 X_2 = 0$. The response function for mutual firms therefore is:

(10.7a) $$E\{Y\} = \beta_0 + \beta_1 X_1 + \beta_2(0) + \beta_3(0) = \beta_0 + \beta_1 X_1 \qquad \text{Mutual firms}$$

This response function is shown in Figure 10.3. Note that the Y intercept is β_0 and the slope is β_1 for the response function for mutual firms.

For stock firms, $X_2 = 1$ and hence $X_1 X_2 = X_1$. The response function for stock firms therefore is:

$$E\{Y\} = \beta_0 + \beta_1 X_1 + \beta_2(1) + \beta_3 X_1$$

or:

(10.7b) $$E\{Y\} = (\beta_0 + \beta_2) + (\beta_1 + \beta_3)X_1 \qquad \text{Stock firms}$$

This response function is also shown in Figure 10.3. Note that the response function for stock firms has Y intercept $\beta_0 + \beta_2$ and slope $\beta_1 + \beta_3$.

Thus, β_2 indicates how much greater (smaller) is the Y intercept for the class coded 1 than that for the class coded 0, and similarly β_3 indicates how much greater (smaller) is the slope for the class coded 1 than that for the class coded 0. Because both the intercept and the slope differ for the two classes in regression model (10.6), it is no longer true that β_2 indicates how much higher (lower) one response line is than the other for any given level of X_1. Figure 10.3 makes it clear that the effect of type of firm with regression model (10.6) depends on X_1, the size of the firm. For smaller firms, according to Figure 10.3, mutual firms tend to innovate more quickly, but for larger firms stock firms tend to innovate more quickly. Thus, when interaction effects are present, the effect of the qualitative variable can only be studied by comparing the regression functions for each class of the qualitative variable.

Figure 10.4 illustrates another possible interaction pattern for the insurance inno-

FIGURE 10.3   Illustration of Meaning of Regression Coefficients for Regression Model (10.6) with Indicator Variable X_2 and Interaction Term— Insurance Innovation Example

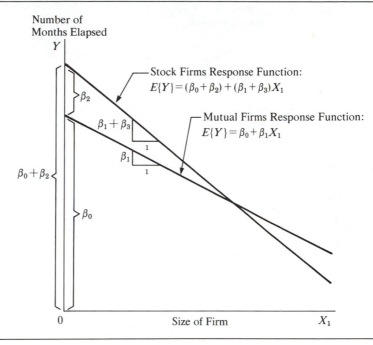

vation example. Here, mutual firms tend to introduce the innovation more quickly than stock firms for all sizes of firms in the scope of the model, but the differential effect is much smaller for large firms than for small ones.

Example

Since the economist anticipated that interaction effects between size and type of firm may be present, he actually first wished to fit regression model (10.6):

$$Y_i = \beta_0 + \beta_1 X_{i1} + \beta_2 X_{i2} + \beta_3 X_{i1} X_{i2} + \varepsilon_i$$

Table 10.4 shows the $\mathbf{X}$ matrix for this model. The $\mathbf{Y}$ matrix is the same as in Table 10.2. Note that the $X_1 X_2$ column in the $\mathbf{X}$ matrix in Table 10.4 contains 0 for mutual companies and X_{i1} for stock companies.

Given the $\mathbf{Y}$ and $\mathbf{X}$ matrices, the regression fit is routine. Basic results from a computer run are shown in Table 10.5. To test for the presence of interaction effects:

$$H_0: \beta_3 = 0$$

$$H_a: \beta_3 \neq 0$$

FIGURE 10.4 Another Illustration of Regression Model (10.6) with Indicator
Variable X_2 and Interaction Term—Insurance Innovation Example

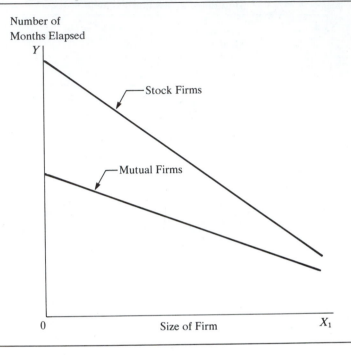

the economist used the t^* statistic from Table 10.5a:

$$t^* = \frac{b_3}{s\{b_3\}} = \frac{-.0004171}{.01833} = -.02$$

For level of significance .05, we require $t(.975; 16) = 2.120$. Since $|t^*| =$
$.02 \leq 2.120$, we conclude H_0, i.e., that $\beta_3 = 0$ or that no interaction effects are
present. The two-sided P-value for the test is very high, namely, .98. It was because
of this result that the economist adopted regression model (10.4) with no interaction
term, which we discussed earlier.

Note

Fitting regression model (10.6) yields the same response functions as fitting separate regressions for stock firms and mutual firms. An advantage of using model (10.6) with an indicator
variable is that one regression run on the computer will yield both fitted regressions.

Another advantage is that tests for comparing the regression functions for the different
classes of the qualitative variable can be clearly seen to be tests of regression coefficients in a
general linear model. For instance, Figure 10.3 makes it clear for the insurance innovation example that the test whether the two regression functions have the same slope involves:

$$H_0: \beta_3 = 0$$

$$H_a: \beta_3 \neq 0$$

TABLE 10.4   **X** Matrix for Fitting Regression Model (10.6) with Interaction Term—
Insurance Innovation Example

$$
\mathbf{X} = \begin{array}{cccc}
X_0 & X_1 & X_2 & X_1X_2 \\
\begin{bmatrix}
1 & 151 & 0 & 0 \\
1 & 92 & 0 & 0 \\
1 & 175 & 0 & 0 \\
1 & 31 & 0 & 0 \\
1 & 104 & 0 & 0 \\
1 & 277 & 0 & 0 \\
1 & 210 & 0 & 0 \\
1 & 120 & 0 & 0 \\
1 & 290 & 0 & 0 \\
1 & 238 & 0 & 0 \\
1 & 164 & 1 & 164 \\
1 & 272 & 1 & 272 \\
1 & 295 & 1 & 295 \\
1 & 68 & 1 & 68 \\
1 & 85 & 1 & 85 \\
1 & 224 & 1 & 224 \\
1 & 166 & 1 & 166 \\
1 & 305 & 1 & 305 \\
1 & 124 & 1 & 124 \\
1 & 246 & 1 & 246
\end{bmatrix}
\end{array}
$$

TABLE 10.5   Regression Results for Fit of Regression Model (10.6) with Interaction
Term—Insurance Innovation Example

(a) Regression Coefficients

Regression Coefficient	Estimated Regression Coefficient	Estimated Standard Deviation	t^*
β_0	33.83837	2.44065	13.86
β_1	−.10153	.01305	−7.78
β_2	8.13125	3.65405	2.23
β_3	−.0004171	.01833	−.02

(b) Analysis of Variance

Source of Variation	SS	df	MS
Regression	1,504.42	3	501.47
Error	176.38	16	11.02
Total	1,680.80	19	

Similarly, the test whether the two regression functions are identical would involve:

$$H_0: \beta_2 = \beta_3 = 0$$

$$H_a: \text{not both } \beta_2 = 0 \text{ and } \beta_3 = 0$$

10.3 MORE COMPLEX MODELS

We now briefly consider more complex models involving qualitative independent variables.

Qualitative Variable with More than Two Classes

If a qualitative independent variable has more than two classes, we require additional indicator variables in the regression model. Consider the regression of tool wear (Y) on tool speed (X_1), where we wish to include also tool model (M1, M2, M3, M4) as an independent variable. Since the qualitative variable (tool model) has four classes, we require three indicator variables. Let us define them as follows:

$$X_2 = \begin{matrix} 1 & \text{if tool model M1} \\ 0 & \text{otherwise} \end{matrix}$$

(10.8)
$$X_3 = \begin{matrix} 1 & \text{if tool model M2} \\ 0 & \text{otherwise} \end{matrix}$$

$$X_4 = \begin{matrix} 1 & \text{if tool model M3} \\ 0 & \text{otherwise} \end{matrix}$$

First-Order Model. A first-order regression model is:

(10.9)
$$Y_i = \beta_0 + \beta_1 X_{i1} + \beta_2 X_{i2} + \beta_3 X_{i3} + \beta_4 X_{i4} + \varepsilon_i$$

For this model, the data input for the **X** matrix would be as follows:

Tool Model	X_0	X_1	X_2	X_3	X_4
M1	1	X_{i1}	1	0	0
M2	1	X_{i1}	0	1	0
M3	1	X_{i1}	0	0	1
M4	1	X_{i1}	0	0	0

The response function for regression model (10.9) is:

(10.10)
$$E\{Y\} = \beta_0 + \beta_1 X_1 + \beta_2 X_2 + \beta_3 X_3 + \beta_4 X_4$$

To see the meaning of the regression coefficients, consider first the response function for tool models M4 for which $X_2 = 0$, $X_3 = 0$, and $X_4 = 0$:

(10.10a) $$E\{Y\} = \beta_0 + \beta_1 X_1 \qquad \text{Tool models M4}$$

For tool models M1, $X_2 = 1$, $X_3 = 0$, and $X_4 = 0$, and the response function is:

(10.10b) $$E\{Y\} = \beta_0 + \beta_1 X_1 + \beta_2 = (\beta_0 + \beta_2) + \beta_1 X_1 \qquad \text{Tool models M1}$$

Similarly, the response functions for tool models M2 and M3 are:

(10.10c) $$E\{Y\} = (\beta_0 + \beta_3) + \beta_1 X_1 \qquad \text{Tool models M2}$$

(10.10d) $$E\{Y\} = (\beta_0 + \beta_4) + \beta_1 X_1 \qquad \text{Tool models M3}$$

Thus, the response function (10.10) implies that the regression of tool wear on tool speed is linear, with the same slope for all types of tool models. The coefficients β_2, β_3, and β_4 indicate, respectively, how much higher (lower) the response functions for tool models M1, M2, and M3 are than the one for tool models M4, for any given level of tool speed. Thus, β_2, β_3, and β_4 measure the differential effects of the qualitative variable classes on the height of the response function for any given level of X_1, always compared with the class for which $X_2 = X_3 = X_4 = 0$. Figure 10.5 illustrates a possible arrangement of the response functions.

When using regression model (10.9), one may wish to estimate differential effects other than against tool models M4. This can be done by estimating differences be-

---

FIGURE 10.5   Illustration of Regression Model (10.9)—Tool Wear Example

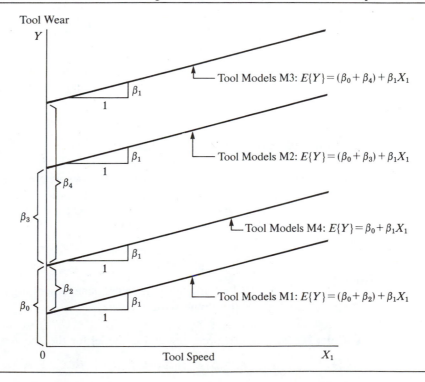

tween regression coefficients. For instance, $\beta_4 - \beta_3$ measures how much higher (lower) the response function for tool models M3 is than the response function for tool models M2 for any given level of tool speed, as may be seen by comparing (10.10c) and (10.10d). The point estimator of this quantity is, of course, $b_4 - b_3$, and the estimated variance of this estimator is:

$$(10.11) \qquad s^2\{b_4 - b_3\} = s^2\{b_4\} + s^2\{b_3\} - 2s\{b_4, b_3\}$$

The needed variances and covariance can be readily obtained from the estimated variance-covariance matrix of the regression coefficients.

First-Order Model with Interactions Added. If interaction effects between tool speed and tool model are present in our illustration, regression model (10.9) would be modified as follows:

$$(10.12) \qquad Y_i = \beta_0 + \beta_1 X_{i1} + \beta_2 X_{i2} + \beta_3 X_{i3} + \beta_4 X_{i4} + \beta_5 X_{i1} X_{i2}$$
$$+ \beta_6 X_{i1} X_{i3} + \beta_7 X_{i1} X_{i4} + \varepsilon_i$$

The response function for tool models M4, for which $X_2 = 0$, $X_3 = 0$, and $X_4 = 0$, is as follows:

$$(10.13a) \qquad E\{Y\} = \beta_0 + \beta_1 X_1 \qquad \text{Tool models M4}$$

Similarly, we find for the other tool models:

$$(10.13b) \qquad E\{Y\} = (\beta_0 + \beta_2) + (\beta_1 + \beta_5)X_1 \qquad \text{Tool models M1}$$

$$(10.13c) \qquad E\{Y\} = (\beta_0 + \beta_3) + (\beta_1 + \beta_6)X_1 \qquad \text{Tool models M2}$$

$$(10.13d) \qquad E\{Y\} = (\beta_0 + \beta_4) + (\beta_1 + \beta_7)X_1 \qquad \text{Tool models M3}$$

Thus, the interaction regression model (10.12) implies that each tool model has its own regression line, with different intercepts and slopes for the different tool models.

More than One Qualitative Independent Variable

Regression models can readily be constructed for cases where two or more of the independent variables are qualitative. Consider the regression of advertising expenditures (Y) on sales (X_1), type of firm (incorporated, not incorporated), and quality of sales management (high, low). We may define:

$$(10.14)$$
$$X_2 = \begin{matrix} 1 & \text{if firm incorporated} \\ 0 & \text{otherwise} \end{matrix}$$

$$X_3 = \begin{matrix} 1 & \text{if quality of sales management high} \\ 0 & \text{otherwise} \end{matrix}$$

First-Order Model. A first-order regression model for the above example is:

$$(10.15) \qquad Y_i = \beta_0 + \beta_1 X_{i1} + \beta_2 X_{i2} + \beta_3 X_{i3} + \varepsilon_i$$

This model implies that the response function of advertising expenditures on sales is linear, with the same slope for all "type of firm—quality of sales management" combinations, and β_2 and β_3 indicate the additive differential effects of type of firm and quality of sales management on the height of the regression line for any given levels of X_1 and the other independent variable.

First-Order Model with Certain Interactions Added.   A first-order regression model with interaction effects between each pair of the independent variables is:

$$(10.16) \qquad Y_i = \beta_0 + \beta_1 X_{i1} + \beta_2 X_{i2} + \beta_3 X_{i3} + \beta_4 X_{i1} X_{i2}$$
$$+ \beta_5 X_{i1} X_{i3} + \beta_6 X_{i2} X_{i3} + \varepsilon_i$$

Note the implications of this model:

Type of Firm	Quality of Sales Management	Response Function
Incorp.	High	$E\{Y\} = (\beta_0 + \beta_2 + \beta_3 + \beta_6) + (\beta_1 + \beta_4 + \beta_5)X_1$
Not incorp.	High	$E\{Y\} = (\beta_0 + \beta_3) + (\beta_1 + \beta_5)X_1$
Incorp.	Low	$E\{Y\} = (\beta_0 + \beta_2) + (\beta_1 + \beta_4)X_1$
Not incorp.	Low	$E\{Y\} = \beta_0 + \beta_1 X_1$

Not only are all response functions different for the various "type of firm—quality of sales management" combinations, but the differential effects of one qualitative variable on the intercept depend on the particular class of the other qualitative variable. For instance, when we move from "not incorporated—low quality" to "incorporated—low quality," the intercept changes by β_2. But if we move from "not incorporated—high quality" to "incorporated—high quality," the intercept changes by $\beta_2 + \beta_6$.

Qualitative Independent Variables Only

Regression models containing only qualitative independent variables can also be constructed. With reference to our previous advertising example, we could regress advertising expenditures only on type of firm and quality of sales management. The first-order regression model then would be:

$$(10.17) \qquad Y_i = \beta_0 + \beta_2 X_{i2} + \beta_3 X_{i3} + \varepsilon_i$$

where X_{i2} and X_{i3} are defined in (10.14).

Comments

1. Models in which all independent variables are qualitative are called *analysis of variance models*.

2. Models containing some quantitative and some qualitative independent variables, where the chief independent variables of interest are qualitative and the quantitative independent variables are introduced primarily to reduce the variance of the error terms, are called *covariance models.*

10.4 COMPARISON OF TWO OR MORE REGRESSION FUNCTIONS

Frequently we encounter regressions for two or more populations and wish to examine their similarities and differences. We present three examples.

1. A company operates two production lines for making soap bars. For each line, the relation between the speed of the line and the amount of scrap for the day was studied. A scatter plot of the data for the two production lines suggests that the regression relation between production line speed and amount of scrap is linear but not the same for the two production lines. The slopes appear to be about the same, but the heights of the regression lines seem to differ. A formal test is desired whether or not the two regression lines are identical. If it is found that the two regression lines are not the same, an investigation is to be made of why the difference in scrap yield exists.

2. An economist is studying the relation between amount of savings and level of income for middle-income families from urban and rural areas, based on independent samples from the two populations. Each of the two relations can be modeled by linear regression. She wishes to compare whether, at given income levels, urban and rural families tend to save the same amount—i.e., whether the two regression lines are the same. If they are not, she wishes to explore whether at least the amounts of savings out of an additional dollar of income are the same for the two groups—i.e., whether the slopes of the two regression lines are the same.

3. A company had two instruments constructed to identical specifications to measure pressure in an industrial process. A study was then made for each instrument of the relation between its gauge readings and actual pressures as determined by an almost exact but slow and costly method. If the two regression lines are the same, a single calibration schedule can be developed for the two instruments; otherwise, two different calibration schedules will be required.

When it is reasonable to assume that the error term variances in the regression models for the different populations are equal, we can use indicator variables to test the equality of the different regression functions. If the error variances are not equal, transformations may equalize them at least approximately.

We have already seen how regression models with indicator variables that contain interaction terms permit testing of the equality of regression functions for the different classes of a qualitative variable. This methodology can be used directly for testing the equality of regression functions for different populations. One simply considers the different populations studied as classes of an independent variable, defines indicator variables for the different populations, and develops a regression model containing appropriate interaction terms. Since no new principles arise in the testing of the equality of regression functions for different populations, we shall immediately utilize two of the earlier examples to illustrate the approach.

Soap Production Lines Example

The data on amount of scrap (Y) and line speed (X_1) for the soap production lines example are presented in Table 10.6. The variable X_2 is a code for the production line. A scatter plot of the data, using different symbols for the two production lines, is shown in Figure 10.6.

Tentative Model. Based on the scatter plot in Figure 10.6, the analyst decided to tentatively fit regression model (10.6) which assumes that the regression relation between amount of scrap and line speed is linear for both production lines and that the variances of the error terms are the same, but permits the two regression lines to have different slopes and intercepts.

$$(10.18) \qquad Y_i = \beta_0 + \beta_1 X_{i1} + \beta_2 X_{i2} + \beta_3 X_{i1} X_{i2} + \varepsilon_i$$

where:

X_{i1} = Line speed

$$X_{i2} = \begin{cases} 1 & \text{if production line 1} \\ 0 & \text{if production line 2} \end{cases}$$

$i = 1, 2, \ldots, 27$

TABLE 10.6 Data for Soap Production Lines Example (all data are coded)

	Production Line 1				Production Line 2		
Case i	Amount of Scrap Y_i	Line Speed X_{i1}	X_{i2}	Case i	Amount of Scrap Y_i	Line Speed X_{i1}	X_{i2}
1	218	100	1	16	140	105	0
2	248	125	1	17	277	215	0
3	360	220	1	18	384	270	0
4	351	205	1	19	341	255	0
5	470	300	1	20	215	175	0
6	394	255	1	21	180	135	0
7	332	225	1	22	260	200	0
8	321	175	1	23	361	275	0
9	410	270	1	24	252	155	0
10	260	170	1	25	422	320	0
11	241	155	1	26	273	190	0
12	331	190	1	27	410	295	0
13	275	140	1				
14	425	290	1				
15	367	265	1				

FIGURE 10.6   Scatter Plot for Soap Production Lines Example

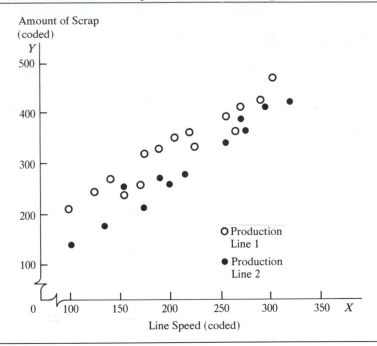

Note that for purposes of this model, the 15 cases for production line 1 and the 12 cases for production line 2 are combined into one group of 27 cases.

Diagnostics.   A fit of regression model (10.18) to the data in Table 10.6 led to the results presented in Table 10.7 and the following fitted regression function:

$$\hat{Y} = 7.57 + 1.322X_1 + 90.39X_2 - .1767X_1X_2$$

Plots of the residuals against $\hat{Y}$ are shown in Figure 10.7 for each production line. There are two plots in order to facilitate the diagnosis of possible differences between the two production lines. Both of the plots in Figure 10.7 are reasonably consistent with regression model (10.18). The splits between positive and negative residuals of 10 to 5 for production line 1 and 4 to 8 for production line 2 can be accounted for by randomness of the outcomes. Plots of the residuals against X_2 and a normal probability plot of the residuals (not shown) also support the appropriateness of the fitted model. For the latter plot, the coefficient of correlation between the ordered residuals and their expected values under normality is .990. This is sufficiently high according to Table 4.3 to indicate normality of the error terms.

Finally, the analyst desired to make a formal test of the equality of the variances of the error terms for the two production lines. To obtain independent estimates of the variances, she fitted separate linear regression models to the data for the two production lines and obtained the following results:

TABLE 10.7 Regression Results for Fit of Regression Model (10.18)—Soap Production Lines Example

(a) Regression Coefficients

Regression Coefficient	Estimated Regression Coefficient	Estimated Standard Deviation	t^*
β_0	7.57	20.87	.36
β_1	1.322	.09262	14.27
β_2	90.39	28.35	3.19
β_3	−.1767	.1288	−1.37

(b) Analysis of Variance

Source of Variation	SS	df	MS
Regression	169,165	3	56,388
X_1	149,661	1	149,661
$X_2 \mid X_1$	18,694	1	18,694
$X_1 X_2 \mid X_1, X_2$	810	1	810
Error	9,904	23	430.6
Total	179,069	26	

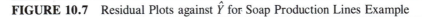

FIGURE 10.7 Residual Plots against $\hat{Y}$ for Soap Production Lines Example

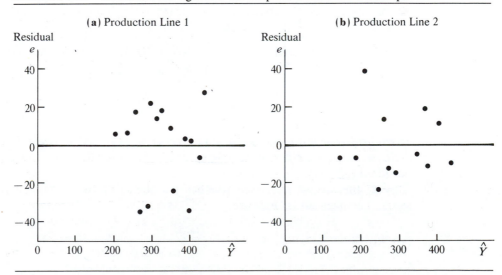

Production Line	Fitted Regression Line	MSE	df
1	$\hat{Y} = 97.965 + 1.145X_1$	492.52	13
2	$\hat{Y} = 7.574 + 1.322X_1$	350.13	10

Using the F test in Table 1.4a, we obtain:

$$F^* = \frac{492.52}{350.13} = 1.41$$

Specifying the level of significance at .05, we require $F(.025; 13, 10) = .308$ and $F(.975; 13, 10) = 3.58$. Since F^* falls between these two limits, we conclude that the two production line regressions have equal error variances.

At this point, the analyst was satisfied about the aptness of regression model (10.18) with normal error terms and was ready to proceed with comparing the regression relation between amount of scrap and line speed for the two production lines.

Inferences about Two Regression Lines.   Identity of the regression functions for the two production lines is tested by considering the alternatives:

(10.19)
$$H_0: \beta_2 = \beta_3 = 0$$
$$H_a: \text{not both } \beta_2 = 0 \text{ and } \beta_3 = 0$$

The appropriate test statistic is given by (8.25):

(10.19a)
$$F^* = \frac{SSR(X_2, X_1X_2 \mid X_1)}{2} \div \frac{SSE(X_1, X_2, X_1X_2)}{n - 4}$$

where n represents the combined sample size for both populations. Using the regression results in Table 10.7, we find:

$$SSR(X_2, X_1X_2 \mid X_1) = SSR(X_2 \mid X_1) + SSR(X_1X_2 \mid X_1, X_2)$$

$$= 18,694 + 810 = 19,504$$

$$F^* = \frac{19,504}{2} \div \frac{9,904}{23} = 22.65$$

To control α at level .01, we require $F(.99; 2, 23) = 5.67$. Since $F^* = 22.65 > 5.67$, we conclude H_a, that the regression functions for the two production lines are not identical.

Next, the analyst examined whether the slopes of the regression lines are the same. The alternatives here are:

(10.20)
$$H_0: \beta_3 = 0$$
$$H_a: \beta_3 \neq 0$$

and the appropriate test statistic is either the t^* statistic (8.23) or the partial F test statistic (8.22):

(10.20a)
$$F^* = \frac{SSR(X_1X_2 \mid X_1, X_2)}{1} \div \frac{SSE(X_1, X_2, X_1X_2)}{n-4}$$

Using the regression results in Table 10.7 and the partial F test statistic, we obtain:

$$F^* = \frac{810}{1} \div \frac{9,904}{23} = 1.88$$

For $\alpha = .01$, we require $F(.99; 1, 23) = 7.88$. Since $F^* = 1.88 \leq 7.88$, we conclude H_0, that the slope of the regression function is the same for the two production lines.

Using the Bonferroni inequality (5.3), the analyst can therefore conclude at family significance level .02 that a given increase in line speed leads to the same amount of increase in expected scrap in each of the two production lines, but that the expected amount of scrap for any given line speed differs by a constant amount for the two production lines.

We can estimate this constant difference in the regression lines by obtaining a confidence interval for β_2. For a 95 percent confidence interval, we require $t(.975; 23) = 2.069$. Using the results in Table 10.7, we obtain the confidence limits $90.39 \pm 2.069(28.35)$. Hence, the confidence interval for β_2 is:

$$31.7 \leq \beta_2 \leq 149.0$$

We thus conclude, with 95 percent confidence, that the mean amount of scrap for production line 1, at any given line speed, exceeds that for production line 2 by somewhere between 32 and 149.

Instrument Calibration Study Example

The engineer making the calibration study believed that the regression functions relating gauge reading (Y) and actual pressure (X_1) for both instruments are second-order polynomials:

$$E\{Y\} = \beta_0 + \beta_1 X_1 + \beta_2 X_1^2$$

but that they might differ for the two instruments. Hence, he employed the model (using a deviation variable for X_1 to reduce multicollinearity problems—see Chapter 9):

(10.21) $Y_i = \beta_0 + \beta_1 x_{i1} + \beta_2 x_{i1}^2 + \beta_3 X_{i2} + \beta_4 x_{i1} X_{i2} + \beta_5 x_{i1}^2 X_{i2} + \varepsilon_i$

where:

$x_{i1} = X_{i1} - \bar{X}_1 =$ deviation of actual pressure

$X_{i2} = \begin{cases} 1 & \text{if instrument B} \\ 0 & \text{otherwise} \end{cases}$

Note that for instrument A, where $X_2 = 0$, the response function is:

$$(10.22a) \qquad E\{Y\} = \beta_0 + \beta_1 x_1 + \beta_2 x_1^2$$

and for instrument B, where $X_2 = 1$, the response function is:

$$(10.22b) \qquad E\{Y\} = (\beta_0 + \beta_3) + (\beta_1 + \beta_4)x_1 + (\beta_2 + \beta_5)x_1^2$$

Hence, the test for equality of the two response functions involves the alternatives:

$$(10.23) \qquad \begin{array}{l} H_0: \beta_3 = \beta_4 = \beta_5 = 0 \\ H_a: \text{not all } \beta_k \text{ in } H_0 \text{ equal zero} \end{array}$$

and the appropriate test statistic is (8.25):

$$(10.23a) \qquad F^* = \frac{SSR(X_2, x_1 X_2, x_1^2 X_2 \mid x_1, x_1^2)}{3} \div \frac{SSE(x_1, x_1^2, X_2, x_1 X_2, x_1^2 X_2)}{n - 6}$$

where n represents the combined sample size for both populations.

Comments

1. The approach just described is completely general. If three or more populations are involved, additional indicator variables are simply added to the model.

2. The use of indicator variables for testing whether two or more regression functions are the same is equivalent to the general linear test approach where fitting the full model involves fitting separate regressions to the data from each population and fitting the reduced model involves fitting one regression to the combined data.

10.5 OTHER USES OF INDICATOR VARIABLES

Piecewise Linear Regression

Sometimes the regression of Y on X follows a particular linear relation in some range of X, but a different linear relation elsewhere. For instance, unit cost (Y) regressed on lot size may follow a certain linear regression up to $X_p = 500$, at which point the slope changes because of some operating efficiencies only possible with lot sizes of more than 500. For example, there may be a break in the unit price when raw materials are purchased for lots greater than 500. Figure 10.8 illustrates this situation.

We consider now how indicator variables may be used to fit piecewise linear regressions consisting of two pieces. We take up the case where X_p, the point where the slope changes, is known.

We return to our lot size illustration, for which it is known that the slope changes at $X_p = 500$. The model for our illustration may be expressed as follows:

$$(10.24) \qquad Y_i = \beta_0 + \beta_1 X_{i1} + \beta_2 (X_{i1} - 500)X_{i2} + \varepsilon_i$$

FIGURE 10.8 Illustration of Piecewise Linear Regression

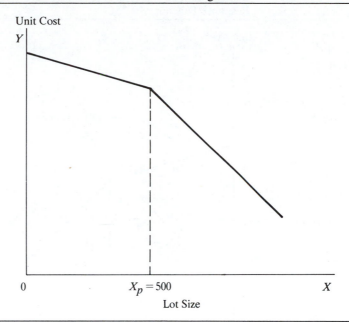

where:

X_{i1} = Lot size

$$X_{i2} = \begin{matrix} 1 & \text{if } X_{i1} > 500 \\ 0 & \text{otherwise} \end{matrix}$$

To check that regression model (10.24) does provide a two-piecewise linear regression, consider the response function for this model:

(10.25) $$E\{Y\} = \beta_0 + \beta_1 X_1 + \beta_2(X_1 - 500)X_2$$

When $X_1 \leq 500$, $X_2 = 0$ so that (10.25) becomes:

(10.25a) $$E\{Y\} = \beta_0 + \beta_1 X_1 \qquad X_1 \leq 500$$

On the other hand, when $X_1 > 500$, $X_2 = 1$ and we obtain:

(10.25b) $$E\{Y\} = (\beta_0 - 500\beta_2) + (\beta_1 + \beta_2)X_1 \qquad X_1 > 500$$

Thus, β_1 and $\beta_1 + \beta_2$ are the slopes of the two regression lines, and β_0 and $\beta_0 - 500\beta_2$ are the two Y intercepts. These parameters are shown in Figure 10.9. The reason for subtracting $500\beta_2$ from β_0 is that β_2 pertains to a unit increase in X_1, but here we are measuring the impact of the differential effect in the slope in the negative direction from $X_p = 500$ to zero.

FIGURE 10.9   Illustration of Meaning of Regression Coefficients of Piecewise Linear
Response Function (10.25)

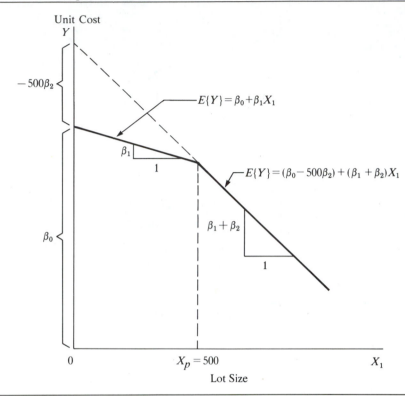

Example.   Table 10.8a contains eight observations on unit costs for given lot sizes.
It is known that the response function slope changes at $X_p = 500$ so that regression
model (10.24) is to be employed. Table 10.8b contains the $\mathbf{X}$ matrix for this example. The left column of $\mathbf{X}$ is a column of 1s as usual. The next column contains X_{i1}.
The final column on the right contains $(X_{i1} - 500)X_{i2}$, which consists of 0s for all lot
sizes up to 500, and of $X_{i1} - 500$ for all lot sizes above 500. The fitting of regression model (10.24) at this point becomes routine. The fitted response function is:

$$\hat{Y} = 5.89545 - .00395X_1 - .00389(X_1 - 500)X_2$$

From this fitted regression function, expected unit cost is estimated to decline by
.00395 for a lot size increase of one when the lot size is less than 500 and by
.00395 + .00389 = .00784 when the lot size is 500 or more.

Note

The extension of regression model (10.24) to more than two-piecewise regression lines is
straightforward. For instance, if the slope of the regression line in the earlier lot size illustra-

TABLE 10.8   Data and **X** Matrix for Piecewise Linear Regression—Lot Size Example

	(a)			(b)		
Lot i	Unit Cost (dollars) Y_i	Lot Size X_i		X_0	X_1	$(X_1 - 500)X_2$
1	2.57	650		1	650	150
2	4.40	340		1	340	0
3	4.52	400		1	400	0
4	1.39	800	$\mathbf{X} =$	1	800	300
5	4.75	300		1	300	0
6	3.55	570		1	570	70
7	2.49	720		1	720	220
8	3.77	480		1	480	0

tion actually changes at both $X = 500$ and $X = 800$, the model would be:

$$(10.26) \qquad Y_i = \beta_0 + \beta_1 X_{i1} + \beta_2(X_{i1} - 500)X_{i2} + \beta_3(X_{i1} - 800)X_{i3} + \varepsilon_i$$

where:

X_{i1} = Lot size

$$X_{i2} = \begin{matrix} 1 \\ 0 \end{matrix} \quad \begin{matrix} \text{if } X_{i1} > 500 \\ \text{otherwise} \end{matrix}$$

$$X_{i3} = \begin{matrix} 1 \\ 0 \end{matrix} \quad \begin{matrix} \text{if } X_{i1} > 800 \\ \text{otherwise} \end{matrix}$$

Discontinuity in Regression Function

Sometimes the linear regression function may not only change its slope at some value X_p but may also have a jump point there. Figure 10.10 illustrates this case. Another indicator variable must now be introduced to take care of the jump. Suppose time required to solve a task successfully (Y) is to be regressed on complexity of task (X), when complexity of task is measured·on a quantitative scale from 0 to 100. It is known that the slope of the response line changes at $X_p = 40$, and it is believed that the regression relation may be discontinuous there. We therefore set up the regression model:

$$(10.27) \qquad Y_i = \beta_0 + \beta_1 X_{i1} + \beta_2(X_{i1} - 40)X_{i2} + \beta_3 X_{i3} + \varepsilon_i$$

where:

X_{i1} = Complexity of task

FIGURE 10.10   Illustration of Response Function (10.28) for Discontinuous Piecewise Linear Regression

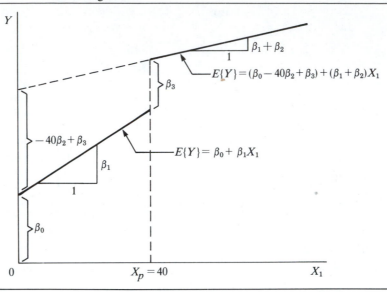

$$X_{i2} = \begin{matrix} 1 \\ 0 \end{matrix} \quad \begin{matrix} \text{if } X_{i1} > 40 \\ \text{otherwise} \end{matrix}$$

$$X_{i3} = \begin{matrix} 1 \\ 0 \end{matrix} \quad \begin{matrix} \text{if } X_{i1} > 40 \\ \text{otherwise} \end{matrix}$$

The response function for regression model (10.27) is:

(10.28) $E\{Y\} = \beta_0 + \beta_1 X_1 + \beta_2(X_1 - 40)X_2 + \beta_3 X_3$

When $X_1 \le 40$, then $X_2 = 0$ and $X_3 = 0$, so (10.28) becomes:

(10.28a) $E\{Y\} = \beta_0 + \beta_1 X_1 \qquad X_1 \le 40$

Similarly, when $X_1 > 40$, then $X_2 = 1$ and $X_3 = 1$, so (10.28) becomes:

(10.28b) $E\{Y\} = (\beta_0 - 40\beta_2 + \beta_3) + (\beta_1 + \beta_2)X_1 \qquad X_1 > 40$

These two response functions are shown in Figure 10.10, together with the parameters involved. Note that β_3 represents the difference in the mean responses for the two regression lines at $X_p = 40$ and β_2 represents the difference in the two slopes.

The estimation of the regression coefficients for model (10.27) presents no new problems. One may test whether or not $\beta_3 = 0$ in the usual manner. If it is concluded that $\beta_3 = 0$, the regression function is continuous at X_p so that the earlier piecewise linear regression model applies.

Time Series Applications

Economists and business analysts frequently use time series data in regression analysis. For instance, savings (Y) may be regressed on income (X), where both the savings and income data pertain to a number of years. The model employed might be:

$$(10.29) \qquad Y_t = \beta_0 + \beta_1 X_t + \varepsilon_t \qquad t = 1, \ldots, n$$

where Y_t and X_t are savings and income, respectively, for time period t. Suppose that the period covered includes both peacetime and wartime years, and that this factor should be recognized since it is anticipated that savings in wartime years tend to be higher. The following model might then be appropriate:

$$(10.30) \qquad Y_t = \beta_0 + \beta_1 X_{t1} + \beta_2 X_{t2} + \varepsilon_t$$

where:

X_{t1} = Income

$X_{t2} = \begin{cases} 1 & \text{if period } t \text{ peacetime} \\ 0 & \text{otherwise} \end{cases}$

Note that regression model (10.30) assumes that the marginal propensity to save (β_1) is constant in both peacetime and wartime years, and that only the height of the response curve is affected by this qualitative variable.

Another use of indicator variables in time series applications occurs when monthly or quarterly data are used. Suppose that quarterly sales (Y) are regressed on quarterly advertising expenditures (X_1) and quarterly disposable personal income (X_2). If seasonal effects also have an influence on quarterly sales, a first-order regression model incorporating seasonal effects would be:

$$(10.31) \qquad Y_t = \beta_0 + \beta_1 X_{t1} + \beta_2 X_{t2} + \beta_3 X_{t3} + \beta_4 X_{t4} + \beta_5 X_{t5} + \varepsilon_t$$

where:

X_{t1} = Quarterly advertising expenditures

X_{t2} = Quarterly disposable personal income

$X_{t3} = \begin{cases} 1 & \text{if first quarter} \\ 0 & \text{otherwise} \end{cases}$

$X_{t4} = \begin{cases} 1 & \text{if second quarter} \\ 0 & \text{otherwise} \end{cases}$

$X_{t5} = \begin{cases} 1 & \text{if third quarter} \\ 0 & \text{otherwise} \end{cases}$

10.6 SOME CONSIDERATIONS IN USING INDICATOR VARIABLES

Indicator Variables versus Allocated Codes

An alternative to the use of indicator variables for a qualitative independent variable is to employ *allocated codes*. Consider, for instance, the independent variable "frequency of product use" which has three classes: frequent user, occasional user, nonuser. With the allocated codes approach, a single independent variable is employed and values are assigned to the classes; for instance:

Class	X_1
Frequent user	3
Occasional user	2
Nonuser	1

The allocated codes are, of course, arbitrary and could be other sets of numbers. The model with allocated codes for our example, assuming no other independent variables, would be:

$$(10.32) \qquad Y_i = \beta_0 + \beta_1 X_{i1} + \varepsilon_i$$

The basic difficulty with allocated codes is that they define a metric for the classes of the qualitative variable which may not be reasonable. To see this concretely, consider the mean responses with regression model (10.32) for the three classes of the qualitative variable:

Class	$E\{Y\}$
Frequent user	$E\{Y\} = \beta_0 + \beta_1(3) = \beta_0 + 3\beta_1$
Occasional user	$E\{Y\} = \beta_0 + \beta_1(2) = \beta_0 + 2\beta_1$
Nonuser	$E\{Y\} = \beta_0 + \beta_1(1) = \beta_0 + \beta_1$

Note the key implication:

$$E\{Y \mid \text{frequent user}\} - E\{Y \mid \text{occasional user}\}$$

$$= E\{Y \mid \text{occasional user}\} - E\{Y \mid \text{nonuser}\} = \beta_1$$

Thus, the coding 1, 2, 3 implies that the mean response changes by the same amount when going from a nonuser to an occasional user as when going from an occasional user to a frequent user. This may not be in accord with reality, and is the result of the coding 1, 2, 3 which assigns equal distances between the three user classes. Other allocated codes may, of course, imply different spacings of the classes of the qualitative variable, but these would ordinarily still be arbitrary.

Indicator variables, in contrast, make no assumptions about the spacing of the classes and rely on the data to show the differential effects that occur. If, for the same example, two indicator variables, say, X_1 and X_2, are employed to represent the qualitative variable, as follows:

Class	X_1	X_2
Frequent user	1	0
Occasional user	0	1
Nonuser	0	0

the regression model would be:

(10.33) $$Y_i = \beta_0 + \beta_1 X_{i1} + \beta_2 X_{i2} + \varepsilon_i$$

Here β_1 measures the differential effect:

$$E\{Y \mid \text{frequent user}\} - E\{Y \mid \text{nonuser}\}$$

and β_2 measures:

$$E\{Y \mid \text{occasional user}\} - E\{Y \mid \text{nonuser}\}$$

Thus, β_2 measures the differential effect between occasional user and nonuser, and $\beta_1 - \beta_2$ measures the differential effect between frequent user and occasional user. Notice that there are no arbitrary restrictions to be satisfied by these two differential effects.

Indicator Variables versus Quantitative Variables

If an independent variable is quantitative, such as age, one can nevertheless use indicator variables instead. For instance, the quantitative variable age may be transformed by grouping ages into classes such as under 21, 21–34, 35–49, etc. Indicator variables may then be used for the classes of this new independent variable. At first sight, this may seem to be a questionable approach because information about the actual ages is thrown away. Furthermore, additional parameters are placed into the model, which leads to a reduction of the degrees of freedom associated with MSE.

Nevertheless, there are occasions when replacement of a quantitative variable by indicator variables may be appropriate. Consider, for example, a large-scale survey in which the relation between liquid assets (Y) and age (X) of head of household is to be studied. Two thousand households will be included in the study, so that the loss of 10 or 20 degrees of freedom is immaterial. The analyst is very much in doubt about the shape of the regression function, which could be highly complex, and hence may prefer the indicator variable approach in order to obtain information about the shape without making any assumptions about the functional form of the regression relation.

Another alternative, also utilizing indicator variables, is available to the analyst in doubt about the functional form of a possibly complex regression relation. He or she could use the quantitative variable age, but employ piecewise linear regression with a number of pieces. Again, this approach loses degrees of freedom for estimating *MSE*, but this is of no concern in large-scale studies. The benefit would be that information about the shape of the regression relation is obtained without making strong assumptions about its functional form.

Other Codings for Indicator Variables

As stated earlier, many different codings of indicator variables are possible. We mention here two alternatives to our 0, 1 coding with $c - 1$ indicator variables for a qualitative variable with c classes.

For the insurance innovation example, where Y is time to adopt an innovation, X_1 is size of insurance firm, and the second independent variable is type of company (stock, mutual), we could use the coding:

$$(10.34) \qquad X_2 = \begin{matrix} 1 & \text{if stock company} \\ -1 & \text{if mutual company} \end{matrix}$$

In that case, the first-order linear regression model would be:

$$(10.35) \qquad Y_i = \beta_0 + \beta_1 X_{i1} + \beta_2 X_{i2} + \varepsilon_i$$

which has the response function:

$$(10.36) \qquad E\{Y\} = \beta_0 + \beta_1 X_1 + \beta_2 X_2$$

This response function is as follows for the two types of companies:

$$(10.36a) \qquad E\{Y\} = (\beta_0 + \beta_2) + \beta_1 X_1 \qquad \text{Stock firms}$$

$$(10.36b) \qquad E\{Y\} = (\beta_0 - \beta_2) + \beta_1 X_1 \qquad \text{Mutual firms}$$

Thus, β_0 here may be viewed as an "average" intercept of the regression line, from which the stock company and mutual company intercepts differ by β_2 in opposite directions. A test whether the regression lines are the same for both types of companies involves $H_0: \beta_2 = 0$, $H_a: \beta_2 \neq 0$.

A second alternative coding scheme is to use a 0, 1 indicator variable for each of the c classes of the qualitative variable and drop the intercept term in the regression model. For the insurance innovation example, we would have:

$$(10.37) \qquad Y_i = \beta_1 X_{i1} + \beta_2 X_{i2} + \beta_3 X_{i3} + \varepsilon_i$$

where:

$X_{i1} = $ Size of firm

$$X_{i2} = \begin{matrix} 1 & \text{if stock company} \\ 0 & \text{otherwise} \end{matrix}$$

$$X_{i3} = \begin{array}{ll} 1 & \text{if mutual company} \\ 0 & \text{otherwise} \end{array}$$

Here, the two response functions would be:

(10.38a)	$E\{Y\} = \beta_2 + \beta_1 X_1$	Stock firms
(10.38b)	$E\{Y\} = \beta_3 + \beta_1 X_1$	Mutual firms

A test of whether or not the two regression lines are the same would involve the alternatives $H_0: \beta_2 = \beta_3$, $H_a: \beta_2 \neq \beta_3$. This type of test is discussed in Section 8.2.

PROBLEMS

10.1. A student who used a regression model that included indicator variables was upset when receiving only the following output on the multiple regression printout: XTRANSPOSE X SINGULAR. What is a likely source of the difficulty?

10.2. Refer to regression model (10.4). Portray graphically the response curves for this model if $\beta_0 = 25.3$, $\beta_1 = .20$, and $\beta_2 = -12.1$.

10.3. In a regression study of factors affecting learning time for a certain task (measured in minutes), gender of learner was included as an independent variable (X_2) that was coded $X_2 = 1$ if male and 0 if female. It was found that $b_2 = 22.3$ and $s\{b_2\} = 3.8$. An observer questioned whether the coding scheme for gender is fair because it results in a positive coefficient, leading to longer learning times for males than females. Comment.

10.4. Refer to **Calculator maintenance** Problem 2.18. The users of the desk calculators are either training institutions that use a student model, or business firms that employ a commercial model. An analyst at Tri-City wishes to fit a regression model including both number of calculators serviced (X_1) and type of calculator (X_2) as independent variables and estimate the effect of calculator model (S—student, C—commercial) on number of minutes spent on the service call. Records show that the models serviced in the 18 calls were:

i:	1	2	3	4	5	6	7	8	9
Model:	C	S	C	C	C	S	S	C	C

i:	10	11	12	13	14	15	16	17	18
Model:	C	C	S	S	C	S	C	C	C

Assume that regression model (10.4) is appropriate, and let $X_2 = 1$ if student model and 0 if commercial model.

a. Explain the meaning of all regression coefficients in the model.

b. Fit the regression model and state the estimated regression function.

c. Estimate the effect of calculator model on mean service time with a 95 percent confidence interval. Interpret your interval estimate.

d. Why would the analyst wish to include X_1, number of calculators, in the regression model when interest is in estimating the effect of type of calculator model on service time?

e. Obtain the residuals and plot them against $X_1 X_2$. Is there any indication that an interaction term in the regression model would be helpful?

10.5. Refer to **Grade point average** Problem 2.17. An assistant to the director of admissions conjectures that the predictive power of the model could be improved by adding information on whether the student had chosen a major field of concentration at the time the application was submitted. Assume that regression model (10.4) is appropriate, where X_1 is entrance test score and $X_2 = 1$ if student had indicated a major field of concentration at the time of application and 0 if the major field was undecided. Students 3, 4, 5, 6, 9, and 17 had indicated a major field at the time of application.

a. Explain how each regression coefficient in model (10.4) is interpreted here.

b. Fit the regression model and state the estimated regression function.

c. Test whether the X_2 variable can be dropped from the regression model; use $\alpha = .01$. State the alternatives, decision rule, and conclusion.

d. Obtain the residuals for regression model (10.4) and plot them against $X_1 X_2$. Is there any evidence in your plot that it would be helpful to include an interaction term in the model?

10.6. Refer to regression models (10.4) and (10.6). Would the conclusion that $\beta_2 = 0$ have the same implication for each of these models? Explain.

10.7. Refer to regression model (10.6). Portray graphically the response curves for this model if $\beta_0 = 25$, $\beta_1 = .30$, $\beta_2 = -12.5$, and $\beta_3 = .05$. Describe the nature of the interaction effect.

10.8. Refer to **Calculator maintenance** Problems 2.18 and 10.4.

a. Fit regression model (10.6) and state the estimated regression function.

b. Test whether the interaction term can be dropped from the model; control the α risk at .10. State the alternatives, decision rule, and conclusion. What is the P-value of the test? If the interaction term cannot be dropped from the model, describe the nature of the interaction effect.

10.9. Refer to **Grade point average** Problems 2.17 and 10.5.

a. Fit regression model (10.6) and state the estimated regression function.

b. Test whether the interaction term can be dropped from the model; use $\alpha = .05$. State the alternatives, decision rule, and conclusion. If the interaction term cannot be dropped from the model, describe the nature of the interaction effect.

10.10. In a regression analysis of on-the-job head injuries of warehouse laborers caused by falling objects, Y is a measure of severity of the injury, X_1 is an index reflecting both the weight of the object and the distance it fell, and X_2 and X_3 are indicator variables for nature of head protection worn at the time of the accident, coded as follows:

Type of Protection	X_2	X_3
Hard hat	1	0
Bump cap	0	1
None	0	0

The response function to be used in the study is $E\{Y\} = \beta_0 + \beta_1 X_1 + \beta_2 X_2 + \beta_3 X_3$.

a. Develop the response function for each type-of-protection category.

b. For each of the following questions, specify the alternatives H_0 and H_a for the appropriate test: (1) With X_1 fixed, does wearing a bump cap reduce the expected severity of injury as compared with wearing no protection? (2) With X_1 fixed, is the expected severity of injury the same when wearing a hard hat as when wearing a bump cap?

10.11. Refer to the tool wear regression model (10.12). Suppose the indicator variables had been defined as follows: $X_2 = 1$ if tool model M2 and 0 otherwise, $X_3 = 1$ if tool model M3 and 0 otherwise, $X_4 = 1$ if tool model M4 and 0 otherwise. Indicate the meaning of each of the following: (1) β_3, (2) $\beta_4 - \beta_3$, (3) β_1, (4) β_7.

10.12. Refer to the advertising expenditures regression model (10.16).

a. How is β_4 interpreted here? What is the meaning of $\beta_3 - \beta_2$ here?

b. State the alternatives for a test of whether the response functions are the same in incorporated and not-incorporated firms with high-quality sales management.

10.13. A marketing research trainee in the national office of a chain of shoe stores used the following response function to study seasonal (winter, spring, summer, fall) effects on sales of a certain line of shoes: $E\{Y\} = \beta_0 + \beta_1 X_1 + \beta_2 X_2 + \beta_3 X_3$. The Xs are indicator variables defined as follows: $X_1 = 1$ if winter and 0 otherwise, $X_2 = 1$ if spring and 0 otherwise, $X_3 = 1$ if fall and 0 otherwise. After fitting the model, she tested the regression coefficients β_k ($k = 0, \ldots, 3$) and came to the following set of conclusions at an .05 family level of significance: $\beta_0 \neq 0$, $\beta_1 = 0$, $\beta_2 \neq 0$, $\beta_3 \neq 0$. In her report she then wrote: "Results of regression analysis show that climatic and other seasonal factors have no influence in determining sales of this shoe line in the winter. Seasonal influences do exist in the other seasons." Do you agree with this interpretation of the test results? Discuss.

10.14. **Assessed valuations.** A tax consultant studied the current relation between selling price and assessed valuation of one-family residential dwellings in a large tax district. He obtained data for a random sample of nine recent "arm's-length" sales transactions of one-family dwellings located on corner lots and also obtained data for a random sample of 14 recent sales of one-family dwellings not located on corner lots. In the data that follow, both selling price (Y) and assessed valuation (X_1) are expressed in thousand dollars. Assume that the error variances in the two populations are equal and that regression model (10.6) is appropriate.

Corner Lots

i:	1	2	3	4	5	6	7	8	9
X_{i1}:	17.5	12.5	20.0	16.0	15.0	14.7	17.5	12.3	11.5
Y_i:	56.2	42.5	68.6	54.8	50.0	47.5	56.9	34.0	39.0

Noncorner Lots

i:	1	2	3	4	5	6	7	8	9	10	11	12	13	14
X_{i1}:	10.0	13.8	15.0	19.5	17.0	12.5	14.5	12.8	12.0	16.0	10.0	17.0	10.8	15.0
Y_i:	31.2	36.9	41.0	51.8	48.0	33.3	38.0	35.9	32.0	44.3	29.0	46.1	30.0	42.0

a. Plot the sample data for the two populations on one graph, using different symbols for the two samples. Does the regression relation appear to be the same for the two populations?

b. Test for identity of the regression functions for dwellings on corner lots and dwellings in other locations; control the risk of Type I error at .10. State the alternatives, decision rule, and conclusion.

c. Plot the estimated regression functions for the two populations and describe the nature of the differences between them.

d. Estimate the difference in the slopes of the two regression functions using a 90 percent confidence interval.

e. Obtain the residuals and plot them separately against $\hat{Y}$ for each sample. Does the assumption of equal error variances appear to be reasonable here?

f. Conduct a formal F test for equality of the error variances for the two types of locations; use $\alpha = .05$. State the alternatives, decision rule, and conclusion. What is the P-value of the test?

g. Plot the residuals against X_1, X_2, and $X_1 X_2$ on separate graphs. Also prepare a normal probability plot of the residuals. Interpret your plots and summarize your findings.

10.15. Tire testing. A testing laboratory with equipment that simulates highway driving studied for two makes (A, B) of a certain type of truck tire the relation between operating cost per mile (Y) and cruising speed (X_1). The observations are shown below (all data are coded). An engineer now wishes to decide whether or not the regression of operating cost on cruising speed is the same for the two makes. Assume that the error variances for the two makes of tires are the same and that regression model (10.6) is appropriate.

Make A

i:	1	2	3	4	5	6	7	8	9	10
X_{i1}:	10	20	20	30	40	40	50	60	60	70
Y_i:	9.8	12.5	14.2	14.9	19.0	16.5	20.9	22.4	24.1	25.8

Make B

i:	1	2	3	4	5	6	7	8	9	10
X_{i1}:	10	20	20	30	40	40	50	60	60	70
Y_i:	15.0	14.5	16.1	16.5	16.4	19.1	20.9	22.3	19.8	21.4

a. Plot the sample data for the two populations on one graph, using different symbols for the two samples. Does the relation between speed and cost appear to be the same for the two makes of tires?

b. Test whether or not the regression functions are the same for the two makes of tires. Control the risk of Type I error at .05. State the alternatives, decision rule, and conclusion.

c. Suppose the question of interest simply had been whether the two regression lines have equal slopes. Answer this question by setting up a 95 percent confidence interval for the difference between the two slopes. What do you find?

d. Obtain the residuals and plot them separately against $\hat{Y}$ for each make of tire. Does the assumption of equal error variances appear to be reasonable here?

e. Conduct a formal test for equality of the error variances for each make of tire; use $\alpha = .01$. State the alternatives, decision rule, and conclusion. What is the P-value of the test?

f. Plot the residuals against X_1, X_2, and $X_1 X_2$ on separate graphs. Also prepare a normal probability plot of the residuals. Summarize the results of your analysis of these plots.

10.16. Refer to **Muscle mass** Problem 2.25. The nutritionist conjectures that the regression of muscle mass on age follows a two-piece linear relation, with the slope changing at age 60 without discontinuity.

    a. State the regression model that applies if the nutritionist's conjecture is correct. What are the respective response functions when age is 60 or less and when age is over 60?

    b. Fit the regression model specified in part (a) and state the estimated regression function.

    c. Test whether a two-piece linear regression function is needed; use $\alpha = .01$. State the alternatives, decision rule, and conclusion. What is the P-value of the test?

10.17. **Shipment handling.** Global Electronics periodically imports shipments of a certain large part used as a component in several of its products. The size of the shipment varies depending upon production schedules. For handling and distribution to assembling plants, shipments of size 250 thousand parts or less are sent to warehouse A; larger shipments are sent to warehouse B since this warehouse has specialized equipment that provides greater economies of scale for large shipments. The data below were collected on the 10 most recent shipments; X is the size of the shipment (in thousand parts), and Y is the direct cost of handling the shipment in the warehouse (in thousand dollars).

i:	1	2	3	4	5	6	7	8	9	10
X_i:	225	350	150	200	175	180	325	290	400	125
Y_i:	11.95	14.13	8.93	10.98	10.03	10.13	13.75	13.30	15.00	7.97

A two-piece linear regression model with a possible discontinuity at $X = 250$ is to be fitted.

    a. Specify the regression model to be used.

    b. Fit this regression model. Plot the fitted response function and the data. Is there any indication that greater economies of scale are obtained in handling relatively large shipments than relatively small ones?

    c. Test whether or not both the two separate slopes and the discontinuity can be dropped from the model. Control the level of significance at .025. State the alternatives, decision rule, and conclusion.

    d. For relatively small shipments, what is the point estimate of the increase in expected handling cost for each increase of 1 thousand in the size of the shipment? What is the corresponding estimate for relatively large shipments?

10.18. In times series analysis, the X variable representing time usually is defined to take on values 1, 2, etc., for the successive time periods. Does this represent an allocated code when the time periods are actually 1989, 1990, etc.?

10.19. An analyst wishes to include number of older siblings in family as an independent variable in a regression analysis of factors affecting maturation in eighth graders. The number of older siblings in the sample observations ranges from 0 to 4. Discuss whether this variable should be placed in the model as an ordinary quantitative variable or by means of four 0,1 indicator variables.

10.20. Refer to regression model (10.2) for the insurance innovation study. Suppose X_0 were dropped to eliminate the linear dependence in the $\mathbf{X}$ matrix so that the model is $Y_i = \beta_1 X_{i1} + \beta_2 X_{i2} + \beta_3 X_{i3} + \varepsilon_i$. What is the meaning here of each of the regression coefficients β_1, β_2, and β_3?

EXERCISES

10.21. Refer to the instrument calibration study example in Section 10.4. Suppose that three instruments (A, B, C) had been developed to identical specifications, that the regression functions relating gauge reading (Y) to actual pressure (X_1) are second-order polynomials for each instrument, that the error variances are the same, and that the polynomial coefficients may differ from one instrument to the next. Let X_3 denote a second indicator variable, where $X_3 = 1$ if instrument C and 0 otherwise.
   a. Expand regression model (10.21) to cover this situation.
   b. State the alternatives, define the test statistic, and give the decision rule for each of the following tests when the level of significance is .01: (1) test whether the second-order regression functions for the three instruments are identical, (2) test whether all three regression functions have the same intercept, (3) test whether both the linear and quadratic effects are the same in all three regression functions.

10.22. Refer to **Muscle mass** Problem 10.16. Specify the regression model for the case where the slope changes at age 40 and again at age 60 with no discontinuities.

10.23. In a regression study, three types of banks were involved, namely, commercial, mutual savings, and savings and loan. Consider the following system of indicator variables for type of bank:

Type of Bank	X_2	X_3
Commercial	1	0
Mutual savings	0	1
Savings and loan	-1	-1

   a. Develop a first-order linear regression model for relating last year's profit or loss (Y) to size of bank (X_1) and type of bank (X_2, X_3).
   b. State the response functions for the three types of banks.
   c. Interpret each of the following quantities: (1) β_2, (2) β_3, (3) $-\beta_2 - \beta_3$.

10.24. Refer to regression model (10.17) and exclude variable X_3.
   a. Obtain the $\mathbf{X'X}$ matrix for this special case of a single qualitative independent variable, for $i = 1, \ldots, n$ and n_1 firms that are not incorporated.
   b. Using (7.21), find $\mathbf{b}$.
   c. Using (7.30) and (7.31), find SSE and SSR.

PROJECTS

10.25. Refer to the **SMSA** data set. The number of active physicians (Y) is to be regressed against total population (X_1), total personal income (X_2), and geographic region.
   a. Fit a first-order regression model. Let $X_3 = 1$ if NE and 0 otherwise, $X_4 = 1$ if NC and 0 otherwise, and $X_5 = 1$ if S and 0 otherwise.
   b. Examine whether the effect for the northeastern region on number of active physicians differs from the effect for the north central region by constructing an appropriate 90 percent confidence interval. Interpret your interval estimate.
   c. Test whether any geographic effects are present; use $\alpha = .10$. State the alternatives, decision rule, and conclusion. What is the P-value of the test?

10.26. Refer to the **SENIC** data set. Infection risk (Y) is to be regressed against length of stay (X_1), age (X_2), routine chest X-ray ratio (X_3), and medical school affiliation (X_4).
   a. Fit a first-order regression model. Let $X_4 = 1$ if hospital has medical school affiliation and 0 if not.
   b. Estimate the effect of medical school affiliation on infection risk using a 98 percent confidence interval. Interpret your interval estimate.
   c. It has been suggested that the effect of medical school affiliation on infection risk may interact with the effects of age and routine chest X-ray ratio. Add appropriate interaction terms to the regression model, fit the revised regression model, and test whether the interaction terms are helpful; use $\alpha = .10$. State the alternatives, decision rule, and conclusion.

10.27. Refer to the **SENIC** data set. Length of stay (Y) is to be regressed on age (X_1), routine culturing ratio (X_2), average daily census (X_3), available facilities and services (X_4), and region (X_5, X_6, X_7).
   a. Fit a first-order regression model. Let $X_5 = 1$ if NE and 0 otherwise, $X_6 = 1$ if NC and 0 otherwise, and $X_7 = 1$ if S and 0 otherwise.
   b. Test whether the routine culturing ratio can be dropped from the model; use a level of significance of .05. State the alternatives, decision rule, and conclusion.
   c. Examine whether the effect on length of stay for hospitals located in the western region differs from that for hospitals located in the other three regions by constructing an appropriate confidence interval for each pairwise comparison. Use the Bonferroni procedure with a 95 percent family confidence coefficient. Summarize your findings.

Chapter 11

Diagnostics and Remedial Measures—II

In this chapter we take up a number of refined diagnostics for checking the adequacy of a regression model. These include methods for detecting model inadequacy for an independent variable, outliers, influential observations, and multicollinearity. We also consider a variety of remedial measures for dealing with these problems, including ridge regression for multicollinearity and weighted least squares for unequal error variances.

11.1 MODEL ADEQUACY FOR AN INDEPENDENT VARIABLE— PARTIAL REGRESSION PLOTS

We discussed in Chapters 4 and 7 how plots of residuals against an independent variable in the regression model can be used to check whether a curvature effect is required in the model for that independent variable and to examine whether the variance of the error terms varies in some systematic fashion with that independent variable. We also described there the plotting of residuals against independent variables not yet in the regression model to determine whether it would be helpful to add one or more of these variables to the model.

A limitation of these residual plots is that, while they show when a linear relation for an independent variable is inadequate, they do not display the nature of the relation for the independent variable that should be represented in the regression model.

Partial regression plots are refined residual plots that do show the proper relation for an independent variable in the regression model, and they are therefore a valuable complement to the regular residual plots. The partial regression plots are *partial* in the sense that they consider the marginal role of an independent variable X_k, given that the other independent variables under consideration are already in the model. Thus, partial regression plots need to be used with caution, primarily for providing information about the functional representation for, and marginal importance of, an independent variable that is to be added to a regression model. This caution is similar to one discussed in Chapter 8 about simultaneous tests on each of the regression coefficients individually, since these tests are also marginal in nature.

In a partial regression plot, both the dependent variable Y and the independent variable X_k under consideration are regressed against the other independent variables in the regression model and the residuals are obtained for each. These residuals reflect the part of each variable that is not linearly associated with the other independent variables already in the regression model. The plot of these residuals against each other reveals (1) the nature of the regression relation for the independent variable X_k under consideration for possible inclusion in the regression model and (2) the marginal importance of this variable in reducing the residual variability.

To make these ideas more specific, let us consider a multiple regression model with two independent variables X_1 and X_2. The extension to more than two independent variables is direct. Suppose we are concerned about the nature of the regression effect for X_1, given that X_2 is already in the model. We regress Y on X_2, and obtain the fitted values and residuals:

(11.1a)
$$\hat{Y}_i(X_2) = b_0 + b_2 X_{i2}$$

(11.1b)
$$e_i(Y \mid X_2) = Y_i - \hat{Y}_i(X_2)$$

The notation here indicates the dependent and independent variables in the fitted model. We also regress X_1 on X_2 and obtain:

(11.2a)
$$\hat{X}_{i1}(X_2) = b_0^* + b_2^* X_{i2}$$

(11.2b)
$$e_i(X_1 \mid X_2) = X_{i1} - \hat{X}_{i1}(X_2)$$

The partial regression plot for independent variable X_1 consists of a plot of the Y residuals $e(Y \mid X_2)$ against the X_1 residuals $e(X_1 \mid X_2)$.

Figure 11.1 contains several prototype partial regression plots in terms of our example, where X_2 is already in the regression model and X_1 is under consideration to be added. Figure 11.1a shows a horizontal band, indicating that X_1 contains no additional regression relation for predicting Y beyond that contained in X_2, so that it is not helpful to add X_1 to the regression model here.

Figure 11.1b shows a linear band with a nonzero slope. This indicates that a linear term in X_1 may be a helpful addition to the regression model already containing X_2. It can be shown that the slope of the least squares line through the origin fitted to the plotted residuals is b_1, the regression coefficient of X_1 if this variable were added to the regression model already containing X_2.

Vertical deviations of the plotted points around the horizontal line $e(Y \mid X_2) = 0$ in Figure 11.1b represent the Y residuals when X_2 is in the regression model. When these deviations are squared and summed, we obtain the error sum of squares $SSE(X_2)$. It can be shown that the vertical deviations of the plotted points taken around the line through the origin with slope b_1 are the residuals $e(Y \mid X_1, X_2)$ when both X_1 and X_2 are in the regression model. Hence, the sum of the squares of these deviations is the error sum of squares $SSE(X_1, X_2)$. The difference between these two sums of squares according to (8.1a) is the extra sum of squares $SSR(X_1 \mid X_2)$.

FIGURE 11.1 Prototype Partial Regression Plots

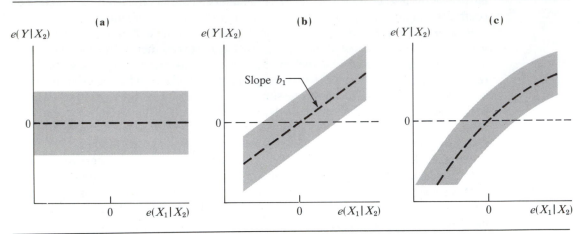

Hence, if the scatter of the points around the line through the origin with slope b_1 is much less than the scatter around the horizontal line, inclusion of the variable X_1 in the regression model will provide a substantial further reduction in the error sum of squares.

Figure 11.1c shows a curvilinear band, indicating that the addition of X_1 to the regression model can be helpful and that it should involve a curvature effect of the type shown by the pattern.

Partial regression plots are also useful for uncovering outlying, influential data points.

Example 1

For a sample of 18 managers in the 30–39 age group, Table 11.1 shows the average annual income during the past two years (X_1), risk aversion score (X_2), and amount of life insurance carried (Y). Risk aversion was measured by a standard questionnaire administered to each manager: the higher the score, the greater the degree of risk aversion. Income and risk aversion are mildly correlated here, the coefficient of correlation being $r_{12} = .254$.

A fit of a first-order regression model yields (calculations not shown):

(11.3)
$$\hat{Y} = -205.72 + 6.2880X_1 + 4.738X_2$$

The residuals for this fitted model are plotted against X_1 in Figure 11.2a. The residual plot clearly shows that a curvilinear effect for X_1 is needed in the regression model.

To study the nature of this effect, we shall use a partial regression plot. We regress Y and X_1 each against X_2. When doing this, we obtain (calculations not shown):

(11.4a)
$$\hat{Y}(X_2) = 50.70 + 15.54X_2$$

TABLE 11.1   Data for Life Insurance Example

Manager i	Average Annual Income (thousand dollars) X_{i1}	Risk Aversion Score X_{i2}	Amount of Life Insurance Carried (thousand dollars) Y_i
1	66.290	7	240
2	40.964	5	73
3	72.996	10	311
4	45.010	6	91
5	57.204	4	162
6	26.852	5	11
7	38.122	4	54
8	35.840	6	53
9	75.796	9	326
10	37.408	5	55
11	54.376	2	130
12	46.186	7	112
13	46.130	4	91
14	30.366	3	14
15	39.060	5	63
16	79.380	1	316
17	52.766	8	154
18	55.916	6	164

FIGURE 11.2   Residual Plot and Partial Regression Plot—Life Insurance Example

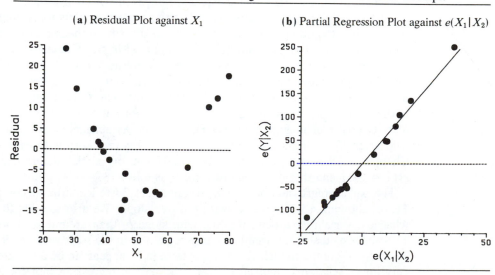

(a) Residual Plot against X_1

(b) Partial Regression Plot against $e(X_1 | X_2)$

(11.4b) $\hat{X}_1(X_2) = 40.779 + 1.718X_2$

The residuals from these two fitted models are plotted against each other in the partial regression plot in Figure 11.2b. This plot also contains the least squares line through the origin with slope $b_1 = 6.2880$. The partial regression plot shows the nature of the curvilinear relation between Y and X_1 when X_2 is already in the regression model, namely that it is convex. This is evident from the vertical deviations around the line through the origin with slope b_1. These deviations are positive at the left, negative in the middle, and positive again at the right. Overall, the curvature effect is modest in the range of the independent variables.

Note also that the vertical deviations of the points around the curvilinear relation, when fitted, will be much smaller than the vertical deviations around the horizontal line, indicating that adding X_1 to the regression model with a curvilinear relation will substantially reduce the error sum of squares. In fact, the coefficient of partial determination for the linear effect of X_1 is $r_{Y1.2}^2 = .984$.

Example 2

For the body fat example in Table 8.1 (page 272), we consider here the regression of body fat (Y) only on triceps skinfold thickness (X_1) and thigh circumference (X_2). We omit the third independent variable $(X_3,$ midarm circumference$)$ in order to focus the discussion of partial regression plots on its essentials. Recall that X_1 and X_2 are highly correlated $(r_{12} = .92)$. The fitted regression function was obtained in Table 8.2c (page 273):

$$\hat{Y} = -19.174 + .2224X_1 + .6594X_2$$

Figures 11.3a and 11.3c contain plots of the residuals against X_1 and X_2, respectively. These plots do not indicate any lack of fit for the linear terms in the regression model or the existence of unequal variances of the error terms.

Figures 11.3b and 11.3d contain the partial regression plots for X_1 and X_2, respectively, when the other independent variable is already in the regression model. Both plots also show the line through the origin with slope equal to the regression coefficient for the independent variable if it were added to the fitted model. These two plots provide some useful additional information. The scatter in Figure 11.3b follows the prototype in Figure 11.1a, suggesting that X_1 is of little additional help in the model when X_2 is already present. This information is not provided by the regular residual plot in Figure 11.3a. The fact that X_1 appears to be of little marginal help when X_2 is already in the regression model is in accord with earlier findings in Chapter 8. We saw there that the coefficient of partial determination is only $r_{Y1.2}^2 = .031$ and that the t^* statistic for b_1 is only .73.

The partial regression plot for X_2 in Figure 11.3d follows the prototype in Figure 11.1b, showing a linear scatter with positive slope. We also see that there is somewhat less variability around the line with slope b_2 than around the horizontal line. This suggests that: (1) Variable X_2 may be helpful in the regression model even when X_1 is already in the model. (2) A linear term in X_2 appears to be adequate because no curvilinear relation is indicated. Thus, the partial regression plot for X_2 in Figure

FIGURE 11.3 Residual Plots and Partial Regression Plots—Body Fat Example with Two Independent Variables

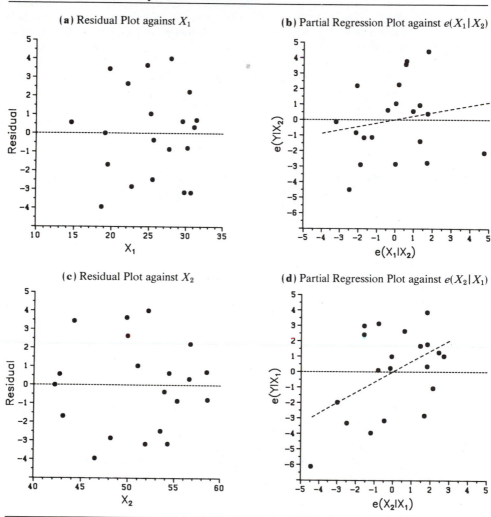

11.3d complements the regular residual plot in Figure 11.3c by indicating the potential usefulness of thigh circumference (X_2) in the regression model when triceps skinfold thickness (X_1) is already in the model. This information is consistent with the $t*$ statistic for b_2 of 2.26 in Table 8.2c and the moderate coefficient of partial determination of $r^2_{Y2.1} = .232$.

Comments

1. A partial regression plot shows graphically the nature of the functional relation in which a variable should be added to the regression model. It does not provide an analytic expression of the relation. Often, a variety of transformations or curvature effect terms are available to

represent the relation shown in the plot. These alternatives will need to be investigated and additional residual plots utilized to identify the best transformation or curvature effect terms.

2. When several partial regression plots are required for a set of independent variables, it is not necessary to fit entirely new regression models each time. Computational procedures are available that economize on the calculations required; these are explained in specialized texts such as Reference 11.1.

3. Any fitted multiple regression function can be obtained from a sequence of fitted partial regressions. To illustrate this, consider again the life insurance example, where the fitted regression of Y on X_2 is given in (11.4a) and the fitted regression of X_1 on X_2 is given in (11.4b). If we now regress the residuals $Y - \hat{Y}(X_2)$ on the residuals $X_1 - \hat{X}_1(X_2)$, using regression through the origin, we obtain (calculations not shown):

$$(11.5) \qquad \widehat{[Y - \hat{Y}(X_2)]} = 6.2880[X_1 - \hat{X}_1(X_2)]$$

By simple substitution, using (11.4a) and (11.4b), we obtain:

$$[Y - (50.70 + 15.54X_2)] = 6.2880[X_1 - (40.779 + 1.718X_2)]$$

or:

$$(11.6) \qquad \hat{Y} = -205.72 + 6.2880X_1 + 4.737X_2$$

where the solution for Y is the fitted value $\hat{Y}$ when X_1 and X_2 are included in the regression model. Note that the result in (11.6) is the same as for the regression model fitted to X_1 and X_2 directly in (11.3), except for a minor difference due to rounding effects.

4. A residual plot closely related to the partial regression plot that is also useful for identifying the nature of the relationship for an independent variable X_k under consideration for addition to the regression model is the *partial residual plot*. This residual plot takes as the starting point the usual residuals $e_i = Y_i - \hat{Y}_i$, to which the regression effect for X_k is added. Specifically, the partial residuals for examining the effect of independent variable X_k, denoted by $p_i(X_k)$, are defined as follows:

$$(11.7) \qquad p_i(X_k) = e_i + b_k X_{ik}$$

Thus for a partial residual, we add the effect of X_k, as reflected by the fitted model term $b_k X_{ik}$, back onto the residual. A plot of these partial residuals against X_k is referred to as a partial residual plot. The reader is referred to Reference 11.2 for more details on partial residual plots.

11.2 IDENTIFYING OUTLYING X OBSERVATIONS— HAT MATRIX LEVERAGE VALUES

Outlying Cases

Frequently in regression analysis applications, the data set contains some cases that are outlying or extreme; that is, the observations for these cases are well separated from the remainder of the data. These outlying cases may involve large residuals and often have dramatic effects on the fitted least squares regression function. It is therefore important to study the outlying cases carefully and decide whether they should be retained or eliminated, and if retained, whether their influence should be reduced in the fitting process and/or the regression model revised.

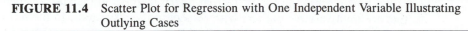

FIGURE 11.4   Scatter Plot for Regression with One Independent Variable Illustrating Outlying Cases

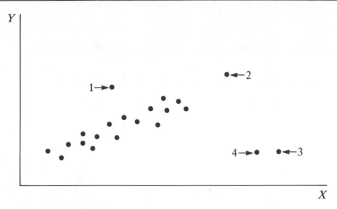

A case may be outlying or extreme with respect to its Y value, its X value(s), or both. Figure 11.4 illustrates this for the case of regression with a single independent variable. In the scatter plot in Figure 11.4, case 1 is outlying with respect to its Y value. Note that this point falls far outside the scatter, although its X value is near the middle of the range of observations on the independent variable. Cases 2, 3, and 4 are outlying with respect to their X values since they have much larger X values than those for the other cases; cases 3 and 4 are also outlying with respect to their Y values.

Not all outlying cases have a strong influence on the fitted regression function. Case 1 in Figure 11.4 may not be too influential because there are a number of other cases that have similar X values, which will keep the fitted regression function from being displaced too far by the outlying case. Likewise, case 2 may not be too influential because its Y value is consistent with the regression relation displayed by the nonextreme cases. Cases 3 and 4, on the other hand, are likely to be very influential in affecting the fit of the regression function. They are outlying with regard to their X values, and their Y values are not consistent with the regression relation for the other cases.

A basic step in any regression analysis is to determine if the regression model under consideration is heavily influenced by one or a few cases in the data set. For regression with one or two independent variables, it is relatively simple to identify outlying cases with respect to their X or Y values by means of box plots, stem-and-leaf plots, scatter plots, and residual plots, and to study whether they are influential in affecting the fitted regression function. When more than two independent variables are included in the regression model, however, the identification of outlying cases by simple graphic means becomes difficult because single-variable or two-variable examinations do not necessarily help find outliers relative to a multivariable regression model. Some univariate outliers may not be extreme in a multiple regression model, and, conversely, some multivariable outliers may not be detectable in single-variable or two-variable analyses.

We now discuss the use of the hat matrix defined in (7.25a) for identifying cases that are multivariable outliers with respect to their X values. In the following section we take up some refined measures for identifying cases with outlying Y observations.

Use of Hat Matrix for Identifying Outlying X Observations

We encountered the hat matrix $\mathbf{H}$ in Chapters 6 and 7:

$$(11.8) \qquad \underset{n \times n}{\mathbf{H}} = \mathbf{X}(\mathbf{X'X})^{-1}\mathbf{X'}$$

We noted in (7.25) that the fitted values $\hat{Y}_i$ can be expressed as linear combinations of the observations Y_i through the hat matrix:

$$(11.9) \qquad \hat{\mathbf{Y}} = \mathbf{HY}$$

and similarly we noted in (7.26) that the least squares residuals can be expressed as linear combinations of the observations Y_i by means of the hat matrix:

$$(11.10) \qquad \mathbf{e} = (\mathbf{I} - \mathbf{H})\mathbf{Y}$$

Further, we noted in (7.27) that the variances and covariances of the residuals involve the hat matrix:

$$(11.11) \qquad \sigma^2\{\mathbf{e}\} = \sigma^2(\mathbf{I} - \mathbf{H})$$

so that the variance of residual e_i, denoted by $\sigma^2\{e_i\}$, is:

$$(11.12) \qquad \sigma^2\{e_i\} = \sigma^2(1 - h_{ii})$$

where h_{ii} is the ith element on the main diagonal of the hat matrix.

The diagonal element h_{ii} of the hat matrix can be obtained directly from:

$$(11.13) \qquad h_{ii} = \mathbf{X}_i'(\mathbf{X'X})^{-1}\mathbf{X}_i$$

where $\mathbf{X}_i$ corresponds to the $\mathbf{X}_h$ vector in (7.48) except that $\mathbf{X}_i$ pertains to the ith case:

$$(11.13a) \qquad \underset{p \times 1}{\mathbf{X}_i} = \begin{bmatrix} 1 \\ X_{i,1} \\ \vdots \\ X_{i,p-1} \end{bmatrix}$$

Note that $\mathbf{X}_i'$ is simply the ith row of the $\mathbf{X}$ matrix, pertaining to the ith case.

The diagonal elements h_{ii} of the hat matrix have some useful properties. In particular, their values are always between 0 and 1 and their sum is p:

$$(11.14) \qquad 0 \le h_{ii} \le 1 \qquad \sum_{i=1}^{n} h_{ii} = p$$

where p is the number of regression parameters in the regression function including the intercept term.

FIGURE 11.5   Illustration of Cases with *X* Values Near and Far from Center

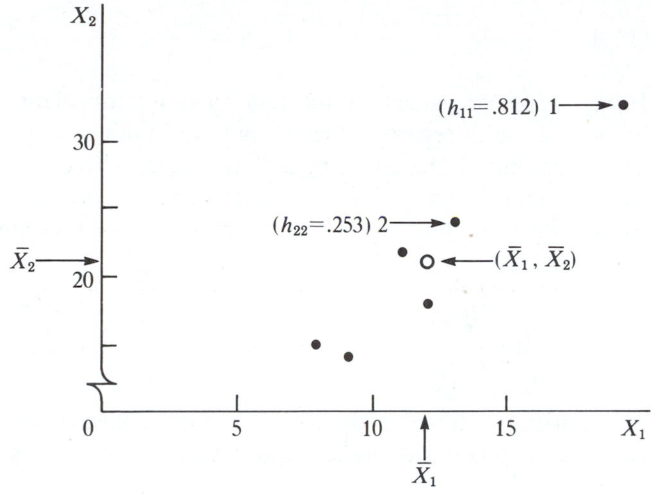

The diagonal element h_{ii} in the hat matrix is a useful indicator in a multivariable setting of whether or not the ith case is outlying with respect to its X values. The diagonal element h_{ii} is called the *leverage* (in terms of the X values) of the ith case. It indicates whether or not the X values for the ith case are outlying because it can be shown that h_{ii} is a measure of the distance between the X values for the ith case and the means of the X values for all n cases. Thus, a large leverage value h_{ii} indicates that the ith case is distant from the center of all X observations. Figure 11.5 illustrates this for two independent variables. Case 1 is distant from the center $(\bar{X}_1, \bar{X}_2)$ and has a large leverage value $h_{11} = .812$, while case 2 is near the center and has a small leverage value $h_{22} = .253$.

If the ith case is outlying in terms of its X observations and therefore has a large leverage value h_{ii}, it exercises substantial leverage in determining the fitted value $\hat{Y}_i$. This is so for the following reasons:

1. The fitted value $\hat{Y}_i$ is a linear combination of the observed Y values, as shown in (11.9), and h_{ii} is the weight of observation Y_i in determining this fitted value. Thus, the larger is h_{ii}, the more important is Y_i in determining $\hat{Y}_i$. Remember that h_{ii} is a function only of the X values, so h_{ii} measures the role of the X values in determining how important Y_i is in affecting the fitted value $\hat{Y}_i$.
2. The larger is h_{ii}, the smaller is the variance of the residual e_i, as may be seen from (11.12). Hence, the larger is h_{ii}, the closer the fitted value $\hat{Y}_i$ will tend to be to the observed value Y_i. In the extreme case where $h_{ii} = 1$, $\sigma^2\{e_i\} = 0$ so that the fitted value $\hat{Y}_i$ is then forced to equal the observed value Y_i. Since cases with high leverage tend to have smaller residuals, it may not be possible to detect them by an examination of the residuals alone.

A leverage value h_{ii} is usually considered to be large if it is more than twice as large as the mean leverage value, denoted by $\bar{h}$, which according to (11.14) is:

$$(11.15) \qquad \bar{h} = \frac{\sum_{i=1}^{n} h_{ii}}{n} = \frac{p}{n}$$

Hence, leverage values greater than $2p/n$ are considered by this rule to indicate outlying cases with regard to their X values. Another suggested guideline is that h_{ii} values exceeding .5 indicate very high leverage, whereas those between .2 and .5 indicate moderate leverage. Additional evidence of an outlying case is the existence of a gap between the leverage values for most of the cases and the unusually large leverage value(s).

Example

We continue with the body fat example of Table 8.1, again using only the two independent variables triceps skinfold thickness (X_1) and thigh circumference (X_2) so that the results using the hat matrix can be compared to simple graphic plots. Figure 11.6 contains a scatter plot of X_2 against X_1, where the data points are identified by their case number. We note from Figure 11.6 that cases 15 and 3 appear to be outlying ones with respect to the pattern of the X values. Case 15 is outlying for X_1 and at the low end of the range for X_2, while case 3 is outlying in terms of the pattern of multi-

FIGURE 11.6   Scatter Plot of Thigh Circumference against Triceps Skinfold Thickness— Body Fat Example with Two Independent Variables

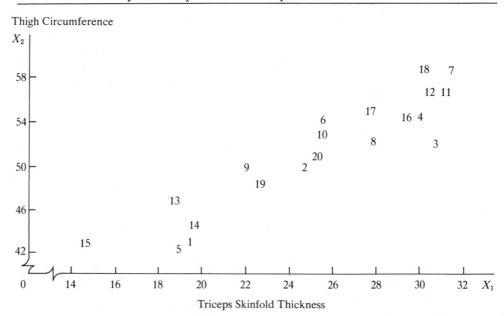

TABLE 11.2 Residuals, Diagonal Elements of the Hat Matrix, and Studentized Deleted Residuals—Body Fat Example with Two Independent Variables

i	(1) e_i	(2) h_{ii}	(3) d_i^*
1	−1.683	.201	−.730
2	3.643	.059	1.534
3	−3.176	.372	−1.656
4	−3.158	.111	−1.348
5	.000	.248	.000
6	−.361	.129	−.148
7	.716	.156	.298
8	4.015	.096	1.760
9	2.655	.115	1.117
10	−2.475	.110	−1.034
11	.336	.120	.137
12	2.226	.109	.923
13	−3.947	.178	−1.825
14	3.447	.148	1.524
15	.571	.333	.267
16	.642	.095	.258
17	−.851	.106	.344
18	−.783	.197	.335
19	−2.857	.067	−1.176
20	1.040	.050	.409

collinearity, though it is not outlying for either of the independent variables separately. Cases 1 and 5 also appear to be somewhat extreme.

Calculation of the hat matrix (11.8) confirms these impressions. Table 11.2, column 2, contains the leverage values h_{ii} for the body fat example. Note that the two largest leverage values are $h_{33} = .372$ and $h_{15,15} = .333$. Both exceed the criterion of twice the mean leverage value, $2p/n = 2(3)/20 = .30$, and are separated by a substantial gap from the next largest leverage values $h_{55} = .248$ and $h_{11} = .201$. Having identified cases 3 and 15 as outlying in terms of their X values, we shall need to ascertain how influential these cases are in the fitting of the regression function. We consider this question after taking up the identification of outlying Y observations.

11.3 IDENTIFYING OUTLYING Y OBSERVATIONS— STUDENTIZED DELETED RESIDUALS

The detection of outlying or extreme Y observations based on an examination of the residuals has been considered in earlier chapters. We utilized there either the residuals e_i:

(11.16)
$$e_i = Y_i - \hat{Y}_i$$

or the standardized residuals:

(11.17)
$$\frac{e_i}{\sqrt{MSE}}$$

We introduce now two refinements to make the analysis of residuals more effective for identifying outlying Y observations.

Internally Studentized Residuals

When the residuals e_i have substantially different variances $\sigma^2\{e_i\}$, as given in (11.12), it is well to consider the magnitude of e_i relative to $\sigma\{e_i\}$ to give recognition to differences in their sampling errors. We noted in (7.28) that an unbiased estimator of this variance is:

(11.18)
$$s^2\{e_i\} = MSE(1 - h_{ii})$$

The ratio of e_i to $s\{e_i\}$ is called the *internally studentized residual* and will be denoted by e_i^*:

(11.19)
$$e_i^* = \frac{e_i}{s\{e_i\}}$$

While the residuals e_i will have substantially different sampling variations if the leverage values h_{ii} differ markedly, the internally studentized residuals have constant variance (when the model is appropriate).

Deleted Residuals

The second refinement is to measure the ith residual $e_i = Y_i - \hat{Y}_i$ when the fitted regression is based on the cases excluding the ith one. The reason for this refinement is that if Y_i is far outlying, the fitted least squares regression function based on all cases may be influenced to come close to Y_i, yielding a fitted value $\hat{Y}_i$ near Y_i. In that event, the residual e_i will be small and will not disclose that Y_i is outlying. On the other hand, if the ith case is deleted before the regression function is fitted, the least squares fitted value $\hat{Y}_i$ is not influenced by the outlying Y_i observation, and the residual will then tend to be larger and therefore more likely to disclose the outlying Y observation.

The procedure then is to delete the ith case, fit the regression function to the remaining $n - 1$ cases, and compare the point estimate of the expected value when the X levels are those of the ith case, to be denoted by $\hat{Y}_{i(i)}$, with the actual observed value Y_i. The notation $\hat{Y}_{i(i)}$ reminds us that the ith case was omitted when fitting the regression function. The residual:

(11.20)
$$d_i = Y_i - \hat{Y}_{i(i)}$$

is called a *deleted residual* and is denoted by d_i. An algebraically equivalent expression for d_i which does not require a recomputation of the fitted regression function omitting the ith case is:

$$(11.20a) \qquad d_i = \frac{e_i}{1 - h_{ii}}$$

where e_i is the ordinary residual for the ith case and h_{ii} is the leverage value (11.13) for this case. Note that the larger is the leverage value h_{ii}, the larger will be the deleted residual as compared to the ordinary residual.

Thus, deleted residuals will at times identify outlying Y observations when ordinary residuals would not identify these; at other times deleted residuals lead to the same identifications as ordinary residuals.

Note that a deleted residual corresponds to the prediction error in the numerator of (3.34) when predicting a new observation from the fitted regression function based on earlier cases, except that in (3.34) the difference considered is $\hat{Y}_{i(i)} - Y_i$ and the notation differs from the present one. Hence, we know from (7.58a) that the estimated variance of d_i is:

$$(11.21) \qquad s^2\{d_i\} = MSE_{(i)}(1 + \mathbf{X}_i'(\mathbf{X}_{(i)}'\mathbf{X}_{(i)})^{-1}\mathbf{X}_i)$$

where $\mathbf{X}_i$ is the X observations vector (11.13a) for the ith case, $MSE_{(i)}$ is the mean square error when the ith case is omitted in fitting the regression function, and $\mathbf{X}_{(i)}$ is the $\mathbf{X}$ matrix with the ith case deleted. An algebraically equivalent expression for $s^2\{d_i\}$ is:

$$(11.21a) \qquad s^2\{d_i\} = \frac{MSE_{(i)}}{1 - h_{ii}}$$

It follows from (7.58) that:

$$(11.22) \qquad \frac{d_i}{s\{d_i\}} \sim t(n - p - 1)$$

Remember that $n - 1$ cases are used here in predicting the ith observation; hence, the degrees of freedom are $(n - 1) - p = n - p - 1$.

Studentized Deleted Residuals

Combining the above two refinements, we obtain for diagnosis of outlying or extreme Y observations the deleted residual d_i in (11.20) studentized by dividing it by its estimated standard deviation given by (11.21). The *studentized deleted residual,* to be denoted by d_i^*, therefore is:

$$(11.23) \qquad d_i^* = \frac{d_i}{s\{d_i\}}$$

It follows from (11.20a) and (11.21a) that an algebraically equivalent expression for d_i^* is:

$$(11.23a) \qquad d_i^* = \frac{e_i}{\sqrt{MSE_{(i)}(1 - h_{ii})}}$$

The studentized residual (11.23) is sometimes called an *externally studentized residual*. We know from (11.22) that each studentized deleted residual d_i^* follows the t distribution with $n - p - 1$ degrees of freedom. The d_i^*, however, are not independent.

Fortunately, the studentized deleted residuals d_i^* in (11.23) can be calculated without having to fit new regression functions each time a different case is omitted. A simple relationship exists between MSE and $MSE_{(i)}$:

$$(11.24) \qquad (n - p)MSE = (n - p - 1)MSE_{(i)} + \frac{e_i^2}{1 - h_{ii}}$$

Using this relationship in (11.23a) yields the following equivalent expression for d_i^*:

$$(11.25) \qquad d_i^* = e_i \left[\frac{n - p - 1}{SSE(1 - h_{ii}) - e_i^2} \right]^{1/2}$$

Thus, the studentized deleted residuals d_i^* can be calculated from the residuals e_i, the error sum of squares SSE, and the leverage values h_{ii}, all for the fitted regression based on the n cases.

To identify outlying Y observations, we examine the studentized deleted residuals for large absolute values and use the appropriate t distribution to ascertain how far in the tails such outlying values fall.

Example.   We illustrate the calculation of studentized deleted residuals for the first case in the body fat example of Table 11.2, based on the two independent variables X_1 and X_2. The X values for this case, given in Table 8.1, are $X_{11} = 19.5$ and $X_{12} = 43.1$. Using the fitted regression function from Table 8.2c, we obtain:

$$\hat{Y}_1 = -19.174 + .2224(19.5) + .6594(43.1) = 13.583$$

Since $Y_1 = 11.9$, the residual for this case is $e_1 = 11.9 - 13.583 = -1.683$. We also know from Table 8.2c that $SSE = 109.95$ and from Table 11.2 that $h_{11} = .201$. Hence, by (11.25), we find:

$$d_1^* = -1.683 \left[\frac{20 - 3 - 1}{109.95(1 - .201) - (-1.683)^2} \right]^{1/2} = -.730$$

The studentized deleted residuals for all 20 cases are shown in column 3 of Table 11.2.

Note that cases 3, 8, and 13 have the largest absolute studentized deleted residuals. If we consider tail areas of .05 on each side to be extreme, we will need to compare the absolute values of the studentized deleted residuals with the t distribution with $n - p - 1 = 16$ degrees of freedom, specifically with $t(.95; 16) = 1.746$. Based on this comparison, we should consider cases 8 and 13 extreme enough to warrant studying whether or not they are influential cases. Incidentally, consideration of the residuals e_i (shown in Table 11.2, column 1) here would also have identified cases 8 and 13 as the most outlying ones.

11.4 IDENTIFYING INFLUENTIAL CASES—*DFFITS*, *DFBETAS*, AND COOK'S DISTANCE MEASURES

After identifying cases that are outlying with respect to their X values and/or their Y values, the next step is to ascertain whether or not these outlying cases are influential. We shall consider a case to be *influential* if its exclusion causes major changes in the fitted regression function. As we noted in Figure 11.4, not all outlying cases need be influential. For example, case 1 in Figure 11.4 may not affect the fitted regression function to any substantial extent.

We shall take up three measures of influence that are widely used in practice, each based on the omission of a single case to measure its influence.

Influence on the Fitted Values—*DFFITS*

A useful measure of the influence that case i has on the fitted value $\hat{Y}_i$ is given by:

$$(11.26) \qquad (DFFITS)_i = \frac{\hat{Y}_i - \hat{Y}_{i(i)}}{\sqrt{MSE_{(i)} h_{ii}}}$$

The letters *DF* stand for the difference between the fitted value $\hat{Y}_i$ for the ith case when all n cases are used in fitting the regression function and the predicted value $\hat{Y}_{i(i)}$ for the ith case obtained when the ith case is omitted in fitting the regression function. The denominator of (11.26) involves the mean square error when the ith case is omitted in fitting the regression function and the leverage value h_{ii} defined in (11.13). The denominator provides a standardization so that the value $(DFFITS)_i$ for the ith data point represents roughly the number of estimated standard deviations that the fitted value $\hat{Y}_i$ changes when the ith case is removed from the data set.

It can be shown that the *DFFITS* values can be computed using only the results from fitting the entire data set, as follows:

$$(11.26a) \quad (DFFITS)_i = e_i \left[\frac{n - p - 1}{SSE(1 - h_{ii}) - e_i^2} \right]^{1/2} \left(\frac{h_{ii}}{1 - h_{ii}} \right)^{1/2} = d_i^* \left(\frac{h_{ii}}{1 - h_{ii}} \right)^{1/2}$$

Note from the last expression that the *DFFITS* values are studentized deleted residuals, as given in (11.25), increased or decreased by a factor that is a function of the leverage value. If case i is an X outlier and has a high leverage value, this factor will be greater than 1 and $(DFFITS)_i$ will tend to be large absolutely.

As a guideline for identifying influential cases, we suggest considering a case influential if the absolute value of *DFFITS* exceeds 1 for small to medium-size data sets and $2\sqrt{p/n}$ for large data sets.

Example. Table 11.3, column 1, lists the *DFFITS* values for the body fat example with two independent variables. To illustrate the calculations, consider the *DFFITS* value for case 1. We found earlier that the studentized deleted residual for this case is

TABLE 11.3 *DFFITS, DFBETAS*, and Cook's Distances—Body Fat Example with Two Independent Variables

	(1)	(2)	(3)	(4)	(5)
			DFBETAS		
i	$(DFFITS)_i$	b_0	b_1	b_2	D_i
1	$-.366$	$-.305$	$-.132$	.232	.046
2	.384	.173	.115	$-.143$	.046
3	-1.273	$-.847$	-1.183	1.067	.490
4	$-.476$	$-.102$	$-.294$	.196	.072
5	.000	.000	.000	.000	.000
6	$-.057$	.040	.040	$-.044$	.001
7	.128	$-.078$	$-.016$	.054	.006
8	.575	.261	.391	$-.333$	.098
9	.402	$-.151$	$-.295$	.247	.053
10	$-.364$	.238	.245	$-.269$	.044
11	.051	$-.009$	.017	$-.003$	.001
12	.323	$-.131$	.023	.070	.035
13	$-.851$	.119	.592	$-.390$	.212
14	.636	.452	.113	$-.298$	.125
15	.189	$-.003$	$-.125$	.069	.013
16	.084	.009	.043	$-.025$	.002
17	$-.118$	.080	.055	$-.076$	.005
18	$-.166$	.132	.075	$-.116$	.010
19	$-.315$	$-.130$	$-.004$	.064	.032
20	.094	.010	.002	$-.003$	.003

$d_1^* = -.730$ and the leverage value is $h_{11} = .201$. Hence, using (11.26a) we obtain:

$$(DFFITS)_1 = -.730 \left(\frac{.201}{1 - .201} \right)^{1/2} = -.366$$

The only *DFFITS* value in Table 11.3 that exceeds our guideline for a medium-size data set is case 3, where $|(DFFITS)_3| = 1.273$. This value is somewhat larger than our guideline of 1. However, the value is close enough to 1 that the case may not be influential enough to require remedial action.

Influence on the Regression Coefficients

DFBETAS. One measure of the influence of the ith case on each regression coefficient b_k ($k = 0, 1, \ldots, p - 1$) is the difference between the estimated regression coefficient b_k based on all n cases and the regression coefficient obtained when the ith case is omitted, to be denoted by $b_{k(i)}$. When this difference is divided by an appropriate standardization, we obtain the measure *DFBETAS*:

(11.27) $\qquad (DFBETAS)_{k(i)} = \dfrac{b_k - b_{k(i)}}{\sqrt{MSE_{(i)} c_{kk}}} \qquad k = 0, 1, \ldots, p - 1$

where c_{kk} is the kth diagonal element of $(\mathbf{X'X})^{-1}$. The denominator is an estimate of the standard error of b_k, with *MSE* based on $n - 1$ cases.

A large absolute value of $(DFBETAS)_{k(i)}$ is indicative of a large impact of the ith case on the kth regression coefficient. As a guideline for identifying influential cases, we recommend considering a case influential if the absolute value of *DFBETAS* exceeds 1 for small to medium-size data sets and $2/\sqrt{n}$ for large data sets.

Example.   For the body fat example with two independent variables, Table 11.3 lists the *DFBETAS* values in columns 2, 3, and 4. Note that case 3 is the only case that exceeds our guideline of 1 for medium-size data sets for both b_1 and b_2. Thus, case 3 is again tagged as potentially influential. Again, however, the *DFBETAS* values do not exceed 1 by very much so that case 3 may not be so influential as to require remedial action.

Cook's Distance.   An overall measure of the combined impact of the ith case on all of the estimated regression coefficients is *Cook's distance measure D_i*. This measure is derived from the concept of a confidence region for all p regression coefficients β_k $(k = 0, 1, \ldots, p - 1)$ simultaneously. It can be shown that the boundary of this joint confidence region for the normal error multiple regression model (7.18) is given by:

$$(11.28) \qquad \frac{(\mathbf{b} - \boldsymbol{\beta})'\mathbf{X'X}(\mathbf{b} - \boldsymbol{\beta})}{pMSE} = F(1 - \alpha; p, n - p)$$

Cook's distance measure D_i uses the same structure for measuring the combined impact of the differences in the estimated regression coefficients when the ith case is deleted:

$$(11.29) \qquad D_i = \frac{(\mathbf{b} - \mathbf{b}_{(i)})'\mathbf{X'X}(\mathbf{b} - \mathbf{b}_{(i)})}{pMSE}$$

where $\mathbf{b}_{(i)}$ is the vector of the estimated regression coefficients obtained when the ith case is omitted and $\mathbf{b}$, as usual, is the vector when all n cases are used.

While D_i does not follow the F distribution, it has been found useful to relate the value D_i to the corresponding F distribution according to (11.28) and ascertain the percentile value. If the percentile value is less than about 10 or 20 percent, the ith case has little apparent influence on the regression coefficients. If, on the other hand, the percentile value is near 50 percent or more, the distance between the vectors $\mathbf{b}$ and $\mathbf{b}_{(i)}$ should be considered large, implying that the ith case has a substantial influence on the fit of the regression function.

Fortunately, Cook's distance measure D_i can be calculated without fitting a new regression function each time a different case is deleted. An algebraically equivalent expression is:

$$(11.29a) \qquad D_i = \frac{e_i^2}{pMSE}\left[\frac{h_{ii}}{(1 - h_{ii})^2}\right]$$

Note from (11.29a) that D_i depends on two factors: (1) the size of the residual e_i and

(2) the leverage value h_{ii}. The larger is either e_i or h_{ii}, the larger is D_i. Thus, the ith case can be influential: (1) by having a large residual e_i and only a moderate leverage value h_{ii}, or (2) by having a large leverage value h_{ii} with only a moderately sized residual e_i, or (3) by having both a large residual e_i and a large leverage value h_{ii}.

Example.   For the body fat example with two independent variables, Table 11.3, column 5, presents the D_i values. To illustrate the calculations, we shall consider case 1. We know from Table 11.2 that $e_1 = -1.683$ and $h_{11} = .201$. Further, $MSE = 6.47$ according to Table 8.2c and $p = 3$ for the model with two independent variables. Hence, we obtain:

$$D_1 = \frac{(-1.683)^2}{3(6.47)} \left[\frac{.201}{(1 - .201)^2} \right] = .046$$

We note from Table 11.3, column 5 that case 3 clearly is the most influential one, having $D_3 = .490$, with the next largest distance measure $D_{13} = .212$ being substantially smaller.

To assess the magnitude of $D_3 = .490$, we refer to the corresponding F distribution, namely, $F(p, n - p) = F(3, 17)$. It can be shown that .490 is about the 31st percentile of this distribution. Hence, it appears that case 3 does influence the regression fit, but the extent of the influence may not be large enough to call for consideration of remedial measures.

Comments

1. Cook's distance measure D_i may be viewed as reflecting in the aggregate the differences for each case between the fitted value when all n cases are used and the fitted value when the ith case is deleted, since it can be shown that an equivalent expression for D_i is:

(11.30)
$$D_i = \frac{(\hat{\mathbf{Y}} - \hat{\mathbf{Y}}_{(i)})'(\hat{\mathbf{Y}} - \hat{\mathbf{Y}}_{(i)})}{pMSE}$$

Here, $\hat{\mathbf{Y}}$ as usual is the vector of the fitted values when all n cases are used for the regression fit and $\hat{\mathbf{Y}}_{(i)}$ is the vector of the fitted values when the ith case is deleted.

2. Analysis of outlying and influential cases is a necessary component of good regression analysis. However, it is neither automatic nor foolproof and requires good judgment by the analyst. The methods which have been described often work well but at other times will be ineffective. For example, if two influential outlying cases are nearly coincident as depicted in Figure 11.4 by cases 3 and 4, an analysis that deletes one case at a time and estimates the change in fit will result in virtually no change for these two outlying cases. The reason is that the retained outlying case will mask the effect of the deleted outlying case.

Influence on Inferences

To round out the determination of influential cases, it is usually a good idea to examine in a direct fashion the inferences from the fitted regression model that would be made with and without the case(s) of concern. If the inferences are not essentially changed, there is little need to think of remedial actions for the cases diagnosed as influential. On the other hand, serious changes in the inferences drawn from the fitted model when a case is omitted will require consideration of remedial measures. Some possible remedial measures will be discussed in the next section.

Example. In the body fat example with two independent variables, cases 3 and 15 were identified as outlying X observations and cases 8 and 13 as outlying Y observations. All three influence measures (*DFFITS, DFBETAS,* and Cook's distance) identified only case 3 as influential, and indeed suggested that its influence may be of marginal importance so that remedial measures might not be required.

The analyst in the body fat example was primarily interested in the fit of the regression model because the model was intended to be used for making predictions within the range of the observations on the independent variables in the data set. Hence, the analyst considered the fitted regression functions with and without case 3:

$$\text{With case 3:} \quad \hat{Y} = -19.174 + .2224X_1 + .6594X_2$$

$$\text{Without case 3:} \quad \hat{Y} = -12.428 + .5641X_1 + .3635X_2$$

Because of the high multicollinearity between X_1 and X_2, the analyst was not surprised by the shifts in the magnitudes of b_1 and b_2 when case 3 is omitted. Remember that the estimated standard deviations of the coefficients, given in Table 8.2c, are very large and that a single case can change the estimated coefficients very substantially when the independent variables are highly correlated.

To examine the effect of case 3 on inferences to be made from the fitted regression function in the range of the X observations in a direct fashion, the analyst calculated for each of the 20 cases the relative difference between the fitted value $\hat{Y}_i$ based on all 20 cases and the fitted value $\hat{Y}_{i(3)}$ obtained when case 3 is omitted. The measure of interest was the average absolute percent difference:

$$\frac{\sum_{i=1}^{n} \left| \dfrac{\hat{Y}_{i(3)} - \hat{Y}_i}{\hat{Y}_i} \right| 100}{n}$$

This mean difference is 3.1 percent; further, 17 of the 20 differences are less than 5 percent (calculations not shown). On the basis of this direct evidence about the effect of case 3 on the inferences to be made, the analyst was satisfied that case 3 does not exercise undue influence so that no remedial action is required for handling this case.

11.5 REMEDIAL MEASURES FOR INFLUENTIAL CASES

After using the hat matrix, studentized deleted residuals, and *DFFITS, DFBETAS,* and Cook's distance measures to identify outlying influential cases that have a substantial impact on the least squares regression fit, one must decide what to do about such cases. Clearly, an outlying influential case should not be automatically discarded, because it may be entirely correct and simply represents an unlikely event.

Discarding of such an outlying case could lead to the undesirable consequence of increased variances of some of the estimated regression coefficients.

If, on the other hand, the circumstances surrounding the data provide an explanation of the unusual case which indicates an exceptional situation not to be covered by the model, the discarding of the case may be appropriate. Thus, when an outlying influential observation can definitely be shown to be the result of a gross measurement error, it would be appropriate to discard that case.

When the outlying influential observation is accurate, it may not represent an unlikely event but rather a failure of the model. The failure may be either the omission of an important independent variable, the choice of an incorrect functional form, such as omission of a curvature effect for an independent variable included in the model, or the omission of an important interaction term. Often, identification of outlying influential cases leads to valuable insights for strengthening the model.

When an outlying influential observation is accurate but no explanation can be found for it, a less severe alternative than discarding the case is to dampen its influence. One possible means of dampening is to use a transformation. If, for example, a case is outlying in terms of one of the X variables, use of a transformation on that variable, such as a logarithmic or square root transformation, may bring the outlying observation closer to the remaining observations and thereby dampen its influence. One would, of course, need to check the continuing appropriateness of the model with the transformed variable to make sure that no new problems have been created by the transformation.

Another possible means of dampening influence is to use a different estimation procedure. We shall now discuss one of these alternative estimation procedures.

Method of Least Absolute Deviations

This method is one of a variety of robust methods that have the property of being insensitive to both outlying data values and inadequacies of the model employed. The method of least absolute deviations estimates the regression coefficients by minimizing the sum of the absolute deviations of the observations from their means. The criterion to be minimized is:

$$(11.31) \qquad \sum_{i=1}^{n} |Y_i - (\beta_0 + \beta_1 X_{i1} + \cdots + \beta_{p-1} X_{i,p-1})|$$

Since absolute deviations rather than squared ones are involved here, the method of least absolute deviations places less emphasis on outlying observations than does the method of least squares.

The estimated regression coefficients according to the method of least absolute deviations can be obtained by linear programming techniques. Details about the computational aspects may be found in specialized texts, such as Reference 11.3.

Example.   In the body fat example with two independent variables, case 3 was identified as having a substantial influence on the fitted regression function. We noted that the fitted regression functions with and without case 3 are:

$$\text{With case 3:} \quad \hat{Y} = -19.174 + .2224X_1 + .6594X_2$$

$$\text{Without case 3:} \quad \hat{Y} = -12.428 + .5641X_1 + .3635X_2$$

Using the method of least absolute deviations, the fitted regression function is:

$$\hat{Y} = -17.027 + .4173X_1 + .5203X_2$$

It is thus seen that the method of least absolute deviations leads to more modest changes than dropping case 3 entirely. An analysis of the residuals shows that the method of least absolute deviations resulted in reductions of those residuals that with the method of least squares are largest absolutely.

Comments

1. The residuals for the method of least absolute deviations ordinarily will not sum to zero.

2. The solution for the estimated regression coefficients with the method of least absolute deviations may not be unique.

3. The method of least absolute deviations is also called *minimum absolute deviations, minimum sum of absolute deviations,* and *minimum L_1-norm.*

4. Numerous other robust procedures besides the method of least absolute deviations have been proposed. Reference 11.4 discusses a number of these procedures.

11.6 MULTICOLLINEARITY DIAGNOSTICS— VARIANCE INFLATION FACTOR

When we discussed multicollinearity in Chapter 8, we noted some key problems that typically arise when the independent variables being considered for the regression model are highly correlated among themselves:

1. Adding or deleting an independent variable changes the regression coefficients.
2. The extra sum of squares associated with an independent variable varies, depending upon which independent variables already are included in the model.
3. The estimated standard deviations of the regression coefficients become large when the independent variables in the regression model are highly correlated with each other.
4. The estimated regression coefficients individually may not be statistically significant even though a definite statistical relation exists between the dependent variable and the set of independent variables.

These problems can also arise without substantial multicollinearity being present, but only under unusual circumstances not likely to be found in practice.

We shall now consider some informal diagnostics for multicollinearity as well as a highly useful formal diagnostic, the variance inflation factor.

Informal Diagnostics

Indications of the presence of serious multicollinearity are given by the following informal diagnostics:

1. Large changes in the estimated regression coefficients when a variable is added or deleted, or when an observation is altered or deleted.
2. Nonsignificant results in individual tests on the regression coefficients for important independent variables.
3. Estimated regression coefficients with an algebraic sign that is the opposite of that expected from theoretical considerations or prior experience.
4. Large coefficients of simple correlation between pairs of independent variables in the correlation matrix $\mathbf{r}_{XX}$.
5. Wide confidence intervals for the regression coefficients representing important independent variables.

Example.   We consider again the body fat example of Chapter 8, this time with all three independent variables—triceps skinfold thickness (X_1), thigh circumference (X_2), and midarm circumference (X_3). The data were given in Table 8.1. We noted there that the independent variables triceps skinfold thickness and thigh circumference are highly correlated with each other. Also, we noted in Chapter 8 large changes in the estimated regression coefficients and their estimated standard errors when a variable was added, nonsignificant results in individual tests on anticipated important variables, and an estimated negative coefficient when a positive coefficient was expected. These are all informal indications that suggest serious multicollinearity among the independent variables.

Note

The informal methods just described have important limitations. They do not provide quantitative measurements of the impact of multicollinearity nor may they identify the nature of the multicollinearity. For instance, if independent variables X_1, X_2, and X_3 have low pairwise correlations, then the examination of simple correlation coefficients will not necessarily disclose the existence of relations among groups of independent variables.

   Another limitation of the informal diagnostic methods is that sometimes the observed behavior may occur without multicollinearity being present.

Variance Inflation Factor

A formal method of detecting the presence of multicollinearity that is widely used is by means of variance inflation factors. These factors measure how much the variances of the estimated regression coefficients are inflated as compared to when the independent variables are not linearly related.

   To understand the signficance of variance inflation factors, we begin with the precision of least squares estimated regression coefficients, which is measured by their variances. We know from (7.41) that the variance-covariance matrix of the estimated regression coefficients is:

(11.32) $$\sigma^2\{\mathbf{b}\} = \sigma^2(\mathbf{X}'\mathbf{X})^{-1}$$

To reduce roundoff errors in calculating $(\mathbf{X}'\mathbf{X})^{-1}$, we noted in Chapter 8 that it is desirable to first transform the variables by means of the correlation transformation (8.41). When the transformed model (8.42) is fitted, the estimated regression coefficients b'_k are standardized coefficients that are related to the estimated regression coefficients for the untransformed variables according to (8.50). The variance-covariance matrix of the estimated standardized regression coefficients is obtained from (11.32) by using the result in (8.47), which states that the $\mathbf{X'X}$ matrix for the transformed variables is $\mathbf{r}_{XX}$:

$$(11.33) \qquad\qquad \sigma^2\{\mathbf{b}\} = (\sigma')^2 \mathbf{r}_{XX}^{-1}$$

where $\mathbf{r}_{XX}$ is the matrix of the pairwise simple correlation coefficients among the X variables, as defined in (8.44), and $(\sigma')^2$ is the error term variance for the transformed model.

Note from (11.33) that the variance of b'_k $(k = 1, \ldots, p - 1)$ is equal to the product of the error term variance $(\sigma')^2$ and the kth diagonal element of the matrix $\mathbf{r}_{XX}^{-1}$. This second factor is called the *variance inflation factor* (*VIF*). It can be shown that the variance inflation factor for b'_k, denoted by $(VIF)_k$, is:

$$(11.34) \qquad\qquad (VIF)_k = (1 - R_k^2)^{-1} \qquad k = 1, 2, \ldots, p - 1$$

where R_k^2 is the coefficient of multiple determination when X_k is regressed on the $p - 2$ other X variables in the model. Hence, we have:

$$(11.35) \qquad\qquad \sigma^2\{b'_k\} = (\sigma')^2(VIF)_k = \frac{(\sigma')^2}{1 - R_k^2}$$

We presented in (8.62) the special results for $\sigma^2\{b'_k\}$ when $p - 1 = 2$, for which $R_k^2 = r_{12}^2$, the coefficient of simple determination between X_1 and X_2.

The variance inflation factor $(VIF)_k$ is equal to 1 when $R_k^2 = 0$, i.e., when X_k is not linearly related to the other X variables. When $R_k^2 \neq 0$, then $(VIF)_k$ is greater than 1, indicating an inflated variance for b'_k. This is evident from (11.35) since the denominator becomes smaller as R_k^2 becomes larger, leading to a larger variance. When X_k has a perfect linear association with the other X variables in the model so that $R_k^2 = 1$, then $(VIF)_k$ and $\sigma^2\{b'_k\}$ are unbounded.

Diagnostic Uses. The largest *VIF* value among all X variables is often used as an indicator of the severity of multicollinearity. A maximum *VIF* value in excess of 10 is often taken as an indication that multicollinearity may be unduly influencing the least squares estimates.

The mean of the *VIF* values also provides information about the severity of the multicollinearity in terms of how far the estimated standardized regression coefficients b'_k are from the true values β'_k. It can be shown that the expected value of the sum of these squared errors $(b'_k - \beta'_k)^2$ is given by:

$$(11.36) \qquad\qquad E\left\{\sum_{k=1}^{p-1} (b'_k - \beta'_k)^2\right\} = (\sigma')^2 \sum_{k=1}^{p-1} (VIF)_k$$

Thus, large *VIF* values result, on the average, in larger differences between the estimated and true standardized regression coefficients.

When no X variable is linearly related to the others in the regression model, $R_k^2 \equiv 0$; hence, $(VIF)_k \equiv 1$ and:

(11.36a) $E\left\{\sum\limits_{k=1}^{p-1} (b_k' - \beta_k')^2\right\} = (\sigma')^2(p - 1)$ when $(VIF)_k \equiv 1$

A ratio of the results in (11.36) and (11.36a) provides useful information about the effect of multicollinearity on the sum of the squared errors:

$$\frac{(\sigma')^2\Sigma(VIF)_k}{(\sigma')^2(p - 1)} = \frac{\Sigma(VIF)_k}{p - 1}$$

Note that this ratio is simply the mean of the VIF values, to be denoted by $(\overline{VIF})$:

(11.37) $$(\overline{VIF}) = \frac{\sum\limits_{i=1}^{p-1}(VIF)_k}{p - 1}$$

Mean VIF values considerably larger than 1 are indicative of serious multicollinearity problems.

Example.   Table 11.4 contains the estimated standardized regression coefficients and the VIF values for the body fat example with three independent variables (calculations not shown). The maximum VIF value is 708.84 and $(\overline{VIF}) = 459.26$. Thus, the expected sum of the squared errors in the least squares standardized regression coefficients is nearly 460 times as large as it would be if the X variables were uncorrelated. In addition, all three VIF values greatly exceed 10, which again indicates that serious multicollinearity problems exist.

It is interesting to note that $(VIF)_3 = 105$ despite the fact that both r_{13}^2 and r_{23}^2 (see Table 8.1b) are not large. Here is an instance where X_3 is strongly related to X_1 and X_2 together ($R_3^2 = .990$), even though the pairwise coefficients of simple determination are not large. Examination of the pairwise correlations does not disclose this multicollinearity.

Comments

1. A number of computer regression programs use the reciprocal of the variance inflation factor to detect instances where an X variable should not be allowed into the fitted regression

---

TABLE 11.4   Variance Inflation Factors for Body Fat Example with Three Independent Variables

Variable	b_k'	$(VIF)_k$
X_1	4.2637	708.84
X_2	−2.9287	564.34
X_3	−1.5614	104.61

Maximum $(VIF)_k = 708.84$ $(\overline{VIF}) = 459.26$

model because of excessively high interdependence between this variable and the other X variables in the model. Tolerance limits for $1/(VIF)_k = 1 - R_k^2$ frequently used are .01, .001, or .0001, below which the variable is not entered into the model.

2. A limitation of variance inflation factors for detecting multicollinearities is that they cannot distinguish between several simultaneous multicollinearities.

3. A number of other formal methods for detecting multicollinearity have been proposed. These are more complex than variance inflation factors and are discussed in specialized texts such as Reference 11.5.

11.7 MULTICOLLINEARITY REMEDIAL MEASURES—RIDGE REGRESSION

We consider now some remedial measures for serious multicollinearity that can be used with ordinary least squares, and then take up ridge regression, a method of overcoming serious multicollinearity problems by modifying the method of least squares.

Remedial Measures with Ordinary Least Squares

1. As we have seen in Chapter 8, the presence of serious multicollinearity often does not affect the usefulness of the fitted model for making inferences about mean responses or making predictions, provided that the values of the independent variables for which inferences are to be made follow the same multicollinearity pattern as the data on which the regression model is based. Hence, one remedial measure is to restrict the use of the fitted regression model to inferences for values of the independent variables that follow the same pattern of multicollinearity.

2. In polynomial regression models, as we noted in Chapter 9, expressing the independent variable(s) in the form of deviations from the mean serves to reduce substantially the multicollinearity among the first-order, second-order, and higher-order terms for any given independent variable.

3. One or several independent variables may be dropped from the model in order to lessen the multicollinearity and thereby reduce the standard errors of the estimated regression coefficients of the independent variables remaining in the model. This remedial measure has two important limitations. First, no direct information is obtained about the dropped independent variables. Second, the magnitudes of the regression coefficients for the independent variables remaining in the model are affected by the correlated independent variables not included in the model.

4. Sometimes it is possible to add some cases which break the pattern of multicollinearity. Often, however, this option is not available. In business and economics, for instance, many independent variables cannot be controlled, so that new cases will tend to show the same intercorrelation patterns as the earlier ones.

5. In some economic studies, it is possible to estimate the regression coefficients for different independent variables from different sets of data to avoid the problems of multicollinearity. Demand studies, for instance, may use both cross-section and time series data to this end. Suppose the independent variables in a demand study are price and income, and the relation to be estimated is:

(11.38) $$Y_i = \beta_0 + \beta_1 X_{i1} + \beta_2 X_{i2} + \varepsilon_i$$

where Y is demand, X_1 is income, and X_2 is price. The income coefficient β_1 may then be estimated from cross-section data. The demand variable Y is thereupon adjusted:

(11.39) $$Y_i' = Y_i - b_1 X_{i1}$$

Finally, the price coefficient β_2 is estimated by regressing the adjusted demand variable Y' on X_2.

Ridge Regression

Biased Estimation.   Ridge regression is one of several methods that have been proposed to remedy multicollinearity problems by modifying the method of least squares to allow biased estimators of the regression coefficients. When an estimator has only a small bias and is substantially more precise than an unbiased estimator, it may well be the preferred estimator since it will have a larger probability of being close to the true parameter value. Figure 11.7 illustrates this situation. Estimator b is unbiased but imprecise, while estimator b^R is much more precise but has a small bias. The probability that b^R falls near the true value β is much greater than that for the unbiased estimator b.

A measure of the combined effect of bias and sampling variation is the expected value of the squared deviation of the biased estimator b^R from the true parameter β. This measure is called the *mean squared error,* and it can be shown to equal:

(11.40). $$E\{b^R - \beta\}^2 = \sigma^2\{b^R\} + (E\{b^R\} - \beta)^2$$

FIGURE 11.7   Biased Estimator with Small Variance May Be Preferable to Unbiased Estimator with Large Variance

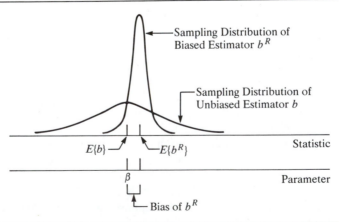

Thus, the mean squared error equals the variance of the estimator plus the squared bias. Note that if the estimator is unbiased, the mean squared error is identical to the variance of the estimator.

Ridge Estimators.   For ordinary least squares, the normal equations are given by (7.20):

$$(11.41) \qquad (\mathbf{X}'\mathbf{X})\mathbf{b} = \mathbf{X}'\mathbf{Y}$$

When all variables are transformed by the correlation transformation (8.41), the transformed regression model is given by (8.42):

$$(11.42) \qquad Y_i' = \beta_1' X_{i1}' + \beta_2' X_{i2}' + \cdots + \beta_{p-1}' X_{i,p-1}' + \varepsilon_i'$$

and the least squares normal equations are given by (8.49a):

$$(11.43) \qquad \mathbf{r}_{XX}\mathbf{b} = \mathbf{r}_{YX}$$

where $\mathbf{r}_{XX}$ is the correlation matrix of the X variables defined in (8.44) and $\mathbf{r}_{YX}$ is the vector of coefficients of simple correlation between Y and each X variable defined in (8.45).

The ridge standardized regression estimators are obtained by introducing into the least squares normal equations (11.43) a biasing constant $c \geq 0$, in the following form:

$$(11.44) \qquad (\mathbf{r}_{XX} + c\mathbf{I})\mathbf{b}^R = \mathbf{r}_{YX}$$

where $\mathbf{b}^R$ is the vector of the standardized ridge regression coefficients b_k^R:

$$(11.45) \qquad \underset{(p-1)\times 1}{\mathbf{b}^R} = \begin{bmatrix} b_1^R \\ b_2^R \\ \vdots \\ b_{p-1}^R \end{bmatrix}$$

and $\mathbf{I}$ is the $(p-1) \times (p-1)$ identity matrix. Solution of the normal equations (11.44) yields the ridge standardized regression coefficients:

$$(11.46) \qquad \mathbf{b}^R = (\mathbf{r}_{XX} + c\mathbf{I})^{-1}\mathbf{r}_{YX}$$

The constant c reflects the amount of bias in the estimators. When $c = 0$, (11.46) reduces to the ordinary least squares regression coefficients in standardized form, as given in (8.49b). When $c > 0$, the ridge regression coefficients are biased but tend to be more stable (i.e., less variable) than ordinary least squares estimators.

Choice of Biasing Constant c.   It can be shown that the bias component of the total mean squared error of the ridge regression estimator $\mathbf{b}^R$ increases as c gets larger (with all b_k^R tending toward zero), while at the same time the variance component becomes smaller. It can further be shown that there always exists some value c for which the ridge regression estimator $\mathbf{b}^R$ has a smaller total mean squared error than the ordinary least squares estimator $\mathbf{b}$. The difficulty is that the optimum value of c varies from one application to another and is unknown.

A commonly used method of determining the biasing constant c is based on the *ridge trace* and the variance inflation factors $(VIF)_k$ in (11.34). The ridge trace is a simultaneous plot of the values of the $p - 1$ estimated ridge standardized regression coefficients for different values of c, usually between 0 and 1. Extensive experience has indicated that an estimated regression coefficient b_k^R may fluctuate widely as c is changed slightly from 0, and may even change signs. Gradually, however, these wide fluctuations cease and the magnitude of the regression coefficient tends to change only slowly as c is increased further. At the same time, the value of $(VIF)_k$ tends to fall rapidly as c is changed from 0, and gradually $(VIF)_k$ also tends to change only moderately as c is increased further. One therefore examines the ridge trace and the VIF values and chooses the smallest value of c where it is deemed that the regression coefficients first become stable in the ridge trace and the VIF values have become sufficiently small. The choice is thus a judgmental one.

Example.   We noted previously several informal indications of severe multicollinearity in the data for the body fat example with three independent variables. Indeed, in the fitted model with three independent variables (Table 8.2d, page 273), the estimated regression coefficient b_2 is negative even though it was expected that amount of body fat is positively related to thigh circumference. Ridge regression calculations were made for the body fat example data in Table 8.1 (calculations not shown). The ridge standardized regression coefficients for selected values of c are presented in Table 11.5, and the variance inflation factors are given in Table 11.6. The coefficients of multiple determination R^2 are also shown in the latter table. Figure 11.8 presents the ridge trace of the estimated standardized regression coefficients. To facilitate the analysis, the horizontal c scale in Figure 11.8 is logarithmic.

Note the instability in Figure 11.8 of the regression coefficients for very small values of c. The estimated regression coefficient b_2^R, in fact, changes signs. Also note the rapid decrease in the VIF values in Table 11.6. It was decided to employ $c = .02$ here because for this value of the biasing constant the ridge regression coefficients have VIF values near 1 and the estimated regression coefficients appear to have become reasonably stable. The resulting fitted model for $c = .02$ is:

$$\hat{Y}' = .5463X_1' + .3774X_2' - .1369X_3'$$

Transforming back to the original variables by (8.50), we obtain:

$$\hat{Y} = -7.3978 + .5553X_1 + .3681X_2 - .1917X_3$$

where $\bar{Y} = 20.195$, $\bar{X}_1 = 25.305$, $\bar{X}_2 = 51.170$, $\bar{X}_3 = 27.620$, $s_Y = 5.106$, $s_1 = 5.023$, $s_2 = 5.235$, and $s_3 = 3.647$.

The improper sign on the estimate for β_2 has now been eliminated, and the estimated regression coefficients are more in line with prior expectations. The sum of the squared residuals for the transformed variables, which increases with c, has only increased from .1986 at $c = 0$ to .2182 at $c = .02$ while R^2 decreased from .8014 to .7818. These changes are relatively modest. The estimated mean body fat when $X_{h1} = 25.0$, $X_{h2} = 50.0$, and $X_{h3} = 29.0$ is 19.33 for the ridge regression at $c = .02$ compared to 19.19 utilizing the ordinary least squares solution. Thus, the ridge solu-

TABLE 11.5 Ridge Estimated Standardized Regression Coefficients for Different Biasing Constants c—Body Fat Example with Three Independent Variables

c	b_1^R	b_2^R	b_3^R
.000	4.264	−2.929	−1.561
.001	2.035	−.9408	−.7087
.002	1.441	−.4113	−.4813
.003	1.165	−.1661	−.3758
.004	1.006	−.0248	−.3149
.005	.9028	.0670	−.2751
.006	.8300	.1314	−.2472
.007	.7760	.1791	−.2264
.008	.7343	.2158	−.2103
.009	.7012	.2448	−.1975
.010	.6742	.2684	−.1870
.020	.5463	.3774	−.1369
.030	.5004	.4134	−.1181
.040	.4760	.4302	−.1076
.050	.4605	.4392	−.1005
.060	.4494	.4443	−.0952
.070	.4409	.4472	−.0909
.080	.4341	.4486	−.0873
.090	.4283	.4491	−.0841
.100	.4234	.4490	−.0812
.200	.3914	.4347	−.0613
.300	.3703	.4154	−.0479
.400	.3529	.3966	−.0376
.500	.3377	.3791	−.0295
.600	.3240	.3629	−.0229
.700	.3116	.3481	−.0174
.800	.3002	.3344	−.0129
.900	.2896	.3218	−.0091
1.000	.2798	.3101	−.0059

tion at $c = .02$ appears to be quite satisfactory here and a reasonable alternative to the ordinary least squares solution.

Comments

1. The normal equations (11.44) for the ridge estimators are as follows:

$$(11.47) \quad \begin{aligned}
(1 + c)b_1^R + \quad r_{12}b_2^R + \cdots + \quad r_{1,p-1}b_{p-1}^R &= r_{Y1} \\
r_{21}b_1^R + (1 + c)b_2^R + \cdots + \quad r_{2,p-1}b_{p-1}^R &= r_{Y2} \\
\vdots \qquad\qquad \vdots \qquad\qquad\qquad \vdots \qquad\quad &\quad \vdots \\
r_{p-1,1}b_1^R + \quad r_{p-1,2}b_2^R + \cdots + (1 + c)b_{p-1}^R &= r_{Y,p-1}
\end{aligned}$$

TABLE 11.6   *VIF* Values for Regression Coefficients and R^2 for Different Biasing Constants c—Body Fat Example with Three Independent Variables

c	$(VIF)_1$	$(VIF)_2$	$(VIF)_3$	R^2
.000	708.84	564.34	104.61	.8014
.001	125.73	100.27	19.28	.7943
.002	50.56	40.45	8.28	.7901
.003	27.18	21.84	4.86	.7878
.004	16.98	13.73	3.36	.7864
.005	11.64	9.48	2.58	.7854
.006	8.50	6.98	2.19	.7847
.007	6.50	5.38	1.82	.7842
.008	5.15	4.30	1.62	.7838
.009	4.19	3.54	1.48	.7834
.010	3.49	2.98	1.38	.7832
.020	1.10	1.08	1.01	.7818
.030	.63	.70	.92	.7812
.040	.45	.56	.88	.7808
.050	.37	.49	.85	.7804
.060	.32	.45	.83	.7801
.070	.30	.42	.81	.7797
.080	.28	.40	.79	.7793
.090	.26	.39	.78	.7789
.100	.25	.37	.76	.7784
.200	.21	.31	.63	.7723
.300	.18	.27	.54	.7638
.400	.17	.24	.46	.7538
.500	.15	.21	.40	.7427
.600	.14	.19	.35	.7310
.700	.13	.18	.31	.7189
.800	.12	.16	.28	.7065
.900	.11	.15	.25	.6941
1.000	.11	.14	.23	.6818

where r_{ij} is the coefficient of simple correlation between the ith and jth X variables and r_{Yj} is the coefficient of simple correlation between the dependent variable Y and the jth X variable.

2. The *VIF* values for the ridge regression coefficients b_k^R are defined analogously to those for the ordinary least squares regression coefficients. Namely, the *VIF* value for b_k^R measures how large is the variance of b_k^R relative to what the variance would be if the independent variables were uncorrelated. It can be shown that the *VIF* values for the ridge regression coefficients b_k^R are the diagonal elements of the following $(p - 1) \times (p - 1)$ matrix:

$$(11.48) \qquad (\mathbf{r}_{XX} + c\mathbf{I})^{-1}\mathbf{r}_{XX}(\mathbf{r}_{XX} + c\mathbf{I})^{-1}$$

3. The coefficient of multiple determination R^2, which for ordinary least squares is given in (7.35):

$$(11.49) \qquad R^2 = 1 - \frac{SSE}{SSTO}$$

FIGURE 11.8 Ridge Trace of Estimated Standardized Regression Coefficients—
Body Fat Example with Three Independent Variables

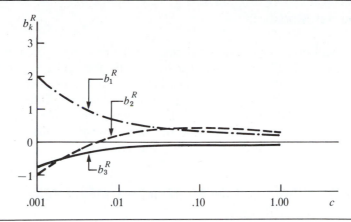

can be defined analogously for ridge regression. A simplification occurs, however, because the total sum of squares for the correlation-transformed dependent variable Y' in (8.41a) is:

$$(11.50) \qquad SSTO_R = \sum (Y'_i - \bar{Y}')^2 = 1$$

The fitted values with ridge regression are:

$$(11.51) \qquad \hat{Y}'_i = b_1^R X'_{i1} + \cdots + b_{p-1}^R X'_{i,p-1}$$

where the X'_{ik} are the X variables transformed according to the correlation transformation (8.41b). The error sum of squares, as usual, is:

$$(11.52) \qquad SSE_R = \sum (Y'_i - \hat{Y}'_i)^2$$

where $\hat{Y}'_i$ is given in (11.51). R^2 for ridge regression then becomes:

$$(11.53) \qquad R_R^2 = 1 - SSE_R$$

4. Ridge regression estimates tend to be stable in the sense that they are usually little affected by small changes in the data on which the fitted regression is based. In contrast, ordinary least squares estimates may be highly unstable under these conditions when the independent variables are highly multicollinear. Also, the ridge estimated regression function at times will provide good estimates of mean responses or predictions of new observations for levels of the independent variables outside the region of the observations on which the regression function is based. In contrast, the estimated regression function based on ordinary least squares may perform quite poorly in such instances. Of course, any estimation or prediction well outside the region of the observations should always be made with great caution.

5. A major limitation of ridge regression is that ordinary inference procedures are not applicable and exact distributional properties are not known. Another limitation is that the choice of the biasing constant c is a judgmental one. While formal methods have been developed for making this choice, these methods have their own limitations.

6. The ridge regression procedures have been generalized to allow for differing biasing constants for the different estimated regression coefficients.

Other Remedial Measures

Still other approaches to remedying the problems of multicollinearity have been developed. These include regression with principal components, where the independent variables are linear combinations of the original independent variables, and Bayesian regression, where prior information about the regression coefficients is incorporated into the estimation procedure. More information about these approaches, as well as about ridge regression and generalized ridge regression, may be obtained from specialized works such as Reference 11.5.

11.8 UNEQUAL ERROR VARIANCES REMEDIAL MEASURES— WEIGHTED LEAST SQUARES

We explained in Chapters 4 and 7 how transformations of Y may be helpful in reducing or eliminating unequal variances of the error terms. A difficulty with transformations of Y is that they may create an inappropriate regression relationship. When an appropriate regression relationship has been found but the variances of the error terms are unequal, an alternative is weighted least squares, a procedure that is frequently effective in these circumstances.

Weighted Least Squares

We begin our explanation of weighted least squares for simple linear regression. The least squares criterion for simple linear regression in (2.8):

$$Q = \sum_{i=1}^{n} (Y_i - \beta_0 - \beta_1 X_i)^2$$

weights each Y observation equally. The weighted least squares criterion for simple linear regression provides for differing weights:

(11.54)
$$Q_w = \sum_{i=1}^{n} w_i(Y_i - \beta_0 - \beta_1 X_i)^2$$

where w_i is the weight of the ith Y observation. Minimizing Q_w with respect to β_0 and β_1 leads to the normal equations:

(11.55)
$$\sum w_i Y_i = b_0 \sum w_i + b_1 \sum w_i X_i$$

$$\sum w_i X_i Y_i = b_0 \sum w_i X_i + b_1 \sum w_i X_i^2$$

In turn, these can be solved for the weighted least squares estimators b_0 and b_1:

$$(11.56a) \qquad b_1 = \frac{\Sigma w_i X_i Y_i - \dfrac{\Sigma w_i X_i \, \Sigma w_i Y_i}{\Sigma w_i}}{\Sigma w_i X_i^2 - \dfrac{(\Sigma w_i X_i)^2}{\Sigma w_i}}$$

$$(11.56b) \qquad b_0 = \frac{\Sigma w_i Y_i - b_1 \Sigma w_i X_i}{\Sigma w_i}$$

Note that if all weights are equal so w_i is identically equal to a constant, the normal equations (11.55) for weighted least squares reduce to the ones for unweighted least squares in (2.9) and the weighted least squares estimators (11.56) reduce to the ones for unweighted least squares in (2.10).

We now generalize weighted least squares to multiple regression and present it more formally. We shall denote the error term variance of ε_i by σ_i^2 and consider it to be made up of a proportionality constant, to be denoted by σ^2, and a component w_i that varies for each error term, as follows:

$$(11.57) \qquad \sigma_i^2 = \sigma^2\{\varepsilon_i\} = \sigma^2\{Y_i\} = \frac{\sigma^2}{w_i}$$

The proportionality constant σ^2 can be any positive value.

The weighted least squares criterion for multiple regression is:

$$(11.58) \qquad Q_w = \sum_{i=1}^{n} w_i(Y_i - \beta_0 - \beta_1 X_{i1} - \cdots - \beta_{p-1} X_{i,p-1})^2$$

Note from (11.57) that the weights w_i are inversely proportional to the variances σ_i^2. Thus, an observation Y_i that has a large variance receives less weight than another observation that has a smaller variance. Intuitively, this is reasonable. The more precise is Y_i (i.e., the smaller is σ_i^2), the more information Y_i provides about $E\{Y_i\}$ and therefore the more weight it should receive in fitting the regression function.

Let the matrix $\mathbf{W}$ be a diagonal matrix containing the weights w_i:

$$(11.59) \qquad \mathop{\mathbf{W}}_{n \times n} = \begin{bmatrix} w_1 & & & 0 \\ & w_2 & & \\ & & \cdot & \\ & & & \cdot \\ 0 & & & w_n \end{bmatrix}$$

The weighted least squares normal equations can then be expressed as follows:

$$(11.60) \qquad (\mathbf{X}'\mathbf{W}\mathbf{X})\mathbf{b} = \mathbf{X}'\mathbf{W}\mathbf{Y}$$

and the weighted least squares estimators of the regression coefficients are:

$$(11.61) \qquad \mathop{\mathbf{b}}_{p \times 1} = (\mathbf{X}'\mathbf{W}\mathbf{X})^{-1}\mathbf{X}'\mathbf{W}\mathbf{Y}$$

The variance-covariance matrix of the weighted least squares estimated regression coefficients is:

$$(11.62) \qquad \mathop{\boldsymbol{\sigma}^2\{\mathbf{b}\}}_{p \times p} = \sigma^2(\mathbf{X}'\mathbf{W}\mathbf{X})^{-1}$$

where σ^2 is the proportionality constant in (11.57). The estimated variance-covariance matrix of the regression coefficients is:

(11.63)
$$\underset{p \times p}{\mathbf{s}^2\{\mathbf{b}\}} = MSE_w(\mathbf{X'WX})^{-1}$$

where MSE_w is based on the weighted squared deviations:

(11.63a)
$$MSE_w = \frac{\Sigma w_i(Y_i - \hat{Y}_i)^2}{n - p}$$

Thus, MSE_w here is an estimator of the propotionality constant σ^2.

Weights When the σ_i^2 Are Unknown

If the variances σ_i^2 were known, or even known up to a proportionality constant, the use of weighted least squares with weights w_i would be straightforward. Unfortunately, one rarely has knowledge of the variances σ_i^2. We are then forced to use estimates of the variances. These can be obtained in a variety of ways. We discuss two methods for obtaining estimates of the variances σ_i^2.

1. Sometimes the error term variances vary with the level of an independent variable in the regression model in a systematic fashion. For simple linear regression, for instance, the relation might be one of the following:

(11.64a)
$$\sigma_i^2 = \sigma^2 X_i$$

(11.64b)
$$\sigma_i^2 = \sigma^2 X_i^2$$

(11.64c)
$$\sigma_i^2 = \sigma^2 \sqrt{X_i}$$

Here, σ^2 is again a proportionality constant.

If analysis of the residuals suggests that σ_i^2 tends to vary directly with X_i, for example, then one would use weights:

$$w_i = \frac{1}{X_i}$$

Similarly, the weights for situations (11.64b) and (11.64c) would be $w_i = 1/X_i^2$ and $w_i = 1/\sqrt{X_i}$, respectively.

2. When the error term variances vary with the level of an independent variable but not in a regular pattern, the cases are placed into a small number of groups according to the level of the independent variable, and the variances of the residuals are calculated for each group. Every Y observation in a group then receives a weight which is the reciprocal of the estimated variance for that group. This same procedure can be used when the error term variance in a multiple regression varies with the level of the fitted value $\hat{Y}$. Here, the cases are grouped according to their fitted values.

These approximate weighting methods can be very helpful if analysis of the residuals indicates major differences in the variances of the error terms. When the differences are only small or modest, weighted least squares with these approximate methods will not be particularly helpful.

Example

A health researcher interested in studying the relationship between diastolic blood pressure and age among healthy adult women 20 to 60 years old collected data on 54 subjects. The data are presented in Table 11.7. The scatter plot of the data in Figure 11.9a strongly suggests a linear relationship between diastolic blood pressure and age but also indicates that the error term variance increases with age. The researcher fitted a linear regression function by unweighted least squares to conduct some preliminary analyses of the residuals. A plot of the residuals against X is presented in Figure 11.9b, which confirms the nonconstant error variance.

To explore whether the error term variance has a simple relation to age, the researcher divided the cases into four groups of approximately equal size according to age. Then, the sample variance of the residuals for the unweighted least squares regression above was calculated for each group. The four age groups and the number of cases in each are presented in columns 1 and 2 of Table 11.8, and the estimated error variances are presented in column 3. The researcher considered the differences in the variances to be substantial, calling for the use of weighted least squares. She

TABLE 11.7   Data for Diastolic Blood Pressure Example

Subject i	Age X_i	Diastolic Blood Pressure Y_i	Subject i	Age X_i	Diastolic Blood Pressure Y_i	Subject i	Age X_i	Diastolic Blood Pressure Y_i
1	27	73	19	37	78	37	42	85
2	21	66	20	38	87	38	44	71
3	22	63	21	33	76	39	46	80
4	26	79	22	35	79	40	47	96
5	25	68	23	30	73	41	45	92
6	28	67	24	37	68	42	55	76
7	24	75	25	31	80	43	54	71
8	25	71	26	39	75	44	57	99
9	23	70	27	46	89	45	52	86
10	20	65	28	49	101	46	53	79
11	29	79	29	40	70	47	56	92
12	24	72	30	42	72	48	52	85
13	20	70	31	43	80	49	57	109
14	38	91	32	46	83	50	50	71
15	32	76	33	43	75	51	59	90
16	33	69	34	49	80	52	50	91
17	31	66	35	40	90	53	52	100
18	34	73	36	48	70	54	58	80

FIGURE 11.9   Scatter and Residual Plots for Blood Pressure Example

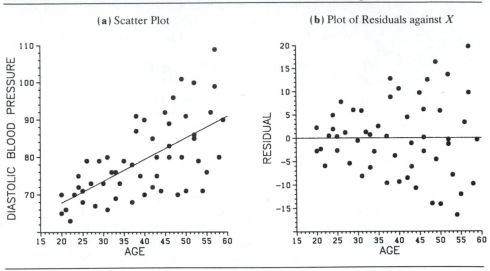

(**a**) Scatter Plot

(**b**) Plot of Residuals against X

TABLE 11.8   Estimated Error Variances and Weights for Age Groups—Blood Pressure Example

	(1)	(2)	(3)	(4)
			Estimated Error	*Estimated*
Group	*Age*	*Sample*	*Variance*	*Weight*
j	X_j	*Size*	s_j^2	$w_j = 1/s_j^2$
1	20—under 30	13	17.74260	.0563615
2	30—under 40	13	42.13678	.0237322
3	40—under 50	15	87.93657	.0113718
4	50—under 60	13	124.14565	.0080551

examined the variances to see if they followed any of the simple relationships in (11.64), using the midpoint of each range of ages as the value of X. She obtained the following results:

Group j	X_j	s_j^2/X_j	s_j^2/X_j^2	$s_j^2/\sqrt{X_j}$
1	25	.71	.028	3.5
2	35	1.20	.034	7.1
3	45	1.95	.043	13.1
4	55	2.26	.041	16.7

She did not consider any of these relations stable enough and therefore decided to use the reciprocals of the variances as weights for each case in the group. These weights are shown in Table 11.8, column 4.

A weighted regression analysis computer program yielded the fitted regression line:

$$(11.65) \qquad \hat{Y} = 56.15693 + .580031X$$

This fitted regression line is shown in Figure 11.9a and appears to be a reasonably good fit to the data.

The estimated regression line fitted to the same data by unweighted least squares is:

$$(11.66) \qquad \hat{Y} = 56.08962 + .589583X$$

This differs somewhat from the weighted least squares line in (11.65), as will generally be the case, but the differences are not great here.

While the estimates obtained by unweighted least squares are unbiased even when the error variances are unequal, as are the estimates obtained by weighted least squares, the unweighted least squares estimates are subject to greater sampling variation. In our example, the estimated standard deviations of the regression coefficients for the two methods are:

Unweighted Least Squares	Weighted Least Squares
$s\{b_0\} = 3.9937$	$s\{b_0\} = 2.7908$
$s\{b_1\} = .09695$	$s\{b_1\} = .08401$

Comments

1. The condition of the error variance not being constant over all cases is called *heteroscedasticity*, in contrast to the condition of equal error variances, called *homoscedasticity*.

2. When heteroscedasticity prevails but the other conditions of regression model (7.18) are met, the estimated regression coefficients obtained by ordinary least squares procedures are still unbiased and consistent, but they are no longer minimum variance unbiased estimators, as illustrated in the previous example.

3. Heteroscedasticity is inherent when the response in regression analysis follows a distribution in which the variance is functionally related to the mean. (Significant nonnormality in Y is encountered as well in most such cases.) Consider, in this connection, a regression analysis where X is the speed of a machine which puts a plastic coating on cable and Y is the number of blemishes in the coating per thousand feet drum of cable. If Y is Poisson distributed with a mean which increases as X increases, the distributions of Y cannot have constant variance at all levels of X since the variance of a Poisson variable equals the mean, which is increasing with X.

4. Replicate observations are very helpful for obtaining information about any pattern in the error variances. Often, however, replicate observations are not available and one then needs to use groupings of approximately equal size, as in the blood pressure example, to obtain information about the error variances.

5. An *iterative weighted least squares procedure* can be employed to get improved weighted least squares estimates. This procedure involves initially estimating the weights w_i from the data, and then obtaining the fitted regression function and residuals by weighted least squares. Using the residuals from this first stage, we reestimate the weights w_i and obtain a new weighted least squares fit. The process can be countinued until no substantial changes in the fitted regression function take place. Often, one or two iterations are sufficient.

6. Weighted least squares estimates can be obtained by employing unweighted least squares on appropriately transformed variables. For example, consider a simple linear regression where σ_i^2 is proportional to X_i^2 so that the weights are $w_i = 1/X_i^2$. The weighted least squares criterion (11.54) then is:

$$(11.67) \qquad Q_w = \sum \frac{1}{X_i^2}(Y_i - \beta_0 - \beta_1 X_i)^2 = \sum \left(\frac{Y_i}{X_i} - \frac{\beta_0}{X_i} - \beta_1\right)^2$$

This can be expressed as:

$$(11.67a) \qquad Q_w = \sum (Y_i' - \beta_0' - \beta_1' X_i')^2$$

where:

$$Y_i' = \frac{Y_i}{X_i} \qquad \beta_0' = \beta_1 \qquad X_i' = \frac{1}{X_i} \qquad \beta_1' = \beta_0$$

Note that (11.67a) is in the form of the unweighted least squares criterion (2.8). Hence, ordinary least squares can be used on the transformed observations Y_i' and X_i' to yield the same estimates as weighted least squares applied to the original observations. It can be shown that the error variance for the transformed variable Y' is constant.

7. Weighted least squares is a special case of *generalized least squares* where the error terms not only may have different variances but pairs of error terms may also be correlated.

CITED REFERENCES

11.1. Atkinson, A. C. *Plots, Transformations, and Regression.* Oxford: Clarendon Press, 1985.

11.2. Mansfield, E. R., and M. D. Conerly. "Diagnostic Value of Residual and Partial Residual Plots." *The American Statistician* 41 (1987), pp. 107–16.

11.3. Kennedy, W. J., Jr., and J. E. Gentle. *Statistical Computing.* New York: Marcel Dekker, 1980.

11.4. Hogg, R. V. "Statistical Robustness: One View of Its Use in Applications Today." *The American Statistician* 33 (1979), pp. 108–15.

11.5. Belsley, D. A.; E. Kuh; and R. E. Welsch. *Regression Diagnostics: Identifying Influential Data and Sources of Collinearity.* New York: John Wiley & Sons, 1980.

PROBLEMS

11.1. A student asked: "Why is it necessary to perform diagnostic checks of the fit when R^2 is large?" Comment.

11.2. A researcher stated: "One good thing about partial regression plots is that they are extremely useful for identifying model adequacy even when the independent variables are highly correlated." Comment.

11.3. A student suggested: "If extremely influential outlying cases are detected in a data set, simply discard these cases from the data set." Comment.

11.4. Describe several informal methods that can be helpful in identifying multicollinearity among the independent variables in a multiple regression model.

11.5. Refer to **Brand preference** Problem 7.8a.
   a. Prepare a partial regression plot for each of the independent variables.
   b. Do your plots in part (a) suggest that the regression relationships in the fitted regression function in Problem 7.8a are inappropriate for any of the independent variables? Explain.
   c. Obtain the fitted regression function in Problem 7.8a by first regressing Y and X_2 each on X_1 and then regressing the residuals in an appropriate fashion.

11.6. Refer to **Chemical shipment** Problem 7.12c.
   a. Prepare a partial regression plot for each of the independent variables.
   b. Do your plots in part (a) suggest that the regression relationships in the fitted regression function in Problem 7.12c are inappropriate for any of the independent variables? Explain.
   c. Obtain the fitted regression function in Problem 7.12c by first regressing Y and X_2 each on X_1 and then regressing the residuals in an appropriate fashion.

11.7. Refer to **Patient satisfaction** Problem 7.17b.
   a. Prepare a partial regression plot for each of the independent variables.
   b. Do your plots in part (a) suggest that the regression relationships in the fitted regression function in Problem 7.17b are inappropriate for any of the independent variables? Explain.

11.8. Refer to **Mathematicians' salaries** Problem 7.20b.
   a. Prepare a partial regression plot for each of the independent variables.
   b. Do your plots in part (a) suggest that the regression relationships in the fitted regression function in Problem 7.20b are inappropriate for any of the independent variables? Explain.

11.9. Refer to **Brand preference** Problem 7.8. The diagonal elements of the hat matrix are: $h_{55} = h_{66} = h_{77} = h_{88} = h_{99} = h_{10,10} = h_{11,11} = h_{12,12} = .137$ and $h_{11} = h_{22} = h_{33} = h_{44} = h_{13,13} = h_{14,14} = h_{15,15} = h_{16,16} = .237$.
   a. Explain the reason for the pattern in the diagonal elements of the hat matrix.
   b. According to the rule of thumb stated in the chapter, are any of the observations outlying with regard to their X values?
   c. Obtain the studentized deleted residuals and identify any outlying Y observations.
   d. Case 14 appears to be a borderline outlying Y observation. Obtain the *DFFITS*, *DFBETAS*, and Cook's distance values for this case to assess its influence. What do you conclude?
   e. Calculate the average absolute percent difference in the fitted values with and without case 14. What does this measure indicate about the influence of case 14?
   f. Calculate Cook's distance D_i for each case. Are any cases influential according to this measure?

11.10. Refer to **Chemical shipment** Problem 7.12. The diagonal elements of the hat matrix are:

i:	1	2	3	4	5	6	7	8	9	10
h_{ii}:	.091	.194	.131	.268	.149	.141	.429	.067	.135	.165

i:	11	12	13	14	15	16	17	18	19	20
h_{ii}:	.179	.051	.110	.156	.095	.128	.097	.230	.112	.073

a. Identify any outlying X observations using the rule of thumb presented in the chapter.

b. Obtain the studentized deleted residuals and identify any outlying Y observations.

c. Case 7 appears to be an outlying X observation and case 12 an outlying Y observation. Obtain the *DFFITS*, *DFBETAS*, and Cook's distance values for each of these cases to assess their influence. What do you conclude?

d. Calculate the average absolute percent difference in the fitted values with and without case 7 and with and without case 12. What does this measure indicate about the influence of each of the two cases?

e. Calculate Cook's distance D_i for each case. Are any cases influential according to this measure?

11.11. Refer to **Patient satisfaction** Problem 7.17. The diagonal elements of the hat matrix are:

i:	1	2	3	4	5	6	7	8	9	10	11	12
h_{ii}:	.134	.193	.070	.235	.204	.319	.060	.174	.339	.137	.245	.057

i:	13	14	15	16	17	18	19	20	21	22	23
h_{ii}:	.230	.072	.313	.104	.209	.143	.078	.231	.158	.238	.059

a. Identify any outlying X observations.

b. Obtain the studentized deleted residuals and identify any outlying Y observations.

c. Case 14 appears to be a borderline outlying Y observation. Obtain the *DFFITS*, *DFBETAS*, and Cook's distance values for this case to assess its influence. What do you conclude?

d. Calculate the average absolute percent difference in the fitted values with and without case 14. What does this measure indicate about the influence of case 14?

e. Calculate Cook's distance D_i for each case. Are any cases influential according to this measure?

11.12. Refer to **Mathematicians' salaries** Problem 7.20. The diagonal elements of the hat matrix are:

i:	1	2	3	4	5	6	7	8	9	10	11	12
h_{ii}:	.184	.059	.132	.071	.214	.146	.115	.179	.241	.288	.083	.128

i:	13	14	15	16	17	18	19	20	21	22	23	24
h_{ii}:	.320	.098	.186	.151	.267	.146	.198	.206	.118	.136	.225	.110

a. Identify any outlying X observations.

b. Obtain the studentized deleted residuals and identify any outlying Y observations.

c. Case 19 appears to be a borderline outlying Y observation. Obtain the *DFFITS*,

DFBETAS, and Cook's distance values for this case to assess its influence. What do you conclude?

d. Calculate the average absolute percent difference in the fitted values with and without case 19. What does this measure indicate about the influence of case 19?

e. Calculate Cook's distance D_i for each case. Are any cases influential according to this measure?

11.13. Cosmetics sales. An assistant in the district sales office of a national cosmetics firm obtained data, shown below, on advertising expenditures and sales last year in the district's 14 territories. X_1 denotes expenditures for point-of-sale displays in beauty salons and department stores (in thousand dollars) while X_2 and X_3 represent the corresponding expenditures for local media advertising and prorated share of national media advertising, respectively. Y denotes sales (in thousand cases). The assistant was instructed to estimate the increase in expected sales when X_1 is increased by 1 thousand dollars and X_2 and X_3 are held constant, and was told to use an ordinary multiple regression model with linear terms for the independent variables and with independent normal error terms.

i:	1	2	3	4	5	6	7
X_{i1}:	4.2	6.5	3.0	2.1	2.9	7.2	4.8
X_{i2}:	4.0	6.5	3.5	2.0	3.0	7.0	5.0
X_{i3}:	3.0	5.0	4.0	3.0	4.0	3.0	4.5
Y_i:	8.26	14.70	9.73	5.62	7.84	12.18	8.56

i:	8	9	10	11	12	13	14
X_{i1}:	4.3	2.6	3.1	6.2	5.5	2.2	3.0
X_{i2}:	4.0	2.5	3.0	6.0	5.5	2.0	2.8
X_{i3}:	5.0	5.0	4.0	4.5	5.0	4.0	3.0
Y_i:	10.77	7.56	8.90	12.51	10.46	7.15	6.74

a. State the regression model to be employed and fit it to the data.

b. Test whether there is a regression relation between sales and the three independent variables. Use a level of significance of .05. State the alternatives, decision rule, and conclusion.

c. Test for each of the regression coefficients β_k ($k = 1, 2, 3$) individually whether or not $\beta_k = 0$. Use a level of significance of .05 each time. Do the conclusions of these tests correspond to that obtained in part (b)?

d. Obtain the correlation matrix of the X variables.

e. What do the results in parts (b), (c), and (d) suggest about the suitability of the data for the research objective?

11.14. Refer to **Cosmetics sales** Problem 11.13.

a. Verify that the variance inflation factor for variable X_1 is $(VIF)_1 = 66.29$. The other variance inflation factors are $(VIF)_2 = 66.99$ and $(VIF)_3 = 1.09$. What do these suggest about the effects of multicollinearity here?

b. The assistant eventually decided to drop variables X_2 and X_3 from the model "to clear up the picture." Fit the assistant's revised model. Is the assistant now in a better position to achieve the research objective?

c. Why would an experiment here be more effective in providing suitable data to meet the research objective? How would you design such an experiment? What regression model would you employ?

11.15. Refer to **Patient satisfaction** Problem 7.17.
  a. Obtain the correlation matrix of the X variables. What does it show about pairwise linear associations among the independent variables?
  b. The variance inflation factors are $(VIF)_1 = 1.35$, $(VIF)_2 = 2.76$, and $(VIF)_3 = 2.87$. What do these results suggest about the effects of multicollinearity here? Are these results more revealing than those in part (a)?

11.16. Refer to **Brand preference** Problem 7.8.
  a. Obtain the correlation matrix of the X variables. What does it show about pairwise linear associations among the independent variables?
  b. Find the two variance inflation factors. Why are they both equal to 1?

11.17. Refer to **Mathematicians' salaries** Problem 7.20.
  a. Obtain the correlation matrix of the X variables. What does it show about pairwise linear associations among the independent variables?
  b. Obtain the variance inflation factors. Do they indicate that a serious multicollinearity problem exists here?

11.18. Refer to **Cosmetics sales** Problem 11.13. Given below are the estimated ridge standardized regression coefficients, the variance inflation factors, and R^2 for selected biasing constants c.

c:	.000	.005	.01	.02	.03	.04	.05	.06
b_1^R:	.273	.327	.349	.368	.376	.380	.382	.383
b_2^R:	.549	.494	.470	.447	.435	.427	.422	.417
b_3^R:	.260	.260	.260	.259	.257	.256	.254	.253
$(VIF)_1$:	66.29	24.11	12.45	5.20	2.92	1.91	1.38	1.07
$(VIF)_2$:	66.99	24.36	12.57	5.25	2.94	1.92	1.39	1.07
$(VIF)_3$:	1.09	1.06	1.04	1.01	.99	.97	.95	.93
R^2:	.8402	.8401	.8401	.8401	.8398	.8397	.8395	.8393

  a. Make a ridge trace plot for the given c values. Do the ridge regression coefficients exhibit substantial changes near $c = 0$?
  b. Suggest a reasonable value for the biasing constant c based on the ridge trace, the VIF values, and R^2.
  c. Transform the estimated standardized regression coefficients selected in part (b) back to the original variables and obtain the fitted values for the 14 cases. How similar are these fitted values to those obtained with the ordinary least squares fit in Problem 11.13a?

11.19. Refer to **Chemical shipment** Problem 7.12. Given below are the estimated ridge standardized regression coefficients, the variance inflation factors, and R^2 for selected biasing constants c.

c:	.000	.005	.01	.05	.07	.09	.10	.20
b_1^R:	.451	.453	.455	.460	.460	.459	.458	.444
b_2^R:	.561	.556	.552	.526	.517	.508	.504	.473
$(VIF)_1 = (VIF)_2$:	7.03	6.20	5.51	2.65	2.03	1.61	1.46	.71
R^2:	.9869	.9869	.9869	.9862	.9856	.9852	.9844	.9780

  a. Make a ridge trace plot for the given c values. Do the regression coefficients exhibit substantial changes near $c = 0$?

b. Why are the $(VIF)_1$ values the same as the $(VIF)_2$ values here?

c. Suggest a reasonable value for the biasing constant c based on the ridge trace in part (a), the VIF values, and R^2.

d. Transform the estimated standardized regression coefficients selected in part (c) back to the original variables and obtain the fitted values for the 20 cases. How similar are these fitted values to those obtained with the ordinary least squares fit in Problem 7.12c?

11.20. Machine speed. The number of defective items produced by a machine (Y) is known to be linearly related to the speed setting of the machine (X). The data below were collected from recent quality control records.

i:	1	2	3	4	5	6	7	8	9	10	11	12
X_i:	200	400	300	400	200	300	300	400	200	400	200	300
Y_i:	28	75	37	53	22	58	40	96	46	52	30	69

a. Fit regression model (3.1), obtain the estimated regression function, and plot the residuals against X. What does the residual plot show?

b. Calculate the sample variance s^2 of the residuals for each of the three machine speeds: $X = 200, 300, 400$. What is suggested by these three sample variances about whether or not the true error variances at the three X levels are equal?

c. For each of the three machine speeds, calculate s^2/X, s^2/X^2, and $s^2/\sqrt{X}$. Do any of these relations appear to be stable?

d. Using weights $w_i = 1/X_i^2$, obtain the weighted least squares estimates of β_0 and β_1. Are these estimates similar to the ones obtained with ordinary least squares in part (a)?

e. Compare the estimated standard deviations of the weighted least squares estimates b_0 and b_1 in part (d) with those for the ordinary least squares estimates in part (a). What do you find?

11.21. Computer-assisted learning. Data from a study of computer-assisted learning by 12 students, showing the total number of responses in completing a lesson (X) and the cost of computer time $(Y$, in cents), follow.

i:	1	2	3	4	5	6	7	8	9	10	11	12
X_i:	16	14	22	10	14	17	10	13	19	12	18	11
Y_i:	77	70	85	50	62	70	55	63	88	57	81	51

a. Fit a linear regression function by ordinary least squares, obtain the residuals, and plot the residuals against X. What does the residual plot suggest?

b. Group cases with $X = 10, 11, 12$ into one group, cases with $X = 13, 14, 16$ into a second group, and cases with $X = 17, 18, 19, 22$ into a third group. For each group, calculate the sample variance of the residuals. Does it appear that the error variances are equal?

c. For each of the three groups, calculate s^2/X, s^2/X^2, and $s^2/\sqrt{X}$, using the midpoint of the group for X. Do any of these relations appear to be stable?

d. Using weights $w_i = 1/X_i^2$, obtain the weighted least squares estimates of β_0 and β_1. Are these estimates similar to the ones obtained with ordinary least squares in part (a)?

e. Compare the estimated standard deviations of the weighted least squares estimates

b_0 and b_1 in part (d) with those for the ordinary least squares estimates in part (a). What do you find?

11.22. Refer to the blood pressure example in Table 11.7. An analyst who reviewed the researcher's findings on page 422 about a possible simple relationship between the error variances and the level of X concluded that s_j^2/X_j^2 is relatively stable and that weights $w_i = 1/X_i^2$ would be appropriate.

   a. Using the suggested weights, obtain the weighted least squares estimates of β_0 and β_1 and their estimated standard deviations.

   b. How do your results in part (a) compare with those obtained by the researcher? Does the choice of weights have important effects here? Discuss.

EXERCISES

11.23. Derive the mean squared error in (11.40).

11.24. Refer to the least absolute deviations estimates for the body fat example on page 407—namely, $b_0 = -17.027$, $b_1 = .4173$, and $b_2 = .5203$.

   a. Find the sum of the absolute deviations from the fitted values based on the least absolute deviations estimates.

   b. For the least squares estimated regression coefficients $b_0 = -19.174$, $b_1 = .2224$, and $b_2 = .6594$, find the sum of the absolute deviations. Is this sum larger than the sum obtained in part (a)?

11.25. (Calculus needed.) Derive the weighted least squares normal equations for fitting a linear regression function when $\sigma_i^2 = \sigma^2 X_i$, where σ^2 is a proportionality constant.

11.26. Express the weighted least squares estimator b_1 in (11.56a) in terms of the deviations $Y_i - \bar{Y}_w$ and $X_i - \bar{X}_w$, where $\bar{Y}_w$ and $\bar{X}_w$ are weighted means.

11.27. Refer to **Machine speed** Problem 11.20. Demonstrate numerically that the weighted least squares estimates obtained in part (d) are identical to those obtained using transformation (11.67a) and ordinary least squares.

11.28. Refer to **Computer-assisted learning** Problem 11.21. Demonstrate numerically that the weighted least squares estimates obtained in part (d) are identical to those obtained using transformation (11.67a) and ordinary least squares.

11.29. Consider the weighted least squares criterion (11.54) with weights given by (11.64b). Set up the variance-covariance matrix for the error terms when $i = 1, \ldots, 4$. Assume $\sigma\{\varepsilon_i, \varepsilon_j\} = 0$ for $i \neq j$.

11.30. Derive the variance-covariance matrix $\sigma^2\{\mathbf{b}\}$ in (11.62) for the weighted least squares estimators when the variance-covariance matrix of the observations Y_i is $\sigma^2 \mathbf{W}^{-1}$, where $\mathbf{W}$ is given in (11.59).

PROJECTS

11.31. Refer to **Patient satisfaction** Problem 7.17.

   a. Obtain the estimated ridge standardized regression coefficients, variance inflation factors, and R^2 for the following biasing constants: $c = .000, .005, .01, .02, .03, .04, .05$.

b. Make a ridge trace plot for the given c values. Do the ridge regression coefficients exhibit substantial changes near $c = 0$?

c. Suggest a reasonable value for the biasing constant c based on the ridge trace, the *VIF* values, and R^2.

d. Transform the estimated standardized regression coefficients selected in part (c) back to the original variables and obtain the fitted values for the 23 cases. How similar are these fitted values to those obtained with the ordinary least squares fit in Problem 7.17b?

11.32. Refer to **Mathematicians' salaries** Problem 7.20.

a. Obtain the estimated ridge standardized regression coefficients, variance inflation factors, and R^2 for the following biasing constants: $c = .000, .005, .01, .02, .03, .04, .05$.

b. Make a ridge trace plot for the given c values. Do the ridge regression coefficients exhibit substantial changes near $c = 0$?

c. Suggest a reasonable value for the biasing constant c based on the ridge trace, the *VIF* values, and R^2.

d. Transform the estimated standardized regression coefficients selected in part (c) back to the original variables and obtain the fitted values for the 24 cases. How similar are these fitted values to those obtained with the ordinary least squares fit in Problem 7.20b?

11.33. Refer to the **SENIC** data set.

a. Regress the logarithm of length of stay (Y') on infection risk (X_1), number of beds (X_2), and average daily census (X_3).

b. Obtain the residuals and identify outliers.

c. Obtain the correlation matrix of the X variables and the variance inflation factors. What do these suggest about the effects of multicollinearity?

d. Obtain the estimated ridge regression coefficients, variance inflation factors, and R^2 for the values of the biasing constant c given in Table 11.6.

e. Make a ridge trace plot and determine a reasonable value for the biasing constant c based on this plot, the *VIF* values, and R^2.

11.34. Refer to the **SMSA** data set.

a. Regress number of active physicians (Y) on number of hospital beds (X_1), total personal income (X_2), and total serious crimes (X_3).

b. Obtain the residuals and identify outliers.

c. Obtain the correlation matrix of the X variables and the variance inflation factors. What do these suggest about the effects of multicollinearity?

d. Obtain the estimated ridge regression coefficients, variance inflation factors, and R^2 for the values of the biasing constant c in Table 11.6.

e. Make a ridge trace plot and determine a reasonable value for the biasing constant c based on this plot, the *VIF* values, and R^2.

11.35. Refer to **Mathematicians' salaries** Problem 7.20.

a. Find the least absolute deviations estimates for the parameters β_0, β_1, β_2, and β_3.

b. Find the sum of the absolute deviations from the fitted values based on the least absolute deviations estimates.

c. For the least squares estimated regression function in Problem 7.20b, find the sum of the absolute deviations. Is this sum larger than the sum obtained in part (b)?

11.36. Five observations on Y are to be taken when $X = 10, 20, 30, 40,$ and 50, respectively. The true regression function is $E\{Y\} = 20 + 10X$. The error terms are independent and normally distributed, with $E\{\varepsilon_i\} = 0$ and $\sigma^2\{\varepsilon_i\} = .8X_i$.

a. Generate a random Y observation for each X level, and calculate both the ordinary and weighted least squares estimates of the regression coefficient β_1 in the linear regression function.

b. Repeat part (a) 200 times, generating new random numbers each time.

c. Calculate the mean and variance of the 200 ordinary least squares estimates of β_1, and do the same for the 200 weighted least squares estimates.

d. Do both the ordinary least squares and weighted least squares estimators appear to be unbiased? Explain. Which estimator appears to be more precise here? Comment.

Chapter 12

Building the Regression Model

In earlier chapters, we considered how to fit simple and multiple regression models, how to make inferences from these models, and how to diagnose a variety of conditions that affect the appropriateness of the fitted regression model.

For pedagogic reasons, we considered these topics in isolation. Now we need to examine how they are interrelated in the process of building a regression model. In this chapter, we first present an overview of the model-building process. Then we consider each major step of the process in more detail. In doing so, we shall take up a few additional procedures. One new set of procedures involves computerized techniques that are helpful in identifying the independent variables to be included in the regression model. Also, we shall present several methods for validating a regression model once it has been developed.

Throughout this chapter, we shall use the same example to illustrate each of the steps of the model-building process, culminating in a validated regression model.

12.1 OVERVIEW OF MODEL-BUILDING PROCESS

At the risk of oversimplifying, we present in Figure 12.1 a strategy for the building of a regression model. This strategy involves four phases:

1. Data collection and preparation
2. Reduction of the number of independent variables
3. Model refinement and selection
4. Model validation

We take up each of these phases in turn now.

Data Collection and Preparation

In some fields, theory can aid in selecting the independent variables to be employed. Often in these fields, controlled experiments can be undertaken to furnish data on the basis of which the regression parameters can be estimated and the theoretical form of the regression function tested.

433

FIGURE 12.1   Strategy for Building a Regression Model

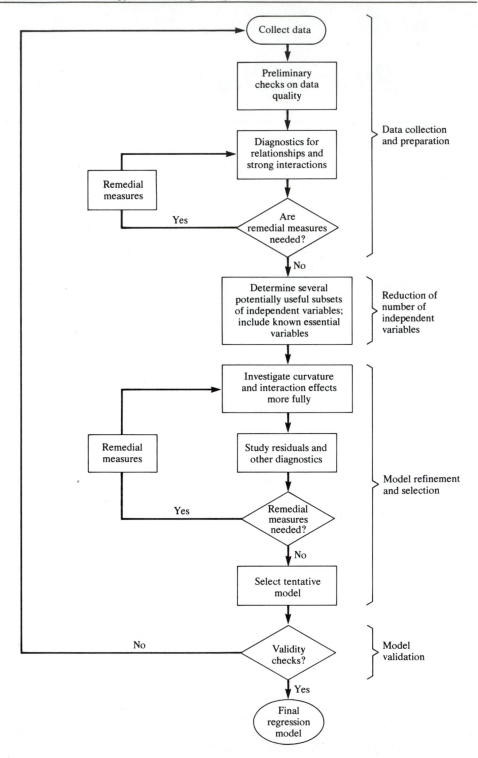

In many other subject matter fields, however, including the social, behavioral and health sciences, and management, serviceable theoretical models are relatively rare. To complicate matters further, the available theoretical models may involve independent variables that are not directly measurable, such as a family's future earnings over the next ten years. Under these conditions, investigators are often forced to prospect for independent variables that could conceivably be related to the dependent variable under study. Obviously, such a set of potentially useful independent variables can be large. For example, a company's sales of portable dishwashers in a district may be affected by population size, per capita income, percent of population in urban areas, percent of population under 50 years of age, percent of families with children at home, etc., etc.!

After a lengthy list of potentially useful independent variables has been compiled, some of the independent variables can be screened out. An independent variable (1) may not be fundamental to the problem, (2) may be subject to large measurement errors, and/or (3) may effectively duplicate another independent variable in the list. Independent variables that cannot be measured may either be deleted or replaced by proxy variables that are highly correlated with them.

The number of cases to be collected for a regression study depends on the size of the pool of independent variables at this stage. When the pool of potentially useful independent variables is large, more cases are required than when the pool is small. A general rule of thumb states that there should be at least 6 to 10 cases for every variable in the pool. This need to have more cases for larger pools of independent variables also affects the model-building strategy. For example, when there is a large pool of potential independent variables, it may be difficult to investigate all two-variable interactions, whereas this might be feasible when the pool is small.

Once the data have been collected, edit checks and plots should be performed to identify gross data errors as well as extreme outliers. Difficulties with data errors are especially prevalent in large data sets and should be corrected or resolved before the model building begins. Whenever possible, the investigator should carefully monitor and control the data collection process to reduce the likelihood of data errors.

Once the data have been properly edited, the formal modeling process can begin. A variety of diagnostics should be employed to identify important independent variables, the functional forms in which the independent variables should enter the regression model, and important interactions. Scatter plots are useful for determining relationships and their strengths. Selected independent variables can be fitted in regression functions to explore relationships, possible strong interactions, and transformations. Residual plots and partial regression plots should play a major role here. Whenever possible, of course, one should also rely on the investigator's prior knowledge and expertise to suggest appropriate transformations and interactions to investigate. All of the diagnostic procedures explained in previous chapters should be used as resources in this phase of model building.

Reduction of Number of Independent Variables

Once the investigator has tentatively decided upon the functional forms of the regression relations (whether given variables are to appear in linear form, quadratic form, etc.) and whether any interaction terms are to be included, the next step is to select a few "good" subsets of X variables. These subsets should include not only the potential independent variables in first-order form but also any needed quadratic and other curvature terms and any necessary interaction terms.

The reason for focusing on subsets of the pool of independent variables is that the number of independent variables that remain after the initial screening typically is still large. Further, many of these variables frequently will be highly intercorrelated. Hence, the investigator usually will wish to reduce the number of independent variables to be used in the final model. There are several reasons for this. A regression model with a large number of independent variables is difficult to maintain. Further, regression models with a limited number of independent variables are easier to work with and understand. Finally, the presence of many highly intercorrelated independent variables may add little to the predictive power of the model while substantially increasing the sampling variation of the regression coefficients, detracting from the model's descriptive abilities, and increasing the problem of roundoff errors (as we have noted in Chapter 8).

The selection of "good" subsets of potentially useful independent variables to be included in the final regression model and the determination of appropriate functional and interaction relations for these variables are usually some of the most difficult problems in regression analysis. Since the uses of regression models vary, no one subset of independent variables is usually always "best." For instance, a descriptive use of a regression model typically will emphasize precise estimation of the regression coefficients, while a predictive use will focus on the prediction errors. Often, different subsets of the pool of potential independent variables will best serve these varying purposes. Even for a given purpose, it is often found that several subsets are about equally "good" according to a given criterion, and the choice among these "good" subsets needs to be made on the basis of additional considerations.

The choice of a few appropriate subsets of independent variables for final consideration needs to be done with particular care for observational data. With such data, elimination of key explanatory variables can seriously damage the explanatory power of the model and lead to biased estimates of regression coefficients, mean responses, and predictions of new observations, as well as biased estimates of the error variance. The bias in these estimates is related to the fact that with observational data, the error terms in an underfitted regression model may reflect nonrandom effects of the independent variables not incorporated in the regression model. Important omitted independent variables are sometimes called *latent predictor variables*.

On the other hand, if too many independent variables are included in the subset, then this overfitted model will often result in variances of estimated parameters that are larger than those for simpler models.

Another danger when the data are observational is that important independent variables may be observed only over narrow ranges. As a result, such important independent variables may be omitted just because they occur in the sample within a narrow range of values and therefore turn out to be statistically nonsignificant.

Another consideration in selecting subsets of independent variables is that these subsets need to be small enough so that maintenance costs are manageable and analysis is facilitated, yet large enough so that adequate description, control, or prediction is possible.

A variety of computerized approaches have been developed to assist the investigator in reducing the number of independent variables to be considered for the regression model when these variables are correlated among themselves. We shall present two of these approaches in this chapter. The first, which is practical for pools of independent variables that are small or moderate in size, considers all possible regression models that can be developed from the pool of potential independent variables and identifies subsets of the independent variables that are "good" according to a criterion specified by the investigator. The second approach employs automatic search procedures to arrive at a single subset of the independent variables. This approach is recommended primarily for reductions involving large pools of independent variables.

While computerized approaches can be very helpful in identifying appropriate subsets for detailed, final consideration, the process of developing a useful regression model must be pragmatic and needs to utilize large doses of subjective judgment. Independent variables that are considered essential should be included in the regression model before any computerized assistance is sought. Further, computerized approaches that provide only a single subset of independent variables as "best" need to be supplemented so that additional subsets are also considered before deciding upon the final regression model.

Comments

1. Carefully designed experiments usually eliminate many of the problems connected with choosing "good" subsets of the independent variables. For example, the effects of latent predictor variables are minimized by using randomization. In addition, adequate ranges of the predictor variables can be selected and correlations among the predictor variables can be eliminated by appropriate choices of their levels. Unfortunately, many regression analyses must be based on observational data where these controls are not possible.

2. All too often, unwary investigators will screen a set of independent variables by fitting the regression model containing the entire set of potential X variables and then simply dropping those for which the $t*$ statistic (8.23):

$$t_k^* = \frac{b_k}{s\{b_k\}}$$

has a small absolute value. As we know from Chapter 8, this procedure can lead to the dropping of important intercorrelated independent variables. Clearly, a good search procedure must be able to handle important intercorrelated independent variables in such a way that not all of them will be dropped.

Model Refinement and Selection

After successfully reducing the number of independent variables, the investigator usually arrives at a small number of potential "good" regression models, each of which contains those independent variables that are known to be essential. At this stage, more detailed checks of curvature and interaction effects are desirable. Residual plots and partial regression plots are helpful in deciding whether one model is to be preferred over another. In addition, the diagnostic checks described in Chapter 11 are useful to identify influential outlying observations, multicollinearity, etc.

Eventually after thorough checking and various remedial actions, such as transformations, the investigator decides to select one regression model as best. At this point, it is good statistical practice to validate the model.

Model Validation

Model validity refers to the stability and reasonableness of the regression coefficients, the plausibility and usability of the regression function, and the ability to generalize inferences drawn from the regression analysis. Validation is a useful and necessary part of the model-building process. Several methods of assessing model validity will be described later in this chapter.

12.2 DATA PREPARATION

In order to illustrate the model-building procedures discussed in this chapter, we shall use a relatively simple example that has four potential independent variables. By limiting the number of potential independent variables, we shall be able to explain the procedures without overwhelming the reader with masses of computer printouts.

We mentioned earlier some of the data preparation steps called for at the beginning of the model-building process. We illustrate these now in terms of the surgical unit example.

Example

A hospital surgical unit was interested in predicting survival in patients undergoing a particular type of liver operation. A random selection of 54 patients was available for analysis. From each patient record, the following information was extracted from the preoperation evaluation:

X_1 blood clotting score
X_2 prognostic index, which includes the age of patient
X_3 enzyme function test score
X_4 liver function test score

These constitute the potential independent variables for a predictive regression model. The dependent variable is survival time, which was ascertained in a follow-up study. The data on the potential independent variables and the dependent variable are presented in Table 12.1. These data have already been screened and properly edited for errors.

Since the pool of independent variables is small, a reasonably full exploration of relationships and possible strong interaction effects is possible at this stage of data

TABLE 12.1   Potential Independent Variables and Dependent Variable—Surgical Unit Example

Case Number i	Blood Clotting Score X_{i1}	Prognostic Index X_{i2}	Enzyme Function Test X_{i3}	Liver Function Test X_{i4}	Survival Time Y_i	$Y_i' = \log_{10} Y_i$
1	6.7	62	81	2.59	200	2.3010
2	5.1	59	66	1.70	101	2.0043
3	7.4	57	83	2.16	204	2.3096
4	6.5	73	41	2.01	101	2.0043
5	7.8	65	115	4.30	509	2.7067
6	5.8	38	72	1.42	80	1.9031
7	5.7	46	63	1.91	80	1.9031
8	3.7	68	81	2.57	127	2.1038
9	6.0	67	93	2.50	202	2.3054
10	3.7	76	94	2.40	203	2.3075
11	6.3	84	83	4.13	329	2.5172
12	6.7	51	43	1.86	65	1.8129
13	5.8	96	114	3.95	830	2.9191
14	5.8	83	88	3.95	330	2.5185
15	7.7	62	67	3.40	168	2.2253
16	7.4	74	68	2.40	217	2.3365
17	6.0	85	28	2.98	87	1.9395
18	3.7	51	41	1.55	34	1.5315
19	7.3	68	74	3.56	215	2.3324
20	5.6	57	87	3.02	172	2.2355
21	5.2	52	76	2.85	109	2.0374
22	3.4	83	53	1.12	136	2.1335
23	6.7	26	68	2.10	70	1.8451
24	5.8	67	86	3.40	220	2.3424
25	6.3	59	100	2.95	276	2.4409
26	5.8	61	73	3.50	144	2.1584
27	5.2	52	86	2.45	181	2.2577
28	11.2	76	90	5.59	574	2.7589
29	5.2	54	56	2.71	72	1.8573
30	5.8	76	59	2.58	178	2.2504
31	3.2	64	65	.74	71	1.8513
32	8.7	45	23	2.52	58	1.7634
33	5.0	59	73	3.50	116	2.0645
34	5.8	72	93	3.30	295	2.4698
35	5.4	58	70	2.64	115	2.0607
36	5.3	51	99	2.60	184	2.2648
37	2.6	74	86	2.05	118	2.0719
38	4.3	8	119	2.85	120	2.0792
39	4.8	61	76	2.45	151	2.1790
40	5.4	52	88	1.81	148	2.1703
41	5.2	49	72	1.84	95	1.9777
42	3.6	28	99	1.30	75	1.8751
43	8.8	86	88	6.40	483	2.6840

TABLE 12.1 *(concluded)*

Case Number i	Blood Clotting Score X_{i1}	Prognostic Index X_{i2}	Enzyme Function Test X_{i3}	Liver Function Test X_{i4}	Survival Time Y_i	$Y_i' = \log_{10} Y_i$
44	6.5	56	77	2.85	153	2.1847
45	3.4	77	93	1.48	191	2.2810
46	6.5	40	84	3.00	123	2.0899
47	4.5	73	106	3.05	311	2.4928
48	4.8	86	101	4.10	398	2.5999
49	5.1	67	77	2.86	158	2.1987
50	3.9	82	103	4.55	310	2.4914
51	6.6	77	46	1.95	124	2.0934
52	6.4	85	40	1.21	125	2.0969
53	6.4	59	85	2.33	198	2.2967
54	8.8	78	72	3.20	313	2.4955

preparation. Stem-and-leaf plots were prepared for each of the independent variables (not shown). These highlighted case 28 as an outlying X_1 observation, case 38 as an outlying X_2 observation, cases 17 and 32 as outlying X_3 observations, and cases 28, 31, and 43 as outlying X_4 observations. The investigator was thereby alerted to examine later the influence of these cases.

A first-order regression model based on all independent variables was fitted. The normal probability plot of the residuals for this fitted model is shown in Figure 12.2a. It suggests some departure from normality. The coefficient of correlation between the ordered residuals and their expected values under normality is only .826; reference to Table 4.3 supports the conclusion of the error terms not being normally distributed.

Various other first-order models were fitted, and many residual plots were prepared. One plot of particular interest was the plot of residuals from the fit of Y on X_2 and X_3 against the interaction term $X_2 X_3$; this plot is shown in Figure 12.2b. Note the suggested strong interaction effects in this plot.

To make the distribution of the error terms more nearly normal and to see if the same transformation would also reduce the $X_2 X_3$ interaction effect, the investigator examined the logarithmic transformation $Y' = \log_{10} Y$. The data for the transformed dependent variable are given in Table 12.1. Figure 12.2c shows the normal probability plot of the residuals when Y' is regressed on all independent variables in a first-order model; the coefficient of correlation between the ordered residuals and their expected values under normality now is .959. It is clear that the transformation was helpful in making the distribution of the error terms more normal. Further, the transformation was also helpful in reducing the interaction effect between X_2 and X_3. Figure 12.2d shows a plot of the residuals, when Y' is regressed on X_2 and X_3, against $X_2 X_3$. This residual plot shows much less indication of a strong interaction effect. None of the other residual plots against interaction terms indicated a presence of strong interaction effects either.

FIGURE 12.2   Some Preliminary Residual Plots for Surgical Unit Example

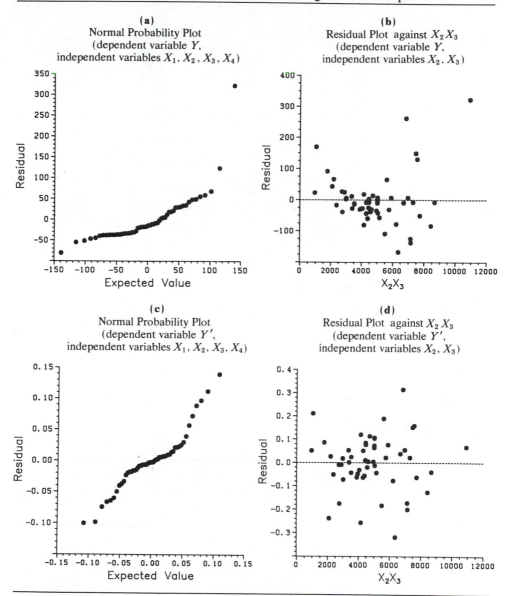

The investigator also obtained the correlation matrix with the transformed Y variable; this is presented in Table 12.2, omitting the duplicate terms. In addition, scatter plots of Y' against each independent variable and scatter plots for each pair of independent variables were obtained. Figure 12.3 is illustrative, showing the scatter plot of Y' against X_1.

Table 12.2 and the various scatter, residual, and partial regression plots (not shown here, other than the scatter plot in Figure 12.3) all indicate that each of the

TABLE 12.2   Correlation Matrix for Surgical Unit Example

	Y'	X_1	X_2	X_3	X_4
Y'	1.000	.346	.593	.665	.726
X_1		1.000	.090	−.150	.502
X_2			1.000	−.024	.369
X_3				1.000	.416
X_4					1.000

independent variables is linearly associated with Y', X_4 showing the highest degree of association and X_1 the lowest. The correlation matrix further shows intercorrelations among the potential independent variables. In particular, X_4 has moderately high pairwise correlation with X_1, X_2, and X_3.

On the basis of these analyses, the investigator concluded to use at this stage of the model-building process $Y' = \log_{10} Y$ as the dependent variable, to represent the independent variables in linear terms, and not to include any interaction terms.

FIGURE 12.3   Scatter Plot of Y' against X_1—Surgical Unit Example

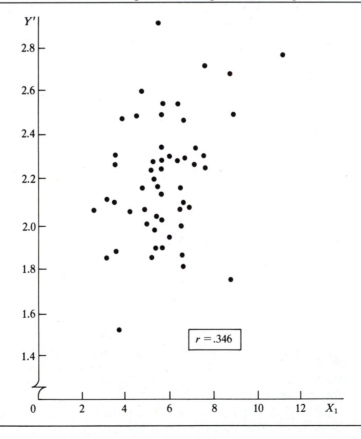

12.3 ALL-POSSIBLE-REGRESSIONS PROCEDURE FOR VARIABLES REDUCTION

The *all-possible-regressions selection procedure* calls for a consideration of all possible regression models involving the potential X variables and identifying a few "good" subsets according to some criterion. When this selection procedure is used with the surgical unit example, for instance, 16 different regression models are to be considered, as shown in Table 12.3. First, there is the regression model with no X variables, i.e., the model $Y_i = \beta_0 + \varepsilon_i$. Then there are the regression models with one X variable (X_1, X_2, X_3, X_4), with two X variables (X_1 and X_2, X_1 and X_3, X_1 and X_4, X_2 and X_3, X_2 and X_4, X_3 and X_4), and so on.

The purpose of the all-possible-regressions approach is to identify a small group of regression models that are "good" according to a specified criterion so that a detailed examination can be made of these models, leading to the selection of the final regression model to be employed. In most circumstances, it would be impossible for an analyst to make a detailed examination of all possible regression models. For instance, when there are 10 potential independent variables in the pool, there would be $2^{10} = 1,024$ possible regression models. (This calculation is based on the fact that each potential independent variable either can be included or excluded.) With the availability of high-speed computers today, running all possible regression models for 10 potential X variables is not too time consuming. Still, the sheer volume of 1,024 candidate models to examine would be an overwhelming task for a data analyst.

Hence, unless the pool of potential X variables is very small, the investigator can concentrate only on a few of all of the possible regression models. This limited number might consist of 5 or 10 "good" subsets according to the criterion specified, so that the investigator can then carefully study these regression models for choosing the final model to be employed.

Different criteria for comparing the regression models may be used with the all-possible-regressions selection procedure. We shall discuss four—R_p^2, MSE_p, C_p, and $PRESS_p$. Before doing so, we need to develop some notation. Let us denote the number of potential X variables in the pool by $P - 1$. We assume throughout this chapter that all regression models contain an intercept term β_0. Hence, the regression function containing all potential X variables contains P parameters, and the function with no X variables contains one parameter (β_0).

The number of X variables in a subset will be denoted by $p - 1$, as always, so that there are p parameters in the regression function for this subset of X variables. Thus, we have:

(12.1)
$$1 \le p \le P$$

The all-possible-regressions approach assumes that the number of observations n exceeds the maximum number of potential parameters:

(12.2)
$$n > P$$

and, indeed, it is highly desirable that n be substantially larger than P, as we noted earlier, so that sound results can be obtained.

R_p^2 Criterion

The R_p^2 criterion calls for an examination of the coefficient of multiple determination R^2, defined in (7.35), in order to select several "good" subsets of X variables. We show the number of parameters in the regression model as a subscript of R^2. Thus R_p^2 indicates that there are p parameters, or $p - 1$ predictor variables, in the regression function on which R_p^2 is based.

Since R_p^2 is a ratio of sums of squares:

$$(12.3) \qquad R_p^2 = \frac{SSR_p}{SSTO} = 1 - \frac{SSE_p}{SSTO}$$

and the denominator is constant for all possible regressions, R_p^2 varies inversely with the error sums of squares SSE_p. But we know that SSE_p can never increase as additional X variables are included in the model. Thus, R_p^2 will be a maximum when all $P - 1$ potential X variables are included in the regression model. The reason for using the R_p^2 criterion with the all-possible-regressions approach therefore cannot be to maximize R_p^2. Rather, the intent is to find the point where adding more X variables is not worthwhile because it leads to a very small increase in R_p^2. Often, this point is reached when only a limited number of X variables is included in the regression model. Clearly, the determination of where diminishing returns set in is a judgmental one.

Example. Table 12.3 shows, for the surgical unit example, in columns 1, 2, and 3,

TABLE 12.3 R_p^2, MSE_p, C_p, and $PRESS_p$ Values for all Possible Regression Models—Surgical Unit Example

X Variables in Model	(1) p	(2) df	(3) SSE_p	(4) R_p^2	(5) MSE_p	(6) C_p	(7) $PRESS_p$
None	1	53	3.9728	0	.0750	1,721.6	4.1241
X_1	2	52	3.4961	.120	.0672	1,510.8	3.8084
X_2	2	52	2.5763	.352	.0495	1,100.1	2.8627
X_3	2	52	2.2153	.442	.0426	939.0	2.4268
X_4	2	52	1.8776	.527	.0361	788.2	2.0292
X_1, X_2	3	51	2.2325	.438	.0438	948.7	2.6388
X_1, X_3	3	51	1.4072	.646	.0276	580.2	1.6095
X_1, X_4	3	51	1.8758	.528	.0368	789.4	2.1203
X_2, X_3	3	51	.7430	.813	.0146	283.7	.8352
X_2, X_4	3	51	1.3922	.650	.0273	573.5	1.5833
X_3, X_4	3	51	1.2453	.687	.0244	507.9	1.4287
X_1, X_2, X_3	4	50	.1099	.972	.00220	3.1	.1405
X_1, X_2, X_4	4	50	1.3905	.650	.0278	574.8	1.6513
X_1, X_3, X_4	4	50	1.1156	.719	.0223	452.0	1.3286
X_2, X_3, X_4	4	50	.4652	.883	.00930	161.7	.5487
X_1, X_2, X_3, X_4	5	49	.1098	.972	.00224	5.0	.1456

respectively, the number of parameters in the regression function, the degrees of freedom associated with the error sum of squares, and the error sum of squares for each possible regression model. In column 4 are given the R_p^2 values. The results were obtained from a series of computer runs. For instance, when X_4 is the only X variable in the regression model, we obtain:

$$R_2^2 = 1 - \frac{SSE(X_4)}{SSTO} = 1 - \frac{1.8776}{3.9728} = .527$$

Note that $SSTO = SSE_1 = 3.9728$.

The R_p^2 values are plotted in Figure 12.4. The maximum R_p^2 value for the possible subsets of $p - 1$ predictor variables, denoted by $\max(R_p^2)$, appears at the top of the graph for each p. These points are connected by dashed lines to show the impact of adding additional X variables. Figure 12.4 makes it clear that little increase in $\max(R_p^2)$ takes place after three X variables are included in the model. Hence, the

FIGURE 12.4 R_p^2 Plot for Surgical Unit Example

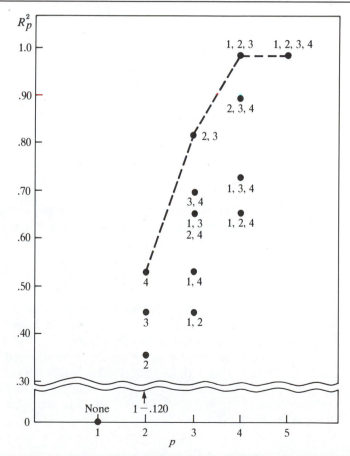

use of the subset (X_1, X_2, X_3) in the regression model appears to be reasonable according to the R_p^2 criterion.

Note that variable X_4, which singly correlates most highly with the dependent variable, is not in the $\max(R_p^2)$ models for $p = 3$ and $p = 4$, indicating that X_1, X_2, and X_3 contain much of the information presented by X_4. If it were desired that X_4 be retained in the model and that the subset model be limited to three X variables, the subset (X_2, X_3, X_4) should then be considered as next best according to the R_p^2 criterion for $p = 4$. The coefficient of multiple determination associated with this subset, $R_4^2 = .883$, would be somewhat smaller than $R_4^2 = .972$ for the subset (X_1, X_2, X_3).

MSE_p or R_a^2 Criterion

Since R_p^2 does not take account of the number of parameters in the regression model, and since $\max(R_p^2)$ can never decrease as p increases, the use of the adjusted coefficient of multiple determination R_a^2 in (7.37):

$$(12.4) \qquad R_a^2 = 1 - \left(\frac{n-1}{n-p}\right)\frac{SSE}{SSTO} = 1 - \frac{MSE}{\dfrac{SSTO}{n-1}}$$

has been suggested as a criterion which takes the number of parameters in the model into account through the degrees of freedom. It can be seen from (12.4) that R_a^2 increases if and only if MSE decreases since $SSTO/(n - 1)$ is fixed for the given Y observations. Hence, R_a^2 and MSE are equivalent criteria. We shall consider here the criterion MSE_p. $\text{Min}(MSE_p)$ can, indeed, increase as p increases when the reduction in SSE_p becomes so small that it is not sufficient to offset the loss of an additional degree of freedom. Users of the MSE_p criterion seek to find a few subsets for which MSE_p is at the minimum or so close to the minimum that adding more variables is not worthwhile.

Example.   The MSE_p values for all possible regression models for the surgical unit example are shown in Table 12.3, column 5. For instance, if the regression model contains only X_4, we have:

$$MSE_2 = \frac{SSE(X_4)}{n-2} = \frac{1.8776}{52} = .0361$$

Figure 12.5 contains the MSE_p plot for the surgical unit example. We have connected the $\min(MSE_p)$ values for each p by dashed lines. The story which Figure 12.5 tells is very similar to that told by Figure 12.4. The subset (X_1, X_2, X_3) appears to be reasonable. Indeed, the mean square error achieved with this subset is practically the same as that with (X_1, X_2, X_3, X_4), which uses all potential X variables.

If X_4 were to be included in the model with $p = 4$, the subset (X_2, X_3, X_4) should be considered, involving $MSE_4 = .009$ which is somewhat higher than $MSE_4 = .002$ for subset (X_1, X_2, X_3).

FIGURE 12.5 MSE_p Plot for Surgical Unit Example

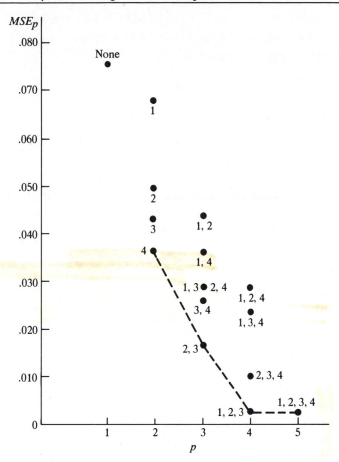

C_p Criterion

This criterion is concerned with the *total mean squared error* of the n fitted values for each subset regression model. The mean squared error concept involves a bias component and a random error component. The mean squared error for an estimated regression coefficient was defined in (11.40). Here, the mean squared error pertains to the fitted values $\hat{Y}_i$ for the regression model employed. The bias component for the ith fitted value $\hat{Y}_i$ is:

$$(12.5) \qquad\qquad E\{\hat{Y}_i\} - \mu_i$$

where $E\{\hat{Y}_i\}$ is the expectation of the ith fitted value for the given regression model and μ_i is the true mean response. The random error component for $\hat{Y}_i$ is simply

$\sigma^2\{\hat{Y}_i\}$, its variance. The mean squared error for $\hat{Y}_i$ is then the sum of the squared bias and the variance:

$$(12.6) \qquad (E\{\hat{Y}_i\} - \mu_i)^2 + \sigma^2\{\hat{Y}_i\}$$

The total mean squared error for all n fitted values $\hat{Y}_i$ is the sum of the n individual mean squared errors in (12.6):

$$(12.7) \qquad \sum_{i=1}^{n} [(E\{\hat{Y}_i\} - \mu_i)^2 + \sigma^2\{\hat{Y}_i\}] = \sum_{i=1}^{n} (E\{\hat{Y}_i\} - \mu_i)^2 + \sum_{i=1}^{n} \sigma^2\{\hat{Y}_i\}$$

The criterion measure, denoted by Γ_p, is simply the total mean squared error in (12.7) divided by σ^2, the true error variance:

$$(12.8) \qquad \Gamma_p = \frac{1}{\sigma^2}\left[\sum_{i=1}^{n} (E\{\hat{Y}_i\} - \mu_i)^2 + \sum_{i=1}^{n} \sigma^2\{\hat{Y}_i\}\right]$$

The model which includes all $P - 1$ potential X variables is assumed to have been carefully chosen so that $MSE(X_1, \ldots, X_{P-1})$ is an unbiased estimator of σ^2. It can then be shown that an estimator of Γ_p is C_p:

$$(12.9) \qquad C_p = \frac{SSE_p}{MSE(X_1, \ldots, X_{P-1})} - (n - 2p)$$

where SSE_p is the error sum of squares for the fitted subset regression model with p parameters (i.e., with $p - 1$ predictor variables).

When there is no bias in the regression model with $p - 1$ predictor variables so that $E\{\hat{Y}_i\} \equiv \mu_i$, the expected value of C_p is approximately p:

$$(12.10) \qquad E\{C_p \mid E\{\hat{Y}_i\} \equiv \mu_i\} \simeq p$$

Thus, when the C_p values for all possible regression models are plotted against p, those models with little bias will tend to fall near the line $C_p = p$. Models with substantial bias will tend to fall considerably above this line. C_p values below the line $C_p = p$ are interpreted as showing no bias; that is, they are below the line due to sampling error.

In using the C_p criterion, one seeks to identify subsets of X variables for which (1) the C_p value is small and (2) the C_p value is near p. Sets of X variables with small C_p values have a small total mean squared error, and when the C_p value is also near p, the bias of the regression model is small. It may sometimes occur that the regression model based on the subset of X variables with the smallest C_p value involves substantial bias. In that case, one may at times prefer a regression model based on a somewhat larger subset of X variables for which the C_p value is slightly larger but which does not involve a substantial bias component. Reference 12.1 contains extended discussions of applications of the C_p criterion.

Example.   Table 12.3, column 6, contains the C_p values for all possible regression models for the surgical unit example. For instance, when X_4 is the only X variable in the regression model, the C_p value is:

$$C_2 = \frac{SSE(X_4)}{MSE(X_1, X_2, X_3, X_4)} - [n - 2(2)]$$

$$= \frac{1.8776}{.00224} - (54 - 4) = 788.2$$

The C_p values for all possible regression models are plotted in Figure 12.6. We find again that subset (X_1, X_2, X_3) is suggested. This subset has the smallest C_p value, with no indication of any bias in the regression model. The fact that the C_p measure for this model, $C_4 = 3.1$, is below $p = 4$ is, as stated earlier, the result of random variation in the C_p estimate.

Note that use of all potential X variables (X_1, X_2, X_3, X_4) would involve a larger total mean squared error. Also, use of subset (X_2, X_3, X_4) with C_p value $C_4 = 161.7$ would be poor because of the substantial bias with that model.

Comments

1. Effective use of the C_p criterion requires careful development of the pool of $P - 1$ potential X variables, with the independent variables expressed in appropriate form (linear,

---

FIGURE 12.6 C_p Plot for Surgical Unit Example

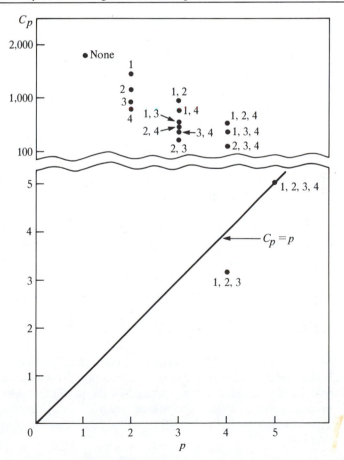

quadratic, transformed) and useless variables excluded so that $MSE(X_1, \ldots, X_{P-1})$ provides an unbiased estimate of the error variance σ^2.

2. The C_p criterion places major emphasis on the fit of the subset model for the n sample observations. At times, a modification of the C_p criterion that emphasizes new observations to be predicted may be preferable.

3. To see why C_p as defined in (12.9) is an estimator of Γ_p, we need to utilize two results that we shall simply state. First, it can be shown that:

$$(12.11) \qquad \sum_{i=1}^{n} \sigma^2\{\hat{Y}_i\} = p\sigma^2$$

Thus, the total random error of the n fitted values $\hat{Y}_i$ increases as the number of variables in the regression model increases.

Further, it can be shown that:

$$(12.12) \qquad E\{SSE_p\} = \sum (E\{\hat{Y}_i\} - \mu_i)^2 + (n - p)\sigma^2$$

Hence, Γ_p in (12.8) can be expressed as follows:

$$(12.13) \qquad \Gamma_p = \frac{1}{\sigma^2}[E\{SSE_p\} - (n - p)\sigma^2 + p\sigma^2]$$

$$= \frac{E\{SSE_p\}}{\sigma^2} - (n - 2p)$$

Replacing $E\{SSE_p\}$ by the estimator SSE_p and using $MSE(X_1, \ldots, X_{P-1})$ as an estimator of σ^2 yields C_p in (12.9).

PRESS$_p$ Criterion

The *PRESS* (prediction sum of squares) selection criterion is based on the deleted residuals d_i, defined in (11.20):

$$(12.14) \qquad d_i = Y_i - \hat{Y}_{i(i)}$$

where $\hat{Y}_{i(i)}$ is the predicted value for the ith case when the regression function is fitted without the ith case.

Each regression model has n deleted residuals associated with it, and the *PRESS* criterion is simply the sum of the squares of these deleted residuals:

$$(12.15) \qquad PRESS_p = \sum_{i=1}^{n} d_i^2 = \sum_{i=1}^{n} (Y_i - \hat{Y}_{i(i)})^2$$

Models with small *PRESS* values are considered good candidate models. The reason is that the deleted residual d_i is the prediction error when a regression model is fitted without the ith case and $\hat{Y}_{i(i)}$ is used as the predicted value. Small prediction errors involve small d_i values and hence small d_i^2 values and a small sum of d_i^2 values. Thus, models with small *PRESS* values fit well in the sense of having small prediction errors.

PRESS values can be calculated without repeated regression runs for each subset. Using the equivalent expression for d_i in (11.20a):

$$d_i = \frac{e_i}{1 - h_{ii}}$$

PRESS$_p$ becomes:

(12.16)
$$PRESS_p = \sum_{i=1}^{n} \left(\frac{e_i}{1 - h_{ii}}\right)^2$$

where e_i is the ordinary residual and h_{ii} is the leverage value in (11.13), both based on all n cases.

Example.   Table 12.3, column 7, contains the *PRESS* values for all possible regression models for the surgical unit example. The *PRESS* values are plotted in Figure 12.7. We have connected the min(*PRESS$_p$*) values by dashed lines. The message which Figure 12.7 tells is very similar to that told by the other criteria. We find that subset (X_1, X_2, X_3) is suggested as reasonable, having the smallest *PRESS* value. Use of all potential X variables (X_1, X_2, X_3, X_4) would involve a slightly larger *PRESS* value. If X_4 were to be included in the model with $p = 4$, the subset (X_2, X_3, X_4) would be suggested, involving a *PRESS* value of .5487, which is moderately larger than the *PRESS* value of .1405 for subset (X_1, X_2, X_3).

FIGURE 12.7   *PRESS$_p$* Plot for Surgical Unit Example

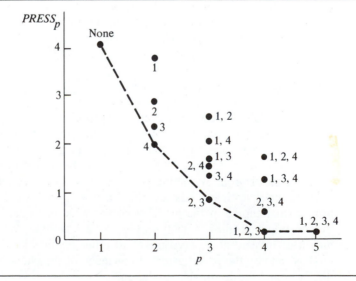

Note

PRESS values can also be useful for model validation, as will be explained in Section 12.6.

"Best" Subsets Algorithms

Time-saving algorithms have been developed in which the "best" subsets according to a specified criterion are identified without requiring the fitting of all of the possible subset regression models. In fact, these algorithms require the calculation of only a small fraction of all possible regression models. Thus, if the C_p criterion is to be employed and the five "best" subsets according to this criterion are to be identified, these algorithms search for the five subsets of X variables with the smallest C_p values using much less computational effort than when all possible subsets are evaluated. These algorithms are called *"best" subsets algorithms*. Not only do these algorithms provide the best subsets according to the specified criterion, but they often will also provide a number of "good" subsets for each possible number of X variables in the model to give the investigator additional helpful information in making the final selection of the subset of X variables to be employed in the regression model.

When the pool of potential independent variables is very large, say greater than 40 to 60, even the "best" subsets algorithms may require excessive computer time. Under these conditions, one of the automatic selection procedures described in Section 12.4 may need to be employed to assist in the selection of independent variables.

Example.   In the surgical unit example, use of one of the "best" subsets algorithms will provide a portion of the information in Table 12.3. When the C_p criterion is to be employed and the "best" three subsets are to be identified, the algorithm will identify subsets (X_1, X_2, X_3), (X_1, X_2, X_3, X_4), and (X_2, X_3, X_4) as the three subsets with the smallest C_p values. In addition, information about three "good" subsets for each level of p may be provided.

Some Final Comments

The all-possible-regressions selection approach leads to the identification of a small number of subsets that are "good" according to a specified criterion. While in the surgical unit example, each of the three criteria pointed to the same "best" subset, this is not always the case. It therefore may be desirable at times to consider more than one criterion in evaluating possible subsets of X variables.

Once the investigator has identified a few "good" subsets for intensive examination, a final choice of the model variables must be made. This choice, as indicated by our model-building strategy in Figure 12.1, is aided by residual analyses, examination of influential observations, and other diagnostics for each of the competing models, and by the investigator's knowledge of the phenomenon under study, and is finally confirmed by model validation.

12.4 FORWARD STEPWISE REGRESSION, OTHER AUTOMATIC SEARCH PROCEDURES, AND USE OF RIDGE REGRESSION FOR VARIABLES REDUCTION

In those occasional cases when the pool of potential X variables contains 40 to 60 or even more variables, use of a "best" subsets algorithm may not be feasible. An automatic search procedure that develops sequentially the subset of X variables to be included in the regression model may be helpful in those cases. The *forward stepwise regression procedure* is probably the most widely used of the automatic search methods. It was developed to economize on computational efforts, as compared with the all-possible-regressions approach, while arriving at the "best" subset of independent variables. Essentially, this search method develops a sequence of regression models, at each step adding or deleting an X variable. The criterion for adding or deleting an X variable can be stated equivalently in terms of error sum of squares reduction, coefficient of partial correlation, or F^* statistic.

An essential difference between automatic search procedures and the all-possible-regressions approach is that the automatic search procedures end with the identification of a *single* regression model as "best." With the all-possible-regressions approach, on the other hand, *several* regression models can be identified as "good" for final consideration.

Forward Stepwise Regression

We shall describe the forward stepwise regression search algorithm in terms of the F^* statistic for the partial F test.

1. The stepwise regression routine first fits a simple linear regression model for each of the $P - 1$ potential X variables. For each simple linear regression model, the F^* statistic (3.59) for testing whether or not the slope is zero is obtained:

$$(12.17) \qquad F_k^* = \frac{MSR(X_k)}{MSE(X_k)}$$

Recall that $MSR(X_k) = SSR(X_k)$ measures the reduction in the total variation of Y associated with the use of the variable X_k. The X variable with the largest F^* value is the candidate for first addition. If this F^* value exceeds a predetermined level, the X variable is added. Otherwise, the program terminates with no X variable considered sufficiently helpful to enter the regression model.

2. Assume X_7 is the variable entered at step 1. The stepwise regression routine now fits all regression models with two X variables, where X_7 is one of the pair. For each such regression model, the partial F test statistic (8.22):

$$(12.18) \qquad F_k^* = \frac{MSR(X_k \mid X_7)}{MSE(X_7, X_k)} = \left(\frac{b_k}{s\{b_k\}}\right)^2$$

is obtained. This is the statistic for testing whether or not $\beta_k = 0$ when X_7 and X_k are the variables in the model. The X variable with the largest F^* value is the candidate for addition at the second stage. If this F^* value exceeds a predetermined level, the second X variable is added. Otherwise the program terminates.

3. Suppose X_3 is added at the second stage. Now the stepwise regression routine examines whether any of the other X variables already in the model should be dropped. For our illustration, there is at this stage only one other X variable in the model, X_7, so that only one partial F test statistic is obtained:

$$(12.19) \qquad\qquad F_7^* = \frac{MSR(X_7 \mid X_3)}{MSE(X_3, X_7)}$$

At later stages, there would be a number of these F^* statistics, for each of the variables in the model besides the one last added. The variable for which this F^* value is smallest is the candidate for deletion. If this F^* value falls below a predetermined limit, the variable is dropped from the model; otherwise, it is retained.

4. Suppose X_7 is retained so that both X_3 and X_7 are now in the model. The stepwise regression routine now examines which X variable is the next candidate for addition, then examines whether any of the variables already in the model should now be dropped, and so on until no further X variables can either be added or deleted, at which point the search terminates.

It should be noted that the stepwise regression algorithm allows an X variable, brought into the model at an earlier stage, to be dropped subsequently if it is no longer helpful in conjunction with variables added at later stages.

Example.   Figure 12.8 shows the computer printout obtained when a particular stepwise regression routine (BMDP2R, Ref. 12.2) was applied to the surgical unit example. The minimum acceptable F limit for adding a variable and the maximum acceptable F limit for removing a variable were specified to be 4.0 and 3.9, respectively, as shown at the top right of Figure 12.8. Since the degrees of freedom associated with MSE vary, depending on the number of X variables in the model, and since repeated tests on the same data are undertaken, fixed F limits for adding or deleting a variable have no precise probabilistic meaning. Note, however, that $F(.95; 1, 50) = 4.03$, so that the specified F limits of 4.0 and 3.9 would correspond roughly to a level of significance of .05 for any single test based on approximately 50 degrees of freedom.

The minimum acceptable tolerance of .01 shown in the upper right of Figure 12.8 is a specification to guard against the entry of a variable that is highly correlated with the other X variables already in the model. As explained in Section 11.6, the tolerance is defined as $1 - R_k^2$, where R_k^2 is the coefficient of multiple determination when X_k is regressed on the other X variables in the regression model. The tolerance specification of .01 in Figure 12.8 provides that no variable is to be added to the model when it has a coefficient of multiple determination with the other X variables already in the model that exceeds $1 - .01 = .99$ or when it would cause the R_k^2 for any variable in the model to exceed .99.

We shall now follow through the steps.

1. At step 0, no X variable is in the model so that the model to be fitted is $Y_i = \beta_0 + \varepsilon_i$. The residual or error sum of squares shown in the ANOVA table in

FIGURE 12.8 Forward Stepwise Regression for Surgical Unit Example (BMDP2R, Ref. 12.2)

```
STEPPING ALGORITHM. . . . . . . . . . . . . .F          MINIMUM ACCEPTABLE F TO ENTER . . . . . . . . .   4.000,   4.000
MAXIMUM NUMBER OF STEPS . . . . . . . . . . .   10        MAXIMUM ACCEPTABLE F TO REMOVE. . . . . . . . .   3.900,   3.900
DEPENDENT VARIABLE. . . . . . . . . . . . . .    5 y      MINIMUM ACCEPTABLE TOLERANCE. . . . . . . . . .   0.01000
                                                          SUBSCRIPTS OF THE INDEPENDENT VARIABLES . . . .   1   2   3   4
STEP NO.   0
----------------
```

STD. ERROR OF EST.   0.2738

ANALYSIS OF VARIANCE

	SUM OF SQUARES	DF	MEAN SQUARE
RESIDUAL	3.9727724	53	0.7495797E-01 $\leftarrow SSTO/(n-1)$

$SSTO$

VARIABLES IN EQUATION FOR y

VARIABLE	COEFFICIENT	STD. ERROR OF COEFF	STD REG COEFF	TOLERANCE	F TO REMOVE	LEVEL
(Y-INTERCEPT	2.20614)					

$\bar{Y}$

VARIABLES NOT IN EQUATION

	VARIABLE	PARTIAL CORR.	TOLERANCE	F TO ENTER	LEVEL
.	x1	1	0.34640	1.00000	7.09	1
.	x2	2	0.59289	1.00000	28.19	1
.	x3	3	0.66512	1.00000	41.25	1
.	x4	4	0.72621	1.00000	58.02	1

r_{Yk} F_k^* $MSR(X_4)$

```
STEP NO.   1
----------------
VARIABLE ENTERED   4 x4
```

MULTIPLE R   0.7262 $\leftarrow R$
MULTIPLE R-SQUARE   0.5274 $\leftarrow R^2$
ADJUSTED R-SQUARE   0.5183 $\leftarrow R_a^2$

ANALYSIS OF VARIANCE $SSR(X_4)$

	SUM OF SQUARES	DF	MEAN SQUARE	F RATIO
REGRESSION	2.0951402	1	2.095140	58.02
RESIDUAL	1.8776320	52	0.3610831E-01	

$SSE(X_4)$ $MSE(X_4)$ F^*

STD. ERROR OF EST.   0.1900 $\leftarrow \sqrt{MSE}$

VARIABLES IN EQUATION FOR y

VARIABLE	COEFFICIENT	STD. ERROR OF COEFF	STD REG COEFF	TOLERANCE	F TO REMOVE	LEVEL
(Y-INTERCEPT	1.69639)					
x4 4	0.18575	0.0244	0.726	1.00000	58.02	1

b_k $s\{b_k\}$ b_k' F_k^*

VARIABLES NOT IN EQUATION

	VARIABLE	PARTIAL CORR.	TOLERANCE	F TO ENTER	LEVEL
.	x1	1	-0.03104	0.74758	0.05	1
.	x2	2	0.50849	0.86382	17.78	1
.	x3	3	0.58031	0.82659	25.89	1

$r_{Yk.4}$ $1-r_{4k}^2$ F_k^*

```
STEP NO.   2
----------------
VARIABLE ENTERED   3 x3
```

MULTIPLE R   0.8286
MULTIPLE R-SQUARE   0.6865
ADJUSTED R-SQUARE   0.6742

ANALYSIS OF VARIANCE

	SUM OF SQUARES	DF	MEAN SQUARE	F RATIO
REGRESSION	2.7274444	2	1.363722	55.85
RESIDUAL	1.2453278	51	0.2441819E-01	

STD. ERROR OF EST.   0.1563

VARIABLES IN EQUATION FOR y

VARIABLE	COEFFICIENT	STD. ERROR OF COEFF	STD REG COEFF	TOLERANCE	F TO REMOVE	LEVEL
(Y-INTERCEPT	1.38878)					
x3 3	0.00565	0.0011	0.439	0.82659	25.89	1
x4 4	0.13901	0.0221	0.543	0.82659	39.72	1

(annotation: MSR(X4|X3) / MSE(X4 X3))

VARIABLES NOT IN EQUATION

	VARIABLE	PARTIAL CORR.	TOLERANCE	F TO ENTER	LEVEL
.	x1	1	0.32276	0.59179	5.81	1
.	x2	2	0.79148	0.82580	83.85	1

(annotation: $1-R_k^2$; MSR(X2|X3 X4) / MSE(X2 X3 X4))

```
STEP NO.   3
----------------
VARIABLE ENTERED   2 x2
```

MULTIPLE R   0.9396
MULTIPLE R-SQUARE   0.8829
ADJUSTED R-SQUARE   0.8759

ANALYSIS OF VARIANCE

	SUM OF SQUARES	DF	MEAN SQUARE	F RATIO
REGRESSION	3.5075624	3	1.169187	125.66
RESIDUAL	0.46520978	50	0.9304196E-02	

STD. ERROR OF EST.   0.0965

VARIABLES IN EQUATION FOR y

VARIABLE	COEFFICIENT	STD. ERROR OF COEFF	STD REG COEFF	TOLERANCE	F TO REMOVE	LEVEL
(Y-INTERCEPT	0.94226)					
x2 2	0.78987E-02	0.8626E-03	0.488	0.82580	83.85	1
x3 3	0.69997E-02	0.7013E-03	-0.543	0.79021	99.63	1
x4 4	0.08185	0.0150	0.320	0.68298	29.86	1

VARIABLES NOT IN EQUATION

	VARIABLE	PARTIAL CORR.	TOLERANCE	F TO ENTER	LEVEL
.	x1	1	0.87409	0.55586	158.66	1

```
STEP NO.   4
----------------
VARIABLE ENTERED   1 x1
```

MULTIPLE R   0.9861
MULTIPLE R-SQUARE   0.9724
ADJUSTED R-SQUARE   0.9701

ANALYSIS OF VARIANCE

	SUM OF SQUARES	DF	MEAN SQUARE	F RATIO
REGRESSION	3.8630006	4	0.9657502	431.09
RESIDUAL	0.10977174	49	0.2240240E-02	

STD. ERROR OF EST.   0.0473

VARIABLES IN EQUATION FOR y

VARIABLE	COEFFICIENT	STD. ERROR OF COEFF	STD REG COEFF	TOLERANCE	F TO REMOVE	LEVEL
(Y-INTERCEPT	0.48876)					
x1 1	0.06852	0.0054	0.401	0.55586	158.66	1
x2 2	0.92541E-02	0.4367E-03	0.571	0.77567	448.98	1
x3 3	0.94745E-02	0.3963E-03	0.736	0.59594	571.70	1
x4 4	0.00193	0.0097	0.008	0.39134	0.04	1

VARIABLES NOT IN EQUATION

```
STEP NO.   5
----------------
VARIABLE REMOVED   4 x4
```

MULTIPLE R   0.9861
MULTIPLE R-SQUARE   0.9723
ADJUSTED R-SQUARE   0.9707

ANALYSIS OF VARIANCE

	SUM OF SQUARES	DF	MEAN SQUARE	F RATIO
REGRESSION	3.8629124	3	1.287637	586.04
RESIDUAL	0.10985984	50	0.2197197E-02	

STD. ERROR OF EST.   0.0469

VARIABLES IN EQUATION FOR y

VARIABLE	COEFFICIENT	STD. ERROR OF COEFF	STD REG COEFF	TOLERANCE	F TO REMOVE	LEVEL
(Y-INTERCEPT	0.48362)					
x1 1	0.06923	0.0041	0.405	0.97011	288.16	1
x2 2	0.92945E-02	0.3825E-03	0.574	0.99177	590.44	1
x3 3	0.95236E-02	0.3064E-03	0.739	0.97751	966.06	1

VARIABLES NOT IN EQUATION

	VARIABLE	PARTIAL CORR.	TOLERANCE	F TO ENTER	LEVEL
.	x4	4	0.02832	0.39134	0.04	1

***** F LEVELS(4.000, 3.900) OR TOLERANCE INSUFFICIENT FOR FURTHER STEPPING

Figure 12.8 for step 0 is therefore $\Sigma(Y_i - \bar{Y})^2 = SSTO = 3.9728$. For each potential X variable, the F^* statistic (12.17) is calculated. In Figure 12.8, these F_k^* values are shown under the heading "Variables not in equation" and are called "F to enter" values. We see that F_4^* is the largest one:

$$F_4^* = \frac{MSR(X_4)}{MSE(X_4)} = \frac{2.095140}{.03610831} = 58.02$$

Since this value exceeds the minimum acceptable F-to-enter value 4.0, X_4 is added to the model.

The column headed "Level" refers to a package option which permits the user to give different priorities to the various potential X variables. Note that in the present example all X variables have the same priority.

2. At this stage, step 1 has been completed. The current regression model contains X_4, and the estimated regression coefficients, the analysis of variance table, and selected other information about the current model are provided.

Next, all regression models containing X_4 and another independent variable are fitted, and the F^* statistics calculated. They are now:

$$F_k^* = \frac{MSR(X_k \mid X_4)}{MSE(X_4, X_k)}$$

These statistics are shown in step 1 under the heading "Variables not in equation." X_3 has the highest F^* value, which exceeds 4.0, so that X_3 now enters the model.

3. Step 2 in Figure 12.8 summarizes the situation at this point. X_3 and X_4 are now in the model, and information about this model is provided. Next, a test whether X_4 should be dropped is undertaken. The F^* statistic is shown under the heading "Variables in equation" and is called "F to remove":

$$F_4^* = \frac{MSR(X_4 \mid X_3)}{MSE(X_3, X_4)} = 39.72$$

Since this F^* value exceeds the maximum acceptable F-to-remove value 3.9, X_4 is not dropped.

4. Next, all regression models containing X_3, X_4, and one of the remaining potential X variables are fitted. The appropriate F^* statistics now are:

$$F_k^* = \frac{MSR(X_k \mid X_3, X_4)}{MSE(X_3, X_4, X_k)}$$

These statistics are shown in step 2 under the heading "Variables not in equation." X_2 has the highest F^* value, which exceeds 4.0, so that X_2 now enters the model.

5. Step 3 in Figure 12.8 summarizes the situation at this point. X_2, X_3, and X_4 are now in the model. Next, a test is undertaken whether X_3 or X_4 should be dropped. The F^* statistics to remove a variable are shown under the heading "Variables in equation" in step 3. F_4^* is smallest:

$$F_4^* = \frac{MSR(X_4 \mid X_2, X_3)}{MSE(X_2, X_3, X_4)} = 29.86$$

Since its value exceeds 3.9, X_4 is not dropped from the model.

6. At this point, only X_1 remains in the potential pool. Its F^* value to enter exceeds 4.0 (see "Variables not in equation" under step 3), so X_1 is entered into the model.

7. Step 4 in Figure 12.8 summarizes the addition of variable X_1 into the model containing variables X_2, X_3, and X_4. Next, a test is undertaken to determine whether either X_2, X_3, or X_4 should be dropped. The F^* statistics are shown under the heading "Variables in equation" in step 4. Note that:

$$F_4^* = \frac{MSR(X_4 \mid X_1, X_2, X_3)}{MSE(X_1, X_2, X_3, X_4)} = .04$$

is smallest, and that its value is less than 3.9; hence, X_4 is deleted.

8. Step 5 summarizes the dropping of X_4 from the model. Since the only potential variable remaining is X_4, which has just been dropped, it cannot enter the regression now. The algorithm therefore next considers the "F to remove" values in step 5 which indicate that F_1^* is smallest. However, since its value exceeds 3.9, X_1 is not dropped from the model and the search process is terminated.

Thus, the stepwise search algorithm identifies (X_1, X_2, X_3) as the "best" subset of X variables, a result that happens to be consistent with our previous analyses based on the all-possible-regressions approach.

Comments

1. In the surgical unit example, the tolerance requirement was always met; hence, no variable was excluded from the model as a result of too high a correlation with the other X variables in the model.

2. Variations of the rules for entering and removing variables illustrated in the example are possible. For instance, different F-to-enter and F-to-remove values can be employed in accordance with the degrees of freedom associated with MSE in the F^* statistic. However, this refinement often is not utilized and fixed values are employed instead, since the repeated testing in the search procedure does not permit precise probabilistic interpretations.

3. The F limits for adding and deleting a variable need not be selected in terms of approximate significance levels, but may be determined descriptively in terms of error reduction. For instance, an F limit of 2.0 for adding a variable may be specified with the thought that the marginal error reduction associated with the added variable should be at least twice as great as the remaining error mean square once that variable has been added.

4. The choice of F-to-enter and F-to-delete values essentially represents a balancing of opposing tendencies. Simulation studies have shown that for large pools of uncorrelated independent variables which have been generated to be uncorrelated with the dependent variable, use of small or moderately small F-to-enter values as the entry criterion results in a procedure which is too liberal, that is, which allows too many independent variables into the model. On the other hand, models produced by an automatic selection procedure with large F-to-enter values are often underspecified, resulting in σ^2 being badly overestimated and the procedure being too conservative (see References 12.3 and 12.4).

5. The minimum acceptable F-to-enter value should never be smaller than the maximum acceptable F-to-remove value; otherwise cycling is possible where a variable is continually entered and removed.

6. The order in which variables enter the regression model does not reflect their importance. In the surgical unit example, for instance, X_4 was the first variable to enter the model, yet it was eventually dropped.

7. The stepwise regression routine we employed prints out the partial correlation coefficients at each stage. These could be used equivalently to the F^* values for screening the X variables, and indeed some routines actually use the partial correlation coefficients for screening.

8. A limitation of the forward stepwise regression search approach is that it presumes there is a single "best" subset of X variables and seeks to identify it. As noted earlier, there is often no unique "best" subset. Another limitation of the forward stepwise regression routine is that it sometimes arrives at an unreasonable "best" subset when the X variables are very highly correlated.

Other Automatic Search Procedures

There are a number of other automatic search procedures that have been proposed to find a "best" subset of independent variables. We mention two of these. Neither of the two methods, however, has gained the acceptance of the forward stepwise search procedure.

Forward Selection. This search procedure is a simplified version of forward stepwise regression, omitting the test whether a variable once entered into the model should be dropped.

Backward Elimination. This search procedure is the opposite of forward selection. It begins with the model containing all potential X variables and identifies the one with the smallest F^* value. For instance, the F^* value for X_1 is:

$$(12.20) \qquad F_1^* = \frac{MSR(X_1 \mid X_2, \ldots, X_{P-1})}{MSE(X_1, \ldots, X_{P-1})}$$

If the minimum F_k^* value is less than a predetermined limit, that independent variable is dropped. The model with the remaining $P - 2$ predictor variables is then fitted, and the next candidate for dropping is identified. This process continues until no further independent variables can be dropped. A stepwise modification can also be adapted which allows variables that were eliminated earlier to be added later; this modification is called the backward stepwise regression procedure.

Note

For small and moderate numbers of variables in the pool of potential X variables, some statisticians argue for backward stepwise search over forward stepwise search (see Reference 12.5). This argument is based primarily on situations where it is useful as a first step to look at each independent variable in the regression function adjusted for all the other independent variables in the pool.

Computer Options and Refinements

Our discussion of the major automatic selection procedures for identifying the "best" subset of X variables has focused on the main conceptual issues and not on options, variations, and refinements available with particular computer packages. It is essential that the specific features of the package to be employed are fully understood so that intelligent use of the package can be made. In some packages, there is an option for regression models through the origin. Some packages permit variables to be brought into the model and tested in pairs or other groupings instead of singly, to save computing time or for other reasons. Some packages, once a "best" regression model is identified, will fit all the possible regression models in the same number of variables and will develop information for each model so that a final choice can be made by the user. Some stepwise programs have options for forcing variables into the regression model; such variables are not removed even if their F^* values become too low.

The diversity of these options and special features serves to emphasize a point made earlier: there is no unique way of searching for "good" subsets of X variables, and subjective elements must play an important role in the search process.

Selection of Variables with Ridge Regression

In Section 11.7, we discussed the use of ridge regression for helping to overcome problems related to multicollinearities among the X variables. The ridge trace mentioned there (Figure 11.8, page 417) can also be used to identify variables that might be dropped from the regression model. It has been suggested that variables be dropped whose ridge trace is unstable, with the coefficient tending toward the value of zero. Also, variables should be dropped whose ridge trace is stable but at a very small value. Finally, variables with unstable ridge traces that do not tend toward zero should be considered as candidates for dropping.

12.5 MODEL REFINEMENT AND SELECTION

The screening of variables, by a computerized selection process or otherwise, usually results in a small number of potentially useful models. These models then need to be studied further for aptness by the diagnostic methods of Chapters 4 and 11. The selection of the ultimate regression model often depends greatly upon these diagnostic results. For example, one fitted model may be very much influenced by a single case, whereas another is not. Again, one fitted model may show correlations among the error terms whereas another one does not.

When repeat observations are available, formal tests for lack of fit can be made. In any case, a variety of residual plots and analyses can be employed to identify any lack of fit, outliers, and influential observations. When the original set of $P - 1$ potential X variables excludes cross-product terms and powers of the independent variables to keep the selection problem within reasonable bounds, residual plots against such "missing" variables, or augmenting each of the "good" subsets of independent variables by adding cross-product and/or power terms, can be useful in identifying ways in which the model fit can be improved further.

When an automatic selection procedure is utilized and only a single model is identified as "best," other models should also be explored. One procedure is to use the number of independent variables in the model identified as "best" as an estimate of the number of independent variables needed in the regression model. Then the investigator explores and identifies other candidate models with approximately the same number of independent variables identified by the automatic procedure.

We will now illustrate the model refinement and selection phase of the model-building process with the surgical unit example.

Example

Since we started with a small pool of potential independent variables in the surgical unit example, it was feasible to generate all possible first-order regression models for the four independent variables in the pool. Recall that all criteria suggested that a model containing variables X_1, X_2, and X_3 was "best." Hence, model refinement and selection here will focus on further study of curvature and interaction effects, multicollinearity, and influential cases for the regression model containing X_1, X_2, and X_3, using residuals and other diagnostics.

To examine interaction effects further, a regression model containing first-order terms in X_1, X_2, and X_3 and all pairwise interaction terms was fitted and the residuals for this model were plotted against the three-variable interaction term $X_1 X_2 X_3$. This plot (not shown) did not suggest any need for a three-variable interaction term in the regression model.

In addition, a regression model containing X_1, X_2, and X_3 in first-order terms and all two-variable interaction terms and the three-variable interaction term was fitted. Adding all of the interaction terms increased R^2 only to .979 as compared to an R^2 of .972 for the first-order model in the three independent variables. Based on these and earlier results, it was decided not to include any interaction terms in the regression model.

Figure 12.9 contains some of the additional diagnostic plots that were generated to check on the adequacy of the first-order model:

$$(12.21) \qquad Y_i' = \beta_0 + \beta_1 X_{i1} + \beta_2 X_{i2} + \beta_3 X_{i3} + \varepsilon_i$$

The following points are worth noting:

1. The residual plot against the fitted values in Figure 12.9a shows no evidence of serious departures from the model.

2. Figure 12.9b shows the plot of the residuals against X_4, the independent variable not in the model. This plot shows no need to include liver function (X_4) in the model to predict survival time.

3. The partial regression plot in Figure 12.9c shows a possible curvature effect for X_1. This was investigated both by adding the squared term X_1^2 to the model and by using the logarithmic transformation of X_1. Neither of these remedial measures improved things appreciably.

4. The normal probability plot of the residuals in Figure 12.9d shows some departure from linearity. Also, the coefficient of correlation between the ordered residuals and their expected values under normality is .959, which is slightly below the critical value for significance level .01 in Table 4.3, thus indicating some departure from normality.

FIGURE 12.9 Residual and Other Diagnostic Plots for Surgical Unit Example—
Regression Model (12.21)

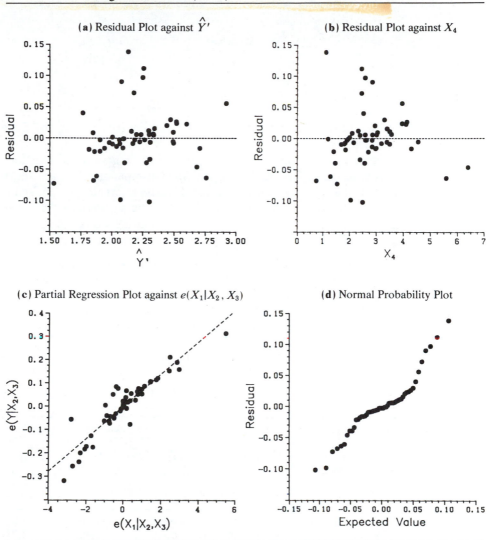

(**a**) Residual Plot against $\hat{Y}'$

(**b**) Residual Plot against X_4

(**c**) Partial Regression Plot against $e(X_1|X_2, X_3)$

(**d**) Normal Probability Plot

However, the problem of nonnormality was not considered to be serious here. An
examination of the residuals in Table 12.4 and of the normal probability plot in Figure 12.9d suggests that the departure from normality is chiefly in the tails of the distribution. There are somewhat more outlying residuals than would be expected under
normality. Still, there is only one studentized residual among the 54 whose absolute
value exceeds 3, and all of the other outlying residuals are much smaller. In view of
this indication that the outlying residuals are only moderately large and because the
sample size of 54 cases is reasonably large, it was felt that the effect of a moderate

departure from normality would not have any serious impact on inferences to be made from the regression model.

Multicollinearity was studied by calculating the variance inflation factors:

Variable	$(VIF)_k$
X_1	1.03
X_2	1.01
X_3	1.02

As may be seen from these results, multicollinearity among the three independent variables is not a problem.

Table 12.4 contains the results of some diagnostic checks for outlying and influential cases. The measures presented in columns 1–7 are the residuals e_i in (11.16), the leverage values h_{ii} in (11.13), the internally studentized residuals e_i^* in (11.19), the deleted residuals d_i in (11.20), the studentized deleted residuals d_i^* in (11.23), the $(DFFITS)_i$ values in (11.26), and the Cook's distance measures D_i in (11.29). The following are noteworthy points about the diagnostics in Table 12.4:

1. Case 22 is outlying with regard to its Y value according to both its studentized residual and studentized deleted residual, outlying by more than 3 standard deviations. Case 27 might also be considered to be marginally outlying.

2. Using $2p/n = 2(4)/54 = .148$ as a guide for identifying outlying X observations, we note that cases 13, 17, 28, 32, and 38 are outlying according to their leverage values. Incidentally, the univariate stem-and-leaf plots did not clearly identify case 13 as outlying. Here we see the value of multivariable outlier identification.

3. To determine the influence of cases 13, 17, 22, 27, 28, 32, and 38, we consider their $DFFITS$ and Cook's distance values. According to each of these mea-

TABLE 12.4   Various Diagnostics for Surgical Unit Example—Regression Model (12.21)

Case Number i	(1) e_i	(2) h_{ii}	(3) e_i^*	(4) d_i	(5) d_i^*	(6) $(DFFITS)_i$	(7) D_i
1	0.0059	0.0264	0.1274	0.0061	0.1262	0.0208	0.0001
2	−0.0093	0.0295	−0.2016	−0.0096	−0.1996	−0.0348	0.0003
3	−0.0065	0.0452	−0.1427	−0.0068	−0.1413	−0.0307	0.0002
4	0.0017	0.0789	0.0388	0.0019	0.0384	0.0112	0.0000
5	−0.0162	0.1234	−0.3700	−0.0185	−0.3668	−0.1376	0.0048
6	−0.0209	0.0622	−0.4609	−0.0223	−0.4572	−0.1178	0.0035
7	−0.0026	0.0472	−0.0578	−0.0028	−0.0572	−0.0127	0.0000
8	−0.0394	0.0534	−0.8639	−0.0416	−0.8617	−0.2047	0.0105
9	−0.1020	0.0311	−2.2108	−0.1053	−2.3041	−0.4131	0.0393
10	−0.0339	0.0720	−0.7499	−0.0365	−0.7466	−0.2080	0.0109
11	0.0263	0.0500	0.5747	0.0276	0.5708	0.1310	0.0043
12	−0.0181	0.0812	−0.4021	−0.0197	−0.3987	−0.1185	0.0036

TABLE 12.4 (*concluded*)

Case Number i	(1) e_i	(2) h_{ii}	(3) e_i^*	(4) d_i	(5) d_i^*	(6) $(DFFITS)_i$	(7) D_i
13	0.0560	0.1495	1.2955	0.0658	1.3046	0.5469	0.0737
14	0.0238	0.0498	0.5219	0.0251	0.5181	0.1186	0.0036
15	−0.0057	0.0478	−0.1246	−0.0060	−0.1234	−0.0277	0.0002
16	0.0052	0.0448	0.1137	0.0055	0.1126	0.0244	0.0002
17	−0.0162	0.1499	−0.3741	−0.0190	−0.3709	−0.1558	0.0062
18	−0.0727	0.1271	−1.6610	−0.0833	−1.6917	−0.6454	0.1004
19	0.0067	0.0361	0.1447	0.0069	0.1433	0.0277	0.0002
20	0.0059	0.0250	0.1269	0.0060	0.1256	0.0201	0.0001
21	−0.0133	0.0289	−0.2880	−0.0137	−0.2854	−0.0492	0.0006
22	0.1383	0.1274	3.1588	0.1585	3.4951	1.3353	0.3641
23	0.0084	0.1240	0.1916	0.0096	0.1897	0.0714	0.0013
24	0.0155	0.0229	0.3347	0.0159	0.3317	0.0508	0.0007
25	0.0204	0.0463	0.4461	0.0214	0.4425	0.0975	0.0024
26	0.0111	0.0196	0.2387	0.0113	0.2365	0.0334	0.0003
27	0.1118	0.0311	2.4222	0.1153	2.5522	0.4571	0.0471
28	−0.0636	0.2619	−1.5781	−0.0861	−1.6027	−0.9546	0.2209
29	−0.0215	0.0474	−0.4704	−0.0226	−0.4667	−0.1042	0.0028
30	0.0970	0.0430	2.1153	0.1014	2.1945	0.4654	0.0503
31	−0.0677	0.0808	−1.5071	−0.0737	−1.5270	−0.4529	0.0499
32	0.0402	0.2113	0.9662	0.0510	0.9656	0.4997	0.0625
33	−0.0089	0.0252	−0.1912	−0.0091	−0.1894	−0.0305	0.0002
34	0.0298	0.0346	0.6463	0.0308	0.6425	0.1216	0.0037
35	−0.0025	0.0239	−0.0534	−0.0025	−0.0529	−0.0083	0.0000
36	−0.0026	0.0479	−0.0563	−0.0027	−0.0558	−0.0125	0.0000
37	−0.0985	0.1059	−2.2231	−0.1102	−2.3183	−0.7979	0.1463
38	0.0902	0.2902	2.2850	0.1271	2.3903	1.5282	0.5336
39	0.0723	0.0261	1.5637	0.0743	1.5873	0.2600	0.0164
40	−0.0085	0.0317	−0.1850	−0.0088	−0.1832	−0.0332	0.0003
41	−0.0070	0.0353	−0.1526	−0.0073	−0.1511	−0.0289	0.0002
42	−0.0608	0.1393	−1.3986	−0.0707	−1.4124	−0.5682	0.0791
43	−0.0462	0.1242	−1.0535	−0.0528	−1.0547	−0.3972	0.0394
44	−0.0027	0.0265	−0.0584	−0.0028	−0.0578	−0.0095	0.0000
45	−0.0394	0.0828	−0.8769	−0.0429	−0.8748	−0.2629	0.0174
46	−0.0155	0.0631	−0.3406	−0.0165	−0.3375	−0.0876	0.0020
47	0.0097	0.0686	0.2135	0.0104	0.2115	0.0574	0.0008
48	0.0228	0.0841	0.5078	0.0249	0.5040	0.1528	0.0059
49	0.0060	0.0233	0.1290	0.0061	0.1277	0.0197	0.0001
50	−0.0053	0.0941	−0.1185	−0.0058	−0.1173	−0.0378	0.0004
51	−0.0009	0.0714	−0.0193	−0.0009	−0.0192	−0.0053	0.0000
52	−0.0007	0.1054	−0.0168	−0.0008	−0.0166	−0.0057	0.0000
53	0.0122	0.0264	0.2627	0.0125	0.2603	0.0428	0.0005
54	−0.0080	0.0948	−0.1789	−0.0088	−0.1771	−0.0573	0.0008

sures, case 38 is the most influential of these, with $(DFFITS)_{38} = 1.5282$ and Cook's distance $D_{38} = .5336$. Referring to the F distribution with 4 and 50 degrees of freedom, we note that the Cook's value corresponds to the 29th percentile. It thus appears that the influence of case 38 is not large enough to warrant remedial measures, and consequently the other outlying cases also do not appear to be overly influential.

A direct check of the influence of case 38 on the inferences of interest was also conducted. Here, the inferences of primary interest are in the fit of the regression model because the model is intended to be used for making predictions in the range of the X observations. Hence, each fitted value $\hat{Y}_i$ based on all 54 observations was compared with the fitted value $\hat{Y}_{i(38)}$ when case 38 is deleted in fitting the regression model. The average of the absolute percent differences:

$$\left| \frac{\hat{Y}_{i(38)} - \hat{Y}_i}{\hat{Y}_i} \right| 100$$

is only .2 percent, and the largest absolute percent difference (which is for case 38) is only 1.85 percent. Thus, case 38 does not have such a disproportionate influence on the fitted values that remedial action would be required.

4. In summary, the diagnostic analysis identified a number of potential problems, but none of these was considered to be serious enough to require further remedial action.

Consequently regression model (12.21) was selected from among the various competing models for final model validation. Results of fitting this model to the data in Table 12.1 are given in Table 12.5 in the column headed "Model-Building Data Set."

TABLE 12.5 Regression Results Based on Model-Building and Validation Data Sets for Surgical Unit Example—Regression Model (12.21)

Statistic	Model-Building Data Set	Validation Data Set
b_0	.484	.501
$s\{b_0\}$	.043	.042
b_1	.0692	.0674
$s\{b_1\}$	.0041	.0050
b_2	.00929	.0101
$s\{b_2\}$	.00038	.00037
b_3	.00952	.00974
$s\{b_3\}$	.00031	.00030
SSE	.1099	.1056
$PRESS$	.1405	—
MSE	.0022	.0021
$MSPR$	—	.0072
R^2	.972	.978

The estimated regression function for the selected model is:

$$(12.22) \qquad \hat{Y}' = .484 + .0692X_1 + .00929X_2 + .00952X_3$$

12.6 MODEL VALIDATION

The final step in the model-building process is the validation of the selected regression model. Model validation usually involves checking the model against independent data. Since regression models are often used for a variety of purposes, several different validation procedures should be employed when practical.

Three basic ways of validating a regression model are:

1. Collection of new data to check the model and its predictive ability.
2. Comparison of results with theoretical expectations, earlier empirical results, and simulation results.
3. Use of a hold-out sample to check the model and its predictive ability.

We shall now discuss each of these procedures in turn.

Collection of New Data to Check Model

The best means of model validation is through the collection of new data. The purpose of collecting new data is to be able to examine whether the regression model developed from the earlier data is still applicable for the new data. If so, one has assurance about the applicability of the model to data beyond those on which the model is based.

There are a variety of methods of examining the validity of the regression model against the new data. One validation method is to reestimate the model form chosen earlier using the new data. The estimated regression coefficients and various characteristics of the fitted model are then compared for consistency to those of the regression model based on the earlier data. If the results are consistent, one has strong support that the chosen regression model is applicable under broader circumstances than those related to the original data.

Another validation method is to reestimate from the new data all of the "good" models that had been considered to see if the selected regression model is still the preferred model according to the new data. If so, one has assurance about the efficacy of the selected model under new conditions.

A third validation method is designed to calibrate the predictive capability of the selected regression model. When a regression model is developed from given data, it is inevitable that the selected model is chosen, at least in large part, because it fits well the data at hand. For different random outcomes in the data set, one may likely have arrived at a different model in terms of the independent variables selected and/ or their functional forms and interaction terms present in the model. A result of this model development process is that the error mean square *MSE* will tend to understate the inherent variability in making future predictions from the selected model.

A means of measuring the actual predictive capability of the selected regression model is to use this model to predict each case in the new data set and then to calculate the mean of the squared prediction errors, to be denoted by *MSPR*, which stands for *mean squared prediction error:*

$$(12.23) \qquad MSPR = \frac{\sum\limits_{i=1}^{n^*}(Y_i - \hat{Y}_i)^2}{n^*}$$

where:

Y_i is the value of the response variable in the ith validation case

$\hat{Y}_i$ is the predicted value for the ith validation case based on the model-building data set

n^* is the number of cases in the validation data set

If the mean squared prediction error *MSPR* is fairly close to *MSE* based on the regression fit to the model-building data set, then the error mean square *MSE* for the selected regression model is not seriously biased and gives an appropriate indication of the predictive ability of the model. If the mean squared prediction error is much larger than *MSE*, one should rely on the mean squared prediction error as an indicator of how well the selected regression model will predict in the future.

Note

When the new data are collected under controlled conditions in an experiment, it is desirable to include data points of major interest to check out the model predictions. If the model is to be used for making predictions over the entire range of the X observations, a possibility is to include data points that are uniformly distributed over the X space.

Comparison with Theory, Empirical Evidence, or Simulation Results

In some cases, theory, simulation results, or previous empirical results may be helpful in determining whether the selected model is reasonable. Comparisons of regression coefficients and predictions with theoretical expectations, previous empirical results, or simulation results should be made. Unfortunately, there is often little theory that can be used to validate regression models.

Data Splitting

By far the preferred method to validate a regression model is through the collection of new data. Often, however, this is neither practical nor feasible. A reasonable alternative when the data set is large enough is to split the data into two sets. The first, called the *model-building set,* is used to develop the model. The second data set, called the *validation* or *prediction set,* is used to evaluate the reasonableness and predictive ability of the selected model. This validation procedure is often called *cross-*

validation. Data splitting in effect simulates partial or complete replication of the study.

The validation data set is used for validation in the same way as when new data are collected. One can reestimate the regression coefficients for the selected model and compare them for consistency with the coefficients obtained from the model-building data set. Also, predictions can be made for the data in the validation data set from the regression model developed from the model-building data set to calibrate the predictive ability of this regression model for new data. When the calibration data set is large enough, one can also study how the "good" models considered in the model selection phase fare with the new data.

Data sets are often split equally into the model-building and validation data sets. It is important, however, that the model-building data set be sufficiently large so that a reliable model can be developed. Recall in this connection that the number of cases should be at least 6 to 10 times the number of variables in the pool of independent variables. Thus, when ten variables are in the pool, the model-building data set should contain at least 60 to 100 cases. If the entire data set is not large enough under these circumstances to enable one to make an equal split, the validation data set will need to be smaller than the model-building data set.

Splits of the data can be made at random. Another possibility is to match cases in pairs and place one of each pair into one of the two split data sets. When data are collected sequentially in time, it is often useful to pick a point in time to divide the data. Generally, the earlier data are selected for the model-building set and the later data for the validation set. When seasonal or cyclical effects are present in the data (e.g., sales data), the split should be made at a point where the cycles are balanced.

Use of time or some other characteristic of the data to split the data set can create problems. Conditions may differ for the two data sets. Data in the validation set may have been created under different causal conditions than those of the model-building set. Alternatively, data in the validation set may represent extrapolations with respect to the data in the model-building set (e.g., sales data collected over time). Such differential conditions may lead to a lack of validity and indicate a need to broaden the regression model so that it is applicable under a broader scope of conditions. One may then need to use part of the validation data to extend the range of the model-building data set, while still retaining to some extent a validation data set.

A possible drawback of data splitting is that the variances of the estimated regression coefficients developed from the model-building data set will usually be larger than those that would have been obtained from the fit to the entire data set. If the model-building data set is reasonably large, however, these variances generally will not be that much larger than those for the entire data set. In any case, once the model has been validated, it is customary practice to use the entire data set for estimating the final regression model even though, in theory, it is more appropriate to retain the fitted model obtained from the model-building data set alone.

Note

For small data sets where data splitting is impractical, the *PRESS* procedure, considered earlier as a criterion for subset selection, can be used as a form of data splitting to assess the precision of model predictions. Recall that with this procedure, each data point is predicted from the least squares fitted regression function developed from the remaining $n - 1$ data points. A fairly close agreement between *PRESS* and *SSE* would suggest that *MSE* may be a reasonably valid indicator of the selected model's predictive capability.

Example

In the surgical unit example, fitted regression model (12.22) was finally selected as the model to be checked for validity. The estimated regression coefficients, their estimated standard deviations, and some other statistics pertaining to the fitted model are shown in Table 12.5.

Some evidence of the internal validity of this fitted model was mentioned earlier. We noted in Table 12.3 the closeness for this model of *PRESS* = .1405 and *SSE* = .1099. The *PRESS* value will always be larger than *SSE* because the regression fit for the *i*th case when this case is deleted in fitting can never be as good as that when the *i*th case is included. A *PRESS* value reasonably close to *SSE*, as here, supports the validity of the fitted regression model and of *MSE* as an indicator of the predictive capability of this model.

To validate the selected regression model externally, 54 additional cases had been held out for a validation data set. The data for these cases are shown in Table 12.6. The correlation matrix for these data (not shown) is quite similar to the one in Table 12.2 for the model-building data set. The estimated regression coefficients, their estimated standard deviations, and some other statistics when regression model (12.21) is fitted to the validation data set are shown in Table 12.5. Note the good agreement between the two sets of estimated regression coefficients, and between the two *MSE* and R^2 values.

TABLE 12.6   Validation Data Set—Surgical Unit Example

Case Number i	Blood Clotting Score X_{i1}	Prognostic Index X_{i2}	Enzyme Function Test X_{i3}	Liver Function Test X_{i4}	Y_i'
1	7.1	23	78	1.93	2.0326
2	4.9	66	91	3.05	2.4086
3	6.4	90	35	1.06	2.2177
4	5.7	35	70	2.13	1.9078
5	6.1	42	69	2.25	2.0035
6	8.0	27	83	2.03	2.0945
7	6.8	34	51	1.27	1.7652
8	4.7	63	36	1.71	1.7925
9	7.0	47	67	1.60	2.1292
10	6.7	69	65	2.91	2.2295
11	6.7	46	78	3.26	2.1524
12	5.8	60	86	3.11	2.3188

TABLE 12.6 (*concluded*)

Case Number i	Blood Clotting Score X_{i1}	Prognostic Index X_{i2}	Enzyme Function Test X_{i3}	Liver Function Test X_{i4}	Y_i'
13	6.7	56	32	1.53	1.9039
14	6.8	51	58	2.18	2.0508
15	7.2	95	82	4.68	2.6525
16	7.4	52	67	3.28	2.2053
17	5.3	53	62	2.42	1.9246
18	3.5	58	84	1.74	2.1541
19	6.8	74	79	2.25	2.4970
20	4.4	47	49	2.42	1.7237
21	7.0	66	118	4.69	2.8339
22	6.7	61	57	3.87	2.1282
23	5.6	75	103	3.11	2.6884
24	6.9	58	88	3.46	2.4284
25	6.2	62	57	1.25	2.0261
26	4.7	97	27	1.77	2.0843
27	6.8	69	60	2.90	2.2826
28	6.0	73	58	1.22	2.2073
29	5.9	50	62	3.19	2.0443
30	5.5	88	74	3.21	2.4863
31	3.8	55	52	1.41	1.9037
32	4.3	99	83	3.93	2.6647
33	6.6	48	54	2.94	1.9071
34	6.2	42	63	1.85	1.9093
35	5.0	60	105	3.17	2.4389
36	5.8	62	82	3.18	2.3343
37	4.7	42	10	.28	1.3379
38	5.7	70	59	2.28	2.1996
39	4.7	64	48	1.30	1.8795
40	7.8	74	40	2.58	2.1504
41	2.9	43	32	.94	1.4330
42	4.9	72	90	3.51	2.4381
43	4.6	73	57	2.82	2.1075
44	5.9	78	70	4.28	2.2843
45	4.6	69	70	3.17	2.1615
46	6.1	53	52	1.84	2.0558
47	5.9	88	98	3.33	2.7249
48	4.7	66	68	1.80	2.0520
49	10.4	62	85	4.65	2.6810
50	5.8	70	64	2.52	2.2604
51	5.4	64	81	1.36	2.2553
52	6.9	90	33	2.78	2.1745
53	7.9	45	55	2.46	2.0224
54	4.5	68	60	2.07	2.1413

To calibrate the predictive ability of the regression model fitted from the model-building data set, the mean squared prediction error *MSPR* in (12.23) was calculated for the cases in the validation data set in Table 12.6; it is *MSPR* = .0072. The mean squared prediction error generally will be larger than *MSE* based on the model-building data set because entirely new data are involved in the validation data set. The fact that *MSPR* here does not differ too greatly from *MSE* implies that the error mean square *MSE* based on the model-building data set is a reasonably valid indicator of the predictive ability of the fitted regression model.

These validation results support the appropriateness of the model selected.

Comments

1. Algorithms are available to split data so that the two data sets have similar statistical properties. The reader is referred to Reference 12.6 for a discussion of this and other issues associated with validation.

2. Refinements of data splitting have been proposed. With the *double cross-validation procedure,* for example, the model is built for each half of the split data and then tested on the other half of the data. Thus, two measures of consistency and predictive ability are obtained from the two fitted models.

3. When the data set is small, variations of *PRESS* have been proposed whereby *m* cases are held out for validation and the remaining $n - m$ cases are used to build the model. Reference 12.7 discusses these procedures, as well as issues dealing with optimal splitting of data sets.

4. When regression models built on observational data do not predict well outside the range of the *X* observations in the data set, the usual reason is the existence of multicollinearity among the independent variables. As mentioned in Chapter 11, possible solutions for this difficulty include using ridge regression or other biased estimation techniques.

12.7 CONCLUDING REMARKS

Building an appropriate and effective regression model is a complex undertaking. The keys to success are proper problem formulation, collection of an adequate amount of data of high quality, selection of important variables and proper model form, good diagnostic checking of assumptions, and use of appropriate validation procedures.

CITED REFERENCES

12.1. Daniel, C., and F. S. Wood. *Fitting Equations to Data.* 2nd ed. New York: Wiley-Interscience, 1980.

12.2. Dixon, W. J., chief editor. *BMDP Statistical Software Manual,* vols. 1 and 2. Berkeley, Calif.: University of California Press, 1988.

12.3. Freedman, D. A. "A Note on Screening Regression Equations." *The American Statistician* 37 (1983), pp. 152–55.

12.4. Pope, P. T., and J. T. Webster. "The Use of an *F*-Statistic in Stepwise Regression." *Technometrics* 14 (1972), pp. 327–40.

12.5. Mantel, N. "Why Stepdown Procedures in Variable Selection." *Technometrics* 12 (1970), pp. 621–25.

12.6. Snee, R. D. "Validation of Regression Models: Methods and Examples." *Technometrics* 19 (1977), pp. 415–28.

12.7. Stone, M. "Cross-validatory Choice and Assessment of Statistical Prediction." *Journal of the Royal Statistical Society B* 36 (1974), pp. 111–47.

PROBLEMS

12.1. A speaker stated: "In well-designed experiments involving quantitative independent variables, a procedure for reducing the number of independent variables after the data are obtained is not necessary." Discuss.

12.2. The dean of a graduate school wishes to predict the grade point average in graduate work for recent applicants. List a dozen variables that might be useful independent variables here.

12.3. Two researchers investigated factors affecting summer attendance at privately operated beaches on Lake Ontario and collected information on attendance and 11 explanatory variables for 42 beaches. Two summers were studied, of relatively hot and relatively cool weather, respectively. A "best" subsets algorithm now is to be used to reduce the independent variables for the final regression model.

    a. Should the variables reduction be done for both summers combined or should it be done separately for each summer? Explain the problems involved and how you might handle them.

    b. Will the "best" subsets selection procedure choose those independent variables that are most important in a causal sense for determining beach attendance?

12.4. In forward stepwise regression, what advantage is there in using a relatively large F-to-enter value for adding variables? What advantage is there in using a smaller F-to-enter value?

12.5. In forward stepwise regression, why should the F-to-remove value for deleting variables never exceed the F-to-enter value for adding variables?

12.6. Draw a flowchart of each of the following selection methods: (1) forward stepwise regression, (2) forward selection, (3) backward elimination.

12.7. An engineer has stated: "Reduction of the independent variables should always be done using the objective forward stepwise regression procedure." Discuss.

12.8. An attendee of a regression modeling short course stated: "I rarely see validation of regression models mentioned in published papers, so it must really not be an important component of model building." Comment.

12.9. Refer to **Patient satisfaction** Problem 7.17. The hospital administrator wishes to determine the best subset of independent variables for predicting patient satisfaction.

    a. Indicate which subset of independent variables you would recommend as best for predicting patient satisfaction according to each of the following criteria: (1) R_p^2, (2) MSE_p, (3) C_p, (4) $PRESS_p$. Support your recommendations with appropriate graphs.

    b. Do the four criteria in part (a) identify the same best subset? Does this always happen?

c. Would forward stepwise regression have any advantages here as a screening procedure over the all-possible-regressions procedure?

12.10. Roofing shingles. Data on sales last year (Y, in thousand squares) in 26 sales districts are given below for a maker of asphalt roofing shingles. Shown also are promotional expenditures (X_1, in thousand dollars), number of active accounts (X_2), number of competing brands (X_3), and district potential (X_4, coded) for each of the districts.

District i	X_{i1}	X_{i2}	X_{i3}	X_{i4}	Y_i
1	5.5	31	10	8	79.3
2	2.5	55	8	6	200.1
3	8.0	67	12	9	163.2
4	3.0	50	7	16	200.1
5	3.0	38	8	15	146.0
6	2.9	71	12	17	177.7
7	8.0	30	12	8	30.9
8	9.0	56	5	10	291.9
9	4.0	42	8	4	160.0
10	6.5	73	5	16	339.4
11	5.5	60	11	7	159.6
12	5.0	44	12	12	86.3
13	6.0	50	6	6	237.5
14	5.0	39	10	4	107.2
15	3.5	55	10	4	155.0
16	8.0	70	6	14	291.4
17	6.0	40	11	6	100.2
18	4.0	50	11	8	135.8
19	7.5	62	9	13	223.3
20	7.0	59	9	11	195.0
21	6.7	53	13	5	73.4
22	6.1	38	13	10	47.7
23	3.6	43	9	17	140.7
24	4.2	26	8	3	93.5
25	4.5	75	8	19	259.0
26	5.6	71	4	9	331.2

a. Prepare separate dot plots for each of the independent variables. Are there any noteworthy features in these plots? Comment.

b. Prepare separate scatter plots of sales (Y) against each of the four independent variables. Do the scatter plots suggest linear or curvilinear relationships between the dependent variable Y and each of the independent variables? Discuss.

c. Obtain the correlation matrix of the X variables. Are any serious multicollinearity problems evident from this matrix? Explain.

d. Fit the multiple regression function containing all four independent variables as first-order terms.

e. Obtain the variance inflation factors for the model fitted in part (d). Are there indications that serious multicollinearity problems exist here? Explain.

f. Obtain the residuals and plot them separately against $\hat{Y}$ and each of the independent variables. Also prepare a normal probability plot of the residuals. On the basis of these plots, should any modifications of the regression model be made?

12.11. Refer to **Roofing shingles** Problem 12.10.

   a. Using only first-order terms for the independent variables, find the three best subset regression models according to the C_p criterion.

   b. Is there relatively little bias in each of these subset models?

12.12. Refer to **Roofing shingles** Problems 12.10 and 12.11. The subset model containing only first-order terms in X_2 and X_3 is to be evaluated in detail.

   a. Obtain the residuals and plot them separately against $\hat{Y}$, each of the four independent variables, and the cross-product term $X_2 X_3$. On the basis of these plots, should any modifications of the regression model be investigated?

   b. Prepare separate partial regression plots against $e(X_2 \mid X_3)$ and $e(X_3 \mid X_2)$. Do these plots suggest that any modifications in the model form are warranted?

   c. Prepare a normal probability plot of the residuals. Also obtain the coefficient of correlation between the ordered residuals and their expected values under normality. Test the reasonableness of the normality assumption using Table 4.3 and $\alpha = .05$. What do you conclude?

   d. Obtain the diagonal elements of the hat matrix. Using the $2p/n$ rule of thumb, identify any outlying X observations. Are the findings consistent with those in Problem 12.10a? Should they be? Discuss.

   e. Obtain the studentized deleted residuals and identify any outlying Y observations.

   f. Case 8 appears to be reasonably far outlying with respect to its Y value. Obtain the *DFFITS*, *DFBETAS*, and Cook's distance values for this case to assess its influence. What do you conclude?

12.13. **Job proficiency.** A personnel officer in a governmental agency administered four newly developed aptitude tests to each of 25 applicants for entry-level clerical positions in the agency. For purposes of the study, all 25 applicants were accepted for positions irrespective of their test scores. After a probationary period, each applicant was rated for proficiency on the job. The scores on the four tests (X_1, X_2, X_3, X_4) and the job proficiency score (Y) for the 25 employees were as follows:

Subject i	Test Score X_{i1}	X_{i2}	X_{i3}	X_{i4}	Job Proficiency Score Y_i
1	86	110	100	87	88
2	62	97	99	100	80
3	110	107	103	103	96
4	101	117	93	95	76
5	100	101	95	88	80
6	78	85	95	84	73
7	120	77	80	74	58
8	105	122	116	102	116
9	112	119	106	105	104
10	120	89	105	97	99
11	87	81	90	88	64
12	133	120	113	108	126
13	140	121	96	89	94
14	84	113	98	78	71

| Subject | Test Score | | | | Job Proficiency Score |
i	X_{i1}	X_{i2}	X_{i3}	X_{i4}	Y_i
15	106	102	109	109	111
16	109	129	102	108	109
17	104	83	100	102	100
18	150	118	107	110	127
19	98	125	108	95	99
20	120	94	95	90	82
21	74	121	91	85	67
22	96	114	114	103	109
23	104	73	93	80	78
24	94	121	115	104	115
25	91	129	97	83	83

a. Prepare separate stem-and-leaf plots for each of the four sets of newly developed aptitude test scores. Are there any noteworthy features in these plots? Comment.

b. Prepare separate scatter plots of job proficiency scores (Y) against each of the four independent variables. What do the scatter plots suggest about the nature of the functional relationship between the dependent variable Y and each of the independent variables?

c. Obtain the correlation matrix of the X variables. Are any serious multicollinearity problems evident from this matrix? Explain.

d. Fit the multiple regression function containing all four independent variables as first-order terms.

e. Obtain the variance inflation factors for the regression model fitted in part (d). Are there indications that serious multicollinearity problems exist here? Explain.

f. Obtain the residuals and plot them separately against $\hat{Y}$ and each of the independent variables. Also prepare a normal probability plot of the residuals. On the basis of these plots, should any modifications of the regression model be made?

12.14. Refer to **Job proficiency** Problem 12.13.

a. Using only first-order terms for the independent variables in the pool of potential X variables, find the four best subset regression models according to the adjusted R^2 criterion.

b. Since there is relatively little difference in R_a^2 for the four best subset models, what other criteria would you use to help in the selection of the best model? Discuss.

12.15. Refer to **Job proficiency** Problems 12.13 and 12.14. The subset model containing only first-order terms in X_1 and X_3 is to be evaluated in detail.

a. Obtain the residuals and plot them separately against $\hat{Y}$, each of the four independent variables, and the cross-product term $X_1 X_3$. On the basis of these plots, should any modifications in the regression model be investigated?

b. Prepare separate partial regression plots against $e(X_1 \mid X_3)$ and $e(X_3 \mid X_1)$. Do these plots suggest that any modifications in the model form are warranted?

c. Prepare a normal probability plot of the residuals. Also obtain the coefficient of correlation between the ordered residuals and their expected values under normality. Test the reasonableness of the normality assumptions using Table 4.3 and $\alpha = .01$. What do you conclude?

d. Obtain the diagonal elements of the hat matrix. Using the $2p/n$ rule of thumb, identify any outlying X observations. Are your findings consistent with those in Problem 12.13a? Should they be? Comment.

e. Obtain the studentized deleted residuals and identify any outlying Y observations.

f. Cases 7 and 18 appear to be moderately outlying with respect to their X values and case 16 is reasonably far outlying with respect to its Y value. Obtain *DFFITS*, *DFBETAS*, and Cook's distance values for these cases to assess their influence. What do you conclude?

12.16. Lung pressure. Increased arterial blood pressure in the lungs frequently leads to the development of heart failure in patients with chronic obstructive pulmonary disease (COPD). The standard method for determining arterial lung pressure is invasive, technically difficult, and involves some risk to the patient. The radionuclide imaging technique is a noninvasive, less risky method for estimating arterial pressure in the lungs. To investigate the predictive ability of this method, a cardiologist collected data on 19 mild to moderate COPD patients. The data that follow include the invasive measure of systolic pulmonary arterial pressure (Y) and three potential noninvasive predictor variables. Two were obtained by using the radionuclide imaging technique—emptying rate of blood into the pumping chamber of the heart (X_1) and ejection rate of blood pumped out of the heart into the lungs (X_2)—and the third independent variable measures a blood gas (X_3).

Subject i	X_{i1}	X_{i2}	X_{i3}	Y_i	Subject i	X_{i1}	X_{i2}	X_{i3}	Y_i
1	45	36	45	49	11	37	37	55	31
2	30	28	40	55	12	29	34	47	49
3	11	16	42	85	13	26	32	28	38
4	30	46	40	32	14	38	45	30	41
5	39	76	43	26	15	38	99	26	12
6	42	78	27	28	16	25	38	47	44
7	17	24	36	95	17	27	51	44	29
8	63	80	42	26	18	37	32	54	40
9	25	12	52	74	19	34	40	36	31
10	32	27	35	37					

Adapted from A. T. Marmor et al., "Improved Radionuclide Method for Assessment of Pulmonary Artery Pressure in COPD," *Chest* 89 (1986), pp. 64–69.

a. Prepare separate dot plots for each of the three independent variables. Are there any noteworthy features in these plots? Comment.

b. Prepare separate scatter plots of Y against each of the three independent variables. What do these scatter plots suggest about the nature of the functional relationship between Y and each of the independent variables?

c. Obtain the correlation matrix of the X variables. Are any serious multicollinearity problems evident from this matrix? Explain.

d. Fit the multiple regression function containing the three independent variables as first-order terms.

e. Obtain the variance inflation factors for the regression model fitted in part (d). Are there indications that serious multicollinearity problems exist here? Explain.

    f. Obtain the residuals and plot them separately against $\hat{Y}$, each of the three independent variables, and against each two-factor interaction term. Also prepare a normal probability plot of the residuals. On the basis of these plots, should the model be modified? Discuss.

12.17. Refer to **Lung pressure** Problem 12.16.

    a. Using first-order and second-order terms for each of the three independent variables (expressed as deviations from the mean) in the pool of potential X variables (including cross-products of the first-order terms), find the three best hierarchical subset regression models according to the R_a^2 criterion. Hierarchical subsets must contain the first-order term for an independent variable (e.g., x_1) if a second-order term (e.g., x_1^2 or $x_1 x_2$) is included in the model.

    b. Is there much difference in R^2 for the three best subset models?

12.18. Refer to **Lung pressure** Problems 12.16 and 12.17. The subset regression model containing first-order terms for X_1 and X_2 and the cross-product term $X_1 X_2$ is to be evaluated in detail.

    a. Obtain the residuals and plot them separately against $\hat{Y}$ and each of the three independent variables. On the basis of these plots, should any further modifications of the regression model be attempted?

    b. Prepare a normal probability plot of the residuals. Also obtain the coefficient of correlation between the ordered residuals and their expected values under normality. Does the normality assumption appear to be reasonable here?

    c. Obtain the variance inflation factors. Are there any indications that serious multicollinearity problems are present? Explain.

    d. Obtain the diagonal elements of the hat matrix. Using the $2p/n$ rule of thumb, identify any outlying X observations. Are your findings consistent with those in Problem 12.16a? Should they be? Discuss.

    e. Obtain the studentized deleted residuals and identify any outlying Y observations.

    f. Cases 3, 8, and 15 are moderately far outlying with respect to their X values and case 7 is relatively far outlying with respect to its Y value. Obtain *DFFITS*, *DFBETAS*, and Cook's distance values for these cases to assess their influence. What do you conclude?

12.19. **Kidney function.** Creatinine clearance (Y) is an important measure of kidney function but is difficult to obtain in a clinical office setting because it requires 24-hour urine collection. To determine whether this measure can be predicted from some data that are easily available, a kidney specialist obtained the data that follow for 33 male subjects. The independent variables are serum creatinine concentration (X_1), age (X_2), and weight (X_3).

Subject i	X_{i1}	X_{i2}	X_{i3}	Y_i	*Subject* i	X_{i1}	X_{i2}	X_{i3}	Y_i
1	.71	38	71	132	8	.92	61	81	92
2	1.48	78	69	53	9	1.55	68	74	60
3	2.21	69	85	50	10	.94	64	87	94
4	1.43	70	100	82	11	1.00	66	79	105
5	.68	45	59	110	12	1.07	49	93	98
6	.76	65	73	100	13	.70	43	60	112
7	1.12	76	63	68	14	.71	42	70	125

Subject i	X_{i1}	X_{i2}	X_{i3}	Y_i	Subject i	X_{i1}	X_{i2}	X_{i3}	Y_i
15	1.00	66	83	108	25	.68	32	80	140
16	2.52	78	70	30	26	1.20	21	67	80
17	1.13	35	73	111	27	2.10	73	72	43
18	1.12	34	85	130	28	1.36	78	67	75
19	1.38	35	68	94	29	1.50	58	60	41
20	1.12	16	65	130	30	.82	62	107	120
21	.97	54	53	59	31	1.53	70	75	52
22	1.61	73	50	38	32	1.58	63	62	73
23	1.58	66	74	65	33	1.37	68	52	57
24	1.40	31	67	85					

Adapted from W. J. Shih and S. Weisberg, "Assessing Influence in Multiple Linear Regression with Incomplete Data," *Technometrics* 28 (1986), pp. 231–40.

a. Prepare separate dot plots for each of the three independent variables. Are there any noteworthy features in these plots? Comment.

b. Prepare separate scatter plots of Y against each of the three independent variables. What do these plots suggest about the nature of the functional relationship between the dependent variable Y and each of the independent variables? Discuss.

c. Obtain the correlation matrix of the X variables. Are any serious multicollinearity problems evident from this matrix? Explain.

d. Fit the multiple regression function containing the three independent variables as first-order terms.

e. Obtain the variance inflation factors for the regression model fitted in part (d). Are there indications that serious multicollinearity problems exist here? Explain.

f. Obtain the residuals and plot them separately against $\hat{Y}$ and each of the independent variables. Also prepare a normal probability plot of the residuals.

g. Prepare separate partial regression plots against $e(X_1 \mid X_2, X_3)$, $e(X_2 \mid X_1, X_3)$, and $e(X_3 \mid X_1, X_2)$.

h. Do the plots in parts (f) and (g) suggest that the regression model should be modified?

i. Theoretical arguments suggest use of the regression function:

$$E\{\log_e Y\} = \beta_0 + \beta_1 \log_e X_1 + \beta_2 \log_e(140 - X_2) + \beta_3 \log_e X_3$$

Are the plots in parts (f) and (g) consistent with the theoretical expectations? Discuss.

12.20. Refer to **Kidney function** Problem 12.19. The regression function based on theoretical considerations:

$$E\{\log_e Y\} = \beta_0 + \beta_1 \log_e X_1 + \beta_2 \log_e(140 - X_2) + \beta_3 \log_e X_3$$

is to be fitted and evaluated in detail.

a. Fit the regression function based on theoretical considerations.

b. Obtain the residuals and plot them separately against $\hat{Y}$ and each of the independent variables in the fitted model. Also prepare a normal probability plot of the residuals. Have the difficulties noted in Problem 12.19 now been largely eliminated?

    c. Obtain the variance inflation factors for the regression model fitted in part (a). Are there indications that serious multicollinearity problems exist here? Explain.

    d. Obtain the diagonal elements of the hat matrix. Using the $2p/n$ rule of thumb, identify any outlying X observations.

    e. Obtain the studentized deleted residuals and identify any outlying Y observations.

    f. Cases 28 and 29 are relatively far outlying with respect to their Y values. Obtain *DFFITS*, *DFBETAS*, and Cook's distance values for these cases to assess their influence. What do you conclude?

12.21. Refer to **Patient satisfaction** Problems 7.17 and 12.9. The hospital administrator was interested to learn how the forward stepwise selection procedure and some of its variations would perform here.

    a. Determine the subset of variables that is selected as best by the forward stepwise regression procedure using F limits of 3.0 and 2.9 to add or delete a variable, respectively. Show your steps.

    b. To what level of significance in any individual test is the F limit of 3.0 for adding a variable approximately equivalent here?

    c. Determine the subset of variables that is selected as best by the forward selection procedure using an F limit of 3.0 to add a variable. Show your steps.

    d. Determine the subset of variables that is selected as best by the backward elimination procedure using an F limit of 2.9 to delete a variable. Show your steps.

    e. Compare the results of the three selection procedures. How consistent are these results? How do the results compare with those for all possible regressions in Problem 12.9?

12.22. Refer to **Roofing shingles** Problems 12.10 and 12.11.

    a. Using forward stepwise regression, find the best subset of independent variables to predict sales. Use F limits for adding or deleting a variable of 4.0 and 3.9, respectively.

    b. How does the best subset according to forward stepwise regression compare with the best subset according to the C_p criterion obtained in Problem 12.11a?

    c. Suppose that variable X_1 is to be forced into the best subset because of its causal importance by entering it first and not removing it even if its F^* value is too low. Which subset of variables (including X_1) is now selected as best by the forward stepwise regression procedure if F limits of 4.0 and 3.9 are used to add or delete a variable, respectively? Did the forced inclusion of X_1 affect the selection of the other variables in the best subset? Will this always happen?

12.23. Refer to **Job proficiency** Problems 12.13 and 12.14.

    a. Using forward stepwise regression, find the best subset of independent variables to predict job proficiency. Use F limits of 4.0 and 3.9 for adding or deleting a variable, respectively.

    b. How does the best subset according to forward stepwise regression compare with the best subset according to the adjusted R^2 criterion obtained in Problem 12.14a?

12.24. Refer to **Roofing shingles** Problems 12.10 and 12.12. To assess internally the predictive ability of the regression model identified in Problem 12.12, calculate the *PRESS* statistic and compare it to *SSE*. What does this comparison suggest about the validity of *MSE* as an indicator of the fitted model's predictive ability?

12.25. Refer to **Roofing shingles** Problems 12.10 and 12.12. To assess externally the validity of the regression model identified in Problem 12.12, data on sales last year for 23 other comparable districts were held out. These data follow.

District i	X_{i1}	X_{i2}	X_{i3}	X_{i4}	Y_i
27	5.3	70	7	10	291.5
28	5.3	83	10	20	277.9
29	7.5	46	14	3	48.0
30	6.2	51	7	20	213.4
31	5.4	52	11	9	135.1
32	5.8	45	9	4	150.0
33	2.5	54	9	12	180.1
34	5.4	70	6	8	295.8
35	5.2	31	9	6	91.3
36	3.7	46	10	11	116.7
37	5.1	49	8	6	190.3
38	6.0	60	12	10	155.3
39	6.7	51	7	9	217.6
40	6.1	55	7	9	236.5
41	3.6	55	8	4	202.2
42	5.8	24	10	8	64.7
43	7.1	63	5	15	287.2
44	6.8	30	8	10	128.0
45	5.0	62	8	4	219.7
46	6.9	75	8	17	272.9
47	4.5	47	9	10	168.0
48	6.2	67	8	17	269.1
49	6.5	48	11	11	115.6

a. Obtain the correlation matrix of the X variables for the validation data set and compare it with that obtained in Problem 12.10c for the model-building data set. What do you conclude?

b. Fit the regression model identified in Problem 12.12 to the validation data set. Compare the estimated regression coefficients and their estimated standard deviations to those obtained in Problem 12.11a. Also compare the error mean squares and coefficients of multiple determination. Do the validation data set estimates appear to be reasonably similar to those obtained using the model-building data set?

c. Calculate the mean squared prediction error in (12.23) and compare it to *MSE* obtained from the model-building data set. Does there appear to be a serious bias problem in *MSE* here? Is this conclusion consistent with your finding in Problem 12.24? Comment.

d. Combine the model-building data set in Problem 12.10 with the validation data set and fit the selected regression model to the combined data. Are the estimated standard deviations of the estimated regression coefficients now reduced appreciably?

12.26. Refer to **Job proficiency** Problems 12.13 and 12.15. To assess internally the predictive ability of the regression model identified in Problem 12.15, compute the *PRESS* statistic and compare it to *SSE*. What does this comparison suggest about the validity of *MSE* as an indicator of the predictive ability of the fitted model?

12.27. Refer to **Job proficiency** Problems 12.13 and 12.15. To assess externally the validity of the regression model identified in Problem 12.15, 25 additional applicants for entry-level clerical positions in the agency were similarly tested and hired irrespective of their test scores. Their data follow.

Subject i	Test Score				Job Proficiency Score
	X_{i1}	X_{i2}	X_{i3}	X_{i4}	Y_i
26	65	109	88	84	58
27	85	90	104	98	92
28	93	73	91	82	71
29	95	57	95	85	77
30	102	139	101	92	92
31	63	101	93	84	66
32	81	129	88	76	61
33	111	102	83	72	57
34	67	98	98	84	66
35	91	111	96	84	75
36	128	99	98	89	98
37	116	103	103	103	100
38	105	102	88	83	67
39	99	132	109	105	111
40	93	95	106	98	97
41	99	113	104	95	99
42	110	114	91	78	74
43	128	134	108	98	117
44	99	110	96	97	92
45	111	113	101	91	95
46	109	120	104	106	104
47	78	125	115	102	100
48	115	119	102	94	95
49	129	70	94	95	81
50	136	104	106	104	109

a. Obtain the correlation matrix of the X variables for the validation data set and compare it with that obtained in Problem 12.13c for the model-building data set. Are the two correlation matrices reasonably similar?

b. Fit the regression model identified in Problem 12.15 to the validation data set. Compare the estimated regression coefficients and their estimated standard deviations to those obtained in Problem 12.14a. Also compare the error mean squares and coefficients of multiple determination. Do the estimates for the validation data set appear to be reasonably similar to those obtained for the model-building data set?

c. Calculate the mean squared prediction error in (12.23) and compare it to *MSE* obtained from the model-building data set. Is there evidence of a substantial bias problem in *MSE* here? Is this conclusion consistent with your finding in Problem 12.26? Discuss.

d. Combine the model-building data set in Problem 12.13 with the validation data set and fit the selected regression model to the combined data. Are the estimated standard deviations of the estimated regression coefficients appreciably reduced now from those obtained for the model-building data set?

12.28. Refer to **Lung pressure** Problems 12.16 and 12.18. The validity of the regression model identified in Problem 12.18 is to be assessed internally.

   a. Calculate the *PRESS* statistic and compare it to *SSE*. What does this comparison suggest about the validity of *MSE* as an indicator of the predictive ability of the fitted model?

   b. Case 8 alone accounts for approximately one-half of the entire *PRESS* statistic. Would you recommend modification of the model because of the strong impact of this case? What are some corrective action options that would lessen the effect of case 8? Discuss.

EXERCISES

12.29. The true quadratic regression function is $E\{Y\} = 15 + 20X + 3X^2$. The fitted linear regression function is $\hat{Y} = 13 + 40X$, for which $E\{b_0\} = 10$ and $E\{b_1\} = 45$. What are the bias and sampling error components of the mean squared error for $X_i = 10$ and for $X_i = 20$?

12.30. Prove (12.11). [*Hint:* Use Exercise 6.31 and (11.14).]

12.31. Refer to (12.18). Show that the same variable X_k that maximizes the test statistic F_k^* also maximizes the coefficient of partial determination $r_{Yk.7}^2$.

PROJECTS

12.32. Refer to the surgical unit example model-building data set in Table 12.1 and the validation data set in Table 12.6. Combine the two data sets and fit regression model (12.21) to the combined data. Compare the estimated regression coefficients and their estimated standard deviations with those obtained for the model-building data set in Table 12.5. Are there appreciable differences in the estimated standard deviations of the estimated regression coefficients? Comment.

12.33. Refer to the **SENIC** data set. Length of stay (Y) is to be predicted, and the pool of potential independent variables includes all other variables in the data set except medical school affiliation and region. It is believed that a model with $\log_{10} Y$ as the dependent variable and the independent variables in first-order terms with no interaction terms will be appropriate. Consider cases 57–113 to constitute the model-building data set to be used for the following analyses.

   a. Prepare separate dot plots for each of the independent variables. Are there any noteworthy features in these plots? Comment.

   b. Obtain the correlation matrix of the X variables. Is there evidence of strong linear pairwise associations here?

   c. Obtain the three best subsets according to the C_p criterion. Which of these subset models appears to have the smallest bias?

12.34. Refer to the **SENIC** data set and Problem 12.33. The regression model containing age, routine chest X-ray ratio, and average daily census in first-order terms is to be evaluated in detail based on the model-building data set.

a. Obtain the residuals and plot them separately against $\hat{Y}$, each of the independent variables in the model, and each of the related cross-product terms. On the basis of these plots, should any modifications of the model be made?

b. Prepare a normal probability plot of the residuals. Also obtain the coefficient of correlation between the ordered residuals and their expected values under normality. Test the reasonableness of the normality assumption using Table 4.3 and $\alpha = .05$. What do you conclude?

c. Obtain the variance inflation factors. Are there any indications that serious multicollinearity problems are present? Explain.

d. Obtain the diagonal elements of the hat matrix. Using the $2p/n$ rule of thumb, identify any outlying X observations.

e. Obtain the studentized deleted residuals and identify any outlying Y observations.

f. Cases 62, 75, 106, and 112 are moderately outlying with respect to their X values and case 87 is reasonably far outlying with respect to its Y value. Obtain *DFFITS*, *DFBETAS*, and Cook's distance values for these cases to assess their influence. What do you conclude?

g. Calculate the *PRESS* statistic and compare it to *SSE*. What does this comparison suggest about the validity of *MSE* as an indicator of the fitted model's predictive ability?

12.35. Refer to the **SENIC** data set and Problems 12.33 and 12.34. The regression model identified in Problem 12.34 is to be validated by means of the validation data set consisting of cases 1–56.

a. Fit the regression model identified in Problem 12.34 to the validation data set. Compare the estimated regression coefficients and their estimated standard deviations with those obtained in Problem 12.33c. Also compare the error mean squares and coefficients of multiple determination. Does the model fitted to the validation data set yield similar estimates as the model fitted to the model-building data set?

b. Calculate the mean squared prediction error in (12.23) and compare it to *MSE* obtained from the model-building data set. Is there evidence of a substantial bias problem in *MSE* here? Is this conclusion consistent with your finding in Problem 12.34g?

c. Combine the model-building and validation data sets and fit the selected regression model to the combined data. Are the estimated regression coefficients and their estimated standard deviations appreciably different from those for the model-building data set? Should you expect any differences in the estimates? Explain.

12.36. Refer to the **SMSA** data set. A public safety official wishes to predict the rate of serious crimes in an **SMSA** (Y, total number of serious crimes per 100,000 population). The pool of potential independent variables includes all other variables in the data set except total population and region. It is believed that a model with independent variables in first-order terms with no interaction terms will be appropriate. Consider the even-numbered cases to constitute the model-building data set to be used for the following analyses.

a. Prepare separate stem-and-leaf plots for each of the independent variables. Are there any noteworthy features in these plots? Comment.

b. Obtain the correlation matrix of the X variables. Is there evidence of strong linear pairwise associations here?

c. Using the C_p criterion, obtain the three best subsets. Which of these subset models appears to have the smallest bias?

12.37. Refer to the **SMSA** data set and Problem 12.36. The regression model containing land area and percent high school graduates in first-order terms is to be evaluated in detail based on the model-building data set.

    a. Obtain the residuals and plot them separately against $\hat{Y}$, each of the independent variables in the model, and the related cross-product term. On the basis of these plots, should any modifications in the model be made?

    b. Prepare a normal probability plot of the residuals. Also obtain the coefficient of correlation between the ordered residuals and their expected values under normality. Test the reasonableness of the normality assumption using Table 4.3 and $\alpha = .01$. What do you conclude?

    c. Obtain the variance inflation factors. Are there any indications that serious multicollinearity problems are present? Explain.

    d. Obtain the diagonal elements of the hat matrix. Using the $2p/n$ rule of thumb, identify any outlying X observations.

    e. Obtain the studentized deleted residuals and identify any outlying Y observations.

    f. Cases 42, 74, 92, 94, 124, and 138 are moderately outlying with respect to their X values, and cases 40 and 54 are reasonably far outlying with respect to their Y values. Obtain *DFFITS*, *DFBETAS*, and Cook's distance values for these cases to assess their influence. What do you conclude?

    g. Calculate the *PRESS* statistic and compare it to *SSE*. What does this comparison suggest about the validity of *MSE* as an indicator of the fitted model's predictive ability?

12.38. Refer to the **SMSA** data set and Problems 12.36 and 12.37. The regression model identified in Problem 12.37 is to be validated by means of the validation data set consisting of the odd-numbered **SMSAs**.

    a. Fit the regression model identified in Problem 12.37 to the validation data set. Compare the estimated regression coefficients and their estimated standard deviations with those obtained in Problem 12.36c. Also compare the error mean squares and coefficients of multiple determination. Does the model fitted to the validation data set yield similar estimates as the model fitted to the model-building data set?

    b. Calculate the mean squared prediction error in (12.23) and compare it to *MSE* obtained from the model-building data set. Is there evidence of a substantial bias problem in *MSE* here? Is this conclusion consistent with your finding in Problem 12.37g?

    c. Fit the regression model identified in Problem 12.37 to the combined model-building and validation data sets. Are the estimated regression coefficients and their estimated standard deviations appreciably different from those for the model fitted to the model-building data set? Should you expect any differences in the estimates? Explain.

Chapter 13

Autocorrelation in Time Series Data

The basic regression models considered so far have assumed that the random error terms ε_i are either uncorrelated random variables or independent normal random variables. In business and economics, many regression applications involve time series data. For such data, the assumption of uncorrelated or independent error terms is often not appropriate; rather, the error terms are frequently correlated positively over time. Error terms correlated over time are said to be *autocorrelated* or *serially correlated*.

A major cause of positively autocorrelated error terms in business and economic regression applications involving time series data is the omission of one or several key variables from the model. When time-ordered effects of such "missing" key variables are positively correlated, the error terms in the regression model will tend to be positively autocorrelated since the error terms include effects of missing variables. Suppose, for example, that annual sales of a product are regressed against average yearly price over a period of 30 years. If population size has an important effect on sales, its omission from the model may lead to the error terms being positively autocorrelated because the effect of population size on sales likely is positively correlated over time.

Another common cause of positively autocorrelated error terms in economic data is systematic coverage errors in the dependent variable time series, which errors often tend to be positively correlated over time.

13.1 PROBLEMS OF AUTOCORRELATION

If the error terms in the regression model are positively autocorrelated, the use of ordinary least squares procedures has a number of important consequences. We summarize these first, and then discuss them in more detail:

1. The estimated regression coefficients are still unbiased, but they no longer have the minimum variance property and may be quite inefficient.

2. *MSE* may seriously underestimate the variance of the error terms.
3. $s\{b_k\}$ calculated according to ordinary least squares procedures may seriously underestimate the true standard deviation of the estimated regression coefficient.
4. The confidence intervals and tests using the t and F distributions, discussed earlier, are no longer strictly applicable.

To illustrate these problems intuitively, we shall consider the simple linear regression model with time series data:

$$Y_t = \beta_0 + \beta_1 X_t + \varepsilon_t$$

Here, Y_t and X_t are observations for period t. Let us assume that the error terms ε_t are positively autocorrelated as follows:

$$\varepsilon_t = \varepsilon_{t-1} + u_t$$

The u_t, called *disturbances,* are independent normal random variables. Thus, any error term ε_t is the sum of the previous error term ε_{t-1} and a new disturbance term u_t. We shall assume here that the u_t have mean 0 and variance 1.

In Table 13.1, column 1, we show 10 random observations on the normal variable u_t with mean 0 and variance 1, obtained from a standard normal random numbers generator. Suppose now that $\varepsilon_0 = 3.0$; we obtain then:

$$\varepsilon_1 = \varepsilon_0 + u_1 = 3.0 + .5 = 3.5$$

$$\varepsilon_2 = \varepsilon_1 + u_2 = 3.5 - .7 = 2.8$$

etc.

The error terms ε_t are shown in Table 13.1, column 2, and they are plotted in Figure 13.1. Note the systematic pattern in these error terms. Their positive relation over time is shown by the fact that adjacent error terms tend to be of the same magnitude.

TABLE 13.1   Example of Positively Autocorrelated Error Terms

t	(1) u_t	(2) $\varepsilon_{t-1} + u_t = \varepsilon_t$
0	—	3.0
1	$+.5$	$3.0 + .5 = 3.5$
2	$-.7$	$3.5 - .7 = 2.8$
3	$+.3$	$2.8 + .3 = 3.1$
4	0	$3.1 + 0 = 3.1$
5	-2.3	$3.1 - 2.3 = .8$
6	-1.9	$.8 - 1.9 = -1.1$
7	$+.2$	$-1.1 + .2 = -.9$
8	$-.3$	$-.9 - .3 = -1.2$
9	$+.2$	$-1.2 + .2 = -1.0$
10	$-.1$	$-1.0 - .1 = -1.1$

FIGURE 13.1 Example of Positively Autocorrelated Error Terms

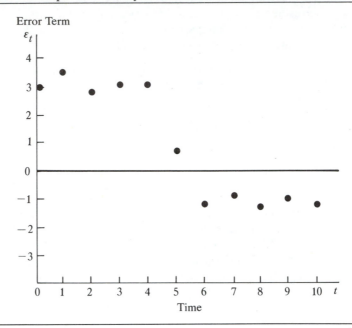

Suppose that X_t in the regression model represents time, such that $X_1 = 1$, $X_2 = 2$, etc. Further, suppose we know that $\beta_0 = 2$ and $\beta_1 = .5$ so that the true regression function is $E\{Y\} = 2 + .5X$. Figure 13.2a contains this true regression line and the observed Y values based on the error terms in Table 13.1. For example, $Y_0 = 2 + .5(0) + 3.0 = 5.0$, and $Y_1 = 2 + .5(1) + 3.5 = 6.0$. Figure 13.2b contains the estimated regression line, fitted by ordinary least squares methods, and repeats the observed Y values. Notice that the fitted regression line differs sharply from the true regression line because the initial ε_0 value was large and the succeeding positively autocorrelated error terms tended to be large for some time. This persistency pattern in the positively autocorrelated error terms leads to a fitted regression line far from the true one. Had the initial ε_0 value been small, say, $\varepsilon_0 = -.2$, and the disturbances different, a sharply different fitted regression line might have been obtained because of the persistency pattern, as shown in Figure 13.2c. This variation from sample to sample in the fitted regression lines due to the positively autocorrelated error terms may be so substantial as to lead to large variances of the estimated regression coefficients when ordinary least squares methods are used.

Another key problem with applying ordinary least squares methods when the error terms are positively autocorrelated, as mentioned before, is that *MSE* may seriously underestimate the variance of the ε_t. Figure 13.2 makes this clear. Note that the variability of the Y values around the fitted regression line in Figure 13.2b is substantially smaller than the variability of the Y values around the true regression line in Figure 13.2a. This is one of the factors leading to an indication of greater

FIGURE 13.2   Regression with Positively Autocorrelated Error Terms

(a) True Regression Line and Observations when $\varepsilon_0 = 3$

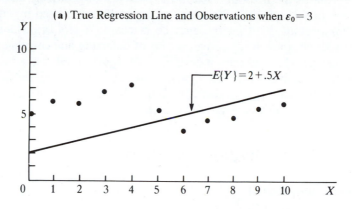

$E\{Y\} = 2 + .5X$

(b) Fitted Regression Line and Observations when $\varepsilon_0 = 3$

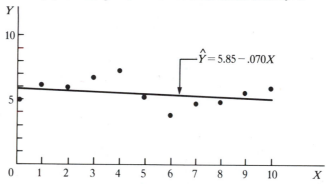

$\hat{Y} = 5.85 - .070X$

(c) Fitted Regression Line and Observations with $\varepsilon_0 = -.2$
and Different Disturbances

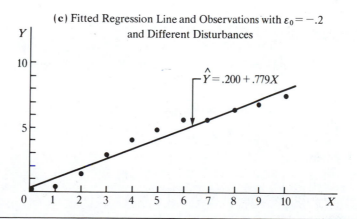

$\hat{Y} = .200 + .779X$

precision of the regression coefficients than is actually the case when ordinary least squares methods are used in the presence of positively autocorrelated errors.

In view of the seriousness of the problems created by autocorrelated errors, it is important that their presence be detected. A plot of residuals against time is an effective, though subjective, means of detecting autocorrelated errors. Formal statistical tests have also been developed. A widely used test is based on the first-order autoregressive error model, which we take up next. This model is a simple one, yet experience suggests that it is frequently applicable in business and economics when the error terms are serially correlated.

13.2 FIRST-ORDER AUTOREGRESSIVE ERROR MODEL

Simple Linear Regression

The simple linear regression model for one independent variable with the random error terms following a first-order autoregressive process is:

$$Y_t = \beta_0 + \beta_1 X_t + \varepsilon_t$$

(13.1)

$$\varepsilon_t = \rho \varepsilon_{t-1} + u_t$$

where:

ρ is a parameter such that $|\rho| < 1$
u_t are independent $N(0, \sigma^2)$

Note that (13.1) is identical to the simple linear regression model (3.1) except for the structure of the error terms. Each error term in model (13.1) consists of a fraction of the previous error term (when $\rho > 0$) plus a new disturbance term u_t. The parameter ρ is called the *autocorrelation parameter*.

Multiple Regression

The multiple regression model with the random error terms following a first-order autoregressive process is:

$$Y_t = \beta_0 + \beta_1 X_{t1} + \beta_2 X_{t2} + \cdots + \beta_{p-1} X_{t,p-1} + \varepsilon_t$$

(13.2)

$$\varepsilon_t = \rho \varepsilon_{t-1} + u_t$$

where:

$|\rho| < 1$
u_t are independent $N(0, \sigma^2)$

Thus, we see that the multiple regression model (13.2) is identical to the earlier multiple regression model (7.7) except for the structure of the error terms.

Properties of Error Terms

It is instructive to expand the definition of the first-order autoregressive error term ε_t:

$$\varepsilon_t = \rho \varepsilon_{t-1} + u_t$$

which holds for all t. Thus, $\varepsilon_{t-1} = \rho \varepsilon_{t-2} + u_{t-1}$. When we substitute this expression above, we obtain:

$$\varepsilon_t = \rho(\rho \varepsilon_{t-2} + u_{t-1}) + u_t = \rho^2 \varepsilon_{t-2} + \rho u_{t-1} + u_t$$

Replacing now ε_{t-2} by $\rho \varepsilon_{t-3} + u_{t-2}$, we obtain:

$$\varepsilon_t = \rho^3 \varepsilon_{t-3} + \rho^2 u_{t-2} + \rho u_{t-1} + u_t$$

Continuing in this fashion, we find:

$$(13.3) \qquad \varepsilon_t = \sum_{s=0}^{\infty} \rho^s u_{t-s}$$

Thus, the error term ε_t in period t is a linear combination of the current and preceding disturbance terms. When $0 < \rho < 1$, (13.3) indicates that the further the period is in the past, the smaller is the weight of that disturbance term in determining ε_t.

It can be shown that the mean and variance of ε_t for the first-order autoregressive error models (13.1) and (13.2) are as follows:

$$(13.4) \qquad E\{\varepsilon_t\} = 0$$

$$(13.5) \qquad \sigma^2\{\varepsilon_t\} = \frac{\sigma^2}{1 - \rho^2}$$

We thus see that the error terms ε_t have mean zero and constant variance, just as for regression models with uncorrelated error terms.

Unlike in earlier regression models, however, the error terms for the first-order autoregressive error models (13.1) and (13.2) are correlated. It can be shown that the covariance between adjacent error terms ε_t and ε_{t-1} is:

$$(13.6) \qquad \sigma\{\varepsilon_t, \varepsilon_{t-1}\} = \rho\left(\frac{\sigma^2}{1 - \rho^2}\right)$$

The coefficient of correlation between ε_t and ε_{t-1}, denoted by $\rho\{\varepsilon_t, \varepsilon_{t-1}\}$, is defined as follows:

$$(13.7) \qquad \rho\{\varepsilon_t, \varepsilon_{t-1}\} = \frac{\sigma\{\varepsilon_t, \varepsilon_{t-1}\}}{\sigma\{\varepsilon_t\}\sigma\{\varepsilon_{t-1}\}}$$

Since the variance of each error term according to (13.5) is $\sigma^2/(1 - \rho^2)$, the coefficient of correlation is:

(13.7a)
$$\rho\{\varepsilon_t, \varepsilon_{t-1}\} = \frac{\rho\left(\dfrac{\sigma^2}{1 - \rho^2}\right)}{\sqrt{\dfrac{\sigma^2}{1 - \rho^2}} \sqrt{\dfrac{\sigma^2}{1 - \rho^2}}} = \rho$$

Thus, the autocorrelation parameter ρ is the coefficient of correlation between adjacent error terms.

The covariance between error terms that are s periods apart can be shown to be:

(13.8)
$$\sigma\{\varepsilon_t, \varepsilon_{t-s}\} = \rho^s\left(\frac{\sigma^2}{1 - \rho^2}\right) \qquad s \neq 0$$

and the coefficient of correlation between ε_t and ε_{t-s} therefore is:

(13.9)
$$\rho\{\varepsilon_t, \varepsilon_{t-s}\} = \rho^s \qquad s \neq 0$$

Thus, when ρ is positive, all error terms are correlated, but the further apart they are, the less is the correlation between them. The only time the error terms for the autoregressive error models (13.1) and (13.2) are uncorrelated is when $\rho = 0$.

Comments

1. The derivation of (13.4), that the error terms have expectation zero, follows directly from taking the expectation of ε_t in (13.3) and using the fact that $E\{u_t\} = 0$ according to models (13.1) and (13.2) for all t.

2. To derive the variance of the error terms, we utilize the assumption of models (13.1) and (13.2) that the u_t are independent with variance σ^2. It follows from (13.3) then that:

$$\sigma^2\{\varepsilon_t\} = \sum_{s=0}^{\infty} \rho^{2s}\sigma^2\{u_{t-s}\} = \sigma^2 \sum_{s=0}^{\infty} \rho^{2s}$$

Now for $|\rho| < 1$, it is known that:

$$\sum_{s=0}^{\infty} \rho^{2s} = \frac{1}{1 - \rho^2}$$

Hence, we have:

$$\sigma^2\{\varepsilon_t\} = \frac{\sigma^2}{1 - \rho^2}$$

3. To derive the covariance of ε_t and ε_{t-1}, we need to recognize that:

$$\sigma^2\{\varepsilon_t\} = E\{\varepsilon_t^2\}$$
$$\sigma\{\varepsilon_t, \varepsilon_{t-1}\} = E\{\varepsilon_t\varepsilon_{t-1}\}$$

These results follow from theorems (1.15a) and (1.21a), respectively, since $E\{\varepsilon_t\} = 0$ by (13.4) for all t.

By (13.3), we have:

$$E\{\varepsilon_t\varepsilon_{t-1}\} = E\{(u_t + \rho u_{t-1} + \rho^2 u_{t-2} + \cdots)(u_{t-1} + \rho u_{t-2} + \rho^2 u_{t-3} + \cdots)\}$$

which can be rewritten:

$$E\{\varepsilon_t \varepsilon_{t-1}\} = E\{[u_t + \rho(u_{t-1} + \rho u_{t-2} + \cdots)][u_{t-1} + \rho u_{t-2} + \rho^2 u_{t-3} + \cdots]\}$$

$$= E\{u_t(u_{t-1} + \rho u_{t-2} + \rho^2 u_{t-3} + \cdots)\}$$

$$+ E\{\rho(u_{t-1} + \rho u_{t-2} + \rho^2 u_{t-3} + \cdots)^2\}$$

Since $E\{u_t u_{t-s}\} = 0$ for all $s \neq 0$ by the assumed independence of the u_t and the fact that $E\{u_t\} = 0$ for all t, the first term drops out and we obtain:

$$E\{\varepsilon_t \varepsilon_{t-1}\} = \rho E\{\varepsilon_{t-1}^2\} = \rho \sigma^2\{\varepsilon_{t-1}\}$$

Hence, by (13.5), which holds for all t, we have:

$$\sigma\{\varepsilon_t, \varepsilon_{t-1}\} = \rho\left(\frac{\sigma^2}{1 - \rho^2}\right)$$

4. The first-order autoregressive error process in models (13.1) and (13.2) is the simplest kind. A second-order process would be:

(13.10) $$\varepsilon_t = \rho_1 \varepsilon_{t-1} + \rho_2 \varepsilon_{t-2} + u_t$$

Still higher-order processes could be postulated. Specialized approaches have been developed for complex autoregressive error processes. These are discussed in treatments of time series procedures and forecasting, such as in Reference 13.1.

13.3 DURBIN-WATSON TEST FOR AUTOCORRELATION

The Durbin-Watson test for autocorrelation assumes the first-order autoregressive error models (13.1) or (13.2), with the values of the independent variable(s) fixed. The test consists of determining whether or not the autocorrelation parameter ρ in (13.1) or (13.2) is zero. Note that if $\rho = 0$, $\varepsilon_t = u_t$. Hence, the error terms ε_t are then independent since the u_t are independent.

Because correlated error terms in business and economic applications tend to show positive serial correlation, the usual test alternatives considered are:

(13.11)
$$H_0: \rho = 0$$
$$H_a: \rho > 0$$

The test statistic D is obtained by using ordinary least squares to fit the regression function, calculating the ordinary residuals:

(13.12) $$e_t = Y_t - \hat{Y}_t$$

and then calculating the statistic:

(13.13) $$D = \frac{\sum_{t=2}^{n} (e_t - e_{t-1})^2}{\sum_{t=1}^{n} e_t^2}$$

where n is the number of cases.

An exact test procedure is not available, but Durbin and Watson have obtained lower and upper bounds d_L and d_U such that a value of D outside these bounds leads to a definite decision. The decision rule for testing between the alternatives in (13.11) is:

If $D > d_U$, conclude H_0

(13.14)   If $D < d_L$, conclude H_a

If $d_L \leq D \leq d_U$, the test is inconclusive

Small values of D lead to the conclusion that $\rho > 0$ because the adjacent error terms ε_t and ε_{t-1} tend to be of the same magnitude when they are positively autocorrelated. Hence, the differences in the residuals, $e_t - e_{t-1}$, would tend to be small when $\rho > 0$, leading to a small numerator in D and hence to a small test statistic D.

Table A.6 contains the bounds d_L and d_U for various sample sizes (n), for two levels of significance (.05 and .01), and for various numbers of X variables ($p - 1$) in the regression model.

Example

The Blaisdell Company wished to predict its sales by using industry sales as a predictor variable. (Accurate predictions of industry sales are available from the industry's trade association.) In Table 13.2, columns 1 and 2 contain seasonally adjusted quarterly data on company sales and industry sales, respectively, for the period 1983–87. A scatter plot (not shown) suggested that a linear regression model is appropriate. The market research analyst was, however, concerned whether or not the error terms are positively autocorrelated.

He used ordinary least squares to fit a regression line to the data in Table 13.2. The results are shown at the bottom of Table 13.2. He then obtained the residuals e_t, which are shown in column 3 of Table 13.2 and which are plotted against time in Figure 13.3. Note how the residuals consistently are above or below the fitted values for extended periods. Positive autocorrelation in the error terms is suggested by such a pattern when an appropriate regression function has been employed.

The analyst wished to confirm this graphic diagnosis by using the Durbin-Watson test for the alternatives:

$$H_0: \rho = 0$$

$$H_a: \rho > 0$$

Columns 4, 5, and 6 of Table 13.2 contain the necessary calculations for the test statistic D. The analyst then obtained:

$$D = \frac{\sum\limits_{t=2}^{20}(e_t - e_{t-1})^2}{\sum\limits_{t=1}^{20} e_t^2} = \frac{.09794}{.13330} = .735$$

Using a level of significance of .01, he found in Table A.6 for $n = 20$ and

TABLE 13.2   Data for Blaisdell Company Example, Regression Results, and Durbin-Watson Test Calculations (company and industry sales data are seasonally adjusted)

Year and Quarter	t	(1) Company Sales ($ millions) Y_t	(2) Industry Sales ($ millions) X_t	(3) Residual e_t	(4) $e_t - e_{t-1}$	(5) $(e_t - e_{t-1})^2$	(6) e_t^2
1983: 1	1	20.96	127.3	−.026052	—	—	.0006787
2	2	21.40	130.0	−.062015	−.035963	.0012933	.0038459
3	3	21.96	132.7	.022021	.084036	.0070620	.0004849
4	4	21.52	129.4	.163754	.141733	.0200882	.0268154
1984: 1	5	22.39	135.0	.046570	−.117184	.0137321	.0021688
2	6	22.76	137.1	.046377	−.000193	.0000000	.0021508
3	7	23.48	141.2	.043617	−.002760	.0000076	.0019024
4	8	23.66	142.8	−.058435	−.102052	.0104146	.0034146
1985: 1	9	24.10	145.5	−.094399	−.035964	.0012934	.0089112
2	10	24.01	145.3	−.149142	−.054743	.0029968	.0222433
3	11	24.54	148.3	−.147991	.001151	.0000013	.0219013
4	12	24.30	146.4	−.053054	.094937	.0090130	.0028147
1986: 1	13	25.00	150.2	−.022928	.030126	.0009076	.0005257
2	14	25.64	153.1	.105852	.128780	.0165843	.0112046
3	15	26.36	157.3	.085464	−.020388	.0004157	.0073041
4	16	26.98	160.7	.106102	.020638	.0004259	.0112576
1987: 1	17	27.52	164.2	.029112	−.076990	.0059275	.0008475
2	18	27.78	165.6	.042316	.013204	.0001743	.0017906
3	19	28.24	168.7	−.044160	−.086476	.0074781	.0019501
4	20	28.78	171.7	−.033009	.011151	.0001243	.0010896
Total						.0979400	.1333018

$$\hat{Y} = -1.4548 + .17628X$$

$$s\{b_0\} = .21415 \qquad s\{b_1\} = .00144$$

$$MSE = .00741$$

FIGURE 13.3   Residuals Plotted against Time—Blaisdell Company Example

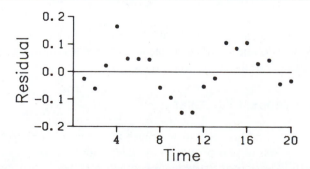

$p - 1 = 1$:

$$d_L = .95$$

$$d_U = 1.15$$

Since $D = .735$ falls below $d_L = .95$, decision rule (13.14) indicates that the appropriate conclusion is H_a, namely, that the error terms are positively autocorrelated.

Comments

1. If a test for negative autocorrelation is required, the test statistic to be used is $4 - D$, where D is defined as above. The test is then conducted in the same manner described for testing for positive autocorrelation. That is, if the quantity $4 - D$ falls below d_L, we conclude $\rho < 0$, that negative autocorrelation exists, and so on.

2. A two-sided test for H_0: $\rho = 0$ versus H_a: $\rho \neq 0$ can be made by employing both one-sided tests separately. The Type I risk with the two-sided test is 2α, where α is the Type I risk with each one-sided test.

3. When the Durbin-Watson test employing the bounds d_L and d_U gives indeterminate results, in principle more cases are required. Of course, with time series data it may be impossible to obtain more cases, or additional cases may lie in the future and be obtainable only with great delay. Durbin and Watson (Ref. 13.2) do give an approximate test which may be used when the bounds test is indeterminate, but the degrees of freedom should be larger than about 40 before this approximate test will give more than a rough indication of whether autocorrelation exists.

A reasonable procedure is to treat indeterminate results as suggesting the presence of autocorrelated errors and employ one of the remedial actions to be discussed next. If such an action does not lead to substantially different regression results, the assumption of uncorrelated error terms would appear to be satisfactory. When the remedial action does lead to substantially different regression results (such as larger estimated standard errors for the regression coefficients or the elimination of autocorrelated errors), the results obtained by means of the remedial action are probably the more useful ones.

4. The Durbin-Watson test is not robust against misspecifications of the model. For example, the Durbin-Watson test may not disclose the presence of autocorrelated errors that follow the second-order autoregressive pattern in (13.10).

5. While the Durbin-Watson test is widely used, other tests for autocorrelation are available. One such alternative test, due to Theil and Nagar, is found in Reference 13.3.

13.4 REMEDIAL MEASURES FOR AUTOCORRELATION

The two principal remedial measures when autocorrelated error terms are present are to add one or more independent variables to the regression model or to use transformed variables.

Addition of Independent Variables

As noted earlier, one major cause of autocorrelated error terms is the omission from the model of one or more key independent variables that have time-ordered effects on the dependent variable. When autocorrelated error terms are found to be present, the first remedial action should always be to search for missing key independent variables. The missing variable, population size, was mentioned previously for a regression of annual sales of a product on average yearly price of the product during a 30-year period. Sometimes, use of a simple linear trend variable or use of indicator variables for seasonal effects can be helpful in eliminating or reducing autocorrelation in the error terms.

Use of Transformed Variables

Only when use of additional independent variables is not helpful in eliminating the problem of autocorrelated errors should a remedial action based on transformed variables be employed. A number of remedial procedures that rely on transformations of the variables have been developed. We shall explain three of these methods. Our explanation will be in terms of simple linear regression, but the extension to multiple regression is direct.

The three methods to be described are each based on an interesting property of the first-order autoregressive error term regression model (13.1). Consider the transformed dependent variable:

$$Y_t' = Y_t - \rho Y_{t-1}$$

Substituting in this expression for Y_t and Y_{t-1} according to regression model (13.1), we obtain:

$$Y_t' = (\beta_0 + \beta_1 X_t + \varepsilon_t) - \rho(\beta_0 + \beta_1 X_{t-1} + \varepsilon_{t-1})$$
$$= \beta_0(1 - \rho) + \beta_1(X_t - \rho X_{t-1}) + (\varepsilon_t - \rho \varepsilon_{t-1})$$

But by (13.1), $\varepsilon_t - \rho \varepsilon_{t-1} = u_t$. Hence:

$$(13.15) \qquad Y_t' = \beta_0(1 - \rho) + \beta_1(X_t - \rho X_{t-1}) + u_t$$

where the u_t are the independent disturbance terms. Thus, when we use the transformed variable Y_t', the regression model contains error terms that are independent. Further, model (13.15) is still a simple linear regression model with new independent variable $X_t' = X_t - \rho X_{t-1}$, as may be seen by rewriting (13.15) as follows:

$$(13.16) \qquad Y_t' = \beta_0' + \beta_1' X_t' + u_t$$

where:

$$Y_t' = Y_t - \rho Y_{t-1}$$
$$X_t' = X_t - \rho X_{t-1}$$
$$\beta_0' = \beta_0(1 - \rho)$$
$$\beta_1' = \beta_1$$

Hence, by use of the transformed variables X_t' and Y_t', we obtain a standard simple linear regression model with independent error terms. This means that ordinary least squares methods have their usual optimum properties with this model.

In order to be able to use the transformed model (13.16), one generally needs to estimate the autocorrelation parameter ρ since its value is usually unknown. The three methods to be described differ in how this is done. Often, however, the results obtained with the three methods are quite similar.

Once an estimate of ρ has been obtained, to be denoted by r, transformed variables are obtained using this estimate of ρ:

(13.17a) $$Y'_t = Y_t - rY_{t-1}$$

(13.17b) $$X'_t = X_t - rX_{t-1}$$

Regression model (13.16) is then fitted to these transformed data, yielding an estimated regression function:

(13.18) $$\hat{Y}' = b'_0 + b'_1 X'$$

If this fitted regression function has eliminated the autocorrelation in the error terms, one can transform back to a fitted regression model in the original variables as follows:

(13.19) $$\hat{Y} = b_0 + b_1 X$$

where:

(13.19a) $$b_0 = \frac{b'_0}{1 - r}$$

(13.19b) $$b_1 = b'_1$$

The estimated standard deviations of the regression coefficients for the original variables can be obtained from those for the regression coefficients for the transformed variables as follows:

(13.20a) $$s\{b_0\} = \frac{s\{b'_0\}}{1 - r}$$

(13.20b) $$s\{b_1\} = s\{b'_1\}$$

Cochrane-Orcutt Procedure

The Cochrane-Orcutt procedure involves an iteration of three steps.

1. *Estimation of ρ.* This is accomplished by noting that the autoregressive error process assumed in model (13.1) can be viewed as a regression through the origin:

$$\varepsilon_t = \rho\varepsilon_{t-1} + u_t$$

where ε_t is the dependent variable, ε_{t-1} the independent variable, u_t the error term, and ρ the slope of the line through the origin. Since the ε_t and ε_{t-1} are unknown we use the residuals e_t and e_{t-1}, obtained by ordinary least squares methods, as the dependent and independent variables, respectively, and estimate ρ by fitting a straight line through the origin. From our previous discussion of regression through the origin, we know by (5.15) that the estimate of the slope ρ, denoted by r, is:

$$(13.21) \qquad r = \frac{\sum\limits_{t=2}^{n} e_{t-1} e_t}{\sum\limits_{t=2}^{n} e_{t-1}^2}$$

2. *Fitting of transformed model (13.16).* Using the estimate of ρ in (13.21), we next obtain the transformed variables Y_t' and X_t' in (13.17) and use ordinary least squares with these transformed variables to yield the fitted regression function (13.18).

3. *Test for need to iterate.* The Durbin-Watson test is then employed to test whether the error terms for the transformed model are uncorrelated. If the test indicates that they are uncorrelated, the procedure terminates. The fitted regression model in the original variables is then obtained by transforming the regression coefficients back according to (13.19).

If the Durbin-Watson test indicates that autocorrelation is still present after the first iteration, the parameter ρ is reestimated from the new residuals for the fitted regression model (13.19) with the original variables, which was derived from the fitted regression model (13.18) with the transformed variables. A new set of transformed variables is then obtained with the new r. This process may be continued for another iteration or two until the Durbin-Watson test suggests that the error terms in the transformed model are uncorrelated. If the process does not terminate after one or two iterations, a different procedure should be employed.

Example. For the Blaisdell Company example, the necessary calculations for estimating the autocorrelation parameter ρ, based on the residuals obtained with ordinary least squares applied to the original variables, appear in Table 13.3. Column 1 repeats the residuals from Table 13.2. Column 2 contains the residuals e_{t-1}, and columns 3 and 4 contain the necessary calculations. Hence, we estimate:

$$r = \frac{.0834478}{.1322122} = .631166$$

We now obtain the transformed variables Y_t' and X_t' in (13.17a) and (13.17b):

$$Y_t' = Y_t - .631166 Y_{t-1}$$

$$X_t' = X_t - .631166 X_{t-1}$$

These are found in Table 13.4. Columns 1 and 2 repeat the original variables Y_t and X_t, and columns 3 and 4 contain the transformed variables Y_t' and X_t'. Ordinary least squares fitting of linear regression is now used with these transformed variables based on the $n - 1$ cases remaining after the transformations. The fitted regression line and other regression results are given at the bottom of Table 13.4. The fitted regression line in the transformed variables is:

$$(13.22) \qquad \hat{Y}' = -.3941 + .17376 X'$$

where:

$$Y_t' = Y_t - .631166 Y_{t-1} \qquad X_t' = X_t - .631166 X_{t-1}$$

TABLE 13.3   Calculations for Estimating ρ with the Cochrane-Orcutt Procedure—Blaisdell Company Example

t	(1) e_t	(2) e_{t-1}	(3) $e_{t-1}e_t$	(4) e_{t-1}^2
1	−.026052	—	—	—
2	−.062015	−.026052	.0016156	.0006787
3	.022021	−.062015	−.0013656	.0038459
4	.163754	.022021	.0036060	.0004849
5	.046570	.163754	.0076260	.0268154
6	.046377	.046570	.0021598	.0021688
7	.043617	.046377	.0020228	.0021508
8	−.058435	.043617	−.0025488	.0019024
9	−.094399	−.058435	.0055162	.0034146
10	−.149142	−.094399	.0140789	.0089112
11	−.147991	−.149142	.0220718	.0222433
12	−.053054	−.147991	.0078515	.0219013
13	−.022928	−.053054	.0012164	.0028147
14	.105852	−.022928	−.0024270	.0005257
15	.085464	.105852	.0090465	.0112046
16	.106102	.085464	.0090679	.0073041
17	.029112	.106102	.0030889	.0112576
18	.042316	.029112	.0012319	.0008475
19	−.044160	.042316	−.0018687	.0017906
20	−.033009	−.044160	.0014577	.0019501
Total			.0834478	.1322122

$$r = \frac{\Sigma e_{t-1}e_t}{\Sigma e_{t-1}^2} = \frac{.0834478}{.1322122} = .631166$$

Since the random term in the transformed regression model (13.16) is the disturbance term u_t, $MSE = .00451$ is an estimate of the variance of this disturbance term; recall that $\sigma^2\{u_t\} = \sigma^2$.

Based on the fitted regression function for the transformed variables in (13.22), residuals were obtained and the Durbin-Watson statistic calculated. The result was (calculations not shown) $D = 1.65$. From Table A.6, we find for $\alpha = .01$, $p - 1 = 1$, and $n = 19$:

$$d_L = .93 \qquad d_U = 1.13$$

Since $D = 1.65 > d_U = 1.13$, we conclude that the autocorrelation coefficient for the error terms in the model with the transformed variables is zero.

Having successfully handled the problem of autocorrelated error terms, we now transform the fitted model in (13.22) back to the original variables, using (13.19):

$$b_0 = \frac{b_0'}{1 - r} = \frac{-.3941}{1 - .631166} = -1.0685$$

$$b_1 = b_1' = .17376$$

TABLE 13.4   Transformed Variables and Regression Results for First Iteration with Cochrane-Orcutt Procedure—Blaisdell Company Example

t	(1) Y_t	(2) X_t	(3) $Y_t' = Y_t - .631166Y_{t-1}$	(4) $X_t' = X_t - .631166X_{t-1}$
1	20.96	127.3	—	—
2	21.40	130.0	8.1708	49.653
3	21.96	132.7	8.4530	50.648
4	21.52	129.4	7.6596	45.644
5	22.39	135.0	8.8073	53.327
6	22.76	137.1	8.6282	51.893
7	23.48	141.2	9.1147	54.667
8	23.66	142.8	8.8402	53.679
9	24.10	145.5	9.1666	55.369
10	24.01	145.3	8.7989	53.465
11	24.54	148.3	9.3857	56.592
12	24.30	146.4	8.8112	52.798
13	25.00	150.2	9.6627	57.797
14	25.64	153.1	9.8608	58.299
15	26.36	157.3	10.1769	60.668
16	26.98	160.7	10.3425	61.418
17	27.52	164.2	10.4911	62.772
18	27.78	165.6	10.4103	61.963
19	28.24	168.7	10.7062	64.179
20	28.78	171.7	10.9559	65.222

$$\hat{Y}' = -.3941 + .17376X'$$

$$s\{b_0'\} = .1672 \qquad s\{b_1'\} = .002957$$

$$MSE = .00451$$

leading to the fitted regression line in the original variables:

(13.23) $$\hat{Y} = -1.0685 + .17376X$$

Finally, we obtain the estimated standard deviations of the regression coefficients for the original variables by using (13.20). Based on the results in Table 13.4, we find:

$$s\{b_0\} = \frac{s\{b_0'\}}{1 - r} = \frac{.1672}{1 - .631166} = .45332$$

$$s\{b_1\} = s\{b_1'\} = .002957$$

Comments

1. The Cochrane-Orcutt approach does not always work properly. A major reason is that when the error terms are positively autocorrelated, the estimate r in (13.21) tends to underestimate the autocorrelation parameter ρ. When this bias is serious, it can significantly reduce the effectiveness of the Cochrane-Orcutt approach.

2. There exists an approximate relation between the Durbin-Watson test statistic D in (13.13) and the estimated autocorrelation parameter r in (13.21):

(13.24)
$$D \simeq 2(1 - r)$$

This relation indicates that the Durbin-Watson statistic ranges approximately between 0 and 4 since r takes on values between -1 and 1, and that D is approximately 2 when $r = 0$. Note that for the Blaisdell Company example ordinary least squares regression fit, $D = .735$, $r = .631$, and $2(1 - r) = .738$.

3. Under certain circumstances, it may be helpful to construct pseudotransformed values for period 1 so that the regression for the transformed variables is based on n cases, rather than on $n - 1$ cases. Procedures for doing this are discussed in specialized texts such as Reference 13.4.

4. The least squares properties of the residuals, such as that the sum of the residuals is zero, apply to the residuals for the fitted regression function with the transformed variables, not to the residuals for the fitted regression function transformed back to the original variables.

Hildreth-Lu Procedure

The Hildreth-Lu procedure takes the same approach to estimating the autocorrelation parameter ρ for use in the transformations (13.17) as the Box-Cox approach to estimating the parameter λ in the power transformation of Y to improve the appropriateness of the standard regression model. That value of ρ is chosen with the Hildreth-Lu procedure, which minimizes the error sum of squares for the transformed regression model (13.16):

(13.25)
$$SSE = \sum (Y_t' - \hat{Y}_t')^2$$

$$= \sum (Y_t' - b_0' - b_1' X_t')^2$$

Computer programs are available to find the value of ρ that minimizes SSE. Alternatively, one can do a numerical search, running repeated regressions with different values of ρ for identifying the approximate magnitude of ρ that minimizes SSE. In the region of ρ that leads to minimum SSE, a finer search can be conducted to obtain a more precise value of ρ.

Once the value of ρ that minimizes SSE is found, the fitted regression function corresponding to that value of ρ is identified. If the transformation has successfully eliminated the autocorrelation, the fitted regression function in the original variables can then be obtained by means of (13.19).

Example. Table 13.5 contains the regression results for the Hildreth-Lu procedure when fitting the transformed regression model (13.16) to the Blaisdell Company data for different values of the autocorrelation parameter ρ. Note that SSE is minimized when ρ is near .96, so we shall let $r = .96$ be the estimate of ρ. The fitted regression function for the transformed variables corresponding to $r = .96$ and other re-

TABLE 13.5 Hildreth-Lu Results for Blaisdell Company Example

ρ	SSE
.10	.1170
.30	.0938
.50	.0805
.70	.0758
.90	.0728
.92	.0723
.94	.0718
.95	.07171
.96	.07167
.97	.07175
.98	.07197

For $\rho = .96$: $\hat{Y}' = .07117 + .16045X'$

$$s\{b_0'\} = .05798 \qquad s\{b_1'\} = .006840$$

$$MSE = .00422$$

gression results are given at the bottom of Table 13.5. The fitted regression function in the transformed variables is:

$$(13.26) \qquad \hat{Y}' = .07117 + .16045X'$$

where:

$$Y_t' = Y_t - .96Y_{t-1}$$
$$X_t' = X_t - .96X_{t-1}$$

The Durbin-Watson test statistic for this fitted model is $D = 1.73$. Since for $n = 19$, $p - 1 = 1$, and $\alpha = .01$ the upper critical value is $d_U = 1.13$, we conclude that no autocorrelation remains in the transformed model.

Therefore, we shall transform regression function (13.26) back to the original variables. Using (13.19), we obtain:

$$(13.27) \qquad \hat{Y} = 1.7793 + .16045X$$

The estimated standard deviations of these regression coefficients are:

$$s\{b_0\} = 1.450 \qquad s\{b_1\} = .006840$$

Comments

1. The Hildreth-Lu procedure, unlike the Cochrane-Orcutt procedure, does not require any iterations once the estimate of the autocorrelation parameter ρ is obtained.

2. Note from Table 13.5 that SSE as a function of ρ is quite stable in a wide region around the minimum. This is often the case, and indicates that the numerical search for finding the best value of ρ need not be too fine unless there is particular interest in the intercept term β_0, since the estimate b_0 is sensitive to the value of r.

First Differences Procedure

Since the autocorrelation parameter ρ is frequently large, and *SSE* as a function of ρ often is quite flat for large values of ρ up to 1.0, as in the Blaisdell example, some economists and statisticians have suggested use of $\rho = 1.0$ in the transformed model (13.16). If $\rho = 1$, $\beta_0' = \beta_0(1 - \rho) = 0$, and the transformed model (13.16) becomes:

(13.28) $$Y_t' = \beta_1' X_t' + u_t$$

where:

(13.28a) $$Y_t' = Y_t - Y_{t-1}$$

(13.28b) $$X_t' = X_t - X_{t-1}$$

Thus again, the regression coefficient $\beta_1' = \beta_1$ can be directly estimated by ordinary least squares methods, this time based on regression through the origin. Note that the transformed variables in (13.28a) and (13.28b) are ordinary first differences. It has been found that this first differences approach is effective in a variety of applications in reducing the autocorrelations of the error terms, and of course it is much simpler than the Cochrane-Orcutt and Hildreth-Lu procedures.

The fitted regression function in the transformed variables:

(13.29) $$\hat{Y}' = b_1' X'$$

can be transformed back to the original variables as follows:

(13.30) $$\hat{Y} = b_0 + b_1 X$$

where:

(13.30a) $$b_0 = \bar{Y} - b_1' \bar{X}$$

(13.30b) $$b_1 = b_1'$$

Example.   Table 13.6 contains the transformed variables Y_t' and X_t', based on the first differences transformations in (13.28a, b) for the Blaisdell Company example. Application of ordinary least squares for estimating a linear regression through the origin leads to the results shown at the bottom of Table 13.6. The fitted regression line in the transformed variables is:

(13.31) $$\hat{Y}' = .16849 X'$$

where:

$$Y_t' = Y_t - Y_{t-1}$$
$$X_t' = X_t - X_{t-1}$$

To examine whether the first differences procedure has removed the autocorrelations, we shall use the Durbin-Watson test. There are two points to note when using the Durbin-Watson test with the first differences procedure. Sometimes the first differences procedure can overcorrect, leading to negative autocorrelations in the error terms. Hence, it may be appropriate to use a two-sided Durbin-Watson test when

TABLE 13.6 First Differences and Regression Results with First Differences Procedure—Blaisdell Company Example

t	(1) Y_t	(2) X_t	(3) $Y_t' = Y_t - Y_{t-1}$	(4) $X_t' = X_t - X_{t-1}$
1	20.96	127.3	—	—
2	21.40	130.0	.44	2.7
3	21.96	132.7	.56	2.7
4	21.52	129.4	−.44	−3.3
5	22.39	135.0	.87	5.6
6	22.76	137.1	.37	2.1
7	23.48	141.2	.72	4.1
8	23.66	142.8	.18	1.6
9	24.10	145.5	.44	2.7
10	24.01	145.3	−.09	−.2
11	24.54	148.3	.53	3.0
12	24.30	146.4	−.24	−1.9
13	25.00	150.2	.70	3.8
14	25.64	153.1	.64	2.9
15	26.36	157.3	.72	4.2
16	26.98	160.7	.62	3.4
17	27.52	164.2	.54	3.5
18	27.78	165.6	.26	1.4
19	28.24	168.7	.46	3.1
20	28.78	171.7	.54	3.0

$$\hat{Y}' = .16849X'$$

$$s\{b_1'\} = .005096 \quad MSE = .00482$$

testing for autocorrelation with first differences data. The second point is that the first differences model (13.28) has no intercept term, but the Durbin-Watson test requires a fitted regression with an intercept term. A valid test for autocorrelation in a no-intercept model can be carried out by fitting for this purpose a regression function with an intercept term. Of course, the fitted no-intercept model is still the model of basic interest.

In the Blaisdell Company example, the Durbin-Watson statistic for the fitted first differences regression model with an intercept term is $D = 1.75$. This indicates uncorrelated error terms for either a one-sided test (with $\alpha = .01$) or a two-sided test (with $\alpha = .02$).

With the first differences procedure successfully eliminating the autocorrelation, we return to a fitted model in the original variables by using (13.30):

(13.32) $$\hat{Y} = -.30349 + .16849X$$

where:

$$b_0 = 24.569 - .16849(147.62) = -.30349$$

We know from Table 13.6 that the estimated standard deviation of b_1 is $s\{b_1\} = .005096$ since $b_1 = b_1'$.

TABLE 13.7   Major Regression Results for Three Transformation Procedures—
Blaisdell Company Example

Procedure	b_1	$s\{b_1\}$	r	Estimate of σ^2 (MSE)
Cochrane-Orcutt	.1738	.0030	.63	.0045
Hildreth-Lu	.1605	.0068	.96	.0042
First differences	.1685	.0051	1.0	.0048
Original variables	.1763	.0014	—	—

Comparison of Three Methods

Table 13.7 contains some of the main regression results for the three transformation methods, as well for the ordinary least squares regression fit to the original variables. A number of key points stand out.

1. All of the estimates of β_1 are quite close to each other.

2. The estimated standard deviations of b_1 based on Hildreth-Lu and first differences transformation methods are quite close to each other; that with the Cochrane-Orcutt procedure is somewhat smaller. The estimated standard deviation of b_1 based on ordinary least squares regression with the original variables is still smaller. This is as expected, since we noted earlier that the estimated standard deviations $s\{b_k\}$ calculated according to ordinary least squares may seriously underestimate the true standard deviations $\sigma\{b_k\}$ when positive autocorrelation is present.

3. All three transformation methods provide essentially the same estimate of σ^2, the variance of the disturbance terms u_t.

The three transformation methods do not always work equally well, as happens to be the case here for the Blaisdell Company example. The Cochrane-Orcutt procedure may not remove autocorrelation in only one or two iterations, in which case the Hildreth-Lu or the first differences procedures may be preferable. When several of the transformation methods are effective in removing autocorrelation, then simplicity of calculations may be considered in choosing from among these procedures.

Note

Further discussions of the Cochrane-Orcutt, Hildreth-Lu, and first differences procedures, as well as other remedial procedures for autocorrelated errors, may be found in specialized texts, such as Reference 13.4.

13.5 FORECASTING WITH AUTOCORRELATED ERROR TERMS

One important use of autoregressive error regression models is to make forecasts. With these models, one can incorporate information about the error term in the most recent period n into the forecast for period $n + 1$. This will provide a more accurate forecast because when autoregressive error regression models are appropriate, the error terms in successive periods are correlated. Thus, if sales in period n are above their expected value and successive error terms are positively correlated, it follows that sales in period $n + 1$ will likely be above their expected value also.

We shall explain the basic ideas underlying the development of forecasts by again using the simple linear autoregressive error term regression model (13.1). The extension to multiple regression model (13.2) is direct. First, we consider forecasting when either the Cochrane-Orcutt or the Hildreth-Lu procedure has been utilized for estimating the regression parameters.

When we express regression model (13.1):

$$Y_t = \beta_0 + \beta_1 X_t + \varepsilon_t$$

by using the structure of the error terms:

$$\varepsilon_t = \rho \varepsilon_{t-1} + u_t$$

we obtain:

$$Y_t = \beta_0 + \beta_1 X_t + \rho \varepsilon_{t-1} + u_t$$

For period $n + 1$, we obtain:

(13.33) $$Y_{n+1} = \beta_0 + \beta_1 X_{n+1} + \rho \varepsilon_n + u_{n+1}$$

Thus, Y_{n+1} is made up of three components:

1. The expected value $\beta_0 + \beta_1 X_{n+1}$.
2. A multiple ρ of the preceding error term ε_n.
3. An independent, random disturbance term with $E\{u_{n+1}\} = 0$.

The forecast for next period $n + 1$, to be denoted by F_{n+1}, is constructed by dealing with each of the three components in (13.33):

1. Given X_{n+1}, we estimate the expected value $\beta_0 + \beta_1 X_{n+1}$ as usual from the fitted regression function:

$$\hat{Y}_{n+1} = b_0 + b_1 X_{n+1}$$

where b_0 and b_1 are the estimated regression coefficients for the original variables obtained from b_0' and b_1' for the transformed variables according to (13.19).

2. ρ is estimated by r, and ε_n is estimated by the residual e_n:

$$e_n = Y_n - (b_0 + b_1 X_n) = Y_n - \hat{Y}_n$$

Thus, $\rho \varepsilon_n$ is estimated by $r e_n$.

3. The disturbance term u_{n+1} has expected value zero and is independent of earlier information. Hence, we use its expected value of zero in the forecast.

Thus, the forecast for period $n + 1$ is:

(13.34) $$F_{n+1} = \hat{Y}_{n+1} + r e_n$$

An approximate $1 - \alpha$ prediction interval for $Y_{n+1(\text{new})}$, the new observation on the dependent variable, may be obtained by employing the usual prediction interval for a new observation in (3.35), but based on the transformed observations. Thus, Y_i and

X_i in formula (3.37a) for the estimated variance $s^2\{Y_{h(\text{new})}\}$ are replaced by Y_t' and X_t' as defined in (13.17a, b).

The approximate $1 - \alpha$ prediction limits for $Y_{n+1(\text{new})}$ with simple linear regression therefore are:

$$(13.35) \qquad\qquad F_{n+1} \pm t(1 - \alpha/2; n - 3)s\{Y_{n+1(\text{new})}\}$$

where $s\{Y_{n+1(\text{new})}\}$, defined in (3.37a), is here based on the transformed observations. Note the use of $n - 3$ degrees of freedom for the t multiple, since there are only $n - 1$ transformed cases and two degrees of freedom are lost for estimating the two parameters in the simple linear regression function.

When forecasts are based on the first differences procedure, the forecast in (13.34) is still applicable, but $r = 1$ now. The estimated standard deviation $s\{Y_{n+1(\text{new})}\}$ now is calculated according to (5.21) for one independent variable, using the transformed variables. Finally, the degrees of freedom for the t multiple in (13.35) will be $n - 2$, since only one parameter has to be estimated in the no-intercept regression model (13.28).

Example

For the Blaisdell Company example, the trade association has projected that deseasonalized industry sales in the first quarter of 1988 (i.e., quarter 21) will be $X_{21} = \$175.3$ million. To forecast Blaisdell Company sales for quarter 21, we shall use the Cochrane-Orcutt fitted regression function (13.23):

$$\hat{Y} = -1.0685 + .17376X$$

First, we need to obtain the residual e_{20}:

$$e_{20} = Y_{20} - \hat{Y}_{20} = 28.78 - [-1.0685 + .17376(171.7)] = .0139$$

The fitted value for $X_{21} = 175.3$ is:

$$\hat{Y}_{21} = -1.0685 + .17376(175.3) = 29.392$$

The forecast for period 21 then is:

$$F_{21} = \hat{Y}_{21} + re_{20} = 29.392 + .631166(.0139) = 29.40$$

Note how the fact that company sales in quarter 20 were slightly above their estimated mean has a small positive influence on the forecast for company sales for quarter 21.

We wish to set up a 95 percent prediction interval for $Y_{21(\text{new})}$. Using the data for the transformed variables in Table 13.4, we calculate $s\{Y_{n+1(\text{new})}\}$ by (3.37) for:

$$X_{n+1}' = X_{n+1} - .631166X_n = 175.3 - .631166(171.7) = 66.929$$

We obtain $s\{Y_{n+1(\text{new})}\} = .0757$ (calculations not shown). We require $t(.975; 17) = 2.110$. We therefore obtain the prediction limits $29.40 \pm 2.110(.0757)$ and the prediction interval:

$$29.24 \leq Y_{21(\text{new})} \leq 29.56$$

Given quarter 20 seasonally adjusted company sales of $28.78 million and other past sales, and given quarter 21 industry sales of $175.3 million, we predict with approximate 95 percent confidence that seasonally adjusted Blaisdell Company sales in quarter 21 will be between $29.24 and $29.56 million.

To obtain a forecast of actual sales including seasonal effects in quarter 21, the Blaisdell Company will still need to incorporate the first quarter seasonal effect into the forecast of seasonally adjusted sales.

The forecasts with the other transformation procedures are very similar to the one with the Cochrane-Orcutt procedure. With the first differences estimated regression function (13.32), the forecast for quarter 21 is:

$$F_{21} = [-.30349 + .16849(175.3)]$$
$$+ 1.0[28.78 + .30349 - .16849(171.70)]$$
$$= 29.39$$

The estimated standard deviation $s\{Y_{n+1(new)}\}$ calculated according to (5.21) with the transformed data in Table 13.6 is $s\{Y_{n+1(new)}\} = .0718$ (calculations not shown). For a 95 percent prediction interval, we require $t(.975; 18) = 2.101$. The prediction limits therefore are $29.39 \pm 2.101(.0718)$ and the approximate 95 percent prediction interval is:

$$29.24 \leq Y_{21(new)} \leq 29.54$$

This forecast is practically the same as that with the Cochrane-Orcutt estimates.

The approximate 95 percent prediction interval with the estimated regression function (13.27) based on the Hildreth-Lu procedure is (calculations not shown):

$$29.24 \leq Y_{21(new)} \leq 29.52$$

This forecast is practically the same as the other two.

Comments

1. Forecasts obtained with autoregressive error regression models (13.1) and (13.2) are conditional on the past observations Y_n, Y_{n-1}, etc. They are also conditional on X_{n+1}, which often has to be projected as in the Blaisdell Company example.

2. Forecasts for two or more periods ahead can also be developed, using the recursive relations of ε_t to earlier error terms developed in Section 13.2. For example, given X_{n+2} the forecast for period $n + 2$, based on either Cochrane-Orcutt or Hildreth-Lu estimates, is:

$$(13.36) \qquad F_{n+2} = \hat{Y}_{n+2} + r^2 e_n$$

For the first differences estimates, the forecast in (13.36) is calculated with $r = 1$.

3. The approximate prediction limits (13.35) assume that the value of r used in the transformations (13.17a, b) is the true value of ρ; that is, $r = \rho$. If that is the case, the standard regression assumptions apply since we are then dealing with the transformed model (13.16). To see that the prediction limits obtained from the transformed model are applicable to the forecast F_{n+1} in (13.34), recall that $\sigma^2\{Y_{h(new)}\}$ in (3.36) is the variance of the difference $\hat{Y}_h - Y$, where Y is the new observation to be predicted. In terms of the situation here for the transformed variables, we have the following correspondences:

$$\hat{Y}_h \text{ corresponds to } \hat{Y}'_{n+1} = b'_0 + b'_1 X'_{n+1} = b_0(1 - r) + b_1(X_{n+1} - rX_n)$$

$$Y \text{ corresponds to } Y'_{n+1} = Y_{n+1} - rY_n$$

The difference $\hat{Y}'_{n+1} - Y'_{n+1}$ therefore is:

$$\hat{Y}'_{n+1} - Y'_{n+1} = b_0(1 - r) + b_1(X_{n+1} - rX_n) - (Y_{n+1} - rY_n)$$

$$= (b_0 + b_1 X_{n+1}) + r(Y_n - b_0 - b_1 X_n) - Y_{n+1}$$

$$= \hat{Y}_{n+1} + re_n - Y_{n+1}$$

$$= F_{n+1} - Y_{n+1}$$

Hence, F_{n+1} plays the role of $\hat{Y}_h$ in (3.36) and Y_{n+1} plays the role of the new observation Y. The prediction limits (13.35) are approximate because r is only an estimate of ρ.

CITED REFERENCES

13.1. Box, G. E. P., and G. M. Jenkins. *Time Series Analysis, Forecasting and Control.* Rev. ed. San Francisco: Holden-Day, 1976.

13.2. Durbin, J., and G. S. Watson. "Testing for Serial Correlation in Least Squares Regression. II." *Biometrika* 38 (1951), pp. 159–78.

13.3. Theil, H., and A. L. Nagar. "Testing the Independence of Regression Disturbances." *Journal of the American Statistical Association* 56 (1961), pp. 793–806.

13.4. Pindyck, R. S., and D. L. Rubinfeld. *Econometric Models and Economic Forecasts.* 2nd ed. New York: McGraw-Hill, 1981.

PROBLEMS

13.1. Refer to Table 13.1.
  a. Plot ε_t against ε_{t-1} for $t = 1, \ldots, 10$ on a graph. How is the positive first-order autocorrelation in the error terms shown by the plot?
  b. If you plotted u_t against ε_{t-1} for $t = 1, \ldots, 10$, what pattern would you expect?

13.2. Refer to **Plastic hardness** Problem 2.20. If the same test item were measured at 12 different points in time, would the error terms in the regression model likely be autocorrelated? Discuss.

13.3. A student stated that the first-order autoregressive error models (13.1) and (13.2) are too simple for business time series data because the error term in period t in such data is also influenced by random effects that occurred more than one period in the past. Comment.

13.4. A student writing a term paper used ordinary least squares in fitting a simple linear regression model to some time series data containing positively autocorrelated errors, and found that the 90 percent confidence interval for β_1 was too wide to be useful. She then decided to employ regression model (13.1) to improve the precision of the estimate. Comment.

13.5. For each of the following tests concerning the autocorrelation parameter ρ in regression model (13.2) with three independent variables, state the appropriate decision rule based on the Durbin-Watson statistic for a sample of size 38: (1) H_0: $\rho = 0$, H_a: $\rho \neq 0$, $\alpha = .02$; (2) H_0: $\rho = 0$, H_a: $\rho < 0$, $\alpha = .05$; (3) H_0: $\rho = 0$, H_a: $\rho > 0$, $\alpha = .01$.

13.6. Refer to **Calculator maintenance** Problem 2.18. The observations are listed in time order. Assume that regression model (13.1) is appropriate. Test whether or not positive autocorrelation is present; use $\alpha = .01$. State the alternatives, decision rule, and conclusion.

13.7. Refer to **Chemical shipment** Problem 7.12. The observations are listed in time order. Assume that regression model (13.2) is appropriate. Test whether or not positive autocorrelation is present; use $\alpha = .05$. State the alternatives, decision rule, and conclusion.

13.8. Refer to **Crop yield** Problem 9.12. The observations are listed in time order. Assume that regression model (13.2) with first- and second-order terms for the two independent variables and no interaction term is appropriate. Test whether or not positive autocorrelation is present; use $\alpha = .01$. State the alternatives, decision rule, and conclusion.

13.9. Microcomputer components. A staff analyst for a manufacturer of microcomputer components has compiled monthly data for the past 16 months on the value of industry production of processing units that use these components (X, in million dollars) and the value of the firm's components used (Y, in thousand dollars). The analyst believes that a simple linear regression relation is appropriate but anticipates positive autocorrelation. The data follow.

t:	1	2	3	4	5	6	7	8
X_t:	2.052	2.026	2.002	1.949	1.942	1.887	1.986	2.053
Y_t:	102.9	101.5	100.8	98.0	97.3	93.5	97.5	102.2

t:	9	10	11	12	13	14	15	16
X_t:	2.102	2.113	2.058	2.060	2.035	2.080	2.102	2.150
Y_t:	105.0	107.2	105.1	103.9	103.0	104.8	105.0	107.2

a. Fit a simple linear regression model by ordinary least squares and obtain the residuals. Also obtain $s\{b_0\}$ and $s\{b_1\}$.

b. Plot the residuals against time and explain whether you find any evidence of positive autocorrelation.

c. Conduct a formal test for positive autocorrelation using a significance level of .05. State the alternatives, decision rule, and conclusion. Is the residual analysis in part (b) in accord with the test result?

13.10. Refer to **Microcomputer components** Problem 13.9. The analyst has decided to employ regression model (13.1) and use the Cochrane-Orcutt procedure to fit the model.

a. Obtain a point estimate of the autocorrelation parameter. How well does the approximate relationship (13.24) hold here between this point estimate and the Durbin-Watson test statistic?

b. Use one iteration to obtain the estimates b_0' and b_1' of the regression coefficients β_0' and β_1' in transformed model (13.16) and state the estimated regression function. Also obtain $s\{b_0'\}$ and $s\{b_1'\}$.

c. Test whether any positive autocorrelation remains after the first iteration using a significance level of .05. State the alternatives, decision rule, and conclusion.

    d. Restate the estimated regression function obtained in part (b) in terms of the original variables. Also obtain $s\{b_0\}$ and $s\{b_1\}$. Compare the estimated regression coefficients obtained with the Cochrane-Orcutt procedure and their estimated standard deviations with those obtained with ordinary least squares in Problem 13.9a.

    e. Based on the results in parts (c) and (d), does the Cochrane-Orcutt procedure appear to have been effective here?

    f. The value of industry production in month 17 will be \$2.210 million. Predict the value of the firm's components used in month 17; employ a 95 percent prediction interval. Interpret your interval.

    g. Estimate β_1 with a 95 percent confidence interval. Interpret your interval.

13.11. Refer to **Microcomputer components** Problem 13.9. Assume that regression model (13.1) is applicable.

    a. Use the Hildreth-Lu procedure to obtain a point estimate of the autocorrelation parameter. Do a search at the values $\rho = .1, .2, \ldots, 1.0$ and select from these the value of ρ that minimizes SSE.

    b. Based on your estimate in part (a), obtain an estimate of the transformed regression function (13.16). Also obtain $s\{b_0'\}$ and $s\{b_1'\}$.

    c. Test whether any positive autocorrelation remains in the transformed regression model; use $\alpha = .05$. State the alternatives, decision rule, and conclusion.

    d. Restate the estimated regression function obtained in part (b) in terms of the original variables. Also obtain $s\{b_0\}$ and $s\{b_1\}$. Compare the estimated regression coefficients obtained with the Hildreth-Lu procedure and their estimated standard deviations with those obtained with ordinary least squares in Problem 13.9a.

    e. Based on the results in parts (c) and (d), has the Hildreth-Lu procedure been effective here?

    f. The value of industry production in month 17 will be \$2.210 million. Predict the value of the firm's components used in month 17; employ a 95 percent prediction interval. Interpret your interval.

    g. Estimate β_1 with a 95 percent confidence interval. Interpret your interval.

13.12. Refer to **Microcomputer components** Problem 13.9. Assume that regression model (13.1) is applicable and that the first differences procedure is to be employed.

    a. Estimate the regression coefficient β_1' in the transformed regression model (13.28), and obtain the estimated standard deviation of this estimate. State the estimated regression function.

    b. Test whether or not the error terms with the first differences procedure are autocorrelated using a two-sided test and a level of significance of .10. State the alternatives, decision rule, and conclusion. Why is a two-sided test meaningful here?

    c. Restate the estimated regression function obtained in part (a) in terms of the original variables. Also obtain $s\{b_1\}$. Compare the estimated regression coefficients obtained with the first differences procedure and the estimated standard deviation $s\{b_1\}$ with the results obtained with ordinary least squares in Problem 13.9a.

    d. Based on the results in parts (b) and (c), has the first differences procedure been effective here?

    e. The value of industry production in month 17 will be \$2.210 million. Predict the value of the firm's components used in month 17; employ a 95 percent prediction interval. Interpret your interval.

    f. Estimate β_1 with a 95 percent confidence interval. Interpret your interval.

13.13. Advertising agency. The managing partner of an advertising agency is interested in the possibility of making accurate predictions of monthly billings. Monthly data on amount of billings (Y, in thousands of constant dollars) and on number of hours of staff time (X, in thousand hours) for the 20 most recent months follow. A simple linear regression model is believed to be appropriate, but positively autocorrelated error terms may be present.

t:	1	2	3	4	5	6	7
X_t:	2.521	2.171	2.234	2.524	2.305	2.523	3.020
Y_t:	220.4	203.9	207.2	221.9	211.3	222.7	247.6

t:	8	9	10	11	12	13	14
X_t:	3.014	3.532	3.461	3.737	3.801	3.576	3.586
Y_t:	247.6	272.9	269.1	283.9	287.0	275.4	275.1

t:	15	16	17	18	19	20
X_t:	3.447	2.723	3.019	3.117	3.623	3.618
Y_t:	269.1	232.8	248.1	252.4	278.6	278.5

    a. Fit a simple linear regression model by ordinary least squares and obtain the residuals. Also obtain $s\{b_0\}$ and $s\{b_1\}$.

    b. Plot the residuals against time and explain whether you find any evidence of positive autocorrelation.

    c. Conduct a formal test for positive autocorrelation using a significance level of .01. State the alternatives, decision rule, and conclusion. Is the residual analysis in part (b) in accord with the test result?

13.14. Refer to **Advertising agency** Problem 13.13. Assume that regression model (13.1) is applicable and that the Cochrane-Orcutt procedure is to be employed.

    a. Obtain a point estimate of the autocorrelation parameter. How well does the approximate relationship (13.24) hold here between the point estimate and the Durbin-Watson test statistic?

    b. Use one iteration to obtain the estimates b_0' and b_1' of the regression coefficients β_0' and β_1' in transformed model (13.16) and state the estimated regression function. Also obtain $s\{b_0'\}$ and $s\{b_1'\}$.

    c. Test whether any positive autocorrelation remains after the first iteration using a significance level of .01. State the alternatives, decision rule, and conclusion.

    d. Restate the estimated regression function obtained in part (b) in terms of the original variables. Also obtain $s\{b_0\}$ and $s\{b_1\}$. Compare the estimated regression coefficients obtained with the Cochrane-Orcutt procedure and their estimated standard deviations with those obtained with ordinary least squares in Problem 13.13a.

    e. Based on the results in parts (c) and (d), does the Cochrane-Orcutt procedure appear to have been effective here?

    f. Staff time in month 21 is expected to be 3.625 thousand hours. Predict the amount of billings in constant dollars for month 21, using a 99 percent prediction interval. Interpret your interval.

    g. Estimate β_1 with a 99 percent confidence interval. Interpret your interval.

13.15. Refer to **Advertising agency** Problem 13.13. Assume that regression model (13.1) is applicable.

a. Use the Hildreth-Lu procedure to obtain a point estimate of the autocorrelation parameter. Do a search at the values $\rho = .1, .2, \ldots, 1.0$ and select from these the value of ρ that minimizes SSE.

b. Based on your estimate in part (a), obtain an estimate of the transformed regression function (13.16). Also obtain $s\{b_0'\}$ and $s\{b_1'\}$.

c. Test whether any positive autocorrelation remains in the transformed regression model; use $\alpha = .01$. State the alternatives, decision rule, and conclusion.

d. Restate the estimated regression function obtained in part (b) in terms of the original variables. Also obtain $s\{b_0\}$ and $s\{b_1\}$. Compare the estimated regression coefficients obtained with the Hildreth-Lu procedure and their estimated standard deviations with those obtained with ordinary least squares in Problem 13.13a.

e. Based on the results in parts (c) and (d), has the Hildreth-Lu procedure been effective here?

f. Staff time in month 21 is expected to be 3.625 thousand hours. Predict the amount of billings in constant dollars for month 21, using a 99 percent prediction interval. Interpret your interval.

g. Estimate β_1 with a 99 percent confidence interval. Interpret your interval.

13.16. Refer to **Advertising agency** Problem 13.13. Assume that regression model (13.1) is applicable and that the first differences procedure is to be employed.

a. Estimate the regression coefficient β_1' in the transformed regression model (13.28), and obtain the estimated standard deviation of this estimate. State the estimated regression function.

b. Test whether or not the error terms with the first differences procedure are autocorrelated using a two-sided test and a level of significance of .02. State the alternatives, decision rule, and conclusion. Why is a two-sided test meaningful here?

c. Restate the estimated regression function obtained in part (a) in terms of the original variables. Also obtain $s\{b_1\}$. Compare the estimated regression coefficients obtained with the first differences procedure and the estimated standard deviation $s\{b_1\}$ with the results obtained with ordinary least squares in Problem 13.13a.

d. Based on the results in parts (b) and (c), has the first differences procedure been effective here?

e. Staff time in month 21 is expected to be 3.625 thousand hours. Predict the amount of billings in constant dollars for month 21, using a 99 percent prediction interval. Interpret your interval.

f. Estimate β_1 with a 99 percent confidence interval. Interpret your interval.

13.17. **McGill Company sales.** The data below show seasonally adjusted quarterly sales for the McGill Company (Y, in million dollars) and for the entire industry (X, in million dollars), for the most recent 20 quarters.

t:	1	2	3	4	5	6	7
X_t:	127.3	130.0	132.7	129.4	135.0	137.1	141.1
Y_t:	20.96	21.40	21.96	21.52	22.39	22.76	23.48

t:	8	9	10	11	12	13	14
X_t:	142.8	145.5	145.3	148.3	146.4	150.2	153.1
Y_t:	23.66	24.10	24.01	24.54	24.28	25.00	25.64

t:	15	16	17	18	19	20
X_t:	157.3	160.7	164.2	165.6	168.7	172.0
Y_t:	26.46	26.98	27.52	27.78	28.24	28.78

a. Would you expect the autocorrelation parameter ρ to be positive, negative, or zero here?

b. Fit a simple linear regression model by ordinary least squares and obtain the residuals. Also obtain $s\{b_0\}$ and $s\{b_1\}$.

c. Plot the residuals against time and explain whether you find any evidence of positive autocorrelation.

d. Conduct a formal test for positive autocorrelation using $\alpha = .01$. State the alternatives, decision rule, and conclusion. Is the residual analysis in part (c) in accord with the test result?

13.18. Refer to **McGill Company sales** Problem 13.17. Assume that regression model (13.1) is applicable and that the Cochrane-Orcutt procedure is to be employed.

a. Obtain a point estimate of the autocorrelation parameter. How well does the approximate relationship (13.24) hold here between the point estimate and the Durbin-Watson test statistic?

b. Use one iteration to obtain the estimates b_0' and b_1' of the regression coefficients β_0' and β_1' in transformed model (13.16) and state the estimated regression function. Also obtain $s\{b_0'\}$ and $s\{b_1'\}$.

c. Test whether any positive autocorrelation remains after the first iteration; use $\alpha = .01$. State the alternatives, decision rule, and conclusion.

d. Restate the estimated regression function obtained in part (b) in terms of the original variables. Also obtain $s\{b_0\}$ and $s\{b_1\}$. Compare the estimated regression coefficients obtained with the Cochrane-Orcutt procedure and their estimated standard deviations with those obtained with ordinary least squares in Problem 13.17b.

e. Based on the results in parts (c) and (d), does the Cochrane-Orcutt procedure appear to have been effective here?

f. Industry sales for quarter 21 are expected to be $181.0 million. Predict the McGill Company sales for quarter 21, using a 90 percent prediction interval. Interpret your interval.

g. Estimate β_1 with a 90 percent confidence interval. Interpret your interval.

13.19. Refer to **McGill Company sales** Problem 13.17. Assume that regression model (13.1) is applicable.

a. Use the Hildreth-Lu procedure to obtain a point estimate of the autocorrelation parameter. Do a search at the values $\rho = .1, .2, \ldots, 1.0$ and select from these the value of ρ that minimizes SSE.

b. Based on your estimate in part (a), obtain an estimate of the transformed regression function (13.16). Also obtain $s\{b_0'\}$ and $s\{b_1'\}$.

c. Test whether any positive autocorrelation remains in the transformed regression model; use $\alpha = .01$. State the alternatives, decision rule, and conclusion.

d. Restate the estimated regression function obtained in part (b) in terms of the original variables. Also obtain $s\{b_0\}$ and $s\{b_1\}$. Compare the estimated regression coefficients obtained with the Hildreth-Lu procedure and their estimated standard deviations with those obtained with ordinary least squares in Problem 13.17b.

e. Based on the results in parts (c) and (d), has the Hildreth-Lu procedure been effective here?

      f. Industry sales for quarter 21 are expected to be \$181.0 million. Predict the McGill Company sales for quarter 21, using a 90 percent prediction interval. Interpret your interval.

      g. Estimate β_1 with a 90 percent confidence interval. Interpret your interval.

13.20. Refer to **McGill Company sales** Problem 13.17. Assume that regression model (13.1) is applicable and that the first differences procedure is to be employed.

      a. Estimate the regression coefficient β_1' in the transformed regression model (13.28) and obtain the estimated standard deviation of this estimate. State the estimated regression function.

      b. Test whether or not the error terms with the first differences procedure are positively autocorrelated using $\alpha = .01$. State the alternatives, decision rule, and conclusion.

      c. Restate the estimated regression function obtained in part (a) in terms of the original variables. Also obtain $s\{b_1\}$. Compare the estimated regression coefficients obtained with the first differences procedure and the estimated standard deviation $s\{b_1\}$ with the results obtained with ordinary least squares in Problem 13.17b.

      d. Based on the results in parts (b) and (c), has the first differences procedure been effective here?

      e. Industry sales for quarter 21 are expected to be \$181.0 million. Predict the McGill Company sales for quarter 21, using a 90 percent prediction interval. Interpret your interval.

      f. Estimate β_1 with a 90 percent confidence interval. Interpret your interval.

13.21. A student applying the first differences transformations in (13.28a, b) found that several X_t' values equaled zero but that the corresponding Y_t' values were nonzero. Does this signify that the first differences transformations are not apt for the data?

EXERCISES

13.22. Derive (13.8) for $s = 2$.

13.23. Refer to the first-order autoregressive error model (13.1). Suppose Y_t is company's percent share of the market, X_t is company's selling price as a percent of average competitive selling price, $\beta_0 = 100$, $\beta_1 = -.35$, $\rho = .6$, $\sigma^2 = 1$, and $\varepsilon_0 = 2.403$. Let X_t and u_t be as follows for $t = 1, \ldots, 10$:

t:	1	2	3	4	5	6	7	8	9	10
X_t:	100	115	120	90	85	75	70	95	105	110
u_t:	.764	.509	−.242	−1.808	−.485	.501	−.539	.434	−.299	.030

      a. Plot the true regression line. Generate the observations Y_t ($t = 1, \ldots, 10$), and plot these on the same graph. Fit a least squares regression line to the generated observations Y_t and plot it also on the same graph. How does your fitted regression line relate to the true line?

      b. Repeat the steps in part (a) but this time let $\rho = 0$. In which of the two cases does the fitted regression line come closer to the true line? Is this the expected outcome?

      c. Generate the observations Y_t for $\rho = -.7$. For each of the cases $\rho = .6$, $\rho = 0$, and $\rho = -.7$, obtain the successive error term differences $\varepsilon_t - \varepsilon_{t-1}$ ($t = 1, \ldots, 10$).

d. For which of the three cases in part (c) is $\Sigma(\varepsilon_t - \varepsilon_{t-1})^2$ smallest? For which is it largest? What generalization does this suggest?

13.24. For multiple regression model (13.2) with $p - 1 = 2$, derive the transformed model in which the random terms are uncorrelated.

13.25. Suppose the autoregressive error process for the model $Y_t = \beta_0 + \beta_1 X_t + \varepsilon_t$ is that given by (13.10).

a. What would be the transformed variables Y_t' and X_t' for which the random terms in the regression model are uncorrelated?

b. How would you estimate the parameters ρ_1 and ρ_2 for use with the Cochrane-Orcutt procedure?

c. How would you estimate the parameters ρ_1 and ρ_2 with the Hildreth-Lu procedure?

13.26. Derive the forecast F_{n+1} for a simple linear regression model with the second-order autogressive error process (13.10).

PROJECTS

13.27. The true regression model is $Y_t = 10 + 24X_t + \varepsilon_t$, where $\varepsilon_t = .8\varepsilon_{t-1} + u_t$ and u_t are independent $N(0, 25)$.

a. Generate 11 independent random numbers from $N(0, 25)$. Use the first random number as ε_0, obtain the 10 error terms $\varepsilon_1, \ldots, \varepsilon_{10}$, and then calculate the 10 observations $Y_1, \ldots, Y_{10}$ corresponding to $X_1 = 1, X_2 = 2, \ldots, X_{10} = 10$. Fit a linear regression function by ordinary least squares and calculate MSE.

b. Repeat part (a) 100 times, using new random numbers each time.

c. Calculate the mean of the 100 estimates b_1. Does it appear that b_1 is an unbiased estimator of β_1 despite the presence of positive autocorrelation?

d. Calculate the mean of the 100 estimates MSE. Does it appear that MSE is a biased estimator of σ^2? If so, does the magnitude of the bias appear to be small or large?

Single-Factor Analysis of Variance

Chapter 14

Single-Factor ANOVA Model and Tests

Analysis of variance (ANOVA) models are versatile statistical tools for studying the relation between a dependent variable and one or more independent variables. They do not require making assumptions about the nature of the statistical relation, nor do they require that the independent variables be quantitative.

In the present chapter, we consider first the relation between analysis of variance and regression. We then take up the basic elements of single-factor analysis of variance models, which are appropriate when one independent variable is being studied. In the remainder of Part III, we continue our discussion of single-factor analysis of variance. In Part IV, we shall take up multifactor analysis of variance models, when two or more independent variables are under investigation.

14.1 RELATION BETWEEN REGRESSION AND ANALYSIS OF VARIANCE

Regression analysis, as we have seen, is concerned with the statistical relation between one or more independent variables and a dependent variable. Both the independent and dependent variables in ordinary regression models are quantitative. (We exclude from this discussion the use of indicator variables in regression models, described in Chapter 10.) The regression function describes the nature of the statistical relation between the mean response and the level(s) of the independent variable(s).

We encountered the analysis of variance in our consideration of regression. It was used there for a variety of tests concerning the regression coefficients, the fit of the regression model, and the like. The analysis of variance is actually much more general than its use with regression models indicated. Analysis of variance models are a basic type of statistical model. They are concerned, like regression models, with the statistical relation between one or more independent variables and a dependent variable. Like regression models, analysis of variance models are appropriate for both observational data and data based on formal experiments. Further, like the usual regression models, the dependent variable for analysis of variance models is a quantitative variable. Analysis of variance models differ from ordinary regression models in two key respects:

1. The independent variables in analysis of variance models may be qualitative (gender, geographic location, plant shift, etc.).
2. If the independent variables are quantitative, no assumption is made in analysis of variance models about the nature of the statistical relation between them and the dependent variable. Thus, the need to specify the nature of the regression function encountered in ordinary regression analysis does not arise in analysis of variance models.

Illustrations

Figure 14.1 illustrates the essential differences between regression and analysis of variance models for the case where the independent variable is quantitative. Shown in Figure 14.1a is the regression model for a pricing study involving three different price levels, $X = \$50, \$60, \$70$. Note that the XY plane has been rotated from its usual position so that the Y axis faces the viewer. For each level of the independent variable, there is a probability distribution of sales volumes. The means of these probability distributions fall on the regression curve, which describes the statistical relation between price level and mean sales volume.

The analysis of variance model for the same study is illustrated in Figure 14.1b. The three price levels are treated as separate populations, each leading to a probability distribution of sales volumes. The quantitative differences in the three price levels and their statistical relation to expected sales volume are not considered by the analysis of variance model.

Figure 14.2 illustrates the analysis of variance model for a study of the effects of four different types of incentive pay systems on employee productivity. Here, each type of incentive pay system corresponds to a different population, and there is associated with each a probability distribution of employee productivities. Since type of incentive pay system is a qualitative variable, Figure 14.2 does not contain a corresponding regression model representation.

Choice between Two Types of Models

When the independent variables are qualitative, there is no fundamental choice available between regression and analysis of variance models. The situation differs, however, when the independent variables are quantitative. Here, the choice involves on one side an analysis not requiring a specification of the nature of the statistical relation (analysis of variance) and on the other an analysis requiring this specification (regression analysis). If there is substantial doubt about the nature of the statistical relation, a strategy sometimes followed is to first employ an analysis of variance model to study the effects of the independent variables on the dependent variable

FIGURE 14.1   Relation between Regression and Analysis of Variance Models

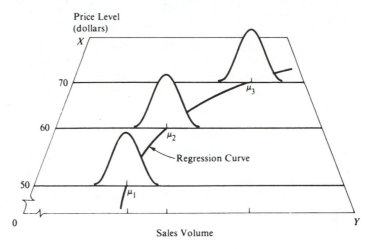

(a) Regression Model

Price Level
(dollars)

X

70

60

50

0

μ_1

μ_2

μ_3

Regression Curve

Sales Volume

Y

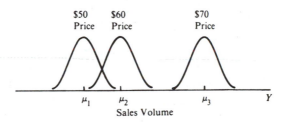

(b) Analysis of Variance Model

$50
Price

$60
Price

$70
Price

μ_1 μ_2 μ_3

Sales Volume

Y

FIGURE 14.2   Analysis of Variance Model Representation for Incentive Pay Example

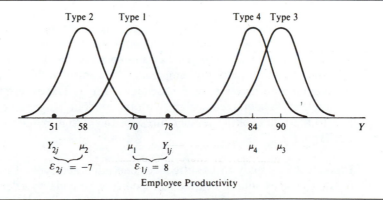

Type 2   Type 1   Type 4   Type 3

51   58   70   78   84   90 Y

Y_{2j} μ_2 μ_1 Y_{1j} μ_4 μ_3

$\varepsilon_{2j} = -7$ $\varepsilon_{1j} = 8$

Employee Productivity

without restrictive assumptions on the nature of the statistical relation. The next step, then, is to turn to regression analysis to exploit the quantitative character of the independent variables.

Note

If you have studied Chapter 10, you will know that regression analysis by means of indicator variables can handle qualitative independent variables, and it can also handle by the same means quantitative independent variables without making assumptions about the nature of the statistical relationship. When indicator variables are so used with regression models, the same results will be obtained as with analysis of variance models. The reason why analysis of variance exists as a distinct statistical methodology is that the structure of the independent indicator variables permits computational simplifications that are explicitly recognized in the statistical procedures for the analysis of variance.

14.2 EXPERIMENTAL AND OBSERVATIONAL STUDIES, FACTORS, AND TREATMENTS

As noted before, analysis of variance models, like regression models, may be employed for experimental data and for observational data. Similarly, analysis of variance models are intended for applications where the effects of one or more independent variables on the dependent variable are of interest. A different terminology is often employed, however, for analysis of variance models. For instance, the independent variables for analysis of variance models are called factors or treatments.

We shall now consider the role of experimental and observational studies in the analysis of variance, and the correspondence of factors and treatments in the analysis of variance to independent variables in regression.

Experimental and Observational Studies

Analysis of variance models are useful for data from both experimental and observational studies.

Example 1. The effect of four cooking temperatures on the fluffiness of omelets prepared from a mix was studied by randomly assigning five packages of mix to each of four cooking temperatures. This is an experimental study because the independent variable of interest (cooking temperature) is controlled by the experimenter. The experimental units here are the 20 packages of mix. The experimental design used was a completely randomized design, with the random assignment of the 20 packages of mix to the four cooking temperatures for this design made in the manner explained in Chapter 2.

Example 2. A study of the effects of education and type of experience of salespeople on the sales volumes of the salespeople was made by selecting a random sample of salespeople currently employed by the company and obtaining information on highest degree obtained, type of experience, and sales volume for each of the selected employees. This is an observational study because the data were obtained without controlling the independent variables of interest (education and type of experience).

Factor

A *factor* is an independent variable to be studied in an investigation. For instance, in an investigation of the effect of price on sales of a luxury item, the factor being studied is price. Similarly, in a study comparing the appeal of four different television programs, the factor under investigation is type of television program. In the omelet fluffiness example, the factor under investigation is cooking temperature.

Factor Level

A *level* of a factor is a particular form of that factor. In the pricing study in Figure 14.1, three prices were used, namely, $50, $60, and $70. Each of these prices is a level of the factor under study, and we say that the price factor has three levels in this study. As another example, in a study of the effect of color of the questionnaire paper on response rate in a mail survey, color of paper is the factor under study, and each different color used is a level of that factor. In the omelet fluffiness example, each of the four cooking temperatures is a factor level.

Single-Factor and Multifactor Studies

Investigations differ as to the number of factors studied. Some are *single-factor studies,* where only one factor is of concern. For instance, the study of the appeal of four different television programs mentioned earlier is an example of a single-factor study. In *multifactor studies,* two or more factors are investigated simultaneously. An example of a multifactor investigation is a study of the effects of temperature and concentration of the solvent upon the yield of a certain chemical process. Here, two factors—temperature and concentration—are studied simultaneously to obtain information about their effects. Similarly, in the earlier sales volume example, two factors were studied simultaneously—education and type of experience. Multifactor investigations will be discussed in Part IV; now we shall focus on single-factor studies.

Experimental and Classification Factors

A factor may be categorized as to whether it is an experimental factor or a classification factor. In any investigation based on observational data, the factors under study are classification factors. A *classification factor* pertains to the characteristic of the units under study and is not under the control of the investigator. For instance, in the sales volume example, education and type of experience are classification factors because they refer to the characteristics of the salespeople in the study and were not manipulated experimentally. On the other hand, an *experimental factor* is one where the level of the factor is assigned at random to the experimental unit. An illustration is the factor cooking temperature in the omelet fluffiness example.

We have pointed out before that experimental data provide firmer foundations for conclusions than do observational data. If experimental factors were only to appear in experimental studies and classification factors only in observational studies, there would be no need to distinguish between these two types of factors. However, classification factors can be found in experimental studies, and therefore it is important to recognize them as such, since the inferences pertaining to these factors will not be as clear as those for experimental factors.

Example.   An appliance manufacturer operates three training centers in the United States for training mechanics to service the company's products. At each center, two different training programs were studied, with the trainees assigned at random to one of the two training programs. One may view this as a two-factor study, the factors being training program and training center. If the same training program is superior to the other in all three centers, the evidence is quite clear as to the effects of the training programs since at each center the trainees were assigned at random to the two programs.

On the other hand, differences between the three training centers cannot be interpreted as clearly because training center is a classification factor. One center may excel for any number of reasons, such as because its staff is doing a better training job, it has better facilities, or because trainees assigned to it come from a geographic region in which better education is provided. External evidence would be required as to whether or not the education of trainees at the three centers is the same, whether or not the facilities are equal, and the like, before a clear understanding of the reasons for differences between training centers could be obtained.

Qualitative and Quantitative Factors

Finally, we need to distinguish between qualitative and quantitative factors. A *qualitative factor* is one where the levels differ by some qualitative attribute. Examples are type of advertisement or brand of rust inhibitor. On the other hand, a *quantitative factor* is one where each level is described by a numerical quantity on a scale. Examples are temperature in degrees centigrade, age in years, or price in dollars.

Treatments

In single-factor studies, a treatment corresponds to a factor level. Thus, in a study of five advertisements, each advertisement is a treatment. In multifactor studies, a treatment corresponds to a combination of factor levels. Thus, in a study of the effects on sales volume of price ($.25, $.29) and package color (red, blue), each combination, such as red package color—$.25 price, is a treatment. This particular study contains four treatments since there are four price-package color combinations.

14.3 DESIGN OF ANALYSIS OF VARIANCE STUDIES

We now briefly discuss a few important considerations in designing analysis of variance studies. Some of the considerations are particularly relevant to experimental studies, while others pertain to both observational and experimental studies.

Choice of Treatments

The choice of treatments to be included in an investigation is basically a matter to be decided by the investigator. A few general comments may, however, be appropriate. In a scientific investigation, the treatments included should be able to provide some insights into the mechanism underlying the phenomenon under study. Initial studies should not attempt to examine this mechanism in full detail, but rather should seek to find the principal factors involved and obtain indications of the magnitudes of their effects. Subsequent studies can then be conducted to provide more detailed findings.

Even if the investigation is not of scientific but rather of practical interest, inclusion of treatments to provide some explanation of the results is often desirable. Consider a company that is planning to purchase a new type of tool machine from a new maker and wishes to compare this new type with the present machines by means of an experiment. Since the new machines are of a much larger size than the present ones, the company wishes to include a third type of machine in the experiment, namely, one corresponding in size to the present ones, but from the new maker. By doing so, the company will get information as to whether any differences that might be observed between the present machines and the new ones are connected with the maker, the size of the machine, or both.

Definition of Treatment. The definition of a treatment can be a difficult problem. Consider an experiment to study whether FORTRAN or PASCAL is a better programming language to teach in an introductory computing course. Some teachers will prefer FORTRAN, others PASCAL. Should the treatments then be defined as the programming language taught by instructors who prefer that language? If so, differences in findings may be due to differences between the two groups of instructors. Should the definition of a treatment not include the instructor, and instructors be randomized, with some being forced to teach a language they do not prefer? Or should instructor preference be a second factor, with each instructor teaching both languages? Problems of this kind need careful resolution so that the results of the study will be useful.

Control Treatment.   A control treatment is needed in some experiments, but not in all. A control treatment consists of applying the identical procedures to experimental units that are used with the other treatments, except for the effects under investigation. In a study of food additives, for instance, a treatment may consist of a portion of a vegetable containing a particular additive that is served to a consumer in a particular experimental setting in the laboratory. A control treatment here would consist of a portion of the same vegetable served to a consumer in the identical experimental setting except that no food additive has been used.

A control treatment is required when the general effectiveness of the treatments under study is not known, or when the general effectiveness of the treatments is known but is not consistent under all conditions. In the food additives example, suppose it is known that food additive A is highly effective in enhancing the tastiness of vegetables and it is desired to see if additives B and C are equally effective or possibly even more effective. In that case, a standard of comparison is available and no control treatment is required. On the other hand, suppose there is no knowledge about the general effectiveness of the three additives, and the following results are obtained (ratings can range between 0 and 60):

Additive	Mean Rating
A	39
B	37
C	41

Assume that the sample sizes are large so that the mean ratings are very precise. In the absence of a standard of comparison, one would not know here whether each of the three additives is effective or whether none of the additives is effective.

It is crucial that the control treatment be conducted in the identical experimental setting as the other treatments. In the food additives example, for instance, a survey of consumers at home, in which persons are asked to rate the general tastiness of the vegetable (without any additive) on the same scale as in the experiment, would not qualify as a control treatment. Such a survey might yield a mean rating of 22, suggesting that the three additives substantially increase the tastiness of the vegetable. This conclusion, however, could be grossly misleading. If the control treatment actually were incorporated into the experiment so that consumers are given portions of the vegetable with no additive in the laboratory setting, the mean rating for the control treatment might be 40. This result would imply that none of the three additives is effective in enhancing the tastiness of the vegetable. The reason for the higher mean rating in the laboratory setting could be a "halo" effect connected with the experimental procedures. Possibly, foods served in the experimental setting taste better than at home, or perhaps consumers try to oblige by giving higher ratings when they participate in an experimental study. Thus, only a control treatment incorporated into the experiment can serve as the proper standard of comparison.

Basic Unit of Study

Another important issue in both experimental and observational studies is how to define the basic unit of the study. Consider, for instance, an experimental study of two incentive pay systems. Should the study unit be an individual employee, a shift, or a plant? Often, technical considerations dictate the choice of the size of the study unit. For instance, morale considerations might preclude the use of different incentive pay systems in the same plant.

A different aspect of defining the basic unit of study occurs in investigations of sales and similar phenomena. Suppose that we are interested in measuring the effectiveness of five different television commercials in terms of sales during a period of time subsequent to their showing. Should the length of time be one week, two weeks, one month, or some other time period? Clearly, the purposes of the study will need to govern the length of time which makes up the basic study unit here.

Representativeness of the study units is another important consideration in the design of analysis of variance studies. Consider a study of management behavior with different communications networks. A university investigator may be tempted to use students as subjects because of their ready availability. If, however, information is desired about the behavior of business people, the students may not be representative experimental units. It hardly needs to be stated that an investigator should make every effort to obtain representative study units. Conversely, one should be cautious in extending results of an investigation to groups for which the study units are not representative. Thus, if the communications network study cited above *did* use students, one should not automatically assume that the findings are relevant to business people.

14.4   USES OF ANALYSIS OF VARIANCE MODELS

Analysis of variance models basically are used to analyze the effects of the independent variable(s) under study on the dependent variable. Specifically, single-factor studies are utilized to compare different factor level effects, to ascertain the "best" factor level, and the like. In multifactor studies, analysis of variance models are employed to determine whether the different factors interact, which factors are the key ones, which factor combinations are "best," and so on. We illustrate these points with three examples.

Example 1

A hospital was using a standard type of therapy to treat a medical problem. Recently two new types of therapy were proposed. The analysis of variance model was then utilized to determine whether either of the two new types of therapy is superior to the present type and, if so, which one is best.

Example 2

Four machines in a plant were studied with respect to the diameters of ball bearings they were turning out. The purpose of using the analysis of variance model in the study was to determine whether substantial differences between the machines existed. If so, the machines would need to be calibrated.

Example 3

In a study of a random sample of employees of an organization, the employees' absentee rates were analyzed according to length of service, gender, marital status, and type of job of the employee. The analysis of variance model for this multifactor investigation was utilized to determine whether the effects of these independent variables interact in important ways and which one of the independent variables shows the greatest statistical relation to the dependent variable.

14.5 ANOVA MODEL I—FIXED FACTOR LEVELS

Distinction between ANOVA Models I and II

We shall consider two single-factor analysis of variance models. For brevity, we shall refer to these as ANOVA models I and II. ANOVA model I, to be taken up here, applies to such cases as a comparison of five different advertisements or a comparison of four different rust inhibitors, where the conclusions will pertain to just those factor levels included in the study. ANOVA model II, to be discussed in Chapter 17, applies to a different type of situation, namely, where the conclusions will extend to a population of factor levels of which the levels in the study are a sample. Consider, for instance, a company that owns several hundred retail stores throughout the country. Seven of these stores are selected at random, and a sample of employees from each store is then chosen and asked in a confidential interview for an evaluation of the management of the store. The seven stores in the study constitute the seven levels of the factor under study, namely, retail stores. In this case, however, management is not just interested in the seven stores included in the study but wishes to generalize the study results to all of the retail stores it owns. Another example when ANOVA model II is applicable is when 3 machines out of 75 in a plant are selected at random and their daily output is studied for a period of 10 days. The three machines constitute the three factor levels in this study, but interest is not just in the three machines in the study but in all machines in the plant.

Thus, the essential difference between situations where ANOVA models I and II are applicable is that model I is the relevant one when the factor levels are chosen because of intrinsic interest in them (e.g., five different advertisements) and they are not considered as a sample from a larger population. ANOVA model II is appropriate when the factor levels constitute a sample from a larger population (e.g., 3 machines out of 75) and interest is in this larger population.

Basic Ideas

The basic elements of ANOVA model I for a single-factor study are quite simple. Corresponding to each factor level, there is a probability distribution of responses. For example, in a study of the effects of four types of incentive pay on employee productivity, there is a probability distribution of employee productivities for each type of incentive pay. ANOVA model I assumes that:

1. Each of the probability distributions is normal.
2. Each probability distribution has the same variance (standard deviation).
3. The observations for each factor level are random observations from the corresponding probability distribution and are independent of the observations for any other factor level.

Figure 14.2 illustrates these conditions. Note the normality of the probability distributions and the constant variability. The probability distributions differ only with respect to their means. Differences in the means hence reflect the essential factor level effects, and it is for this reason that the analysis of variance focuses on the mean responses for the different factor levels. The analysis of the sample data from the factor level probability distributions therefore usually proceeds in two steps:

1. Determine whether or not the factor level means are the same.
2. If the factor level means are not the same, examine how they differ and what the implications of the differences are.

In this chapter, we consider step 1, the testing procedure for determining whether or not the factor level means are equal. In the next chapter, we take up the analysis of the factor level effects when the means are not equal.

ANOVA Model I—Cell Means Model

Before stating ANOVA model I for single-factor studies, we need to develop some notation. We shall denote by r the number of levels of the factor under study (e.g., $r = 4$ types of incentive pay), and we shall denote any one of these levels by the index i ($i = 1, \ldots, r$). The number of cases for the ith factor level is denoted by n_i, and the total number of cases in the study is denoted by n_T, where:

$$(14.1) \qquad n_T = \sum_{i=1}^{r} n_i$$

This notation differs from that used earlier for regression models, where the subscript i identifies the case or trial.

For analysis of variance models we shall always use the last subscript to represent the case or trial for a given factor level or treatment. Here, the index j will be used to identify the given case or trial for a particular factor level. Further, we shall let Y_{ij} denote the jth observation on the dependent or response variable for the ith factor level. For instance, Y_{ij} is the productivity of the jth employee in the ith plant, or the sales volume of the jth store featuring the ith type of shelf display. Since the number of cases or trials for the ith factor level is denoted by n_i, we have $j = 1, \ldots, n_i$.

ANOVA model I can now be stated as follows:

(14.2)
$$Y_{ij} = \mu_i + \varepsilon_{ij}$$

where:

Y_{ij} is the value of the response variable in the jth trial for the ith factor level or treatment

μ_i are parameters

ε_{ij} are independent $N(0, \sigma^2)$

$i = 1, \ldots, r; j = 1, \ldots, n_i$

This model is called the *cell means model* for reasons to be explained shortly. This model may be used for data from observational studies or for data from experimental studies based on a completely randomized design.

Important Features of Model

1. The observed value of Y in the jth trial for the ith factor level or treatment is the sum of two components: (a) a constant term μ_i and (b) a random error term ε_{ij}.

2. Since $E\{\varepsilon_{ij}\} = 0$, it follows that:

(14.3)
$$E\{Y_{ij}\} = \mu_i$$

Thus, all observations for the ith factor level have the same expectation μ_i.

3. Since μ_i is a constant, it follows from (1.16a) that:

(14.4)
$$\sigma^2\{Y_{ij}\} = \sigma^2\{\varepsilon_{ij}\} = \sigma^2$$

Thus, all observations have the same variance, regardless of factor level.

4. Since each ε_{ij} is normally distributed, so is each Y_{ij}. This follows from (1.33) because Y_{ij} is a linear function of ε_{ij}.

5. The error terms are assumed to be independent. Hence, the error term outcome on any one trial has no effect on the error term for the outcome of any other trial for the same factor level or for a different factor level. Since the ε_{ij} are independent, so are the observations Y_{ij}.

6. In view of these features, ANOVA model (14.2) can be restated as follows:

(14.5)
$$Y_{ij} \text{ are independent } N(\mu_i, \sigma^2)$$

Comments

1. ANOVA model (14.2) is a linear model because it can be expressed in matrix terms in the form (7.18), i.e., as $\mathbf{Y} = \mathbf{X}\boldsymbol{\beta} + \boldsymbol{\varepsilon}$. We illustrate this for a study involving $r = 3$ treatments for each of which two observations are made. Thus, $n_1 = n_2 = n_3 = 2$. $\mathbf{Y}$, $\mathbf{X}$, $\boldsymbol{\beta}$, and $\boldsymbol{\varepsilon}$ are defined as follows here:

$$
(14.6) \quad \mathbf{Y} = \begin{bmatrix} Y_{11} \\ Y_{12} \\ Y_{21} \\ Y_{22} \\ Y_{31} \\ Y_{32} \end{bmatrix} \quad \mathbf{X} = \begin{bmatrix} 1 & 0 & 0 \\ 1 & 0 & 0 \\ 0 & 1 & 0 \\ 0 & 1 & 0 \\ 0 & 0 & 1 \\ 0 & 0 & 1 \end{bmatrix} \quad \boldsymbol{\beta} = \begin{bmatrix} \mu_1 \\ \mu_2 \\ \mu_3 \end{bmatrix} \quad \boldsymbol{\varepsilon} = \begin{bmatrix} \varepsilon_{11} \\ \varepsilon_{12} \\ \varepsilon_{21} \\ \varepsilon_{22} \\ \varepsilon_{31} \\ \varepsilon_{32} \end{bmatrix}
$$

Note the simple structure of the $\mathbf{X}$ matrix and that the $\boldsymbol{\beta}$ vector consists of the means μ_i.

To see that these matrices yield ANOVA model (14.2), recall from (7.18a) that the vector of expected values $E\{Y_{ij}\}$ is given by $E\{\mathbf{Y}\} = \mathbf{X}\boldsymbol{\beta}$. We thus obtain:

$$
(14.7) \quad E\{\mathbf{Y}\} = \begin{bmatrix} E\{Y_{11}\} \\ E\{Y_{12}\} \\ E\{Y_{21}\} \\ E\{Y_{22}\} \\ E\{Y_{31}\} \\ E\{Y_{32}\} \end{bmatrix} = \mathbf{X}\boldsymbol{\beta} = \begin{bmatrix} 1 & 0 & 0 \\ 1 & 0 & 0 \\ 0 & 1 & 0 \\ 0 & 1 & 0 \\ 0 & 0 & 1 \\ 0 & 0 & 1 \end{bmatrix} \begin{bmatrix} \mu_1 \\ \mu_2 \\ \mu_3 \end{bmatrix} = \begin{bmatrix} \mu_1 \\ \mu_1 \\ \mu_2 \\ \mu_2 \\ \mu_3 \\ \mu_3 \end{bmatrix}
$$

which indicates properly that $E\{Y_{ij}\} = \mu_i$. Hence, ANOVA model (14.2)—$Y_{ij} = \mu_i + \varepsilon_{ij}$—in matrix form is given by $\mathbf{Y} = \mathbf{X}\boldsymbol{\beta} + \boldsymbol{\varepsilon}$:

$$
(14.8) \quad \mathbf{Y} = \begin{bmatrix} Y_{11} \\ Y_{12} \\ Y_{21} \\ Y_{22} \\ Y_{31} \\ Y_{32} \end{bmatrix} = \mathbf{X}\boldsymbol{\beta} + \boldsymbol{\varepsilon} = \begin{bmatrix} \mu_1 \\ \mu_1 \\ \mu_2 \\ \mu_2 \\ \mu_3 \\ \mu_3 \end{bmatrix} + \begin{bmatrix} \varepsilon_{11} \\ \varepsilon_{12} \\ \varepsilon_{21} \\ \varepsilon_{22} \\ \varepsilon_{31} \\ \varepsilon_{32} \end{bmatrix}
$$

2. ANOVA model (14.2) is called a cell means model because the $\boldsymbol{\beta}$ vector contains the means of the "cells"—here factor levels. In Section 14.10 we discuss an equivalent ANOVA model called the factor effects model, where the $\boldsymbol{\beta}$ vector contains components of the factor level means.

Example

Suppose that ANOVA model (14.2) is applicable to the earlier incentive pay study illustration and that the parameters are as follows:

$$
\mu_1 = 70 \qquad \mu_2 = 58 \qquad \mu_3 = 90 \qquad \mu_4 = 84 \qquad \sigma = 4
$$

Figure 14.2 contains a representation of this model. Note that employee productivities for incentive pay type 1 according to this model are normally distributed with mean $\mu_1 = 70$ and standard deviation $\sigma = 4$.

Suppose that in the jth trial of incentive pay type 1, the observed productivity is $Y_{1j} = 78$. In that case, the error term value is $\varepsilon_{1j} = 8$, for we have:

$$
\varepsilon_{1j} = Y_{1j} - \mu_1 = 78 - 70 = 8
$$

Figure 14.2 shows this observation Y_{1j}. Note that the deviation of Y_{1j} from the mean μ_1 represents the error term ε_{1j}. This figure also shows the observation $Y_{2j} = 51$, for which the error term value is $\varepsilon_{2j} = -7$.

Interpretation of Factor Level Means

Observational Data.   In an observational study, the factor level means μ_i correspond to the means for the different factor level populations. For instance, in a study of the productivity of employees in each of three shifts operated in a plant, the populations consist of the employee productivities for each of the three shifts. The population mean μ_1 is the mean productivity for employees in shift 1; μ_2 and μ_3 are interpreted similarly. The variance σ^2 refers to the variability of employee productivities within a shift.

Experimental Data.   In an experimental study, the factor level mean μ_i stands for the mean response that would be obtained if the ith treatment were applied to all units in the population of experimental units about which inferences are to be drawn. Similarly, the variance σ^2 refers to the variability of responses if any given experimental treatment were applied to the entire population of experimental units. For instance, in a completely randomized design to study the effects of three different training programs on employee productivity, in which 90 employees participate, a third of these employees is assigned at random to each of the three programs. The mean μ_1 here denotes the mean productivity if training program 1 were given to each employee in the population of experimental units; the means μ_2 and μ_3 are interpreted correspondingly. The variance σ^2 denotes the variability in productivities if any one training program were given to each employee in the population of experimental units.

Comments

1. ANOVA model I, like any other statistical model, is not likely to be met exactly by any real-world situation. However, it will be met approximately in many cases. As we shall note later, the statistical procedures based on ANOVA model I are quite robust so that even if the actual conditions of the situation under study differ substantially from those of ANOVA model I, the statistical analysis may still be an appropriate approximation.

2. At times, all treatments under study are given to each of the units in the study. For instance, a person may be asked to use toothpaste A for a week and then give it a rating, which is followed by a week's use of toothpastes B and C each. In this type of case, ANOVA model I is usually not appropriate since the several responses by the same person for the different treatments under study are likely to be correlated. For instance, if a person prefers tooth powder to toothpaste, all of the ratings for the different toothpastes are likely to be low. In Chapter 28, we take up models for this situation.

14.6 FITTING OF ANOVA MODEL

The parameters of ANOVA model (14.2) are ordinarily unknown and must be estimated from sample data. As with regression, we use the method of least squares to fit the ANOVA model and provide estimators of the model parameters. Before turning to these estimators, we shall describe an example to be used throughout the remainder of this chapter and shall develop needed additional notation.

Example

The Kenton Food Company wished to test four different package designs for a new breakfast cereal. Ten stores, with approximately equal sales volumes, were selected as the experimental units. Each store was randomly assigned one of the package designs, with two of the package designs assigned to three stores each and the other two designs to two stores each. Other relevant conditions besides package design, such as price, amount and location of shelf space, and special promotional efforts, were kept the same for all stores in the experiment. Sales, in number of cases, were observed for the study period, and the results are recorded in Table 14.1a.

The sample sizes here are very small to keep the illustrative computations to a minimum. In actual practice, larger sample sizes would usually be employed to obtain more precise and powerful results.

Also, in an application of this type, it would often be more reasonable to assign an equal number of stores to each package design since there is frequently an equal interest in each of the treatments. We are utilizing unequal sample sizes to demonstrate the analytical procedures in full generality.

TABLE 14.1   Number of Cases Sold by Stores for Each of Four Package Designs—Kenton Food Company Example

(a) Sample Data

| | Store | | | | | Number of |
Package Design	1	2	3	Total	Mean	Stores
1	12	18		30	15	2
2	14	12	13	39	13	3
3	19	17	21	57	19	3
4	24	30		54	27	2
All designs				180	18	10

These are different stores

(b) Symbolic Notation

| Factor Level | Sample Unit (j) | | | | | Number of |
i	1	2	3	Total	Mean	Sample Units
1	Y_{11}	Y_{12}		$Y_{1.}$	$\bar{Y}_{1.}$	n_1
2	Y_{21}	Y_{22}	Y_{23}	$Y_{2.}$	$\bar{Y}_{2.}$	n_2
3	Y_{31}	Y_{32}	Y_{33}	$Y_{3.}$	$\bar{Y}_{3.}$	n_3
4	Y_{41}	Y_{42}		$Y_{4.}$	$\bar{Y}_{4.}$	n_4
All factor levels				$Y_{..}$	$\bar{Y}_{..}$	n_T

Notation

Table 14.1b shows the symbolic notation for the Kenton Food Company data in Table 14.1a. Y_{ij}, as explained earlier, represents the observation on the jth sample unit for the ith factor level; here, Y_{ij} is the number of cases sold by the jth store assigned to the ith package design. For instance, Y_{11} represents the sales of the first store assigned package design 1. For our example, $Y_{11} = 12$ cases. Similarly, sales of the second store assigned package design 3 are $Y_{32} = 17$ cases.

The total of the observations for the ith factor level is denoted by $Y_{i.}$:

$$(14.9) \qquad Y_{i.} = \sum_{j=1}^{n_i} Y_{ij}$$

Thus, the dot in $Y_{i.}$ indicates an aggregation over the j index; in our example, the aggregation is over all stores assigned to the ith package design. For instance, the total sales for all stores assigned package design 1 are, according to Table 14.1, $Y_{1.} = 30$ cases. Similarly, total sales for all stores assigned package design 4 are $Y_{4.} = 54$ cases.

The sample mean for the ith factor level is denoted by $\bar{Y}_{i.}$:

$$(14.10) \qquad \bar{Y}_{i.} = \frac{\sum_{j=1}^{n_i} Y_{ij}}{n_i} = \frac{Y_{i.}}{n_i}$$

In our example, the mean number of cases sold by stores assigned package design 1 is $\bar{Y}_{1.} = 30/2 = 15$. Thus, the dot in the subscript indicates that the averaging was done over j (stores).

The total of all observations in the study is denoted by $Y_{..}$:

$$(14.11) \qquad Y_{..} = \sum_{i=1}^{r} \sum_{j=1}^{n_i} Y_{ij}$$

where the two dots indicate aggregation over both the j and i indexes (in our example, over all stores for any one package design and then over all package designs).

Finally, the overall mean for all observations is denoted by $\bar{Y}_{..}$:

$$(14.12) \qquad \bar{Y}_{..} = \frac{\sum_{i} \sum_{j} Y_{ij}}{n_T} = \frac{Y_{..}}{n_T}$$

The dots here indicate that the averaging was done over both i and j. For our example, we have from Table 14.1 that $\bar{Y}_{..} = 180/10 = 18$.

Least Squares Estimators

According to the least squares criterion, the sum of the squared deviations of the observations around their expected values must be minimized with respect to the parameters. For ANOVA model (14.2), we have by (14.3) that:

$$E\{Y_{ij}\} = \mu_i$$

Hence, the quantity to be minimized is:

$$(14.13) \qquad Q = \sum_i \sum_j (Y_{ij} - \mu_i)^2$$

Now (14.13) can be written as follows:

$$(14.13a) \quad Q = \sum_j (Y_{1j} - \mu_1)^2 + \sum_j (Y_{2j} - \mu_2)^2 + \cdots + \sum_j (Y_{rj} - \mu_r)^2$$

Note that each of the parameters appears in only one of the component sums in (14.13a). Hence, Q can be minimized by minimizing each of the component sums separately. It is well known that the sample mean minimizes a sum of squared deviations. Hence, the least squares estimator of μ_i, denoted by $\hat{\mu}_i$, is:

$$(14.14) \qquad\qquad \hat{\mu}_i = \bar{Y}_{i.}$$

Thus, the *fitted value* for observation Y_{ij}, denoted as for regression models by $\hat{Y}_{ij}$, is simply the corresponding factor level sample mean here:

$$(14.15) \qquad\qquad \hat{Y}_{ij} = \bar{Y}_{i.}$$

Example.   For the Kenton Food Company example, the least squares estimates of the model parameters are as follows according to Table 14.1:

Parameter	Least Squares Estimate
μ_1	$\hat{\mu}_1 = \bar{Y}_{1.} = 15$
μ_2	$\hat{\mu}_2 = \bar{Y}_{2.} = 13$
μ_3	$\hat{\mu}_3 = \bar{Y}_{3.} = 19$
μ_4	$\hat{\mu}_4 = \bar{Y}_{4.} = 27$

Thus, the mean sales per store with package design 1 are estimated to be 15 cases for the population of stores under study, and the fitted values for the two observations for package design 1 are $\hat{Y}_{11} = \hat{Y}_{12} = \bar{Y}_{1.} = 15$. Similarly, the mean sales for package design 2 are estimated to be 13 cases per store, and the fitted values for the three observations for this package design are $\hat{Y}_{2j} = \bar{Y}_{2.} = 13$.

Comments

1. The least squares estimators in (14.14) are also maximum likelihood estimators for the normal error ANOVA model (14.2). Hence, they have all of the desirable properties mentioned in Chapter 2 for the regression estimators. For example, they are minimum variance unbiased estimators.

2. To derive the least squares estimator of μ_i, we need to minimize, with respect to μ_i, the ith component sum of squares in (14.13a):

$$(14.16) \qquad\qquad Q_i = \sum_j (Y_{ij} - \mu_i)^2$$

Differentiating with respect to μ_i, we obtain:

$$\frac{dQ_i}{d\mu_i} = \sum_j - 2(Y_{ij} - \mu_i)$$

When we set this derivative equal to zero and replace the parameter μ_i by the least squares estimator $\hat{\mu}_i$, we obtain the result in (14.14):

$$-2 \sum_{j=1}^{n_i} (Y_{ij} - \hat{\mu}_i) = 0$$

$$\sum_j Y_{ij} = n_i \hat{\mu}_i$$

$$\hat{\mu}_i = \bar{Y}_{i.}$$

Residuals

Residuals are highly useful for examining the aptness of ANOVA models. The residual e_{ij} is again defined, as for regression models, as the difference between the observed and fitted values:

(14.17)
$$e_{ij} = Y_{ij} - \hat{Y}_{ij} = Y_{ij} - \bar{Y}_{i.}$$ *group mean.*

Thus, a residual here represents the deviation of an observation from the respective estimated factor level mean.

Example.   Table 14.2 contains the residuals for the Kenton Food Company example. For instance, from Table 14.1, we find:

$$e_{11} = Y_{11} - \bar{Y}_{1.} = 12 - 15 = -3$$
$$e_{21} = Y_{21} - \bar{Y}_{2.} = 14 - 13 = +1$$

Note from Table 14.2 that the residuals sum to zero for each factor level. This illustrates an important property of the residuals for ANOVA model (14.2)—they sum to zero for each factor level i:

(14.18)
$$\sum_j e_{ij} = 0   \quad i = 1, \ldots, r$$

TABLE 14.2   Residuals for Kenton Food Company Example

Package Design i	Store (j) 1	2	3	Total
1	−3	+3		0
2	+1	−1	0	0
3	0	−2	+2	0
4	−3	+3		0
All designs				0

The use of residuals for examining the aptness of an ANOVA model will be discussed in Chapter 16.

14.7 ANALYSIS OF VARIANCE

Just as the analysis of variance for a regression model partitions the total sum of squares into the regression sum of squares and the error sum of squares, so a corresponding partitioning exists for ANOVA model (14.2).

Partitioning of *SSTO*

The total variability of the Y_{ij} observations, not using any information about factor levels, is measured in terms of the deviation of each observation Y_{ij} around the overall mean $\bar{Y}_{..}$:

$$(14.19) \qquad Y_{ij} - \bar{Y}_{..}$$

When we utilize information about the factor levels, the deviations reflecting the uncertainty remaining in the data are those of each observation Y_{ij} around its respective estimated factor level mean $\bar{Y}_{i.}$:

$$(14.20) \qquad Y_{ij} - \bar{Y}_{i.}$$

The difference between the deviations (14.19) and (14.20) reflects the difference between the estimated factor level mean and the overall mean:

$$(14.21) \qquad (Y_{ij} - \bar{Y}_{..}) - (Y_{ij} - \bar{Y}_{i.}) = \bar{Y}_{i.} - \bar{Y}_{..}$$

Note from (14.21) that we can decompose the total deviation $Y_{ij} - \bar{Y}_{..}$ into two components:

$$(14.22) \qquad \underbrace{Y_{ij} - \bar{Y}_{..}}_{\substack{\text{Total} \\ \text{deviation}}} = \underbrace{\left(\bar{Y}_{i.} - \bar{Y}_{..} \right)}_{\substack{\text{Deviation of} \\ \text{estimated} \\ \text{factor level} \\ \text{mean around} \\ \text{overall mean}}} + \underbrace{\left(Y_{ij} - \bar{Y}_{i.} \right)}_{\substack{\text{Deviation} \\ \text{around} \\ \text{estimated} \\ \text{factor} \\ \text{level mean}}}$$

Thus, the total deviation $Y_{ij} - \bar{Y}_{..}$ can be viewed as the sum of two components:

1. The deviation of the estimated factor level mean around the overall mean.
2. The deviation of Y_{ij} around its estimated factor level mean. This deviation is simply the residual e_{ij} according to (14.17).

Figure 14.3 illustrates this decomposition for the Kenton Food Company example.
When we square (14.22) and then sum, the cross products on the right drop out and we obtain:

$$(14.23) \qquad \underbrace{\sum_i \sum_j (Y_{ij} - \bar{Y}_{..})^2}_{SSTO} = \underbrace{\sum_i n_i (\bar{Y}_{i.} - \bar{Y}_{..})^2}_{SSTR} + \underbrace{\sum_i \sum_j (Y_{ij} - \bar{Y}_{i.})^2}_{SSE}$$

FIGURE 14.3   Partitioning of Total Deviations $Y_{ij} - \bar{Y}_{..}$—Kenton Food Company Example

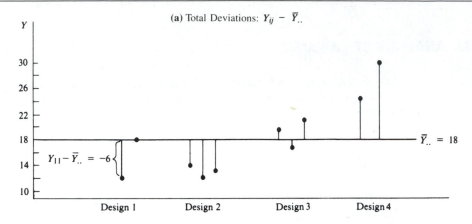

(a) Total Deviations: $Y_{ij} - \bar{Y}_{..}$

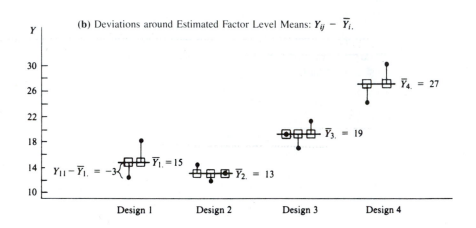

(b) Deviations around Estimated Factor Level Means: $Y_{ij} - \bar{Y}_{i.}$

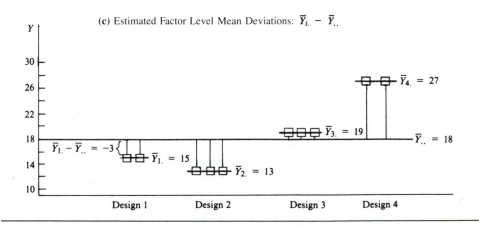

(c) Estimated Factor Level Mean Deviations: $\bar{Y}_{i.} - \bar{Y}_{..}$

The term on the left measures the total variability of the Y_{ij} observations and is denoted, as for regression, by $SSTO$ for *total sum of squares*:

(14.24)
$$SSTO = \sum_i \sum_j (Y_{ij} - \bar{Y}_{..})^2$$

The first term on the right in (14.23) will be denoted by $SSTR$, standing for *treatment sum of squares*:

(14.25)
$$SSTR = \sum_i n_i(\bar{Y}_{i.} - \bar{Y}_{..})^2$$

The second term on the right in (14.23) will be denoted by SSE, standing for *error sum of squares*:

(14.26)
$$SSE = \sum_i \sum_j (Y_{ij} - \bar{Y}_{i.})^2 = \sum_i \sum_j e_{ij}^2$$

Thus, (14.23) can be written equivalently:

(14.27)
$$SSTO = SSTR + SSE$$

The correspondence to the regression decomposition in (3.48a) is readily apparent.

The total sum of squares for the analysis of variance model is therefore made up of these two components:

1. *SSE*: A measure of the random variation of the observations around the respective estimated factor level means. The less variation among the observations for each factor level, the smaller is SSE. If $SSE = 0$, all observations for a factor level are the same, and this holds for all factor levels. The more the observations for a factor level differ among themselves, the larger will be SSE.

2. *SSTR*: A measure of the extent of differences between the estimated factor level means, based on the deviations of the estimated factor level means $\bar{Y}_{i.}$ around the overall mean $\bar{Y}_{..}$. If all estimated factor level means $\bar{Y}_{i.}$ are the same, then $SSTR = 0$. The more the estimated factor level means differ, the larger will be $SSTR$.

Comments

1. To prove (14.23), we begin by considering (14.22):

$$Y_{ij} - \bar{Y}_{..} = (\bar{Y}_{i.} - \bar{Y}_{..}) + (Y_{ij} - \bar{Y}_{i.})$$

Squaring both sides we obtain:

$$(Y_{ij} - \bar{Y}_{..})^2 = (\bar{Y}_{i.} - \bar{Y}_{..})^2 + (Y_{ij} - \bar{Y}_{i.})^2 + 2(\bar{Y}_{i.} - \bar{Y}_{..})(Y_{ij} - \bar{Y}_{i.})$$

If we sum over all sample observations in the study (i.e., over both i and j), we obtain:

(14.28)
$$\sum_i \sum_j (Y_{ij} - \bar{Y}_{..})^2 = \sum_i \sum_j (\bar{Y}_{i.} - \bar{Y}_{..})^2 + \sum_i \sum_j (Y_{ij} - \bar{Y}_{i.})^2$$

$$+ \sum_i \sum_j 2(\bar{Y}_{i.} - \bar{Y}_{..})(Y_{ij} - \bar{Y}_{i.})$$

The first term on the right in (14.28) equals:

$$(14.29) \qquad \sum_i \sum_j (\bar{Y}_{i.} - \bar{Y}_{..})^2 = \sum_i n_i(\bar{Y}_{i.} - \bar{Y}_{..})^2$$

since $(\bar{Y}_{i.} - \bar{Y}_{..})^2$ is constant when summed over j; hence, n_i such terms are picked up for the summation over j.

The third term on the right in (14.28) equals zero:

$$(14.30) \qquad \sum_i \sum_j 2(\bar{Y}_{i.} - \bar{Y}_{..})(Y_{ij} - \bar{Y}_{i.}) = 2 \sum_i (\bar{Y}_{i.} - \bar{Y}_{..}) \sum_j (Y_{ij} - \bar{Y}_{i.}) = 0$$

This follows because $\bar{Y}_{i.} - \bar{Y}_{..}$ is constant for the summation over j; hence, it can be brought in front of the summation sign over j. Further, $\sum_j (Y_{ij} - \bar{Y}_{i.}) = 0$, since the sum of the deviations around the arithmetic mean is zero.

Thus, (14.28) reduces to (14.23).

2. The squared estimated factor level mean deviations $(\bar{Y}_{i.} - \bar{Y}_{..})^2$ in $SSTR$ in (14.25) are weighted by the number of observations n_i at that factor level. The reason is that for each of the observations at factor level i, the deviation component $\bar{Y}_{i.} - \bar{Y}_{..}$ is the same, as Figure 14.3c makes clear.

Computational Formulas. For hand computing, the definitional formulas for $SSTO$, $SSTR$, and SSE given earlier are usually not too convenient. Useful computational formulas for hand computing, which are algebraically identical to the definitional formulas, are:

$$(14.31a) \qquad SSTO = \sum_i \sum_j Y_{ij}^2 - \frac{Y_{..}^2}{n_T}$$

$$(14.31b) \qquad SSTR = \sum_i \frac{Y_{i.}^2}{n_i} - \frac{Y_{..}^2}{n_T}$$

$$(14.31c) \qquad SSE = \sum_i \sum_j Y_{ij}^2 - \sum_i \frac{Y_{i.}^2}{n_i}$$

Example. The analysis of variance breakdown of the total sum of squares for the Kenton Food Company example in Table 14.1a is obtained as follows, using the computational formulas in (14.31):

$$SSTO = (12)^2 + (18)^2 + (14)^2 + \cdots + (30)^2 - \frac{(180)^2}{10}$$

$$= 3{,}544 - 3{,}240$$

$$= 304$$

$$SSTR = \frac{(30)^2}{2} + \frac{(39)^2}{3} + \frac{(57)^2}{3} + \frac{(54)^2}{2} - \frac{(180)^2}{10}$$

$$= 3{,}498 - 3{,}240$$

$$= 258$$

$$SSE = 3{,}544 - 3{,}498$$

$$= 46$$

Thus, the decomposition of *SSTO* is:

$$304 = 258 + 46$$

$$SSTO = SSTR + SSE$$

Note that much of the total variation in the observations is associated with variation between the estimated factor level means.

Breakdown of Degrees of Freedom

Corresponding to the decomposition of the total sum of squares, we can also obtain a breakdown of the associated degrees of freedom.

SSTO has $n_T - 1$ degrees of freedom associated with it. There are altogether n_T deviations $Y_{ij} - \bar{Y}_{..}$, but one degree of freedom is lost because the deviations are not independent in that they must sum to zero, i.e., $\Sigma\Sigma(Y_{ij} - \bar{Y}_{..}) = 0$.

SSTR has $r - 1$ degrees of freedom associated with it. There are r estimated factor level mean deviations $\bar{Y}_{i.} - \bar{Y}_{..}$, but one degree of freedom is lost because the deviations are not independent in that the weighted sum must equal zero, i.e., $\Sigma n_i(\bar{Y}_{i.} - \bar{Y}_{..}) = 0$.

SSE has $n_T - r$ degrees of freedom associated with it. This can be readily seen by considering the component of *SSE* for the *i*th factor level:

(14.32)
$$\sum_{j=1}^{n_i} (Y_{ij} - \bar{Y}_{i.})^2 \quad - \quad \textit{total sum of squares considering ith factor level.}$$

The expression in (14.32) is the equivalent of a total sum of squares considering only the *i*th factor level. Hence, there are $n_i - 1$ degrees of freedom associated with this sum of squares. Since *SSE* is a sum of component sums of squares such as the one in (14.32), the degrees of freedom associated with *SSE* are the sum of the component degrees of freedom:

(14.33)
$$(n_1 - 1) + (n_2 - 1) + \cdots + (n_r - 1) = n_T - r$$

Example. For the Kenton Food Company example, for which $n_T = 10$ and $r = 4$, the degrees of freedom associated with the three sums of squares are as follows:

SS	df
SSTO	10 − 1 = 9
SSTR	4 − 1 = 3
SSE	10 − 4 = 6

Note that degrees of freedom, like sums of squares, are additive:

$$9 = 3 + 6$$

Mean Squares

The mean squares, as usual, are obtained by dividing each sum of squares by its associated degrees of freedom. We therefore have:

(14.34a)
$$MSTR = \frac{SSTR}{r - 1}$$

(14.34b)
$$MSE = \frac{SSE}{n_T - r}$$

Here, *MSTR* stands for *treatment mean square* and *MSE*, as before, stands for *error mean square*.

Example.   For the Kenton Food Company example, we obtain from earlier results:

$$MSTR = \frac{258}{3} = 86$$

$$MSE = \frac{46}{6} = 7.67$$

Note that the two mean squares do not add to $SSTO/(n_T - 1) = 304/9 = 33.8$. Thus, the mean squares here, as in regression, are not additive.

Analysis of Variance Table

The breakdowns of the total sum of squares and degrees of freedom, together with the resulting mean squares, are presented in an ANOVA table such as Table 14.3. The ANOVA table for the Kenton Food Company example is presented in Table 14.4.

TABLE 14.3   ANOVA Table for Single-Factor Study

Source of Variation	SS	df	MS	E{MS}
Between treatments	$SSTR = \Sigma n_i(\bar{Y}_{i.} - \bar{Y}_{..})^2$	$r - 1$	$MSTR = \dfrac{SSTR}{r - 1}$	$\sigma^2 + \dfrac{1}{r - 1}\Sigma n_i(\mu_i - \mu.)^2$
Error (within treatments)	$SSE = \Sigma\Sigma(Y_{ij} - \bar{Y}_{i.})^2$	$n_T - r$	$MSE = \dfrac{SSE}{n_T - r}$	σ^2
Total	$SSTO = \Sigma\Sigma(Y_{ij} - \bar{Y}_{..})^2$	$n_T - 1$		

TABLE 14.4   ANOVA Table for Kenton Food Company Example

Source of Variation	SS	df	MS
Between designs	258	3	86
Error	46	6	7.67
Total	304	9	

Expected Mean Squares

The expected values of *MSE* and *MSTR* can be shown to be as follows:

(14.35a) $$E\{MSE\} = \sigma^2$$

(14.35b) $$E\{MSTR\} = \sigma^2 + \frac{\Sigma n_i(\mu_i - \mu_.)^2}{r - 1}$$

where:

(14.35c) $$\mu_. = \frac{\Sigma n_i \mu_i}{n_T}$$

These expected values are shown in the $E\{MS\}$ column of Table 14.3.

Two important features of the expected mean squares deserve attention:

1. *MSE* is an unbiased estimator of the variance of the error terms ε_{ij}, whether or not the factor level means μ_i are equal. This is intuitively reasonable since the variability of the observations within each factor level is not affected by the magnitudes of the estimated factor level means for normal populations.

2. When all factor level means μ_i are equal and hence equal to the weighted mean $\mu_.$, then $E\{MSTR\} = \sigma^2$ since the second term on the right in (14.35b) becomes zero. Hence, *MSTR* and *MSE* both estimate the error variance σ^2 when all factor level means μ_i are equal. When, however, the factor level means are not equal, *MSTR* tends on the average to be larger than *MSE*, since the second term in (14.35b) will then be positive. This is intuitively reasonable, as illustrated in Figure 14.4 for four treatments. The situation portrayed there assumes that all sample sizes are equal, i.e., $n_i \equiv n$. When all μ_i are equal, then all $\bar{Y}_{i.}$ follow the same sampling distribution, with common mean μ_c and variance σ^2/n; this is portrayed in Figure 14.4a. When the μ_i are not equal, on the other hand, the $\bar{Y}_{i.}$ follow different sampling distributions, each with the same variability σ^2/n but centered on different means μ_i. One such possibility is shown in Figure 14.4b. Hence, the $\bar{Y}_{i.}$ will tend to differ more from each other if the μ_i differ than if the μ_i are equal, and consequently *MSTR* will tend to be larger when the factor level means are not the same than when they are equal. This property of *MSTR* is utilized in constructing the statistical test discussed in the next section to determine whether or not the factor level means μ_i are the same. If *MSTR* and *MSE* are of the same order of magnitude, this is taken to

$$MSTR = \frac{SSTR}{r-1} = \frac{\Sigma n_i(\bar{Y}_{i.} - \bar{Y}_{..})^2}{r-1}$$

FIGURE 14.4   Sampling Distributions of $\bar{Y}_{i.}$ for Four Treatments $(n_i \equiv n)$

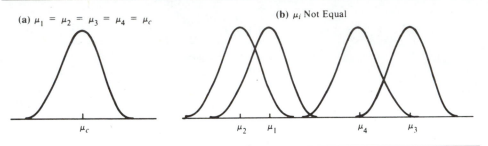

suggest that the factor level means μ_i are equal. If $MSTR$ is substantially larger than MSE, this is taken to suggest that the μ_i are not equal.

Comments

1. To find the expected value of MSE, we first note that MSE can be expressed as follows:

$$(14.36) \qquad MSE = \frac{1}{n_T - r} \sum_i \sum_j (Y_{ij} - \bar{Y}_{i.})^2$$

$$= \frac{1}{n_T - r} \sum_i \left[(n_i - 1) \frac{\sum_j (Y_{ij} - \bar{Y}_{i.})^2}{n_i - 1} \right]$$

Now let us denote the ordinary sample variance of the observations for the ith factor level by s_i^2:

$$(14.37) \qquad s_i^2 = \frac{\sum_j (Y_{ij} - \bar{Y}_{i.})^2}{n_i - 1}$$

Hence, (14.36) can be expressed as follows:

$$(14.38) \qquad MSE = \frac{1}{n_T - r} \sum_i (n_i - 1) s_i^2$$

Since it is well known that the sample variance (14.37) is an unbiased estimator of the population variance, which in our case is σ^2 for all factor levels, we obtain:

$$E\{MSE\} = \frac{1}{n_T - r} \sum_i (n_i - 1) E\{s_i^2\}$$

$$= \frac{1}{n_T - r} \sum_i (n_i - 1) \sigma^2$$

$$= \sigma^2$$

2. We shall derive the expected value of $MSTR$ for the special case when all sample sizes n_i are the same, namely, $n_i \equiv n$. The general result in (14.35b) becomes for this special case:

$$(14.39) \qquad E\{MSTR\} = \sigma^2 + \frac{n\sum(\mu_i - \mu_.)^2}{r - 1} \qquad \text{when } n_i \equiv n$$

Further, when all factor level sample sizes are n, $MSTR$ as defined in (14.25) and (14.34a) becomes:

(14.40)
$$MSTR = \frac{n \Sigma (\bar{Y}_{i.} - \bar{Y}_{..})^2}{r - 1} \qquad \text{when } n_i \equiv n$$

To derive (14.39), consider the model formulation for Y_{ij} in (14.2):

$$Y_{ij} = \mu_i + \varepsilon_{ij}$$

Averaging the Y_{ij} for the ith factor level, we obtain:

(14.41)
$$\bar{Y}_{i.} = \mu_i + \bar{\varepsilon}_{i.}$$

where $\bar{\varepsilon}_{i.}$ is the average of the ε_{ij} for the ith factor level:

(14.42)
$$\bar{\varepsilon}_{i.} = \frac{\sum_j \varepsilon_{ij}}{n}$$

Averaging the Y_{ij} over all factor levels, we obtain:

(14.43)
$$\bar{Y}_{..} = \mu_. + \bar{\varepsilon}_{..}$$

where $\mu_.$, which is defined in (14.35c), becomes for $n_i \equiv n$:

(14.44)
$$\mu_. = \frac{n \Sigma \mu_i}{nr} = \frac{\Sigma \mu_i}{r} \qquad \text{when } n_i \equiv n$$

and $\bar{\varepsilon}_{..}$ is the average of all ε_{ij}:

(14.45)
$$\bar{\varepsilon}_{..} = \frac{\Sigma \Sigma \varepsilon_{ij}}{nr}$$

Since the sample sizes are equal, we also have:

(14.46)
$$\bar{Y}_{..} = \frac{\Sigma \bar{Y}_{i.}}{r} \qquad \bar{\varepsilon}_{..} = \frac{\Sigma \bar{\varepsilon}_{i.}}{r}$$

Using (14.41) and (14.43), we obtain:

(14.47)
$$\bar{Y}_{i.} - \bar{Y}_{..} = (\mu_i + \bar{\varepsilon}_{i.}) - (\mu_. + \bar{\varepsilon}_{..}) = (\mu_i - \mu_.) + (\bar{\varepsilon}_{i.} - \bar{\varepsilon}_{..})$$

If we square $\bar{Y}_{i.} - \bar{Y}_{..}$ and sum over the factor levels, we obtain:

(14.48)
$$\Sigma (\bar{Y}_{i.} - \bar{Y}_{..})^2 = \Sigma (\mu_i - \mu_.)^2 + \Sigma (\bar{\varepsilon}_{i.} - \bar{\varepsilon}_{..})^2 + 2 \Sigma (\mu_i - \mu_.)(\bar{\varepsilon}_{i.} - \bar{\varepsilon}_{..})$$

We now wish to find $E\{\Sigma(\bar{Y}_{i.} - \bar{Y}_{..})^2\}$, and therefore need to find the expected value of each term on the right in (14.48):

a. Since $\Sigma(\mu_i - \mu_.)^2$ is a constant, its expectation is:

(14.49)
$$E\left\{ \Sigma (\mu_i - \mu_.)^2 \right\} = \Sigma (\mu_i - \mu_.)^2$$

b. Before finding the expectation of the second term on the right, consider first the expression:

$$\frac{\Sigma(\bar{\varepsilon}_{i.} - \bar{\varepsilon}_{..})^2}{r - 1}$$

This is an ordinary sample variance, since $\bar{\varepsilon}_{..}$ is the sample mean of the r terms $\bar{\varepsilon}_{i.}$ per (14.46). We further know that the sample variance is an unbiased estimator of the variance of the variable, in this case the variable being $\bar{\varepsilon}_{i.}$. But $\bar{\varepsilon}_{i.}$ is just the mean of n independent error terms ε_{ij} by (14.42). Hence:

$$\sigma^2\{\bar{\varepsilon}_{i.}\} = \frac{\sigma^2\{\varepsilon_{ij}\}}{n} = \frac{\sigma^2}{n}$$

Therefore:

$$E\left\{\frac{\Sigma(\bar{\varepsilon}_{i.} - \bar{\varepsilon}_{..})^2}{r - 1}\right\} = \frac{\sigma^2}{n}$$

so that:

(14.50)
$$E\left\{\sum (\bar{\varepsilon}_{i.} - \bar{\varepsilon}_{..})^2\right\} = \frac{(r - 1)\sigma^2}{n}$$

c. Since both $\bar{\varepsilon}_{i.}$ and $\bar{\varepsilon}_{..}$ are means of ε_{ij} terms, all of which have expectation 0, it follows that:

$$E\{\bar{\varepsilon}_{i.}\} = 0 \qquad E\{\bar{\varepsilon}_{..}\} = 0$$

Hence:

(14.51) $$E\left\{2 \sum (\mu_i - \mu_.)(\bar{\varepsilon}_{i.} - \bar{\varepsilon}_{..})\right\} = 2 \sum (\mu_i - \mu_.)E\{\bar{\varepsilon}_{i.} - \bar{\varepsilon}_{..}\} = 0$$

We have thus shown, by (14.49), (14.50), and (14.51), that:

$$E\left\{\sum (\bar{Y}_{i.} - \bar{Y}_{..})^2\right\} = \sum (\mu_i - \mu_.)^2 + \frac{(r - 1)\sigma^2}{n}$$

But then (14.39) follows at once:

$$E\{MSTR\} = E\left\{\frac{n\Sigma(\bar{Y}_{i.} - \bar{Y}_{..})^2}{r - 1}\right\} = \frac{n}{r - 1}\left[\sum (\mu_i - \mu_.)^2 + \frac{(r - 1)\sigma^2}{n}\right]$$

$$= \sigma^2 + \frac{n\Sigma(\mu_i - \mu_.)^2}{r - 1}$$

14.8 *F* TEST FOR EQUALITY OF FACTOR LEVEL MEANS

It is customary to begin the analysis of a single-factor study by determining whether or not the factor level means μ_i are equal. If, for instance, the four package designs in the Kenton Food Company example lead to the same mean sales volumes, there is no need for further analysis, such as to determine which design is best or how two particular designs compare in stimulating sales.

Thus, the alternative conclusions we wish to consider are:

(14.52)
$$H_0: \mu_1 = \mu_2 = \cdots = \mu_r$$
$$H_a: \text{not all } \mu_i \text{ are equal}$$

Test Statistic

The test statistic to be used for choosing between the alternatives in (14.52) is:

(14.53)
$$F^* = \frac{MSTR}{MSE}$$

Note that *MSTR* here plays the role corresponding to *MSR* for a regression model.

Large values of F^* support H_a, since *MSTR* will tend to exceed *MSE* when H_a holds, as we saw from (14.35). Values of F^* near 1 support H_0, since both *MSTR* and *MSE* have the same expected value when H_0 holds. Hence, the appropriate test is an upper-tail one.

Distribution of F^*

When all treatment means μ_i are equal, each observation Y_{ij} has the same expected value. In view of the additivity of sums of squares and degrees of freedom, Cochran's theorem (3.60) then implies:

When H_0 holds, $\dfrac{SSE}{\sigma^2}$ and $\dfrac{SSTR}{\sigma^2}$ are independent χ^2 variables

It follows in the same fashion as for regression: *SSTR SSE.*

When H_0 holds, F^* is distributed as $F(r - 1, n_T - r)$

If H_a holds, that is, if the μ_i are not all equal, F^* does *not* follow the F distribution. Rather, it follows a complex distribution called the *noncentral F distribution*. We shall make use of the noncentral F distribution when we discuss the power of the F test in Chapter 17.

Note

SSTR and *SSE* are independent even if all μ_i are not equal. Intuitively, this can be seen because *SSE* reflects the variability within the factor level samples, and this within-sample variability is not affected by the magnitudes of the estimated factor level means when the error terms are normally distributed. *SSTR*, on the other hand, is solely based on the estimated factor level means $Y_{i.}$.

Construction of Decision Rule

Usually, the risk of making a Type I error is controlled in constructing the decision rule. This provides protection against making further, more detailed, analyses of the factor effects when in fact there are no differences in the factor level means. The Type II error can also be controlled, as we shall see later, through sample size determination.

Since we know that F^* is distributed as $F(r - 1, n_T - r)$ when H_0 holds and that large values of F^* lead to conclusion H_a, the appropriate decision rule to control the level of significance at α is:

(14.54)
$$\text{If } F^* \le F(1 - \alpha; r - 1, n_T - r), \text{ conclude } H_0$$
$$\text{If } F^* > F(1 - \alpha; r - 1, n_T - r), \text{ conclude } H_a$$

where $F(1 - \alpha; r - 1, n_T - r)$ is the $(1 - \alpha)100$ percentile of the appropriate F distribution.

Example

For the Kenton Food Company example, we wish to test whether or not mean sales are the same for the four package designs:

$$H_0: \mu_1 = \mu_2 = \mu_3 = \mu_4$$

$$H_a: \text{not all } \mu_i \text{ are equal}$$

Management wishes to control the risk of making a Type I error at $\alpha = .05$. We therefore require $F(.95; 3, 6)$, where the degrees of freedom are those shown in Table 14.4. From Table A.4 in the Appendix, we find $F(.95; 3, 6) = 4.76$. Hence, the decision rule is:

$$\text{If } F^* \le 4.76, \text{ conclude } H_0$$

$$\text{If } F^* > 4.76, \text{ conclude } H_a$$

Using the data from Table 14.4, the test statistic is:

$$F^* = \frac{MSTR}{MSE} = \frac{86}{7.67} = 11.2$$

Since $F^* = 11.2 > 4.76$, we conclude H_a, that the factor level means μ_i are not equal, or that the four different package designs do not lead to the same mean sales volume. Thus, we conclude that there is a relation between package design and sales volume.

The P-value for the test statistic is the probability $P\{F(3, 6) > F^* = 11.2\}$. We see from Table A.4 that the P-value is between .005 and .01. The P-value actually is .007.

The conclusion of a relation between package design and sales volume did not surprise the sales manager of the Kenton Food Company. He conducted the study in the first place because he expected the four package designs to have different effects on sales volume and was interested in finding out the nature of these differences. In the next chapter, we discuss the second stage of the analysis, namely, how to study the nature of the factor level effects when differences exist.

Comments

1. If there are only two factor levels so that $r = 2$, it can easily be shown that the test employing F^* in (14.53) is the equivalent of the two-population, two-sided t test in Table 1.2a. The F test here has $(1, n_T - 2)$ degrees of freedom, and the t test has $n_1 + n_2 - 2$ or $n_T - 2$ degrees of freedom; thus both tests lead to equivalent critical regions. For comparing two population means, the t test generally is to be preferred since it can be used to conduct both two-sided and one-sided tests (Table 1.2); the F test can be used only for two-sided tests.

2. Since the F test for testing the alternatives (14.52) is a test of a linear statistical model, it can be obtained by the general method explained in Chapter 3:

a. The full model is ANOVA model (14.2):

$$(14.55) \qquad\qquad Y_{ij} = \mu_i + \varepsilon_{ij} \qquad\qquad \text{Full model}$$

Fitting the full model by the method of least squares leads to the fitted values $\hat{Y}_{ij} = \bar{Y}_{i.}$, per (14.15), and the resulting error sum of squares:

$$SSE(F) = \sum\sum (Y_{ij} - \hat{Y}_{ij})^2 = \sum\sum (Y_{ij} - \bar{Y}_{i.})^2$$

$SSE(F)$ has $df_F = n_T - r$ degrees of freedom associated with it because r parameter values $(\mu_1, \ldots, \mu_r)$ had to be estimated.

b. The reduced model under H_0 is:

$$(14.56) \qquad\qquad Y_{ij} = \mu_c + \varepsilon_{ij} \qquad\qquad \text{Reduced model}$$

where μ_c is the common mean for all factor levels. Fitting the reduced model leads to the least squares estimator $\hat{\mu}_c = \bar{Y}_{..}$ so that all fitted values are $\hat{Y}_{ij} \equiv \bar{Y}_{..}$, and the resulting error sum of squares is:

$$SSE(R) = \sum\sum (Y_{ij} - \hat{Y}_{ij})^2 = \sum\sum (Y_{ij} - \bar{Y}_{..})^2$$

The degrees of freedom associated with $SSE(R)$ are $df_R = n_T - 1$ because one parameter (μ_c) had to be estimated.

c. Since, according to (14.24) and (14.26), respectively:

$$SSE(R) = SSTO$$

$$SSE(F) = SSE$$

and since by (14.27) $SSTO - SSE = SSTR$, test statistic (3.69) becomes:

$$F^* = \frac{SSE(R) - SSE(F)}{df_R - df_F} \div \frac{SSE(F)}{df_F}$$

$$= \frac{SSTO - SSE}{(n_T - 1) - (n_T - r)} \div \frac{SSE}{n_T - r} = \frac{SSR}{r - 1} \div \frac{SSE}{n_T - r}$$

$$= \frac{MSTR}{MSE}$$

14.9 COMPUTER INPUT AND OUTPUT FOR ANOVA PACKAGES

For analysis of variance as for regression analysis, packaged computer programs are readily available for handling calculations. We will assume, as in our regression analysis discussions, that computers or programmable calculators are used in handling analysis of variance calculations for all but the simplest data sets.

For regression analysis, a single "multiple regression" packaged program frequently suffices for a variety of applications such as simple linear regression, polynomial regression, and multiple regression. Packaged analysis of variance programs, on the other hand, are often keyed to specific analyses. Thus, a library may contain one program for single-factor analysis of variance and other programs for multifactor analyses, which we will discuss in later chapters. Formats for input of observations and output of results often differ substantially from one library to another, and may even differ somewhat among the different analysis of variance programs in a given library.

One format for input of the Kenton Food Company data in Table 14.1a requires two columns:

Treatment	Observation
1	12
1	18
2	14
2	12
2	13
·	·
·	·
·	·
4	30

Here the treatment identifications are entered in one column and the observations are entered in another. Other formats will also be encountered when using different ANOVA packages.

Figure 14.5 illustrates a typical output format when the single-factor analysis of variance model is used. The output illustrated in Figure 14.5 is for the Kenton Food Company data in Table 14.1a and was produced by the BMDP computer program (Ref. 14.1). We have added some annotations to facilitate understanding of the output. Note that the data format is listed at the top of Figure 14.5, followed by the number of cases (stores) for each package design, estimates of the factor level means, and the analysis of variance table.

The results in Figure 14.5 coincide with those obtained by hand calculations in Tables 14.1a and 14.4. This is because roundoff errors are no problem in calculations involving the simple data set of the Kenton Food Company example. With more complex data sets, roundoff effects can arise, so that somewhat differing results could be obtained then.

FIGURE 14.5   Segment of Computer Output for Single-Factor Analysis of Variance— Kenton Food Company Example (BMDP, Ref. 14.1)

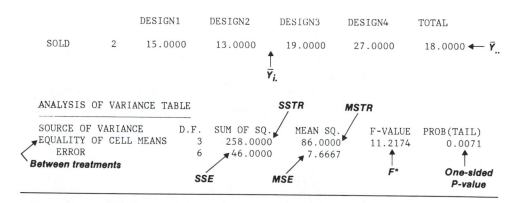

C A S E NO. LABEL	1 DESIGN	2 SOLD Y_{ij}
1	1	12
2	1	18
3	2	14
4	2	12
5	2	13
6	3	19
7	3	17
8	3	21
9	4	24
10	4	30

NUMBER OF CASES PER GROUP

DESIGN1	2.
DESIGN2	3. ← n_i
DESIGN3	3.
DESIGN4	2.
TOTAL	10. ← n_T

ESTIMATES OF MEANS

		DESIGN1	DESIGN2	DESIGN3	DESIGN4	TOTAL
SOLD	2	15.0000	13.0000	19.0000	27.0000	18.0000 ← $\bar{Y}_{..}$

$\bar{Y}_{i.}$

ANALYSIS OF VARIANCE TABLE   **SSTR**   **MSTR**

SOURCE OF VARIANCE	D.F.	SUM OF SQ.	MEAN SQ.	F-VALUE	PROB(TAIL)
EQUALITY OF CELL MEANS	3	258.0000	86.0000	11.2174	0.0071
ERROR	6	46.0000	7.6667		

Between treatments   **SSE**   **MSE**   **F***   **One-sided P-value**

We noted, in discussing roundoff errors in regression analysis, that least squares results tend to be highly sensitive to roundoff effects and that different regression packages do not control such effects equally well. Similarly, different analysis of variance packages differ in quality, and it is a good idea to check a library program before using it for the first time. One procedure is to test the program on a complex data set for which accurate results are already known.

14.10 ALTERNATIVE FORMULATION OF MODEL I

ANOVA Model I—Factor Effects Model

At times, an alternative but completely equivalent formulation of the single-factor ANOVA model I in (14.2) is used. This alternative formulation is called the *factor effects model*. With this alternative formulation, the treatment means μ_i are expressed in an equivalent fashion by means of the identity:

$$(14.57) \qquad \mu_i \equiv \mu_. + (\mu_i - \mu_.)$$

where $\mu_.$ is a constant that can be defined to fit the purposes of the study. We shall denote the difference $\mu_i - \mu_.$ by τ_i:

$$(14.58) \qquad \tau_i = \mu_i - \mu_.$$

so that (14.57) can be expressed in equivalent fashion as:

$$(14.59) \qquad \mu_i \equiv \mu_. + \tau_i$$

The difference $\tau_i = \mu_i - \mu_.$ is called the *effect of the ith factor level*.

ANOVA model I in (14.2) can now be stated as follows:

$$(14.60) \qquad Y_{ij} = \mu_. + \tau_i + \varepsilon_{ij}$$

where:

$\mu_.$ is a constant component common to all observations
τ_i is the effect of the ith factor level (a constant for each factor level)
ε_{ij} are independent $N(0, \sigma^2)$
$i = 1, \ldots, r; j = 1, \ldots, n_i$

ANOVA model (14.60) is called the factor effects model because it is expressed in terms of the factor effects τ_i, in distinction to the cell means model (14.2), which is expressed in terms of the treatment means μ_i.

The factor effects model (14.60) is a linear model, like the equivalent cell means model (14.2). We shall demonstrate this in the next section.

Definition of $\mu_.$

The splitting up of the factor level mean μ_i into two components, an overall constant $\mu_.$ and a factor level effect τ_i, depends on the definition of $\mu_.$ and can be done in many ways. We shall now explain two basic ways to define $\mu_.$.

Unweighted Mean.   Often, a definition of $\mu_.$ as the unweighted average of all factor level means μ_i is found to be useful:

$$(14.61) \qquad \mu_. = \frac{\sum\limits_{i=1}^{r} \mu_i}{r}$$

This definition implies that:

$$(14.62) \qquad \sum_{i=1}^{r} \tau_i = 0$$

because by (14.58) we have:

$$\sum \tau_i = \sum (\mu_i - \mu_.) = \sum \mu_i - r\mu_.$$

and:

$$\sum \mu_i = r\mu_.$$

by (14.61). Thus, the definition of the overall constant $\mu_.$ in (14.61) implies a restriction on the τ_i, in this case that their sum must be zero.

Example. For the earlier incentive pay example in Figure 14.2, we have $\mu_1 = 70$, $\mu_2 = 58$, $\mu_3 = 90$, and $\mu_4 = 84$. When $\mu_.$ is defined according to (14.61), we obtain:

$$\mu_. = \frac{70 + 58 + 90 + 84}{4} = 75.5$$

Hence:

$$\tau_1 = 70 - 75.5 = -5.5$$
$$\tau_2 = 58 - 75.5 = -17.5$$
$$\tau_3 = 90 - 75.5 = 14.5$$
$$\tau_4 = 84 - 75.5 = 8.5$$

The first factor level effect $\tau_1 = -5.5$, for instance, indicates that the mean employee productivity for incentive pay type 1 is 5.5 units less than the average productivity for all four types of incentive pay.

Weighted Mean. The constant $\mu_.$ can also be defined as some weighted average of the factor level means μ_i:

$$(14.63) \qquad \mu_. = \sum_{i=1}^{r} w_i \mu_i$$

where the w_i are weights defined so that $\Sigma w_i = 1$. The restriction on the τ_i then is:

$$(14.64) \qquad \sum_{i=1}^{r} w_i \tau_i = 0$$

This follows in the same fashion as (14.62).

The choice of weights w_i should depend on the meaningfulness of the resulting measures of the factor level effects. We present now two examples where different weightings are appropriate: (1) weighting according to a known measure of importance and (2) weighting according to sample size.

Example 1. A car rental firm wanted to estimate the average fuel consumption (in miles per gallon) for its large fleet of cars which consists of 50 percent compacts, 30 percent sedans, and 20 percent station wagons. Here, a meaningful measure of $\mu_.$ might be in terms of the overall mean fuel consumption:

$$(14.65) \qquad \mu_. = .5\mu_1 + .3\mu_2 + .2\mu_3$$

where μ_1, μ_2, and μ_3 are the mean fuel consumptions for the three types of cars in the fleet.

An estimate of $\mu_.$ here is:

$$(14.66) \qquad \hat{\mu}_. = .5\bar{Y}_{1.} + .3\bar{Y}_{2.} + .2\bar{Y}_{3.}$$

Example 2. If the car rental firm in Example 1 used sample sizes for the three types of cars in its fleet that are approximately in the same ratios as the numbers of cars of each type in the fleet, use of the proportions n_1/n_T, n_2/n_T, and n_3/n_T, respectively, as weights might be meaningful. The resulting definition of the overall fuel consumption constant $\mu_.$ would then be:

$$(14.67) \qquad \mu_. = \frac{n_1}{n_T}\mu_1 + \frac{n_2}{n_T}\mu_2 + \frac{n_3}{n_T}\mu_3$$

This quantity would be estimated by $\bar{Y}_{..}$:

$$(14.68) \qquad \hat{\mu}_. = \frac{n_1}{n_T}\bar{Y}_{1.} + \frac{n_2}{n_T}\bar{Y}_{2.} + \frac{n_3}{n_T}\bar{Y}_{3.} = \bar{Y}_{..}$$

When all sample sizes are equal, $\mu_.$ as defined in (14.67) reduces to the unweighted mean (14.61).

Test for Equality of Factor Level Means

Since the factor effects model (14.60) is equivalent to the cell means model (14.2), the test for equality of factor level means uses the same test statistic F^* in (14.53). The only difference is in the statement of the alternatives. For the cell means model (14.2), the alternatives are as specified in (14.52):

$$H_0: \mu_1 = \mu_2 = \cdots = \mu_r$$

$$H_a: \text{not all } \mu_i \text{ are equal}$$

For the factor effects model (14.60), these same alternatives in terms of the factor effects are:

$$(14.69) \qquad H_0: \tau_1 = \tau_2 = \cdots = \tau_r = 0$$

$$H_a: \text{not all } \tau_i \text{ equal zero}$$

The equivalence of the two forms can be readily established. The equality of the factor level means $\mu_1 = \mu_2 = \cdots = \mu_r$ implies that all τ_i are equal. This follows from (14.59) since the constant term $\mu_.$ is common to all factor level effects τ_i. Further, the equality of the factor level means implies that all $\tau_i = 0$, whether the restriction on the τ_i is of the form in (14.62) or (14.64). In either case, the restriction can be satisfied in only one way given the equality of the τ_i, namely, that $\tau_i \equiv 0$. Thus, it is equivalent to state that all factor level means μ_i are equal or that all factor level effects τ_i equal zero.

14.11 REGRESSION APPROACH TO SINGLE-FACTOR ANALYSIS OF VARIANCE

We noted earlier that the cell means model (14.2) is a linear model, as is also the equivalent factor effects model (14.60). Thus, we can obtain the test statistic F^* for testing the equality of the factor level means μ_i by means of the matrix formulation in Section 8.6. In fact, we can obtain the test statistic F^* without matrix manipulations by use of a multiple regression program. We shall now explain the regression approach to single-factor analysis of variance. For this purpose, we shall utilize the factor effects model (14.60):

$$Y_{ij} = \mu_. + \tau_i + \varepsilon_{ij}$$

and shall assume that equal weightings of the factor level means are appropriate for defining the overall constant $\mu_.$.

To state ANOVA model (14.60) as a linear model, the parameters $\mu_., \tau_1, \ldots, \tau_r$ need to be represented in the model. However, constraint (14.62) for the case of equal weightings:

$$\sum_{i=1}^{r} \tau_i = 0$$

implies that:

(14.70) $$\tau_r = -\tau_1 - \tau_2 - \cdots - \tau_{r-1}$$

Thus, we shall need only the parameters $\mu_., \tau_1, \ldots, \tau_{r-1}$ for the linear model since τ_r is a function of $\tau_1, \ldots, \tau_{r-1}$.

To illustrate how a linear model is developed with this approach, consider a single-factor study with $r = 3$ factor levels when $n_1 = n_2 = n_3 = 2$. The $\mathbf{Y}$, $\mathbf{X}$, $\boldsymbol{\beta}$, and $\boldsymbol{\varepsilon}$ matrices for this case are as follows:

(14.71) $$\mathbf{Y} = \begin{bmatrix} Y_{11} \\ Y_{12} \\ Y_{21} \\ Y_{22} \\ Y_{31} \\ Y_{32} \end{bmatrix} \quad \mathbf{X} = \begin{bmatrix} 1 & 1 & 0 \\ 1 & 1 & 0 \\ 1 & 0 & 1 \\ 1 & 0 & 1 \\ 1 & -1 & -1 \\ 1 & -1 & -1 \end{bmatrix} \quad \boldsymbol{\beta} = \begin{bmatrix} \mu_. \\ \tau_1 \\ \tau_2 \end{bmatrix} \quad \boldsymbol{\varepsilon} = \begin{bmatrix} \varepsilon_{11} \\ \varepsilon_{12} \\ \varepsilon_{21} \\ \varepsilon_{22} \\ \varepsilon_{31} \\ \varepsilon_{32} \end{bmatrix}$$

Note that the vector of expected values, $E\{\mathbf{Y}\} = \mathbf{X}\boldsymbol{\beta}$, yields the following:

$$(14.72) \quad E\{\mathbf{Y}\} = \begin{bmatrix} E\{Y_{11}\} \\ E\{Y_{12}\} \\ E\{Y_{21}\} \\ E\{Y_{22}\} \\ E\{Y_{31}\} \\ E\{Y_{32}\} \end{bmatrix} = \mathbf{X}\boldsymbol{\beta} = \begin{bmatrix} 1 & 1 & 0 \\ 1 & 1 & 0 \\ 1 & 0 & 1 \\ 1 & 0 & 1 \\ 1 & -1 & -1 \\ 1 & -1 & -1 \end{bmatrix} \begin{bmatrix} \mu. \\ \tau_1 \\ \tau_2 \end{bmatrix} = \begin{bmatrix} \mu. + \tau_1 \\ \mu. + \tau_1 \\ \mu. + \tau_2 \\ \mu. + \tau_2 \\ \mu. - \tau_1 - \tau_2 \\ \mu. - \tau_1 - \tau_2 \end{bmatrix}$$

Since $\tau_3 = -\tau_1 - \tau_2$ according to (14.70), we see that $E\{Y_{31}\} = E\{Y_{32}\} = \mu. + \tau_3$. Thus, the above $\mathbf{X}$ matrix and $\boldsymbol{\beta}$ vector representation provides in all cases the appropriate expected values:

$$E\{Y_{ij}\} = \mu. + \tau_i$$

The illustration in (14.71) indicates how we need to define in general the multiple regression model so that it is the equivalent of the single-factor ANOVA model (14.60). Note that we shall need indicator variables that take on values 0, 1, or -1. This coding was discussed in Section 10.6. While this coding is not as simple as a 0,1 coding for an indicator variable, it is desirable here because it leads to regression coefficients in the $\boldsymbol{\beta}$ vector that are the parameters in the factor effects ANOVA model, i.e., $\mu., \tau_1, \ldots, \tau_r$.

We shall let X_{ij1} denote the value of indicator variable X_1 for the jth case from the ith factor level, X_{ij2} the value of indicator variable X_2 for this same case, and so on, using altogether $r - 1$ indicator variables in the model. The multiple regression model then is as follows:

$$(14.73) \qquad Y_{ij} = \mu. + \tau_1 X_{ij1} + \tau_2 X_{ij2} + \cdots + \tau_{r-1} X_{ij,r-1} + \varepsilon_{ij} \qquad \text{Full model}$$

where:

$$X_{ij1} = \begin{array}{l} 1 \text{ if case from factor level } 1 \\ -1 \text{ if case from factor level } r \\ 0 \text{ otherwise} \end{array}$$

$$\vdots \qquad\qquad \vdots$$

$$X_{ij,r-1} = \begin{array}{l} 1 \text{ if case from factor level } r - 1 \\ -1 \text{ if case from factor level } r \\ 0 \text{ otherwise} \end{array}$$

Note how the ANOVA model parameters play the role of regression function parameters in (14.73); the intercept term is $\mu.$, and the regression coefficients are $\tau_1, \tau_2, \ldots, \tau_{r-1}$.

To test the equality of the treatment means μ_i by means of the regression approach, we state the alternatives in the equivalent formulation (14.69), noting that τ_r must equal zero when $\tau_1 = \tau_2 = \cdots = \tau_{r-1} = 0$ according to (14.70):

$$(14.74) \qquad \begin{array}{l} H_0: \tau_1 = \tau_2 = \cdots = \tau_{r-1} = 0 \\ \\ H_a: \text{ not all } \tau_i \text{ equal zero} \end{array}$$

Note that H_0 states that all regression coefficients in regression model (14.73) are zero. Thus, we employ the usual test statistic (7.34b) for testing whether or not there is a regression relation:

$$(14.75) \qquad\qquad F^* = \frac{MSR}{MSE}$$

Example

To test the equality of mean sales for the four cereal package designs in the Kenton Food Company example by means of the regression approach, we shall employ the regression model:

$$(14.76) \qquad\qquad Y_{ij} = \mu_. + \tau_1 X_{ij1} + \tau_2 X_{ij2} + \tau_3 X_{ij3} + \varepsilon_{ij}$$

where:

$$X_{ij1} = \begin{array}{l} 1 \text{ if case from factor level 1} \\ -1 \text{ if case from factor level 4} \\ 0 \text{ otherwise} \end{array}$$

$$X_{ij2} = \begin{array}{l} 1 \text{ if case from factor level 2} \\ -1 \text{ if case from factor level 4} \\ 0 \text{ otherwise} \end{array}$$

$$X_{ij3} = \begin{array}{l} 1 \text{ if case from factor level 3} \\ -1 \text{ if case from factor level 4} \\ 0 \text{ otherwise} \end{array}$$

The observations vector **Y** and the **X** matrix for the data in Table 14.1a are shown in Table 14.5a. For observation Y_{11}, for instance, note that $X_1 = 1$, $X_2 = 0$, and $X_3 = 0$; hence, we obtain from (14.76):

$$E\{Y_{11}\} = \mu_. + \tau_1$$

Similarly, for observation Y_{42} we have $X_1 = -1$, $X_2 = -1$, and $X_3 = -1$; hence:

$$E\{Y_{42}\} = \mu_. - \tau_1 - \tau_2 - \tau_3 = \mu_. + \tau_4$$

since $\tau_4 = -\tau_1 - \tau_2 - \tau_3$.

Note that we employ the following codings in the indicator variables for cases from each of the four factor levels:

	Coding		
Factor Level	X_1	X_2	X_3
1	1	0	0
2	0	1	0
3	0	0	1
4	-1	-1	-1

TABLE 14.5 Regression Approach to the Analysis of Variance—
Kenton Food Company Example

(a) Data Matrices for Regression Model (14.76)

$$
\mathbf{Y} = \begin{bmatrix} 12 \\ 18 \\ 14 \\ 12 \\ 13 \\ 19 \\ 17 \\ 21 \\ 24 \\ 30 \end{bmatrix}
\qquad
\mathbf{X} = \begin{bmatrix}
 & X_1 & X_2 & X_3 \\
1 & 1 & 0 & 0 \\
1 & 1 & 0 & 0 \\
1 & 0 & 1 & 0 \\
1 & 0 & 1 & 0 \\
1 & 0 & 1 & 0 \\
1 & 0 & 0 & 1 \\
1 & 0 & 0 & 1 \\
1 & 0 & 0 & 1 \\
1 & -1 & -1 & -1 \\
1 & -1 & -1 & -1
\end{bmatrix}
$$

(b) Fitted Regression Function

$$\hat{Y} = 18.5 - 3.5X_1 - 5.5X_2 + .5X_3$$

(c) Regression Analysis of Variance Table

Source of Variation	SS	df	MS
Regression	$SSR = 258$	3	$MSR = 86$
Error	$SSE = 46$	6	$MSE = 7.67$
Total	$SSTO = 304$	9	

A computer run of a multiple regression package for the data in Table 14.5a yielded the fitted regression function and analysis of variance table presented in Tables 14.5b and 14.5c. Test statistic (14.75) therefore is:

$$F^* = \frac{MSR}{MSE} = \frac{86}{7.67} = 11.2$$

This is the same test statistic obtained earlier based on the analysis of variance calculations. Indeed, the analysis of variance table in Table 14.5c obtained with the regression approach is the same as the one in Table 14.4 obtained with the analysis of variance approach except that the treatment sum of squares and mean square in Table 14.4 are called the regression sum of squares and mean square in Table 14.5c.

From this point on, the test procedure based on the regression approach parallels the analysis of variance test procedure explained earlier.

Comments

1. In the fitted regression function in Table 14.5b, the intercept term 18.5 is the unweighted average of the estimated factor level means $\bar{Y}_{i.}$ because $\mu_.$ was defined as the unweighted average of the factor level means μ_i.

2. The regression approach is not generally utilized for ordinary analysis of variance problems. The reason is that the **X** matrix for analysis of variance problems usually is of a very simple structure, as we have seen in Table 14.5a for the Kenton Food Company example. This simple structure permits computational simplifications that are explicitly recognized in the statistical procedures for analysis of variance. We take up the regression approach to analysis of variance here, and in later chapters, for two principal reasons. First, we see that analysis of variance models are encompassed by the general linear statistical model (7.18) considered in Chapter 7. Second, the regression approach is very useful for analyzing some multifactor studies when the structure of the **X** matrix is not simple.

3. When the analysis of variance test is to be conducted by means of the regression approach based on the cell means model (14.2):

$$Y_{ij} = \mu_i + \varepsilon_{ij}$$

the $\boldsymbol{\beta}$ vector is defined to contain all r treatment means μ_i:

$$(14.77) \qquad\qquad \boldsymbol{\beta} = \begin{bmatrix} \mu_1 \\ \vdots \\ \mu_r \end{bmatrix}$$

and r indicator variables $X_1, X_2, \ldots, X_r$ are utilized, each defined as a 0,1 variable as illustrated in Chapter 10:

$$(14.78)$$

$$X_1 = \begin{array}{l} 1 \text{ if case from factor level 1} \\ 0 \text{ otherwise} \end{array}$$

$$\vdots \qquad\qquad \vdots$$

$$X_r = \begin{array}{l} 1 \text{ if case from factor level } r \\ 0 \text{ otherwise} \end{array}$$

The regression model therefore is:

$$(14.79) \qquad Y_{ij} = \mu_1 X_{ij1} + \mu_2 X_{ij2} + \cdots + \mu_r X_{ijr} + \varepsilon_{ij} \qquad\qquad \text{Full model}$$

with the μ_i playing the role of regression coefficients.

The **X** matrix with this approach contains only 0 and 1 entries. For example, for $r = 3$ factor levels with $n_1 = n_2 = n_3 = 2$ cases, the **X** matrix (observations in order Y_{11}, Y_{12}, Y_{21}, etc.) and $\boldsymbol{\beta}$ vector would be as follows:

$$\mathbf{X} = \begin{bmatrix} 1 & 0 & 0 \\ 1 & 0 & 0 \\ 0 & 1 & 0 \\ 0 & 1 & 0 \\ 0 & 0 & 1 \\ 0 & 0 & 1 \end{bmatrix} \qquad \boldsymbol{\beta} = \begin{bmatrix} \mu_1 \\ \mu_2 \\ \mu_3 \end{bmatrix}$$

Note that regression model (14.79) has no intercept term. When a computer regression package is to be employed for this case, it is important that a fit with no intercept term be specified.

The test of whether the factor level means are equal, i.e., $\mu_1 = \mu_2 = \cdots = \mu_r$, asks only whether or not the regression coefficients in (14.79) are equal, not whether or not they equal zero. Hence, we need to fit the full model and then the reduced model to conduct this test. The reduced model when H_0: $\mu_1 = \cdots = \mu_r$ holds is:

(14.80)
$$Y_{ij} = \mu_c + \varepsilon_{ij}$$
Reduced model

where μ_c is the common value of all μ_i under H_0. The $\mathbf{X}$ matrix here consists simply of a column of 1s. The $\mathbf{X}$ matrix and $\boldsymbol{\beta}$ vector for the reduced model in our example would be:

$$\mathbf{X} = \begin{bmatrix} 1 \\ 1 \\ 1 \\ 1 \\ 1 \\ 1 \end{bmatrix} \qquad \boldsymbol{\beta} = [\mu_c]$$

After the full and reduced models are fitted and the error sums of squares are obtained for each fit, the usual general linear test statistic (3.69) is then calculated.

CITED REFERENCE

14.1. Dixon, W. J., chief editor. *BMDP Statistical Software Manual,* vols. 1 and 2. Berkeley, Calif.: University of California Press, 1988.

PROBLEMS

14.1. Refer to Figure 14.1a. Could you determine the mean sales level when the price level is 68 dollars if you knew the true regression function? Could you make this determination from Figure 14.1b if you only knew the values of the parameters μ_1, μ_2, and μ_3 of the ANOVA model (14.2)? What distinction between regression models and ANOVA models is demonstrated by your answers?

14.2. A market researcher, having collected data on breakfast cereal expenditures by families with 1, 2, 3, 4, and 5 children living at home, plans to use an ordinary regression model to estimate the mean expenditures at each of these five family size levels. However, she is undecided between fitting a linear or a quadratic regression model, and the data do not give clear evidence in favor of one model or the other. A colleague suggests: "For your purposes you might simply use an ANOVA model." Is this a useful suggestion? Explain.

14.3. Refer to the **SENIC** data set. An analyst has set up four age groupings for variable 3 (age) and wishes to apply ANOVA model (14.2) to determine whether the mean infection risk (variable 4) is the same in the four age groups.
a. What is the dependent variable here?
b. Identify the factor studied. What are the factor levels?
c. Is the factor quantitative or qualitative?
d. Is the factor an experimental or a classification factor?

14.4. Thirty trainees are randomly divided into three groups of 10 and each group is given instruction in the use of a different word-processing system. At the end of the training period, each trainee is given the same "benchmark" word-processing project to complete and the time required for completion is recorded. ANOVA model (14.2) will be used to test whether or not the mean time is the same for the three systems.
   a. What is the dependent variable here?
   b. Identify the factor studied and the factor levels.
   c. Is the factor an experimental or a classification factor? Would your answer differ if each trainee had been allowed to select the word-processing system of his or her choice?

14.5. In a study of intentions to get flu-vaccine shots in an area threatened by an epidemic, 90 persons were classified into three groups of 30 according to the degree of risk of getting flu. Each group was together when the persons were asked about the likelihood of getting the shots, on a probability scale ranging from 0 to 1.0. Unavoidably, most persons overheard the answers of nearby respondents. An analyst wishes to test whether the mean intent scores are the same for the three risk groups. Consider each assumption for ANOVA model (14.2) and explain whether this assumption is likely to hold in the present situation.

14.6. In a study of the effectiveness of subliminal advertising, subjects are brought to a studio and shown a film with one of three types of subliminal advertising for a product. Each subject's attitudes toward the product before and after the film viewing are measured.
   a. In this type of situation, should a control treatment be employed?
   b. Explain precisely what would be the nature of the control treatment for this study.

14.7. A company, studying the relation between job satisfaction and length of service of employees, classified employees into three length-of-service groups (less than 5 years, 5–10 years, more than 10 years). Suppose $\mu_1 = 65$, $\mu_2 = 80$, $\mu_3 = 95$, and $\sigma = 3$, and that ANOVA model (14.2) is applicable.
   a. Draw a representation of this model in the format of Figure 14.2.
   b. Find $E\{MSTR\}$ and $E\{MSE\}$ if 25 employees from each group are selected at random for intensive interviewing about job satisfaction. Is $E\{MSTR\}$ substantially larger than $E\{MSE\}$ here? What is the implication of this?

14.8. In a study of length of hospital stay (in number of days) of persons in four income groups, the parameters are as follows: $\mu_1 = 5.1$, $\mu_2 = 6.3$, $\mu_3 = 7.9$, $\mu_4 = 9.5$, $\sigma = 2.8$. Assume that ANOVA model (14.2) is appropriate.
   a. Draw a representation of this model in the format of Figure 14.2.
   b. Suppose 100 persons from each income group are randomly selected for the study. Find $E\{MSTR\}$ and $E\{MSE\}$. Is $E\{MSTR\}$ substantially larger than $E\{MSE\}$ here? What is the implication of this?
   c. If $\mu_2 = 5.6$ and $\mu_3 = 9.0$, everything else remaining the same, what would $E\{MSTR\}$ be? Why is $E\{MSTR\}$ larger here than in part (b) even though the range of the factor level means is the same?

14.9. A student asks: "Why is the F test for equality of factor level means not a two-tail test since any differences among the factor level means can occur in either direction?" Explain, utilizing the expressions for the expected mean squares in (14.35).

14.10. Productivity improvement.   An economist compiled data on productivity improvements last year for a sample of firms producing electronic computing equipment. The firms were classified according to the level of their average expenditures for research and development in the past three years (low, moderate, high). The results of the study follow (productivity improvement is measured on a scale from 0 to 100). Assume that ANOVA model (14.2) is appropriate.

	i	1	2	3	4	5	6	7	8	9	10	11	12
							j						
1	Low	7.6	8.2	6.8	5.8	6.9	6.6	6.3	7.7	6.0			
2	Moderate	6.7	8.1	9.4	8.6	7.8	7.7	8.9	7.9	8.3	8.7	7.1	8.4
3	High	8.5	9.7	10.1	7.8	9.6	9.5						

Summary calculational results are: $SSTR = 20.125$, $SSE = 15.362$.
a. Obtain the fitted values.
b. Obtain the residuals. Do they sum to zero in accord with (14.18)?
c. Obtain the analysis of variance table.
d. Test whether or not the mean productivity improvement differs according to the level of research and development expenditures. Control the α risk at .05. State the alternatives, decision rule, and conclusion.
e. What is the P-value of the test in part (d)? How does it support the conclusion reached in part (d)?
f. What appears to be the nature of the relationship between research and development expenditures and productivity improvement?

14.11. Questionnaire color. In an experiment to investigate the effect of color of paper (blue, green, orange) on response rates for questionnaires distributed by the "windshield method" in supermarket parking lots, 15 representative supermarket parking lots were chosen in a metropolitan area and each color was assigned at random to five of the lots. The response rates (in percent) follow. Assume that ANOVA model (14.2) is appropriate.

	i	1	2	3	4	5
				j		
1	Blue	28	26	31	27	35
2	Green	34	29	25	31	29
3	Orange	31	25	27	29	28

Summary calculational results are: $SSTR = 7.60$, $SSE = 116.40$.
a. Obtain the fitted values.
b. Obtain the residuals.
c. Obtain the analysis of variance table.
d. Conduct a test to determine whether or not the mean response rates for the three colors differ. Use a level of significance of $\alpha = .10$. State the alternatives, decision rule, and conclusion. What is the P-value of the test?
e. When informed of the findings, an executive said: "See? I was right all along. We might as well print the questionnaires on plain white paper, which is cheaper." Does this conclusion follow from the findings of the study? Discuss.

14.12. Rehabilitation therapy. A rehabilitation center researcher was interested in examining the relationship between physical fitness prior to surgery of persons undergoing corrective knee surgery and time required in physical therapy until successful rehabilitation. Patient records in the rehabilitation center were examined, and 24 male subjects ranging in age from 18 to 30 years who had undergone similar corrective knee surgery during the past year were selected for the study. The number of days required for successful completion of physical therapy and the prior physical fitness status (below average, average, above average) for each patient follow.

	i	1	2	3	4	5	6	7	8	9	10
1	Below average	29	42	38	40	43	40	30	42		
2	Average	30	35	39	28	31	31	29	35	29	33
3	Above average	26	32	21	20	23	22				

Assume that ANOVA model (14.2) is appropriate.
a. Obtain the fitted values.
b. Obtain the residuals. Do they sum to zero in accord with (14.18)?
c. Obtain the analysis of variance table.
d. Test whether or not the mean number of days required for successful rehabilitation is the same for the three fitness groups. Control the α risk at .01. State the alternatives, decision rule, and conclusion.
e. Obtain the P-value for the test in part (d). Explain how the same conclusion reached in part (d) can be obtained by knowing the P-value.
f. What appears to be the nature of the relationship between physical fitness status and duration of required physical therapy?

14.13. Cash offers.   A consumer organization studied the effect of age of automobile owner on size of cash offer for a used car by utilizing 12 persons in each of three age groups (young, middle, elderly) who acted as the owner of a used car. A medium price, six-year-old car was selected for the experiment, and the "owners" solicited cash offers for this car from 36 dealers selected at random from the dealers in the region. Randomization was used in assigning the dealers to the "owners." The offers (in hundred dollars) follow. Assume that ANOVA model (14.2) is applicable.

	i	1	2	3	4	5	6	7	8	9	10	11	12
1	Young	23	25	21	22	21	22	20	23	19	22	19	21
2	Middle	28	27	27	29	26	29	27	30	28	27	26	29
3	Elderly	23	20	25	21	22	23	21	20	19	20	22	21

a. Obtain the fitted values.
b. Obtain the residuals.
c. Obtain the analysis of variance table.
d. Conduct the F test for equality of factor level means; use $\alpha = .01$. State the alternatives, decision rule, and conclusion. What is the P-value of the test?
e. What appears to be the nature of the relationship between age of owner and mean cash offer?

14.14. Filling machines.   A company uses six filling machines of the same make and model to place detergent into cartons that show a label weight of 32 ounces. The production manager has complained that the six machines do not place the same amount of fill into the cartons. A consultant requested that 20 filled cartons be selected randomly from each of the six machines and the content of each carton carefully weighed. The observations (stated for convenience as deviations from 32.00 ounces) follow. Assume that ANOVA model (14.2) is applicable.

						j				
i	1	2	3	4	5	6	7	8	9	10
1	−.14	.20	.07	.18	.38	.10	−.04	−.27	.27	−.21
2	.46	.11	.12	.47	.24	.06	−.12	.33	.06	−.03
3	.21	.78	.32	.45	.22	.35	.54	.24	.47	.62
4	.49	.58	.52	.29	.27	.55	.40	.14	.48	.34
5	−.19	.27	.06	.11	.23	.15	.01	.22	.29	.14
6	.05	−.05	.28	.47	.12	.27	.08	.17	.43	−.07

						j				
i	11	12	13	14	15	16	17	18	19	20
1	.39	−.07	−.02	.28	.09	.13	.26	.07	−.01	−.19
2	.05	.53	.42	.29	.36	.04	.17	.02	.11	.12
3	.47	.55	.59	.71	.45	.48	.44	.50	.20	.61
4	.01	.33	.18	.13	.48	.54	.51	.42	.45	.20
5	.20	.30	−.11	.27	−.20	.24	.20	.14	.35	−.18
6	.20	.01	.10	.16	−.06	.13	.43	.35	−.09	.05

a. Obtain the fitted values.
b. Obtain the residuals. Do they sum to zero in accord with (14.18)?
c. Obtain the analysis of variance table.
d. Test whether or not the mean fill differs among the six machines; control the α risk at .05. State the alternatives, decision rule, and conclusion. Does your conclusion support the production manager's complaint?
e. What is the P-value of the test in part (d)? Is this value consistent with your conclusion in part (d)? Explain.
f. Does the variation between the mean fills for the six machines appear to be large relative to the variability in fills between cartons for any given machine? Explain.

14.15. **Premium distribution.**   A soft-drink manufacturer uses five agents (1, 2, 3, 4, 5) to handle premium distributions for its various products. The marketing director desired to study the timeliness with which the premiums are distributed. Twenty transactions for each agent were selected at random, and the time lapse (in days) for handling each transaction was determined. The results follow. Assume that ANOVA model (14.2) is appropriate.

																	j			
i	1	2	3	4	5	6	7	8	9	10	11	12	13	14	15	16	17	18	19	20
1	24	24	29	20	21	25	28	27	23	21	24	26	23	24	28	23	23	27	26	25
2	18	20	20	24	22	29	23	24	28	19	24	25	21	20	24	22	19	26	22	21
3	10	11	8	12	12	10	14	9	8	11	16	12	18	14	13	11	14	9	11	12
4	15	13	18	16	12	19	10	18	11	17	15	12	13	13	14	17	16	17	14	16
5	33	22	28	35	29	28	30	31	29	28	33	30	32	33	29	35	32	26	30	29

a. Obtain the fitted values.

b. Obtain the residuals. Do they sum to zero in accord with (14.18)?

c. Obtain the analysis of variance table.

d. Test whether or not the mean time lapse differs for the five agents; use $\alpha = .10$. State the alternatives, decision rule, and conclusion.

e. What is the P-value of the test in part (d)? Explain how the same conclusion as in part (d) can be reached by knowing the P-value.

f. Based on the estimated treatment means, does there appear to be much variation in the mean time lapse for the five agents? Is this variation necessarily the result of differences in the efficiency of operations of the five agents? Discuss.

14.16. Four treatments are to be studied in a completely randomized design, and 16 experimental units are available. Each treatment is to be assigned to four experimental units selected at random. There are four observers; each is to be assigned at random to one experimental unit for each treatment. Make all appropriate randomizations.

14.17. Five treatments are to be studied in a completely randomized design, and 30 experimental units are available. Each treatment is to be assigned to six experimental units selected at random. The study is to be conducted over a period of six days, with each treatment included on each day. The order of the five treatments within each day is to be randomized. Make all appropriate randomizations.

14.18. Refer to **Questionnaire color** Problem 14.11. Explain how you would make the random assignments of supermarket parking lots to colors in this single-factor study. Make all appropriate randomizations.

14.19. Refer to **Cash offers** Problem 14.13. Explain how you would make the random assignments of dealers to "owners" in this single-factor study. Make all appropriate randomizations.

14.20. Refer to Problem 14.7. What are the values of τ_1, τ_2, and τ_3 if the ANOVA model is expressed in the alternative formulation (14.60) in terms of factor effects and $\mu.$ is defined by (14.61)?

14.21. Refer to Problem 14.8. What are the values of τ_i if the ANOVA model is expressed in the alternative formulation (14.60) in terms of factor effects and $\mu.$ is defined by (14.61)?

14.22. Refer to **Premium distribution** Problem 14.15. Suppose that 25 percent of all premium distributions are handled by agent 1, 20 percent by agent 2, 20 percent by agent 3, 20 percent by agent 4, and 15 percent by agent 5.

a. Obtain a point estimate of $\mu.$ when the ANOVA model is expressed in the alternative formulation (14.60) and $\mu.$ is defined by (14.63), with the weights being the proportions of premium distribution handled by each agent.

b. State the alternatives for the test of equality of factor level means in terms of factor effects model (14.60) for the present case. Would this statement be affected if $\mu.$ were defined according to (14.61)? Explain.

14.23. Refer to **Productivity improvement** Problem 14.10. Regression model (14.73) is to be employed for testing the equality of the factor level means.

a. Set up the $\mathbf{Y}$, $\mathbf{X}$, and $\boldsymbol{\beta}$ matrices.

b. Obtain $\mathbf{X}\boldsymbol{\beta}$. Develop equivalent expressions of the elements of this vector in terms of μ_i.

c. Obtain the fitted regression function. What is estimated by the intercept term?

d. Obtain the regression analysis of variance table.

    e. Conduct the test for equality of factor level means; use $\alpha = .05$. State the alternatives, decision rule, and conclusion.

14.24. Refer to **Questionnaire color** Problem 14.11. Regression model (14.73) is to be employed for testing the equality of the factor level means.

    a. Set up the **Y**, **X**, and **β** matrices.

    b. Obtain **Xβ**. Develop equivalent expressions of the elements of this vector in terms of μ_i.

    c. Obtain the fitted regression function. What is estimated by the intercept term?

    d. Obtain the regression analysis of variance table.

    e. Conduct the test for equality of factor level means; use $\alpha = .10$. State the alternatives, decision rule, and conclusion.

14.25. Refer to **Cash offers** Problem 14.13.

    a. Fit regression model (14.73) to the data. What is estimated by the intercept term?

    b. Obtain the regression analysis of variance table and test whether or not the factor level means are equal; use $\alpha = .01$. State the alternatives, decision rule, and conclusion.

14.26. Refer to **Rehabilitation therapy** Problem 14.12.

    a. Fit the full regression model (14.79) to the data. Would a fitted regression model containing an intercept term be proper here?

    b. Fit the reduced model (14.80) to the data.

    c. Use test statistic (3.69) for testing the equality of the factor level means; employ a level of significance of $\alpha = .01$.

EXERCISES

14.27. (Calculus needed.) State the likelihood function for ANOVA model (14.2) when $r = 3$ and $n_i \equiv 2$. Find the maximum likelihood estimators. Are they the same as the least squares estimators (14.14)?

14.28. Show that the result in (14.31b) is the algebraic equivalent of (14.25).

14.29. Show that when test statistic t^* in Table 1.2a is squared, it is equivalent to the F^* test statistic (14.53) for $r = 2$.

14.30. Derive the restriction in (14.64) when the constant $\mu.$ is defined according to (14.63).

14.31. a. Obtain the least squares estimators of the regression coefficients in the full regression model (14.79). What is $SSE(F)$ here?

    b. Obtain the least squares estimator of μ_c in the reduced regression model (14.80). What is $SSE(R)$ here?

14.32. Consider the illustration in (14.6)–(14.8) demonstrating that ANOVA model (14.2) is a linear model. To test the equality of the three treatment means, we could therefore employ the general matrix approach and use (8.70) for constructing the test statistic. Show that if:

$$\mathbf{C} = \begin{bmatrix} 1 & -1 & 0 \\ 1 & 0 & -1 \end{bmatrix} \qquad \mathbf{h} = \begin{bmatrix} 0 \\ 0 \end{bmatrix}$$

formula (8.70) is equivalent to *SSTR*.

PROJECTS

14.33. Refer to the **SENIC** data set. Test whether or not the mean infection risk (variable 4) is the same in the four geographic regions (variable 9); use a level of significance of $\alpha = .05$. Assume that ANOVA model (14.2) is applicable. State the alternatives, decision rule, and conclusion.

14.34. Refer to the **SENIC** data set. The effect of average age of patient (variable 3) on mean infection risk (variable 4) is to be studied. For purposes of this ANOVA study, average age is to be classified into four categories: Under 50.0, 50.0–54.9, 55.0–59.9, 60.0 and over. Assume that ANOVA model (14.2) is applicable. Test whether or not the mean infection risk differs for the four age groups. Control the α risk at .10. State the alternatives, decision rule, and conclusion.

14.35. Refer to the SMSA data set. The effect of geographic region (variable 12) on the crime rate (variable 11 ÷ variable 3) is to be studied. Assume that ANOVA model (14.2) is applicable. Test whether or not the mean crime rates for the four geographic regions differ; use $\alpha = .05$. State the alternatives, decision rule, and conclusion.

14.36. Consider a test involving H_0: $\mu_1 = \mu_2 = \mu_3$. Five observations are to be taken for each factor level, and a level of significance of $\alpha = .05$ is to be employed in the test.
   a. Generate five random normal observations when $\mu_1 = 100$ and $\sigma = 12$ to represent the observations for treatment 1. Repeat this for the other two treatments when $\mu_2 = \mu_3 = 100$ and $\sigma = 12$. Finally, calculate the F^* test statistic (14.53).
   b. Repeat part (a) 100 times.
   c. Calculate the mean of the 100 F^* statistics.
   d. What proportion of the F^* statistics lead to conclusion H_0? Is this consistent with theoretical expectations?
   e. Repeat parts (a) and (b) when $\mu_1 = 80$, $\mu_2 = 60$, $\mu_3 = 160$, and $\sigma = 12$. Calculate the mean of the 100 F^* statistics. How does this mean compare with the mean obtained in part (c) when $\mu_1 = \mu_2 = \mu_3 = 100$? Is this result consistent with use of decision rule (14.54)?
   f. What proportion of the 100 test statistics obtained in part (e) lead to conclusion H_a? Does it appear that the test has satisfactory power when $\mu_1 = 80$, $\mu_2 = 60$, and $\mu_3 = 160$?

Chapter 15

Analysis of Factor Level Effects

The F test for determining whether or not the factor level means μ_i differ, discussed in the previous chapter, is a preliminary test to establish whether detailed analysis of the factor level effects is warranted. If the F test leads to the conclusion that the factor level means μ_i are equal, the implication is that there is no relation between the factor and the dependent variable. On the other hand, if the F test leads to the conclusion that the factor level means μ_i differ, the implication is that there is a relation between the factor and the dependent variable. In that case, a thorough analysis of the nature of the factor level effects is usually undertaken. This is done in two principal ways:

1. A direct analysis of the factor level effects of interest using estimation techniques.
2. Statistical tests in regard to the factor level effects of interest.

We shall illustrate each of these two avenues of approach in turn, but will concentrate on the estimation approach in view of its greater usefulness. Throughout this chapter, we continue to assume ANOVA model I. The cell means version of this model was given in (14.2):

$$(15.1) \qquad\qquad Y_{ij} = \mu_i + \varepsilon_{ij}$$

where:

μ_i are parameters
ε_{ij} are independent $N(0, \sigma^2)$

15.1 GRAPHIC PLOTS OF ESTIMATED FACTOR LEVEL MEANS

It is usually helpful, before undertaking formal analysis of the nature of the factor level effects, to examine these factor effects informally from a plot of the estimated factor level means $\overline{Y}_{i.}$. We shall take up two types of plots: (1) a line plot, which is

568

appropriate whether the sample sizes n_i are equal or not; and (2) a normal probability plot, which is appropriate when the sample sizes n_i are equal.

Line Plot

A line plot of the estimated factor level means simply shows the positions of the $\bar{Y}_{i.}$ on a line scale. It is a very simple, but effective, device for indicating when one or several factor level means may differ substantially from the others.

Example.   We return to the Kenton Food Company example of Chapter 14. The basic results of the study are repeated in summary form in Table 15.1. In Figure 15.1 we present a line plot of the estimated factor level means $\bar{Y}_{i.}$. It is clear from Figure 15.1 that design 4 led by far to the highest mean sales in the study, and that package designs 1 and 2 led to mean sales which did not differ much from each other. The purpose of the formal inference procedures to be taken up shortly is to provide information whether the pattern noted here is simply the result of random variation or reflects underlying differences in the factor level means μ_i.

TABLE 15.1   Summary of Results for Kenton Food Company Example Obtained in Chapter 14

	Package Design (i)				
	1	2	3	4	Total
n_i	2	3	3	2	10
$Y_{i.}$	30	39	57	54	180
$\bar{Y}_{i.}$	15	13	19	27	18

Source of Variation	SS	df	MS
Between designs	258	3	86
Error	46	6	7.67
Total	304	9	

Package Design	Characteristics
1	3 colors, with cartoons
2	3 colors, without cartoons
3	5 colors, with cartoons
4	5 colors, without cartoons

FIGURE 15.1   Line Plot of Estimated Factor Level Means—Kenton Food Company
Example

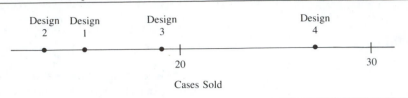

Cases Sold

Normal Probability Plot

A normal probability plot of the estimated factor level means recognizes that the $\overline{Y}_{i\cdot}$ are subject to random variation and permits one to assess in an informal fashion whether the differences in the estimated means reflect real effects. The use of normal probability plots of the estimated factor level means requires that *all sample sizes be equal,* in other words, that $n_i \equiv n$.

We considered normal probability plots for residuals in Chapter 4. There, the residuals were plotted against their expected values under normality, given in (4.6). Here, normality of the observations Y_{ij}, with constant variance σ^2, is assumed. The condition being tested is whether the factor level means μ_i are equal. If they are, the estimated means $\overline{Y}_{i\cdot}$ have the same expected value and the same variance (because of the constant variance of the Y_{ij} and the equal sample sizes). Hence, the estimated means $\overline{Y}_{i\cdot}$ should behave like random observations from the same normal distribution when shown in a normal probability plot, if the factor level means are equal. A substantial departure from a linear pattern therefore indicates that the factor level means are not equal, and the nature of the plot may suggest which of the factor level means differ from the others.

The expected value of the ith ordered estimated factor level mean, if all factor level means μ_i are equal, is approximately as follows:

$$(15.2) \qquad \text{Expected value} = \overline{Y}_{\cdot\cdot} + z\left(\frac{i - .375}{r + .25}\right)\sqrt{\frac{MSE}{n}}$$

where the square root term is the estimated standard deviation of $\overline{Y}_{i\cdot}$, $i = 1, \ldots, r$, as will be explained shortly. Rather than plotting the estimated factor level means $\overline{Y}_{i\cdot}$ against their expected values, as we did for residual plots, it will be more effective here to plot the $\overline{Y}_{i\cdot}$ against the normal percentiles. Since the expected values are a linear function of the percentiles, a linear pattern of the points is preserved whether we plot against the percentiles or the expected values. We shall also plot the line:

$$y = \text{expected value}$$

on the same graph to serve as a reference line to assist in judging whether any estimated mean $\overline{Y}_{i\cdot}$ is far from its expected value.

FIGURE 15.2   Prototype Normal Probability Plots of Estimated Factor Level Means

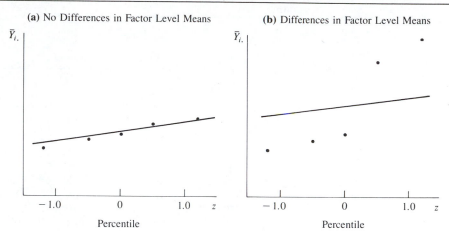

Figure 15.2a shows a prototype pattern suggesting that all treatment means μ_i are equal. Each of the points is close to the reference line y = expected value, and the pattern of the points therefore is reasonably linear.

Figure 15.2b shows a prototype pattern where the jump in the points suggests that the two largest factor level means differ from the other three factor level means. The fact that the points for the three smallest $\bar{Y}_{i.}$ follow a linear pattern approximately parallel to the reference line y = expected value suggests that these factor level means do not differ from each other. The substantial deviations of these points from the reference line is the result of a major difference between the two groups of factor level means.

Example.   In a study of the effectiveness of different rust inhibitors, four brands (A, B, C, D) were tested. Altogether, 40 experimental units were randomly assigned to the four brands, with 10 units assigned to each brand. The results after exposing the experimental units to severe weather conditions are given in coded form in Table 15.2a. The higher is the coded value, the more effective is the rust inhibitor. Note in Table 15.2a that the treatments are shown in columns because of layout consider-ations.

The analysis of variance is shown in Table 15.2b. Using a level of significance of α = .05 for testing whether or not the four rust inhibitors differ in effectiveness, we require $F(.95; 3, 36) = 2.87$. Using the mean squares from Table 15.2b, the test statistic is:

$$F^* = \frac{MSTR}{MSE} = \frac{5,317.82}{6.140} = 866.1$$

Since $F^* = 866.1 > 2.87$, we conclude that the four rust inhibitors differ in effec-tiveness. The P-value of the test is $0+$.

TABLE 15.2   Data and Analysis of Variance Results for Rust Inhibitor Example (data are coded)

(a) Data

Rust Inhibitor Brand

j	A $i = 1$	B $i = 2$	C $i = 3$	D $i = 4$
1	43.9	89.8	68.4	36.2
2	39.0	87.1	69.3	45.2
3	46.7	92.7	68.5	40.7
4	43.8	90.6	66.4	40.5
5	44.2	87.7	70.0	39.3
6	47.7	92.4	68.1	40.3
7	43.6	86.1	70.6	43.2
8	38.9	88.1	65.2	38.7
9	43.6	90.8	63.8	40.9
10	40.0	89.1	69.2	39.7
$\bar{Y}_{i.}$	43.14	89.44	67.95	40.47

$$\bar{Y}_{..} = 60.25$$

(b) Analysis of Variance

Source of Variation	SS	df	MS
Between brands	15,953.47	3	5,317.82
Error	221.03	36	6.140
Total	16,174.50	39	

Figure 15.3a contains a line plot of the estimated factor level means $\bar{Y}_{i.}$. This plot suggests that brands B and C performed much better, on the average, than the other two brands.

Before preparing a normal probability plot of the estimated factor level means, we list them in ascending order in Table 15.3, column 1. Column 2 shows the standard normal percentiles, and column 3 shows the expected values of the ordered estimated means under the assumption that the factor level means are equal. We illustrate the calculations for the smallest estimated factor level mean, for which the ordering index is $i = 1$. The standard normal percentile required is:

$$z\left(\frac{i - .375}{r + .25}\right) = z\left(\frac{1 - .375}{4 + .25}\right) = z(.147) = -1.049$$

We know from Table 15.2 that $\bar{Y}_{..} = 60.25$ and $MSE = 6.140$. Hence, using (15.2) we obtain the approximate expected value:

$$60.25 + (-1.049)\sqrt{\frac{6.140}{10}} = 59.4$$

FIGURE 15.3   Plots of Estimated Factor Level Means—Rust Inhibitor Example

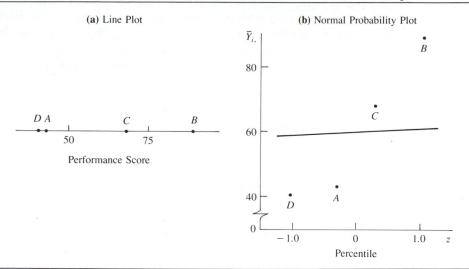

(a) Line Plot

(b) Normal Probability Plot

The expected values for the other estimated factor level means are obtained in similar fashion and are shown in Table 15.3, column 3.

Figure 15.3b contains the normal probability plot of the estimated factor level means for the rust inhibitor example. It suggests strongly, like the line plot in Figure 15.3a, that the four rust inhibitors differ in effectiveness because the points deviate substantially from the reference line. In addition, the normal probability plot suggests that rust inhibitors B and C performed better than rust inhibitors A and D. Fur-

TABLE 15.3   Estimated Factor Level Means $\bar{Y}_{i.}$ and Expected Values Under Assumption of Equality of μ_i—Rust Inhibitor Example

Ascending Order i	(1) Ordered Means $\bar{Y}_{i.}$	(2) $z\left(\dfrac{i - .375}{4 + .25}\right)$	(3) Expected Value under Equality
1	40.5	−1.049	59.4
2	43.1	−.299	60.0
3	68.0	.299	60.5
4	89.4	1.049	61.1

$$\bar{Y}_{..} = 60.25$$

$$\sqrt{\frac{MSE}{n}} = \sqrt{\frac{6.140}{10}} = .784$$

ther, the normal probability plot suggests that the mean performance for brands A and D may be the same since these two points form a line that is fairly parallel to the reference line. On the other hand, the mean performance for rust inhibitor B appears to be greater than that for rust inhibitor C because these two points form a line with a slope far greater than that of the reference line.

Formal inference procedures are now required to confirm the suggestions made by the graphic plots.

15.2 ESTIMATION OF FACTOR LEVEL EFFECTS

Estimates of factor level effects usually employed include:

1. Estimation of a factor level mean μ_i.
2. Estimation of the difference between two factor level means.
3. Estimation of a contrast among factor level means.
4. Estimation of a linear combination of factor level means.

We shall discuss each of these four types of estimation in turn.

Estimation of Factor Level Mean

An unbiased point estimator of the factor level mean μ_i was obtained in (14.14):

$$(15.3) \qquad \hat{\mu}_i = \bar{Y}_{i.}$$

This estimator has mean and variance:

$$(15.4a) \qquad E\{\bar{Y}_{i.}\} = \mu_i$$

$$(15.4b) \qquad \sigma^2\{\bar{Y}_{i.}\} = \frac{\sigma^2}{n_i}$$

The latter result follows because (14.41) indicates that $\bar{Y}_{i.} = \mu_i + \bar{\varepsilon}_{i.}$, the sum of a constant plus a mean of n_i independent ε_{ij} terms, each of which has variance σ^2. Further, $\bar{Y}_{i.}$ is normally distributed because the error terms ε_{ij} are independent normal random variables.

The estimated variance of $\bar{Y}_{i.}$ is denoted by $s^2\{\bar{Y}_{i.}\}$ and is obtained as usual by replacing σ^2 in (15.4b) by the unbiased point estimator MSE:

$$(15.5) \qquad s^2\{\bar{Y}_{i.}\} = \frac{MSE}{n_i}$$

The estimated standard deviation $s\{\bar{Y}_{i.}\}$ is the positive square root of (15.5).

It can be shown that:

$$(15.6) \qquad \frac{\bar{Y}_{i.} - \mu_i}{s\{\bar{Y}_{i.}\}} \text{ is distributed as } t(n_T - r) \text{ for ANOVA model (15.1)}$$

where the degrees of freedom are those associated with MSE. The result (15.6) follows from the definition of t in (1.41) since: (1) $\bar{Y}_{i.}$ is normally distributed and (2)

MSE/σ^2 is distributed independently of $\bar{Y}_{i.}$ as $\chi^2(n_T - r)/(n_T - r)$ according to the following theorem:

(15.7)   For ANOVA model (15.1), SSE/σ^2 is distributed as χ^2 with $n_T - r$ degrees of freedom, and is independent of $\bar{Y}_{1.}, \ldots, \bar{Y}_{r.}$.

It follows directly from (15.6) that the confidence limits for μ_i with confidence coefficient $1 - \alpha$ are:

(15.8) $$\bar{Y}_{i.} \pm t(1 - \alpha/2; n_T - r)s\{\bar{Y}_{i.}\}$$

Example. In the Kenton Food Company example, the sales manager wished to estimate mean sales for package design 1 with a 95 percent confidence coefficient.

Using the results from Table 15.1, we have:

$$\bar{Y}_{1.} = 15 \qquad n_1 = 2 \qquad MSE = 7.67$$

We require $t(.975; 6)$. From Table A.2 in the Appendix, we obtain $t(.975; 6) = 2.447$. Finally, we need $s\{\bar{Y}_{1.}\}$. We have:

$$s^2\{\bar{Y}_{1.}\} = \frac{MSE}{n_1} = \frac{7.67}{2} = 3.835$$

so that $s\{\bar{Y}_{1.}\} = 1.958$. Hence, we obtain the confidence limits $15 \pm 2.447(1.958)$ and the confidence interval:

$$10.2 \le \mu_1 \le 19.8$$

Thus, we estimate with confidence coefficient .95 that the mean sales per store for package design 1 are between 10.2 and 19.8 cases.

Estimation of Difference between Two Factor Level Means

Frequently two treatments or factor levels are to be compared by estimating the difference D between the two factor level means, say, μ_i and $\mu_{i'}$: → *another gr. mean.*

(15.9) $$D = \mu_i - \mu_{i'}$$ *one gr. mean*

Such a difference between two factor level means will be called a *pairwise comparison*. A point estimator of (15.9), denoted by $\hat{D}$, is:

(15.10) $$\hat{D} = \bar{Y}_{i.} - \bar{Y}_{i'.}$$

This point estimator is unbiased:

(15.11) $$E\{\hat{D}\} = \mu_i - \mu_{i'}$$

Since $\bar{Y}_{i.}$ and $\bar{Y}_{i'.}$ are independent, the variance of $\hat{D}$ follows from (1.28b):

(15.12) $$\sigma^2\{\hat{D}\} = \sigma^2\{\bar{Y}_{i.}\} + \sigma^2\{\bar{Y}_{i'.}\} = \sigma^2\left(\frac{1}{n_i} + \frac{1}{n_{i'}}\right)$$

The estimated variance of $\hat{D}$, denoted by $s^2\{\hat{D}\}$, is given by:

$$(15.13) \qquad\qquad s^2\{\hat{D}\} = MSE\left(\frac{1}{n_i} + \frac{1}{n_{i'}}\right)$$

Finally, $\hat{D}$ is normally distributed by (1.37) because $\hat{D}$ is a linear combination of independent normal variables.

It follows from these characteristics, theorem (15.7), and the definition of t in (1.41) that:

$$(15.14) \qquad\qquad \frac{\hat{D} - D}{s\{\hat{D}\}} \text{ is distributed as } t(n_T - r) \text{ for ANOVA model (15.1)}$$

Hence, the $1 - \alpha$ confidence limits for D are:

$$(15.15) \qquad\qquad \hat{D} \pm t(1 - \alpha/2; n_T - r)s\{\hat{D}\}$$

Example. For the Kenton Food Company example, package designs 1 and 2 used 3-color printing and designs 3 and 4 used 5-color printing, as shown in Table 15.1. We wish to estimate the difference in mean sales for 5-color designs 3 and 4 using a 95 percent confidence interval. That is, we wish to estimate $D = \mu_3 - \mu_4$. From Table 15.1, we have:

$$\bar{Y}_{3.} = 19 \qquad n_3 = 3 \qquad MSE = 7.67$$
$$\bar{Y}_{4.} = 27 \qquad n_4 = 2$$

Hence:

$$\hat{D} = \bar{Y}_{3.} - \bar{Y}_{4.} = 19 - 27 = -8$$

The estimated variance of $\hat{D}$ is:

$$s^2\{\hat{D}\} = MSE\left(\frac{1}{n_3} + \frac{1}{n_4}\right) = 7.67\left(\frac{1}{3} + \frac{1}{2}\right) = 6.392$$

so that the estimated standard deviation of $\hat{D}$ is $s\{\hat{D}\} = 2.528$. We require $t(.975; 6) = 2.447$. The confidence limits therefore are $-8 \pm 2.447(2.528)$, and the desired 95 percent confidence interval is:

$$-14.2 \leq \mu_3 - \mu_4 \leq -1.8$$

Thus, we estimate with confidence coefficient .95 that the mean sales for package design 3 fall short of those for package design 4 by somewhere between 1.8 and 14.2 cases per store.

Estimation of Contrast

A *contrast* is a comparison involving two or more factor level means and includes the previous case of a pairwise difference between two factor level means in (15.9). A contrast will be denoted by L, and is defined as a linear combination of the factor level means μ_i where the coefficients c_i sum to zero:

(15.16) $$L = \sum_{i=1}^{r} c_i \mu_i \qquad \text{where } \sum_{i=1}^{r} c_i = 0$$

Illustrations of Contrasts. We illustrate some contrasts by reference to the Kenton Food Company example. Recall that package designs 1 and 2 used 3-color printing and designs 3 and 4 used 5-color printing. In addition, as shown in Table 15.1, package designs 1 and 3 utilized cartoons while no cartoons were utilized in designs 2 and 4.

1. Comparison of the mean sales for the two 3-color designs:

$$L = \mu_1 - \mu_2$$

Here, $c_1 = 1$, $c_2 = -1$, $c_3 = 0$, $c_4 = 0$, and $\Sigma c_i = 0$.

2. Comparison of the mean sales for the 3-color and 5-color designs:

$$L = \frac{\mu_1 + \mu_2}{2} - \frac{\mu_3 + \mu_4}{2}$$

Here, $c_1 = 1/2$, $c_2 = 1/2$, $c_3 = -1/2$, $c_4 = -1/2$, and $\Sigma c_i = 0$.

3. Comparison of the mean sales for designs with and without cartoons:

$$L = \frac{\mu_1 + \mu_3}{2} - \frac{\mu_2 + \mu_4}{2}$$

Here, $c_1 = 1/2$, $c_2 = -1/2$, $c_3 = 1/2$, $c_4 = -1/2$, and $\Sigma c_i = 0$.

Note that the first contrast is simply a pairwise comparison. In the second and third contrasts, we compare averages of several factor level means. The averages used here are unweighted averages of the means μ_i; these are ordinarily the averages of interest. In special cases one might be interested in weighted averages of the μ_i to describe the mean response for a group of several factor levels. For example, if both 3-color and 5-color designs were to be employed, with 3-color printing used three times as often as 5-color printing, the comparison of the effect of cartoons versus no cartoons might be based on the contrast:

$$L = \frac{3\mu_1 + \mu_3}{4} - \frac{3\mu_2 + \mu_4}{4}$$

Here, $c_1 = 3/4$, $c_2 = -3/4$, $c_3 = 1/4$, $c_4 = -1/4$, and $\Sigma c_i = 0$.

Confidence Interval for Contrast. An unbiased estimator of a contrast L is:

(15.17) $$\hat{L} = \sum_{i=1}^{r} c_i \bar{Y}_{i.}$$

Since the $\bar{Y}_{i.}$ are independent, the variance of $\hat{L}$ according to (1.28) is:

(15.18) $$\sigma^2\{\hat{L}\} = \sum_{i=1}^{r} c_i^2 \sigma^2\{\bar{Y}_{i.}\} = \sum_{i=1}^{r} c_i^2 \left(\frac{\sigma^2}{n_i}\right) = \sigma^2 \sum_{i=1}^{r} \frac{c_i^2}{n_i}$$

An unbiased estimator of this variance is:

(15.19)
$$s^2\{\hat{L}\} = MSE \sum_{i=1}^{r} \frac{c_i^2}{n_i}$$

$\hat{L}$ is normally distributed by (1.37) because it is a linear combination of independent normal random variables. It can be shown by theorem (15.7), the characteristics of $\hat{L}$ just mentioned, and the definition of t that:

(15.20) $\dfrac{\hat{L} - L}{s\{\hat{L}\}}$ is distributed as $t(n_T - r)$ for ANOVA model (15.1)

Consequently, the $1 - \alpha$ confidence limits for L are:

(15.21)
$$\hat{L} \pm t(1 - \alpha/2; n_T - r)s\{\hat{L}\}$$

Example.   In the Kenton Food Company example, the mean sales for the 3-color designs are to be compared to the mean sales for the 5-color designs. Let us estimate this effect using a 95 percent confidence interval. We wish to estimate:

$$L = \frac{\mu_1 + \mu_2}{2} - \frac{\mu_3 + \mu_4}{2}$$

The point estimate is (see data in Table 15.1):

$$\hat{L} = \frac{\bar{Y}_{1.} + \bar{Y}_{2.}}{2} - \frac{\bar{Y}_{3.} + \bar{Y}_{4.}}{2} = \frac{15 + 13}{2} - \frac{19 + 27}{2} = -9$$

Since $c_1 = 1/2$, $c_2 = 1/2$, $c_3 = -1/2$, and $c_4 = -1/2$, we obtain:

$$\sum \frac{c_i^2}{n_i} = \frac{(1/2)^2}{2} + \frac{(1/2)^2}{3} + \frac{(-1/2)^2}{3} + \frac{(-1/2)^2}{2} = \frac{5}{12} = .4167$$

and:

$$s^2\{\hat{L}\} = MSE \sum \frac{c_i^2}{n_i} = 7.67(.4167) = 3.196$$

so that $s\{\hat{L}\} = 1.79$.

For a confidence coefficient of 95 percent, we require $t(.975; 6) = 2.447$. The confidence limits for L therefore are $-9 \pm 2.447(1.79)$, and the desired 95 percent confidence interval is:

$$-13.4 \le L \le -4.6$$

Therefore, we conclude with confidence coefficient .95 that mean sales for the 3-color designs fall below those for the 5-color designs by somewhere between 4.6 and 13.4 cases per store.

Note

Theorem (15.20) enables us to test any hypothesis concerning a contrast L by means of a t test. Such tests are called single degree of freedom tests. These are discussed in Section 15.6.

Estimation of Linear Combination

Occasionally, we are interested in a linear combination of the factor level means that is not a contrast. For example, suppose that the Kenton Food Company will use all four package designs, one in each of its four major marketing regions, and that these marketing regions account for 35, 28, 12, and 25 percent of sales, respectively. In that case, there might be interest in the overall mean sales per store for all regions:

$$L = .35\mu_1 + .28\mu_2 + .12\mu_3 + .25\mu_4$$

Note that this linear combination is of the form $L = \Sigma c_i \mu_i$ but that the coefficients c_i sum to 1.0, not to zero as they must for a contrast.

We define a *linear combination of the factor level means* μ_i as:

$$(15.22) \qquad\qquad L = \sum_{i=1}^{r} c_i \mu_i$$

with no restrictions on the coefficients c_i.

Confidence limits for a linear combination L are obtained in exactly the same way as those for a contrast by means of (15.21), using the point estimator (15.17) and the estimated variance (15.19).

Need for Multiple Comparison Procedures

The procedures for estimating the factor level effects discussed up to this point have two important limitations:

1. The confidence coefficient $1 - \alpha$ applies only to a particular estimate, not to a series of estimates.
2. The confidence coefficient $1 - \alpha$ is appropriate only if the estimate was not suggested by the data.

The first limitation is familiar from regression analysis. It is particularly serious for analysis of variance models because frequently many different comparisons are of interest here, and one needs to piece the different findings together. Consider the very simple case where three different advertisements are being compared for their effectiveness in stimulating sales. The following estimates of their comparative effectiveness have been obtained, each with a 95 percent statement confidence coefficient:

$$59 \leq \mu_2 - \mu_1 \leq 62$$

$$-2 \leq \mu_3 - \mu_1 \leq 3$$

$$58 \leq \mu_2 - \mu_3 \leq 64$$

It would be natural here to piece the different comparisons together and conclude that advertisement 2 leads to highest mean sales, while advertisements 1 and 3 are substantially less effective and do not differ much among themselves. One would therefore like a family confidence coefficient for this family of statements, to provide known assurance that all statements in the family are correct.

The second limitation of the procedures for estimating factor level effects, namely, that the estimate must not be suggested by the data, is an important one in exploratory investigations where many new questions may be suggested once the data are being analyzed. Sometimes, the process of studying effects suggested by the data is called *data snooping*. Analysts often have the tendency to investigate comparisons where the effect appears to be large from the sample data. Now, effects may appear large because in fact they are, or because a random occurrence made them appear large even though they are not. Consequently, investigating only comparisons for which the effect appears to be large implies a smaller confidence coefficient than the specified one if in fact the effect is small or nonexistent. Thus, it can be shown that if six factor levels are being studied and the analyst will always compare the smallest and largest estimated factor level means by using the confidence limits (15.15) with a 95 percent confidence coefficient, the interval estimate will suggest a real effect 40 percent of the time when indeed there is no difference between any of the factor level means (Ref. 15.1). With a greater number of factor levels, the likelihood of an erroneous indication of a real effect would be even greater.

One solution to this problem of making comparisons that are suggested by initial analysis of the data is to use a multiple comparison procedure where the family of statements includes all the possible statements one anticipates might be made after the data are examined. For instance, in an investigation where five factor level means are being studied, it is decided in advance that principal interest is in three pairwise comparisons. However, it is also agreed that other pairwise comparisons that will appear interesting should be studied as well. In this case, one can use the family of *all* pairwise comparisons as the basis for obtaining an appropriate family confidence coefficient for the comparisons suggested by the data.

In the next three sections, we shall discuss three multiple comparison procedures for analysis of variance models that permit the family confidence coefficient to be controlled. Two of these procedures allow data snooping to be undertaken naturally without affecting the confidence coefficient. Of the three methods, the Scheffé and Bonferroni methods have been encountered before. The third, the Tukey method, is new and will be discussed first.

15.3 TUKEY METHOD OF MULTIPLE COMPARISONS

The Tukey method of multiple comparisons that we will consider here applies when:

The family of interest is the set of all pairwise comparisons of factor level means; in other words, the family consists of estimates of all pairs $D = \mu_i - \mu_{i'}$.

When all sample sizes are equal, the family confidence coefficient for the Tukey method is exactly $1 - \alpha$. When the sample sizes are not equal, the family confidence coefficient is greater than $1 - \alpha$; in other words, the Tukey method is then conservative.

Studentized Range Distribution

The Tukey method utilizes the *studentized range distribution*. Suppose that we have r independent observations $Y_1, \ldots, Y_r$ from a normal distribution with mean μ and variance σ^2. Let w be the range for this set of observations; thus:

$$(15.23) \qquad w = \max(Y_i) - \min(Y_i)$$

Suppose further that we have an estimate s^2 of the variance σ^2 which is based on ν degrees of freedom and is independent of the Y_i. Then, the ratio w/s is called the *studentized range*. It is denoted by:

$$(15.24) \qquad q(r, \nu) = \frac{w}{s}$$

where the arguments in parentheses remind us that the distribution of q depends on r and ν. The distribution of q has been tabulated, and selected percentiles are presented in Table A.9.

This table is simple to use. Suppose that $r = 5$ and $\nu = 10$. The 95th percentile is then $q(.95; 5, 10) = 4.65$, which means:

$$P\left\{\frac{w}{s} = q(5, 10) \le 4.65\right\} = .95$$

Thus, with five normal Y observations, the probability is .95 that their range is not more than 4.65 times as great as an independent sample standard deviation based on 10 degrees of freedom.

Multiple Comparison Confidence Intervals

The Tukey multiple comparison confidence limits for all pairwise comparisons $D = \mu_i - \mu_{i'}$ with family confidence coefficient of at least $1 - \alpha$ are as follows:

$$(15.25) \qquad \hat{D} \pm Ts\{\hat{D}\}$$

where:

$$(15.25a) \qquad \hat{D} = \bar{Y}_{i.} - \bar{Y}_{i'.}$$

$$(15.25b) \qquad s^2\{\hat{D}\} = s^2\{\bar{Y}_{i.}\} + s^2\{\bar{Y}_{i'.}\} = MSE\left(\frac{1}{n_i} + \frac{1}{n_{i'}}\right)$$

$$(15.25c) \qquad T = \frac{1}{\sqrt{2}}q(1 - \alpha; r, n_T - r)$$

Note that the point estimator $\hat{D}$ in (15.25a) and the estimated variance in (15.25b) are the same as those in (15.10) and (15.13) for a single pairwise comparison. Thus, the only difference between the confidence limits (15.25) for simultaneous comparisons and those in (15.15) for a single comparison is the multiple of the estimated standard deviation.

The family confidence coefficient $1 - \alpha$ pertaining to the multiple pairwise comparisons refers to the proportion of correct families of pairwise comparisons when repeated sets of samples are selected and all pairwise confidence intervals are calculated each time. A family of pairwise comparisons is considered to be correct if every pairwise comparison in the family is correct. Thus, when the family confidence coefficient is $1 - \alpha$, all pairwise comparisons in the family will be correct in $(1 - \alpha)100$ percent of the families.

Example 1—Equal Sample Sizes

In the rust inhibitor example in Table 15.2, it was desired to estimate all pairwise comparisons by means of the Tukey procedure, using a family confidence coefficient of 95 percent. Since $r = 4$ and $n_T - r = 36$, the required percentile of the studentized range distribution is $q(.95; 4, 36)$. From Table A.9, we find $q(.95; 4, 36) = 3.814$. Hence, by (15.25c), we obtain:

$$T = \frac{1}{\sqrt{2}}(3.814) = 2.70$$

Further, we need $s\{\hat{D}\}$. Using (15.25b), we find for any pairwise comparison since equal sample sizes were employed:

$$s^2\{\hat{D}\} = MSE\left(\frac{1}{n_i} + \frac{1}{n_{i'}}\right) = 6.140\left(\frac{1}{10} + \frac{1}{10}\right) = 1.23$$

so that $s\{\hat{D}\} = 1.11$. Hence, we obtain for each pairwise comparison:

$$Ts\{\hat{D}\} = 2.70(1.11) = 3.0$$

The pairwise confidence intervals with 95 percent family confidence coefficient therefore are:

$$43.3 = (89.44 - 43.14) - 3.0 \leq \mu_2 - \mu_1 \leq (89.44 - 43.14) + 3.0 = 49.3$$

$$21.8 = (67.95 - 43.14) - 3.0 \leq \mu_3 - \mu_1 \leq (67.95 - 43.14) + 3.0 = 27.8$$

$$-.3 = (43.14 - 40.47) - 3.0 \leq \mu_1 - \mu_4 \leq (43.14 - 40.47) + 3.0 = 5.7$$

$$18.5 = (89.44 - 67.95) - 3.0 \leq \mu_2 - \mu_3 \leq (89.44 - 67.95) + 3.0 = 24.5$$

$$46.0 = (89.44 - 40.47) - 3.0 \leq \mu_2 - \mu_4 \leq (89.44 - 40.47) + 3.0 = 52.0$$

$$24.5 = (67.95 - 40.47) - 3.0 \leq \mu_3 - \mu_4 \leq (67.95 - 40.47) + 3.0 = 30.5$$

The pairwise comparisons indicate that all but one of the differences (D and A) are statistically significant (confidence interval does not cover zero). We incorporate this information in a line plot of the estimated factor level means by underlining non-significant comparisons.

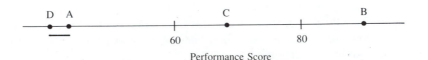

Performance Score

The line between D and A indicates that there is no clear evidence whether D or A is the better rust inhibitor. The absence of a line signifies that a difference in performance has been found. Thus, the multiple comparison procedure permits us to infer with a 95 percent family confidence coefficient for the chain of conclusions that B is the best inhibitor (better by somewhere between 18.5 and 24.5 units than the second best), C is second best, and A and D follow substantially behind with little or no difference between them. Thus, the effects suggested by the normal probability plot in Figure 15.3b are confirmed by this analysis.

Example 2—Unequal Sample Sizes

In the Kenton Food Company example in Table 15.1, the sales manager was interested in the comparative performance of the four package designs. The analyst developed all pairwise comparisons by means of the Tukey procedure with a family confidence coefficient of at least 90 percent. Since the sample sizes are not equal here, the estimated standard deviation $s\{\hat{D}\}$ must be recalculated for each pairwise comparison. To compare designs 1 and 2, for instance, we obtain:

$$\hat{D} = \bar{Y}_{1.} - \bar{Y}_{2.} = 15 - 13 = 2$$

$$s^2\{\hat{D}\} = MSE\left(\frac{1}{n_1} + \frac{1}{n_2}\right) = 7.67\left(\frac{1}{2} + \frac{1}{3}\right) = 6.39$$

$$s\{\hat{D}\} = 2.53$$

For a 90 percent family confidence coefficient, we require $q(.90; 4, 6) = 4.07$ so that we obtain:

$$T = \frac{1}{\sqrt{2}}(4.07) = 2.88$$

Hence, the confidence limits are $2 \pm 2.88(2.53)$ and the confidence interval for $\mu_1 - \mu_2$ is:

$$-5.3 \le \mu_1 - \mu_2 \le 9.3$$

In the same way, we obtain the other five confidence intervals:

$$-3.3 = (19 - 15) - 2.88(2.53) \le \mu_3 - \mu_1 \le (19 - 15) + 2.88(2.53) = 11.3$$

$$4.0 = (27 - 15) - 2.88(2.77) \le \mu_4 - \mu_1 \le (27 - 15) + 2.88(2.77) = 20.0$$

$$-.5 = (19 - 13) - 2.88(2.26) \le \mu_3 - \mu_2 \le (19 - 13) + 2.88(2.26) = 12.5$$

$$6.7 = (27 - 13) - 2.88(2.53) \le \mu_4 - \mu_2 \le (27 - 13) + 2.88(2.53) = 21.3$$

$$.7 = (27 - 19) - 2.88(2.53) \le \mu_4 - \mu_3 \le (27 - 19) + 2.88(2.53) = 15.3$$

We summarize the comparative performance by a line plot, indicating each nonsignificant difference by a rule.

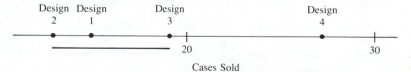

Cases Sold

Thus, we can conclude with at least 90 percent family confidence coefficient that design 4 is clearly the most effective design. However, the small-scale study does not permit any ordering among the other three designs because each of those pairwise confidence intervals encompasses the possibility of either design having larger mean sales per store.

Comments

1. The Tukey method when used with unequal sample sizes is sometimes called the *Tukey-Kramer method*.

2. If not all pairwise comparisons are of interest, the confidence coefficient for the family of comparisons being considered will be greater than the specification $1 - \alpha$ used in setting up the Tukey intervals. Thus, the confidence coefficient $1 - \alpha$ with the Tukey method serves as a guaranteed minimum level when not all pairwise comparisons are of interest.

3. The Tukey method can be used for data snooping as long as the effects to be studied on the basis of preliminary data analysis are pairwise comparisons.

4. The Tukey method can be modified to handle general contrasts between factor level means. We do not discuss this modification since the Scheffé method (to be discussed next) is to be preferred for this situation.

5. To derive the Tukey simultaneous confidence intervals for the case when all sample sizes are equal, i.e., $n_i \equiv n$ so that $n_T = rn$, consider the deviations:

$$(15.26) \qquad (\bar{Y}_{1.} - \mu_1), \ldots, (\bar{Y}_{r.} - \mu_r)$$

and assume that ANOVA model (15.1) applies. The deviations in (15.26) are then independent variables (because the error terms are independent), they are normally distributed (because the error terms are independent normal variables), they have the same expectation zero (because μ_i is subtracted from $\bar{Y}_{i.}$), and they have the same variance σ^2/n. Further, MSE/n is an estimator of σ^2/n that is independent of the deviations $(\bar{Y}_{i.} - \mu_i)$ per theorem (15.7). Thus, it follows from the definition of the studentized range q in (15.24) that:

$$(15.27) \qquad \frac{\max(\bar{Y}_{i.} - \mu_i) - \min(\bar{Y}_{i.} - \mu_i)}{\sqrt{\dfrac{MSE}{n}}} \sim q(r, n_T - r)$$

where $n_T - r$ is the number of degrees of freedom associated with SSE, $\max(\bar{Y}_{i.} - \mu_i)$ is the largest deviation, and $\min(\bar{Y}_{i.} - \mu_i)$ is the smallest deviation.

In view of (15.27), we can write the following probability statement:

$$(15.28) \qquad P\left\{ \frac{\max(\bar{Y}_{i.} - \mu_i) - \min(\bar{Y}_{i.} - \mu_i)}{\sqrt{\dfrac{MSE}{n}}} \leq q(1 - \alpha; r, n_T - r) \right\} = 1 - \alpha$$

Note now that the following inequality holds for *all* pairs of factor levels i and i':

$$(15.29) \qquad |(\bar{Y}_{i.} - \mu_i) - (\bar{Y}_{i'.} - \mu_{i'})| \leq \max(\bar{Y}_{i.} - \mu_i) - \min(\bar{Y}_{i.} - \mu_i)$$

The absolute value at the left is needed since the factor levels i and i' are not ordered so that we may be subtracting the larger deviation from the smaller. To put this another way, we are merely concerned here with the difference between the two factor level deviations regardless of direction.

Since the inequality (15.29) holds for all pairs of factor levels i and i', it follows from (15.28) that the probability:

$$(15.30) \qquad P\left\{ \left| \frac{(\bar{Y}_{i.} - \mu_i) - (\bar{Y}_{i'.} - \mu_{i'})}{\sqrt{\dfrac{MSE}{n}}} \right| \leq q(1 - \alpha; r, n_T - r) \right\} = 1 - \alpha$$

holds for all $r(r - 1)/2$ pairwise comparisons among the r factor levels. By rearranging the inequality in (15.30), using the definitions of $s^2\{\hat{D}\}$ in (15.25b) and of T in (15.25c), and noting that for the equal sample size case $s^2\{\hat{D}\}$ becomes:

$$s^2\{\hat{D}\} = MSE\left(\frac{1}{n} + \frac{1}{n}\right) = \frac{2MSE}{n}$$

we obtain the Tukey multiple comparison confidence limits in (15.25).

15.4 SCHEFFÉ METHOD OF MULTIPLE COMPARISONS

The Scheffé method of multiple comparisons was encountered previously for regression models. It is also applicable for analysis of variance models. It applies for analysis of variance models when:

The family of interest is the set of estimates of all possible contrasts among the factor level means:

$$(15.31) \qquad L = \sum c_i \mu_i \qquad \text{where } \sum c_i = 0$$

Thus, infinitely many statements belong to this family. The Scheffé method family confidence coefficient is exactly $1 - \alpha$ whether the factor level sample sizes are equal or unequal.

We noted earlier that an unbiased estimator of L is:

$$(15.32) \qquad \hat{L} = \sum c_i \bar{Y}_{i.}$$

for which the estimated variance is:

$$(15.33) \qquad s^2\{\hat{L}\} = MSE \sum \frac{c_i^2}{n_i}$$

It can be shown that the probability is $1 - \alpha$ that *all* confidence limits of the type:

$$(15.34) \qquad \hat{L} \pm Ss\{\hat{L}\}$$

are correct simultaneously, where $\hat{L}$ and $s\{\hat{L}\}$ are given by (15.32) and (15.33), respectively, and S is given by:

$$(15.34a) \qquad S^2 = (r - 1)F(1 - \alpha; r - 1, n_T - r)$$

Thus, if we were to calculate the confidence intervals for all conceivable contrasts by use of (15.34), then in $(1 - \alpha)100$ percent of repetitions of the experiment, the entire set of confidence intervals in the family would be correct.

Note that the simultaneous confidence limits in (15.34) differ from those for a single confidence limit in (15.21) only with respect to the multiple of the estimated standard deviation.

Example

In the Kenton Food Company example, we wish to estimate the following four contrasts with family confidence coefficient of 90 percent:

Comparison of 3-color and 5-color designs:

$$L_1 = \frac{\mu_1 + \mu_2}{2} - \frac{\mu_3 + \mu_4}{2}$$

Comparison of designs with and without cartoons:

$$L_2 = \frac{\mu_1 + \mu_3}{2} - \frac{\mu_2 + \mu_4}{2}$$

Comparison of the two 3-color designs:

$$L_3 = \mu_1 - \mu_2$$

Comparison of the two 5-color designs:

$$L_4 = \mu_3 - \mu_4$$

Consider the estimation of L_1. Earlier, we found:

$$\hat{L}_1 = -9$$

$$s\{\hat{L}_1\} = 1.79$$

Since $r = 4$ and $n_T - r = 6$ (Table 15.1), we have:

$$S^2 = (r - 1)F(1 - \alpha; r - 1, n_T - r) = 3F(.90; 3, 6) = 3(3.29) = 9.87$$

so that $S = 3.14$. Hence, the confidence limits for L_1 by the Scheffé multiple comparison method are $-9 \pm 3.14(1.79)$ and the desired confidence interval is:

$$-14.6 \leq L_1 \leq -3.4$$

In similar fashion, we obtain the other desired confidence intervals, and the entire set is:

$$-14.6 \leq L_1 \leq -3.4$$
$$-8.6 \leq L_2 \leq 2.6$$
$$-5.9 \leq L_3 \leq 9.9$$
$$-15.9 \leq L_4 \leq -.1$$

This set of confidence intervals has a family confidence coefficient of 90 percent, so that any chain of conclusions derived from the intervals has associated with it this confidence coefficient. The principal conclusions drawn by the sales manager from the above set of estimates were as follows: 5-color designs lead to higher mean sales than 3-color designs, the increase being somewhere between 3 and 15 cases per store. No overall effect of cartoons in the package design is indicated, although for 5-color designs the use of a cartoon leads to mean sales that are lower than those when no cartoon is used.

Comments

1. If in the Kenton Food Company example we wished to estimate a single contrast with statement confidence coefficient .90, the required t value would be $t(.95; 6) = 1.943$. This t value is smaller than the Scheffé multiple $S = 3.14$, so that the single confidence interval would be somewhat narrower. The increased width of the interval with the Scheffé method is the price paid for a known confidence coefficient for a family of statements and a chain of conclusions drawn from them, and for the possibility of making comparisons not specified in advance of the data analysis.

2. Since applications of the Scheffé method never involve all conceivable contrasts, the confidence coefficient for the finite family of statements actually considered will be greater than $1 - \alpha$. Thus, when we state the confidence coefficient is $1 - \alpha$ with the Scheffé method, we really mean it is guaranteed to be at least $1 - \alpha$. For this reason, it has been suggested that the confidence coefficient $1 - \alpha$ used with the Scheffé method be below the level ordinarily used, since $1 - \alpha$ is a lower bound and the actual confidence coefficient will be greater. Confidence coefficients of 90 percent and 95 percent with the Scheffé method are frequently mentioned.

3. The Scheffé method can be used for a wide variety of data snooping since the family of statements contains all possible contrasts.

Comparison of Scheffé Method with Tukey Method

1. If only pairwise comparisons are to be made, the Tukey method gives narrower confidence limits and is therefore the preferred method.
2. In the case of general contrasts, the Scheffé method tends to give narrower confidence limits and is therefore the preferred method.
3. The Scheffé method has the property that if the test based on F^* indicates that the factor level means μ_i are not equal, the corresponding Scheffé multiple comparison procedure will find at least one contrast (out of all possible contrasts) that differs significantly from zero (the confidence interval does not cover zero). It may be, though, that this contrast is not one of those estimated by the analyst.

15.5 BONFERRONI MULTIPLE COMPARISON METHOD

The Bonferroni method of multiple comparisons was encountered earlier for regression models. It is also applicable for analysis of variance models when:

The family of interest is the particular set of pairwise comparisons, contrasts, or linear combinations specified by the user.

The Bonferroni method is applicable whether the factor level sample sizes are equal or unequal and whether pairwise comparisons, contrasts, linear combinations, or a mixture of these are to be estimated.

We shall denote the number of statements in the family by g and treat them all as linear combinations since pairwise comparisons and contrasts are special cases of linear combinations. The Bonferroni inequality (5.5) then implies that the confidence coefficient is at least $1 - \alpha$ that the following confidence limits for the g linear combinations L_i are all correct:

(15.35)
$$\hat{L}_i \pm Bs\{\hat{L}_i\} \qquad i = 1, \ldots, g$$

where:

(15.35a)
$$B = t(1 - \alpha/2g; n_T - r)$$

Example

The sales manager of the Kenton Food Company is interested in estimating the following two contrasts with family confidence coefficient .975:

Comparison of 3-color and 5-color designs:

$$L_1 = \frac{\mu_1 + \mu_2}{2} - \frac{\mu_3 + \mu_4}{2}$$

Comparison of designs with and without cartoons:

$$L_2 = \frac{\mu_1 + \mu_3}{2} - \frac{\mu_2 + \mu_4}{2}$$

Earlier we found:

$$\hat{L}_1 = -9 \qquad s\{\hat{L}_1\} = 1.79$$
$$\hat{L}_2 = -3 \qquad s\{\hat{L}_2\} = 1.79$$

For a 97.5 percent family confidence coefficient with the Bonferroni method, we require:

$$B = t[1 - .025/2(2); 6] = t(.99375; 6) = 3.57$$

We can now complete the confidence intervals for the two contrasts. For L_1, we have confidence limits $-9 \pm 3.57(1.79)$, which lead to the confidence interval:

$$-15.4 \leq L_1 \leq -2.6$$

Similarly, we obtain the other confidence interval:

$$-9.4 \leq L_2 \leq 3.4$$

These confidence intervals have a guaranteed family confidence coefficient of 97.5 percent, which means that in at least 97.5 percent of repetitions of the experiment, both intervals will be correct.

Again, we would conclude from this family of estimates that mean sales for 5-color designs are higher than those for 3-color designs (by somewhere between 3 and 15 cases per store), and that no overall effect of cartoons in the package design is indicated.

The Scheffé multiple for a 97.5 percent family confidence coefficient in this case would have been:

$$S^2 = 3F(.975; 3, 6) = 3(6.60) = 19.8$$

or $S = 4.45$, as compared to the Bonferroni multiple $B = 3.57$. Thus, the Scheffé method here would have led to wider confidence intervals than the Bonferroni method.

Note

It is not necessary that all comparisons be estimated with statement confidence coefficients $1 - (\alpha/g)$ for the Bonferroni family confidence coefficient to be $1 - \alpha$. Different statement confidence coefficients $1 - \alpha_i$ may be used, depending upon the importance of each statement, provided that $\alpha_1 + \alpha_2 + \cdots + \alpha_g = \alpha$.

Comparison of Bonferroni Method with Scheffé and Tukey Methods

1. If all pairwise comparisons are of interest, the Tukey method is superior to the Bonferroni method in the sense of leading to narrower confidence intervals. If not all pairwise comparisons are to be considered, however, the Bonferroni method may be the better at times.
2. The Bonferroni method will be better than the Scheffé method when the number of contrasts to be estimated is about the same as the number of factor levels, or less. Indeed, the number of statements to be made must exceed the number of factor levels by a considerable amount before the Scheffé method becomes better.
3. In any given problem, one may compute the Bonferroni multiple as well as the Scheffé multiple and, when appropriate, the Tukey multiple, and select the one that is smallest. This choice is proper since it does not depend on the observed data.
4. The Bonferroni multiple comparison method does not lend itself to data snooping unless one can specify in advance the family of statements one may be interested in, and provided this family is not large. On the other hand, the Tukey and Scheffé methods involve families of statements that lend themselves naturally to data snooping.
5. There are still other methods of making multiple comparisons. Many of these are designed for special cases, such as for comparing experimental treatments with a control treatment. A good reference book on multiple comparisons is that by Miller (Ref. 15.2).

15.6 SINGLE DEGREE OF FREEDOM TESTS

When the overall F test statistic (14.53) leads to the conclusion that the factor level means μ_i in a single-factor study are not all equal, the investigation of the nature of the factor level effects sometimes is conducted by means of tests dealing with specific questions rather than by estimation of pairwise comparisons, contrasts, or linear combinations. For example, the sales manager of the Kenton Food Company wished to know whether or not mean sales per store for the 3-color designs are the same as those for the 5-color designs. Since all factor level means μ_i are considered to be of the same importance, this question involves the alternatives:

$$H_0: \frac{\mu_1 + \mu_2}{2} = \frac{\mu_3 + \mu_4}{2}$$

$$H_a: \frac{\mu_1 + \mu_2}{2} \neq \frac{\mu_3 + \mu_4}{2}$$

These alternatives can be stated equivalently as:

$$H_0: \frac{\mu_1 + \mu_2}{2} - \frac{\mu_3 + \mu_4}{2} = 0$$

$$H_a: \frac{\mu_1 + \mu_2}{2} - \frac{\mu_3 + \mu_4}{2} \neq 0$$

A test that involves a linear combination of the factor level means μ_i is called a *single degree of freedom test*. In general, the two-sided alternatives for a single degree of freedom test are stated as follows:

(15.36)
$$H_0: \sum c_i \mu_i = c$$
$$H_a: \sum c_i \mu_i \neq c$$

where the c_i and c are appropriate constants. In the earlier example, for instance, we have $c_1 = 1/2$, $c_2 = 1/2$, $c_3 = -1/2$, $c_4 = -1/2$, and $c = 0$.

To test the alternatives (15.36), we utilize theorem (15.20) yielding the t^* test statistic:

(15.37)
$$t^* = \frac{\sum c_i \bar{Y}_{i.} - c}{\sqrt{MSE \sum \frac{c_i^2}{n_i}}}$$

which follows the t distribution with $n_T - r$ degrees of freedom when H_0 holds. An equivalent test statistic is $(t^*)^2$, denoted by F^*:

(15.38)
$$F^* = (t^*)^2 = \frac{(\sum c_i \bar{Y}_{i.} - c)^2}{MSE \sum \frac{c_i^2}{n_i}}$$

which follows the F distribution with 1 and $n_T - r$ degrees of freedom when H_0 holds. Remember from (1.47a) that $[t(n_T - r)]^2 = F(1, n_T - r)$.

Example

To test in the Kenton Food Company example whether or not the mean sales per store for the 3-color designs and the 5-color designs are equal, we shall use test statistic (15.38) with $\alpha = .05$. The alternatives are:

$$H_0: \frac{\mu_1 + \mu_2}{2} - \frac{\mu_3 + \mu_4}{2} = 0$$

$$H_a: \frac{\mu_1 + \mu_2}{2} - \frac{\mu_3 + \mu_4}{2} \neq 0$$

so that $c_1 = c_2 = 1/2$, $c_3 = c_4 = -1/2$, and $c = 0$. Using the sample results in Table 15.1, we obtain the test statistic:

$$F^* = \frac{\left(\dfrac{\bar{Y}_{1.} + \bar{Y}_{2.}}{2} - \dfrac{\bar{Y}_{3.} + \bar{Y}_{4.}}{2} - 0\right)^2}{MSE\left[\dfrac{(1/2)^2}{n_1} + \dfrac{(1/2)^2}{n_2} + \dfrac{(-1/2)^2}{n_3} + \dfrac{(-1/2)^2}{n_4}\right]}$$

$$= \frac{\left(\dfrac{15 + 13}{2} - \dfrac{19 + 27}{2}\right)^2}{7.67\left(\dfrac{.25}{2} + \dfrac{.25}{3} + \dfrac{.25}{3} + \dfrac{.25}{2}\right)} = 25.3$$

For $\alpha = .05$, we require $F(.95; 1, 6) = 5.99$. Hence, the decision rule is:

If $F^* \leq 5.99$, conclude H_0

If $F^* > 5.99$, conclude H_a

Since $F^* = 25.3 > 5.99$, we conclude H_a, that mean sales for the 3-color designs differ from the mean sales for the 5-color designs. The P-value of this test is $P\{F(1, 6) > 25.3\} = .0024$.

If we had utilized test statistic (15.37), the value of the test statistic would have been:

$$t^* = \frac{\dfrac{\bar{Y}_{1.} + \bar{Y}_{2.}}{2} - \dfrac{\bar{Y}_{3.} + \bar{Y}_{4.}}{2} - 0}{\sqrt{MSE\left[\dfrac{(1/2)^2}{n_1} + \dfrac{(1/2)^2}{n_2} + \dfrac{(-1/2)^2}{n_3} + \dfrac{(-1/2)^2}{n_4}\right]}}$$

$$= \frac{\dfrac{15 + 13}{2} - \dfrac{19 + 27}{2}}{\sqrt{7.67\left(\dfrac{.25}{2} + \dfrac{.25}{3} + \dfrac{.25}{3} + \dfrac{.25}{2}\right)}} = -5.03$$

For $\alpha = .05$, we require $t(.975; 6) = 2.447$ and the decision rule is:

If $|t*| \leq 2.447$, conclude H_0

If $|t*| > 2.447$, conclude H_a

Since $|t*| = 5.03 > 2.447$, we conclude H_a, as before with the $F*$ test statistic.

Comments

1. Any single degree of freedom test can also be conducted by setting up a confidence interval for the appropriate pairwise comparison, contrast, or linear combination and noting whether or not the confidence interval includes the value c specified in the alternatives. For instance, we obtained earlier the 95 percent confidence interval for the difference in mean sales between the 3-color and 5-color designs:

$$-13.4 \leq \frac{\mu_1 + \mu_2}{2} - \frac{\mu_3 + \mu_4}{2} \leq -4.6$$

Since this 95 percent confidence interval does not include $c = 0$, a test of the difference between the two group means for $\alpha = .05$ leads to the conclusion that the two group means are not equal.

2. The test statistic $t*$ can also be used when one-sided tests are to be conducted, but the test statistic $F*$ is only appropriate for two-sided tests.

3. A number of single-factor analysis of variance computer packages permit the user to specify a contrast of interest and then will furnish either the $t*$ or the $F*$ test statistic.

Multiple Single Degree of Freedom Tests

When the analysis of factor effects is carried out by single degree of freedom tests, usually several single degree of freedom tests are conducted to answer related questions. For instance, the sales manager of the Kenton Food Company might wish to know not only whether the number of colors has an effect on mean sales but also whether use of cartoons has an effect. Whenever several single degree of freedom tests are conducted, both the level of signficance and the power, insofar as the family of tests is concerned, are affected. For example, if three different single degree of freedom F or t tests are conducted, each with $\alpha = .05$, then the probability that each of the tests will lead to conclusion H_0 when indeed H_0 is correct in each case, assuming independence of the tests, is $(.95)^3 = .857$. Thus, the level of significance that at least one of the three tests leads to conclusion H_a when H_0 holds in each case would be $1 - .857 = .143$ and not .05. We see then that the level of significance and power for a *family* of tests is not the same as that for an *individual* test. Actually, the $F*$ or $t*$ statistics are dependent since they all are based on the same sample data and use the same MSE value. It often therefore becomes difficult to determine the actual level of significance and power for a family of tests.

Control of the family level of significance when conducting a number of single degree of freedom tests can be achieved by using one of the simultaneous estimation procedures—either the Bonferroni method, the Scheffé method (when contrasts are involved), or the Tukey method (when pairwise comparisons are involved). One simply constructs the appropriate confidence intervals and draws the proper test conclusion from each of the confidence intervals. Not only does this approach permit simultaneous single degree of freedom tests with control over the family level of significance, but it also provides information about the magnitude of any effects that are present.

Example. In the Kenton Food Company example, the analyst wished to test whether or not any two factor level means μ_i and $\mu_{i'}$ differ. Thus, the analyst was interested in conducting tests involving the alternatives:

Test 1: $H_0: \mu_1 = \mu_2$ Test 2: $H_0: \mu_1 = \mu_3$ Test 3: $H_0: \mu_1 = \mu_4$
      $H_a: \mu_1 \neq \mu_2$ $H_a: \mu_1 \neq \mu_3$ $H_a: \mu_1 \neq \mu_4$

Test 4: $H_0: \mu_2 = \mu_3$ Test 5: $H_0: \mu_2 = \mu_4$ Test 6: $H_0: \mu_3 = \mu_4$
      $H_a: \mu_2 \neq \mu_3$ $H_a: \mu_2 \neq \mu_4$ $H_a: \mu_3 \neq \mu_4$

The family level of significance is to be controlled at $\alpha = .10$.

The analyst utilized the Tukey multiple comparison procedure for obtaining all pairwise comparisons and used the decision rule:

If the confidence interval includes 0, conclude H_0

Otherwise conclude H_a

We obtained these confidence intervals earlier and repeat the results here, together with the appropriate conclusions:

Test 1: $-5.3 \leq \mu_1 - \mu_2 \leq 9.3$ Test 2: $-3.3 \leq \mu_3 - \mu_1 \leq 11.3$

Conclude H_0 Conclude H_0

Test 3: $4.0 \leq \mu_4 - \mu_1 \leq 20.0$ Test 4: $-.5 \leq \mu_3 - \mu_2 \leq 12.5$

Conclude H_a Conclude H_0

Test 5: $6.7 \leq \mu_4 - \mu_2 \leq 21.3$ Test 6: $.7 \leq \mu_4 - \mu_3 \leq 15.3$

Conclude H_a Conclude H_a

Thus, all pairwise differences between package design 4 and the other designs are statistically significant (i.e., the differences between the means are not zero), while all other pairwise differences are not statistically significant.

Often, the results of these pairwise tests are summarized by setting up groups of factor levels whose means do not differ according to the single degree of freedom tests. Here, there are two such groups:

Group 1		Group 2	
Package design 1	$\bar{Y}_{1.} = 15$	Package design 4	$\bar{Y}_{4.} = 27$
Package design 2	$\bar{Y}_{2.} = 13$		
Package design 3	$\bar{Y}_{3.} = 19$		

Comments

1. The Bonferroni procedure can be used for multiple single degree of freedom tests not only by means of confidence intervals but also with either test statistic t^* or F^*. If, say, four single degree of freedom tests are to be conducted with family level of significance $\alpha = .10$, each test with t^* or F^* is simply conducted with individual significance level $.10/4 = .025$.

2. When the Tukey multiple comparison method is used for testing pairwise differences, the tests are sometimes called *honestly significant differences tests*.

15.7 ANALYSIS OF FACTOR EFFECTS WHEN FACTOR QUANTITATIVE

When the factor under investigation is quantitative, the analysis of factor effects can be carried beyond the point of multiple comparisons to include a study of the nature of the response function. Consider an experimental study undertaken to investigate the effect of the price of a product on sales. Five different price levels are investigated (28 cents, 29 cents, 30 cents, 31 cents, and 32 cents), and the experimental unit is a store. After a preliminary test whether mean sales differ for the price levels studied, the analyst might use multiple comparisons to examine whether "odd pricing" at 29 cents actually leads to higher sales than "even pricing" at 28 cents, as well as other questions of interest. In addition, the analyst may wish to study whether mean sales are a specified function of price, in the range of prices studied in the experiment. Further, once the relation has been established, the analyst may wish to use it for estimating sales volumes at various price levels not studied.

The methods of regression analysis discussed in Parts I and II are, of course, appropriate for the analysis of the response function. It should only be noted that the single-factor studies discussed in this chapter almost always involve replications at the different factor levels, so that the lack of fit of a specified response function can be tested. For this purpose, it should be remembered that the analysis of variance error sum of squares in (14.26) is identical to the pure error sum of squares of Chapter 4 in (4.11). We illustrate this relation in the following example.

Example

In a study to reduce raw material costs in a glass working firm, an operations analyst collected the experimental data in Table 15.4 on the number of acceptable units produced from equal amounts of raw material by 28 entry-level piecework employees who had received special training as part of the experiment. Four training levels were used (6, 8, 10, and 12 hours), with seven of the employees being assigned at random to each level. The higher the number of acceptable pieces, the more efficient is the employee in utilizing the raw material.

Preliminary Analysis. The analyst first tested whether or not the mean number of acceptable pieces is the same for the four training levels. ANOVA model (15.1) was employed:

$$(15.39) \qquad\qquad Y_{ij} = \mu_i + \varepsilon_{ij}$$

TABLE 15.4   Data for Piecework Trainees Example

Treatment (hours of training) i		j						
		1	2	3	4	5	6	7
1	6 hours	40	39	39	36	42	43	41
2	8 hours	53	48	49	50	51	50	48
3	10 hours	53	58	56	59	53	59	58
4	12 hours	63	62	59	61	62	62	61

The alternative conclusions and appropriate test statistic are:

$$H_0: \mu_1 = \mu_2 = \mu_3 = \mu_4$$

$$H_a: \text{not all } \mu_i \text{ are equal}$$

$$F^* = \frac{MSTR}{MSE}$$

A computer program for single-factor ANOVA gave the output shown in Figure 15.4. Residual analysis (to be discussed in Chapter 16) showed ANOVA model (15.1) to be apt. Therefore, the analyst proceeded with the test, using $\alpha = .05$. The decision rule is:

If $F^* \leq F(.95; 3, 24) = 3.01$, conclude H_0

If $F^* > 3.01$, conclude H_a

From the printout in Figure 15.4, we have:

$$F^* = \frac{MSTR}{MSE} = \frac{602.8926}{4.2619} = 141.5$$

Since $F^* = 141.5 > 3.01$, the analyst concluded H_a, that training level effects differed and that further analysis of them is warranted. The P-value for the test statistic is $0+$, as shown in Figure 15.4.

Investigation of Treatment Effects.   The analyst's interest next centered on multiple comparisons of all pairs of treatment means. A Tukey multiple comparison option in the analyst's computer package was used. It gave the output shown in the lower portion of Figure 15.4. This output presents the results of single degree of freedom tests conducted by means of the Tukey multiple comparison procedure for all pairwise comparisons. (The confidence intervals for the pairwise comparisons are not shown in the output.) All factor levels for which the test concludes that the pairwise means are equal are placed in the same subset. This form of summary of single degree of freedom tests was illustrated earlier for the Kenton Food Company example. When a subset contains only one factor level, as is the case for all groups in the output of Figure 15.4, the implication is that all single degree of freedom tests involving this factor level and each of the other factor levels led to conclusion H_a, that the two factor level means being compared are not equal.

FIGURE 15.4   Segment of Computer Output for Piecework Trainees Example (SPSS^X, Ref. 15.3)

		n_i	$\bar{Y}_{i\cdot}$	
	GROUP	COUNT	MEAN	STANDARD DEVIATION
Treatment →	GRP01	7	40.0000	2.3094
	GRP02	7	49.8571	1.7728
	GRP03	7	56.5714	2.6367
	GRP04	7	61.4286	1.2724
	TOTAL	28	51.9643	8.4129

ANALYSIS OF VARIANCE

SOURCE	D.F.	SUM OF SQUARES	MEAN SQUARES
BETWEEN GROUPS	3	SSTR → 1808.6778	602.8926 ← MSTR
WITHIN GROUPS	24	SSE → 102.2856	4.2619 ← MSE
TOTAL	27	SSTO → 1910.9634	

F RATIO   F PROB.

141.461   0.0000

F*   One-sided P-value

MULTIPLE RANGE TEST

TUKEY-HSD PROCEDURE
RANGES FOR THE 0.050 LEVEL -

3.90 ← q(.95; 4, 24)

HOMOGENEOUS SUBSETS

SUBSET 1		SUBSET 3	
GROUP	GRP01	GROUP	GRP03
MEAN	40.0000	MEAN	56.5714
- - - - - - - - - -		- - - - - - - - - -	
SUBSET 2		SUBSET 4	
GROUP	GRP02	GROUP	GRP04
MEAN	49.8571	MEAN	61.4286
- - - - - - - - - -		- - - - - - - - - -	

Two points are to be noted in particular from the results in Figure 15.4: (1) All pairwise factor level differences are statistically significant. (2) There is some indication that differences between the means for adjoining factor levels diminish as the number of hours of training increases; that is, diminishing returns appear to be obtained as the length of training is increased.

Estimation of Response Function.   These findings were in accord with the analyst's expectations. She had surmised that the treatment means μ_i would most likely follow a quadratic response function with respect to training level. The computer plot of the data in Figure 15.5 supported this expectation. She now wished to investigate this point further by fitting a quadratic regression model. The model to be fitted and tested is:

$$(15.40) \qquad Y_{ij} = \beta_0 + \beta_1 x_i + \beta_{11} x_i^2 + \varepsilon_{ij}$$

where Y_{ij} and ε_{ij} are defined as earlier, the βs are regression parameters, and x_i denotes the number of hours of training in the ith training level (X_i) expressed as a deviation around the mean of all training levels ($\overline{X} = 9$), i.e., $x_i = X_i - 9$.

The **X** and **Y** matrices for the regression analysis are given in Table 15.5. A computer run of a multiple regression package yielded the estimated regression function:

$$(15.41) \qquad \hat{Y} = 53.52679 + 3.55000x - .31250x^2$$

The analysis of variance for regression model (15.40) is shown in Table 15.6a. For completeness, we repeat in Table 15.6b the analysis of variance for ANOVA model (15.39).

Since the data contain replicates, the analyst could test the regression model

---

TABLE 15.5   X and Y Matrices for Piecework Trainees Example

$$
\mathbf{Y} =
\begin{bmatrix}
40 \\ 39 \\ 39 \\ 36 \\ 42 \\ 43 \\ 41 \\ 53 \\ 48 \\ 49 \\ 50 \\ 51 \\ 50 \\ 48 \\ 53 \\ 58 \\ 56 \\ 59 \\ 53 \\ 59 \\ 58 \\ 63 \\ 62 \\ 59 \\ 61 \\ 62 \\ 62 \\ 61
\end{bmatrix}
\qquad
\mathbf{X} =
\begin{bmatrix}
1 & -3 & 9 \\
1 & -3 & 9 \\
1 & -3 & 9 \\
1 & -3 & 9 \\
1 & -3 & 9 \\
1 & -3 & 9 \\
1 & -3 & 9 \\
1 & -1 & 1 \\
1 & -1 & 1 \\
1 & -1 & 1 \\
1 & -1 & 1 \\
1 & -1 & 1 \\
1 & -1 & 1 \\
1 & -1 & 1 \\
1 & 1 & 1 \\
1 & 1 & 1 \\
1 & 1 & 1 \\
1 & 1 & 1 \\
1 & 1 & 1 \\
1 & 1 & 1 \\
1 & 1 & 1 \\
1 & 3 & 9 \\
1 & 3 & 9 \\
1 & 3 & 9 \\
1 & 3 & 9 \\
1 & 3 & 9 \\
1 & 3 & 9 \\
1 & 3 & 9
\end{bmatrix}
\begin{matrix} x & x^2 \end{matrix}
$$

TABLE 15.6   Analyses of Variance for Piecework Trainees Example

(a) Regression Model (15.40)

Source of Variation	SS	df	MS
Regression	1,808.100	2	904.05
Error	102.864	25	4.11
Total	1,910.964	27	

(b) Analysis of Variance Model (15.39)

Source of Variation	SS	df	MS
Treatments	1,808.678	3	602.89
Error	102.286	24	4.26
Total	1,910.964	27	

(c) ANOVA for Lack of Fit Test

Source of Variation	SS	df	MS
Regression	1,808.100	2	904.05
Error	102.864	25	4.11
Lack of fit	.578	1	.58
Pure error	102.286	24	4.26
Total	1,910.964	27	

(15.40) for lack of fit. She utilized the fact that the ANOVA error sum of squares in (14.26) is identical to the regression pure error sum of squares in (4.11). Both measure variation around the mean at any given level of X (i.e., around the estimated treatment mean $\bar{Y}_{i.}$). Hence, the lack of fit sum of squares can be readily obtained from previous results:

$$(15.42) \quad SSLF = \underset{\text{(Table 15.6a)}}{SSE} - \underset{\text{(Table 15.6b)}}{SSPE} = 102.864 - 102.286 = .578$$

Since there are $c = r = 4$ levels of X here and $p = 3$ parameters in the regression model, $SSLF$ has associated with it $c - p = 4 - 3 = 1$ degree of freedom. Hence, we obtain $MSLF = .578/1 = .578$. Table 15.6c contains the analysis of variance for the regression model, with the error sum of squares and degrees of freedom broken down into lack of fit and pure error components.

The alternative conclusions (7.57a) for the test of lack of fit here are:

$$H_0: E\{Y\} = \beta_0 + \beta_1 x + \beta_{11} x^2$$

$$H_a: E\{Y\} \neq \beta_0 + \beta_1 x + \beta_{11} x^2$$

FIGURE 15.5   Scatter Plot and Fitted Quadratic Response Function—Piecework Trainees Example

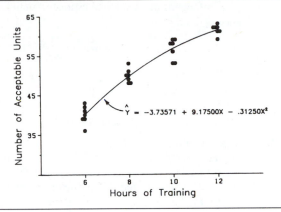

and test statistic (7.57b) is:

$$F^* = \frac{MSLF}{MSPE}$$

For $\alpha = .05$, decision rule (7.57c) becomes:

If $F^* \leq F(.95; 1, 24) = 4.26$, conclude H_0

If $F^* > 4.26$, conclude H_a

We calculate the test statistic from Table 15.6c:

$$F^* = \frac{.58}{4.26} = .136$$

Since $F^* = .136 \leq 4.26$, the analyst concluded that the quadratic response function is a good fit. Consequently, she used the fitted regression function in (15.41) in further evaluation of the relation between mean number of acceptable pieces produced and level of training, after expressing the fitted response function in the original independent variable X (number of hours of training):

$$\hat{Y} = -3.73571 + 9.17500X - .31250X^2$$

Figure 15.5 displays this fitted response function.

CITED REFERENCES

15.1. Cochran, W. G., and G. M. Cox. *Experimental Designs*. 2nd ed. New York: John Wiley & Sons, 1957, p. 74.

15.2. Miller, R. G., Jr. *Simultaneous Statistical Inference*. 2nd ed. New York: Springer-Verlag, 1981.

15.3. *SPSS^X User's Guide*. 2nd ed. Chicago: SPSS, 1986.

PROBLEMS

15.1. Refer to Figure 15.4 for the piecework trainees example. To estimate the treatment mean μ_1 by means of an interval estimate, would it be valid to use for the estimated variance of $\bar{Y}_1.$:

$$s^2\{\bar{Y}_1.\} = \frac{s_1^2}{n_1} = \frac{(2.3094)^2}{7}$$

instead of (15.5):

$$s^2\{\bar{Y}_1.\} = \frac{MSE}{n_1} = \frac{4.2619}{7}$$

What is the advantage of using (15.5)? What is a disadvantage?

15.2. Refer to **Premium distribution** Problem 14.15. A student, asked to give a class demonstration of the use of a confidence interval for comparing two treatment means, proposed to construct a 99 percent confidence interval for the pairwise comparison $D = \mu_5 - \mu_3$. He selected this particular comparison because the sample means $\bar{Y}_5.$ and $\bar{Y}_3.$ are the largest and smallest, respectively. The student stated: "This confidence interval is particularly useful. If it does not straddle zero, it indicates, with significance level $\alpha = .01$, that the factor level means are not equal."
   a. Explain why the student's assertion is not correct.
   b. How should the confidence interval be constructed so that the assertion can be made with significance level $\alpha = .01$?

15.3. A trainee examined a set of experimental data to find comparisons that "look promising" and calculated a family of Bonferroni confidence intervals for these comparisons with a 90 percent family confidence coefficient. Upon being informed that the Bonferroni procedure is not applicable in this case because the comparisons had been suggested by the data, the trainee stated: "This makes no difference. I would use the same formulas for the point estimates and the estimated standard errors even if the comparisons were not suggested by the data." Respond.

15.4. Consider the following linear combinations of interest in a single-factor study involving four factor levels:

$$\text{(i)} \quad \mu_1 + 3\mu_2 - 4\mu_3$$

$$\text{(ii)} \quad .3\mu_1 + .5\mu_2 + .1\mu_3 + .1\mu_4$$

$$\text{(iii)} \quad \frac{\mu_1 + \mu_2 + \mu_3}{3} - \mu_4$$

   a. Which of the linear combinations are contrasts? State the coefficients for each of the contrasts.
   b. Give an unbiased estimator for each of the linear combinations. Also give the estimated variance of each estimator assuming that $n_i \equiv n$.

15.5. A single-factor ANOVA study consists of $r = 6$ treatments with sample sizes $n_i \equiv 10$.
   a. Assuming that pairwise comparisons of the treatment means are to be made with a 90 percent family confidence coefficient, find the T, S, and B multiples for the following numbers of pairwise comparisons in the family: $g = 2, 5, 15$. What generalization is suggested by your results?
   b. Assuming that contrasts of the treatment means are to be estimated with a 90 percent family confidence coefficient, find the S and B multiples for the following

numbers of contrasts in the family: $g = 2, 5, 15$. What generalization is suggested by your results?

15.6. Consider a single-factor study with $r = 5$ treatments and sample sizes $n_i \equiv 5$.

   a. Find the T, S, and B multiples if $g = 2, 5$, and 10 pairwise comparisons are to be made with a 95 percent family confidence coefficient. What generalization is suggested by your results?

   b. What would be the T, S, and B multiples for sample sizes $n_i \equiv 20$? Does the generalization obtained in part (a) still hold?

15.7. In making multiple comparisons, why is it appropriate to use the multiple comparison procedure that leads to the tightest confidence intervals for the sample data obtained? Discuss.

15.8. For a single-factor study with $r = 2$ treatments and sample sizes $n_i \equiv 10$, find the T, S, and B multiples for $g = 1$ pairwise comparison with a 99 percent family confidence coefficient. What generalization is suggested by your results?

15.9. Refer to **Productivity improvement** Problem 14.10. Some additional calculational results are:

i	Research and Development Expenditures Level	n_i	$\bar{Y}_{i.}$	
1	Low	9	6.878	
2	Moderate	12	8.133	$MSE = .6401$
3	High	6	9.200	

   a. Prepare a line plot of the estimated factor level means $\bar{Y}_{i.}$. What does this plot suggest regarding the effect of the level of research and development expenditures on mean productivity improvement?

   b. Estimate the mean productivity improvement for firms with high research and development expenditures levels; use a 95 percent confidence interval.

   c. Obtain a 95 percent confidence interval for $D = \mu_2 - \mu_1$. Interpret your interval estimate.

   d. Obtain confidence intervals for all pairwise comparisons of the treatment means; use the Tukey procedure and a 90 percent family confidence coefficient. State your findings and prepare a graphic summary by underlining nonsignificant comparisons in your line plot in part (a).

   e. Is the Tukey procedure employed in part (d) the most efficient one that could be used here? Explain.

15.10. Refer to **Questionnaire color** Problem 14.11. Some additional calculational results are:

i	Color	n_i	$\bar{Y}_{i.}$	
1	Blue	5	29.4	
2	Green	5	29.6	$MSE = 9.70$
3	Orange	5	28.0	

   a. Prepare a normal probability plot of the estimated factor level means $\bar{Y}_{i.}$. What does this plot suggest about the effect of color on the response rate? Is your conclusion in accord with the test result in Problem 14.11d?

    b. Estimate the mean response rate for blue questionnaires; use a 90 percent confidence interval.

    c. Obtain a 90 percent confidence interval for $D = \mu_3 - \mu_2$. Interpret your interval estimate. In light of the result for the ANOVA test in Problem 14.11d, is it surprising that the confidence interval straddles zero? Explain.

15.11. Refer to **Rehabilitation therapy** Problem 14.12.

    a. Prepare a line plot of the estimated factor level means $\bar{Y}_{i\cdot}$. What does this plot suggest about the effect of prior physical fitness on the mean time required in therapy?

    b. Estimate with a 99 percent confidence interval the mean number of days required in therapy for persons of average physical fitness.

    c. Obtain confidence intervals for $D_1 = \mu_2 - \mu_3$ and $D_2 = \mu_1 - \mu_2$; use the Bonferroni procedure with a 95 percent family confidence coefficient. Interpret your results.

    d. Would the Tukey procedure have been more efficient to use in part (c)? Explain.

    e. If the researcher also wished to estimate $D_3 = \mu_1 - \mu_3$, still with a 95 percent family confidence coefficient, would the B multiple in part (c) need to be modified? Would this also be the case if the Tukey procedure had been employed?

15.12. Refer to **Cash offers** Problem 14.13.

    a. Prepare a normal probability plot of the estimated factor level means $\bar{Y}_{i\cdot}$. What does this plot suggest regarding the effect of the owner's age on the mean cash offer?

    b. Estimate the mean cash offer for young owners; use a 99 percent confidence interval.

    c. Construct a 99 percent confidence interval for $D = \mu_3 - \mu_1$. Interpret your interval estimate.

    d. Obtain confidence intervals for all pairwise comparisons between the treatment means; use the Tukey procedure and a 90 percent family confidence coefficient. Interpret your results and provide a graphic summary by preparing a line plot of the estimated factor level means with nonsignificant comparisons underlined. Are your conclusions in accord with those in part (a)?

    e. Would the Bonferroni procedure have been more efficient to use in part (d) than the Tukey procedure? Explain.

15.13. Refer to **Filling machines** Problem 14.14.

    a. Prepare a normal probability plot of the estimated factor level means $\bar{Y}_{i\cdot}$. What does this plot suggest regarding the variation in the mean fills for the six machines?

    b. Construct a 95 percent confidence interval for the mean fill for machine 1.

    c. Obtain a 95 percent confidence interval for $D = \mu_2 - \mu_1$. Interpret your interval estimate.

    d. The consultant is particularly interested in the mean fills for machines 1, 4, and 5. Construct confidence intervals for all pairwise comparisons among these three treatment means; use the Bonferroni method with a 90 percent family confidence coefficient. Interpret your results and provide a graphic summary by preparing a line plot of the estimated factor level means with nonsignificant differences underlined. Do your conclusions agree with those in part (a)?

    e. Would the Tukey procedure have been more efficient to use in part (d) than the Bonferroni procedure? Explain.

15.14. Refer to **Premium distribution** Problem 14.15.

    a. Prepare a normal probability plot of the estimated factor level means $\bar{Y}_{i\cdot}$. What does this plot suggest about the variation in the mean time lapses for the five agents?

b. Construct a 90 percent confidence interval for the mean time lapse for agent 1.

c. Obtain a 90 percent confidence interval for $D = \mu_2 - \mu_1$. Interpret your interval estimate.

d. The marketing director wishes to compare the mean time lapses for agents 1, 3, and 5. Obtain confidence intervals for all pairwise comparisons among these three treatment means; use the Bonferroni procedure with a 90 percent family confidence coefficient. Interpret your results and present a graphic summary by preparing a line plot of the estimated factor level means with nonsignificant differences underlined. Do your conclusions agree with those in part (a)?

e. Would the Tukey procedure have been more efficient to use in part (d) than the Bonferroni procedure? Explain.

15.15. Refer to **Productivity improvement** Problems 14.10 and 15.9.

a. Estimate the difference in mean productivity improvement between firms with low or moderate research and development expenditures and firms with high expenditures; use a 95 percent confidence interval. Employ an unweighted mean for the low and moderate expenditures groups. Interpret your interval estimate.

b. The sample sizes for the three factor levels are proportional to the population sizes. The economist wishes to estimate the mean productivity gain last year for all firms in the population. Estimate this overall mean productivity improvement with a 95 percent confidence interval.

c. Using the Scheffé procedure, obtain confidence intervals for the following comparisons with 90 percent family confidence coefficient:

$$D_1 = \mu_3 - \mu_2 \qquad D_3 = \mu_2 - \mu_1$$

$$D_2 = \mu_3 - \mu_1 \qquad L_1 = \frac{\mu_1 + \mu_2}{2} - \mu_3$$

Analyze your results and describe your findings.

d. Would the Bonferroni procedure have been more efficient to use in part (c) than the Scheffé procedure? Explain.

15.16. Refer to **Rehabilitation therapy** Problem 14.12.

a. Estimate the contrast $L = (\mu_1 - \mu_2) - (\mu_2 - \mu_3)$ with a 99 percent confidence interval. Interpret your interval estimate.

b. Estimate the following comparisons using the Bonferroni procedure with a 95 percent family confidence coefficient:

$$D_1 = \mu_1 - \mu_2 \qquad D_3 = \mu_2 - \mu_3$$

$$D_2 = \mu_1 - \mu_3 \qquad L_1 = D_1 - D_3$$

Analyze your results and describe your findings.

c. Would the Scheffé procedure have been more efficient to use in part (b) than the Bonferroni procedure? Explain.

15.17. Refer to **Cash offers** Problem 14.13.

a. Estimate the contrast $L = (\mu_3 - \mu_2) - (\mu_2 - \mu_1)$ with a 99 percent confidence interval. Interpret your interval estimate.

b. Estimate the following comparisons with a 90 percent family confidence coefficient; employ the most efficient multiple comparison procedure:

$$D_1 = \mu_2 - \mu_1 \qquad D_3 = \mu_3 - \mu_1$$

$$D_2 = \mu_3 - \mu_2 \qquad L_1 = D_2 - D_1$$

Interpret your results.

15.18. Refer to **Filling machines** Problem 14.14. Machines 1 and 2 were bought new five years ago, machines 3 and 4 were bought in a reconditioned state five years ago, and machines 5 and 6 were bought new last year.

a. Estimate the contrast:

$$L = \frac{\mu_1 + \mu_2}{2} - \frac{\mu_3 + \mu_4}{2}$$

with a 95 percent confidence interval. Interpret your interval estimate.

b. Estimate the following comparisons with a 90 percent family confidence coefficient; use the most efficient multiple comparison procedure:

$$D_1 = \mu_1 - \mu_2 \qquad\qquad L_2 = \frac{\mu_1 + \mu_2}{2} - \frac{\mu_5 + \mu_6}{2}$$

$$D_2 = \mu_3 - \mu_4 \qquad\qquad L_3 = \frac{\mu_1 + \mu_2 + \mu_5 + \mu_6}{4} - \frac{\mu_3 + \mu_4}{2}$$

$$D_3 = \mu_5 - \mu_6 \qquad\qquad L_4 = \frac{\mu_1 + \mu_2 + \mu_3 + \mu_4}{4} - \frac{\mu_5 + \mu_6}{2}$$

$$L_1 = \frac{\mu_1 + \mu_2}{2} - \frac{\mu_3 + \mu_4}{2}$$

Interpret your results. What can the consultant learn from these results about the differences between the six filling machines?

15.19. Refer to **Premium distribution** Problem 14.15. Agents 1 and 2 distribute merchandise only, agents 3 and 4 distribute cash-value coupons only, and agent 5 distributes both merchandise and coupons.

a. Estimate the contrast:

$$L = \frac{\mu_1 + \mu_2}{2} - \frac{\mu_3 + \mu_4}{2}$$

with a 90 percent confidence interval. Interpret your interval estimate.

b. Estimate the following comparisons with 90 percent family confidence coefficient; use the Scheffé procedure:

$$D_1 = \mu_1 - \mu_2 \qquad\qquad L_2 = \frac{\mu_3 + \mu_4}{2} - \mu_5$$

$$D_2 = \mu_3 - \mu_4 \qquad\qquad L_3 = \frac{\mu_1 + \mu_2}{2} - \frac{\mu_3 + \mu_4}{2}$$

$$L_1 = \frac{\mu_1 + \mu_2}{2} - \mu_5$$

Interpret your results.

c. Would the Bonferroni procedure have been more efficient to use in part (b) than the Scheffé procedure? Explain.

d. Of all premium distributions, 25 percent are handled by agent 1, 20 percent by agent 2, 20 percent by agent 3, 20 percent by agent 4, and 15 percent by agent 5. Estimate the overall mean time lapse for premium distributions with a 90 percent confidence interval.

15.20. Refer to **Productivity improvement** Problems 14.10 and 15.9.

a. Use a single degree of freedom test to determine whether or not $(\mu_1 + \mu_2)/2 = \mu_3$; control the α risk at .05 and employ test statistic (15.37). State the alternatives, decision rule, and conclusion.

b. Test for all pairs of factor level means whether or not they differ; use single degree of freedom tests based on the Tukey procedure with $\alpha = .05$. Set up groups of factor levels whose means do not differ.

15.21. Refer to **Cash offers** Problem 14.13.

a. Use a single degree of freedom test to determine whether or not $\mu_2 - \mu_1 = \mu_3 - \mu_2$; control the α risk at .01 and employ test statistic (15.38). State the alternatives, decision rule, and conclusion.

b. Test for all pairs of factor level means whether or not they differ; use single degree of freedom tests based on the Tukey procedure with $\alpha = .01$. Set up groups of factor levels whose means do not differ.

15.22. Refer to **Premium distribution** Problem 14.15.

a. Use a single degree of freedom test to determine whether or not $(\mu_1 + \mu_2)/2 = (\mu_3 + \mu_4)/2$; control the α risk at .10 and employ test statistic (15.37). State the alternatives, decision rule, and conclusion.

b. Test for all pairs of factor level means whether or not they differ; use single degree of freedom tests based on the Tukey procedure with $\alpha = .10$. Set up groups of factor levels whose means do not differ.

15.23. Refer to **Solution concentration** Problem 4.15. Suppose the chemist initially wished to employ ANOVA model (14.2) to determine whether or not the concentration of the solution is affected by the amount of time that has elapsed since preparation.

a. State the analysis of variance model.

b. Prepare a normal probability plot of the estimated factor level means $\bar{Y}_{i\cdot}$. What does this plot suggest about a relation between the solution concentration and time?

c. Obtain the analysis of variance table.

d. Test whether or not the factor level means are equal; use a level of significance of $\alpha = .025$. State the alternatives, decision rule, and conclusion.

e. Make pairwise comparisons of factor level means between all adjacent lengths of time; use the Bonferroni procedure with a 95 percent family confidence coefficient. Do your results suggest that the regression relation is not linear? Are your conclusions in accord with those in part (b)?

15.24. Refer to **Rehabilitation therapy** Problem 14.12. A biometrician has developed a scale for physical fitness status, as follows:

Physical Fitness Status	Scale Value
Below average	83
Average	100
Above average	121

a. Using this physical fitness status scale, fit the first-order regression model (2.1) for regressing number of days required for therapy (Y) on physical fitness status (X).

b. Obtain the residuals and plot them against X. Does a linear regression model appear to fit the data?

c. Perform an F test to determine whether or not there is lack of fit of a linear regression function; use $\alpha = .05$. State the alternatives, decision rule, and conclusion.

d. Could you test for lack of fit of a quadratic regression function here? Explain.

15.25. Refer to **Filling machines** Problem 14.14. A maintenance engineer has suggested that the differences in mean fills for the six machines are largely related to the length of time since a machine last received major servicing. Service records indicate these lengths of time to be as follows (in months):

Filling Machine	Number of Months	Filling Machine	Number of Months
1	.4	4	5.3
2	3.7	5	1.4
3	6.1	6	2.1

a. Fit the second-order polynomial regression model (9.1) for regressing amount of fill (Y) on number of months since major servicing (X).
b. Obtain the residuals and plot them against X. Does a quadratic regression function appear to fit the data?
c. Perform an F test to determine whether or not there is lack of fit of a quadratic regression function; use $\alpha = .01$. State the alternatives, decision rule, and conclusion.
d. Test whether or not the quadratic term in the response function can be dropped from the model; use $\alpha = .01$. State the alternatives, decision rule, and conclusion.

EXERCISES

15.26. Show that when $r = 2$ and $n_i \equiv n$, q defined in (15.27) is equivalent to $\sqrt{2}|t^*|$, where t^* is defined in (1.62).

15.27. Starting with (15.30), complete the derivation of (15.25).

15.28. Show that when $r = 2$, S^2 defined in (15.34a) is equivalent to $[t(1 - \alpha/2; n_T - r)]^2$.

15.29. Set up the test alternatives (15.36) in the matrix format $\mathbf{C\beta} = \mathbf{h}$ of (8.66) and show that test statistic (8.71) reduces to F^* as defined in (15.38).

PROJECTS

15.30. Refer to the **SENIC** data set and Project 14.33. Obtain confidence intervals for all pairwise comparisons between the four regions; use the Tukey procedure and a 90 percent family confidence coefficient. Interpret your results and state your findings. Prepare a line plot of the estimated factor level means and underline all nonsignificant comparisons.

15.31. Refer to the **SMSA** data set and Project 14.35. Obtain confidence intervals for all pairwise comparisons between the four regions; use the Tukey procedure and a 90 percent family confidence coefficient. Interpret your results and state your findings. Prepare a line plot of the estimated factor level means and underline all nonsignificant comparisons.

15.32. Refer to Project 14.36d.
a. For each replication, construct confidence intervals for all pairwise comparisons among the three treatment means; use the Tukey procedure with a 95 percent family confidence coefficient. Then determine whether all confidence intervals for the replication are correct, given that $\mu_1 = 80$, $\mu_2 = 60$, and $\mu_3 = 160$.
b. For what proportion of the 100 replications are all confidence intervals correct? Is this proportion close to theoretical expectations? Discuss.

Chapter 16

Diagnostics and Remedial Measures—III

When discussing regression analysis, we emphasized the importance of examining the aptness of the regression model under consideration, and pointed out the effectiveness of residual plots and other diagnostics for spotting major departures from the tentative model. Examination of aptness is no less important for analysis of variance models.

In this chapter, we take up the use of residual plots for diagnosing the appropriateness of analysis of variance models, as well as formal tests for the equality of the error variances. We also discuss the use of transformations as a remedial measure to improve the aptness of the analysis of variance model and consider the influence of departures from the ANOVA model on estimation and test inferences.

For pedagogic reasons, as in regression analysis, we have discussed inference procedures before diagnostics and remedial measures. The actual sequence of developing and using any statistical model is, of course, the reverse:

1. Examine whether the proposed model is appropriate for the set of data at hand.
2. If the proposed model is not appropriate, take remedial measures such as transformations of the data, or modify the model.
3. After review of the aptness of the model and completion of any necessary remedial measures and an evaluation of their effectiveness, inferences based on the model can be undertaken.

It is not necessary, nor is it usually possible, that an ANOVA model fits perfectly. As will be noted later, ANOVA models are reasonably robust against certain types of departures from the model, such as the error terms not being exactly normally distributed. The major purpose of the examination of the aptness of the model is therefore to detect serious departures from the conditions assumed by the model.

16.1 RESIDUAL ANALYSIS

Residual analysis for ANOVA models corresponds closely to that for regression models. We therefore discuss only briefly some key issues in the use of residual analysis for ANOVA models.

Residuals

The residuals e_{ij} for the ANOVA cell means model (14.2) were defined in (14.17):

$$(16.1) \qquad\qquad e_{ij} = Y_{ij} - \hat{Y}_{ij} = Y_{ij} - \bar{Y}_{i.}$$

As in regression, standardized residuals:

$$(16.2) \qquad\qquad \frac{e_{ij}}{\sqrt{MSE}}$$

are sometimes helpful. An alternative standardized residual is useful at times, as will be seen shortly:

$$(16.3) \qquad\qquad \frac{e_{ij}}{s_i}$$

Here, s_i is the sample standard deviation of the observations from the ith factor level, as defined in (14.37).

Residual Plots

Residual plots useful for analysis of variance models include: (1) plots against the fitted values, (2) time or other sequence plots, (3) dot plots, and (4) normal probability plots. All of these plots have been encountered previously. We illustrate their application for evaluating the aptness of analysis of variance models by an example.

Example.   Table 16.1 contains the residuals for the rust inhibitor example of Chapter 15. For ease of presentation, the treatments are shown in the columns of the table. The residuals were obtained from the data in Table 15.2a. For instance, the residual for the first experimental unit treated with brand A rust inhibitor is:

$$e_{11} = Y_{11} - \hat{Y}_{11} = Y_{11} - \bar{Y}_{1.} = 43.9 - 43.14 = .76$$

Figure 16.1a contains a *residual plot against the fitted values*. This plot, prepared with the Minitab computer package (Ref. 16.1), differs in appearance from similar plots for regression analysis, because for the ANOVA model the fitted values $\hat{Y}_{ij}$ are the same for all observations for a given factor level. Recall from (14.15) that $\hat{Y}_{ij} = \bar{Y}_{i.}$.

Figure 16.1b contains *dot plots* of the residuals for each factor level. These plots are similar to the residual plot against the fitted values in Figure 16.1a, except here

TABLE 16.1   Residuals for Rust Inhibitor Example

| | Brand | | | |
| | A
$i = 1$ | B
$i = 2$ | C
$i = 3$ | D
$i = 4$ |
j				
1	.76	.36	.45	−4.27
2	−4.14	−2.34	1.35	4.73
3	3.56	3.26	.55	.23
4	.66	1.16	−1.55	.03
5	1.06	−1.74	2.05	−1.17
6	4.56	2.96	.15	−.17
7	.46	−3.34	2.65	2.73
8	−4.24	−1.34	−2.75	−1.77
9	.46	1.36	−4.15	.43
10	−3.14	−.34	1.25	−.77

the residual scale is the horizontal one. An advantage of the plot of the residuals against the fitted values in Figure 16.1a is that it facilitates an assessment of the relation between the magnitudes of the error variances and the factor level means. A disadvantage is that some of the estimated factor level means may be far apart, making a comparison of the factor levels more difficult. This difficulty is remedied in Figure 16.1b since dot plots can be placed close together to facilitate comparisons between factor levels.

Figure 16.1c contains a *normal probability plot* of the residuals. This plot is exactly the same as for regression models.

No sequence plot of the residuals is presented here because the data for the rust inhibitor example were not ordered according to time or in some other logical sequence.

All of the plots in Figure 16.1, as we shall see, suggest that ANOVA model (14.2) is appropriate for the rust inhibitor data.

Diagnosis of Departures from ANOVA Model

We consider now how residual plots can be helpful in diagnosing the following departures from ANOVA model (14.2):

1. Nonconstancy of error variance
2. Nonindependence of error terms
3. Outliers
4. Omission of important independent variables
5. Nonnormality of error terms

Nonconstancy of Error Variance.   ANOVA model (14.2) requires that the error terms ε_{ij} have constant variance for all factor levels. The appropriateness of this assump-

FIGURE 16.1   Diagnostic Residual Plots—Rust Inhibitor Example (Minitab, Ref. 16.1)

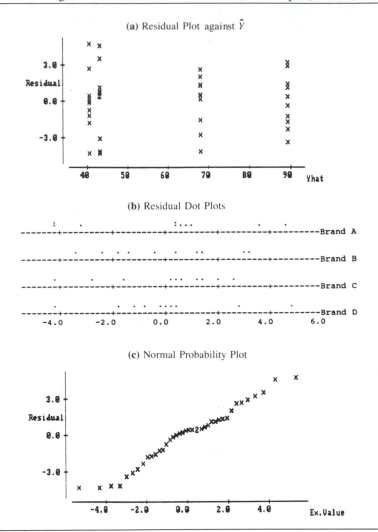

(a) Residual Plot against $\hat{Y}$

(b) Residual Dot Plots

(c) Normal Probability Plot

tion when the sample sizes are not large can best be studied from *residual plots against fitted values* or from *residual dot plots*. When the error variance is constant, these residual plots should show about the same extent of scatter of the residuals around 0 for each factor level. This is the case for the rust inhibitor example in Figures 16.1a and 16.1b.

Figure 16.2 is a prototype residual plot against the fitted values when the error variances are not constant. This plot portrays the case where the error terms for factor level 3 have a larger variance than those for the other two factor levels.

FIGURE 16.2   Prototype Plot of Residuals against Fitted Values when Error Term Variance Is Not Constant for All Factor Levels

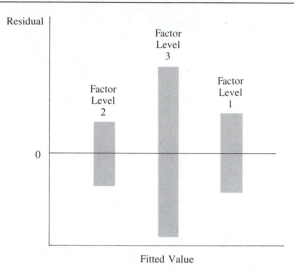

When the sample sizes for the different factor levels are large, *histograms* of the residuals for each treatment—arranged vertically and using the same scale, like the dot plots in Figure 16.1b—are an effective means for examining the constancy of the error variance, as well as for assessing whether the error terms are normally distributed.

A number of statistical tests have been developed for formally examining the equality of r variances; two of these tests will be discussed in Section 16.2.

Nonindependence of Error Terms.   Whenever data are obtained in a time sequence, a *residual sequence plot* should be prepared to examine if the error terms are serially correlated. Figure 16.3 contains the residuals for an experiment on group interactions. Three different treatments were applied, and the group interactions were recorded on video tapes. Seven replications were made for each treatment. Afterwards, the experimenter measured the number of interactions by viewing the tapes in randomized order. Figure 16.3 strongly suggests that the experimenter discerned a larger number of interactions as he gained experience in viewing the tapes, as a result of which the residuals in Figure 16.3 appear to be serially correlated. In this instance, an inclusion in the model of a linear term for the time effect might be sufficient to assure independence of the error terms in the revised model.

Time-related effects may also lead to increases or decreases in the error variance over time. For instance, an experimenter may make more precise measurements over time. Figure 16.4 portrays residual sequence plots where the error variance decreases over time.

FIGURE 16.3   Residual Sequence Plots for Group Interaction Study Illustrating Time-Related Effect

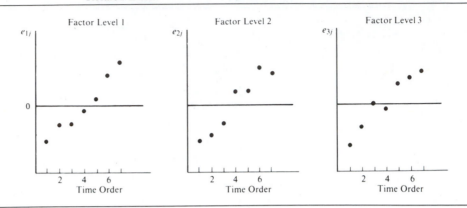

When the data are ordered in some other logical sequence, such as in a geographic sequence, a plot of the residuals against this ordering is helpful for ascertaining whether the error terms are serially correlated according to this ordering.

Outliers.   The detection of outliers is facilitated by *residual plots against fitted values, residual dot plots, box plots,* and *stem-and-leaf plots*. These plots easily show an outlier observation, that is, an observation that differs from the fitted value by far more than do other observations. As noted in Chapter 4, it is wise practice to discard outlier observations only if they can be identified as being due to such specific causes as instrumentation malfunctioning, observer measurement blunder, or recording error.

FIGURE 16.4   Residual Sequence Plots Illustrating Decreasing Error Variance over Time

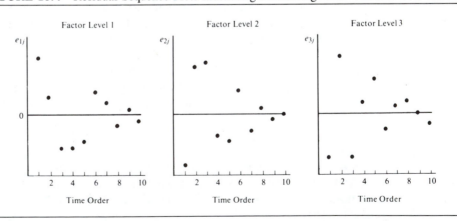

Omission of Important Independent Variables. Residual analysis may also be used to study whether or not the single-factor ANOVA model is an adequate model. In a learning experiment involving three motivational treatments, the residuals shown in Figure 16.5 were obtained. The residual plot against the fitted values in Figure 16.5 shows no unusual overall pattern. The experimenter wondered, however, whether the treatment effects differ according to the gender of the subject. In Figure 16.5 the residuals for male subjects are shown by a box, and those for females by dots. The results in Figure 16.5 suggest strongly that for each of the motivational treatments studied, the treatment effects do differ according to gender. Hence, a multifactor model recognizing both motivational treatment and gender of subject as independent variables might be more useful.

Note that residual analysis here does not invalidate the original single-factor model. Rather, the residual analysis points out that the original model overlooks differences in treatment effects that may be important to recognize. Since there are usually many independent variables that have some effect on the dependent variable, the analyst needs to identify for residual analysis those independent variables that most likely have an important effect on the dependent variable.

Nonnormality of Error Terms. The normality of the error terms can be studied from *histograms, dot plots, box plots,* and *normal probability plots* of the residuals. In addition, comparisons can be made of observed frequencies with expected frequencies if normality holds, and formal chi-square goodness of fit or related tests can be utilized. The discussion in Chapter 4 about these methods for assessing the normality of the error terms is entirely applicable here.

When the factor level sample sizes are large, the study of normality can be made separately for each treatment. When the factor level sample sizes are small, one can

---

FIGURE 16.5   Residual Plot against Fitted Values Illustrating Omission of Important Independent Variable

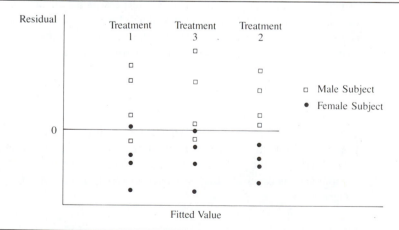

combine the residuals e_{ij} for all treatments into one group, provided that the evidence suggests that there are no major differences in the error variances for the treatments studied. We have done this in the rust inhibitor example in Figure 16.1c. This figure does not indicate any serious departures from normality. The pattern of the points is reasonably linear except possibly in the tails. The coefficient of correlation between the ordered residuals and their expected values under normality is .987, which also supports the reasonableness of the normality assumption.

When the factor level sample sizes are small and unequal variances of the error terms for the different factor levels are indicated, the standardized residuals (16.3) should be used before combining all the residuals for a study of normality. Otherwise, nonnormality may be indicated solely because of the failure of the error terms to have equal variances.

Note

As we pointed out for regression models, the residuals e_{ij} are not independent random variables. For ANOVA model (14.2), they are subject to the restrictions stated in (14.18). Consequently, statistical tests that require independent observations are not exactly appropriate for residuals. If, however, the number of residuals for each factor level is not small, the effect of the correlations will only be modest. It has been noted that graphic plots of residuals are less subject to the effects of correlation than are statistical tests because graphic plots contain the individual residuals and not simply functions of them.

16.2 TESTS FOR EQUALITY OF VARIANCES

Several formal tests are available for studying whether or not r populations have equal variances, as required by the ANOVA model. We shall consider two of these, the Bartlett test and the Hartley test. Both assume that each of the r populations is normal. Also, it is assumed by both tests that independent random samples are obtained from each population. The Bartlett test is an all-purpose test and can be used for equal and unequal sample sizes. The Hartley test is applicable only if the sample sizes are equal, and it is designed to be sensitive against substantial differences between the largest and the smallest population variances.

Bartlett Test

The basic idea underlying the Bartlett test is simple. Let $s_1^2, \ldots, s_r^2$ denote the sample variances from r normal populations, and let df_i denote the degrees of freedom associated with the sample variance s_i^2. Then the weighted arithmetic average of the sample variances, using the associated degrees of freedom df_i as weights, is the mean square error:

(16.4)
$$MSE = \frac{1}{df_T} \sum_{i=1}^{r} df_i s_i^2$$

where:

(16.4a)
$$df_T = \sum_{i=1}^{r} df_i$$

Similarly, the weighted geometric average of the s_i^2, denoted by *GMSE*, is:

(16.5)
$$GMSE = [(s_1^2)^{df_1}(s_2^2)^{df_2} \cdots (s_r^2)^{df_r}]^{1/df_T}$$

It can be shown that for any given set of s_i^2 values, the following relation holds between these two averages:

(16.6)
$$GMSE \leq MSE$$

The two averages are equal when all s_i^2 are equal; the greater the variation between the s_i^2, the further apart the two averages will be. Hence, if the ratio $MSE/GMSE$ is close to 1, we have evidence that the population variances are equal. If the ratio $MSE/GMSE$ is large, it indicates that the population variances are unequal. The same conclusions follow if we consider $\log(MSE/GMSE) = \log MSE - \log GMSE$.

Bartlett has shown that a function of $\log MSE - \log GMSE$ follows, for large sample sizes, approximately the χ^2 distribution with $r - 1$ degrees of freedom when the population variances are equal. This test statistic is:

(16.7)
$$B = \frac{df_T}{C} (\log_e MSE - \log_e GMSE)$$

where:

(16.7a)
$$C = 1 + \frac{1}{3(r-1)} \left[\left(\sum_{i=1}^{r} \frac{1}{df_i} \right) - \frac{1}{df_T} \right]$$

The term C is always greater than 1.

Test statistic (16.7) reduces to:

(16.8)
$$B = \frac{1}{C} \left[(df_T) \log_e MSE - \sum_{i=1}^{r} (df_i) \log_e s_i^2 \right]$$

For deciding between:

(16.9a)
$$H_0: \sigma_1^2 = \sigma_2^2 = \cdots = \sigma_r^2$$
$$H_a: \text{not all } \sigma_i^2 \text{ are equal}$$

we calculate test statistic B. Since B is approximately distributed as χ^2 with $r - 1$ degrees of freedom when H_0 holds, and since we saw that large values of B lead to conclusion H_a, the appropriate decision rule for controlling the risk of a Type I error at α is:

(16.9b)
$$\text{If } B \leq \chi^2(1 - \alpha; r - 1), \text{ conclude } H_0$$
$$\text{If } B > \chi^2(1 - \alpha; r - 1), \text{ conclude } H_a$$

where $\chi^2(1 - \alpha; r - 1)$ is the $(1 - \alpha)100$ percentile of the χ^2 distribution with $r - 1$ degrees of freedom. Percentiles of the χ^2 distribution are given in Table A.3.

The chi-square approximation may be considered to be appropriate when all degrees of freedom df_i are four or more.

When the Bartlett test is used for the single-factor analysis of variance model (14.2), we have:

$$df_i = n_i - 1 \qquad df_T = \sum_{i=1}^{r} (n_i - 1) = n_T - r$$

Example.   Table 16.2 contains data from a study on the time required to complete a certain production operation for each of the three plant shifts. Twenty cycles of the operation were performed by shift 1, 17 by shift 2, and 21 by shift 3. It has been ascertained that the three populations are close to normal ones. We now wish to determine by the Bartlett test whether the shift variances σ_i^2 are the same for all three shifts.

The needed calculations for the Bartlett test are shown in Table 16.2. Substituting into (16.7a) for C and (16.8) for B, we obtain:

$$C = 1 + \frac{1}{3(3-1)} \left[\left(\frac{1}{19} + \frac{1}{16} + \frac{1}{20} \right) - \frac{1}{55} \right] = 1.02449$$

and:

$$B = \frac{1}{1.02449} [55(6.18631) - 338.32164] = \frac{1.92541}{1.02449} = 1.8794$$

Assume that we are to control the risk of making a Type I error at $\alpha = .05$. We therefore require $\chi^2(.95; 3 - 1)$. From Table A.3, we find $\chi^2(.95; 2) = 5.99$. Hence, the decision rule is:

$$\text{If } B \leq 5.99, \text{ conclude } H_0$$

$$\text{If } B > 5.99, \text{ conclude } H_a$$

Since $B = 1.88 \leq 5.99$, we conclude H_0, that the three population variances are equal. The P-value of the test is $P\{\chi^2(2) > 1.8794\} = .39$.

TABLE 16.2   Calculations for Bartlett Test for Equality of Three Population Variances

Population i	s_i^2	$df_i = n_i - 1$	$(df_i)s_i^2$	$\log_e s_i^2$	$(df_i)\log_e s_i^2$
1	415	19	7,885	6.02828	114.53732
2	698	16	11,168	6.54822	104.77152
3	384	20	7,680	5.95064	119.01280
Total		$df_T = 55$	26,733		338.32164

$$MSE = 26,733 \div 55 = 486.05$$
$$\log_e MSE = 6.18631$$

Note

In the above example, we could have avoided calculating the denominator C. Even before dividing by C, we can see that the numerator of the test statistic, 1.92541, falls below the action limit 5.99. Since $C > 1$ always, the effect of dividing by C is to make the test statistic B still smaller. Thus, one may compute the numerator of B first, and only calculate the denominator C if it can affect the outcome.

Box Transformation.   As noted earlier, the chi-square approximation to the distribution of the Bartlett test statistic (16.7) when the population variances are equal should not be used when any of the df_i are less than four. An approximation that can be used when some of the degrees of freedom are small and that also is appropriate for larger numbers of degrees of freedom utilizes the F distribution and is based on the modified Bartlett test statistic B' developed by Box:

$$(16.10) \qquad B' = \frac{f_2 BC}{f_1(A - BC)}$$

where:

$$(16.10a) \qquad f_1 = r - 1$$

$$(16.10b) \qquad f_2 = \frac{r + 1}{(C - 1)^2}$$

$$(16.10c) \qquad A = \frac{f_2}{2 - C + \dfrac{2}{f_2}}$$

For testing the alternatives (16.9a), the appropriate decision rule for controlling the risk of a Type I error at α is:

$$(16.11) \qquad \begin{aligned} &\text{If } B' \le F(1 - \alpha; f_1, f_2), \text{ conclude } H_0 \\ &\text{If } B' > F(1 - \alpha; f_1, f_2), \text{ conclude } H_a \end{aligned}$$

where $F(1 - \alpha; f_1, f_2)$ is the $(1 - \alpha)100$ percentile of the F distribution with f_1 and f_2 degrees of freedom. Percentiles of the F distribution are given in Table A.4. The value of f_2 will usually not be an integer so that it will then be necessary to interpolate in the F table. Reciprocal interpolation should be employed, as illustrated in the following example.

Example.   We shall employ again the example in Table 16.2. We found earlier:

$$C = 1.02449 \qquad B = 1.8794$$

We now require:

$$f_1 = 3 - 1 = 2$$

$$f_2 = \frac{3 + 1}{(1.02449 - 1.0)^2} = 6,669.3$$

$$A = \frac{6{,}669.3}{2 - 1.02449 + \dfrac{2}{6{,}669.3}} = 6{,}834.63$$

$$B' = \frac{6{,}669.3(1.8794)(1.02449)}{2[6{,}834.63 - 1.8794(1.02449)]} = .94$$

For controlling the level of significance at $\alpha = .05$, we require $F(.95; 2, 6{,}669.3)$. From Table A.4, we find:

$$F(.95; 2, 120) = 3.07 \qquad F(.95; 2, \infty) = 3.00$$

Reciprocal interpolation is similar to linear interpolation except the reciprocals are used for determining the fraction of the difference between 3.00 and 3.07, as follows:

$$F(.95; 2, 6{,}669.3) = 3.07 + \frac{\dfrac{1}{6{,}669.3} - \dfrac{1}{120}}{\dfrac{1}{\infty} - \dfrac{1}{120}}(3.00 - 3.07)$$

$$= 3.001$$

Hence, the decision rule is:

$$\text{If } B' \leq 3.001, \text{ conclude } H_0$$

$$\text{If } B' > 3.001, \text{ conclude } H_a$$

Since $B' = .94 \leq 3.001$, we conclude H_0, that the three population variances are equal. This is the same conclusion as we obtained with the Bartlett test statistic B and the chi-square approximation. The P-value of the modified Bartlett test, $P\{F(2, 6{,}669.3) > .94\} = .39$, is the same as that for the Bartlett test statistic with the chi-square approximation.

Comments

1. The Bartlett test is quite sensitive to departures from normality. That is, if the populations in fact are not normal, the actual level of significance may differ substantially from the specified one. Hence, if the populations depart substantially from normality, it is not recommended that the Bartlett test be employed for testing equality of variances. Instead, a robust nonparametric test should be employed. Reference 16.2 cites several such tests.

2. As will be noted in Section 16.4, the F test for equality of factor level means is not much affected by unequal variances when the factor level sample sizes are approximately equal, as long as the differences in the variances are not unusually large. Hence, if the populations are reasonably normal so that the Bartlett test can be employed and the sample sizes do not differ greatly, a fairly low α level may be justified in testing the equality of variances for determining the aptness of the ANOVA model, since only large differences between variances need be detected.

Hartley Test

If each of the r sample variances s_i^2 has the same number of degrees of freedom so that $df_i \equiv df$, a simple test for deciding between:

(16.12)
$$H_0: \sigma_1^2 = \sigma_2^2 = \cdots = \sigma_r^2$$

$$H_a: \text{not all } \sigma_i^2 \text{ are equal}$$

is due to Hartley. It is based solely on the largest sample variance, denoted by $\max(s_i^2)$, and the smallest sample variance, denoted by $\min(s_i^2)$. The test statistic is:

(16.13)
$$H = \frac{\max(s_i^2)}{\min(s_i^2)}$$

Clearly, values of H near 1 support H_0, and large values of H support H_a. The distribution of H when H_0 holds has been tabulated, and selected percentiles are presented in Table A.12. The distribution of H depends on the number of populations r and the common number of degrees of freedom df. As we noted earlier, the Hartley test, like the Bartlett test, assumes normal populations.

The appropriate decision rule for controlling the risk of making a Type I error at α is:

(16.14)
$$\text{If } H \leq H(1 - \alpha; r, df), \text{ conclude } H_0$$

$$\text{If } H > H(1 - \alpha; r, df), \text{ conclude } H_a$$

where $H(1 - \alpha; r, df)$ is the $(1 - \alpha)100$ percentile of the distribution of H when H_0 holds, for r populations and df degrees of freedom for each sample variance.

When the Hartley test is used for the single-factor ANOVA model (14.2) with equal sample sizes $n_i \equiv n$, we have $df = n - 1$.

Example. In a study of the appeal of four different television commercials, 10 observations were made for each commercial. The sample variances were as follows:

$$s_1^2 = 293 \qquad s_2^2 = 146 \qquad s_3^2 = 985 \qquad s_4^2 = 528$$

Before proceeding with the analysis of variance, it was determined that the populations are close to normal and it is now desired to use the Hartley test to ascertain whether or not the four treatment variances are equal:

$$H_0: \sigma_1^2 = \sigma_2^2 = \sigma_3^2 = \sigma_4^2$$

$$H_a: \text{not all } \sigma_i^2 \text{ are equal}$$

The level of significance is to be controlled at $\alpha = .05$. For $r = 4$ and $df = n - 1 = 9$, we require from Table A.12 $H(.95; 4, 9) = 6.31$. Hence, the appropriate decision rule is:

$$\text{If } H \leq 6.31, \text{ conclude } H_0$$

$$\text{If } H > 6.31, \text{ conclude } H_a$$

We have $\max(s_i^2) = 985$ and $\min(s_i^2) = 146$. Hence:

$$H = \frac{985}{146} = 6.75$$

Since $H = 6.75 > 6.31$, we conclude H_a, that the four treatment variances are not equal.

Comments

1. The Hartley test strictly requires equal sample sizes. If the sample sizes are unequal but do not differ greatly, the Hartley test may still be used as an approximate test. For this purpose, the average number of degrees of freedom would be used for entering Table A.12.

2. The Hartley test, like the Bartlett test, is quite sensitive to departures from the assumption of normal populations and should not be used when substantial departures from normality exist.

3. For the same reasons noted for the Bartlett test, a low α level may be justified when the Hartley test is used for determining the aptness of the ANOVA model with respect to equal treatment variances.

16.3 TRANSFORMATIONS

When residual plots or other diagnostics indicate that the ANOVA model is not appropriate for the data at hand, a choice of possible corrective measures is available. One is to modify the model. This approach may have the disadvantage at times of leading to fairly complex analysis. Another approach is to use transformations on the data. A third approach, useful when serious lack of normality is the basic difficulty, is to employ nonparametric tests such as the median test or the Kruskal-Wallis test based on ranks (discussed in Chapter 17).

The use of transformations is the principal topic of this section. Since transformations were discussed in Chapter 4 in connection with regression analysis, our discussion here will be brief.

Transformations to Stabilize Variances

There are several types of situations in which the variances of the error terms are not constant, each of which requires a different type of transformation to stabilize the variance.

Variance Proportional to μ_i.   When the variance of the error terms for any factor level (denoted by σ_i^2) varies proportionally to the factor level mean μ_i, the sample statistics $s_i^2/\bar{Y}_{i\cdot}$ will tend to be constant. Here s_i^2 is the sample variance for the ith factor level observations, as defined in (14.37). This type of situation is often found when the observed variable Y is a count, such as the number of attempts by a subject before the correct solution is found. For this case, a square root transformation is helpful for stabilizing the variances:

(16.15)   If σ_i^2 proportional to μ_i:

$$Y' = \sqrt{Y} \quad \text{or} \quad Y' = \sqrt{Y} + \sqrt{Y+1}$$

Standard Deviation Proportional to μ_i. When the standard deviation of the error terms for any factor level is proportional to the mean, $s_i/\bar{Y}_{i.}$ tends to be constant for the different factor levels. A helpful transformation in this case for stabilizing the variances is the logarithmic transformation:

(16.16)   If σ_i proportional to μ_i: $Y' = \log Y$

Standard Deviation Proportional to μ_i^2. When the error term standard deviation is proportional to the square of the factor level mean, $s_i/\bar{Y}_{i.}^2$ tends to be constant. An appropriate transformation here is the reciprocal transformation:

(16.17)   If σ_i proportional to μ_i^2: $Y' = \dfrac{1}{Y}$

Dependent Variable Is a Proportion. At times, the observed variable Y_{ij} is a proportion p_{ij}. For instance, the treatments may be different training procedures, the unit of observation is a company training class, and the observed variable Y_{ij} is the proportion of employees in the jth class for the ith training procedure who benefited substantially by the training. Note that n_i here refers to the number of classes receiving the ith training procedure, not to the number of students.

It is well known that in the binomial case the variance of the sample proportion depends on the true proportion. When the number of cases on which each sample proportion is based is the same, this variance is:

(16.18) $\sigma^2\{p_{ij}\} = \dfrac{\pi_i(1 - \pi_i)}{m}$

Here π_i is the population proportion for the ith treatment and m is the common number of cases on which each sample proportion is based. Since $\sigma^2\{p_{ij}\}$ depends on the treatment proportion π_i, the variances of the error terms will not be stable if the treatment proportions π_i differ. An appropriate transformation for this case is the arc sine transformation:

(16.19)   If observation is a proportion: $Y' = 2 \arcsin \sqrt{Y}$

When the proportions p_{ij} are based on different numbers of cases (for instance, in our earlier illustration there may be different numbers of employees in each training class), transformation (16.19) should be employed together with a weighted least squares analysis as described in Section 11.8.

Example. In the television commercials example, the Hartley test indicated that the error variances for the four commercials are not equal. To study which transformation may be helpful here to stabilize the variances, we need to examine the relation between the sample variances s_i^2 and the estimated factor level means $\bar{Y}_{i.}$. The needed summary data are (the original Y_{ij} observations are not given):

i:	1	2	3	4
s_i^2:	293	146	985	528
$\bar{Y}_{i.}$:	65.2	31.9	211.8	117.4

We calculate now the ratios $s_i^2/\bar{Y}_{i.}$, $s_i/\bar{Y}_{i.}$, and $s_i/\bar{Y}_{i.}^2$. The results are as follows:

i	$\dfrac{s_i^2}{\bar{Y}_{i.}}$	$\dfrac{s_i}{\bar{Y}_{i.}}$	$\dfrac{s_i}{\bar{Y}_{i.}^2}$
1	4.5	.26	.0040
2	4.6	.38	.0119
3	4.7	.15	.0007
4	4.5	.20	.0017

The relation $s_i^2/\bar{Y}_{i.}$ is the most stable, hence the square root transformation (16.15) may be helpful in stabilizing the error variances here.

Transformations to Correct Lack of Normality

When the error terms are normally distributed but with unequal variances, a transformation of the observations to stabilize the variances will destroy the normality. Fortunately, in practice, lack of normality and unequal variances tend to go hand in hand. Further, the transformation that helps to correct the lack of constancy of error variances usually also is effective in making the distribution of the error terms more normal. It is wise policy, however, as mentioned in Chapter 4 to check the residuals after a transformation has been applied to make sure that the transformation has been effective in both stabilizing the variances and making the distribution of the error terms reasonably normal.

Comments

1. When a transformation of the observations is required, one can work completely with the transformed data for testing the equality of factor level means. On the other hand, it is often desirable when making estimates of factor level effects to change a confidence interval based on the transformed variable back to an interval in the original variable for easier understanding of the significance of the results.

2. The transformations suggested earlier are obtained by the following argument. Suppose that Y is a random variable with mean μ and variance σ_Y^2 that is a function of μ, say, $\sigma_Y^2 = g(\mu)$. For a transformation of Y, say, $Y' = h(Y)$, it can be shown using a Taylor series expansion that:

$$(16.20) \qquad \sigma_{Y'}^2 \cong [h'(\mu)]^2 g(\mu)$$

where $h'(\mu)$ is the first derivative of $h(Y)$ at μ. We seek that transformation h for which $\sigma_{Y'}^2$ is a constant, for convenience, say, 1. Hence, we wish:

$$[h'(\mu)]^2 g(\mu) = 1$$

or:

(16.21)
$$h'(\mu) = \frac{1}{\sqrt{g(\mu)}}$$

Formula (16.21) is a differential equation whose solution is (neglecting the arbitrary constant):

(16.22)
$$h(\mu) = \int \frac{d\mu}{\sqrt{g(\mu)}}$$

To illustrate the use of (16.22), suppose σ_Y^2 is proportional to μ, say, $\sigma_Y^2 = k\mu = g(\mu)$. We obtain from (16.22):

$$h(\mu) = \int \frac{d\mu}{\sqrt{k\mu}} = \frac{2}{\sqrt{k}}\sqrt{\mu}$$

Thus, a square root transformation $Y' = \sqrt{Y}$ will yield a constant variance of Y', and $Y' = (2/\sqrt{k})\sqrt{Y}$ will yield a constant variance equal to 1.

Formula (16.20) makes it clear that the transformations obtained by this procedure only approximately stabilize the variances. It is therefore important to inspect the residuals for the transformed variable to see how effectively the variances have actually been stabilized.

16.4 EFFECTS OF DEPARTURES FROM MODEL

In preceding sections, we considered how residual analysis and other diagnostic techniques can be helpful in assessing the aptness of the ANOVA model for the data at hand. We also discussed the use of transformations, chiefly for stabilizing variances but also for obtaining as a by-product error distributions more nearly normal. The question now arises what are the effects of any remaining departures from the model on the inferences made. A thorough review of the many studies investigating these effects has been made by Scheffé (Ref. 16.3). Here, we shall summarize the findings.

Nonnormality

For the fixed ANOVA model I, lack of normality is not an important matter, provided the departure from normality is not of extreme form. It may be noted in this connection that *kurtosis* of the error distribution (either more or less peaked than a normal distribution) is more important than skewness of the distribution in terms of the effects on inferences.

The point estimators of factor level means and contrasts are unbiased whether or not the populations are normal. The F test for the equality of factor level means is but little affected by lack of normality, either in terms of the level of significance or power of the test. Hence, the F test is a *robust* test against departures from normality. For instance, the specified level of significance might be .05, whereas the actual level for a nonnormal error distribution might be .04 or .065. Typically, the achieved level of significance in the presence of nonnormality is slightly higher than the specified one, and the achieved power of the test is slightly less than the calculated one. Single interval estimates of factor level means and contrasts and the Scheffé multiple comparison method also are not much affected by lack of normality provided the sample sizes are not extremely small.

For the random ANOVA model II (to be discussed in the next chapter), lack of normality has more serious implications. The estimators of the variance components are still unbiased, but the actual confidence coefficient for interval estimates may be substantially different from the specified one.

Unequal Error Variances

When the error variances are unequal, the F test for the equality of means with the fixed ANOVA model is only slightly affected if all factor level sample sizes are equal or do not differ greatly. Specifically, unequal error variances then raise the actual level of significance only slightly higher than the specified level. Similarly, the Scheffé multiple comparison procedure based on the F distribution is not affected to any substantial extent by unequal variances when the sample sizes are equal or are approximately the same. Thus, the F test and related analyses are robust against unequal variances when the sample sizes are approximately equal. Single comparisons between factor level means, on the other hand, can be substantially affected by unequal variances, so that the actual and specified confidence coefficients may differ markedly in these cases.

The use of equal sample sizes for all factor levels not only tends to minimize the effects of unequal variances on inferences with the F distribution but also simplifies calculational procedures. Thus, here at least, simplicity and robustness go hand in hand.

For the random ANOVA model, unequal error variances can have pronounced effects on inferences about the variance components, even with equal sample sizes.

Nonindependence of Error Terms

Lack of independence of the error terms can have serious effects on inferences in the analysis of variance, for both fixed and random ANOVA models. Since this defect is often difficult to correct, it is important to prevent it in the first place whenever feasible. The use of randomization in those stages of a study that are likely to lead to correlated error terms can be a most important insurance policy. In the case of observational data, however, randomization may not be possible. Here, in the presence of correlated error terms, one may be able to modify the model. For instance, in the earlier discussion based on Figure 16.3, it was noted that inclusion in the model of a linear term for the learning effect of the analyst might remove the correlation of the error terms.

Modification of the model because of correlated error terms may also be necessary in experimental studies. In one case, the experimenter asked each of 10 subjects to give ratings to four new flavors of a fruit syrup and to the standard flavor, on a scale from 0 to 100. She applied the single-factor analysis of variance model but found high degrees of correlation in the residuals for each subject. She thereupon modified her model to a repeated measures design model (Chapter 28). These models are intended for situations when the same subject is given each of the different treatments and differences between subjects are expected.

CITED REFERENCES

16.1. *MINITAB Reference Manual, Release 7*. State College, Pa.: Minitab, Inc., 1989.

16.2. Glaser, R. E. "Bartlett's Test of Homogeneity of Variances." In *Encyclopedia of Statistical Sciences,* vol. 1, ed. S. Kotz and N. L. Johnson. New York: John Wiley & Sons, 1982, pp. 189–91.

16.3. Scheffé, H. *The Analysis of Variance*. New York: John Wiley & Sons, 1959.

PROBLEMS

16.1. Refer to Figures 16.3 and 16.4. What feature of the residual sequence plots enables you to diagnose that in one case the error variance changes over time whereas in the other case the effect is of a different nature? Could you make a diagnosis about time effects from a residual dot plot?

16.2. A student proposed in class that deviations of the observations Y_{ij} around the estimated overall mean $\bar{Y}_{..}$ be plotted to assist in evaluating the aptness of ANOVA model (14.2). Would these deviations be helpful in studying the independence of the error terms? The constancy of the variance of the error terms? The normality of the error terms? Discuss.

16.3. A consultant discussing ANOVA applications stated: "Sometimes I find that treatment effects in an experiment do not show up through differences in the treatment means. Hence, it is important to compare the residual plots for the treatments." Later a member of the audience said: "I don't think I understood the reference to residual plots." Explain.

16.4. Refer to **Productivity improvement** Problem 14.10. The residuals are as follows:

					j							
i	1	2	3	4	5	6	7	8	9	10	11	12
1 Low	.72	1.32	−.08	−1.08	.02	−.28	−.58	.82	−.88			
2 Moderate	−1.43	−.03	1.27	.47	−.33	−.43	.77	−.23	.17	.57	−1.03	.27
3 High	−.70	.50	.90	−1.40	.40	.30						

a. Prepare aligned residual dot plots by factor level. What departures from ANOVA model (14.2) can be studied from these plots? What are your findings?

b. Prepare a normal probability plot of the residuals. Also obtain the coefficient of correlation between the ordered residuals and their expected values under normality. Does the normality assumption appear to be reasonable here?

c. The economist wishes to investigate whether location of the firm's home office is related to productivity improvement. The home office locations are as follows (U: U.S.; E: Europe):

	j											
i	1	2	3	4	5	6	7	8	9	10	11	12
1	U	E	E	E	E	U	U	U	U			
2	E	E	E	E	U	U	U	U	U	E	E	E
3	E	U	E	U	U	E						

Make residual dot plots in which the location of the home office is identified. Does it appear that ANOVA model (14.2) could be improved by adding location of home office as a second factor? Explain.

16.5. Refer to **Questionnaire color** Problem 14.11. The residuals are as follows:

	j				
i	1	2	3	4	5
1	−1.4	−3.4	1.6	−2.4	5.6
2	4.4	−.6	−4.6	1.4	−.6
3	3.0	−3.0	−1.0	1.0	0.0

a. Prepare aligned residual dot plots by color. What departures from ANOVA model (14.2) can be studied from these plots? What are your findings?

b. Prepare a normal probability plot of the residuals. Also obtain the coefficient of correlation between the ordered residuals and their expected values under normality. Does the normality assumption appear to be reasonable here?

c. The observations within each factor level are in geographic sequence. Prepare residual sequence plots. What can be studied from these plots? What are your findings?

d. Calculate the standardized residuals (16.2). Obtain the centered intervals within which approximately 50 percent and 90 percent of the standardized residuals should fall if the error terms are normally distributed with constant variance. What are the actual percents of residuals within these intervals? Are these results consistent with those in part (b)?

16.6. Refer to **Rehabilitation therapy** Problem 14.12.

a. Obtain the residuals and prepare aligned residual dot plots by factor level. What departures from ANOVA model (14.2) can be studied from these plots? What are your findings?

b. Prepare a normal probability plot of the residuals. Also obtain the coefficient of correlation between the ordered residuals and their expected values under normality. Does the normality assumption appear to be reasonable here?

c. The observations within each factor level are in time order. Prepare residual sequence plots and analyze them. What are your findings?

d. Calculate the standardized residuals (16.2). Obtain the centered intervals within which approximately 50 percent and 90 percent of the standardized residuals should fall if the error terms are normally distributed with constant variance. What are the actual percents of the residuals within these intervals? Are these results consistent with those in part (b)?

16.7. Refer to **Cash offers** Problem 14.13.

a. Obtain the residuals and prepare aligned residual dot plots by factor level. What departures from ANOVA model (14.2) can be studied from these plots? What are your findings?

b. Prepare a normal probability plot of the residuals. Also obtain the coefficient of correlation between the ordered residuals and their expected values under normality. Does the normality assumption appear to be reasonable here?

c. The observations within each factor level are in time order. Prepare residual sequence plots and analyze them. What are your findings?

d. An executive in the consumer organization has been told that used-car dealers in the region tend to make lower cash offers during weekends (Friday evening through Sunday) than at other times. The times when offers were obtained are as follows (W: weekend; O: other time):

							j					
i	1	2	3	4	5	6	7	8	9	10	11	12
1	O	O	W	O	W	O	W	O	W	O	W	W
2	O	W	W	O	W	O	W	O	O	W	W	O
3	O	W	O	W	O	O	O	W	W	W	O	W

Make residual dot plots in which the time of the offer is identified. Does it appear that ANOVA model (14.2) could be improved by adding time of offer as a second factor? Explain.

16.8. Refer to **Filling machines** Problem 14.14.

a. Obtain the residuals and prepare aligned residual dot plots by machine. What departures from ANOVA model (14.2) can be studied from these plots? What are your findings?

b. Prepare a normal probability plot of the residuals. Also obtain the coefficient of correlation between the ordered residuals and their expected values under normality. Does the normality assumption appear to be reasonable here?

c. The observations within each factor level are in time order. Prepare residual sequence plots and analyze them. What are your findings?

16.9. Refer to **Premium distribution** Problem 14.15.

a. Obtain the residuals and prepare aligned residual dot plots by agent. What departures from ANOVA model (14.2) can be studied from these plots? What are your findings?

b. Prepare a normal probability plot of the residuals. Also obtain the coefficient of correlation between the ordered residuals and their expected values under normality. Does the normality assumption appear to be reasonable here?

c. The observations within each factor level are in time order. Prepare residual sequence plots and analyze them. What are your findings?

16.10. Computerized game. Four teams competed in 20 trials of a computerized business game. Each trial involved a new game, the objective for each team being to maximize profits in the given trial. A researcher fitted ANOVA model (14.2) to determine whether or not the mean profits for the four teams are the same and obtained the following residuals:

					j					
i	1	2	3	4	5	6	7	8	9	10
1	.10	.28	.10	.47	.83	.65	.28	−.08	.10	−.26
2	−1.44	−1.44	−1.12	−1.28	−.95	−.62	−.29	−.46	−.29	.03
3	−.93	−.70	−.81	−.59	−.25	−.36	−.14	.00	−.14	.09
4	−.15	.11	.25	−.02	−.15	−.29	−.02	.11	.38	.25

					j					
i	11	12	13	14	15	16	17	18	19	20
1	−.45	−.63	−1.00	−.63	−.45	−.08	.10	.10	.28	.28
2	.20	.20	.36	.69	.85	1.02	.85	1.02	1.18	1.51
3	.20	.43	.31	.09	.20	.43	.54	.54	.43	.65
4	−.02	−.42	−.29	−.42	−.15	−.15	.25	.11	.25	.38

The residuals for each team are given in time order. Construct appropriate residual plots to study whether the error terms are independent from trial to trial for each team. What are your findings?

16.11. Helicopter service. An operations analyst in a sheriff's department studied how frequently their emergency helicopter was used during a recent 20-day period, by time of day (shift 1: 2 A.M.–8 A.M.; shift 2: 8 A.M.–2 P.M.; shift 3: 2 P.M.–8 P.M.; shift 4: 8 P.M.–2 A.M.). The data are as follows (in time order):

												j								
i	1	2	3	4	5	6	7	8	9	10	11	12	13	14	15	16	17	18	19	20
1	4	3	5	4	6	3	2	5	7	1	2	5	4	7	4	5	0	4	1	6
2	0	2	0	3	2	1	0	3	1	0	0	1	1	0	1	3	1	2	2	0
3	2	1	0	3	4	1	3	4	2	0	1	3	2	4	0	1	3	0	2	4
4	5	2	4	4	6	5	3	5	7	3	1	0	2	3	3	4	1	5	2	3

Since the data involved are counts, the analyst was concerned about the normality and equal variances assumptions of ANOVA model (14.2).

a. Obtain suitable residual plots to study whether or not the error variances are equal for the four shifts. What are your findings?

b. For each shift, calculate $\bar{Y}_{i.}$ and s_i. Examine whether relation (16.15), (16.16), or (16.17) is most appropriate here. What do you conclude?

c. Obtain the standardized residuals (16.3) and construct a normal probability plot. Also obtain the coefficient of correlation between the ordered standardized residuals and their expected values under normality. Does the normality assumption appear to be reasonable here?

d. The analyst decided to apply the square root transformation (16.15). Obtain the transformed data $Y' = \sqrt{Y}$, and then calculate the residuals.

e. Prepare suitable plots of the residuals obtained in part (d) to study the equality of the error variances for the four shifts. Also obtain a normal probability plot and the coefficient of correlation between the ordered residuals and their expected values under normality. What are your findings?

16.12. Winding speeds. In an experiment to study the effect of the speed of winding thread (1: slow; 2: normal; 3: fast; 4: maximum) onto 75-yard spools, 16 runs of 10,000 spools each were made at each of the four winding speeds. The dependent variable is the number of thread breaks during the production run. The results (in time order) are as follows:

								j								
i	1	2	3	4	5	6	7	8	9	10	11	12	13	14	15	16
1	4	3	2	3	4	4	3	6	5	4	2	4	4	2	3	4
2	7	6	4	6	7	2	9	5	5	9	3	8	6	4	7	6
3	12	6	14	12	10	9	12	17	7	6	12	11	6	13	10	14
4	17	15	7	20	13	11	16	25	11	24	18	21	16	19	9	23

Since the data involved are counts, the researcher was concerned about the normality and equal variances assumptions of ANOVA model (14.2).

a. Obtain suitable residual plots to study whether or not the error variances are equal for the four winding speeds. What are your findings?

b. For each winding speed, calculate $\bar{Y}_{i.}$ and s_i. Examine whether relation (16.15), (16.16), or (16.17) is most appropriate here. What do you conclude?

c. Obtain the standardized residuals (16.3) and construct a normal probability plot. Also obtain the coefficient of correlation between the ordered standardized residuals and their expected values under normality. Does the normality assumption appear to be reasonable here?

d. The researcher decided to apply the logarithmic transformation (16.16). Obtain the transformed data $Y' = \log_{10} Y$, and then calculate the residuals.

e. Prepare suitable plots of the residuals obtained in part (d) to study the equality of the error variances for the four winding speeds. Also obtain a normal probability plot and the coefficient of correlation between the ordered residuals and their expected values under normality. What are your findings?

16.13. Use reciprocal interpolation to find the following percentiles:
   a. $F(.95; 3, 360)$.
   b. $F(.99; 200, 4)$.
   c. $F(.90; 400, 500)$.

16.14. Refer to **Productivity improvement** Problem 14.10. Some additional calculational results are:

i:	1	2	3
s_i:	.8136	.7572	.8672

Assume that the error terms are approximately normally distributed.

a. Examine by means of the Bartlett test whether or not the treatment error variances are equal; use $\alpha = .05$. State the alternatives, decision rule, and conclusion. What is the P-value of the test?

b. Would you reach the same conclusion as in part (a) with the modified Bartlett test?

c. Would the tests in parts (a) and (b) be appropriate if the distributions of the error terms were far from normal?

16.15. Refer to **Rehabilitation therapy** Problem 14.12. Assume that the distributions of the error terms are approximately normal.

a. Examine by means of the Bartlett test whether or not the treatment error variances are equal; use $\alpha = .10$. State the alternatives, decision rule, and conclusion. What is the P-value of the test?

b. Would you reach the same conclusion as in part (a) with the modified Bartlett test?

c. Would the tests in parts (a) and (b) be appropriate if the distributions of the error terms were far from normal?

16.16. Refer to **Cash offers** Problem 14.13. Assume that the error terms are approximately normally distributed.

a. Examine by means of the Bartlett test whether or not the treatment error variances are equal; use $\alpha = .01$. State the alternatives, decision rule, and conclusion. What is the P-value of the test?

b. Would you reach the same conclusion as in part (a) with the Hartley test?

16.17. Refer to **Filling machines** Problem 14.14. Assume that the error terms are approximately normally distributed.

a. Examine by means of the Bartlett test whether or not the treatment error variances are equal; use $\alpha = .01$. State the alternatives, decision rule, and conclusion. What is the P-value of the test?

b. Would you reach the same conclusion as in part (a) with the Hartley test statistic?

16.18. Refer to **Helicopter service** Problem 16.11. Assume that the error terms are approximately normally distributed.

a. For the untransformed data, test by means of the Bartlett test whether or not the treatment error variances are equal; use $\alpha = .10$. What is the P-value of the test? Are your results consistent with the diagnosis in Problem 16.11a?

b. Repeat the Bartlett test for the transformed data in Problem 16.11d. What are your findings now?

16.19. Refer to **Winding speeds** Problem 16.12. Assume that the error terms are approximately normally distributed.

a. For the untransformed data, test by means of the Hartley test whether or not the treatment error variances are equal; use $\alpha = .05$. What is the P-value of the test? Are your results consistent with the diagnosis in Problem 16.12a?

b. Repeat the Hartley test for the transformed data in Problem 16.12d. What are your findings now?

EXERCISES

16.20. Refer to Figure 16.3. Modify ANOVA model (14.2) to include a linear trend term for the time effect. Is this modified model still an ANOVA model? A linear model?

16.21. (Calculus needed.) Use (16.22) to find the appropriate transformation when: (1) $\sigma_i = k\mu_i$, (2) $\sigma_i = k\mu_i^2$.

PROJECTS

16.22. Refer to the **SENIC** data set and Project 14.33.
   a. Obtain the residuals and prepare aligned residual dot plots by region. Are any serious departures from ANOVA model (14.2) suggested by your plots?
   b. Obtain a normal probability plot of the residuals and calculate the coefficient of correlation between the ordered residuals and their expected values under normality. Is the normality assumption reasonable here?
   c. Assume that the error terms are approximately normally distributed. Examine by means of the Bartlett test whether or not the geographic region error variances are equal; use $\alpha = .05$. State the alternatives, decision rule, and conclusion. What is the P-value of the test?

16.23. Refer to the **SENIC** data set. A test of whether or not mean length of stay (variable 2) is the same in the four geographic regions (variable 9) is desired but concern exists about the normality and equal variances assumptions of ANOVA model (14.2).
   a. Obtain the residuals and plot them against the fitted values to study whether or not the error variances are equal for the four geographic regions. What are your findings?
   b. For each geographic region, calculate $\bar{Y}_{i.}$ and s_i. Examine whether relation (16.15), (16.16), or (16.17) is the most appropriate one here. What do you conclude?
   c. Use the reciprocal transformation (16.17) to obtain transformed data $Y' = 1/Y$.
   d. Obtain the residuals when ANOVA model (14.2) is fitted to the transformed data. Plot these residuals against the fitted values to study the equality of the error variances for the four regions. Also obtain a normal probability plot of the residuals and the coefficient of correlation between the ordered residuals and their expected values under normality. What are your findings?
   e. Assume that the transformed error terms are approximately normally distributed. Examine by means of the Bartlett test whether or not the geographic region variances are equal; use $\alpha = .01$. State the alternatives, decision rule, and conclusion. What is the P-value of the test?
   f. Assume that ANOVA model (14.2) is appropriate for the transformed data Y'. Test whether or not the mean length of stay is the same in the four geographic regions. Control the α risk at .01. State the alternatives, decision rule, and conclusion. What is the P-value of the test?

16.24. Refer to the **SMSA** data set and Project 14.35.
   a. Obtain the residuals and prepare aligned residual dot plots by region. Are any serious departures from ANOVA model (14.2) suggested by your plots?

    b. Obtain a normal probability plot of the residuals and calculate the coefficient of correlation between the ordered residuals and their expected values under normality. Is the normality assumption reasonable here?

    c. Assume that the error terms are approximately normally distributed. Examine by means of the Bartlett test whether or not the geographic region error variances are equal; use $\alpha = .01$. State the alternatives, decision rule, and conclusion. What is the P-value of the test?

Chapter 17

Planning Sample Sizes, Nonparametric Tests, and Random ANOVA Model

In this chapter, we discuss the planning of sample sizes for analysis of variance studies, which is an integral part of the design of such studies. We also discuss some alternative tests to the F test for deciding whether or not the treatment means are equal. Finally, we present ANOVA model II for single-factor analysis of variance, which is appropriate when the treatment levels are a random sample from a larger population.

17.1 PLANNING OF SAMPLE SIZES WITH POWER APPROACH

Design of ANOVA Studies

For analysis of variance studies, as for other statistical studies, it is important to plan the sample sizes so that needed protection against both Type I and Type II errors can be obtained, or so that the estimates of interest have sufficient precision to be useful. This planning is necessary to ensure that the sample sizes are large enough to detect important differences with high probability. At the same time, the sample sizes should not be so large that the cost of the study becomes excessive and that unimportant differences become statistically significant with high probability. Planning of sample sizes is therefore an integral part of the design of an analysis of variance study.

We shall generally assume in this discussion that all factor levels are to have equal sample sizes, reflecting that they are about equally important. Indeed, if major interest lies in pairwise comparisons of all factor level means, it can be shown that equal sample sizes maximize the precision of the various comparisons. Another reason for equal sample sizes is that certain departures from the assumed ANOVA model are not troublesome if all factor levels have the same sample size, as was noted in Section 16.4. There will be times, however, when unequal sample sizes are appropriate. For instance, when four experimental treatments are each to be compared to a control, it may be reasonable to make the sample size for the control larger. We shall comment later on the planning of sample sizes for such a case.

Planning of sample sizes can be approached in terms of controlling (1) the risks of making Type I and Type II errors, (2) the widths of desired confidence intervals, or (3) a combination of these two. In this section, we consider planning of sample sizes with the power approach, which permits controlling the risks of making Type I and Type II errors. First, we need to consider the power of the F test.

Power of F Test

By the power of the F test, we refer to the probability that the decision rule will lead to conclusion H_a when in fact H_a holds. Specifically, the power is given by the following expression:

(17.1) $$\text{Power} = P\{F^* > F(1 - \alpha; r - 1, n_T - r) | \phi\}$$

where ϕ is the *noncentrality parameter*, that is, a measure of how unequal the μ_i are:

(17.1a) $$\phi = \frac{1}{\sigma} \sqrt{\frac{\Sigma n_i(\mu_i - \mu_.)^2}{r}}$$

and:

(17.1b) $$\mu_. = \frac{\Sigma n_i \mu_i}{n_T}$$

When all factor level samples are of equal size n, the parameter ϕ becomes:

(17.2) $$\phi = \frac{1}{\sigma} \sqrt{\frac{n}{r} \Sigma (\mu_i - \mu_.)^2} \qquad \text{when } n_i \equiv n$$

where:

(17.2a) $$\mu_. = \frac{\Sigma \mu_i}{r}$$

To determine power probabilities, we need to utilize the noncentral F distribution since this is the sampling distribution of F^* when H_a holds. The resulting calculations are quite complex but charts have been prepared that make the determination of power probabilities relatively simple. Table A.8 contains the Pearson-Hartley charts of the power of the F test. The proper curve to utilize depends on the number of factor levels, the sample sizes, and the level of significance employed in the decision rule. The Pearson-Hartley charts are used as follows:

1. Each page refers to a different ν_1, the number of degrees of freedom for the numerator of F^*. For ANOVA model (14.2), $\nu_1 = r - 1$, or the number of factor

levels minus one. Table A.8 contains charts for $\nu_1 = 2, 3, 4, 5,$ and 6 as shown in the upper left-hand corner of each chart.

2. Two levels of significance, denoted by α, are used in the charts, namely, $\alpha = .05$ and $\alpha = .01$. There are two X scales, depending on which level of significance is employed. Further, the left set of curves on each chart refers to $\alpha = .05$ and the right set to $\alpha = .01$.

3. There are separate curves for different values of ν_2, the degrees of freedom for the denominator of F^*. For ANOVA model (14.2), $\nu_2 = n_T - r$. The curves are indexed according to the value of ν_2 at the top of the chart. Since only selected values of ν_2 are used in the charts, one needs to interpolate for intermediate values of ν_2.

4. The X scale represents ϕ, the noncentrality parameter defined in (17.1a).

5. Finally, the Y scale gives the power $1 - \beta$, where β is the risk of making a Type II error.

Examples

1. Consider the case where $\nu_1 = 2$, $\nu_2 = 10$, $\phi = 3$, and $\alpha = .05$. We then find from Table A.8 (p. 1143) that the power is $1 - \beta = .983$ approximately.

2. Suppose that for the Kenton Food Company example in Chapter 14, the analyst wishes to know the power of the decision rule on page 548 when there are substantial differences between the factor level means. Specifically, she wishes to consider the case when $\mu_1 = 12.5$, $\mu_2 = 13$, $\mu_3 = 18$, and $\mu_4 = 21$. The weighted mean in (17.1b) therefore is:

$$\mu_. = \frac{2(12.5) + 3(13) + 3(18) + 2(21)}{10} = 16$$

Thus, the specified value of ϕ is:

$$\phi = \frac{1}{\sigma}\left[\frac{2(-3.5)^2 + 3(-3)^2 + 3(2)^2 + 2(5)^2}{4}\right]^{1/2}$$

$$= \frac{1}{\sigma}(5.33)$$

Note that we still need to know σ, the standard deviation of the error terms ε_{ij} in the model. Suppose that from past experience it is known that $\sigma = 2.5$ cases approximately. Then we have:

$$\phi = \frac{1}{2.5}(5.33) = 2.13$$

Further, we have for this example:

$$\nu_1 = r - 1 = 3$$

$$\nu_2 = n_T - r = 6$$

$$\alpha = .05$$

Hence, from the chart on page 1144 we find that the power is $1 - \beta = .72$ approximately. In other words, there are about 72 chances in 100 that the decision rule will lead to the detection of differences in the mean sales volumes for the four package designs when the differences are the ones specified earlier.

Comments

1. Any given value of ϕ encompasses many different combinations of factor level means μ_i. Thus, in the Kenton Food example, $\mu_1 = 12.5$, $\mu_2 = 13$, $\mu_3 = 18$, $\mu_4 = 21$ and $\mu_1 = 21$, $\mu_2 = 18$, $\mu_3 = 13$, $\mu_4 = 12.5$ lead to the same value of $\phi = 2.13$ and hence to the same power.

2. The larger ϕ, that is, the larger the differences between the factor level means, the higher the power and hence the smaller the probability of making a Type II error for a given risk α of making a Type I error. Also, the smaller the specified α risk, the smaller is the power for any given ϕ, and hence the larger the risk of a Type II error.

3. Since many single-factor studies are undertaken because of the expectation that the factor level means differ and it is desired to investigate these differences, the α risk used in constructing the decision rule for determining whether or not the factor level means are equal is often set relatively high (e.g., .05 or .10 instead of .01) so as to increase the power of the test.

4. The Pearson-Hartley chart for $\nu_1 = 1$ is not reproduced in Table A.8 since this case corresponds to the comparison of two population means. As was noted earlier, the F test is the equivalent of the two-sided t test for this case, and the power charts for the two-sided t test presented in Table A.5 can then be used, with noncentrality parameter:

$$(17.3) \qquad \delta = \frac{|\mu_1 - \mu_2|}{\sigma\sqrt{\dfrac{1}{n_1} + \dfrac{1}{n_2}}}$$

and degrees of freedom $n_1 + n_2 - 2$.

Use of Table A.10

One procedure for implementing the power approach in planning sample sizes is to use the power charts for the F test presented in Table A.8. A trial-and-error process is required, however, with these charts. Alternatively, tables are available that furnish the appropriate sample sizes directly. In Table A.10, we present sample size determinations that are applicable when *all factor levels are to have equal sample sizes*—that is, when $n_i \equiv n$.

The planning of sample sizes using Table A.10 is done in terms of the noncentrality parameter (17.2) for equal sample sizes. However, instead of requiring a direct specification of the levels of μ_i for which it is important to control the risk of making a Type II error, Table A.10 only requires a specification of the minimum range of factor level means for which it is important to detect differences between the μ_i with high probability. This minimum range is denoted by Δ:

$$(17.4) \qquad \Delta = \max(\mu_i) - \min(\mu_i)$$

The following three specifications need to be made in using Table A.10:

1. The level α at which the risk of making a Type I error is to be controlled.

2. The magnitude of the minimum range Δ of the μ_i which it is important to detect with high probability. The magnitude of σ, the standard deviation of the probability distributions of Y, must also be specified since entry into Table A.10 is in terms of the ratio:

$$(17.5) \qquad\qquad \frac{\Delta}{\sigma}$$

3. The level β at which the risk of making a Type II error is to be controlled for the specification given in 2. Entry into Table A.10 is in terms of the power $1 - \beta$.

When using Table A.10, four α levels are available at which the risk of making a Type I error can be controlled ($\alpha = .2, .1, .05, .01$). The Type II error risk can be controlled at one of four β levels ($\beta = .3, .2, .1, .05$) through the specification on the power $1 - \beta$. Table A.10 provides necessary sample sizes for studies consisting of $r = 2, \ldots, 10$ factor levels or treatments.

Example. A company owning a large fleet of trucks wishes to determine whether or not four different brands of snow tires have the same mean tread life (in thousands of miles). It is important to conclude that the four brands of snow tires have different mean tread lives when the difference between the means of the best and worst brands is 3 (thousand miles) or more. Thus, the minimum range specification is $\Delta = 3$. It is known from past experience that the standard deviation of the tread lives of these tires is 2 (thousand miles), approximately. Management would like to control the risks of making incorrect decisions at the following levels:

$$\alpha = .05$$

$$\beta = .10 \quad \text{or} \quad \text{Power} = 1 - \beta = .90$$

Entering Table A.10 for $\Delta/\sigma = 3/2 = 1.5$, $\alpha = .05$, $1 - \beta = .90$, and $r = 4$, we find $n = 14$. Hence, 14 snow tires of each brand need to be tested in order to control the risks of making incorrect decisions at the desired levels.

Specification of Δ/σ Directly. Table A.10 can also be used when the minimum range is specified directly in units of the standard deviation σ. Let the specification of Δ in this case be $k\sigma$ so that we have by (17.5):

$$\frac{\Delta}{\sigma} = \frac{k\sigma}{\sigma} = k$$

Hence, Table A.10 is entered directly for the specified value k with this approach.

Example. Suppose it is specified in the snow tires example that it is important to detect differences between the mean tread lives if the range of the mean tread lives is $k = 2$ standard deviations or more. Suppose also that the other specifications are:

$$\alpha = .10$$

$$\beta = .05 \quad \text{or} \quad \text{Power} = 1 - \beta = .95$$

From Table A.10, we find for $k = 2$ and $r = 4$ that $n = 9$ tires will need to be tested for each brand in order that the specified risk protection will be achieved.

Note

While specifying Δ/σ directly does not require an advance planning value of the standard deviation σ, this is not of as much advantage as it might seem because a meaningful specification of Δ in units of σ will frequently require knowledge of the magnitude of the standard deviation.

Some Further Comments

1. The exact specification of Δ/σ has great effect on the sample sizes n when Δ/σ is small, but it has much less effect when Δ/σ is large. For instance, when $r = 3$, $\alpha = .05$, and $\beta = .10$, we have from Table A.10:

Δ/σ	n
1.0	27
1.5	13
2.0	8
2.5	6

Thus, unless Δ/σ is quite small, one need not be too concerned about some imprecision in specifying Δ/σ.

2. Reducing either the specified α or β risks or both increases the required sample sizes. For instance, when $r = 4$, $\alpha = .10$, and $\Delta/\sigma = 1.25$, we have:

β	$1 - \beta$	n
.20	.80	13
.10	.90	16
.05	.95	20

3. A moderate error in the advance planning value of σ can cause a substantial miscalculation of needed sample sizes, although the needed sample sizes will still be "in the same ball park." For instance, when $r = 5$, $\alpha = .05$, $\beta = .10$, and $\Delta = 3$, we have:

σ	Δ/σ	n
1	3.0	5
2	1.5	15
3	1.0	32

In view of the usual approximate nature of the advance planning value of σ, it is generally desirable to investigate the needed sample sizes for a range of likely values of σ before deciding on the sample sizes to be employed.

4. Table A.10 is based on the noncentrality parameter ϕ in (17.2) even though no specification is made of the individual factor level means μ_i for which it is important to conclude that the factor level means differ. Consider again the snow tires example where $r = 4$ brands are to be tested and a minimum range of $\Delta = 3$ (thousand miles) of the four mean tread lives μ_i is to be detected with high probability. The following are possible sets of values of the μ_i, each of which has range $\Delta = 3$:

Case	μ_1	μ_2	μ_3	μ_4	$\Sigma(\mu_i - \mu_.)^2$
1	24	27	25	26	5.00
2	25	25	26	23	4.75
3	25	25	25	28	6.75
4	25	25	26.5	23.5	4.50

The term $\Sigma(\mu_i - \mu_.)^2$ of the noncentrality parameter ϕ in (17.2) differs for each of these four possibilities and hence the power differs, even though the range is the same in all cases. Note that the term $\Sigma(\mu_i - \mu_.)^2$ is smallest for case 4, where two factor level means are at $\mu_.$ and the other two are equally spaced around $\mu_.$. It can be shown that for a given range Δ, the term $\Sigma(\mu_i - \mu_.)^2$ is minimized when all but two factor level means are at $\mu_.$ and the two remaining factor level means are equally spaced around $\mu_.$. Thus, we have:

$$(17.6) \qquad \min \sum_{i=1}^{r} (\mu_i - \mu_.)^2 = \left(\frac{\Delta}{2}\right)^2 + \left(-\frac{\Delta}{2}\right)^2 + 0 + \cdots + 0 = \frac{\Delta^2}{2}$$

Since the power of the test varies directly with $\Sigma(\mu_i - \mu_.)^2$, use of (17.6) in calculating Table A.10 assures that the power is at least $1 - \beta$ for any combination of μ_i values with range Δ.

17.2 PLANNING OF SAMPLE SIZES WITH ESTIMATION APPROACH

The estimation approach to planning sample sizes may be used either in conjunction with the control of Type I and Type II errors or by itself. The essence of the approach is to specify the major comparisons of interest and to determine the expected widths of the confidence intervals for various sample sizes, given an advance planning value for the standard deviation σ. The approach is iterative, starting with an initial judgment of needed sample sizes. This initial judgment may be based on the needed sample sizes to control the risks of Type I and Type II errors when these have been previously obtained. If the anticipated widths of the confidence intervals based on the initial sample sizes are satisfactory, the iteration process is terminated. If one or more widths are too great, larger sample sizes should be tried next. If the widths are unnecessarily tight, smaller sample sizes should be tried next. This process is continued until those sample sizes are found that yield satisfactory anticipated widths.

Example

We are to plan sample sizes for the snow tires example by means of the estimation approach, given that the sample sizes for each tire brand are to be equal, that is, $n_i \equiv n$. Management has indicated it wishes three types of estimates:

1. A comparison of the mean tread lives for each pair of brands:

$$\mu_i - \mu_{i'}$$

2. A comparison of the mean tread lives for the two high-priced brands (1 and 4) and the two low-priced brands (2 and 3):

$$\frac{\mu_1 + \mu_4}{2} - \frac{\mu_2 + \mu_3}{2}$$

3. A comparison of the mean tread lives for the national brands (1, 2, and 4) and the local brand (3):

$$\frac{\mu_1 + \mu_2 + \mu_4}{3} - \mu_3$$

Management further has indicated that it wishes a family confidence coefficient of .95 for the entire set of comparisons.

We first need a planning value for the standard deviation of the tread lives of tires. Suppose that from past experience we judge the standard deviation to be approximately $\sigma = 2$ (thousand miles). Next, we require an initial judgment of needed sample sizes and shall consider $n = 10$ as a starting point.

We know from (15.18) that the variance of an estimated contrast $\hat{L}$ when $n_i \equiv n$ is:

$$\sigma^2\{\hat{L}\} = \frac{\sigma^2}{n} \sum c_i^2 \quad \text{when } n_i \equiv n$$

Hence, given $\sigma = 2$ and $n = 10$, our anticipations of the standard deviations of the required estimators are:

Contrast	Anticipated Variance	Anticipated Standard Deviation
Pairwise comparisons	$\frac{(2)^2}{10}[(1)^2 + (-1)^2] = .80$	.89
High- and low-priced brands	$\frac{(2)^2}{10}\left[\left(\frac{1}{2}\right)^2 + \left(\frac{1}{2}\right)^2 + \left(-\frac{1}{2}\right)^2 + \left(-\frac{1}{2}\right)^2\right] = .40$	.63
National and local brands	$\frac{(2)^2}{10}\left[\left(\frac{1}{3}\right)^2 + \left(\frac{1}{3}\right)^2 + \left(\frac{1}{3}\right)^2 + (-1)^2\right] = .53$	.73

We shall employ the Scheffé multiple comparison procedure and therefore require the Scheffé multiple S in (15.34a) for $r = 4$, $n_T = nr = 10(4) = 40$, and $1 - \alpha = .95$:

$$S^2 = (r - 1)F(1 - \alpha; r - 1, n_T - r) = 3F(.95; 3, 36) = 3(2.87) = 8.61$$

or $S = 2.93$. Hence, the anticipated widths of the confidence intervals are:

Contrast	Anticipated Width of Confidence Interval $= \pm S\sigma\{\hat{L}\}$
Pairwise comparisons	$\pm 2.93(.89) = \pm 2.61$ (thousand miles)
High- and low-priced brands	$\pm 2.93(.63) = \pm 1.85$ (thousand miles)
National and local brands	$\pm 2.93(.73) = \pm 2.14$ (thousand miles)

Management was satisfied with these anticipated widths. However, it was decided to increase the sample sizes from 10 to 15 in case the actual standard deviation of the tread lives of tires is somewhat greater than the anticipated value $\sigma = 2$ (thousand miles).

Comments

1. Since one cannot usually be certain that the planning value for the standard deviation is correct, it is advisable to study a range of values for the standard deviation before making a decision on sample size.

2. If the sample sizes for the factor levels are to be unequal, one can still use the iterative procedure with the estimation approach just described. For instance, suppose that four new flavors of a fruit syrup are each to be compared with a control, the present flavor. One may therefore wish to make the control sample size larger in order to increase the precision of these key comparisons. Suppose the control sample size is to be twice as large as the other factor level sample sizes. Then we can represent the control sample size as $2n$ and the other sample sizes as n. We then proceed as before with an initial specification for n, and utilize formula (15.18) for the variance of an estimated contrast.

17.3 PLANNING OF SAMPLE SIZES TO FIND "BEST" TREATMENT

There are occasions when the chief purpose of the study is to ascertain the factor level or treatment with the highest or lowest mean μ_i. In the snow tires example, for instance, it may be desired to determine which of the four brands has the longest mean tread life.

Bechhofer has developed a procedure, on which Table A.11 is based, that enables us to determine the necessary sample sizes so that with probability $1 - \alpha$ the highest (lowest) estimated factor level mean $\bar{Y}_{i\cdot}$ is from the factor level with the highest (lowest) population mean μ_i. We need to specify the probability $1 - \alpha$, the standard deviation σ, and the smallest difference λ between the highest (lowest) and second highest (second lowest) factor level means that it is important to recognize. Table A.11 assumes that *equal sample sizes are to be used for all factor levels*, and that analysis of variance model (14.2) is appropriate.

Example

Suppose that in the snow tires example, the chief objective is to identify the brand with the longest mean tread life. There are $r = 4$ brands. We anticipate, as before, that $\sigma = 2$ (thousand miles). Further, we are informed that a difference $\lambda = 1$ (thousand miles) between the highest and second highest brand means is important to recognize, and that the probability is to be $1 - \alpha = .90$ or greater that we identify correctly the brand with the highest mean tread life when $\lambda \geq 1$.

The entry in Table A.11 is $\lambda \sqrt{n}/\sigma$. For $r = 4$ and probability $1 - \alpha = .90$, we find from Table A.11 that $\lambda \sqrt{n}/\sigma = 2.4516$. Hence, since the λ specification is $\lambda = 1$, we obtain:

$$\frac{(1)\sqrt{n}}{2} = 2.4516$$

$$\sqrt{n} = 4.9032 \quad \text{or} \quad n = 25$$

Thus, when the mean tread life for the best brand exceeds that of the second best by at least 1 (thousand miles) and when $\sigma = 2$ (thousand miles), sample sizes of 25 tires for each brand provide an assurance of at least .90 that the brand with the highest estimated mean $\overline{Y}_{i.}$ is the brand with the highest population mean.

Note

If the planning value for the standard deviation is not accurate, the probability of identifying the population with the highest (lowest) mean correctly is, of course, affected. This is no different from the other approaches, where a misjudgment of the standard deviation affects the risks of making a Type II error or the widths of the confidence intervals actually obtained.

17.4 KRUSKAL-WALLIS RANK TEST

We have noted that the F test for the analysis of variance is robust against departures from normality, provided the departures are not extreme. In the occasional instances where the F test would not be appropriate because of major departures from normality, a nonparametric test may be employed. We shall now discuss two such tests, the Kruskal-Wallis test in this section and the median test in the next section.

Kruskal-Wallis Test Statistic

The Kruskal-Wallis test is based on the ranks of the observations. To test whether the treatment means are equal with this test, the only assumption required about the population distributions is that *they are continuous and of the same shape*. Thus, the population distributions must have the same variability, skewness, etc., but may differ as to the location of the mean. It is also assumed that the samples from the different populations are independent random ones.

First, all n_T observations Y_{ij} are ranked from 1 to n_T. Let $\bar{R}_{i.}$ be the mean of the ranks for the ith factor level and $\bar{R}_{..}$ the overall mean rank. The test statistic is then simply:

$$(17.7) \qquad X_{KW}^2 = \frac{SSTR}{\dfrac{SSTO}{n_T - 1}} = \frac{\displaystyle\sum_{i=1}^{r} n_i(\bar{R}_{i.} - \bar{R}_{..})^2}{\dfrac{n_T(n_T + 1)}{12}}$$

The numerator is the usual treatment sum of squares, but with the data expressed in ranks, and the denominator is the variance of the ranks 1, 2, 3, . . . , n_T. The test statistic X_{KW}^2 can be expressed equivalently as follows:

$$(17.7a) \qquad X_{KW}^2 = \left[\frac{12}{n_T(n_T + 1)} \sum_{i=1}^{r} n_i \bar{R}_{i.}^2 \right] - 3(n_T + 1)$$

If the n_i are reasonably large (5 or more is the usual advice), X_{KW}^2 is approximately a χ^2 random variable with $r - 1$ degrees of freedom when H_0 (all μ_i are equal) holds. Large values of X_{KW}^2, as expected, lead to H_a (not all μ_i are equal). Thus, in choosing between:

$$(17.8a) \qquad \begin{aligned} &H_0: \mu_1 = \mu_2 = \cdots = \mu_r \\ &H_a: \text{not all } \mu_i \text{ are equal} \end{aligned}$$

the appropriate decision rule for controlling the risk of making a Type I error at α is:

$$(17.8b) \qquad \begin{aligned} &\text{If } X_{KW}^2 \le \chi^2(1 - \alpha; r - 1), \text{ conclude } H_0 \\ &\text{If } X_{KW}^2 > \chi^2(1 - \alpha; r - 1), \text{ conclude } H_a \end{aligned}$$

Example. Servo-Data, Inc., operates mainframe computers at three different locations. The computers are identical as to make and model, but are subject to different degrees of voltage fluctuation in the power lines serving the respective installations. Management wishes to test whether or not the mean lengths of operating time between computer failures are the same for the three locations. The alternative conclusions are:

$$H_0: \mu_1 = \mu_2 = \mu_3$$

$$H_a: \text{not all } \mu_i \text{ are equal}$$

Table 17.1 contains the lengths of time between computer failures for the three locations, for five failure intervals each. Even though the sample sizes are small, the data suggest highly skewed distributions. In the same table, the data are ranked from 1 to 15 and the mean ranks are shown.

Using the computational formula (17.7a), we obtain for the test statistic:

$$X_{KW}^2 = \left\{ \frac{12}{15(16)} 5[(8.2)^2 + (4.8)^2 + (11.0)^2] \right\} - 3(16) = 4.8$$

TABLE 17.1   Times between Computer Failures at Three Locations (in hours)— Servo-Data Example

Location i		Failure Interval (j)					Mean $\bar{R}_{i.}$
		1	2	3	4	5	
A							
Time	Y_{1j}	105	3	90	217	22	
Rank	R_{1j}	11	2	10	14	4	8.2
B							
Time	Y_{2j}	56	43	1	37	14	
Rank	R_{2j}	8	7	1	5	3	4.8
C							
Time	Y_{3j}	183	144	219	86	39	
Rank	R_{3j}	13	12	15	9	6	11.0

$$\bar{R}_{..} = 8.0$$

A level of significance of $\alpha = .10$ has been specified. Since $r = 3$, we require $\chi^2(.90; 2)$. From Table A.3 we find $\chi^2(.90; 2) = 4.61$. The decision rule for choosing between H_0 and H_a therefore is:

$$\text{If } X^2_{KW} \leq 4.61, \text{ conclude } H_0$$

$$\text{If } X^2_{KW} > 4.61, \text{ conclude } H_a$$

Since $X^2_{KW} = 4.8 > 4.61$, we conclude H_a, that the mean times between computer failures differ for the three locations. The P-value of the test is $P\{\chi^2(2) > 4.8\} = .09$.

Comments

1. The Kruskal-Wallis test, like the ordinary F test, does not require equal sample sizes.

2. If the n_i are small so that the χ^2 approximation is not appropriate, special tables should be used for conducting the Kruskal-Wallis test; see, for instance, the tables by Owen in Reference 17.1.

3. In case of ties among some observations, each of the tied observations is given the mean of the ranks involved. Thus, if two observations are tied for what would otherwise have been the third and fourth ranked positions, each would be given the mean value 3.5. If a large number of ties exist, test statistic (17.7) needs to be modified.

4. The Kruskal-Wallis test can also be used to choose among the alternatives:

(17.9)

H_0: all populations are identical

H_a: all populations are not identical

This statement of the decision problem avoids the earlier assumption that all populations are identical except for the location of the means. If conclusion H_a in (17.9) is reached, however, one cannot identify the reason for the differences. For example, the means might differ, or the variances, or the nature of the skewness, or some combination of these.

F^* Test Statistic

Instead of the Kruskal-Wallis test statistic X_{KW}^2 which utilizes the chi-square distribution, another test statistic also based on ranked data that utilizes the F distribution is often applied. This alternative test statistic is:

$$(17.10) \qquad F^* = \frac{MSTR}{MSE}$$

When H_0 holds, this test statistic follows approximately the $F(r - 1, n_T - r)$ distribution for large sample sizes.

Let R_{ij} denote the rank of Y_{ij} when all observations are ranked from 1 to n_T. As defined before, $\bar{R}_{i.}$ is the mean of the ranks for the ith factor level, and $\bar{R}_{..}$ is the overall mean rank. For ranked data, we have:

$$(17.11) \qquad \bar{R}_{..} = \frac{n_T + 1}{2}$$

Test statistic (17.10) then can be expressed as follows:

$$(17.12) \qquad F^* = \frac{\sum n_i (\bar{R}_{i.} - \bar{R}_{..})^2}{r - 1} \div \frac{\sum\sum (R_{ij} - \bar{R}_{i.})^2}{n_T - r}$$

The decision rule to control the Type I error at α is the usual one:

$$(17.13)$$

If $F^* \leq F(1 - \alpha; r - 1, n_T - r)$, conclude H_0

If $F^* > F(1 - \alpha; r - 1, n_T - r)$, conclude H_a

Example. In the Servo-Data example, we calculate $SSTR$ and SSE as follows based on the rank data in Table 17.1:

$$SSTR = 5[(8.2 - 8.0)^2 + (4.8 - 8.0)^2 + (11.0 - 8.0)^2]$$

$$= 96.4$$

$$SSE = (11 - 8.2)^2 + (2 - 8.2)^2 + \cdots + (6 - 11.0)^2$$

$$= 183.6$$

The overall mean $\bar{R}_{..}$ here is $(15 + 1)/2 = 8.0$.

The F^* test statistic therefore is:

$$F^* = \frac{96.4}{3 - 1} \div \frac{183.6}{15 - 3}$$

$$= 3.15$$

For $\alpha = .10$, we require $F(.90; 2, 12) = 2.81$. Since $F^* = 3.15 > 2.81$, we conclude H_a, just as we did with the X_{KW}^2 test statistic. The P-value of the test here is .08, which is very similar to the P-value of .09 with the X_{KW}^2 test statistic.

Note

It can be shown that test statistic (17.10) is a simple function of the Kruskal-Wallis test statistic, namely:

$$F^* = \frac{(n_T - r)X^2_{KW}}{(r - 1)(n_T - 1 - X^2_{KW})}$$

Multiple Pairwise Testing Procedure

If the Kruskal-Wallis test leads to the conclusion that the factor level means μ_i are not equal, it is frequently desired to obtain information about the comparative magnitudes of these means. A large-sample testing analogue of the Bonferroni pairwise comparison procedure discussed in Section 15.5 based on the ranks of the observations may be employed for this purpose, provided that the sample sizes are not too small. Testing limits for all $g = r(r - 1)/2$ pairwise tests using the mean ranks $\bar{R}_{i.}$ are set up as follows for family level of significance α:

$$(17.14) \qquad (\bar{R}_{i.} - \bar{R}_{i'.}) \pm B\left[\frac{n_T(n_T + 1)}{12}\left(\frac{1}{n_i} + \frac{1}{n_{i'}}\right)\right]^{1/2}$$

where:

$$(17.14a) \qquad\qquad B = z(1 - \alpha/2g)$$

$$(17.14b) \qquad\qquad g = \frac{r(r - 1)}{2}$$

If the testing limits include zero, we conclude that the corresponding treatment means μ_i and $\mu_{i'}$ do not differ. If the testing limits do not include zero, we conclude that the two corresponding treatment means differ. Based on all pairwise tests, we then set up groups of treatment means whose members do not differ according to the simultaneous testing procedure. In this way, we obtain information about the comparative magnitudes of the treatment means μ_i.

Example.   For the Servo-Data study results in Table 17.1, we wish to ascertain, if possible, which location has the longest mean time between computer failures. We shall base the analysis on ranks in view of the serious departures from normality of the distributions of times between failures. For a family significance level of $\alpha = .10$ and $g = r(r - 1)/2 = 3(2)/2 = 3$ pairwise tests, we require $B = z(.9833) = 2.13$. Since all treatment sample sizes are equal, we need calculate the right term in (17.14) only once:

$$B\left[\frac{n_T(n_T + 1)}{12}\left(\frac{1}{n_i} + \frac{1}{n_{i'}}\right)\right]^{1/2} = 2.13\left[\frac{15(16)}{12}\left(\frac{1}{5} + \frac{1}{5}\right)\right]^{1/2} = 6.02$$

Hence, the testing limits for the three pairwise tests are:

$$\text{A and B:} \qquad (8.2 - 4.8) \pm 6.02 \qquad \text{or} \qquad -2.6 \text{ and } 9.4$$

$$\text{C and B:} \quad (11.0 - 4.8) \pm 6.02 \quad \text{or} \quad .2 \text{ and } 12.2$$

$$\text{C and A:} \quad (11.0 - 8.2) \pm 6.02 \quad \text{or} \quad -3.2 \text{ and } 8.8$$

The only test showing a significant difference is between locations B and C. Hence, we obtain two groupings:

Group 1	Group 2
Location A	Location A
Location B	Location C

These groupings indicate that the mean times between computer failures differ for locations B and C, but not for any other pairs of locations. Thus, the only conclusion about comparative magnitudes of the means possible from this analysis is that the mean time between computer failures for location C is larger than that for location B.

17.5 MEDIAN TEST

The median test is another test that may be utilized when populations are far from normally distributed. With this test, one is again interested in choosing between the two alternatives:

(17.15a)
$$H_0: \mu_1 = \mu_2 = \cdots = \mu_r$$
$$H_a: \text{not all } \mu_i \text{ are equal}$$

The median test assumes only that *all populations are of the same shape,* but they may differ as to the location of the mean. Also it is assumed that the samples from the different populations are independent random ones.

Median Test Statistic

All sample data are combined to determine the median value for the combined sample. For each treatment i, the number of observations above this median value (O_{i1}) and the number not above it (O_{i2}) are then ascertained. Finally, a test for homogeneity is conducted using the test statistic:

(17.15b)
$$X^2 = \sum_{i=1}^{r} \sum_{j=1}^{2} \frac{(O_{ij} - E_{ij})^2}{E_{ij}}$$

where O_{ij} $(i = 1, \ldots, r; j = 1, 2)$ is the observed frequency in a cell and E_{ij} is the expected frequency under H_0, when all populations are identical.

When the sample sizes are reasonably large, the test statistic X^2 is distributed approximately as χ^2 with $r - 1$ degrees of freedom when H_0 holds. Large values of X^2

lead to conclusion H_a. Hence, the appropriate decision rule for controlling the risk of making a Type I error at α is:

(17.15c)

$$\text{If } X^2 \leq \chi^2(1 - \alpha; r - 1), \text{ conclude } H_0$$

$$\text{If } X^2 > \chi^2(1 - \alpha; r - 1), \text{ conclude } H_a$$

Example

We consider again the Servo-Data example dealing with the times between computer failures for three different locations. Fifteen additional observations on times between failures were made for each location to provide more precise information. The median number of hours between failures for the combined sample of 60 observations was 64 hours. Table 17.2 summarizes the results for the three samples (the original data are not shown). The expected frequencies when all populations are identical are shown in parentheses in Table 17.2. These are obtained by allocating the total frequencies in each column to the three computers in proportion to the total number of observations for each computer. In our example, 20 observations were made on each computer; hence, the 30 frequencies above the median are allocated equally to each computer. Similarly, the 30 frequencies not above the median are allocated equally to each computer. These then are the frequencies that are expected if the three populations have the same mean and hence are identical.

The test statistic is calculated using (17.15b):

$$X^2 = \frac{(13 - 10)^2}{10} + \frac{(7 - 10)^2}{10} + \cdots + \frac{(6 - 10)^2}{10} = 14.8$$

It has been specified that the level of significance is to be $\alpha = .05$; hence we require $\chi^2(.95; 2) = 5.99$. The decision rule then is:

$$\text{If } X^2 \leq 5.99, \text{ conclude } H_0$$

$$\text{If } X^2 > 5.99, \text{ conclude } H_a$$

TABLE 17.2   Frequency of Times between Computer Failures at Three Locations—Servo-Data Example with Enlarged Samples

Location	Number of Observations				Total
	Above Median		Not above Median		
i	O_{i1}	(E_{i1})	O_{i2}	(E_{i2})	Total
A	13	(10)	7	(10)	20
B	3	(10)	17	(10)	20
C	14	(10)	6	(10)	20
Total	30		30		60

Median for combined sample = 64 hours

Since $X^2 = 14.8 > 5.99$, we conclude H_a, that the mean number of hours between computer failures is not the same for the three locations. The P-value of the test is $P\{\chi^2(2) > 14.8\} = .001$.

Note

Like the Kruskal-Wallis test, the median test does not require equal sample sizes.

17.6 ANOVA MODEL II—RANDOM FACTOR LEVELS

As we noted earlier, there are occasions when the employed factor levels or treatments are not of intrinsic interest in themselves but constitute a sample from a larger population of factor levels. ANOVA model II is designed for this type of situation. Consider, for instance, Apex Enterprises, a company that builds roadside restaurants carrying one of several promoted trade names, leases franchises to individuals to operate the restaurants, and provides management services. This company employs a large number of personnel officers who interview applicants for jobs in the restaurants. At the end of an interview, the personnel officer assigns a subjective rating between 0 and 100 to indicate the applicant's potential value on the job. Suppose now that five personnel officers were selected at random and each was assigned four candidates at random. In this case, the company would not wish to make inferences concerning the five personnel officers who happened to be selected but rather about the population of all personnel officers. Questions of interest might include: How great is the variation in ratings between all personnel officers? What is the mean rating by all personnel officers?

The distinction between this situation, for which ANOVA model II is designed, and one where fixed ANOVA model I is appropriate can be seen readily by modifying our example slightly. If a smaller company had only five personnel officers who were all included in the study and interest is limited to these five officers, ANOVA model I would be relevant since the factor levels (the five personnel officers) are then not considered a sample from a larger population. A repetition of the experiment for the smaller company would involve the same five personnel officers, but in the case of the large company, a repetition would involve a new random sample of five personnel officers that likely would consist of different officers.

Random Cell Means Model

ANOVA model II for single-factor analysis of variance is as follows:

(17.16) $$Y_{ij} = \mu_i + \varepsilon_{ij}$$

where:

μ_i are independent $N(\mu_., \sigma_\mu^2)$
ε_{ij} are independent $N(0, \sigma^2)$
μ_i and ε_{ij} are independent random variables
$i = 1, \ldots, r; j = 1, \ldots, n_i$

ANOVA model (17.16) is similar in appearance to the fixed ANOVA model (14.2). The main distinction is that the factor level means μ_i are constants for ANOVA model I but are random variables for ANOVA model II. Hence, ANOVA model II is often called a *random* ANOVA model. Note that ANOVA model (17.16) is the cell means version.

Meaning of Model Terms.   We shall explain the meaning of the model terms with reference to the personnel officers in the Apex Enterprises example. The term μ_i corresponds in this example to the mean of all ratings by the ith personnel officer if he or she interviewed all prospective employees. The expected value of μ_i is $\mu_.$. Thus, $\mu_.$ represents in this example the mean rating for all prospective employees by all personnel officers. The variability of the μ_i is measured by the variance σ_μ^2. The more the different personnel officers vary in their mean ratings (for instance, some may rate consistently higher than others), the greater will be σ_μ^2. On the other hand, if all personnel officers rate at the same mean level, all μ_i will be equal to $\mu_.$ and then $\sigma_\mu^2 = 0$.

The term ε_{ij} represents the variation associated with the differing potential values of different prospective employees. Note that ANOVA model (17.16) assumes that all ε_{ij} have the same variance σ^2. This means that the distributions of ratings for prospective employees by the different personnel officers are assumed to have the same variability. The distributions for the different personnel officers may differ, however, with respect to the mean levels of the distributions.

Figure 17.1 illustrates ANOVA model II. On the top is shown the distribution of the μ_i, which is normal. A number of μ_i (two in the illustration) are selected at random from this distribution. Each in turn leads to a distribution of $Y_{ij} = \mu_i + \varepsilon_{ij}$, which are all normal distributions with the same variance. A number of Y_{ij} observations (two each in the illustration) are then selected from each of these distributions.

Important Features of Model

1. The expected value of an observation Y_{ij} is:

$$(17.17a) \qquad E\{Y_{ij}\} = \mu_.$$

because we have by (17.16):

$$E\{Y_{ij}\} = E\{\mu_i\} + E\{\varepsilon_{ij}\}$$
$$= \mu_. + 0$$
$$= \mu_.$$

Note that this expectation averages over the selections of both μ_i and ε_{ij}.

2. The variance of Y_{ij}, to be denoted by σ_Y^2, is:

$$(17.17b) \qquad \sigma^2\{Y_{ij}\} = \sigma_Y^2 = \sigma_\mu^2 + \sigma^2$$

Thus, all observations Y_{ij} have the same variance. The result in (17.17b) follows because ANOVA model II assumes that μ_i and ε_{ij} are independent random variables, and $\sigma^2\{\mu_i\} = \sigma_\mu^2$ and $\sigma^2\{\varepsilon_{ij}\} = \sigma^2$ according to ANOVA model (17.16). Because the

FIGURE 17.1 Representation of ANOVA Model II

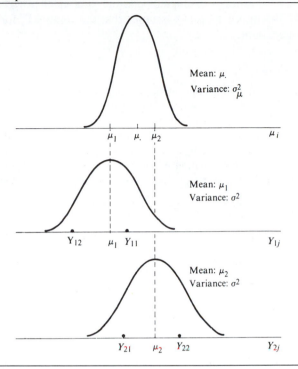

variance of Y in this model is the sum of two components, σ_μ^2 and σ^2, this model is sometimes called a *components of variance* model.

3. The Y_{ij} are normally distributed because they are linear combinations of the independent normal variables μ_i and ε_{ij}.

4. Unlike for the fixed ANOVA model I where all observations Y_{ij} are independent, the Y_{ij} for random ANOVA model II are only independent if they pertain to different factor levels. It can be shown that the covariance of any two observations Y_{ij} and $Y_{ij'}$ for the same factor level i with random ANOVA model (17.16) is:

$$(17.17c) \qquad\qquad \sigma\{Y_{ij}, Y_{ij'}\} = \sigma_\mu^2 \qquad j \neq j'$$

Thus, random ANOVA model (17.16) assumes that the covariance between any two observations for the same factor level is constant for all factor levels.

The reason why any two observations for the same factor level are correlated is that, in advance of the random trials, the observations are expected to be similar because they will both have the same random component μ_i and will differ only because of the error terms ε_{ij}.

Once the factor levels have been selected, however, random ANOVA model (17.16) assumes that any two observations for the same factor level are independent because the factor level mean μ_i is then fixed and the two observations differ only because of the error terms ε_{ij} which are assumed to be independent. Thus, in the Apex Enterprises example, once the personnel officers have been selected, random ANOVA model (17.16) assumes that the ratings Y_{ij} for a given personnel officer are independent.

Note

At times, the population of the μ_i may be relatively small and should be treated as a finite population. This can be done, but we do not discuss this case here. If the population of the μ_i is finite but large, little is lost in treating it as an infinite population. We did this, in fact, in our illustration of the personnel officers. The number of officers is finite, but since there are many, we treated the population of the μ_i as an infinite one. Thus, there are two basic situations when the population of the μ_i is treated as infinite—when the population is finite but large, and when interest centers in the underlying *process* generating the μ_i.

Questions of Interest

When ANOVA model II is appropriate, one is usually not interested in inferences about the particular μ_i included in the study, such as which is the largest or smallest, but rather in inferences about the entire population of the μ_i. Specifically, interest often centers on the mean of the μ_i, $\mu.$, and on the variability of the μ_i, measured by σ_μ^2. In the Apex Enterprises example, for instance, management would not ordinarily be as interested in the mean ratings of the five personnel officers who happened to be included in the study as in the mean rating by all personnel officers and in the effect of variability among all personnel officers.

While σ_μ^2 is a direct measure of the variability of the μ_i, the effect of this variability is often measured more meaningfully by the ratio:

$$(17.18) \qquad \frac{\sigma_\mu^2}{\sigma_\mu^2 + \sigma^2}$$

Note the following characteristics of this ratio:

1. The ratio takes on values between 0 (when $\sigma^2 = \infty$) and 1 (when $\sigma^2 = 0$).
2. The denominator is σ_Y^2 according to (17.17b).
3. In view of properties 1 and 2, the ratio measures the proportion of the total variability of the Y_{ij} that is accounted for by the variability of the μ_i.

With reference to the Apex Enterprises example, the denominator of the ratio measures the variability of ratings for all candidates by all personnel officers, and the numerator measures the variability of the mean ratings by all personnel officers. The ratio then measures the proportion of the total variability of ratings that is accounted for by differences among the personnel officers. If the ratio is near zero, differences among personnel officers are relatively insignificant. On the other hand, if the ratio is large, say, .5 or more, then much of the total variability is accounted for by differences between personnel officers, and management may wish to study the advisability of giving the personnel officers more training to improve uniformity of ratings between officers.

Note

It can be shown that the coefficient of correlation between any two observations for the same factor level with random ANOVA model (17.16) is:

$$(17.19) \qquad \rho\{Y_{ij}, Y_{ij'}\} = \frac{\sigma_\mu^2}{\sigma_\mu^2 + \sigma^2}$$

Thus, the measure in (17.18) is actually the coefficient of correlation between any two observations for the same factor level, which here indicates the proportion of the total variability of the Y_{ij} that is accounted for by the variability of the μ_i.

The result in (17.19) follows from the definition of the coefficient of correlation in (13.7):

$$\rho\{Y_{ij}, Y_{ij'}\} = \frac{\sigma\{Y_{ij}, Y_{ij'}\}}{\sigma\{Y_{ij}\}\sigma\{Y_{ij'}\}}$$

The covariance is given in (17.17c), and $\sigma\{Y_{ij}\} = \sigma\{Y_{ij'}\} = \sigma_Y$ is given in (17.17b).

Test Whether $\sigma_\mu^2 = 0$

We first consider how to decide between:

$$(17.20) \qquad \begin{aligned} H_0: \sigma_\mu^2 &= 0 \\ H_a: \sigma_\mu^2 &> 0 \end{aligned}$$

H_0 implies that all μ_i are equal, that is, $\mu_i \equiv \mu_.$. H_a implies that the μ_i differ. For the personnel officers example, H_0 implies that the mean ratings for all personnel officers are equal, while H_a implies that they differ.

Despite the fact that ANOVA model II differs from ANOVA model I, the analysis of variance for a single-factor study is conducted in identical fashion. (This is not always the case in more complex situations.) The difference between the two models appears in the expected mean squares. It can be shown in a similar manner to that employed in our derivation for ANOVA model I that, for ANOVA model II:

$$(17.21) \qquad E\{MSE\} = \sigma^2$$

$$(17.22) \qquad E\{MSTR\} = \sigma^2 + n'\sigma_\mu^2$$

where:

$$(17.22a) \qquad n' = \frac{1}{r-1}\left[\left(\sum n_i\right) - \frac{\sum n_i^2}{\sum n_i}\right]$$

If all $n_i = n$, then $n' = n$.

It is clear from (17.21) and (17.22) that if $\sigma_\mu^2 = 0$, *MSE* and *MSTR* have the same expectation σ^2. Otherwise, $E\{MSTR\} > E\{MSE\}$ since $n' > 0$ always. Hence, large values of the test statistic:

$$(17.23) \qquad F^* = \frac{MSTR}{MSE}$$

will lead to conclusion H_a in (17.20). Since F^* again follows the F distribution when H_0 holds, the decision rule for controlling the risk of making a Type I error at α is the same one as for ANOVA model I:

$$(17.24) \qquad \begin{aligned} &\text{If } F^* \leq F(1 - \alpha; r - 1, n_T - r), \text{ conclude } H_0 \\ &\text{If } F^* > F(1 - \alpha; r - 1, n_T - r), \text{ conclude } H_a \end{aligned}$$

Example. Table 17.3 contains the results of a study by Apex Enterprises on the evaluation ratings of potential employees by its personnel officers. Five personnel officers were selected at random, and each was assigned at random four prospective employees. The ANOVA calculations are routine and were done by a computer program package. The results are shown in Table 17.4, which also shows the expected mean squares in general and for this particular example.

Using the data from Table 17.4, the appropriate test statistic is:

$$F^* = \frac{370}{75.6} = 4.89$$

Assuming that we are to control the risk of making a Type I error at $\alpha = .05$, we require $F(.95; 4, 15) = 3.06$. Hence, the decision rule is:

If $F^* \leq 3.06$, conclude H_0

If $F^* > 3.06$, conclude H_a

Since $F^* = 4.89 > 3.06$, we conclude H_a, that $\sigma_\mu^2 > 0$ or that the mean ratings of the personnel officers differ. The *P*-value of the test is .01.

TABLE 17.3 Ratings by Five Personnel Officers—Apex Enterprises Example

Officer	Candidate (j)				
i	1	2	3	4	*Mean*
A	76	64	85	75	$\bar{Y}_{1.} = 75$
B	58	75	81	66	$\bar{Y}_{2.} = 70$
C	49	63	62	46	$\bar{Y}_{3.} = 55$
D	74	71	85	90	$\bar{Y}_{4.} = 80$
E	66	74	81	79	$\bar{Y}_{5.} = 75$
Mean					$\bar{Y}_{..} = 71$

TABLE 17.4 ANOVA Table for Single-Factor ANOVA Model II—Apex Enterprises Example

Source of Variation	SS	df	MS	E{MS} General	E{MS} Example
Between personnel officers	$SSTR = 1,480$	4	$MSTR = 370$	$\sigma^2 + n'\sigma_\mu^2$	$\sigma^2 + 4\sigma_\mu^2$
Error (within personnel officers)	$SSE = 1,134$	15	$MSE = 75.6$	σ^2	σ^2
Total	$SSTO = 2,614$	19			

$$n' = \frac{1}{r-1}\left[\left(\sum n_i\right) - \frac{\sum n_i^2}{\sum n_i}\right]$$
$$(n' = n \text{ if all } n_i = n)$$

Note

We shall illustrate the derivation of an expected mean square for ANOVA model II by sketching the development for deriving $E\{MSTR\}$ in (17.22) when $n_i \equiv n$. The proof parallels that for ANOVA model I. By model (17.16), we can write:

$$\bar{Y}_{i.} = \mu_i + \bar{\varepsilon}_{i.}$$
$$\bar{Y}_{..} = \bar{\mu}_. + \bar{\varepsilon}_{..}$$

where $\bar{\varepsilon}_{i.}$ and $\bar{\varepsilon}_{..}$ are defined in (14.42) and (14.45), respectively, and:

$$\bar{\mu}_. = \frac{\sum\mu_i}{r}$$

(Note the use of a different notation for the mean of the μ_i here than for ANOVA model I to emphasize the random nature of the mean for ANOVA model II.) Corresponding to (14.47), we obtain:

$$\bar{Y}_{i.} - \bar{Y}_{..} = (\mu_i - \bar{\mu}_.) + (\bar{\varepsilon}_{i.} - \bar{\varepsilon}_{..})$$

so that:

$$\sum(\bar{Y}_{i.} - \bar{Y}_{..})^2 = \sum(\mu_i - \bar{\mu}_.)^2 + \sum(\bar{\varepsilon}_{i.} - \bar{\varepsilon}_{..})^2 + 2\sum(\mu_i - \bar{\mu}_.)(\bar{\varepsilon}_{i.} - \bar{\varepsilon}_{..})$$

When we take the expectation, the cross-product term drops out because of the independence of the μ_i and ε_{ij} and because the deviations $\mu_i - \bar{\mu}_.$ and $\bar{\varepsilon}_{i.} - \bar{\varepsilon}_{..}$ all have expectation zero. From (14.50) we know that:

$$E\left\{\sum(\bar{\varepsilon}_{i.} - \bar{\varepsilon}_{..})^2\right\} = \frac{(r-1)\sigma^2}{n}$$

Lastly, since $\Sigma(\mu_i - \bar{\mu}_.)^2$ is the numerator of an ordinary sample variance for r independent μ_i observations, it follows from the unbiasedness of the sample variance that:

$$E\left\{\sum (\mu_i - \bar{\mu}_.)^2\right\} = (r - 1)\sigma_\mu^2$$

Hence, we obtain:

$$E\left\{\frac{n}{r - 1} \sum (\bar{Y}_{i.} - \bar{Y}_{..})^2\right\} = \frac{n}{r - 1}\left[(r - 1)\sigma_\mu^2 + \frac{r - 1}{n}\sigma^2\right] = n\sigma_\mu^2 + \sigma^2$$

which is the result in (17.22) for the case $n_i \equiv n$.

Estimation of $\mu_.$

When ANOVA model II is applicable, one is frequently interested in estimating the overall mean $\mu_.$. We shall assume in developing an interval estimate for $\mu_.$ that *all factor level sample sizes are equal*, that is, $n_i \equiv n$.

We know from (17.17a) that:

$$E\{Y_{ij}\} = \mu_.$$

Hence, an unbiased estimator of $\mu_.$ is:

(17.25) $$\hat{\mu}_. = \bar{Y}_{..}$$

It can be shown that the variance of this estimator is:

(17.26) $$\sigma^2\{\bar{Y}_{..}\} = \frac{\sigma_\mu^2}{r} + \frac{\sigma^2}{n_T} = \frac{n\sigma_\mu^2 + \sigma^2}{n_T}$$

(Remember that $n_T = rn$ here.)

Formula (17.26) shows that the variance of $\bar{Y}_{..}$ is made up of two components. The first corresponds to the variance of a sample mean based on r observations when sampling from the population of the μ_i, and it reflects the contribution due to sampling the factor levels. The second component corresponds to the variance of a sample mean based on n_T observations when sampling from the populations of the Y_{ij}, given the μ_i, and it reflects the contribution due to variation within factor levels.

An unbiased estimator of $\sigma^2\{\bar{Y}_{..}\}$ is:

(17.27) $$s^2\{\bar{Y}_{..}\} = \frac{MSTR}{n_T}$$

This estimator is unbiased because we know from (17.22) that when $n_i \equiv n$:

(17.28) $$E\{MSTR\} = n\sigma_\mu^2 + \sigma^2$$

(Remember that $n' = n$ when $n_i \equiv n$.) Dividing the result in (17.28) by n_T yields (17.26).

It can be shown that:

(17.29) $\dfrac{\bar{Y}_{..} - \mu_.}{s\{\bar{Y}_{..}\}}$ is distributed as $t(r-1)$ for ANOVA model (17.16) when $n_i \equiv n$

Hence, we obtain in usual fashion the confidence limits for $\mu_.$:

(17.30) $$\bar{Y}_{..} \pm t(1-\alpha/2; r-1)s\{\bar{Y}_{..}\}$$

Example. Management of Apex Enterprises wishes to estimate the mean rating for all prospective employees by all personnel officers with a 90 percent confidence interval. We have from Tables 17.3 and 17.4:

$$\bar{Y}_{..} = 71 \qquad MSTR = 370 \qquad n_T = 20$$

We require $t(.95; 4) = 2.132$ and:

$$s^2\{\bar{Y}_{..}\} = \frac{370}{20} = 18.5$$

Hence, $s\{\bar{Y}_{..}\} = 4.301$, the confidence limits are $71 \pm 2.132(4.301)$, and the desired 90 percent confidence interval is:

$$62 \le \mu_. \le 80$$

Thus, with a 90 percent confidence coefficient, we conclude that the mean rating assigned by all personnel officers to all prospective employees is between 62 and 80. The interval estimate is not too precise because of the relatively small sample sizes of personnel officers and potential employees.

Note

The variance of $\bar{Y}_{..}$ in (17.26) can be derived readily. First, we consider:

$$\bar{Y}_{i.} = \mu_i + \bar{\varepsilon}_{i.}$$

where $\bar{\varepsilon}_{i.}$ is defined in (14.42). Because of the independence of μ_i and the ε_{ij}, we have:

$$\sigma^2\{\bar{Y}_{i.}\} = \sigma_\mu^2 + \frac{\sigma^2}{n}$$

Remember that $\bar{\varepsilon}_{i.}$ is just an ordinary mean of n independent ε_{ij} observations.

For the case $n_i \equiv n$ which we are considering here, we have:

$$\bar{Y}_{..} = \frac{\sum\limits_{i=1}^{r} \bar{Y}_{i.}}{r}$$

In view of the independence of the μ_i and the ε_{ij} among themselves and between each other, it follows that the $\bar{Y}_{i.}$ are independent so that:

$$\sigma^2\{\bar{Y}_{..}\} = \frac{\sigma^2\{\bar{Y}_{i.}\}}{r} = \frac{\sigma_\mu^2}{r} + \frac{\sigma^2}{rn} = \frac{n\sigma_\mu^2 + \sigma^2}{n_T}$$

Estimation of $\sigma_\mu^2/(\sigma_\mu^2 + \sigma^2)$

As noted earlier, the ratio $\sigma_\mu^2/(\sigma_\mu^2 + \sigma^2)$ reveals meaningfully the effect of the extent of variation between the μ_i. To develop an interval estimate for this ratio, we shall assume that *all factor level sample sizes are equal*, that is, $n_i \equiv n$.

We begin by obtaining confidence limits for the ratio σ_μ^2/σ^2. First, we need to note that *MSTR* and *MSE* are independent random variables for ANOVA model II, just as for ANOVA model I. When $n_i \equiv n$, it can be shown further that:

$$(17.31) \qquad \frac{MSTR}{n\sigma_\mu^2 + \sigma^2} \div \frac{MSE}{\sigma^2} \sim F(r - 1, n_T - r) \qquad \text{when } n_i \equiv n$$

Hence, we can write the probability statement:

$$(17.32) \qquad P\left\{F(\alpha/2; r - 1, n_T - r) \le \frac{MSTR}{n\sigma_\mu^2 + \sigma^2} \div \frac{MSE}{\sigma^2}\right.$$

$$\left. \le F(1 - \alpha/2; r - 1, n_T - r)\right\} = 1 - \alpha$$

Rearranging the inequalities, we obtain the following confidence limits L and U for σ_μ^2/σ^2:

$$(17.33a) \qquad L = \frac{1}{n}\left[\frac{MSTR}{MSE}\left(\frac{1}{F(1 - \alpha/2; r - 1, n_T - r)}\right) - 1\right]$$

$$(17.33b) \qquad U = \frac{1}{n}\left[\frac{MSTR}{MSE}\left(\frac{1}{F(\alpha/2; r - 1, n_T - r)}\right) - 1\right]$$

where L is the lower confidence limit and U the upper.

The confidence limits L^* and U^* for $\sigma_\mu^2/(\sigma_\mu^2 + \sigma^2)$ can now be readily obtained and are as follows:

$$(17.34) \qquad L^* = \frac{L}{1 + L} \qquad U^* = \frac{U}{1 + U}$$

Example. Management of Apex Enterprises wishes a 90 percent confidence interval for $\sigma_\mu^2/(\sigma_\mu^2 + \sigma^2)$. From previous work, we have:

$$MSTR = 370 \qquad MSE = 75.6 \qquad n = 4 \qquad r = 5 \qquad n_T = 20$$

For a 90 percent confidence interval, we require:

$$F(.05; 4, 15) = .170 \qquad F(.95; 4, 15) = 3.06$$

Hence, the 90 percent confidence limits for σ_μ^2/σ^2 are by (17.33):

$$L = \frac{1}{4}\left[\frac{370}{75.6}\left(\frac{1}{3.06}\right) - 1\right] = .15 \qquad U = \frac{1}{4}\left[\frac{370}{75.6}\left(\frac{1}{.170}\right) - 1\right] = 6.9$$

and the confidence interval for σ_μ^2/σ^2 is:

$$.15 \leq \frac{\sigma_\mu^2}{\sigma^2} \leq 6.9$$

Finally, the confidence limits for $\sigma_\mu^2/(\sigma_\mu^2 + \sigma^2)$ are by (17.34) $L^* = .15/1.15 = .13$ and $U^* = 6.9/7.9 = .87$ so that the 90 percent confidence interval is:

$$.13 \leq \frac{\sigma_\mu^2}{\sigma_\mu^2 + \sigma^2} \leq .87$$

Hence, with confidence coefficient .90, we conclude that the variability of the mean ratings for the different personnel officers accounts for somewhere between 13 and 87 percent of the total variance of the ratings. Note that this interval estimate is not very precise. The reason is the relatively small sample sizes. The confidence interval does indicate, though, that the variability among personnel officers is not negligible since it accounts for at least 13 percent of the total variability.

Comments

1. It may happen occasionally that the lower limit of the confidence interval for σ_μ^2/σ^2 is negative. Since this ratio cannot be negative, the usual practice is to consider the lower limit L in (17.33a) to be zero in that case.

2. If one-sided or two-sided tests concerning the relative magnitudes of σ_μ^2 and σ^2 are desired, such as the following (where c is a specified constant):

$$H_0: \sigma_\mu^2 \leq c\sigma^2 \qquad H_0: \sigma_\mu^2 = c\sigma^2$$

$$H_a: \sigma_\mu^2 > c\sigma^2 \qquad H_a: \sigma_\mu^2 \neq c\sigma^2$$

the decision rule can be constructed by utilizing (17.31). Alternatively, one-sided or two-sided confidence intervals can be set up from which the appropriate conclusion can be drawn. For instance, suppose that for the Apex Enterprises example we are considering:

$$H_0: \sigma_\mu^2 = \frac{1}{2}\sigma^2$$

$$H_a: \sigma_\mu^2 \neq \frac{1}{2}\sigma^2$$

Since the 90 percent confidence interval for σ_μ^2/σ^2 above (corresponding to a level of significance of .10) includes .5, the conclusion to be reached is H_0.

3. The ratio σ_μ^2/σ^2 is of relevance when planning investigations. In the Apex Enterprises example dealing with the personnel officers, suppose that the mean rating $\mu.$ is to be estimated, and that the costs of including in the study a personnel officer and a candidate are c_1 and c_2 respectively. For a given total budget C, the ratio σ_μ^2/σ^2 is the determining variable for finding the optimum balance between the number of personnel officers and the number of candidates to include in the study so as to minimize the variance of the estimator. If the populations are not large, the model will need to take account of their finite nature.

Estimation of σ^2 and σ_μ^2

At times, interest exists in estimating σ^2 and σ_μ^2 separately. According to (17.21), an unbiased estimator of σ^2 is:

$$(17.35) \qquad \hat{\sigma}^2 = MSE$$

A confidence interval for σ^2 can be obtained in the usual fashion by means of (1.68); here, the degrees of freedom will be $n_T - r$.

An unbiased point estimator of σ_μ^2 is also available. Since by (17.21) and (17.22), we have:

$$E\{MSE\} = \sigma^2$$

$$E\{MSTR\} = \sigma^2 + n'\sigma_\mu^2$$

it follows that:

$$(17.36) \qquad \hat{\sigma}_\mu^2 = \frac{MSTR - MSE}{n'}$$

is an unbiased estimator of σ_μ^2. Occasionally, this point estimator will turn out to be negative. Since a variance cannot be negative, the usual practice is to consider the point estimator to be zero in that event. Only approximate confidence intervals for σ_μ^2 are available. These are discussed in Reference 17.2.

Example.   For the Apex Enterprises example, a 90 percent confidence interval for σ^2 requires:

$$MSE = 75.6 \qquad \chi^2(.05; 15) = 7.26 \qquad \chi^2(.95; 15) = 25.0$$

Using (1.68), we find:

$$45.4 = \frac{15(75.6)}{25.0} \le \sigma^2 \le \frac{15(75.6)}{7.26} = 156.2$$

An unbiased point estimate of σ_μ^2 requires:

$$MSE = 75.6 \qquad MSTR = 370 \qquad n' = n = 4$$

Hence, by (17.36) we find:

$$\hat{\sigma}_\mu^2 = \frac{370 - 75.6}{4} = 73.6$$

Random Factor Effects Model

We can express the single-factor random cell means model (17.16) in an equivalent random factor effects fashion, just as we did for fixed factor levels in Chapter 14. We do this by expressing the factor level mean μ_i as a deviation from its expected value,

$E\{\mu_i\} = \mu_.$, as follows:

(17.37) $$\tau_i = \mu_i - \mu_.$$

Then, we simply replace μ_i in ANOVA model (17.16) by its equivalent expression from (17.37):

(17.38) $$\mu_i = \mu_. + \tau_i$$

The random factor effects model therefore is expressed as follows:

(17.39) $$Y_{ij} = \mu_. + \tau_i + \varepsilon_{ij}$$

where:

$\mu_.$ is a constant component common to all observations
τ_i are independent $N(0, \sigma_\mu^2)$
ε_{ij} are independent $N(0, \sigma^2)$
τ_i and ε_{ij} are independent
$i = 1, \ldots, r; j = 1, \ldots, n_i$

Note that the τ_i are random variables in ANOVA model (17.39). With reference to the personnel officers in the Apex Enterprises example, τ_i represents the effect of the ith personnel officer who is selected at random. Specifically, τ_i measures by how much the mean evaluation of all potential employees by the ith personnel officer differs from the mean evaluation by all personnel officers.

CITED REFERENCES

17.1. Owen, D. B. *Handbook of Statistical Tables*. Reading, Mass.: Addison-Wesley Publishing, 1962.

17.2. Scheffé, H. *The Analysis of Variance*. New York: John Wiley & Sons, 1959.

PROBLEMS

17.1. Refer to Example 1 on page 635. Find the power of the test if $\alpha = .01$, everything else remaining unchanged. How does this power compare with that in Example 1?

17.2. Refer to Example 2 on page 635. The analyst is also interested in the power of the test when $\mu_1 = \mu_2 = 13$ and $\mu_3 = \mu_4 = 18$. Assume that $\sigma = 2.5$.
a. Obtain the power of the test if $\alpha = .05$.
b. What would be the power of the test if $\alpha = .01$?

17.3. Refer to **Productivity improvement** Problem 14.10. Obtain the power of the test in Problem 14.10d if $\mu_1 = 7.0$, $\mu_2 = 8.0$, and $\mu_3 = 9.0$. Assume that $\sigma = .9$.

17.4. Refer to **Rehabilitation therapy** Problem 14.12. Obtain the power of the test in Problem 14.12d if $\mu_1 = 37$, $\mu_2 = 35$, and $\mu_3 = 28$. Assume that $\sigma = 4.5$.

17.5. Refer to **Cash offers** Problem 14.13. Obtain the power of the test in Problem 14.13d if the mean cash offers are $\mu_1 = 22$, $\mu_2 = 28$, and $\mu_3 = 22$. Assume that $\sigma = 1.6$.

17.6. A market researcher stated in a seminar: "The power approach to determining sample sizes for analysis of variance problems is not meaningful; only the estimation approach should be used. We never conduct a study where all treatment means are expected to be equal, so we are always interested in a variety of estimates." Discuss.

17.7. Why do you think that the approach to planning sample sizes to find the best treatment by means of Table A.11 does not consider the risk of an incorrect identification when the best two treatment means are the same or practically the same?

17.8. Consider a single-factor study where $r = 5$, $\alpha = .01$, $\beta = .05$, and $\sigma = 10$, and equal treatment sample sizes are desired by means of the approach in Table A.10.
   a. What are the required sample sizes if $\Delta = 10, 15, 20, 30$? What generalization is suggested by your results?
   b. What are the required sample sizes for the same values of Δ as in part (a) if $\alpha = .05$, all other specifications remaining the same? How do these sample sizes compare with those in part (a)?

17.9. Consider a single-factor study where $r = 6$, $\alpha = .05$, $\beta = .10$, and $\Delta = 50$, and equal treatment sample sizes are desired by means of the approach in Table A.10.
   a. What are the required sample sizes if $\sigma = 50, 25, 20$? What generalization is suggested by your results?
   b. What are the required sample sizes for the same values of σ as in part (a) if $r = 4$, all other specifications remaining the same? How do these sample sizes compare with those in part (a)?

17.10. Consider a single-factor study where $r = 5$, $1 - \alpha = .95$, and $\sigma = 20$, and equal sample sizes are desired by means of the approach in Table A.11.
   a. What are the required sample sizes if $\lambda = 20, 10, 5$? What generalization is suggested by your results?
   b. What are the required sample sizes for the same values of λ as in part (a) if $\sigma = 30$, all other specifications remaining the same? How do these sample sizes compare with those in part (a)?

17.11. Refer to **Questionnaire color** Problem 14.11. Suppose that the sample sizes have not yet been determined but it has been decided to sample the same number of supermarket parking lots for each questionnaire color. A reasonable planning value for the error standard deviation is $\sigma = 3.0$.
   a. What would be the required sample sizes if: (1) differences in the response rates are to be detected with probability .90 or more when the range of the treatment means is 4.5, and (2) the α risk is to be controlled at .05?
   b. If the sample sizes determined in part (a) were employed, what would be the minimum power of the test for treatment mean differences (using $\alpha = .05$) when the range of the treatment means is 6.0?
   c. Suppose pairwise comparisons are of primary importance. What would be the required sample sizes if the precision of all pairwise comparisons is to be ± 3.0, using the Tukey procedure with a 95 percent family confidence coefficient?
   d. Suppose the chief objective is to identify the color with the highest mean response rate. The probability should be at least .99 that the best color is recognized correctly when the difference between the response rates for the best and second best colors is 1.5 percent points or more. What are the required sample sizes?

17.12. Refer to **Rehabilitation therapy** Problem 14.12. Suppose that the sample sizes have not yet been determined but it has been decided to use the same number of patients for each physical fitness group. Assume that a reasonable planning value for the error standard deviation is $\sigma = 4.5$ days.

a. What would be the required sample sizes if: (1) differences in the mean times for the three physical fitness categories are to be detected with probability .80 or more when the range of the treatment means is 5.63 days, and (2) the α risk is to be controlled at .01?

b. If the sample sizes determined in part (a) were employed, what would be the power of the test for treatment mean differences when $\mu_1 = 37$, $\mu_2 = 32$, and $\mu_3 = 28$?

c. Suppose primary interest is in estimating the two pairwise comparisons:

$$D_1 = \mu_1 - \mu_2 \qquad D_2 = \mu_3 - \mu_2$$

What would be the required sample sizes if the precision of each comparison is to be ± 3.0 days, using the most efficient multiple comparison procedure with a 95 percent family confidence coefficient?

d. Suppose the chief objective is to identify the physical fitness group with the smallest mean required time for therapy. The probability should be at least .90 that the correct group is identified when the mean required time for the second best group differs by 2.0 days or more. What are the required sample sizes?

17.13 Refer to **Filling machines** Problem 14.14. Suppose that the sample sizes have not yet been determined but it has been decided to sample the same number of cartons for each filling machine. Assume that a reasonable planning value for the error standard deviation is $\sigma = .15$ ounce.

a. What would be the required sample sizes if: (1) differences in the mean amount of fill for the six filling machines are to be detected with probability .70 or more when the range of the treatment means is .15 ounce, and (2) the α risk is to be controlled at .05?

b. For the sample sizes determined in part (a), what would be the power of the test if $\mu_1 = .09$, $\mu_2 = .18$, $\mu_3 = .30$, $\mu_4 = .20$, $\mu_5 = .10$, and $\mu_6 = .20$?

c. Suppose primary interest is in estimating the following comparisons:

$$D_1 = \mu_1 - \mu_2 \qquad L_1 = \frac{\mu_1 + \mu_2}{2} - \frac{\mu_3 + \mu_4}{2}$$

$$D_2 = \mu_3 - \mu_4 \qquad L_2 = \frac{\mu_1 + \mu_2 + \mu_3 + \mu_4}{4} - \frac{\mu_5 + \mu_6}{2}$$

What would be the required sample sizes if the precision of each of these comparisons is not to exceed $\pm .08$ ounce, using the best multiple comparison procedure with a 95 percent family confidence coefficient?

d. Suppose the chief objective is to identify the filling machine with the smallest mean fill. The probability should be at least .95 that the filling machine with the smallest mean fill is recognized correctly when the filling machine with the next smallest amount of mean fill differs by .10 ounce or more. What are the required sample sizes?

17.14. Refer to **Premium distribution** Problem 14.15. Suppose that the sample sizes have not yet been determined but it has been decided to sample the same number of premium distributions for each agent. Assume that a reasonable planning value for the error standard deviation is $\sigma = 3.0$ days.

a. What would be the required sample sizes if: (1) differences in the mean time lapse for the five agents are to be detected with probability .95 or more when the range of the treatment means is 3.75 days, and (2) the α risk is to be controlled at .10?

b. Suppose primary interest is in estimating the following comparisons:

$$D_1 = \mu_1 - \mu_2 \qquad L_1 = \frac{\mu_1 + \mu_2}{2} - \mu_5$$

$$D_2 = \mu_3 - \mu_4 \qquad L_2 = \frac{\mu_1 + \mu_2}{2} - \frac{\mu_3 + \mu_4}{2}$$

What would be the required sample sizes if the precision of each of the estimated comparisons is not to exceed ± 1.0 day, using the most efficient multiple comparison procedure with a 90 percent family confidence coefficient?

c. Suppose the chief objective is to identify the best agent, i.e., the one with the smallest mean time lapse. The probability should be at least .90 that the best agent is recognized correctly when the mean time lapse for the second best agent differs by 1.0 day or more. What are the required sample sizes?

17.15. Refer to **Rehabilitation therapy** Problem 14.12. Suppose that primary interest is in comparing the below-average and above-average physical fitness groups, respectively, with the average physical fitness group. Thus, two comparisons are of interest:

$$D_1 = \mu_1 - \mu_2 \qquad D_2 = \mu_3 - \mu_2$$

Assume that a reasonable planning value for the error standard deviation is 4.5 days.

a. It has been decided to use equal sample sizes (n) for the below-average and above-average groups. If twice this sample size ($2n$) were to be used for the average physical fitness group, what would be the required sample sizes if the precision of each pairwise comparison is to be ± 2.5 days, using the Bonferroni procedure and a 90 percent family confidence coefficient?

b. Repeat the calculations in part (a) if the sample size for the average physical fitness group is to be: (1) n and (2) $3n$, all other specifications remaining the same.

c. Compare your results in parts (a) and (b). Which design leads to the smallest total sample size here?

17.16. Why are the Kruskal-Wallis rank test and the median test nonparametric tests?

17.17. Is there a basic distinction in the assumptions for the Kruskal-Wallis rank test and those for the median test? If so, what is it? If not, how should one make a choice between the two tests?

17.18. Explain why the limits in (17.14) are testing limits and not confidence limits.

17.19. Refer to **Productivity improvement** Problem 14.10.

a. Conduct the Kruskal-Wallis rank test; use $\alpha = .05$. State the alternatives, decision rule, and conclusion. Were ties a source of difficulty here?

b. What is the P-value of the test in part (a)?

c. Does the conclusion in part (a) differ from the one in Problem 14.10d?

d. Do the data suggest that a nonparametric test is needed here?

e. Conduct multiple pairwise tests based on the ranked data to group the three types of firms according to mean productivity improvement. Use a family level of significance of $\alpha = .10$. Describe your findings.

f. Conduct the rank test in part (a) by means of the F^* test statistic (17.10). Is the P-value for this test similar to that for the Kruskal-Wallis test in part (b)?

17.20. Telephone communications. A management consultant was engaged by a firm to improve the cost effectiveness of its communications. As part of his study, the consultant selected 10 home-office executives at random from each of the (1) sales, (2) production, and (3) research and development divisions, and studied the communications of these executives during the past 10 weeks in great detail. Among other data, he obtained the following information on weekly dollar costs of long-distance telephone calls to branch offices by the executives:

					j					
i	1	2	3	4	5	6	7	8	9	10
1	666	920	495	602	1,499	960	796	343	894	813
2	488	362	156	546	216	542	345	291	516	126
3	391	450	609	910	705	472	645	496	763	1,309

The consultant decided to employ a nonparametric approach to test whether or not the mean telephone expenses for the three divisions are equal.

a. What feature of the data may have suggested the use of a nonparametric test?

b. Conduct the Kruskal-Wallis rank test, controlling the risk of Type I error at $\alpha = .05$. State the alternatives, decision rule, and conclusion. What is the P-value of the test?

c. Conduct multiple pairwise tests based on the ranked data to group the three divisions according to mean telephone expenditures; use a family level of significance of $\alpha = .05$. Describe your findings.

d. Conduct the rank test in part (b) by means of the F^* test statistic (17.10). Is the P-value for this test similar to that for the Kruskal-Wallis test in part (b)?

17.21. Refer to **Telephone communications** Problem 17.20. Suppose that in conducting the Kruskal-Wallis rank test the alternatives had been:

$$H_0: \text{all populations are identical}$$

$$H_a: \text{all populations are not identical}$$

a. Would the same test assumptions be involved as in Problem 17.20?

b. If H_a were concluded, would it necessarily imply that the mean telephone expenses are not the same in the three divisions? Explain.

17.22. Battery life. A special field version of a laboratory instrument was developed, powered by a battery. Four different designs of the battery were tested. Data on the number of operating hours in the field for 20 batteries of each type follow.

					j					
i	1	2	3	4	5	6	7	8	9	10
1	7.48	10.08	3.81	13.22	10.19	8.31	13.27	4.85	5.71	11.71
2	10.86	23.41	6.45	17.42	12.36	11.52	14.08	7.07	14.82	8.41
3	6.70	8.12	9.68	5.35	16.40	8.60	7.14	16.70	6.35	22.41
4	12.40	4.99	6.74	4.59	9.09	6.31	4.21	7.34	11.74	3.57

					j					
i	11	12	13	14	15	16	17	18	19	20
1	6.52	2.66	2.25	19.37	4.19	8.03	6.71	3.08	3.37	6.24
2	9.06	11.00	6.53	7.53	7.10	9.21	15.37	17.07	10.17	8.85
3	10.52	6.01	13.28	9.14	11.30	5.66	14.63	13.80	7.58	11.14
4	8.36	4.07	11.76	7.86	12.38	5.40	8.22	14.76	20.17	5.35

    a. Obtain the standardized residuals (16.2) when fitting ANOVA model (14.2) and prepare aligned residual dot plots by treatment. Also prepare a normal probability plot of the residuals and calculate the coefficient of correlation between the ordered residuals and their expected values under normality. Does it appear that the distribution of the error terms is not normal?

    b. Conduct the Kruskal-Wallis rank test; use $\alpha = .10$. State the alternatives, decision rule, and conclusion. What is the P-value of the test? Were ties a source of difficulty here?

    c. Conduct multiple pairwise tests based on the ranked data to group the four types of batteries according to mean operating life; use a family level of significance of $\alpha = .10$. Describe your findings.

    d. Conduct the rank test in part (b) by means of the F^* test statistic (17.10). Is the P-value for this test similar to that for the Kruskal-Wallis test in part (b)?

17.23. Refer to **Cash offers** Problem 14.13.

    a. Conduct the median test for equality of factor level means; use $\alpha = .01$. State the alternatives, decision rule, and conclusion. What is the P-value of the test?

    b. Is the conclusion in part (a) the same as the one in Problem 14.13d?

17.24. Refer to **Telephone communications** Problem 17.20. Conduct the median test for equality of treatment means; control the α risk at .05. State the alternatives, decision rule, and conclusion. What is the P-value of the test?

17.25. Refer to **Battery life** Problem 17.22. Conduct the median test for equality of treatment means; use $\alpha = .10$. State the alternatives, decision rule, and conclusion. What is the P-value of the test?

17.26. A student asks why ε_{ij} is shown as a separate term in the random effects model (17.16) in view of μ_i being a random variable in this model. Respond.

17.27. Refer to Figure 17.1. Here, the situation portrayed is one where the variance σ^2 is larger than the variance σ_μ^2. Is this always the case? Explain.

17.28. In each of the following cases, indicate whether ANOVA model I or model II is more appropriate and state your reasons:

    (1) In a study of absenteeism at a plant, the treatments are the three eight-hour shifts.

    (2) In a study of employee productivity, the treatments are 10 production employees selected at random from all production employees in a large company.

    (3) In a study of anticipated annual income at retirement, the treatments are the four types of retirement plans available to employees.

    (4) In a study of tire wear in 18-wheel trucks, the treatments are four tire locations selected at random.

17.29. Refer to the Apex Enterprises personnel officers example on page 654. Explain with reference to this example over what the expectation in (17.17a) is taken. Over what is the variance in (17.17b) taken? Over what is the covariance in (17.17c) taken?

17.30. Refer to **Filling machines** Problem 14.14. Suppose that the company uses a large number of filling machines and the six machines studied were selected randomly. Assume that ANOVA model (17.16) is applicable.

a. Interpret the following with reference to this example:
   (1) $\mu_.$, (2) σ_μ^2, (3) σ^2, (4) $\sigma^2\{Y_{ij}\}$.

b. Test whether or not all machines in the population have the same mean fill; use $\alpha = .05$. State the alternatives, decision rule, and conclusion. What is the P-value of the test?

c. Estimate the mean fill for all machines in the population with a 95 percent confidence interval.

17.31. Refer to **Filling machines** Problems 14.14 and 17.30.

a. Estimate the proportion of the total variability in carton fills that reflects the differences in mean fills between machines; use a 95 percent confidence interval.

b. Estimate σ^2 with a 95 percent confidence interval. Interpret your interval estimate.

c. Obtain a point estimate of σ_μ^2.

17.32. **Sodium content.** A researcher studied the sodium content in lager beer by selecting at random six brands from the large number of brands of U.S. and Canadian beers sold in a metropolitan area. She then chose eight 12-ounce cans or bottles of each selected brand at random from retail outlets in the area, and measured the sodium content (in milligrams) of each can or bottle. The observations follow.

				j				
i	1	2	3	4	5	6	7	8
1	24.4	22.6	23.8	22.0	24.5	22.3	25.0	24.5
2	10.2	12.1	10.3	10.2	9.9	11.2	12.0	9.5
3	19.2	19.4	19.8	19.0	19.6	18.3	20.0	19.4
4	17.4	18.1	16.7	18.3	17.6	17.5	18.0	16.4
5	13.4	15.0	14.1	13.1	14.9	15.0	13.4	14.8
6	21.3	20.2	20.7	20.8	20.1	18.8	21.1	20.3

Assume that ANOVA model (17.16) is applicable.

a. Test whether or not the mean sodium content is the same in all brands sold in the metropolitan area; use $\alpha = .01$. State the alternatives, decision rule, and conclusion. What is the P-value of the test?

b. Estimate the mean sodium content for all brands; use a 99 percent confidence interval.

17.33. Refer to **Sodium content** Problem 17.32.

a. Estimate $\sigma_\mu^2/(\sigma_\mu^2 + \sigma^2)$ with a 99 percent confidence interval. Interpret your interval estimate.

b. Obtain point estimates of σ^2 and σ_μ^2.

c. Estimate σ^2 with a 99 percent confidence interval.

d. It has been conjectured that the variance of sodium content between brands is more than twice as great as that within brands. Conduct an appropriate test using $\alpha = .01$. State the alternatives, decision rule, and conclusion.

17.34. **Coil winding machines.** A plant contains a large number of coil winding machines. A production analyst studied a certain characteristic of the wound coils produced by these machines by selecting four machines at random and then choosing 10 coils at random from the day's output of each selected machine. The results follow.

						j				
i	1	2	3	4	5	6	7	8	9	10
1	205	204	207	202	208	206	209	205	207	206
2	201	204	198	203	209	207	199	206	205	204
3	198	204	196	201	199	203	202	198	202	197
4	210	209	214	215	211	208	210	209	211	210

Assume that ANOVA model (17.16) is appropriate.
a. Test whether or not the mean coil characteristic is the same for all machines in the plant. Use a level of significance of $\alpha = .10$. State the alternatives, decision rule, and conclusion. What is the P-value of the test?
b. Estimate the mean coil characteristic for all coil winding machines in the plant; use a 90 percent confidence interval.

17.35. Refer to **Coil winding machines** Problem 17.34.
a. Estimate $\sigma_\mu^2/(\sigma_\mu^2 + \sigma^2)$ with a 90 percent confidence interval. Interpret your interval estimate.
b. Estimate σ^2 with a 90 percent confidence interval. Interpret your interval estimate.
c. Obtain a point estimate of σ_μ^2.
d. Test whether or not σ_μ^2 and σ^2 are equal; use $\alpha = .10$. State the alternatives, decision rule, and conclusion.

EXERCISES

17.36. (Calculus needed.) Given $\mu_1 = 0$, $\mu_3 = 1$, and $0 \leq \mu_2 \leq 1$, show that $\Sigma(\mu_i - \mu.)^2$ is minimized when $\mu_2 = .5$, where $\mu. = (\mu_1 + \mu_2 + \mu_3)/3$.

17.37. (Calculus needed.) Refer to **Rehabilitation therapy** Problem 14.12. The sample sizes for the below-average, average, and above-average physical fitness groups are to be n, kn, and n, respectively. Assuming that ANOVA model (14.2) is appropriate, find the optimal value of k to minimize the variances of $\hat{D}_1 = \bar{Y}_1. - \bar{Y}_2.$ and $\hat{D}_2 = \bar{Y}_3. - \bar{Y}_2.$ for a given total sample size n_T.

17.38. Show that $n_T(n_T + 1)/12$ is the sample variance of the consecutive integers 1 to n_T.

17.39. Show that test statistic (17.10) can be expressed as the simple function of X_{KW}^2 on page 646.

17.40. Show that n' defined in (17.22a) equals n when $n_i \equiv n$.

17.41. What are the values r and n that minimize $\sigma^2\{\bar{Y}..\}$ in (17.26) for a given total sample size n_T? Ignore any cost considerations.

17.42. Derive the confidence limits in (17.34) from those in (17.33).

PROJECTS

17.43. Refer to the **SENIC** data set and Project 14.33.

    a. Use the Kruskal-Wallis test to determine whether or not the mean infection risk is the same in the four regions; control the level of significance at $\alpha = .05$. State the alternatives, decision rule, and conclusion. What is the P-value of the test?

    b. Is your conclusion in part (a) the same as that obtained in Project 14.33? Are the assumptions for ANOVA model (14.2) or those underlying the Kruskal-Wallis test more reasonable here?

    c. Use the multiple pairwise testing procedure (17.14) to group the regions; employ family significance level $\alpha = .10$. What are your findings?

    d. Conduct the rank test in part (a) by means of the F^* test statistic (17.10). Is the P-value for this test similar to that for the Kruskal-Wallis test in part (a)?

17.44. Refer to the **SMSA** data set and Project 14.35.

    a. Using the Kruskal-Wallis test, determine whether or not the mean crime rate is the same in the four regions; control the level of significance at $\alpha = .05$. State the alternatives, decision rule, and conclusion. What is the P-value of the test?

    b. Is your conclusion in part (a) the same as that obtained in Project 14.35? Are the assumptions for ANOVA model (14.2) or those underlying the Kruskal-Wallis test more reasonable here?

    c. Use the multiple pairwise testing procedure (17.14) to group the regions; employ family significance level $\alpha = .05$. What are your findings?

    d. Conduct the rank test in part (a) by means of the F^* test statistic (17.10). Is the P-value for this test similar to that for the Kruskal-Wallis test in part (a)?

17.45. Obtain the exact sampling distribution of X^2_{KW} when H_0 holds, for the case $r = 2$ and $n_i \equiv 2$. (*Hint:* What does the equality of the treatment means imply about the arrangements of the ranks 1, 2, 3, and 4?)

17.46. Three populations are being studied; each is uniform between 300 and 800.

    a. Generate 10 random observations from each of the three uniform populations and calculate the X^2_{KW} test statistic (17.7).

    b. Repeat part (a) 100 times.

    c. Calculate the mean and standard deviation of the 100 test statistics. How do these values compare with the characteristics of the relevant chi-square distribution?

    d. What proportion of the 100 test statistics obtained in part (b) is less than 4.61? What proportion is less than 9.21? How do these proportions agree with theoretical expectations?

Part IV

Multifactor Analysis of Variance

Chapter 18

Two-Factor Analysis of Variance—Equal Sample Sizes

In Part III, we considered studies in which the effect of one factor is investigated. Now we are concerned with investigations of the simultaneous effects of two or more factors. In this chapter, we take up the analysis of variance for two-factor studies when all sample sizes are equal. In Chapters 19, 20, and 21, we continue the discussion of two-factor studies by taking up the analysis of factor effects, the planning of sample sizes, the case of unequal sample sizes, and a number of other topics. In Chapter 22, we consider the analysis of variance for studies in which three or more factors are being investigated. Finally in Chapter 23, we take up the analysis of covariance for factorial studies.

18.1 MULTIFACTOR STUDIES

Before focusing specifically on two-factor studies, we shall first make some general remarks about multifactor studies, which encompass investigations of two or more factors. Multifactor studies, like single-factor studies, can be based on experimental or observational data.

Examples of Two-Factor Studies

Example 1. A company investigated the effects of selling price and type of promotional campaign on sales of one of its products. Three selling prices (59 cents, 60 cents, 64 cents) were studied, as were two types of promotional campaigns (radio advertising, newspaper advertising). Let us consider selling price to be factor A and promotional campaign to be factor B. Factor A here was studied at three price levels;

in general, we use the symbol a to denote the number of levels of factor A investigated. Factor B was here studied at two levels; we shall use the symbol b to denote the number of levels of factor B investigated. Each combination of price and promotional campaign was studied as follows:

Treatment	Description
1	59¢ price, radio advertising
2	60¢ price, radio advertising
3	64¢ price, radio advertising
4	59¢ price, newspaper advertising
5	60¢ price, newspaper advertising
6	64¢ price, newspaper advertising

Each of the combinations of a factor level of A and a factor level of B is a *treatment*. Thus, there are $3 \times 2 = 6$ treatments here altogether. In general, the total number of possible treatments in a two-factor study is ab.

Twelve communities throughout the United States, of approximately equal size and similar socioeconomic characteristics, were selected and assigned at random to the treatments such that each treatment was given to two experimental units. As before, we shall use the symbol n for the number of units receiving a given treatment when all treatment sample sizes are the same. In the two communities assigned to treatment 1, for instance, the product price was fixed at 59 cents and radio advertising was employed, and so on for the other communities in the study.

This is an experimental study because control was exercised in assigning the factor A and factor B levels to the experimental units by means of random assignments of the treatments to the communities. The design used was a completely randomized design.

Example 2. A steel company studied the effects of carbon content and tempering temperature on the strength of steel. Carbon content was investigated at a high level and at a low level (the precise definitions of these levels is not important here). Tempering temperature was also studied at a high level and at a low level. Altogether, $2 \times 2 = 4$ treatments were defined for this study:

Treatment	Description
1	High carbon level, high tempering temperature
2	High carbon level, low tempering temperature
3	Low carbon level, high tempering temperature
4	Low carbon level, low tempering temperature

These four treatments were then assigned to 12 production batches in randomized fashion, with each treatment assigned to three batches.

This again is an experimental study because control was exercised over carbon content and temperature by means of the random assignment of treatments to the production batches. The design used again was a completely randomized design.

Example 3. An analyst studied the effects of family income (under $10,000, $10,000–$29,999, $30,000–$49,999, $50,000 and more) and stage in the life cycle of the family (stages 1, 2, 3, 4) on appliance purchases. Here, $4 \times 4 = 16$ treatments are defined. These are in part:

Treatment	Description
1	Under $10,000 income, stage 1
2	Under $10,000 income, stage 2
.	.
.	.
.	.
16	$50,000 and more income, stage 4

The analyst selected 20 families with the required income and life-cycle characteristics for each of the "treatment" classes for this study, yielding 320 families for the entire study.

This study is an observational one because the data were obtained without assigning income and life-cycle stage to the families. Rather, the families were selected because they had the specified characteristics.

Comments

1. When we considered single-factor studies, we did not place any restrictions on the nature of the r factor levels under study. Formally, the ab treatments in a two-factor investigation could be considered as the r factor levels in a single-factor investigation and analyzed according to the methods discussed in Part III. The reason why new methods of analysis are required is that we wish to analyze the ab treatments in special ways that recognize two factors are involved and enable us to obtain information about the effects of each of the two factors as well as about any special joint effects.

2. When a completely randomized design is used in a multifactor study, the random assignments of treatments to the experimental units are made in the manner explained in Chapter 2. No new problems are encountered once the treatments are defined in terms of the factor levels of the various factors under study.

Complete and Fractional Factorial Studies

The three examples cited above are *complete factorial studies* because all possible combinations of factor levels for the different factors were included. At times, it is not feasible or desirable to include all possible combinations of factor levels for the different factors. For instance, suppose the steel company mentioned in Example 2 wished to study six temperatures, five levels of carbon content, and four methods of cooling the steel. A complete factorial study would then involve $6 \times 5 \times 4 = 120$ treatments. Such a study might be extremely costly and time-consuming. Under these conditions, it may be possible to design a *fractional factorial study* containing only a fraction of the 120 factor level combinations, which will still provide information about the effects of each of the three factors as well as about any important special joint effects of these factors.

The discussion of multifactor investigations in Part IV deals solely with complete factorial studies.

Advantages of Multifactor Studies

Efficiency. Multifactor studies are more efficient than the traditional experimental approach of manipulating only one factor at a time and keeping all other conditions constant. With reference to Example 1, the traditional approach to studying the effect of promotional campaign would have been to keep price constant at a given level and vary only the promotional campaign. An important problem with this approach is the choice of the price level to be held constant. This choice is especially difficult when one is not sure whether the promotional effect is the same at different price levels. Even though the traditional approach devotes all resources to studying the effect of only one factor, it does not yield any more precise information about that factor than a multifactor experiment of the same size. With reference to Example 1 again, suppose that 12 communities were to be utilized in a traditional study, six assigned to radio advertising and the other six to newspaper advertising, and that the price would be kept constant at 59 cents. For this traditional study, the comparison between the two types of promotional campaigns would be based on two samples of six communities each. The same is true for the two-factor study in Example 1, since each promotional campaign occurs there in three treatments and each treatment has two communities assigned to it.

Amount of Information. The traditional study provides less information than the two-factor study. Specifically in our previous illustration, it does not provide any information about the effect of price, nor about any special joint effects of price and promotional campaign. Information about price effects would require an additional traditional experiment for which promotional campaign would be kept constant at a given level and price varied. Thus, the traditional approach would require a larger sample to provide information about both price and promotional campaign effects, and unless the traditional study were yet further enlarged, it would still not provide full information about any special joint effects of the two factors. Such special joint effects are called *interactions*. Interaction effects were encountered in regression models and will be discussed in the context of analysis of variance models in the next section. Here it suffices to point out that interaction effects may be very important. For instance, it might be that the price effect is not large when the promotional campaign is in newspapers, but it is large with radio advertising. Such interaction effects can be readily investigated from factorial studies.

Validity of Findings.   In addition to being more efficient and readily providing information about interaction effects, multifactor studies also can strengthen the validity of the findings. Suppose that in Example 1, management was principally interested in investigating the effect of price on sales. If the promotional campaign used in the price study had been newspaper advertising, doubts might exist whether or not the price effect differs for other promotional vehicles. By including type of promotional campaign as another factor in the study, management can get information about the persistence of the price effect with different promotional vehicles, without increasing the number of experimental units in the study. Thus, multifactor studies can include some factors of secondary importance to permit inferences about the primary factors with a greater range of validity.

Comments

1. In studies based on observational data, as for ones utilizing experimental data, multifactor analysis of the data permits a ready evaluation of interaction effects and economizes on the number of cases required for analysis.

2. The advantages of multifactor experiments just described should not lead one to think that the more factors are included in the study, the better. Experiments involving many factors, each at numerous levels, become complex, costly, and time-consuming. It is often better research strategy to begin with a few factors, investigate the effects of these, and extend the investigation in accordance with the results obtained to date. In this way, resources can be devoted principally to the most promising avenues of investigation and a better understanding of the working of the factors can be obtained.

18.2   MEANING OF MODEL ELEMENTS

Before presenting a formal statement of the analysis of variance model for two-factor studies, we shall develop the model elements and discuss their meaning. This will not only be helpful in understanding the ANOVA model but will also provide insights into how the analysis of two-factor studies should proceed. *Throughout this section, we assume that all population means are known and are of equal importance when averages of these means are required.*

Illustration

To illustrate the meaning of the model elements, we shall consider a simple two-factor study in which the effects of gender and age on learning of a task are of interest. For simplicity, the age factor has been defined in terms of only three factor levels (young, middle, old), as shown in Table 18.1a.

Treatment Means

The mean response for a given treatment in a two-factor study is denoted by μ_{ij}, where i refers to the level of factor A ($i = 1, \ldots, a$) and j refers to the level of fac-

---

TABLE 18.1 Age Effect but No Gender Effect, with No Interactions—Learning Example

(a) Mean Learning Times (in minutes)

	Factor B—Age			
Factor A—Gender	$j = 1$ Young	$j = 2$ Middle	$j = 3$ Old	Row Average
$i = 1$ Male	$9\ (\mu_{11})$	$11\ (\mu_{12})$	$16\ (\mu_{13})$	$12\ (\mu_{1.})$
$i = 2$ Female	$9\ (\mu_{21})$	$11\ (\mu_{22})$	$16\ (\mu_{23})$	$12\ (\mu_{2.})$
Column average	$9\ (\mu_{.1})$	$11\ (\mu_{.2})$	$16\ (\mu_{.3})$	$12\ (\mu_{..})$

(b) Main Gender Effects (in minutes)

$$\alpha_1 = \mu_{1.} - \mu_{..} = 12 - 12 = 0$$
$$\alpha_2 = \mu_{2.} - \mu_{..} = 12 - 12 = 0$$

(c) Main Age Effects (in minutes)

$$\beta_1 = \mu_{.1} - \mu_{..} = 9 - 12 = -3$$
$$\beta_2 = \mu_{.2} - \mu_{..} = 11 - 12 = -1$$
$$\beta_3 = \mu_{.3} - \mu_{..} = 16 - 12 = 4$$

---

tor $B(j = 1, \ldots, b)$. Table 18.1a contains the true treatment means μ_{ij} for the learning example. Note, for instance, that $\mu_{11} = 9$, which indicates that the mean learning time for young males is nine minutes. Similarly, we see that $\mu_{22} = 11$, so that the mean learning time for middle-aged females is 11 minutes.

Note

The interpretation of a treatment mean μ_{ij} depends on whether the study is an observational or an experimental one. In an observational study, the treatment mean μ_{ij} corresponds to the population mean for the elements having the characteristics of the ith level of factor A and the jth level of factor B. For instance, in the learning example, the treatment mean μ_{11} is the mean learning time for the population of young males.

In an experimental study, the treatment mean μ_{ij} stands for the mean response that would be obtained if the treatment consisting of the ith level of factor A and the jth level of factor B were applied to all units in the population of experimental units about which inferences are to be drawn. For instance, in a study where factor A is type of training program (structured, partially structured, unstructured) and factor B is time of training (during work, after work), $6n$ employees are selected and n are assigned at random to each of the six treatments. The mean μ_{ij} here represents the mean response, say, mean gain in productivity, if the ith training program administered during the jth time were given to all employees in the population of experimental units.

Factor Level Means

The treatment means in Table 18.1a for the learning example indicate that the mean learning times for men and women are the same for each age group. On the other hand, the mean learning time increases with age for each gender. Thus, gender has no effect on mean learning time, but age does. This can also be seen quickly from the row averages and column averages shown in Table 18.1a, which in this case tell the complete story. The row averages are the gender *factor level means,* and the column averages are the age factor level means. We denote the column average for the first column by $\mu_{.1}$, which is the average of μ_{11} and μ_{21}. In general, the column average for the *j*th column is denoted by $\mu_{.j}$:

(18.1)
$$\mu_{.j} = \frac{\sum_{i=1}^{a} \mu_{ij}}{a}$$ — *average for the jth column*

and the row average for the *i*th row is denoted by $\mu_{i.}$:

(18.2)
$$\mu_{i.} = \frac{\sum_{j=1}^{b} \mu_{ij}}{b}$$, *average for the ith row.*

The overall mean learning time for all ages and both genders is denoted by $\mu_{..}$, and is defined in the following equivalent fashions:

(18.3a)
$$\mu_{..} = \frac{\sum_{i}\sum_{j} \mu_{ij}}{ab}$$

(18.3b)
$$\mu_{..} = \frac{\sum_{i} \mu_{i.}}{a}$$

(18.3c)
$$\mu_{..} = \frac{\sum_{j} \mu_{.j}}{b}$$

Main Effects

Main Age Effects. To summarize the main age effects, we shall consider the differences between each factor level mean and the overall mean. For instance, the main effect for young persons in Table 18.1a is the difference between $\mu_{.1}$, the mean learning time for young persons, and $\mu_{..}$, the overall mean. This difference is denoted by β_1:

$$\beta_1 = \mu_{.1} - \mu_{..} = 9 - 12 = -3$$

β_1 is called the *main effect* for factor B at the first level. This and the other main effects for factor B are shown in Table 18.1c.

Main Gender Effects. The main gender effects are defined in corresponding fashion, and denoted by α_i. Thus, we have:

$$\alpha_1 = \mu_{1.} - \mu_{..} = 12 - 12 = 0$$

α_1 is called the main effect for factor A at the first level. The main effects for factor A are shown in Table 18.1b. They are both zero, indicating that gender does not affect mean learning time.

General Definitions. In general, we define the main effect of factor A at the ith level as follows:

(18.4) $$\alpha_i = \mu_{i.} - \mu_{..}$$

Similarly, the main effect of the jth level of factor B is defined:

(18.5) $$\beta_j = \mu_{.j} - \mu_{..}$$

It follows from (18.3b) and (18.3c) that:

(18.6) $$\sum_i \alpha_i = 0 \qquad \sum_j \beta_j = 0$$

Thus, the sum of the main effects for each factor is zero.

Additive Factor Effects

The factor effects in Table 18.1 have an interesting property. Each mean response μ_{ij} can be obtained by adding the respective gender and age main effects to the overall mean $\mu_{..}$. For instance, we have:

$$\mu_{11} = \mu_{..} + \alpha_1 + \beta_1 = 12 + 0 + (-3) = 9$$
$$\mu_{23} = \mu_{..} + \alpha_2 + \beta_3 = 12 + 0 + 4 = 16$$

In general, we have for Table 18.1a:

(18.7) $$\mu_{ij} = \mu_{..} + \alpha_i + \beta_j \qquad \text{Additive factor effects}$$

which can be expressed equivalently, using the definitions of α_i in (18.4) and of β_j in (18.5), as:

(18.7a) $$\mu_{ij} = \mu_{i.} + \mu_{.j} - \mu_{..} \qquad \text{Additive factor effects}$$

It can also be shown that each treatment mean μ_{ij} in Table 18.1a can be expressed in terms of three other treatment means:

(18.7b) $$\mu_{ij} = \mu_{ij'} + \mu_{i'j} - \mu_{i'j'} \qquad \text{Additive factor effects}$$
$$i \neq i', j \neq j'$$

For instance, we have:

$$\mu_{11} = \mu_{12} + \mu_{21} - \mu_{22} = 11 + 9 - 11 = 9$$

or:

$$\mu_{11} = \mu_{13} + \mu_{21} - \mu_{23} = 16 + 9 - 16 = 9$$

When all treatment means can be expressed in the form of (18.7), (18.7a), or (18.7b), we say that the *factors do not interact,* or that *no factor interactions are present,* or that the *factor effects are additive.* The significance of no factor interactions is that the effects of the two factors can be described separately merely by analyzing the factor level means or the factor main effects. Thus, in the learning example in Table 18.1a, the two gender means signify that gender has no influence regardless of age, and the three age means portray the influence of age regardless of gender. The analysis of factor effects is therefore quite simple when there are no factor interactions.

Graphic Presentation

Figure 18.1 presents the mean learning times of Table 18.1a in graphic form. The X axis contains the gender factor levels (denoted by A_1 and A_2), and the Y axis contains learning time. Separate curves are drawn for each of the age factor levels (denoted by B_1, B_2, and B_3). The zero slope of each curve indicates that gender has no effect. The differences in the heights of the three curves show the age effects on learning time.

The points on each curve are conventionally connected by straight lines even though the variable on the X axis (gender, in our example) is not a continuous variable. When the variable on the X axis is qualitative, the slopes of the curves have no meaning, except when the slope is zero, which implies there are no factor level effects. If one of the two factors is a quantitative variable, it is ordinarily advisable to place that factor on the X scale.

FIGURE 18.1   Age Effect but No Gender Effect, with No Interactions—
Learning Example

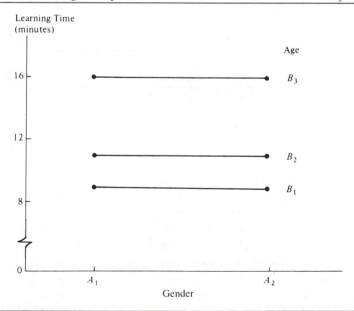

A Second Example with Additive Factor Effects

Table 18.2a contains another illustration of factor effects that do not interact, for the same gender-age learning example as before. The situation here differs from that of Table 18.1a in that not only age but also gender affect the learning time. This is evident from the fact that the mean learning times for men and women are not the same for any age group.

In Table 18.2a, as in Table 18.1a, every mean response can be decomposed according to (18.7):

$$\mu_{ij} = \mu_{..} + \alpha_i + \beta_j$$

For instance:

$$\mu_{11} = \mu_{..} + \alpha_1 + \beta_1 = 12 + 2 + (-3) = 11$$

Hence, the two factors do not interact and the factor effects can be analyzed separately by examining the factor level means $\mu_{i.}$ and $\mu_{.j}$, respectively.

Figure 18.2 presents the data from Table 18.2a in graphic form. This time we have placed age on the X axis and used different curves for each gender. Note that the difference in the heights of the two curves reflects the gender difference and the departure from horizontal for each of the curves reflects the age effect.

Equivalent Statements of Additive Factor Effects

We have said that two factors do not interact if *all* treatment means μ_{ij} can be expressed according to (18.7), (18.7a), or (18.7b). There are a number of other, equivalent, methods of recognizing when two factors do not interact. These are:

1. The difference between the mean responses for any two levels of factor B is the same for all levels of factor A. (Thus, in Table 18.2a, going from young to mid-

---

TABLE 18.2 Age and Gender Effects, with No Interactions—Learning Example

(a) Mean Learning Times (in minutes)

Factor A—Gender	Factor B—Age $j = 1$ Young	$j = 2$ Middle	$j = 3$ Old	Row Average
$i = 1$ Male	11 (μ_{11})	13 (μ_{12})	18 (μ_{13})	14 ($\mu_{1.}$)
$i = 2$ Female	7 (μ_{21})	9 (μ_{22})	14 (μ_{23})	10 ($\mu_{2.}$)
Column average	9 ($\mu_{.1}$)	11 ($\mu_{.2}$)	16 ($\mu_{.3}$)	12 ($\mu_{..}$)

(b) Main Gender Effects (in minutes)

$$\alpha_1 = \mu_{1.} - \mu_{..} = 14 - 12 = 2$$
$$\alpha_2 = \mu_{2.} - \mu_{..} = 10 - 12 = -2$$

(c) Main Age Effects (in minutes)

$$\beta_1 = \mu_{.1} - \mu_{..} = 9 - 12 = -3$$
$$\beta_2 = \mu_{.2} - \mu_{..} = 11 - 12 = -1$$
$$\beta_3 = \mu_{.3} - \mu_{..} = 16 - 12 = 4$$

FIGURE 18.2   Age and Gender Effects, with No Interactions—Learning Example

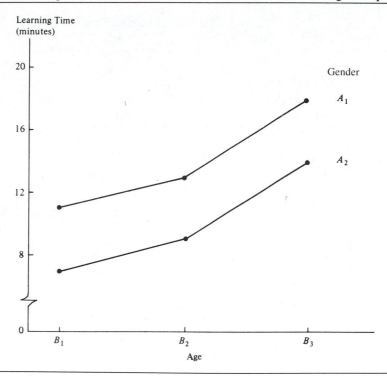

dle age leads to an increase of 2 minutes for both males and females, and going from middle age to old leads to an increase of 5 minutes for both males and fe-males.) Note that it is *not* required that the changes, say, between levels 1 and 2 and between levels 2 and 3 of factor B are the same. These, of course, may differ depending upon the nature of the factor B effect.

2. The difference between the mean responses for any two levels of factor A is the same for all levels of factor B. (Thus, in Table 18.2a, going from male to female leads to a decrease of 4 minutes for all three age groups.)

3. The curves of the mean responses for the different levels of a factor are all paral-lel (such as the two gender curves in Figure 18.2).

All of these conditions are equivalent, implying that the two factors do not inter-act.

Interacting Factor Effects

Table 18.3a contains an illustration for the learning example where the factor effects do interact. The mean learning times for the different gender-age combinations in Table 18.3a indicate that gender has no effect on learning time for young persons but

TABLE 18.3 Age and Gender Effects, with Interactions—Learning Example

(a) Mean Learning Times (in minutes)

Factor A—Gender	Factor B—Age			Row Average	Main Gender Effect
	$j = 1$ Young	$j = 2$ Middle	$j = 3$ Old		
$i = 1$ Male	$9\ (\mu_{11})$	$12\ (\mu_{12})$	$18\ (\mu_{13})$	$13\ (\mu_{1.})$	$1\ (\alpha_1)$
$i = 2$ Female	$9\ (\mu_{21})$	$10\ (\mu_{22})$	$14\ (\mu_{23})$	$11\ (\mu_{2.})$	$-1\ (\alpha_2)$
Column average	$9\ (\mu_{.1})$	$11\ (\mu_{.2})$	$16\ (\mu_{.3})$	$12\ (\mu_{..})$	
Main age effect	$-3\ (\beta_1)$	$-1\ (\beta_2)$	$4\ (\beta_3)$		

(b) Interactions (in minutes)

	$j = 1$	$j = 2$	$j = 3$	Row Average
$i = 1$	-1	0	1	0
$i = 2$	1	0	-1	0
Column average	0	0	0	0

has a substantial effect for old persons. This differential influence of gender, which depends on the age of the person, implies that the age and gender factors interact in their effect on learning time.

Definition of Interaction. We can study the existence of interacting factor effects formally by examining whether or not all treatment means μ_{ij} can be expressed according to (18.7):

$$\mu_{ij} = \mu_{..} + \alpha_i + \beta_j$$

If they can, the factor effects are additive; otherwise, the factor effects are interacting.

For the learning example in Table 18.3a, the main factor effects α_i and β_j are shown in the margins of the table. It is clear that the factors interact. For instance:

$$\mu_{..} + \alpha_1 + \beta_1 = 12 + 1 + (-3) = 10$$

whereas $\mu_{11} = 9$. If the two factors were additive, these would be the same.

The difference between the treatment mean μ_{ij} and the value $\mu_{..} + \alpha_i + \beta_j$ that would be expected if the two factors were additive is called the *interaction effect*, or more simply the *interaction,* of the ith level of factor A with the jth level of factor B, and is denoted by $(\alpha\beta)_{ij}$. Thus, we define $(\alpha\beta)_{ij}$ as follows:

(18.8) $$(\alpha\beta)_{ij} = \mu_{ij} - (\mu_{..} + \alpha_i + \beta_j) = \mu_{ij} - \mu_{..} - \mu_{i.} + \mu_{..} - \mu_{.j} + \mu$$

Replacing α_i and β_j by their definitions in (18.4) and (18.5), respectively, we obtain an alternative definition:

(18.8a) $$(\alpha\beta)_{ij} = \mu_{ij} - \mu_{i.} - \mu_{.j} + \mu_{..}$$ $= 0 \ \&\ no\ interaction$

Utilizing (18.7b), still another alternative definition is:

(18.8b) $$(\alpha\beta)_{ij} = \mu_{ij} - \mu_{ij'} - \mu_{i'j} + \mu_{i'j'}$$

To repeat, the interaction of the ith level of A with the jth level of B, denoted by $(\alpha\beta)_{ij}$, is simply the difference between μ_{ij} and the value that would be expected if the factors were additive. If in fact the two factors are additive, all interactions equal zero, i.e., $(\alpha\beta)_{ij} \equiv 0$.

The interactions for the learning example in Table 18.3a are shown in Table 18.3b. We have, for instance:

$$(\alpha\beta)_{13} = \mu_{13} - (\mu_{..} + \alpha_1 + \beta_3)$$
$$= 18 - (12 + 1 + 4)$$
$$= 1$$

[handwritten marginal notes:]
$\alpha_i = \mu_{i.} - \mu_{..}$
effect of factor A at the ith level
$\beta_j = \mu_{.j} - \mu_{..}$
effect of B at j^{th} level.

Recognition of Interactions. We may recognize whether or not interactions are present in one of the following equivalent fashions:

1. By examining whether all μ_{ij} can be expressed as the sums $\mu_{..} + \alpha_i + \beta_j$.
2. By examining whether the difference between the mean responses for any two levels of factor B is the same for all levels of factor A. (Note in Table 18.3a that the mean learning time increases when going from young to middle-aged persons by 3 minutes for men but only by 1 minute for women.)
3. By examining whether the difference between the mean responses for any two levels of factor A is the same for all levels of factor B. (Note in Table 18.3a that there is no difference between genders for young persons, but there is a difference of 4 minutes for old persons.)
4. By examining whether the treatment mean curves for the different factor levels in a graph are parallel. (Figure 18.3 presents the treatment means in Table 18.3a with age on the X axis. Note that the treatment mean curves for the two genders are not parallel.)

Comments

1. Note from Table 18.3b that some interactions are zero even though the two factors are interacting. *All* interactions must equal zero in order for the two factors to be additive.

2. Table 18.3b illustrates that interactions sum to zero when added over either rows or columns:

(18.9a) $$\sum_i (\alpha\beta)_{ij} = 0 \qquad j = 1, \ldots, b$$

(18.9b) $$\sum_j (\alpha\beta)_{ij} = 0 \qquad i = 1, \ldots, a$$

Consequently, the sum of all interactions is also zero:

(18.9c) $$\sum_i \sum_j (\alpha\beta)_{ij} = 0$$

FIGURE 18.3   Age and Gender Effects, with Important Interactions—Learning Example

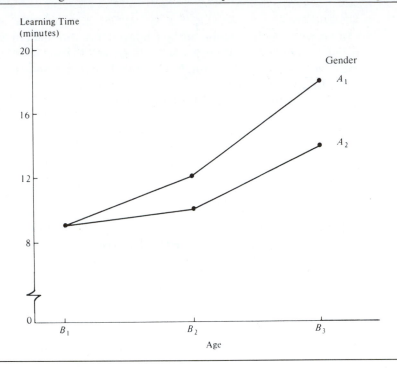

We show this for (18.9a):

$$\sum_i (\alpha\beta)_{ij} = \sum_{i=1}^{a} (\mu_{ij} - \mu_{..} - \alpha_i - \beta_j)$$

$$= \sum_i \mu_{ij} - a\mu_{..} - \sum_i \alpha_i - a\beta_j$$

Now $\sum_i \mu_{ij} = a\mu_{.j}$ by (18.1) and $\Sigma\alpha_i = 0$ by (18.6). Finally, $\beta_j = \mu_{.j} - \mu_{..}$ by (18.5). Hence, we obtain:

$$\sum_i (\alpha\beta)_{ij} = a\mu_{.j} - a\mu_{..} - a(\mu_{.j} - \mu_{..}) = 0$$

Important and Unimportant Interactions

When two factors interact, the question arises whether the factor level means are meaningful measures. In Table 18.3a, for instance, it may well be argued that the gender factor level means 13 and 11 are misleading measures. They indicate that some difference exists in learning time for men and women, but that this difference is not too great. These factor level means hide the fact that there is no difference in mean learning time between genders for young persons, but there is a relatively large difference for old persons. The interactions in Table 18.3a would therefore be considered *important interactions,* implying that one should not ordinarily discuss the effects of each factor separately in terms of the factor level means. A graph, such as Figure 18.3, presents effectively a description of the nature of the interacting effects of the two factors.

Sometimes when two factors interact, the interaction effects are so small that they are considered to be *unimportant interactions.* Table 18.4 and Figure 18.4 present such a case. Note from Figure 18.4 that the curves are *almost* parallel. Perfectly parallel curves, we know, would indicate there are no interactions. For practical purposes, one may say that the mean learning time for women is 2 minutes less than that for men, and this statement is approximately true for all age groups. Alternatively, statements based on average learning time for different age groups will hold approximately for both genders.

Thus, in the case of unimportant interactions, the analysis of factor effects can proceed as for the case of no interactions. Each factor can be studied separately, based on the factor level means $\mu_{i.}$ and $\mu_{.j}$, respectively. This separate analysis of factor effects is, of course, much simpler than a joint analysis for the two factors based on the treatment means μ_{ij}, which is required when the interactions are important.

Comments

1. The determination of whether interactions are important or unimportant is admittedly sometimes difficult. This decision is not a statistical decision and should be made by the subject area specialist (researcher). The advantage of unimportant (or no) interactions, namely, that one is then able to analyze the factor effects separately, is especially great when the study contains more than two factors.

2. Occasionally, it is meaningful to consider the effects of each factor in terms of the factor level means even when important interactions are present. For example, two methods of teaching college mathematics (abstract and standard) were used in teaching students of excellent, good, and moderate quantitative ability. Important interactions between teaching method and student's quantitative ability were found to be present. Students with excellent quantitative ability tended to perform equally well with the two teaching methods whereas

TABLE 18.4 Age and Gender Effects, with Unimportant Interactions— Learning Example

| | Factor B—Age | | | |
Factor A—Gender	j = 1 Young	j = 2 Middle	j = 3 Old	Row Average
i = 1 Male	9.75	12.00	17.25	13.0
i = 2 Female	8.25	10.00	14.75	11.0
Column average	9.0	11.0	16.0	12.0

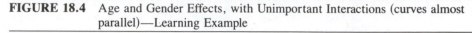

FIGURE 18.4   Age and Gender Effects, with Unimportant Interactions (curves almost parallel)—Learning Example

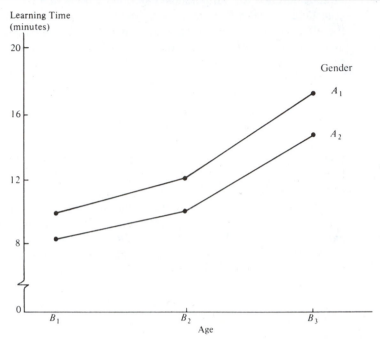

students of moderate or good quantitative ability tended to perform better when taught by the standard method. If equal numbers of students with moderate, good, and excellent quantitative ability are to be taught by one of the two teaching methods, then the method that produces the best average result for all students might be of interest even in the presence of important interactions. A comparison of the teaching method factor level means would then be relevant, even though important interactions are present.

Transformable and Nontransformable Interactions

When important interactions exist, they are sometimes the result of the dependent variable being measured on an inappropriate scale. Consider, for instance, the case where the factor main effects act multiplicatively, rather than additively as in (18.7):

$$(18.10) \qquad \mu_{ij} = \mu_{..} \alpha_i \beta_j \qquad \text{Multiplicative factor effects}$$

If we were to assume in this case that the factor effects are additive, we would find that condition (18.7) does not hold and therefore that interactions are present. These interactions can be removed, however, by applying a logarithmic transformation to (18.10):

$$(18.11) \qquad \log \mu_{ij} = \log \mu_{..} + \log \alpha_i + \log \beta_j$$

This result can be restated equivalently as follows:

(18.11a)
$$\mu'_{ij} = \mu'_{..} + \alpha'_i + \beta'_j$$

where:

$$\mu'_{ij} = \log \mu_{ij}$$
$$\mu'_{..} = \log \mu_{..}$$
$$\alpha'_i = \log \alpha_i$$
$$\beta'_j = \log \beta_j$$

The result in (18.11a) suggests that the original measurement scale for the dependent variable Y may not be the most appropriate one in the sense of leading to easily understood results. Rather, use of $Y' = \log Y$ for the response variable may be better, making the additive model (18.7) then more appropriate.

We say that the interactions present when the factor effects are actually multiplicative are *transformable interactions* because a simple transformation of Y will remove most of these interaction effects and thus make them unimportant.

Another example of transformable interactions occurs when each interaction effect equals the product of functions of the main effects:

(18.12) $$\mu_{ij} = \alpha_i + \beta_j + 2\sqrt{\alpha_i}\sqrt{\beta_j} \qquad \text{Multiplicative interactions}$$

An equivalent form of (18.12) is:

(18.12a) $$\mu_{ij} = \left(\sqrt{\alpha_i} + \sqrt{\beta_j}\right)^2$$

If we now apply the square root transformation, we obtain an additive effects model:

(18.13) $$\mu'_{ij} = \alpha'_i + \beta'_j$$

where:

$$\mu'_{ij} = \sqrt{\mu_{ij}}$$
$$\alpha'_i = \sqrt{\alpha_i}$$
$$\beta'_j = \sqrt{\beta_j}$$

Some simple transformations that may be helpful in making important interactions unimportant are the square, square root, logarithmic, and reciprocal transformations. When interactions cannot be largely removed by a simple transformation, they are called *nontransformable interactions*.

Table 18.5a contains an example of important interactions that are transformable. When a square root transformation is applied to these means, the resulting treatment means in Table 18.5b show no interacting effects. Ordinarily, of course, one cannot hope that a simple transformation of scale removes all interactions as in Table 18.5, but only that interactions become unimportant after the transformation.

Interpretation of Interactions

The interpretation of interactions can be quite difficult when the interacting effects are complex. There are many occasions, however, when the interactions have a simple structure, such as in Table 18.3a, so that the joint factor effects can be described

---

TABLE 18.5   Illustration of a Transformable Interaction

(a) Treatment Means—Original Scale			(b) Treatment Means after Square Root Transformation		
	Factor B			Factor B	
Factor A	$j = 1$	$j = 2$	Factor A	$j = 1$	$j = 2$
$i = 1$	16	64	$i = 1$	4	8
$i = 2$	49	121	$i = 2$	7	11
$i = 3$	64	144	$i = 3$	8	12

---

in a straightforward manner. Table 18.6 provides several additional illustrations.

In Table 18.6a, we have a situation where either raising the pay or increasing the authority of low-paid executives with small authority leads to increased productivity. However, combining both higher pay and greater authority does not lead to further improvements in productivity than increasing either one alone. Table 18.6b represents a case where both higher pay and greater authority are required before any substantial increase in productivity takes place. Table 18.6c portrays a situation where size of crew and personality of crew chief do not interact on the productivity per person when the crew size is 6, 8, or 10 persons.

TABLE 18.6   Examples of Different Types of Interactions

(a) Productivity of Executives

	Factor B—Authority	
Factor A—Pay	Small	Great
Low	50	76
High	74	75

(b) Productivity of Executives

	Factor B—Authority	
Factor A—Pay	Small	Great
Low	50	52
High	53	75

(c) Productivity per Person in Crew

	Factor B—Personality of Crew Chief	
Factor A—Crew Size	Extrovert	Introvert
4 persons	28	20
6 persons	22	20
8 persons	20	18
10 persons	17	15

Note

It is possible that two factors interact, yet the main effects for one (or both) factors are zero. This would be the result of interactions in opposite directions that balance out over one (or both) factors. Thus, there would be definite factor effects, but these would not be disclosed by the factor level means. The case of interacting factors with no main effects for one (or both) factors fortunately is unusual. Typically, interaction effects are smaller than main effects.

18.3 MODEL I (FIXED FACTOR LEVELS) FOR TWO-FACTOR STUDIES

Having explained the model elements, we are now ready to develop ANOVA model I with fixed factor levels for two-factor studies *when all treatment sample sizes are equal and all treatment means are of equal importance.* This ANOVA model is applicable to observational studies and to experimental studies based on a completely randomized design. In Part V we shall consider ANOVA models for some other experimental designs.

The basic situation is as follows: Factor A is studied at a levels, and these are of intrinsic interest in themselves; in other words, the a levels are not considered a sample from a larger population of factor A levels. Similarly, factor B is studied at b levels that are of intrinsic interest in themselves. All ab factor level combinations are included in the study. The number of cases for each of the ab treatments is the same, denoted by n, and it is required that $n > 1$. Thus, the total number of cases for the study is:

$$(18.14) \qquad n_T = abn$$

The kth observation ($k = 1, \ldots, n$) for the treatment where A is at the ith level and B at the jth level is denoted by Y_{ijk} ($i = 1, \ldots, a; j = 1, \ldots, b$). Table 18.7 on page 695 illustrates this notation for an example where A is at three levels, B is at two levels, and two replications have been made for each treatment.

We shall state the fixed ANOVA model for two-factor studies in two equivalent versions—the cell means version and the factor effects version—and later will use one or the other as convenience dictates.

Cell Means Model

When we regard the ab treatments without explicitly considering the factorial structure of the study, we express the analysis of variance model in terms of the cell means (here treatment means) μ_{ij}:

$$(18.15) \qquad Y_{ijk} = \mu_{ij} + \varepsilon_{ijk}$$

where:

μ_{ij} are parameters
ε_{ijk} are independent $N(0, \sigma^2)$
$i = 1, \ldots, a; j = 1, \ldots, b; k = 1, \ldots, n$

Important Features of Model

1. The parameter μ_{ij} is the mean response for the treatment in which factor A is at the ith level and factor B is at the jth level. This follows because $E\{\varepsilon_{ijk}\} = 0$:

(18.16)
$$E\{Y_{ijk}\} = \mu_{ij}$$

2. Since μ_{ij} is a constant, the variance of Y_{ijk} is:

(18.17)
$$\sigma^2\{Y_{ijk}\} = \sigma^2\{\varepsilon_{ijk}\} = \sigma^2$$

3. Since the error terms ε_{ijk} are independent and normally distributed, so are the observations Y_{ijk}. Hence, we can state ANOVA model (18.15) also as follows:

(18.18)
$$Y_{ijk} \text{ are independent } N(\mu_{ij}, \sigma^2)$$

4. ANOVA model (18.15) is a linear model because it can be expressed in the form $\mathbf{Y} = \mathbf{X\beta} + \boldsymbol{\varepsilon}$. Consider a two-factor study with each factor having two levels, i.e., $a = b = 2$, and each treatment having two trials (i.e., $n = 2$). Then, $\mathbf{Y}$, $\mathbf{X}$, $\boldsymbol{\beta}$, and $\boldsymbol{\varepsilon}$ are defined as follows:

(18.19)
$$\mathbf{Y} = \begin{bmatrix} Y_{111} \\ Y_{112} \\ Y_{121} \\ Y_{122} \\ Y_{211} \\ Y_{212} \\ Y_{221} \\ Y_{222} \end{bmatrix} \quad \mathbf{X} = \begin{bmatrix} 1 & 0 & 0 & 0 \\ 1 & 0 & 0 & 0 \\ 0 & 1 & 0 & 0 \\ 0 & 1 & 0 & 0 \\ 0 & 0 & 1 & 0 \\ 0 & 0 & 1 & 0 \\ 0 & 0 & 0 & 1 \\ 0 & 0 & 0 & 1 \end{bmatrix} \quad \boldsymbol{\beta} = \begin{bmatrix} \mu_{11} \\ \mu_{12} \\ \mu_{21} \\ \mu_{22} \end{bmatrix} \quad \boldsymbol{\varepsilon} = \begin{bmatrix} \varepsilon_{111} \\ \varepsilon_{112} \\ \varepsilon_{121} \\ \varepsilon_{122} \\ \varepsilon_{211} \\ \varepsilon_{212} \\ \varepsilon_{221} \\ \varepsilon_{222} \end{bmatrix}$$

Recall that the $\mathbf{E\{Y\}}$ vector, which consists of the elements $E\{Y_{ijk}\}$, equals $\mathbf{X\beta}$ according to (7.18a). Hence, this vector here is:

(18.20)
$$\mathbf{E\{Y\}} = \mathbf{X\beta} = \begin{bmatrix} 1 & 0 & 0 & 0 \\ 1 & 0 & 0 & 0 \\ 0 & 1 & 0 & 0 \\ 0 & 1 & 0 & 0 \\ 0 & 0 & 1 & 0 \\ 0 & 0 & 1 & 0 \\ 0 & 0 & 0 & 1 \\ 0 & 0 & 0 & 1 \end{bmatrix} \begin{bmatrix} \mu_{11} \\ \mu_{12} \\ \mu_{21} \\ \mu_{22} \end{bmatrix} = \begin{bmatrix} \mu_{11} \\ \mu_{11} \\ \mu_{12} \\ \mu_{12} \\ \mu_{21} \\ \mu_{21} \\ \mu_{22} \\ \mu_{22} \end{bmatrix}$$

Thus, $E\{Y_{ijk}\} = \mu_{ij}$, as it must according to (18.16), and we have the proper matrix representation for the two-factor ANOVA model (18.15):

$$(18.21) \qquad \mathbf{Y} = \begin{bmatrix} Y_{111} \\ Y_{112} \\ Y_{121} \\ Y_{122} \\ Y_{211} \\ Y_{212} \\ Y_{221} \\ Y_{222} \end{bmatrix} = \mathbf{X\beta} + \mathbf{\varepsilon} = \begin{bmatrix} \mu_{11} \\ \mu_{11} \\ \mu_{12} \\ \mu_{12} \\ \mu_{21} \\ \mu_{21} \\ \mu_{22} \\ \mu_{22} \end{bmatrix} + \begin{bmatrix} \varepsilon_{111} \\ \varepsilon_{112} \\ \varepsilon_{121} \\ \varepsilon_{122} \\ \varepsilon_{211} \\ \varepsilon_{212} \\ \varepsilon_{221} \\ \varepsilon_{222} \end{bmatrix}$$

5. ANOVA model (18.15) is similar to the single-factor ANOVA model (14.2), except for the two subscripts now needed to identify the treatment. Normality, independent error terms, and constant variances for the error terms are properties of the ANOVA models for both single-factor and two-factor studies.

Factor Effects Model

An equivalent version of the cell means model (18.15) can be obtained by utilizing an identical expression for the treatment means μ_{ij} in terms of factor effects based on the definition of an interaction in (18.8):

$$(\alpha\beta)_{ij} = \mu_{ij} - (\mu_{..} + \alpha_i + \beta_j)$$

Rearranging terms, we obtain the identity:

$$(18.22) \qquad \mu_{ij} \equiv \mu_{..} + \alpha_i + \beta_j + (\alpha\beta)_{ij}$$

where:

$$\mu_{..} = \frac{\sum\limits_i \sum\limits_j \mu_{ij}}{ab} \qquad \begin{array}{l} \text{→ the sum of means across all cells.} \\ \text{→ this is equivalent to the \# of cells.} \end{array}$$

$$\alpha_i = \mu_{i.} - \mu_{..} \qquad \text{→ deviations of the row means from the}$$
$$\beta_j = \mu_{.j} - \mu_{..} \qquad \text{→ deviation of the column means. grand.}$$
$$(\alpha\beta)_{ij} = \mu_{ij} - \mu_{i.} - \mu_{.j} + \mu_{..} \qquad \text{→ interaction}$$

This formulation indicates that the cell mean μ_{ij} for any treatment can be viewed as the sum of four component factor effects. Specifically, (18.22) states that the mean response for the treatment where factor A is at the ith level and factor B is at the jth level is the sum of:

1. An overall constant $\mu_{...}$.
2. The main effect α_i for factor A at the ith level.
3. The main effect β_j for factor B at the jth level.
4. The interaction effect $(\alpha\beta)_{ij}$ when factor A is at the ith level and factor B is at the jth level.

Replacing μ_{ij} in ANOVA model (18.15) by the equivalent expression in (18.22), we obtain an equivalent factor effects ANOVA model for two-factor studies:

$$(18.23) \qquad Y_{ijk} = \mu_{..} + \alpha_i + \beta_j + (\alpha\beta)_{ij} + \varepsilon_{ijk}$$

where:

equivalent to T_j of a single factor Anova model.

$\mu_{..}$ is a constant
α_i are constants subject to the restriction $\Sigma\alpha_i = 0$
β_j are constants subject to the restriction $\Sigma\beta_j = 0$
$(\alpha\beta)_{ij}$ are constants subject to the restrictions

$$\sum_i (\alpha\beta)_{ij} = 0 \qquad \sum_j (\alpha\beta)_{ij} = 0$$

$$j = 1, \ldots, b \qquad i = 1, \ldots, a$$

ε_{ijk} are independent $N(0, \sigma^2)$
$i = 1, \ldots, a; j = 1, \ldots, b; k = 1, \ldots, n$

Important Features of Model

1. ANOVA model (18.23) corresponds to the fixed factor effects ANOVA model (14.60) for a single-factor study except that the single-factor treatment effect is here replaced by the sum of a factor A effect, a factor B effect, and an interaction effect.

2. The properties of the observations Y_{ijk} for ANOVA model (18.23) are the same as those for the equivalent cell means model (18.15). Since $E\{\varepsilon_{ijk}\} = 0$, we have:

$$(18.24) \qquad E\{Y_{ijk}\} = \mu_{..} + \alpha_i + \beta_j + (\alpha\beta)_{ij} = \mu_{ij}$$

The second equality follows from identity (18.22). Further, we have:

$$(18.25) \qquad \sigma^2\{Y_{ijk}\} = \sigma^2$$

because the error term ε_{ijk} is the only random term on the right-hand side in (18.23) and $\sigma^2\{\varepsilon_{ijk}\} = \sigma^2$. Finally, the Y_{ijk} are independent normal random variables because the error terms are independent normal random variables. Hence, we can also state ANOVA model (18.23) as follows:

$$(18.26) \qquad Y_{ijk} \text{ are independent } N[\mu_{..} + \alpha_i + \beta_j + (\alpha\beta)_{ij}, \sigma^2]$$

3. ANOVA model (18.23) is a linear model because it can be stated in the form $\mathbf{Y} = \mathbf{X}\boldsymbol{\beta} + \boldsymbol{\varepsilon}$. We shall show this explicitly in Section 18.8.

18.4 ANALYSIS OF VARIANCE

Illustration

Table 18.7 contains an illustration that we shall employ both in this chapter and the next. The Castle Bakery Company supplies wrapped Italian bread to a large number of supermarkets in a metropolitan area. An experimental study was made of the effects of height of the shelf display (bottom, middle, top) and the width of the shelf display (regular, wide) on sales of this bakery's bread (measured in cases) during the

TABLE 18.7 Sample Data and Notation for Two-Factor Study—Castle Bakery Example (sales in cases)

Factor A (display height) i	Factor B (display width) j		Row Total	Display Height Average
	B_1 (regular)	B_2 (wide)		
A_1 (Bottom)	47 (Y_{111})	46 (Y_{121})		
	43 (Y_{112})	40 (Y_{122})		
Total	90 ($Y_{11.}$)	86 ($Y_{12.}$)	176 ($Y_{1..}$)	
Average	45 ($\bar{Y}_{11.}$)	43 ($\bar{Y}_{12.}$)		44 ($\bar{Y}_{1..}$)
A_2 (Middle)	62 (Y_{211})	67 (Y_{221})		
	68 (Y_{212})	71 (Y_{222})		
Total	130 ($Y_{21.}$)	138 ($Y_{22.}$)	268 ($Y_{2..}$)	
Average	65 ($\bar{Y}_{21.}$)	69 ($\bar{Y}_{22.}$)		67 ($\bar{Y}_{2..}$)
A_3 (Top)	41 (Y_{311})	42 (Y_{321})		
	39 (Y_{312})	46 (Y_{322})		
Total	80 ($Y_{31.}$)	88 ($Y_{32.}$)	168 ($Y_{3..}$)	
Average	40 ($\bar{Y}_{31.}$)	44 ($\bar{Y}_{32.}$)		42 ($\bar{Y}_{3..}$)
Column total	300 ($Y_{.1.}$)	312 ($Y_{.2.}$)	612 ($Y_{...}$)	
Display width average	50 ($\bar{Y}_{.1.}$)	52 ($\bar{Y}_{.2.}$)		51 ($\bar{Y}_{...}$)

experimental period. Twelve supermarkets, similar in terms of sales volumes and clientele, were utilized in the study. Two stores were assigned at random to each of the six treatments according to a completely randomized design, and the display of the bread in each store followed the treatment specifications for that store. Sales of the bread were recorded, and these results are presented in Table 18.7.

Notation

Table 18.7 illustrates the notation we shall use for two-factor studies. It is a straight-forward extension of the notation for single-factor studies. An observation is denoted by Y_{ijk}. The subscripts i and j specify the levels of factors A and B, respectively, and the subscript k refers to the given case or trial for a particular treatment (i.e., factor level combination).

A dot in the subscript indicates aggregation or averaging over the variable represented by the index. For instance, the sum of the observations for the treatment corresponding to the ith level of factor A and the jth level of factor B is:

(18.27a)
$$Y_{ij.} = \sum_{k=1}^{n} Y_{ijk}$$

The corresponding mean is:

$$(18.27b) \qquad \bar{Y}_{ij.} = \frac{Y_{ij.}}{n}$$

The total of all observations for the ith factor level of A is:

$$(18.27c) \qquad Y_{i..} = \sum_{j}^{b} \sum_{k}^{n} Y_{ijk}$$

and the corresponding mean is:

$$(18.27d) \qquad \bar{Y}_{i..} = \frac{Y_{i..}}{bn}$$

Similarly, for the jth factor level of B the sum of all observations and their mean are denoted by:

$$(18.27e) \qquad Y_{.j.} = \sum_{i}^{a} \sum_{k}^{n} Y_{ijk}$$

$$(18.27f) \qquad \bar{Y}_{.j.} = \frac{Y_{.j.}}{an}$$

Finally, the sum of all observations in the study is:

$$(18.27g) \qquad Y_{...} = \sum_{i}^{a} \sum_{j}^{b} \sum_{k}^{n} Y_{ijk}$$

and the overall mean is:

$$(18.27h) \qquad \bar{Y}_{...} = \frac{Y_{...}}{nab}$$

Fitting of ANOVA Model

Cell Means Model (18.15).   We fit the two-factor cell means ANOVA model (18.15) to the sample data by the method of least squares. The least squares criterion here is:

$$(18.28) \qquad Q = \sum_{i} \sum_{j} \sum_{k} (Y_{ijk} - \mu_{ij})^2$$

When we perform the minimization of Q, we obtain the least squares estimators:

$$(18.29) \qquad \hat{\mu}_{ij} = \bar{Y}_{ij.}$$

Thus, the *fitted values* are the estimated treatment means:

$$(18.30) \qquad \hat{Y}_{ijk} = \hat{\mu}_{ij} = \bar{Y}_{ij.}$$

The *residuals,* as usual, are defined as the difference between the observed and fitted values:

$$(18.31) \qquad e_{ijk} = Y_{ijk} - \hat{Y}_{ijk} = Y_{ijk} - \bar{Y}_{ij.}$$

Residuals are highly useful for assessing the aptness of the two-factor ANOVA model (18.15), as they also are for the other statistical models considered earlier.

Factor Effects Model (18.23). For the equivalent factor effects ANOVA model (18.23), we minimize the least squares criterion:

$$(18.32) \qquad Q = \sum_i \sum_j \sum_k [Y_{ijk} - \mu_{..} - \alpha_i - \beta_j - (\alpha\beta)_{ij}]^2$$

subject to the restrictions:

$$\sum_i \alpha_i = 0 \qquad \sum_j \beta_j = 0 \qquad \sum_i (\alpha\beta)_{ij} = 0 \qquad \sum_j (\alpha\beta)_{ij} = 0$$

When we perform this minimization, we obtain the following least squares estimators of the parameters:

	Parameter	Estimator
(18.33a)	$\mu_{..}$	$\hat{\mu}_{..} = \bar{Y}_{...}$
(18.33b)	$\alpha_i = \mu_{i.} - \mu_{..}$	$\hat{\alpha}_i = \bar{Y}_{i..} - \bar{Y}_{...}$
(18.33c)	$\beta_j = \mu_{.j} - \mu_{..}$	$\hat{\beta}_j = \bar{Y}_{.j.} - \bar{Y}_{...}$
(18.33d)	$(\alpha\beta)_{ij} = \mu_{ij} - \mu_{i.} - \mu_{.j} + \mu_{..}$	$(\widehat{\alpha\beta})_{ij} = \bar{Y}_{ij.} - \bar{Y}_{i..} - \bar{Y}_{.j.} + \bar{Y}_{...}$

The correspondences of the least squares estimators to the definitions of the parameters are readily apparent.

The fitted values and residuals for the factor effects model (18.23) are exactly the same as those for the cell means model (18.15). Specifically, the fitted values for ANOVA model (18.23) are:

$$(18.34) \qquad \hat{Y}_{ijk} = \hat{\mu}_{..} + \hat{\alpha}_i + \hat{\beta}_j + (\widehat{\alpha\beta})_{ij}$$
$$= \bar{Y}_{...} + (\bar{Y}_{i..} - \bar{Y}_{...}) + (\bar{Y}_{.j.} - \bar{Y}_{...}) + (\bar{Y}_{ij.} - \bar{Y}_{i..} - \bar{Y}_{.j.} + \bar{Y}_{...})$$
$$= \bar{Y}_{ij.}$$

so that the residuals are again:

$$(18.35) \qquad e_{ijk} = Y_{ijk} - \bar{Y}_{ij.}$$

Note

The least squares estimators in (18.29) and (18.33) are the same as those obtained by the method of maximum likelihood.

Example. For the Castle Bakery example, the fitted values are the estimated treatment means $\bar{Y}_{ij.}$ shown in Table 18.7. These estimated treatment means are presented graphically in Figure 18.5. We see from this figure that for both regular and wide display widths, mean sales for the middle display height are substantially larger than those for the other two display heights. The effect of display width does not ap-

FIGURE 18.5   Plot of Estimated Treatment Means—Castle Bakery Example

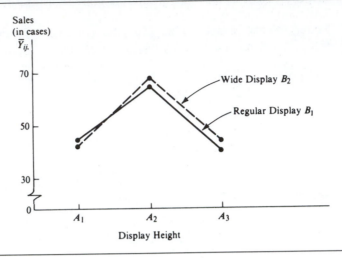

pear to be large at all. Indeed, there may be no effect of display width, with the variations between the estimated treatment means for any given display height being solely of a random nature. In that event, there would be no interactions between display height and display width.

Figure 18.5 differs from the earlier figures portraying factor effects because the earlier figures presented the true treatment means μ_{ij} while Figure 18.5 presents sample estimates. We therefore need to test whether or not the effects shown in Figure 18.5 are real effects or represent only random variations. To conduct such tests, we require a breakdown of the total sum of squares, to be discussed next.

Breakdown of Total Sum of Squares

Breakdown of Total Deviation.   We shall break down the deviation of an observation Y_{ijk} from the overall mean $\bar{Y}_{...}$ in two stages. First, we shall obtain a decomposition of the total deviation $Y_{ijk} - \bar{Y}_{...}$ by viewing the study as consisting of ab treatments:

$$(18.36) \qquad \underbrace{Y_{ijk} - \bar{Y}_{...}}_{\substack{\text{Total} \\ \text{deviation}}} = \underbrace{\bar{Y}_{ij.} - \bar{Y}_{...}}_{\substack{\text{Deviation of estimated} \\ \text{treatment mean} \\ \text{around overall} \\ \text{mean}}} + \underbrace{Y_{ijk} - \bar{Y}_{ij.}}_{\substack{\text{Deviation} \\ \text{around estimated} \\ \text{treatment} \\ \text{mean}}}$$

Note that the deviation around the estimated treatment mean is simply the residual e_{ijk}:

$$e_{ijk} = Y_{ijk} - \hat{Y}_{ijk} = Y_{ijk} - \bar{Y}_{ij.}$$

Next, we shall decompose the estimated treatment mean deviation $\bar{Y}_{ij.} - \bar{Y}_{...}$ in terms of components reflecting the factor A main effect, the factor B main effect, and the AB interaction:

(18.37) $$\underbrace{\bar{Y}_{ij.} - \bar{Y}_{...}}_{\substack{\text{Deviation of} \\ \text{estimated treatment} \\ \text{mean around} \\ \text{overall mean}}} = \underbrace{\bar{Y}_{i..} - \bar{Y}_{...}}_{\substack{A \text{ main} \\ \text{effect}}} + \underbrace{\bar{Y}_{.j.} - \bar{Y}_{...}}_{\substack{B \text{ main} \\ \text{effect}}} + \underbrace{\bar{Y}_{ij.} - \bar{Y}_{i..} - \bar{Y}_{.j.} + \bar{Y}_{...}}_{AB \text{ interaction effect}}$$

Treatment and Error Sums of Squares.   When we square (18.36) and sum over all cases, the cross-product term drops out and we obtain:

(18.38) $$SSTO = SSTR + SSE$$

where:

(18.38a) $$SSTO = \sum_i \sum_j \sum_k (Y_{ijk} - \bar{Y}_{...})^2$$

(18.38b) $$SSTR = n \sum_i \sum_j (\bar{Y}_{ij.} - \bar{Y}_{...})^2$$

(18.38c) $$SSE = \sum_i \sum_j \sum_k (Y_{ijk} - \bar{Y}_{ij.})^2 = \sum_i \sum_j \sum_k e_{ijk}^2$$

$SSTR$ reflects the variability between the ab estimated treatment means and is the ordinary *treatment sum of squares,* and SSE reflects the variability within treatments and is the usual *error sum of squares.* The only difference between these formulas and those for the single-factor case is the use of the two subscripts i and j to designate a treatment.

   Example.   For the Castle Bakery example, Table 18.8 contains the decomposition of the total sum of squares in (18.38). This is the ordinary ANOVA table treating the study as a single-factor one with $ab = r = 6$ treatments. The sums of squares are obtained as follows:

$$SSTO = (47 - 51)^2 + (43 - 51)^2 + (46 - 51)^2 + \cdots + (46 - 51)^2 = 1{,}642$$

$$SSTR = 2[(45 - 51)^2 + (43 - 51)^2 + (65 - 51)^2 + \cdots + (44 - 51)^2] = 1{,}580$$

$$SSE = (47 - 45)^2 + (43 - 45)^2 + (46 - 43)^2 + \cdots + (46 - 44)^2 = 62$$

TABLE 18.8   ANOVA Table Neglecting Factorial Structure—Castle Bakery Example

Source of Variation	SS	df	MS
Between treatments	1,580	5	316
Error	62	6	10.3
Total	1,642	11	

One could test at this point, by means of test statistic (14.53), whether or not the six treatment means are equal. If they are, neither of the two factors has any effect. Ordinarily, however, testing of factor effects is not made until a further decomposition of the treatment sum of squares has been carried out to reflect the factorial nature of the study.

Breakdown of Treatment Sum of Squares.   When we square (18.37) and sum over all treatments and over the n cases associated with each estimated treatment mean $\bar{Y}_{ij.}$, all cross-product terms drop out and we obtain:

$$(18.39) \qquad\qquad SSTR = SSA + SSB + SSAB$$

where:

$$(18.39a) \qquad\qquad SSA = nb \sum_i (\bar{Y}_{i..} - \bar{Y}_{...})^2$$

$$(18.39b) \qquad\qquad SSB = na \sum_j (\bar{Y}_{.j.} - \bar{Y}_{...})^2$$

$$(18.39c) \qquad\qquad SSAB = n \sum_i \sum_j (\bar{Y}_{ij.} - \bar{Y}_{i..} - \bar{Y}_{.j.} + \bar{Y}_{...})^2$$

SSA, called the *factor A sum of squares*, measures the variability of the estimated factor A level means $\bar{Y}_{i..}$. The more variable they are, the bigger will be SSA. Similarly SSB, called the *factor B sum of squares*, measures the variability of the estimated factor B level means $\bar{Y}_{.j.}$. Finally, $SSAB$, called the *AB interaction sum of squares*, measures the variability of the estimated interactions $\bar{Y}_{ij.} - \bar{Y}_{i..} - \bar{Y}_{.j.} + \bar{Y}_{...}$ for the ab treatments. Remember that the mean of all estimated interactions is zero, so that the deviations of the estimated interactions around their mean is not explicitly shown as it was in SSA and SSB. The larger are the estimated interactions (sign disregarded), the larger will be $SSAB$.

The breakdown of $SSTR$ into the components SSA, SSB, and $SSAB$ is called an *orthogonal decomposition*. An orthogonal decomposition is one where the component sums of squares add to the total sum of squares ($SSTR$ here), and likewise for the degrees of freedom. Thus, the decompositions of $SSTO$ into $SSTR$ and SSE for single-factor and two-factor studies are also orthogonal decompositions. While many different orthogonal decompositions of $SSTR$ are possible here, the one into the SSA, SSB, and $SSAB$ components is of interest because these three components provide information about the factor A main effects, the factor B main effects, and the AB interactions, respectively, as will be seen shortly.

Computational Forms.   For hand computing purposes, we ordinarily use the following formulas that are algebraically identical to the definitional formulas previously given:

$$(18.40a) \qquad\qquad SSTO = \sum_i \sum_j \sum_k Y_{ijk}^2 - \frac{Y_{...}^2}{nab}$$

(18.40b)
$$SSE = \sum_i \sum_j \sum_k Y_{ijk}^2 - \frac{\sum_i \sum_j Y_{ij\cdot}^2}{n}$$

(18.40c)
$$SSA = \frac{\sum_i Y_{i\cdot\cdot}^2}{nb} - \frac{Y_{\cdot\cdot\cdot}^2}{nab}$$

(18.40d)
$$SSB = \frac{\sum_j Y_{\cdot j\cdot}^2}{na} - \frac{Y_{\cdot\cdot\cdot}^2}{nab}$$

Ordinarily, the interaction sum of squares is obtained as a remainder:

(18.40e)
$$SSAB = SSTO - SSE - SSA - SSB$$

or from:

(18.40f)
$$SSAB = SSTR - SSA - SSB$$

where:

(18.40g)
$$SSTR = \frac{\sum_i \sum_j Y_{ij\cdot}^2}{n} - \frac{Y_{\cdot\cdot\cdot}^2}{nab}$$

Example. For the Castle Bakery example, we obtain the following decomposition of $SSTR$, using the data in Table 18.7 and the computational formulas in (18.40):

$$SSA = \frac{(176)^2 + (268)^2 + (168)^2}{2(2)} - \frac{(612)^2}{2(3)(2)} = 1{,}544$$

$$SSB = \frac{(300)^2 + (312)^2}{2(3)} - \frac{(612)^2}{2(3)(2)} = 12$$

$$SSAB = 1{,}580 - 1{,}544 - 12 = 24$$

Hence, we have:

$$1{,}580 = 1{,}544 + 12 + 24$$
$$SSTR = SSA + SSB + SSAB$$

Recapitulation. Combining the decompositions in (18.38) and (18.39), we have established that:

(18.41)
$$SSTO = SSA + SSB + SSAB + SSE$$

where the component sums of squares are defined in (18.40).
  For the Castle Bakery example, we have found:

$$1{,}642 = 1{,}544 + 12 + 24 + 62$$
$$SSTO = SSA + SSB + SSAB + SSE$$

Thus, much of the total variability in this instance is associated with the factor A (display height) effects.

Breakdown of Degrees of Freedom

We are familiar from single-factor analysis of variance as to how the degrees of freedom are divided between the treatment and error components. For two-factor studies with n cases for each treatment, there are a total of $n_T = nab$ cases and $r = ab$ treatments; hence, the degrees of freedom associated with $SSTO$, $SSTR$, and SSE are $nab - 1$, $ab - 1$, and $nab - ab = (n - 1)ab$, respectively. These degrees of freedom for the Castle Bakery example are $2(3)(2) - 1 = 11$, $3(2) - 1 = 5$, and $(2 - 1)(3)(2) = 6$, respectively, as shown in Table 18.8.

Corresponding to the further breakdown of the treatment sum of squares, we can also obtain a breakdown of the associated $ab - 1$ degrees of freedom. SSA has $a - 1$ degrees of freedom associated with it. There are a factor level deviations $\bar{Y}_{i..} - \bar{Y}_{...}$, but one degree of freedom is lost because the deviations are subject to one restriction, i.e., $\Sigma(\bar{Y}_{i..} - \bar{Y}_{...}) = 0$. Similarly, SSB has $b - 1$ degrees of freedom associated with it. The degrees of freedom associated with $SSAB$, the interaction sum of squares, is the remainder:

$$(ab - 1) - (a - 1) - (b - 1) = (a - 1)(b - 1)$$

The degrees of freedom associated with $SSAB$ may be understood as follows: There are ab interaction terms. These are subject to b restrictions since:

$$\sum_i (\bar{Y}_{ij.} - \bar{Y}_{i..} - \bar{Y}_{.j.} + \bar{Y}_{...}) = 0 \qquad j = 1, \dots, b$$

There are a additional restrictions since:

$$\sum_j (\bar{Y}_{ij.} - \bar{Y}_{i..} - \bar{Y}_{.j.} + \bar{Y}_{...}) = 0 \qquad i = 1, \dots, a$$

However, only $a - 1$ of these latter restrictions are independent since the last one is implied by the previous b restrictions. Altogether, therefore, there are $b + (a - 1)$ independent restrictions. Hence, the degrees of freedom are:

$$ab - (b + a - 1) = (a - 1)(b - 1)$$

Example.   For the Castle Bakery example, SSA has $3 - 1 = 2$ degrees of freedom associated with it, SSB has $2 - 1 = 1$ degree of freedom, and $SSAB$ has $(3 - 1)(2 - 1) = 2$ degrees of freedom.

Mean Squares

Mean squares are obtained in the usual way by dividing the sums of squares by their associated degrees of freedom. We thus obtain:

(18.42a)
$$MSA = \frac{SSA}{a - 1}$$

(18.42b)
$$MSB = \frac{SSB}{b - 1}$$

(18.42c)
$$MSAB = \frac{SSAB}{(a - 1)(b - 1)}$$

Example. For the Castle Bakery example, these mean squares are:

$$MSA = \frac{1,544}{2} = 772$$

$$MSB = \frac{12}{1} = 12$$

$$MSAB = \frac{24}{2} = 12$$

Note that these mean squares do not add to the treatment mean square, $MSTR = 1,580/5 = 316$. Here again, we see that mean squares are not additive.

Expected Mean Squares

It can be shown, along the same lines used for the single-factor case, that the mean squares for two-factor ANOVA model (18.23) have the following expectations:

(18.43a)
$$E\{MSE\} = \sigma^2$$

(18.43b)
$$E\{MSA\} = \sigma^2 + nb\frac{\Sigma\alpha_i^2}{a - 1} = \sigma^2 + nb\frac{\Sigma(\mu_{i.} - \mu_{..})^2}{a - 1}$$

(18.43c)
$$E\{MSB\} = \sigma^2 + na\frac{\Sigma\beta_j^2}{b - 1} = \sigma^2 + na\frac{\Sigma(\mu_{.j} - \mu_{..})^2}{b - 1}$$

(18.43d)
$$E\{MSAB\} = \sigma^2 + n\frac{\Sigma\Sigma(\alpha\beta)_{ij}^2}{(a - 1)(b - 1)}$$

$$= \sigma^2 + n\frac{\Sigma\Sigma(\mu_{ij} - \mu_{i.} - \mu_{.j} + \mu_{..})^2}{(a - 1)(b - 1)}$$

These expectations show that if there are no factor A main effects (i.e., if all $\mu_{i.}$ are equal, or all $\alpha_i = 0$), MSA and MSE have the same expectation; otherwise MSA tends to be larger than MSE. Similarly, if there are no factor B main effects, MSB and MSE have the same expectation; otherwise MSB tends to be larger than MSE. Finally, if there are no interactions [i.e., if all $(\alpha\beta)_{ij} = 0$] so that the factor effects are additive, $MSAB$ has the same expectation as MSE; otherwise, $MSAB$ tends to be larger than MSE. This suggests that F^* test statistics based on the ratios MSA/MSE, MSB/MSE, and $MSAB/MSE$ will provide information about the main effects and in-

TABLE 18.9 ANOVA Table for Two-Factor Study with Fixed Factor Levels

Source of Variation	SS	df	MS	$E\{MS\}$
Between treatments	$SSTR = n\sum\sum(\bar{Y}_{ij.} - \bar{Y}_{...})^2$	$ab - 1$	$MSTR = \dfrac{SSTR}{ab - 1}$	$\sigma^2 + \dfrac{n}{ab - 1}\sum\sum(\mu_{ij} - \mu_{..})^2$
Factor A	$SSA = nb\sum(\bar{Y}_{i..} - \bar{Y}_{...})^2$	$a - 1$	$MSA = \dfrac{SSA}{a - 1}$	$\sigma^2 + \dfrac{bn}{a - 1}\sum(\mu_{i.} - \mu_{..})^2$
Factor B	$SSB = na\sum(\bar{Y}_{.j.} - \bar{Y}_{...})^2$	$b - 1$	$MSB = \dfrac{SSB}{b - 1}$	$\sigma^2 + \dfrac{an}{b - 1}\sum(\mu_{.j} - \mu_{..})^2$
AB Interactions	$SSAB = n\sum\sum(\bar{Y}_{ij.} - \bar{Y}_{i..} - \bar{Y}_{.j.} + \bar{Y}_{...})^2$	$(a - 1)(b - 1)$	$MSAB = \dfrac{SSAB}{(a - 1)(b - 1)}$	$\sigma^2 + \dfrac{n}{(a - 1)(b - 1)}\sum\sum(\mu_{ij} - \mu_{i.} - \mu_{.j} + \mu_{..})^2$
Error	$SSE = \sum\sum\sum(Y_{ijk} - \bar{Y}_{ij.})^2$	$ab(n - 1)$	$MSE = \dfrac{SSE}{ab(n - 1)}$	σ^2
Total	$SSTO = \sum\sum\sum(Y_{ijk} - \bar{Y}_{...})^2$	$nab - 1$		

TABLE 18.10   ANOVA Table for Two-Factor Study—Castle Bakery Example

Source of Variation	SS	df	MS
Between treatments	1,580	5	316
Factor A (display height)	1,544	2	772
Factor B (display width)	12	1	12
AB Interactions	24	2	12
Error	62	6	10.3
Total	1,642	11	

teractions of the two factors, respectively, with large values of the test statistics indicating the presence of factor effects. We shall see shortly that tests based on these statistics are regular F tests.

Analysis of Variance Table

The breakdown of the total sum of squares into the treatment and error components is shown in Table 18.9, as well as the further breakdown of the treatment sum of squares into the several factor components. Also shown there are the associated degrees of freedom, the mean squares, and the expected mean squares. Table 18.10 contains the two-factor analysis of variance for the Castle Bakery example.

18.5 EVALUATION OF APTNESS OF ANOVA MODEL

Before undertaking formal inference procedures, it is desirable to evaluate the aptness of the two-factor ANOVA model (18.23). No new problems arise here. The residuals in (18.35):

$$e_{ijk} = Y_{ijk} - \bar{Y}_{ij.}$$

may be examined for normality, constancy of error variance, and independence of error terms in the same fashion as for a single-factor study.

Transformations may be employed to stabilize the error variance and to make the error distributions more normal.

Our discussion of this topic in Chapter 16 for the single-factor case applies completely to the two-factor case.

Finally, the earlier discussion on effects of departures from the model applies fully to the two-factor case. In particular, the employment of equal sample sizes for each treatment minimizes the effect of unequal variances.

Example

In the Castle Bakery example, there are only two replications for each treatment. Also, the data are rounded to keep the illustrative computations as simple as possible. As a result, the analysis of the residuals will only be of limited value here. The residuals are obtained according to (18.35). Using the data in Table 18.7, we have, for instance:

$$e_{111} = 47 - 45 = 2$$

$$e_{121} = 46 - 43 = 3$$

These residuals are plotted against the fitted values $\hat{Y}_{ijk} = \bar{Y}_{ij.}$ in Figure 18.6a and against the expected values under normality in Figure 18.6b. There is no strong evidence of unequal error variances for the different treatments in Figure 18.6a. The plot in Figure 18.6b is moderately linear, the flat steps being due to the rounded nature of the data. The coefficient of correlation between the ordered residuals and their expected values under normality is .940, which tends to support the reasonableness of approximate normality.

FIGURE 18.6   Diagnostic Residual Plots—Castle Bakery Example (Minitab, Ref. 18.1)

(a) Residual Plot against $\hat{Y}$

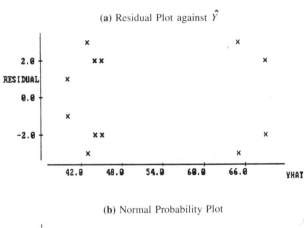

(b) Normal Probability Plot

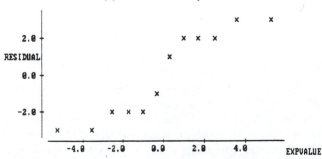

In view of the robustness of the inference procedures for the fixed ANOVA model, it appears to be reasonable to proceed for the Castle Bakery example with tests for factor effects and other inference procedures.

18.6 *F* TESTS

In view of the additivity of sums of squares and degrees of freedom, Cochran's theorem (3.60) applies when no factor effects are present. Hence, the test statistics F^* based on the appropriate mean squares then follow the F distribution, leading to the usual type of F tests for factor effects.

Test for Interactions

Ordinarily, the analysis of a two-factor study begins with a test to determine whether or not the two factors interact:

(18.44)
$$H_0: \mu_{ij} - \mu_{i.} - \mu_{.j} + \mu_{..} = 0 \qquad \text{for all } i, j$$
$$H_a: \mu_{ij} - \mu_{i.} - \mu_{.j} + \mu_{..} \neq 0 \qquad \text{for some } i, j$$

or equivalently:

(18.44a)
$$H_0: \text{all } (\alpha\beta)_{ij} = 0$$
$$H_a: \text{not all } (\alpha\beta)_{ij} \text{ equal zero}$$

As we noted from an examination of the expected mean squares in Table 18.9, the appropriate test statistic is:

(18.45)
$$F^* = \frac{MSAB}{MSE}$$

Large values of F^* indicate the existence of interactions. When H_0 holds, F^* is distributed as $F[(a - 1)(b - 1), (n - 1)ab]$. Hence, the appropriate decision rule to control the Type I error at α is: *⊘t* $_{MSAB}$ *DR* *mse .*

(18.46)
If $F^* \leq F[1 - \alpha; (a - 1)(b - 1), (n - 1)ab]$, conclude H_0

If $F^* > F[1 - \alpha; (a - 1)(b - 1), (n - 1)ab]$, conclude H_a

where $F[1 - \alpha; (a - 1)(b - 1), (n - 1)ab]$ is the $(1 - \alpha)100$ percentile of the appropriate F distribution.

Test for Factor *A* Main Effects

Tests for factor A main effects and for factor B main effects ordinarily follow the test for interactions when no strong interactions exist. To test whether or not A main effects are present:

$$(18.47) \qquad \begin{array}{c} H_0: \mu_{1.} = \mu_{2.} = \cdots = \mu_{a.} \\ H_a: \text{not all } \mu_{i.} \text{ are equal} \end{array}$$

or equivalently:

$$(18.47a) \qquad \begin{array}{c} H_0: \alpha_1 = \alpha_2 = \cdots = \alpha_a = 0 \\ H_a: \text{not all } \alpha_i \text{ equal zero} \end{array}$$

we use the test statistic:

$$(18.48) \qquad F^* = \frac{MSA}{MSE}$$

Again, large values of F^* indicate the existence of factor A main effects. Since F^* is distributed as $F[a - 1, (n - 1)ab]$ when H_0 holds, the appropriate decision rule for controlling the risk of making a Type I error at α is:

$$(18.49) \qquad \begin{array}{c} \text{If } F^* \leq F[1 - \alpha; a - 1, (n - 1)ab], \text{ conclude } H_0 \\ \text{If } F^* > F[1 - \alpha; a - 1, (n - 1)ab], \text{ conclude } H_a \end{array}$$

Test for Factor *B* Main Effects

This test is similar to the one for factor A main effects. The alternatives are:

$$(18.50) \qquad \begin{array}{c} H_0: \mu_{.1} = \mu_{.2} = \cdots = \mu_{.b} \\ H_a: \text{not all } \mu_{.j} \text{ are equal} \end{array}$$

or equivalently:

$$(18.50a) \qquad \begin{array}{c} H_0: \beta_1 = \beta_2 = \cdots = \beta_b = 0 \\ H_a: \text{not all } \beta_j \text{ equal zero} \end{array}$$

The test statistic is:

$$(18.51) \qquad F^* = \frac{MSB}{MSE}$$

and the appropriate decision rule for controlling the risk of a Type I error at α is:

$$(18.52) \qquad \begin{array}{c} \text{If } F^* \leq F[1 - \alpha; b - 1, (n - 1)ab], \text{ conclude } H_0 \\ \text{If } F^* > F[1 - \alpha; b - 1, (n - 1)ab], \text{ conclude } H_a \end{array}$$

Example

We shall investigate in the Castle Bakery example the presence of display height and display width effects, using a level of significance of $\alpha = .05$ for each test. First, we begin by testing whether or not interaction effects are present:

$$H_0: \text{all } (\alpha\beta)_{ij} = 0$$

$$H_a: \text{not all } (\alpha\beta)_{ij} \text{ equal zero}$$

Using the data from Table 18.10 in test statistic (18.45), we obtain:

$$F^* = \frac{12}{10.3} = 1.17$$

Since we are to control the risk of making a Type I error at $\alpha = .05$, we require $F(.95; 2, 6) = 5.14$, so that the decision rule is:

If $F^* \leq 5.14$, conclude H_0

If $F^* > 5.14$, conclude H_a

Since $F^* = 1.17 \leq 5.14$, we conclude H_0, that display height and display width do not interact in their effects on sales. The P-value of this test is $P\{F(2, 6) > 1.17\} = .37$.

Since the two factors do not interact, we turn to testing for display height (factor A) main effects; the alternative conclusions are given in (18.47). The test statistic (18.48) for our example becomes:

$$F^* = \frac{772}{10.3} = 75.0$$

For $\alpha = .05$, we require $F(.95; 2, 6) = 5.14$. Since $F^* = 75.0 > 5.14$, we conclude H_a, that not all factor A level means $\mu_{i.}$ are equal, or that some definite effects associated with height of display level exist. The P-value of this test is $P\{F(2, 6) > 75.0\} = .0001$.

Next, we test for display width (factor B) main effects; the alternative conclusions are given in (18.50). The test statistic (18.51) becomes for our example:

$$F^* = \frac{12}{10.3} = 1.17$$

For $\alpha = .05$, we require $F(.95; 1, 6) = 5.99$. Since $F^* = 1.17 \leq 5.99$, we conclude H_0, that all $\mu_{.j}$ are equal, or that display width has no effect on sales. The P-value of this test is $P\{F(1, 6) > 1.17\} = .32$.

Thus, the analysis of variance tests confirm the impressions from the plot of the estimated treatment means $\overline{Y}_{ij.}$ in Figure 18.5—that only display height has an effect on sales for the treatments studied. At this point, it is clearly desirable to conduct further analyses of the nature of the display height effects. We shall discuss such analyses of the factor effects in Chapter 19.

Comments

1. If the test for interactions is conducted with a level of significance of α_1, that for factor A main effects with a level of significance of α_2, and that for factor B main effects with a level of significance of α_3, the level of significance α for the *family* of three tests is greater than the individual levels of significance. From the Bonferroni inequality in (5.5), we can derive the inequality:

$$(18.53) \qquad \alpha \leq \alpha_1 + \alpha_2 + \alpha_3$$

For the case considered here, a somewhat tighter inequality can be used, the *Kimball inequality,* which utilizes the fact that the numerators of the three test statistics are independent and the denominator is the same in each case. This inequality states:

$$(18.54) \qquad \alpha \leq 1 - (1 - \alpha_1)(1 - \alpha_2)(1 - \alpha_3)$$

For the Castle Bakery example, where $\alpha_1 = \alpha_2 = \alpha_3 = .05$, the Bonferroni inequality yields as the bound for the family level of significance:

$$\alpha \leq .05 + .05 + .05 = .15$$

while the Kimball inequality yields the bound:

$$\alpha \leq 1 - (.95)(.95)(.95) = .143$$

This illustration makes it clear that the level of significance for the family of three tests may be substantially higher than the levels of significance for the individual tests.

2. The F^* test statistics in (18.45), (18.48), and (18.51) can be obtained by the general linear test method explained in Chapter 3. For example, in testing for the presence of interaction effects, the alternatives are those given in (18.44) and the full model is the ANOVA model in (18.23):

$$(18.55) \qquad Y_{ijk} = \mu_{..} + \alpha_i + \beta_j + (\alpha\beta)_{ij} + \varepsilon_{ijk} \qquad \text{Full model}$$

Fitting this full model leads to the fitted values $\hat{Y}_{ijk} = \bar{Y}_{ij.}$ and the error sum of squares:

$$(18.56) \qquad SSE(F) = \Sigma\Sigma\Sigma(Y_{ijk} - \hat{Y}_{ijk})^2 = \Sigma\Sigma\Sigma(Y_{ijk} - \bar{Y}_{ij.})^2 = SSE$$

which is the usual ANOVA error sum of squares in (18.38c). This error sum of squares has $ab(n - 1)$ degrees of freedom associated with it.

The reduced model under H_0: $(\alpha\beta)_{ij} \equiv 0$ is:

$$(18.57) \qquad Y_{ijk} = \mu_{..} + \alpha_i + \beta_j + \varepsilon_{ijk} \qquad \text{Reduced model}$$

It can be shown that the fitted values for the reduced model are $\hat{Y}_{ijk} = \bar{Y}_{i..} + \bar{Y}_{.j.} - \bar{Y}_{...}$ so that the error sum of squares for the reduced model is:

$$(18.58) \qquad SSE(R) = \Sigma\Sigma\Sigma(Y_{ijk} - \hat{Y}_{ijk})^2 = \Sigma\Sigma\Sigma(Y_{ijk} - \bar{Y}_{i..} - \bar{Y}_{.j.} + \bar{Y}_{...})^2$$

This error sum of squares can be shown to have $nab - a - b + 1$ degrees of freedom associated with it. Test statistic (3.69) then simplifies to $F^* = MSAB/MSE$ in (18.45).

Similarly, in testing for the presence of factor A main effects, the full model is the ANOVA model in (18.23), the alternatives are those in (18.47), and the reduced model is:

$$(18.59) \qquad Y_{ijk} = \mu_{..} + \beta_j + (\alpha\beta)_{ij} + \varepsilon_{ijk} \qquad \text{Reduced model}$$

18.7 COMPUTER INPUT AND OUTPUT

Computer input and output formats for two-factor ANOVA vary considerably from one program to another. Figure 18.7 shows an example of computer printout for the Castle Bakery case, produced by the BMDP computer program (Ref. 18.2).

FIGURE 18.7   Segment of Computer Output for Two-Factor Analysis of Variance—
Castle Bakery Example (BMDP8V, Ref. 18.2)

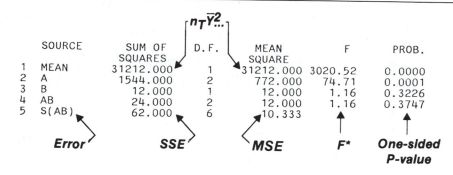

	SOURCE	SUM OF SQUARES	D.F.	MEAN SQUARE	F	PROB.
1	MEAN	31212.000	1	31212.000	3020.52	0.0000
2	A	1544.000	2	772.000	74.71	0.0001
3	B	12.000	1	12.000	1.16	0.3226
4	AB	24.000	2	12.000	1.16	0.3747
5	S(AB)	62.000	6	10.333		

GRAND MEAN   51.00000 ← $\bar{Y}_{...}$

CELL AND MARGINAL MEANS

A =	1	2	3	
	44.00000	67.00000	42.00000	← $\bar{Y}_{i..}$

B =	1	2	
	50.00000	52.00000	← $\bar{Y}_{.j.}$

	B =	1	2	
A =	1	45.00000	43.00000	
	2	65.00000	69.00000	← $\bar{Y}_{ij.}$
	3	40.00000	44.00000	

CELL DEVIATIONS

A =	1	2	3	
	-7.00000	16.00000	-9.00000	← $\bar{Y}_{i..} - \bar{Y}_{...}$

B =	1	2	
	-1.00000	1.00000	← $\bar{Y}_{.j.} - \bar{Y}_{...}$

	B =	1	2	
A =	1	2.00000	-2.00000	
	2	-1.00000	1.00000	← $\bar{Y}_{ij.} - \bar{Y}_{i..} - \bar{Y}_{.j.} + \bar{Y}_{...}$
	3	-1.00000	1.00000	

The first output block shows ANOVA results similar to those presented in Table 18.10. The treatment sum of squares $SSTR$ is not shown separately; it can be obtained as the sum of SSA, SSB, and $SSAB$. Instead, a source of variation attributable to the mean is given. This line in the ANOVA table can be used to test whether $\mu_{..} = 0$ and is usually not of interest (see Table 3.4 for an analogous ANOVA table in regression). The correction for mean sum of squares here is $n_T \bar{Y}^2_{...}$, which has one degree of freedom associated with it. For the Castle Bakery case we have $n_T \bar{Y}^2_{...} = 12(51)^2 = 31,212$.

The second block presents various estimated means while the final block shows the point estimates of the effects (labeled cell deviations) $\alpha_i = \mu_{i.} - \mu_{..}$, $\beta_j = \mu_{.j} - \mu_{..}$, and $(\alpha\beta)_{ij} = \mu_{ij} - \mu_{i.} - \mu_{.j} + \mu_{..}$, for $i = 1, 2, 3$ and $j = 1, 2$.

18.8 REGRESSION APPROACH TO TWO-FACTOR ANALYSIS OF VARIANCE

We shall explain the regression approach to two-factor analysis of variance in terms of the factor effects model (18.23):

[handwritten: $\cdot \alpha_i = \mu_{i.} - \mu_{..}$]

$$（18.60）\qquad Y_{ijk} = \mu_{..} + \alpha_i + \beta_j + (\alpha\beta)_{ij} + \varepsilon_{ijk}$$

[handwritten: $\cdot \beta_j = \mu_{.j} - \mu_{..}$]

As we know from (18.24), the mean responses for this model are given by:

$$（18.61）\qquad E\{Y_{ijk}\} = \mu_{..} + \alpha_i + \beta_j + (\alpha\beta)_{ij}$$

[handwritten: $\cdot \alpha\beta_{ij} = \mu_{ij} - (\mu_{..} + \alpha_i + \beta_j)$]

To represent this model in matrix terms, we proceed in the same fashion as in the regression approach to single-factor ANOVA. Since $\Sigma\alpha_i = 0$, we need only $a - 1$ parameters α_i in the regression model and shall represent the parameter α_a as follows:

[handwritten: this is the last α. It is equal to - "sum of all α-s.]

$$（18.62）\qquad \alpha_a = -\alpha_1 - \alpha_2 - \cdots - \alpha_{a-1}$$

Hence, we shall utilize $a - 1$ indicator variables that can take on values 1, -1, or 0 for the α_i parameters, as in the single-factor ANOVA representation. Similarly, we need only $b - 1$ parameters β_j in the regression model and shall represent the parameter β_b as follows:

$$（18.63）\qquad \beta_b = -\beta_1 - \beta_2 - \cdots - \beta_{b-1}$$

Hence, we shall utilize $b - 1$ indicator variables that can take on values 1, -1, or 0 for the β_j parameters.

For the interaction parameters, we need to recognize that:

$$（18.64）\qquad \sum_i (\alpha\beta)_{ij} = 0 \qquad \sum_j (\alpha\beta)_{ij} = 0$$

$$j = 1, \ldots, b \qquad i = 1, \ldots, a$$

Therefore, we represent the parameters $(\alpha\beta)_{ib}$ and $(\alpha\beta)_{aj}$ as follows:

$$（18.65）\qquad (\alpha\beta)_{ib} = -(\alpha\beta)_{i1} - (\alpha\beta)_{i2} - \cdots - (\alpha\beta)_{i,b-1}$$

$$(18.66) \qquad (\alpha\beta)_{aj} = -(\alpha\beta)_{1j} - (\alpha\beta)_{2j} - \cdots - (\alpha\beta)_{a-1,j}$$

Indeed, because of the interrelations in the constraints in (18.64), only $(a-1)(b-1)$ terms $(\alpha\beta)_{ij}$ are needed in the regression model. These are the terms associated with the cross-products between the indicator variables for the factor A and factor B main effects, as we will now demonstrate.

Example

We present in Table 18.11 the $\mathbf{Y}$, $\boldsymbol{\beta}$, and $\boldsymbol{\varepsilon}$ vectors and the $\mathbf{X}$ matrix for the Castle Bakery example. The X variables are defined as follows:

$$(18.67) \qquad X_1 = \begin{array}{l} 1 \text{ if case from level 1 for factor } A \\ -1 \text{ if case from level 3 for factor } A \\ 0 \text{ otherwise} \end{array}$$

$$X_2 = \begin{array}{l} 1 \text{ if case from level 2 for factor } A \\ -1 \text{ if case from level 3 for factor } A \\ 0 \text{ otherwise} \end{array}$$

$$X_3 = \begin{array}{l} 1 \text{ if case from level 1 for factor } B \\ -1 \text{ if case from level 2 for factor } B \end{array}$$

Hence, the α_i parameters included in the $\boldsymbol{\beta}$ vector are α_1 and α_2, and β_1 is the only β_j parameter included here. The interaction terms in the $\boldsymbol{\beta}$ vector are those associated with independent variables $X_4 = X_1 X_3$ and $X_5 = X_2 X_3$. The X variable $X_1 X_3$ refers to level $i = 1$ for factor A and $j = 1$ for factor B; hence, the associated interaction parameter is $(\alpha\beta)_{11}$. Correspondingly, the X variable $X_2 X_3$ refers to levels $i = 2$, $j = 1$; hence, the associated interaction term is $(\alpha\beta)_{21}$.

TABLE 18.11   Regression Representation for Two-Factor ANOVA Model— Castle Bakery Example   *Codes for factor B (both levels)*

$$\mathbf{Y} = \begin{bmatrix} Y_{111} \\ Y_{112} \\ Y_{121} \\ Y_{122} \\ Y_{211} \\ Y_{212} \\ Y_{221} \\ Y_{222} \\ Y_{311} \\ Y_{312} \\ Y_{321} \\ Y_{322} \end{bmatrix}$$

	X_1	X_2	X_3	X_1X_3	X_2X_3
	1	1	0	1	1	0
	1	1	0	1	1	0
	1	1	0	-1	-1	0
	1	1	0	-1	-1	0
	1	0	1	1	0	1
	1	0	1	1	0	1
	1	0	1	-1	0	-1
	1	0	1	-1	0	-1
	1	-1	-1	1	-1	-1
	1	-1	-1	1	-1	-1
	1	-1	-1	-1	1	1
	1	-1	-1	-1	1	1

$$\boldsymbol{\beta} = \begin{bmatrix} \mu_{..} \\ \alpha_1 \\ \alpha_2 \\ \beta_1 \\ (\alpha\beta)_{11} \\ (\alpha\beta)_{21} \end{bmatrix} \qquad \boldsymbol{\varepsilon} = \begin{bmatrix} \varepsilon_{111} \\ \varepsilon_{112} \\ \varepsilon_{121} \\ \varepsilon_{122} \\ \varepsilon_{211} \\ \varepsilon_{212} \\ \varepsilon_{221} \\ \varepsilon_{222} \\ \varepsilon_{311} \\ \varepsilon_{312} \\ \varepsilon_{321} \\ \varepsilon_{322} \end{bmatrix}$$

TABLE 18.12   Mean Vector for Castle Bakery Example

$$
E\{Y\} =
\begin{bmatrix}
E\{Y_{111}\} \\
E\{Y_{112}\} \\
E\{Y_{121}\} \\
E\{Y_{122}\} \\
E\{Y_{211}\} \\
E\{Y_{212}\} \\
E\{Y_{221}\} \\
E\{Y_{222}\} \\
E\{Y_{311}\} \\
E\{Y_{312}\} \\
E\{Y_{321}\} \\
E\{Y_{322}\}
\end{bmatrix}
= X\beta =
\begin{bmatrix}
\mu_{..} + \alpha_1 + \beta_1 + (\alpha\beta)_{11} \\
\mu_{..} + \alpha_1 + \beta_1 + (\alpha\beta)_{11} \\
\mu_{..} + \alpha_1 - \beta_1 - (\alpha\beta)_{11} \\
\mu_{..} + \alpha_1 - \beta_1 - (\alpha\beta)_{11} \\
\mu_{..} + \alpha_2 + \beta_1 + (\alpha\beta)_{21} \\
\mu_{..} + \alpha_2 + \beta_1 + (\alpha\beta)_{21} \\
\mu_{..} + \alpha_2 - \beta_1 - (\alpha\beta)_{21} \\
\mu_{..} + \alpha_2 - \beta_1 - (\alpha\beta)_{21} \\
\mu_{..} - \alpha_1 - \alpha_2 + \beta_1 - (\alpha\beta)_{11} - (\alpha\beta)_{21} \\
\mu_{..} - \alpha_1 - \alpha_2 + \beta_1 - (\alpha\beta)_{11} - (\alpha\beta)_{21} \\
\mu_{..} - \alpha_1 - \alpha_2 - \beta_1 + (\alpha\beta)_{11} + (\alpha\beta)_{21} \\
\mu_{..} - \alpha_1 - \alpha_2 - \beta_1 + (\alpha\beta)_{11} + (\alpha\beta)_{21}
\end{bmatrix}
=
\begin{bmatrix}
\mu_{..} + \alpha_1 + \beta_1 + (\alpha\beta)_{11} \\
\mu_{..} + \alpha_1 + \beta_1 + (\alpha\beta)_{11} \\
\mu_{..} + \alpha_1 + \beta_2 + (\alpha\beta)_{12} \\
\mu_{..} + \alpha_1 + \beta_2 + (\alpha\beta)_{12} \\
\mu_{..} + \alpha_2 + \beta_1 + (\alpha\beta)_{21} \\
\mu_{..} + \alpha_2 + \beta_1 + (\alpha\beta)_{21} \\
\mu_{..} + \alpha_2 + \beta_2 + (\alpha\beta)_{22} \\
\mu_{..} + \alpha_2 + \beta_2 + (\alpha\beta)_{22} \\
\mu_{..} + \alpha_3 + \beta_1 + (\alpha\beta)_{31} \\
\mu_{..} + \alpha_3 + \beta_1 + (\alpha\beta)_{31} \\
\mu_{..} + \alpha_3 + \beta_2 + (\alpha\beta)_{32} \\
\mu_{..} + \alpha_3 + \beta_2 + (\alpha\beta)_{32}
\end{bmatrix}
$$

To verify that the **X** matrix representation yields the proper model, we present in Table 18.12 the mean vector $E\{Y\} = X\beta$. We see, for instance, that:

$$E\{Y_{111}\} = \mu_{..} + \alpha_1 + \beta_1 + (\alpha\beta)_{11}$$

Also, we see that:

$$E\{Y_{121}\} = \mu_{..} + \alpha_1 - \beta_1 - (\alpha\beta)_{11}$$
$$= \mu_{..} + \alpha_1 + \beta_2 + (\alpha\beta)_{12}$$

since $\beta_2 = -\beta_1$ by (18.63) and $(\alpha\beta)_{12} = -(\alpha\beta)_{11}$ by (18.65). Similarly, we see that:

$$E\{Y_{322}\} = \mu_{..} - \alpha_1 - \alpha_2 - \beta_1 + (\alpha\beta)_{11} + (\alpha\beta)_{21}$$
$$= \mu_{..} + \alpha_3 + \beta_2 + (\alpha\beta)_{32}$$

since:

$$\alpha_3 = -\alpha_1 - \alpha_2 \qquad \text{by (18.62)}$$
$$\beta_2 = -\beta_1 \qquad \text{by (18.63)}$$
$$-(\alpha\beta)_{12} = (\alpha\beta)_{11} \qquad \text{and} \qquad -(\alpha\beta)_{22} = (\alpha\beta)_{21} \qquad \text{by (18.65)}$$
$$-(\alpha\beta)_{12} - (\alpha\beta)_{22} = (\alpha\beta)_{32} \qquad \text{by (18.66)}$$

Thus, the linear model representation in Table 18.11 yields the proper mean response for each observation.

The multiple regression model that is the equivalent of the two-factor ANOVA model for the Castle Bakery example therefore is:

$$(18.68) \qquad Y_{ijk} = \mu_{..} + \underbrace{\alpha_1 X_{ijk1} + \alpha_2 X_{ijk2}}_{A \text{ main effect}} + \underbrace{\beta_1 X_{ijk3}}_{B \text{ main effect}}$$

$$+ \underbrace{(\alpha\beta)_{11} X_{ijk1} X_{ijk3} + (\alpha\beta)_{21} X_{ijk2} X_{ijk3}}_{AB \text{ interaction effect}} + \varepsilon_{ijk} \qquad \text{Full model}$$

$$i = 1, 2, 3; \ j = 1, 2; \ k = 1, 2$$

Here X_{ijk1} denotes the value of independent variable X_1 for the kth case from the treatment for which factor A is at the ith level and factor B is at the jth level, and X_{ijk2} and X_{ijk3} have corresponding meanings. [These independent variables were defined in (18.67).] Finally, the regression parameters are the ANOVA model parameters $\mu_{..}$ (the intercept term) and α_1, α_2, β_1, $(\alpha\beta)_{11}$, and $(\alpha\beta)_{21}$ (the regression coefficients).

The tests for interaction effects, factor A main effects, and factor B main effects for the Castle Bakery example involve testing whether certain of the regression coefficients in regression model (18.68) are zero, as follows:

Test for interaction effects

$$(18.69) \qquad \begin{aligned} &H_0: (\alpha\beta)_{11} = (\alpha\beta)_{21} = 0 \\ &H_a: \text{not both } (\alpha\beta)_{11} \text{ and } (\alpha\beta)_{21} \text{ equal zero} \end{aligned}$$

Test for factor A main effects

$$(18.70) \qquad \begin{aligned} &H_0: \alpha_1 = \alpha_2 = 0 \\ &H_a: \text{not both } \alpha_1 \text{ and } \alpha_2 \text{ equal zero} \end{aligned}$$

Test for factor B main effects

$$(18.71) \qquad \begin{aligned} &H_0: \beta_1 = 0 \\ &H_a: \beta_1 \neq 0 \end{aligned}$$

Thus, the test statistic in each case is the F^* statistic in (8.25), and only the appropriate extra sum of squares and numerator degrees of freedom vary for the three tests.

To conduct the analysis of variance tests for the Castle Bakery example by means of the regression approach, we utilize regression model (18.68). The observations Y_{ijk} are given in Table 18.7, and the **X** matrix is given in Table 18.11. A computer run of a multiple regression package for these data provided the results shown in Table 18.13, including the extra sums of squares for each X variable in the order of the X variables in the regression model.

It can be shown that when all treatment sample sizes are equal, as they are for the Castle Bakery example, the total of the extra sums of squares for the factor A regression terms equals SSA. Similarly, the totals of the extra sums of squares for the factor B regression terms and the interaction regression terms equal, respectively, SSB and $SSAB$. Because of the balance in the **X** matrix when all treatment sample sizes

TABLE 18.13   Regression Approach to Two-Factor Analysis of Variance—
Castle Bakery Example

(a) Fitted Regression Function

$$\hat{Y} = 51.0 - 7.0X_1 + 16.0X_2 - 1.0X_3 + 2.0X_1X_3 - 1.0X_2X_3$$

(b) ANOVA Table

Source of Variation	SS		df		MS
Regression	1,580		5		$MSTR = 316$
X_1	$8\}$		$1\}$		
$X_2\|X_1$	$1,536\}$ $SSA = 1,544$		$1\}$ 2		$MSA = 772$
$X_3\|X_1, X_2$	$12\}$ $SSB = 12$		$1\}$ 1		$MSB = 12$
$X_1X_3\|X_1, X_2, X_3$	$18\}$		$1\}$		
$X_2X_3\|X_1, X_2, X_3, X_1X_3$	$6\}$ $SSAB = 24$		$1\}$ 2		$MSAB = 12$
Error	62		6		$MSE = 10.3$
Total	1,642		11		

are equal, these totals of the extra sums of squares are the same regardless of the order of the factor A, factor B, and interaction extra sums of squares.

We see from Table 18.13 that the totals SSA, SSB, and $SSAB$ and also the error sum of squares SSE obtained with the regression approach are the same as in Table 18.10 where they were calculated by means of the ANOVA formulas.

From this point on, the test procedures based on the regression approach are identical to the analysis of variance tests explained earlier.

18.9   OTHER APPROACHES TO ANOVA ANALYSIS

We now briefly discuss two other approaches to the ANOVA analysis considered in this chapter.

Revision of ANOVA Model

The approach presented in this chapter assumes that the ANOVA model in (18.23) is the full model for all tests of factor effects, regardless of the conclusions reached in any of these tests. The rationale for this approach is that ANOVA model (18.23) is based on the identity (18.22) for the treatment means μ_{ij}. Once the analysis of residuals and other diagnostics demonstrate that this model is appropriate, it is used for all tests.

Some statisticians take the view that ANOVA model (18.23) should be revised if the test for interaction effects leads to the conclusion that no interactions are present. The full model considered with this approach in testing for factor A and factor B main effects when the test for interaction effects leads to the conclusion that no interactions are present is:

$$(18.72) \qquad Y_{ijk} = \mu_{..} + \alpha_i + \beta_j + \varepsilon_{ijk} \qquad \text{Full model}$$

As we just noted with the regression approach for the Castle Bakery example, the extra sums of squares for factor A and factor B main effects do not depend on the order of the extra sums of squares for factor effects when all treatment sample sizes are equal. Hence, the numerator of the test statistic F^* is not affected by this revision in the full model when the treatment sample sizes are equal. The denominator of the F^* test statistic is affected, however, leading to the following error sum of squares for the full model:

$$(18.73) \qquad SSE(F) = SSE + SSAB$$

Thus, the error sum of squares for the full model with this approach involves the *pooling* of the interaction and error sums of squares. Likewise, the degrees of freedom are pooled; the degrees of freedom associated with $SSE(F)$ are:

$$df_F = (a - 1)(b - 1) + (n - 1)ab = nab - a - b + 1$$

For the Castle Bakery example, the pooled error sum of squares for testing factor A and factor B main effects would be (Table 18.10):

$$SSE(F) = 62 + 24 = 86$$

and the pooled degrees of freedom would be:

$$df_F = 6 + 2 = 8$$

Hence, the error mean square for testing factor A or factor B main effects with the model revision approach here would be $86/8 = 10.75$.

This pooling procedure affects both the level of significance and the power of the tests for factor A and factor B main effects, in ways not yet fully understood. It has been suggested therefore by some statisticians that pooling should not be considered unless: (1) the degrees of freedom associated with MSE are small, perhaps 5 or less, and (2) the test statistic $MSAB/MSE$ falls substantially below the action limit of the decision rule, perhaps when $MSAB/MSE < 2$ for $\alpha = .05$. Part (1) of this rule is designed to limit pooling to cases where the gains may be substantial, while part (2) is designed to give reasonable assurance that there are indeed no interactions.

Treatment Means of Unequal Importance

When the treatment means are of unequal importance, the test for interaction effects is not affected by this. However, the analysis of variance formulas for factor A and factor B main effects presented in this chapter are no longer appropriate. Instead, the general linear test approach in matrix terms described in Section 8.6 will usually be required.

In the Castle Bakery example, suppose that the middle and top shelf heights are twice as frequently used for the Italian bread as the bottom shelf height. For testing factor B (shelf width) main effects, the mean response of interest for the regular shelf width would then be:

$$\mu_R = \frac{\mu_{11} + 2\mu_{21} + 2\mu_{31}}{5}$$

and the mean response of interest for the wide shelf width would be:

$$\mu_W = \frac{\mu_{12} + 2\mu_{22} + 2\mu_{32}}{5}$$

The test for display width effect would therefore involve the alternatives:

$$H_0: \mu_R = \mu_W \qquad \qquad H_0: \mu_R - \mu_W = 0$$
$$\text{or}$$
$$H_a: \mu_R \neq \mu_W \qquad \qquad H_a: \mu_R - \mu_W \neq 0$$

Note that these alternatives consist of linear combinations of the cell means μ_{ij}:

(18.74)
$$H_0: \frac{\mu_{11} + 2\mu_{21} + 2\mu_{31}}{5} - \frac{\mu_{12} + 2\mu_{22} + 2\mu_{32}}{5} = 0$$

$$H_a: \frac{\mu_{11} + 2\mu_{21} + 2\mu_{31}}{5} - \frac{\mu_{12} + 2\mu_{22} + 2\mu_{32}}{5} \neq 0$$

Hence, the procedures for the general linear test in matrix terms would usually be required. We shall illustrate this testing approach when the treatment means are of unequal importance in Section 21.3.

In contrast, estimation procedures when the treatment means are of unequal importance are basically the same as those used when all treatment means are of equal importance. We shall illustrate the estimation of factor effects when the treatment means are of unequal importance in Section 19.4.

CITED REFERENCES

18.1. *MINITAB Reference Manual, Release 7*. State College, Pa.: Minitab, Inc., 1989.

18.2. Dixon, W. J., chief editor. *BMDP Statistical Software Manual*, vols. 1 and 2. Berkeley, Calif.: University of California Press, 1988.

PROBLEMS

18.1. Refer to the **SENIC** data set. An analyst wishes to investigate the effects of medical school affiliation (factor A) and geographic region (factor B) on infection risk. All factor level combinations will be included in the study.

a. How many treatments are being studied?

b. What is the dependent variable here?

18.2. A student in a class discussion stated: "A treatment is a treatment, whether the study involves a single factor or multiple factors. The number of factors only affects the analysis of the results." Discuss.

18.3. Verify the interactions in Table 18.3b.

18.4. In a two-factor study, the treatment means μ_{ij} are as follows:

	Factor B		
Factor A	B_1	B_2	B_3
A_1	34	23	36
A_2	40	29	42

a. Obtain the factor A level means.

b. Obtain the main effects of factor A.

c. Does the fact that $\mu_{12} - \mu_{11} = -11$ while $\mu_{13} - \mu_{12} = 13$ imply that factors A and B interact? Explain.

d. Plot the mean responses μ_{ij} in the format of Figure 18.2 and determine whether the two factors interact. What do you find?

18.5. In a two-factor study, the treatment means μ_{ij} are as follows:

		Factor B		
Factor A	B_1	B_2	B_3	B_4
A_1	250	265	268	269
A_2	288	273	270	269

a. Obtain the factor B main effects. What do your results imply about factor B?

b. Plot the treatment means μ_{ij} in the format of Figure 18.2 and determine whether the two factors interact. How can you tell that interactions are present? Are the interactions important or unimportant?

c. Make a logarithmic transformation of the μ_{ij} and plot the transformed values to explore whether this transformation is helpful in reducing the interactions. What are your findings?

18.6. Three sets of treatment means μ_{ij} for students' grades in a course follow, where factor A is student's major (A_1: computer science; A_2: mathematics) and factor B is student's class affiliation (B_1: junior; B_2: senior; B_3: graduate).

	Set 1				Set 2				Set 3		
	B_1	B_2	B_3		B_1	B_2	B_3		B_1	B_2	B_3
A_1	80	80	80	A_1	75	80	90	A_1	75	80	85
A_2	90	90	90	A_2	80	86	97	A_2	75	85	100

Plot each set of μ_{ij} in the format of Figure 18.2 to study interaction effects. Analyze each plot and state your findings. If interactions are present, describe their nature and indicate whether they are important or unimportant.

18.7. Refer to Problem 18.4. Assume that $\sigma = 1.4$ and $n = 10$.
   a. Obtain $E\{MSE\}$ and $E\{MSA\}$.
   b. Is $E\{MSA\}$ substantially larger than $E\{MSE\}$? What is the implication of this?

18.8. Refer to Problem 18.5. Assume that $\sigma = 4$ and $n = 6$.
   a. Obtain $E\{MSE\}$ and $E\{MSAB\}$.
   b. Is $E\{MSAB\}$ substantially larger than $E\{MSE\}$? What is the implication of this?

18.9. A psychologist stated: "I feel uncomfortable about deciding in a research study whether the interactions are important or unimportant. I would rather have the statistician make that decision." Comment.

18.10. Refer to **Cash offers** Problem 14.13. Six male and six female volunteers were used in each age group. The observations (in hundred dollars), classified by age (factor A) and gender of owner (factor B), follow.

	Factor B (gender of owner)	
Factor A (age)	$j = 1$ Male	$j = 2$ Female
$i = 1$ Young	21	21
	23	22
	19	20
	22	21
	22	19
	23	25
$i = 2$ Middle	30	26
	29	29
	26	27
	28	28
	27	27
	27	29
$i = 3$ Elderly	25	23
	22	19
	23	20
	21	21
	22	20
	21	20

Summary calculational results are: $SSA = 316.722$, $SSB = 5.444$, $SSAB = 5.056$, $SSE = 71.667$.

   a. Obtain the fitted values for ANOVA model (18.23).
   b. Obtain the residuals. Do they sum to zero for each treatment?
   c. Prepare aligned residual dot plots for the treatments. What departures from ANOVA model (18.23) can be studied from these plots? What are your findings?
   d. Prepare a normal probability plot of the residuals. Also obtain the coefficient of correlation between the ordered residuals and their expected values under normality. Does the normality assumption appear to be reasonable here?
   e. The observations for each treatment were obtained in the order shown. Prepare residual sequence plots and analyze them. What are your findings?

18.11. Refer to **Cash offers** Problems 14.13 and 18.10. Assume that ANOVA model (18.23) is applicable.

a. Plot the estimated treatment means $\bar{Y}_{ij.}$ in the format of Figure 18.5. Does it appear that any factor effects are present? Explain.

b. Set up the analysis of variance table. Does any one source account for most of the total variability in cash offers in the study? Explain.

c. Test whether or not interaction effects are present; use $\alpha = .05$. State the alternatives, decision rule, and conclusion. What is the P-value of the test?

d. Test whether or not age and gender main effects are present. In each case, use $\alpha = .05$ and state the alternatives, decision rule, and conclusion. What is the P-value of each test? Is it meaningful here to test for main factor effects? Explain.

e. Obtain an upper bound on the family level of significance for the tests in parts (c) and (d); use the Kimball inequality.

f. Do the results in parts (c) and (d) confirm your graphic analysis in part (a)?

g. What are the relations between the sums of squares in the two-factor analysis of variance in part (b) and the sums of squares in the single-factor analysis of variance in Problem 14.13c? Do the same relations hold for the degrees of freedom?

18.12. Eye contact effect.   In a study of the effect of eye contact (factor A) and gender of personnel officer (factor B) on a personnel officer's assessment of likely job success of an applicant, 10 male and 10 female personnel officers were shown a frontview photograph of an applicant's face and were asked to give the person in the photograph a success rating on a scale of 0 (total failure) to 20 (outstanding success). Half of the officers in each gender group were chosen at random to receive a version of the photograph in which the subject made eye contact with the camera lens. The other half received a version in which there was no eye contact. The success ratings follow.

	Factor B (gender of officer)	
Factor A (eye contact)	$j = 1$ Male	$j = 2$ Female
$i = 1$ Present	11	15
	7	12
	12	14
	6	11
	10	16
$i = 2$ Absent	12	14
	16	17
	10	13
	13	20
	14	18

Summary calculational results are: $SSA = 54.45$, $SSB = 76.05$, $SSAB = 1.25$, $SSE = 97.2$.

a. Obtain the fitted values for ANOVA model (18.23).

b. Obtain the residuals. Do they sum to zero for each treatment?

c. Prepare aligned residual dot plots for the treatments. What departures from ANOVA model (18.23) can be studied from these plots? What are your findings?

d. Prepare a normal probability plot of the residuals. Also obtain the coefficient of correlation between the ordered residuals and their expected values under normality. Does the normality assumption appear to be reasonable here?

e. The observations for each treatment were obtained in the order shown. Prepare residual sequence plots and analyze them. What are your findings?

18.13. Refer to **Eye contact effect** Problem 18.12. Assume that ANOVA model (18.23) is applicable.

a. Plot the estimated treatment means $\bar{Y}_{ij}$ in the format of Figure 18.5. Does it appear that any factor effects are present? Explain.

b. Set up the analysis of variance table. Does any one source account for most of the total variability in the success ratings in the study? Explain.

c. Test whether or not interaction effects are present; use $\alpha = .01$. State the alternatives, decision rule, and conclusion. What is the P-value of the test?

d. Test whether or not eye contact and gender main effects are present. In each case, use $\alpha = .01$ and state the alternatives, decision rule, and conclusion. What is the P-value of each test? Is it meaningful here to test for main factor effects? Explain.

e. Obtain an upper bound on the family level of significance for the tests in parts (c) and (d); use the Kimball inequality.

f. Do the results in parts (c) and (d) confirm your graphic analysis in part (a)?

18.14. **Hay fever relief.**   A research laboratory was developing a new compound for the relief of severe cases of hay fever. In an experiment with 36 volunteers, the amounts of the two active ingredients (factors A and B) in the compound were varied at three levels each. Randomization was used in assigning four volunteers to each of the nine treatments. The data on hours of relief follow.

Factor A (ingredient 1)	Factor B (ingredient 2) $j = 1$ Low	$j = 2$ Medium	$j = 3$ High
$i = 1$ Low	2.4	4.6	4.8
	2.7	4.2	4.5
	2.3	4.9	4.4
	2.5	4.7	4.6
$i = 2$ Medium	5.8	8.9	9.1
	5.2	9.1	9.3
	5.5	8.7	8.7
	5.3	9.0	9.4
$i = 3$ High	6.1	9.9	13.5
	5.7	10.5	13.0
	5.9	10.6	13.3
	6.2	10.1	13.2

a. Obtain the fitted values for ANOVA model (18.23).

b. Obtain the residuals.

c. Plot the residuals against the fitted values. What departures from ANOVA model (18.23) can be studied from this plot? What are your findings?

d. Prepare a normal probability plot of the residuals. Also obtain the coefficient of correlation between the ordered residuals and their expected values under normality. Does the normality assumption appear to be reasonable here?

18.15. Refer to **Hay fever relief** Problem 18.14. Assume that ANOVA model (18.23) is applicable.

    a. Plot the estimated treatment means $\bar{Y}_{ij}$ in the format of Figure 18.5. Does your graph suggest that any factor effects are present? Explain.

    b. Obtain the analysis of variance table. Does any one source account for most of the total variability in hours of relief in the study? Explain.

    c. Test whether or not the two factors interact; use $\alpha = .05$. State the alternatives, decision rule, and conclusion. What is the P-value of the test?

    d. Test whether or not main effects for the two ingredients are present. Use $\alpha = .05$ in each case and state the alternatives, decision rule, and conclusion. What is the P-value of each test? Is it meaningful here to test for main factor effects? Explain.

    e. Obtain an upper bound on the family level of significance for the tests in parts (c) and (d); use the Kimball inequality.

    f. Do the results in parts (c) and (d) confirm your graphic analysis in part (a)?

18.16. **Disk drive service.** The staff of a service center for electronic equipment includes three technicians who specialize in repairing three widely used makes of disk drives for desk-top computers. It was desired to study the effects of technician (factor A) and make of disk drive (factor B) on the service time. The data that follow show the number of minutes required to complete the repair job in a study where each technician was randomly assigned to five jobs on each make of disk drive.

Factor A (technician)	Factor B (make of drive)		
	$j = 1$ Make 1	$j = 2$ Make 2	$j = 3$ Make 3
$i = 1$ Technician 1	62	57	59
	48	45	53
	63	39	67
	57	54	66
	69	44	47
$i = 2$ Technician 2	51	61	55
	57	58	58
	45	70	50
	50	66	69
	39	51	49
$i = 3$ Technician 3	59	58	47
	65	63	56
	55	70	51
	52	53	44
	70	60	50

    a. Obtain the fitted values for ANOVA model (18.23).

    b. Obtain the residuals.

    c. Plot the residuals against the fitted values. What departures from ANOVA model (18.23) can be studied from this plot? What are your findings?

    d. Prepare a normal probability plot of the residuals. Also obtain the coefficient of correlation between the ordered residuals and their expected values under normality. Does the normality assumption appear to be reasonable here?

e. The observations for each treatment were obtained in the order shown. Prepare residual sequence plots and analyze them. What are your findings?

18.17. Refer to **Disk drive service** Problem 18.16. Assume that ANOVA model (18.23) is applicable.

a. Plot the estimated treatment means $\bar{Y}_{ij}$ in the format of Figure 18.5. Does your graph suggest that any factor effects are present? Explain.

b. Obtain the analysis of variance table. Does any one source account for most of the total variability?

c. Test whether or not the two factors interact; use $\alpha = .01$. State the alternatives, decision rule, and conclusion. What is the P-value of the test?

d. Test whether or not main effects for technician and make of drive are present. Use $\alpha = .01$ in each case and state the alternatives, decision rule, and conclusion. What is the P-value of each test? Is it meaningful here to test for main factor effects? Explain.

e. Obtain an upper bound on the family level of significance for the tests in parts (c) and (d); use the Kimball inequality.

f. Do the results in parts (c) and (d) confirm your graphic analysis in part (a)?

18.18. **Kidney failure hospitalization.** Kidney failure patients are commonly treated on dialysis machines that filter toxic substances from the blood. The appropriate "dose" for effective treatment depends, among other things, on duration of treatment and weight gain between treatments as a result of fluid buildup. To study the effects of these two factors on the number of days hospitalized (attributable to the disease) during a year, a random sample of 10 patients per group who had undergone treatment at a large dialysis facility was obtained. Treatment duration (factor A) was categorized into two groups: short duration (average dialyzing time for the year under four hours) and long duration (average dialyzing time for the year equal to or greater than four hours). Average weight gain between treatments (factor B) during the year was categorized into three groups: slight, moderate, and severe. The data on number of days hospitalized follow.

Factor A (duration)	Factor B (weight gain)					
	$j = 1$ Mild		$j = 2$ Moderate		$j = 3$ Severe	
$i = 1$ Short	0	2	2	4	15	16
	2	0	4	3	10	7
	1	5	7	1	8	30
	3	6	12	5	5	3
	0	8	15	20	25	27
$i = 2$ Long	0	2	5	1	10	15
	1	7	3	3	8	4
	1	4	2	6	12	9
	0	0	0	7	3	6
	4	3	1	9	7	1

The transformed data $Y' = \log_{10}(Y + 1)$ are to be used for the analysis.

a. Transform the data.

b. Obtain the fitted values and residuals for ANOVA model (18.23) with the transformed data.

c. Prepare aligned residual dot plots for the treatments. What departures from ANOVA model (18.23) can be studied from these plots? What are your findings?

d. Prepare a normal probability plot of the residuals. Also obtain the coefficient of correlation between the ordered residuals and their expected values under normality. Does the normality assumption appear to be reasonable here?

18.19. Refer to **Kidney failure hospitalization** Problem 18.18. Assume that ANOVA model (18.23) is appropriate for the transformed data.

   a. Plot the estimated treatment means $\bar{Y}'_{ij.}$ in the format of Figure 18.5. Does your graph suggest that any factor effects are present? Explain.

   b. Obtain the analysis of variance table. Does any one source account for most of the total variability?

   c. Test whether or not the two factors interact; use $\alpha = .05$. State the alternatives, decision rule, and conclusion. What is the P-value of the test?

   d. Test whether or not main effects for duration and weight gain are present. Use $\alpha = .05$ in each case and state the alternatives, decision rule, and conclusion. What is the P-value of each test? Is it meaningful here to test for main factor effects? Explain.

   e. Obtain an upper bound on the family level of significance for the tests in parts (c) and (d); use the Kimball inequality.

   f. Do the results in parts (c) and (d) confirm your graphic analysis in part (a)?

18.20. **Programmer requirements.**   A computer software firm was encountering difficulties in forecasting the programmer requirements for large-scale programming projects. As part of a study to remedy the difficulties, 24 programmers, classified into equal groups by type of experience (factor A) and amount of experience (factor B), were asked to predict the number of programmer-days required to complete a large project about to be initiated. After this project was completed, the prediction errors (actual minus predicted programmer-days) were determined. The data on prediction errors follow.

	Factor B (years of experience)		
Factor A (type of experience)	$j = 1$ Under 5	$j = 2$ 5–under 10	$j = 3$ 10 or more
$i = 1$ Small systems only	240	110	56
	206	118	60
	217	103	68
	225	95	58
$i = 2$ Small and large systems	71	47	37
	53	52	33
	68	31	40
	57	49	45

   a. Obtain the fitted values for ANOVA model (18.23).

   b. Obtain the residuals.

   c. Prepare aligned residual dot plots for the treatments. What departures from ANOVA model (18.23) can be studied from these plots? What are your findings?

      d. Prepare a normal probability plot of the residuals. Also obtain the coefficient of correlation between the ordered residuals and their expected values under normality. Does the normality assumption appear to be reasonable here?

18.21. Refer to **Programmer requirements** Problem 18.20. Assume that ANOVA model (18.23) is applicable.

      a. Plot the estimated treatment means $\bar{Y}_{ij.}$ in the format of Figure 18.5. Does your graph suggest that any factor effects are present? Explain.

      b. Obtain the analysis of variance table. Does any one source account for most of the total variability?

      c. Test whether or not the two factors interact; use $\alpha = .01$. State the alternatives, decision rule, and conclusion. What is the P-value of the test?

      d. Test whether or not main effects for type of experience and years of experience are present. Use $\alpha = .01$ in each case and state the alternatives, decision rule, and conclusion. What is the P-value of each test? Is it meaningful here to test for main factor effects? Explain.

      e. Obtain an upper bound on the family level of significance for the tests in parts (c) and (d); use the Kimball inequality.

      f. Do the results in parts (c) and (d) confirm your graphic analysis in part (a)?

18.22. How does the randomization of treatment assignments in a two-factor study differ when both factors are experimental factors and when only one factor is an experimental factor?

18.23. Refer to **Eye contact effect** Problem 18.12.

      a. Explain how you would make the assignments of personnel officers to treatments in this two-factor study. Make all appropriate randomizations.

      b. Did you randomize the officers to the factor levels of each factor?

18.24. Refer to **Hay fever relief** Problem 18.14.

      a. Explain how you would make the assignments of volunteers to treatments in this study. Make all appropriate randomizations.

      b. Did you randomize the volunteers to the factor levels of each factor?

18.25. Refer to **Disk drive service** Problem 18.16.

      a. Is any randomization of treatment assignments called for in this study? Is any randomization utilized? Explain.

      b. Would you consider this study to be experimental in nature? Discuss.

18.26. Refer to **Eye contact effect** Problem 18.12.

      a. Modify regression model (18.68) to apply to this two-factor study with $a = 2$ and $b = 2$.

      b. Set up the **Y**, **X**, and $\boldsymbol{\beta}$ matrices for the regression model in part (a).

      c. Obtain $\mathbf{X}\boldsymbol{\beta}$. Verify the correctness of the expected values.

      d. Obtain the fitted regression function. What is estimated by the intercept term?

      e. Obtain the regression analysis of variance table based on appropriate extra sums of squares.

      f. Test separately for interaction effects, factor A main effects, and factor B main effects. Use $\alpha = .01$ for each test and state the alternatives, decision rule, and conclusion.

18.27. Refer to **Hay fever relief** Problem 18.14.

      a. Modify regression model (18.68) to apply to this two-factor study with $a = 3$ and $b = 3$.

b. Set up the **Y**, **X**, and **β** matrices for the regression model in part (a).

c. Obtain **Xβ**. Verify the correctness of the expected values.

d. Obtain the fitted regression function. What is estimated by $\hat{\alpha}_1$?

e. Obtain the regression analysis of variance table based on appropriate extra sums of squares.

f. Test separately for interaction effects, factor A main effects, and factor B main effects. Use $\alpha = .05$ for each test and state the alternatives, decision rule, and conclusion.

18.28. Refer to **Disk drive service** Problem 18.16.

a. Modify regression model (18.68) to apply to this two-factor study with $a = 3$ and $b = 3$.

b. Obtain the fitted regression function. What is estimated by $\hat{\beta}_1$?

c. Obtain the regression analysis of variance table based on appropriate extra sums of squares.

d. Test separately for interaction effects, factor A main effects, and factor B main effects. Use $\alpha = .01$ for each test and state the alternatives, decision rule, and conclusion.

18.29. Refer to **Cash offers** Problem 18.10. It is desired to test for factor B main effects by the general linear test approach and matrix methods.

a. Obtain the **Y**, **X**, and **β** matrices for full model (18.15), as illustrated in (18.19).

b. Obtain the estimated regression coefficients for the full model using (8.64).

c. Express the alternatives in matrix form (8.66).

d. Obtain $SSE(R) - SSE(F)$ using (8.70). Does your result equal $SSB = 5.444$, as it should?

EXERCISES

18.30. Derive (18.7a) from (18.7).

18.31. Prove the result in (18.9b).

18.32. (Calculus needed.) State the likelihood function for ANOVA model (18.15) when $a = 2$, $b = 2$, and $n = 2$. Find the maximum likelihood estimators. Are they the same as the least squares estimators in (18.29)?

18.33. (Calculus needed.) Derive (18.29).

18.34. Derive (18.39) from (18.37).

18.35. Show the equivalence between the expressions in (18.40c) and (18.39a).

PROJECTS

18.36. Refer to the **SENIC** data set. The following hospitals are to be considered in a study of the effects of region (factor A: variable 9) and average age of patients (factor B: variable 3) on the mean length of hospital stay of patients (variable 2):

1–44	46	48	51	53	57	58	60	63	66	74
76	79	80	83	84	88	94	101	103	111	

For purposes of this ANOVA study, average age is to be classified into two categories: less than or equal to 53.9 years, 54.0 years or more.

a. Assemble the required data and obtain the fitted values for ANOVA model (18.23).

b. Obtain the residuals.

c. Plot the residuals against the fitted values. What departures from ANOVA model (18.23) can be studied from this plot? What are your findings?

d. Prepare a normal probability plot of the residuals. Also obtain the coefficient of correlation between the ordered residuals and their expected values under normality. Does the normality assumption appear to be reasonable here?

18.37. Refer to the **SENIC** data set and Project 18.36. Assume that ANOVA model (18.23) is applicable.

a. Plot the estimated treatment means $\bar{Y}_{ij.}$ in the format of Figure 18.5. Does it appear that any factor effects are present? Explain.

b. Obtain the analysis of variance table. Does any one source account for most of the total variability in the study? Explain.

c. Test whether or not interaction effects are present; use $\alpha = .05$. State the alternatives, decision rule, and conclusion. What is the P-value of the test?

d. Test whether or not region and age main effects are present. In each case, use $\alpha = .05$ and state the alternatives, decision rule, and conclusion. What is the P-value of each test? Is it meaningful here to test for main factor effects? Explain.

e. Obtain an upper bound on the family level of significance for the tests in parts (c) and (d); use the Kimball inequality.

f. Do the results in parts (c) and (d) confirm your graphic analysis in part (a)?

18.38. Refer to the **SMSA** data set. The following metropolitan areas are to be considered in a study of the effects of region (factor A: variable 12) and percent of population in central cities (factor B: variable 4) on the crime rate (variable 11 ÷ variable 3):

$$1-36 \quad 38 \quad 39 \quad 45 \quad 46 \quad 49 \quad 53 \quad 62 \quad 71 \quad 73 \quad 101 \quad 123 \quad 130$$

For purposes of this ANOVA study, percent of population in central cities is to be classified into two categories: less than or equal to 38.9 percent, 39.0 percent or more.

a. Assemble the required data and obtain the fitted values for ANOVA model (18.23).

b. Obtain the residuals.

c. Prepare aligned residual dot plots for the treatments. What departures from ANOVA model (18.23) can be studied from these plots? What are your findings?

d. Prepare a normal probability plot of the residuals. Also obtain the coefficient of correlation between the ordered residuals and their expected values under normality. Does the normality assumption appear to be reasonable here?

18.39. Refer to the **SMSA** data set and Project 18.38. Assume that ANOVA model (18.23) is applicable.

a. Plot the estimated treatment means $\bar{Y}_{ij.}$ in the format of Figure 18.5. Does it appear that any factor effects are present? Explain.

b. Set up the analysis of variance table. Does any one source account for most of the total variability in the study? Explain.

c. Test whether or not interaction effects are present; use $\alpha = .01$. State the alternatives, decision rule, and conclusion. What is the P-value of the test?

d. Test whether or not region and percent of population in central cities main effects are present. In each case, use $\alpha = .01$ and state the alternatives, decision rule, and conclusion. What is the P-value of each test? Is it meaningful here to test for main factor effects? Explain.

e. Obtain an upper bound on the family level of significance for the tests in parts (c) and (d); use the Kimball inequality.

f. Do the results in parts (c) and (d) confirm your graphic analysis in part (a)?

Chapter 19

Analysis and Planning of Two-Factor Studies— Equal Sample Sizes

When the analysis of variance tests described in Chapter 18 indicate the presence of factor effects in two-factor studies, the next step is to analyze the nature of the factor effects. In this chapter, we discuss how to conduct such analyses. We continue to consider the two-factor fixed ANOVA model (18.23), *when there are n observations for each treatment and all treatment means are deemed to be of equal importance*. We take up first the analysis of factor effects when both factors are qualitative and then the analysis when one or both factors are quantitative.

We conclude this chapter by discussing briefly the planning of sample sizes for two-factor studies, a key element in the design of such studies.

19.1 STRATEGY FOR ANALYSIS

Our consideration in Chapter 18 of the meaning of the model elements suggests the following basic strategy for analyzing factor effects in two-factor studies:

1. Examine whether the two factors interact.
2. If they do not interact, examine whether the main effects for factors A and B are important. For important A or B main effects, describe the nature of these effects in terms of the factor level means $\mu_{i.}$ or $\mu_{.j}$, respectively. In some special cases, there may also be interest in the treatment means μ_{ij}.
3. If the factors do interact, examine if the interactions are important or unimportant.
4. If the interactions are unimportant, proceed as in step 2.
5. If the interactions are important, consider whether they can be made unimportant by a meaningful simple transformation of scale. If so, make the transformation and proceed as in step 2.

6. For important interactions that cannot be made unimportant by a simple transformation, analyze the two factor effects jointly in terms of the treatment means μ_{ij}. In some special cases, there may also be interest in the factor level means $\mu_{i\cdot}$ and $\mu_{\cdot j}$.

A flowchart of this strategy is presented in Figure 19.1.

Step 1 of this strategy, testing for interaction effects, was discussed in Chapter 18. Also discussed there was step 5, the possible diminution of important interactions by a meaningful simple transformation, as well as how to test for the presence of factor main effects. Now we turn to steps 2 and 6 of the strategy for analysis, namely, how to compare factor level means $\mu_{i\cdot}$ or $\mu_{\cdot j}$ when there are no interactions or only unim-

FIGURE 19.1   Strategy for Analysis of Two-Factor Studies

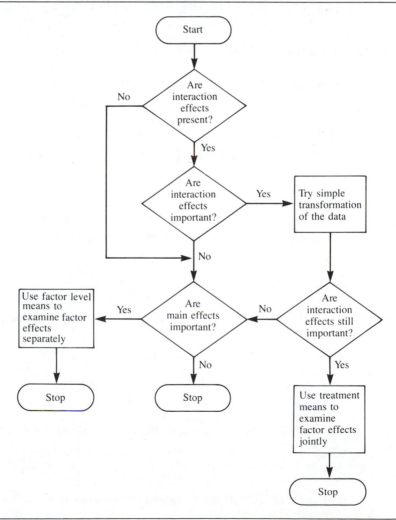

portant ones, and how to compare treatment means μ_{ij} when there are important interactions. We begin with a discussion of the analysis of factor effects when the factors do not interact or interact only in unimportant fashion.

19.2 ANALYSIS OF FACTOR EFFECTS WHEN FACTORS DO NOT INTERACT

As just noted, the analysis of factor effects usually only involves the factor level means $\mu_{i.}$ and $\mu_{.j}$ when the two factors do not interact, or when they interact only in unimportant fashion. Before proceeding with formal estimation procedures, it is often useful to plot the estimated factor level means in a normal probability plot, as explained in Section 15.1 for single-factor studies.

Estimation of Factor Level Mean

Unbiased point estimators of $\mu_{i.}$ and $\mu_{.j}$ are:

(19.1a)
$$\hat{\mu}_{i.} = \bar{Y}_{i..}$$

(19.1b)
$$\hat{\mu}_{.j} = \bar{Y}_{.j.}$$

where $\bar{Y}_{i..}$ and $\bar{Y}_{.j.}$ are defined in (18.27d) and (18.27f), respectively. The variance of $\bar{Y}_{i..}$ is:

(19.2a)
$$\sigma^2\{\bar{Y}_{i..}\} = \frac{\sigma^2}{bn}$$

since $\bar{Y}_{i..}$ contains bn independent observations, each with variance σ^2. Similarly, we have:

(19.2b)
$$\sigma^2\{\bar{Y}_{.j.}\} = \frac{\sigma^2}{an}$$

Unbiased estimators of these variances are obtained by replacing σ^2 with *MSE*:

(19.3a)
$$s^2\{\bar{Y}_{i..}\} = \frac{MSE}{bn}$$

(19.3b)
$$s^2\{\bar{Y}_{.j.}\} = \frac{MSE}{an}$$

Confidence limits for $\mu_{i.}$ and $\mu_{.j}$ utilize, as usual, the t distribution:

(19.4a)
$$\bar{Y}_{i..} \pm t[1 - \alpha/2; (n - 1)ab]s\{\bar{Y}_{i..}\}$$

(19.4b)
$$\bar{Y}_{.j.} \pm t[1 - \alpha/2; (n - 1)ab]s\{\bar{Y}_{.j.}\}$$

The degrees of freedom $(n - 1)ab$ are those associated with *MSE*.

Estimation of Contrast of Factor Level Means

A contrast among the factor level means $\mu_{i.}$:

$$(19.5) \qquad L = \sum c_i \mu_{i.} \qquad \text{where } \sum c_i = 0$$

is estimated unbiasedly by:

$$(19.6) \qquad \hat{L} = \sum c_i \bar{Y}_{i..}$$

Because of the independence of the $\bar{Y}_{i..}$, the variance of this estimator is:

$$(19.7) \qquad \sigma^2\{\hat{L}\} = \sum c_i^2 \sigma^2\{\bar{Y}_{i..}\} = \frac{\sigma^2}{bn} \sum c_i^2$$

An unbiased estimator of this variance is:

$$(19.8) \qquad s^2\{\hat{L}\} = \frac{MSE}{bn} \sum c_i^2$$

Finally, the appropriate $1 - \alpha$ confidence limits for L are:

$$(19.9) \qquad \hat{L} \pm t[1 - \alpha/2; (n - 1)ab]s\{\hat{L}\}$$

To estimate a contrast among the factor level means $\mu_{.j}$:

$$(19.10) \qquad L = \sum c_j \mu_{.j} \qquad \text{where } \sum c_j = 0$$

we use the estimator:

$$(19.11) \qquad \hat{L} = \sum c_j \bar{Y}_{.j.}$$

whose estimated variance is:

$$(19.12) \qquad s^2\{\hat{L}\} = \frac{MSE}{an} \sum c_j^2$$

The $1 - \alpha$ confidence limits for L in (19.9) are still appropriate, with $\hat{L}$ and $s\{\hat{L}\}$ now defined in (19.11) and (19.12), respectively.

Estimation of Linear Combination of Factor Level Means

A linear combination of the factor level means $\mu_{i.}$:

$$(19.13) \qquad L = \sum c_i \mu_{i.}$$

is estimated unbiasedly by $\hat{L}$ in (19.6). The variance of this estimator is given in

(19.7), and an unbiased estimator of this variance is given in (19.8). The appropriate $1 - \alpha$ confidence limits for L are given in (19.9).

Analogous results follow for a linear combination of the factor level means $\mu_{.j}$:

(19.14)
$$L = \sum c_j \mu_{.j}$$

Multiple Pairwise Comparisons of Factor Level Means

Usually, more than one comparison is of interest, and the multiple comparison procedures discussed in Chapter 15 can be employed with only minor modifications. If all or a large number of pairwise comparisons among the factor level means $\mu_{i.}$ are to be made, of the form:

(19.15)
$$D = \mu_{i.} - \mu_{i'.}$$

the Tukey procedure of (15.25) is appropriate. The formulas are as follows, reflecting the case of equal sample sizes considered here:

(19.16a)
$$\hat{D} = \bar{Y}_{i..} - \bar{Y}_{i'..}$$

(19.16b)
$$s^2\{\hat{D}\} = \frac{2MSE}{bn}$$

(19.16c)
$$T = \frac{1}{\sqrt{2}} q[1 - \alpha; a, (n - 1)ab]$$

The confidence limits as usual are:

(19.17)
$$\hat{D} \pm Ts\{\hat{D}\}$$

The probability then is $1 - \alpha$ that all statements in the family are correct.

For pairwise comparisons of the factor level means $\mu_{.j}$, the only changes are:

(19.18a)
$$D = \mu_{.j} - \mu_{.j'}$$

(19.18b)
$$\hat{D} = \bar{Y}_{.j.} - \bar{Y}_{.j'.}$$

(19.18c)
$$s^2\{\hat{D}\} = \frac{2MSE}{an}$$

(19.18d)
$$T = \frac{1}{\sqrt{2}} q[1 - \alpha; b, (n - 1)ab]$$

If only a few pairwise comparisons are to be made, the Bonferroni method may be best. All of the above formulas still apply, but the Tukey multiple T is replaced by the Bonferroni multiple B:

(19.19)
$$B = t[1 - \alpha/2g; (n - 1)ab]$$

where g is the number of statements in the family.

If it is desired to have a family confidence coefficient $1 - \alpha$ for the joint set of pairwise comparisons involving *both* factor A and factor B means, the Bonferroni method can be used directly, with g representing the total number of statements in the joint set. Alternatively, the Bonferroni method can be used in conjunction with the Tukey method. To illustrate this use, suppose the pairwise comparisons for factor A are made with the Tukey procedure with a family confidence coefficient of .95, and likewise for the pairwise comparisons for factor B. The Bonferroni inequality then assures us that the family confidence coefficient for the joint set of comparisons for both factors is at least .90.

Multiple Contrasts of Factor Level Means

When a large number of contrasts among the factor level means $\mu_{i.}$ or $\mu_{.j}$ are of interest, the Scheffé method should be used. If the contrasts involve the $\mu_{i.}$, as in (19.5), the unbiased estimator is given in (19.6) and its estimated variance in (19.8). For this case, the Scheffé multiple is defined by:

$$(19.20) \qquad S^2 = (a - 1)F[1 - \alpha; a - 1, (n - 1)ab]$$

leading to the confidence limits for the contrast L:

$$(19.21) \qquad \hat{L} \pm Ss\{\hat{L}\}$$

The probability is then $1 - \alpha$ that every confidence interval (19.21) in the family of all possible contrasts is correct.

If contrasts among the factor level means $\mu_{.j}$ are desired, the unbiased point estimator is given in (19.11), its estimated variance is given in (19.12), and the Scheffé confidence limits (19.21) are appropriate with:

$$(19.22) \qquad S^2 = (b - 1)F[1 - \alpha; b - 1, (n - 1)ab]$$

When the number of contrasts of interest is small, the Bonferroni method may be best. Confidence limits (19.21) would need to be modified by replacing the Scheffé multiple S with the Bonferroni multiple B:

$$(19.23) \qquad B = t[1 - \alpha/2g; (n - 1)ab]$$

where g is the number of statements in the family.

When it is desired to obtain a family confidence coefficient for the joint set of contrasts for both factors, several possibilities exist:

1. The Bonferroni method may be used directly, with g representing the total number of statements in the joint set.

2. The Bonferroni method can be used to join the two sets of Scheffé multiple comparison families in the same way explained earlier for joining two Tukey sets.

3. The Scheffé method can be modified to use the S multiple defined by:

$$(19.24) \qquad S^2 = (a + b - 2)F[1 - \alpha; a + b - 2, (n - 1)ab]$$

When this S multiple is used in both sets of multiple comparisons, the probability is $1 - \alpha$ that all statements in the combined family are correct.

One may try each of these three approaches and see which leads to the narrowest confidence intervals, without affecting the validity of the procedure.

Estimates Based on Treatment Means

Occasionally in analyzing the factor effects in a two-factor study when no interactions are present, one is interested in particular treatment means μ_{ij}. For example, in a two-factor study of the effects of price and type of advertisement on sales, interest may exist in estimating the mean sales for two different price levels when a particular advertisement is used. In such cases, the methods of analysis for single-factor studies discussed in Chapter 15 are appropriate. The number of treatments now is simply $r = ab$, the degrees of freedom associated with MSE are $n_T - r = nab - ab = (n - 1)ab$, and the estimated treatment means are $\bar{Y}_{ij\cdot}$, based on n observations each.

Example 1—Pairwise Comparisons of Factor Level Means

In the Castle Bakery example of Chapter 18, the plot of the estimated treatment means in Figure 18.5 suggested that no interaction effects are present. The formal analysis of variance based on Table 18.10 supported this conclusion. Suppose now that no tests of main factor effects had been conducted, and that we wish to analyze the effects of shelf width and shelf height by means of estimation procedures.

We shall perform the analysis in terms of the factor level means since no interaction effects are present. First, we plot the estimated factor level means given in Table 18.7 in normal probability plots. Figure 19.2a contains the normal probability plot of the estimated factor A level means $\bar{Y}_{i\cdot\cdot}$, and Figure 19.2b contains a similar plot for the estimated factor B level means $\bar{Y}_{\cdot j\cdot}$. These plots are prepared in the same fashion as the one in Figure 15.3b for a single-factor study. Again, we show in each plot the line $y =$ expected value, representing the expected values if all factor level means are equal. The expected values are obtained as follows:

$$\text{Factor } A: \quad \bar{Y}_{\cdots} + z\left(\frac{i - .375}{a + .25}\right)\sqrt{\frac{MSE}{bn}}$$

$$\text{Factor } B: \quad \bar{Y}_{\cdots} + z\left(\frac{i - .375}{b + .25}\right)\sqrt{\frac{MSE}{an}}$$

Figure 19.2a suggests that level 2 of factor A (middle shelf height) leads to significantly larger sales than the other two factor levels. Figure 19.2b follows the prototype pattern in Figure 15.2a, indicating no factor level effects for display width.

Turning now to formal estimation procedures, we shall make pairwise comparisons between the factor level means for each factor, with a combined confidence

FIGURE 19.2 Normal Probability Plots of Estimated Factor Level Means— Castle Bakery Example

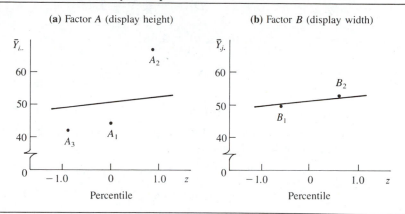

(a) Factor A (display height)   (b) Factor B (display width)

coefficient of 90 percent for all comparisons. The Tukey multiple comparison procedure will be used, with the comparisons for display height assigned a family confidence coefficient of 95 percent, and similarly for the display width comparison. The Bonferroni inequality then guarantees a family confidence coefficient of at least 90 percent for the joint set of comparisons.

For the comparison of display height means ($i = 1$—bottom, 2—middle, 3—top), we have from Tables 18.7 and 18.8:

$$\bar{Y}_{2..} - \bar{Y}_{1..} = 67 - 44 = 23 \qquad MSE = 10.3$$

$$a = 3$$

$$\bar{Y}_{1..} - \bar{Y}_{3..} = 44 - 42 = 2 \qquad b = 2$$

$$n = 2$$

$$\bar{Y}_{2..} - \bar{Y}_{3..} = 67 - 42 = 25 \qquad (n - 1)ab = 6$$

Hence, by (19.16), we obtain:

$$s^2\{\hat{D}\} = \frac{2(10.3)}{2(2)} = 5.15$$

$$q(.95; 3, 6) = 4.34$$

$$T = \frac{4.34}{\sqrt{2}} = 3.07$$

$$Ts\{\hat{D}\} = 3.07\sqrt{5.15} = 7.0$$

Similarly, for the comparison of display widths ($j = 1$—regular, 2—wide), we have by (19.18):

$$\bar{Y}_{.2.} - \bar{Y}_{.1.} = 52 - 50 = 2$$

$$s^2\{\hat{D}\} = \frac{2(10.3)}{3(2)} = 3.43$$

$$q(.95; 2, 6) = 3.46$$

$$T = \frac{3.46}{\sqrt{2}} = 2.45$$

$$Ts\{\hat{D}\} = 2.45\sqrt{3.43} = 4.5$$

We obtain therefore the following confidence intervals for all pairwise comparisons of factor level means:

$$16 = 23 - 7.0 \leq \mu_{2.} - \mu_{1.} \leq 23 + 7.0 = 30$$

$$-5 = 2 - 7.0 \leq \mu_{1.} - \mu_{3.} \leq 2 + 7.0 = 9$$

$$18 = 25 - 7.0 \leq \mu_{2.} - \mu_{3.} \leq 25 + 7.0 = 32$$

$$-2.5 = 2 - 4.5 \leq \mu_{.2} - \mu_{.1} \leq 2 + 4.5 = 6.5$$

It can be concluded from these confidence intervals with family confidence coefficient of 90 percent for the joint set that for the product studied and the types of stores in the experiment, the middle shelf height is far better than either the bottom or the top heights, the latter two do not differ significantly in sales effectiveness, and widening the display has no significant effect on sales. All these conclusions are covered by the family confidence coefficient of 90 percent. The effects of shelf height can be summarized graphically as follows:

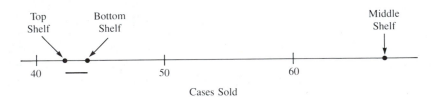

Example 2—Estimation of Treatment Means

The manager of a supermarket similar in terms of sales volume and clientele to the supermarkets included in the Castle Bakery study has room only for the regular shelf display width, and wishes to obtain estimates of mean sales for the middle and top shelf heights. We shall obtain interval estimates with a 90 percent family confidence coefficient using the Bonferroni procedure.

From Tables 18.7 and 18.8, we have:

$$\bar{Y}_{21.} = 65 \qquad \bar{Y}_{31.} = 40 \qquad MSE = 10.3$$

Hence, we obtain:

$$s^2\{\bar{Y}_{21.}\} = s^2\{\bar{Y}_{31.}\} = \frac{MSE}{n} = \frac{10.3}{2} = 5.15$$

$$s\{\bar{Y}_{21.}\} = s\{\bar{Y}_{31.}\} = 2.27$$

For $g = 2$, we require $B = t[1 - \alpha/2g; (n - 1)ab] = t(.975; 6) = 2.447$. Thus, we obtain the confidence limits:

$$65 \pm 2.447(2.27) \qquad 40 \pm 2.447(2.27)$$

and the desired confidence intervals are:

$$59.4 \leq \mu_{21} \leq 70.6 \qquad 34.4 \leq \mu_{31} \leq 45.6$$

19.3 ANALYSIS OF FACTOR EFFECTS WHEN INTERACTIONS IMPORTANT

When important interactions exist that cannot be made unimportant by a simple transformation, the analysis of factor effects generally must be based on the treatment means μ_{ij}. Typically, this analysis will involve multiple comparisons or single degree of freedom tests of treatment means.

Multiple Pairwise Comparisons of Treatment Means

If pairs of treatment means μ_{ij} are to be compared, either the Tukey or the Bonferroni multiple comparison procedure may be used, depending on which is more advantageous. In effect, the analysis is equivalent to the single-factor case, with the total number of treatments here equal to $r = ab$, the degrees of freedom associated with MSE here equal to $n_T - r = (n - 1)ab$, and the estimated treatment mean, now denoted by $\bar{Y}_{ij.}$, based on n cases. Formula (15.25) for the Tukey multiple comparison procedure for $D = \mu_{ij} - \mu_{i'j'}$ with equal sample sizes becomes:

(19.25) $\hat{D} \pm Ts\{\hat{D}\} \qquad i, j \neq i', j'$

where:

(19.25a) $\hat{D} = \bar{Y}_{ij.} - \bar{Y}_{i'j'.}$

(19.25b) $s^2\{\hat{D}\} = \frac{2MSE}{n}$

(19.25c) $T = \frac{1}{\sqrt{2}}q[1 - \alpha; ab, (n - 1)ab]$

If the Bonferroni method is employed, the multiple in the confidence interval is:

(19.26) $B = t[1 - \alpha/2g; (n - 1)ab]$

where g is the number of statements in the family.

Multiple Contrasts of Treatment Means

The Scheffé multiple comparison procedure for single-factor studies is directly applicable to the estimation of contrasts involving the treatment means μ_{ij}. We denote these contrasts as follows:

$$(19.27) \qquad L = \sum\sum c_{ij}\mu_{ij} \qquad \text{where } \sum\sum c_{ij} = 0$$

The point estimator of L is:

$$(19.28) \qquad \hat{L} = \sum\sum c_{ij}\bar{Y}_{ij.}$$

and the estimated variance of this estimator is:

$$(19.29) \qquad s^2\{\hat{L}\} = \frac{MSE}{n}\sum\sum c_{ij}^2$$

The Scheffé multiple S is given by:

$$(19.30) \qquad S^2 = (ab - 1)F[1 - \alpha; ab - 1, (n - 1)ab]$$

and the joint confidence limits are as usual:

$$(19.31) \qquad \hat{L} \pm Ss\{\hat{L}\}$$

When the number of contrasts is small, the Bonferroni procedure may be preferable. The confidence intervals (19.31) would simply be modified by replacing S with B as defined in (19.26).

Single Degree of Freedom Tests of Treatment Means

When important interactions exist, a single degree of freedom test of treatment means μ_{ij} may sometimes be of interest. The two-sided alternatives for a single degree of freedom test here are as follows:

$$(19.32) \qquad \begin{aligned} H_0&: \sum\sum c_{ij}\mu_{ij} = c \\ H_a&: \sum\sum c_{ij}\mu_{ij} \neq c \end{aligned}$$

where the c_{ij} and c are appropriate constants.

To test alternatives (19.32) we can use the t^* test statistic:

$$(19.33) \qquad t^* = \frac{\sum\sum c_{ij}\bar{Y}_{ij.} - c}{\sqrt{\dfrac{MSE}{n}\sum\sum c_{ij}^2}}$$

which follows the t distribution with $(n-1)ab$ degrees of freedom when H_0 holds. Alternatively, we can use test statistic $F^* = (t^*)^2$, which follows the F distribution with 1 and $(n-1)ab$ degrees of freedom when H_0 holds.

When multiple single degree of freedom tests are to be conducted with a specified family level of significance, an appropriate multiple comparison procedure (Tukey, Scheffé, Bonferroni) should be used for setting up the relevant confidence intervals. The confidence intervals will show which of the two alternatives should be concluded in each case, as explained in Chapter 15 for single-factor studies.

Example 1—Pairwise Comparisons of Treatment Means

A junior college system studied the effects of teaching method (factor A) and student's quantitative ability (factor B) on learning of college mathematics. Two teaching methods were studied—the standard method of teaching (to be called the standard method) and a method that emphasizes teaching of concepts in the abstract before going into drill routines (to be called the abstract method). The quantitative ability of a student was determined by a standard aptitude test, on the basis of which the student was classified as having excellent, good, or moderate quantitative ability. Thus, factor A in this study (teaching method) has $a = 2$ levels, and factor B (student's quantitative ability) has $b = 3$ levels.

For each of the quantitative ability groups, 42 students were selected and randomly placed into classes according to the designated teaching method, with each class containing equal numbers of students of each quantitative ability level. For simplicity, it is assumed that any effects associated with the classes are negligible.

The dependent variable was the amount of learning of college mathematics, as measured by a standard mathematics achievement test. The results of the study are summarized in Table 19.1 (the original data are not shown). The estimated treatment means are shown in Table 19.1a, and the analysis of variance table is presented in Table 19.1b.

Figure 19.3 contains two plots of the estimated treatment means $\bar{Y}_{ij.}$. In Figure 19.3a, the two curves represent the different factor A levels; and in Figure 19.3b, the three curves represent the different factor B levels. The clear lack of parallelism of the curves suggests the presence of interaction effects between teaching method and student's quantitative ability on amount of mathematics learning. A formal test for interactions confirms this. From Table 19.1b, we have $F^* = MSAB/MSE = 325.5/28 = 11.625$. For $\alpha = .01$ we require $F(.99; 2, 120) = 4.79$. Since $F^* = 11.625 > 4.79$, we conclude that interaction effects are present. The P-value of this test is $0+$.

Figure 19.3 suggests that the interactions are important: students with excellent quantitative ability are but little affected by teaching method (perhaps doing slightly better with the abstract method); students with good or moderate abilities learn much better with the standard teaching method. Hence, we shall first investigate whether some simple transformation can make the interactions unimportant. We do

TABLE 19.1   Results of Study on Mathematics Learning

(a) Mean Learning Scores ($n = 21$)

Teaching Method i	Quantitative Ability (j)		
	Excellent	Good	Moderate
Abstract	92 ($\bar{Y}_{11.}$)	81 ($\bar{Y}_{12.}$)	73 ($\bar{Y}_{13.}$)
Standard	90 ($\bar{Y}_{21.}$)	86 ($\bar{Y}_{22.}$)	82 ($\bar{Y}_{23.}$)

(b) ANOVA Table

Source of Variation	SS	df	MS
Between treatments	4,998	5	999.6
Factor A (teaching methods)	504	1	504
Factor B (quantitative ability)	3,843	2	1,921.5
AB Interactions	651	2	325.5
Error	3,360	120	28
Total	8,358	125	

FIGURE 19.3   Plots of Estimated Treatment Means—Mathematics Learning Example

(a) Teaching Method Curves

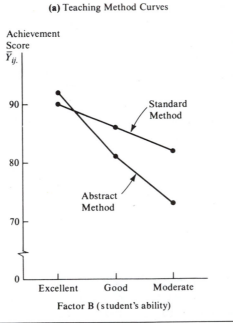

(b) Student Ability Curves

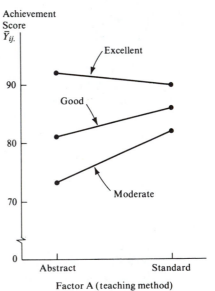

this in an approximate fashion by considering some transformations of $\bar{Y}_{ij.}$. In Figure 19.4a we present the teaching method curves for $\bar{Y}'_{ij.} = \sqrt{\bar{Y}_{ij.}}$, and in Figure 19.4b we present the teaching method curves for $\bar{Y}'_{ij.} = \log_{10} \bar{Y}_{ij.}$. In neither case have the interactions become unimportant, so it appears that the interactions here may be nontransformable.

We now proceed to investigate the nature of the interaction effects in Figure 19.3. We do this by estimating separately for students with excellent, good, and moderate quantitative abilities how large is the difference in mean learning for the two teaching methods. Thus, we wish to estimate:

$$D_1 = \mu_{11} - \mu_{21}$$

(19.34)
$$D_2 = \mu_{12} - \mu_{22}$$

$$D_3 = \mu_{13} - \mu_{23}$$

We shall employ the Bonferroni multiple comparison procedure with family confidence coefficient .95. (Since only three pairwise comparisons are of interest, the Bonferroni method yields more precise estimates here than the Tukey method.)

Using the data in Table 19.1a, the point estimates of the pairwise comparisons are:

$$\hat{D}_1 = 92 - 90 = 2$$

(19.35)
$$\hat{D}_2 = 81 - 86 = -5$$

$$\hat{D}_3 = 73 - 82 = -9$$

FIGURE 19.4   Plots of Teaching Method Curves with Transformed Estimated Means—Mathematics Learning Example

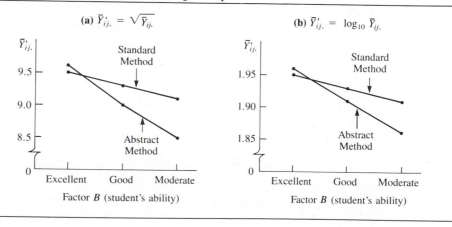

Since $n = 21$, we find the estimated variances of these estimates by (19.25b) to be:

$$s^2\{\hat{D}_1\} = s^2\{\hat{D}_2\} = s^2\{\hat{D}_3\} = \frac{2(28)}{21} = 2.667$$

so that:

$$s\{\hat{D}_1\} = s\{\hat{D}_2\} = s\{\hat{D}_3\} = 1.633$$

Finally, for family confidence coefficient $1 - \alpha = .95$ and $g = 3$, we require $B = t[1 - .05/2(3); 120] = t(.99167; 120) = 2.428$. Hence, the confidence limits are by (19.25):

$$2 \pm 2.428(1.633) \qquad -5 \pm 2.428(1.633) \qquad -9 \pm 2.428(1.633)$$

and the 95 percent confidence intervals for the family of comparisons are:

$$-1.96 \le \mu_{11} - \mu_{21} \le 5.96$$

$$-8.96 \le \mu_{12} - \mu_{22} \le -1.04$$

$$-12.96 \le \mu_{13} - \mu_{23} \le -5.04$$

From this family of confidence intervals the following conclusions may be drawn with family confidence coefficient of 95 percent: (1) The mean learning scores with the two teaching methods for students with excellent quantitative ability do not differ. (2) The mean learning score with the abstract teaching method is lower than that with the standard method for both students with good and moderate quantitative abilities. The superiority of the standard teaching method may be particularly strong for students with moderate quantitative abilities.

Example 2—Contrasts of Treatment Means

A school administrator also wished to know in the mathematics learning example whether the amount of gain in learning with the standard teaching method over the abstract method for students with moderate quantitative ability is greater than the gain for students with good quantitative ability. This question had been raised before the study began. We shall estimate the single contrast:

$$(19.36) \qquad L = (\mu_{23} - \mu_{13}) - (\mu_{22} - \mu_{12})$$

by means of a one-sided lower confidence interval. Utilizing the results in Table 19.1a, the point estimate of L is $\hat{L} = (82 - 73) - (86 - 81) = 4$. The estimated variance by (19.29) is:

$$s^2\{\hat{L}\} = \frac{28}{21}[(1)^2 + (-1)^2 + (-1)^2 + (1)^2] = 5.333$$

so that the estimated standard deviation is $s\{\hat{L}\} = 2.309$. For a 95 percent confidence coefficient, we require $t(.05; 120) = -1.658$. Hence, the lower confidence limit is $4 - 1.658(2.309)$ and the desired confidence interval is:

$$L \geq .17$$

We conclude, therefore, with 95 percent confidence coefficient that the gain in learning with the standard teaching method over the abstract method is greater for students with moderate ability than for students with good ability, the difference in the mean gain being at least .17 point.

19.4 ANALYSIS WHEN TREATMENT MEANS OF UNEQUAL IMPORTANCE

We mentioned in Chapter 18 that the usual ANOVA formulas for *testing* factor effects are not appropriate for testing factor A and factor B main effects when the treatment means are of unequal importance, and that the general linear test approach needs to be used instead.

For purposes of *estimating* factor effects, however, no additional complexities arise when the treatment means are of unequal importance. The formulas in Section 19.3 for contrasts of treatment means and single degree of freedom tests still apply in this case.

We illustrate the analysis of factor effects when the treatment means are of unequal importance by returning to the mathematics learning example.

Example

A school administrator in the mathematics learning example had requested information about which teaching method leads to better learning of college mathematics when 20 percent of the students in the class have excellent quantitative ability, 50 percent have good ability, and 30 percent have moderate ability. The mean learning scores for such a class mix with the two teaching methods are the following linear combinations of the treatment means:

(19.37)
$$\text{Abstract method:} \quad L_1 = .2\mu_{11} + .5\mu_{12} + .3\mu_{13}$$
$$\text{Standard method:} \quad L_2 = .2\mu_{21} + .5\mu_{22} + .3\mu_{23}$$

This assumes that the mean learning scores for students with different quantitative abilities will not be affected by a class mix that is somewhat different from the one in the experimental study.

We could conduct a single degree of freedom test for the effect of teaching method here or could develop an interval estimate. We shall do the latter because the confidence interval will provide information not only about the direction of any difference between the two teaching methods but also about the magnitude of the difference.

Point estimates of the mean scores in (19.37) are:

$$\hat{L}_1 = .2(92) + .5(81) + .3(73) = 80.8$$
$$\hat{L}_2 = .2(90) + .5(86) + .3(82) = 85.6$$

The difference between the two mean scores in (19.37) is a contrast:

(19.38)
$$L = L_1 - L_2$$

This contrast is estimated to be:

$$\hat{L} = \hat{L}_1 - \hat{L}_2 = 80.8 - 85.6 = -4.8$$

The estimated variance of $\hat{L}$ is by (19.29):

$$s^2\{\hat{L}\} = \frac{28}{21}[(.2)^2 + (.5)^2 + (.3)^2 + (-.2)^2 + (-.5)^2 + (-.3)^2] = 1.013$$

so that the estimated standard deviation is $s\{\hat{L}\} = 1.006$. For a 95 percent confidence coefficient, we require $t(.975; 120) = 1.980$. Hence, the confidence limits are $-4.8 \pm 1.980(1.006)$ and the desired confidence interval is:

$$-6.79 \leq L \leq -2.81$$

With 95 percent confidence we conclude that the standard teaching method is better for the specified class mix, leading to a mean learning score that is at least 2.8 points greater than that for the abstract teaching method and may be as much as 6.8 points greater.

If it were desired to test formally whether or not for the class mix specified by the school administrator, the mean learning score (L_2) for the standard teaching method exceeds that for the abstract method (L_1), the alternatives would be:

$$H_0: L_2 \leq L_1 \qquad H_0: L_1 - L_2 \geq 0$$
$$\text{or}$$
$$H_a: L_2 > L_1 \qquad H_a: L_1 - L_2 < 0$$

where L_1 and L_2 are defined in (19.37). In view of the one-sided nature of the alternatives, we shall use the t^* test statistic (19.33):

$$t^* = \frac{\hat{L} - 0}{s\{\hat{L}\}} = \frac{-4.8}{1.006} = -4.77$$

Assuming that the level of significance is to be controlled at $\alpha = .05$, we require $t(.05; 120) = -1.658$. The decision rule therefore is:

If $t^* \geq -1.658$, conclude H_0

If $t^* < -1.658$, conclude H_a

Since $t^* = -4.77 < -1.658$, we conclude H_a, that the mean score for the standard teaching method exceeds that for the abstract method when the class mix is as specified. The one-sided P-value of this test is $0+$.

19.5 ANALYSIS WHEN ONE OR BOTH FACTORS QUANTITATIVE

When one or both of the factors in a two-factor study are quantitative, the analysis of factor effects can be carried beyond the point of multiple comparisons to include a study of the nature of the response function. Since the familiar methods of regression analysis, discussed earlier, then come into use, we shall only briefly discuss this extension of the analysis. The initial study of factor effects by multiple comparison techniques can be very helpful in choosing an appropriate functional form for the regression relation.

Analysis of Response Function when One Factor Quantitative

(This section assumes previous study of Chapter 10.)

Consider an experiment in which the effect of type of cake mix (factor A) and temperature (factor B) on the lightness of the cake texture, suitably measured, is to be investigated. Two kinds of cake mixes (G, H) and four temperatures ($300°$, $315°$, $330°$, $345°$) are studied. The analyst in this case might be interested in extending the investigation of factor effects into the nature of the response function relating cake texture to baking temperature. Since factor A is qualitative, indicator variables will be used to represent it in the response function. If kind of cake mix and baking temperature do not interact, a first-order model might be appropriate:

$$(19.39) \qquad Y_{ijk} = \beta_0 + \beta_1 X_{ijk1} + \beta_2 X_{ijk2} + \varepsilon_{ijk}$$

where:

$$X_{ijk1} = \frac{1 \text{ if case from first level of factor } A}{0 \text{ otherwise}}$$

X_{ijk2} is the baking temperature for the case

The βs denote regression parameters here, and X_{ijk1} and X_{ijk2} are the respective values of X_1 and X_2 for the kth case or trial for the treatment corresponding to the ith level of factor A and the jth level of factor B.

We know from Chapter 10 that regression model (19.39) implies a linear relation between cake texture and temperature that has constant slope for both kinds of cake mixes but different heights. Figure 10.1 provides an illustration of this model.

If cake mix and baking temperature interact, an appropriate model might be:

$$(19.40) \qquad Y_{ijk} = \beta_0 + \beta_1 X_{ijk1} + \beta_2 X_{ijk2} + \beta_3 X_{ijk1} X_{ijk2} + \varepsilon_{ijk}$$

From Chapter 10, we know that this model implies a linear relation between cake texture and temperature with different slopes and intercepts for the two kinds of cake mixes. Figure 10.3 provides an illustration of this model.

If the regression relation is quadratic and no interactions between the two factors exist, a suitable model would be:

$$(19.41) \qquad Y_{ijk} = \beta_0 + \beta_1 X_{ijk1} + \beta_2 x_{ijk2} + \beta_3 x_{ijk2}^2 + \varepsilon_{ijk}$$

where:

$$x_{ijk2} = X_{ijk2} - \bar{X}_2$$

Analysis of Response Function when Both Factors Quantitative

When both factors are quantitative, the analysis of the nature of the response function involves ordinary multiple regression. We shall denote the two factor variables as X_1 and X_2. A first-order model would then be:

$$(19.42) \qquad Y_{ijk} = \beta_0 + \beta_1 X_{ijk1} + \beta_2 X_{ijk2} + \varepsilon_{ijk}$$

where X_{ijk1} and X_{ijk2} are the respective values of X_1 and X_2 for the kth case or trial for the treatment corresponding to the ith level of factor A and the jth level of factor B. A second-order model, including interactions between the two factors, would be:

$$(19.43) \quad Y_{ijk} = \beta_0 + \beta_1 x_{ijk1} + \beta_2 x_{ijk2} + \beta_3 x_{ijk1}^2 + \beta_4 x_{ijk2}^2 + \beta_5 x_{ijk1} x_{ijk2} + \varepsilon_{ijk}$$

where:

$$x_{ijk1} = X_{ijk1} - \bar{X}_1$$
$$x_{ijk2} = X_{ijk2} - \bar{X}_2$$

The discussion of response surfaces in Chapter 9 is completely applicable here.

Example (adapted from Reference 19.1).   A study was conducted to improve the efficiency of a burr remover in a wool textile carding machine. The rollers in the burr remover are adjustable as to speed and spacing. Four spacings and three speeds, as shown in Table 19.2, were used in the study. Four replications were made for each treatment, but only the estimated treatment means are presented in Table 19.2. The observed variable is a measure of the efficiency of the carding. A plot of the treatment means is presented in Figure 19.5.

An analysis of variance was conducted by a computer package for this two-factor study. The results are summarized in Table 19.3a. The test for interactions utilizes the test statistic:

$$F^* = \frac{MSAB}{MSE} = \frac{4.92}{1.32} = 3.73$$

TABLE 19.2   Estimated Treatment Means in Two-Factor Burr Remover Study with Both Factors Quantitative ($n = 4$)

	Speed		
Spacing	*300 rpm*	*400 rpm*	*500 rpm*
1.0 unit	21.6	22.3	22.9
1.2 units	18.7	19.1	21.6
1.4 units	15.8	17.9	19.4
1.6 units	13.2	16.7	19.5

Source: Reprinted, with permission, from D. R. Cox, *Planning of Experiments* (New York: John Wiley & Sons, 1958), p. 124.

FIGURE 19.5   Plots of Estimated Treatment Means—Burr Remover Example

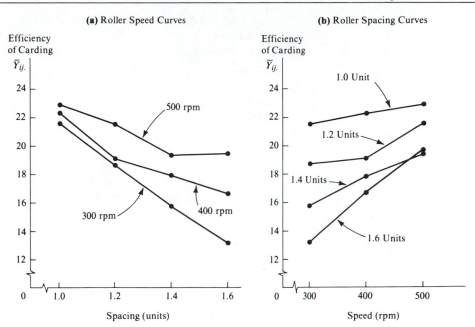

For level of significance $\alpha = .05$, we require $F(.95; 6, 36) = 2.36$. Since $F^* = 3.73 > 2.36$, it was concluded that spacing and speed interact. The P-value of the test is .0055.

Figure 19.5, along with further analysis, suggested that a first-order model with interaction effects added might be appropriate. The response function for this model is:

(19.44) $$E\{Y\} = \beta_0 + \beta_1 X_1 + \beta_2 X_2 + \beta_3 X_1 X_2$$

where X_1 represents spacing and X_2 speed. This model was fitted by a multiple regression computer package. The results are summarized in Table 19.3b.

Since replications were used in the study, a test of the appropriateness of the response function can be conducted. The lack of fit sum of squares is readily obtainable because SSE for the ANOVA model is the same as $SSPE$ for the regression model. Hence, we find:

$$SSLF = \underset{\text{(Table 19.3b)}}{SSE} - \underset{\text{(Table 19.3a)}}{SSPE} = 57.29 - 47.61 = 9.68$$

Table 19.3c contains the decomposition needed for testing lack of fit. The appropriate test statistic is:

$$F^* = \frac{MSLF}{MSE} = \frac{1.21}{1.32} = .92$$

TABLE 19.3 Analysis of Variance Tables for Burr Remover Study

(a) Analysis of Variance Model

Source of Variation	SS	df	MS
Spacing (A)	232.86	3	77.62
Speed (B)	99.49	2	49.74
AB Interactions	29.53	6	4.92
Error	47.61	36	1.32
Total	409.49	47	

(b) Regression Model

$$\hat{Y} = 45.09333 - 25.45000X_1$$
$$- .03340X_2 + .03925X_1X_2$$

Source of Variation	SS	df	MS
Regression	352.20	3	117.40
Error	57.29	44	1.30
Total	409.49	47	

(c) ANOVA for Lack of Fit Test

Source of Variation	SS	df	MS
Regression	352.20	3	117.40
Error	57.29	44	1.30
Lack of fit	9.68	8	1.21
Pure error	47.61	36	1.32
Total	409.49	47	

Assuming a level of significance of $\alpha = .05$ is to be used, we require $F(.95; 8, 36) = 2.21$. Since $F^* = .92 \leq 2.21$, we conclude that the response function (19.44) is appropriate. The P-value of this test is .51.

Let us take a closer look at the estimated response function:

$$\hat{Y} = 45.09333 - 25.45000X_1 - .03340X_2 + .03925X_1X_2$$

Note that b_3 is positive. It implies here, as a glance at Figure 19.5 will confirm, that efficiency declines with increased spacing for all speeds (e.g., if $X_2 = 300$, $\hat{Y} = 35.073 - 13.675X_1$), but the decline is smaller for high speeds. Correspondingly, the efficiency increases with speed for all spacings (e.g., if $X_1 = 1.0$, $\hat{Y} = 19.643 + .00585X_2$), but the increase is larger for large spacings.

19.6 PLANNING OF SAMPLE SIZES

Planning of sample sizes for two-factor studies is handled in essentially the same fashion as discussed in Chapter 17 for single-factor studies. Hence, we make only a few brief comments. We consider first the power of F tests for two-factor studies, since one main approach to planning sample sizes is by means of controlling the power of the test.

Power of F Tests

The power of the F tests for interactions, factor A main effects, and factor B main effects can be ascertained in similar fashion to the single-factor case by using the Pearson-Hartley charts in Table A.8. The noncentrality parameter ϕ and the appropriate degrees of freedom for each of these cases are as follows:

(19.45a)   Test for interactions:

$$\phi = \frac{1}{\sigma}\sqrt{\frac{n\Sigma\Sigma(\alpha\beta)_{ij}^2}{(a-1)(b-1)+1}} = \frac{1}{\sigma}\sqrt{\frac{n\Sigma\Sigma(\mu_{ij} - \mu_{i.} - \mu_{.j} + \mu_{..})^2}{(a-1)(b-1)+1}}$$

$$\nu_1 = (a-1)(b-1) \qquad \nu_2 = ab(n-1)$$

(19.45b)   Test for A main effects:

$$\phi = \frac{1}{\sigma}\sqrt{\frac{nb\Sigma\alpha_i^2}{a}} = \frac{1}{\sigma}\sqrt{\frac{nb\Sigma(\mu_{i.} - \mu_{..})^2}{a}}$$

$$\nu_1 = a-1 \qquad \nu_2 = ab(n-1)$$

(19.45c)   Test for B main effects:

$$\phi = \frac{1}{\sigma}\sqrt{\frac{na\Sigma\beta_j^2}{b}} = \frac{1}{\sigma}\sqrt{\frac{na\Sigma(\mu_{.j} - \mu_{..})^2}{b}}$$

$$\nu_1 = b-1 \qquad \nu_2 = ab(n-1)$$

Example.   For the test of A main effects (display height) in the Castle Bakery example of Chapter 18, we wish to find the power of the test when $\mu_{1.} = 50$, $\mu_{2.} = 55$, and $\mu_{3.} = 45$, or equivalently when $\alpha_1 = 0$, $\alpha_2 = +5$, and $\alpha_3 = -5$. Assume that we know from past experience that $\sigma = 3$ cases. We have for this test from before:

$$n = 2 \qquad a = 3 \qquad b = 2 \qquad \alpha = .05$$

so that:

$$\phi = \frac{1}{3}\sqrt{\frac{2(2)[(0)^2 + (5)^2 + (-5)^2]}{3}} = 2.7$$

For $\nu_1 = 2$, $\nu_2 = 6$, $\alpha = .05$, and $\phi = 2.7$, we find from Table A.8 that the power is about .89. Thus, if $\mu_1. = 50$, $\mu_2. = 55$, and $\mu_3. = 45$ (and $\sigma = 3$), the probability is about .89 that the F test will detect the differences in the display height means.

Planning Approaches

One can plan the sample sizes for two-factor studies using either the power approach or the estimation approach discussed in Chapter 17. In most cases, equal sample sizes will be desired for each treatment.

With the power approach, one would be concerned typically with both the power of detecting factor A main effects and the power of detecting factor B main effects. One can first specify the minimum range of factor A level means for which it is important to detect factor A main effects, and obtain the needed sample sizes from Table A.10, with $r = a$. The resulting sample size is bn, from which n can be readily obtained. The use of Table A.10 for this purpose is appropriate provided the resulting sample size is not small, specifically provided $a(bn - 1) \geq 20$. If this condition is not met, one should use the Pearson-Hartley power charts in Table A.8. These charts, as we noted earlier, require an iterative approach for determining needed sample sizes.

In the same way, one can then specify the minimum range of factor B level means for which it is important to detect factor B main effects, and find the needed sample sizes. If the sample sizes obtained from the factor A and factor B power specifications differ substantially, a judgment will need to be made as to the final sample sizes.

Alternatively, or in conjunction with the power approach, one can specify the important contrasts to be estimated and then find the sample sizes that are expected to provide the needed precisions for the desired family confidence coefficient. Frequently this approach is more useful than the power approach, although both of these approaches can be used jointly for arriving at a determination of needed sample sizes.

If the purpose of the factorial study is to identify the best of the ab factor combinations, Table A.11 can be used for finding the needed sample sizes, as described in Section 17.3. For this purpose, $r = ab$.

CITED REFERENCE

19.1. Cox, D. R. *Planning of Experiments*. New York: John Wiley & Sons, 1958.

PROBLEMS

19.1. Why is it suggested in the flowchart in Figure 19.1 that a test for interactions should be conducted before tests for main factor effects? Explain.

19.2. A two-factor study was conducted with $a = 5$, $b = 5$, and $n = 4$. No interactions between factors A and B were noted, and the analyst now wishes to estimate all pairwise comparisons among the factor A level means and all pairwise comparisons among the factor B level means. The family confidence coefficient for the joint set of interval estimates is to be 90 percent.

    a. Is it more efficient to use the Bonferroni procedure for the entire family or to use the Tukey procedure for each family of factor level mean comparisons and then to join the two families by means of the Bonferroni procedure?

    b. Would your answer differ if each factor had three levels, everything else remaining the same?

19.3. A two-factor study was conducted with $a = 6$, $b = 6$, and $n = 10$. No interactions between factors A and B were found, and it is now desired to estimate five contrasts of factor A level means and four contrasts of factor B level means. The family confidence coefficient for the joint set of estimates is to be 95 percent. Which of the three procedures on page 735 will be most efficient here?

19.4. Refer to the Castle Bakery example on page 736, where multiple pairwise comparisons were made by means of the Tukey procedure and then combined by the Bonferroni procedure to yield a 90 percent family confidence coefficient for the entire set of estimates. Would it have been more efficient here to use the Bonferroni procedure entirely for the family of four estimates? Explain. Would the Scheffé method of (19.24) have been more efficient here?

19.5. Refer to **Cash offers** Problems 18.10 and 18.11. Some additional calculational results are:

i	$\bar{Y}_{i..}$	j	$\bar{Y}_{.j.}$	
1	21.50	1	23.94	
2	27.75	2	23.17	$MSE = 2.389$
3	21.42			

    a. Estimate μ_{11} with a 95 percent confidence interval. Interpret your interval estimate.

    b. Prepare a normal probability of the estimated factor B level means. What does this plot suggest about the equality of the factor B level means?

    c. Estimate $D = \mu_{.1} - \mu_{.2}$ by means of a 95 percent confidence interval. Is your confidence interval consistent with the test result in Problem 18.11d? Is your confidence interval consistent with your finding in part (b)? Explain.

    d. Prepare a normal probability plot of the estimated factor A level means. What does this plot suggest about the factor A main effects?

    e. Obtain all pairwise comparisons among the factor A level means; use the Tukey procedure with a 90 percent family confidence coefficient. Present your findings graphically and summarize your results. Are your conclusions consistent with those in part (d)?

    f. Is the Tukey procedure used in part (e) the most efficient one that could be used here? Explain.

    g. Estimate the contrast:

$$L = \frac{\mu_{1.} + \mu_{3.}}{2} - \mu_{2.}$$

with a 95 percent confidence interval. Interpret your interval estimate.

h. Suppose that in the population of female owners, 30 percent are young, 60 percent are middle-aged, and 10 percent are elderly. Obtain a 95 percent confidence interval for the mean cash offer in the population of female owners.

19.6. Refer to **Eye contact effect** Problems 18.12 and 18.13. Some additional calculational results are:

i	$\bar{Y}_{i..}$	j	$\bar{Y}_{.j.}$	
1	11.4	1	11.1	$MSE = 6.075$
2	14.7	2	15.0	

a. Estimate μ_{21} with a 99 percent confidence interval. Interpret your interval estimate.

b. Estimate $\mu_{1.}$ with a 99 percent confidence interval. Interpret your interval estimate.

c. Prepare a normal probability plot of the estimated factor B level means. What does this plot suggest about the factor B main effects?

d. Obtain confidence intervals for $\mu_{.1}$ and $\mu_{.2}$, each with a 99 percent confidence coefficient. Interpret your interval estimates. What is the family confidence coefficient for the set of two estimates?

e. Prepare a normal probability plot of the estimated factor A level means. What does this plot suggest about the factor A main effects?

f. Obtain confidence intervals for $D_1 = \mu_{2.} - \mu_{1.}$ and $D_2 = \mu_{.2} - \mu_{.1}$; use the Bonferroni procedure and a 95 percent family confidence coefficient. Summarize your findings. Are your findings consistent with those in parts (c) and (e)?

g. Is the Bonferroni procedure used in part (f) the most efficient one that could be used here? Explain.

19.7. Refer to **Hay fever relief** Problems 18.14 and 18.15.

a. Estimate μ_{23} with a 95 percent confidence interval. Interpret your interval estimate.

b. Estimate $D = \mu_{12} - \mu_{11}$ with a 95 percent confidence interval. Interpret your interval estimate.

c. The analyst decided to study the nature of the interacting factor effects by means of the following contrasts:

$$L_1 = \frac{\mu_{12} + \mu_{13}}{2} - \mu_{11} \qquad L_4 = L_2 - L_1$$

$$L_2 = \frac{\mu_{22} + \mu_{23}}{2} - \mu_{21} \qquad L_5 = L_3 - L_1$$

$$L_3 = \frac{\mu_{32} + \mu_{33}}{2} - \mu_{31} \qquad L_6 = L_3 - L_2$$

Obtain confidence intervals for these contrasts; use the Scheffé multiple comparison procedure with a 90 percent family confidence coefficient. Interpret your findings.

d. The analyst also wished to identify the treatment(s) yielding the longest mean relief. Using the Tukey procedure with a 90 percent family confidence coefficient, identify the treatment(s) providing the longest mean relief.

e. Use the single degree of freedom test statistic (19.33) to choose between the following two alternatives:

$$H_0: \mu_{32} - \mu_{31} \le \mu_{33} - \mu_{32}$$

$$H_a: \mu_{32} - \mu_{31} > \mu_{33} - \mu_{32}$$

Control the level of significance at $\alpha = .05$. State the decision rule and conclusion.

f. To examine whether a transformation of the data would make the interactions unimportant, plot separately the transformed estimated treatment means for the reciprocal and square root transformations in the format of Figure 19.4. Would either of these transformations have made the interaction effects unimportant? Explain.

19.8. Refer to **Disk drive service** Problems 18.16 and 18.17.

a. Estimate μ_{11} with a 99 percent confidence interval. Interpret your interval estimate.

b. Estimate $D = \mu_{22} - \mu_{21}$ with a 99 percent confidence interval. Interpret your interval estimate.

c. The nature of the interaction effects is to be studied by making, for each technician, all three pairwise comparisons among the disk drive makes in order to identify, if possible, the make of disk drive for which the technician's mean service time is lowest. The family confidence coefficient for each set of three pairwise comparisons is to be 95 percent. Use the Bonferroni procedure to make all required pairwise comparisons. Summarize your findings.

d. The service center currently services 30 disk drives of each of the three makes per week, with each technician servicing 10 machines of each make. Estimate the expected total amount of service time required per week to service the 90 disk drives; use a 99 percent confidence interval.

e. How much time could be saved per week, on the average, if technician 1 services only make 2, technician 2 services only make 1, and technician 3 services only make 3? Use a 99 percent confidence interval.

f. To examine whether a transformation of the data would make the interactions unimportant, plot separately the transformed estimated treatment means for the reciprocal and logarithmic transformations in the format of Figure 19.4. Would either of these transformations have made the interaction effects unimportant? Explain.

19.9. Refer to **Kidney failure hospitalization** Problems 18.18 and 18.19. Continue to work with the transformed observations $Y' = \log_{10}(Y + 1)$.

a. Estimate μ_{22} with a 95 percent confidence interval. Interpret your interval estimate.

b. Estimate $D = \mu_{23} - \mu_{21}$ with a 95 percent confidence interval. Interpret your interval estimate.

c. Prepare separate normal probability plots of the estimated factor A and factor B level means. What do these plots suggest about the factor main effects?

d. The researcher wishes to study the main effects of each of the two factors by making all pairwise comparisons of factor level means with a 90 percent family confidence coefficient for the entire set of comparisons. Which mutiple comparison procedure is most efficient here?

   e. Using the Bonferroni procedure, make all pairwise comparisons called for in part (d). State your findings and prepare a graphic summary. Are your findings consistent with those in part (c)?

   f. It is known from past experience that 30 percent of patients have mild weight gains, 40 percent have moderate weight gains, and 30 percent have severe weight gains, and that these proportions are the same for the two duration groups. Estimate the mean number of days hospitalized (in transformed units) in the entire population with a 95 percent confidence interval. Convert your confidence limits to the original units. Does it appear that the mean number of days is less than 7?

19.10. Refer to **Programmer requirements** Problems 18.20 and 18.21.

   a. Estimate μ_{23} with a 99 percent confidence interval. Interpret your interval estimate.

   b. Estimate $D = \mu_{12} - \mu_{13}$ with a 99 percent confidence interval. Interpret your interval estimate.

   c. The nature of the interaction effects is to be studied by comparing the effect of type of experience for each years-of-experience group. Specifically, the following comparisons are to be estimated:

$$D_1 = \mu_{11} - \mu_{21} \qquad L_1 = D_1 - D_2$$
$$D_2 = \mu_{12} - \mu_{22} \qquad L_2 = D_1 - D_3$$
$$D_3 = \mu_{13} - \mu_{23} \qquad L_3 = D_2 - D_3$$

The family confidence coefficient is to be 95 percent. Which multiple comparison procedure is most efficient here?

   d. Use the most efficient procedure to estimate the comparisons specified in part (c). State your findings.

   e. Use the Tukey procedure with a 95 percent family confidence coefficient to identify the type of experience-years of experience group(s) with the smallest mean prediction errors.

   f. For each of the groups identified in part (e), obtain a confidence interval for the mean prediction error. Use the Bonferroni procedure with a 95 percent family confidence coefficient. Does any group have a mean prediction error that could be zero? Explain.

   g. Use the single degree of freedom test statistic (19.33) to choose between the two alternatives:

$$H_0: \frac{\mu_{21} + \mu_{22} + \mu_{23}}{3} \leq 40$$

$$H_a: \frac{\mu_{21} + \mu_{22} + \mu_{23}}{3} > 40$$

Control the level of significance at $\alpha = .05$. State the decision rule and conclusion.

   h. To examine whether a transformation of the data would make the interactions unimportant, plot separately the transformed estimated treatment means for the reciprocal and logarithmic transformations in the format of Figure 19.4. Would either of these transformations have made the interaction effects unimportant? Explain.

19.11. Refer to **Brand preference** Problem 7.8. Suppose the market researcher first wished to employ analysis of variance model (18.23) to determine whether or not moisture content (factor A) and sweetness (factor B) affect the degree of brand liking.
  a. State the analysis of variance model for this case.
  b. Obtain the analysis of variance table.
  c. Test whether or not the two factors interact; use $\alpha = .01$. State the alternatives, decision rule, and conclusion.
  d. Study possible curvilinearity of the moisture content effect by estimating the following contrast:

$$L = (\mu_{4.} - \mu_{3.}) - (\mu_{2.} - \mu_{1.})$$

  Use a 95 percent confidence interval. What do you conclude?
  e. Test whether or not sweetness affects brand liking; use $\alpha = .01$. State the alternatives, decision rule, and conclusion.

19.12. Refer to **Cash offers** Problem 18.10. The mean ages of the "owners" in the three age classes were:

$$
\begin{array}{ll}
\text{Young:} & 24.8 \\
\text{Middle:} & 45.3 \\
\text{Elderly:} & 66.7
\end{array}
$$

Since the actual ages of the "owners" in each age group for both genders were close to the mean age, each mean age can be used to represent the ages (X_1) of the "owners" in that group.
  a. Fit the regression model:

$$Y_{ijk} = \beta_0 + \beta_1 x_{ijk1} + \beta_2 x_{ijk1}^2 + \beta_3 X_{ijk2} + \varepsilon_{ijk}$$

  where $x_{ijk1} = X_{ijk1} - \bar{X}_1$ and $X_{ijk2} = 1$ if owner is male and 0 if female.
  b. Obtain the residuals and plot them against the fitted values. What does your plot show?
  c. Conduct a formal test for lack of fit using a level of significance of $\alpha = .01$. State the alternatives, decision rule, and conclusion.
  d. Test whether or not the quadratic term in the regression model in part (a) can be dropped; use $\alpha = .01$. State the alternatives, decision rule, and conclusion.

19.13. Refer to **Hay fever relief** Problem 18.14. The researcher now wishes to study the nature of the relationship between the amounts of the two active ingredients and the duration of relief. The amounts of the ingredients used in the study were as follows:

| Factor Level | Quantity (in milligrams) | |
	X_1 (ingredient 1)	X_2 (ingredient 2)
Low	5.0	7.5
Medium	10.0	10.0
High	15.0	12.5

  a. Fit regression model (19.43).

  b. Estimate the mean duration of relief when $X_1 = 7.50$ and $X_2 = 8.75$; use a 95 percent confidence interval. Could this estimate have been obtained from the ANOVA model? Explain.

  c. Obtain the residuals and plot them against the fitted values. What does your plot show?

  d. Conduct a formal test for lack of fit; use $\alpha = .005$. State the alternatives, decision rule, and conclusion.

  e. Test whether or not the interaction term can be dropped from the model; use $\alpha = .05$. State the alternatives, decision rule, and conclusion.

19.14. In a two-factor study, factor A has four levels, factor B has three levels, and $n = 6$. Separate tests were conducted for factor A and factor B main effects, each with significance level $\alpha = .05$. The researcher now wishes to investigate the power of the two tests. Assume that $\sigma = 20$.

  a. What is the power of the test for factor A main effects when $\alpha_1 = -10$, $\alpha_2 = 7$, $\alpha_3 = 3$, and $\alpha_4 = 0$?

  b. What is the power of the test for factor B main effects when $\beta_1 = -4$, $\beta_2 = 8$, and $\beta_3 = -4$?

19.15. Refer to **Cash offers** Problems 18.10 and 18.11. Assume that $\sigma = 2.0$.

  a. What is the power of the test for interaction effects in Problem 18.11c if $(\alpha\beta)_{11} = -.2$, $(\alpha\beta)_{12} = .2$, $(\alpha\beta)_{21} = -.8$, $(\alpha\beta)_{22} = .8$, $(\alpha\beta)_{31} = 1.0$, and $(\alpha\beta)_{32} = -1.0$?

  b. What is the power of the test for factor A main effects in Problem 18.11d if $\mu_{1.} = 23$, $\mu_{2.} = 25$, and $\mu_{3.} = 21$?

  c. What is the power of the test for factor B main effects in Problem 18.11d if $\mu_{.1} = 24$ and $\mu_{.2} = 20$? (*Hint:* Use Table A.5.)

19.16. Refer to **Hay fever relief** Problems 18.14 and 18.15. Assume that $\sigma = .28$.

  a. What is the power of the test for factor A main effects in Problem 18.15d if $\mu_{1.} = 6.6$, $\mu_{2.} = 7.0$, and $\mu_{3.} = 7.4$?

  b. How is the power of the test in part (a) affected if $\mu_{2.} = 7.3$, everything else remaining the same?

  c. What is the power of the test for factor B main effects in Problem 18.15d if $\beta_1 = -.25$, $\beta_2 = -.25$, and $\beta_3 = .50$?

19.17. Refer to **Programmer requirements** Problems 18.20 and 18.21. Obtain the power of the test for factor B main effects in Problem 18.21d if $\beta_1 = 15$, $\beta_2 = -5$, and $\beta_3 = -10$. Assume that $\sigma = 9.0$.

19.18. A market research manager is planning to study the effects of duration of advertising (factor A) and price level (factor B) on sales. Each factor has three levels. No important interactions are expected, and the primary analysis is to consist of pairwise comparisons of factor level means for each factor. Equal sample sizes are to be used for each treatment. The precision of each comparison is to be ± 3 thousand dollars. The family confidence coefficient for the joint set of comparisons is to be 90 percent, the Tukey procedure is to be used in making the comparisons for each factor, and the Bonferroni procedure is then to be used to join the two sets of comparisons. Assume that $\sigma = 7$ thousand dollars is a reasonable planning value for the error standard deviation. What sample sizes do you recommend?

19.19. Refer to **Cash offers** Problem 18.10. Suppose that the sample sizes have not yet been determined but it has been decided to use the same number of "owners" in each age-gender group. What are the required sample sizes if: (1) differences in the age factor level means are to be detected with probability .90 or more when the range of the factor level means is 3 (hundred dollars), and (2) the α risk is to be controlled at .05? Assume that a reasonable planning value for the error standard deviation is $\sigma = 1.5$ (hundred dollars).

19.20. Refer to **Eye contact effect** Problem 18.12. Suppose that the sample sizes have not yet been determined but it has been decided to use equal sample sizes for each treatment. Primary interest is in the two comparisons $D_1 = \mu_{1.} - \mu_{2.}$ and $D_2 = \mu_{.1} - \mu_{.2}$. What are the required sample sizes if each of these comparisons is to be estimated with precision not to exceed ± 1.2 with a 95 percent family confidence coefficient, using the most efficient multiple comparison procedure? Assume that a reasonable planning value for the error standard deviation is $\sigma = 2.4$.

19.21. Refer to **Hay fever relief** Problem 18.14. Suppose that the sample sizes have not yet been determined but it has been decided to use equal sample sizes for each treatment. The chief objective is to identify the dosage combination that yields the longest mean relief. The probability should be at least .99 that the correct dosage combination is identified when the mean relief duration for the second best combination differs by .5 hour or more. What are the required sample sizes? Assume that a reasonable planning value for the error standard deviation is $\sigma = .29$ hour.

19.22. Refer to **Kidney failure hospitalization** Problem 18.18. Suppose that the sample sizes have not yet been determined but it has been decided to use equal sample sizes for each treatment. The chief objective is to estimate the pairwise comparisons:

$$D_1 = \mu_{1.} - \mu_{2.} \qquad D_3 = \mu_{.1} - \mu_{.3}$$
$$D_2 = \mu_{.1} - \mu_{.2} \qquad D_4 = \mu_{.2} - \mu_{.3}$$

What are the required sample sizes if the precision of each of the estimates should not exceed $\pm .20$ (in transformed units), using the Bonferroni procedure with a family confidence coefficient of 90 percent for the joint set of comparisons? A reasonable planning value for the error standard deviation is $\sigma = .32$ (in transformed units).

19.23. Refer to **Programmer requirements** Problem 18.20. Suppose that the sample sizes have not yet been determined but it has been decided to use equal sample sizes for each treatment. Primary interest is in identifying the type of experience-years of experience combination for which the mean prediction error is smallest. The probability should be at least .95 that the correct combination is identified when the mean prediction error for the second best combination differs by 8 programmer-days or more. Assume that a reasonable planning value for the error standard deviation is $\sigma = 9.1$ days. What are the required sample sizes?

EXERCISES

19.24. Show that the point estimator (19.11) is unbiased. Find the variance of this estimator.

19.25. Find the variance of the estimator (19.28).

19.26. Consider a two-factor study with $a = 2$ and $b = 2$. Use (18.8b) to develop a contrast for each of the interactions $(\alpha\beta)_{12}$ and $(\alpha\beta)_{21}$. Use (6.20) to show that the two contrasts are linearly dependent.

PROJECTS

19.27. Refer to the **SENIC** data set and Projects 18.36 and 18.37.

  a. Prepare a normal probability plot of the estimated factor level means $\bar{Y}_{i..}$. What does this plot suggest regarding the region main effects?

  b. Analyze the effect of region on mean length of hospital stay by making all pairwise comparisons between regions; use the Tukey procedure and a 90 percent family confidence coefficient. State your findings and present a graphic summary. Are your findings consistent with those in part (a)?

19.28. Refer to the **SMSA** data set and Projects 18.38 and 18.39.

  a. Prepare a normal probability plot of the estimated factor level means $\bar{Y}_{i..}$. What does this plot suggest regarding the region main effects?

  b. Analyze the effect of region on crime rate by making all pairwise comparisons between regions; use the Tukey procedure and a 95 percent family confidence coefficient. State your findings and present a graphic summary. Are your findings consistent with those in part (a)?

Chapter 20

Unequal Sample Sizes in Two-Factor Studies

Up to this point in our discussion of two-factor studies we have restricted ourselves to the case of equal treatment sample sizes for the two-factor ANOVA model (18.23). In this chapter, we take up procedures for handling cases when the treatment sample sizes are not equal. We continue to assume that *all treatment means are of equal importance*.

20.1 UNEQUAL SAMPLE SIZES

Unequal sample sizes for the treatments are common with observational data. For instance, a market research analyst wished to study the effects of temperature and precipitation on sales of a product from data for the 30 largest metropolitan areas in the United States. In this type of uncontrolled situation, it is unlikely that each temperature-precipitation category will contain the same number of cities.

One may also encounter unequal treatment sample sizes in experimental studies. For instance, an experimenter may seek to have the same number of cases for each treatment, but for a variety of reasons (e.g., illness of subject, incomplete records, technical problems) ends up with unequal sample sizes. In addition, some designed experiments require variable precision among treatment comparisons, and thus unequal sample sizes are specified by design.

In our discussion of unequal sample sizes in Sections 20.2 and 20.3, we shall assume that there is at least one case for each treatment. We relax this restriction in Section 20.4.

Notation

Our notation remains the same as before, except that the sample size for the treatment consisting of the *i*th level of factor *A* and *j*th level of factor *B* will now be denoted by n_{ij}. The total number of cases for the ith level of factor A will be denoted by:

$$(20.1a) \qquad n_{i.} = \sum_j n_{ij}$$

for the jth level of factor B by:

$$(20.1b) \qquad n_{.j} = \sum_i n_{ij}$$

and for the total study by:

$$(20.1c) \qquad n_T = \sum_i \sum_j n_{ij}$$

The estimated treatment mean when factor A is at the ith level and factor B is at the jth level is defined as usual:

$$(20.2) \qquad \bar{Y}_{ij.} = \frac{Y_{ij.}}{n_{ij}}$$

where:

$$(20.2a) \qquad Y_{ij.} = \sum_{k=1}^{n_{ij}} Y_{ijk}$$

20.2 USE OF REGRESSION APPROACH FOR TESTING FACTOR EFFECTS WHEN SAMPLE SIZES UNEQUAL

When the treatment sample sizes are unequal, the analysis of variance for two-factor studies becomes more complex. The least squares equations are no longer of a simple structure, yielding direct and easy solutions, and the regular analysis of variance formulas in (18.38) and (18.39) are now inappropriate. Furthermore, the factor effect component sums of squares are no longer orthogonal—that is, they do not sum to $SSTR$.

An easy way to obtain the proper sums of squares for testing factor interactions and main effects is through the regression approach described in Section 18.8. The only difference when sample sizes are unequal is that a reduced model needs to be fitted for each test of factor interactions and main effects. Since no new principles are involved, we shall turn directly to an example to illustrate how ANOVA tests are conducted by means of the regression approach when the treatment sample sizes are unequal.

Example

Human growth hormone was administered at a clinical research center to growth hormone deficient, short children who had not yet reached puberty. The investigator was interested in the effects of gender (factor A) and bone development (factor B) on the rate of growth induced by the hormone administration. A child's bone development was classified into one of three categories—severely depressed, moderately depressed, mildly depressed. Three children were randomly selected for each gender-bone development group. The dependent variable (Y) utilized was the difference between the growth rate during growth hormone treatment and the normal growth rate prior to the treatment, expressed in centimeters per month. Four of the 18 children were unable to complete the year-long study, thus creating unequal treatment sample sizes.

Table 20.1 presents the study data. Plots of the estimated treatment means are shown in Figure 20.1. The plots in Figure 20.1 clearly suggest that bone development has a major impact on the change in growth rate. The plots also raise the questions whether some interaction effects are present and whether the gender of a child affects the growth rate.

To test formally whether or not these factor effects are present, we utilize the regression approach because of the unequal sample sizes.

Development of Regression Model. The two-factor ANOVA model (18.23) here is:

$$(20.3) \qquad Y_{ijk} = \mu_{..} + \alpha_i + \beta_j + (\alpha\beta)_{ij} + \varepsilon_{ijk} \qquad i = 1, 2; j = 1, 2, 3$$

To express this model in regression terms, we utilize indicator variables that take on the values 1, -1, or 0, as explained in Section 18.8. Specifically, we shall need $a - 1 = 2 - 1 = 1$ indicator variable for the factor A main effects and $b - 1 = 3 - 1 = 2$ indicator variables for the factor B main effects. The interaction terms will correspond to the cross-products of the indicator variables for factor A and factor B main effects. Specifically, the regression model equivalent to ANOVA model (20.3) is:

TABLE 20.1 Sample Data and Notation for Growth Hormone Example (growth rate difference in centimeters per month)

Gender (factor A) i	Bone Development (factor B) j		
	Severely Depressed (B_1)	Moderately Depressed (B_2)	Mildly Depressed (B_3)
Male (A_1)	1.4 (Y_{111})	2.1 (Y_{121})	.7 (Y_{131})
	2.4 (Y_{112})	1.7 (Y_{122})	1.1 (Y_{132})
	2.2 (Y_{113})		
Mean	2.0 ($\bar{Y}_{11.}$)	1.9 ($\bar{Y}_{12.}$)	.9 ($\bar{Y}_{13.}$)
Female (A_2)	2.4 (Y_{211})	2.5 (Y_{221})	.5 (Y_{231})
		1.8 (Y_{222})	.9 (Y_{232})
		2.0 (Y_{223})	1.3 (Y_{233})
Mean	2.4 ($\bar{Y}_{21.}$)	2.1 ($\bar{Y}_{22.}$)	.9 ($\bar{Y}_{23.}$)

FIGURE 20.1 Plots of Estimated Treatment Means—Growth Hormone Example

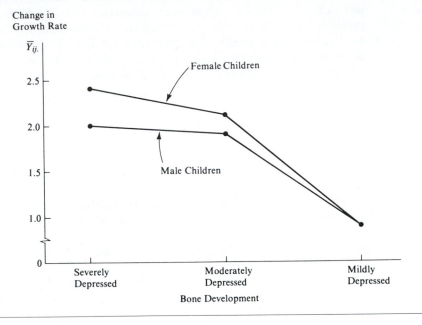

$$(20.4) \quad Y_{ijk} = \mu_{..} + \underbrace{\alpha_1 X_{ijk1}}_{\substack{A \text{ main} \\ \text{effect}}} + \underbrace{\beta_1 X_{ijk2} + \beta_2 X_{ijk3}}_{\substack{B \text{ main} \\ \text{effect}}}$$

$$+ \underbrace{(\alpha\beta)_{11} X_{ijk1} X_{ijk2} + (\alpha\beta)_{12} X_{ijk1} X_{ijk3}}_{AB \text{ interaction effect}} + \varepsilon_{ijk} \qquad \text{Full model}$$

where:

$$X_{ijk1} = \begin{array}{l} 1 \text{ if case from level 1 for factor } A \\ -1 \text{ if case from level 2 for factor } A \end{array}$$

$$X_{ijk2} = \begin{array}{l} 1 \text{ if case from level 1 for factor } B \\ -1 \text{ if case from level 3 for factor } B \\ 0 \text{ otherwise} \end{array}$$

$$X_{ijk3} = \begin{array}{l} 1 \text{ if case from level 2 for factor } B \\ -1 \text{ if case from level 3 for factor } B \\ 0 \text{ otherwise} \end{array}$$

The regression coefficients in (20.4) are the ANOVA model parameters:

$$\mu_{..}$$
$$(20.5) \qquad \alpha_1 = \mu_{1.} - \mu_{..}$$
$$\beta_1 = \mu_{.1} - \mu_{..}$$

$$\beta_2 = \mu_{.2} - \mu_{..}$$

(20.5)
$$(\alpha\beta)_{11} = \mu_{11} - \mu_{1.} - \mu_{.1} + \mu_{..}$$

$$(\alpha\beta)_{12} = \mu_{12} - \mu_{1.} - \mu_{.2} + \mu_{..}$$

The remaining ANOVA model parameters are not required in the regression model because of the constraints in (18.23). Thus, for instance:

$$\alpha_2 = -\alpha_1$$

(20.6)
$$\beta_3 = -\beta_1 - \beta_2$$

$$(\alpha\beta)_{13} = -(\alpha\beta)_{11} - (\alpha\beta)_{12}$$

$$(\alpha\beta)_{21} = -(\alpha\beta)_{11}$$

Table 20.2 presents the **Y** and **X** data matrices for the growth hormone study regression model (20.4). Table 20.3a presents the fitted regression function and regression ANOVA table when the full regression model (20.4) is fitted to the data. Note that the fitted values for the full model are the estimated treatment means $\bar{Y}_{ij.}$, just as when all treatment sample sizes are equal. For instance, we have for the cases from treatment $i = 1, j = 1$ for which $X_1 = X_2 = 1$ and $X_3 = 0$:

$$\hat{Y}_{111} = \hat{Y}_{112} = \hat{Y}_{113} = 1.7 - .1(1) + .5(1) + .3(0) - .1(1)(1) - 0(1)(0)$$

$$= 2.0 = \bar{Y}_{11.}$$

and for the single case from treatment $i = 2, j = 1$ for which $X_1 = -1$, $X_2 = 1$, and $X_3 = 0$:

$$\hat{Y}_{211} = 1.7 - .1(-1) + .5(1) + .3(0) - .1(-1)(1) - 0(-1)(0)$$

$$= 2.4 = \bar{Y}_{21.}$$

TABLE 20.2   Y and X Data Matrices for Growth Hormone Study
Regression Model (20.4)

$$
\mathbf{Y} =
\begin{bmatrix}
Y_{111} \\
Y_{112} \\
Y_{113} \\
Y_{121} \\
Y_{122} \\
Y_{131} \\
Y_{132} \\
Y_{211} \\
Y_{221} \\
Y_{222} \\
Y_{223} \\
Y_{231} \\
Y_{232} \\
Y_{233}
\end{bmatrix}
=
\begin{bmatrix}
1.4 \\
2.4 \\
2.2 \\
2.1 \\
1.7 \\
.7 \\
1.1 \\
2.4 \\
2.5 \\
1.8 \\
2.0 \\
.5 \\
.9 \\
1.3
\end{bmatrix}
\qquad
\mathbf{X} =
\begin{matrix}
& X_1 & X_2 & X_3 & X_1X_2 & X_1X_3 \\
\begin{bmatrix}
1 & 1 & 1 & 0 & 1 & 0 \\
1 & 1 & 1 & 0 & 1 & 0 \\
1 & 1 & 1 & 0 & 1 & 0 \\
1 & 1 & 0 & 1 & 0 & 1 \\
1 & 1 & 0 & 1 & 0 & 1 \\
1 & 1 & -1 & -1 & -1 & -1 \\
1 & 1 & -1 & -1 & -1 & -1 \\
1 & -1 & 1 & 0 & -1 & 0 \\
1 & -1 & 0 & 1 & 0 & -1 \\
1 & -1 & 0 & 1 & 0 & -1 \\
1 & -1 & 0 & 1 & 0 & -1 \\
1 & -1 & -1 & -1 & 1 & 1 \\
1 & -1 & -1 & -1 & 1 & 1 \\
1 & -1 & -1 & -1 & 1 & 1
\end{bmatrix}
\end{matrix}
$$

TABLE 20.3   Regression Results for Growth Hormone Example

(a) Fit of Full Model (20.4)

Source of Variation	SS	df
Regression	4.4743	5
Error	1.3000	8
Total	5.7743	13

$$\hat{Y} = 1.7 - .1X_1 + .5X_2 + .3X_3 - .1X_1X_2 - 0.0X_1X_3$$

(b) Fit of Reduced Model (20.8)

Source of Variation	SS	df
Regression	4.3989	3
Error	1.3754	10
Total	5.7743	13

$$\hat{Y} = 1.68 - .0857X_1 + .467X_2 + .327X_3$$

$$SSE(R) - SSE(F) = 1.3754 - 1.3000 = .0754$$

(c) Fit of Reduced Model (20.10)

Source of Variation	SS	df
Regression	4.3543	4
Error	1.4200	9
Total	5.7743	13

$$\hat{Y} = 1.69 + .444X_2 + .328X_3 - .0667X_1X_2 - .0167X_1X_3$$

$$SSE(R) - SSE(F) = 1.4200 - 1.3000 = .1200$$

(d) Fit of Reduced Model (20.11)

Source of Variation	SS	df
Regression	.2846	3
Error	5.4897	10
Total	5.7743	13

$$\hat{Y} = 1.63 + .0190X_1 + .0667X_1X_2 - .193X_1X_3$$

$$SSE(R) - SSE(F) = 5.4897 - 1.3000 = 4.1897$$

Test for Interaction Effects.   To test whether or not interaction effects are present, the ANOVA model alternatives:

(20.7)
$$H_0: \text{all } (\alpha\beta)_{ij} = 0$$
$$H_a: \text{not all } (\alpha\beta)_{ij} \text{ equal zero}$$

become for regression model (20.4):

(20.7a)
$$H_0: (\alpha\beta)_{11} = (\alpha\beta)_{12} = 0$$
$$H_a: \text{not both } (\alpha\beta)_{11} \text{ and } (\alpha\beta)_{12} \text{ equal zero}$$

Thus, we are simply testing whether or not two regression coefficients equal zero. The reduced regression model therefore is:

(20.8) $$Y_{ijk} = \mu_{..} + \alpha_1 X_{ijk1} + \beta_1 X_{ijk2} + \beta_2 X_{ijk3} + \varepsilon_{ijk}$$ Reduced model

When this reduced model is fitted, the results presented in Table 20.3b are obtained. The general linear test statistic (3.69) therefore is:

$$F^* = \frac{SSE(R) - SSE(F)}{df_R - df_F} \div \frac{SSE(F)}{df_F}$$

$$= \frac{1.3754 - 1.3000}{10 - 8} \div \frac{1.3000}{8} = \frac{.0377}{.1625} = .23$$

To control the risk of making a Type I error at $\alpha = .05$, we require $F(.95; 2, 8) = 4.46$. Since $F^* = .23 \leq 4.46$, we conclude H_0, that no interaction effects are present. The P-value for this test statistic is .80.

Tests for Factor Main Effects.   We now proceed to test whether or not factor A and factor B main effects are present. The ANOVA model alternatives:

(20.9) H_0: $\alpha_1 = \alpha_2 = 0$ H_0: $\beta_1 = \beta_2 = \beta_3 = 0$

H_a: not both α_i equal zero H_a: not all β_j equal zero

become for regression model (20.4):

(20.9a) H_0: $\alpha_1 = 0$ H_0: $\beta_1 = \beta_2 = 0$

H_a: $\alpha_1 \neq 0$ H_a: not both β_j equal zero

The reduced regression models for testing for factor A main effects and factor B main effects therefore are:

Test for factor A main effects

(20.10) $Y_{ijk} = \mu_{..} + \beta_1 X_{ijk2} + \beta_2 X_{ijk3} + (\alpha\beta)_{11} X_{ijk1} X_{ijk2} + (\alpha\beta)_{12} X_{ijk1} X_{ijk3} + \varepsilon_{ijk}$

Reduced model

Test for factor B main effects

(20.11) $Y_{ijk} = \mu_{..} + \alpha_1 X_{ijk1} + (\alpha\beta)_{11} X_{ijk1} X_{ijk2} + (\alpha\beta)_{12} X_{ijk1} X_{ijk3} + \varepsilon_{ijk}$

Reduced model

Tables 20.3c and 20.3d present the results of fitting these reduced models. The two test statistics therefore are:

$$F_1^* = \frac{1.4200 - 1.3000}{9 - 8} \div \frac{1.3000}{8} = \frac{.1200}{.1625} = .74$$

$$F_2^* = \frac{5.4897 - 1.3000}{10 - 8} \div \frac{1.3000}{8} = \frac{2.0949}{.1625} = 12.89$$

For $\alpha = .05$, we require $F(.95; 1, 8) = 5.32$ and $F(.95; 2, 8) = 4.46$ for the two tests. Since $F_1^* = .74 \leq 5.32$ and $F_2^* = 12.89 > 4.46$, we conclude that there are no factor A main effects but that factor B main effects are present. The respective P-values for these two test statistics are .41 and .003.

TABLE 20.4   ANOVA Table for Growth Hormone Example

Source of Variation	SS	df	MS	F*
Gender (A)	.1200	1	.1200	.74
Bone development (B)	4.1897	2	2.0949	12.89
AB Interactions	.0754	2	.0377	.23
Error	1.3000	8	.1625	

Thus, these tests support the evident effect of bone development on the change in growth rate during growth hormone treatment noted earlier from Figure 20.1 and also indicate that the gender and interaction variations in this figure can be accounted for as random behavior. The family level of significance for the set of three tests just conducted, according to the Bonferroni inequality (5.5), is .15.

At this point, it would clearly be desirable to conduct further analyses of the nature of the bone development effects. We shall discuss such analyses in the next section.

Table 20.4 contains a consolidated ANOVA table presenting the results from fitting the four regression models in Table 20.3. The sums of squares for the factor effects in each case are the differences between the error sums of squares for the reduced and full models, and the associated degrees of freedom are the differences between the respective degrees of freedom for these error sums of squares. Note that a total sum of squares is not shown in Table 20.4 because the sums of squares for the three types of factor effects and the error sum of squares do not add to *SSTO* when the treatment sample sizes are unequal.

Revision of ANOVA Model

If it is desired to revise the ANOVA model for testing factor main effects when the test for interactions leads to the conclusion that no interaction effects are present, the regression approach just described needs to be modified. Specifically with reference to the growth hormone example, the full model in (20.4) would no longer be the full model for testing for factor A and factor B main effects when interactions are not present. Instead, the full model for these tests would exclude the interaction effects and would be as follows:

$$(20.12) \qquad Y_{ijk} = \mu_{..} + \alpha_1 X_{ijk1} + \beta_1 X_{ijk2} + \beta_2 X_{ijk3} + \varepsilon_{ijk}$$

Note

If the sample sizes n_{ij} do not differ too much (some statisticians say not more than by the ratio 2 to 1, with most n_{ij} agreeing more closely) and if no n_{ij} is zero, an approximate analysis of variance, called the *method of unweighted means*, may be utilized. This approximate procedure is also used sometimes when the n_{ij} do differ substantially but only a quick first approximation to the factor effects is desired.

The procedure is simple. An analysis of variance is conducted using the $\bar{Y}_{ij.}$ as if they were single observations for each treatment. *SSA*, *SSB*, and *SSAB* are thus calculated in the usual way, except that each treatment has only one "observation" $\bar{Y}_{ij.}$. We know that the "observation" $\bar{Y}_{ij.}$ has variance σ^2/n_{ij}. Hence, the average variance of the "observations" $\bar{Y}_{ij.}$ is:

(20.13)
$$\frac{\sum_i \sum_j \dfrac{\sigma^2}{n_{ij}}}{ab} = \frac{\sigma^2}{ab} \sum_i \sum_j \frac{1}{n_{ij}}$$

The variance σ^2 is estimated by *MSE*, whether frequencies are equal or unequal:

(20.14)
$$MSE = \frac{\sum_i \sum_j \sum_k (Y_{ijk} - \bar{Y}_{ij.})^2}{n_T - ab}$$

Hence, the estimated average variance of the "observations" is:

(20.15)
$$\frac{MSE}{ab} \sum_i \sum_j \frac{1}{n_{ij}}$$

This estimate is then used for the error variance in the analysis of variance of the "observations" $\bar{Y}_{ij.}$.

Another method of approximation has been developed by Federer and Zelen (Ref. 20.1). It is more exact, though somewhat more complicated, than the method of unweighted means.

20.3 ESTIMATION OF FACTOR EFFECTS WHEN SAMPLE SIZES UNEQUAL

No new problems arise in estimating factor effects when the treatment sample sizes are unequal. The nature of the analysis, as for the case of equal sample sizes, depends on whether or not strong interactions are present. When no important interactions are present, the analysis generally is concerned with the factor level means $\mu_{i.}$ and $\mu_{.j}$. On the other hand, when important interactions are present, the analysis usually focuses on the treatment means μ_{ij}.

The estimators and estimated variances presented in Chapter 19 for equal sample sizes must, of course, be modified when the treatment sample sizes are unequal. For instance, if interest is in estimating the factor level means $\mu_{i.}$ as defined in (18.2):

$$\mu_{i.} = \frac{\sum_j \mu_{ij}}{b}$$

the appropriate estimator is simply the unweighted average of the estimated treatment means $\bar{Y}_{ij.}$:

$$\hat{\mu}_{i.} = \frac{\sum_j \bar{Y}_{ij.}}{b}$$

Since the $\bar{Y}_{ij.}$ are independent, the variance of this estimator is:

$$\sigma^2\{\hat{\mu}_{i.}\} = \frac{1}{b^2} \sum_j \sigma^2\{\bar{Y}_{ij.}\} = \frac{1}{b^2} \sum_j \frac{\sigma^2}{n_{ij}} = \frac{\sigma^2}{b^2} \sum_j \frac{1}{n_{ij}}$$

TABLE 20.5   Point Estimators and Estimated Variances for Two-Factor Analyses
when Sample Sizes Are Unequal

(a) Factor Level Mean

$$\mu_{i.} = \frac{\sum_j \mu_{ij}}{b} \qquad\qquad \mu_{.j} = \frac{\sum_i \mu_{ij}}{a}$$

$$(20.16) \qquad \hat{\mu}_{i.} = \frac{\sum_j \bar{Y}_{ij.}}{b} \qquad\qquad \hat{\mu}_{.j} = \frac{\sum_i \bar{Y}_{ij.}}{a}$$

$$s^2\{\hat{\mu}_{i.}\} = \frac{MSE}{b^2} \sum_j \frac{1}{n_{ij}} \qquad\qquad s^2\{\hat{\mu}_{.j}\} = \frac{MSE}{a^2} \sum_i \frac{1}{n_{ij}}$$

(b) Pairwise Comparison of Factor Level Means

$$D = \mu_{i.} - \mu_{i'.} \qquad\qquad D = \mu_{.j} - \mu_{.j'}$$

$$(20.17) \qquad \hat{D} = \hat{\mu}_{i.} - \hat{\mu}_{i'.} \qquad\qquad \hat{D} = \hat{\mu}_{.j} - \hat{\mu}_{.j'}$$

$$s^2\{\hat{D}\} = \frac{MSE}{b^2} \sum_j \left(\frac{1}{n_{ij}} + \frac{1}{n_{i'j}}\right) \qquad\qquad s^2\{\hat{D}\} = \frac{MSE}{a^2} \sum_i \left(\frac{1}{n_{ij}} + \frac{1}{n_{ij'}}\right)$$

(c) Contrast or Linear Combination of Factor Level Means

$$L = \sum_i c_i \mu_{i.} \qquad\qquad L = \sum_j c_j \mu_{.j}$$

$$(20.18) \qquad \hat{L} = \sum_i c_i \hat{\mu}_{i.} \qquad\qquad \hat{L} = \sum_j c_j \hat{\mu}_{.j}$$

$$s^2\{\hat{L}\} = \frac{MSE}{b^2} \sum_i c_i^2 \sum_j \frac{1}{n_{ij}} \qquad\qquad s^2\{\hat{L}\} = \frac{MSE}{a^2} \sum_j c_j^2 \sum_i \frac{1}{n_{ij}}$$

(d) Confidence Interval Multiple

Single Estimate

$$t(1 - \alpha/2; n_T - ab) \qquad\qquad t(1 - \alpha/2; n_T - ab)$$

Multiple Comparisons

$$(20.19) \qquad B = t(1 - \alpha/2g; n_T - ab) \qquad\qquad B = t(1 - \alpha/2g; n_T - ab)$$

$$T = \frac{1}{\sqrt{2}} q(1 - \alpha; a, n_T - ab) \qquad\qquad T = \frac{1}{\sqrt{2}} q(1 - \alpha; b, n_T - ab)$$

$$S^2 = (a - 1)F(1 - \alpha; a - 1, n_T - ab) \qquad S^2 = (b - 1)F(1 - \alpha; b - 1, n_T - ab)$$

and the estimated variance is:

$$s^2\{\hat{\mu}_{i.}\} = \frac{MSE}{b^2} \sum_j \frac{1}{n_{ij}}$$

Table 20.5 presents the formulas for the point estimator and estimated variance

TABLE 20.5   *(concluded)*

(e) Treatment Mean

$$\mu_{ij}$$

(20.20)

$$\hat{\mu}_{ij} = \bar{Y}_{ij.}$$

$$s^2\{\hat{\mu}_{ij}\} = \frac{MSE}{n_{ij}}$$

(f) Pairwise Comparison of Treatment Means

$$D = \mu_{ij} - \mu_{i'j'}$$

(20.21)

$$\hat{D} = \bar{Y}_{ij.} - \bar{Y}_{i'j'.}$$

$$s^2\{\hat{D}\} = MSE\left(\frac{1}{n_{ij}} + \frac{1}{n_{i'j'}}\right)$$

(g) Contrast or Linear Combination of Treatment Means

$$L = \sum \sum c_{ij}\mu_{ij}$$

(20.22)

$$\hat{L} = \sum \sum c_{ij}\bar{Y}_{ij.}$$

$$s^2\{\hat{L}\} = MSE \sum \sum \frac{c_{ij}^2}{n_{ij}}$$

(h) Confidence Interval Multiple

Single Estimate

$$t(1 - \alpha/2; n_T - ab)$$

Multiple Comparisons

(20.23)

$$B = t(1 - \alpha/2g; n_T - ab)$$

$$T = \frac{1}{\sqrt{2}}q(1 - \alpha; ab, n_T - ab)$$

$$S^2 = (ab - 1)F(1 - \alpha; ab - 1, n_T - ab)$$

when estimating factor level means, pairwise comparisons of factor level means, and contrasts or linear combinations of factor level means with unequal sample sizes. The corresponding formulas for treatment means, pairwise comparisons of treatment means, and contrasts or linear combinations of treatment means are also presented in this table.

All multiple comparison procedures that are applicable for the equal sample size case are appropriate when the treatment sample sizes are unequal. The Tukey pairwise comparison procedure now is conservative. The degrees of freedom associated with MSE are $n_T - ab$, as before. For equal sample sizes, recall that $n_T = nab$; hence, $n_T - ab = (n - 1)ab$ then. Table 20.5 also presents the appropriate simultaneous comparison multiples for making inferences about factor level means or treatment means.

Since no new issues are involved in estimating factor effects when the sample sizes are unequal, we proceed directly to two examples.

Example 1—Pairwise Comparisons of Factor Level Means

We continue with the growth hormone example. We found earlier that a child's gender and bone development do not interact in their effects on the change in the growth rate when growth hormone is administered. We further found no main gender (factor A) effects, but concluded that a child's bone development (factor B) does affect the change in growth rate. We shall now analyze the nature of the bone development effects by means of pairwise comparisons among the three bone development groups. The Tukey multiple comparison method is to be used. This method is conservative when sample sizes are unequal and use of the Bonferroni method would lead to wider confidence intervals here. The family confidence coefficient has been specified to be .90.

We use formulas (20.17) for the point estimates and estimated variances. The estimated treatment means are given in Table 20.1, and MSE is found in Table 20.4. For the pairwise comparisons of the bone development factor level means ($j = 1$: severely depressed; $j = 2$: moderately depressed; $j = 3$: mildly depressed), we obtain:

$$\hat{\mu}_{.1} = \frac{\bar{Y}_{11.} + \bar{Y}_{21.}}{2} = \frac{2.0 + 2.4}{2} = 2.2$$

$$\hat{\mu}_{.2} = \frac{\bar{Y}_{12.} + \bar{Y}_{22.}}{2} = \frac{1.9 + 2.1}{2} = 2.0$$

$$\hat{\mu}_{.3} = \frac{\bar{Y}_{13.} + \bar{Y}_{23.}}{2} = \frac{.9 + .9}{2} = .9$$

$$\hat{D}_1 = \hat{\mu}_{.1} - \hat{\mu}_{.2} = 2.2 - 2.0 = .2$$

$$\hat{D}_2 = \hat{\mu}_{.1} - \hat{\mu}_{.3} = 2.2 - .9 = 1.3$$

$$\hat{D}_3 = \hat{\mu}_{.2} - \hat{\mu}_{.3} = 2.0 - .9 = 1.1$$

$$s^2\{\hat{D}_1\} = \frac{.1625}{(2)^2}\left(\frac{1}{3} + \frac{1}{2} + \frac{1}{1} + \frac{1}{3}\right) = .0880 \qquad s\{\hat{D}_1\} = .297$$

$$s^2\{\hat{D}_2\} = \frac{.1625}{(2)^2}\left(\frac{1}{3} + \frac{1}{2} + \frac{1}{1} + \frac{1}{3}\right) = .0880 \qquad s\{\hat{D}_2\} = .297$$

$$s^2\{\hat{D}_3\} = \frac{.1625}{(2)^2}\left(\frac{1}{2} + \frac{1}{2} + \frac{1}{3} + \frac{1}{3}\right) = .0677 \qquad s\{\hat{D}_3\} = .260$$

For a 90 percent family confidence coefficient, we require:

$$T = \frac{1}{\sqrt{2}}q(.90; 3, 8) = \frac{1}{\sqrt{2}}(3.37) = 2.38$$

Hence, we obtain the following confidence intervals:

$$-.51 = .2 - 2.38(.297) \le \mu_{.1} - \mu_{.2} \le .2 + 2.38(.297) = .91$$

$$.59 = 1.3 - 2.38(.297) \le \mu_{.1} - \mu_{.3} \le 1.3 + 2.38(.297) = 2.01$$

$$.48 = 1.1 - 2.38(.260) \le \mu_{.2} - \mu_{.3} \le 1.1 + 2.38(.260) = 1.72$$

We conclude from these confidence intervals with 90 percent family confidence coefficient that growth hormone deficient short children with mildly depressed bone development on the average have a substantially smaller increase in the growth rate than children with either moderately depressed or severely depressed bone development. Further, the latter two groups of children do not show significantly different mean changes in the growth rate. We summarize these findings in the following line plot of the estimated factor level means:

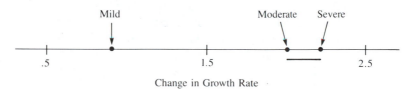

Change in Growth Rate

Example 2—Single Degree of Freedom Test

In the growth hormone example, a researcher wanted to know whether children with only mildly depressed bone development obtain, on the average, any increase in the growth rate with administration of growth hormone. Thus, the alternatives to be considered are those for a one-sided test:

$$H_0: \mu_{.3} \le 0$$

$$H_a: \mu_{.3} > 0$$

The level of significance is to be controlled at $\alpha = .05$.

The test statistic to be employed is:

$$t^* = \frac{\hat{\mu}_{.3} - 0}{s\{\hat{\mu}_{.3}\}}$$

We found earlier that $\hat{\mu}_{.3} = .9$ and $MSE = .1625$. Hence, using (20.16) we obtain:

$$s^2\{\hat{\mu}_{.3}\} = \frac{.1625}{(2)^2}\left(\frac{1}{2} + \frac{1}{3}\right) = .0339 \qquad s\{\hat{\mu}_{.3}\} = .184$$

Hence, the test statistic is:

$$t^* = \frac{.9 - 0}{.184} = 4.89$$

For $\alpha = .05$ we require $t(.95; 8) = 1.860$. Therefore the one-sided decision rule is:

If $t^* \leq 1.860$, conclude H_0

If $t^* > 1.860$, conclude H_a

Since $t^* = 4.89 > 1.860$, we conclude H_a, that the mean change in the growth rate for children with mildly depressed bone development is greater than zero. The one-sided P-value for this test statistic is .0006.

20.4 EMPTY CELLS IN TWO-FACTOR STUDIES

Occasionally one finds after a two-factor study has been completed that there are no cases in one or more treatment cells. Not only are the treatment sample sizes unequal then, but there is no sample information about the treatment means for the empty cells. Consider again Table 20.1 for the growth hormone study. Note that two female children with severely depressed bone condition dropped out of the study before its completion so that only one case ($n_{21} = 1$) is present for that treatment. We can easily imagine that all three of these children could have dropped out of the study. Then we would have had $n_{21} = 0$, and no sample information would be available about the treatment mean μ_{21}.

Partial Analysis of Factor Effects

When one or several treatment cells are empty, the usual analysis of variance for unequal sample sizes by means of the regression approach, as explained earlier, cannot be conducted. This does not mean, however, that the entire two-factor study has become useless. Usually, a variety of analyses can be conducted that will provide at least partial information about the nature of the factor effects. The analyses that can be undertaken depend on the particular cells for which no sample information is available. We shall illustrate by means of an example how partial information can be obtained from two-factor studies with empty cells.

Example.   In the growth hormone example, suppose that there were no observations for female children with severely depressed bone development; i.e., $n_{21} = 0$. In that case no sample information would be available about the treatment mean μ_{21}.

Partial information about interactions could still be obtained by restricting attention to children with moderately depressed and mildly depressed bone development. For these children, interactions are present if the differences between the treatment means for the two genders are not the same for the two bone development groups. The two differences are:

$$\mu_{12} - \mu_{22} \qquad \mu_{13} - \mu_{23}$$

Thus, we would consider the following contrast among the treatment means:

$$L = \mu_{12} - \mu_{22} - \mu_{13} + \mu_{23}$$

We could either estimate L by means of a confidence interval and note whether or not the interval includes zero, or we could conduct a single degree of freedom test to establish whether or not interactions are present. With either approach, we would use MSE based on all sample observations so that the associated degrees of freedom for MSE would be $n_T - (ab - 1) = 13 - 5 = 8$ (remember that $n_{21} = 0$ now).

If the partial analysis of interactions were to suggest that no interactions are present, the effect of gender could be studied by comparing the factor level means excluding children with severely depressed bone development:

$$\mu_{1.} = \frac{\mu_{12} + \mu_{13}}{2} \qquad \mu_{2.} = \frac{\mu_{22} + \mu_{23}}{2}$$

Further, the effect of bone development could be studied for male children by comparing the treatment means μ_{11}, μ_{12}, and μ_{13}, or it could be studied for children of both genders by excluding those with severely depressed bone development:

$$\mu_{.2} = \frac{\mu_{12} + \mu_{22}}{2} \qquad \mu_{.3} = \frac{\mu_{13} + \mu_{23}}{2}$$

Analysis If Model with No Interactions Can Be Employed

Occasionally, information is available from previous studies that the two factors in a two-factor study do not interact. In that case, a simpler model than ANOVA model (18.23) can be employed. The two-factor no-interaction model for fixed factor levels is:

$$(20.24) \qquad Y_{ijk} = \mu_{..} + \alpha_i + \beta_j + \varepsilon_{ijk} \qquad \text{No-interaction model}$$

We have not discussed this model previously because information about the appropriateness of this model (i.e., whether or not the two factors interact) is usually not available in advance.

However, if the no-interaction model (20.24) is appropriate, the analysis of variance and the analysis of factor main effects can be conducted by means of the regression approach even when one or several cells are empty, as long as the empty cell means can be estimated from nonempty cells by means of (18.7b).

For example, suppose again in the growth hormone example that the cell for female children with severely depressed bone development is empty but that from past knowledge the researcher is able to assume that there are no interactions between gender and bone development. In that case, regression model (20.4) reduces to that in (20.12):

$$(20.25) \qquad Y_{ijk} = \mu_{..} + \alpha_1 X_{ijk1} + \beta_1 X_{ijk2} + \beta_2 X_{ijk3} + \varepsilon_{ijk} \qquad \text{Full model}$$

To test for, say, gender main effects, we first fit this full model and obtain $SSE(F)$. The alternatives to be tested are:

$$H_0: \alpha_1 = 0$$

$$H_a: \alpha_1 \neq 0$$

Hence, the reduced model is:

$$(20.26) \qquad Y_{ijk} = \mu_{..} + \beta_1 X_{ijk2} + \beta_2 X_{ijk3} + \varepsilon_{ijk} \qquad \text{Reduced model}$$

We then fit this reduced model, obtain $SSE(R)$, and calculate the general linear test statistic (3.69) in the usual fashion. In this particular case, we could also use the t^* test statistic (8.23) since the test is simply for whether a single regression coefficient equals zero. A test for bone development effects would be carried out similarly.

The reason why the usual analysis of variance by means of the regression approach can be conducted here even though $n_{21} = 0$ is that the assumption of no interactions permits us in effect to estimate μ_{21}. Conceptually, this estimate of μ_{21} requires two steps. First, we need to estimate the treatment means μ_{ij} for the nonempty cells. These estimates are more complicated than simply using the estimated treatment means $\bar{Y}_{ij}$ because we need to utilize the model assumption of no interactions. We obtain these estimates by using the matrix methods described in Section 8.6. (We shall illustrate how to estimate a treatment mean μ_{ij} for the no-interaction model in Section 21.1.) Once we have estimates of the treatment means μ_{ij} for the nonempty cells, the second step in estimating μ_{21} is to utilize relation (18.7b) for the no-interaction case, whereby we can express μ_{21} in terms of three other treatment means; for instance, we have:

$$\mu_{21} = \mu_{22} + \mu_{11} - \mu_{12}$$

Estimates of μ_{22}, μ_{11}, and μ_{12}, for which sample data are available, thus enable us to estimate μ_{21} when no interactions are present.

We need to caution that it is not appropriate to use a no-interaction model as the full model when no prior information about the absence of interactions is available. Only partial analyses of factor effects can then be undertaken when some treatment cells are empty, as explained earlier.

20.5 STATISTICAL COMPUTING PACKAGES

Extreme care must be exercised when using packaged analysis of variance programs with unequal sample sizes because the default option of the package may not necessarily assign equal importance to each treatment mean. The user should read the package documentation carefully and make sure that the package generates the appropriate sums of squares for the tests of interest.

For the BMDP, SAS, and SPSS^X statistical packages, the outputs that are the equivalents of the regression results obtained in Section 20.2 for the case of treatment means with equal importance and no empty cells are as follows at the time of this writing:

BMDP2V—Default option
SAS PROC GLM—Type III or Type IV sums of squares
SPSS^X ANOVA—Option 9

Extreme caution should also be used with ANOVA computer packages that provide results when some treatment cells are empty. The package may make assumptions about interactions that the researcher is unwilling to make. In the absence of a clear description of how the package handles empty cells, it is preferable that appropriate analyses be conducted without the assistance of the package other than for obtaining the estimated treatment means and *MSE*.

CITED REFERENCE

20.1. Federer, W. T., and M. Zelen. "Analysis of Multifactor Classifications with Unequal Numbers of Observations." *Biometrics* 22 (1966), pp. 525–52.

PROBLEMS

20.1. A market research intern selected a random sample of 400 communities and classified them according to population size (four levels) and geographic region (five levels) to study the effects of these factors on sales of the company's products. When he found that the treatment sample sizes were unequal, the smallest cell frequency being four, he generated random numbers to reduce the number of communities in each cell to four. He then proceeded to analyze the effects of population size and region on the basis of the 80 communities remaining.
   a. Does the method of randomly discarding cases lead to any biases? Explain.
   b. Was it wise for the intern to discard 320 cases randomly in order to obtain equal treatment sample sizes?

20.2. A student asked: "If two-factor studies with unequal sample sizes must be analyzed by the regression approach, why bother with the two-factor analysis of variance model at all?" Comment.

20.3. Refer to **Cash offers** Problem 18.10. Assume that observations $Y_{214} = 28$ and $Y_{323} = 20$ are missing because the offer received in each of these cases was a trade-in offer, not a cash offer.
   a. State the ANOVA model for this case. Also state the equivalent regression model; use 1, -1, 0 indicator variables.
   b. Present the **X** and **β** matrices for the regression model in part (a).
   c. Obtain **Xβ** and show that the proper treatment means are obtained by your model.
   d. What is the reduced model for testing for interaction effects?
   e. Test whether or not interaction effects are present by fitting the full and reduced models; use $\alpha = .05$. State the alternatives, decision rule, and conclusion. What is the *P*-value of the test?

f. State the reduced models for testing for age and gender main effects, respectively. Conduct each of the tests. Use $\alpha = .05$ each time and state the alternatives, decision rule, and conclusion. What is the P-value for each test?

g. To study the nature of the age main effects, estimate the following pairwise comparisons:

$$D_1 = \mu_{1.} - \mu_{2.} \qquad D_3 = \mu_{2.} - \mu_{3.}$$

$$D_2 = \mu_{1.} - \mu_{3.}$$

Use the most efficient multiple comparison procedure with a 90 percent family confidence coefficient.

h. In the population of female owners, 30 percent are young, 60 percent are middle-aged, and 10 percent are elderly. Estimate the mean cash offer for this population with a 95 percent confidence interval.

20.4. Refer to **Hay fever relief** Problem 18.14. Assume that observations $Y_{113} = 2.3$, $Y_{221} = 8.9$, and $Y_{224} = 9.0$ are missing because the subjects did not immediately record the time when they began to suffer again from hay fever.

a. State the ANOVA model for this case. Also state the equivalent regression model; use $1, -1, 0$ indicator variables.

b. Present the **X** and $\boldsymbol{\beta}$ matrices for the regression model in part (a).

c. Obtain $\mathbf{X}\boldsymbol{\beta}$ and show that the proper treatment means are obtained by your model.

d. What is the reduced model for testing for interaction effects?

e. Test whether or not interaction effects are present by fitting the full and reduced models; use $\alpha = .05$. State the alternatives, decision rule, and conclusion. What is the P-value of the test?

f. The nature of the interaction effects is to be studied by means of the following contrasts:

$$L_1 = \frac{\mu_{12} + \mu_{13}}{2} - \mu_{11} \qquad L_4 = L_2 - L_1$$

$$L_2 = \frac{\mu_{22} + \mu_{23}}{2} - \mu_{21} \qquad L_5 = L_3 - L_1$$

$$L_3 = \frac{\mu_{32} + \mu_{33}}{2} - \mu_{31} \qquad L_6 = L_3 - L_2$$

Obtain confidence intervals for these contrasts; use the Scheffé multiple comparison procedure with a 90 percent family confidence coefficient. Interpret your findings.

20.5. Refer to **Kidney failure hospitalization** Problem 18.18. Assume that observations $Y_{124} = 12$, $Y_{216} = 2$, and $Y_{238} = 9$ are missing because the hospitalization records for these patients were not complete. Continue to work with the transformed data $Y' = \log_{10}(Y + 1)$.

a. State the ANOVA model for this case. Also state the equivalent regression model; use $1, -1, 0$ indicator variables.

b. Present the **X** and $\boldsymbol{\beta}$ matrices for the regression model in part (a).

c. Obtain $\mathbf{X}\boldsymbol{\beta}$ and show that the proper treatment means are obtained by your model.

d. What is the reduced model for testing for interaction effects?

e. Test whether or not interaction effects are present by fitting the full and reduced models; use $\alpha = .05$. State the alternatives, decision rule, and conclusion. What is the P-value of the test?

f. State the reduced models for testing for treatment duration and weight gain main effects, respectively. Conduct each of the tests. Use $\alpha = .05$ each time and state the alternatives, decision rule, and conclusion. What is the P-value for each test?

g. Use the single degree of freedom t^* statistic for testing whether or not the mean number of days hospitalized (in transformed units) for persons with mild weight gains exceeds .5; use $\alpha = .05$. State the alternatives, decision rule, and conclusion. What is the P-value of the test?

h. To analyze the nature of the factor main effects, estimate the following pairwise comparisons:

$$D_1 = \mu_{1.} - \mu_{2.} \qquad D_3 = \mu_{.3} - \mu_{.1}$$

$$D_2 = \mu_{.2} - \mu_{.1} \qquad D_4 = \mu_{.3} - \mu_{.2}$$

Use the Bonferroni procedure with a 90 percent family confidence coefficient. State your findings.

20.6. Adjunct professors.

Factor A (subject matter)	Factor B (highest degree)		
	$j = 1$ Bachelor's	$j = 2$ Master's	$j = 3$ Doctorate
$i = 1$ Humanities	1.7	1.8	2.5
	1.9	2.1	2.7
			2.9
			2.5
			2.6
			2.8
			2.7
			2.9
$i = 2$ Social sciences	2.5	2.7	3.5
	2.3	2.4	3.3
	2.6	2.6	3.6
	2.4	2.4	3.4
		2.5	
$i = 3$ Engineering	2.7	2.9	3.7
	2.8	3.0	3.6
		2.8	3.7
		2.7	3.8
			3.9
$i = 4$ Management	2.5	2.3	3.3
	2.6	2.8	3.4
			3.3
			3.5
			3.6

A sociologist selected a random sample of 45 adjunct professors who teach in the evening division of a large metropolitan university for a study of special problems associated with teaching in the evening division. The data collected include the amount of payment received by the faculty member for teaching a course during the past semester. The sociologist classified the faculty members by subject matter of course (factor A) and highest degree earned (factor B). The earnings per course (in thousand dollars) were given at the beginning of this problem.

a. State the ANOVA model for this case. Also state the equivalent regression model; use $1, -1, 0$ indicator variables.

b. Present the $\mathbf{X}$ and $\boldsymbol{\beta}$ matrices for the regression model in part (a).

c. Obtain $\mathbf{X}\boldsymbol{\beta}$ and show that the proper treatment means are obtained by your model.

d. Obtain the residuals, and prepare aligned residual dot plots for the treatments. What are your findings?

e. Prepare a normal probability plot of the residuals. Also obtain the coefficient of correlation between the ordered residuals and their expected values under normality. Does the normality assumption appear to be reasonable here?

20.7. Refer to **Adjunct professors** Problem 20.6. Assume that ANOVA model (18.23) is appropriate, except that now $k = 1, \ldots, n_{ij}$.

a. Plot the estimated treatment means $\bar{Y}_{ij.}$ in the format of Figure 20.1. Does it appear that any factor effects are present? Explain.

b. What is the reduced model for testing for interaction effects?

c. Test whether or not interaction effects are present by fitting the full and reduced models; use $\alpha = .01$. State the alternatives, decision rule, and conclusion. What is the P-value of the test?

d. State the reduced models for testing for subject matter and highest degree main effects, respectively. Conduct each of the tests. Use $\alpha = .01$ each time and state the alternatives, decision rule, and conclusion. What is the P-value for each test?

e. Make all pairwise comparisons between the subject matter means; use the Tukey procedure with a 95 percent family confidence coefficient. State your findings and present a graphic summary.

f. Make all pairwise comparisons between the highest degree means; use the Tukey procedure with a 95 percent family confidence coefficient. State your findings and present a graphic summary.

20.8. Refer to **Adjunct professors** Problem 20.6. Suppose that the sociologist had prior information indicating that the two factors do not interact and that no-interaction model (20.24) is therefore appropriate.

a. State the equivalent full regression model for this case. Also state the reduced models for testing for factor A and factor B main effects. Use $1, -1, 0$ indicator variables.

b. Fit the full and reduced models and test for factor A and factor B main effects; use $\alpha = .05$ for each test. State the alternatives, decision rule, and conclusion for each test. What is the P-value for each test?

20.9. Refer to **Hay fever relief** Problem 18.14. Suppose that the data for the treatment when each of the two active ingredients is at the medium level were lost and immediate analyses of the available data are required; i.e., assume that $n_T = 32$ and $n_{22} = 0$.

a. To study whether or not interaction effects are present, estimate the following comparisons:

$$D_1 = \mu_{13} - \mu_{11} \qquad L_1 = D_1 - D_2$$

$$D_2 = \mu_{23} - \mu_{21} \qquad L_2 = D_1 - D_3$$

$$D_3 = \mu_{33} - \mu_{31}$$

Use the Bonferroni procedure with a 90 percent family confidence coefficient. State your findings.

b. To further explore the nature of possible interaction effects, conduct separate single degree of freedom tests of whether $\mu_{12} = \mu_{13}$ and whether $\mu_{32} = \mu_{33}$. Use $\alpha = .02$ for each test and state the alternatives, decision rule, and conclusion. What is the family level of significance, using the Bonferroni inequality?

20.10. Refer to **Kidney failure hospitalization** Problem 18.18. Suppose that there were no patients who received the dialysis treatment for long duration and had mild weight gains; i.e., assume that $n_T = 50$ and $n_{21} = 0$. Continue to work with the transformed data $Y' = \log_{10}(Y + 1)$.

   Based on related research, the analyst believes it is reasonable to assume that the two factors do not interact and that no-interaction model (20.24) is appropriate.

a. State the equivalent full regression model for this case. Also state the reduced models for testing for factor A and factor B main effects. Use $1, -1, 0$ indicator variables in the regression model.

b. Fit the full and reduced models. Test for factor A and factor B main effects; use $\alpha = .05$ for each test. State the alternatives, decision rule, and conclusion for each test. What is the P-value for each test?

20.11. Refer to **Programmer requirements** Problem 18.20. Suppose that there were no programmers with experience on both small and large systems who had less than five years' experience; i.e., assume that $n_T = 20$ and $n_{21} = 0$.

a. To study whether or not interaction effects are present, estimate the following comparisons:

$$D_1 = \mu_{12} - \mu_{13} \qquad L_1 = D_1 - D_2$$

$$D_2 = \mu_{22} - \mu_{23}$$

Use the Bonferroni procedure with a 95 percent family confidence coefficient. State your findings.

b. To study further the nature of possible interaction effects, test whether or not μ_{22} exceeds μ_{23}; use $\alpha = .05$. State the alternatives, decision rule, and conclusion. What is the P-value of the test?

20.12. Refer to **Adjunct professors** Problem 20.6. Suppose that there were no professors teaching humanities courses who had only a bachelor's degree, so that the study consists of $n_T = 43$ adjunct professors and $n_{11} = 0$. On the basis of previous research, the sociologist believes it is reasonable to assume that the two factors do not interact and that no-interaction model (20.24) is appropriate here.

a. State the equivalent full regression model for this case. Also state the reduced models for testing for factor A and factor B main effects. Use $1, -1, 0$ indicator variables in the regression model.

b. Fit the full and reduced models and test for factor A and factor B main effects. Use $\alpha = .01$ for each test and state the alternatives, decision rule, and conclusion. What is the P-value of each test?

EXERCISES

20.13. Give expressions for SSA, SSB, and $SSAB$ for the method of unweighted means.

20.14. Derive $\sigma^2\{\hat{L}\}$ for the estimated contrast involving $\hat{\mu}_{i\cdot}$ in (20.18).

20.15. Show that $s^2\{\hat{L}\}$ in (20.22) is an unbiased estimator of $\sigma^2\{\hat{L}\}$.

20.16. Consider a two-factor study where $a = 2$, $b = 2$, $n_{11} = n_{12} = n_{21} = 2$, $n_{22} = 1$, and no-interaction model (20.24) applies. Use the matrix methods of Section 8.6 to estimate μ_{22}. (*Hint:* Consider the development in Section 21.1.)

20.17. Refer to **Kidney failure hospitalization** Problem 20.10. Suppose that you are going to use the general matrix approach of Section 8.6, rather than the regression approach, to test for factor A main effects.
   a. State the $\mathbf{X}$ and $\boldsymbol{\beta}$ matrices to be used in the full model (8.63).
   b. State the test hypothesis (8.66) in matrix form.

PROJECTS

20.18. Refer to the **SENIC** data set. The effects of region (factor A: variable 9) and average age of patients (factor B: variable 3) on mean length of hospital stay (variable 2) are to be studied. For purposes of this ANOVA study, average age is to be classified into three categories: under 52.0 years, 52.0–under 55.0 years, 55.0 years or more.
   a. State the ANOVA model for this case. Also state the equivalent regression model; use 1, -1, 0 indicator variables.
   b. Obtain the residuals and prepare aligned residual dot plots for the treatments. What are your findings?
   c. Prepare a normal probability plot of the residuals. Also obtain the coefficient of correlation between the ordered residuals and their expected values under normality. Does the normality assumption appear to be reasonable here?

20.19. Refer to the **SENIC** data set and Project 20.18. Assume that ANOVA model (18.23), with $k = 1, \ldots , n_{ij}$, is appropriate.
   a. Plot the estimated treatment means $\bar{Y}_{ij\cdot}$ in the format of Figure 20.1. Does it appear that any factor effects are present? Explain.
   b. State the reduced model for testing for interaction effects.
   c. Test whether or not interaction effects are present by fitting the full and reduced models; use $\alpha = .01$. State the alternatives, decision rule, and conclusion. What is the P-value of the test?
   d. State the reduced model for testing for factor A main effects. Conduct this test using $\alpha = .01$. State the alternatives, decision rule, and conclusion. What is the P-value of the test?
   e. State the reduced model for testing for factor B main effects. Conduct this test using $\alpha = .01$. State the alternatives, decision rule, and conclusion. What is the P-value of the test?
   f. Make all pairwise comparisons between regions; use the Tukey procedure and a 95 percent family confidence coefficient. State your findings and present a graphic summary.

20.20. Refer to the **SMSA** data set. The effects of region (factor A: variable 12) and percent of population in central cities (factor B: variable 4) on the crime rate (variable 11 ÷ variable 3) are to be studied. For purposes of this ANOVA study, percent of population in central cities is to be classified into three categories: under 30.0 percent, 30.0–under 50.0 percent, 50.0 percent or more.

    a. State the ANOVA model for this case. Also state the equivalent regression model; use 1, −1, 0 indicator variables.

    b. Obtain the residuals and prepare aligned residual dot plots for the treatments. What are your findings?

    c. Prepare a normal probability plot of the residuals. Also obtain the coefficient of correlation between the ordered residuals and their expected values under normality. Does the normality assumption appear to be reasonable here?

20.21. Refer to the **SMSA** data set and Project 20.20. Assume that ANOVA model (18.23) with $k = 1, \ldots, n_{ij}$, is appropriate.

    a. Plot the estimated treatment means $\bar{Y}_{ij.}$ in the format of Figure 20.1. Does it appear that any factor effects are present? Explain.

    b. State the reduced model for testing for interaction effects.

    c. Test whether or not interaction effects are present by fitting the full and reduced models; use $\alpha = .005$. State the alternatives, decision rule, and conclusion. What is the P-value of the test?

    d. State the reduced model for testing for factor A main effects. Conduct this test using $\alpha = .005$. State the alternatives, decision rule, and conclusion. What is the P-value of the test?

    e. State the reduced model for testing for factor B main effects. Conduct this test using $\alpha = .005$. State the alternatives, decision rule, and conclusion. What is the P-value of the test?

    f. Make all pairwise comparisons between regions; use the Tukey procedure and a 95 percent family confidence coefficient. State your findings and present a graphic summary.

Chapter 21

Random and Mixed Effects Models for Two-Factor Studies and Other Topics in ANOVA

In this chapter, we discuss a number of selected topics in the analysis of variance for two-factor studies. We first take up the special case where there is only one case for each treatment. In this connection, we discuss the Tukey test for additivity, a test that is important not only when there is just a single case for each treatment in a two-factor study but that is also useful for a variety of experimental designs to be discussed in later chapters. We then illustrate how to perform the ANOVA tests when the treatment means are of unequal importance. Finally, we consider two types of models for two-factor studies that are appropriate when the factor levels of one or both factors may be viewed as random.

21.1 ONE CASE PER TREATMENT

When there is only one case for each treatment, we no longer can work with the two-factor ANOVA model (18.23) because no estimate of the error variance σ^2 will be available. Recall from (18.38c) that SSE is a sum of squares made up of components measuring the variability within each treatment, $\sum_k (Y_{ijk} - \bar{Y}_{ij.})^2$. With only one case per treatment, there is no variability within a treatment, and SSE will then always be zero.

A way out of this difficulty is to change the model. A glance at Table 18.9 indicates that if the two factors do not interact, the interaction mean square $MSAB$ has expectation σ^2. Thus, if it is possible to assume that the two factors do not interact, we may use $MSAB$ as the estimator of the error variance σ^2 and proceed with the analysis of factor effects as usual. If it is unreasonable to assume that the two factors do not interact, transformations may be tried to remove the interaction effects. We shall say more about this in the next section.

No-Interaction Model

The two-factor ANOVA model with fixed factor levels and no interactions was given in (20.24). For the case $n = 1$ considered here, this model is:

$$(21.1) \qquad\qquad Y_{ij} = \mu_{..} + \alpha_i + \beta_j + \varepsilon_{ij}$$

where the model terms are as in the two-factor ANOVA model (18.23). Note that the third subscript has been dropped from the Y and ε terms because there is now only one case per treatment.

SSA and SSB are calculated as before from (18.39a) and (18.39b), respectively, with $n = 1$. The interaction sum of squares in (18.39c) with $n = 1$ now is expressed as follows:

$$(21.2) \qquad\qquad SSAB = \sum_i \sum_j (Y_{ij} - \bar{Y}_{i.} - \bar{Y}_{.j} + \bar{Y}_{..})^2$$

Note that SSAB in (21.2) is identical to SSAB in (18.39c) with $n = 1$, the third subscript dropped because there is only one case per treatment, and the mean $\bar{Y}_{ij.}$ replaced by the observation Y_{ij} for the same reason. The number of degrees of freedom associated with SSAB in (21.2) is the same as the number associated with SSAB in (18.39c), namely, $(a - 1)(b - 1)$.

The analysis of variance table for the case $n = 1$ for the no-interaction model (21.1) is shown in Table 21.1. No new problems arise in the tests for factor A and factor B main effects, nor in estimating these effects. Since the expected value of MSAB is σ^2 for the no-interaction model (21.1), as shown in the last column of Table 21.1, the F^* test statistics for testing factor A and factor B main effects will now utilize MSAB in the denominator, instead of MSE as before:

$$(21.3a) \qquad\qquad \text{Factor } A \text{ main effects:} \quad F^* = \frac{MSA}{MSAB}$$

$$(21.3b) \qquad\qquad \text{Factor } B \text{ main effects:} \quad F^* = \frac{MSB}{MSAB}$$

Similarly, for estimating comparisons of factor A and factor B level means, we simply replace MSE in all of the earlier results with MSAB as the estimator of the error variance σ^2 and modify the degrees of freedom accordingly.

A special problem does exist, however, in estimating treatment means. We shall explain how to handle this problem after presenting an example.

Example

Table 21.2a shows the amounts of three-month premiums charged by an automobile insurance firm for a specific type and amount of coverage in a given risk category for six cities, classified by size of city (factor A) and geographic region (factor B). Note there is only one observation per cell, namely, the amount of the premium for one city in each factor level combination. This firm adjusts its premiums periodically to

TABLE 21.1 ANOVA Table for No-Interaction Two-Factor Model with Fixed Factor Levels, $n = 1$

Source of Variation	SS	df	MS	$E\{MS\}$
Factor A	$SSA = b \sum (\bar{Y}_{i.} - \bar{Y}_{..})^2$	$a - 1$	$MSA = \dfrac{SSA}{a - 1}$	$\sigma^2 + \dfrac{b}{a - 1} \sum (\mu_{i.} - \mu_{..})^2$
Factor B	$SSB = a \sum (\bar{Y}_{.j} - \bar{Y}_{..})^2$	$b - 1$	$MSB = \dfrac{SSB}{b - 1}$	$\sigma^2 + \dfrac{a}{b - 1} \sum (\mu_{.j} - \mu_{..})^2$
Error	$SSAB = \sum\sum (Y_{ij} - \bar{Y}_{i.} - \bar{Y}_{.j} + \bar{Y}_{..})^2$	$(a - 1)(b - 1)$	$MSAB = \dfrac{SSAB}{(a - 1)(b - 1)}$	σ^2
Total	$SSTO = \sum\sum (Y_{ij} - \bar{Y}_{..})^2$	$ab - 1$		

TABLE 21.2   Two-Factor Insurance Premium Study with $n = 1$

(a) Premiums for Automobile Insurance Policy (in dollars)

Size of City (factor A)	Region (factor B)		
	East ($j = 1$)	West ($j = 2$)	Average
Small ($i = 1$)	140	100	120
Medium ($i = 2$)	210	180	195
Large ($i = 3$)	220	200	210
Average	190	160	175

(b) ANOVA Table

Source of Variation	SS	df	MS
Size of city (A)	9,300	2	4,650
Region (B)	1,350	1	1,350
Error	100	2	50
Total	10,750	5	

reflect comparative loss experiences in different localities. An insurance analyst wished to evaluate the effects of the city sizes and geographic regions shown in Table 21.2 on the amount of the premium. From experience in other cases, he conjectured that interaction effects would not be present in the premiums under analysis. A test for interactions (to be discussed in Section 21.2) did indeed lead to the conclusion that interaction effects are not present, so the analyst adopted ANOVA model (21.1).

The required sums of squares are obtained as follows [using the definitional formulas (18.38) and (18.39) for $n = 1$]:

$$SSA = 2[(120 - 175)^2 + (195 - 175)^2 + (210 - 175)^2] = 9,300$$

$$SSB = 3[(190 - 175)^2 + (160 - 175)^2] = 1,350$$

$$SSAB = (140 - 120 - 190 + 175)^2 + \cdots + (200 - 210 - 160 + 175)^2 = 100$$

$$SSTO = (140 - 175)^2 + \cdots + (200 - 175)^2 = 10,750$$

The ANOVA table is given in Table 21.2b. In the analyst's test of city size (factor A) effects, the alternative conclusions are:

$$H_0: \alpha_1 = \alpha_2 = \alpha_3 = 0$$

$$H_a: \text{not all } \alpha_i \text{ equal zero}$$

The F^* test statistic here is given by (21.3a):

$$F^* = \frac{MSA}{MSAB}$$

and the decision rule is [remember that the denominator of F^* here involves $(a - 1)(b - 1)$ degrees of freedom]:

$$\text{If } F^* \leq F[1 - \alpha; a - 1, (a - 1)(b - 1)], \text{ conclude } H_0$$

$$\text{If } F^* > F[1 - \alpha; a - 1, (a - 1)(b - 1)], \text{ conclude } H_a$$

Let $\alpha = .05$. Hence, we need $F(.95; 2, 2) = 19.0$. Table 21.2b provides the data for the test statistic:

$$F^* = \frac{4{,}650}{50} = 93$$

Since $F^* = 93 > 19.0$, we conclude H_a, that city size effects are present. The *P*-value of the test is .011.

The test for geographic region (factor B) effects proceeds similarly, the alternative conclusions being:

$$H_0: \beta_1 = \beta_2 = 0$$

$$H_a: \text{not all } \beta_j \text{ equal zero}$$

For $\alpha = .05$ the decision rule is:

$$\text{If } F^* \leq F(.95; 1, 2) = 18.5, \text{ conclude } H_0$$

$$\text{If } F^* > F(.95; 1, 2) = 18.5, \text{ conclude } H_a$$

Test statistic (21.3b) here is:

$$F^* = \frac{MSB}{MSAB} = \frac{1{,}350}{50} = 27$$

Since $F^* = 27 > 18.5$, we conclude H_a, that geographic region effects are present. The *P*-value of this test is .035.

Given the findings of both factor A and factor B main effects, the analyst next examined the factor level means $\mu_{i.}$ and $\mu_{.j}$ by methods discussed in Section 19.2.

Estimation of Treatment Mean

When there is only a single case for each treatment in a two-factor study and the no-interaction model (21.1) is employed, the usual method of estimation of a treatment mean μ_{ij} by the sample mean $\bar{Y}_{ij.}$, here simply the single observation Y_{ij}, needs to be modified. The reason is that this estimator does not make use of the model assumption of no interactions. We shall use the general matrix methods described in Section 8.6 whereby we can utilize the assumption of no interactions.

To illustrate the procedure for the insurance premium example, we begin with the cell means ANOVA model (18.15) with no restrictions about interactions (remember $n = 1$ here):

$$Y_{ij} = \mu_{ij} + \varepsilon_{ij}$$

We express this model in matrix terms in the form $\mathbf{Y} = \mathbf{X}\boldsymbol{\beta} + \boldsymbol{\varepsilon}$ with the matrices illustrated in (18.19). For our example, where we have $n = 1$, $ab = 6$, and $n_T = 6$, $\mathbf{X}$ and $\boldsymbol{\beta}$ are defined as follows:

$$(21.4) \qquad \mathbf{X} = \mathop{\mathbf{I}}_{6\times6} \qquad \boldsymbol{\beta} = \begin{bmatrix} \mu_{11} \\ \mu_{12} \\ \mu_{21} \\ \mu_{22} \\ \mu_{31} \\ \mu_{32} \end{bmatrix}$$

The least squares estimators of the treatment means μ_{ij} in $\boldsymbol{\beta}$ as usual are the sample means $\bar{Y}_{ij.}$; since $n = 1$, each sample mean is simply the individual observation Y_{ij}. Hence, the least squares estimator of $\boldsymbol{\beta}$, to be denoted by $\mathbf{b}_F$ to tie in with the notation of Chapter 8, is the observations vector $\mathbf{Y}$:

$$(21.5) \qquad \mathbf{b}_F = \mathbf{Y}$$

We shall express the constraint of no interactions in the form (8.66):

$$(21.6) \qquad \mathbf{C}\boldsymbol{\beta} = \mathbf{h}$$

utilizing the relation (18.7b) for expressing a treatment mean μ_{ij} in terms of three other treatment means when no interactions are present. We require $(a - 1)(b - 1) = 2(1) = 2$ such relations here. We shall use the relations:

$$\mu_{11} = \mu_{12} + \mu_{21} - \mu_{22} \qquad \qquad \mu_{11} - \mu_{12} - \mu_{21} + \mu_{22} = 0$$
$$\text{or}$$
$$\mu_{11} = \mu_{12} + \mu_{31} - \mu_{32} \qquad \qquad \mu_{11} - \mu_{12} - \mu_{31} + \mu_{32} = 0$$

Hence, the matrices $\mathbf{C}$ and $\mathbf{h}$ in constraint (21.6) are as follows for our example:

$$\mathop{\mathbf{C}}_{2\times6} = \begin{bmatrix} 1 & -1 & -1 & 1 & 0 & 0 \\ 1 & -1 & 0 & 0 & -1 & 1 \end{bmatrix} \qquad \mathop{\mathbf{h}}_{2\times1} = \begin{bmatrix} 0 \\ 0 \end{bmatrix}$$

We now utilize (8.68) to obtain the least squares estimators of the treatment means μ_{ij} under the constraint of no interactions. These constrained estimators are denoted by $\mathbf{b}_R$ in (8.68). Since $\mathbf{X} = \mathbf{I}$, $\mathbf{h} = \mathbf{0}$, and $\mathbf{b}_F = \mathbf{Y}$ here, (8.68) can be simplified to become:

$$(21.7) \qquad \mathbf{b}_R = \mathbf{AY} \qquad \text{No-interaction model, } n = 1$$

where:

$$(21.7a) \qquad \mathbf{A} = \mathbf{I} - \mathbf{C}'(\mathbf{CC}')^{-1}\mathbf{C}$$

For the insurance premium example, substitution into (21.7) yields:

$$\mathbf{b}_R = \begin{bmatrix} 135 \\ 105 \\ 210 \\ 180 \\ 225 \\ 195 \end{bmatrix}$$

Thus, for instance, the least squares estimate of the mean premium for a small city in the East is $\hat{\mu}_{11} = \$135$.

It is of interest to note that the least squares estimators $\hat{\mu}_{ij}$ in (21.7) turn out to be simple in structure, reflecting the constraint of no interactions:

(21.7b) $$\hat{\mu}_{ij} = \bar{Y}_{i.} + \bar{Y}_{.j} - \bar{Y}_{..}$$

Thus, using the data in Table 21.2a, we obtain:

$$\hat{\mu}_{11} = 120 + 190 - 175 = 135$$

$$\hat{\mu}_{12} = 120 + 160 - 175 = 105$$

etc.   etc.

To set up a confidence interval for a treatment mean μ_{ij}, we require the estimated variance of $\hat{\mu}_{ij}$. This can be obtained from (6.47), where $\mathbf{A}$ is given by (21.7a) and $\sigma^2\{\mathbf{Y}\}$ is estimated by $(MSAB)\mathbf{I}$.

Comments

1. The analysis of two-factor studies with $n = 1$ just outlined depends on the assumption that the two factors do not interact. If one utilizes this analysis when in fact interactions are present, the result is that the actual level of significance for testing factor A and factor B main effects is below the specified level and the actual power of the tests is lower than the expected power. Correspondingly, confidence intervals for contrasts of factor level means will tend to be too wide. This means that when interactions are present, the analysis is more likely to fail to disclose real effects than anticipated. However, when the analysis based on the no-interaction model does indicate the presence of factor A main effects or of factor B main effects, they may be taken as real effects even though interactions are actually present.

2. Sometimes, the case $n = 1$ is encountered when the observations Y_{ij} are proportions. For instance, the data may consist of the proportion of employees in a plant absent during the past week, with the plants classified by size and geographic area. As noted earlier, the arc sine transformation can be used for such data to stabilize the variances. The transformed data then can be analyzed through the no-interaction model (21.1), provided that each proportion is based on roughly the same number of cases. If the number of cases differs greatly, a weighted least squares approach should be utilized.

21.2 TUKEY TEST FOR ADDITIVITY

We describe now a test devised by Tukey that may be used for examining, when $n = 1$, whether or not the two factors in a two-factor study interact. This test is also useful for a variety of experimental designs to be discussed in later chapters.

Development of Test Statistic

As noted in Section 21.1, we considered the no-interaction model (21.1) when $n = 1$ to enable us to obtain an estimate of the error variance in this case. It would have been possible, however, to impose less severe restrictions on the $(\alpha\beta)_{ij}$ and include the restricted interaction effects in the ANOVA model. Suppose we assume that:

$$(\alpha\beta)_{ij} = D\alpha_i\beta_j \qquad (21.8)$$

where D is some constant. One motivation for this restriction is that if $(\alpha\beta)_{ij}$ is any second-degree polynomial function of α_i and β_j, then it must be of the form (21.8) because of the restrictions in (18.23) on the α_i, β_j, and $(\alpha\beta)_{ij}$ that the sums over each subscript be 0.

Using (21.8) in a regular two-factor ANOVA model with interactions for the case $n = 1$, we obtain:

$$(21.9) \qquad Y_{ij} = \mu_{..} + \alpha_i + \beta_j + D\alpha_i\beta_j + \varepsilon_{ij}$$

where each of the terms has the usual meaning. Remember there is no third subscript here because $n = 1$. The interaction sum of squares $\Sigma\Sigma D^2\alpha_i^2\beta_j^2$ now needs to be obtained. Assuming the other parameters are known, the least squares estimator of D turns out to be:

$$(21.10) \qquad \hat{D} = \frac{\sum_i \sum_j \alpha_i\beta_j Y_{ij}}{\sum_i \alpha_i^2 \sum_j \beta_j^2}$$

The usual estimator of α_i is $\bar{Y}_{i.} - \bar{Y}_{..}$, and that of β_j is $\bar{Y}_{.j} - \bar{Y}_{..}$. Replacing the parameters in $\hat{D}$ by these estimators, we obtain:

$$(21.10a) \qquad \hat{D} = \frac{\sum_i \sum_j (\bar{Y}_{i.} - \bar{Y}_{..})(\bar{Y}_{.j} - \bar{Y}_{..})Y_{ij}}{\sum_i (\bar{Y}_{i.} - \bar{Y}_{..})^2 \sum_j (\bar{Y}_{.j} - \bar{Y}_{..})^2}$$

The sample counterpart of the interaction sum of squares $\Sigma\Sigma D^2\alpha_i^2\beta_j^2$ will be denoted by $SSAB^*$ to remind us that this interaction sum of squares is for the special form of interaction in model (21.9). Substituting the sample estimates into $\Sigma\Sigma D^2\alpha_i^2\beta_j^2$, we obtain the interaction sum of squares:

$$(21.11) \qquad SSAB^* = \sum_i \sum_j \hat{D}^2(\bar{Y}_{i.} - \bar{Y}_{..})^2(\bar{Y}_{.j} - \bar{Y}_{..})^2$$

$$= \frac{\left[\sum_i \sum_j (\bar{Y}_{i.} - \bar{Y}_{..})(\bar{Y}_{.j} - \bar{Y}_{..})Y_{ij}\right]^2}{\sum_i (\bar{Y}_{i.} - \bar{Y}_{..})^2 \sum_j (\bar{Y}_{.j} - \bar{Y}_{..})^2}$$

This expression can be expanded for computational simplifications.

The analysis of variance decomposition for the special interaction model (21.9) therefore is:

$$(21.12) \qquad SSTO = SSA + SSB + SSAB^* + SSRem^*$$

where $SSRem^*$ is the *remainder sum of squares:*

$$(21.12a) \qquad SSRem^* = SSTO - SSA - SSB - SSAB^*$$

It can be shown that if $D = 0$, that is, if no interactions of the type $D\alpha_i\beta_j$ exist, $SSAB^*$ and $SSRem^*$ are independently distributed as chi-square random variables with 1 and $ab - a - b$ degrees of freedom, respectively. Hence, if $D = 0$, the test statistic:

(21.13)
$$F^* = \frac{SSAB^*}{1} \div \frac{SSRem^*}{ab - a - b}$$

is distributed as $F(1, ab - a - b)$.

Thus, for testing:

(21.14a)
$$H_0: D = 0 \text{ (no interactions present)}$$
$$H_a: D \neq 0 \text{ (interactions } D\alpha_i\beta_j \text{ present)}$$

we use the test statistic F^* defined in (21.13). Large values of F^* lead to conclusion H_a. The appropriate decision rule for controlling the risk of a Type I error at α is:

(21.14b)
$$\text{If } F^* \leq F(1 - \alpha; 1, ab - a - b), \text{ conclude } H_0$$
$$\text{If } F^* > F(1 - \alpha; 1, ab - a - b), \text{ conclude } H_a$$

The power of this test has been studied, and it appears that if interactions of approximately the type postulated in (21.8) are present and the factor A and factor B main effects are large, the test is effective in detecting the interactions. The test is usually called the *Tukey one degree of freedom test*. This test also may be used for testing for the presence of general interactions.

Example

We shall conduct the Tukey test for the insurance premium example. The data are presented in Table 21.2a. First, we obtain the elements of $SSAB^*$:

$$\sum\sum (\bar{Y}_{i.} - \bar{Y}_{..})(\bar{Y}_{.j} - \bar{Y}_{..})Y_{ij} = (120 - 175)(190 - 175)(140) + \cdots$$

$$+ (210 - 175)(160 - 175)(200) = -13,500$$

$$\sum (\bar{Y}_{i.} - \bar{Y}_{..})^2 = \frac{SSA}{2} = \frac{9,300}{2} = 4,650$$

$$\sum (\bar{Y}_{.j} - \bar{Y}_{..})^2 = \frac{SSB}{3} = \frac{1,350}{3} = 450$$

Hence, the special interaction sum of squares is:

$$SSAB^* = \frac{(-13,500)^2}{4,650(450)} = 87.1$$

We found earlier in Table 21.2b that $SSTO = 10,750$, $SSA = 9,300$, and $SSB = 1,350$; hence, we have by (21.12a):

$$SSRem^* = 10,750 - 9,300 - 1,350 - 87.1 = 12.9$$

Finally, we obtain the test statistic by (21.13):

$$F^* = \frac{87.1}{1} \div \frac{12.9}{3(2) - 3 - 2} = 6.8$$

Assuming the level of significance is to be controlled at $\alpha = .10$, we require $F(.90; 1, 1) = 39.9$. Since $F^* = 6.8 \leq 39.9$, we conclude that region and size of city do not interact. The P-value of this test is .23.

The use of the no-interaction model for the data in Table 21.2a therefore appears to be reasonable.

Remedial Actions If Interaction Effects Are Present

If the Tukey test indicates the presence of interaction effects in an analysis of variance application where $n = 1$, efforts should be made to remove the interactions so that the analysis described in Section 21.1 can be utilized. As we described in Chapter 18, transformations of the data can often be used to remove interaction effects or to make them unimportant.

One can try simple transformations such as a square root or logarithmic transformation. Alternatively, one can search in the family of power transformations on Y described in Chapter 4 in connection with the Box-Cox transformations. The procedure is to make transformations on Y according to (4.29) for selected values of λ. For each value of λ, the Tukey test statistic (21.13) is then obtained. If a λ value leads to a nonsignificant F^* test statistic, a transformation will then have been found that removes the interaction effect. Frequently, a range of λ values will yield non-significant test statistics, in which case a simple λ value in this range, such as $\lambda = .5$, may be chosen.

If no transformation can be found to make the interactions unimportant, an approximate method of analysis can be employed; see, for instance, Reference 21.1.

Note

If one or both factors are quantitative, a test for interaction effects can also be obtained by regression methods. These regression tests for interaction effects were discussed earlier.

21.3 ANOVA TESTS WHEN TREATMENT MEANS OF UNEQUAL IMPORTANCE

General Linear Test Approach

We mentioned in earlier chapters that the same basic estimation procedures are applicable whether the treatment means are of equal or unequal importance, but that the ordinary ANOVA formulas for testing are not appropriate when the treatment means are of unequal importance. Instead, analysis of variance tests when the treatment means have unequal importance must usually be conducted by means of the general linear test procedure in matrix terms described in Section 8.6.

We now consider how to use this general linear test approach when the treatment means have unequal importance. First, however, we need to stress that the test for interaction effects is not affected by unequal importance of treatment means since this test is concerned with the parallelism, or lack of it, of the treatment mean curves. This was illustrated in Figures 18.1, 18.2, and 18.3. These treatment mean curves are based solely on the individual treatment means μ_{ij} and hence do not involve averages of the treatment means. Thus, the test for interactions is conducted as explained in Section 18.6 when the sample sizes are equal and as explained in Section 20.2 when the sample sizes are unequal, whether the treatment means are of equal or unequal importance.

Tests for factor main effects when the treatment means are of unequal importance can be carried out most easily by working with the cell means model (18.15). Since no new principles are involved, we shall illustrate the tests for main effects by turning to an example.

Example.   In the growth hormone example of Chapter 20 (Table 20.1), it is known that twice as many male as female children undergo growth hormone treatment therapy, and that this ratio is the same for children who have severe, moderate, and mild depression in bone development. Inferences are desired about the target population of children undergoing therapy. Specifically, we are to test whether or not the state of bone development affects the change in growth rate. The alternatives therefore are:

(21.15)
$$H_0: \frac{2\mu_{11} + \mu_{21}}{3} = \frac{2\mu_{12} + \mu_{22}}{3} = \frac{2\mu_{13} + \mu_{23}}{3}$$

$$H_a: \text{not all equalities hold}$$

We shall restate the alternative H_0 in the following equivalent fashion:

(21.15a)
$$H_0: \begin{aligned} \frac{2\mu_{11} + \mu_{21}}{3} - \frac{2\mu_{12} + \mu_{22}}{3} = 0 \\ \frac{2\mu_{11} + \mu_{21}}{3} - \frac{2\mu_{13} + \mu_{23}}{3} = 0 \end{aligned}$$

Since H_0 is expressed in terms of the treatment means μ_{ij}, we shall use the two-factor cell means model (18.15):

(21.16)
$$Y_{ijk} = \mu_{ij} + \varepsilon_{ijk}$$

To state this linear model in matrix terms in the form $\mathbf{Y} = \mathbf{X}\boldsymbol{\beta} + \boldsymbol{\varepsilon}$, we define the $\mathbf{X}$ matrix as illustrated in (18.19). Table 21.3 contains the $\mathbf{X}$ matrix and $\boldsymbol{\beta}$ vector for the growth hormone example data in Table 20.1. The $\mathbf{Y}$ vector was given earlier in Table 20.2. The mean vector $\mathbf{X}\boldsymbol{\beta}$ is also shown in Table 21.3. We see that $E\{Y_{ijk}\} \equiv \mu_{ij}$, as ANOVA model (21.16) requires.

The hypothesis H_0 in (21.15a) can now be stated as follows in the matrix form (8.66):

(21.17)
$$H_0: \underset{s \times p}{\mathbf{C}} \ \underset{p \times 1}{\boldsymbol{\beta}} = \underset{s \times 1}{\mathbf{h}}$$

where:

$$\underset{2 \times 6}{\mathbf{C}} = \begin{bmatrix} \frac{2}{3} & -\frac{2}{3} & 0 & \frac{1}{3} & -\frac{1}{3} & 0 \\ \frac{2}{3} & 0 & -\frac{2}{3} & \frac{1}{3} & 0 & -\frac{1}{3} \end{bmatrix}$$

TABLE 21.3 X and β Matrices for ANOVA Model (21.16)—Growth Hormone Example

$$
X = \begin{bmatrix}
1 & 0 & 0 & 0 & 0 & 0 \\
1 & 0 & 0 & 0 & 0 & 0 \\
1 & 0 & 0 & 0 & 0 & 0 \\
0 & 1 & 0 & 0 & 0 & 0 \\
0 & 1 & 0 & 0 & 0 & 0 \\
0 & 0 & 1 & 0 & 0 & 0 \\
0 & 0 & 1 & 0 & 0 & 0 \\
0 & 0 & 0 & 1 & 0 & 0 \\
0 & 0 & 0 & 0 & 1 & 0 \\
0 & 0 & 0 & 0 & 1 & 0 \\
0 & 0 & 0 & 0 & 1 & 0 \\
0 & 0 & 0 & 0 & 0 & 1 \\
0 & 0 & 0 & 0 & 0 & 1 \\
0 & 0 & 0 & 0 & 0 & 1
\end{bmatrix}
\qquad
\beta = \begin{bmatrix}
\mu_{11} \\
\mu_{12} \\
\mu_{13} \\
\mu_{21} \\
\mu_{22} \\
\mu_{23}
\end{bmatrix}
\qquad
X\beta = \begin{bmatrix}
\mu_{11} \\
\mu_{11} \\
\mu_{11} \\
\mu_{12} \\
\mu_{12} \\
\mu_{13} \\
\mu_{13} \\
\mu_{21} \\
\mu_{22} \\
\mu_{22} \\
\mu_{22} \\
\mu_{23} \\
\mu_{23} \\
\mu_{23}
\end{bmatrix}
$$

$$
\underset{6\times 1}{\beta} = \begin{bmatrix}
\mu_{11} \\
\mu_{12} \\
\mu_{13} \\
\mu_{21} \\
\mu_{22} \\
\mu_{23}
\end{bmatrix}
\qquad
\underset{2\times 1}{h} = \begin{bmatrix} 0 \\ 0 \end{bmatrix}
$$

Note that this formulation yields (21.15a):

$$
\underset{2\times 1}{C\beta} = \begin{bmatrix}
(\tfrac{2}{3})\mu_{11} - (\tfrac{2}{3})\mu_{12} + (\tfrac{1}{3})\mu_{21} - (\tfrac{1}{3})\mu_{22} \\
(\tfrac{2}{3})\mu_{11} - (\tfrac{2}{3})\mu_{13} + (\tfrac{1}{3})\mu_{21} - (\tfrac{1}{3})\mu_{23}
\end{bmatrix} = \begin{bmatrix} 0 \\ 0 \end{bmatrix} = h
$$

To calculate $SSE(R) - SSE(F)$, we utilize (8.70):

(21.18) $$(Cb_F - h)'(C(X'X)^{-1}C')^{-1}(Cb_F - h)$$

The vector of parameter estimates b_F needs to be obtained first. Since we know from (18.29) that the estimated treatment means $\bar{Y}_{ij.}$ are the least squares estimators, we have:

$$
b_F = \begin{bmatrix}
\bar{Y}_{11.} \\
\bar{Y}_{12.} \\
\bar{Y}_{13.} \\
\bar{Y}_{21.} \\
\bar{Y}_{22.} \\
\bar{Y}_{23.}
\end{bmatrix} = \begin{bmatrix}
2.0 \\
1.9 \\
.9 \\
2.4 \\
2.1 \\
.9
\end{bmatrix}
$$

We can now substitute into (21.18) to obtain $SSE(R) - SSE(F)$. We do not show the matrix calculations since they are routine. We find:

$$SSE(R) - SSE(F) = 3.454$$

The error sum of squares for the full model was obtained earlier in Table 20.3a; it is $SSE(F) = 1.3000$. The degrees of freedom associated with SSE for the full model are $df_F = 8$, as shown in Table 20.3a. The degrees of freedom associated with $SSE(R) - SSE(F)$ are $df_R - df_F = s = 2$. Hence, the general linear test statistic (8.71) is:

$$F* = \frac{SSE(R) - SSE(F)}{s} \div \frac{SSE(F)}{n - p}$$

$$= \frac{3.454}{2} \div \frac{1.3000}{8} = 10.63$$

If H_0 holds, $F*$ follows the F distribution with 2 and 8 degrees of freedom. To control the level of significance at $\alpha = .05$, we require $F(.95; 2, 8) = 4.46$. Since $F* = 10.63 > 4.46$, we conclude H_a, that the weighted factor level means for the different bone development groups are not equal. The P-value of this test is .006.

Note

The test for the alternatives (21.15a) could also have been conducted by estimating the two contrasts:

$$L_1 = \frac{2\mu_{11} + \mu_{21}}{3} - \frac{2\mu_{12} + \mu_{22}}{3} \qquad L_2 = \frac{2\mu_{11} + \mu_{21}}{3} - \frac{2\mu_{13} + \mu_{23}}{3}$$

with a multiple comparison procedure (e.g., the Bonferroni procedure) and noting whether or not both confidence intervals include zero.

Weights Proportional to Sample Sizes

Simplifications in determining the term $SSE(R) - SSE(F)$ in the general linear test statistic for testing factor A and factor B main effects occur when the weights for the treatment means μ_{ij} are proportional to the treatment sample sizes n_{ij}. Such weights will be appropriate in some circumstances.

Consider a study of retail stores. The effects of size of store (factor A) and location of store within the city (factor B) on shoplifting losses are to be studied. Inferences about all retail stores in the population of interest are to be made. A random sample of n_T retail stores is selected from the population of all stores, and the selected stores are then classified by size and location. We shall denote the resulting cell sample sizes by n_{ij}. If the proportions of stores in the different size-location groups in the population were known, these proportions would serve as the appropriate weights in making inferences about size and location main effects, and the general linear test procedures just discussed would be employed. When these proportions are not known, however, the cell sample sizes n_{ij} may be used to estimate these proportions and therefore may serve as reasonable weights.

To illustrate this, suppose that $a = 2$ size groups and $b = 3$ location groups are employed in the study of retail stores, and that a random sample of $n_T = 60$ stores resulted in the following cell sample sizes n_{ij}:

	$j = 1$	$j = 2$	$j = 3$	Total
$i = 1$	20	5	4	29
$i = 2$	10	15	6	31
Total	30	20	10	60

Thus $n_{11} = 20$, $n_{21} = 10$, and so on. Further, denoting by $n_{i.}$ and $n_{.j}$ the total factor A and factor B level sample sizes as defined in (20.1a) and (20.1b), respectively, we have $n_{1.} = 29$, $n_{.1} = 30$, and so on.

The test for factor A main effects, when the sample sizes n_{ij} reflect the importance of the treatment means, would then involve a comparison of the weighted mean for factor A level $i = 1$:

$$\frac{20\mu_{11} + 5\mu_{12} + 4\mu_{13}}{29}$$

and the weighted mean for factor A level $i = 2$:

$$\frac{10\mu_{21} + 15\mu_{22} + 6\mu_{23}}{31}$$

Expressed in symbolic notation, the alternatives would be:

$$H_0: \left(\frac{n_{11}}{n_{1.}}\right)\mu_{11} + \left(\frac{n_{12}}{n_{1.}}\right)\mu_{12} + \left(\frac{n_{13}}{n_{1.}}\right)\mu_{13} = \left(\frac{n_{21}}{n_{2.}}\right)\mu_{21} + \left(\frac{n_{22}}{n_{2.}}\right)\mu_{22} + \left(\frac{n_{23}}{n_{2.}}\right)\mu_{23}$$

H_a: equality does not hold

Similarly, the alternatives for testing factor B main effects would be as follows for the growth hormone example when the sample sizes reflect the importance of the treatment means:

$$H_0: \left(\frac{n_{11}}{n_{.1}}\right)\mu_{11} + \left(\frac{n_{21}}{n_{.1}}\right)\mu_{21} = \left(\frac{n_{12}}{n_{.2}}\right)\mu_{12} + \left(\frac{n_{22}}{n_{.2}}\right)\mu_{22} = \left(\frac{n_{13}}{n_{.3}}\right)\mu_{13} + \left(\frac{n_{23}}{n_{.3}}\right)\mu_{23}$$

H_a: not all equalities hold

In general, then, the alternatives for testing for factor A main effects when the weights of the treatment means are proportional to the sample sizes are:

(21.19)
$$H_0: \sum_j \left(\frac{n_{1j}}{n_{1.}}\right)\mu_{1j} = \cdots = \sum_j \left(\frac{n_{aj}}{n_{a.}}\right)\mu_{aj}$$

H_a: not all equalities hold

and the alternatives for testing for factor B main effects are:

(21.20)
$$H_0: \sum_i \left(\frac{n_{i1}}{n_{.1}}\right)\mu_{i1} = \cdots = \sum_i \left(\frac{n_{ib}}{n_{.b}}\right)\mu_{ib}$$

H_a: not all equalities hold

It can be shown that the term $SSE(R) - SSE(F)$ for testing factor A main effects involving the alternatives in (21.19) simplifies to the ordinary single-factor treatment sum of squares in (14.25), with the factor A levels considered to be the treatments:

(21.21)
$$SSA = \sum_i n_{i.} (\bar{Y}_{i..} - \bar{Y}_{...})^2$$

where:

(21.21a)
$$\bar{Y}_{i..} = \frac{Y_{i..}}{n_{i.}}$$

(21.21b)
$$\bar{Y}_{...} = \frac{Y_{...}}{n_T}$$

and:

(21.21c)
$$Y_{i..} = \sum_{j=1}^{b} \sum_{k=1}^{n_{ij}} Y_{ijk}$$

(21.21d)
$$Y_{...} = \sum_{i=1}^{a} \sum_{j=1}^{b} \sum_{k=1}^{n_{ij}} Y_{ijk}$$

Similarly, the term $SSE(R) - SSE(F)$ for testing factor B main effects involving the alternatives in (21.20) simplifies to the single-factor treatment sum of squares in (14.25) when the factor B levels are considered to be the treatments:

(21.22)
$$SSB = \sum_j n_{.j} (\bar{Y}_{.j.} - \bar{Y}_{...})^2$$

where:

(21.22a)
$$\bar{Y}_{.j.} = \frac{Y_{.j.}}{n_{.j}}$$

and:

(21.22b)
$$Y_{.j.} = \sum_{i=1}^{a} \sum_{k=1}^{n_{ij}} Y_{ijk}$$

Example.   In the growth hormone example of Table 20.1, suppose that the treatment sample sizes n_{ij} reflect the relative importance of the treatment means. We saw in Chapter 20 that gender (factor A) and bone development (factor B) do not interact. We now wish to test whether gender affects the mean change in the growth rate. The alternatives (21.19) here are:

$$H_0: \left(\frac{3}{7}\right)\mu_{11} + \left(\frac{2}{7}\right)\mu_{12} + \left(\frac{2}{7}\right)\mu_{13} = \left(\frac{1}{7}\right)\mu_{21} + \left(\frac{3}{7}\right)\mu_{22} + \left(\frac{3}{7}\right)\mu_{23}$$

H_a: equality does not hold

To calculate SSA in (21.21), we require from Table 20.1:

$$Y_{1..} = 11.6 \quad n_{1.} = 7 \quad \bar{Y}_{1..} = 1.65714$$

$$Y_{2..} = 11.4 \quad n_{2.} = 7 \quad \bar{Y}_{2..} = 1.62857$$

$$Y_{...} = 23.0 \quad n_T = 14 \quad \bar{Y}_{...} = 1.64286$$

We then obtain:

$$SSA = 7(1.65714 - 1.64286)^2 + 7(1.62857 - 1.64286)^2$$

$$= .002857$$

There is $a - 1 = 2 - 1 = 1$ degree of freedom associated with SSA.

We found earlier in Table 20.3a that the error sum of squares for the full model is $SSE(F) = 1.3000$, with 8 degrees of freedom associated with it. Hence, the general linear test statistic here is:

$$F^* = \frac{SSE(R) - SSE(F)}{df_R - df_F} \div \frac{SSE(F)}{df_F} = \frac{SSA}{1} \div MSE(F)$$

$$= \frac{.002857}{1} \div \frac{1.3000}{8}$$

$$= .018$$

For $\alpha = .05$, we require $F(.95; 1, 8) = 5.32$. Since $F^* = .018 \leq 5.32$, we conclude H_0, that the mean change in the growth rate is the same for male and female children. The P-value of the test is .897.

The test for factor B main effects would be carried out in similar fashion.

Comments

1. The cell sample sizes in alternatives (21.19) and (21.20) are considered to be fixed, not random variables. Thus, the relevance of the alternatives depends on the reasonableness of the actual cell sample sizes as indicators of the importance of the treatment means.

2. The numerator sums of squares SSA and SSB in (21.21) and (21.22) can also be obtained by utilizing the regression approach explained in Chapter 20 for the case of treatment means with equal importance and placing the relevant factor effect variables as the *initial* variables for obtaining extra sums of squares.

For instance, we could fit regression model (20.4) for the growth hormone example so as to obtain $SSR(X_1)$. We would then have:

$$SSA = SSR(X_1)$$

Similarly, we could obtain SSB by:

$$SSB = SSR(X_2) + SSR(X_3 \mid X_2)$$

3. For the SAS and SPSS[X] packages, the sums of squares in (21.21) and (21.22) may be obtained using PROC GLM—Type I sum of squares and ANOVA Option 10, respectively. The BMDP2V program does not provide these results directly.

Proportional Sample Sizes. A special case of weights proportional to the sample sizes occurs when the sample sizes themselves follow a proportional pattern. Suppose that a chain of diet establishments is experimenting with two diets that are of equal importance. The establishments cater to three times as many women as men. One hundred men and 300 women are selected, and half of each group is randomly assigned to each diet. Hence, the treatment sample sizes are as follows:

Diet	Men	Women	Total
1	50	150	200
2	50	150	200
Total	100	300	400

Note that these treatment sample sizes follow the relation:

$$(21.23) \qquad n_{ij} = \frac{n_i. n_{.j}}{n_T}$$

Condition (21.23) implies that the sample sizes in any two rows (or columns) are proportional. This is called a case of *proportional frequencies*.

In this special case of weights proportional to sample sizes (i.e., when the sample sizes are also proportional themselves), not only are SSA and SSB given by the simple formulas (21.21) and (21.22), respectively, but the interaction sum of squares as well is given by a simple formula:

$$(21.24) \qquad SSAB = \sum_i \sum_j n_{ij}(\bar{Y}_{ij.} - \bar{Y}_{i..} - \bar{Y}_{.j.} + \bar{Y}_{...})^2$$

Furthermore, the sums of squares in this special case are orthogonal so that SSA, SSB, SSAB, and SSE sum to SSTO.

Comments

1. Formula (21.24) is applicable whenever the sample sizes are proportional, whether the treatment means are of equal or unequal importance since, as we noted earlier, the interaction sum of squares does not depend on the weights of the treatment means.

2. When proportional sample sizes are employed but the sample sizes do not reflect the importance of the treatment means (e.g., when the sample sizes are unequal but the treatment means are of equal importance), the regression approach or the general linear test approach explained earlier must be employed.

21.4 MODELS II (RANDOM FACTOR LEVELS) AND III (MIXED FACTOR LEVELS) FOR TWO-FACTOR STUDIES

Random ANOVA Model

Consider an investigation of the effects of machine operators (factor A) and machines (factor B) on the number of pieces produced in a day. Five operators and three machines are used in the study. Yet the inferences are not to be confined to the particular five operators and three machines participating in the study, but rather they are to pertain to all operators and all machines available to the company. Here a random ANOVA model (model II) would be appropriate for the two-factor study, since each of the two sets of factor levels may be considered as a sample from a population (all operators, all machines) about which inferences are to be drawn.

In a random ANOVA model for a two-factor study, we assume analogously to the random ANOVA model for a single-factor study that both the factor A main effects α_i and the factor B main effects β_j are independent random variables. Further, it is assumed that the interaction effects $(\alpha\beta)_{ij}$ are independent random variables. Thus, the random ANOVA model for a two-factor study with equal sample sizes n is:

$$(21.25) \qquad\qquad Y_{ijk} = \mu_{..} + \alpha_i + \beta_j + (\alpha\beta)_{ij} + \varepsilon_{ijk}$$

where:

$\mu_{..}$ is a constant

α_i, β_j, $(\alpha\beta)_{ij}$ are independent normal random variables with expectations zero and respective variances σ_α^2, σ_β^2, $\sigma_{\alpha\beta}^2$

ε_{ijk} are independent $N(0, \sigma^2)$

α_i, β_j, $(\alpha\beta)_{ij}$, and ε_{ijk} are pairwise independent

$i = 1, \ldots, a; j = 1, \ldots, b; k = 1, \ldots, n$

For this ANOVA model, the expected value of observation Y_{ijk} is:

$$(21.25a) \qquad\qquad E\{Y_{ijk}\} = \mu_{..}$$

and the variance of Y_{ijk}, denoted by σ_Y^2, is:

$$(21.25b) \qquad\qquad \sigma^2\{Y_{ijk}\} = \sigma_Y^2 = \sigma_\alpha^2 + \sigma_\beta^2 + \sigma_{\alpha\beta}^2 + \sigma^2$$

The Y_{ijk} thus have constant variance. They are normally distributed because they are linear combinations of independent normal random variables. Further, in advance of the random trials, different observations are independent except for observations from the same factor A level and/or from the same factor B level, which are correlated because they contain some common random terms.

Meaning of Model.   We shall explain the meaning of the terms in the random ANOVA model (21.25) with reference to the production example involving the two factors machine operators and machines. The main effect of operator i in the study (selected at random from the population of operators) is α_i. Similarly, the main effect of machine j in the study (selected at random from the population of machines) is β_j. Further, the interaction between operator i and machine j is $(\alpha\beta)_{ij}$. ANOVA model (21.25) assumes that the main effects of operators on output per day are normally distributed with zero mean and variance σ_α^2. Similarly, the main effects of machines are normally distributed with zero mean and variance σ_β^2. Finally, the operator-machine interactions are normally distributed with mean zero and variance $\sigma_{\alpha\beta}^2$. Since random ANOVA model (21.25) assumes these three effects to be independent random variables, the mean output for the operator i–machine j combination, namely, $\mu_{ij} = \mu_{..} + \alpha_i + \beta_j + (\alpha\beta)_{ij}$, may be viewed in the random ANOVA model as the sum of independent selections of α_i, β_j, and $(\alpha\beta)_{ij}$ from three different normal distributions.

Note

We caution that a random ANOVA model should only be used if the factor levels of the different factors do indeed represent random samples from populations of interest.

Mixed ANOVA Model

When one of the two factors involves fixed factor levels while the other involves random factor levels, a mixed ANOVA model (model III) is appropriate. An instance where this model may be appropriate is an investigation of the effects of four different training materials (factor A) and five instructors (factor B) upon learning in a company training program. The four levels for training materials may be considered fixed, since interest centers in these particular training materials. On the other hand, the levels for instructors may be viewed as random since inferences are to be made about a population of instructors of which the five used in the study are a sample.

When factor A has fixed factor levels and factor B has random factor levels, the α_i effects are constants and the β_j effects are random variables. The interaction effects $(\alpha\beta)_{ij}$ are also random variables because the factor B levels are random. For this case where A is the fixed effects factor, B the random effects factor, and equal sample sizes n are selected for each treatment, a relatively simple mixed ANOVA model is:

$$(21.26) \qquad Y_{ijk} = \mu_{..} + \alpha_i + \beta_j + (\alpha\beta)_{ij} + \varepsilon_{ijk}$$

where:

$\mu_{..}$ is a constant
α_i are constants subject to the restriction $\Sigma\alpha_i = 0$
β_j are independent $N(0, \sigma_\beta^2)$
$(\alpha\beta)_{ij}$ are $N\left(0, \dfrac{a-1}{a}\sigma_{\alpha\beta}^2\right)$, subject to the restrictions:
$\quad \sum_i (\alpha\beta)_{ij} = 0$ for all j
$\quad \sigma\{(\alpha\beta)_{ij}, (\alpha\beta)_{i'j}\} = -\dfrac{1}{a}\sigma_{\alpha\beta}^2 \quad i \neq i'$
ε_{ijk} are independent $N(0, \sigma^2)$
$\beta_j, (\alpha\beta)_{ij}$, and ε_{ijk} are pairwise independent
$i = 1, \ldots, a; j = 1, \ldots, b; k = 1, \ldots, n$

Note in this mixed ANOVA model that any two interaction terms $(\alpha\beta)_{ij}$ and $(\alpha\beta)_{i'j'}$ are assumed to be independent unless both refer to the same random factor B level, in which case they are negatively correlated.

For mixed ANOVA model (21.26), the expected value of observation Y_{ijk} is:

$$(21.26a) \qquad E\{Y_{ijk}\} = \mu_{..} + \alpha_i$$

and the variance of Y_{ijk} is:

$$(21.26b) \qquad \sigma^2\{Y_{ijk}\} = \sigma_Y^2 = \sigma_\beta^2 + \frac{a-1}{a}\sigma_{\alpha\beta}^2 + \sigma^2$$

The Y_{ijk} thus have constant variance. Further, they are normally distributed because they are linear combinations of independent normal random variables. Finally, in advance of the random trials, different observations are independent except for observations from the same random factor B level, which are correlated because they contain a common and some correlated random terms. A more detailed discussion of the correlations among the observations Y_{ijk} in mixed ANOVA models will be presented in Chapter 25.

Comments

1. The reason why the variance of the interaction terms in ANOVA model (21.26) is expressed as $(a - 1)\sigma_{\alpha\beta}^2/a$ rather than simply as $\sigma_{\alpha\beta}^2$ is so that the expected mean squares will be expressed relatively simply. This facilitates the making of inferences for this model.

2. We noted that model (21.26) assumes that $\sum_i(\alpha\beta)_{ij} = 0$ for all j, since all factor A levels are included in the study. However, the model places no constraints on the sums $\sum_j(\alpha\beta)_{ij}$, and they will ordinarily not equal zero.

3. More complex mixed ANOVA models have also been formulated.

21.5 ANALYSIS OF VARIANCE TESTS FOR MODELS II AND III

For both the mixed and random ANOVA models for a two-factor study, the analysis of variance calculations for sums of squares are identical to those for a fixed ANOVA model. Thus, formulas (18.38)–(18.40) are entirely applicable for ANOVA models II and III. Similarly, the degrees of freedom and mean square determinations are exactly the same as for the fixed ANOVA model, as shown in Table 18.9. The random and mixed ANOVA models depart from the fixed ANOVA model only in the expected mean squares and the consequent choice of the appropriate test statistic.

Expected Mean Squares

The expected mean squares for the random and mixed ANOVA models can be worked out by utilizing the properties of the model and applying the usual expectation theorems. They are shown in Table 21.4, together with those for the fixed ANOVA model. The derivations are tedious, but simple rules have been developed for finding the expected mean squares for random and mixed ANOVA models. We take up these rules in Chapter 27.

Note

To illustrate the derivation of an expected mean square by use of expectation theorems, we shall find $E\{MSA\}$ for the two-factor random ANOVA model (21.25). We wish to find:

$$E\{MSA\} = E\left\{\frac{nb\sum(\bar{Y}_{i..} - \bar{Y}_{...})^2}{a - 1}\right\}$$

TABLE 21.4 Expected Mean Squares in Two-Factor Studies

Mean Square	df	Fixed Factor Levels (A and B fixed)	Random Factor Levels (A and B random)	Mixed Factor Levels (A fixed, B random)
MSA	$a - 1$	$\sigma^2 + nb\dfrac{\Sigma\alpha_i^2}{a-1}$	$\sigma^2 + nb\sigma_\alpha^2 + n\sigma_{\alpha\beta}^2$	$\sigma^2 + nb\dfrac{\Sigma\alpha_i^2}{a-1} + n\sigma_{\alpha\beta}^2$
MSB	$b - 1$	$\sigma^2 + na\dfrac{\Sigma\beta_j^2}{b-1}$	$\sigma^2 + na\sigma_\beta^2 + n\sigma_{\alpha\beta}^2$	$\sigma^2 + na\sigma_\beta^2$
MSAB	$(a-1)(b-1)$	$\sigma^2 + n\dfrac{\Sigma\Sigma(\alpha\beta)_{ij}^2}{(a-1)(b-1)}$	$\sigma^2 + n\sigma_{\alpha\beta}^2$	$\sigma^2 + n\sigma_{\alpha\beta}^2$
MSE	$(n-1)ab$	σ^2	σ^2	σ^2

Now:

$$Y_{i..} = \sum_j \sum_k [\mu_{..} + \alpha_i + \beta_j + (\alpha\beta)_{ij} + \varepsilon_{ijk}]$$

$$= nb\mu_{..} + nb\alpha_i + n\sum_j \beta_j + n\sum_j (\alpha\beta)_{ij} + \sum_j \sum_k \varepsilon_{ijk}$$

We obtain then:

(21.27) $$\bar{Y}_{i..} = \frac{Y_{i..}}{nb} = \mu_{..} + \alpha_i + \bar{\beta}_{.} + \overline{(\alpha\beta)}_{i.} + \bar{\varepsilon}_{i..}$$

where the bars as usual indicate means over the subscripts shown by dots. Similarly, we find:

(21.27a) $$\bar{Y}_{...} = \mu_{..} + \bar{\alpha}_{.} + \bar{\beta}_{.} + \overline{(\alpha\beta)}_{..} + \bar{\varepsilon}_{...}$$

Hence, we have:

(21.28) $$\bar{Y}_{i..} - \bar{Y}_{...} = (\alpha_i - \bar{\alpha}_{.}) + [\overline{(\alpha\beta)}_{i.} - \overline{(\alpha\beta)}_{..}] + (\bar{\varepsilon}_{i..} - \bar{\varepsilon}_{...})$$

Squaring each side of (21.28) and summing, we obtain:

(21.29) $$\sum_i (\bar{Y}_{i..} - \bar{Y}_{...})^2 = \sum_i (\alpha_i - \bar{\alpha}_{.})^2 + \sum_i [\overline{(\alpha\beta)}_{i.} - \overline{(\alpha\beta)}_{..}]^2$$

$$+ \sum_i (\bar{\varepsilon}_{i..} - \bar{\varepsilon}_{...})^2 + \text{cross-product terms}$$

To find $E\{\sum(\bar{Y}_{i..} - \bar{Y}_{...})^2\}$, we need to take the expectation of each term on the right. The cross-product terms drop out because of the independence of α_i, $(\alpha\beta)_{ij}$, and ε_{ijk} and the fact that each of these random variables has zero expectation. Each of the remaining terms can be thought of as the numerator of a sample variance of a observations. We know from the unbiasedness of a sample variance that:

(21.30) $$E\left\{\sum_{i=1}^a (Y_i - \bar{Y})^2\right\} = (a - 1)\sigma^2(Y_i)$$

Hence:

(21.31) $$E\left\{\sum_i (\alpha_i - \bar{\alpha}_{.})^2\right\} = (a - 1)\sigma_\alpha^2$$

because $\sigma^2\{\alpha_i\} = \sigma_\alpha^2$. Similarly, we find:

(21.32) $$E\left\{\sum_i (\bar{\varepsilon}_{i..} - \bar{\varepsilon}_{...})^2\right\} = (a - 1)\frac{\sigma^2}{bn}$$

since $\sigma^2\{\bar{\varepsilon}_{i..}\} = \sigma^2/bn$, and:

(21.33) $$E\left\{\sum_i [\overline{(\alpha\beta)}_{i.} - \overline{(\alpha\beta)}_{..}]^2\right\} = (a - 1)\frac{\sigma_{\alpha\beta}^2}{b}$$

Using (21.31)–(21.33), we find:

(21.34) $$E\left\{\sum (\bar{Y}_{i..} - \bar{Y}_{...})^2\right\} = (a - 1)\sigma_\alpha^2 + (a - 1)\frac{\sigma_{\alpha\beta}^2}{b} + (a - 1)\frac{\sigma^2}{bn}$$

and:

$$(21.35) \qquad E\{MSA\} = \frac{nb}{a-1} E\left\{ \sum (\bar{Y}_{i..} - \bar{Y}_{...})^2 \right\} = nb\sigma_\alpha^2 + n\sigma_{\alpha\beta}^2 + \sigma^2$$

as shown in Table 21.4.

Construction of Test Statistics

As usual, each statistic for testing factor effects is constructed by comparing two mean squares that have the properties:

1. Under H_0, both have the same expectation.
2. Under H_a, the numerator mean square has a larger expectation than the denominator mean square.

It can be shown that such a test statistic follows the F distribution if H_0 holds. The decision rule is constructed in the ordinary fashion, with large values of the test statistic leading to H_a.

For instance, to test for the presence of factor A main effects in the random ANOVA model (21.25), namely:

$$(21.36) \qquad \begin{aligned} H_0&: \sigma_\alpha^2 = 0 \\ H_a&: \sigma_\alpha^2 > 0 \end{aligned}$$

we see from Table 21.4 that MSA and $MSAB$ both have the same expectation if $\sigma_\alpha^2 = 0$, that is, if factor A has no main effects. If $\sigma_\alpha^2 > 0$, $E\{MSA\}$ is greater than $E\{MSAB\}$. Hence, the appropriate test statistic is:

$$(21.37) \qquad F^* = \frac{MSA}{MSAB}$$

and the decision rule for controlling the Type I error at α is:

$$(21.38) \qquad \begin{aligned} &\text{If } F^* \le F[1 - \alpha; a - 1, (a - 1)(b - 1)], \text{ conclude } H_0 \\ &\text{If } F^* > F[1 - \alpha; a - 1, (a - 1)(b - 1)], \text{ conclude } H_a \end{aligned}$$

Note that the denominator for testing for factor A main effects in the random ANOVA model is $MSAB$, whereas it is MSE in the fixed ANOVA model.

We summarize the appropriate test statistics for mixed and random ANOVA models in Table 21.5. For comparison purposes, we also present the test statistics for the fixed ANOVA model there. As may be seen from Table 21.5, in a number of instances the denominator of the test statistic for mixed and random ANOVA models differs from that for the fixed ANOVA model. Hence, it is important that the expected mean squares be known when random or mixed models are utilized so that the appropriate test statistics can be determined.

TABLE 21.5 Test Statistics for Mixed and Random ANOVA Models

Test for Presence of Effects of—	Fixed ANOVA Model (A and B fixed)	Random ANOVA Model (A and B random)	Mixed ANOVA Model (A fixed, B random)
Factor A	MSA/MSE	MSA/MSAB	MSA/MSAB
Factor B	MSB/MSE	MSB/MSAB	MSB/MSE
AB interactions	MSAB/MSE	MSAB/MSE	MSAB/MSE

Example

We return to our earlier mixed ANOVA example of four different training materials (factor A, fixed) and five instructors (factor B, random). Four classes were tested for each training material–instructor combination. The analysis of variance as obtained from a computer package for two-factor studies is shown in Table 21.6; the original data are not presented. To test whether or not training materials and instructors interact:

$$H_0: \sigma_{\alpha\beta}^2 = 0$$

$$H_a: \sigma_{\alpha\beta}^2 > 0$$

we utilize according to Table 21.5 the test statistic:

$$F^* = \frac{MSAB}{MSE}$$

Using the results from Table 21.6, we obtain:

$$F^* = \frac{3.9}{2.1} = 1.86$$

TABLE 21.6 ANOVA Table for Mixed ANOVA Model Training Example (A fixed, B random, a = 4, b = 5, n = 4)

Source of Variation	SS	df	MS	F*
Factor A (training materials–fixed)	42	3	14.0	14.0/3.9 = 3.59
Factor B (instructors–random)	54	4	13.5	13.5/2.1 = 6.43
AB interactions	47	12	3.9	3.9/2.1 = 1.86
Error	126	60	2.1	
Total	269	79		

$$F(.95; 3, 12) = 3.49 \qquad F(.95; 4, 60) = 2.53$$
$$F(.95; 12, 60) = 1.92$$

For a level of significance of $\alpha = .05$, we require $F(.95; 12, 60) = 1.92$. Since $F^* = 1.86 \le 1.92$, we conclude that training materials and instructors do not interact. The P-value of this test is .06.

The test statistics for testing training material main effects and instructor main effects are shown in Table 21.6. By comparing the test statistics with the appropriate percentiles of the F distribution shown at the bottom of Table 21.6 for level of significance $\alpha = .05$, we find that both training materials and instructors differ in effectiveness.

Note

If there is only one case per treatment ($n = 1$), it will be recalled from Section 21.1 that no exact tests are possible with the fixed two-factor ANOVA model unless one can modify the model. The reason is that $MSE = 0$ always in that case so that no estimate of σ^2 can be obtained. Table 21.4 indicates that exact tests for both factor A and factor B main effects are possible with the random ANOVA model when $n = 1$ without any restrictive assumptions about the interactions. This is because $MSAB$ is the appropriate denominator of the test statistic here, and $MSAB$ can be determined regardless of sample size. With the mixed ANOVA model where factor A is the fixed factor, the presence of factor A main effects can also be tested when $n = 1$ without the need for restrictive assumptions about the interactions. However, an exact test for factor B main effects would require the assumption that all interactions are zero or some other modification of the ANOVA model.

21.6 ESTIMATION OF FACTOR EFFECTS FOR MODELS II AND III

Estimation of Variance Components

For random factors that have significant main effects, we often would like to estimate the magnitude of the variance component. Unbiased point estimators can readily be derived, using linear combinations of the expected mean squares in Table 21.4. With the random ANOVA model, for instance, σ_α^2 can be estimated by noting that:

$$E\{MSA\} - E\{MSAB\} = \sigma^2 + nb\sigma_\alpha^2 + n\sigma_{\alpha\beta}^2 - \sigma^2 - n\sigma_{\alpha\beta}^2 = nb\sigma_\alpha^2$$

Hence, an unbiased point estimator of σ_α^2 is:

$$(21.39) \qquad s_\alpha^2 = \frac{MSA - MSAB}{nb}$$

Example.   In the training example of Table 21.6, random factor B (instructors) had significant effects. To estimate σ_β^2, we utilize the expected mean squares in Table 21.4 for the mixed model with factor A fixed and factor B random, and determine an unbiased point estimator to be:

$$(21.40) \qquad s_\beta^2 = \frac{MSB - MSE}{na}$$

Substituting, we obtain:

$$s_\beta^2 = \frac{13.5 - 2.1}{16} = .71$$

Estimation of Fixed Effects in Mixed Model

When the main effects of the fixed factor in a mixed two-factor model are of interest, one usually desires to obtain estimates of these effects in the form of contrasts. Our earlier discussion for the two-factor fixed ANOVA model in Chapter 19 is applicable here, with the principal change occurring in the estimated variance of the contrast. For a fixed ANOVA model, this estimated variance involves MSE, as shown, for instance, in (19.8). When dealing with a mixed model, however, the appropriate mean square to be used in the estimated variance formula is no longer MSE. A simple rule tells us which mean square is appropriate, namely, the one that is used in the denominator of the test statistic for testing the presence of main effects of the fixed factor under consideration. For instance, with the mixed model (21.26) where A is the fixed factor, $MSAB$ is the appropriate mean square (Table 21.5). The degrees of freedom in constructing the confidence interval are those associated with the mean square utilized for estimating the variance of the contrast.

Example. In the training example of Table 21.6, no interaction effects were found. It is now desired that we estimate the difference in the mean amount of learning with training materials 1 and 2 using a 95 percent confidence interval. The relevant sample results are:

$$\bar{Y}_{1..} = 43.1 \qquad \bar{Y}_{2..} = 40.8$$

Hence, our point estimate of $D = \mu_{1.} - \mu_{2.}$ is:

(21.41) $$\hat{D} = \bar{Y}_{1..} - \bar{Y}_{2..} = 43.1 - 40.8 = 2.3$$

Using formula (19.16b) with MSE replaced by $MSAB$, the estimated variance is:

(21.42) $$s^2\{\hat{D}\} = \frac{2MSAB}{bn}$$

For our example, we have:

$$s^2\{\hat{D}\} = \frac{2(3.9)}{20} = .39$$

or $s\{\hat{D}\} = .62$. There are 12 degrees of freedom associated with $MSAB$; hence, we require $t(.975; 12) = 2.179$. The confidence limits therefore are $2.3 \pm 2.179(.62)$ and the desired confidence interval is:

$$.9 \le \mu_{1.} - \mu_{2.} \le 3.7$$

Thus, we conclude with confidence coefficient .95 that training material 1 is more effective than training material 2, its mean being somewhere between .9 and 3.7 units larger.

Multiple Comparison Procedures. Multiple comparison procedures can be utilized for the main effects of the fixed factor in a mixed ANOVA model in the same way as for the fixed ANOVA model. In the estimated variance of the contrast, *MSE* will simply need to be replaced by the appropriate mean square. For example, suppose we wish to obtain all pairwise comparisons between the different training materials in the training example in Table 21.6 by means of the Tukey method. We would calculate $s^2\{\hat{D}\}$ as in the previous example. The T multiple in (19.16c) now would be:

$$(21.43) \qquad T = \frac{1}{\sqrt{2}} q[1 - \alpha; a, (a - 1)(b - 1)]$$

where $(a - 1)(b - 1)$ replaces $(n - 1)ab$ as the degrees of freedom associated with the mean square utilized. With specific reference to the training example in Table 21.6, for constructing 95 percent family confidence coefficient intervals for all pairwise comparisons between training materials we would require:

$$q(.95; 4, 12) = 4.20$$

and:

$$T = \frac{1}{\sqrt{2}}(4.20) = 2.97$$

CITED REFERENCE

21.1. Johnson, D. E., and F. A. Graybill. "Estimation of σ^2 in a Two-Way Classification Model with Interaction." *Journal of the American Statistical Association* 67 (1972), pp. 388–94.

PROBLEMS

21.1. Suppose that two-factor analysis of variance model (18.23) were to be employed with $n = 1$ for each factor level combination. How many degrees of freedom would be associated with *SSE* in (18.38c)? What does this imply?

21.2. **Coin-operated terminals.** A university computer service conducted an experiment in which one coin-operated computer terminal was placed at each of four different locations on the campus last semester during the midterm week and again during the final week of classes. The data that follow show the number of hours each terminal was *not* in use during the week at the four locations (factor A) and for the two different weeks (factor B).

Factor A (location)	Factor B (week)	
	$j = 1$ Midterm	$j = 2$ Final
$i = 1$	16.5	21.4
$i = 2$	11.8	17.3
$i = 3$	12.3	16.9
$i = 4$	16.6	21.0

Assume that no-interaction ANOVA model (21.1) is appropriate.

a. Plot the data in the format of Figure 18.5. Does it appear that interaction effects are present? Does it appear that factor A and factor B main effects are present? Discuss.

b. Conduct separate tests for location and week main effects. In each test, use a level of significance of $\alpha = .05$ and state the alternatives, decision rule, and conclusion. Give an upper bound for the family level of significance; use the Kimball inequality. What is the P-value for each test?

c. Make all pairwise comparisons among location means and estimate the difference between the means for the two weeks; use the Bonferroni procedure with a 90 percent family confidence coefficient. State your findings.

21.3. Refer to **Coin-operated terminals** Problem 21.2. It is desired to estimate μ_{32}.

a. Obtain a point estimate of μ_{32} using (21.7b).

b. Verify the point estimate $\hat{\mu}_{32}$ obtained in part (a) using (21.7) and (21.7a). State the **C** and **A** matrices utilized.

c. Obtain the estimated variance of $\hat{\mu}_{32}$. [*Hint:* Use (6.47) and note that $\mathbf{A}[(MSE)\mathbf{I}]\mathbf{A}' = (MSE)\mathbf{A}.$]

d. Construct a 95 percent confidence interval for μ_{32}. Interpret your interval estimate. Is your interval estimate applicable if next year three terminals will be placed at location 3? Explain.

21.4. Refer to **Coin-operated terminals** Problem 21.2. Conduct the Tukey test for additivity; use $\alpha = .025$. State the alternatives, decision rule, and conclusion. If the additive model is not appropriate, what might you do?

21.5. **Brainstorming.** A researcher investigated whether brainstorming is more effective for larger groups than for smaller ones by setting up four groups of agribusiness executives, the group sizes being two, three, four, and five, respectively. He also set up four groups of agribusiness scientists, the group sizes being the same as for the agribusiness executives. The researcher gave each group the same problem: "How can Canada increase the value of its agricultural exports?" Each group was allowed 30 minutes to generate ideas. The variable of interest was the number of different ideas proposed by the group. The results, classified by type of group (factor A) and size of group (factor B), were:

Factor A (type of group)	Factor B (size of group)			
	$j = 1$ Two	$j = 2$ Three	$j = 3$ Four	$j = 4$ Five
$i = 1$ Agribusiness executives	18	22	31	32
$i = 2$ Agribusiness scientists	15	23	29	33

Assume that no-interaction ANOVA model (21.1) is appropriate.

a. Plot the data in the format of Figure 18.5. Does it appear that interaction effects are present? Does it appear that factor A and factor B main effects are present? Discuss.

b. Conduct separate tests for type of group and size of group main effects. In each test, use a level of significance of $\alpha = .01$ and state the alternatives, decision rule, and conclusion. Give an upper bound for the family level of significance; use the Kimball inequality. What is the P-value for each test?

c. Obtain confidence intervals for $D_1 = \mu_{.2} - \mu_{.1}$, $D_2 = \mu_{.3} - \mu_{.2}$, and $D_3 = \mu_{.4} - \mu_{.3}$; use the Bonferroni procedure with a 95 percent family confidence coefficient. State your findings.

d. Is the Bonferroni procedure used in part (c) the most efficient one here? Explain.

21.6. Refer to **Brainstorming** Problem 21.5. It is desired to estimate μ_{14}.
a. Obtain a point estimate of μ_{14} using (21.7b).
b. Verify the point estimate $\hat{\mu}_{14}$ obtained in part (a) using (21.7) and (21.7a). State the $\mathbf{C}$ and $\mathbf{A}$ matrices utilized.
c. Obtain the estimated variance of $\hat{\mu}_{14}$. [*Hint:* Use (6.47) and note that $\mathbf{A}[(MSE)\mathbf{I}]\mathbf{A}' = (MSE)\mathbf{A}.]$
d. Construct a 99 percent confidence interval for μ_{14}. Interpret your interval estimate. Is your interval estimate applicable if the two factors interact?

21.7. Refer to **Brainstorming** Problem 21.5. Conduct the Tukey test for additivity; use $\alpha = .01$. State the alternatives, decision rule, and conclusion. If the additive model is not appropriate, what might you do?

21.8. **Soybean sausage.** A food technologist, testing storage capabilities for a newly developed type of imitation sausage made from soybeans, conducted an experiment to test the effects of temperature level (factor A) and humidity level (factor B) in the freezer compartment on color change in the sausage. Three humidity levels and four temperature levels were considered. Five hundred sausages were stored at each of the 12 temperature-humidity combinations for 90 days. At the end of the storage period, the researcher determined the proportion of sausages for each temperature-humidity combination that exhibited color changes. The researcher transformed the data by means of the arc sine transformation (16.19) to stabilize the variances. The transformed data $Y' = 2 \arcsin \sqrt{Y}$ follow.

Factor A	Factor B (temperature level)			
(humidity level)	$j = 1$	$j = 2$	$j = 3$	$j = 4$
$i = 1$	13.9	14.2	20.5	24.8
$i = 2$	15.7	16.3	21.7	23.6
$i = 3$	15.1	15.4	19.9	26.1

Assume that no-interaction ANOVA model (21.1) is appropriate.
a. Plot the data in the format of Figure 18.5. Does it appear that interaction effects are present? Does it appear that factor A and factor B main effects are present? Discuss.
b. Conduct separate tests for humidity and temperature main effects. In each test, use a level of significance of $\alpha = .025$ and state the alternatives, decision rule, and conclusion. What is the P-value for each test?
c. Obtain confidence intervals for $D_1 = \mu_{.2} - \mu_{.1}$, $D_2 = \mu_{.3} - \mu_{.2}$, and $D_3 = \mu_{.4} - \mu_{.3}$; use the Bonferroni procedure with a 95 percent family confidence coefficient. State your findings.
d. Is the Bonferroni procedure used in part (c) the most efficient one here? Explain.

21.9. Refer to **Soybean sausage** Problem 21.8. It is desired to estimate μ_{23}.
a. Obtain a point estimate of μ_{23} using (21.7b).
b. Verify the point estimate $\hat{\mu}_{23}$ obtained in part (a) using (21.7) and (21.7a). State the $\mathbf{C}$ and $\mathbf{A}$ matrices utilized.
c. Obtain the estimated variance of $\hat{\mu}_{23}$. [*Hint:* Use (6.47) and note that $\mathbf{A}[(MSE)\mathbf{I}]\mathbf{A}' = (MSE)\mathbf{A}.]$

d. Construct a 98 percent confidence interval for μ_{23} and transform it back to the original units. Interpret your interval estimate. Is your interval estimate applicable if the two factors interact?

21.10. Refer to **Soybean sausage** Problem 21.8. Conduct the Tukey test for additivity; use $\alpha = .005$. State the alternatives, decision rule, and conclusion. If the additive model is not appropriate, what might you do?

21.11. Refer to **Cash offers** Problem 18.10. It is known that in both populations of male and female owners, 30 percent are young, 60 percent are middle-aged, and 10 percent are elderly. Test by means of the single degree of freedom t^* test statistic whether or not the mean cash offers for male and female owners are equal; use $\alpha = .05$. State the alternatives, decision rule, and conclusion. What is the P-value of the test?

21.12. Refer to **Kidney failure hospitalization** Problem 18.18. Continue to work with the transformed data $Y' = \log_{10}(Y + 1)$. It is known that 75 percent of patients in each weight gain group receive the short duration treatment.
a. Use the cell means model (18.15) for expressing the two alternatives for testing whether or not factor B main effects are present.
b. Define the $\mathbf{X}$ matrix and $\boldsymbol{\beta}$ vector for expressing full model (18.15) in matrix form for this case.
c. Express the two alternatives in part (a) in matrix form (8.66).
d. Use (8.70) to calculate $SSE(R) - SSE(F)$.
e. Test whether or not factor B main effects are present; use $\alpha = .05$. State the decision rule and conclusion. What is the P-value of the test?
f. Compare the mean number of days of hospitalization (in transformed units) for patients with severe and mild weight gains; use a 95 percent confidence interval.

21.13. Refer to **Adjunct professors** Problem 20.6. It is known that 10 percent of professors in each subject matter area have a bachelor's degree, 20 percent have a master's degree, and 70 percent have a doctorate.
a. Use the cell means model (18.15) for expressing the two alternatives for testing whether or not factor A main effects are present.
b. Define the $\mathbf{X}$ matrix and $\boldsymbol{\beta}$ vector for expressing full model (18.15) in matrix form for this case.
c. Express the two alternatives in part (a) in matrix form (8.66).
d. Use (8.70) to calculate $SSE(R) - SSE(F)$.
e. Test whether or not factor A main effects are present; use $\alpha = .01$. State the decision rule and conclusion. What is the P-value of the test?
f. Compare the mean amounts of payment received by faculty members teaching humanities and engineering courses; use a 99 percent confidence interval. Interpret your interval estimate.

21.14. Refer to **Programmer requirements** Problem 18.20. Assume that the following observations did not exist: $Y_{133}, Y_{134}, Y_{234}$. Further assume that the sample sizes reflect the importance of the treatment means. Test whether or not type of experience main effects are present; control the level of significance at $\alpha = .01$. State the alternatives, decision rule, and conclusion. What is the P-value of the test?

21.15. Refer to **Adjunct professors** Problem 20.6. Assume that the sample sizes reflect the importance of the treatment means. Test whether or not subject matter main effects are present; control the level of significance at $\alpha = .05$. State the alternatives, decision rule, and conclusion. What is the P-value of the test?

21.16. For the mixed effects model (21.26), why is $\sum_i(\alpha\beta)_{ij} = 0$ whereas usually $\sum_j(\alpha\beta)_{ij} \neq 0$?

21.17. A marketing consultant is designing several experiments involving a newly developed low-cost food processor. The initial experiment has the objectives (1) to compare the effects on unit sales of three possible prices recommended by the sales department ($23.99, $25.49, $25.95) and (2) to determine whether the color scheme used for the appliance affects unit sales. A great many color schemes are feasible; three (white, green, pink) have been selected for the initial experiment to represent the range of possible colors. If the experiment suggests that color scheme does have an effect, this aspect of the product design will be investigated in detail in a follow-up study. Which ANOVA model would you employ for analyzing the initial experiment? Discuss.

21.18. In a two-factor ANOVA study with $a = 3$, $b = 2$, and $n = 5$, the two factor effects are both random with $\sigma^2 = 5.0$, $\sigma_\alpha^2 = 8.0$, $\sigma_\beta^2 = 10.0$, and $\sigma_{\alpha\beta}^2 = 6.0$. Assume that ANOVA model (21.25) is applicable.
a. Obtain $E\{MSA\}$, $E\{MSB\}$, and $E\{MSAB\}$.
b. What would be the expected mean squares if $\sigma_{\alpha\beta}^2 = 0$, all other parameters remaining the same?

21.19. A survey statistician has commented: "I am rather suspicious of uses of random effects and mixed effects ANOVA models. Seldom are the factor levels chosen by a random mechanism from a known population." Discuss.

21.20. Miles per gallon. An automobile manufacturer wished to study the effects of differences between drivers (factor A) and differences between cars (factor B) on gasoline consumption. Four drivers were selected at random; also five cars of the same model with manual transmission were randomly selected from the assembly line. Each driver drove each car twice over a 40-mile test course and the miles per gallon were recorded. The data follow.

Factor A	Factor B (car)				
(driver)	$j = 1$	$j = 2$	$j = 3$	$j = 4$	$j = 5$
$i = 1$	25.3	28.9	24.8	28.4	27.1
	25.2	30.0	25.1	27.9	26.6
$i = 2$	33.6	36.7	31.7	35.6	33.7
	32.9	36.5	31.9	35.0	33.9
$i = 3$	27.7	30.7	26.9	29.7	29.2
	28.5	30.4	26.3	30.2	28.9
$i = 4$	29.2	32.4	27.7	31.8	30.3
	29.3	32.4	28.9	30.7	29.9

Assume that random ANOVA model (21.25) is applicable.
a. Test whether or not the two factors interact; use $\alpha = .05$. State the alternatives, decision rule, and conclusion. What is the P-value of the test?
b. Test separately whether or not factor A and factor B main effects are present. For each test, use $\alpha = .05$ and state the alternatives, decision rule, and conclusion. What is the P-value for each test?
c. Obtain point estimates of σ_α^2 and σ_β^2. Which factor appears to have the greater effect on gasoline consumption?

21.21. Refer to **Disk drive service** Problem 18.16. Suppose that the service center employs a large number of technicians and that the three included in the study were selected at random. Assume that the conditions of mixed ANOVA model (21.26) are applicable, except that here factor A effects are random and factor B effects are fixed. Under current conditions, all technicians service each of the three makes with approximately equal frequency.

  a. Test whether or not the two factors interact; use $\alpha = .01$. State the alternatives, decision rule, and conclusion. What is the P-value of the test?

  b. Obtain a point estimate of $\sigma_{\alpha\beta}^2$. Does $\sigma_{\alpha\beta}^2$ appear to be large relative to σ^2? Explain.

  c. Test whether or not factor A main effects are present; use $\alpha = .01$. State the alternatives, decision rule, and conclusion. Why is it meaningful here to test for factor A main effects?

  d. Test whether or not factor B main effects are present; use $\alpha = .01$. State the alternatives, decision rule, and conclusion. Why is it meaningful here to test for factor B main effects?

  e. It is desired to obtain all pairwise comparisons between the means for the three disk drive makes. Use the Tukey procedure and a 95 percent family confidence coefficient to make these comparisons. State your findings.

  f. Obtain a point estimate of σ_α^2. Does the variability between technicians appear to be large? Explain.

21.22. **Imitation pearls.** Preliminary research on the production of imitation pearls entailed studying the effect of the number of coats of a special lacquer (factor A) applied to an opalescent plastic bead used as the base of the pearl on the market value of the pearl. Four batches of 12 beads (factor B) were used in the study, and it is desired to also consider their effect on the market value. The three levels of factor A (6, 8, and 10 coats) were fixed in advance, while the four batches can be regarded as a random sample of batches from the bead production process. The market value of each pearl was determined by a panel of experts. The data (coded) follow.

Factor A		Factor B (batch)			
(number of coats)		$j = 1$	$j = 2$	$j = 3$	$j = 4$
$i = 1$	6	72.0	72.1	75.2	70.4
		74.6	76.9	73.8	68.1
		67.4	74.8	75.7	72.4
		72.8	73.3	77.8	72.4
$i = 2$	8	76.9	80.3	80.2	74.3
		78.1	79.3	76.6	77.6
		72.9	76.6	77.3	74.4
		74.2	77.2	79.9	72.9
$i = 3$	10	76.3	80.9	79.2	71.6
		74.1	73.7	78.0	77.7
		77.1	78.6	77.6	75.2
		75.0	80.2	81.2	74.4

Assume that mixed ANOVA model (21.26) is applicable.

    a. Test for interaction effects; use $\alpha = .05$. State the alternatives, decision rule, and conclusion. What is the P-value of the test?

    b. Test for factor A and factor B main effects. For each test, use $\alpha = .05$ and state the alternatives, decision rule, and conclusion. What is the P-value for each test?

    c. Estimate $D_1 = \mu_2. - \mu_1.$ and $D_2 = \mu_3. - \mu_2.$ by means of the Bonferroni procedure with a 90 percent family confidence coefficient. State your findings.

    d. Obtain a point estimate of σ_β^2. Does σ_β^2 appear to be large compared to σ^2? Discuss.

21.23. Refer to **Coin-operated terminals** Problem 21.2. Suppose that the weeks (factor B) had been selected intentionally but the locations (factor A) had been selected at random from a large number of possible locations. Assume that the conditions of mixed ANOVA model (21.26) are appropriate, except that here factor A effects are random and factor B effects are fixed.

    a. Test for factor B main effects; use $\alpha = .05$. State the alternatives, decision rule, and conclusion. What is the P-value of the test?

    b. Why can you not test for factor A main effects here?

EXERCISES

21.24. Modify formulas (18.39a) and (18.39b) to apply to ANOVA model (21.1), where $n = 1$.

21.25. Consider a two-factor study with $a = 2$, $b = 2$, and $n = 1$. Obtain the **C** matrix to be used in (21.7a) for the constraint $\mu_{11} = \mu_{12} + \mu_{21} - \mu_{22}$. Show that $\mathbf{b}_R$ in (21.7) has as its elements the point estimators in (21.7b).

21.26. Show that (21.8) is the only second-degree polynomial function of α_i and β_j such that $\sum_i (\alpha\beta)_{ij} = \sum_j (\alpha\beta)_{ij} = 0$.

21.27. For random ANOVA model (21.25), derive $\sigma^2\{\bar{Y}_{i..}\}$. [*Hint:* Use (21.27).]

PROJECTS

21.28. Refer to the **SENIC** data set and Projects 20.18 and 20.19. Assume that the sample sizes reflect the importance of the treatment means.

    a. Test for region (factor A) main effects; use $\alpha = .01$. State the alternatives, decision rule, and conclusion. What is the P-value of the test?

    b. Test for average age of patients (factor B) main effects; use $\alpha = .01$. State the alternatives, decision rule, and conclusion. What is the P-value of the test?

21.29. Refer to the **SMSA** data set and Projects 20.20 and 20.21. Assume that the sample sizes reflect the importance of the treatment means.

    a. Test for region (factor A) main effects; use $\alpha = .005$. State the alternatives, decision rule, and conclusion. What is the P-value of the test?

    b. Test for percent of population in central cities (factor B) main effects; use $\alpha = .005$. State the alternatives, decision rule, and conclusion. What is the P-value of the test?

21.30. Consider a two-factor study with $a = 3$, $b = 2$, and $n = 5$. Random ANOVA model (21.25) is applicable with $\mu_{..} = 92$, $\sigma_\alpha^2 = 24$, $\sigma_\beta^2 = 11$, $\sigma_{\alpha\beta}^2 = .1$, and $\sigma^2 = 8$.

  a. Using a normal random number generator, obtain a value for each of the main effects α_i ($i = 1, 2, 3$) and β_j ($j = 1, 2$) and for each interaction effect $(\alpha\beta)_{ij}$.

  b. Generate five error terms for each treatment.

  c. Combine the parameter values obtained in part (a) and the error terms obtained in part (b) to yield five observations Y_{ijk} for each treatment.

  d. Based on the observations obtained in part (c), calculate the F^* test statistic for testing whether or not factor A main effects are present. What is your conclusion using $\alpha = .05$?

  e. Repeat the steps in parts (a)–(d) 100 times. Calculate the mean of the 100 numerator mean squares and the mean of the 100 denominator mean squares. Are these means close to theoretical expectations?

  f. In what proportion of the 100 trials did the test lead to the conclusion of the presence of factor A main effects? Does the test have good power for the case considered here?

Chapter 22

Multifactor Studies

When three or more factors are studied simultaneously, the model and analysis employed are straightforward extensions of the two-factor case. We shall illustrate the nature of the extensions with reference to three-factor studies. Ordinarily, computer packages will be utilized for performing the needed calculations for multifactor studies involving three or more factors. For completeness, however, we shall present the necessary computational formulas for three-factor studies.

22.1 MODEL I (FIXED FACTOR LEVELS) FOR THREE-FACTOR STUDIES

We shall develop here ANOVA model I with fixed factor levels for three-factor studies *when all treatment sample sizes are equal and all treatment means are of equal importance*. This ANOVA model will be applicable to observational studies and to experimental studies based on a completely randomized design.

Notation

Three factors, A, B, and C, are investigated at a, b, and c levels, respectively. The mean response for the treatment when factor A is at the ith level ($i = 1, \ldots, a$), factor B is at the jth level ($j = 1, \ldots, b$), and factor C is at the kth level ($k = 1, \ldots, c$) is denoted by μ_{ijk}. The number of cases for each treatment is assumed to be constant, denoted by n. We assume $n \geq 2$.

The mean response when A is at the ith level and B is at the jth level is denoted by $\mu_{ij.}$, and similar notation is used for other pairs of factor levels. Since all treatment means are assumed to have equal importance, we define:

$$(22.1a) \qquad \mu_{ij.} = \frac{\sum\limits_{k} \mu_{ijk}}{c}$$

(22.1b)
$$\mu_{i.k} = \frac{\sum\limits_{j} \mu_{ijk}}{b}$$

(22.1c)
$$\mu_{.jk} = \frac{\sum\limits_{i} \mu_{ijk}}{a}$$

The mean response when A is at the ith level is denoted by $\mu_{i..}$, and similar notation is used for the other factor level means. We define:

(22.2a)
$$\mu_{i..} = \frac{\sum\limits_{j}\sum\limits_{k} \mu_{ijk}}{bc}$$

(22.2b)
$$\mu_{.j.} = \frac{\sum\limits_{i}\sum\limits_{k} \mu_{ijk}}{ac}$$

(22.2c)
$$\mu_{..k} = \frac{\sum\limits_{i}\sum\limits_{j} \mu_{ijk}}{ab}$$

Finally, the overall mean response is denoted by $\mu_{...}$ and is defined:

(22.3)
$$\mu_{...} = \frac{\sum\limits_{i}\sum\limits_{j}\sum\limits_{k} \mu_{ijk}}{abc}$$

Illustration

To illustrate the meaning of the model terms for a three-factor analysis of variance model, we shall consider a study of the effects of gender, age, and intelligence level of college graduates on learning of a complex task. Gender is factor A and has $a = 2$ levels (male, female). Age is factor B and is defined in terms of $b = 3$ levels (young, middle, old). Finally, intelligence is factor C and is defined in terms of $c = 2$ levels (high IQ, normal IQ). Table 22.1 shows the treatment means μ_{ijk} for all factor level combinations, as well as their notational representations. Also shown in Table 22.1 are the various means of the μ_{ijk}. We shall refer repeatedly to this learning example as we explain the model terms for a three-factor study.

Main Effects

The main effects in a three-factor study are defined analogously to those for a two-factor study. Thus, the *main effect* of the ith level of factor A is defined:

(22.4a)
$$\alpha_i = \mu_{i..} - \mu_{...}$$

For the learning example in Table 22.1, we have, for instance:

$$\alpha_1 = \mu_{1..} - \mu_{...} = 16.5 - 16 = .5$$

TABLE 22.1 Mean Learning Times according to Gender, Age, and Intelligence (in minutes)—Learning Example

Intelligence (factor C) and Age (factor B)

Factor A—Gender	k = 1 High IQ				k = 2 Normal IQ				Average			
	j = 1 Young	j = 2 Middle	j = 3 Old	Average	j = 1 Young	j = 2 Middle	j = 3 Old	Average	j = 1 Young	j = 2 Middle	j = 3 Old	Average
i = 1 Male	9 (μ_{111})	12 (μ_{121})	18 (μ_{131})	13 ($\mu_{1.1}$)	19 (μ_{112})	20 (μ_{122})	21 (μ_{132})	20 ($\mu_{1.2}$)	14 ($\mu_{11.}$)	16 ($\mu_{12.}$)	19.5 ($\mu_{13.}$)	16.5 ($\mu_{1..}$)
i = 2 Female	9 (μ_{211})	10 (μ_{221})	14 (μ_{231})	11 ($\mu_{2.1}$)	19 (μ_{212})	20 (μ_{222})	21 (μ_{232})	20 ($\mu_{2.2}$)	14 ($\mu_{21.}$)	15 ($\mu_{22.}$)	17.5 ($\mu_{23.}$)	15.5 ($\mu_{2..}$)
Average	9 ($\mu_{.11}$)	11 ($\mu_{.21}$)	16 ($\mu_{.31}$)	12 ($\mu_{..1}$)	19 ($\mu_{.12}$)	20 ($\mu_{.22}$)	21 ($\mu_{.32}$)	20 ($\mu_{..2}$)	14 ($\mu_{.1.}$)	15.5 ($\mu_{.2.}$)	18.5 ($\mu_{.3.}$)	16 ($\mu_{...}$)

Similarly, we define the main effect of the *j*th level of factor *B*:

(22.4b)
$$\beta_j = \mu_{.j.} - \mu_{...}$$

and the main effect of the *k*th level of *C*:

(22.4c)
$$\gamma_k = \mu_{..k} - \mu_{...}$$

It follows from these definitions that the sums of the main effects are zero:

(22.5)
$$\sum_i \alpha_i = \sum_j \beta_j = \sum_k \gamma_k = 0$$

Two-Factor Interactions

The two-factor interaction effects in a three-factor study are defined in the same fashion as for a two-factor study, except that all means are averaged over the third factor. Thus, following (18.8a) we define the two-factor interaction between factor *A* at the *i*th level and factor *B* at the *j*th level, denoted as before by $(\alpha\beta)_{ij}$, as follows:

(22.6a)
$$(\alpha\beta)_{ij} = \mu_{ij.} - \mu_{i..} - \mu_{.j.} + \mu_{...}$$

For the learning example in Table 22.1, we have for instance:

$$(\alpha\beta)_{11} = 14 - 16.5 - 14 + 16 = -.5$$

In corresponding fashion, we define the *AC* and *BC* two-factor interactions:

(22.6b)
$$(\alpha\gamma)_{ik} = \mu_{i.k} - \mu_{i..} - \mu_{..k} + \mu_{...}$$

(22.6c)
$$(\beta\gamma)_{jk} = \mu_{.jk} - \mu_{.j.} - \mu_{..k} + \mu_{...}$$

The two-factor interactions $(\alpha\beta)_{ij}$, $(\alpha\gamma)_{ik}$, and $(\beta\gamma)_{jk}$ are often called *first-order interactions*. It can readily be shown that the sums of the first-order interactions over each subscript are zero:

(22.7a) $\sum_i (\alpha\beta)_{ij} = 0$ for all *j* $\sum_j (\alpha\beta)_{ij} = 0$ for all *i*

(22.7b) $\sum_i (\alpha\gamma)_{ik} = 0$ for all *k* $\sum_k (\alpha\gamma)_{ik} = 0$ for all *i*

(22.7c) $\sum_j (\beta\gamma)_{jk} = 0$ for all *k* $\sum_k (\beta\gamma)_{jk} = 0$ for all *j*

Three-Factor Interactions

Just as in a two-factor study, where the interaction between the *i*th level of factor *A* and the *j*th level of factor *B* is defined as the difference between the treatment mean μ_{ij} and the value that would be expected if the factor effects were additive, so in a three-factor study the three-factor interaction $(\alpha\beta\gamma)_{ijk}$ is defined as the difference between the treatment mean μ_{ijk} and the value that would be expected if main effects and first-order interactions were sufficient to account for all factor effects. The value that would be expected from main effects and first-order interactions when A is at the ith level, B at the jth level, and C at the kth level is:

$$(22.8) \qquad \mu_{...} + \alpha_i + \beta_j + \gamma_k + (\alpha\beta)_{ij} + (\alpha\gamma)_{ik} + (\beta\gamma)_{jk}$$

Hence, the *three-factor interaction* $(\alpha\beta\gamma)_{ijk}$, also called the *second-order interaction*, is defined as:

$$(22.9a) \qquad (\alpha\beta\gamma)_{ijk} = \mu_{ijk} - [\mu_{...} + \alpha_i + \beta_j + \gamma_k + (\alpha\beta)_{ij} + (\alpha\gamma)_{ik} + (\beta\gamma)_{jk}]$$

or equivalently:

$$(22.9b) \qquad (\alpha\beta\gamma)_{ijk} = \mu_{ijk} - \mu_{ij.} - \mu_{i.k} - \mu_{.jk} + \mu_{i..} + \mu_{.j.} + \mu_{..k} - \mu_{...}$$

From the definition of the three-factor interactions, it follows that they sum to zero when added over any index:

$$(22.10) \qquad \sum_i (\alpha\beta\gamma)_{ijk} = 0 \qquad \sum_j (\alpha\beta\gamma)_{ijk} = 0 \qquad \sum_k (\alpha\beta\gamma)_{ijk} = 0$$

$$\text{for all } j, k \qquad\qquad \text{for all } i, k \qquad\qquad \text{for all } i, j$$

If *all* three-factor interactions $(\alpha\beta\gamma)_{ijk}$ are zero, we say that there are no three-factor interactions between factors A, B, and C. If some $(\alpha\beta\gamma)_{ijk}$ are not zero, we say that three-factor interactions are present.

Let us find the three-factor interaction $(\alpha\beta\gamma)_{111}$ for the learning example in Table 22.1. We require the following terms:

$$\mu_{...} = 16 \qquad\qquad (\alpha\beta)_{11} = 14 - 16.5 - 14 + 16 = -.5$$
$$\alpha_1 = 16.5 - 16 = .5 \qquad (\alpha\gamma)_{11} = 13 - 16.5 - 12 + 16 = .5$$
$$\beta_1 = 14 - 16 = -2 \qquad (\beta\gamma)_{11} = 9 - 14 - 12 + 16 = -1$$
$$\gamma_1 = 12 - 16 = -4 \qquad \mu_{111} = 9$$

Hence, we have:

$$(\alpha\beta\gamma)_{111} = 9 - (16 + .5 - 2 - 4 - .5 + .5 - 1) = -.5$$

Since $(\alpha\beta\gamma)_{111}$ is not zero, we know at once that three-factor interactions are present in this example.

Interpretation of Interactions in Three-Factor Studies

To shed light on the nature of interactions in three-factor studies, we shall examine some examples by means of tables and graphs, beginning with very simple factor effects and building up to complex three-factor interaction effects. In each example, we present the true treatment means μ_{ijk}.

Example 1—Main Effects Only. Figure 22.1 illustrates a case where there are A, B, and C main effects but no interactions of any kind. The AB response curves in Figure 22.1a are plots of the treatment means μ_{ijk} against C. Here, the main A effects are reflected by the slopes of the AB curves not being zero. The main B effects are reflected by the differences in the heights of the two AB curves within each panel, and the main C effects are reflected by the corresponding curves in the two panels being at different heights.

FIGURE 22.1 A, B, and C Main Effects and No Interactions

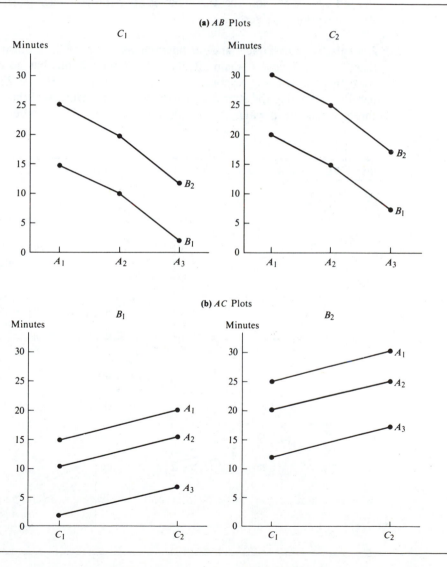

The absence of AB interactions is shown by the parallel response curves in each panel. We know from our discussion of two-factor analysis that parallel response curves imply absence of interactions. Here, the parallel response curves within each panel imply:

1. The AB interactions $(\alpha\beta)_{ij}$ equal zero.
2. The ABC interactions $(\alpha\beta\gamma)_{ijk}$ equal zero.

The absences of BC and AC interactions in this example become evident when the BC and AC response curves are plotted against A and B, respectively. Thus, in Figure 22.1b, the same treatment means μ_{ijk} are shown with the AC response curves plotted against B. Note that these curves are parallel in each panel, implying that the AC interactions are zero.

Example 2—Main Effects and *AB* Interactions.   Figure 22.2 illustrates a case where there are A, B, and C main effects and AB interactions but no other interactions. Note that the two AB response curves in each panel of Figure 22.2a are no longer parallel, reflecting the presence of AB interactions. However, the upper curves in the three panels are parallel, as are the lower curves. This implies that when the AC

FIGURE 22.2 A, B, and C Main Effects and AB Interactions

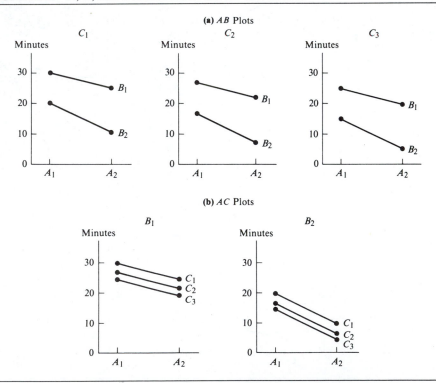

curves are plotted against B, the AC response curves within each panel will be parallel. This is shown in Figure 22.2b, which contains the same treatment means as in Figure 22.2a. The parallelism of the AC response curves within each panel in Figure 22.2b in turn implies that there are no AC interactions, as well as no ABC interactions, that is, all $(\alpha\gamma)_{ik} = 0$ and all $(\alpha\beta\gamma)_{ijk} = 0$.

Thus, if the AB (AC, BC) response curves corresponding to any given level of factor C (B, A) are parallel for all levels of factor C (B, A), as in Figure 22.2a, even though the response curves within a panel are not parallel, it follows that there are no three-factor interactions.

Example 3—Main Effects and *AB* and *AC* Interactions.

It does not follow, however, that lack of parallelism either within or between panels implies the presence of three-factor interactions. Figure 22.3 portrays a case where main A, B, and C effects and AB and AC two-factor interactions are present but no three-factor interactions exist. Yet there are no parallel response curves either within or between panels. The fact that the AB response curves when factor B is at level B_1 are not parallel for the three levels of factor C reflects the presence of AC interactions. The lack of parallelism of the AB response curves within each panel reflects the presence of AB interactions. Note, however, that the differences $\mu_{11k} - \mu_{12k}$ are the same for all levels of factor C, as are the differences $\mu_{21k} - \mu_{22k}$. Hence, no three-factor interactions are present.

Example 4—Main Effects and *AB, AC, BC,* and *ABC* Interactions.

Figure 22.4 shows the AB response plots for the different levels of factor C for the learning example in Table 22.1 of the effects of gender, age, and intelligence on learning time. The plots of the treatment means in Figure 22.4 reflect, as we have seen earlier, the presence of main effects and first-order and second-order interactions. Specifically, Figure 22.4 shows that for persons with normal IQ, gender has no effect on mean learning time, and age has only a small effect leading to slightly longer learning times for

FIGURE 22.3 A, B, and C Main Effects and AB and AC Interactions—AB Plots

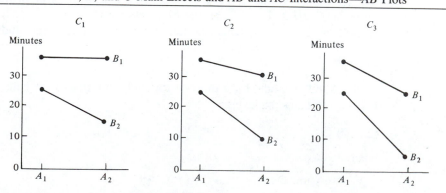

FIGURE 22.4   *A*, *B*, and *C* Main Effects and *AB*, *AC*, *BC*, and *ABC* Interactions—
Table 22.1 Learning Example

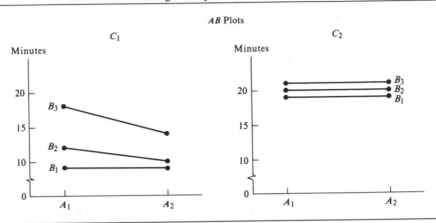

older persons. For persons with high IQ, on the other hand, females tend to learn more quickly than males for older persons but not for young persons, and older persons tend to require substantially longer learning times than young persons.

Note

If three-factor interactions are difficult to understand, higher-order interactions such as four-factor interactions are yet more abstruse. Fortunately, it is often found in practice that these higher-order interactions are quite small or nonexistent. When this is the case, they can be disregarded in the analysis of factor effects.

Cell Means Model

Let Y_{ijkm} be the mth observation ($m = 1, \ldots, n$) for the treatment consisting of the ith level of A ($i = 1, \ldots, a$), the jth level of B ($j = 1, \ldots, b$), and the kth level of C ($k = 1, \ldots, c$). Thus, the total number of cases in the study is:

$$(22.11) \qquad\qquad n_T = nabc$$

The ANOVA model for a three-factor study in terms of the cell (treatment) means μ_{ijk} with fixed factor levels is:

$$(22.12) \qquad\qquad Y_{ijkm} = \mu_{ijk} + \varepsilon_{ijkm}$$

where:

μ_{ijk} are parameters

ε_{ijkm} are independent $N(0, \sigma^2)$

$i = 1, \ldots, a; j = 1, \ldots, b; k = 1, \ldots, c; m = 1, \ldots, n$

Factor Effects Model

An equivalent factor effects model can be developed that incorporates the factorial structure by expressing the treatment mean μ_{ijk} in terms of the various factor effects. From the three-factor interaction definition (22.9a), we have the identity:

$$(22.13) \quad \mu_{ijk} \equiv \mu_{...} + \alpha_i + \beta_j + \gamma_k + (\alpha\beta)_{ij} + (\alpha\gamma)_{ik} + (\beta\gamma)_{jk} + (\alpha\beta\gamma)_{ijk}$$

where:

$$\mu_{...} = \frac{\Sigma\Sigma\Sigma\mu_{ijk}}{abc}$$

$$\alpha_i = \mu_{i..} - \mu_{...}$$
$$\beta_j = \mu_{.j.} - \mu_{...}$$
$$\gamma_k = \mu_{..k} - \mu_{...}$$
$$(\alpha\beta)_{ij} = \mu_{ij.} - \mu_{i..} - \mu_{.j.} + \mu_{...}$$
$$(\alpha\gamma)_{ik} = \mu_{i.k} - \mu_{i..} - \mu_{..k} + \mu_{...}$$
$$(\beta\gamma)_{jk} = \mu_{.jk} - \mu_{.j.} - \mu_{..k} + \mu_{...}$$
$$(\alpha\beta\gamma)_{ijk} = \mu_{ijk} - \mu_{ij.} - \mu_{i.k} - \mu_{.jk} + \mu_{i..} + \mu_{.j.} + \mu_{..k} - \mu_{...}$$

Hence, the equivalent factor effects ANOVA model for a three-factor study is:

$$(22.14) \qquad Y_{ijkm} = \mu_{...} + \alpha_i + \beta_j + \gamma_k + (\alpha\beta)_{ij}$$
$$+ (\alpha\gamma)_{ik} + (\beta\gamma)_{jk} + (\alpha\beta\gamma)_{ijk} + \varepsilon_{ijkm}$$

where:

ε_{ijkm} are independent $N(0, \sigma^2)$

$\alpha_i, \beta_j, \gamma_k, (\alpha\beta)_{ij}, (\alpha\gamma)_{ik}, (\beta\gamma)_{jk}, (\alpha\beta\gamma)_{ijk}$ are constants subject to the restrictions:

$$\sum_i \alpha_i = \sum_j \beta_j = \sum_k \gamma_k = 0$$

$$\sum_i (\alpha\beta)_{ij} = \sum_j (\alpha\beta)_{ij} = \sum_i (\alpha\gamma)_{ik} = 0$$

$$\sum_k (\alpha\gamma)_{ik} = \sum_j (\beta\gamma)_{jk} = \sum_k (\beta\gamma)_{jk} = 0$$

$$\sum_i (\alpha\beta\gamma)_{ijk} = \sum_j (\alpha\beta\gamma)_{ijk} = \sum_k (\alpha\beta\gamma)_{ijk} = 0$$

The cell means model (22.12) and the equivalent factor effects model (22.14) are linear models, just as in the two-factor case. We shall illustrate this for an example later in the chapter.

22.2 ANALYSIS OF VARIANCE

Notation

The notation for sample totals and means is a straightforward extension of that for two-factor studies. As usual, a dot in the subscript indicates aggregation or averaging over the index represented by the dot. We have:

$$(22.15a) \qquad Y_{ijk.} = \sum_m Y_{ijkm} \qquad\qquad \bar{Y}_{ijk.} = \frac{Y_{ijk.}}{n}$$

$$(22.15b) \qquad Y_{ij..} = \sum_k \sum_m Y_{ijkm} \qquad\qquad \bar{Y}_{ij..} = \frac{Y_{ij..}}{cn}$$

$$(22.15c) \qquad Y_{i.k.} = \sum_j \sum_m Y_{ijkm} \qquad\qquad \bar{Y}_{i.k.} = \frac{Y_{i.k.}}{bn}$$

$$(22.15d) \qquad Y_{.jk.} = \sum_i \sum_m Y_{ijkm} \qquad\qquad \bar{Y}_{.jk.} = \frac{Y_{.jk.}}{an}$$

$$(22.15e) \qquad Y_{i...} = \sum_j \sum_k \sum_m Y_{ijkm} \qquad\qquad \bar{Y}_{i...} = \frac{Y_{i...}}{bcn}$$

$$(22.15f) \qquad Y_{.j..} = \sum_i \sum_k \sum_m Y_{ijkm} \qquad\qquad \bar{Y}_{.j..} = \frac{Y_{.j..}}{acn}$$

$$(22.15g) \qquad Y_{..k.} = \sum_i \sum_j \sum_m Y_{ijkm} \qquad\qquad \bar{Y}_{..k.} = \frac{Y_{..k.}}{abn}$$

$$(22.15h) \qquad Y_{....} = \sum_i \sum_j \sum_k \sum_m Y_{ijkm} \qquad\qquad \bar{Y}_{....} = \frac{Y_{....}}{abcn}$$

Example.   Table 22.2 illustrates this notation for a study of the effects of gender, body fat, and smoking history on exercise tolerance in stress testing of persons 25 to 35 years old. Each of the three factors has two levels, and there are three replications for each treatment. Tables 22.2a, b, and c show, respectively, the data, totals, and means, together with the corresponding notation. We shall fully analyze these data later on.

Fitting of ANOVA Model

When the cell means model (22.12) is fitted by the method of least squares, the estimators as usual turn out to be the estimated treatment means:

$$(22.16) \qquad\qquad \hat{\mu}_{ijk} = \bar{Y}_{ijk.}$$

TABLE 22.2 Sample Data, Totals, and Means for Three-Factor Study—Stress Test Example

(a) Data

| | Smoking History | |
| | k = 1 Light | k = 2 Heavy |

$j = 1$ Low fat:

$i = 1$ Male

	k = 1 Light	k = 2 Heavy
	24 (Y_{1111})	18 (Y_{1121})
	29 (Y_{1112})	19 (Y_{1122})
	25 (Y_{1113})	23 (Y_{1123})

$i = 2$ Female

	20 (Y_{2111})	15 (Y_{2121})
	22 (Y_{2112})	10 (Y_{2122})
	18 (Y_{2113})	11 (Y_{2123})

$j = 2$ High fat:

$i = 1$ Male

	15 (Y_{1211})	15 (Y_{1221})
	15 (Y_{1212})	20 (Y_{1222})
	12 (Y_{1213})	13 (Y_{1223})

$i = 2$ Female

	16 (Y_{2211})	10 (Y_{2221})
	9 (Y_{2212})	14 (Y_{2222})
	11 (Y_{2213})	6 (Y_{2223})

(b) Totals

	k = 1	k = 2	All k
$j = 1$:			
$i = 1$	78 ($Y_{111.}$)	60 ($Y_{112.}$)	138 ($Y_{11..}$)
$i = 2$	60 ($Y_{211.}$)	36 ($Y_{212.}$)	96 ($Y_{21..}$)
All i	138 ($Y_{.11.}$)	96 ($Y_{.12.}$)	234 ($Y_{.1..}$)
$j = 2$:			
$i = 1$	42 ($Y_{121.}$)	48 ($Y_{122.}$)	90 ($Y_{12..}$)
$i = 2$	36 ($Y_{221.}$)	30 ($Y_{222.}$)	66 ($Y_{22..}$)
All i	78 ($Y_{.21.}$)	78 ($Y_{.22.}$)	156 ($Y_{.2..}$)
All j:			
$i = 1$	120 ($Y_{1.1.}$)	108 ($Y_{1.2.}$)	228 ($Y_{1...}$)
$i = 2$	96 ($Y_{2.1.}$)	66 ($Y_{2.2.}$)	162 ($Y_{2...}$)
All i	216 ($Y_{..1.}$)	174 ($Y_{..2.}$)	390 ($Y_{....}$)

(c) Averages

	k = 1	k = 2	All k
$j = 1$:			
$i = 1$	26 ($\bar{Y}_{111.}$)	20 ($\bar{Y}_{112.}$)	23 ($\bar{Y}_{11..}$)
$i = 2$	20 ($\bar{Y}_{211.}$)	12 ($\bar{Y}_{212.}$)	16 ($\bar{Y}_{21..}$)
All i	23 ($\bar{Y}_{.11.}$)	16 ($\bar{Y}_{.12.}$)	19.5 ($\bar{Y}_{.1..}$)
$j = 2$:			
$i = 1$	14 ($\bar{Y}_{121.}$)	16 ($\bar{Y}_{122.}$)	15 ($\bar{Y}_{12..}$)
$i = 2$	12 ($\bar{Y}_{221.}$)	10 ($\bar{Y}_{222.}$)	11 ($\bar{Y}_{22..}$)
All i	13 ($\bar{Y}_{.21.}$)	13 ($\bar{Y}_{.22.}$)	13 ($\bar{Y}_{.2..}$)
All j:			
$i = 1$	20 ($\bar{Y}_{1.1.}$)	18 ($\bar{Y}_{1.2.}$)	19 ($\bar{Y}_{1...}$)
$i = 2$	16 ($\bar{Y}_{2.1.}$)	11 ($\bar{Y}_{2.2.}$)	13.5 ($\bar{Y}_{2...}$)
All i	18 ($\bar{Y}_{..1.}$)	14.5 ($\bar{Y}_{..2.}$)	16.25 ($\bar{Y}_{....}$)

Thus, the *fitted values* for the observations are the estimated treatment means:

(22.17)
$$\hat{Y}_{ijkm} = \hat{\mu}_{ijk} = \bar{Y}_{ijk.}$$

and the *residuals* are the deviations of the observed values from the estimated treatment means:

(22.18)
$$e_{ijkm} = Y_{ijkm} - \hat{Y}_{ijkm} = Y_{ijkm} - \bar{Y}_{ijk.}$$

For the equivalent factor effects model (22.14), the least squares estimators of the parameters are as follows:

	Parameter	Estimator
(22.19a)	$\mu_{...}$	$\hat{\mu}_{...} = \bar{Y}_{....}$
(22.19b)	α_i	$\hat{\alpha}_i = \bar{Y}_{i...} - \bar{Y}_{....}$
(22.19c)	β_j	$\hat{\beta}_j = \bar{Y}_{.j..} - \bar{Y}_{....}$
(22.19d)	γ_k	$\hat{\gamma}_k = \bar{Y}_{..k.} - \bar{Y}_{....}$
(22.19e)	$(\alpha\beta)_{ij}$	$\widehat{(\alpha\beta)}_{ij} = \bar{Y}_{ij..} - \bar{Y}_{i...} - \bar{Y}_{.j..} + \bar{Y}_{....}$
(22.19f)	$(\alpha\gamma)_{ik}$	$\widehat{(\alpha\gamma)}_{ik} = \bar{Y}_{i.k.} - \bar{Y}_{i...} - \bar{Y}_{..k.} + \bar{Y}_{....}$
(22.19g)	$(\beta\gamma)_{jk}$	$\widehat{(\beta\gamma)}_{jk} = \bar{Y}_{.jk.} - \bar{Y}_{.j..} - \bar{Y}_{..k.} + \bar{Y}_{....}$
(22.19h)	$(\alpha\beta\gamma)_{ijk}$	$\widehat{(\alpha\beta\gamma)}_{ijk} = \bar{Y}_{ijk.} - \bar{Y}_{ij..} - \bar{Y}_{i.k.} - \bar{Y}_{.jk.}$ $+ \bar{Y}_{i...} + \bar{Y}_{.j..} + \bar{Y}_{..k.} - \bar{Y}_{....}$

The fitted values and residuals for factor effects model (22.14) are exactly the same as those for cell means model (22.12), as was also the case for two-factor studies.

Breakdown of Total Sum of Squares

Neglecting the factorial structure of the study and simply considering it to contain abc treatments, we obtain the usual breakdown of the total sum of squares:

$$(22.20) \qquad SSTO = SSTR + SSE$$

where:

$$(22.20a) \qquad SSTO = \sum_i \sum_j \sum_k \sum_m (Y_{ijkm} - \bar{Y}_{....})^2$$

$$(22.20b) \qquad SSTR = n \sum_i \sum_j \sum_k (\bar{Y}_{ijk.} - \bar{Y}_{....})^2$$

$$(22.20c) \qquad SSE = \sum_i \sum_j \sum_k \sum_m (Y_{ijkm} - \bar{Y}_{ijk.})^2 = \sum_i \sum_j \sum_k \sum_m e_{ijkm}^2$$

Consider now the estimated treatment mean deviation $\bar{Y}_{ijk.} - \bar{Y}_{....}$, which appears in $SSTR$. This can be decomposed in terms of the least squares estimators (22.19) of the main effects, two-factor interactions, and three-factor interaction:

$$\underbrace{\bar{Y}_{ijk.} - \bar{Y}_{....}}_{\substack{\text{Estimated} \\ \text{treatment} \\ \text{mean deviation}}} = \underbrace{\bar{Y}_{i...} - \bar{Y}_{....}}_{A \text{ main effect}} + \underbrace{\bar{Y}_{.j..} - \bar{Y}_{....}}_{B \text{ main effect}} + \underbrace{\bar{Y}_{..k.} - \bar{Y}_{....}}_{C \text{ main effect}} + \underbrace{\bar{Y}_{ij..} - \bar{Y}_{i...} - \bar{Y}_{.j..} + \bar{Y}_{....}}_{AB \text{ interaction effect}}$$

$$+ \underbrace{\bar{Y}_{i.k.} - \bar{Y}_{i...} - \bar{Y}_{..k.} + \bar{Y}_{....}}_{AC \text{ interaction effect}} + \underbrace{\bar{Y}_{.jk.} - \bar{Y}_{.j..} - \bar{Y}_{..k.} + \bar{Y}_{....}}_{BC \text{ interaction effect}}$$

$$+ \underbrace{\bar{Y}_{ijk.} - \bar{Y}_{ij..} - \bar{Y}_{i.k.} - \bar{Y}_{.jk.} + \bar{Y}_{i...} + \bar{Y}_{.j..} + \bar{Y}_{..k.} - \bar{Y}_{....}}_{ABC \text{ interaction effect}}$$

When we square each side and sum over i, j, k, and m, all cross-product terms drop out and we obtain:

(22.21) $SSTR = SSA + SSB + SSC + SSAB + SSAC + SSBC + SSABC$

where:

(22.21a) $SSA = nbc \sum_i (\bar{Y}_{i...} - \bar{Y}_{....})^2$

(22.21b) $SSB = nac \sum_j (\bar{Y}_{.j..} - \bar{Y}_{....})^2$

(22.21c) $SSC = nab \sum_k (\bar{Y}_{..k.} - \bar{Y}_{....})^2$

(22.21d) $SSAB = nc \sum_i \sum_j (\bar{Y}_{ij..} - \bar{Y}_{i...} - \bar{Y}_{.j..} + \bar{Y}_{....})^2$

(22.21e) $SSAC = nb \sum_i \sum_k (\bar{Y}_{i.k.} - \bar{Y}_{i...} - \bar{Y}_{..k.} + \bar{Y}_{....})^2$

(22.21f) $SSBC = na \sum_j \sum_k (\bar{Y}_{.jk.} - \bar{Y}_{.j..} - \bar{Y}_{..k.} + \bar{Y}_{....})^2$

(22.21g) $SSABC = n \sum_i \sum_j \sum_k (\bar{Y}_{ijk.} - \bar{Y}_{ij..} - \bar{Y}_{i.k.} - \bar{Y}_{.jk.} + \bar{Y}_{i...}$

$$+ \bar{Y}_{.j..} + \bar{Y}_{..k.} - \bar{Y}_{....})^2$$

Combining (22.20) and (22.21), we have thus established the orthogonal decomposition:

(22.22) $SSTO = SSA + SSB + SSC + SSAB + SSAC$

$$+ SSBC + SSABC + SSE$$

SSA, SSB, and SSC are the usual main effect sums of squares. For instance, the larger (absolutely) are the estimated main B effects $\bar{Y}_{.j..} - \bar{Y}_{....}$, the larger will be SSB.

$SSAB$, $SSAC$, and $SSBC$ are the usual two-factor interaction sums of squares. For instance, the larger (absolutely) are the estimated AB interactions $\bar{Y}_{ij..} - \bar{Y}_{i...} - \bar{Y}_{.j..} + \bar{Y}_{....}$, the larger will be $SSAB$.

Finally, *SSABC* is the three-factor interaction sum of squares. The larger (absolutely) are these estimated three-factor interactions, the larger will be *SSABC*.

Computational Formulas. For the occasional instance when the calculations are to be done by hand, the use of the definitional formulas previously given is cumbersome. The computational formulas for the three-factor case follow the pattern of the two-factor formulas:

$$(22.23a) \qquad SSTO = \sum_i \sum_j \sum_k \sum_m Y_{ijkm}^2 - \frac{Y_{....}^2}{nabc}$$

$$(22.23b) \qquad SSE = \sum_i \sum_j \sum_k \sum_m Y_{ijkm}^2 - \sum_i \sum_j \sum_k \frac{Y_{ijk.}^2}{n}$$

$$(22.23c) \qquad SSA = \frac{\sum_i Y_{i...}^2}{nbc} - \frac{Y_{....}^2}{nabc}$$

$$(22.23d) \qquad SSB = \frac{\sum_j Y_{.j..}^2}{nac} - \frac{Y_{....}^2}{nabc}$$

$$(22.23e) \qquad SSC = \frac{\sum_k Y_{..k.}^2}{nab} - \frac{Y_{....}^2}{nabc}$$

The two-factor interaction sum of squares *SSAB* can be obtained by working with the means $\bar{Y}_{ij..}$ and treating these *ab* means as constituting a two-factor study. The "treatment sum of squares" for this two-factor study, which we will denote by *SSTR*(*A*, *B*), is as usual:

$$(22.24) \qquad SSTR(A, B) = nc \sum_i \sum_j (\bar{Y}_{ij..} - \bar{Y}_{....})^2 = \frac{\sum_i \sum_j Y_{ij..}^2}{nc} - \frac{Y_{....}^2}{nabc}$$

It can be shown that:

$$(22.25) \qquad SSTR(A, B) = SSA + SSB + SSAB$$

Hence, we can find *SSAB* by subtraction:

$$(22.26a) \qquad SSAB = SSTR(A, B) - SSA - SSB$$

Similarly, we find *SSAC* and *SSBC* as follows:

$$(22.26b) \qquad SSAC = SSTR(A, C) - SSA - SSC$$

$$(22.26c) \qquad SSBC = SSTR(B, C) - SSB - SSC$$

where:

$$(22.26d) \qquad SSTR(A, C) = \frac{\sum_i \sum_k Y_{i.k.}^2}{nb} - \frac{Y_{....}^2}{nabc}$$

$$(22.26e) \qquad SSTR(B, C) = \frac{\sum_j \sum_k Y_{.jk.}^2}{na} - \frac{Y_{....}^2}{nabc}$$

The three-factor interaction sum of squares is obtained by subtraction:

$$(22.27) \quad SSABC = SSTO - SSE - SSA - SSB - SSC - SSAB - SSAC - SSBC$$

Note

The computational formulas above can be extended readily if more than three factors are studied simultaneously. Usually, however, a computer package will be utilized under these circumstances.

Degrees of Freedom and Mean Squares

Table 22.3 contains the general ANOVA table for the three-factor model (22.14). The degrees of freedom for main effect and two-factor interaction sums of squares correspond to those for two-factor studies. The number of degrees of freedom associated with $SSABC$ is obtained by subtraction and corresponds to the number of independent linear relations among all the interaction terms $(\alpha\beta\gamma)_{ijk}$.

The expected mean squares are also given in Table 22.3. Note that MSA, MSB, MSC, $MSAB$, $MSAC$, $MSBC$, and $MSABC$ all have expectations equal to σ^2 if there are no factor effects of the type reflected by the mean square. If such effects are present, each mean square has an expectation exceeding σ^2. As usual, $E\{MSE\} = \sigma^2$ always. Hence, the test for factor effects consists of comparing the appropriate mean square against MSE by means of an F^* test statistic, with large values of F^* indicating the presence of factor effects.

Table 22.5 (p. 842) contains the ANOVA table for the three-factor stress test example (calculational details are not shown).

Tests for Factor Effects

The various tests for factor effects all follow the same pattern; we illustrate them with the test for three-factor interactions. The alternatives are:

$$(22.28a) \qquad \begin{aligned} &H_0\text{: all } (\alpha\beta\gamma)_{ijk} = 0 \\ &H_a\text{: not all } (\alpha\beta\gamma)_{ijk} \text{ equal zero} \end{aligned}$$

The appropriate test statistic is:

$$(22.28b) \qquad F^* = \frac{MSABC}{MSE}$$

If H_0 holds, F^* follows the F distribution with $(a - 1)(b - 1)(c - 1)$ degrees of freedom for the numerator and $abc(n - 1)$ degrees of freedom for the denominator. Hence, the decision rule to control the Type I error at α is:

TABLE 22.3 General ANOVA Table for Three-Factor Study with Fixed Factor Levels

Source of Variation	SS	df	MS	$E\{MS\}$
Between treatments	SSTR	$abc - 1$	MSTR	$\sigma^2 + \dfrac{n\sum\sum\sum(\mu_{ijk} - \mu_{....})^2}{abc - 1}$
Factor A	SSA	$a - 1$	MSA	$\sigma^2 + \dfrac{bcn}{a - 1}\sum \alpha_i^2$
Factor B	SSB	$b - 1$	MSB	$\sigma^2 + \dfrac{acn}{b - 1}\sum \beta_j^2$
Factor C	SSC	$c - 1$	MSC	$\sigma^2 + \dfrac{abn}{c - 1}\sum \gamma_k^2$
AB interactions	SSAB	$(a - 1)(b - 1)$	MSAB	$\sigma^2 + \dfrac{cn}{(a - 1)(b - 1)}\sum\sum (\alpha\beta)_{ij}^2$
AC interactions	SSAC	$(a - 1)(c - 1)$	MSAC	$\sigma^2 + \dfrac{bn}{(a - 1)(c - 1)}\sum\sum (\alpha\gamma)_{ik}^2$
BC interactions	SSBC	$(b - 1)(c - 1)$	MSBC	$\sigma^2 + \dfrac{an}{(b - 1)(c - 1)}\sum\sum (\beta\gamma)_{jk}^2$
ABC interactions	SSABC	$(a - 1)(b - 1)(c - 1)$	MSABC	$\sigma^2 + \dfrac{n}{(a - 1)(b - 1)(c - 1)}\sum\sum\sum (\alpha\beta\gamma)_{ijk}^2$
Error	SSE	$abc(n - 1)$	MSE	σ^2
Total	SSTO	$abcn - 1$		

Note: $\mu_{....}$, α_i, β_j, γ_k, $(\alpha\beta)_{ij}$, $(\alpha\gamma)_{ik}$, $(\beta\gamma)_{jk}$, and $(\alpha\beta\gamma)_{ijk}$ are defined in (22.13).

(22.28c)

$$\text{If } F^* \leq F[1 - \alpha; (a - 1)(b - 1)(c - 1), (n - 1)abc], \text{ conclude } H_0$$

$$\text{If } F^* > F[1 - \alpha; (a - 1)(b - 1)(c - 1), (n - 1)abc], \text{ conclude } H_a$$

Table 22.4 contains the appropriate test statistics and percentiles of the F distribution for the various possible tests for a three-factor study.

Comments

1. The Kimball inequality for the family level of significance α in a three-factor study when the family consists of the combined set of seven tests, including three on main effects, three on two-factor interactions, and one on three-factor interactions, is:

(22.29)

$$\alpha < 1 - (1 - \alpha_1)(1 - \alpha_2) \cdots (1 - \alpha_7)$$

where α_i is the level of significance for the ith test.

2. If the three-factor interactions (and also perhaps some sets of two-factor interactions) equal zero, the question sometimes arises whether the corresponding sums of squares should be pooled with the error sum of squares. Our earlier discussion on revising the ANOVA model (p. 716) is applicable here also.

3. If there is only one case per treatment in a three-factor study with fixed factor levels, analysis of variance tests can only be conducted if it is possible to assume that some interactions equal zero. Usually, the interactions most likely to equal zero are the three-factor interactions. If it is possible to assume that all three-factor interactions equal zero, $MSABC$ has expectation σ^2 and plays the role of the error mean square MSE. All mean squares are calculated in the usual manner, except that $n = 1$.

TABLE 22.4   Test Statistics for Three-Factor Study with Fixed Factor Levels

Alternatives	Test Statistic	Percentile
H_0: all $\alpha_i = 0$ H_a: not all $\alpha_i = 0$	$F^* = \dfrac{MSA}{MSE}$	$F[1 - \alpha; a - 1, (n - 1)abc]$
H_0: all $\beta_j = 0$ H_a: not all $\beta_j = 0$	$F^* = \dfrac{MSB}{MSE}$	$F[1 - \alpha; b - 1, (n - 1)abc]$
H_0: all $\gamma_k = 0$ H_a: not all $\gamma_k = 0$	$F^* = \dfrac{MSC}{MSE}$	$F[1 - \alpha; c - 1, (n - 1)abc]$
H_0: all $(\alpha\beta)_{ij} = 0$ H_a: not all $(\alpha\beta)_{ij} = 0$	$F^* = \dfrac{MSAB}{MSE}$	$F[1 - \alpha; (a - 1)(b - 1), (n - 1)abc]$
H_0: all $(\alpha\gamma)_{ik} = 0$ H_a: not all $(\alpha\gamma)_{ik} = 0$	$F^* = \dfrac{MSAC}{MSE}$	$F[1 - \alpha; (a - 1)(c - 1), (n - 1)abc]$
H_0: all $(\beta\gamma)_{jk} = 0$ H_a: not all $(\beta\gamma)_{jk} = 0$	$F^* = \dfrac{MSBC}{MSE}$	$F[1 - \alpha; (b - 1)(c - 1), (n - 1)abc]$
H_0: all $(\alpha\beta\gamma)_{ijk} = 0$ H_a: not all $(\alpha\beta\gamma)_{ijk} = 0$	$F^* = \dfrac{MSABC}{MSE}$	$F[1 - \alpha; (a - 1)(b - 1)(c - 1), (n - 1)abc]$

4. The F^* test statistics in Table 22.4 can be obtained by the general linear test method explained in Chapter 3. For example, for testing whether all three-factor interactions are zero, the alternatives are those in (22.28a), the full model is that in (22.14), and the reduced model under H_0: $(\alpha\beta\gamma)_{ijk} \equiv 0$ is:

$$(22.30) \qquad Y_{ijkm} = \mu_{...} + \alpha_i + \beta_j + \gamma_k + (\alpha\beta)_{ij} + (\alpha\gamma)_{ik} + (\beta\gamma)_{jk} + \varepsilon_{ijkm}$$

22.3 EVALUATION OF APTNESS OF ANOVA MODEL

No new problems arise in examining the aptness of the analysis of variance three-factor model. The residuals (22.18):

$$(22.31) \qquad e_{ijkm} = Y_{ijkm} - \bar{Y}_{ijk.}$$

may be examined for normality, constancy of error variance, and independence of error terms in the same fashion as for single-factor and two-factor studies.

Transformations may be employed to stabilize the error variances, to make the error distributions more normal, and/or to make important interactions unimportant. Our earlier discussions of this topic apply completely to the three-factor case.

Finally, the earlier discussion on effects of departures from the ANOVA model applies fully to the three-factor case. In particular, the employment of equal sample sizes for each treatment minimizes the effect of unequal variances.

22.4 ANALYSIS OF FACTOR EFFECTS

No new problems are encountered in the analysis of factor effects for three-factor studies with fixed factor levels. As for two-factor studies, the focus of the analysis is usually on factor level means when no important interactions are present, and on treatment means when there are important interactions. We shall now present some selected results for estimating factor effects. Other analyses follow the same pattern.

Analysis of Factor Effects when Factors Do Not Interact

Estimation of Factor Level Mean. The factor A level mean $\mu_{i..}$ is estimated by:

$$(22.32) \qquad \hat{\mu}_{i..} = \bar{Y}_{i...}$$

The estimated variance of this estimator is:

$$(22.33) \qquad s^2\{\bar{Y}_{i...}\} = \frac{MSE}{nbc}$$

Confidence limits for $\mu_{i..}$ are obtained by means of the t distribution with $(n-1)abc$ degrees of freedom:

$$(22.34) \qquad \bar{Y}_{i...} \pm t[1 - \alpha/2; (n-1)abc]s\{\bar{Y}_{i...}\}$$

Estimation of factor level means for factors B or C is done in similar fashion.

Estimation of Contrast of Factor Level Means. When a contrast involving the factor A level means $\mu_{i..}$ is to be estimated:

$$(22.35) \qquad L = \sum c_i \mu_{i..} \quad \text{where } \sum c_i = 0$$

the unbiased estimator of L we shall employ is:

$$(22.36) \qquad \hat{L} = \sum c_i \bar{Y}_{i...}$$

The estimated variance of $\hat{L}$ is:

$$(22.37) \qquad s^2\{\hat{L}\} = \frac{MSE}{nbc} \sum c_i^2$$

and the $1 - \alpha$ confidence limits for L are:

$$(22.38) \qquad \hat{L} \pm t[1 - \alpha/2; (n - 1)abc]s\{\hat{L}\}$$

Contrasts of factor level means for factors B or C are estimated in similar fashion.

Multiple Contrasts of Factor Level Means. If a number of contrasts of the factor A level means $\mu_{i..}$ are to be estimated and assurance is to be provided by a family confidence coefficient, the t multiple in (22.38) is simply replaced by the T, S, or B multiple defined as follows:

$$(22.39a) \qquad \text{Tukey procedure (for pairwise comparisons)} \qquad T = \frac{1}{\sqrt{2}}q[1 - \alpha; a, (n - 1)abc]$$

$$(22.39b) \qquad \text{Scheffé procedure} \qquad S^2 = (a - 1)F[1 - \alpha; a - 1, (n - 1)abc]$$

$$(22.39c) \qquad \text{Bonferroni procedure} \qquad B = t[1 - \alpha/2g; (n - 1)abc]$$

Multiple contrasts based on the factor level means $\mu_{.j.}$ or $\mu_{..k}$ are estimated in corresponding fashion.

Analysis of Factor Effects when Interactions Important

As explained earlier for two-factor studies, one usually attempts to make important interactions unimportant by use of some simple transformation of the data. If that effort is not successful, the important interactions are then typically analyzed in terms of the treatment means μ_{ijk}.

Estimation of Treatment Mean. The treatment mean μ_{ijk} is estimated by:

$$(22.40) \qquad \hat{\mu}_{ijk} = \bar{Y}_{ijk.}$$

The estimated variance of $\bar{Y}_{ijk.}$ is:

(22.41)
$$s^2\{\bar{Y}_{ijk.}\} = \frac{MSE}{n}$$

Confidence limits for μ_{ijk} are:

(22.42)
$$\bar{Y}_{ijk.} \pm t[1 - \alpha/2; (n - 1)abc]s\{\bar{Y}_{ijk.}\}$$

Estimation of Contrast of Treatment Means.   When interactions are present, contrasts among the treatment means μ_{ijk} are ordinarily desired. Let, as usual, L denote such a contrast:

(22.43)
$$L = \sum\sum\sum c_{ijk}\mu_{ijk} \quad \text{where} \sum\sum\sum c_{ijk} = 0$$

An unbiased estimator of L is:

(22.44)
$$\hat{L} = \sum\sum\sum c_{ijk}\bar{Y}_{ijk.}$$

for which the estimated variance is:

(22.45)
$$s^2\{\hat{L}\} = \frac{MSE}{n}\sum\sum\sum c_{ijk}^2$$

As usual, confidence limits for L are:

(22.46)
$$\hat{L} \pm t[1 - \alpha/2; (n - 1)abc]s\{\hat{L}\}$$

At times, not all types of interaction effects are present. In such a case, the desired contrasts may involve means of the μ_{ijk} taken over one of the factors. For example, when the only interactions present are the BC interactions, one may be interested in contrasts of the means $\mu_{.jk}$:

(22.47)
$$L = \sum\sum c_{jk}\mu_{.jk} \quad \text{where} \sum\sum c_{jk} = 0$$

Such contrasts are, of course, special cases of contrasts of the treatment means μ_{ijk} in (22.43). The estimator of the contrast in (22.47) can be readily obtained from (22.44) and the estimated variance from (22.45); they are:

(22.48)
$$\hat{L} = \sum\sum c_{jk}\bar{Y}_{.jk.}$$

(22.49)
$$s^2\{\hat{L}\} = \frac{MSE}{na}\sum\sum c_{jk}^2$$

Multiple Contrasts of Treatment Means.   For multiple comparisons, the t multiple in (22.46) is replaced by the T, S, or B multiple defined as follows:

(22.50a)   Tukey procedure (for pairwise comparisons) $T = \dfrac{1}{\sqrt{2}}q[1 - \alpha; abc, (n - 1)abc]$

(22.50b)   Scheffé procedure $S^2 = (abc - 1)F[1 - \alpha; abc - 1,$

$$(n - 1)abc]$$

(22.50c)   Bonferroni procedure $B = t[1 - \alpha/2g; (n - 1)abc]$

Single Degree of Freedom Tests.   When interactions are present, either one or several single degree of freedom tests on the treatment means μ_{ijk} are sometimes used instead of estimation of contrasts. The two-sided alternatives for a single degree of freedom test are:

(22.51)

$$H_0: \sum\sum\sum c_{ijk}\mu_{ijk} = c$$

$$H_a: \sum\sum\sum c_{ijk}\mu_{ijk} \neq c$$

where the c_{ijk} and c are appropriate constants.

To test the two-sided alternatives (22.51), we can use the test statistic:

(22.52)

$$t^* = \frac{\sum\sum\sum c_{ijk}\bar{Y}_{ijk.} - c}{\sqrt{\frac{MSE}{n}\sum\sum\sum c_{ijk}^2}}$$

which follows the t distribution with $(n - 1)abc$ degrees of freedom when H_0 holds. Alternatively, $(t^*)^2 = F^*$ can be used as the test statistic. When H_0 holds, F^* follows the F distribution with 1 and $(n - 1)abc$ degrees of freedom.

22.5 EXAMPLE OF THREE-FACTOR STUDY

The exercise tolerance stress test study data presented in Table 22.2 (p. 829) will now be analyzed. Exercise tolerance is measured in minutes until fatigue occurs while the subject is performing on a bicycle apparatus. It will be recalled that the three factors in the study are gender of subject (A), body fat of subject measured in percent (B), and smoking history of subject (C). Each of the factors has two levels. Figure 22.5 presents the estimated treatment means $\bar{Y}_{ijk.}$ shown in Table 22.2c in a simple format. The same estimated treatment means are presented in Figure 22.6 in a graphic form. It appears from Figures 22.5 and 22.6 that some factors may interact in their effects on exercise tolerance and that gender, in particular, may affect the endurance in stress testing. The researcher now wishes to analyze the nature of the factor effects in detail.

Residual Analysis

The researcher first prepared aligned residual dot plots for the eight treatments. These plots (not shown), though based on only three observations for each treatment, did not suggest any gross differences in the error variances for the eight treatments.

FIGURE 22.5   Schematic Presentation of Estimated Treatment Means—
Stress Test Example

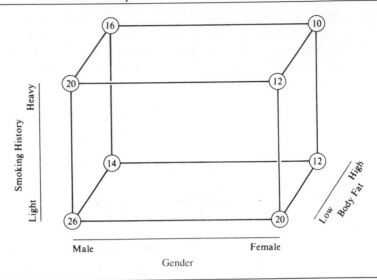

FIGURE 22.6   Plots of Estimated Treatment Means—Stress Test Example

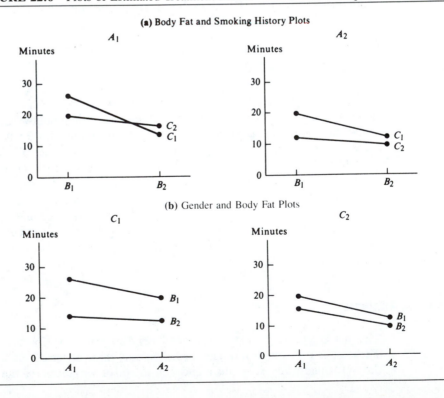

The researcher also obtained a normal probability plot of the residuals, shown in Figure 22.7. The points in this plot form a reasonably linear pattern, even though there are a considerable number of ties among the residuals because of the rounded nature of the data. This impression of linearity is supported by the high coefficient of correlation between the ordered residuals and their expected values under normality, namely, .969. The researcher was therefore satisfied that the three-factor ANOVA model (22.14) is applicable here.

Tests for Factor Effects

The researcher first wished to test for the various factor effects. She desired a family level of significance of $\alpha = .10$ for the seven potential tests. This will assure that if in fact no factor effects are present, there will be only 1 chance in 10 for one or more of the seven tests to lead to the conclusion of the presence of factor effects. Using the Kimball inequality (22.29), she solved the equation:

$$\alpha = .10 = 1 - (1 - \alpha_i)^7$$

and found $\alpha_i = .015$. Thus, use of significance level $\alpha_i = .015$ for each test assures that the family level of significance will not exceed .10.

Table 22.5 contains the results of a run of a multifactor ANOVA computer package. The ANOVA table is shown, as well as the seven test statistics and their *P*-values. Each test statistic has in the numerator the appropriate factor effect mean square, and the denominator of each test statistic is *MSE*.

Test for Three-Factor Interactions. The first test was conducted for three-factor interactions. The alternatives are:

FIGURE 22.7 Normal Probability Plot of Residuals—Stress Test Example

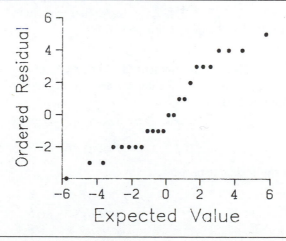

TABLE 22.5   ANOVA Table for Three-Factor Study—Stress Test Example

Source of Variation	SS	df	MS	F*	P-Value
Between treatments	610.50	7	87.21		
Factor A (gender)	181.50	1	181.50	20.74	0+
Factor B (body fat)	253.50	1	253.50	28.97	0+
Factor C (smoking)	73.50	1	73.50	8.40	.01
AB interactions	13.50	1	13.50	1.54	.23
AC interactions	13.50	1	13.50	1.54	.23
BC interactions	73.50	1	73.50	8.40	.01
ABC interactions	1.50	1	1.50	.17	.69
Error	140.00	16	8.75		
Total	750.50	23			

$$F(.985; 1, 16) = 7.42$$

$$H_0: \text{all } (\alpha\beta\gamma)_{ijk} = 0$$

$$H_a: \text{not all } (\alpha\beta\gamma)_{ijk} \text{ equal zero}$$

The decision rule is:

$$\text{If } F^* \leq F(.985; 1, 16) = 7.42, \text{ conclude } H_0$$

$$\text{If } F^* > F(.985; 1, 16) = 7.42, \text{ conclude } H_a$$

The F^* test statistic obtained from Table 22.5 is:

$$F^* = \frac{MSABC}{MSE} = \frac{1.50}{8.75} = .17$$

Since $F^* = .17 \leq 7.42$, the researcher concluded that no ABC interactions are present. The P-value of this test is .69.

Tests for Two-Factor Interactions.   The researcher next tested for two-factor interactions. In the test for AB interactions, the decision rule is (the alternatives are given in Table 22.4):

$$\text{If } F^* \leq F(.985; 1, 16) = 7.42, \text{ conclude } H_0$$

$$\text{If } F^* > F(.985; 1, 16) = 7.42, \text{ conclude } H_a$$

and the test statistic is:

$$F^* = \frac{MSAB}{MSE} = \frac{13.50}{8.75} = 1.54$$

Since $F^* = 1.54 \leq 7.42$, the researcher concluded that no AB interactions are present. The P-value of this test is .23.

The tests for AC and BC interactions proceeded similarly. We obtain:

1. $F^* = \dfrac{MSAC}{MSE} = \dfrac{13.50}{8.75} = 1.54 \leq F(.985; 1, 16) = 7.42 \qquad P\text{-value} = .23$

    Conclusion: No AC interactions are present.

2. $F^* = \dfrac{MSBC}{MSE} = \dfrac{73.50}{8.75} = 8.40 > F(.985; 1, 16) = 7.42 \qquad P\text{-value} = .01$

    Conclusion: Some BC interactions are present.

At this point, the researcher considered several simple transformations of the data to see if the BC interactions could be removed. However, she was not successful in this endeavor.

Tests for Main Effects. Since factor A (gender) did not interact with the other two factors, attention next turned to testing for factor A main effects. In testing for factor A main effects, the decision rule is (the alternatives are given in Table 22.4):

$$\text{If } F^* \leq F(.985; 1, 16) = 7.42, \text{ conclude } H_0$$

$$\text{If } F^* > F(.985; 1, 16) = 7.42, \text{ conclude } H_a$$

The test statistic is:

$$F^* = \frac{MSA}{MSE} = \frac{181.50}{8.75} = 20.74$$

Since $F^* = 20.74 > 7.42$, the conclusion was reached that factor A main effects are present. The P-value of this test is $0+$.

The factor B and factor C main effects were not tested at this point because BC interactions were found to be present. The researcher first wished to study the nature of the BC interaction effects before determining whether the factor B and factor C main effects are of any practical interest under the circumstances.

Family of Conclusions. The five separate F tests for factor effects led the researcher to conclude (with family level of significance $\leq .10$):

1. There are no three-factor interactions.
2. There are no two-factor interactions between gender (factor A) and either of the other two factors—body fat (factor B) and smoking history (factor C). Body fat and smoking history interactions do exist, however.
3. Main effects for gender (factor A) are present.

This set of test results was most useful to the researcher. The next step in her analysis was to examine the nature of the BC interaction effects and of the factor A main effects.

Estimation of Factor Effects

To study the nature of the *BC* interaction effects, the researcher wished to estimate separately, for persons with high and low percent body fat, the difference in mean fatigue time for light smokers and heavy smokers. The desired contrasts are:

$$L_1 = \mu_{.11} - \mu_{.12}$$

$$L_2 = \mu_{.21} - \mu_{.22}$$

In addition, a single comparison between the factor level means for factor *A* is sufficient to analyze the factor *A* main effects since factor *A* has only two levels. The contrast of interest (here a pairwise comparison of factor level means) is:

$$L_3 = \mu_{1..} - \mu_{2..}$$

These three contrasts are estimated by:

$$\hat{L}_1 = \bar{Y}_{.11.} - \bar{Y}_{.12.}$$

$$\hat{L}_2 = \bar{Y}_{.21.} - \bar{Y}_{.22.}$$

$$\hat{L}_3 = \bar{Y}_{1...} - \bar{Y}_{2...}$$

From Table 22.2c, we obtain:

$$\hat{L}_1 = 23 - 16 = 7$$

$$\hat{L}_2 = 13 - 13 = 0$$

$$\hat{L}_3 = 19 - 13.5 = 5.5$$

The researcher used estimated variances (22.49) and (22.37) and confidence limits (22.46) with a 95 percent family confidence coefficient based on the Bonferroni procedure. Hence, she required for the three comparisons:

$$B = t(1 - .05/6; 16) = 2.673$$

$$s^2\{\hat{L}_1\} = s^2\{\hat{L}_2\} = \frac{MSE}{na}[(1)^2 + (-1)^2] = \frac{8.75}{6}(2) = 2.917$$

$$s^2\{\hat{L}_3\} = \frac{MSE}{nbc}[(1)^2 + (-1)^2] = \frac{8.75}{12}(2) = 1.458$$

$$s\{\hat{L}_1\} = s\{\hat{L}_2\} = 1.708 \qquad s\{\hat{L}_3\} = 1.207$$

The desired confidence intervals therefore are:

$$2.4 = 7.0 - 2.673(1.708) \le \mu_{.11} - \mu_{.12} \le 7.0 + 2.673(1.708) = 11.6$$

$$-4.6 = 0 - 2.673(1.708) \le \mu_{.21} - \mu_{.22} \le 0 + 2.673(1.708) = 4.6$$

$$2.3 = 5.5 - 2.673(1.207) \le \mu_{1..} - \mu_{2..} \le 5.5 + 2.673(1.207) = 8.7$$

The researcher therefore concluded with family confidence coefficient .95: (1) Among people with low percent body fat, those who have a light smoking history have a mean stress test endurance that is 2.4 to 11.6 minutes longer than the mean endurance for people with a heavy smoking history. (2) People with high percent body fat do not differ in mean stress test endurance whether they have a light or a heavy smoking history. (3) The mean stress test endurance for men is 2.3 to 8.7 minutes longer than the mean endurance for women.

In view of the important interaction effects noted between body fat and smoking history on stress test endurance, the researcher concluded that factor B and factor C main effects are of no interest, and therefore terminated her analysis at this point. The principal findings are presented graphically in Figure 22.8. Figure 22.8a shows the magnitude of the effect of gender on stress test endurance, and Figure 22.8b shows the nature of the interaction effects between body fat and smoking history.

22.6 PLANNING OF SAMPLE SIZES

Planning of sample sizes is handled in essentially the same fashion for three-factor studies as for single-factor and two-factor studies. Hence, we make only a few brief remarks. We consider first the power of the F tests for three-factor studies.

FIGURE 22.8 Key Findings from Stress Test Endurance Study

(a) Effect of Gender

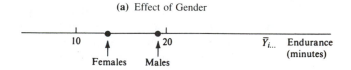

(b) Effects of Body Fat and Smoking History

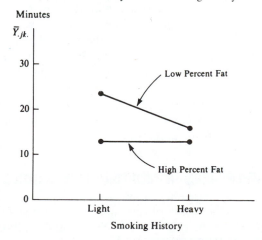

Power of *F* Tests

The power of the factor effects tests for three-factor studies can be obtained from Table A.8 in the manner described for one-factor and two-factor studies. The noncentrality parameter ϕ for a given test is defined as follows:

$$(22.53) \qquad \phi = \frac{1}{\sigma}\left[\frac{\text{numerator of second term in } E\{MS\} \text{ in Table 22.3}}{\text{denominator of second term in } E\{MS\} \text{ plus } 1}\right]^{1/2}$$

Thus, for testing the existence of three-factor interactions, we have:

$$\phi = \frac{1}{\sigma}\left[\frac{n\Sigma\Sigma\Sigma(\alpha\beta\gamma)_{ijk}^2}{(a-1)(b-1)(c-1)+1}\right]^{1/2}$$

Planning Approaches

In most cases, equal replications will be desired for each treatment. When planning sample sizes with the power approach, one is typically concerned with the power of detecting factor *A* main effects, the power of detecting factor *B* main effects, and the power of detecting factor *C* main effects. One can first specify the minimum range of factor *A* level means for which it is important to detect factor *A* main effects and obtain the needed sample sizes from Table A.10, with $r = a$. The resulting sample size is bcn, from which n can readily be obtained. The use of Table A.10 for this purpose is appropriate provided the resulting sample sizes are not small, specifically provided $a(bcn - 1) \geq 20$. If this condition is not met, one should use the Pearson-Hartley power charts in Table A.8 with an iterative approach.

In the same way one can specify values for the minimum range of factor level means for factors *B* and *C* for which it is important to detect the factor main effects and find the needed sample sizes. If the sample sizes obtained from the factor *A*, factor *B*, and factor *C* power specifications differ substantially, a judgment will need to be made as to the final sample sizes.

Alternatively, or in conjunction with the power approach, one can specify the important contrasts to be estimated and then find the sample sizes that are expected to provide the needed precisions for the desired family confidence coefficient. Frequently this approach is more useful than the power approach, although both of these approaches can be used jointly for arriving at a determination of needed sample sizes.

If the purpose of the factorial study is to identify the best of the abc factor combinations, Table A.11 can be used for finding the needed sample sizes, as described in Section 17.3. For this purpose, $r = abc$.

22.7 UNEQUAL SAMPLE SIZES IN MULTIFACTOR STUDIES

When the treatment sample sizes in a multifactor study are not equal, the procedures explained in Chapter 20 for two-factor studies should be followed with routine modifications. We continue to assume that *all treatment means are of equal importance and that there are no empty cells.*

Tests for Factor Effects

Tests for factor effects in multifactor studies with unequal sample sizes can be conducted by means of the regression approach. Indicator variables taking on the values $1, -1, 0$, are designated for each factor, the number of such variables being one less than the number of factor levels. Interaction effects are represented by cross-product terms, as usual. Since the sums of squares are no longer orthogonal when the treatment sample sizes are unequal, different reduced models need to be fitted for the tests of interest.

Example.   Suppose that in the stress test example of Table 22.2, observations Y_{1113} and Y_{2212} were missing. To develop a regression model for this example, we note that each of the three factors is at two levels. Hence, one indicator variable is required for each factor. The full regression model therefore is:

$$(22.54) \qquad Y_{ijkm} = \mu_{...} + \alpha_1 X_{ijkm1} + \beta_1 X_{ijkm2} + \gamma_1 X_{ijkm3} + (\alpha\beta)_{11} X_{ijkm1} X_{ijkm2}$$
$$+ (\alpha\gamma)_{11} X_{ijkm1} X_{ijkm3} + (\beta\gamma)_{11} X_{ijkm2} X_{ijkm3}$$
$$+ (\alpha\beta\gamma)_{111} X_{ijkm1} X_{ijkm2} X_{ijkm3} + \varepsilon_{ijkm} \qquad \text{Full model}$$

where:

$$X_{ijkm1} = \begin{array}{l} 1 \text{ if case from level 1 for factor } A \\ -1 \text{ if case from level 2 for factor } A \end{array}$$

$$X_{ijkm2} = \begin{array}{l} 1 \text{ if case from level 1 for factor } B \\ -1 \text{ if case from level 2 for factor } B \end{array}$$

$$X_{ijkm3} = \begin{array}{l} 1 \text{ if case from level 1 for factor } C \\ -1 \text{ if case from level 2 for factor } C \end{array}$$

The regression parameters in model (22.54) are the ANOVA model parameters as defined in (22.13).

Table 22.6 contains the **Y** vector and **X** matrix for the full regression model (22.54) for the stress test example with two observations missing. The reduced models for testing different factor effects are obtained by dropping appropriate columns from the **X** matrix in Table 22.6.

Note

The discussion in Section 20.5 on the use of statistical packages for analysis of variance with unequal sample sizes and/or empty cells is applicable for multifactor studies in its entirety.

Estimation of Factor Effects

Estimation of factor effects in multifactor studies with unequal sample sizes is conducted in similar fashion as for two-factor studies. The formulas in Table 20.5 need simply be extended to three or more factors.

TABLE 22.6   Data Matrices for Regression Model (22.54)—Stress Test Example with Y_{1113} and Y_{2212} Missing

		X_1	X_2	X_3	X_1X_2	X_1X_3	X_2X_3	$X_1X_2X_3$
Y_{1111}	24	1	1	1	1	1	1	1
Y_{1112}	29	1	1	1	1	1	1	1
Y_{1121}	18	1	1	1	−1	1	−1	−1
Y_{1122}	19	1	1	1	−1	1	−1	−1
Y_{1123}	23	1	1	1	−1	1	−1	−1
Y_{1211}	15	1	1	−1	1	−1	1	−1
Y_{1212}	15	1	1	−1	1	−1	1	−1
Y_{1213}	12	1	1	−1	1	−1	1	−1
Y_{1221}	15	1	1	−1	−1	−1	−1	1
Y_{1222}	20	1	1	−1	−1	−1	−1	1
Y_{1223}	13	1	1	−1	−1	−1	−1	1
Y_{2111}	20	1	−1	1	1	−1	−1	1
Y_{2112}	22	1	−1	1	1	−1	−1	1
Y_{2113}	18	1	−1	1	1	−1	−1	1
Y_{2121}	15	1	−1	1	−1	−1	1	−1
Y_{2122}	10	1	−1	1	−1	−1	1	−1
Y_{2123}	11	1	−1	1	−1	−1	1	−1
Y_{2211}	16	1	−1	−1	1	1	−1	−1
Y_{2213}	11	1	−1	−1	1	1	−1	−1
Y_{2221}	10	1	−1	−1	−1	1	1	−1
Y_{2222}	14	1	−1	−1	−1	1	1	−1
Y_{2223}	6	1	−1	−1	−1	1	1	−1

$$\mathbf{Y} = \qquad = \qquad \mathbf{X} =$$

To illustrate these extensions, consider pairwise comparisons of factor A level means in a three-factor study. Such a comparison, its estimator, and the estimated variance are:

(22.55a)
$$D = \mu_{i..} - \mu_{i'..}$$

(22.55b)
$$\hat{D} = \hat{\mu}_{i..} - \hat{\mu}_{i'..} \qquad \text{where } \hat{\mu}_{i..} = \frac{\sum_j \sum_k \bar{Y}_{ijk.}}{bc}$$

(22.55c)
$$s^2\{\hat{D}\} = \frac{MSE}{b^2 c^2} \sum_j \sum_k \left(\frac{1}{n_{ijk}} + \frac{1}{n_{i'jk}} \right)$$

The appropriate degrees of freedom associated with MSE are $n_T - abc$.

22.8 MODELS II AND III FOR THREE-FACTOR STUDIES

Just as for single-factor and two-factor studies, the analysis of variance sums of squares and degrees of freedom for random and mixed multifactor models are the same as those for the fixed ANOVA model. The principal problem with random and mixed multifactor models, as we saw for two-factor models, is the determination of the expected mean squares. Once these are known, the proper test statistics and confidence intervals can be constructed. Rules for finding expected mean squares for random and mixed models for any number of factors will be presented later in Chapter 27. We now present model II (random factor levels) and model III (mixed factor levels) for three-factor studies and show how appropriate tests are conducted. We consider again the case where *all treatment sample sizes are equal.*

Model II (Random Factor Levels)

In a study of the effects of operators, machines, and batches of raw material on daily output, all three factors may be considered to have random factor levels. The random ANOVA model for such a three-factor study is:

$$(22.56) \quad Y_{ijkm} = \mu_{...} + \alpha_i + \beta_j + \gamma_k + (\alpha\beta)_{ij} + (\alpha\gamma)_{ik} + (\beta\gamma)_{jk} + (\alpha\beta\gamma)_{ijk} + \varepsilon_{ijkm}$$

where:

$\mu_{...}$ is a constant

$\alpha_i, \beta_j, \gamma_k, (\alpha\beta)_{ij}, (\alpha\gamma)_{ik}, (\beta\gamma)_{jk}, (\alpha\beta\gamma)_{ijk}, \varepsilon_{ijkm}$ are independent normal random variables with expectations zero and respective variances $\sigma_\alpha^2, \sigma_\beta^2, \sigma_\gamma^2, \sigma_{\alpha\beta}^2, \sigma_{\alpha\gamma}^2, \sigma_{\beta\gamma}^2, \sigma_{\alpha\beta\gamma}^2, \sigma^2$

$i = 1, \ldots, a; j = 1, \ldots, b; k = 1, \ldots, c; m = 1, \ldots, n$

Just as for the two-factor random ANOVA model (21.25), the observations Y_{ijkm} for the three-factor random ANOVA model (22.56) are normally distributed with constant variance. The expected value and variance of observation Y_{ijkm} are:

$$(22.57a) \quad E\{Y_{ijkm}\} = \mu_{...}$$

$$(22.57b) \quad \sigma^2\{Y_{ijkm}\} = \sigma_Y^2 = \sigma_\alpha^2 + \sigma_\beta^2 + \sigma_\gamma^2 + \sigma_{\alpha\beta}^2 + \sigma_{\alpha\gamma}^2 + \sigma_{\beta\gamma}^2 + \sigma_{\alpha\beta\gamma}^2 + \sigma^2$$

Any two observations are independent except for observations that have one or more common factor levels, which are correlated because they contain some common random terms.

Table 22.7 contains the expected mean squares for all components of the ANOVA table for the random ANOVA model (22.56).

TABLE 22.7 Expected Mean Squares in Random Three-Factor Study

Mean Square	df	Expected Mean Square
MSA	$a - 1$	$\sigma^2 + nbc\sigma_\alpha^2 + nc\sigma_{\alpha\beta}^2 + nb\sigma_{\alpha\gamma}^2 + n\sigma_{\alpha\beta\gamma}^2$
MSB	$b - 1$	$\sigma^2 + nac\sigma_\beta^2 + nc\sigma_{\alpha\beta}^2 + na\sigma_{\beta\gamma}^2 + n\sigma_{\alpha\beta\gamma}^2$
MSC	$c - 1$	$\sigma^2 + nab\sigma_\gamma^2 + nb\sigma_{\alpha\gamma}^2 + na\sigma_{\beta\gamma}^2 + n\sigma_{\alpha\beta\gamma}^2$
MSAB	$(a - 1)(b - 1)$	$\sigma^2 + nc\sigma_{\alpha\beta}^2 + n\sigma_{\alpha\beta\gamma}^2$
MSAC	$(a - 1)(c - 1)$	$\sigma^2 + nb\sigma_{\alpha\gamma}^2 + n\sigma_{\alpha\beta\gamma}^2$
MSBC	$(b - 1)(c - 1)$	$\sigma^2 + na\sigma_{\beta\gamma}^2 + n\sigma_{\alpha\beta\gamma}^2$
MSABC	$(a - 1)(b - 1)(c - 1)$	$\sigma^2 + n\sigma_{\alpha\beta\gamma}^2$
MSE	$(n - 1)abc$	σ^2

Model III (Mixed Factor Levels)

Suppose that in a three-factor study, factors B and C have random factor levels while factor A has fixed factor levels. A mixed ANOVA model for this three-factor study is:

$$(22.58) \quad Y_{ijkm} = \mu_{...} + \alpha_i + \beta_j + \gamma_k + (\alpha\beta)_{ij} + (\alpha\gamma)_{ik} + (\beta\gamma)_{jk} + (\alpha\beta\gamma)_{ijk} + \varepsilon_{ijkm}$$

where:

$\mu_{...}$ is a constant

α_i are constants

β_j, γ_k, $(\alpha\beta)_{ij}$, $(\alpha\gamma)_{ik}$, $(\beta\gamma)_{jk}$, $(\alpha\beta\gamma)_{ijk}$ are pairwise independent normal random variables with expectations zero and constant variances

ε_{ijkm} are independent $N(0, \sigma^2)$, and are independent of the other random components

$\sum_i \alpha_i = \sum_i (\alpha\beta)_{ij} = \sum_i (\alpha\gamma)_{ik} = \sum_i (\alpha\beta\gamma)_{ijk} = 0$

$i = 1, \ldots, a; j = 1, \ldots, b; k = 1, \ldots, c; m = 1, \ldots, n$

Note that all interaction terms in this model are random, since at least one of the factors contained in each has random factor levels. Note also that all sums of effects involving the fixed factor are zero over the fixed factor levels. Various correlations exist between the random effects terms, which we shall not detail.

The observations Y_{ijkm} for the three-factor mixed ANOVA model (22.58) are normally distributed with constant variance. The expected value of observation Y_{ijkm} is:

$$(22.59) \qquad E\{Y_{ijkm}\} = \mu_{...} + \alpha_i$$

In advance of the random trials, any two observations are independent except for observations that contain common and/or correlated random effects terms; these observations are correlated.

Table 22.8 contains all the expected mean squares for the mixed ANOVA model (22.58).

TABLE 22.8 Expected Mean Squares in Mixed Three-Factor Study (A fixed, B and C random)

Mean Square	df	Expected Mean Square
MSA	$a - 1$	$\sigma^2 + nbc\dfrac{\sum \alpha_i^2}{a-1} + nc\sigma_{\alpha\beta}^2 + nb\sigma_{\alpha\gamma}^2 + n\sigma_{\alpha\beta\gamma}^2$
MSB	$b - 1$	$\sigma^2 + nac\sigma_{\beta}^2 + na\sigma_{\beta\gamma}^2$
MSC	$c - 1$	$\sigma^2 + nab\sigma_{\gamma}^2 + na\sigma_{\beta\gamma}^2$
MSAB	$(a-1)(b-1)$	$\sigma^2 + nc\sigma_{\alpha\beta}^2 + n\sigma_{\alpha\beta\gamma}^2$
MSAC	$(a-1)(c-1)$	$\sigma^2 + nb\sigma_{\alpha\gamma}^2 + n\sigma_{\alpha\beta\gamma}^2$
MSBC	$(b-1)(c-1)$	$\sigma^2 + na\sigma_{\beta\gamma}^2$
MSABC	$(a-1)(b-1)(c-1)$	$\sigma^2 + n\sigma_{\alpha\beta\gamma}^2$
MSE	$(n-1)abc$	σ^2

Other mixed ANOVA models can be developed in similar fashion. The expected mean squares for these mixed models can be found by employing the rules to be presented in Chapter 27.

Appropriate Test Statistics

From the expected mean squares, we seek to determine the appropriate F^* statistic for a given test. An exact test statistic can often be found for random and mixed multifactor models, but not always.

Exact F Test. Suppose we wish to determine whether or not BC interactions are present in the random ANOVA model for Table 22.7. We see easily from the expected mean squares column that the appropriate test statistic is $MSBC/MSABC$. If we wish to study the same question for the mixed ANOVA model for Table 22.8, we are again able to find an appropriate test statistic, but this time it is $MSBC/MSE$. We thus see that the two test statistics are not the same, even though the same factor effects are being studied, because of the differences between the two models.

It is not always possible to find an exact F test for mixed and random multifactor ANOVA models. For instance, one cannot directly test for the presence of factor A main effects in the random ANOVA model for Table 22.7. Note from this table that there is no expected mean square that consists of the components of $E\{MSA\}$ except for the $nbc\sigma_\alpha^2$ term.

Sometimes it is possible to assume that certain interactions are zero, and then proceed in the usual way with an exact F test. For example, to test for factor A main effects in the random ANOVA model for Table 22.7, it may be possible to assume that $\sigma_{\alpha\gamma}^2 = 0$ (indeed, this can be tested with $MSAC/MSABC$). If this assumption is appropriate one can use the test statistic $MSA/MSAB$ to test for factor A main effects.

Satterthwaite Approximate F Test. Often, one may not know whether certain interactions are zero. In that case, an approximate F test may be employed which utilizes a *pseudo F* or *quasi F* test statistic. This approximate test, called the *Satterthwaite test,* involves developing a linear combination of mean squares which has the same expectation when H_0 holds as the factor effects mean square. Let us express this linear combination as follows:

$$(22.60) \qquad a_1 MS_1 + a_2 MS_2 + \cdots + a_h MS_h$$

where the a_i are constants. It can be shown that the approximate number of degrees of freedom associated with the linear combination (22.60) is:

$$(22.61) \qquad df \cong \frac{(a_1 MS_1 + a_2 MS_2 + \cdots + a_h MS_h)^2}{\dfrac{(a_1 MS_1)^2}{df_1} + \dfrac{(a_2 MS_2)^2}{df_2} + \cdots + \dfrac{(a_h MS_h)^2}{df_h}}$$

where df_i denotes the degrees of freedom associated with MS_i. The test statistic is then set up in the usual way, and follows approximately the F distribution when H_0 holds.

We illustrate this procedure for testing factor A main effects in the random ANOVA model for Table 22.7:

(22.62)
$$H_0: \sigma_\alpha^2 = 0$$
$$H_a: \sigma_\alpha^2 > 0$$

Note from Table 22.7 that:

(22.63) $E\{MSAB\} + E\{MSAC\} - E\{MSABC\} = \sigma^2 + nc\sigma_{\alpha\beta}^2 + nb\sigma_{\alpha\gamma}^2 + n\sigma_{\alpha\beta\gamma}^2$

This equals precisely $E\{MSA\}$ when $\sigma_\alpha^2 = 0$. Hence, the suggested test statistic is:

(22.64)
$$F^{**} = \frac{MSA}{MSAB + MSAC - MSABC}$$

where we denote the test statistic as F^{**} as a reminder that a pseudo F test is involved.

Example.   Table 22.9 contains the analysis of variance for a study of the effects of operators, machines, and batches on the daily output of a highly automated process. Each factor is assumed to be a random factor. To test whether operators (factor A) have a main effect on output, we use the test statistic (22.64):

$$F^{**} = \frac{8.5}{2.5 + 4.0 - 1.5} = \frac{8.5}{5.0} = 1.7$$

The approximate number of degrees of freedom associated with the denominator is:

$$df \cong \frac{(5.0)^2}{\dfrac{(2.5)^2}{2} + \dfrac{(4.0)^2}{8} + \dfrac{(-1.5)^2}{8}} = 4.6$$

TABLE 22.9   ANOVA Table for Random Three-Factor Study
($a = 3, b = 2, c = 5, n = 3$)

Source of Variation	SS	df	MS
Factor A (operators)	17	2	8.5
Factor B (machines)	4	1	4.0
Factor C (batches)	25	4	6.2
AB interactions	5	2	2.5
AC interactions	32	8	4.0
BC interactions	12	4	3.0
ABC interactions	12	8	1.5
Error	138	60	2.3
Total	245	89	

Usually, the degrees of freedom will not turn out to be an integer. One can then either interpolate in the F distribution table, or round to the nearest integer. We shall round. For a level of significance of $\alpha = .05$, we require $F(.95; 2, 5) = 5.79$. Since $F^{**} = 1.7 \leq 5.79$, we conclude H_0, that operators do not have a main effect on daily output. This test, it should be remembered, is an approximate one but can be quite useful if employed with caution.

Estimation of Effects

No new problems arise in the development of unbiased estimators of variance components for random factors or in the estimation of contrasts for fixed factors in mixed models, when three or more factors are studied at one time. Confidence limits for contrasts of the factor level means of a fixed factor are obtained by using the mean square utilized in the denominator of the test statistic for examining the presence of main effects for that factor. The degrees of freedom are those associated with the mean square utilized.

PROBLEMS

22.1. Refer to Table 22.1 containing the mean responses μ_{ijk} for a three-factor study.
  a. Find the main effects of age.
  b. Find the main interaction effect of young age and normal IQ.
  c. Find the interaction effect of young age, normal IQ, and female gender.

22.2. Prepare AC plots of the mean responses μ_{ijk} in Table 22.1 in the format of Figure 22.2b. Do your plots convey the same information as Figure 22.4? Discuss.

22.3. Prepare BC plots of the mean responses μ_{ijk} in Figure 22.3. Do your plots bring out any information on main effects and interactions not readily seen from Figure 22.3? Discuss.

22.4. In a three-factor study, the mean responses μ_{ijk} are as follows:

	$k = 1$		$k = 2$	
	$j = 1$	$j = 2$	$j = 1$	$j = 2$
$i = 1$	130	138	140	144
$i = 2$	126	130	134	136
$i = 3$	122	125	122	131

  a. Find α_1, α_2, and α_3.
  b. Find β_2 and γ_1.
  c. Find $(\alpha\beta)_{12}$, $(\alpha\gamma)_{21}$, and $(\beta\gamma)_{12}$.
  d. Find $(\alpha\beta\gamma)_{111}$ and $(\alpha\beta\gamma)_{322}$.

22.5. Refer to Problem 22.4. Prepare AB plots of the mean responses μ_{ijk} in the format of Figure 22.3. What do these plots show about factor main effects and interactions?

22.6. Case hardening. An experiment involving the case hardening of lightweight shafts machined from bars of an alloy was run to study the effects of the amount of a chemical agent added to the alloy in a molten state (factor A), the temperature of the hardening process (factor B), and the time duration of the hardening process (factor C) on the outside hardness of the shaft. All factors were at two levels (1: low; 2: high), and the number of rods tested for each treatment was $n = 3$. The data on hardness (in Brinell units) follow.

	$k = 1$		$k = 2$	
	$j = 1$	$j = 2$	$j = 1$	$j = 2$
$i = 1$	39.9	53.5	56.0	70.9
	32.2	50.7	56.9	73.3
	36.3	52.8	56.6	71.6
$i = 2$	45.2	63.3	69.4	82.9
	48.0	65.5	66.6	85.2
	47.5	63.6	68.8	82.3

a. Obtain the residuals for ANOVA model (22.14) and prepare aligned residual dot plots for each level of factor A. Do the same for each of the other two factors. What do your plots suggest about the appropriateness of ANOVA model (22.14)?

b. Prepare a normal probability plot of the residuals. Also obtain the coefficient of correlation between the ordered residuals and their expected values under normality. Does the normality assumption appear to be reasonable here?

22.7. Refer to **Case hardening** Problem 22.6. Assume that fixed ANOVA model (22.14) is appropriate.

a. Prepare AB plots of the estimated treatment means $\bar{Y}_{ijk}$ in the format of Figure 22.6b. Does it appear that any interactions are present? Any main effects?

b. Obtain the analysis of variance table.

c. Test for three-factor interactions; use $\alpha = .025$. State the alternatives, decision rule, and conclusion. What is the P-value of the test?

d. Test for AB, AC, and BC interactions. For each test, use $\alpha = .025$ and state the alternatives, decision rule, and conclusion. What is the P-value of each test?

e. Test for A, B, and C main effects. For each test, use $\alpha = .025$ and state the alternatives, decision rule, and conclusion. What is the P-value of each test?

f. State the set of conclusions that can be arrived at from the tests in parts (c), (d), and (e). Obtain an upper bound for the family level of significance for the set of tests; use the Kimball inequality.

g. Do the results in part (f) confirm your graphic analysis in part (a)?

22.8. Refer to **Case hardening** Problems 22.6 and 22.7.

a. To study the nature of the main factor effects, estimate the following pairwise comparisons:

$$D_1 = \mu_{2..} - \mu_{1..} \qquad D_2 = \mu_{.2.} - \mu_{.1.} \qquad D_3 = \mu_{..2} - \mu_{..1}$$

Use the Bonferroni procedure with a 95 percent family confidence coefficient. State your findings.

b. Estimate μ_{222} with a 95 percent confidence interval.

22.9. **Marketing research contractors.** A marketing research consultant evaluated the effects of fee schedule (factor A), scope of work (factor B), and type of supervisory control (factor C) on the quality of work performed under contract by independent marketing research agencies. The factor levels in the study were as follows:

Factor		Factor Levels
A Fee level	$i = 1$:	High
	$i = 2$:	Average
	$i = 3$:	Low
B Scope	$j = 1$:	All contract work performed in house
	$j = 2$:	Some work subcontracted out
C Supervision	$k = 1$:	Local supervisors
	$k = 2$:	Traveling supervisors only

The quality of work performed was measured by an index taking into account several characteristics of quality. Four agencies were chosen for each factor level combination and the quality of their work evaluated. The data on quality follow.

		$k = 1$		$k = 2$	
		$j = 1$	$j = 2$	$j = 1$	$j = 2$
$i = 1$		124.3	115.1	112.7	88.2
		120.6	119.9	110.2	96.0
		120.7	115.4	113.5	96.4
		122.6	117.3	108.6	90.1
$i = 2$		119.3	117.2	113.6	92.7
		118.9	114.4	109.1	91.1
		125.3	113.4	108.9	90.7
		121.4	120.0	112.3	87.9
$i = 3$		90.9	89.9	78.6	58.6
		95.3	83.0	80.6	63.5
		88.8	86.5	83.5	59.8
		92.0	82.7	77.1	62.3

a. Obtain the residuals for ANOVA model (22.14) and plot them against the fitted values. What does your plot suggest about the appropriateness of ANOVA model (22.14)?

b. Prepare a normal probability plot of the residuals. Also obtain the coefficient of correlation between the ordered residuals and their expected values under normality. Does the normality assumption appear to be reasonable here?

22.10. Refer to **Marketing research contractors** Problem 22.9. Assume that fixed ANOVA model (22.14) is appropriate.

a. Prepare AB plots of the estimated treatment means $\bar{Y}_{ijk.}$ in the format of Figure 22.6b. Does it appear that any interactions are present? Any main effects?

b. Prepare a normal probability plot of the estimated factor A level means. What does this plot suggest about the nature of the factor A main effects?

c. Obtain the analysis of variance table.

d. Test for three-factor interactions; use $\alpha = .01$. State the alternatives, decision rule, and conclusion. What is the P-value of the test?

e. Test for AB, AC, and BC interactions. For each test, use $\alpha = .01$ and state the alternatives, decision rule, and conclusion. What is the P-value of each test?

f. Test for factor A main effects; use $\alpha = .01$. State the alternatives, decision rule, and conclusions. What is the P-value of the test?

g. State the set of conclusions that can be arrived at from the tests in parts (d), (e), and (f). Obtain an upper bound for the family level of significance for the set of tests; use the Kimball inequality.

h. Do the results in part (g) confirm your graphic analysis in parts (a) and (b)?

22.11. Refer to **Marketing research contractors** Problems 22.9 and 22.10.

a. To study the nature of the factor A main effects and the BC interactions, it is desired to estimate the following comparisons:

$$D_1 = \mu_{1..} - \mu_{2..} \qquad D_4 = \mu_{.11} - \mu_{.12}$$

$$D_2 = \mu_{2..} - \mu_{3..} \qquad D_5 = \mu_{.21} - \mu_{.22}$$

$$D_3 = \mu_{1..} - \mu_{3..} \qquad L_1 = D_4 - D_5$$

Use the Bonferroni procedure with a 90 percent family confidence coefficient to make the desired comparisons. State your findings.

b. Estimate $D = \mu_{121} - \mu_{221}$ with a 95 percent confidence interval.

c. The consultant wishes to identify the type(s) of independent marketing research agencies that provide the highest quality of work. Use the Tukey procedure with a 90 percent family confidence coefficient to make the desired identifications.

22.12. **Electronics assembly.** Assemblers in an electronics firm will attach 12 components to a newly developed "board" that will be used in automatic-control equipment in manufacturing plants. An operations analyst conducted an experiment to study the effects of three factors on the mean time to assemble a board. Factor A was the gender of the assembler ($i = 1$: male; $i = 2$: female), factor B was the sequence of assembling the components ($j = 1, 2, 3$), and factor C was the amount of experience by the assembler ($k = 1$: under 18 months; $k = 2$: 18 months or more). Randomization was used to assign 15 assemblers of each gender with a given amount of experience to each of the three assembly sequences, with each sequence assigned to five assemblers. After a learning period, the total time (in minutes) to assemble 50 boards was observed. The data follow.

		$k = 1$			$k = 2$	
	$j = 1$	$j = 2$	$j = 3$	$j = 1$	$j = 2$	$j = 3$
$i = 1$	1,250	1,319	1,217	1,021	1,119	1,033
	1,175	1,251	1,190	1,099	1,110	1,067
	1,236	1,241	1,201	1,069	1,123	1,057
	1,239	1,295	1,232	996	1,097	1,077
	1,193	1,265	1,251	1,070	1,163	1,022
$i = 2$	1,066	1,105	1,021	864	927	841
	1,076	1,043	1,020	848	944	865
	1,004	1,051	1,035	881	957	817
	1,002	1,128	1,000	892	897	911
	1,034	1,060	1,026	868	933	868

a. Obtain the residuals for ANOVA model (22.14) and plot them against the fitted values. What does your plot suggest about the appropriateness of ANOVA model (22.14)?

b. Prepare a normal probability plot of the residuals. Also obtain the coefficient of correlation between the ordered residuals and their expected values under normality. Does the normality assumption appear to be reasonable here?

22.13. Refer to **Electronics assembly** Problem 22.12. Assume that fixed ANOVA model (22.14) is appropriate.

a. Prepare AB plots of the estimated treatment means $\bar{Y}_{ijk.}$ in the format of Figure 22.6b. Does it appear that any interactions are present? Any main effects?

b. For each factor, prepare a normal probability plot of the estimated factor level means. What do these plots suggest about the nature of the factor effects?

c. Obtain the analysis of variance table.

d. Test for three-factor interactions; use $\alpha = .05$. State the alternatives, decision rule, and conclusion. What is the P-value of the test?

e. Test for AB, AC, and BC interactions. For each test, use $\alpha = .05$ and state the alternatives, decision rule, and conclusion. What is the P-value of each test?

f. Test for A, B, and C main effects. For each test, use $\alpha = .05$ and state the alternatives, decision rule, and conclusion. What is the P-value of each test?

g. State the set of conclusions that can be arrived at from the tests in parts (d), (e), and (f). Obtain an upper bound for the family level of significance for the set of tests; use the Kimball inequality.

h. Do the results in part (g) confirm your graphic analysis in parts (a) and (b)?

22.14. Refer to **Electronics assembly** Problems 22.12 and 22.13.

a. To study the nature of the factor main effects, estimate the following pairwise comparisons:

$$D_1 = \mu_{1..} - \mu_{2..} \qquad D_4 = \mu_{.2.} - \mu_{.3.}$$

$$D_2 = \mu_{.1.} - \mu_{.2.} \qquad D_5 = \mu_{..1} - \mu_{..2}$$

$$D_3 = \mu_{.1.} - \mu_{.3.}$$

Use the Bonferroni procedure with a 90 percent family confidence coefficient. State your findings.

b. Estimate μ_{231} with a 95 percent confidence interval.

22.15. For the three-factor fixed ANOVA model (22.14), what is the noncentrality parameter ϕ for testing for factor A main effects? For testing for AB interactions?

22.16. Refer to **Marketing research contractors** Problems 22.9 and 22.10. Assume that $\sigma = 3.0$. What is the power of the test for factor A main effects in Problem 22.10f if $\mu_{1..} = 97$, $\mu_{2..} = 95$, and $\mu_{3..} = 90$?

22.17. Refer to **Electronics assembly** Problems 22.12 and 22.13. Assume that $\sigma = 29$. What is the power of the test for factor B main effects in Problem 22.13f if $\mu_{.1.} = 1,050$, $\mu_{.2.} = 1,100$, and $\mu_{.3.} = 1,060$?

22.18. Refer to **Case hardening** Problem 22.6. Suppose that the sample sizes have not yet been determined but it has been decided to use equal sample sizes for all treatments. The chief objective is to identify the treatment that leads to the highest mean hardness. The probability should be at least .99 that the correct treatment is identified when the mean hardness for the second best treatment differs by 2.0 or more Brinell units. Assume that a reasonable planning value for the error standard deviation is $\sigma = 1.8$. What are the required sample sizes?

22.19. Refer to **Electronics assembly** Problem 22.12. Suppose that the sample sizes have not yet been determined but it has been decided to use equal sample sizes for all treatments. The chief objective is to estimate the following pairwise comparisons:

$$D_1 = \mu_{1..} - \mu_{2..} \qquad D_4 = \mu_{.2.} - \mu_{.3.}$$

$$D_2 = \mu_{.1.} - \mu_{.2.} \qquad D_5 = \mu_{..1} - \mu_{..2}$$

$$D_3 = \mu_{.1.} - \mu_{.3.}$$

What are the required sample sizes if the precision of each of the estimates should not exceed ± 20, using the Bonferroni procedure with a 90 percent family confidence coefficient for the joint set of comparisons? A reasonable planning value for the error standard deviation is $\sigma = 29$.

22.20. Refer to **Case hardening** Problem 22.6. Suppose that observations $Y_{1211} = 53.5$ and $Y_{1212} = 50.7$ are missing.
  a. State the full regression model equivalent to ANOVA model (22.14); use $1, -1, 0$ indicator variables.
  b. What is the reduced model for testing for factor A main effects?
  c. Test whether or not factor A main effects are present by fitting the full and reduced models; use $\alpha = .025$. State the alternatives, decision rule, and conclusion. What is the P-value of the test?
  d. Estimate $D = \mu_{2..} - \mu_{1..}$ with a 95 percent confidence interval.

22.21. Refer to **Electronics assembly** Problem 22.12. Suppose that observations $Y_{1224} = 1,097$, $Y_{2213} = 1,051$, and $Y_{2125} = 868$ are missing.
  a. State the full regression model equivalent to ANOVA model (22.14); use $1, -1, 0$ indicator variables.
  b. What is the reduced model for testing for factor C main effects?
  c. Test whether or not factor C main effects are present by fitting the full and reduced models; use $\alpha = .05$. State the alternatives, decision rule, and conclusion. What is the P-value of the test?
  d. Estimate $D = \mu_{..1} - \mu_{..2}$ with a 95 percent confidence interval.

22.22. Refer to Table 22.9. All three factors in this study have random effects.
  a. Test whether or not AB interactions are present. Use significance level $\alpha = .01$. State the alternatives, decision rule, and conclusion.
  b. Test whether machines (factor B) have main effects. Use significance level $\alpha = .01$. State the alternatives, decision rule, and conclusion.

22.23. Refer to **Electronics assembly** Problem 22.12. Suppose that the number of feasible sequences in which the components can be attached to the board is very large and that the three sequences studied were selected randomly from the set of operationally feasible sequences. Assume that a normal error ANOVA model is applicable where factors A and C have fixed effects and factor B has random effects. Some relevant expected mean squares for this model are:

$$E\{MSA\} = \sigma^2 + bcn\frac{\Sigma\alpha_i^2}{a - 1} + cn\sigma_{\alpha\beta}^2$$

$$E\{MSB\} = \sigma^2 + acn\sigma_\beta^2$$

$$E\{MSAC\} = \sigma^2 + bn\frac{\Sigma\Sigma(\alpha\gamma)_{ik}^2}{(a - 1)(c - 1)} + n\sigma_{\alpha\beta\gamma}^2$$

$$E\{MSABC\} = \sigma^2 + n\sigma_{\alpha\beta\gamma}^2$$

$$E\{MSE\} = \sigma^2$$

a. What is the appropriate test statistic for testing for AC interactions? For testing for factor B main effects?

b. Test whether or not AC interactions are present; use $\alpha = .05$. State the alternatives, decision rule, and conclusion.

c. Test whether or not factor B main effects are present; use $\alpha = .05$. State the alternatives, decision rule, and conclusion.

d. Obtain a point estimate of σ_β^2.

22.24. Consider the mixed ANOVA model (22.58) where factor A has fixed effects and the other two factors have random effects. Find a pseudo F test statistic F^{**} for testing for factor A main effects. What is the approximate number of degrees of freedom associated with the denominator of this test statistic?

EXERCISES

22.25. Show that for fixed ANOVA model (22.14) $\sum_i (\alpha\beta\gamma)_{ijk} = 0$.

22.26. Derive (22.23c) from (22.21a).

22.27. Derive the sum of squares breakdown in (22.25).

22.28. State the fixed ANOVA model for a three-factor study with $n = 1$ when all three-factor interactions are zero. Show the ANOVA table for this case.

22.29. For fixed ANOVA model (22.14), derive the variance of the estimated contrast $\hat{L} = \sum\sum c_{ij} \bar{Y}_{ij..}$.

22.30. For random ANOVA model (22.56), find the variance of the estimated mean $\bar{Y}_{i...}$.

PROJECTS

22.31. Refer to the **SENIC** data set. The following hospitals are to be considered in a study of the effects of average age of patients (factor A: variable 3), available facilities and services (factor B: variable 12), and region (factor C: variable 9) on the mean length of hospital stay of patients (variable 2):

1–14	16–28	31	32	34	35	37–39	41	44	46	50
52	53	57	58	63	66	76	77	83	111	

For purposes of this ANOVA study, average age is to be classified into two categories (less than 53.0 years, 53.0 years or more) and available facilities and services are to be classified into two categories (less than 40.2 percent, 40.2 percent or more).

a. Assemble the required data and obtain the residuals for ANOVA model (22.14).

b. Plot the residuals against the fitted values. What does your plot suggest about the appropriateness of ANOVA model (22.14)?

c. Prepare a normal probability plot of the residuals. Also obtain the coefficient of correlation between the ordered residuals and their expected values under normality. Does the normality assumption appear reasonable here?

22.32. Refer to the **SENIC** data set and Project 22.31. Assume that fixed ANOVA model (22.14) is appropriate.

   a. Plot the estimated treatment means $\bar{Y}_{ijk.}$ in the format of Figure 22.6a. Does it appear that any factor effects are present? Explain.

   b. Prepare a separate normal probability plot of the estimated factor level means for each factor. What do these plots suggest about the nature of the factor main effects?

   c. Obtain the analysis of variance table. Does any one source account for most of the total variability in the study? Explain.

   d. Test for three-factor interactions; use $\alpha = .01$. State the alternatives, decision rule, and conclusion. What is the P-value of the test?

   e. Test for AB, AC, and BC interactions. For each test, use $\alpha = .01$ and state the alternatives, decision rule, and conclusion. What is the P-value of each test?

   f. Test for A, B, and C main effects. For each test, use $\alpha = .01$ and state the alternatives, decision rule, and conclusion. What is the P-value of each test?

   g. To study the nature of the available facilities and region main effects, make all pairwise comparisons for each of these two factors. Use the Bonferroni procedure with a 90 percent family confidence coefficient. State your findings.

22.33. Refer to the **SMSA** data set. The effects of region (factor A: variable 12), percent of population in central cities (factor B: variable 4), and percent of population 65 or older (factor C: variable 5) on the crime rate (variable 11 $\div$ variable 3) are to be studied. For purposes of this ANOVA study, percent of population in central cities is to be classified into two categories (40.0 percent or less, 40.1 percent or more) and percent of population 65 or older is to be classified into two categories (9.9 percent or less, 10.0 percent or more).

   a. Assemble the required data and obtain the residuals for ANOVA model (22.14).

   b. Plot the residuals against the fitted values. What does your plot suggest about the appropriateness of ANOVA model (22.14)?

   c. Prepare a normal probability plot of the residuals. Also, obtain the coefficient of correlation between the ordered residuals and their expected values under normality. Does the normality assumption appear reasonable here?

22.34. Refer to the **SMSA** data set and Project 22.33. Assume that fixed effects ANOVA model (22.14) with $m = 1, \ldots, n_{ijk}$ is appropriate.

   a. Plot the estimated treatment means $\bar{Y}_{ijk.}$ in the format of Figure (22.6b). Does it appear that any factor effects are present?

   b. State the equivalent regression model for this case; use 1, -1, 0 indicator variables, and fit this full model.

   c. Test for three-factor interactions and AB, AC, and BC interactions. For each test, use $\alpha = .025$ and state the alternatives, reduced model, decision rule, and conclusion. What is the P-value of each test?

   d. Test for A, B, and C main effects. For each test, use $\alpha = .025$ and state the alternatives, reduced model, decision rule, and conclusion. What is the P-value of each test?

   e. To study the nature of the region main effects, make all pairwise comparisons between the region means. Use the Tukey procedure with a 95 percent family confidence coefficient. State your findings.

Chapter 23

Analysis of Covariance

Analysis of covariance is a technique that combines features of analysis of variance and regression. It can be used for either observational studies or designed experiments. The basic idea is to augment the analysis of variance model containing the factor effects with one or more additional quantitative variables that are related to the dependent variable. This augmentation is intended to reduce the variance of the error terms in the model, i.e., to make the analysis more precise. We considered covariance models briefly in Chapter 10, and noted there that they are linear models containing both qualitative and quantitative independent variables. Thus, covariance models are just a special type of regression model.

In this chapter, we shall first consider how a covariance model can be more effective than an ordinary ANOVA model. Then we shall discuss how to use a single-factor covariance model for making inferences in terms of the regression approach. This is followed by a demonstration that analysis of covariance can also be viewed as a modification of analysis of variance. We conclude the chapter by taking up analysis of covariance models for multifactor studies and some additional considerations for the use of covariance analysis.

23.1 BASIC IDEAS

How Covariance Analysis Reduces Error Variability

Covariance analysis may be helpful in reducing large error term variances that sometimes are present in analysis of variance models. Consider a study in which the effects of three different films promoting travel in a state are studied. A subject receives an initial questionnaire to elicit information about his or her attitudes toward the state. The subject is then shown a five-minute film, and immediately afterwards is questioned about the film, about desire to travel in the state, and so on.

In this type of situation, covariance analysis can be utilized. To see why it might be highly effective, consider Figure 23.1a. Here are plotted the desire-to-travel scores, obtained after each of the three promotional films was shown to a different group of five subjects. Three different symbols are used to distinguish the different treatments. It is evident from Figure 23.1a that the error terms, as shown by the scatter around the treatment means μ_i, are fairly large, indicating a large error term variance.

Suppose now that we were to utilize also the subjects' initial attitude scores. We plot in Figure 23.1b the desire-to-travel score (obtained after exposure to the film) against the initial attitude score for each of the 15 subjects. Note that the three treatment regression relations happen to be linear (this need not be so). Also note that the scatter around the treatment regression lines is much less than the scatter in Figure 23.1a around the treatment means μ_i, as a result of the desire-to-travel scores being highly linearly related to the initial attitude scores. The relatively large scatter in Figure 23.1a reflects the large error term variability that would be encountered with an analysis of variance model for this single-factor study. The smaller scatter in Figure 23.1b reflects the smaller error term variability that would be involved in an analysis of covariance model.

Covariance analysis, it is thus seen, utilizes the relationship between the dependent variable (desire-to-travel score in our example) and one or more independent quantitative variables for which observations are available (pre-study attitude score in our example) in order to reduce the error term variability and make the study a more powerful one for comparing treatment effects.

Concomitant Variables

In covariance analysis terminology, each independent quantitative variable added to the study is called a *concomitant variable*. Clearly, the choice of concomitant variables is an important one. If such variables have no relation to the dependent variable, nothing is to be gained by covariance analysis, and one might as well use a simpler analysis of variance model. Concomitant variables frequently used with human subjects include pre-study attitudes, age, IQ, and aptitude. When stores are used as study units, concomitant variables might be last period's sales and number of employees.

Choice of Concomitant Variables. There is one problem in selecting concomitant variables that is unique to covariance analysis. For clear interpretations of the results, a concomitant variable should be observed before the study; or if observed during the study, it should not be influenced by the treatments in any way. A pre-study attitude score meets this requirement. Also, if a subject's age is ascertained during the study, it would be reasonable in many instances to expect that the response about age will not be affected by the treatment. The reason for this requirement can be readily seen from the following example. A company was conducting a training school for engineers to teach them accounting and budgeting principles.

FIGURE 23.1   Illustration of Error Variability Reduction by Covariance Analysis

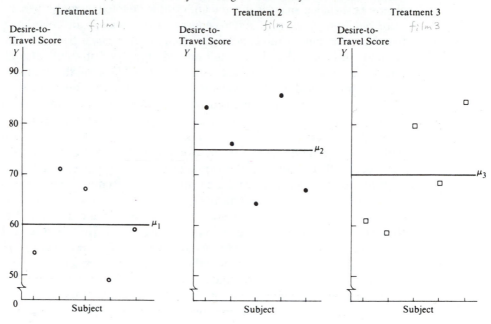

(a) Error Variability with Single-Factor Analysis of Variance Model

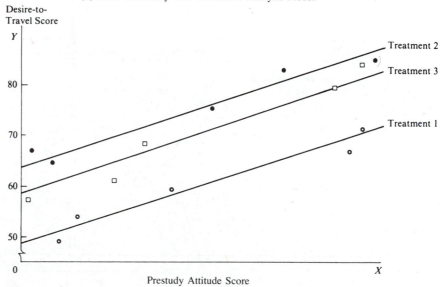

(b) Error Variability with Covariance Analysis Model

Two treatments

Two teaching methods were used, and engineers were assigned at random to one of the two. At the end of the program, a score was obtained for each engineer reflecting the amount of learning. The analyst decided to use as a concomitant variable in covariance analysis the amount of time devoted to study (which the engineers were required to record). After conducting the analysis of covariance, the analyst found that training method had virtually no effect. He was baffled by this until it was pointed out to him that the amount of study time probably was also affected by the treatments, and analysis indeed confirmed this. One of the training methods involved computer-assisted learning which appealed to the engineers so that they spent more study time and also learned more. In other words, both learning score and amount of study time were dependent upon the treatment in this case.

When a concomitant variable is affected by the treatments, covariance analysis will remove some (or much) of the effect that the treatments had on the dependent variable, so that an uncritical analysis may be badly misleading. Great care should be taken in the analysis when the covariance approach is employed with a concomitant variable affected by treatments.

Figure 23.2 shows a scatter plot of learning score and amount of study time for the experiment involving the training of engineers. Treatment 1 is the one using computer-assisted learning. Note that most persons with this treatment devoted large amounts of time to study. On the other hand, persons receiving treatment 2 tended to devote smaller amounts of time to study. As a result, the observations for the two treatments tend to be bunched over different intervals on the X scale.

Contrast this situation with the one seen in Figure 23.1b for the study on promotional films. Figure 23.1b illustrates how the concomitant variable observations should be scattered if treatments have no effect on the concomitant variable. Here,

FIGURE 23.2   Illustration of Treatments Affecting the Concomitant Variable

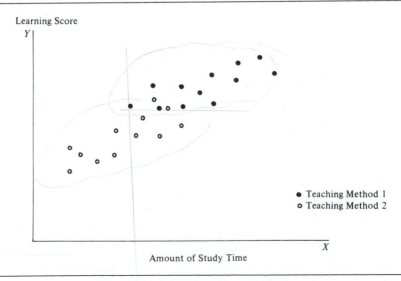

learning score and time measurement (concomitant variable) were both affected by treatments.

the distribution of subjects along the X scale by pre-study attitude scores is roughly similar for all treatments, subject only to chance variation.

23.2 SINGLE-FACTOR COVARIANCE MODEL

The covariance models to be presented in this chapter are applicable to observational studies and to experimental studies based on a completely randomized design. In the earlier example on learning by engineers, the 24 engineers participating in the study were randomly assigned to the two teaching methods, with 12 engineers assigned to each teaching method. Thus, this experimental study was based on a completely randomized design.

The covariance models to be taken up in this chapter are also applicable to observational studies, such as an investigation of the salary increases of a company's employees in the accounting department by gender, where age is utilized as a concomitant variable.

Notation

We shall employ the notation for single-factor analysis of variance. The number of cases for the ith factor level is denoted by n_i, the total number of cases by $n_T = \Sigma n_i$, and the jth observation on the dependent variable for the ith factor level is denoted by Y_{ij}. We shall initially consider a single-factor covariance model with only one concomitant variable. Later we shall take up models with more than one concomitant variable. We shall denote the value of the concomitant variable associated with the jth case for the ith factor level by X_{ij}.

Development of Covariance Model

The single-factor ANOVA model in terms of fixed factor effects was given in (14.60):

(23.1)
$$Y_{ij} = \mu_. + \tau_i + \varepsilon_{ij}$$

The covariance model starts with this model and simply adds another term (or several), reflecting the relationship between the concomitant and dependent variables. Usually, a linear relation is utilized as a first approximation:

(23.2)
$$Y_{ij} = \mu_. + \tau_i + \gamma X_{ij} + \varepsilon_{ij}$$

Here γ is a regression coefficient for the relation between Y and X. The constant $\mu_.$ now is no longer an overall mean. We can, however, make this constant an overall mean, and incidentally simplify some computations, if we express the concomitant variable as a deviation from the overall mean $\bar{X}_{..}$. The resulting model is the usual covariance model for a single-factor study with fixed factor levels:

(23.3)
$$Y_{ij} = \mu_. + \tau_i + \gamma(X_{ij} - \bar{X}_{..}) + \varepsilon_{ij}$$

where:

> $\mu_.$ is an overall mean
> τ_i are the fixed treatment effects subject to the restriction $\Sigma\tau_i = 0$
> γ is a regression coefficient for the relation between Y and X
> X_{ij} are constants
> ε_{ij} are independent $N(0, \sigma^2)$
> $i = 1, \ldots, r; j = 1, \ldots, n_i$ ~ r treatments , $\sum_{i=1}^{r} n_i$ observations

Covariance model (23.3) corresponds to ANOVA model (23.1) except for the added term $\gamma(X_{ij} - \bar{X}_{..})$ to reflect the relationship between Y and X. Note that the concomitant observations X_{ij} are assumed to be constants. Since ε_{ij} is the only random variable on the right side of (23.3), it follows at once that:

(23.4a) $$E\{Y_{ij}\} = \mu_. + \tau_i + \gamma(X_{ij} - \bar{X}_{..})$$

(23.4b) $$\sigma^2\{Y_{ij}\} = \sigma^2$$

In view of the independence of the ε_{ij}, the Y_{ij} are also independent. Hence, an alternative statement of covariance model (23.3) is:

(23.5) $$Y_{ij} \text{ are independent } N(\mu_{ij}, \sigma^2)$$

where:

> $\mu_{ij} = \mu_. + \tau_i + \gamma(X_{ij} - \bar{X}_{..})$
> $\Sigma\tau_i = 0$

Properties of Covariance Model

Some of the properties of covariance model (23.3) are identical to those of ANOVA model (23.1). For instance, the error terms ε_{ij} are independent and have constant variance. There are also some new properties, and we discuss these now.

Comparisons of Treatment Effects. With the analysis of variance model, all observations for the ith treatment have the same mean response (μ_i). This is not so with the covariance model, since the mean response here depends not only on the treatment τ_i but also on the value of the concomitant variable X_{ij} for the study unit. Thus, the expected response for the ith treatment with covariance model (23.3) is given by a regression line:

(23.6) $$E(Y_{ij}) = \mu_{ij} = \mu_. + \tau_i + \gamma(X_{ij} - \bar{X}_{..}) \neq \mu_i \text{ in ANOVA}$$

which indicates, for any value of X, the mean response with treatment i. Figure 23.3 illustrates for a study with three treatments how these treatment regression lines might appear. Note that $\mu_. + \tau_i$ is the ordinate of the line for the ith treatment when $X - \bar{X}_{..} = 0$, that is, when $X = \bar{X}_{..}$, and that γ is the slope of each line. Since all treatment regression lines have the same slope, they are parallel.

While we no longer can speak of *the* mean response with the ith treatment since it varies with X, we can still measure the effect of any treatment compared with any

FIGURE 23.3 Treatment Regression Lines with Covariance Model (23.3)

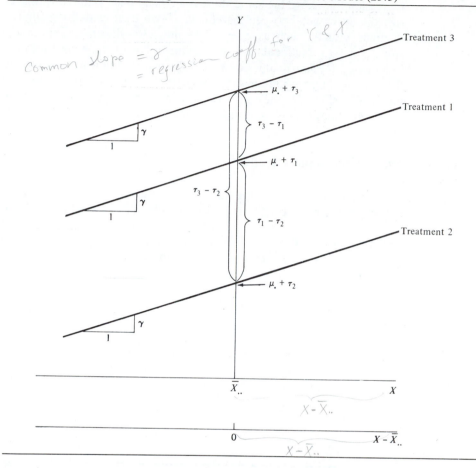

other by a single number. In Figure 23.3, for instance, treatment 1 leads to a higher mean response than treatment 2, no matter what is the value of X. The difference between the two mean responses is the same for all values of X, since the slopes of the regression lines are equal. Hence, we can measure the difference at any convenient X, say, at $X = \bar{X}_{..}$:

$$(23.7) \qquad \mu_{.} + \tau_1 - (\mu_{.} + \tau_2) = \tau_1 - \tau_2$$

Thus, $\tau_1 - \tau_2$ measures how much higher the mean response is with treatment 1 than with treatment 2 for any value of X. We can compare any other two treatments similarly. It follows directly from this discussion that when all treatments have the same mean responses for each X (i.e., the treatments have no differential effects), the treatment regression lines must be identical; and hence, $\tau_1 - \tau_2 = 0$, $\tau_1 - \tau_3 = 0$, etc. Indeed, all τ_i equal zero in that case.

Constancy of Slopes.   The assumption in covariance model (23.3) that all treatment regression lines have the same slope is a crucial one. Without it, one cannot summarize the difference between the effects of two treatments by a single number based on main effects, such as $\tau_2 - \tau_1$. Figure 23.4 illustrates the case of nonparallel slopes for two treatments. Here, treatment 1 leads to higher mean responses than treatment 2 for some values of X, and the reverse holds for other values of X. When the treatment regression lines interact with the concomitant variable X in the form of nonparallel slopes, covariance analysis is not appropriate. Instead, separate treatment regression lines should be estimated and then compared.

Generalizations of Covariance Model

Covariance model (23.3) for single-factor studies can be generalized in several respects. We mention briefly three ways in which this model can be generalized.

Nonconstant Xs.   Covariance model (23.3) assumes that the observations X_{ij} on the concomitant variable are constants. At times, it might be more reasonable to consider the concomitant observations as random variables. In that case, if one is willing to interpret covariance model (23.3) as a conditional one, applying for any X values that might be observed, the covariance analysis to be presented is still appropriate.

Nonlinearity of Relation.   The linear relation between Y and X assumed in covariance model (23.3) is not essential to covariance analysis. Any other relation could be used. For instance, the model would be as follows for a quadratic relation:

$$(23.8) \qquad Y_{ij} = \mu_. + \tau_i + \gamma_1(X_{ij} - \bar{X}_{..}) + \gamma_2(X_{ij} - \bar{X}_{..})^2 + \varepsilon_{ij}$$

FIGURE 23.4   Nonparallel Treatment Regression Lines

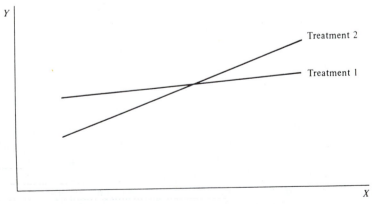

Linearity of the relation leads to simpler analysis and is often a sufficiently good approximation to provide meaningful results. If a linear relation is not a good approximation, however, a more adequate description of the relation should be utilized in the covariance model.

Several Concomitant Variables.   Covariance model (23.3) uses a single concomitant variable. This is often sufficient to reduce the error variability substantially. However, the model can be extended in a straightforward fashion to include two or more concomitant variables. The single-factor covariance model for two concomitant variables, X_1 and X_2, to the first order would be:

$$(23.9) \qquad Y_{ij} = \mu_. + \tau_i + \gamma_1(X_{ij1} - \bar{X}_{..1}) + \gamma_2(X_{ij2} - \bar{X}_{..2}) + \varepsilon_{ij}$$

X_{ij} associated with jth case for the ith factor level.

$\underline{x}'\underline{x}$ where $\underline{x} = \begin{pmatrix} \delta_1 \\ \delta_2 \end{pmatrix}$

$\underline{x} = \begin{pmatrix} X_{ij1} - \bar{X}_{..1} \\ X_{ij2} - \bar{X}_{..2} \end{pmatrix}$

Regression Formulation of Covariance Model

An easy way to estimate the parameters of covariance model (23.3) and make inferences is through the regression approach. (If the reader has not yet read Sections 10.1–10.4, he or she should do so before proceeding further.) *p. 349*

As for analysis of variance models, we shall employ $r - 1$ indicator variables taking on the values 1, -1, or 0 to represent the r treatments:

$$(23.10)$$

$$I_1 = \begin{cases} 1 & \text{if case from treatment 1} \\ - 1 & \text{if case from treatment } r \\ 0 & \text{otherwise} \end{cases}$$

$$\vdots \qquad \qquad \vdots$$

$$I_{r-1} = \begin{cases} 1 & \text{if case from treatment } r - 1 \\ - 1 & \text{if case from treatment } r \\ 0 & \text{otherwise} \end{cases}$$

Note that we now denote the indicator variables by the symbol I to clearly distinguish the treatment effects from the concomitant variable X.

In expressing covariance model (23.3) in regression form, we shall, as in the regression chapters, denote the deviation $X_{ij} - \bar{X}_{..}$ by x_{ij}. Covariance model (23.3) can then be expressed as follows:

$$(23.11) \qquad Y_{ij} = \mu_. + \tau_1 I_{ij1} + \cdots + \tau_{r-1} I_{ij,r-1} + \gamma x_{ij} + \varepsilon_{ij}$$

for $i = 1$

$I_1 = I_2 = \cdots = I_{r-1} = -1$

where:

$$x_{ij} = X_{ij} - \bar{X}_{..}$$

$\sum \tau_i = 0 \;\Rightarrow\; \tau_r = -\sum_{i=1}^{r-1} \tau_i$

$\sum_{i=1}^{r-1} \tau_i I_i = -\sum_{i=1}^{r-1} \tau_i$

$= \tau_r$

Here, I_{ij1} is the value of indicator variable I_1 for the jth case from treatment i, and similarly for the other indicator variables. Note that the treatment effects $\tau_1, \ldots, \tau_{r-1}$ are the regression coefficients for the indicator variables.

Now that we have formulated covariance model (23.3) as a regression model, the earlier discussion of regression analysis applies. We shall therefore consider only briefly how to examine the aptness of the covariance model and how to make relevant inferences and then turn to an example to illustrate the procedures.

Aptness of Covariance Model

Some of the key issues concerning the aptness of covariance model (23.3) and the equivalent regression model (23.11) deal with:

1. Normality of error terms.
2. Equality of error variances for different treatments.
3. Equality of slopes of the different treatment regression lines.
4. Linearity of regression relation with concomitant variable.
5. Uncorrelatedness of error terms.

Only the issue about the equality of the slopes of the different treatment regression lines is new in evaluating the aptness of the model. We did discuss in Section 10.4 how to compare several regression lines, and this discussion is applicable for testing whether the condition of equal slopes in the covariance model is met. We shall illustrate this test in the example in Section 23.3.

Inferences of Interest

The key statistical inferences of interest in covariance analysis are the same as with analysis of variance models, namely, whether the treatments have any effects, and if so what these effects are. Testing for fixed treatment effects involves the same alternatives as for analysis of variance models:

(23.12)
$$H_0: \tau_1 = \tau_2 = \cdots = \tau_r = 0$$
$$H_a: \text{not all } \tau_i \text{ equal zero}$$

As we can see by referring to the equivalent regression model (23.11), this test involves simply testing whether several regression coefficients equal zero. The appropriate test statistic therefore is (8.25).

If the treatment effects differ, one usually wishes to investigate the nature of these effects. Pairwise comparisons of treatment effects $\tau_i - \tau_{i'}$ (the vertical distance between the two treatment regression lines) may be of interest, or more general contrasts of the τ_i may be relevant. In either case, linear combinations of the regression coefficients $\tau_1, \ldots, \tau_{r-1}$ are to be estimated.

Occasionally, the nature of the regression relationship between Y and X is of interest, but usually the concomitant variable X is only employed to help reduce the error variability.

Note

One is usually not concerned in covariance analysis with whether the regression coefficient γ is zero, that is, whether there is indeed a regression relation between Y and X. If there is no relation, no bias results in the covariance analysis. The error mean square would simply be the same as for the analysis of variance model (allowing for sampling variation), and one degree of freedom for the error mean square would be lost.

When one is concerned equally with the concomitant variable and the treatment effects, then the methods presented in Chapter 10 should be employed for the analysis pertaining to the concomitant variable.

23.3 EXAMPLE OF SINGLE-FACTOR COVARIANCE ANALYSIS

A company wished to study the effects of three different types of promotions on sales of its crackers. The three promotions were:

Treatment 1—Sampling of product by customers in store and regular shelf space
Treatment 2—Additional shelf space in regular location
Treatment 3—Special display shelves at ends of aisle in addition to regular shelf space

Fifteen stores were selected for the study and a completely randomized experimental design was utilized. Each store was randomly assigned one of the promotion types, with five stores assigned to each type of promotion. Other relevant conditions under the control of the company, such as price and advertising, were kept the same for all stores in the study. Data on the number of cases of the product sold during the promotional period, denoted by Y, are presented in Table 23.1a, as are also data on

TABLE 23.1 Data for Cracker Promotion Example (number of cases sold)

(a) Data

Treatment i	Store (j) 1		2		3		4		5	
	Y_{i1}	X_{i1}	Y_{i2}	X_{i2}	Y_{i3}	X_{i3}	Y_{i4}	X_{i4}	Y_{i5}	X_{i5}
1	38	21	39	26	36	22	45	28	33	19
2	43	34	38	26	38	29	27	18	34	25
3	24	23	32	29	31	30	21	16	28	29

(b) Means, Totals, and Sums of Squares and Products

$\bar{Y}_{1.} = 38.2$ $\bar{Y}_{2.} = 36.0$ $\bar{Y}_{3.} = 27.2$ $\bar{Y}_{..} = 33.8$
$\bar{X}_{1.} = 23.2$ $\bar{X}_{2.} = 26.4$ $\bar{X}_{3.} = 25.4$ $\bar{X}_{..} = 25.0$

Treatment i	$\sum_j Y_{ij}$	$\sum_j Y_{ij}^2$	$\sum_j X_{ij}$	$\sum_j X_{ij}^2$	$\sum_j X_{ij} Y_{ij}$
1	191	7,375	116	2,746	4,491
2	180	6,622	132	3,622	4,888
3	136	3,786	127	3,367	3,558
Total	507	17,783	375	9,735	12,937

the sales of the product in the preceding period, denoted by X. Sales of the preceding period are to be used as the concomitant variable.

Development of Model

Figure 23.5 presents the data of Table 23.1a in the form of a scatter plot. Linear regression and parallel slopes for the treatment regression lines appear to be reasonable. Therefore, the following regression model was tentatively selected:

$$(23.13) \qquad Y_{ij} = \mu. + \tau_1 I_{ij1} + \tau_2 I_{ij2} + \gamma x_{ij} + \varepsilon_{ij} \qquad \text{Full model}$$

where:

$$I_{ij1} = \begin{cases} 1 & \text{if store received treatment 1} \\ -1 & \text{if store received treatment 3} \\ 0 & \text{otherwise} \end{cases}$$

$$I_{ij2} = \begin{cases} 1 & \text{if store received treatment 2} \\ -1 & \text{if store received treatment 3} \\ 0 & \text{otherwise} \end{cases}$$

$$x_{ij} = X_{ij} - \bar{X}_{..}$$

FIGURE 23.5   Scatter Plot of Cracker Sales—Cracker Promotion Example

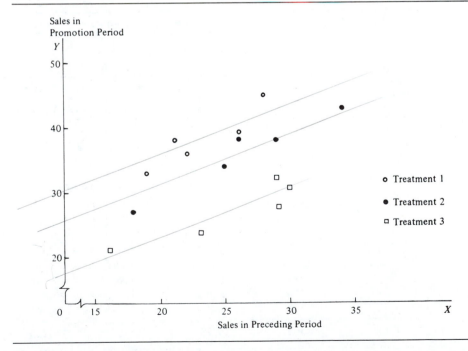

The **Y** observations vector and the **X** matrix for the data in Table 23.1a are given in Table 23.2. A computer run of a multiple regression package led to the results summarized in Table 23.3.

Various residual plots were then obtained to examine the aptness of regression model (23.13). Figure 23.6 contains two of these. Figure 23.6a contains residual dot plots for the three treatments. These do not suggest any major differences in the variances of the error terms. Figure 23.6b contains a normal probability plot of the residuals, which shows some modest departure from linearity. However, the coefficient of correlation between the ordered residuals and their expected values under normality is .958, which does not suggest any significant departure from normality. The analyst also conducted a test for the equality of the slopes of the three treatment regression lines. This test will be described shortly. On the basis of these analyses, the analyst concluded that regression model (23.13) is appropriate here.

Test for Treatment Effects

To test whether or not the three cracker promotions differ in effectiveness, we can either follow the general linear test approach of fitting full and reduced models and using test statistic (3.69) or use extra sums of squares and test statistic (8.25). In either case, the alternatives are:

$$p.283$$

(23.14)
$$H_0: \tau_1 = \tau_2 = 0$$

$$H_a: \text{not both } \tau_1 \text{ and } \tau_2 \text{ equal zero}$$

Note that $\tau_3 = -\tau_1 - \tau_2$ must equal zero when $\tau_1 = \tau_2 = 0$.

TABLE 23.2 Data Matrices for Cracker Promotion Example—Covariance Model (23.13)

$$
\mathbf{Y} = \begin{bmatrix} 38 \\ 39 \\ 36 \\ 45 \\ 33 \\ 43 \\ 38 \\ 38 \\ 27 \\ 34 \\ 24 \\ 32 \\ 31 \\ 21 \\ 28 \end{bmatrix}
\quad
\mathbf{X} = \begin{bmatrix}
\mu & I_1 & I_2 & & x & \\
1 & 1 & 0 & 21 - 25 = & -4 \\
1 & 1 & 0 & 26 - 25 = & 1 \\
1 & 1 & 0 & 22 - 25 = & -3 \\
1 & 1 & 0 & 28 - 25 = & 3 \\
1 & 1 & 0 & 19 - 25 = & -6 \\
1 & 0 & 1 & 34 - 25 = & 9 \\
1 & 0 & 1 & 26 - 25 = & 1 \\
1 & 0 & 1 & 29 - 25 = & 4 \\
1 & 0 & 1 & 18 - 25 = & -7 \\
1 & 0 & 1 & 25 - 25 = & 0 \\
1 & -1 & -1 & 23 - 25 = & -2 \\
1 & -1 & -1 & 29 - 25 = & 4 \\
1 & -1 & -1 & 30 - 25 = & 5 \\
1 & -1 & -1 & 16 - 25 = & -9 \\
1 & -1 & -1 & 29 - 25 = & 4
\end{bmatrix}
$$

TABLE 23.3 Computer Output for Cracker Promotion Example— Covariance Model (23.13)

(a) Regression Coefficients

$\hat{\mu}. = 33.800 \qquad \hat{\tau}_2 = .942$

$\hat{\tau}_1 = 6.017 \qquad \hat{\gamma} = .899$

Full model

(b) Analysis of Variance

$1+1-1 = 3+1-1 = 3 = r$

Source of Variation	SS	df	MS
Regression	$SSR = 607.829$	3	$MSR = 202.610$
Error	$SSE = 38.571$	11	$MSE = 3.506$
Total	$SSTO = 646.400$	$14 = \sum_{i=1}^{r} n_i - 1$	

(c) Estimated Variance-Covariance Matrix of Regression Coefficients

$$\begin{array}{c} \hat{\mu}. \\ \hat{\tau}_1 \\ \hat{\tau}_2 \\ \hat{\gamma} \end{array} \begin{bmatrix} .2338 & & & \\ 0 & .5016 & & \\ 0 & -.2603 & .4882 & \\ 0 & .0189 & -.0147 & .0105 \end{bmatrix}$$

with column headers $\hat{\mu}. \quad \hat{\tau}_1 \quad \hat{\tau}_2 \quad \hat{\gamma}$

FIGURE 23.6 Diagnostic Residual Plots—Cracker Promotion Example (Minitab, Ref. 23.1)

(a) Residual Dot Plots

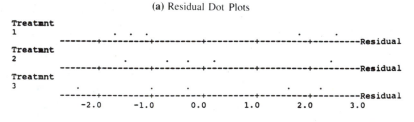

(b) Normal Probability Plot

Obs.

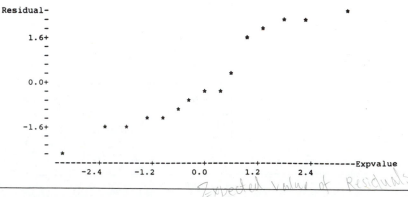

Expected value of Residuals

We shall conduct the test by means of the general linear test approach. First, we develop the reduced model under H_0:

(23.15)
$$Y_{ij} = \mu_. + \gamma x_{ij} + \varepsilon_{ij}$$
Reduced model

Model (23.15) is just a simple linear regression model where none of the parameters vary for the different treatments. The data matrices for this reduced model are shown in Table 23.4a and the analysis of variance results are presented in Table 23.4b.

We see from Table 23.4b that $SSE(R) = 455.722$ and from Table 23.3b that $SSE(F) = 38.571$. Hence, test statistic (3.69) here is:

$$F^* = \frac{SSE(R) - SSE(F)}{(n_T - 2) - [n_T - (r + 1)]} \div \frac{SSE(F)}{n_T - r - 1}$$

$$= \frac{455.722 - 38.571}{13 - 11} \div \frac{38.571}{11} = 59.5$$

The level of significance is to be controlled at $\alpha = .05$; hence, we require $F(.95; 2, 11) = 3.98$. The decision rule therefore is:

TABLE 23.4   Data Matrices and Regression Results for Cracker Promotion Example—Reduced Model (23.15)

(a) Data Matrices

$$
Y = \begin{bmatrix} 38 \\ 39 \\ 36 \\ 45 \\ 33 \\ 43 \\ 38 \\ 38 \\ 27 \\ 34 \\ 24 \\ 32 \\ 31 \\ 21 \\ 28 \end{bmatrix}
\qquad
X = \begin{bmatrix} 1 & -4 \\ 1 & 1 \\ 1 & -3 \\ 1 & 3 \\ 1 & -6 \\ 1 & 9 \\ 1 & 1 \\ 1 & 4 \\ 1 & -7 \\ 1 & 0 \\ 1 & -2 \\ 1 & 4 \\ 1 & 5 \\ 1 & -9 \\ 1 & 4 \end{bmatrix}
$$

(b) Analysis of Variance

Source of Variation	SS	df
Regression	$SSR = 190.678$	1
Error	$SSE = 455.722$	13
Total	$SSTO = 646.400$	14

$$\text{If } F^* \leq 3.98, \text{ conclude } H_0$$

$$\text{If } F^* > 3.98, \text{ conclude } H_a$$

Since $F^* = 59.5 > 3.98$, we conclude H_a, that the three cracker promotions differ in sales effectiveness. The P-value of the test is $0+$.

Note

Occasionally, a test whether or not $\gamma = 0$ is of interest. This is simply the ordinary test whether or not a single regression coefficient equals zero. It can be conducted by means of the t^* test statistic (8.23) or by means of the F^* test statistic (8.22).

Estimation of Treatment Effects

Since treatment effects were found to be present in the cracker promotion study, the analyst wished to investigate them further. We noted earlier that a comparison of two treatments involves $\tau_i - \tau_{i'}$, the vertical distance between the two treatment regression lines. Using the fact that $\tau_3 = -\tau_1 - \tau_2$ and theorem (1.27b) for variances, we know at once that the estimators of all pairwise comparisons and their variances are as follows:

Comparison	Estimator	Variance
$\tau_1 - \tau_2$	$\hat{\tau}_1 - \hat{\tau}_2$	$\sigma^2\{\hat{\tau}_1\} + \sigma^2\{\hat{\tau}_2\} - 2\sigma\{\hat{\tau}_1, \hat{\tau}_2\}$
(23.16) $\tau_1 - \tau_3 = 2\tau_1 + \tau_2$	$2\hat{\tau}_1 + \hat{\tau}_2$	$4\sigma^2\{\hat{\tau}_1\} + \sigma^2\{\hat{\tau}_2\} + 4\sigma\{\hat{\tau}_1, \hat{\tau}_2\}$
$\tau_2 - \tau_3 = \tau_1 + 2\tau_2$	$\hat{\tau}_1 + 2\hat{\tau}_2$	$\sigma^2\{\hat{\tau}_1\} + 4\sigma^2\{\hat{\tau}_2\} + 4\sigma\{\hat{\tau}_1, \hat{\tau}_2\}$

Table 23.3a furnishes us the needed estimated regression coefficients and Table 23.3c provides their estimated variances and covariances. We obtain from there:

Comparison	Estimate	Estimated Variance
$\tau_1 - \tau_2$	$6.017 - .942$ $= 5.075$	$.5016 + .4882 - 2(-.2603)$ $= 1.5104$
(23.16a) $\tau_1 - \tau_3$	$2(6.017) + .942$ $= 12.976$	$4(.5016) + .4882 + 4(-.2603)$ $= 1.4534$
$\tau_2 - \tau_3$	$6.017 + 2(.942)$ $= 7.901$	$.5016 + 4(.4882) + 4(-.2603)$ $= 1.4132$

When a single interval estimate is to be constructed, the t distribution with $n_T - r - 1$ degrees of freedom is used. (The degrees of freedom are those associated with MSE in the full covariance model.) Usually, however, a family of interval estimates is desired. In that case, the Scheffé multiple comparison procedure may be employed with the S multiple defined by:

$$(23.17) \qquad S^2 = (r - 1)F(1 - \alpha; r - 1, n_T - r - 1)$$

or the Bonferroni method may be employed with the B multiple:

(23.18) $$B = t(1 - \alpha/2g; n_T - r - 1)$$

where g is the number of statements in the family. The Tukey method is not appropriate for covariance analysis.

In the case at hand, the analyst wished to obtain all pairwise comparisons with a 95 percent family confidence coefficient. The analyst used the Scheffé procedure because he anticipated making some additional estimates of contrasts. He required therefore:

$$S^2 = (3 - 1)F(.95; 2, 11) = 2(3.98) = 7.96 \qquad S = 2.82$$

Using the results in (23.16a), the confidence intervals for all pairwise treatment comparisons with a 95 percent family confidence coefficient then are:

$$1.61 = 5.075 - 2.82\sqrt{1.5104} \le \tau_1 - \tau_2 \le 5.075 + 2.82\sqrt{1.5104} = 8.54$$

$$9.58 = 12.976 - 2.82\sqrt{1.4534} \le \tau_1 - \tau_3 \le 12.976 + 2.82\sqrt{1.4534} = 16.38$$

$$4.55 = 7.901 - 2.82\sqrt{1.4132} \le \tau_2 - \tau_3 \le 7.901 + 2.82\sqrt{1.4132} = 11.25$$

These results indicate clearly that sampling in store (treatment 1) is significantly better for stimulating cracker sales than either of the two shelf promotions, and that increasing the regular shelf space (treatment 2) is superior to additional displays at the end of the aisle (treatment 3).

Comments

1. Occasionally, more general contrasts among treatment effects than pairwise comparisons are desired. No new problems arise either in the use of the t distribution for a single contrast or in the use of the Scheffé or Bonferroni procedures for multiple comparisons. For instance, if the analyst had desired in the cracker promotion example to compare the treatment effect for sampling in the store (treatment 1) with the two treatments involving shelf displays (treatments 2 and 3), he would have been interested in the contrast:

(23.19) $$L = \tau_1 - \frac{\tau_2 + \tau_3}{2}$$

The appropriate estimator is:

(23.20) $$\hat{L} = \hat{\tau}_1 - \frac{\hat{\tau}_2 + (-\hat{\tau}_1 - \hat{\tau}_2)}{2} = \frac{3}{2}\hat{\tau}_1$$

The variance of this estimator is by (1.16b):

(23.21) $$\sigma^2\{\hat{L}\} = \frac{9}{4}\sigma^2\{\hat{\tau}_1\}$$

C.I.

$$\hat{L} \pm t_{.025;11} \sqrt{\sigma^2\{\hat{L}\}}$$

$E\{Y_{ij}\}$

2. It may be of interest to estimate the mean response with the ith treatment for some value of X. Frequently $X = \bar{X}_{..}$ is considered to be a "typical" value of X. We know from Figure 23.3 that at $X = \bar{X}_{..}$, the mean response for the ith treatment is the intercept of the treatment regression line, $\mu_. + \tau_i$. An estimator of $\mu_. + \tau_i$ can be readily developed. For the cracker promotion example, we obtain the following estimators and their variances:

p.867

	Mean Response at $X = \bar{X}_{..}$	Estimator	Variance
(23.22)	$\mu_. + \tau_1$	$\hat{\mu}_. + \hat{\tau}_1$	$\sigma^2\{\hat{\mu}_.\} + \sigma^2\{\hat{\tau}_1\} + 2\sigma\{\hat{\mu}_., \hat{\tau}_1\}$
	$\mu_. + \tau_2$	$\hat{\mu}_. + \hat{\tau}_2$	$\sigma^2\{\hat{\mu}_.\} + \sigma^2\{\hat{\tau}_2\} + 2\sigma\{\hat{\mu}_., \hat{\tau}_2\}$
	$\mu_. + \tau_3$	$\hat{\mu}_. - \hat{\tau}_1 - \hat{\tau}_2$	$\sigma^2\{\hat{\mu}_.\} + \sigma^2\{\hat{\tau}_1\} + \sigma^2\{\hat{\tau}_2\}$ $- 2\sigma\{\hat{\mu}_., \hat{\tau}_1\} - 2\sigma\{\hat{\mu}_., \hat{\tau}_2\}$ $+ 2\sigma\{\hat{\tau}_1, \hat{\tau}_2\}$

Use of the results in Table 23.3 leads to the following estimates:

Treatment	Estimated Mean Response at $\bar{X}_{..}$	Estimated Variance
1	$33.800 + 6.017 = 39.817$	$.2338 + .5016 + 2(0) = .7354$
2	$33.800 + .942 = 34.742$	$.2338 + .4882 + 2(0) = .7220$
3	$33.800 - 6.017 - .942$ $= 26.841$	$.2338 + .5016 + .4882 - 2(0) - 2(0)$ $+ 2(-.2603) = .7030$

Test for Parallel Slopes

An important assumption made in covariance analysis is that all treatment regression lines have the same slope γ. The analyst who conducted the cracker promotion study, indeed, tested this assumption before proceeding with the analysis discussed earlier. We know from Chapter 10 that regression model (23.13) can be generalized to allow for different slopes for the treatments by introducing cross-product interaction terms. Specifically, interaction variables I_1x and I_2x will be required here. We shall denote the corresponding regression coefficients by β_1 and β_2. Thus, the generalized model is:

$$(23.23) \qquad Y_{ij} = \mu_. + \tau_1 I_{ij1} + \tau_2 I_{ij2} + \gamma x_{ij} + \beta_1 I_{ij1} x_{ij}$$
$$+ \beta_2 I_{ij2} x_{ij} + \varepsilon_{ij} \qquad \text{Generalized model}$$

Table 23.5a contains the data matrices for this generalized model for the cracker promotion example. Note that the $\mathbf{X}$ matrix for the generalized model differs from the $\mathbf{X}$ matrix for regression model (23.13) simply by the addition of the I_1x and I_2x columns. A fit of regression model (23.23) by a computer multiple regression package yielded the ANOVA results in Table 23.5b. The error sum of squares SSE obtained by fitting the generalized model (23.23) is the equivalent of fitting separate regression lines for each treatment and summing these error sums of squares.

To test for parallel slopes is equivalent to testing for no interactions in the generalized model (23.23):

TABLE 23.5 Data Matrices and Regression Results for Cracker Promotion Example—Generalized Model (23.23)

(a) Data Matrices

$$
Y = \begin{bmatrix} 38 \\ 39 \\ 36 \\ 45 \\ 33 \\ 43 \\ 38 \\ 38 \\ 27 \\ 34 \\ 24 \\ 32 \\ 31 \\ 21 \\ 28 \end{bmatrix}
\qquad
X = \begin{array}{c} \begin{array}{ccccc} I_1 & I_2 & x & I_1x & I_2x \end{array} \\ \begin{bmatrix} 1 & 1 & 0 & -4 & -4 & 0 \\ 1 & 1 & 0 & 1 & 1 & 0 \\ 1 & 1 & 0 & -3 & -3 & 0 \\ 1 & 1 & 0 & 3 & 3 & 0 \\ 1 & 1 & 0 & -6 & -6 & 0 \\ 1 & 0 & 1 & 9 & 0 & 9 \\ 1 & 0 & 1 & 1 & 0 & 1 \\ 1 & 0 & 1 & 4 & 0 & 4 \\ 1 & 0 & 1 & -7 & 0 & -7 \\ 1 & 0 & 1 & 0 & 0 & 0 \\ 1 & -1 & -1 & -2 & 2 & 2 \\ 1 & -1 & -1 & 4 & -4 & -4 \\ 1 & -1 & -1 & 5 & -5 & -5 \\ 1 & -1 & -1 & -9 & 9 & 9 \\ 1 & -1 & -1 & 4 & -4 & -4 \end{bmatrix} \end{array}
$$

(b) Analysis of Variance

Source of Variation	SS	df
Regression	$SSR = 614.879$	5
Error	$SSE = 31.521$	9
Total	$SSTO = 646.400$	14

$$H_0: \beta_1 = \beta_2 = 0$$

(23.24)

$$H_a: \text{not both } \beta_1 \text{ and } \beta_2 \text{ equal zero}$$

We now need to recognize that generalized model (23.23) here is the "full" model and covariance model (23.13) is now the "reduced" model. Hence, we have from Tables 23.3b and 23.5b:

$$SSE(F) = 31.521 \qquad SSE(R) = 38.571$$

Thus, test statistic (3.69) becomes here:

$$F^* = \frac{38.571 - 31.521}{11 - 9} \div \frac{31.521}{9} = 1.01$$

For a level of significance of $\alpha = .05$, we require $F(.95; 2, 9) = 4.26$. Since $F^* = 1.01 \leq 4.26$, we conclude H_0, that the three treatment regression lines have the same slope. The P-value of the test is .40.

23.4 SINGLE-FACTOR COVARIANCE ANALYSIS AS MODIFICATION OF ANALYSIS OF VARIANCE

In previous sections, we explained the fundamental ideas underlying covariance analysis by viewing the model as a regression model. We saw that the statistical questions that arise in covariance analysis fit readily in the regression framework. When analysis of covariance was first developed, however, computers did not exist and it was not feasible to do the covariance analysis computations by means of the regression approach. Instead, computational formulas were developed that exploit the fact that the indicator variables take on the values 1, -1, or 0 in certain structures. These computational formulas may appear formidable, but computations by hand via these formulas are much simpler than hand computations via the regression approach. To explain the rationale of these computational formulas, analysis of covariance is usually viewed as a process that begins with the regular analysis of variance and *adjusts* this analysis for the concomitant variables. We shall now examine this adjustment approach for single-factor covariance analysis with one concomitant variable and demonstrate its equivalence to the regression approach.

Analysis of Covariance Table

Analysis of Variance for X and XY.   The first idea to be considered when viewing covariance analysis as an adjustment of regular analysis of variance is that the decomposition of the total sum of squares in (14.27):

$$(23.25) \qquad SSTO_Y = SSTR_Y + SSE_Y$$

can also be carried out for the variable X and the product XY. Note that the subscript Y has been added in (23.25) to make clear that this decomposition refers to variable Y. It will be recalled that:

$$(23.26a) \qquad SSTO_Y = \sum_i \sum_j (Y_{ij} - \bar{Y}_{..})^2 = \sum_i \sum_j Y_{ij}^2 - \frac{Y_{..}^2}{n_T}$$

$$(23.26b) \qquad SSTR_Y = \sum_i n_i(\bar{Y}_{i.} - \bar{Y}_{..})^2 = \sum_i \frac{Y_{i.}^2}{n_i} - \frac{Y_{..}^2}{n_T}$$

$$(23.26c) \qquad SSE_Y = \sum_i \sum_j (Y_{ij} - \bar{Y}_{i.})^2 = \sum_i \sum_j Y_{ij}^2 - \sum_i \frac{Y_{i.}^2}{n_i}$$

The same analysis of variance for the variable X is:

$$(23.27a) \qquad SSTO_X = \sum_i \sum_j (X_{ij} - \bar{X}_{..})^2 = \sum_i \sum_j X_{ij}^2 - \frac{X_{..}^2}{n_T}$$

$$(23.27b) \qquad SSTR_X = \sum_i n_i(\bar{X}_{i.} - \bar{X}_{..})^2 = \sum_i \frac{X_{i.}^2}{n_i} - \frac{X_{..}^2}{n_T}$$

$$(23.27c) \qquad SSE_X = \sum_i \sum_j (X_{ij} - \bar{X}_{i.})^2 = \sum_i \sum_j X_{ij}^2 - \sum_i \frac{X_{i.}^2}{n_i}$$

where the notation for the Xs corresponds exactly to that for the Ys.

The analysis of variance for the products XY starts with the definition of *total sum of products (SPTO)*:

$$(23.28a) \quad SPTO = \sum_i \sum_j (X_{ij} - \bar{X}_{..})(Y_{ij} - \bar{Y}_{..}) = \sum_i \sum_j X_{ij} Y_{ij} - \frac{X_{..} Y_{..}}{n_T}$$

To see that the total sum of products is related to the total sum of squares for Y or X, note that if X_{ij} is replaced by Y_{ij} in (23.28a), we obtain $SSTO_Y$, and if Y_{ij} is replaced by X_{ij}, we obtain $SSTO_X$. The two components of the total sum of products are the *treatment sum of products (SPTR)* and the *error sum of products (SPE)*:

$$(23.28b) \quad SPTR = \sum_i n_i(\bar{X}_{i.} - \bar{X}_{..})(\bar{Y}_{i.} - \bar{Y}_{..}) = \sum_i \frac{X_{i.} Y_{i.}}{n_i} - \frac{X_{..} Y_{..}}{n_T}$$

$$(23.28c) \quad SPE = \sum_i \sum_j (X_{ij} - \bar{X}_{i.})(Y_{ij} - \bar{Y}_{i.}) = \sum_i \sum_j X_{ij} Y_{ij} - \sum_i \frac{X_{i.} Y_{i.}}{n_i}$$

Unlike sums of squares, *sums of products may be negative.*

Example.   Table 23.6 contains the analyses of variance for Y, X, and XY for the cracker promotion example in Table 23.1a. We illustrate two of the calculations, using the results presented in Table 23.1b:

$$SPTR = \frac{1}{5}[116(191) + 132(180) + 127(136)] - \frac{375(507)}{15}$$

$$= 12{,}637.6 - 12{,}675.0 = -37.4$$

$$SSE_X = 9{,}735 - \frac{1}{5}[(116)^2 + (132)^2 + (127)^2] = 9{,}735 - 9{,}401.8 = 333.2$$

Adjusted Analysis of Variance for Y.   We are now ready to adjust the analysis of variance for Y to obtain the analysis of covariance. The rationale of the adjustment

---

TABLE 23.6   Analyses of Variance for Y, X, and XY for Cracker Promotion Example

Source of Variation	Sums of Squares or Products			df
	Y	X	XY	
Treatments	338.8	26.8	−37.4	2
Error	307.6	333.2	299.4	12
Total	646.4	360.0	262.0	14

can best be seen from simple linear regression analysis. There we found that the error sum of squares (2.21):

(23.29) $$SSE = \sum (Y_i - \hat{Y}_i)^2 = \sum (Y_i - b_0 - b_1 X_i)^2$$

could be expressed in the algebraically equivalent form (2.24b):

(23.29a) $$SSE = \sum (Y_i - \bar{Y})^2 - \frac{[\sum (X_i - \bar{X})(Y_i - \bar{Y})]^2}{\sum (X_i - \bar{X})^2}$$

which can be rewritten as follows:

(23.29b) $$SSE = \sum (Y_i - \bar{Y})^2 - b_1 \sum (X_i - \bar{X})(Y_i - \bar{Y})$$

In terms of our analysis of variance notation (which uses double subscripts and $\bar{X}_{..}$ and $\bar{Y}_{..}$ for $\bar{X}$ and $\bar{Y}$), the regression SSE can therefore be expressed in either of the following ways:

(23.30a) $$SSE = SSTO_Y - \frac{(SPTO)^2}{SSTO_X}$$

(23.30b) $$SSE = SSTO_Y - b_1(SPTO)$$

Hence, in the analysis of covariance, the total sum of squares for Y, adjusted for the linear relation to X, would be obtained as follows:

(23.31a) $$SSTO(\text{adj.}) = SSTO_Y - \frac{(SPTO)^2}{SSTO_X}$$

where $SSTO(\text{adj.})$ stands for the *adjusted total sum of squares for Y.*
    It may then be argued by analogy that:

(23.31b) $$SSE(\text{adj.}) = SSE_Y - \frac{(SPE)^2}{SSE_X}$$

where $SSE(\text{adj.})$ stands for the *adjusted error sum of squares for Y.*
    Finally, we obtain by subtraction:

(23.31c) $$SSTR(\text{adj.}) = SSTO(\text{adj.}) - SSE(\text{adj.})$$

where $SSTR(\text{adj.})$ stands for the *adjusted treatment sum of squares for Y.* Note carefully that $SSTR(\text{adj.})$ is obtained by subtraction and not by an analogous adjustment. The reason for this will become clear later.
    Table 23.7 contains the general covariance analysis table for a single-factor study with one concomitant variable. First are presented the sums of squares and products. Then the adjusted sums of squares are given. When these are divided by the degrees of freedom, the adjusted mean squares are obtained. Note that there is one less degree of freedom for $SSTO(\text{adj.})$ and for $SSE(\text{adj.})$ than with the analysis of variance model. The reason is that the regression coefficient γ for the concomitant variable had to be estimated.

TABLE 23.7 Covariance Analysis for Single-Factor Study with One Concomitant Variable

Source of Variation	Sums of Squares or Products			df
	Y	*X*	*XY*	
Treatments	$SSTR_Y$	$SSTR_X$	$SPTR$	$r - 1$
Error	SSE_Y	SSE_X	SPE	$n_T - r$
Total	$SSTO_Y$	$SSTO_X$	$SPTO$	$n_T - 1$

Source of Variation	Adjusted SS	Adjusted df	Adjusted MS
Treatments	$SSTR$(adj.)	$r - 1$	$MSTR$(adj.)
Error	SSE(adj.)	$n_T - r - 1$	MSE(adj.)
Total	$SSTO$(adj.)	$n_T - 2$	

Example. For the cracker promotion example, we find, using (23.31) and the results in Table 23.6:

$$SSTO\text{(adj.)} = 646.4 - \frac{(262)^2}{360} = 455.722$$

$$SSE\text{(adj.)} = 307.6 - \frac{(299.4)^2}{333.2} = 38.571$$

$$SSTR\text{(adj.)} = 455.722 - 38.571 = 417.151$$

These results are presented in an analysis of covariance table in Table 23.8, which also contains the adjusted degrees of freedom and the adjusted mean squares. Note there is one less degree of freedom for the adjusted total sum of squares and for the adjusted error sum of squares than in the analysis of variance (Table 23.6).

TABLE 23.8 Covariance Analysis Table for Cracker Promotion Example

Source of Variation	Adjusted SS	Adjusted df	Adjusted MS
Treatments	417.151	2	208.576
Error	38.571	11	3.506
Total	455.722	13	

Test for Treatment Effects

The test for treatment effects:

(23.32a)

$$H_0: \tau_1 = \tau_2 = \cdots = \tau_r = 0$$

$$H_a: \text{not all } \tau_i \text{ equal zero}$$

is then based on the usual test statistic:

(23.32b)

$$F^* = \frac{MSTR(\text{adj.})}{MSE(\text{adj.})}$$

If H_0 holds, F^* follows the $F(r - 1, n_T - r - 1)$ distribution. Hence, the decision rule for controlling the level of significance at α is:

(23.32c)

If $F^* \leq F(1 - \alpha; r - 1, n_T - r - 1)$, conclude H_0

If $F^* > F(1 - \alpha; r - 1, n_T - r - 1)$, conclude H_a

Example.   For the cracker promotion example, we have from Table 23.8:

$$F^* = \frac{MSTR(\text{adj.})}{MSE(\text{adj.})} = \frac{208.576}{3.506} = 59.5$$

which, of course, is the same value as we obtained on page 875 with the regression approach. If $\alpha = .05$, we require $F(.95; 2, 11) = 3.98$. Since $F^* = 59.5 > 3.98$, we conclude that the three promotions have different effects on sales of crackers.

Reconciliation of Two Approaches

Figure 23.7 summarizes the relations between the regression and adjustment approaches to covariance analysis. We shall now explain these relations in three steps.

1. We consider first the equivalence of $SSE(R)$ with the regression approach and $SSTO(\text{adj.})$ with the adjustment approach. When we fit the reduced model (23.15) with the regression approach, noting that $x_{ij} = X_{ij} - \bar{X}_{..}$:

$$Y_{ij} = \mu_. + \gamma(X_{ij} - \bar{X}_{..}) + \varepsilon_{ij}$$

we obtain the usual least squares estimator (2.10a) for the slope γ:

(23.33)

$$g = \frac{\sum_i \sum_j (X_{ij} - \bar{X}_{..})(Y_{ij} - \bar{Y}_{..})}{\sum_i \sum_j (X_{ij} - \bar{X}_{..})^2} = \frac{SPTO}{SSTO_X}$$

and the usual error sum of squares (2.24b):

(23.34)

$$SSE(R) = \sum_i \sum_j (Y_{ij} - \bar{Y}_{..})^2 - \frac{\left[\sum_i \sum_j (X_{ij} - \bar{X}_{..})(Y_{ij} - \bar{Y}_{..})\right]^2}{\sum_i \sum_j (X_{ij} - \bar{X}_{..})^2}$$

FIGURE 23.7   Reconciliation between Regression and Adjustment Approaches
to Covariance Analysis

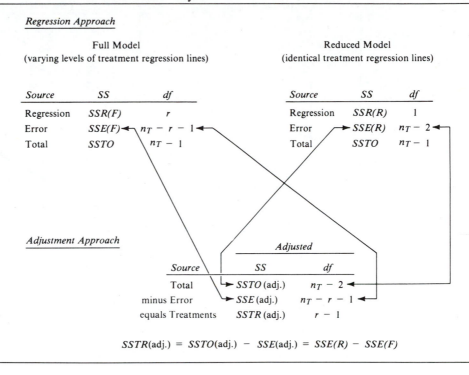

Regression Approach

	Full Model		Reduced Model	
	(varying levels of treatment regression lines)		(identical treatment regression lines)	

Source	SS	df
Regression	SSR(F)	r
Error	SSE(F)	$n_T - r - 1$
Total	SSTO	$n_T - 1$

Source	SS	df
Regression	SSR(R)	1
Error	SSE(R)	$n_T - 2$
Total	SSTO	$n_T - 1$

Adjustment Approach

	Adjusted	
Source	SS	df
Total	SSTO (adj.)	$n_T - 2$
minus Error	SSE (adj.)	$n_T - r - 1$
equals Treatments	SSTR (adj.)	$r - 1$

$$SSTR(\text{adj.}) = SSTO(\text{adj.}) - SSE(\text{adj.}) = SSE(R) - SSE(F)$$

Figure 23.8a schematically illustrates the residuals entering into $SSE(R)$ when two
treatments are in a study.

Expressed in our notation, $SSE(R)$ becomes:

$$(23.34a) \qquad SSE(R) = SSTO_Y - \frac{(SPTO)^2}{SSTO_X} = SSTO(\text{adj.})$$

according to the definition of $SSTO$ (adj.) in formula (23.31a). Thus, $SSTO$ (adj.) is
simply the error sum of squares when a linear regression is fitted to the entire set of
data.

2. We next consider the equivalence of $SSE(F)$ with the regression approach and
SSE (adj.) with the adjustment approach. When the full covariance model (23.3):

$$Y_{ij} = \mu_. + \tau_i + \gamma(X_{ij} - \bar{X}_{..}) + \varepsilon_{ij}$$

is fitted to the data, allowing different intercepts $\mu_. + \tau_i$ for the treatment regression
lines but requiring a common slope γ, it can be shown that the least squares estimator of this common slope is:

$$(23.35) \qquad g_w = \frac{\underset{i\ j}{\sum\sum}(X_{ij} - \bar{X}_{i.})(Y_{ij} - \bar{Y}_{i.})}{\underset{i\ j}{\sum\sum}(X_{ij} - \bar{X}_{i.})^2} = \frac{SPE}{SSE_X}$$

FIGURE 23.8   Schematic Representation of Residuals for $SSE(R)$ and $SSE(F)$

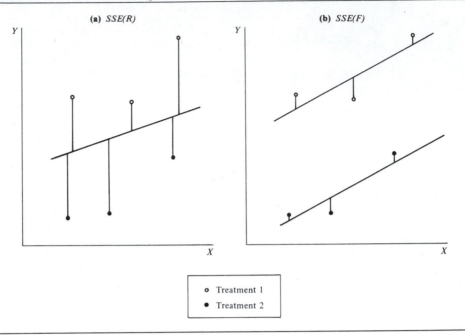

and that the least squares estimator of $\mu. + \tau_i$ is:

(23.36) $$\bar{Y}_{i.} - g_w(\bar{X}_{i.} - \bar{X}_{..})$$

Hence, the error sum of squares for the ith treatment is:

(23.37) $$\sum_j \{Y_{ij} - [\bar{Y}_{i.} - g_w(\bar{X}_{i.} - \bar{X}_{..})] - g_w(X_{ij} - \bar{X}_{..})\}^2$$

$$= \sum_j [(Y_{ij} - \bar{Y}_{i.}) - g_w(X_{ij} - \bar{X}_{i.})]^2$$

Summing these error sums of squares over all treatments, we obtain $SSE(F)$:

(23.38) $$SSE(F) = \sum_i \sum_j [(Y_{ij} - \bar{Y}_{i.}) - g_w(X_{ij} - \bar{X}_{i.})]^2$$

Figure 23.8b illustrates the residuals entering into $SSE(F)$ for the case of two treatments.

Expanding out the expression for $SSE(F)$ in (23.38) and simplifying, we obtain:

(23.38a) $$SSE(F) = \sum_i \sum_j (Y_{ij} - \bar{Y}_{i.})^2 - g_w \sum_i \sum_j (X_{ij} - \bar{X}_{i.})(Y_{ij} - \bar{Y}_{i.})$$

But in our new notation, this expression becomes:

(23.38b) $SSE(F) = SSE_Y - \dfrac{SPE}{SSE_X} SPE = SSE_Y - \dfrac{(SPE)^2}{SSE_X} = SSE(\text{adj.})$

according to the definition of $SSE(\text{adj.})$ in (23.31b). Thus, $SSE(\text{adj.})$ is simply the error sum of squares when separate regression lines, each with the same slope, are fitted to the treatments.

3. $SSTR(\text{adj.})$ is obtained as the difference $SSTO(\text{adj.}) - SSE(\text{adj.})$, just as the numerator of the test statistic with the general linear test approach is based on the difference $SSE(R) - SSE(F)$.

Comments

1. One can obtain an indication of the effectiveness of the analysis of covariance in reducing error variability by comparing $MSE(\text{adj.})$ for covariance analysis with MSE for regular analysis of variance. For the cracker promotion example, we know from Table 23.8 that $MSE(\text{adj.}) = 3.51$. We can also see from Table 23.6 that the error mean square for regular analysis of variance would have been:

$$MSE = \frac{307.6}{12} = 26.63$$

Hence, in this case covariance analysis reduced the residual variance by about 87 percent, a substantial reduction.

2. Covariance analysis and analysis of variance need not lead to the same conclusions about the treatment effects. For instance, analysis of variance might not indicate any treatment effects, whereas covariance analysis with smaller error variance could show significant treatment effects. Ordinarily, of course, one should decide in advance which of the two analyses is to be used.

3. The estimator g_w of the common slope γ can be considered as an average of the separately estimated treatment regression line slopes g_i. If we were fitting a separate regression line for each treatment, the estimated slope g_i for the ith treatment would be given by the method of least squares as follows:

(23.39)
$$g_i = \frac{\sum\limits_j (X_{ij} - \bar{X}_{i.})(Y_{ij} - \bar{Y}_{i.})}{\sum\limits_j (X_{ij} - \bar{X}_{i.})^2}$$

Using $\sum\limits_j (X_{ij} - \bar{X}_{i.})^2$ as weights, a weighted average of the g_i gives us precisely g_w as defined in (23.35):

$$\frac{\sum\limits_i \left[\sum\limits_j (X_{ij} - \bar{X}_{i.})^2 \right] g_i}{\sum\limits_i \left[\sum\limits_j (X_{ij} - \bar{X}_{i.})^2 \right]} = \frac{\sum\limits_i \sum\limits_j (X_{ij} - \bar{X}_{i.})(Y_{ij} - \bar{Y}_{i.})}{\sum\limits_i \sum\limits_j (X_{ij} - \bar{X}_{i.})^2} = g_w$$

Thus, g_w may be thought of as an average within-treatments regression slope.

For the cracker promotion example, the average within-treatments regression slope is:

$$g_w = \frac{SPE}{SSE_X} = \frac{299.4}{333.2} = .8986$$

and the slope when a single regression line is fitted to all data is:

$$g = \frac{SPTO}{SSTO_X} = \frac{262}{360} = .7278$$

Adjusted Treatment Means

In analysis of variance, the estimated treatment mean $\bar{Y}_{i.}$ is an estimate of the mean response with the ith treatment. In analysis of covariance, many writers speak of the need to *adjust* the $\bar{Y}_{i.}$ to make them comparable with respect to the concomitant variable since the X values usually will not be the same for all treatments. The adjustment takes the form:

(23.40) $$\bar{Y}_{i.}(\text{adj.}) = \bar{Y}_{i.} - g_w(\bar{X}_{i.} - \bar{X}_{..})$$

The rationale of the adjustment can be seen from Figure 23.9. This figure contains the points $(\bar{X}_{i.}, \bar{Y}_{i.})$ for the three treatments in the cracker promotion example. Through each of these points, a regression line with the average within-treatment slope $g_w = .8986$ is drawn. The adjusted treatment mean $\bar{Y}_{i.}(\text{adj.})$ is simply the ordinate of its regression line at $X = \bar{X}_{..}$. In this way, the treatments are said to be made

---

FIGURE 23.9   Representation of Adjusted Treatment Means for Cracker Promotion Example

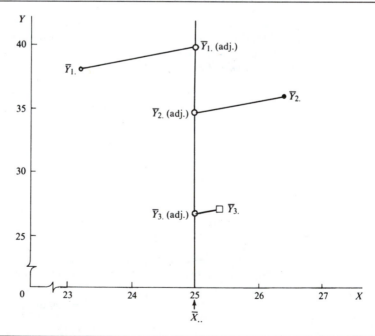

comparable with respect to X. For the cracker promotion example, the adjusted treatment means are obtained as follows:

Treatment	$\bar{Y}_{i.}$	$\bar{X}_{i.}$	$\bar{X}_{..}$	g_w	$g_w(\bar{X}_{i.} - \bar{X}_{..})$	$\bar{Y}_{i.}(\text{adj.})$
1	38.2	23.2	25	.8986	−1.62	39.82
2	36.0	26.4	25	.8986	1.26	34.74
3	27.2	25.4	25	.8986	.36	26.84

A comparison with the results on page 878 indicates that these adjusted treatment means are simply estimates of the intercepts of the treatment regression lines. In other words, $\bar{Y}_{i.}(\text{adj.})$ is simply an estimator of $\mu_. + \tau_i$.

The variance of $\bar{Y}_{i.}(\text{adj.})$ can be shown to be:

$$(23.41) \qquad \sigma^2\{\bar{Y}_{i.}(\text{adj.})\} = \sigma^2\left[\frac{1}{n_i} + \frac{(\bar{X}_{i.} - \bar{X}_{..})^2}{SSE_X}\right]$$

and an unbiased estimator of this variance is:

$$(23.42) \qquad s^2\{\bar{Y}_{i.}(\text{adj.})\} = MSE(\text{adj.})\left[\frac{1}{n_i} + \frac{(\bar{X}_{i.} - \bar{X}_{..})^2}{SSE_X}\right]$$

For instance, the estimated variance of the adjusted mean for treatment 1 in the cracker promotion example is:

$$s^2\{\bar{Y}_{1.}(\text{adj.})\} = 3.506\left[\frac{1}{5} + \frac{(23.2 - 25)^2}{333.2}\right] = .735$$

which is the same result obtained on page 878 by the regression approach.

Comparisons among Adjusted Treatment Means. Since $\bar{Y}_{i.}(\text{adj.})$ is an estimator of $\mu_. + \tau_i$, a pairwise comparison $\bar{Y}_{i.}(\text{adj.}) - \bar{Y}_{i'.}(\text{adj.})$ is an estimator of $\tau_i - \tau_{i'}$. For the cracker promotion example, we would thus estimate $\tau_1 - \tau_3$ from:

$$\bar{Y}_{1.}(\text{adj.}) - \bar{Y}_{3.}(\text{adj.}) = 39.82 - 26.84 = 12.98$$

which is the same result we obtained on page 876. The variance of the difference of two adjusted treatment means is:

$$(23.43) \qquad \sigma^2\{\bar{Y}_{i.}(\text{adj.}) - \bar{Y}_{i'.}(\text{adj.})\} = \sigma^2\left[\frac{1}{n_i} + \frac{1}{n_{i'}} + \frac{(\bar{X}_{i.} - \bar{X}_{i'.})^2}{SSE_X}\right]$$

and this is estimated by:

$$(23.44) \quad s^2\{\bar{Y}_{i.}(\text{adj.}) - \bar{Y}_{i'.}(\text{adj.})\} = MSE(\text{adj.})\left[\frac{1}{n_i} + \frac{1}{n_{i'}} + \frac{(\bar{X}_{i.} - \bar{X}_{i'.})^2}{SSE_X}\right]$$

For the above example, we obtain:

$$s^2\{\bar{Y}_{1.}(\text{adj.}) - \bar{Y}_{3.}(\text{adj.})\} = 3.506\left[\frac{1}{5} + \frac{1}{5} + \frac{(23.2 - 25.4)^2}{333.2}\right] = 1.453$$

which, of course, is the same estimated variance as was obtained via the regression approach on page 876.

The use of the t distribution for a single interval estimate of a contrast among the adjusted treatment means and of the Scheffé and Bonferroni methods for multiple comparisons is identical to their use with the regression approach. An estimator of a contrast among the adjusted treatment means is defined as follows:

$$(23.45) \quad \hat{L} = \sum c_i[\bar{Y}_{i.}(\text{adj.})] = \sum c_i \bar{Y}_{i.} - g_w \sum c_i \bar{X}_{i.} \quad \text{where } \sum c_i = 0$$

Its estimated variance is:

$$(23.46) \qquad s^2\{\hat{L}\} = MSE(\text{adj.}) \left[\sum \frac{c_i^2}{n_i} + \frac{(\sum c_i \bar{X}_{i.})^2}{SSE_X} \right]$$

Confidence limits for a single contrast L are:

$$(23.47) \qquad \hat{L} \pm t(1 - \alpha/2; n_T - r - 1)s\{\hat{L}\}$$

For multiple comparisons, the t multiple is replaced by either the S or B multiples defined in (23.17) and (23.18), respectively.

23.5 MULTIFACTOR STUDIES

We have until now considered the case of covariance analysis for single-factor studies with r treatments. Covariance analysis can also be employed with two-factor and multifactor studies. We illustrate now the use of covariance analysis for a two-factor study with one concomitant variable.

Covariance Model for Two-Factor Study

The fixed ANOVA model for a two-factor study was given in (18.23):

$$(23.48) \qquad Y_{ijk} = \mu_{..} + \alpha_i + \beta_j + (\alpha\beta)_{ij} + \varepsilon_{ijk}$$

$$i = 1, \ldots, a; j = 1, \ldots, b; k = 1, \ldots, n$$

where α_i is the main effect of factor A at the ith level, β_j is the main effect of factor B at the jth level, and $(\alpha\beta)_{ij}$ is the interaction effect when factor A is at the ith level and factor B is at the jth level. The covariance model for a two-factor study with a single concomitant variable, assuming the relation between Y and the concomitant variable X is linear, is:

$$(23.49) \qquad Y_{ijk} = \mu_{..} + \alpha_i + \beta_j + (\alpha\beta)_{ij} + \gamma(X_{ijk} - \bar{X}_{...}) + \varepsilon_{ijk}$$

$$i = 1, \ldots, a; j = 1, \ldots, b; k = 1, \ldots, n$$

Regression Approach

To illustrate the regression approach to covariance analysis for a two-factor study with one concomitant variable, suppose that both factors A and B are at two levels. The regression model counterpart to covariance model (23.49) then is:

$$(23.50) \qquad Y_{ijk} = \mu_{..} + \alpha_1 I_{ijk1} + \beta_1 I_{ijk2} + (\alpha\beta)_{11} I_{ijk1} I_{ijk2} + \gamma x_{ijk} + \varepsilon_{ijk}$$

where:

$$I_{ijk1} = \begin{array}{l} 1 \text{ if case from level 1 for factor } A \\ -1 \text{ if case from level 2 for factor } A \end{array}$$

$$I_{ijk2} = \begin{array}{l} 1 \text{ if case from level 1 for factor } B \\ -1 \text{ if case from level 2 for factor } B \end{array}$$

$$x_{ijk} = X_{ijk} - \bar{X}_{...}$$

Note that the regression coefficients in (23.50) are the analysis of variance factor effects α_1, β_1, and $(\alpha\beta)_{11}$ and the concomitant variable coefficient γ.

Testing for factor A main effects requires that $\alpha_1 = 0$ in the reduced model. Correspondingly, we have $\beta_1 = 0$ in the reduced model when testing for factor B main effects, and $(\alpha\beta)_{11} = 0$ in the reduced model when testing for AB interactions.

Estimation of factor A and factor B main effects can easily be done in terms of comparisons among the regression coefficients. The use of the Scheffé and Bonferroni multiple comparison procedures presents no new problems. For instance, the S multiple for multiple comparisons among the factor A level means is defined as follows:

$$(23.51) \qquad S^2 = (a - 1)F(1 - \alpha; a - 1, n_T - ab - 1)$$

and the B multiple would be given by (23.18) with $r = ab$.

Adjustment Approach

In applying the adjustment approach to covariance analysis for a two-factor study with one concomitant variable, we again require analyses of variance for Y, X, and XY. The analysis of variance for Y is given in (18.38) and (18.39). The analysis of variance for X is identical, with X_{ij} replacing Y_{ij}. Table 23.9 contains these analyses of variance for Y and X. Note that subscripts for Y and X are used for identification.

The analysis of variance for products XY is as follows:

$$(23.52a) \quad SPTO = \sum_i \sum_j \sum_k (X_{ijk} - \bar{X}_{...})(Y_{ijk} - \bar{Y}_{...}) = \sum_i \sum_j \sum_k X_{ijk} Y_{ijk} - \frac{X_{...} Y_{...}}{abn}$$

$$(23.52b) \quad SPA = bn \sum_i (\bar{X}_{i..} - \bar{X}_{...})(\bar{Y}_{i..} - \bar{Y}_{...}) = \frac{\sum_i X_{i..} Y_{i..}}{bn} - \frac{X_{...} Y_{...}}{abn}$$

TABLE 23.9   Analyses of Variance for Y, X, and XY—Two-Factor Study with One Concomitant Variable

Source of Variation	Sums of Squares or Products			df
	Y	X	XY	
Factor A	SSA_Y	SSA_X	SPA	$a - 1$
Factor B	SSB_Y	SSB_X	SPB	$b - 1$
AB interactions	$SSAB_Y$	$SSAB_X$	$SPAB$	$(a - 1)(b - 1)$
Error	SSE_Y	SSE_X	SPE	$ab(n - 1)$
Total	$SSTO_Y$	$SSTO_X$	$SPTO$	$abn - 1$

$$(23.52c) \qquad SPB = an \sum_j (\bar{X}_{.j.} - \bar{X}_{...})(\bar{Y}_{.j.} - \bar{Y}_{...}) = \frac{\sum_j X_{.j.} Y_{.j.}}{an} - \frac{X_{...} Y_{...}}{abn}$$

$$(23.52d) \qquad SPE = \sum_i \sum_j \sum_k (X_{ijk} - \bar{X}_{ij.})(Y_{ijk} - \bar{Y}_{ij.}) = SPTO - SPTR$$

$$(23.52e) \qquad SPAB = n \sum_i \sum_j (\bar{X}_{ij.} - \bar{X}_{i..} - \bar{X}_{.j.} + \bar{X}_{...})(\bar{Y}_{ij.} - \bar{Y}_{i..} - \bar{Y}_{.j.} + \bar{Y}_{...})$$

$$= SPTR - SPA - SPB$$

where:

$$(23.52f) \qquad SPTR = n \sum_i \sum_j (\bar{X}_{ij.} - \bar{X}_{...})(\bar{Y}_{ij.} - \bar{Y}_{...}) = \frac{\sum_i \sum_j X_{ij.} Y_{ij.}}{n} - \frac{X_{...} Y_{...}}{abn}$$

Table 23.9 also contains this analysis of variance for products XY.

The adjustment process ignores all components other than the error component and the component for which the test is to be made, and then proceeds to make adjustments in a fashion analogous to a single-factor study. Thus, to test for factor A main effects:

$$(23.53) \qquad \begin{array}{l} H_0: \alpha_1 = \alpha_2 = \cdots = \alpha_a = 0 \\ H_a: \text{not all } \alpha_i \text{ equal zero} \end{array}$$

we extract the lines for factor A and error in Table 23.9. This is done in Table 23.10a. We then derive the adjusted sums of squares in the usual fashion:

$$(23.53a) \qquad SS(A + E; \text{adj.}) = (SSA_Y + SSE_Y) - \frac{(SPA + SPE)^2}{SSA_X + SSE_X}$$

$$(23.53b) \qquad SSE(\text{adj.}) = SSE_Y - \frac{(SPE)^2}{SSE_X}$$

$$(23.53c) \qquad SSA(\text{adj.}) = SS(A + E; \text{adj.}) - SSE(\text{adj.})$$

TABLE 23.10 Covariance Analysis for Testing Factor A Main Effects in a Two-Factor Study with One Concomitant Variable

(a) Analyses of Variance

Source of Variation	Sums of Squares or Products			df
	Y	X	XY	
Factor A	SSA_Y	SSA_X	SPA	$a - 1$
Error	SSE_Y	SSE_X	SPE	$ab(n - 1)$
Sum	$SSA_Y + SSE_Y$	$SSA_X + SSE_X$	$SPA + SPE$	$a + ab(n - 1) - 1$

(b) Analysis of Covariance

Source of Variation	Adjusted SS	Adjusted df	Adjusted MS
Factor A	$SSA(\text{adj.})$	$a - 1$	$MSA(\text{adj.})$
Error	$SSE(\text{adj.})$	$ab(n - 1) - 1$	$MSE(\text{adj.})$
Sum	$SS(A + E; \text{adj.})$	$a + ab(n - 1) - 2$	

Table 23.10b contains the covariance analysis for testing factor A main effects. The degrees of freedom for error and the sum are reduced by one to take account of the concomitant variable X, and the adjusted degrees of freedom for factor A effects are obtained as a remainder.

As usual, the statistic for testing the alternatives in (23.53) is:

$$(23.54) \qquad F^* = \frac{MSA(\text{adj.})}{MSE(\text{adj.})}$$

If H_0 holds, F^* follows the $F[a - 1, ab(n - 1) - 1]$ distribution.

Tests for other factor effects are developed in similar fashion.

Example

A horticulturist conducted an experiment to study the effects of flower variety (factor A: varieties LP, WB) and moisture level (factor B: low, high) on yield of salable flowers (Y). Because the plots were not of the same size, the horticulturist wished to use plot size (X) as the concomitant variable. Six replications were made for each treatment. The data are presented in Table 23.11.

Regression model (23.50) was fitted to the data by a computer regression package. The fitted regression function is shown in Table 23.12a. The analyst plotted the data and fitted regression lines (not shown), and made a variety of residual plots and tests. On the basis of these diagnostics, she was satisfied that regression model (23.50), which assumes parallel linear regression functions and constant error variances, is suitable here.

TABLE 23.11   Data for Salable Flowers Example

Factor A (flower variety) i	Factor B (moisture level) j			
	B_1 (low)		B_2 (high)	
	Y_{i1k}	X_{i1k}	Y_{i2k}	X_{i2k}
A_1 (variety LP)	98	15	71	10
	60	4	80	12
	77	7	86	14
	80	9	82	13
	95	14	46	2
	64	5	55	3
A_2 (variety WB)	55	4	76	11
	60	5	68	10
	75	8	43	2
	65	7	47	3
	87	13	62	7
	78	11	70	9

TABLE 23.12   Computer Output of Regression Runs for Salable Flowers Example—Regression Model (23.50)

(a) Fitted Regression Function for Model (23.50)

$$\hat{Y} = 70.0 + 2.04234I_1 + 3.68078I_2 + .81922I_1I_2 + 3.27688x$$

Regression Coefficient	Estimated Regression Coefficient	Estimated Standard Deviation
α_1	2.04234	.52108
β_1	3.68078	.51291
$(\alpha\beta)_{11}$	.81922	.51291
γ	3.27688	.13002

(b) Extra Sums of Squares

Effect	Source of Variation	SS	df	MS
Concomitant variable	$x \mid I_1, I_2, I_1I_2$	3,994.52	1	3,994.52
A	$I_1 \mid x, I_2, I_1I_2$	96.60	1	96.60
B	$I_2 \mid x, I_1, I_1I_2$	323.85	1	323.85
AB	$I_1I_2 \mid x, I_1, I_2$	16.04	1	16.04
	Error	119.48	19	6.2884

The fitted regression lines for the four treatments based on the full regression model (23.50) are presented in Figure 23.10a. To study the nature of the factor effects, we show in Figure 23.10b the usual plots of estimated treatment means. These estimated means all correspond to plot size $X = \bar{X}_{...} = 8.25$ or $x = 0$. Any other plot size would yield exactly the same relationships as those in Figure 23.10b. It appears from Figure 23.10b that there are no important interactions between flower variety and moisture level, and that there may be main effects for both factors, particularly for moisture level.

To study formally the factor effects, reduced models were formed by deleting from regression model (23.50) one independent variable at a time (since both factors have two levels), and the reduced models were then fitted. The extra sums of squares so obtained, as well as the error sum of squares for the full model, are presented in Table 23.12b, together with the degrees of freedom and mean squares. No total sum of squares is shown because the factor effects components are not orthogonal.

We test first for the presence of interactions by means of the usual test statistic F^*, using the results in Table 23.12b:

$$F^* = \frac{SSR(I_1 I_2 \mid x, I_1, I_2)}{1} \div MSE = \frac{16.04}{6.2884} = 2.55$$

FIGURE 23.10   Fitted Regression Lines and Estimated Treatment Mean Plots—
Salable Flowers Example

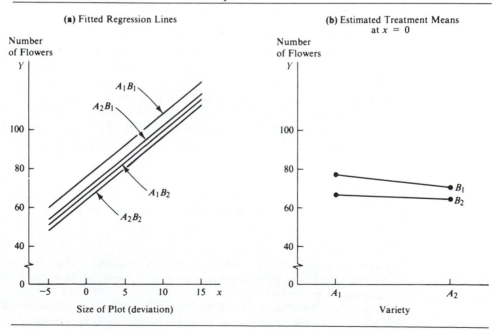

(a) Fitted Regression Lines

(b) Estimated Treatment Means at $x = 0$

For $\alpha = .01$, we require $F(.99; 1, 19) = 8.18$. Since $F^* = 2.55 \le 8.18$, we conclude that no interactions are present. The P-value of the test is .13.

We now wish to compare the factor A main effects and the factor B main effects by means of confidence intervals with a 95 percent family confidence coefficient. Since $\alpha_2 = -\alpha_1$, we have for our example:

$$D_1 = \alpha_1 - \alpha_2 = \alpha_1 - (-\alpha_1) = 2\alpha_1$$

Similarly, we obtain for the comparison of factor B main effects:

$$D_2 = 2\beta_1$$

Point estimates are readily obtained from the results in Table 23.12a:

$$\hat{D}_1 = 2\hat{\alpha}_1 = 2(2.04234) = 4.08$$

$$\hat{D}_2 = 2\hat{\beta}_1 = 2(3.68078) = 7.36$$

The estimated standard deviations also follow easily, using (1.16b):

$$s\{\hat{D}_1\} = 2s\{\hat{\alpha}_1\} = 2(.52108) = 1.042$$

$$s\{\hat{D}_2\} = 2s\{\hat{\beta}_1\} = 2(.51291) = 1.026$$

We shall utilize the Bonferroni simultaneous estimation procedure for $g = 2$ comparisons. For a 95 percent family confidence coefficient, we require $t[1 - .05/2(2); 19] = t(.9875; 19) = 2.433$. The two desired confidence intervals therefore are:

$$1.5 = 4.08 - 2.433(1.042) \le \alpha_1 - \alpha_2 \le 4.08 + 2.433(1.042) = 6.6$$

$$4.9 = 7.36 - 2.433(1.026) \le \beta_1 - \beta_2 \le 7.36 + 2.433(1.026) = 9.9$$

With family confidence coefficient .95, we conclude that variety LP yields, on the average, between 1.5 and 6.6 more salable flowers for any given plot size than variety WB. Also, for any given plot size, the mean number of salable flowers is between 4.9 and 9.9 flowers greater for the low moisture level than for the high one, thus indicating a substantial effect of moisture level on yield.

If interactions had been present, we could have studied the nature of the interaction effects by, for instance, comparing the effect of the moisture level for each of the two flower varieties. It can be shown that this comparison is given by:

$$L = (\alpha\beta)_{12} = -(\alpha\beta)_{11}$$

Hence, we could estimate the desired interaction effect by using the estimated regression coefficient $\widehat{(\alpha\beta)}_{11}$ and its estimated standard deviation in Table 23.12a.

Note

When the cell sample sizes in multifactor covariance studies are unequal, the regression approach is still applicable for testing treatment effects. However, the adjusted sums of squares formulas are no longer appropriate.

Computer packages for covariance analysis with unequal cell sample sizes should be used with great care to make sure that the package conducts the tests of interest.

23.6 ADDITIONAL CONSIDERATIONS FOR THE USE OF COVARIANCE ANALYSIS

Use of Differences

In a variety of studies, a pre-study observation X and a post-study observation Y on the same variable are available for each unit. For instance, X may be the score for a subject's attitude toward a company prior to reading its annual report, and Y may be the score after reading the report. In this situation, an obvious alternative to covariance analysis is to do an analysis of variance on the difference $Y - X$. Sometimes, $Y - X$ is called an *index of response* because it makes one observation out of two separate ones.

If the slope of the treatment regression lines is $\gamma = 1$, analysis of covariance and analysis of variance on $Y - X$ are essentially equivalent. When $\gamma = 1$, covariance model (23.2) becomes:

$$(23.55) \qquad Y_{ij} = \mu_. + \tau_i + X_{ij} + \varepsilon_{ij}$$

which can be written as a regular analysis of variance model:

$$(23.55a) \qquad Y_{ij} - X_{ij} = \mu_. + \tau_i + \varepsilon_{ij}$$

Thus, if a unit change in X leads to about the same change in Y, it makes sense to perform an analysis of variance on $Y - X$, rather than to use covariance analysis, because analysis of variance is much easier. If the regression slope is not near 1, however, covariance analysis may be substantially more effective than use of the difference $Y - X$.

In the earlier cracker promotion example, use of $Y - X$ would have been effective. It would have involved an error mean square (see Table 23.6):

$$\frac{SSE_Y + SSE_X - 2SPE}{12} = \frac{307.6 + 333.2 - 2(299.4)}{12} = 3.50$$

which is practically the same as the error mean square for covariance analysis, $MSE(\text{adj.}) = 3.51$. Recall that the regression slope in our example was close to 1 ($g_w = .8986$); hence, the approximate equivalence of the two procedures.

Correction for Bias

The suggestion is sometimes made that analysis of covariance can be helpful in correcting for bias with observational data. With such data, the groups under study may differ substantially with respect to a concomitant variable, and this may bias the comparisons of the groups. Consider, for instance, a study in which attitudes toward no-fault automobile insurance were compared for persons who are risk averse and persons who are risk seeking. It was found that many persons in the risk averse group tended to be older (50 to 70 years old), while many persons in the risk seeking group tended to be younger (20 to 40 years old). In this type of situation, some would advise that covariance analysis, with age as the concomitant variable, be employed to help remove any bias that may be in the observational data because the two age groups differ so much.

Even though there is great appeal in the idea of removing bias in observational data, covariance analysis should be used with caution for this purpose. In the first place, the adjusted means may require substantial extrapolation of the regression lines to a region where there are no or only few data points (in our example, to near 45 years). It may well be that the regression relationship used in the covariance analysis is not appropriate for substantial extrapolation. In the second place, the treatment variable may be dependent on the concomitant variable (or vice versa) which could affect the proper conclusions to be drawn.

Interest in Nature of Treatment Effects

Covariance analysis is sometimes employed for the principal purpose of shedding more light on the nature of the treatment effects, rather than merely for increasing the precision of the analysis. For instance, a market researcher in a study of the effects of three different advertisements on the maximum price consumers are willing to pay for a new type of home siding may use covariance analysis, with value of the consumer's home as the concomitant variable. The reason is because she is truly interested in the relation for each advertisement between home value and maximum price. Reduction of error variance in this instance may be a secondary consideration.

As in all regression analyses, care must be used in drawing inferences about the causal nature of the relation between the concomitant variable and the dependent variable. In the advertising example, it might well be that value of a consumer's home is largely influenced by income. If this were so, the relation between value of the consumer's home and maximum price the consumer is willing to pay may actually be largely a reflection of a more underlying relation between income and maximum price.

CITED REFERENCE

23.1. *MINITAB Reference Manual, Release 7*. State College, Pa.: Minitab, Inc., 1989.

PROBLEMS

23.1. A student's reaction to the instructor's statement that covariance analysis is inappropriate when the treatment regression lines do not have the same slope was as follows: "It seems to me that this is ducking a real-world problem. If the treatment slopes are different, just use a covariance model that allows for different treatment slopes." Evaluate this reaction.

23.2. A survey analyst remarked: "When covariance analysis is used with survey data, there is the danger that the treatments may be related to the concomitant variable." What is the nature of the problem? Does this same problem exist when the treatments are randomly assigned to the experimental units?

23.3. Portray, analogous to the format of Figure 2.7 on page 33 for a regression model, the nature of covariance model (23.3) when there are three treatments and the parameter values are: $\mu_{.} = 150$, $\tau_1 = 15$, $\tau_2 = -5$, $\tau_3 = -10$, $\gamma = 6$, $\bar{X}_{..} = 70$, $\sigma = 5$. Show several distributions of Y for each treatment.

23.4. Refer to the cracker promotion example on page 871. A student stated, in discussing this case: "Strictly speaking, you cannot conclude anything about whether the three promotions differ in effectiveness because there was no control. The preceding period does not qualify as a control because it might have differed from the promotion period due to seasonal factors or other unique circumstances." Comment.

23.5. Refer to the cracker promotion example on page 877, where three pairwise comparisons of treatment effects were made by the Scheffé procedure.

a. What would be the value of the Bonferroni multiple here for estimating the three comparisons?

b. Did the analyst obtain substantially less precise interval estimates using the Scheffé procedure, which permits him to make additional estimates without modifying the present ones?

23.6. State the analysis of covariance model for a single-factor study with four treatments when there are two concomitant variables, each with linear and quadratic terms in the model.

23.7. Refer to **Productivity improvement** Problem 14.10. The economist also has information on annual productivity improvement in the prior year and wishes to use this information as a concomitant variable. The data on the prior year's productivity improvement (X_{ij}) follow.

	j											
i	1	2	3	4	5	6	7	8	9	10	11	12
1	8.2	7.9	7.0	5.7	7.2	7.0	6.5	7.9	6.3			
2	8.8	10.0	10.7	10.0	9.7	9.4	10.6	9.8	10.0	10.3	8.9	10.0
3	11.5	12.2	12.8	11.0	12.3	12.1						

a. Obtain the residuals for covariance model (23.3).

b. For each treatment, plot the residuals against the fitted values. Also prepare a normal probability plot of the residuals and calculate the coefficient of correlation between the ordered residuals and their expected values under normality. What do you conclude from your analysis?

c. State the generalized regression model to be employed for testing whether or not the treatment regression lines have the same slope. Conduct this test using $\alpha = .01$. State the alternatives, decision rule, and conclusion. What is the P-value of the test?

d. Could you conduct a formal test here as to whether the regression functions are linear? If so, how many degrees of freedom are there for the denominator mean square in the test statistic?

23.8. Refer to **Productivity improvement** Problems 14.10 and 23.7. Assume that covariance model (23.3) is appropriate.

    a. Plot the data in the format of Figure 23.5. Does it appear that there are research and development expenditures level effects on mean productivity improvement? Discuss.

    b. State the regression model equivalent to covariance model (23.3) for this case; use $1, -1, 0$ indicator variables. Also state the reduced model for testing for treatment effects.

    c. Fit the full and reduced models and test for treatment effects; use $\alpha = .05$. State the alternatives, decision rule, and conclusion. What is the P-value of the test?

    d. Is $MSE(F)$ for the covariance model substantially smaller than MSE for the analysis of variance model in Problem 14.10c? Does this affect the conclusion reached about treatment effects? Does it affect the P-value?

    e. Estimate the mean productivity improvement for firms with moderate research and development expenditures who had a prior productivity improvement of $X = 9.0$; use a 95 percent confidence interval.

    f. Make all pairwise comparisons between the treatment effects; use either the Bonferroni or Scheffé procedure with a 90 percent family confidence coefficient, whichever is more efficient. State your findings.

23.9. Refer to **Questionnaire color** Problem 14.11. It has been suggested to the investigator that size of parking lot might be a useful concomitant variable. The number of spaces (X_{ij}) in each parking lot utilized in the study follow.

			j		
i	1	2	3	4	5
1	300	381	226	350	100
2	153	334	473	264	325
3	144	359	296	243	252

    a. Obtain the residuals for covariance model (23.3).

    b. For each treatment, plot the residuals against the fitted values. Also prepare a normal probability plot of the residuals and calculate the coefficient of correlation between the ordered residuals and their expected values under normality. What do you conclude from your analysis?

    c. State the generalized regression model to be employed for testing whether or not the treatment regression lines have the same slope. Conduct this test using $\alpha = .005$. State the alternatives, decision rule, and conclusion. What is the P-value of the test?

    d. Could you conduct a formal test here as to whether the regression functions are linear? Explain.

23.10. Refer to **Questionnaire color** Problems 14.11 and 23.9. Assume that covariance model (23.3) is applicable.

    a. Plot the data in the format of Figure 23.5. Does it appear that there are color effects on the mean response rate? Discuss.

    b. State the regression model equivalent to covariance model (23.3) for this case; use $1, -1, 0$ indicator variables. Also state the reduced model for testing for treatment effects.

c. Fit the full and reduced models and test for treatment effects; use $\alpha = .10$. State the alternatives, decision rule, and conclusion. What is the P-value of the test?

d. Is $MSE(F)$ for the covariance model substantially smaller than MSE for the analysis of variance model in Problem 14.11c? How does this affect the conclusion reached about treatment effects?

e. Estimate the mean response rate for blue questionnaires in parking lots of size $X = 280$; use a 90 percent confidence interval.

f. Make all pairwise comparisons between the treatment effects; use either the Bonferroni or Scheffé procedure with a 90 percent family confidence coefficient, whichever is more efficient. State your findings.

23.11. Refer to **Rehabilitation therapy** Problem 14.12. The rehabilitation researcher wishes to use age of patient as a concomitant variable. The ages (X_{ij}) of patients in the study follow.

i	1	2	3	4	5	6	7	8	9	10
					j					
1	18.3	30.0	26.5	28.1	29.7	27.8	19.8	29.3		
2	20.8	25.2	29.2	20.0	21.5	22.1	19.7	24.7	20.2	22.9
3	22.7	28.7	18.9	18.0	21.7	20.0				

a. Obtain the residuals for covariance model (23.3).

b. For each treatment, plot the residuals against the fitted values. Also prepare a normal probability plot of the residuals and calculate the coefficient of correlation between the ordered residuals and their expected values under normality. What do you conclude from your analysis?

c. State the generalized regression model to be employed for testing whether or not the treatment regression lines have the same slope. Conduct this test using $\alpha = .05$. State the alternatives, decision rule, and conclusion. What is the P-value of the test?

d. Could you conduct a formal test here as to whether the regression functions are linear? Explain.

23.12. Refer to **Rehabilitation therapy** Problems 14.12 and 23.11. Assume that covariance model (23.3) is applicable.

a. Plot the data in the format of Figure 23.5. Does it appear that there are effects of physical fitness status on the mean number of days required for therapy? Discuss.

b. State the regression model equivalent to covariance model (23.3) for this case; use $1, -1, 0$ indicator variables. Also state the reduced model for testing for treatment effects.

c. Fit the full and reduced models and test for treatment effects; use $\alpha = .01$. State the alternatives, decision rule, and conclusion. What is the P-value of the test?

d. Is $MSE(F)$ for the covariance model substantially smaller than MSE for the analysis of variance model in Problem 14.12c? Does this affect the conclusion reached about treatment effects? Does it affect the P-value?

e. Estimate the mean number of days required for therapy for patients of average physical fitness and age 24 years; use a 99 percent confidence interval.

f. Make all pairwise comparisons between the treatment effects; use either the Bonferroni or Scheffé procedure with a 95 percent family confidence coefficient, whichever is more efficient. State your findings.

23.13. **Product display.** A manufacturer of felt-tip markers investigated by an experiment whether a proposed new display, featuring a picture of a physician, is more effective in drugstores than the present counter display, featuring a picture of an athlete and designed to be located in the stationery area. Fifteen drugstores of similar characteristics were chosen for the study. They were assigned at random in equal numbers to one of the following three treatments: (1) present counter display in stationery area, (2) new display in stationery area, (3) new display in checkout area. Sales with the present display (X_{ij}) were recorded in all 15 stores for a three-week period. Then the new display was set up in the 10 stores receiving it, and sales for the next three-week period (Y_{ij}) were recorded in all 15 stores. The data on sales (in dollars) follow.

			j		
i	1	2	3	4	5
Treatment 1					
1st 3 weeks	92	68	74	52	65
2d 3 weeks	69	44	58	38	54
Treatment 2					
1st 3 weeks	77	80	70	73	79
2d 3 weeks	74	75	73	78	82
Treatment 3					
1st 3 weeks	64	43	81	68	71
2d 3 weeks	66	49	84	75	77

The analyst wishes to analyze the effects of the three different display treatments by means of covariance analysis.

a. Obtain the residuals for covariance model (23.3).

b. For each treatment, plot the residuals against the fitted values. Also prepare a normal probability plot of the residuals and calculate the coefficient of correlation between the ordered residuals and their expected values under normality. What do you conclude from your analysis?

c. State the generalized regression model to be employed for testing whether or not the treatment regression lines have the same slope. Conduct this test using $\alpha = .05$. State the alternatives, decision rule, and conclusion. What is the P-value of the test?

d. Could you conduct a formal test here as to whether the regression functions are linear? Explain.

23.14. Refer to **Product display** Problem 23.13. Assume that covariance model (23.3) is applicable.

a. Plot the data in the format of Figure 23.5. Does it appear that there are display effects on mean sales? Discuss.

b. State the regression model equivalent to covariance model (23.3) for this case; use $1, -1, 0$ indicator variables. Also state the reduced model for testing for treatment effects.

c. Fit the full and reduced models and test for treatment effects; use $\alpha = .05$. State the alternatives, decision rule, and conclusion. What is the P-value of the test?

d. Is $MSE(F)$ for the covariance model substantially smaller than the mean square error if analysis of variance model (14.2) had been employed?

e. Estimate the mean sales with display treatment 2 for stores whose sales in the preceding three-week period were $75; use a 95 percent confidence interval.

f. Make all pairwise comparisons between the treatment effects; use either the Bonferroni or Scheffé procedure with a 90 percent family confidence coefficient, whichever is more efficient. State your findings.

23.15. Refer to **Cash offers** Problem 18.10. An analyst wishes to use each dealer's sales volume as a concomitant variable. The sales data (X_{ijk}, in hundred thousand dollars) follow.

i = 1		i = 2		i = 3	
j = 1	j = 2	j = 1	j = 2	j = 1	j = 2
3.0	3.5	6.5	2.2	5.0	4.0
5.1	4.2	4.1	5.4	3.1	.8
1.0	2.2	2.2	3.1	3.2	1.9
4.4	3.1	3.7	4.5	3.2	2.8
2.7	1.3	3.4	3.6	3.0	2.2
4.9	6.6	3.0	5.0	2.9	1.9

a. Obtain the residuals for covariance model (23.49).

b. For each treatment, plot the residuals against the fitted values. Also prepare a normal probability plot of the residuals and calculate the coefficient of correlation between the ordered residuals and their expected values under normality. What do you conclude from your analysis?

c. State the generalized regression model to be employed for testing whether or not the treatment regression lines have the same slope. Conduct this test using $\alpha = .01$. State the alternatives, decision rule, and conclusion. What is the P-value of the test?

23.16. Refer to **Cash offers** Problems 18.10 and 23.15. Assume that covariance model (23.49) is applicable.

a. State the regression model equivalent to covariance model (23.49) for this case; use $1, -1, 0$ indicator variables. Fit this full model.

b. State the reduced models for testing for interaction and factor A and factor B main effects, respectively. Fit these reduced models.

c. Test for interaction effects; use $\alpha = .05$. State the alternatives, decision rule, and conclusion. What is the P-value of the test?

d. Test for factor A main effects; use $\alpha = .05$. State the alternatives, decision rule, and conclusion. What is the P-value of the test?

e. Test for factor B main effects; use $\alpha = .05$. State the alternatives, decision rule, and conclusion. What is the P-value of the test?

f. For each factor, make all pairwise comparisons between the factor level main effects. Use the Bonferroni procedure with a 90 percent family confidence coefficient. State your findings.

23.17. Refer to **Eye contact effect** Problem 18.12. Age of personnel officer is to be used as a concomitant variable. The ages (X_{ijk}) of the personnel officers follow.

i = 1		i = 2	
j = 1	j = 2	j = 1	j = 2
42	51	43	42
30	35	53	47
47	48	40	46
31	38	50	59
35	49	49	56

   a. Obtain the residuals for covariance model (23.49).

   b. For each treatment, plot the residuals against the fitted values. Also prepare a normal probability plot of the residuals and calculate the coefficient of correlation between the ordered residuals and their expected values under normality. What do you conclude from your analysis?

   c. State the generalized regression model to be employed for testing whether or not the treatment regression lines have the same slope. Conduct this test using $\alpha = .005$. State the alternatives, decision rule, and conclusion. What is the P-value of the test?

23.18. Refer to **Eye contact effect** Problems 18.12 and 23.17. Assume that covariance model (23.49) is applicable.

   a. State the regression model equivalent to covariance model (23.49) for this case; use $1, -1, 0$ indicator variables. Fit this full model.

   b. State the reduced models for testing for interaction and factor A and factor B main effects, respectively. Fit these reduced models.

   c. Test for interaction effects; use $\alpha = .01$. State the alternatives, decision rule, and conclusion. What is the P-value of the test?

   d. Test for factor A main effects; use $\alpha = .01$. State the alternatives, decision rule, and conclusion. What is the P-value of the test?

   e. Test for factor B main effects; use $\alpha = .01$. State the alternatives, decision rule, and conclusion. What is the P-value of the test?

   f. Compare the gender main effects by means of a 99 percent confidence interval. Interpret your interval estimate.

   g. Estimate the mean success rating for female personnel officers aged 40 when eye contact is present; use a 99 percent confidence interval.

23.19. Refer to **Productivity improvement** Problems 23.7 and 23.8. The analyst is considering the use of the difference between the productivity improvements in the two years $(Y_{ij} - X_{ij})$ as the dependent variable with the regular analysis of variance model (23.55a).

   a. Obtain the analysis of variance table.

   b. How effective is the use of differences here with the regular ANOVA model compared to the use of covariance model (23.3)? Discuss.

23.20. Refer to **Product display** Problems 23.13 and 23.14. The analyst is considering the use of the difference in sales between the two periods $(Y_{ij} - X_{ij})$ as the dependent variable with the regular analysis of variance model (23.55a).

   a. Obtain the analysis of variance table.

   b. How effective is the use of differences here with the regular ANOVA model compared to the use of covariance model (23.3)? Discuss.

EXERCISES

23.21. (Calculus needed.) Denote $\mu. + \tau_i$ in covariance model (23.3) by Δ_i. Derive the least squares estimators for Δ_i and γ in covariance model (23.3).

23.22. Show that $\sigma^2\{\bar{Y}_{i.}(\text{adj.})\}$ is given by (23.41).

23.23. Show that the variance of a contrast among the estimated adjusted treatment means, $\hat{L} = \Sigma c_i[\bar{Y}_{i.}(\text{adj.})]$, is given by:

$$\sigma^2\{\hat{L}\} = \sigma^2 \left[\sum \frac{c_i^2}{n_i} + \frac{(\Sigma c_i \bar{X}_{i.})^2}{SSE_X} \right]$$

PROJECTS

23.24. Refer to the **SENIC** data set. The following hospitals are to be considered in a study of the effects of region (variable 9) on the mean length of hospital stay of patients (variable 2), with available facilities and services (variable 12) as a concomitant variable:

$$1-52 \quad 54 \quad 55 \quad 57 \quad 58 \quad 63 \quad 76 \quad 83 \quad 84 \quad 94 \quad 101 \quad 103 \quad 111$$

a. Obtain the residuals for covariance model (23.3).

b. For each region, plot the residuals against the fitted values. Also prepare a normal probability plot of the residuals and calculate the coefficient of correlation between the ordered residuals and their expected values under normality. What do you conclude from your analysis?

c. State the generalized regression model to be employed for testing whether or not the treatment regression lines have the same slope. Conduct this test using $\alpha = .005$. State the alternatives, decision rule, and conclusion. What is the P-value of the test?

23.25. Refer to the **SENIC** data set and Project 23.24. Assume that covariance model (23.3) is applicable.

a. Plot the data in the format of Figure 23.5. Does it appear that there are region effects on the mean length of hospital stay? Discuss.

b. State the regression model equivalent to covariance model (23.3) for this case; use $1, -1, 0$ indicator variables. Also state the reduced model for testing for treatment effects.

c. Fit the full and reduced models and test for treatment effects; use $\alpha = .05$. State the alternatives, decision rule, and conclusion. What is the P-value of the test?

d. Make all pairwise comparisons between the region effects; use either the Bonferroni or Scheffé procedure with a 90 percent family confidence coefficient, whichever is more efficient. State your findings.

23.26. Refer to the **SMSA** data set. The following metropolitan areas are to be considered in a study of the effects of region (factor A: variable 12) and percent of population in central cities (factor B: variable 4) on crime rate (variable 11 $\div$ variable 3), with percent of population 65 or older (variable 5) as a concomitant variable:

$$1-45 \quad 49 \quad 51-54 \quad 58 \quad 60-62 \quad 64 \quad 66 \quad 71$$

$$73 \quad 80 \quad 92 \quad 101 \quad 123 \quad 130 \quad 131$$

For purposes of this analysis of covariance study, percent of population in central cities is to be classified into two categories: 37.0 percent or less, 37.1 percent or more.

a. Obtain the residuals for covariance model (23.49).

b. For each treatment, plot the residuals against the fitted values. Also prepare a normal probability plot of the residuals and calculate the coefficient of correlation between the ordered residuals and their expected values under normality. What do you conclude from your analysis?

c. State the generalized regression model to be employed for testing whether or not the treatment regression lines have the same slope. Conduct this test using $\alpha = .001$. State the alternatives, decision rule, and conclusion. What is the P-value of the test?

23.27. Refer to the **SMSA** data set and Project 23.26. Assume that covariance model (23.49) is applicable.

a. State the regression model equivalent to covariance model (23.49) for this case; use $1, -1, 0$ indicator variables. Fit this full model.

b. State the reduced models for testing for interaction and factor A and factor B main effects, respectively. Fit these reduced models.

c. Test for interaction effects; use $\alpha = .01$. State the alternatives, decision rule, and conclusion. What is the P-value of the test?

d. Test for factor A main effects; use $\alpha = .01$. State the alternatives, decision rule, and conclusion. What is the P-value of the test?

e. Test for factor B main effects; use $\alpha = .01$. State the alternatives, decision rule, and conclusion. What is the P-value of the test?

f. Make all pairwise comparisons between the region main effects; use either the Bonferroni or Scheffé procedure with a 95 percent family confidence coefficient, whichever is more efficient. State your findings.

Part V

Experimental Designs

Chapter 24

Randomized Block Designs—I

Formal experimentation is widely employed in the biological and social sciences. It has come of age somewhat more recently in business and economics, but a wide variety of uses now is found in these fields also. One example is an experiment to investigate the best level of aggregation of company data furnished by a management information system to middle management. Another is an experiment on the effect of a guaranteed annual income on the consumption behavior of families. The latter experiment was designed by dividing a group of low-income families at random into two halves, one of which received income supplements up to a guaranteed annual income, while the other half received no supplements. The consumption behavior of each group of families was then observed.

Up to this point we have considered the analysis of experimental data from studies based on a completely randomized design. There are many other types of experimental designs, however, that are widely used, and we shall examine the most important of these in this part.

We begin this chapter by considering the major elements of any experimental design and the statistical contributions to effective experimental designs. Then we shall take up in the remainder of this chapter and in the following one randomized block designs. In succeeding chapters, we shall discuss other widely used experimental designs.

24.1 DESIGN OF EXPERIMENTS

The *design of an experiment* refers to the structure of the experiment, with particular reference to:

1. The set of treatments included in the study.
2. The set of experimental units included in the study.
3. The rules and procedures by which the treatments are assigned to the experimental units (or vice versa).

4. The measurements that are made on the experimental units after the treatments have been applied.

Statistical designs for experiments are concerned with the rules and procedures whereby treatments are assigned to the experimental units. Statistical methodology also makes contributions to the other elements of experimental design, but we shall dwell chiefly on how to assign the treatments to experimental units so that efficient use is made of the experimental units.

Use of improper rules and procedures in assigning the treatments to experimental units can lead to serious difficulties. For instance, in a medical study designed to compare a standard treatment to a potentially risky new treatment, the group of patients undergoing the new treatment consisted of volunteers. These volunteers were less healthy at the beginning of the study than the patients receiving the standard treatment. Despite the fact that both the standard and new treatments were equally effective, the analysis of the health conditions of the patients at the end of the study showed a significant difference between the two groups, namely, that the new treatment group was in poorer health. A conclusion that the new treatment is therefore less effective would be biased. The source of bias here is called *selection bias* because the experimental units for the two treatment groups were not similar. Selection bias can be minimized by randomization. The use of randomization tends to balance the experimental units in each treatment group with regard to factors, other than the treatment, that affect the outcome.

The measurement process is another important element of experimental designs. Ideally, the measurement process should produce measurements that are unbiased and precise. *Measurement bias* can cause serious difficulties in the analysis of a study. An important source of measurement bias is due to unrecognized differences in the evaluation process. For example, a group of plants randomly assigned to a new fungicide treatment might unintentionally be evaluated by the investigators to be responding better to the treatment than actually is the case because of a desire to show the new treatment to be effective. When the experimental unit is a person, knowledge of the treatment by the person may also influence the measurement obtained. For instance, a person who knows that the food additive is salt may respond differently in the evaluation of the tastiness of a vegetable than if the additive were unknown. This source of measurement bias can be minimized by concealing the treatment assignment to both the experimental subject and the evaluator. A study using this kind of concealment is called a *double-blind* study. When knowledge of the assignment is withheld only from the experimental subject or the evaluator, the study is called a *single-blind* study.

24.2 CONTRIBUTIONS OF STATISTICS TO EXPERIMENTATION

Statistics has made a number of important contributions to experimentation. We consider briefly four major ones.

Factorial Experiments

This contribution was considered in Chapter 18. There we noted that multifactor investigations permit the analysis of a number of factors with the same precision as if the entire experiment had been devoted to the study of only one factor. In addition, a single factorial experiment provides information on interaction effects while the classical one-factor-at-a-time approach requires a series of experiments for doing so.

Replication

Replication refers to the repetition of an experiment. Consider an experiment consisting of three treatments. The assignment of three experimental units at random, one to each treatment, constitutes one replication of the experiment. The assignment of an additional three experimental units at random to the three treatments constitutes a second replication, and so on.

Not all repetitions are replications. Suppose that two incentive pay plans are being investigated and two plants are used in the study, with one plant assigned to each incentive plan. Assume now that 10 employees in each plant are selected and their productivity measured. For purposes of comparing the incentive pay plans, the plants are the experimental units so that there is only one replication (one plant for each plan), not 10 (the number of employees studied in each plant). Indeed, with only one replication here, plant effects and incentive pay plan effects cannot be disentangled and are said to be confounded. Selecting additional employees will not enable one to disentangle the incentive pay plan effects from the plant effects; only repetition of the experiment in additional plants (i.e., replicating the experiment) will permit this.

Replication makes it possible to assess the mean square error required for testing the presence of treatment effects or for establishing confidence interval estimates of these effects, as we have seen in earlier chapters. Replication also plays a second role, namely, it permits control over the precision of the estimates or the power of the tests through manipulation of the replication (sample) size. Again, we have observed this in earlier chapters.

Randomization

Randomization in experiments is a relatively recent idea, first introduced by the famous British statistician Sir R. A. Fisher. In the past, treatments had been assigned to experimental units either on a systematic or on a subjective basis. We noted earlier how serious biases can arise when self-selection is employed to assign experimental units to the treatments. The same dangers exist with systematic and subjective selection. To illustrate potential selection biases with systematic assignments, consider an experiment using ten employees and two treatments, where the first five employees on the payroll listing are assigned treatment 1 and the next five treatment 2. Suppose that the payroll listing is by seniority and that this variable is related to the phenomenon under study, say, productivity. A comparison of treatments 1 and 2 then reflects not only differences between the two treatments but also differences in the amount of experience between the two groups of employees. This potential bias may be so transparent that no good experimenter would use the type of systematic assignment just described. Nevertheless, there may be many other sources of bias that are not so apparent.

Subjective assignments of treatments to experimental units can also lead to selection bias, as when an experimenter subconsciously tends to assign one treatment to highly extrovert subjects and the other treatment to less extrovert subjects.

With randomization, the treatments are assigned to experimental units at random. Randomization tends to average out between the treatments whatever systematic effects may be present, apparent or hidden, so that comparisons between treatments measure only the pure treatment effects. Thus, randomization tends to eliminate the influence of extraneous factors not under the direct control of the experimenter and thereby precludes the presence of selection bias. Cochran and Cox (Ref. 24.1, p. 8) have likened randomization to an insurance policy in that it is a precaution against biases that may or may not occur.

Randomization is appropriate not only for the assignment of treatments to experimental units but also for any other phases of the experiment where systematic effects not under the control of the experimenter may be present. For instance, consider an experiment in which five treatments (alternative methods of measuring subjective probability) and 20 subjects are used. Only one subject can be run per day; thus, four weeks are required to complete the experiment. In this type of situation, it usually is highly desirable to determine the order of the treatments randomly since a variety of systematic time effects could be present. The experimenter may with time improve the explanation of the methods of measuring subjective probability, there may be a streak of extremely hot weather during a week, and the like. With these possible time effects, a systematic assignment of one treatment per week could lead to seriously biased results. Randomization, on the other hand, will tend to average out whatever systematic effects are present, whether anticipated or not.

Advice on when randomization is needed in addition to the assignment of treatments to experimental units can only be general. Certainly, randomization should be employed whenever the consequences of systematic effects could be serious. In our illustration on methods of measuring subjective probability, suppose that two observers conduct the experiment. Randomization of observers to treatment-experimental unit combinations would then be highly desirable, since large differences between observers are known to occur in this kind of situation. If the seriousness of the consequences is not known, the safe course is to randomize when feasible and not too costly. If randomization cannot be easily carried out and no serious consequences of systematic effects are anticipated, the experimenter may be willing to forgo randomization. He or she must then realize, however, that the validity of the treatment comparisons depends on the absence of serious systematic effects.

Comments

1. One may view the implications of randomization in a somewhat different fashion than that presented so far. The random errors of experimental units that are adjacent in time or space are often correlated, not independent, as a result of various systematic effects over time or space. Randomization does not erase this correlation pattern but, by making it equally likely that any two treatments are adjacent, tends to eliminate the correlations between treatments with increasing replications. Thus, randomization makes it reasonable to analyze the data as though the model random error terms are independent, an assumption that has been made in almost all models discussed so far.

2. Once in a while, randomization may provide a pattern that makes the experimenter uneasy, such as running the four experimental units for treatment 1 first and then running the four experimental units for treatment 2. This is not a likely occurrence, but one that can take place. Some solutions have been suggested for this problem, but none provides a final answer. In practice, the experimenter typically will discard a randomization sequence that has apparent dangers of systematic effects for the particular experiment and select another randomization.

3. Randomization also can provide the basis for making inferences without requiring that the error terms ε be independent $N(0, \sigma^2)$. We illustrate this for a single-factor experiment, consisting of two treatments and three replications. In this experiment, the treatments were assigned to the experimental units at random. Suppose the data are:

Treatment 1	Treatment 2
3	6
9	2
4	10

We assume now the following simple model (it can be generalized):

$$(24.1) \qquad Y_{ij} = \left(\begin{matrix} \text{A quantity depending on} \\ \text{the experimental unit} \end{matrix} \right) + \left(\begin{matrix} \text{A quantity depending} \\ \text{on the treatment} \end{matrix} \right)$$

Both the quantities for the experimental units and the quantities for the treatments are viewed as fixed. The randomness in the model arises solely from the random assignment of treatments to experimental units. Suppose now that the two treatment effects are equal. In that case, it would have been just as likely that we observed the numbers 3, 9, 4 for treatment 2, and 6, 2, 10 for treatment 1, since the treatments are assigned to experimental units at random. In fact, if the two treatment effects are equal, any division of the six observations into two groups of three is equally likely. Thus, in the list of all possible arrangements, all are equally likely if no treatment effects are present:

Treatment 1	Treatment 2
3, 9, 4	6, 2, 10
3, 9, 6	4, 2, 10
3, 9, 2	6, 4, 10
etc.	etc.

We then view the problem of comparing treatments as a single-factor analysis of variance, and calculate $F^* = MSTR/MSE$ for each arrangement. We thereby obtain the exact sampling distribution of F^* when the two treatment effects are equal. Both empirical and theoretical studies have shown that the sampling distribution so obtained is distributed approximately as the F distribution, provided the sample sizes are not very small. Thus, randomization alone can justify the F test as a good approximate test, without requiring any assumption of independent, normal error terms.

Local Control

The fourth contribution of statistics to experimental design is the concept of local control, which often is considered statistical design proper. Local control is intended to reduce experimental errors and make the experiment more powerful by suitable restrictions on the randomization of treatments to experimental units. Consider again the study of five methods of measuring subjective probability, which is to be conducted over a period of four weeks. The thought may have occurred in our earlier discussion that complete randomization might not provide full balance of all treatments within the four-week period. Would it not be better if we required that each week contain each of the five treatments once? If a substantial time effect is likely, it would indeed be desirable to use this form of restricted randomization, called *blocking*. Thus, we would randomize the order of treatments subject to the restriction that each treatment occur once in each week. It will be seen later that if a time effect is present, blocking will lead to more precise results than complete randomization.

Let us consider this same example from a slightly different point of view. With unrestricted randomization, the five observations in a single replication of the experiment will differ among themselves because of treatment effects, because of time effects (since the treatments may end up in any of the four weeks), and so on. If we require that each of the five treatments be conducted in each week, then a week's observations constitute a replication. Within such a replication, the observations will differ again because of treatment effects and a variety of other causes, but not because of any time effects from one week to another. The only effect of time that is left is that within a week, which may be anticipated to be substantially smaller than that between weeks. Thus, blocking by week will reduce the experimental error variability when a time effect is present, and in this way make the experiment more powerful.

Another benefit of blocking is that it can increase the range of validity for the conclusions from the experiment. In general, experimental errors can be made smaller (i.e., the variance of the random component ε can be made smaller) by using similar experimental units, leading to more precise experimental results. Thus, in a learning experiment, the use of persons of the same age, intelligence, and economic and social background will tend to lead to smaller experimental errors than if a more heterogeneous group of subjects was used. However the more homogeneous are the experimental units, the smaller is the range for which the experimental results are valid. For instance, findings for persons in one age group may not be valid for persons in other age groups. Thus, to make the conclusions broadly valid, one should vary the characteristics of experimental units, but the cost of this is less precision in the experimental results. Blocking of experimental units according to their characteristics can be employed to have one's cake and eat it too, namely, to have sufficient variability between experimental units for a wide range of validity and yet achieve high precision due to small experimental errors.

Blocking by characteristics of the experimental units can be particularly helpful in business, economics, and the life sciences. The experimental units utilized in these areas are frequently highly heterogeneous—for example, persons, families, towns, metropolitan areas. Blocking by a person's age or income or blocking towns by population size may be highly effective in reducing the experimental error variability.

24.3 ELEMENTS OF RANDOMIZED BLOCK DESIGNS

Description of Designs

A randomized block design is a restricted randomization design in which the experimental units are first sorted into homogeneous groups, called *blocks,* and the treatments are then assigned at random within the blocks. We illustrate randomized block designs by considering three examples.

1. In an experiment on the effects of four levels of newspaper advertising saturation on sales volume, the experimental unit is a city, and 16 cities are available for the study. Size of city usually is highly correlated with the dependent variable, sales volume. Hence, it is desirable to block the 16 cities into four groups of four cities each, according to population size. Thus, the four largest cities will constitute block 1, and so on. Within each block, the four treatments are then assigned at random to the four cities, and the assignments from one block to another are made independently.

2. In an experiment on the effects of three different incentive pay schemes on employee productivity of electronic assemblies, the experimental unit is an employee, and 30 employees are available for the study. Since productivity in this situation is highly correlated with manual dexterity, it is desirable to block the 30 employees into 10 groups of three according to their manual dexterity. Thus, the three employees with the highest manual dexterity ratings are grouped into one block, and so on for the other employees. Within each block, the three incentive pay schemes are then assigned randomly to the three employees.

3. A chemist is studying the reaction rate of five chemical agents. Only five agents can be effectively analyzed per day. Since day-to-day differences may affect the reaction rate, each day is used as a block, and all five chemical agents are tested each day in independently randomized orders.

As these examples imply, the key objective in blocking the experimental units is to make them as homogeneous as possible within blocks with respect to the dependent variable under study, and to make the different blocks as heterogeneous as possible with respect to the dependent variable. The design in which each treatment is included in each block is called a *randomized complete block design*. Often, we shall drop the term *complete* because the context makes it clear that all treatments are included in each block.

Comments

1. In a complete block design, each block constitutes a replication of the experiment. For that reason, it is highly desirable that the experimental units within a block be processed together whenever this will help to reduce experimental error variability. As an example, an experimenter may tend to make changes in experimental techniques over time (e.g., in the administration of the experiment to subjects) without being aware of it. Consecutive processing of the experimental units block by block will tend to keep such sources of variation out of the variation within blocks, and thereby make the experimental results more precise.

2. In factorial experiments, some of the factors of interest are often characteristics of the experimental units, such as gender, age, and amount of experience on the job. Even though these factors are not introduced to reduce experimental error variability but rather are included for their intrinsic interest, we shall nevertheless consider such experiments to be randomized block designs since the randomization of treatments to experimental units is restricted by the nature of the classification factors considered.

Criteria for Blocking

As noted earlier, the purpose of blocking is to sort experimental units into groups within each of which the elements are homogeneous with respect to the dependent variable, such that the differences between groups are as great as possible. To help recognize some of the characteristics of experimental units that are fruitful criteria for blocking, we need a precise definition of an experimental unit. All elements of the experimental situation that are not included in the definition of a treatment need to be assigned to the definition of an experimental unit. Suppose the treatment in an experiment consists of a portion of a vegetable containing a particular additive served in the laboratory. The experimental unit might then be defined as a homemaker of a given age, processed by a given observer on a specified day during a particular part of the day, and served food from a given batch of cooked vegetable. Still other elements of the experimental setting might be included in the definition of the experimental unit, and should be if they could be the cause of material variability in the observations.

A full definition of the experimental unit such as the one just given suggests two types of blocking criteria:

1. Characteristics associated with the unit—for persons: gender, age, income, intelligence, education, job experience, attitudes, etc.; for geographic areas: population size, average income, etc.
2. Characteristics associated with the experimental setting—observer, time of processing, machine, batch of material, measuring instrument, etc.

Use of time as a blocking variable frequently captures a number of different sources of variability, such as learning by observer, changes in equipment, and drifts in environmental conditions (e.g., weather). Blocking by observers often eliminates a substantial amount of interobserver variability; similarly, blocking by batches of material frequently is very effective.

There is no need to use only a single blocking criterion; several may be employed if the experimental error can be reduced substantially thereby. We shall consider in Chapter 25 the use of more than one blocking criterion, such as when a block consists of subjects in a given age group processed by a particular observer.

To design effective randomized block experiments requires the ability to select blocking variables that will reduce the experimental error variability. Often, past experience in the subject matter field enables the experimenter to select good blocking variables. If some experiments have been run in the past in which blocking has been employed, these results can be analyzed to determine the effectiveness of the blocking variables. We shall discuss an appropriate method of analysis for doing this in Section 24.9. In the absence of any information on potential blocking variables, uniformity trials can be run where all experimental units are assigned the same treatment. From these trials, information can be obtained on the effectiveness of different variables for blocking.

Note

There is another blocking variable often used in social science research that has not been mentioned yet, namely, the subject. With the subject as a complete block, all treatments are given to every subject. Such designs are often called *repeated measures designs*. Since they involve some special problems, we will discuss them separately in Chapter 28.

Advantages and Disadvantages

The advantages of a randomized complete block design are:

1. It can, with effective grouping, provide substantially more precise results than a completely randomized design of comparable size.
2. It can accommodate any number of treatments and replications.
3. Different treatments need not have equal sample sizes. For instance, if the control is to have twice as large a sample size as each of three treatments, blocks of size five would be used; three units in a block are then assigned at random to the three treatments and two to the control.
4. The statistical analysis is relatively simple.
5. If an entire treatment or a block needs to be dropped from the analysis for some reason, such as spoiled results, the analysis is not complicated thereby.
6. Variability in experimental units can be deliberately introduced to widen the range of validity of the experimental results without sacrificing the precision of the results.

Disadvantages include:

1. Missing observations within a block require more complex calculations.
2. The degrees of freedom for experimental error are not as large as with a com-

pletely randomized design. One degree of freedom is lost for each block after the first.

3. More assumptions are required for the model (e.g., no interactions between treatments and blocks, constant variance from block to block) than for a completely randomized design model.

How to Randomize

The randomization procedure for a randomized block design is straightforward. Within each block a random permutation is used to assign treatments to experimental units, just as in a completely randomized design. Independent permutations are selected for the several blocks.

Illustration

In an experiment on decision making, executives were exposed to one of three methods of quantifying the maximum risk premium they would be willing to pay to avoid uncertainty. The three methods are the utility method, the worry method, and the comparison method. After using the assigned method, each subject was asked to state the degree of confidence in the method of quantifying the risk premium on a scale from 0 (no confidence) to 20 (highest confidence).

Fifteen subjects were used in the study. They were grouped into five blocks of three executives, according to age. Block 1 contained the three oldest executives, and so on. The design layout, after five independent random permutations of 3 were employed, is shown in Table 24.1. Table 24.2 contains the results of the experiment,

TABLE 24.1   Layout for Randomized Block Design—Risk Premium Example

	Experimental Unit		
	1	2	3
Block 1 (oldest executives)	C	W	U
2	C	U	W
3	U	W	C
4	W	U	C
5 (youngest executives)	W	C	U

C: Comparison method
W: Worry method
U: Utility method

TABLE 24.2   Risk Premium Experiment Results (confidence ratings on scale from 0 to 20)

| Block | Method (j) | | | |
i	Utility	Worry	Comparison	Average
1 (oldest)	1	5	8	4.7
2	2	8	14	8.0
3	7	9	16	10.7
4	6	13	18	12.3
5 (youngest)	12	14	17	14.3
Average	5.6	9.8	14.6	10.0

and Figure 24.1 presents graphically the confidence ratings for each method by block. It appears from Figure 24.1 that there is much variation between blocks, but that in all blocks the comparison method leads to the highest confidence rating and the utility method to the lowest rating. We discuss next a widely used model for randomized block designs and the analysis of variance for this model before undertaking a formal analysis of the results in our example.

FIGURE 24.1   Risk Premium Experiment—Plot of Confidence Ratings by Blocks

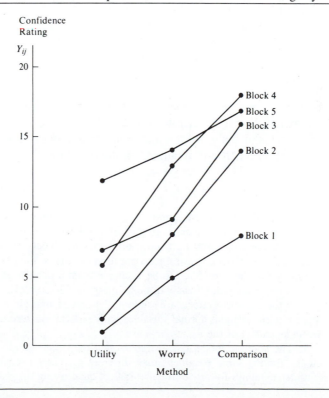

24.4 MODEL FOR RANDOMIZED BLOCK DESIGNS

Table 24.2 is similar in appearance to Table 21.2a, which shows the data for a two-factor study with one observation in each cell. In fact, one may think of a randomized complete block design as corresponding to a two-factor study (blocks and treatments are the factors), with one observation in each cell. As we noted in Section 21.1, if one can assume that there are no interactions between the two factors, an analysis of factor effects can be undertaken when there is only one observation in each cell and the factors have fixed effects.

Thus, the model for a randomized complete block design, when both the block and treatment effects are fixed and there are n blocks (replications) and r treatments, is:

$$(24.2) \qquad Y_{ij} = \mu_{..} + \rho_i + \tau_j + \varepsilon_{ij}$$

where:

$\mu_{..}$ is a constant

ρ_i are constants for the block (row) effects, subject to the restriction $\Sigma \rho_i = 0$

τ_j are constants for the treatment effects, subject to the restriction $\Sigma \tau_j = 0$

ε_{ij} are independent $N(0, \sigma^2)$

$i = 1, \ldots, n; j = 1, \ldots, r$

The observations Y_{ij} with randomized block model (24.2) are independent and normally distributed, with mean:

$$(24.2a) \qquad E\{Y_{ij}\} = \mu_{..} + \rho_i + \tau_j$$

and constant variance:

$$(24.2b) \qquad \sigma^2\{Y_{ij}\} = \sigma^2$$

Randomized block model (24.2) is identical to the two-factor, no-interaction model (21.1), except that we now use ρ_i for the block effect, τ_j for the treatment effect, and n to designate the total number of blocks. Note that Y_{ij} here stands for the observation for the jth treatment in the ith block.

Comments

1. When the experimental units are grouped according to specified categories, such as into particular age groups, income groups, and order-of-processing groups, the block effects ρ_i are usually considered to be fixed. Sometimes the block effects are viewed as random. For instance, when observers or subjects are used as blocks, the particular observers or subjects in the study may be considered to be a sample from a population of observers or subjects. The case of random block effects will be taken up in Chapter 25.

2. If the treatment effects are random, the only changes in model (24.2) are that the τ_j now represent independent normal variables with expectation zero and variance σ_τ^2, and that the τ_j are independent of the ε_{ij}.

3. The additive model (24.2) implies that the expected values of observations in different blocks for the same treatment may differ (e.g., older executives may tend to have lower confidence ratings for any of the methods of quantifying the risk premium than younger executives), but the treatment effects (e.g., how much higher the confidence rating for one method is over that for another) are the same for all blocks. We shall consider the possibility of interactions between blocks and treatments later in this chapter.

24.5 ANALYSIS OF VARIANCE AND TESTS

Fitting of Randomized Block Model

The least squares estimators of the parameters in randomized block model (24.2) are obtained in the customary fashion. Employing our usual notation, they are:

	Parameter	Estimator
(24.3a)	$\mu_{..}$	$\hat{\mu}_{..} = \bar{Y}_{..}$
(24.3b)	ρ_i	$\hat{\rho}_i = \bar{Y}_{i.} - \bar{Y}_{..}$
(24.3c)	τ_j	$\hat{\tau}_j = \bar{Y}_{.j} - \bar{Y}_{..}$

The fitted values therefore are:

$$(24.4) \qquad \hat{Y}_{ij} = \bar{Y}_{..} + (\bar{Y}_{i.} - \bar{Y}_{..}) + (\bar{Y}_{.j} - \bar{Y}_{..}) = \bar{Y}_{i.} + \bar{Y}_{.j} - \bar{Y}_{..}$$

and the residuals are:

$$(24.5) \qquad e_{ij} = Y_{ij} - \hat{Y}_{ij} = Y_{ij} - \bar{Y}_{i.} - \bar{Y}_{.j} + \bar{Y}_{..}$$

Analysis of Variance

The analysis of variance for a randomized complete block design is identical to that for a two-factor, no-interaction model with one observation per cell, as described in Section 21.1:

$$(24.6a) \qquad SSBL = r \sum_i (\bar{Y}_{i.} - \bar{Y}_{..})^2 = \sum_i \frac{Y_{i.}^2}{r} - \frac{Y_{..}^2}{rn}$$

$$(24.6b) \qquad SSTR = n \sum_j (\bar{Y}_{.j} - \bar{Y}_{..})^2 = \sum_j \frac{Y_{.j}^2}{n} - \frac{Y_{..}^2}{rn}$$

$$(24.6c) \qquad SSBL.TR = \sum_i \sum_j (Y_{ij} - \bar{Y}_{i.} - \bar{Y}_{.j} + \bar{Y}_{..})^2$$

$$= \sum_i \sum_j Y_{ij}^2 - \sum_i \frac{Y_{i.}^2}{r} - \sum_j \frac{Y_{.j}^2}{n} + \frac{Y_{..}^2}{rn}$$

$$= \sum_i \sum_j e_{ij}^2$$

Here, SSBL is the *sum of squares for blocks* and SSTR is, as usual, the treatment sum of squares. SSBL.TR denotes the *interaction sum of squares between blocks and treatments;* note from (24.5) that this sum of squares here is the same as the sum of the squared residuals. Finally, *rn* is the total number of experimental units in the study.

A summary of the analysis of variance, including the expected mean squares for both fixed and random treatment effects, is given in Table 24.3. Note that since there are no interaction terms in the model, the expected mean squares contain only σ^2 and, as appropriate, the treatment or block main effects term. Also note from the $E\{MS\}$ columns in Table 24.3 that the appropriate denominator in the $F*$ test statistic for testing treatment effects is the interaction mean square, here denoted by *MSBL.TR*, whether the treatment effects are fixed or random. This is the same as in Section 21.1 for the two-factor no-interaction model with $n = 1$. Hence, to test for treatment effects:

	Fixed Treatment Effects	Random Treatment Effects
(24.7a)	H_0: all $\tau_j = 0$	H_0: $\sigma_\tau^2 = 0$
	H_a: not all τ_j equal zero	H_a: $\sigma_\tau^2 > 0$

we use the same test statistic whether the treatment effects are fixed or random:

$$(24.7b) \qquad F* = \frac{MSTR}{MSBL.TR}$$

and the decision rule for controlling the Type I error at α is:

$$(24.7c) \qquad \begin{array}{l} \text{If } F* \le F[1 - \alpha; r - 1, (n - 1)(r - 1)], \text{ conclude } H_0 \\ \text{If } F* > F[1 - \alpha; r - 1, (n - 1)(r - 1)], \text{ conclude } H_a \end{array}$$

TABLE 24.3   ANOVA Table for Randomized Complete Block Design, Block Effects Fixed

Source of Variation	SS	df	MS	E{MS} Treatments Fixed	E{MS} Treatments Random
Blocks	SSBL	$n - 1$	MSBL	$\sigma^2 + \dfrac{r\Sigma\rho_i^2}{n - 1}$	$\sigma^2 + \dfrac{r\Sigma\rho_i^2}{n - 1}$
Treatments	SSTR	$r - 1$	MSTR	$\sigma^2 + \dfrac{n\Sigma\tau_j^2}{r - 1}$	$\sigma^2 + n\sigma_\tau^2$
Error	SSBL.TR	$(n - 1)(r - 1)$	MSBL.TR	σ^2	σ^2
Total	SSTO	$nr - 1$			

Example

Table 24.4 contains the analysis of variance for the risk premium example in Table 24.2. The calculations are straightforward and were carried out by a computer package. To test for treatment effects:

$$H_0: \tau_1 = \tau_2 = \tau_3 = 0$$

$$H_a: \text{not all } \tau_j \text{ equal zero}$$

we use the results in Table 24.4:

$$F^* = \frac{MSTR}{MSBL.TR} = \frac{101.4}{2.99} = 33.9$$

For a level of significance of $\alpha = .01$, we require $F(.99; 2, 8) = 8.65$. Since $F^* = 33.9 > 8.65$, we conclude H_a, that the mean confidence ratings for the three methods differ. The P-value of the test is .0001.

Comments

1. Sometimes one may also wish to conduct a test for block effects:

(24.8a)
$$H_0: \text{all } \rho_i = 0$$

$$H_a: \text{not all } \rho_i \text{ equal zero}$$

Usually, however, the treatments are of primary interest, and blocks are chiefly the means for reducing experimental error variability. Table 24.3 indicates that the test for fixed block effects uses the test statistic:

(24.8b)
$$F^* = \frac{MSBL}{MSBL.TR}$$

For the risk premium example, this test statistic is:

$$F^* = \frac{42.8}{2.99} = 14.3$$

TABLE 24.4   ANOVA Table for Randomized Complete Block Design—Risk Premium Example of Table 24.2

Source of Variation	SS	df	MS
Blocks	171.3	4	42.8
Methods for risk premium specification	202.8	2	101.4
Error	23.9	8	2.99
Total	398.0	14	

For a level of significance of $\alpha = .01$, we require $F(.99; 4, 8) = 7.01$. Since $F^* = 14.3 > 7.01$, we conclude that the mean confidence ratings (averaged over treatments) differ for the various blocks.

Since blocks correspond to a classification factor, one needs to be careful in interpreting the implications of block effects. In our example, for instance, the block effects might not be due to age, even though age was the grouping variable. Education could be the pivotal independent variable, the effect by age arising if older executives have less formal education than younger ones.

2. The power of the F test for treatment effects for a randomized complete block design involves the same noncentrality parameter as for a completely randomized design. Formula (17.2) gives the appropriate measure. Despite the same form of the noncentrality parameter, the two designs generally lead to different power levels even when based on the same sample sizes, for two reasons. First, the experimental error variance σ^2 will differ for the two designs. Second, the degrees of freedom associated with the denominator of the F^* statistic differ for the two designs.

3. If only two treatments are investigated in a randomized complete block design, it can readily be shown that the F test for treatment effects based on test statistic (24.7b) is equivalent to the two-sided t test for paired observations based on test statistic (1.66).

24.6 EVALUATION OF APTNESS OF RANDOMIZED BLOCK MODEL

Since the importance of examining the aptness of a statistical model for a given set of data has been mentioned many times earlier and since the techniques of examination are similar, we shall make only a few points of special relevance to randomized block designs here.

Diagnostic Plots

Some of the chief ways in which the data may not fit the randomized block model (24.2) are:

1. Unequal error variability by blocks.
2. Unequal error variability by treatments.
3. Time effects.
4. Block-treatment interactions.

Use of residual plots in connection with points 2 and 3 has been considered in Section 16.1 with reference to a completely randomized design, but the discussion there applies also to the residuals of a randomized block design, given in (24.5):

$$e_{ij} = Y_{ij} - \bar{Y}_{i.} - \bar{Y}_{.j} + \bar{Y}_{..}$$

We simply add here that if treatments do have unequal error variability in a randomized complete block design, one can always estimate differences between any two treatments by working with the differences between the paired observations, $Y_{ij} - Y_{ij'}$, which are unaffected by the unequal treatment variances.

FIGURE 24.2   Residual Dot Plots Suggesting Unequal Error Variances by Blocks

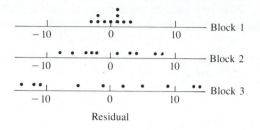

Unequal error variability by blocks can be studied by aligned residual dot plots for each block, as shown in Figure 24.2. The residual dot plots in Figure 24.2 are suggestive of increasing error variability with increasing block number. If, for instance, the blocks were processed in block number order, some modifications in procedures may have taken place leading to larger experimental error variability over time. Tests concerning the equality of variances, such as those described in Section 16.2, may be employed for a more formal determination, provided that the sample sizes are reasonably large so that the residuals can be treated as if they were independent.

Interactions between treatments and blocks are somewhat more difficult to detect from residual plots. Figure 24.3 contains the residuals for an experiment with two treatments run in four blocks. The reversal in pattern of the residuals is suggestive of an interaction effect. There are, however, many other possible types of interaction patterns that would appear very much different from that in Figure 24.3.

Another diagnostic plot that may be helpful to detect interaction effects is a plot of the residuals e_{ij} against the fitted values $\hat{Y}_{ij}$. A curvilinear pattern of the residuals in such a plot often suggests the presence of interaction effects between blocks and treatments. This plot also provides information about the constancy of the error term variance.

FIGURE 24.3   Residual Dot Plots Suggesting Block-Treatment Interactions

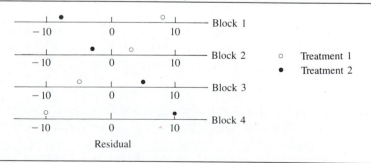

Still another diagnostic plot for interactions, which is often more effective than a residual plot, is a plot of the responses Y_{ij} by blocks. Figure 24.1 illustrates this type of plot. A severe lack of parallelism in such a plot is a strong indication that blocks and treatments interact in their effects on the response.

Example.   Figure 24.1 for the risk premium example does not exhibit a severe lack of parallelism, thus suggesting that blocks and treatments do not interact in any major fashion. Figure 24.4a, which presents a computer plot of the residuals against the fitted values, leads to a similar conclusion. There is no strong evidence of a curvilinear pattern here. In addition, Figure 24.4a does not indicate the existence of substantially unequal error variances.

Figure 24.4b contains a normal probability plot of the residuals. This plot does not suggest any strong departures from a normal error distribution. The coefficient of correlation between the ordered residuals and their expected values under normality is .959 and supports this conclusion. Residual dot plots for each treatment and for

FIGURE 24.4   Diagnostic Residual Plots—Risk Premium Example (Minitab, Ref. 24.2)

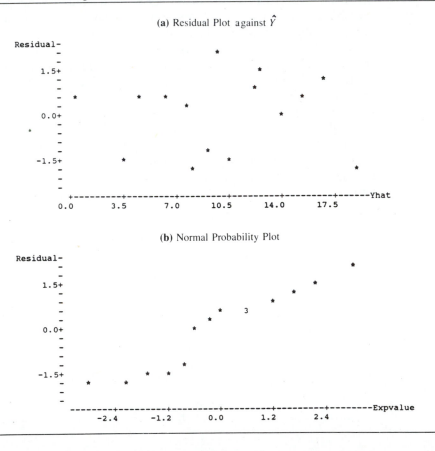

(a) Residual Plot against $\hat{Y}$

(b) Normal Probability Plot

each block were also prepared (they are not shown here). They suggested that the error variances did not differ substantially between treatments and between blocks. These results, in addition to a formal test that found no interactions between block and treatment effects (to be discussed next), led the analyst to conclude that randomized block model (24.2) is appropriate for the data.

Tukey Test for Additivity

The Tukey test for additivity, discussed in Section 21.2, may be employed for a formal test of possible interaction effects between blocks and treatments. We illustrate this test for the risk premium example data in Table 24.2. The special interaction sum of squares, denoted here by $SSBL.TR^*$, is given by (21.11):

$$SSBL.TR^* = \frac{[\sum_i \sum_j (\bar{Y}_{i.} - \bar{Y}_{..})(\bar{Y}_{.j} - \bar{Y}_{..})Y_{ij}]^2}{\sum_i (\bar{Y}_{i.} - \bar{Y}_{..})^2 \sum_j (\bar{Y}_{.j} - \bar{Y}_{..})^2}$$

Using the data in Table 24.2, we find the numerator:

$$[(4.7 - 10)(5.6 - 10)(1) + \cdots + (14.3 - 10)(14.6 - 10)(17)]^2 = 615.04$$

and from the results in Table 24.4 and formulas (24.6a) and (24.6b), we obtain the two terms in the denominator:

$$\sum_i (\bar{Y}_{i.} - \bar{Y}_{..})^2 = \frac{SSBL}{r} = \frac{171.3}{3} = 57.1$$

$$\sum_j (\bar{Y}_{.j} - \bar{Y}_{..})^2 = \frac{SSTR}{n} = \frac{202.8}{5} = 40.56$$

Hence:

$$SSBL.TR^* = \frac{615.04}{57.1(40.56)} = .27$$

Using the results from Table 24.4, we can now obtain the remainder sum of squares (21.12a) for the special interaction model (21.9):

$$SSRem^* = SSTO - SSBL - SSTR - SSBL.TR^*$$

$$= 398.0 - 171.3 - 202.8 - .27$$

$$= 23.63$$

Hence, test statistic (21.13) is:

$$F^* = \frac{SSBL.TR^*}{1} \div \frac{SSRem^*}{rn - r - n}$$

$$= \frac{.27}{1} \div \frac{23.63}{7} = .08$$

For a level of significance of $\alpha = .05$, we need $F(.95; 1, 7) = 5.59$. Since $F^* = .08 \leq 5.59$, we conclude that no block–treatment interaction effects are present. The P-value of this test is .79.

Note

If interaction effects are present, transformations of the data should be attempted to remove at least the important interaction effects. The discussion in Section 21.2 is relevant to this point.

24.7 ANALYSIS OF TREATMENT EFFECTS

Once the existence of fixed treatment effects has been established through the analysis of variance, the analysis of these effects proceeds as described in Chapter 15 for single-factor studies. Often, a useful preliminary view of the treatment effects can be obtained from a normal probability plot of the estimated treatment means $\bar{Y}_{.j}$. The formal analysis of the treatment effects usually involves estimation of one or more contrasts of the treatment means $\mu_{.j}$, where $\mu_{.j}$ is the mean response for treatment j averaged over all blocks. The formulas in Chapter 15 for estimating contrasts of the treatment means apply here, with the treatment means now denoted by $\mu_{.j}$ and the estimated treatment means denoted by $\bar{Y}_{.j}$. The appropriate mean square term to be used in the estimated variance of the contrast is $MSBL.TR$, obtained from (24.6c), since it is the denominator of the F^* statistic for testing fixed treatment effects. The multiples for the estimated standard deviation of the contrast are now as follows:

(24.9a) Single comparison $\qquad\qquad t[1 - \alpha/2; (n - 1)(r - 1)]$

(24.9b) Tukey procedure (for pairwise comparisons) $\qquad T = \dfrac{1}{\sqrt{2}} q[1 - \alpha; r, (n - 1)(r - 1)]$

(24.9c) Scheffé procedure $\qquad\qquad S^2 = (r - 1)F[1 - \alpha; r - 1,$
$$(n - 1)(r - 1)]$$

(24.9d) Bonferroni procedure $\qquad B = t[1 - \alpha/2g; (n - 1)(r - 1)]$

Example

The researcher who conducted the risk premium study was satisfied, on the basis of the residual analyses and tests, that the randomized block model (24.2) is apt for the experiment. He therefore began the analysis of the treatment effects by preparing a normal probability plot of the estimated treatment means $\bar{Y}_{.j}$. This plot is shown in Figure 24.5. The reference line shown in Figure 24.5 is given in (15.2) and here is:

FIGURE 24.5 Normal Probability Plot of Estimated Treatment Means—
Risk Premium Example

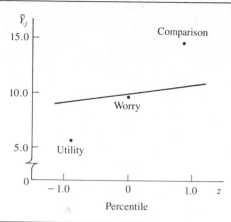

$$\text{Expected value} = \bar{Y}_{..} + z\left(\frac{i - .375}{r + .25}\right) \sqrt{\frac{MSBL.TR}{n}} = 10.0 + z\left(\frac{i - .375}{3.25}\right) \sqrt{\frac{2.99}{5}}$$

where $\bar{Y}_{..} = 10.0$ (Table 24.2), $MSBL.TR = 2.99$ (Table 24.4), and $n = 5$ because each treatment occurred once in each of five blocks.

The plot in Figure 24.5 clearly shows that treatment effects are present. It also suggests, by the fact that the slopes connecting each pair of treatments differs so substantially from that of the reference line, that significant differences are present between each pair of methods of quantifying the maximum risk premium.

To analyze the treatment effects formally, the researcher wished to obtain all pairwise comparisons with a 95 percent family confidence coefficient, utilizing the Tukey procedure. Using (15.25) with MSE replaced by $MSBL.TR$ and the results in Table 24.4, we obtain:

$$s^2\{\hat{D}\} = MSBL.TR\left(\frac{1}{n} + \frac{1}{n}\right) = \frac{2MSBL.TR}{n} = \frac{2(2.99)}{5} = 1.20$$

Remember that each estimated treatment mean $\bar{Y}_{.j}$ consists of n observations (one from each of n blocks). Using (24.9b), we find for a 95 percent family confidence coefficient:

$$T = \frac{1}{\sqrt{2}}q(.95; 3, 8) = \frac{1}{\sqrt{2}}(4.04) = 2.86$$

Hence:

$$Ts\{\hat{D}\} = 2.86\sqrt{1.20} = 3.1$$

Thus, we obtain for the pairwise comparisons (see Table 24.2 for the $\bar{Y}_{.j}$):

$$1.7 = (14.6 - 9.8) - 3.1 \leq \mu_{.3} - \mu_{.2} \leq (14.6 - 9.8) + 3.1 = 7.9$$

$$5.9 = (14.6 - 5.6) - 3.1 \leq \mu_{.3} - \mu_{.1} \leq (14.6 - 5.6) + 3.1 = 12.1$$

$$1.1 = (9.8 - 5.6) - 3.1 \leq \mu_{.2} - \mu_{.1} \leq (9.8 - 5.6) + 3.1 = 7.3$$

Here $\mu_{.1}$ is the mean confidence rating, averaged over all blocks, for the utility method, and $\mu_{.2}$ and $\mu_{.3}$ are the mean confidence ratings for the worry and comparison methods, respectively.

We conclude, just as Figure 24.5 had suggested, that the comparison method has a higher mean confidence rating than the worry method, which in turn has a higher mean confidence rating than the utility method. The family confidence coefficient of .95 applies to this entire set of comparisons. A line plot of the estimated treatment means summarizes the results:

24.8 FACTORIAL TREATMENTS

When the treatments in a randomized block design are combinations of different factor levels, one can simply write the ANOVA model showing the factor effects in place of the treatment effect. For a two-factor study, we have:

$$(24.10) \qquad Y_{ijk} = \mu_{...} + \rho_i + \alpha_j + \beta_k + (\alpha\beta)_{jk} + \varepsilon_{ijk}$$

where the terms in the model have the usual meaning and (j, k) identifies the treatment.

In the analysis of variance, we proceed as always by decomposing the treatment sum of squares $SSTR$ into sums of squares for factor main effects and interactions. This is shown in Table 24.5 for a two-factor study, the factors having a and b levels, respectively. Thus, the total number of treatments r here equals ab. The decomposition is done in the usual fashion, as explained in Section 18.4, utilizing the relation in (18.39):

$$SSTR = SSA + SSB + SSAB$$

Formulas (18.39a, b, c) and their alternative versions (18.40c, d, f) are appropriate for calculating the component sums of squares, remembering that (i, j) subscripts are there used to identify the treatments in terms of the factor combinations. Tests for factor effects are conducted as usual, and no new problems are encountered in the estimation of fixed factor effects.

TABLE 24.5 ANOVA Table for a Two-Factor Study in a Randomized Complete Block Design—Randomized Block Model (24.10)

Source of Variation	SS	df	MS
Blocks	SSBL	$n - 1$	MSBL
Treatments	SSTR	$r - 1$	MSTR
Factor A	SSA	$a - 1$	MSA
Factor B	SSB	$b - 1$	MSB
AB interactions	SSAB	$(a - 1)(b - 1)$	MSAB
Error	SSBL.TR	$(n - 1)(r - 1)$	MSBL.TR
Total	SSTO	$nr - 1$	

Note: $r = ab$

Note

Randomized block model (24.10) assumes no interactions between treatments and blocks. Specifically, it implies that all block-factor A interactions (denoted by $BL.A$) equal zero, and similarly that all $BL.B$ and $BL.AB$ interactions equal zero. One can make a less restrictive analysis by assuming only that the $BL.AB$ interactions equal zero. To see this, consider the layout for a two-factor study ($a = 2$, $b = 2$) in $n = 3$ blocks, as shown in Table 24.6. Here Y_{ijk} denotes the observation in the ith block for the (j, k) factor combination. Note that this layout corresponds to the three-factor layout in Table 22.2 but with only one observation per cell. Table 22.3 contains the general analysis of variance for a three-factor study. It can be seen from there that if all $BL.AB$ interactions ($BL.AB$ corresponds to ABC in Table 22.3) equal zero, the $BL.AB$ interaction mean square is an unbiased estimator of σ^2, the experimental error variance, and is the appropriate denominator mean square in the F^* test statistic for testing all factor effects. Hence, we can conduct all tests and make all desired estimates with a factorial experiment in a complete block design by only assuming that the $BL.AB$ interactions are zero. The price of the less restrictive assumptions is fewer degrees of freedom for the experimental error.

TABLE 24.6 Layout for a Two-Factor Study in a Randomized Complete Block Design

	A_1		A_2	
	B_1	B_2	B_1	B_2
Block 1	Y_{111}	Y_{112}	Y_{121}	Y_{122}
2	Y_{211}	Y_{212}	Y_{221}	Y_{222}
3	Y_{311}	Y_{312}	Y_{321}	Y_{322}

TABLE 24.7 ANOVA Table for a Two-Factor Study in a Randomized Complete Block Design—Randomized Block Model (24.11)

Source of Variation	SS	df	MS
Blocks (*BL*)	*SSBL*	$n - 1$	*MSBL*
Factor *A*	*SSA*	$a - 1$	*MSA*
Factor *B*	*SSB*	$b - 1$	*MSB*
AB interactions	*SSAB*	$(a - 1)(b - 1)$	*MSAB*
BL.A interactions	*SSBL.A*	$(n - 1)(a - 1)$	*MSBL.A*
BL.B interactions	*SSBL.B*	$(n - 1)(b - 1)$	*MSBL.B*
Error	*SSBL.AB*	$(n - 1)(a - 1)(b - 1)$	*MSBL.AB*
Total	*SSTO*	$nab - 1$	

The ANOVA model for this less restrictive case is:

$$(24.11) \qquad Y_{ijk} = \mu_{...} + \rho_i + \alpha_j + \beta_k + (\alpha\beta)_{jk} + (\rho\alpha)_{ij} + (\rho\beta)_{ik} + \varepsilon_{ijk}$$

where the model terms have the usual meaning. The analysis of variance for this model is given in Table 24.7. The sums of squares may be calculated using formulas (22.21a–g) or their alternative computational versions. In using these formulas, remember that n in these formulas (number of observations per cell) now is 1, and that the number of levels for blocks (corresponding to factor C) is n.

24.9 PLANNING RANDOMIZED BLOCK EXPERIMENTS

Necessary Number of Blocks

The planning of sample size for a randomized complete block design is very similar to that for a completely randomized design. One may determine the needed number of blocks n either to obtain specified protection against making Type I and Type II errors or to obtain specified precision for key contrasts of the treatment means. With either approach, it is necessary to assess in advance the magnitude of the experimental error variance σ^2.

Power Approach. The same table as for the completely randomized design (Table A.10) may be used, provided the number of treatments and blocks are not very small, specifically provided that $r(n - 1) \geq 20$. The development in Section 17.1 may be followed directly.

Example. In the experiment on confidence ratings for three methods of measuring the risk premium, suppose that the number of blocks had not yet been determined and the experimenter desired the following risk protections:

1. Type I error is to be controlled at $\alpha = .05$.

2. If any two treatment means differ by 3 or more rating points, i.e., if the minimum range of the treatment means is $\Delta = 3$, the risk of concluding that there are no treatment effects should not exceed $\beta = .20$.

The experimenter anticipates that the experimental error standard deviation when executives are grouped by age will be approximately $\sigma = 2$.

Thus, the specifications can be summarized as follows:

$$r = 3 \qquad\qquad \alpha = .05 \quad \Delta = 3$$

$$\beta = .20 \text{ or Power} = .80 \quad \sigma = 2$$

Using (17.5) we find:

$$\frac{\Delta}{\sigma} = \frac{3}{2} = 1.5$$

Entering Table A.10 for $1 - \beta = .80$, $r = 3$, $\Delta/\sigma = 1.5$, and $\alpha = .05$, we find $n = 10$. Thus, the experimenter requires approximately 10 blocks of three executives each in order to obtain the desired protection against incorrect decisions.

Estimation Approach.   If the experimenter wishes to determine the number of blocks n by means of the estimation approach, one need simply calculate the anticipated standard deviations of key contrasts and modify the replication size iteratively until the desired precision is attained. Often, the experimenter will use a multiple comparison procedure for encompassing the different estimates under a family confidence coefficient.

Example.   For the risk premium illustration, the Tukey procedure is to be used for all pairwise comparisons with a 95 percent family confidence coefficient. Using $n = 10$ as a starting point and assuming that $\sigma = 2$ approximately, the anticipated variance of any pairwise difference is:

$$\sigma^2\{\hat{D}\} = \sigma^2\left(\frac{1}{n} + \frac{1}{n}\right) = (2)^2\left(\frac{1}{10} + \frac{1}{10}\right) = .8$$

or $\sigma\{\hat{D}\} = .89$. Further:

$$T = \frac{1}{\sqrt{2}}q[.95; r, (n - 1)(r - 1)] = \frac{1}{\sqrt{2}}q(.95; 3, 18) = \frac{1}{\sqrt{2}}(3.61) = 2.55$$

Thus, the anticipated half-width of the confidence interval is $T\sigma\{\hat{D}\} = 2.55(.89) = 2.3$. If this precision is not adequate, a larger number of blocks should be tried next. If the precision is greater than necessary, a smaller number of blocks should be used in the next iteration.

Efficiency of Blocking Variable

Once a randomized complete block experiment has been run, it is often desired to estimate the efficiency of the blocking variable employed for guidance in future experimentation.

Let σ_b^2 stand for the experimental error variance for the randomized block design. Up to this point, we have used σ^2 for this error variance, but now that we will compare two designs we need to be more specific. Let σ_r^2 denote the experimental error variance for a completely randomized design. The relative efficiency of blocking, compared to a completely randomized design, is then defined as follows:

$$(24.12) \qquad\qquad E = \frac{\sigma_r^2}{\sigma_b^2}$$

The measure E indicates how much larger the replications need be with a completely randomized design as compared to a randomized block design in order that the variance of any estimated treatment contrast be the same for the two designs.

We know that $MSBL.TR$ for the randomized block design is an unbiased estimator of σ_b^2. The question is how to estimate σ_r^2 from the data for the randomized block design. Since the same experimental units are involved in either case and there are assumed to be no interactions between treatments and blocks, it can be shown that an unbiased estimator of σ_r^2 is:

$$(24.13) \qquad\qquad s_r^2 = \frac{(n-1)MSBL + n(r-1)MSBL.TR}{nr-1}$$

Hence, we estimate E as follows:

$$(24.14) \qquad \hat{E} = \frac{s_r^2}{MSBL.TR} = \frac{(n-1)MSBL + n(r-1)MSBL.TR}{(nr-1)MSBL.TR}$$

Since the number of degrees of freedom for experimental error for a randomized block design is not as great as for a completely randomized one, E overstates the efficiency a little because it considers only the error variances. Several modified measures of efficiency have been suggested to take this overstatement into account. Unless the degrees of freedom for experimental error with both designs are very small, these modifications have little effect. One frequently used modification, applicable for assessing any design relative to another, is:

$$(24.15) \qquad \hat{E}' = \frac{(df_2 + 1)(df_1 + 3)}{(df_2 + 3)(df_1 + 1)}\hat{E}$$

where df_1 denotes the degrees of freedom for the experimental error in the base design (completely randomized design, in our case) and df_2 denotes the degrees of freedom for the experimental error in the design whose efficiency is being assessed (randomized block design, in our case).

Example.   We shall evaluate the efficiency of blocking by age of executives in the risk premium example. Placing the appropriate results from Table 24.4 in the efficiency measure (24.14), we obtain:

$$\hat{E} = \frac{4(42.8) + 5(2)(2.99)}{14(2.99)} = 4.8$$

Thus, we would have required almost five times as many replications per treatment with a completely randomized design to achieve the same variance of any estimated contrast as is obtained with blocking by age. Clearly, blocking by age was highly effective here.

If we had used the modified efficiency measure (24.15), we would have found:

$$\hat{E}' = \frac{(8 + 1)(12 + 3)}{(8 + 3)(12 + 1)}(4.8) = 4.5$$

This result, of course, does not differ greatly from that obtained by using (24.14).

Note

The efficiency measure $\hat{E}$ in (24.14) equals 1 if $MSBL = MSBL.TR$, is greater than 1 if $MSBL > MSBL.TR$, and is less than 1 if $MSBL < MSBL.TR$. Since the test statistic for block effects in (24.8b) is $F^* = MSBL/MSBL.TR$, it follows that good blocking is achieved when F^* exceeds 1 by a considerable margin.

Covariance Analysis as Alternative to Blocking

At times, a choice exists between: (1) a completely randomized design with covariance analysis used to reduce the experimental errors, and (2) a randomized block design, with the blocks formed by means of the concomitant variable. Generally, the latter alternative is to be preferred. There are several reasons for this:

1. If the regression is linear, randomized block designs and covariance analysis are about equally efficient. If the regression is not linear but covariance analysis with a linear relationship is utilized, covariance analysis with a completely randomized design will tend to be not as effective as a randomized block design.
2. Computations with randomized block designs are simpler than those with covariance analysis.
3. Randomized block designs are essentially free of assumptions about the nature of the relationship between the blocking variable and the dependent variable, while covariance analysis assumes a definite form of relationship.

A drawback of randomized block designs is that somewhat fewer degrees of freedom are available for experimental error than with covariance analysis for a completely randomized design. However, in all but small-scale experiments, this difference in degrees of freedom has little effect on the precision of the estimates.

24.10 REGRESSION APPROACH TO RANDOMIZED BLOCK DESIGNS

ANOVA model (24.2) for a randomized block design with both block and treatment effects fixed:

(24.16)
$$Y_{ij} = \mu_{..} + \rho_i + \tau_j + \varepsilon_{ij}$$

$$i = 1, \ldots, n; j = 1, \ldots, r$$

can be expressed in straightforward fashion in the form of a regression model with indicator variables taking on values 1, -1, or 0. The regression model for the risk premium example in Table 24.2, with $r = 3$ treatments and $n = 5$ blocks, can be written as follows:

$$(24.17) \quad Y_{ij} = \mu_{..} + \underbrace{\rho_1 X_{ij1} + \rho_2 X_{ij2} + \rho_3 X_{ij3} + \rho_4 X_{ij4}}_{\text{Block effect}} + \underbrace{\tau_1 X_{ij5} + \tau_2 X_{ij6}}_{\text{Treatment effect}} + \varepsilon_{ij}$$

where:

$$X_{ij1} = \begin{array}{l} 1 \text{ if experimental unit from block 1} \\ -1 \text{ if experimental unit from block 5} \\ 0 \text{ otherwise} \end{array}$$

X_{ij2}, X_{ij3}, X_{ij4} are defined similarly

$$X_{ij5} = \begin{array}{l} 1 \text{ if experimental unit received treatment 1} \\ -1 \text{ if experimental unit received treatment 3} \\ 0 \text{ otherwise} \end{array}$$

$$X_{ij6} = \begin{array}{l} 1 \text{ if experimental unit received treatment 2} \\ -1 \text{ if experimental unit received treatment 3} \\ 0 \text{ otherwise} \end{array}$$

The **Y** observations vector and **X** matrix for the risk premium example are shown in Table 24.8. Note, for instance, that for observation Y_{13}, the indicator variables have the values:

$$X_1 = 1 \quad X_2 = 0 \quad X_3 = 0 \quad X_4 = 0 \quad X_5 = -1 \quad X_6 = -1$$

TABLE 24.8   Data Matrices for Regression Model (24.17) Based on Risk Premium Data in Table 24.2

	X_1	X_2	X_3	X_4	X_5	X_6
$Y_{11} = 1$	1	1	0	0	0	1	0
$Y_{12} = 5$	1	1	0	0	0	0	1
$Y_{13} = 8$	1	1	0	0	0	-1	-1
$Y_{21} = 2$	1	0	1	0	0	1	0
$Y_{22} = 8$	1	0	1	0	0	0	1
$Y_{23} = 14$	1	0	1	0	0	-1	-1
$Y_{31} = 7$	1	0	0	1	0	1	0
$Y_{32} = 9$	1	0	0	1	0	0	1
$Y_{33} = 16$	1	0	0	1	0	-1	-1
$Y_{41} = 6$	1	0	0	0	1	1	0
$Y_{42} = 13$	1	0	0	0	1	0	1
$Y_{43} = 18$	1	0	0	0	1	-1	-1
$Y_{51} = 12$	1	-1	-1	-1	-1	1	0
$Y_{52} = 14$	1	-1	-1	-1	-1	0	1
$Y_{53} = 17$	1	-1	-1	-1	-1	-1	-1

$\mathbf{Y} =$ (left column of observations) $\mathbf{X} =$ (right matrix)

Hence:

$$Y_{13} = \mu_{..} + \rho_1 - \tau_1 - \tau_2 + \varepsilon_{13}$$

$$= \mu_{..} + \rho_1 + \tau_3 + \varepsilon_{13}$$

because $\tau_3 = -\tau_1 - \tau_2$.

Tests and estimation of treatment effects with the regression approach follow easily and will not be discussed further here.

24.11 COVARIANCE ANALYSIS FOR RANDOMIZED BLOCK DESIGNS

Covariance analysis can be employed to further reduce the experimental error variability in a randomized block design. The extension is a straightforward one from covariance analysis for a completely randomized design.

Covariance Model

The usual randomized block design model was given in (24.2):

$$(24.18) \qquad Y_{ij} = \mu_{..} + \rho_i + \tau_j + \varepsilon_{ij}$$

$$i = 1, \ldots, n; j = 1, \ldots, r$$

The covariance model for a randomized block design with one concomitant variable is obtained by simply adding a term (or several terms) for the relation between the dependent variable Y and the concomitant variable X. Assuming this relation can be described by a linear function, we obtain:

$$(24.19) \qquad Y_{ij} = \mu_{..} + \rho_i + \tau_j + \gamma(X_{ij} - \bar{X}_{..}) + \varepsilon_{ij}$$

$$i = 1, \ldots, n; j = 1, \ldots, r$$

Regression Approach

The regression approach to covariance model (24.19) involves no new principles. As in Chapter 23, we shall denote $X_{ij} - \bar{X}_{..}$ in covariance model (24.19) by x_{ij}:

$$(24.20) \qquad x_{ij} = X_{ij} - \bar{X}_{..}$$

Further, we shall again use $1, -1, 0$ indicator variables for the block and treatment effects.

Suppose in a randomized complete block design study, $n = 4$ blocks and $r = 3$ treatments are utilized. The regression model counterpart to covariance model (24.19) then is:

$$(24.21) \qquad Y_{ij} = \mu_{..} + \rho_1 I_{ij1} + \rho_2 I_{ij2} + \rho_3 I_{ij3} + \tau_1 I_{ij4}$$

$$+ \tau_2 I_{ij5} + \gamma x_{ij} + \varepsilon_{ij} \qquad \text{Full model}$$

where:

$$I_{ij1} = \begin{array}{l} 1 \text{ if experimental unit from block 1} \\ -1 \text{ if experimental unit from block 4} \\ 0 \text{ otherwise} \end{array}$$

$$I_{ij2} = \begin{array}{l} 1 \text{ if experimental unit from block 2} \\ -1 \text{ if experimental unit from block 4} \\ 0 \text{ otherwise} \end{array}$$

$$I_{ij3} = \begin{array}{l} 1 \text{ if experimental unit from block 3} \\ -1 \text{ if experimental unit from block 4} \\ 0 \text{ otherwise} \end{array}$$

$$I_{ij4} = \begin{array}{l} 1 \text{ if experimental unit received treatment 1} \\ -1 \text{ if experimental unit received treatment 3} \\ 0 \text{ otherwise} \end{array}$$

$$I_{ij5} = \begin{array}{l} 1 \text{ if experimental unit received treatment 2} \\ -1 \text{ if experimental unit received treatment 3} \\ 0 \text{ otherwise} \end{array}$$

To test for treatment effects:

(24.22)
$$H_0: \tau_1 = \tau_2 = \tau_3 = 0$$
$$H_a: \text{not all } \tau_j \text{ equal zero}$$

we would either need to fit the reduced model under H_0:

(24.23)
$$Y_{ij} = \mu_{..} + \rho_1 I_{ij1} + \rho_2 I_{ij2} + \rho_3 I_{ij3} + \gamma x_{ij} + \varepsilon_{ij} \qquad \text{Reduced model}$$

or else use the appropriate extra sum of squares. The test for treatment effects would then be conducted in the usual way.

Comparisons of two treatment effects by the regression approach are straightforward. For estimating $\tau_1 - \tau_2$, for instance, we use the unbiased estimator $\hat{\tau}_1 - \hat{\tau}_2$ based on the estimated regression coefficients obtained when fitting the full model (24.21). The estimated variance of this estimator is:

(24.24)
$$s^2\{\hat{\tau}_1 - \hat{\tau}_2\} = s^2\{\hat{\tau}_1\} + s^2\{\hat{\tau}_2\} - 2s\{\hat{\tau}_1, \hat{\tau}_2\}$$

The estimated variance-covariance matrix of the regression coefficients from the computer printout can then be used to provide the required estimated variances and covariances.

CITED REFERENCES

24.1. Cochran, W. G., and G. M. Cox. *Experimental Designs*. 2nd ed. New York: John Wiley & Sons, 1957.

24.2. *MINITAB Reference Manual, Release 7*. State College, Pa.: Minitab, Inc., 1989.

PROBLEMS

24.1. Give an example of an experimental study where a repetition is not a replication.

24.2. In an experiment to study the effect of the location of a product display in drugstores of a chain, the manager of one of the drugstores rearranged the displays of other products so as to increase the traffic flow at the experimental display. Does this action potentially lead to either selection bias or measurement bias? Discuss.

24.3. In a study of the effect of size of team on the volume of communications within the team, can a double-blind procedure be utilized? A single-blind procedure? Discuss.

24.4. A student commented in a discussion group: "Random permutations are used to assign treatments to experimental units with a randomized block design just as with a completely randomized design. Hence, there is no basic difference between these two designs." Comment.

24.5. a. What might be some useful blocking variables for an experiment about the effects of different price levels on sales of a product, using stores as experimental units?

   b. What might be some useful blocking variables for an experiment about the effects of different flight crew schedules on the morale of crews, using flight crews as experimental units?

   c. What might be some useful blocking variables for an experiment about the effects of different drugs on the speed of a response to a stimulus, using laboratory animals as experimental units?

24.6. Five treatments are studied in an experiment with a randomized complete block design using four blocks. Obtain randomized assignments of treatments to experimental units.

24.7. Two treatments and a control are studied in an experiment with a randomized block design. Five blocks are employed, each containing four experimental units. In each block, each treatment is to be assigned to one experimental unit, and the control is to be assigned to two experimental units. Obtain randomized assignments of treatments to experimental units.

24.8. Auditor training. An accounting firm, prior to introducing in the firm widespread training in statistical sampling for auditing, tested three training methods: (1) study at home with programmed training materials, (2) training sessions at local offices conducted by local staff, and (3) training session in Chicago conducted by national staff. Thirty auditors were grouped into 10 blocks of three, according to time elapsed since college graduation, and the auditors in each block were randomly assigned to the three training methods. At the end of the training, each auditor was asked to analyze a complex case involving statistical applications; a proficiency measure based on this analysis was obtained for each auditor. The results were (block 1 consists of auditors graduated most recently, block 10 consists of those graduated most distantly):

Block	Training Method (j)			Block	Training Method (j)		
i	1	2	3	i	1	2	3
1	73	81	92	6	73	75	86
2	76	78	89	7	68	72	88
3	75	76	87	8	64	74	82
4	74	77	90	9	65	73	81
5	76	71	88	10	62	69	78

a. Why do you think the blocking variable "time elapsed since college graduation" was employed?

b. Obtain the residuals for randomized block model (24.2) and plot them against the fitted values. Also prepare a normal probability plot of the residuals. What are your findings?

c. Plot the responses Y_{ij} by blocks in the format of Figure 24.1. What does this plot suggest about the appropriateness of the no-interaction assumption here?

d. Conduct the Tukey test for additivity of block and treatment effects; use $\alpha = .01$. State the alternatives, decision rule, and conclusion. What is the P-value of the test?

24.9. Refer to **Auditor training** Problem 24.8. Assume that randomized block model (24.2) is appropriate.

a. Obtain the analysis of variance table.

b. Prepare a normal probability plot of the estimated treatment means. Does it appear that the treatment means differ substantially here?

c. Test whether or not the mean proficiency is the same for the three training methods. Use a level of significance of $\alpha = .05$. State the alternatives, decision rule, and conclusion. What is the P-value of the test?

d. Make all pairwise comparisons between the training method means; use the Tukey procedure with a 90 percent family confidence coefficient. State your findings.

e. Test whether or not blocking effects are present; use $\alpha = .05$. State the alternatives, decision rule, and conclusion. What is the P-value of the test?

24.10. **Fat in diets.** A researcher studied the effects of three experimental diets with varying fat contents on the total lipid (fat) level in plasma. Total lipid level is a widely used predictor of coronary heart disease. Fifteen male subjects who were within 20 percent of their ideal body weight were grouped into five blocks according to age. Within each block, the three experimental diets were randomly assigned to the three subjects. Data on reduction in lipid level (in grams per liter) after the subjects were on the diet for a fixed period of time follow.

| Block i | Fat Content of Diet | | |
	$j = 1$ Extremely Low	$j = 2$ Fairly Low	$j = 3$ Moderately Low
1 Ages 15–24	.73	.67	.15
2 Ages 25–34	.86	.75	.21
3 Ages 35–44	.94	.81	.26
4 Ages 45–54	1.40	1.32	.75
5 Ages 55–64	1.62	1.41	.78

a. Why do you think that age of subject was used as a blocking variable?

b. Obtain the residuals for randomized block model (24.2) and plot them against the fitted values. Also prepare a normal probability plot of the residuals. What are your findings?

c. Plot the responses Y_{ij} by blocks in the format of Figure 24.1. What does this plot suggest about the appropriateness of the no-interaction assumption here?

d. Conduct the Tukey test for additivity of block and treatment effects; use $\alpha = .01$. State the alternatives, decision rule, and conclusion. What is the P-value of the test?

24.11. Refer to **Fat in diets** Problem 24.10. Assume that randomized block model (24.2) is appropriate.

a. Obtain the analysis of variance table.

b. Prepare a normal probability plot of the estimated treatment means. Does it appear that the treatment means differ substantially here?

c. Test whether or not the mean reductions in lipid level differ for the three diets; use $\alpha = .05$. State the alternatives, decision rule, and conclusion. What is the P-value of the test?

d. Estimate $D_1 = \mu_{.1} - \mu_{.2}$ and $D_2 = \mu_{.2} - \mu_{.3}$ using the Bonferroni procedure with a 95 percent family confidence coefficient. State your findings.

e. Test whether or not blocking effects are present; use $\alpha = .05$. State the alternatives, decision rule, and conclusion. What is the P-value of the test?

f. A standard diet was not used in this experiment as a control. What justification do you think the experimenters might give for not having a control treatment here for comparative purposes?

24.12. Dental pain. An anesthesiologist made a comparative study of the effects of acupuncture and codeine on postoperative dental pain in male subjects. The four treatments were: (1) placebo treatment—a sugar capsule and two inactive acupuncture points $(A_1 B_1)$, (2) codeine treatment only—a codeine capsule and two inactive acupuncture points $(A_2 B_1)$, (3) acupuncture treatment only—a sugar capsule and two active acupuncture points $(A_1 B_2)$, and (4) codeine and acupuncture treatment—a codeine capsule and two active acupuncture points $(A_2 B_2)$. Thirty-two subjects were grouped into eight blocks of four according to an initial evaluation of their level of pain tolerance. The subjects in each block were then randomly assigned to the four treatments. Pain relief scores were obtained for all subjects two hours after dental treatment. Data were collected on a double-blind basis. The data on pain relief scores follow (the higher the pain relief score, the more effective the treatment).

Block	Treatment (j, k)			
i	$A_1 B_1$	$A_2 B_1$	$A_1 B_2$	$A_2 B_2$
1 (Lowest)	0.0	.5	.6	1.2
2	.3	.6	.7	1.3
3	.4	.8	.8	1.6
4	.4	.7	.9	1.5
5	.6	1.0	1.5	1.9
6	.9	1.4	1.6	2.3
7	1.0	1.8	1.7	2.1
8 (Highest)	1.2	1.7	1.6	2.4

a. Why do you think that pain tolerance of the subjects was used as a blocking variable?

b. Which of the assumptions involved in randomized block model (24.10) are you most concerned with here?

c. Obtain the residuals for randomized block model (24.10) and plot them against the fitted values. Also prepare a normal probability plot of the residuals. What are your findings?

d. Plot the responses Y_{ijk} by blocks in the format of Figure 24.1, ignoring the factorial structure of the treatments. What does this plot suggest about the appropriateness of the no-interaction assumption here?

   e. Conduct the Tukey test for additivity of block and treatment effects, ignoring the factorial structure of the treatments; use $\alpha = .01$. State the alternatives, decision rule, and conclusion. What is the P-value of the test?

24.13. Refer to **Dental pain** Problem 24.12. Assume that randomized block model (24.10) is appropriate.
   a. Obtain the analysis of variance table.
   b. Test whether or not the two factors interact; use $\alpha = .01$. State the alternatives, decision rule, and conclusion. What is the P-value of the test?
   c. Prepare separate normal probability plots for each set of estimated factor level means. Does it appear that substantial main effects are present here?
   d. Test separately whether main effects are present for each of the factors; use $\alpha = .01$ for each test. State the alternatives, decision rule, and conclusion for each test. What is the P-value of each test?
   e. Estimate:

$$D_1 = \mu_{.1.} - \mu_{.2.} = \alpha_1 - \alpha_2$$
$$D_2 = \mu_{..1} - \mu_{..2} = \beta_1 - \beta_2$$

   Use the Bonferroni procedure with a 95 percent family confidence coefficient. State your findings.
   f. Test whether or not blocking effects are present; use $\alpha = .01$. State the alternatives, decision rule, and conclusion. What is the P-value of the test?

24.14. Refer to **Dental pain** Problem 24.12. Assume that randomized block model (24.11) with fixed effects is to be employed.
   a. Obtain the analysis of variance table.
   b. Test whether or not the two factors interact; use $\alpha = .01$. State the alternatives, decision rule, and conclusion. What is the P-value of the test?
   c. Test separately whether or not blocks interact with each of the factors. For each test, use $\alpha = .01$ and state the alternatives, decision rule, and conclusion. What is the P-value for each test? What do your conclusions imply about the choice between randomized block models (24.10) and (24.11)? Discuss.
   d. Test separately whether main effects are present for each of the factors; use $\alpha = .01$. State the alternatives, decision rule, and conclusion for each test. What is the P-value of each test? Did the choice of models affect your conclusions here?

24.15. Refer to **Auditor training** Problems 24.8 and 24.9. Assume that $\sigma = 2.5$. What is the power of the test for training method effects in Problem 24.9c if $\mu_{.1} = 70$, $\mu_{.2} = 73$, and $\mu_{.3} = 76$?

24.16. Refer to **Fat in diets** Problems 24.10 and 24.11. Assume that $\sigma = .04$. What is the power of the test for diet effects in Problem 24.11c if $\mu_{.1} = 1.1$, $\mu_{.2} = 1.0$, and $\mu_{.3} = .9$?

24.17. Refer to **Auditor training** Problem 24.8. Another accounting firm wishes to conduct the same experiment with some of its auditors, using the same design and model. How many blocks would you recommend that this firm employ if it wishes to make all pairwise treatment comparisons with precision ± 1.5 with a 99 percent family confidence coefficient? Assume that a reasonable planning value for the error standard deviation in model (24.2) is $\sigma = 2.5$.

24.18. Refer to **Fat in diets** Problem 24.10. Suppose that the number of blocks to be used in the study, to consist of male subjects of similar ages, has not yet been determined.

Assume that a reasonable planning value for the error standard deviation in model (24.2) is $\sigma = .04$.

    a. What would be the required number of blocks if it is desired to make all pairwise diet comparisons with precision $\pm.03$ with a 95 percent family confidence coefficient?

    b. What would be the required number of blocks if: (1) differences in lipid level reduction means for the three diets are to be detected with probability .95 or more when the range of the treatment means is .12, and (2) the α risk is to be controlled at .01?

24.19. Refer to **Auditor training** Problems 24.8 and 24.9. Based on the estimated efficiency measure (24.14), how effective was the use of the blocking variable as compared to a completely randomized design?

24.20. Refer to **Fat in diets** Problems 24.10 and 24.11. Based on the estimated efficiency measure (24.15), how effective was the use of the blocking variable as compared to a completely randomized design?

24.21. Refer to **Dental pain** Problems 24.12 and 24.13. Based on the estimated efficiency measure (24.14), how effective was the use of the blocking variable as compared to a completely randomized design?

24.22. Refer to **Auditor training** Problem 24.8.

    a. State the regression model equivalent to randomized block model (24.2); use 1, -1, 0 indicator variables.

    b. Fit the regression model to the data.

    c. Obtain the regression analysis of variance table based on appropriate extra sums of squares.

    d. Test for treatment main effects; use $\alpha = .05$. State the alternatives, decision rule, and conclusion.

24.23. Refer to **Fat in diets** Problem 24.10.

    a. State the regression model equivalent to randomized block model (24.2); use 1, -1, 0 indicator variables.

    b. Fit the regression model to the data.

    c. Obtain the regression analysis of variance table based on appropriate extra sums of squares.

    d. Test for treatment main effects; use $\alpha = .05$. State the alternatives, decision rule, and conclusion.

24.24. Refer to **Auditor training** Problem 24.8. The analyst wishes to examine whether use of pretraining statistical proficiency scores as a concomitant variable would help to reduce the experimental error variability significantly. The pretraining statistical proficiency scores for the auditors are as follows:

Block i	Training Method (j) 1	2	3	Block i	Training Method (j) 1	2	3
1	93	98	91	6	75	74	78
2	94	93	94	7	79	76	72
3	89	91	92	8	71	69	64
4	86	84	90	9	74	71	70
5	78	76	84	10	63	68	64

a. Would you expect the auditor's age to have been a better concomitant variable here than the pretraining statistical proficiency score? Discuss.

b. State the regression model equivalent to covariance model (24.19); use 1, −1, 0 indicator variables.

c. Fit the full regression model.

d. State the reduced regression model for testing treatment effects. Fit the reduced model.

e. Test whether or not the training methods differ in mean effectiveness. Use a level of significance of $\alpha = .05$. State the alternatives, decision rule, and conclusion. What is the P-value of the test?

f. Obtain a 95 percent confidence interval for $D = \tau_1 - \tau_2$. Interpret your interval estimate.

g. Has the error variance been reduced substantially by adding the concomitant variable? Explain.

24.25. Refer to **Fat in diets** Problem 24.10. The researcher wishes to examine whether each subject's body weight expressed as a percent of the ideal weight for that person would be a useful concomitant variable. The body weights as percents of the ideal weights for the 15 male subjects are as follows:

Block	Fat Content of Diet		
i	$j = 1$	$j = 2$	$j = 3$
1	94	96	101
2	97	102	99
3	105	100	106
4	108	107	112
5	118	115	107

a. State the regression model equivalent to covariance model (24.19); use 1, −1, 0 indicator variables.

b. Fit the full regression model.

c. State the reduced regression model for testing treatment effects. Fit the reduced model.

d. Test whether or not the mean reductions in lipid level differ for the three diets; use $\alpha = .05$. State the alternatives, decision rule, and conclusion. What is the P-value of the test?

e. Obtain confidence intervals for $D_1 = \tau_1 - \tau_2$ and $D_2 = \tau_2 - \tau_3$, using the Bonferroni procedure with a 95 percent family confidence coefficient. Interpret your interval estimates.

f. Has the error variance been reduced substantially by adding the concomitant variable? Explain.

EXERCISES

24.26. Consider randomized block model (24.2), but with random treatment effects. Derive $\sigma^2\{Y_{ij}\}$ and $\sigma^2\{\bar{Y}_{.j}\}$.

24.27. (Calculus needed.) State the likelihood function for the randomized block fixed effects model (24.2) when $n = 3$ and $r = 2$. Find the maximum likelihood estimators of the parameters. Are they the same as the least squares estimators in (24.3)?

24.28. For randomized block fixed effects model (24.2), derive $E\{MSTR\}$.

24.29. Show that when two treatments are studied in a randomized complete block design, the F^* test statistic (24.7b) for treatment effects is equivalent to the square of the two-sided t test statistic for paired observations based on (1.66).

24.30. Refer to regression model (24.17), the equivalent to ANOVA model (24.16) when $n = 5$ and $r = 3$. Suppose that the indicator variables in model (24.17) were coded as follows:

$$X_{ij1} = \begin{array}{l} 1 \text{ if experimental unit from block 1} \\ 0 \text{ otherwise} \end{array}$$

$X_{ij2}, X_{ij3}, X_{ij4}$ are defined similarly

$$X_{ij5} = \begin{array}{l} 1 \text{ if experimental unit received treatment 1} \\ 0 \text{ otherwise} \end{array}$$

$$X_{ij6} = \begin{array}{l} 1 \text{ if experimental unit received treatment 2} \\ 0 \text{ otherwise} \end{array}$$

and that the regression coefficients are denoted by $\beta_0, \beta_1, \beta_2, \beta_3, \beta_4, \beta_5,$ and β_6.
  a. Exhibit the **X** matrix for this regression model.
  b. Find the correspondences between the regression coefficients $\beta_0, \beta_1, \ldots, \beta_6$ and the parameters in ANOVA model (24.16).
  c. Discuss the advantages and disadvantages of using 1, 0 indicator variables and 1, -1, 0 indicator variables here.

PROJECT

24.31. Refer to comment 3 on page 913 and model (24.1). Suppose that the quantities depending on the experimental unit are as follows for eight subjects in a psychological reinforcement experiment:

j:	1	2	3	4	5	6	7	8
Experimental unit quantity:	16	14	18	16	12	15	13	12

An experimental treatment is to be compared with a standard treatment, with four subjects assigned at random to each treatment.
  a. Suppose that there are no differential treatment effects, with the quantity depending on the treatment being 4 for each treatment. Generate all possible randomizations of the eight experimental units to the two treatments and obtain the observed values for each treatment sample.
  b. For each of the 70 randomizations obtained in part (a), calculate F^* for testing the alternatives $H_0: \mu_1 = \mu_2$ versus $H_a: \mu_1 \neq \mu_2$. Determine the proportion of F^* values that exceed $F(.90; 1, 6)$, the proportion that exceed $F(.95; 1, 6)$, and the proportion that exceed $F(.99; 1, 6)$.
  c. How do the proportions obtained in part (b) compare with the probabilities for the normal error model? Discuss.
  d. Repeat parts (a) and (b) for the case where the treatment quantities for the experimental and standard treatments are, respectively, 15 and 4. Does the test appear to have reasonable power in this case?

Chapter 25

Randomized Block Designs—II

In this chapter, we continue our discussion of randomized block designs by first considering binary responses for the dependent variable and a nonparametric test for treatment effects. Then we take up missing observations and random block effects. We conclude the chapter by discussing generalized randomized block designs and the use of more than one blocking variable.

25.1 BINARY RESPONSES FOR DEPENDENT VARIABLE

In some randomized block experiments, the responses are binary. For instance, the responses in an experiment involving shopping behavior may be buy or not buy. In an experiment on task motivation, the responses may be success in task or failure in task. In each of these cases, the responses can be coded 1 or 0.

Cochran Test

When the responses in a randomized block experiment are binary, coded 1 or 0, a chi-square test may be used for assessing the presence of treatment effects. To test whether treatment effects are present:

(25.1)
$$H_0: \text{all } \tau_j = 0$$
$$H_a: \text{not all } \tau_j \text{ equal zero}$$

the following test statistic due to Cochran may be used:

(25.2)
$$X_C^2 = SSTR \div \frac{SSTR + SSBL.TR}{n(r - 1)}$$

which reduces to:

$$(25.2a) \qquad X_C^2 = \frac{(r-1)\left(r\sum_j Y_{.j}^2 - Y_{..}^2\right)}{rY_{..} - \sum_i Y_{i.}^2}$$

where the usual notation is employed.

The number of 1s in each block may differ because of block-to-block differences. Given the number of 1s in each block, if all treatments have the same effect, all permutations of 0s and 1s within a block are equally likely. It can then be shown that when H_0 holds, X_C^2 is distributed approximately as χ^2 with $r - 1$ degrees of freedom provided the number of blocks is not very small. Large values of X_C^2 lead to conclusion H_a.

Example. Table 25.1 contains data for an experiment in which 15 teams were grouped into $n = 5$ blocks of three teams each, according to a criterion on the creativity of each team. Within each block the teams were assigned at random to one of $r = 3$ instruction methods, and upon completion of training each team was given the same complex task to perform. Success in the task was coded 1, failure 0.

To test whether the instruction methods have differential effects on successful performance, we use test statistic (25.2a) and obtain:

$$X_C^2 = \frac{2[3(29) - (9)^2]}{3(9) - 19} = 1.5$$

The level of significance is to be controlled at $\alpha = .05$. We require then $\chi^2(.95; 2) = 5.99$. Since $X_C^2 = 1.5 \le 5.99$, we conclude that the instruction methods do not differ in effectiveness. The P-value of this test is .47.

TABLE 25.1   Binary Response Experiment in Randomized Block Design

Block	Instruction Method (j)			
i	1	2	3	Total
1 (high creativity)	1	1	1	3
2	1	0	1	2
3	1	0	0	1
4	0	1	0	1
5 (low creativity)	1	0	1	2
Total	4	2	3	9

$$\sum Y_{.j}^2 = 4^2 + 2^2 + 3^2 = 29$$

$$\sum Y_{i.}^2 = 3^2 + 2^2 + 1^2 + 1^2 + 2^2 = 19$$

Multiple Pairwise Testing Procedure

When the Cochran test leads to the conclusion that treatment effects are present, pairwise tests based on the treatment means $\bar{Y}_{.j}$ can be set up in a similar fashion to that for the Kruskal-Wallis test described in Section 17.4, provided the number of blocks is not too small. Testing limits for all $g = r(r-1)/2$ pairwise tests with family significance level α are set up as follows:

$$(25.3) \qquad \bar{Y}_{.j} - \bar{Y}_{.j'} \pm B\left[\left(\frac{rY_{..} - \Sigma Y_{i.}^2}{nr(r-1)}\right)\left(\frac{2}{n}\right)\right]^{1/2}$$

where:

$$(25.3a) \qquad B = z(1 - \alpha/2g)$$

$$(25.3b) \qquad g = \frac{r(r-1)}{2}$$

If the testing limits include zero, we conclude that the corresponding treatment means $\mu_{.j}$ and $\mu_{.j'}$ do not differ. If the testing limits do not include zero, we conclude that the two corresponding treatment means differ.

25.2 FRIEDMAN RANK TEST

When the observations Y_{ij} in a randomized complete block design are far from normally distributed and transformations of the data are of little help, a nonparametric test of treatment effects may be utilized. This test, called the Friedman test, is based on the ranks of the data in each block.

Friedman Test Statistic

The observations Y_{ij} for each block are first ranked. Let R_{ij} denote the rank of Y_{ij} when the observations in the ith block are ranked from 1 to r. The Friedman test statistic then is:

$$(25.4) \qquad X_F^2 = SSTR \div \frac{SSTR + SSBL.TR}{n(r-1)}$$

which can be reduced to (when no ties are present):

$$(25.4a) \qquad X_F^2 = \left[\frac{12}{nr(r+1)}\sum_j R_{.j}^2\right] - 3n(r+1)$$

where $R_{.j}$ is the sum of the ranks for the jth treatment.

If there are no treatment differences, all ranking permutations for a block are assumed to be equally likely and the statistic X_F^2 will be distributed approximately as χ^2 with $r - 1$ degrees of freedom, provided the number of blocks is not too small.

Large values of the test statistic lead to the conclusion that the treatments have un-equal effects. If the number of blocks is small, tables such as those in Reference 25.1 may be used for an exact test.

Example. Table 25.2a contains data for a randomized complete block design exper-iment in which 15 stores were grouped into 5 blocks according to size of the store, and three different layouts of the produce section were assigned at random to the 3 stores in each block. The data in Table 25.2a are produce sales (in thousands of dollars) for each of the 15 stores. Because these sales data appear to depart substan-tially from normality, it was decided to use the Friedman test to examine whether mean sales differ for the three layouts. The ranks of the sales data in Table 25.2a are shown in Table 25.2b. Using the results from there, we obtain from (25.4a):

$$X_F^2 = \left[\frac{12}{5(3)(4)} (324) \right] - 3(5)(4) = 4.80$$

To test the alternatives:

$$H_0: \mu_{.1} = \mu_{.2} = \mu_{.3}$$

$$H_a: \text{not all } \mu_{.j} \text{ are equal}$$

where $\mu_{.j}$ is the mean for treatment j averaged over all blocks, we use the decision rule:

$$\text{If } X_F^2 \leq \chi^2(1 - \alpha; r - 1), \text{ conclude } H_0$$

$$\text{If } X_F^2 > \chi^2(1 - \alpha; r - 1), \text{ conclude } H_a$$

For level of significance $\alpha = .10$, we need $\chi^2(.90; 2) = 4.61$. Since $X_F^2 = 4.80 > 4.61$, we conclude H_a, that the mean sales for the three layouts are not equal. The P-value of this test is .09.

TABLE 25.2   Sales Data and Ranks for Produce Layout Example

(a) Sales ($ thousands)				**(b) Ranks**			
Block	Layout (j)			Block	Layout (j)		
i	1	2	3	i	1	2	3
1	75.3	69.8	87.7	1	2	1	3
2	64.3	70.0	71.1	2	1	2	3
3	59.0	45.4	71.8	3	2	1	3
4	44.2	35.1	61.0	4	2	1	3
5	21.7	59.9	25.3	5	1	3	2
				$R_{.j}$	8	8	14
				$\bar{R}_{.j}$	1.6	1.6	2.8
				$\sum R_{.j}^2 = (8)^2 + (8)^2 + (14)^2 = 324$			

Comments

1. If ties are present in the data for a treatment, the mean rank may be assigned to the tied observations and test statistic (25.4) still used, provided that there are not too many ties in the entire data set.

2. Test statistic (25.4) is of the same form as the Cochran test statistic (25.2) but applied to ranks instead of 0, 1 data.

Multiple Pairwise Testing Procedure

Just like in the case of the Kruskal-Wallis test for single-factor studies (Section 17.4), we can use a large-sample testing analogue of the Bonferroni pairwise comparison procedure to obtain information about the comparative magnitudes of the treatment means for randomized block designs when the Friedman test indicates that the treatment means differ. Testing limits for all $g = r(r - 1)/2$ pairwise comparisons using the mean ranks $\bar{R}_{.j}$ are set up as follows for family level of significance α:

$$(25.5) \qquad \bar{R}_{.j} - \bar{R}_{.j'} \pm B\left[\frac{r(r + 1)}{6n}\right]^{1/2}$$

where:

$$(25.5a) \qquad B = z(1 - \alpha/2g)$$

$$(25.5b) \qquad g = \frac{r(r - 1)}{2}$$

If the testing limits include zero, we conclude that the corresponding treatment means $\mu_{.j}$ and $\mu_{.j'}$ do not differ. If the testing limits do not include zero, we conclude that the two corresponding treatment means differ. We can then set up groups of treatments whose means do not differ according to this simultaneous testing procedure.

Example.   For the produce layout example, we wish to make all pairwise tests with family level of significance $\alpha = .20$. The mean ranks $\bar{R}_{.j}$ are presented in Table 25.2b. For $r = 3$, we have $g = 3(2)/2 = 3$. Hence, we obtain for $n = 5$ that $B = z[1 - .20/2(3)] = z(.9667) = 1.834$. Thus, the right term in (25.5) is:

$$B\left[\frac{r(r + 1)}{6n}\right]^{1/2} = 1.834\left[\frac{3(4)}{6(5)}\right]^{1/2} = 1.16$$

We note from Table 25.2b that the differences between the mean ranks for layout 3 and each of the other two layouts exceed 1.16 and are therefore significant. Hence, we can set up two groups, within which the treatment means do not differ:

Group 1		Group 2	
Layout 1	$\bar{R}_{.1} = 1.6$	Layout 3	$\bar{R}_{.3} = 2.8$
Layout 2	$\bar{R}_{.2} = 1.6$		

Thus, we can conclude with family level of significance of .20 that layout 3 leads to larger mean sales than do layouts 1 and 2.

25.3 MISSING OBSERVATIONS

There are occasions when one or several observations in a randomized complete block design are "missing"—a subject may have been sick, a record may have been mislaid, a treatment may have been applied incorrectly in one instance. Such missing observations destroy the symmetry (orthogonality) of the complete block design and make the usual ANOVA calculations inappropriate. However, the regression approach to the analysis of randomized block designs, discussed in Section 24.10, is ordinarily still appropriate when there are missing observations. The reason is that the no-interaction randomized block design model (24.2) enables us, in effect, to estimate the mean response for a missing cell. We explained earlier how this is done for a two-factor no-interaction model (Section 21.1), and the same reasoning applies here.

Since no new principles are involved, we turn to an example to illustrate the use of the regression approach when observations are missing in a randomized block design experiment.

Example

Table 25.3 contains the data for a simple randomized block design experiment with $r = 3$ treatments and $n = 3$ blocks, where observation Y_{11} is missing. We set up the regression model equivalent of the randomized block design model (24.2) as follows:

$$(25.6) \qquad Y_{ij} = \mu_{..} + \underbrace{\rho_1 X_{ij1} + \rho_2 X_{ij2}}_{\text{Block effect}} + \underbrace{\tau_1 X_{ij3} + \tau_2 X_{ij4}}_{\text{Treatment effect}} + \varepsilon_{ij} \qquad \text{Full Model}$$

where:

$$X_{ij1} = \begin{array}{l} 1 \text{ if experimental unit from block 1} \\ -1 \text{ if experimental unit from block 3} \\ 0 \text{ otherwise} \end{array}$$

$$X_{ij2} = \begin{array}{l} 1 \text{ if experimental unit from block 2} \\ -1 \text{ if experimental unit from block 3} \\ 0 \text{ otherwise} \end{array}$$

$$X_{ij3} = \begin{array}{l} 1 \text{ if experimental unit received treatment 1} \\ -1 \text{ if experimental unit received treatment 3} \\ 0 \text{ otherwise} \end{array}$$

$$X_{ij4} = \begin{array}{l} 1 \text{ if experimental unit received treatment 2} \\ -1 \text{ if experimental unit received treatment 3} \\ 0 \text{ otherwise} \end{array}$$

TABLE 25.3   Example of Missing Observation in Randomized Block Design
($r = 3$, $n = 3$)

Block	Treatment (j)		
i	1	2	3
1	Missing	10	9
2	11	10	7
3	6	4	3

Table 25.4 exhibits the **Y**, **X**, and **β** matrices for the full model (25.6) for this example where observation Y_{11} is missing.

The analysis of variance for testing treatment effects and block effects is carried out in the usual manner by first fitting the full model (25.6) and then fitting each of the following reduced models:

Test for block effects

$$(25.7) \qquad Y_{ij} = \mu_{..} + \tau_1 X_{ij3} + \tau_2 X_{ij4} + \varepsilon_{ij} \qquad \text{Reduced model}$$

Test for treatment effects

$$(25.8) \qquad Y_{ij} = \mu_{..} + \rho_1 X_{ij1} + \rho_2 X_{ij2} + \varepsilon_{ij} \qquad \text{Reduced model}$$

The extra sums of squares $SSR(X_1, X_2 \mid X_3, X_4)$ for blocks and $SSR(X_3, X_4 \mid X_1, X_2)$ for treatments are then calculated as always. Table 25.5a presents these extra sums of squares for our example obtained from computer runs, as well as the error sum of squares for the full model. No total sum of squares is shown because of lack of orthogonality as a result of the missing observation.

The test for treatment effects is conducted as usual. From Table 25.5a we find:

$$F^* = \frac{MSR(X_3, X_4 \mid X_1, X_2)}{MSE} = \frac{6.25}{.44} = 14.2$$

TABLE 25.4   **Y**, **X**, and **β** Matrices for Missing Observation Example of Table 25.3

$$
\mathbf{Y} = \begin{bmatrix} Y_{12} \\ Y_{13} \\ Y_{21} \\ Y_{22} \\ Y_{23} \\ Y_{31} \\ Y_{32} \\ Y_{33} \end{bmatrix} = \begin{bmatrix} 10 \\ 9 \\ 11 \\ 10 \\ 7 \\ 6 \\ 4 \\ 3 \end{bmatrix} \qquad
\mathbf{X} = \begin{bmatrix} & X_1 & X_2 & X_3 & X_4 \\ 1 & 1 & 0 & 0 & 1 \\ 1 & 1 & 0 & -1 & -1 \\ 1 & 0 & 1 & 1 & 0 \\ 1 & 0 & 1 & 0 & 1 \\ 1 & 0 & 1 & -1 & -1 \\ 1 & -1 & -1 & 1 & 0 \\ 1 & -1 & -1 & 0 & 1 \\ 1 & -1 & -1 & -1 & -1 \end{bmatrix} \qquad
\mathbf{\beta} = \begin{bmatrix} \mu_{..} \\ \rho_1 \\ \rho_2 \\ \tau_1 \\ \tau_2 \end{bmatrix}
$$

TABLE 25.5 ANOVA Table and Other Regression Output for Missing Data Example of Table 25.3

(a) ANOVA Table

Source of Variation	SS	df	MS
Blocks	53.83	2	26.92
Treatments	12.50	2	6.25
Error	1.33	3	.44

(b) Estimated Regression Coefficients for Full Model (25.6)

Regression Coefficient	Estimated Regression Coefficient
$\mu_{..}$	$\hat{\mu}_{..} = 8.000$
ρ_1	$\hat{\rho}_1 = 2.333$
ρ_2	$\hat{\rho}_2 = 1.333$
τ_1	$\hat{\tau}_1 = 1.667$
τ_2	$\hat{\tau}_2 = 0.0$

(c) Estimated Variance-Covariance Matrix of Regression Coefficients

	$\hat{\mu}_{..}$	$\hat{\rho}_1$	$\hat{\rho}_2$	$\hat{\tau}_1$	$\hat{\tau}_2$
$\hat{\mu}_{..}$	.06173				
$\hat{\rho}_1$	.02469	.14815			
$\hat{\rho}_2$	−.01235	−.07407	.11111		
$\hat{\tau}_1$	.02469	.04938	−.02469	.14815	
$\hat{\tau}_2$	−.01235	−.02469	.01235	−.07407	.11111

For $\alpha = .05$, we need $F(.95; 2, 3) = 9.55$. Since $F^* = 14.2 > 9.55$, we conclude that differential treatment effects are present. The P-value of this test is .03. The test for block effects can be carried out along similar lines when it is of interest.

No new problems are encountered with the regression approach in analyzing fixed treatment effects when there are missing observations. To estimate in our example the pairwise comparison $D = \mu_{.1} - \mu_{.3} = \tau_1 - \tau_3$, for instance, we utilize the fact that $\tau_3 = -\tau_1 - \tau_2$ so that we have:

$$(25.9) \qquad D = \mu_{.1} - \mu_{.3} = \tau_1 - \tau_3 = \tau_1 - (-\tau_1 - \tau_2) = 2\tau_1 + \tau_2$$

An unbiased estimator of (25.9) is:

$$(25.10) \qquad \hat{D} = 2\hat{\tau}_1 + \hat{\tau}_2$$

whose estimated variance is, using (1.27b):

$$(25.11) \qquad s^2\{\hat{D}\} = 4s^2\{\hat{\tau}_1\} + s^2\{\hat{\tau}_2\} + 4s\{\hat{\tau}_1, \hat{\tau}_2\}$$

Table 25.5b contains the estimated regression coefficients for the full model, and Table 25.5c contains the estimated variance-covariance matrix for the regression coefficients. We therefore obtain the following estimates:

$$\hat{D} = 2(1.667) + 0.0 = 3.334$$

$$s^2\{\hat{D}\} = 4(.14815) + .11111 + 4(-.07407) = .4074$$

so that the estimated standard deviation is $s\{\hat{D}\} = .638$. A 95 percent confidence interval for D requires $t(.975; 3) = 3.182$, yielding the confidence limits $3.334 \pm 3.182(.638)$ and the confidence interval:

$$1.3 \leq \mu_{.1} - \mu_{.3} \leq 5.4$$

Note

An alternative hand calculation procedure due to Yates is sometimes used when one or two observations are missing. Pseudo-observations for the missing values are obtained, and then the usual ANOVA calculations are carried out as if all observations were at hand. Finally, adjustments to these ANOVA calculations are made. See Reference 25.2 for complete details of this method. With the widespread availability of computer regression packages, the need for the Yates hand calculation procedure has greatly diminished.

25.4 RANDOM BLOCK EFFECTS

Sometimes the blocks can be considered a random sample from a population so that the block effects in the randomized block model should be considered to be random variables.

1. A researcher investigated the improvement in learning in third-grade classes by augmenting the teacher with one or two teaching assistants. Ten schools were selected at random, and three third-grade classes in each school were utilized in the study. In each school, one class was randomly chosen to have no teaching assistant, one class was randomly chosen to have one teaching assistant, and the third class was assigned two teaching assistants. The amount of learning by the class at the end of the school year, suitably measured, was the dependent variable.

Here the blocks are schools, which may be viewed as a random sample from the population of all schools eligible for the study.

2. In a study of the effectiveness of four different dosages of a drug, 20 litters of mice, each consisting of four mice, were utilized. The 20 litters (blocks) here may be viewed as a random sample from the population of all litters that could have been used for the study.

When blocks can be considered to be a random sample from a population of blocks, either an additive (i.e., no-interaction) or a nonadditive (i.e., interaction) model can be employed. The choice can be assisted by the diagnostics discussed in Chapter 24. In particular, plots of the responses Y_{ij} for each block, such as in Figure 24.1, can be helpful in examining whether blocks and treatments interact. A severe lack of parallelism in such a plot would be a clear indication that the interaction model may be preferable. The Tukey test statistic for interactions in (21.13) may also be utilized, with the interpretation here that the test applies to the given blocks that have been selected.

When the primary emphasis of the analysis is on testing and estimating treatment effects, which is the usual case, the choice between the two models actually is not critical because the inference procedures for fixed treatment effects, as we shall see, are exactly the same for the two models.

We first explain the additive, no-interaction model for randomized block designs with random block effects, and then we will take up the nonadditive model. In particular, we shall explore the nature of the correlations between experimental units within a block that is assumed by each model, because the usefulness of a model can often best be judged in terms of these assumed correlations.

A special case of random blocks occurs when the blocks are experimental units such as persons, stores, or cities, each receiving all of the treatments over time or having the effect of a given treatment (e.g., advertising) evaluated at different points of time. These designs are called *repeated measures designs* and will be discussed in Chapter 28.

Additive Model

The additive model for random block effects and fixed treatment effects is analogous to fixed effects model (24.2):

$$(25.12) \qquad Y_{ij} = \mu_{..} + \rho_i + \tau_j + \varepsilon_{ij}$$

where:

> $\mu_{..}$ is a constant
> ρ_i are independent $N(0, \sigma_\rho^2)$
> τ_j are constants subject to the restriction $\Sigma \tau_j = 0$
> ε_{ij} are independent $N(0, \sigma^2)$, and independent of the ρ_i
> $i = 1, \ldots, n; j = 1, \ldots, r$

Properties of Model.   The observations Y_{ij} for additive model (25.12) are normally distributed because each is a linear combination of independent normal variables. The mean and variance of Y_{ij} are:

$$(25.13a) \qquad E\{Y_{ij}\} = \mu_{..} + \tau_j$$

$$(25.13b) \qquad \sigma^2\{Y_{ij}\} = \sigma_Y^2 = \sigma_\rho^2 + \sigma^2$$

Thus, the variance of Y_{ij}, denoted by σ_Y^2, is a constant for all observations, but it is here made up of two components: (1) the variability between blocks σ_ρ^2, and (2) the error variance σ^2.

The additive model (25.12) assumes that observations from different blocks are independent. However, any two observations from the same block, Y_{ij} and $Y_{ij'}$, are correlated for this model. Their covariance can be shown to be:

$$(25.14) \qquad \sigma\{Y_{ij}, Y_{ij'}\} = \sigma_\rho^2 \qquad j \neq j'$$

Thus, any two observations from the same block are positively correlated, in advance of the random trials, the covariance being the same for all blocks. The reason for the correlation is that any two observations from the same random block will have the same component ρ_i, which will tend to make the two observations more alike. This positive covariance is reasonable for many applications. For example, class learning in different classes in the same school will tend to be more similar than for classes in different schools because of similar facilities, similar quality of teachers, and the like.

The coefficient of correlation between any two observations from the same block for model (25.12) is constant for all blocks, to be denoted by ω:

$$(25.15) \qquad \omega = \frac{\sigma_\rho^2}{\sigma_Y \sigma_Y} = \frac{\sigma_\rho^2}{\sigma_Y^2}$$

This follows from the definition of a coefficient of correlation in (13.7) and the fact that $\sigma\{Y_{ij}\} = \sigma\{Y_{ij'}\} = \sigma_Y$. Note also that the covariance in (25.14) can be expressed as follows, using (25.15):

$$(25.16) \qquad \sigma\{Y_{ij}, Y_{ij'}\} = \omega \sigma_Y^2 \qquad j \neq j'$$

An important property of additive model (25.12), as shown by (25.14) and (25.15), is that any two Y_{ij} observations within a given block, in advance of the random trials, are correlated in the same fashion. The variance-covariance matrix of the observations in a given block therefore is of a particular form. We illustrate this variance-covariance matrix for the observations in a block for a randomized block study with $r = 3$ treatments, using the covariance expression in (25.16):

$$(25.17) \qquad \boldsymbol{\sigma}^2\{\mathbf{Y}\} = \begin{bmatrix} \sigma_Y^2 & \omega\sigma_Y^2 & \omega\sigma_Y^2 \\ \omega\sigma_Y^2 & \sigma_Y^2 & \omega\sigma_Y^2 \\ \omega\sigma_Y^2 & \omega\sigma_Y^2 & \sigma_Y^2 \end{bmatrix} = \sigma_Y^2 \begin{bmatrix} 1 & \omega & \omega \\ \omega & 1 & \omega \\ \omega & \omega & 1 \end{bmatrix}$$

where:

$$\mathbf{Y} = \begin{bmatrix} Y_{i1} \\ Y_{i2} \\ Y_{i3} \end{bmatrix}$$

Note that the main diagonal of the matrix contains the variances of the Y_{ij}, σ_Y^2, and off the main diagonal are the covariances, $\omega\sigma_Y^2$.

The particular pattern of the variance-covariance matrix in (25.17) is called *compound symmetry*.

While any two observations in a given block are correlated in advance of the random trials, once a block has been selected, additive model (25.12) assumes that the observations in that block are independent. The only remaining random variation in an observation Y_{ij} then is the error term ε_{ij}, and additive model (25.12) assumes that these are independent. Thus, in the teacher assistant study, model (25.12) assumes that once the schools have been selected, any one class performance is independent of that of another class in each selected school, given all of the common conditions for the classes in that school as reflected in the block effect ρ_i.

Comments

1. The variance of Y_{ij} in (25.13b) can be expressed as follows using (25.15):

$$\sigma_Y^2 = \omega \sigma_Y^2 + \sigma^2$$

Hence, we obtain:

(25.18)
$$\sigma_Y^2 = \frac{\sigma^2}{1 - \omega}$$

2. The assumption of compound symmetry in additive model (25.12) is restrictive. While this assumption is sufficient so that the F^* statistic for testing treatment effects will follow the F distribution when H_0 holds (i.e., when no treatment effects are present), the assumption is not necessary. For this purpose, it would suffice that the condition of *sphericity* be met. This condition requires that the variance of the difference between any two estimated treatment means be constant; that is:

(25.19)
$$\sigma^2\{\bar{Y}_{.j} - \bar{Y}_{.j'}\} = \text{constant} \qquad j \neq j'$$

This condition can be met without the compound symmetry requirement. For example, consider the following variance-covariance matrix for the Y_{ij} observations in any block for a randomized complete block study with $r = 3$ treatments:

$$\sigma^2\{\mathbf{Y}\} = \begin{bmatrix} 2 & 2 & 4 \\ 2 & 4 & 5 \\ 4 & 5 & 8 \end{bmatrix}$$

This matrix does not exhibit compound symmetry. Yet the requirement for sphericity in (25.19) is met because $\sigma^2\{\bar{Y}_{.j} - \bar{Y}_{.j'}\} = 2/n$ always. For example, we have:

$$\sigma^2\{\bar{Y}_{.1} - \bar{Y}_{.3}\} = \frac{2}{n} + \frac{8}{n} - 2\left(\frac{4}{n}\right) = \frac{2}{n}$$

Analysis of Variance.   Table 25.6 contains the analysis of variance for additive model (25.12). The sums of squares are the same as in (24.6) for the fixed effects

TABLE 25.6 ANOVA for Randomized Complete Block Design—Block Effects Random, Treatment Effects Fixed

Source of Variation	SS	df	MS	$E\{MS\}$ Additive Model (25.12)	$E\{MS\}$ Interaction Model (25.20)
Blocks	SSBL	$n - 1$	MSBL	$\sigma^2 + r\sigma_\rho^2$	$\sigma^2 + r\sigma_\rho^2$
Treatments	SSTR	$r - 1$	MSTR	$\sigma^2 + \dfrac{n}{r-1}\sum \tau_j^2$	$\sigma^2 + \sigma_{\rho\tau}^2 + \dfrac{n}{r-1}\sum \tau_j^2$
Error	SSBL.TR	$(n-1)(r-1)$	MSBL.TR	σ^2	$\sigma^2 + \sigma_{\rho\tau}^2$
Total	SSTO	$nr - 1$			

model. Table 25.6 also contains the expected mean squares for model (25.12). The statistic for testing for treatment effects is $F* = MSTR/MSBL.TR$, as may be seen from the $E\{MS\}$ column in Table 25.6. Thus, the test statistic is the same whether block effects are fixed or random. Confidence intervals for treatment contrasts also present no new problems. Again, $MSBL.TR$ will be used as the mean square in the estimated variance of the contrast.

Interaction Model

When the diagnostics discussed in Chapter 24 indicate that block-treatment interactions are present in the case where the blocks are a random sample from a population of blocks, one may use the following randomized block model, which allows for interactions between blocks and treatments:

$$(25.20) \qquad Y_{ij} = \mu_{..} + \rho_i + \tau_j + (\rho\tau)_{ij} + \varepsilon_{ij}$$

where:

$\mu_{..}$ is a constant
ρ_i are independent $N(0, \sigma_\rho^2)$
τ_j are constants subject to the restriction $\Sigma\tau_j = 0$
$(\rho\tau)_{ij}$ are $N\left(0, \dfrac{r-1}{r}\sigma_{\rho\tau}^2\right)$, subject to the restrictions:

$\displaystyle\sum_j (\rho\tau)_{ij} = 0$ for all i

$\sigma\{(\rho\tau)_{ij}, (\rho\tau)_{ij'}\} = -\dfrac{1}{r}\sigma_{\rho\tau}^2$ for $j \neq j'$

$(\rho\tau)_{ij}$ are independent of the ρ_i
ε_{ij} are independent $N(0, \sigma^2)$ and independent of the ρ_i and of the $(\rho\tau)_{ij}$
$i = 1, \ldots, n; j = 1, \ldots, r$

Properties of Model.   The observations Y_{ij} for interaction model (25.20) are normally distributed, with the following expected values and variance:

$$(25.21a) \qquad E\{Y_{ij}\} = \mu_{..} + \tau_j$$

$$(25.21b) \qquad \sigma^2\{Y_{ij}\} = \sigma_Y^2 = \sigma_\rho^2 + \dfrac{r-1}{r}\sigma_{\rho\tau}^2 + \sigma^2$$

Here again, the Y_{ij} have constant variance, but now the variance is made up of three components.

As for no-interaction model (25.12), observations from different blocks are assumed to be independent according to interaction model (25.20). Similarly, any two observations Y_{ij} and $Y_{ij'}$ from the same block are correlated for interaction model (25.20). The covariance can be shown to be:

$$(25.22) \qquad \sigma\{Y_{ij}, Y_{ij'}\} = \sigma_\rho^2 - \dfrac{1}{r}\sigma_{\rho\tau}^2 \qquad j \neq j'$$

Thus any two observations from the same block with interaction model (25.20) have constant covariance, and this holds for all blocks. The coefficient of correlation between any two observations in the same block, to be denoted by ω^*, is:

$$(25.23) \qquad \omega^* = \frac{\sigma_\rho^2 - \dfrac{1}{r}\sigma_{\rho\tau}^2}{\sigma_Y^2}$$

Comments

1. The reason why the variance of the interaction terms in interaction model (25.20) is expressed as $(r - 1)\sigma_{\rho\tau}^2/r$ rather than simply as $\sigma_{\rho\tau}^2$ is so that the expected mean squares will be expressed relatively simply.

2. The covariance result in (25.22) follows from the fact that it can be shown for interaction model (25.20) that:

$$(25.24) \qquad \sigma\{Y_{ij}, Y_{ij'}\} = \sigma\{\rho_i, \rho_i\} + \sigma\{(\rho\tau)_{ij}, (\rho\tau)_{ij'}\}$$

The first term on the right is the variance of ρ_i, which is σ_ρ^2 according to the model, and the second term on the right is the covariance of two interaction terms from the same block, which is given by the model assumptions to be $-\sigma_{\rho\tau}^2/r$.

3. Interaction model (25.20) assumes, just like no-interaction model (25.12), that once the blocks have been selected any two observations from a given block are uncorrelated.

Analysis of Variance. The sums of squares and degrees of freedom for interaction model (25.20) are the same as those for no-interaction model (25.12). The principal difference in the use of the two models occurs in the expected mean squares, as shown in Table 25.6. No exact test for block effects is possible with the interaction model, whereas an exact test is possible with the no-interaction model. This distinction is unimportant whenever blocks are used primarily to reduce the experimental error variability and are not of intrinsic interest themselves.

The F^* test statistic for treatment effects is the same for the two models, namely $F^* = MSTR/MSBL.TR$, which is exactly the same as test statistic (24.7b) for randomized block model (24.2) with fixed block effects. Similarly, estimation of fixed treatment effects for both models with random block effects is carried out in the manner described in Chapter 24 for fixed block effects.

Some Final Comments

1. Table 25.6 indicates that when the block effects are random, $MSBL.TR$ estimates σ^2 for the additive model (25.12). For the nonadditive model (25.20), however, $MSBL.TR$ estimates the sum of the error term variance σ^2 and the interaction variance $\sigma_{\rho\tau}^2$. Separate estimation of these two components is not possible for this latter model, and the two components are said to be confounded.

2. When the assumption of compound symmetry, which underlies both the no-interaction model (25.12) and the interaction model (25.20), or the less restrictive requirement of sphericity is not met, the usual F test becomes biased. Some computer packages provide the user with the option of formally testing for compound symmetry or sphericity.

When these conditions are violated, an approximate conservative test procedure is as follows:

a. Conduct the usual F test; if it leads to conclusion H_0, accept this conclusion.
b. If the usual F test leads to H_a, replace $F[1 - \alpha; r - 1, (n - 1)(r - 1)]$ in decision rule (24.7c) by $F(1 - \alpha; 1, n - 1)$. If this modified decision rule leads to H_a, accept this conclusion.
c. If the modified decision rule leads to H_0, revise the degrees of freedom in the modified decision rule by one of the *epsilon adjustment procedures,* as described in References 25.3 and 25.4.

Alternatively, multivariate analysis of variance techniques may be employed provided that $n > r$. See Reference 25.5 for further discussions of these issues.

3. Mixed models based on less restrictive assumptions regarding the variance-covariance matrix and the parameters in the ANOVA model have also been proposed. See Reference 25.6 for a discussion of these models.

25.5 GENERALIZED RANDOMIZED BLOCK DESIGNS

When block effects are fixed, use of a no-interaction model in the presence of interactions between blocks and treatments has the effect of reducing the power of the test and increasing the width of interval estimates of treatment effects, thus making the experiment less sensitive. In addition, there are occasions when one is interested in the nature of the interactions between blocks and treatments and would like to obtain estimates of these. It is possible to use a design that permits an interaction term in the model even when the block effects are fixed, and that allows the nature of the interaction effects to be investigated. This design is called a *generalized randomized block design.* It is the same as a randomized block design except that d experimental units are assigned to each treatment within a block. This design increases the size of a block from r units for a randomized block design to dr units. This increase in block size often has the effect of increasing experimental error variability when the number of experimental units is fixed. In the social sciences, however, increasing the size of the block moderately may cause little loss in efficiency. For instance, having one block of 10 persons aged 20–29 instead of two blocks of five persons of ages 20–24 and 25–29, respectively, will for many types of experiments involve little loss of efficiency.

As we shall demonstrate by an example, a generalized randomized block design is analyzed like an ordinary multifactor study where blocks are one factor. Hence, no new problems are encountered with a generalized randomized block design in testing for treatment effects or in estimating them. In particular, we shall now be able to calculate *MSE* and use it as an estimator of the error variance σ^2.

Example

Table 25.7 contains the data for a two-factor experiment in which the effects of motivation (factor A: high level, low level) and distraction (factor B: high level, low level) on the time required to complete a task were studied, using eight men and

TABLE 25.7 Two-Factor Study in Generalized Randomized Block Design with $d = 2$ (observations are times for task completion)

	Block (gender)	
	Male	Female
High motivation:		
High distraction	12	3
	8	9
Low distraction	7	5
	5	9
Low motivation:		
High distraction	14	11
	16	9
Low distraction	15	10
	13	14

eight women. Two men were assigned at random to each treatment, and independently two women were assigned at random to each treatment. Here gender is the blocking variable. Each block contains eight persons, with two randomly assigned to each treatment within the block. The layout in Table 25.7 corresponds to the layout in Table 22.2a for a three-factor study; to stress the correspondence, we have placed the blocks in columns rather than in rows as usual. Since blocks, motivation levels, and distraction levels are considered to be fixed, we utilize the fixed effects three-factor model (22.14), with notation modified to fit the present context:

$$(25.25) \quad Y_{ijkm} = \mu_{...} + \rho_i + \alpha_j + \beta_k + (\rho\alpha)_{ij} + (\rho\beta)_{ik}$$
$$+ (\alpha\beta)_{jk} + (\rho\alpha\beta)_{ijk} + \varepsilon_{ijkm}$$

where:

$\mu_{...}$ is a constant

ρ_i, α_j, and β_k are constants subject to the restrictions $\Sigma\rho_i = \Sigma\alpha_j = \Sigma\beta_k = 0$

$(\rho\alpha)_{ij}$, $(\rho\beta)_{ik}$, $(\alpha\beta)_{jk}$, $(\rho\alpha\beta)_{ijk}$ are constants subject to the restrictions that the sums over any subscript are zero

ε_{ijkm} are independent $N(0, \sigma^2)$

$i = 1, \ldots, n; j = 1, \ldots, a; k = 1, \ldots, b; m = 1, \ldots, d$

The analysis of variance for generalized randomized block model (25.25) is the ordinary three-factor ANOVA of Table 22.3, with slight modifications in notation. A computer package was employed to obtain the analysis of variance for the data in Table 25.7, and the results are shown in Table 25.8. We know from Table 22.4 that all test statistics use *MSE* in the denominator. These F^* statistics are shown in Table 25.8. For $\alpha = .01$, we require $F(.99; 1, 8) = 11.3$ for each of the tests. It is evident from the results in Table 25.8 (see also the *P*-values given there) that blocks do not interact with the factorial treatments and that among the two factors, only moti-

TABLE 25.8 Analysis of Variance for Task Completion Example of Table 25.7 ($n = 2$, $a = 2$, $b = 2$, $d = 2$)

Source of Variation	SS	df	MS	F*	P-value
Blocks (gender)	$SSBL = 25.00$	$n - 1 = 1$	$MSBL = 25.00$	4.00	.08
Factor A (motivation)	$SSA = 121.00$	$a - 1 = 1$	$MSA = 121.00$	19.36	.002
Factor B (distraction)	$SSB = 1.00$	$b - 1 = 1$	$MSB = 1.00$	.16	.70
BL.A interactions	$SSBL.A = 4.00$	$(n - 1)(a - 1) = 1$	$MSBL.A = 4.00$	.64	.45
BL.B interactions	$SSBL.B = 16.00$	$(n - 1)(b - 1) = 1$	$MSBL.B = 16.00$	2.56	.15
AB interactions	$SSAB = 4.00$	$(a - 1)(b - 1) = 1$	$MSAB = 4.00$	.64	.45
BL.AB interactions	$SSBL.AB = 1.00$	$(n - 1)(a - 1)(b - 1) = 1$	$MSBL.AB = 1.00$	.16	.70
Error	$SSE = 50.00$	$(d - 1)nab = 8$	$MSE = 6.25$		
Total	$SSTO = 222.00$	$dnab - 1 = 15$			

$$F(.99; 1, 8) = 11.3$$

vation affects the time required to complete the task. At this point, further analysis of the motivation effects would be indicated.

25.6 USE OF MORE THAN ONE BLOCKING VARIABLE

Sometimes, substantial reduction in the experimental error variability can only be obtained by utilizing more than one variable for determining blocks. For instance, both age and gender might be needed for designating blocks:

Block	Characteristics of Experimental Units
1	Male, aged 20–29
2	Female, aged 20–29
3	Male, aged 30–39
etc.	etc.

As another example, both observer and day of treatment application may be helpful as blocking variables:

Block	Characteristics of Experimental Units
1	Observer 1, day 1
2	Observer 2, day 1
3	Observer 1, day 2
etc.	etc.

Unless one wishes to study the separate effects of each of the blocking variables, no new problems arise when the blocks are defined by two or more variables. The n blocks are simply treated as ordinary blocks, and the usual block sum of squares is calculated.

If the effect of each of the blocking variables is to be isolated and the blocks are defined in a complete factorial fashion (e.g., nine blocks are used when three observers and three days are employed for blocking), the analysis simply treats each of the blocking variables as a factor and utilizes the methods developed in Chapter 22.

A problem that sometimes arises when two or more blocking variables are to be used is the large number of blocks called for. Suppose an experiment is to be conducted where the experimental units are stores. In order to reduce the experimental error variability to a reasonable level, it would be desirable to group the stores into six sales volume classes and also into six location classes (suburban shopping center, suburban other, etc.). Thirty-six blocks result from combining these two blocking variables. If six treatments are to be studied, 216 stores would be required for the experiment. Often, use of this many stores would be much too costly. Latin square designs, to be discussed in Chapter 29, permit in this type of study the use of a much smaller number of replications while still preserving the full benefits of error variance reduction by using both blocking variables in six classes each.

CITED REFERENCES

25.1. Owen, D. B. *Handbook of Statistical Tables.* Reading, Mass.: Addision-Wesley Publishing, 1962.

25.2. Cochran, W. G., and G. M. Cox. *Experimental Designs.* 2nd ed. New York: John Wiley & Sons, 1957, pp. 110–12.

25.3. Greenhouse, S. W., and S. Geisser. "On Methods in the Analysis of Profile Data." *Psychometrika* 24 (1959), pp. 95–112.

25.4. Huynh, H., and L. Feldt. "Estimation of the Box Correction for Degrees of Freedom from Sample Data in the Randomized Block and Split-Plot Designs." *Journal of Educational Statistics* 1 (1976), pp. 69–82.

25.5. Winer, B. J. *Statistical Principles in Experimental Design.* 2nd ed. New York: McGraw-Hill, 1971.

25.6. Hocking, R. R. "A Discussion of the Two-way Mixed Model." *The American Statistician* 27 (1973), pp. 148–52.

PROBLEMS

25.1. Antinausea therapy. Cancer patients undergoing chemotherapy commonly suffer from episodes of nausea that are uncontrolled by conventional antinausea drugs. To evaluate the comparative effectiveness of two experimental antinausea drugs, 36 cancer patients were blocked in groups of size three based on their previous history of severe episodes of nausea and were randomly assigned to one of three treatments in a double-blind study while undergoing chemotherapy. Treatment 1 was a conventional antinausea drug while treatments 2 and 3 were the experimental drugs. The effectiveness of each drug was evaluated from reports of the patients, coded 1 for improvement and 0 for no improvement. The data follow.

Block i	Treatment (j) 1	2	3	Block i	Treatment (j) 1	2	3
1	0	1	1	7	1	0	1
2	0	1	1	8	0	1	1
3	1	1	1	9	0	1	1
4	0	0	1	10	0	0	1
5	0	1	0	11	1	0	1
6	0	0	0	12	0	0	1

a. Employ the Cochran test to determine whether or not the three drugs differ in mean effectiveness; use $\alpha = .05$. State the alternatives, decision rule, and conclusion. What is the P-value of the test?

b. Conduct multiple pairwise tests to group the three drugs according to mean effectiveness; use a family level of significance of $\alpha = .10$. Describe your findings.

25.2. Product coupons. An advertising agency designed an experiment to evaluate the effectiveness of four different coupon offers for a household product. Fifty-two households participated. The households were blocked in groups of size four according to income level, and assigned randomly to one of the coupon offers. The data to follow show whether or not each household used the coupon to purchase the product during the following month (1: used coupon; 0: did not use coupon).

Block	Coupon Offer (j)				Block	Coupon Offer (j)			
i	1	2	3	4	i	1	2	3	4
1	0	1	1	1	8	0	0	0	0
2	0	1	0	1	9	1	1	1	1
3	1	1	1	1	10	1	1	1	1
4	0	0	0	0	11	1	1	1	1
5	0	0	1	1	12	0	0	1	1
6	1	0	0	1	13	0	1	1	1
7	0	0	1	1					

a. Employ the Cochran test to determine whether or not mean coupon usage differs for the four offers; use $\alpha = .05$. State the alternatives, decision rule, and conclusion. What is the P-value of the test?

b. Conduct multiple pairwise tests to group the offers according to mean coupon usage; employ a family level of significance of $\alpha = .10$. State your findings.

25.3. Refer to **Auditor training** Problems 24.8 and 24.9. It has been suggested that the nonparametric Friedman test should be used here.

a. Rank the data within each block and perform the Friedman test; use $\alpha = .05$. State the alternatives, decision rule, and conclusion. What is the P-value of the test? Is your conclusion the same as that obtained in Problem 24.9c?

b. Use the multiple pairwise testing procedure (25.5) to group the three training methods according to mean proficiency; employ a family significance level of $\alpha = .10$. Summarize your findings.

25.4. Refer to **Dental pain** Problem 24.12; ignore the factorial treatment structure. A consultant is concerned about the validity of the model assumptions and suggested that the study be analyzed by means of the Friedman test.

a. Rank the data within each block and perform the Friedman test; use $\alpha = .025$. State the alternatives, decision rule, and conclusion. What is the P-value of the test?

b. Use the multiple pairwise testing procedure (25.5) to group the four treatments according to mean pain relief; employ a family significance level of $\alpha = .05$. State your findings.

25.5. Refer to **Auditor training** Problem 24.8. Assume that observation $Y_{23} = 89$ is missing because the auditor became ill and dropped out from the study.

a. State the ANOVA model for this case. Also state the equivalent regression model; use 1, −1, 0 indicator variables.

b. State the reduced regression model for testing for differences in the mean proficiency scores for the three training methods.

c. Test whether or not the mean proficiency scores for the three training methods differ by fitting the full and reduced models; use $\alpha = .05$. State the alternatives, decision rule, and conclusion.

d. Compare the mean proficiency scores for training methods 2 and 3 by means of the regression approach; use a 95 percent confidence interval.

25.6. Refer to **Fat in diets** Problem 24.10. Assume that observations $Y_{13} = .15$ and $Y_{51} = 1.62$ are missing because the subjects did not stay on the prescribed diet.

a. State the ANOVA model for this case. Also state the equivalent regression model; use 1, −1, 0 indicator variables.

b. State the reduced regression model for testing for differences in the mean reductions in lipid level for the three diets.

c. Test whether or not the mean reductions in lipid level differ for the three diets by fitting the full and reduced models; use $\alpha = .05$. State the alternatives, decision rule, and conclusion.

d. Compare the mean reductions in lipid level for diets 1 and 3 by means of the regression approach; use a 98 percent confidence interval.

25.7. Road paint wear. A state highway department studied the wear characteristics of five different paints at eight locations in the state. The standard, currently used paint (paint 1) and four experimental paints (paints 2, 3, 4, 5) were included in the study. The eight locations were randomly selected, thus reflecting variations in traffic densities throughout the state. At each location, a random ordering of the paints to the chosen road surface was employed. After a suitable period of exposure to weather and traffic, a combined measure of wear, considering both durability and visibility, was obtained. The data on wear follow (the higher the score, the better the wearing characteristics).

Location	Paint (j)					Location	Paint (j)				
i	1	2	3	4	5	i	1	2	3	4	5
1	11	13	10	18	15	5	14	16	13	22	16
2	20	28	15	30	18	6	25	27	26	33	25
3	8	10	8	16	12	7	43	46	41	55	42
4	30	35	27	41	28	8	13	14	12	20	13

a. Obtain the residuals for additive randomized block model (25.12) and plot them against the fitted values. Also prepare a normal probability plot of the residuals. Summarize your findings about the appropriateness of model (25.12).

b. Plot the responses by location in the format of Figure 24.1. What does this plot suggest about the appropriateness of the no-interaction assumption here?

c. Conduct the Tukey test for additivity of location and treatment effects, conditional on the locations selected; use $\alpha = .005$. State the alternatives, decision rule, and conclusion.

25.8. Refer to **Road paint wear** Problem 25.7. Assume that additive randomized block model (25.12) is appropriate.

a. Obtain the analysis of variance table.

b. Test whether or not the mean wear differs for the five paints; use significance level $\alpha = .05$. State the alternatives, decision rule, and conclusion. What is the P-value of the test?

c. Compare the mean wear of each experimental paint against that of the standard paint; use the most efficient multiple comparison procedure with a 90 percent family confidence coefficient. Summarize your findings.

d. Paints 1, 3, and 5 are white colored whereas paints 2 and 4 are yellow. Estimate the difference in the mean wear for the two groups of paints with a 95 percent confidence interval. Interpret your findings.

25.9. Muscle tissue. A physiologist studied the effects of three reagents on muscle tissue in dogs. Ten litters of three dogs each were randomly selected and the three reagents were randomly assigned to the three dogs in each litter. The data on the effects of the reagents follow (the higher the value, the higher the activity level):

Litter	Reagent (j)			Litter	Reagent(j)		
i	1	2	3	i	1	2	3
1	10	15	14	6	7	9	10
2	8	12	13	7	24	30	27
3	21	27	25	8	16	18	20
4	14	17	17	9	23	29	32
5	12	18	16	10	18	22	21

a. Obtain the residuals for additive randomized block model (25.12) and plot them against the fitted values. Also prepare a normal probability plot of the residuals. Summarize your findings.

b. Plot the responses by litter in the format of Figure 24.1. What does this plot suggest about the appropriateness of the no-interaction assumption here?

c. Conduct the Tukey test for additivity of litter and reagent effects, conditional on the litters selected; use $\alpha = .025$. State the alternatives, decision rule, and conclusion.

d. Based on parts (b) and (c), would interaction randomized block model (25.20) be more appropriate here? What practical differences exist in using models (25.12) and (25.20)?

25.10. Refer to **Muscle tissue** Problem 25.9. Assume that additive randomized block model (25.12) is applicable.

a. Obtain the analysis of variance table.

b. Test whether or not the mean activity level differs for the three reagents; use significance level $\alpha = .025$. State the alternatives, decision rule, and conclusion. What is the P-value of the test?

c. Reagents 2 and 3 were expected to be similar to each other but to differ from reagent 1. Use the most efficient multiple comparison procedure with a 95 percent family confidence coefficient to estimate:

$$D_1 = \mu_{.2} - \mu_{.3}$$

$$L_1 = \frac{\mu_{.2} + \mu_{.3}}{2} - \mu_{.1}$$

Summarize your findings.

d. Test whether or not σ_p^2 equals zero; use $\alpha = .025$. State the alternatives, decision rule, and conclusion. What is the P-value of the test?

25.11. A social scientist, after learning about generalized randomized block designs, asked: "Why would anyone use a randomized complete block design that requires the assumption that block and treatment effects do not interact when this assumption can be avoided with a generalized randomized block design?" Comment.

25.12. Refer to the task completion example on page 960.

a. Using the data in Table 25.7, verify the analysis of variance in Table 25.8.

b. Estimate the difference in mean effects for the two motivation levels using a 99 percent confidence interval.

25.13. Refer to **Auditor training** Problem 24.8. The accounting firm repeated the experiment with another group of 30 auditors, but this time grouped them into five blocks of six each. In each block, each treatment was randomly assigned to two auditors. The results were:

Block	Training Method (j)			Block	Training Method (j)		
i	1	2	3	i	1	2	3
1	74	84	91	4	65	73	84
	71	78	95		70	78	87
2	73	75	93	5	64	71	81
	69	83	98		61	74	74
3	75	81	89				
	67	74	86				

Assume that generalized randomized block model (25.25), modified for a single-factor study, is appropriate.

a. State the generalized randomized block model for this application.

b. Obtain the analysis of variance table.

c. Test whether or not the mean proficiency scores for the three training methods differ; use $\alpha = .01$. State the alternatives, decision rule, and conclusion. What is the P-value of the test?

d. Make all pairwise comparisons between the three training methods; use the Tukey procedure with a 95 percent family confidence coefficient. Summarize your findings.

e. Obtain the residuals and plot them against the fitted values. Also prepare a normal probability plot of the residuals. State your findings.

f. Test whether or not blocks interact with treatments; use $\alpha = .01$. State the alternatives, decision rule, and conclusion. What is the P-value of the test?

EXERCISES

25.14. Derive (25.2a) from (25.2).

25.15. Derive (25.4a) from (25.4) when no ties are present.

25.16. Consider a randomized complete block design study where $n = 4$ and $r = 2$, randomized block model (24.2) applies, and Y_{31} is missing. Use the matrix methods of

Section 8.6 to obtain an estimator of μ_{31}. (*Hint:* Consider the development in Section 21.1.)

25.17. Refer to **Dental pain** Problem 24.12. Suppose that the subjects in the study had been randomly selected from 8 towns (blocks), and that the towns were randomly selected from a population of towns. Assume that additive randomized block model (25.12) is applicable, except that the factorial structure of the fixed treatment effects needs to be recognized.

    a. State the randomized block model for this case.

    b. What is the appropriate test statistic for testing whether or not the two factors interact? What are the appropriate test statistics for testing for main effects? [*Hint:* Consider the test for treatment effects in model (25.12).]

25.18. Derive (25.14).

PROJECTS

25.19. Refer to **Road paint wear** Problem 25.7.

    a. Estimate the variance-covariance matrix of the treatment observations in a block; use (28.8) to obtain the entries in the matrix.

    b. Does the compound symmetry property of (25.17) appear to be reasonable here? Explain.

    c. Does the sphericity property of (25.19) appear to be reasonable here? Explain.

25.20. Refer to **Muscle tissue** Problem 25.9.

    a. Estimate the variance-covariance matrix of the treatment observations in a block; use (28.8) to obtain the entries in the matrix.

    b. Does the compound symmetry property of (25.17) appear to be reasonable here? Explain.

    c. Does the sphericity property of (25.19) appear to be reasonable here? Explain.

Chapter 26

Nested Designs and Subsampling

In this chapter, we take up the basic elements of nested designs, including the use of subsampling. We begin this chapter by considering the general concept of nested designs and describe how these designs differ from crossed designs. We then take up in detail two-factor nested designs and their analysis. We conclude this chapter by considering subsampling designs.

26.1 DISTINCTION BETWEEN NESTED AND CROSSED FACTORS

In the factorial studies considered so far, where every level of one factor appears with each level of every other factor, the factors are said to be crossed. A different situation occurs when factors are nested. The distinction between nested and crossed factors will now be illustrated by some examples involving two-factor studies.

Example 1

A large manufacturing company operates three regional training schools for mechanics, one in each of its operating districts. The schools have two instructors each who teach classes of about 15 mechanics in three-week sessions. The company was concerned about the effect of school (factor A) and instructor (factor B) on the learning achieved. To investigate these effects, classes in each district were formed in the usual way and then randomly assigned to one of the two instructors in the school. This was done for two sessions, and at the end of each session a suitable measure of learning for the class was obtained. The results are presented in Table 26.1.

The layout of Table 26.1 appears identical to an ordinary two-factor investigation, with two observations per cell (see, e.g., Table 18.7). In fact, the study is not an or-

TABLE 26.1 Sample Data for Nested Two-Factor Study—Training School Example (class learning scores, coded)

Factor A (school) i		Factor B (instructor) j		Total
		1	2	
Atlanta		25	14	
		29	11	
	Total	$Y_{11.} = 54$	$Y_{12.} = 25$	$Y_{1..} = 79$
Chicago		11	22	
		6	18	
	Total	$Y_{21.} = 17$	$Y_{22.} = 40$	$Y_{2..} = 57$
San Francisco		17	5	
		20	2	
	Total	$Y_{31.} = 37$	$Y_{32.} = 7$	$Y_{3..} = 44$
			Total	$Y_{...} = 180$

dinary two-factor study. The reason is that the instructors in the Atlanta school did not also teach in the other two schools, and similarly for the other instructors. Thus, six different instructors were involved. An ordinary two-factor investigation with six different instructors would have consisted of 18 treatments, as shown in Table 26.2a. In the training school example, however, only six treatments were included, as shown in Table 26.2b, where the crossed-out cells represent treatments not studied. Figure 26.1 contains a graphic representation of the nested design for the training school example, including the two replications of the study.

It is clear from Table 26.2b that the experimental design for the training school example involves an incomplete factorial arrangement of a special type, where each level of factor B (instructor) occurs with only one level of factor A (school). Specifically here, each instructor teaches in only one school. Factor B is therefore said to be *nested* within factor A. As noted earlier, in an ordinary factorial study where every factor level of A appears with every factor level of B, factors A and B are said to be *crossed*.

There is another way to look at the distinction between nested and crossed designs. Let μ_{ij} denote the mean response when factor A is at the ith level and factor B is at the jth level. If the factors are crossed, the jth level of B is the same for all levels of A. If, on the other hand, factor B is nested within factor A, the jth level of B when A is at level 1 has nothing in common with the jth level of B when A is at level 2, and so on. For instance, in a crossed factorial study of the effects of price ($1.99, $2.49) and advertising level (high, low), a particular advertising level is the same no matter with which price it appears, and similarly for the price levels. On the other hand, in the nested design training school example, the first instructor in school 1 is not the same as the first instructor in school 2, and so on.

TABLE 26.2 Illustration of Crossed and Nested Factors—Training School Example

(a) Crossed Factors

School (factor A)	Instructor (factor B)					
	1	2	3	4	5	6
Atlanta						
Chicago						
San Francisco						

(b) Nested Factors

School (factor A)	Instructor (factor B)					
	1	2	3	4	5	6
Atlanta			╳	╳	╳	╳
Chicago	╳	╳			╳	╳
San Francisco	╳	╳	╳	╳		

Example 2

An analyst was interested in the effects of community (factor A) and neighborhood (factor B) on the spread of information about new products. Information was obtained from samples of families in various neighborhoods within selected communities. Since the neighborhood designated 1 in a given community is not the same as the neighborhoods designated 1 in the other communities, and similarly for the other neighborhoods, neighborhoods here are nested within communities.

FIGURE 26.1 Graphic Representation of Two-Factor Nested Design—Training School Example

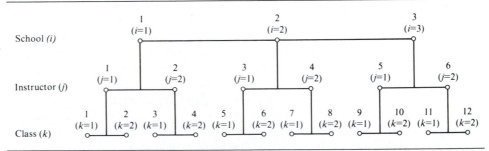

Comments

1. The distinction between crossed and nested factors is often a fine one. In Example 2, if the neighborhoods of each community represented specified average income levels so that, say, the first neighborhoods in each community had an average income of $5,000–$9,999, the second neighborhoods an average income of $10,000–$19,999 and so on for the other neighborhoods, one could view the design as a crossed one. The factors would be community and economic level of neighborhood, and these would be crossed since a given economic level is the same for all communities, and vice versa.

2. Nested factors are frequently encountered in observational studies where the researcher cannot manipulate the factors under study, or in experiments where only some factors can be manipulated. Factors that cannot be manipulated, it will be recalled, are designated classification factors, in distinction to experimental factors that can be assigned at will to the experimental units. Example 2 is an observational study where both community and neighborhood are classification factors since families (the study units) were not randomly assigned to either community or neighborhood. In Example 1, school is a classification factor because the classes of a school (the experimental units) are made up of mechanics from the district in which the school is located. Instructors in this example are an experimental factor since they are assigned randomly to a class, but a nested design results because the randomization is restricted to within a school.

26.2 TWO-FACTOR NESTED DESIGNS

We now consider nested designs involving two factors, one of which is nested inside the other. For consistency, we shall always consider the case where factor B is nested within factor A. We initially assume that both factor effects are fixed, but we shall later also consider the case of random effects. We assume throughout that *all treatment means are of equal importance*.

Development of Model Elements

We shall use the customary notation for a two-factor study and let μ_{ij} denote the mean response when factor A is at the ith level ($i = 1, \ldots, a$) and factor B is at the jth level ($j = 1, \ldots, b$). As usual, when all mean responses are of equal importance we define:

$$(26.1) \qquad \mu_{i.} = \frac{\sum_j \mu_{ij}}{b}$$

For the training school example of Table 26.1, $\mu_{1.}$ represents the mean learning score for the Atlanta school, averaged over the instructors of that school, and $\mu_{2.}$ and $\mu_{3.}$ are interpreted similarly. Note once more that the $\mu_{i.}$ here represent mean learning scores that have been averaged over *different* instructors.

We define the main effect of the ith level of factor A as usual:

$$(26.2) \qquad \alpha_i = \mu_{i.} - \mu_{..}$$

where:

(26.3)
$$\mu_{..} = \frac{\sum_i \sum_j \mu_{ij}}{ab} = \frac{\sum_i \mu_{i.}}{a}$$

is the overall mean response. It follows from (26.3) that:

(26.4)
$$\sum_i \alpha_i = 0$$

In a nested design, it is not meaningful to employ a model component for the main effect of the jth level of factor B. To see why, consider again the training school example. Since each school employs different instructors and the jth instructors in the various schools are not the same, it would be meaningless to consider the effect of the jth instructor, averaged over all schools. Instead, one needs to consider the individual effects of each instructor in each school. We denote these individual effects by $\beta_{j(i)}$, where the subscript $j(i)$ indicates that the jth factor level of B is nested within the ith factor level of A. $\beta_{j(i)}$ is defined as follows:

(26.5)
$$\beta_{j(i)} = \mu_{ij} - \mu_{i.}$$

which can be rewritten, utilizing (26.2):

(26.5a)
$$\beta_{j(i)} = \mu_{ij} - \alpha_i - \mu_{..}$$

It follows from (26.5) and (26.1) that:

(26.6)
$$\sum_j \beta_{j(i)} = 0 \qquad i = 1, \ldots, a$$

The meaning of $\beta_{j(i)}$ can be seen most clearly from (26.5). With reference to the training school example, $\beta_{j(i)}$ is simply the difference in the mean learning score for the jth instructor of school i and the average of the mean learning scores for all instructors in that school. Thus, the effect of the jth instructor in the ith school is measured with respect to the overall mean learning score for the school in which the instructor teaches. We shall call $\beta_{j(i)}$ the *specific effect* of the jth level of factor B nested within the ith level of factor A.

We have now expressed the mean response μ_{ij} in terms of the overall mean, the main effect of the ith level of factor A, and the specific effect of the jth level of factor B nested within the ith level of factor A, as can be seen from (26.5a):

(26.7)
$$\mu_{ij} \equiv \mu_{..} + \alpha_i + \beta_{j(i)} \equiv \mu_{..} + (\mu_{i.} - \mu_{..}) + (\mu_{ij} - \mu_{i.})$$

For the training school example, the mean learning score for the jth instructor in school i has been expressed in terms of the overall mean, the main effect of school i, and the specific effect of instructor j within school i.

To complete the model, we need only add a random error term. This will be denoted by $\varepsilon_{k(ij)}$, where the first subscript represents the kth replication and (i, j) identifies the factor combination for which it is the kth replication. We now use the subscript notation $k(ij)$ since the replications for any (i, j) factor combination can be viewed as being nested within that (i, j) combination. In our training school example, the replications are the several classes of mechanics taught by the same instructor in a given school. Since the first class for school i–instructor j is not the same as the first class for any other school-instructor combination, and similarly for the second class, we can regard the classes as being nested within the school-instructor combinations. Figure 26.1 illustrates this nesting.

We have not pointed out this nesting of the replications within the factor combinations up to now because it has not been necessary. With the current consideration of nested factors, however, it will be useful to recognize the nesting of the replications variable within the factor combinations.

Nested Design Model

Let Y_{ijk} denote the kth observation when factor A is at the ith level and factor B is at the jth level. We assume that there are n replications for each factor combination, i.e., $k = 1, \ldots, n$, and that $i = 1, \ldots, a$ and $j = 1, \ldots, b$. When both factors A and B have fixed effects, an appropriate nested design model is:

$$(26.8) \qquad Y_{ijk} = \mu_{..} + \alpha_i + \beta_{j(i)} + \varepsilon_{k(ij)}$$

where:

$\mu_{..}$ is a constant

α_i are constants subject to the restriction $\Sigma \alpha_i = 0$

$\beta_{j(i)}$ are constants subject to the restrictions $\Sigma_j \beta_{j(i)} = 0$ for all i

$\varepsilon_{k(ij)}$ are independent $N(0, \sigma^2)$

$i = 1, \ldots, a; j = 1, \ldots, b; k = 1, \ldots, n$

The expected value and variance of observation Y_{ijk} for nested design model (26.8) with fixed factor effects are:

$$(26.8a) \qquad E\{Y_{ijk}\} = \mu_{..} + \alpha_i + \beta_{j(i)}$$

$$(26.8b) \qquad \sigma^2\{Y_{ijk}\} = \sigma^2$$

Thus, all observations have a constant variance. Further, the observations Y_{ijk} are independent and normally distributed for this model.

Comments

1. It is not necessary, as in (26.8), that the number of replications be equal for all factor combinations, or that the number of levels of nested factor B (number of instructors in the training school example) be the same for each level of factor A (school in this example). We shall discuss the removal of some of these restrictions later. We only point out now that the computations become more complex when these restrictions are relaxed.

2. There is no interaction term in nested design model (26.8). There is no need for it since factor B is nested within factor A, not crossed with it. To put this somewhat differently, with reference to the training school example, it is not possible to estimate a school-instructor interaction when each instructor teaches in only one school. The teacher effect $\beta_{j(i)}$, since it is specific to a given school i, in a sense incorporates the interaction effect between the particular teacher j (in the ith school) and the ith school, but it is not possible in a nested design to disentangle this interaction effect.

3. The factor level means $\mu_{i.}$ in a nested design are not generally the same as the corresponding means in a crossed design. Remember that in a nested design, the $\mu_{i.}$ are obtained by averaging over only some of the distinctive levels of factor B. With reference to the training school example, the $\mu_{i.}$ are obtained by averaging over only those teachers who instruct in the ith school. In a crossed design, on the other hand, the $\mu_{i.}$ would be obtained by averaging over all instructors included in the study.

Random Factor Effects

If both factors A and B have random factor levels, nested design model (26.8) is modified with α_i, $\beta_{j(i)}$, and $\varepsilon_{k(ij)}$ being independent normal random variables with expectations 0 and variances σ_α^2, σ_β^2, and σ^2, respectively. Thus, it is assumed that all $\beta_{j(i)}$ have the same variance σ_β^2. The assumption that all $\beta_{j(i)}$ have the same variance also is made if only factor B is random. It is important to check whether this assumption is appropriate, since it may well be that the mean responses μ_{i1}, μ_{i2}, . . . , in one factor A level (plant, school, city, etc.) differ in variability from those in other factor A levels (other plants, schools, cities, etc.). Tests for equality of variances were discussed in Section 16.2.

26.3 ANALYSIS OF VARIANCE FOR TWO-FACTOR NESTED DESIGNS

Fitting of Model

The least squares estimators of the parameters in nested design model (26.8) are obtained in the usual fashion. Employing our customary notation for sample data in factorial studies, the least squares estimators are:

	Parameter	Estimator
(26.9a)	$\mu_{..}$	$\hat{\mu}_{..} = \bar{Y}_{...}$
(26.9b)	α_i	$\hat{\alpha}_i = \bar{Y}_{i..} - \bar{Y}_{...}$
(26.9c)	$\beta_{j(i)}$	$\hat{\beta}_{j(i)} = \bar{Y}_{ij.} - \bar{Y}_{i..}$

The fitted values therefore are:

$$(26.10) \qquad \hat{Y}_{ijk} = \bar{Y}_{...} + (\bar{Y}_{i..} - \bar{Y}_{...}) + (\bar{Y}_{ij.} - \bar{Y}_{i..}) = \bar{Y}_{ij.}$$

and the residuals are:

$$(26.11) \qquad e_{ijk} = Y_{ijk} - \hat{Y}_{ijk} = Y_{ijk} - \bar{Y}_{ij.}$$

Sums of Squares

The analysis of variance for nested design model (26.8) is obtained by decomposing the total deviation $Y_{ijk} - \bar{Y}_{...}$ as follows:

(26.12)
$$\underbrace{Y_{ijk} - \bar{Y}_{...}}_{\text{Total deviation}} = \underbrace{\bar{Y}_{i..} - \bar{Y}_{...}}_{\text{A main effect}} + \underbrace{\bar{Y}_{ij.} - \bar{Y}_{i..}}_{\substack{\text{Specific } B \\ \text{effect when } A \\ \text{at } i\text{th level}}} + \underbrace{Y_{ijk} - \bar{Y}_{ij.}}_{\text{Residual}}$$

When we square (26.12) and sum over all cases, all cross-product terms drop out and we obtain:

(26.13)
$$SSTO = SSA + SSB(A) + SSE$$

where:

(26.13a)
$$SSTO = \sum_i \sum_j \sum_k (Y_{ijk} - \bar{Y}_{...})^2$$

(26.13b)
$$SSA = bn \sum_i (\bar{Y}_{i..} - \bar{Y}_{...})^2$$

(26.13c)
$$SSB(A) = n \sum_i \sum_j (\bar{Y}_{ij.} - \bar{Y}_{i..})^2$$

(26.13d)
$$SSE = \sum_i \sum_j \sum_k (Y_{ijk} - \bar{Y}_{ij.})^2 = \sum_i \sum_j \sum_k e_{ijk}^2$$

$SSTO$ is the usual total sum of squares, and SSA is the ordinary factor A sum of squares, reflecting the variability of the estimated factor level means $\bar{Y}_{i..}$.

$SSB(A)$ is the factor B sum of squares, with the notation reflecting that factor B is nested within factor A. $SSB(A)$ is made up of terms such as:

(26.14)
$$n \sum_j (\bar{Y}_{ij.} - \bar{Y}_{i..})^2$$

which is simply the ordinary factor B sum of squares when factor A is at level i. These terms are then summed over all levels of factor A.

Finally, the error sum of squares SSE is, as usual, the sum of the squared residuals and reflects the variability of each observation Y_{ijk} around the corresponding estimated treatment mean $\bar{Y}_{ij.}$. Alternatively, we can view SSE as being made up of terms such as:

(26.15)
$$\sum_j \sum_k (Y_{ijk} - \bar{Y}_{ij.})^2$$

which is simply the ordinary error sum of squares within the ith level of factor A. These terms are then summed over all levels of factor A.

Thus, a nested two-factor design can be viewed as a series of single-factor investigations at the successive levels of the other factor. In terms of the training school example, a study of the effects of instructors (B) within any given school (A_i) leads to the usual sums of squares for instructors and errors in a single-factor analysis of variance within school A_i, denoted by $SSB(A_i)$ and $SSE(A_i)$:

$$SSB(A_i) = n \sum_j (\bar{Y}_{ij.} - \bar{Y}_{i..})^2 \qquad SSE(A_i) = \sum_j \sum_k (Y_{ijk} - \bar{Y}_{ij.})^2$$

These are then aggregated to yield $SSB(A)$ and SSE, respectively. It is only the between-schools sum of squares SSA that introduces explicitly the other factor. Table 26.3 demonstrates this relation between the single-factor analyses of variance for each school and the two-factor analysis of variance for the nested design.

Computational Formulas.   For hand computations, the following formulas may be used:

(26.16a)
$$SSTO = \sum_i \sum_j \sum_k Y_{ijk}^2 - \frac{Y_{...}^2}{abn}$$

(26.16b)
$$SSA = \frac{\sum_i Y_{i..}^2}{bn} - \frac{Y_{...}^2}{abn}$$

(26.16c)
$$SSB(A) = \frac{\sum_i \sum_j Y_{ij.}^2}{n} - \frac{\sum_i Y_{i..}^2}{bn}$$

(26.16d)
$$SSE = \sum_i \sum_j \sum_k Y_{ijk}^2 - \frac{\sum_i \sum_j Y_{ij.}^2}{n}$$

Degrees of Freedom

The degrees of freedom associated with the various sums of squares can be deduced directly from the known relationships already studied. Since there is a total of abn cases, the degrees of freedom associated with $SSTO$ are $abn - 1$. For any level of factor A, there are $b(n - 1)$ degrees of freedom associated with the error sum of squares. Aggregating over all levels of factor A, there must be $ab(n - 1)$ degrees of freedom associated with SSE. Similarly, for any level of factor A, there are $b - 1$ degrees of freedom associated with the factor B sum of squares. Hence, by aggregating over all levels of factor A, we find that there must be $a(b - 1)$ degrees of freedom associated with $SSB(A)$. Finally, since there are a levels of factor A, there must be $a - 1$ degrees of freedom associated with SSA.

Table 26.3 shows this aggregation of the degrees of freedom for the training school example, and Table 26.4 presents the general analysis of variance table for the two-factor nested design model (26.8) where factor B is nested within factor A.

TABLE 26.3 Relation between Nested Two-Factor ANOVA and Single-Factor ANOVAs—Training School Example

Source of Variation	Single-Factor ANOVAs						Nested Two-Factor ANOVA	
	School 1		School 2		School 3			
	SS	df	SS	df	SS	df	SS	df
Between instructors (within schools)	$SSB(A_1)$	$2-1$	$+\ SSB(A_2)$	$2-1$	$+\ SSB(A_3)$	$2-1$	$=\ SSB(A)$	$3(2-1)$
Error	$SSE(A_1)$	$2(2-1)$	$+\ SSE(A_2)$	$2(2-1)$	$+\ SSE(A_3)$	$2(2-1)$	$=\ SSE$	$3(2)(2-1)$
Total within schools	$SSTO(A_1)$	$2(2)-1$	$SSTO(A_2)$	$2(2)-1$	$SSTO(A_3)$	$2(2)-1$		
Between schools							SSA	$3-1$
Total							$SSTO$	$3(2)(2)-1$

TABLE 26.4   ANOVA Table for Nested Two-Factor Fixed Effects Model (26.8) (B nested within A)

Source of Variation	SS	df	MS	E{MS}
Factor A	$SSA = bn \sum (\bar{Y}_{i..} - \bar{Y}_{...})^2$	$a - 1$	MSA	$\sigma^2 + bn\dfrac{\Sigma\alpha_i^2}{a - 1}$
Factor B (within A)	$SSB(A) = n \sum\sum (\bar{Y}_{ij.} - \bar{Y}_{i..})^2$	$a(b - 1)$	MSB(A)	$\sigma^2 + n\dfrac{\Sigma\Sigma\beta_{j(i)}^2}{a(b - 1)}$
Error	$SSE = \sum\sum\sum (Y_{ijk} - \bar{Y}_{ij.})^2$	$ab(n - 1)$	MSE	σ^2
Total	$SSTO = \sum\sum\sum (Y_{ijk} - \bar{Y}_{...})^2$	$abn - 1$		

Example

In the training school example of Table 26.1, both schools and instructors were re-garded as fixed effects factors; hence, model (26.8) was deemed appropriate. Figure 26.2 presents a line plot of the estimated treatment means $\bar{Y}_{ij.}$ for the training school example. Note that this plot is not in the format for a crossed two-factor study be-cause different instructors were used in the different schools. Figure 26.2 suggests strong differences between instructors within a school and also possible differences in mean learning between schools.

To analyze these effects formally, we begin by obtaining the analysis of variance. The sums of squares were obtained as follows using the computational formulas (26.16):

$$SSTO = (25)^2 + (29)^2 + (14)^2 + \cdots + (2)^2 - \frac{(180)^2}{12}$$

$$= 3,466 - 2,700 = 766$$

$$SSA = \frac{1}{4}[(79)^2 + (57)^2 + (44)^2] - 2,700$$

$$= 2,856.5 - 2,700 = 156.5$$

FIGURE 26.2   Line Plot of Estimated Treatment Means—Training School Example

Learning Score

○ Atlanta
□ Chicago
△ San Francisco

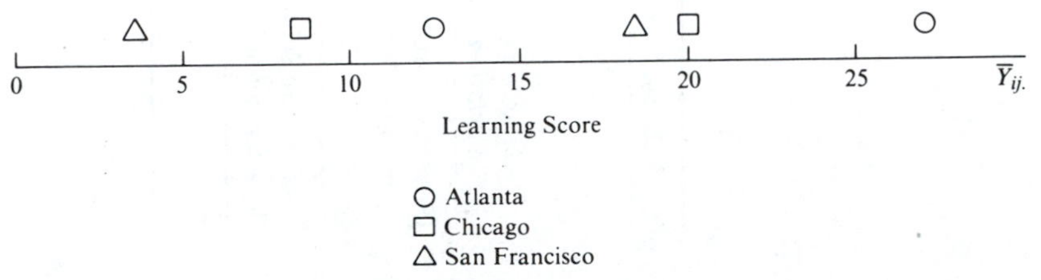

$$SSB(A) = \frac{1}{2}[(54)^2 + (25)^2 + \cdots + (7)^2] - 2,856.5$$

$$= 3,424 - 2,856.5 = 567.5$$

$$SSE = 3,466 - 3,424 = 42$$

Table 26.5a contains the analysis of variance.

Relation between Crossed and Nested Sums of Squares

If a computer program for the analysis of variance for nested designs is unavailable but one for crossed factors is at hand, the latter can be used with only slight inconvenience when the number of levels of nested factor B is the same for each level of factor A and the number of replications is the same for all factor combinations. Such nested designs are said to be *balanced*. Table 26.6 contains the (incorrect) analysis of variance obtained for the training school example in Table 26.1 from a computer run that treated the two factors as crossed. When we compare this incorrect analysis of variance with the correct one in Table 26.5a, we note that $SSTO$, SSA, and SSE are the same in each case, as are the associated degrees of freedom. The difference between the two analyses of variance is that the nested analysis has no interaction sum of squares. However, if we use the relation:

(26.17) $$\underbrace{SSB(A)}_{\text{Nested}} = \underbrace{SSB + SSAB}_{\text{Crossed}}$$

and do likewise for the associated degrees of freedom, we obtain the correct nested

---

TABLE 26.5 ANOVA for Two-Factor Nested Design—Training School Example

(a) ANOVA Table

Source of Variation	SS	df	MS
Schools (A)	$SSA = 156.5$	2	78.25
Instructors, within schools [$B(A)$]	$SSB(A) = 567.5$	3	189.17
Error (E)	$SSE = 42.0$	6	7.00
Total	$SSTO = 766.0$	11	

(b) Decomposition of $SSB(A)$

Source of Variation	$SSB(A_i)$	df	$MSB(A_i)$
Instructors, Atlanta	210.25	1	210.25
Instructors, Chicago	132.25	1	132.25
Instructors, San Francisco	225.00	1	225.00
Total	567.5	3	

TABLE 26.6 Incorrect Crossed Factor ANOVA for Two-Factor Nested Design—Training School Example

Source of Variation	SS	df	MS
Schools (A)	$SSA = 156.5$	2	78.25
Instructors (B)	$SSB = 108.0$	1	108.00
School-instructor interactions (AB)	$SSAB = 459.5$	2	229.75
Error (E)	$SSE = 42.0$	6	7.00
Total	$SSTO = 766.0$	11	

factor B sum of squares and degrees of freedom:

$$SSB(A) = 108.0 + 459.5 = 567.5$$

$$df = 1 + 2 = 3$$

By using the relation (26.17), one can easily obtain the proper sums of squares and degrees of freedom for a balanced nested two-factor design from a computer package for crossed factors.

Tests for Factor Effects

The tests for factor effects in a nested two-factor study are straightforward. The appropriate test statistics are determined, as for a crossed two-factor study, by comparing the expected values of the ANOVA mean squares. The expected mean squares for the nested fixed effects model (26.8) are shown in Table 26.4. They can be obtained by somewhat tedious derivations. We do not illustrate these derivations because we will present in Chapter 27 a relatively simple method of finding expected mean squares for any balanced nested design.

The $E\{MS\}$ column in Table 26.4 indicates that for the fixed effects model (26.8), the test for factor A main effects:

(26.18a)
$$H_0: \text{all } \alpha_i = 0$$
$$H_a: \text{not all } \alpha_i \text{ equal zero}$$

is based on the test statistic:

(26.18b)
$$F^* = \frac{MSA}{MSE}$$

and the decision rule to control the level of significance at α is:

(26.18c)
If $F^* \leq F[1 - \alpha; a - 1, (n - 1)ab]$, conclude H_0

If $F^* > F[1 - \alpha; a - 1, (n - 1)ab]$, conclude H_a

Similarly, to test for factor B specific effects:

(26.19a)
$$H_0: \text{all } \beta_{j(i)} = 0$$
$$H_a: \text{not all } \beta_{j(i)} \text{ equal zero}$$

the appropriate test statistic is:

(26.19b)
$$F^* = \frac{MSB(A)}{MSE}$$

and the appropriate decision rule is:

(26.19c)
$$\text{If } F^* \leq F[1 - \alpha; a(b - 1), (n - 1)ab], \text{ conclude } H_0$$
$$\text{If } F^* > F[1 - \alpha; a(b - 1), (n - 1)ab], \text{ conclude } H_a$$

Example.   We return to the training school example. Based on the analysis of variance in Table 26.5a, the first test conducted was one to determine whether or not main school effects exist. The alternatives are given in (26.18a), and test statistic (26.18b) here is:

$$F^* = \frac{78.25}{7.00} = 11.2$$

The level of significance was to be controlled at $\alpha = .05$. Hence, we require $F(.95; 2, 6) = 5.14$. Since $F^* = 11.2 > 5.14$, it was concluded that the three schools differ in mean learning effects. The P-value of the test is .0094.

Next, a test for differences in mean learning effects between instructors within each school was conducted. The alternatives are given in (26.19a), and test statistic (26.19b) here is:

$$F^* = \frac{189.17}{7.00} = 27.0$$

For $\alpha = .05$, we require $F(.95; 3, 6) = 4.76$. Since $F^* = 27.0 > 4.76$, it was concluded that instructors within at least one school differ in terms of mean learning effects. The P-value of this test is .0007.

Comments

1. The alternative H_0 in (26.19a) can also be expressed in terms of the treatment means μ_{ij}:

(26.20)
$$H_0: \mu_{11} = \mu_{12} = \cdots = \mu_{1b}; \mu_{21} = \mu_{22} = \cdots = \mu_{2b}; \ldots$$

Thus, H_0 states in terms of the training school example that the mean learning scores for all instructors in Atlanta are the same, and similarly for the other schools. It does *not* state that the mean learning scores for all instructors in the different schools are the same.

2. If it is concluded that factor B effects are present, it is often desired to ascertain whether they are present in all levels of factor A or only in some. (In some cases, indeed, one may wish to proceed immediately to this analysis.) With reference to the training school example, the question would be whether the instructor effects differ in all schools or only in some

schools. As noted earlier, $SSB(A)$ in Table 26.5a is made up of the instructor sums of squares within the individual schools, and these component sums of squares can be used for testing instructor effects within each school. Table 26.5b contains the relevant component sums of squares. To test for instructor differences within the Atlanta school, for instance, we use test statistic $F^* = MSB(A_1)/MSE = 210.25/7.00 = 30.0$. For a level of significance of $\alpha = .05$, we need $F(.95; 1, 6) = 5.99$. Since $F^* = 30.0 > 5.99$, it was concluded that the two instructors in Atlanta have different mean learning effects. Using the same level of significance each time, similar conclusions were reached for the other two schools. The family level of significance for the three tests according to the Bonferroni inequality is .15.

3. If the assumption of constant error variance were violated in the training school example through unequal variances for the different schools, it would still be possible to study instructor effects within each school by separate analyses of variance for each school.

Random Factor Effects

Test statistic (26.18b) for factor A main effects is not appropriate if either or both factor effects are random. Table 26.7 gives the expected mean squares for these cases and also the appropriate test statistics.

26.4 EVALUATION OF APTNESS OF NESTED DESIGN MODEL

The diagnostic procedures described earlier are entirely applicable for examining whether the nested design model (26.8) is appropriate. The residuals in (26.11):

$$(26.21) \qquad e_{ijk} = Y_{ijk} - \bar{Y}_{ij.}$$

may be examined as usual for normality, constancy of the error variance, and independence of the error terms. In particular, aligned dot plots of the residuals for each

---

TABLE 26.7 Expected Mean Squares for Nested Two-Factor Designs with Random Factor Effects (B nested within A)

	Expected Mean Square	
Mean Square	A Fixed, B Random	A Random, B Random
MSA	$\sigma^2 + \dfrac{bn}{a-1}\sum \alpha_i^2 + n\sigma_\beta^2$	$\sigma^2 + bn\sigma_\alpha^2 + n\sigma_\beta^2$
$MSB(A)$	$\sigma^2 + n\sigma_\beta^2$	$\sigma^2 + n\sigma_\beta^2$
MSE	σ^2	σ^2
	Appropriate Test Statistic	
Test for	A Fixed, B Random	A Random, B Random
Factor A	$MSA/MSB(A)$	$MSA/MSB(A)$
Factor $B(A)$	$MSB(A)/MSE$	$MSB(A)/MSE$

factor A level may be helpful in examining whether the variance of the error terms is constant for the different factor A levels within which factor B is nested.

Example

Figure 26.3a contains aligned dot plots of the residuals for each school for the training school example. These plots are affected by the rounded nature of the data, but they support the appropriateness of the assumption of constancy of the error variance. Figure 26.3b presents a normal probability plot of the residuals. This plot is also affected by the rounded nature of the observations, but does not indicate any gross departure from normality. This conclusion is supported by the coefficient of correlation between the ordered residuals and their expected values under normality, which is .927. These and other diagnostics (not shown here) support the appropriateness of nested design model (26.8) for the training school example.

Note

In previous normal probability plots, we have plotted tied residuals against expected values that were obtained from (4.6) as if the residuals had distinct values. Figure 4.2d provides an example of this plotting.

FIGURE 26.3   Diagnostic Residual Plots—Training School Example (Minitab, Ref. 26.1)

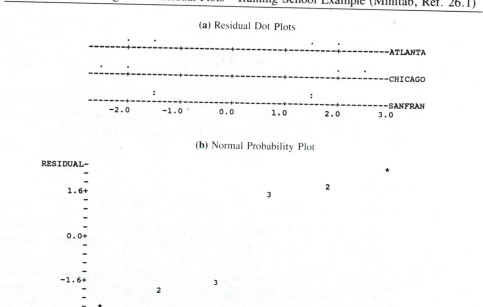

When there are numerous ties, as in the training school example, a more effective normal probability plot is obtained by plotting each of the tied residuals against the expected value for the mean of the tied order positions. The plot then should show the number of tied residuals at that position. This is done in Figure 26.3b.

26.5 ANALYSIS OF FACTOR EFFECTS IN TWO-FACTOR NESTED DESIGNS

When factor effects are present in a nested design, one usually desires to obtain estimates of these and/or make comparisons.

Estimation of Factor Level Means $\mu_{i\cdot}$

When factor A (fixed effects factor) has significant main effects, there is frequent interest in estimating the factor level means $\mu_{i\cdot}$. The estimated factor level mean $\bar{Y}_{i\cdot\cdot}$ is an unbiased point estimator of $\mu_{i\cdot}$. As usual for a fixed effects factor, the estimated variance of $\bar{Y}_{i\cdot\cdot}$ is based on the mean square in the denominator of the statistic used for testing for factor A main effects, and on the number of observations in $\bar{Y}_{i\cdot\cdot}$. Confidence limits for $\mu_{i\cdot}$ are of the customary form:

$$(26.22) \qquad \bar{Y}_{i\cdot\cdot} \pm t(1 - \alpha/2; df)s\{\bar{Y}_{i\cdot\cdot}\}$$

where:

$$(26.22a) \qquad s^2\{\bar{Y}_{i\cdot\cdot}\} = \frac{MSE}{bn} \qquad df = ab(n - 1) \qquad A \text{ and } B \text{ fixed}$$

$$(26.22b) \qquad s^2\{\bar{Y}_{i\cdot\cdot}\} = \frac{MSB(A)}{bn} \qquad df = a(b - 1) \qquad A \text{ fixed, } B \text{ random}$$

Confidence limits for differences $D = \mu_{i\cdot} - \mu_{i'\cdot}$ are set up in the usual way, utilizing the point estimator $\hat{D} = \bar{Y}_{i\cdot\cdot} - \bar{Y}_{i'\cdot\cdot}$ and the t distribution with degrees of freedom those associated with the appropriate mean square:

$$(26.23) \qquad \hat{D} \pm t(1 - \alpha/2; df)s\{\hat{D}\}$$

where:

$$(26.23a) \qquad s^2\{\hat{D}\} = s^2\{\bar{Y}_{i\cdot\cdot}\} + s^2\{\bar{Y}_{i'\cdot\cdot}\} \qquad \text{as given by (26.22a) or (26.22b)}$$

The Tukey and Bonferroni simultaneous comparison procedures can be utilized in the usual way for making pairwise comparisons with family confidence coefficient $1 - \alpha$.

Finally, no new problems arise in estimating contrasts of factor level means. The Scheffé or the Bonferroni procedures can be employed when estimating several contrasts.

Example.   For the training school example in Table 26.1, it was desired to estimate the mean learning score for the Atlanta school with a 95 percent confidence coefficient. Using our earlier results in Tables 26.1 and 26.5, we obtain for the fixed effects model:

$$\bar{Y}_{1..} = \frac{79}{4} = 19.75$$

$$s^2\{\bar{Y}_{1..}\} = \frac{MSE}{bn} = \frac{7.00}{4} = 1.75$$

$$s\{\bar{Y}_{1..}\} = 1.32$$

$$t(.975; 6) = 2.447$$

$$16.5 = 19.75 - 2.447(1.32) \leq \mu_{1.} \leq 19.75 + 2.447(1.32) = 23.0$$

In addition, pairwise comparisons of the three schools were to be made with family confidence coefficient .90. We shall utilize the Tukey method and require:

$$T = \frac{1}{\sqrt{2}} q[1 - \alpha; a, ab(n - 1)] = \frac{1}{\sqrt{2}} q(.90; 3, 6)$$

$$= \frac{1}{\sqrt{2}} (3.56) = 2.52$$

The estimated variance is the same for all pairwise comparisons:

$$s^2\{\hat{D}\} = \frac{MSE}{bn} + \frac{MSE}{bn} = \frac{2(7.00)}{4} = 3.5$$

so that the estimated standard deviation is $s\{\hat{D}\} = 1.87$ and the precision term is $2.52(1.87) = 4.71$.

Using the results in Table 26.1, we find:

$$\bar{Y}_{1..} = 19.75 \qquad \bar{Y}_{2..} = 14.25 \qquad \bar{Y}_{3..} = 11$$

Hence, the 90 percent family of confidence intervals is:

$$.8 = (19.75 - 14.25) - 4.71 \leq \mu_{1.} - \mu_{2.} \leq (19.75 - 14.25) + 4.71 = 10.2$$

$$4.0 = (19.75 - 11) - 4.71 \leq \mu_{1.} - \mu_{3.} \leq (19.75 - 11) + 4.71 = 13.5$$

$$-1.5 = (14.25 - 11) - 4.71 \leq \mu_{2.} - \mu_{3.} \leq (14.25 - 11) + 4.71 = 8.0$$

We conclude with 90 percent family confidence coefficient that the mean learning score is highest in Atlanta and that the difference in the observed mean scores for Chicago and San Francisco is not statistically significant. We summarize these results by the following line plot:

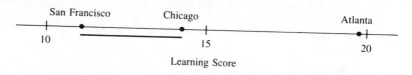

Estimation of Treatment Means μ_{ij}

Confidence limits for μ_{ij} are set up in the usual fashion using the t distribution when both factors A and B have fixed effects:

(26.24) $\qquad\qquad\qquad \bar{Y}_{ij.} \pm t[1 - \alpha/2; (n - 1)ab]s\{\bar{Y}_{ij.}\}$

where:

(26.24a) $\qquad\qquad\qquad s^2\{\bar{Y}_{ij.}\} = \dfrac{MSE}{n}$

To make a pairwise comparison within a factor A level, we estimate the difference $D = \mu_{ij} - \mu_{ij'}$ with the point estimator $\hat{D} = \bar{Y}_{ij.} - \bar{Y}_{ij'.}$ and employ the confidence limits:

(26.25) $\qquad\qquad\qquad \hat{D} \pm t[1 - \alpha/2; (n - 1)ab]s\{\hat{D}\}$

where:

(26.25a) $\qquad\qquad\qquad s^2\{\hat{D}\} = \dfrac{2MSE}{n}$

The Bonferroni procedure may be used when several comparisons are to be made and the family confidence level is to be controlled. The Tukey procedure is also applicable but often will not be efficient since ordinarily only comparisons within each factor level are of interest, whereas the Tukey family is based on all pairwise comparisons among the ab treatments.

Example.   It is desired in the training school example to compare the mean scores for the two instructors in each school, using the Bonferroni procedure with a 90 percent family confidence coefficient. For $g = 3$ comparisons, we require $B = t[1 - .10/2(3); 6] = t(.983; 6) = 2.748$. The estimated variance in each case is:

$$s^2\{\bar{Y}_{i1.} - \bar{Y}_{i2.}\} = \frac{2(7.00)}{2} = 7.0$$

Hence, the precision term in each comparison is $2.748\sqrt{7.0} = 7.27$. Obtaining the estimated treatment means $\bar{Y}_{ij.}$ from Table 26.1, we find:

$$7.2 = (27 - 12.5) - 7.27 \le \mu_{11} - \mu_{12} \le (27 - 12.5) + 7.27 = 21.8$$

$$-18.8 = (8.5 - 20) - 7.27 \le \mu_{21} - \mu_{22} \le (8.5 - 20) + 7.27 = -4.2$$

$$7.7 = (18.5 - 3.5) - 7.27 \le \mu_{31} - \mu_{32} \le (18.5 - 3.5) + 7.27 = 22.3$$

It is evident that substantial differences between the two instructors exist at each school.

Estimation of Overall Mean $\mu_{..}$

Sometimes there is interest in estimating the overall mean $\mu_{..}$. For the training school example, $\mu_{..}$ is the overall mean learning score for all training schools and all instructors in these schools. The point estimator is $\bar{Y}_{...}$. The confidence limits are constructed utilizing the t distribution as follows:

$$(26.26) \qquad \bar{Y}_{...} \pm t(1 - \alpha/2; df)s\{\bar{Y}_{...}\}$$

where:

$$(26.26\text{a}) \qquad s^2\{\bar{Y}_{...}\} = \frac{MSE}{abn} \qquad df = ab(n - 1) \qquad A \text{ and } B \text{ fixed}$$

$$(26.26\text{b}) \qquad s^2\{\bar{Y}_{...}\} = \frac{MSA}{abn} \qquad df = a - 1 \qquad A \text{ and } B \text{ random}$$

$$(26.26\text{c}) \qquad s^2\{\bar{Y}_{...}\} = \frac{MSB(A)}{abn} \qquad df = a(b - 1) \qquad A \text{ fixed}, B \text{ random}$$

Example. For the training school example, we wish to estimate the overall mean $\mu_{..}$ with a 95 percent confidence interval. The estimated variance (26.26a) is appropriate here since the model involves fixed factor effects. Hence, we obtain:

$$s^2\{\bar{Y}_{...}\} = \frac{7.00}{12} = .583 \qquad s\{\bar{Y}_{...}\} = .764$$

For confidence coefficient .95, we require $t(.975; 6) = 2.447$. From Table 26.1, we find $\bar{Y}_{...} = 15$. The desired confidence interval therefore is:

$$13.1 = 15 - 2.447(.764) \leq \mu_{..} \leq 15 + 2.447(.764) = 16.9$$

Estimation of Variance Components

With random factor effects, one may wish to estimate the variance components. No new problems arise for nested designs. For instance, an unbiased estimator of σ_α^2 when both factors A and B have random effects is (Table 26.7):

$$(26.27) \qquad s_\alpha^2 = \frac{MSA - MSB(A)}{bn}$$

26.6 UNEQUAL NESTING AND REPLICATIONS IN NESTED TWO-FACTOR DESIGNS

Up to this point, we have assumed that the same number of levels of factor B is nested within each of the levels of factor A, and that the same number of replications is made for each factor combination. There are occasions, however, when the number of levels of the nested B factor will vary for the different levels of factor A, and when the number of replications for the different factor combinations will be unequal. For instance, in our earlier example dealing with the effects of school (factor A) and instructor (factor B) on the learning achieved by classes of mechanics, there might have been b_i instructors in the ith school and n_{ij} classes taught by the jth instructor in school i.

The ANOVA sums of squares formulas given earlier are not appropriate for unequal nestings and replications. Ordinarily, it is best to use the regression approach for this case, but the general matrix approach described in Section 8.6 can also be employed. Since no new principles are involved when the factor effects are fixed, we proceed directly to an example.

Example

The manufacturing company that conducted the training school study subsequently made a follow-up study involving only Atlanta and Chicago. Three instructors were used in Atlanta and two in Chicago. All instructors were to train two classes, but one class for one of the instructors in Atlanta had to be canceled. The data for this follow-up study are presented in Table 26.8a. Again we assume that the fixed effects

TABLE 26.8 Nested Two-Factor Study with Unequal Nestings and Replications—Follow-up Training School Study

(a) Data

Replication k	Atlanta (A_1)			Chicago (A_2)	
	B_1	B_2	B_3	B_1	B_2
1	20	8	9	4	16
2	22		13	8	20

(b) $\mathbf{Y}$ and $\mathbf{X}$ Matrices for Full Model (26.31)

$$\mathbf{Y} = \begin{bmatrix} Y_{111} \\ Y_{112} \\ Y_{121} \\ Y_{131} \\ Y_{132} \\ Y_{211} \\ Y_{212} \\ Y_{221} \\ Y_{222} \end{bmatrix} = \begin{bmatrix} 20 \\ 22 \\ 8 \\ 9 \\ 13 \\ 4 \\ 8 \\ 16 \\ 20 \end{bmatrix} \qquad \mathbf{X} = \begin{bmatrix} & X_1 & X_2 & X_3 & X_4 \\ 1 & 1 & 1 & 0 & 0 \\ 1 & 1 & 1 & 0 & 0 \\ 1 & 1 & 0 & 1 & 0 \\ 1 & 1 & -1 & -1 & 0 \\ 1 & 1 & -1 & -1 & 0 \\ 1 & -1 & 0 & 0 & 1 \\ 1 & -1 & 0 & 0 & 1 \\ 1 & -1 & 0 & 0 & -1 \\ 1 & -1 & 0 & 0 & -1 \end{bmatrix}$$

(c) Fitted Full Model

$$\hat{Y} = 12.667 + .667X_1 + 7.667X_2 - 5.333X_3 - 6.0X_4$$

nested design model (26.8) is appropriate:

(26.28) $$Y_{ijk} = \mu_{..} + \alpha_i + \beta_{j(i)} + \varepsilon_{k(ij)}$$

$$i = 1, 2; j = 1, \ldots, b_i; k = 1, \ldots, n_{ij}$$

$$b_1 = 3, b_2 = 2$$

$$n_{11} = n_{13} = 2, n_{12} = 1, n_{21} = n_{22} = 2$$

In developing the equivalent regression model, we need to recognize the constraints in (26.8):

(26.29) $$\sum_{i=1}^{2} \alpha_i = 0 \qquad \sum_{j=1}^{3} \beta_{j(1)} = 0 \qquad \sum_{j=1}^{2} \beta_{j(2)} = 0$$

Proceeding as usual, we shall incorporate the parameters α_1, $\beta_{1(1)}$, $\beta_{2(1)}$, and $\beta_{1(2)}$ into the regression model. The other parameters are not required since according to the constraints (26.29) we have:

(26.30) $$\alpha_2 = -\alpha_1 \qquad \beta_{3(1)} = -\beta_{1(1)} - \beta_{2(1)} \qquad \beta_{2(2)} = -\beta_{1(2)}$$

Thus, we require four indicator variables for our example, each taking on values 1, -1, or 0.

The equivalent regression model therefore is:

(26.31) $$Y_{ijk} = \mu_{..} + \underbrace{\alpha_1 X_{ijk1}}_{\substack{\text{School main} \\ \text{effect}}}$$

$$+ \underbrace{\beta_{1(1)} X_{ijk2} + \beta_{2(1)} X_{ijk3} + \beta_{1(2)} X_{ijk4}}_{\substack{\text{Specific instructor within} \\ \text{school effect}}} + \varepsilon_{ijk} \qquad \text{Full model}$$

where:

$$X_{ijk1} = \begin{array}{l} 1 \text{ if class from school 1} \\ -1 \text{ if class from school 2} \end{array}$$

$$X_{ijk2} = \begin{array}{l} 1 \text{ if class for instructor 1 in school 1} \\ -1 \text{ if class for instructor 3 in school 1} \\ 0 \text{ otherwise} \end{array}$$

$$X_{ijk3} = \begin{array}{l} 1 \text{ if class for instructor 2 in school 1} \\ -1 \text{ if class for instructor 3 in school 1} \\ 0 \text{ otherwise} \end{array}$$

$$X_{ijk4} = \begin{array}{l} 1 \text{ if class for instructor 1 in school 2} \\ -1 \text{ if class for instructor 2 in school 2} \\ 0 \text{ otherwise} \end{array}$$

The **Y** vector and **X** matrix for our example are shown in Table 26.8b.

To test for school main effects, we first fit the full model (26.31). The fitted full

TABLE 26.9   ANOVA Table for Nested Two-Factor Study with Unequal Nestings and Replications—Follow-up Training School Study

Source of Variation	SS	df	MS	F*
Schools (A)	3.76	1	3.76	$3.76/6.5 = .58$
Instructors [B(A)]	295.20	3	98.4	$98.4/6.5 = 15.1$
Error (E)	26.00	4	6.5	

model is shown in Table 26.8c. We then fit the reduced model for $H_0: \alpha_1 = 0$:

$$(26.32) \qquad Y_{ijk} = \mu_{..} + \beta_{1(1)} X_{ijk2} + \beta_{2(1)} X_{ijk3} + \beta_{1(2)} X_{ijk4} + \varepsilon_{ijk} \qquad \text{Reduced model}$$

The difference $SSE(R) - SSE(F)$ equals SSA. Test statistic (3.69) is then obtained in the usual fashion.

To test for specific instructor effects, we employ the reduced model for $H_0: \beta_{1(1)} = \beta_{2(1)} = \beta_{1(2)} = 0$:

$$(26.33) \qquad\qquad Y_{ijk} = \mu_{..} + \alpha_1 X_{ijk1} + \varepsilon_{ijk} \qquad\qquad \text{Reduced model}$$

The difference $SSE(R) - SSE(F)$ equals $SSB(A)$.

Table 26.9 contains the ANOVA table for the follow-up training school study. No total sum of squares is shown because the component sums of squares are not orthogonal.

The tests for school and instructor effects are carried out as before. Estimation of factor effects is done by means of the regression parameters. For instance, a comparison of the mean scores for the two schools involves:

$$\mu_{1.} - \mu_{2.} = \alpha_1 - \alpha_2$$

Since $\alpha_2 = -\alpha_1$ by (26.30), we need to estimate:

$$\mu_{1.} - \mu_{2.} = \alpha_1 - (-\alpha_1) = 2\alpha_1$$

A point estimator is $2\hat{\alpha}_1$. Other desired estimates are obtained in similar fashion.

26.7 SUBSAMPLING IN SINGLE-FACTOR STUDY WITH COMPLETELY RANDOMIZED DESIGN

Up to this point in our discussion of experimental designs, we have considered only designs in which one observation of the dependent variable is made on an experimental unit. There are occasions, however, when more than one observation is desirable. Consider an experiment to study the effect of oven temperature on crustiness of bread. Three temperatures were utilized, and two experimental units (batches of flour mix) were randomly assigned to each treatment. It was not economical to use the entire batch to bake breads, nor was it technically feasible to use a batch as a block. Hence, three subsamples were selected from each batch to make three loaves, which were baked at a given temperature. Here then, three observations (subsamples) were made on each experimental unit (batch).

Another instance of several observations being made on the dependent variable for each experimental unit occurred in an experiment on the effectiveness of three different training methods. The experimental units here were persons, and the experiment sought to measure the length of time required to perform a certain engine assembly operation after the given training program was completed. Ten consecutive assemblies were timed, and these constituted the subsamples of the experimental unit (person).

Formally, subsampling (i.e., repeated observations on the same experimental unit) is completely analogous to nested factors. We shall demonstrate this for a completely randomized design.

Model

Consider again the experiment to study the effect of oven temperature on the crustiness of bread. The model for this study can be written as follows:

$$(26.34) \qquad Y_{ijk} = \mu_{..} + \tau_i + \varepsilon_{j(i)} + \eta_{k(ij)}$$

The meaning of the symbols is as follows:

1. $\mu_{..}$ is an overall constant.
2. τ_i is the temperature (i.e., treatment) effect (fixed effect, here).
3. $\varepsilon_{j(i)}$ is the experimental error associated with the particular batch (random effect, here). As usual, the experimental error is nested within the treatment, since the jth batch for treatment i was not used with any other treatment.
4. $\eta_{k(ij)}$ is the error associated with the kth subsample or observation on the jth experimental unit for the ith treatment (random effect, here). This observation error is nested within the experimental unit, and consequently also within the treatment.

Note that subsampling model (26.34) appears the same as nested design model (26.8) for a nested two-factor design, except for changes in notation to reflect the fact that subsampling model (26.34) is a single-factor model and contains both experimental error and observation error. Specifically, the treatment effect τ_i here corresponds to α_i in the nested two-factor model, the batch effect $\varepsilon_{j(i)}$ corresponds to $\beta_{j(i)}$, and the observation error term $\eta_{k(ij)}$ corresponds to $\varepsilon_{k(ij)}$. Consequently, the analysis of variance for the case of subsampling in a single-factor study with a completely randomized design parallels that for a nested two-factor study.

In general, the model for subsampling in a single-factor study with a completely randomized design where the treatment effects are fixed and there are equal numbers of replications and subsamples is:

$$(26.35) \qquad Y_{ijk} = \mu_{..} + \tau_i + \varepsilon_{j(i)} + \eta_{k(ij)}$$

where:

$\mu_{..}$ is a constant

τ_i are constants subject to the restriction $\Sigma \tau_i = 0$

$\varepsilon_{j(i)}$ are independent $N(0, \sigma^2)$

$\eta_{k(ij)}$ are independent $N(0, \sigma_\eta^2)$

$\varepsilon_{j(i)}$ and $\eta_{k(ij)}$ are independent

$i = 1, \ldots, r; j = 1, \ldots, n; k = 1, \ldots, m$

The mean and variance of observation Y_{ijk} for this model are:

(26.35a)
$$E\{Y_{ijk}\} = \mu_{..} + \tau_i$$

(26.35b)
$$\sigma^2\{Y_{ijk}\} = \sigma_Y^2 = \sigma^2 + \sigma_\eta^2$$

Further, the observations Y_{ijk} are normally distributed for this model, and observations from different replications (i.e., from different subsamples) are independent. However, any two observations from the same replication are correlated in advance of the random trials because they contain the same random term $\varepsilon_{j(i)}$:

(26.35c)
$$\sigma\{Y_{ijk}, Y_{ijk'}\} = \sigma^2 \quad k \neq k'$$

Analysis of Variance and Tests of Effects

The appropriate sums of squares for the analysis of variance for subsampling model (26.35) are as follows:

(26.36a)
$$SSTO = \sum_i \sum_j \sum_k (Y_{ijk} - \bar{Y}_{...})^2 = \sum_i \sum_j \sum_k Y_{ijk}^2 - \frac{Y_{...}^2}{rnm}$$

(26.36b)
$$SSTR = nm \sum_i (\bar{Y}_{i..} - \bar{Y}_{...})^2 = \frac{\sum_i Y_{i..}^2}{nm} - \frac{Y_{...}^2}{rnm}$$

(26.36c)
$$SSEE = m \sum_i \sum_j (\bar{Y}_{ij.} - \bar{Y}_{i..})^2 = \frac{\sum_i \sum_j Y_{ij.}^2}{m} - \frac{\sum_i Y_{i..}^2}{nm}$$

(26.36d)
$$SSOE = \sum_i \sum_j \sum_k (Y_{ijk} - \bar{Y}_{ij.})^2 = \sum_i \sum_j \sum_k Y_{ijk}^2 - \frac{\sum_i \sum_j Y_{ij.}^2}{m}$$

Here, *SSEE* stands for the *experimental error sum of squares*, and *SSOE* stands for the *observation error sum of squares*. Note the correspondence of formulas (26.36) to formulas (26.13) and (26.16) for nested two-factor designs. The only difference is that we now have $i = 1, \ldots, r, j = 1, \ldots, n$, and $k = 1, \ldots, m$, whereas before i, j, and k ran to a, b, and n, respectively.

Table 26.10 contains the ANOVA for a single-factor completely randomized experiment with subsampling. Also shown there are the expected mean squares for both fixed and random treatment effects. Note that regardless of whether treatment

TABLE 26.10   ANOVA for Single-Factor Completely Randomized Experiment with Subsampling

Source of Variation	SS	df	MS	$E\{MS\}$ τ_i Fixed	τ_i Random
Treatments	SSTR	$r - 1$	MSTR	$\sigma_\eta^2 + m\sigma^2 + nm\dfrac{\Sigma\tau_i^2}{r-1}$	$\sigma_\eta^2 + m\sigma^2 + nm\sigma_\tau^2$
Experimental error	SSEE	$r(n-1)$	MSEE	$\sigma_\eta^2 + m\sigma^2$	$\sigma_\eta^2 + m\sigma^2$
Observation error	SSOE	$rn(m-1)$	MSOE	σ_η^2	σ_η^2
Total	SSTO	$rnm - 1$			

effects are fixed or random, the appropriate statistic for testing treatment effects is:

$$(26.37a) \qquad\qquad F* = \frac{MSTR}{MSEE}$$

A test for the presence of experimental error effects, i.e., $\sigma^2 > 0$, uses the same test statistic for both fixed and random treatment effects:

$$(26.37b) \qquad\qquad F* = \frac{MSEE}{MSOE}$$

Example.   The data for the study of the effect of baking temperature on the crustiness of bread are contained in Table 26.11. The data are scores on a scale from 1 to 20. The appropriate analysis of variance was obtained from a computer run and is presented in Table 26.12. To test the effect of temperature:

$$H_0: \tau_1 = \tau_2 = \tau_3 = 0$$

$$H_a: \text{not all } \tau_i \text{ equal zero}$$

TABLE 26.11   Data for Single-Factor Completely Randomized Experiment with Subsampling—Bread Crustiness Example

Observation Unit k	Low $(i=1)$ Batch 1 $j=1$	Batch 2 $j=2$	Medium $(i=2)$ Batch 3 $j=1$	Batch 4 $j=2$	High $(i=3)$ Batch 5 $j=1$	Batch 6 $j=2$
1	4	12	14	9	14	16
2	7	8	13	10	17	19
3	5	10	11	12	15	18
	$Y_{11.} = 16$	$Y_{12.} = 30$	$Y_{21.} = 38$	$Y_{22.} = 31$	$Y_{31.} = 46$	$Y_{32.} = 53$
		$Y_{1..} = 46$		$Y_{2..} = 69$		$Y_{3..} = 99$
				$Y_{...} = 214$		

TABLE 26.12 ANOVA for Bread Crustiness Example

Source of Variation	SS	df	MS
Temperatures (*TR*)	235.44	2	117.72
Mix batches (*EE*)	49.00	3	16.33
Observation units (*OE*)	31.33	12	2.61
Total	315.78	17	

Note: Component sums of squares do not add to *SSTO* because of rounding.

---

we use test statistic (26.37a):

$$F^* = \frac{117.72}{16.33} = 7.21$$

A level of significance of $\alpha = .10$ was specified. Hence, we need $F(.90; 2, 3) = 5.46$. Since $F^* = 7.21 > 5.46$, we conclude H_a, that baking temperature does have an effect on the crustiness of the bread. The *P*-value of the test is .07.

To test for batch differences:

$$H_0: \sigma^2 = 0$$

$$H_a: \sigma^2 > 0$$

we employ test statistic (26.37b):

$$F^* = \frac{16.33}{2.61} = 6.26$$

For a level of significance of $\alpha = .10$, we need $F(.90; 3, 12) = 2.61$. Since $F^* = 6.26 > 2.61$, we conclude H_a, that there are batch effects on the crustiness of bread. The *P*-value of this test is .01. Thus, both the particular batch of flour mix and the temperature at which the bread is baked affect the crustiness of the loaf.

Estimation of Treatment Effects

When the treatment effects are fixed, there is usually interest in confidence limits for treatment means $\mu_{i.} = \mu_{..} + \tau_i$ and for pairwise comparisons and contrasts of treatment means. These can be obtained in the usual manner, using *MSEE* as the error variance since this is the quantity in the denominator of the test statistic for fixed treatment effects. The degrees of freedom are those associated with *MSEE*, namely, $(n - 1)r$. For instance, the confidence limits for treatment mean $\mu_{i.}$ are:

(26.38) $$\bar{Y}_{i..} \pm t[1 - \alpha/2; (n - 1)r]s\{\bar{Y}_{i..}\}$$

where:

(26.38a) $$s^2\{\bar{Y}_{i..}\} = \frac{MSEE}{nm}$$

Similarly, confidence limits for a pairwise comparison of treatment means, $D = \mu_{i.} - \mu_{i'.}$, are obtained as follows:

(26.39) $$\hat{D} \pm t[1 - \alpha/2; (n - 1)r]s\{\hat{D}\}$$

where:

(26.39a) $$\hat{D} = \bar{Y}_{i..} - \bar{Y}_{i'..}$$

(26.39b) $$s^2\{\hat{D}\} = \frac{2MSEE}{nm}$$

The Bonferroni and Tukey simultaneous comparison procedures can be utilized in the usual manner.

Example.   To estimate the mean crustiness of bread baked at a low temperature with a 95 percent confidence coefficient, we need:

$$\bar{Y}_{1..} = 7.67$$

$$s^2\{\bar{Y}_{1..}\} = \frac{16.33}{6} = 2.722 \qquad s\{\bar{Y}_{1..}\} = 1.65$$

$$t(.975; 3) = 3.182$$

Hence, the 95 percent confidence interval is:

$$2.4 = 7.67 - 3.182(1.65) \leq \mu_{1.} \leq 7.67 + 3.182(1.65) = 12.9$$

It was also desired to estimate the difference in mean crustiness of bread baked at high and low temperatures with a 95 percent confidence interval. Utilizing (26.39), we require:

$$\bar{Y}_{1..} = 7.67 \qquad \bar{Y}_{3..} = 16.5$$

$$\hat{D} = \bar{Y}_{3..} - \bar{Y}_{1..} = 16.5 - 7.67 = 8.83$$

$$s^2\{\hat{D}\} = \frac{2(16.33)}{6} = 5.443 \qquad s\{\hat{D}\} = 2.33$$

Hence, the desired confidence interval is:

$$1.4 = 8.83 - 3.182(2.33) \leq \mu_{3.} - \mu_{1.} \leq 8.83 + 3.182(2.33) = 16.2$$

Estimation of Variances

At times, one is interested in estimating σ^2, the variance of experimental units, and σ_η^2, the variance of the observation units. It is evident from either of the $E\{MS\}$

columns in Table 26.10 that the following are unbiased estimators:

Parameter	Unbiased Estimator

(26.40a) $\qquad\qquad\qquad\sigma^2 \qquad\qquad\qquad s^2 = \dfrac{(MSEE - MSOE)}{m}$

(26.40b) $\qquad\qquad\qquad\sigma_\eta^2 \qquad\qquad\qquad s_\eta^2 = MSOE$

Example.   For the bread crustiness example, we obtain from Table 26.12 the follow-ing estimates of the two variances:

$$s^2 = \frac{16.33 - 2.61}{3} = 4.57$$

$$s_\eta^2 = 2.61$$

Thus, the estimated variability between batches is somewhat greater here than that between observation units within a batch.

Comments

1. Frequently, the units for subsampling are called *observation units,* to distinguish them from the *experimental units.* Thus, in the bread crustiness example, the batches of flour mix are the experimental units and the portions selected from a batch for making loaves of bread are the observation units.

2. Observation units may be different physical entities, as in the bread crustiness example where they were portions of a batch of flour mix. Observation units also may refer to repeated observations on the entire experimental unit. An example of the latter is the earlier illustration where a person is timed for 10 consecutive assembly operations after receiving a given type of training.

3. Note that subsampling model (26.35) contains no interaction terms. This is because the experimental error terms $\varepsilon_{j(i)}$ are nested within treatments, and the observation error terms $\eta_{k(ij)}$ are nested within experimental units. When one variable is nested within another, we saw earlier that interaction terms are inapplicable.

4. We have considered only the case where an equal number of experimental units (n) are applied to each treatment, and a constant number of observations (m) are made on each exper-imental unit. Serious complications are encountered in the unbalanced case, and no exact test for treatment effects can be made. See an advanced text, such as Reference 26.2, for a discus-sion.

Design Considerations

A problem that arises in designing an experiment with repeated observations is the choice of the number of experimental units and the number of observation units. Suppose that we are to estimate the treatment means $\mu_{i.}$ in a balanced study with fixed treatment effects. It can be shown that:

(26.41)
$$\sigma^2\{\bar{Y}_{i..}\} = \frac{\sigma_\eta^2 + m\sigma^2}{nm}$$

Formula (26.41) makes it clear that if nm is fixed (in the bread crustiness example, nm is the total number of loaves baked in the experiment), $\sigma^2\{\bar{Y}_{i..}\}$ is minimized if m is made as small as possible, namely, $m = 1$. Thus, when nm is fixed, optimum estimation of the treatment means requires that only one observation unit be selected for each experimental unit, so that the total sample is spread among as many experimental units as possible.

The justification for subsampling lies in cost considerations. Suppose that it costs c_1 to include one experimental unit in the study and c_2 to make one observation on an experimental unit. Also, suppose the total cost C is given by:

(26.42)
$$C = c_1 n + c_2 nm$$

It can then be shown that for a given total cost C_0, the variance $\sigma^2\{\bar{Y}_{i..}\}$ is minimized when:

(26.43a)
$$m_{\text{opt}} = \frac{\sigma_\eta}{\sigma}\sqrt{\frac{c_1}{c_2}}$$

(26.43b)
$$n_{\text{opt}} = \frac{C_0}{c_1 + c_2 m_{\text{opt}}}$$

Example. With reference to the bread crustiness example, suppose $c_1 = \$30$ and $c_2 = \$5$, and that the total cost of observations in the experiment is limited to $C_0 = \$400$. Advance assessments indicate that $\sigma = 2.2$ and $\sigma_\eta = 1.5$, approximately. The optimum sample sizes then are obtained as follows:

$$m_{\text{opt}} = \frac{1.5}{2.2}\sqrt{\frac{30}{5}} = 1.67$$

$$n_{\text{opt}} = \frac{400}{30 + 5(1.67)} = 10.4$$

Thus, 10 batches might be used for each treatment, with two observations on each batch.

Comments

1. Note that the optimum number of observation units (m_{opt}) is not affected by the total allowable cost C_0. Only the optimum number of experimental units is affected by C_0.

2. Result (26.43) is obtained by minimizing:

$$\sigma^2\{\bar{Y}_{i..}\} = \frac{\sigma_\eta^2 + m\sigma^2}{nm}$$

subject to the constraint:

$$c_1 n + c_2 nm - C_0 = 0$$

Setting up the Lagrangian function:

$$L = \frac{\sigma_\eta^2 + m\sigma^2}{nm} + \lambda(c_1 n + c_2 nm - C_0)$$

we differentiate L with respect to m, n, and λ and set the partial derivatives equal to zero. When the three equations are solved simultaneously, the results in (26.43) are obtained.

26.8 PURE SUBSAMPLING IN THREE STAGES

Sometimes an investigation does not involve a comparison of treatments, but only subsampling at several levels. Consider, for instance, a quality control engineer who wishes to investigate a certain quality characteristic of a computer assembly. These assemblies are produced in lots of 2,000. The engineer will select a random sample of r lots; from each lot she will select n assemblies; finally, she will obtain m observations on the quality characteristic for each assembly.

Model

Assuming that all random variables are normally distributed and that equal sample sizes are employed at each stage, the model for subsampling in three stages is:

(26.44) $$Y_{ijk} = \mu_{..} + \tau_i + \varepsilon_{j(i)} + \eta_{k(ij)}$$

where:

$\mu_{..}$ is a constant
τ_i, $\varepsilon_{j(i)}$, and $\eta_{k(ij)}$ are independent normal random variables with expectations 0 and variances σ_τ^2, σ^2, and σ_η^2, respectively
$i = 1, \ldots r; j = 1, \ldots, n; k = 1, \ldots, m$

For our illustration, τ_i represents the lot effect, $\varepsilon_{j(i)}$ represents the assembly effect that is nested within the lot, and $\eta_{k(ij)}$ represents the observation effect that is nested within the assembly and therefore also within the lot.

The observations Y_{ijk} for subsampling model (26.44) are normally distributed, with mean and variance:

(26.44a) $$E\{Y_{ijk}\} = \mu_{..}$$

(26.44b) $$\sigma^2\{Y_{ijk}\} = \sigma_Y^2 = \sigma_\tau^2 + \sigma^2 + \sigma_\eta^2$$

Various correlations exist between two observations from the same lot.

Subsampling model (26.44) corresponds to subsampling model (26.35) for a single-factor study except that we assume here that the τ_i are independent $N(0, \sigma_\tau^2)$ and are independent of the $\varepsilon_{j(i)}$ and $\eta_{k(ij)}$. Formally, then, the only difference between models (26.35) and (26.44) is that the τ_i are fixed in one case and random in the other. Subsampling model (26.44) also corresponds to nested model (26.8) with both factor A effects and factor B effects random.

Analysis of Variance

The analysis of variance for pure subsampling model (26.44) uses the same sums of squares as before, namely, those in (26.36). The ANOVA table is the same as that in Table 26.10. The applicable expected mean squares are those for random τ_i effects.

Estimation of $\mu_{..}$

In the case of pure subsampling, one is often interested in estimating the overall mean $\mu_{..}$ (the process mean for the computer assembly quality characteristic in our earlier example). A point estimator of $\mu_{..}$ in model (26.44) is $\bar{Y}_{...}$, and it can be shown that its variance is:

$$(26.45) \qquad \sigma^2\{\bar{Y}_{...}\} = \frac{\sigma_\tau^2}{r} + \frac{\sigma^2}{rn} + \frac{\sigma_\eta^2}{rnm} = \frac{nm\sigma_\tau^2 + m\sigma^2 + \sigma_\eta^2}{rnm}$$

An unbiased estimator of this variance is:

$$(26.46) \qquad s^2\{\bar{Y}_{...}\} = \frac{MSTR}{rnm}$$

and the $1 - \alpha$ confidence limits for $\mu_{..}$ are:

$$(26.47) \qquad \bar{Y}_{...} \pm t(1 - \alpha/2; r - 1)s\{\bar{Y}_{...}\}$$

Subsampling Extensions

Our discussion of subsampling has been confined to completely randomized designs and to three stages of sampling in the case of pure subsampling. Clearly, repeated observations can be used with any experimental design, and pure subsampling can be carried out with any number of stages. In the following chapter, we shall take up procedures for easily handling these more complex situations.

CITED REFERENCES

26.1. *MINITAB Reference Manual, Release 7*, State College, Pa.: Minitab, Inc., 1989.

26.2. Searle, S. R. *Linear Models for Unbalanced Data*. New York: John Wiley & Sons, 1987.

PROBLEMS

26.1. A student asked: "Since the mean squares in the analysis of variance table for a two-factor nested design are the same whether the factor effects are assumed to be random or fixed, what difference does it make whether we assume the factors to have fixed effects or random effects?" Comment.

26.2. A researcher declared: "I prefer analyzing a nested two-factor study as a crossed factor study because I can isolate more sources of variation." Comment on the researcher's strategy.

26.3. Consider a three-factor study where factor C is nested within factor B, and factor B in turn is nested within factor A, and $a = b = c = 2$. Illustrate in the format of Table 26.2 the distinction between this nested design and the corresponding crossed design.

26.4. Bottling plant production. A production engineer studied the effects of machine model (factor A) and operator (factor B) on the output in a bottling plant. Three bottling machines were used, each a different model. Twelve operators were employed. Four operators were assigned to a machine and worked six-hour shifts each. Data on the number of cases produced by each machine and operator were collected for a week. The data that follow represent the number of cases produced per hour for each day during the week.

Machine i:	1				2				3			
Operator j:	1	2	3	4	1	2	3	4	1	2	3	4
Day $k = 1$:	65	68	56	45	74	69	52	73	69	63	81	67
$k = 2$:	58	62	65	56	81	76	56	78	83	70	72	79
$k = 3$:	63	75	58	54	76	80	62	83	74	72	73	73
$k = 4$:	57	64	70	48	80	78	58	75	78	68	76	77
$k = 5$:	66	70	64	60	68	73	51	76	80	75	70	71

    a. Obtain the residuals for nested design model (26.8) with fixed factor effects and plot them against the fitted values. Also prepare a normal probability plot of the residuals. What are your findings about the appropriateness of model (26.8)?

    b. Prepare aligned residual dot plots by machine. Do these plots support the assumption of constancy of the error variance? Discuss.

26.5. Refer to **Bottling plant production** Problem 26.4. Assume that nested design model (26.8) with fixed factor effects is appropriate.

    a. Can the operator effects be distinguished from the effects of shifts in this study? Discuss.

    b. Plot the estimated treatment means $\bar{Y}_{ij.}$ in the format of Figure 26.2. Does it appear that any factor effects are present?

    c. Obtain the analysis of variance table.

    d. Test whether or not the mean outputs differ for the three machine models; use $\alpha = .01$. State the alternatives, decision rule, and conclusion. What is the P-value of the test?

    e. Test whether or not the mean outputs differ for the operators assigned to each machine; use $\alpha = .01$. State the alternatives, decision rule, and conclusion. What is the P-value of the test? What does your conclusion imply about the mean outputs for the four operators assigned to machine 3? Explain.

    f. Test for each machine separately whether or not the mean outputs for the four operators differ. For each test, use $\alpha = .01$ and state the alternatives, decision rule, and conclusion.

g. What is the family level of significance for the combined tests in parts (d), (e), and (f) using the Bonferroni inequality? Summarize the set of conclusions reached in your tests.

26.6. Refer to **Bottling plant production** Problems 26.4 and 26.5.

a. Make all pairwise comparisons among the mean outputs for the three machines. Use the Tukey procedure with a 95 percent family confidence coefficient. State your findings.

b. Make all pairwise comparisons among the mean outputs for the four operators assigned to machine 1. Use the Bonferroni procedure with a 95 percent family confidence coefficient. State your findings.

c. Operator 4 assigned to machine 1 has relatively little experience compared to the other three operators. Estimate the contrast:

$$L = \frac{\mu_{11} + \mu_{12} + \mu_{13}}{3} - \mu_{14}$$

using a 99 percent confidence interval. Interpret your interval estimate.

26.7. Refer to **Bottling plant production** Problem 26.4. Assume that the four operators assigned to each machine were selected at random from a large number of operators.

a. How is nested design model (26.8) modified to fit this case?

b. Obtain a point estimate of the operator variance σ_β^2.

c. Test whether or not σ_β^2 equals zero; use $\alpha = .10$. State the alternatives, decision rule, and conclusion. What is the P-value of the test?

d. Test whether or not the mean outputs differ for the three machine models; use $\alpha = .10$. State the alternatives, decision rule, and conclusion. What is the P-value of the test?

e. Make all pairwise comparisons among the mean outputs for the three machines. Use the Tukey procedure with a 90 percent family confidence coefficient. State your findings.

f. Test the assumption that the $\beta_{j(i)}$ for all machines have the same variance σ_β^2. Use the Hartley test (Section 16.2) with significance level $\alpha = .01$. State the alternatives, decision rule, and conclusion.

26.8. Refer to **Bottling plant production** Problem 26.4. Assume that the four operators assigned to each machine were selected at random from a large number of operators and that the three machines were chosen at random from a large number of machines.

a. How is nested design model (26.8) modified to fit this case?

b. Obtain point estimates of the operator and machine variances σ_β^2 and σ_α^2, respectively.

c. Test whether or not σ_α^2 equals zero; use $\alpha = .05$. State the alternatives, decision rule, and conclusion. What is the P-value of the test?

d. The production engineer is interested in estimating the overall mean $\mu_{..}$ with a 95 percent confidence interval. Obtain the desired confidence interval and interpret your interval estimate.

26.9. Health awareness. Three states (factor A) participated in a health awareness study. Each state independently devised a health awareness program. Three cities (factor B) within each state were selected for participation and five households within each city were randomly selected to evaluate the effectiveness of the program. All members of the selected households were interviewed before and after participation in the program and a composite index was formed for each household measuring the impact of the health awareness program. The data on health awareness follow (the larger the index, the greater the awareness).

State i:	1			2			3		
City j:	1	2	3	1	2	3	1	2	3
Household $k = 1$:	42	26	34	47	56	68	19	18	16
$k = 2$:	56	38	51	58	43	51	36	40	28
$k = 3$:	35	42	60	39	65	49	24	27	45
$k = 4$:	40	35	29	62	70	71	12	31	30
$k = 5$:	28	53	44	65	59	57	33	23	21

a. Obtain the residuals for nested design model (26.8) with fixed factor effects and plot them against the fitted values. Also prepare a normal probability plot of the residuals. What are your findings about the appropriateness of model (26.8)?

b. Prepare aligned residual dot plots by state. Do these plots support the assumption of constancy of the error variance? Discuss.

26.10. Refer to **Health awareness** Problem 26.9. Assume that nested design model (26.8) with fixed factor effects is appropriate.

a. Plot the estimated treatment means $\bar{Y}_{ij.}$ in the format of Figure 26.2. Does it appear that any factor effects are present?

b. Obtain the analysis of variance table.

c. Test whether or not the mean awareness differs for the three states; use $\alpha = .05$. State the alternatives, decision rule, and conclusion. What is the P-value of the test?

d. Test whether or not the mean awareness differs for the three cities within each state; use $\alpha = .05$. State the alternatives, decision rule, and conclusion. What is the P-value for the test? What does your conclusion imply about the awareness means for the three cities in state 1? Explain.

e. What is the family level of significance for the combined tests in parts (c) and (d) using the Bonferroni inequality? Summarize the set of conclusions reached in your tests.

26.11. Refer to **Health awareness** Problems 26.9 and 26.10.

a. Estimate μ_{11} with a 95 percent confidence interval. Interpret your interval estimate.

b. Obtain separate confidence intervals for $\mu_{1.}$, $\mu_{2.}$, and $\mu_{3.}$, each with a 99 percent confidence coefficient. Interpret your interval estimates.

c. Obtain confidence intervals for all pairwise comparisons between the state means. Use the Tukey procedure and a 90 percent family confidence coefficient. Summarize your findings.

d. It is desired to obtain a 95 percent confidence interval for $D = \mu_{11} - \mu_{32}$, since these two cities are of comparable size. Interpret your interval estimate.

26.12. Refer to **Health awareness** Problem 26.9. Assume that the three cities in each state were chosen at random from all the cities in the state.

a. How is nested design model (26.8) modified to fit this case?

b. Obtain a point estimate of the city variance σ_β^2. Is there anything peculiar about the estimate here?

c. Test whether or not σ_β^2 equals zero; use $\sigma = .10$. State the alternatives, decision rule, and conclusion. What is the P-value of the test?

d. Test whether or not the mean awareness differs for the three states; use $\alpha = .10$. State the alternatives, decision rule, and conclusion. What is the P-value of the test?

e. Obtain confidence intervals for all pairwise comparisons between the state means. Use the Tukey procedure and a 90 percent family confidence coefficient. Summarize your findings.

f. Test the assumption that the $\beta_{j(i)}$ for all states have the same variance σ_β^2. Use the Hartley test (Section 16.2) with significance level $\alpha = .05$. State the alternatives, decision rule, and conclusion.

26.13. Refer to **Health awareness** Problem 26.9. Assume that the three cities within each state and the three states were selected at random.

a. How is nested design model (26.8) modified to fit this case?

b. Obtain point estimates of the city and state variances σ_β^2 and σ_α^2, respectively.

c. Test whether or not σ_α^2 equals zero; use $\alpha = .01$. State the alternatives, decision rule, and conclusion. What is the P-value of the test?

d. Estimate the overall mean health awareness index $\mu_{..}$ using a 99 percent confidence interval. Interpret your interval estimate.

26.14. Internal control. A large retailer operates three regional accounting centers (factor A). Center 1 employs three audit teams while the other two centers employ two audit teams each. One function of each center is to review whether a certain internal control operates properly in the processing of payroll. Data on the percent of transactions where the internal control was found to be operating properly were requested for each team in each region for the previous two months. Three months' data were received in one case, and data for only one month in another. The arc sine transformation $Y' = 2 \arcsin \sqrt{p}$ was employed to stabilize the error variances. The transformed data follow.

Region i:	1			2		3	
Team j:	1	2	3	1	2	1	2
Month $k = 1$:	151.6	143.2	131.4	163.8	151.6	157.0	160.0
$k = 2$:	141.2	139.4	136.0	154.2		147.2	151.6
$k = 3$:	149.4						

a. Set up the full regression model for this case, analogous to the illustrative full model (26.31), using $1, -1, 0$ indicator variables.

b. Fit this model and obtain the residuals. Plot the residuals against the fitted values. Also prepare a normal probability plot of the residuals. What are your findings about the appropriateness of the model?

26.15. Refer to **Internal control** Problem 26.14. Assume that nested design model (26.8) with fixed factor effects, modified for unequal nestings and replications, is appropriate.

a. Test for region main effects using test statistic (8.71) and significance level $\alpha = .025$. State the alternatives, reduced model, decision rule, and conclusion. What is the P-value of the test?

b. Test for effects of audit teams within region using test statistic (8.71) and significance level $\alpha = .025$. State the alternatives, reduced model, decision rule, and conclusion.

c. Estimate $D = \mu_{1.} - \mu_{2.}$ (in transformed units) with a 98 percent confidence interval.

26.16. A student asked in class why all experiments do not make use of repeated observations since all measurement procedures are inexact to some degree. Comment.

26.17. Refer to **Questionnaire color** Problem 14.11. Suppose that the experiment was conducted by distributing the fliers to the assigned parking lots in two different weeks and noting the response rates for each week. The complete data on response rates follow.

Color i:	1 (Blue)					2 (Green)					3 (Orange)				
Lot j:	1	2	3	4	5	1	2	3	4	5	1	2	3	4	5
Week $k = 1$:	28	26	31	27	35	34	29	25	31	29	31	25	27	29	28
$k = 2$:	32	23	29	24	37	33	27	22	34	25	35	28	25	25	31

a. Obtain the residuals for subsampling model (26.35) with fixed treatment effects and plot them against the fitted values. Also prepare a normal probability plot of the residuals. What are your findings about the appropriateness of model (26.35)?

b. Test the assumption that the $\varepsilon_{j(i)}$ have the same variance σ^2 for all colors. Use the Hartley test (Section 16.2) with significance level $\alpha = .01$. State the alternatives, decision rule, and conclusion.

26.18. Refer to **Questionnaire color** Problem 26.17. Assume that subsampling model (26.35) with fixed treatment effects is appropriate.

a. Obtain the analysis of variance table.

b. Test whether or not questionnaire color effects are present; use $\alpha = .05$. State the alternatives, decision rule, and conclusion. What is the P-value of the test?

c. Test whether or not lot differences within colors are present; use $\alpha = .05$. State the alternatives, decision rule, and conclusion. What is the P-value of the test?

d. Estimate the mean response rate for blue questionnaires with a 95 percent confidence interval.

e. Obtain point estimates of σ^2 and σ_η^2. Which variance appears to be larger here?

26.19. An economist has $20,000 available for a study to compare the amounts of installment debt owed by urban families with two or fewer children with those of urban families with more than two children in a state. The cost of including a city in the study is $1,000 and the cost of including a family is $50. The number of families with two or fewer children in the study is to be the same as the number of families with more than two children. Assume that the cost function is given by (26.42). The primary purpose of the study is to estimate the mean debt for each of the two types of families as precisely as possible.

a. If reasonable planning values for the standard deviations are $\sigma = 150$ and $\sigma_\eta = 300$, how many cities and families should be included in the study?

b. How would the sample sizes change if $\sigma = 400$ and $\sigma_\eta = 200$?

26.20. Plant acid levels. Four plants of the same variety were randomly selected in an experiment to investigate the concentration of a particular acid. Three leaves per plant were randomly selected and three separate determinations of the acid concentration were obtained per leaf. The data follow.

Plant i:	1			2			3			4		
Leaf j:	1	2	3	1	2	3	1	2	3	1	2	3
Determination												
$k = 1$:	11.2	16.5	18.3	14.1	19.0	11.9	15.3	19.5	16.5	7.3	8.9	11.3
$k = 2$:	11.6	16.8	18.7	13.8	18.5	12.4	15.9	20.1	17.2	7.8	9.4	10.9
$k = 3$:	12.0	16.1	19.0	14.2	18.2	12.0	16.0	19.3	16.9	7.0	9.3	10.5

Obtain the residuals for three-stage subsampling model (26.44) and plot them against the fitted values. Also prepare a normal probability plot of the residuals. What are your findings about the appropriateness of model (26.44)?

26.21. Refer to **Plant acid levels** Problem 26.20. Assume that three-stage subsampling model (26.44) is appropriate.
a. Obtain the analysis of variance table.
b. Test whether or not there are variations in mean concentration levels between plants; use $\alpha = .05$. State the alternatives, decision rule, and conclusion. What is the P-value of the test?
c. Test whether or not there are variations in mean concentration levels between leaves of the same plant; use $\alpha = .05$. State the alternatives, decision rule, and conclusion. What is the P-value of the test?
d. Estimate the overall mean concentration in all plants of the variety; use a 95 percent confidence interval.
e. Obtain point estimates of σ_τ^2, σ^2, and σ_η^2. Which component of variance appears to be most important in the total variance σ_Y^2?

26.22. Chemical consistency. A chemical company wished to study the consistency of the strength of one of its liquid chemical products. The product is made in batches in large vats and then is barreled. The barrels are subsequently stored for a period of time in a warehouse. To examine the consistency of the strength of the chemical, an analyst randomly selected five different batches of the product from the warehouse and then selected four barrels per batch at random. Three determinations per barrel were made. The data on strength follow.

Batch i:	1				2				3			
Barrel j:	1	2	3	4	1	2	3	4	1	2	3	4
Determination												
$k = 1$:	2.3	2.5	2.6	2.4	2.8	2.7	2.6	2.4	3.0	3.4	2.9	3.1
$k = 2$:	2.1	2.3	2.4	2.6	2.9	2.5	2.6	2.8	3.1	3.3	3.0	2.8
$k = 3$:	2.0	2.5	2.7	2.3	2.6	2.8	2.8	2.6	2.9	3.0	3.2	3.2

Batch i:		4				5		
Barrel j:	1	2	3	4	1	2	3	4
Determination								
$k = 1$:	2.5	2.8	3.1	2.7	3.6	3.8	3.7	3.9
$k = 2$:	2.8	3.0	2.8	2.9	3.7	3.8	3.5	3.5
$k = 3$:	2.6	2.7	2.9	2.6	3.4	3.5	3.5	3.7

a. Obtain the residuals for three-stage subsampling model (26.44) and plot them against the fitted values. Also prepare a normal probability plot of the residuals. What are your findings about the appropriateness of model (26.44)?

b. Test the assumption that the $\varepsilon_{j(i)}$ have the same variance σ^2 for all batches. Use the Hartley test (Section 16.2) with significance level $\alpha = .01$. State the alternatives, decision rule, and conclusion.

26.23. Refer to **Chemical consistency** Problem 26.22. Assume that three-stage subsampling model (26.44) is appropriate.

a. Obtain the analysis of variance table.

b. Test whether or not there are variations in mean strength between batches; use $\alpha = .01$. State the alternatives, decision rule, and conclusion. What is the P-value of the test?

c. Test whether or not there are variations in mean strength between barrels within batches; use $\alpha = .01$. State the alternatives, decision rule, and conclusion. What is the P-value of the test?

d. Estimate the overall mean strength of the chemical using a 99 percent confidence interval.

e. Obtain point estimates of σ_τ^2, σ^2, and σ_η^2. Which component of variance appears to be most important in the total variance σ_Y^2?

EXERCISES

26.24. Derive (26.13) by squaring (26.12) and summing over all observations.

26.25. Derive (26.16b) from (26.13b).

26.26. Derive (26.17) for a balanced nested two-factor design.

26.27. Consider a balanced nested two-factor design with factor A having fixed effects and factor B (nested within factor A) having random effects.
a. Derive $\sigma^2\{\bar{Y}_{i..}\}$ and $\sigma^2\{\bar{Y}_{...}\}$.
b. Find an unbiased point estimator of σ_β^2.

26.28. Derive variance (26.41) for subsampling model (26.35) with fixed treatment effects.

26.29. (Calculus needed.) Derive the optimal sample sizes given in (26.43). (*Hint:* See Comment 2 on p. 999.)

26.30. Derive variance (26.45) for three-stage subsampling model (26.44). Using the expected mean squares in Table 26.10, show that the estimated variance (26.46) is an unbiased estimator of variance (26.45).

PROJECTS

26.31. Refer to the **Drug effect experiment** data set. Consider only Part I of the study and dosage level 4; i.e., include only observations for which variable 2 equals 1 and variable 5 equals 4. Assume that initial lever press rate (factor A) has fixed effects and that rats are a second factor (factor D) with random effects.

    a. State the appropriate model for this nested two-factor study.

    b. Obtain the residuals and plot them against the fitted values. Also prepare a normal probability plot of the residuals. What are your findings about the appropriateness of your model?

26.32. Refer to the **Drug effect experiment** data set and Project 26.31. Assume that nested design model (26.8), with $\beta_{j(i)}$ and $\varepsilon_{k(ij)}$ random, is appropriate.

    a. Obtain the analysis of variance table.

    b. Test whether or not the mean lever press rate differs for the three initial rate groups; use $\alpha = .05$. State the alternatives, decision rule, and conclusion. What is the P-value of the test?

    c. Test whether or not the mean lever press rate differs for the rats within the initial rate groups; use $\alpha = .05$. State the alternatives, decision rule, and conclusion. What is the P-value of the test? What does your conclusion imply about the four rats in the slow initial rate group?

    d. Make all pairwise comparisons between the mean lever press rates for the three initial rate groups. Use the Tukey procedure with a 90 percent family confidence coefficient.

    e. Obtain a point estimate of the between-rats variance.

26.33. Refer to the **Drug effect experiment** data set. Consider only Part II of the study and dosage level 3; i.e., include only observations for which variable 2 equals 2 and variable 5 equals 3. Assume that the initial lever press rate groups are the treatments with fixed effects, and that the rats are the experimental units with two observations for each experimental unit.

    a. State the appropriate model for this single-factor study with subsampling.

    b. Obtain the residuals and plot them against the fitted values. Also prepare a normal probability plot of the residuals. What are your findings about the appropriateness of your model?

    c. Test the assumption that the $\varepsilon_{j(i)}$ have the same variance σ^2 for all lever press rates. Use the Hartley test (Section 16.2) with significance level $\alpha = .01$. State the alternatives, decision rule, and conclusion.

26.34. Refer to the **Drug effect experiment** data set and Project 26.33. Assume that the single-factor subsampling model (26.35) with fixed treatment effects is appropriate.

    a. Obtain the analysis of variance table.

    b. Test whether or not the mean lever press rate differs for the three initial rate groups; use $\alpha = .01$. State the alternatives, decision rule, and conclusion. What is the P-value of the test?

    c. Test whether or not differences in the mean lever press rate between rats are present; use $\alpha = .01$. State the alternatives, decision rule, and conclusion. What is the P-value of the test?

    d. Make all pairwise comparisons between the mean lever press rates for the three initial rate groups. Use the Tukey procedure with a 95 percent family confidence coefficient. Summarize your findings.

    e. Obtain point estimates of σ^2 and σ_η^2.

Chapter 27

Rules for Developing ANOVA Models and Tables for Balanced Designs

In this chapter, we present and illustrate rules for developing models for nested and/or crossed factor designs, for finding the appropriate sums of squares and degrees of freedom for the needed mean squares, and for finding the expected values of the mean squares. These rules apply to all balanced designs with two or more replications and with no interactions assumed to equal zero.

In the nested case, as noted earlier, a design is *balanced* when (1) the number of factor levels of a nested factor is the same for each level of the factor in which the nesting takes place, and (2) the number of replications is constant for the different factor combinations. In the crossed case, a design is balanced whenever the number of replications is constant for all factor combinations. In a subsampling design, balance requires that the subsample sizes at each stage of sampling be constant.

In Section 27.6, we shall show that a slight modification of the rules makes them applicable to balanced designs with no replications and/or with some interaction terms assumed to equal zero.

27.1 RULE FOR MODEL DEVELOPMENT

We begin by presenting a rule for the development of a nested and/or crossed factor design model. *This rule is applicable when no interactions are assumed to equal zero.* We shall utilize as an illustration the training school example of Table 26.1, where the effects of three schools (factor *A*) and two instructors within each school (factor *B*) were studied and two replications were made in each instance.

Rule (27.1)

Step 1. *Include an overall constant and a main effect term for each factor, taking into account when one factor is nested within another.*
  Example.   For the training school example, we include:

$$\mu_{..} \qquad \alpha_i \qquad \beta_{j(i)}$$

Note that factor B is nested within factor A.

Step 2. *Include all interaction terms except those containing both a nested factor and the factor within which it is nested.*
  Example.   Since factor B is nested within factor A, the AB interaction (the only possible interaction term here) is not included.

Step 3. *Interactions between a nested factor and another factor with which the nested factor is crossed are always themselves nested.*
  Example.   For the training school example, this situation does not arise.

Step 4. *Include the error term, which is nested within all factors.*
  Example.   For the training school example the error term is $\varepsilon_{k(ij)}$, and the appropriate model therefore is:

$$(27.2) \qquad Y_{ijk} = \mu_{..} + \alpha_i + \beta_{j(i)} + \varepsilon_{k(ij)}$$

$$i = 1, 2, 3; j = 1, 2; k = 1, 2$$

27.2 RULE FOR FINDING SUMS OF SQUARES AND DEGREES OF FREEDOM

Since nested factors and subsampling designs may require sums of squares not discussed so far, we shall now consider a rule for finding sums of squares and associated degrees of freedom. *This rule is applicable to all balanced designs with two or more replications and with no interaction terms assumed to equal zero.*

Illustration

The rule for finding sums of squares and associated degrees of freedom can best be explained in terms of an illustration. We shall continue to consider the training school example where factor B is nested within factor A. It does not matter for this rule whether the factor effects are fixed or random.

Rule (27.3) for Definitional Forms of Sums of Squares

Step 1. *Write the model equation.*

Example. The model equation for the training school example was given earlier. We show this model now in its general form, where factor A has a levels, factor B has b levels, and there are n replications:

$$(27.2a) \qquad Y_{ijk} = \mu_{..} + \alpha_i + \beta_{j(i)} + \varepsilon_{k(ij)}$$

$$i = 1, \ldots, a; j = 1, \ldots, b; k = 1, \ldots, n$$

Step 2. *For each model term other than the overall constant, write the associated SS notation.*

Example. We do this for the training school example in columns 1 and 2 of Table 27.1 for α_i, $\beta_{j(i)}$, and $\varepsilon_{k(ij)}$. The line for Total will not be completed until step 9.

Step 3. *Each sum of squares will have as coefficient the product of the limits of the subscripts not appearing in the model term. The coefficient is taken to be 1 if all subscripts appear in the model term.*

Example. The coefficients for our example are shown in column 3 of Table 27.1. For instance, α_i does not contain j and k. These subscripts have limits of b and n, respectively. The coefficient for the SSA term is therefore bn. Since the model term $\varepsilon_{k(ij)}$ contains all subscripts, the coefficient is taken to be 1 here.

Step 4. *Each sum of squares is summed over all of the subscripts of the model term, whether in parentheses or not.*

Example. The summations for our example are shown in column 4. For instance, the sum of squares term corresponding to α_i is summed over i, the only subscript in that model term. Similarly, the sum of squares term corresponding to $\varepsilon_{k(ij)}$ is summed over i, j, and k since all of these appear in the model term.

Step 5. *Form a symbolic product from the subscripts of the model term, using the subscript if it is in parentheses, and the subscript minus 1 if it is not in parentheses. Expand the product.*

Example. The symbolic products for our example are shown in column 5. For instance, for α_i the symbolic product is $i - 1$. For $\beta_{j(i)}$, the symbolic product is $i(j - 1) = ij - i$. For $\varepsilon_{k(ij)}$, the symbolic product is $(k - 1)ij = ijk - ij$.

Step 6. *The typical term to be squared consists of means of the observations with the subscripts consisting of the symbolic product term and dots elsewhere. The sign of each mean is that of the symbolic product. A 1 refers to the overall mean.*

Example. The terms to be squared for our example are shown in column 6. Note that for α_i, the symbolic product is $i - 1$, and the typical term to be squared therefore is:

$$\bar{Y}_{i..} - \bar{Y}_{...}$$

TABLE 27.1 Derivation of Definitional Sums of Squares Formulas for Nested Two-Factor Experiment (B nested within A)

(1) Model Term	(2) SS	(3) Coefficient	(4) $\sum$	(5) Symbolic Product	(6) Term to Be Squared	(7) Sum of Squares	(8) Degrees of Freedom
α_i	SSA	bn	$\sum_i$	$i-1$	$\bar{Y}_{i..} - \bar{Y}_{...}$	$bn \sum_i (\bar{Y}_{i..} - \bar{Y}_{...})^2$	$a-1$
$\beta_{j(i)}$	$SSB(A)$	n	$\sum_i \sum_j$	$i(j-1)$ $= ij - i$	$\bar{Y}_{ij.} - \bar{Y}_{i..}$	$n \sum_i \sum_j (\bar{Y}_{ij.} - \bar{Y}_{i..})^2$	$a(b-1)$
$\varepsilon_{k(ij)}$	SSE	1	$\sum_i \sum_j \sum_k$	$(k-1)ij$ $= ijk - ij$	$Y_{ijk} - \bar{Y}_{ij.}$	$\sum_i \sum_j \sum_k (Y_{ijk} - \bar{Y}_{ij.})^2$	$ab(n-1)$
Total	$SSTO$				$Y_{ijk} - \bar{Y}_{...}$	$\sum_i \sum_j \sum_k (Y_{ijk} - \bar{Y}_{...})^2$	$abn - 1$

For $\beta_{j(i)}$ the symbolic product is $ij - i$, and hence the typical term to be squared is:

$$\bar{Y}_{ij.} - \bar{Y}_{i..}$$

Similarly, for $\varepsilon_{k(ij)}$, the symbolic product is $ijk - ij$. Hence the typical term to be squared is:

$$Y_{ijk} - \bar{Y}_{ij.}$$

Note that we write the first term as Y_{ijk} since it is not averaged over any subscript.

Step 7. *Combining the steps of squaring, summing, and multiplying by the coefficient yields the appropriate sums of squares.*
  Example.   The sums of squares for our example are shown in column 7.

Step 8. *The degrees of freedom are obtained by replacing in each symbolic product the subscript variable by its limit.*
  Example.   For our example, the degrees of freedom are shown in column 8. For instance, for α_i the symbolic product is $i - 1$; hence $df = a - 1$. Similarly for $\varepsilon_{k(ij)}$, the symbolic product is $ijk - ij$; hence $df = abn - ab = ab(n - 1)$.

Step 9. *The total sum of squares is always defined as the sum, over all observations, of the squared deviations of the observations from the overall mean. The total degrees of freedom are always defined as one less than the total number of observations.*

The results in Table 27.1 are, of course, the same as those given earlier in Table 26.4.

Rule (27.3a) for Computational Forms of Sums of Squares

If the computational forms of the sums of squares are desired, the procedure is modified as follows:

Step 1a. *Obtain the symbolic products as before.*

Step 2a. *For each term in the symbolic product, there corresponds a total of the observations with the subscripts consisting of the symbolic product term and dots elsewhere. A 1 denotes the grand total.*

Step 3a. *The total is squared and summed over the subscript(s) it has.*

Step 4a. *The sum has the sign of the corresponding term in the symbolic product, and is divided by the product of the limits of the subscripts not present.*

TABLE 27.2   Derivation of Computational Sums of Squares Formulas for Nested Two-Factor Experiment (B nested within A)

Model Term	Symbolic Product	Sum of Squares	Degrees of Freedom
α_i	$i - 1$	$SSA = \dfrac{\sum\limits_i Y_{i..}^2}{bn} - \dfrac{Y_{...}^2}{abn}$	$a - 1$
$\beta_{j(i)}$	$(j - 1)i$ $= ij - i$	$SSB(A) = \dfrac{\sum\limits_i \sum\limits_j Y_{ij.}^2}{n} - \dfrac{\sum\limits_i Y_{i..}^2}{bn}$	$(b - 1)a = ab - a$
$\varepsilon_{k(ij)}$	$(k - 1)ij$ $= ijk - ij$	$SSE = \sum\limits_i \sum\limits_j \sum\limits_k Y_{ijk}^2 - \dfrac{\sum\limits_i \sum\limits_j Y_{ij.}^2}{n}$	$ab(n - 1)$
Total		$SSTO = \sum\limits_i \sum\limits_j \sum\limits_k Y_{ijk}^2 - \dfrac{Y_{...}^2}{abn}$	$abn - 1$

Step 5a.   *SSTO always equals the sum of the squared observations minus the square of the sum of all observations divided by the total number of observations.*

The degrees of freedom are obtained as before.

Example.   Table 27.2 contains the derivation of the computational formulas for our nested two-factor example. To illustrate the derivations, consider model term $\beta_{j(i)}$. For this term, the symbolic product is $ij - i$. Hence, we obtain:

$$\frac{\sum\limits_i \sum\limits_j Y_{ij.}^2}{n} - \frac{\sum\limits_i Y_{i..}^2}{bn}$$

The results in Table 27.2 are the same as, or equivalent to, those given earlier in (26.16).

27.3   RULE FOR FINDING EXPECTED MEAN SQUARES

The rule for finding expected mean squares that we shall now present enables us to avoid tedious derivations. The rule applies to both nested factors and crossed factors. *The rule is applicable to all balanced designs with two or more replications and with no interaction terms assumed to equal zero.*

Illustration

We shall again use the training school example of Table 26.1. Here factor A (school) and factor B (instructor) are both fixed factors, factor B is nested within factor A, factor B has b levels within each level of factor A, factor A has a levels, and there are n replications.

Rule (27.4)

The rule for finding expected mean squares to be presented may appear to be a bit complex on first reading. However, with a little practice the desired expected mean squares can be obtained very quickly and easily.

Step 1. *List the model equation.*
   Example.   The model equation is that of (27.2a):

$$Y_{ijk} = \mu_{..} + \alpha_i + \beta_{j(i)} + \varepsilon_{k(ij)}$$

Step 2. *For each term other than the overall constant, write the associated random effects variance term.*
   Example

$$\begin{array}{ccc} \alpha_i & \beta_{j(i)} & \varepsilon_{k(ij)} \\ \sigma_\alpha^2 & \sigma_\beta^2 & \sigma^2 \end{array}$$

If factors have fixed effects, as in this example, we shall at the end replace these variance terms by sums of squared effects divided by degrees of freedom. Thus, for the training school example, the term σ_α^2 later will be replaced by $\Sigma\alpha_i^2/(a-1)$, and likewise σ_β^2 will be replaced by $\Sigma\Sigma\beta_{j(i)}^2/a(b-1)$. In the meantime, however, it is easier to write the variance term rather than a sum of squared effects divided by degrees of freedom.

Step 3. *Set up a table, with the rows consisting of the model elements other than the overall constant.*
   Example

α_i
$\beta_{j(i)}$
$\varepsilon_{k(ij)}$

Step 4. *The column headings for the table are the subscripts in the model. Under each heading, write F if the factor indexed by the subscript is fixed, and write R if it is random. Also write the number of levels for that factor.*

Example

	i	j	k
	F	F	R
	a	b	n

α_i
$\beta_{j(i)}$
$\varepsilon_{k(ij)}$

For instance, i refers to school, a fixed factor that occurs at a levels. Note that the subscript k refers to replication, which is a random "factor" and occurs at n levels.

Step 5. *In each row where one or more subscripts are in parentheses, enter a 1 in the column(s) corresponding to the subscript(s) in parentheses.*
Example

	i	j	k
	F	F	R
	a	b	n
α_i			
$\beta_{j(i)}$	1		
$\varepsilon_{k(ij)}$	1	1	

Thus, in the $\beta_{j(i)}$ row, we enter a 1 in the i column, and so on.

Step 6. *In each row where one or more subscripts are not in parentheses, enter in the column(s) corresponding to the subscript(s) not in parentheses a 1 if the subscript refers to a random factor, and a 0 if the factor is fixed.*
Example

	i	j	k
	F	F	R
	a	b	n
α_i	0		
$\beta_{j(i)}$	1	0	
$\varepsilon_{k(ij)}$	1	1	1

Thus, for the $\beta_{j(i)}$ row, the subscript not in parentheses is j, which refers to factor B, a fixed factor. Hence, a 0 is entered in the j column.

Step 7. *Fill in all remaining empty cells with the number of levels appearing in the column heading.*

Example

	i	j	k	
	F	F	R	
	a	b	n	
α_i	0	b	n
$\beta_{j(i)}$	1	0	n
$\varepsilon_{k(ij)}$	1	1	1

Each $E\{MS\}$ will consist of a linear combination of the variance terms enumerated in step 2, with the coefficients obtained by taking additional steps in the table just completed. Some of the coefficients may be zero, which means that the corresponding variance term is not present in the $E\{MS\}$.

Step 8. *Adjoin on the right of the table just completed the variance term associated with the effect in that row. In addition, adjoin a column for each expected mean square to be found. Under each expected mean square, indicate all of the subscripts (including any parentheses) associated with the corresponding model term.*
  Example

	i	j	k		$E\{MSA\}$	$E\{MSB(A)\}$	$E\{MSE\}$
	F	F	R				
	a	b	n	Variance	i	$(i)j$	$(ij)k$
α_i	0	b	n	σ_α^2			
$\beta_{j(i)}$	1	0	n	σ_β^2			
$\varepsilon_{k(ij)}$	1	1	1	σ^2			

Note that all of the subscripts of the associated model term, whether in parentheses or not, are shown under the expected mean square. For example, $E\{MSB(A)\}$ has associated with it the model term $\beta_{j(i)}$, so that the subscripts shown are (i) and j. Similarly, $E\{MSE\}$ has associated with it the model term $\varepsilon_{k(ij)}$, so that (ij) and k are shown.

Step 9. *For each expected mean square column, the coefficient of any variance term is zero if the subscript(s) of the model term in that row (whether in parentheses or not) do not include all of the subscript(s) in the heading of that $E\{MS\}$ column (whether in parentheses or not).*

Example

	i	j	k				
	F	F	R		$E\{MSA\}$	$E\{MSB(A)\}$	$E\{MSE\}$
	a	b	n	Variance	i	$(i)j$	$(ij)k$
α_i	0	b	n	σ_α^2		0	0
$\beta_{j(i)}$	1	0	n	σ_β^2			0
$\varepsilon_{k(ij)}$	1	1	1	σ^2			

For the $E\{MSA\}$ column, it will be noted that the model terms in all rows contain the subscript i. Hence, none of the variances receives a zero coefficient as a result of this step.

For the $E\{MSB(A)\}$ column, note that the first row has a model term not containing both i and j. Hence, σ_α^2 receives a zero coefficient in the $E\{MSB(A)\}$ column.

Finally, for the $E\{MSE\}$ column, the first and second rows have model terms that do not contain the three subscripts i, j, and k. Hence, both σ_α^2 and σ_β^2 receive zero coefficients in the $E\{MSE\}$ column.

Step 10. *The coefficients of the variance terms that have not been assigned a zero coefficient as a result of step 9 are found as follows:*

a. *For each expected mean square column, delete (e.g., mask or cover) the column(s) on the left corresponding to the subscripts not in parentheses in the heading of the $E\{MS\}$ column.*

b. *Multiply the entries in the remaining columns for each row being considered.*

Step 11. *The expected mean square equals the sum of the products of each coefficient times the associated variance term, with the variance terms for fixed effects replaced by sums of squared effects divided by degrees of freedom.*

Example

	i	j	k				
	F	F	R		$E\{MSA\}$	$E\{MSB(A)\}$	$E\{MSE\}$
	a	b	n	Variance	i	$(i)j$	$(ij)k$
α_i	0	b	n	σ_α^2	bn	0 (step 9)	0 (step 9)
$\beta_{j(i)}$	1	0	n	σ_β^2	0	n	0 (step 9)
$\varepsilon_{k(ij)}$	1	1	1	σ^2	1	1	1

To find the coefficients for the $E\{MSA\}$ column, for example, we noted earlier that no zero coefficient is assigned as a result of step 9. Step 10a calls for column i on the left to be deleted. Hence, we obtain by multiplying the terms in the j and k columns:

	j	k		$E\{MSA\}$
	F	R		i
	b	n	Variance	
α_i	b	n	σ_α^2	bn
$\beta_{j(i)}$	0	n	σ_β^2	0
$\varepsilon_{k(ij)}$	1	1	σ^2	1

Thus:

$$E\{MSA\} = bn\sigma_\alpha^2 + (0)\sigma_\beta^2 + (1)\sigma^2 = bn\sigma_\alpha^2 + \sigma^2$$

Since factor A has fixed effects, we finally obtain:

$$E\{MSA\} = bn\frac{\Sigma\alpha_i^2}{a-1} + \sigma^2$$

We find the remaining coefficients for $E\{MSB(A)\}$ in similar fashion. We delete column j on the left, the subscript not in parentheses, and obtain:

	i	k		$E\{MSB(A)\}$
	F	R		$(i)j$
	a	n	Variance	
α_i	0	n	σ_α^2	0 (step 9)
$\beta_{j(i)}$	1	n	σ_β^2	n
$\varepsilon_{k(ij)}$	1	1	σ^2	1

Thus:

$$E\{MSB(A)\} = (0)\sigma_\alpha^2 + n\sigma_\beta^2 + (1)\sigma^2 = n\sigma_\beta^2 + \sigma^2$$

Since factor B has fixed effects, we finally obtain:

$$E\{MSB(A)\} = n\frac{\Sigma\Sigma\beta_{j(i)}^2}{a(b-1)} + \sigma^2$$

To find the remaining coefficient in the $E\{MSE\}$ column, we delete column k, and the product on the σ^2 line is $1 \cdot 1 = 1$. Thus:

$$E\{MSE\} = (0)\sigma_\alpha^2 + (0)\sigma_\beta^2 + (1)\sigma^2 = \sigma^2$$

Assembling our results, we have:

(27.5a)
$$E\{MSA\} = bn\frac{\Sigma\alpha_i^2}{a-1} + \sigma^2$$

(27.5b)
$$E\{MSB(A)\} = n\frac{\Sigma\Sigma\beta_{j(i)}^2}{a(b-1)} + \sigma^2$$

(27.5c)
$$E\{MSE\} = \sigma^2$$

Of course, these results are identical to those given earlier in Table 26.4.

Note

Some computer packages provide the expected mean squares for the ANOVA study that is being analyzed. An example is shown in Figure 28.3.

27.4 CROSSED TWO-FACTOR STUDY—MIXED FACTOR EFFECTS

In the previous sections of this chapter, we have presented rules for developing the model and finding sums of squares, degrees of freedom, and expected mean squares. We shall now provide in this and the following sections additional illustrations of the use of these rules for designs involving crossed and nested factors.

In this section we shall consider the case of a two-factor experiment in a completely randomized design, where factors A and B are crossed, factor A has fixed effects and factor B has random effects, and n replications are obtained for each factor combination. The model equation is that of (21.26):

$$Y_{ijk} = \mu_{..} + \alpha_i + \beta_j + (\alpha\beta)_{ij} + \varepsilon_{k(ij)}$$

except that we now recognize the nesting of the error term ε.

Table 27.3 contains the derivation of the definitional sums of squares. Table 27.4a contains the preliminary tabulations for finding the expected mean squares, while Table 27.4b presents the results of steps 9 and 10 of rule (27.4). The random effects variance terms corresponding to the model terms are:

$$
\begin{array}{cccc}
\alpha_i & \beta_j & (\alpha\beta)_{ij} & \varepsilon_{k(ij)} \\
\sigma_\alpha^2 & \sigma_\beta^2 & \sigma_{\alpha\beta}^2 & \sigma^2
\end{array}
$$

Here, only the α_i are fixed effects, so at the end σ_α^2 will need to be replaced by a sum of squared effects divided by degrees of freedom. Note in Table 27.4b that for finding $E\{MSA\}$, σ_β^2 receives a zero coefficient as a result of step 9 since the subscript in the β_j model term does not contain the subscript i in the $E\{MSA\}$ column. Column i is deleted for step 10 for finding the coefficients in the $E\{MSA\}$ column since it is the only subscript in the column heading and is not in parentheses. The other expected mean squares coefficients are found in similar fashion. Table 27.4b indicates for each expected mean square whether the zero coefficients are obtained from step 9, and also which columns are deleted. The final expected mean squares, presented in Table 27.4c, are identical, of course, to those shown in Table 21.4.

27.5 CROSSED-NESTED THREE-FACTOR STUDY— MIXED FACTOR EFFECTS

We consider here a situation where some but not all of the factors are nested. Such designs are called *partially nested, partially hierarchical,* or *crossed-nested* designs. An experiment studied the effect of cultural background on group decision making.

TABLE 27.3 Derivation of Definitional Sums of Squares Formulas for Crossed Two-Factor Experiment in Completely Randomized Design

(1) Model Term	(2) SS	(3) Coefficient	(4) $\sum$	(5) Symbolic Product	(6) Term to Be Squared	(7) Sum of Squares	(8) Degrees of Freedom
α_i	SSA	bn	$\sum_i$	$i-1$	$\bar{Y}_{i..} - \bar{Y}_{...}$	$bn \sum_i (\bar{Y}_{i..} - \bar{Y}_{...})^2$	$a-1$
β_j	SSB	an	$\sum_j$	$j-1$	$\bar{Y}_{.j.} - \bar{Y}_{...}$	$an \sum_j (\bar{Y}_{.j.} - \bar{Y}_{...})^2$	$b-1$
$(\alpha\beta)_{ij}$	SSAB	n	$\sum_i \sum_j$	$(i-1)(j-1)$ $= ij - i - j + 1$	$\bar{Y}_{ij.} - \bar{Y}_{i..} - \bar{Y}_{.j.} + \bar{Y}_{...}$	$n \sum_i \sum_j (\bar{Y}_{ij.} - \bar{Y}_{i..} - \bar{Y}_{.j.} + \bar{Y}_{...})^2$	$(a-1)(b-1)$
$\varepsilon_{k(ij)}$	SSE	1	$\sum_i \sum_j \sum_k$	$(k-1)ij$ $= ijk - ij$	$Y_{ijk} - \bar{Y}_{ij.}$	$\sum_i \sum_j \sum_k (Y_{ijk} - \bar{Y}_{ij.})^2$	$ab(n-1)$
Total	SSTO				$Y_{ijk} - \bar{Y}_{...}$	$\sum_i \sum_j \sum_k (Y_{ijk} - \bar{Y}_{...})^2$	$abn-1$

1022

TABLE 27.4 $E\{MS\}$ Derivations for Crossed Two-Factor Experiment (A fixed, B random)

(a) Table

	i	j	k
	F	R	R
	a	b	n
α_i	0	b	n
β_j	a	1	n
$(\alpha\beta)_{ij}$	0	1	n
$\varepsilon_{k(ij)}$	1	1	1

(b) Coefficients

Variance	$E\{MSA\}$ i	$E\{MSB\}$ j	$E\{MSAB\}$ ij	$E\{MSE\}$ $(ij)k$
σ_α^2	$b \cdot n$	0 (step 9)	0 (step 9)	0 (step 9)
σ_β^2	0 (step 9)	$a \cdot n$	0 (step 9)	0 (step 9)
$\sigma_{\alpha\beta}^2$	$1 \cdot n$	$0 \cdot n$	n	0 (step 9)
σ^2	$1 \cdot 1$	$1 \cdot 1$	1	$1 \cdot 1$
	(i col. deleted)	(j col. deleted)	(i, j cols. deleted)	(k col. deleted)

(c) $E\{MS\}$

$$E\{MSA\} = bn(\textstyle\sum \alpha_i^2)/(a - 1) + n\sigma_{\alpha\beta}^2 + \sigma^2$$
$$E\{MSB\} = an\sigma_\beta^2 + \sigma^2$$
$$E\{MSAB\} = n\sigma_{\alpha\beta}^2 + \sigma^2$$
$$E\{MSE\} = \sigma^2$$

Teams of students were formed and assigned a task. One of the dependent variables was the number of questions raised prior to the final group decision. Some teams consisted of foreign students, others of U.S. students. Half of the teams consisted of eight members, the other half of four members. Two foreign observers were used for the foreign teams, and two U.S. observers for the U.S. teams. Thus, the design may be represented as follows:

	U.S. Teams (A_1)		Foreign Teams (A_2)	
	Observer 1 (C_1)	Observer 2 (C_2)	Observer 3 (C_1)	Observer 4 (C_2)
Small team (B_1)	Replication 1 Replication 2	Replication 1 Replication 2	Replication 1 Replication 2	Replication 1 Replication 2
Large team (B_2)	Replication 1 Replication 2	Replication 1 Replication 2	Replication 1 Replication 2	Replication 1 Replication 2

For simplicity, we assume that only two replications (teams) were used in each cell.

Development of Model

Let nationality of team be factor A, size of team factor B, and observer factor C. Note that factor C is nested within factor A since the two observers for the U.S. teams were different from the two observers for the foreign teams. Also note that factors A and B are crossed since each level of factor A appears with every level of factor B, and vice versa. Similarly, factors B and C are crossed. In this example, factors A and B were considered to have fixed effects, while the factor C (observer) effects were considered to be random. In using rule (27.1) to develop an appropriate model, we need to recognize according to step 2 that AC and ABC interactions are to be excluded because factor C is nested within factor A. Further, step 3 tells us that the BC interaction is nested within factor A since factor C is nested within factor A; thus, the BC interaction is a $BC(A)$ interaction. Hence, the appropriate model is:

$$(27.6) \qquad Y_{ijkm} = \mu_{...} + \alpha_i + \beta_j + \gamma_{k(i)} + (\alpha\beta)_{ij} + (\beta\gamma)_{jk(i)} + \varepsilon_{m(ijk)}$$

where:

$\mu_{...}$ is an overall constant
α_i are the fixed nationality effects
β_j are the fixed team size effects
$\gamma_{k(i)}$ are the random observer (within nationality) effects
$(\alpha\beta)_{ij}$ are the fixed nationality–team size interaction effects
$(\beta\gamma)_{jk(i)}$ are the random team size–observer interaction effects (within nationality)
$\varepsilon_{m(ijk)}$ are the random error terms
$i = 1, \ldots, a; j = 1, \ldots, b; k = 1, \ldots, c; m = 1, \ldots, n$

We shall assume as usual that $\gamma_{k(i)}$, $(\beta\gamma)_{jk(i)}$, and $\varepsilon_{m(ijk)}$ are normally distributed with expectations 0 and with constant variances, and that the three groups of random variables are pairwise independent. The interaction effects $(\beta\gamma)_{jk(i)}$ for any given observer are correlated, however, as may be seen from the following restrictions on the model:

$$\sum_i \alpha_i = 0 \qquad\qquad \sum_j \beta_j = 0 \qquad\qquad \sum_i (\alpha\beta)_{ij} = 0 \quad \text{for all } j$$

$$(27.6a)$$

$$\sum_j (\alpha\beta)_{ij} = 0 \quad \text{for all } i \qquad\qquad \sum_j (\beta\gamma)_{jk(i)} = 0 \quad \text{for all } k(i)$$

Analysis of Variance

Table 27.5 contains the development of the computational sums of squares and degrees of freedom for ANOVA model (27.6), and Table 27.6 contains the development of the expected mean squares. In Table 27.6b, we have, as usual, replaced variance terms for fixed effects by sums of squared effects divided by degrees of freedom. Table 27.6b indicates directly how to form test statistics for a variety of tests.

TABLE 27.5 Derivation of Computational Sums of Squares Formulas for Crossed-Nested Model (27.6)

Model Term	Symbolic Product	Sum of Squares	Degrees of Freedom
α_i	$i-1$	$SSA = \dfrac{\sum_i Y_{i...}^2}{bcn} - \dfrac{Y_{....}^2}{abcn}$	$a-1$
β_j	$j-1$	$SSB = \dfrac{\sum_j Y_{.j..}^2}{acn} - \dfrac{Y_{....}^2}{abcn}$	$b-1$
$\gamma_{k(i)}$	$\begin{aligned}(k-1)i \\ = ki - i\end{aligned}$	$SSC(A) = \dfrac{\sum_i\sum_k Y_{i.k.}^2}{bn} - \dfrac{\sum_i Y_{i...}^2}{bcn}$	$a(c-1)$
$(\alpha\beta)_{ij}$	$\begin{aligned}(i-1)(j-1) \\ = ij - i - j + 1\end{aligned}$	$SSAB = \dfrac{\sum_i\sum_j Y_{ij..}^2}{cn} - \dfrac{\sum_i Y_{i...}^2}{bcn} - \dfrac{\sum_j Y_{.j..}^2}{acn} + \dfrac{Y_{....}^2}{abcn}$	$(a-1)(b-1)$
$(\beta\gamma)_{jk(i)}$	$\begin{aligned}(j-1)(k-1)i \\ = ijk - ij - ik + i\end{aligned}$	$SSBC(A) = \dfrac{\sum_i\sum_j\sum_k Y_{ijk.}^2}{n} - \dfrac{\sum_i\sum_j Y_{ij..}^2}{cn} - \dfrac{\sum_i\sum_k Y_{i.k.}^2}{bn} + \dfrac{\sum_i Y_{i...}^2}{bcn}$	$a(b-1)(c-1)$
$\varepsilon_{m(ijk)}$	$\begin{aligned}(m-1)ijk \\ = ijkm - ijk\end{aligned}$	$SSE = \displaystyle\sum_i\sum_j\sum_k\sum_m Y_{ijkm}^2 - \dfrac{\sum_i\sum_j\sum_k Y_{ijk.}^2}{n}$	$abc(n-1)$
Total		$SSTO = \displaystyle\sum_i\sum_j\sum_k\sum_m Y_{ijkm}^2 - \dfrac{Y_{....}^2}{abcn}$	$abcn-1$

TABLE 27.6   Derivation of Expected Mean Squares for Crossed-Nested Model (27.6)

(a) Table

	i	j	k	m	Variance	A	B	C(A)	AB	BC(A)	E
	F	F	R	R							
	a	b	c	n		i	j	$(i)k$	ij	$(i)jk$	$(ijk)m$
α_i	0	b	c	n	σ_α^2	bcn	0	0	0	0	0
β_j	a	0	c	n	σ_β^2	0	acn	0	0	0	0
$\gamma_{k(i)}$	1	b	1	n	σ_γ^2	bn	0	bn	0	0	0
$(\alpha\beta)_{ij}$	0	0	c	n	$\sigma_{\alpha\beta}^2$	0	0	0	cn	0	0
$(\beta\gamma)_{jk(i)}$	1	0	1	n	$\sigma_{\beta\gamma}^2$	0	n	0	n	n	0
$\varepsilon_{m(ijk)}$	1	1	1	1	σ^2	1	1	1	1	1	1

Expected Mean Square of—

(b) Expected Mean Squares

$$E\{MSA\} = bcn\frac{\Sigma\alpha_i^2}{a-1} + bn\sigma_\gamma^2 + \sigma^2$$

$$E\{MSB\} = acn\frac{\Sigma\beta_j^2}{b-1} + n\sigma_{\beta\gamma}^2 + \sigma^2$$

$$E\{MSC(A)\} = bn\sigma_\gamma^2 + \sigma^2$$

$$E\{MSAB\} = cn\frac{\Sigma\Sigma(\alpha\beta)_{ij}^2}{(a-1)(b-1)} + n\sigma_{\beta\gamma}^2 + \sigma^2$$

$$E\{MSBC(A)\} = n\sigma_{\beta\gamma}^2 + \sigma^2$$

$$E\{MSE\} = \sigma^2$$

Example

Table 27.7 contains the results of the group decision-making experiment described earlier, based on $n = 2$ replications. The analysis of variance was obtained by means of a computer program and is presented in Table 27.8.

To test for nationality effects, the alternatives are:

(27.7a)

$$H_0: \alpha_1 = \alpha_2 = 0$$

$$H_a: \text{not both } \alpha_i \text{ equal zero}$$

Table 27.6b indicates that the appropriate test statistic is:

(27.7b)

$$F^* = \frac{MSA}{MSC(A)}$$

We have for our example:

$$F^* = \frac{420.25}{.25} = 1,681$$

TABLE 27.7 Data for Crossed-Nested Three-Factor Study—Group Decision-Making Example

	U.S. Teams ($i = 1$)		Foreign Teams ($i = 2$)	
Size of Team	Observer 1 ($k = 1$)	Observer 2 ($k = 2$)	Observer 3 ($k = 1$)	Observer 4 ($k = 2$)
4 members ($j = 1$)	16 20	14 19	7 5	4 9
8 members ($j = 2$)	21 25	28 19	11 17	12 15

$$Y_{111.} = 36 \quad Y_{112.} = 33 \quad Y_{211.} = 12 \quad Y_{212.} = 13$$
$$Y_{121.} = 46 \quad Y_{122.} = 47 \quad Y_{221.} = 28 \quad Y_{222.} = 27$$
$$Y_{11..} = 69 \quad Y_{1.1.} = 82 \quad Y_{21..} = 25 \quad Y_{2.1.} = 40$$
$$Y_{12..} = 93 \quad Y_{1.2.} = 80 \quad Y_{22..} = 55 \quad Y_{2.2.} = 40$$
$$Y_{1...} = 162 \qquad\qquad Y_{2...} = 80$$
$$Y_{.1..} = 94 \qquad\qquad Y_{.2..} = 148$$
$$Y_{....} = 242$$

For level of significance $\alpha = .05$, we require $F(.95; 1, 2) = 18.5$. Since $F^* = 1,681 > 18.5$, we conclude H_a, that nationality does have an effect on the group behavior. The P-value of the test is .0006. Other tests would be conducted in a similar fashion.

Confidence intervals for contrasts of main factor effects are set up in the usual way when the factor effects are fixed. For example, to estimate the difference between U.S. and foreign teams in the mean number of questions raised prior to a decision, we require $MSC(A)$ since this is the mean square in the denominator of the test statistic for examining nationality effects. Specifically, the confidence limits for

TABLE 27.8 ANOVA for Crossed-Nested Three-Factor Study—Group Decision-Making Example

	Source of Variation	SS	df	MS
A	Nationality	420.25	1	420.25
B	Size of team	182.25	1	182.25
$C(A)$	Observer (within nationality)	.50	2	.25
AB	Nationality–size of team interactions	2.25	1	2.25
$BC(A)$	Size of team–observer interactions (within nationality)	2.50	2	1.25
E	Error	106.00	8	13.25
	Total	713.75	15	

$D = \mu_{1..} - \mu_{2..}$ are:

(27.8)
$$\hat{D} \pm t[1 - \alpha/2; (c - 1)a]s\{\hat{D}\}$$

where:

(27.8a)
$$s^2\{\hat{D}\} = \frac{2MSC(A)}{nbc}$$

For our example, we obtain:

$$\bar{Y}_{1...} = \frac{162}{8} = 20.25 \quad \bar{Y}_{2...} = \frac{80}{8} = 10.00 \quad \hat{D} = 20.25 - 10.00 = 10.25$$

$$s^2\{\hat{D}\} = \frac{2(.25)}{8} = .063 \quad s\{\hat{D}\} = .25$$

For a confidence coefficient of .95, we require $t(.975; 2) = 4.303$. The confidence limits then are $10.25 \pm 4.303(.25)$, and the desired 95 percent confidence interval for D is:

$$9.2 \leq D \leq 11.3$$

Comments

1. The sums of squares SSA, SSB, and $SSAB$ in Table 27.5 for the analysis of a crossed-nested experimental design are the usual sums of squares for factor A main effects, factor B main effects, and AB interactions. $SSC(A)$ is a typical nested sum of squares for factor C effects. This can be seen by writing out the sum of squares in definitional form from the symbolic product $ki - i$:

(27.9)
$$SSC(A) = bn \sum_i \sum_k (\bar{Y}_{i.k.} - \bar{Y}_{i...})^2$$

Thus, $SSC(A)$ simply measures the variability of the factor C level estimated means for any given level of factor A, and then aggregates these sums of squares over factor A.

The definitional form of $SSBC(A)$ can be obtained from the symbolic product $ijk - ij - ik + i$:

(27.10)
$$SSBC(A) = n \sum_i \sum_j \sum_k (\bar{Y}_{ijk.} - \bar{Y}_{ij..} - \bar{Y}_{i.k.} + \bar{Y}_{i...})^2$$

Thus, $SSBC(A)$ contains the usual BC interaction sum of squares for a given level of factor A, and then aggregates these sums of squares over factor A.

2. If the only available computer program provides sums of squares for crossed factors, the two nested sums of squares can be obtained by utilizing the following relationships when the design is balanced:

(27.11a)
$$SSC(A) = SSC + SSAC$$

(27.11b)
$$SSBC(A) = SSBC + SSABC$$

These relationships generalize in a straightforward way. For example, if B is nested within

A and *C* within *B*, we have:

(27.12a) $SSB(A) = SSB + SSAB$

(27.12b) $SSC(AB) = SSC + SSAC + SSBC + SSABC$

3. If important *AB* interactions are present, analysis should usually focus on the means $\mu_{ij.}$ when the factors have fixed effects, rather than on the factor level means $\mu_{i..}$ and $\mu_{.j.}$. It can be shown that the estimated variance for comparing the two team sizes for any given nationality is:

(27.13) $$s^2\{\bar{Y}_{i1..} - \bar{Y}_{i2..}\} = \frac{2MSBC(A)}{cn}$$

This variance has associated with it $a(b-1)(c-1)$ degrees of freedom, as is evident from Table 27.5.

No exact confidence interval exists for comparing the two nationalities for any given team size. An unbiased variance estimator that can be utilized is:

(27.14) $$s^2\{\bar{Y}_{1j..} - \bar{Y}_{2j..}\} = \frac{2}{cn}\left[MSBC(A) + \frac{MSC(A) - MSE}{b}\right]$$

The approximate number of degrees of freedom associated with this variance is obtained from (22.61).

The reason for the different variances in (27.13) and (27.14) is that the observers are the same when the two team sizes for a given nationality are compared, while the observers differ when the two nationalities for a given team size are compared.

27.6 NO REPLICATIONS AND/OR SOME INTERACTIONS EQUAL ZERO

Modification of Rules

When a balanced design includes no replications and/or some interactions are assumed to equal zero—as, for instance, in a randomized complete block design with fixed block effects—rules (27.1) and (27.3) can be modified slightly to apply under these conditions as well. Rule (27.4) requires no modification.

The modification of rule (27.1) is very slight. Step 2 now becomes:

(27.15)   *Rule (27.1) modification: Step 2. Include all interaction terms except those assumed to equal zero and those containing both a nested factor and the factor within which it is nested.*

The modification of rule (27.3) is also a simple one:

(27.16)   *Rule (27.3) modification: Steps 2 through 8 do not apply to the model error term ε. Instead, the sum of squares associated with the model error term ε is obtained as a remainder from the total sum of squares. Likewise, the degrees of freedom associated with this remainder sum of squares are obtained as a remainder from the total degrees of freedom.*

The sum of squares associated with the model error term ε in balanced designs where there are no replications and/or where some interaction terms are assumed to equal zero will be denoted by *SSRem*, which stands for the *remainder sum of squares*. Frequently, the remainder sum of squares will turn out to be an interaction sum of squares for the interaction term in the model that is assumed to equal zero. The *remainder mean square* will be denoted by *MSRem*.

We shall illustrate the use of these modified rules for subsampling in a randomized block design where no replications are present.

Subsampling in Randomized Block Design

The model usually employed for a randomized block design when only a single observation is made on an experimental unit is ANOVA model (24.2) in the case of fixed treatment and block effects:

$$(27.17) \qquad Y_{ij} = \mu_{..} + \rho_i + \tau_j + \varepsilon_{ij}$$

We know from Chapter 24 that the error sum of squares *SSE* always equals zero for this design because there are no replications for any block-treatment combination. However, since model (27.17) assumes that all block-treatment interactions are zero, the interaction mean square *MSBL.TR* is an unbiased estimator of the experimental error variance σ^2 and is used as the denominator of the F^* test statistic for testing treatment effects when blocks and treatments have fixed factor levels.

We shall now show that modification (27.16) to rule (27.3) leads to the remainder mean square *MSRem* which equals *MSBL.TR*. However, we shall illustrate the use of modification (27.16) for a slightly more complex case, namely when subsampling is used in a randomized block design—that is, when more than one observation is made on each experimental unit. Consider, for instance, an experiment to study how three different motivational stimuli affect the length of time a person requires to perform a task. The persons in the experiment are blocked into groups of three, according to age, and each person is assigned at random one of the three motivational stimuli. Three observations are then made on the time required to complete the task; that is, the subject is asked to perform the task three times.

In this type of situation, we simply add a random observation error component to ANOVA model (27.17). Assuming that the treatment and block effects (motivational stimuli and age groups in our example) are fixed, an appropriate model is:

$$(27.18) \qquad Y_{ijk} = \mu_{..} + \rho_i + \tau_j + \varepsilon_{(ij)} + \eta_{k(ij)}$$

where:

$$\sum \rho_i = 0 \qquad \sum \tau_j = 0$$

$\varepsilon_{(ij)}$ and $\eta_{k(ij)}$ are independent normal random variables with expectations 0 and variances σ^2 and σ_η^2, respectively

$i = 1, \ldots, n; j = 1, \ldots, r; k = 1, \ldots, m$

Here ρ_i is the block effect, τ_j the treatment effect, $\varepsilon_{(ij)}$ the random effect associated with the experimental unit, and $\eta_{k(ij)}$ the random effect associated with the kth observation on the experimental unit. Note that the experimental error ε is nested within the (ij) block-treatment combination; there is no additional subscript since only one experimental unit is assigned to a treatment within a block. Thus, there are no replications for experimental units. Also note that the observation error η is nested within the (ij) block-treatment combination.

Table 27.9 contains the derivation of the computational sums of squares for ANOVA model (27.18), and Table 27.10 contains the derivation of the expected mean squares. Note that the sum of squares for experimental units is obtained as a remainder in Table 27.9 because there is only one experimental unit assigned to a treatment within a block. As expected, $SSRem$ turns out to be the block-treatment interaction sum of squares, as for a randomized block design without subsampling.

Table 27.10b indicates that for ANOVA model (27.18) with fixed treatment and block effects, the test statistic for examining the presence of treatment effects is $F^* = MSTR/MSRem$, as is also the case when no subsampling occurs in a randomized complete block design—see (24.7b). Remember that $MSRem$ is simply the interaction mean square $MSBL.TR$ here.

TABLE 27.9   Derivation of Computational Sums of Squares Formulas for Randomized Block Design with Subsampling—ANOVA Model (27.18)

Model Term	Symbolic Product	Sum of Squares	Degrees of Freedom
ρ_i	$i-1$	$SSBL = \dfrac{\sum_i Y_{i..}^2}{rm} - \dfrac{Y_{...}^2}{nrm}$	$n-1$
τ_j	$j-1$	$SSTR = \dfrac{\sum_j Y_{.j.}^2}{nm} - \dfrac{Y_{...}^2}{nrm}$	$r-1$
$\varepsilon_{(ij)}$		$SSRem = SSBL.TR$	Remainder $= (n-1)(r-1)$
		$= \dfrac{\sum_i\sum_j Y_{ij.}^2}{m} - \dfrac{\sum_i Y_{i..}^2}{rm} - \dfrac{\sum_j Y_{.j.}^2}{nm} + \dfrac{Y_{...}^2}{nrm}$	
$\eta_{k(ij)}$	$(k-1)ij$ $= ijk - ij$	$SSOE = \sum_i\sum_j\sum_k Y_{ijk}^2 - \dfrac{\sum_i\sum_j Y_{ij.}^2}{m}$	$nr(m-1)$
Total		$SSTO = \sum_i\sum_j\sum_k Y_{ijk}^2 - \dfrac{Y_{...}^2}{nrm}$	$nrm-1$

TABLE 27.10   Derivation of Expected Mean Squares for Randomized Block Design with Subsampling—ANOVA Model (27.18)

(a) Table

	i	j	k	Variance	Expected Mean Square of— BL	TR	Rem	OE
	F	F	R					
	n	r	m		i	j	(ij)	$(ij)k$
ρ_i	0	r	m	σ_ρ^2	rm	0	0	0
τ_j	n	0	m	σ_τ^2	0	nm	0	0
$\varepsilon_{(ij)}$	1	1	m	σ^2	m	m	m	0
$\eta_{k(ij)}$	1	1	1	σ_η^2	1	1	1	1

(b) Expected Mean Squares

$$E\{MSBL\} = rm\frac{\Sigma\rho_i^2}{n-1} + m\sigma^2 + \sigma_\eta^2$$

$$E\{MSTR\} = nm\frac{\Sigma\tau_j^2}{r-1} + m\sigma^2 + \sigma_\eta^2$$

$$E\{MSRem\} = m\sigma^2 + \sigma_\eta^2$$

$$E\{MSOE\} = \sigma_\eta^2$$

PROBLEMS

27.1. Refer to ANOVA model (21.25) on page 801. Use rule (27.4) to obtain the expected mean squares in Table 21.4 for this model.

27.2. Refer to ANOVA model (22.56) on page 849.
   a. Use rule (27.3) to obtain the definitional sums of squares formulas in (22.21) and the associated degrees of freedom.
   b. Use rule (27.4) to obtain the expected mean squares in Table 22.7.

27.3. Refer to ANOVA model (22.58) on page 850.
   a. Use rule (27.3a) to obtain the computational sums of squares formulas in (22.23) and (22.26) and the associated degrees of freedom.
   b. Use rule (27.4) to obtain the expected mean squares in Table 22.8.

27.4. Refer to nested design model (26.8) on page 975, but assume that factor A is nested within factor B, factor A effects are random, and factor B effects are fixed. (See also the paragraph "Random factor effects" on page 976.)
   a. Use rule (27.3a) to obtain the computational sums of squares formulas and the associated degrees of freedom.
   b. Use rule (27.4) to obtain the expected mean squares.
   c. What is the appropriate mean square to be used in constructing a confidence interval for $\mu_{.j}$?

27.5. Refer to randomized block model (24.2) on page 920.

a. Use rule (27.3) and modification (27.16) to obtain the definitional sums of squares formulas in (24.6) and the associated degrees of freedom.

b. Use rule (27.4) to obtain the expected mean squares in Table 24.3 for this model.

27.6. Refer to randomized block model (24.2) on page 920, but assume that treatment effects are random. (See also Comment 2 on page 920.)

a. Use rule (27.3a) and modification (27.16) to obtain the computational sums of squares formulas in (24.6) and the associated degrees of freedom.

b. Use rule (27.4) to obtain the expected mean squares in Table 24.3 for this model.

27.7. Refer to randomized block model (25.12) on page 955.

a. Use rule (27.3) and modification (27.16) to obtain the definitional sums of squares formulas in (24.6) and the associated degrees of freedom.

b. Use rule (27.4) to obtain the expected mean squares in Table 25.6 for this model.

27.8. Refer to randomized block model (27.18), but assume that block effects are random.

a. Use rule (27.3) and modification (27.16) to obtain the definitional sums of squares formulas and the associated degrees of freedom.

b. Use rule (27.4) to obtain the expected mean squares.

27.9. In a balanced three-factor study, factors A and C are crossed and factor B is nested within factor C. Factor A has fixed effects, and factors B and C have random effects. There are n replications for each treatment.

a. Use rule (27.3a) to obtain the computational sums of squares formulas and the associated degrees of freedom.

b. Use rule (27.4) to obtain the expected mean squares.

c. What is the appropriate denominator mean square for testing for factor A main effects?

27.10. Swimmer motivation. A large metropolitan swim club for youths studied the effects of three motivational stimuli on performance. The three motivational stimuli were: (1) presentation of merit award, (2) granting of team leadership privileges, and (3) publicity in the club newsletter. Since age is known to be related to performance, the nine female swimmers included in the study were grouped according to age into three blocks of three each. Within each age block, the three swimmers were randomly assigned to one of the motivation treatments. After a suitable amount of training, each swimmer was timed on three separate occasions while swimming a fixed distance. The coded data on the time for each of the three trials follow.

| | | Motivation Treatment | | |
| | Obser- | $j = 1$ | $j = 2$ | $j = 3$ |
Block	vation	Merit Award	Leadership	Publicity
$i = 1$	$k = 1$:	28	26	27
(7–8 years)	$k = 2$:	32	24	29
	$k = 3$:	31	27	30
$i = 2$	$k = 1$:	24	22	20
(9–10 years)	$k = 2$:	26	19	21
	$k = 3$:	23	18	22
$i = 3$	$k = 1$:	18	13	17
(11–12 years)	$k = 2$:	21	16	19
	$k = 3$:	20	15	19

Obtain the residuals for randomized block model (27.18) and plot them against the fitted values. Also prepare a normal probability plot of the residuals. What are your findings about the appropriateness of model (27.18)?

27.11. Refer to **Swimmer motivation** Problem 27.10. Assume that randomized block model (27.18) with fixed block and treatment effects is appropriate.
   a. Obtain the analysis of variance table.
   b. Test whether or not the mean times are the same for the three motivational stimuli; use $\alpha = .05$. State the alternatives, decision rule, and conclusion. What is the P-value of the test?
   c. Make all pairwise comparisons among the three treatment means; use the Tukey procedure with a 90 percent family confidence coefficient. State your findings.
   d. Obtain point estimates of σ^2 and σ_η^2. Does one variance appear to be much larger than the other? Discuss.

EXERCISES

27.12. Derive (27.11a) for a balanced three-factor study.

27.13. Use (27.14) and the fact that this estimated variance is unbiased to find $\sigma^2\{\bar{Y}_{1j..} - \bar{Y}_{2j..}\}$ for ANOVA model (27.6). What is the approximate number of degrees of freedom associated with the estimated variance?

Chapter 28

Repeated Measures and Related Designs

In this chapter we take up repeated measures designs—designs that are widely used in the behavioral and life sciences. We begin by considering some basic elements of repeated measures designs. We then take up single-factor repeated measures designs, after which we consider two-factor experiments with repeated measures on both factors and on only one factor. We conclude this chapter by introducing split-plot designs, which may be viewed as special types of repeated measures designs.

28.1 ELEMENTS OF REPEATED MEASURES DESIGNS

Description of Designs

Repeated measures designs utilize the same subject (person, store, plant, test market, etc.) for each of the treatments under study. Thus, the subject serves as a block, and the experimental units within a block may be viewed as the different occasions when a treatment is applied to the subject. A repeated measures study may involve several treatments or only a single treatment that is evaluated at different points in time. Subjects used in repeated measures studies in the behavioral and life sciences include persons, households, observers, and experimental animals. At other times the subjects in repeated measures designs are stores, test markets, cities, and plants. We shall refer to all of these study units that are used in repeated measures designs as *subjects*.

Three examples of repeated measures designs follow.

1. Fifteen test markets are to be used to study each of two different advertising campaigns. In each test market, the order of the two campaigns will be randomized, with a sufficient time lapse between the two campaigns so that the effects of the initial campaign will not carry over into the second campaign. The subjects in this study are the test markets.

2. Two hundred persons who have persistent migraine headaches are each to be given two different drugs and a placebo, for two weeks each, with the order of the drugs randomized for each person. The subjects in the study are the persons with migraine headaches.

3. In a weight loss study, 100 overweight persons are to be given the same diet and their weights measured at the end of each week for 12 weeks to assess the weight loss over time. Here the subjects are the overweight persons, who are observed repeatedly to provide information about the effects of a single treatment over time.

Each of these studies is referred to as a *repeated measures design* because the same subject is measured repeatedly. This key characteristic distinguishes this type of design from the designs considered earlier.

Advantages and Disadvantages

A principal advantage of repeated measures designs is that they provide good precision for comparing treatments because all sources of variability between subjects are excluded from the experimental error. Only variation within subjects enters the experimental error, since any two treatments can be compared directly for each subject. Thus, one may view the subjects as serving as their own controls. Another advantage of a repeated measures design is that it economizes on subjects. This is particularly important when only a few subjects (e.g., stores, plants, test markets) can be utilized for the experiment. Also, when interest is in the effects of a treatment over time, as when one is concerned with the shape of the learning curve for a new process operation, it is usually desirable to observe the same subject at different points in time rather than observing different subjects at the specified points in time.

Repeated measures designs have a serious potential disadvantage, however, namely, that there may be several types of interference. One type of interference is connected with the position in the treatment order. For instance, in evaluating five different advertisements, subjects may tend to give higher (or lower) ratings for advertisements shown toward the end of the sequence than at the beginning. Another type of interference is connected with the preceding treatment or treatments. For instance, in evaluating five different soup recipes, a bland recipe may get a higher (or lower) rating when preceded by a highly spiced recipe than when preceded by a blander recipe. This type of interference is called a *carry-over effect*.

Various steps can be taken to minimize the danger of interference. Randomization of the treatment orders for each subject independently will make it more reasonable to analyze the data as if the error terms are independent. Allowing sufficient time between treatments is often an effective means of reducing carry-over effects. It may be desirable at times to balance the order of treatment presentations and sometimes even the number of times each treatment is preceded by any other treatment. Latin square designs and cross-over designs (discussed in Chapter 29) are helpful to this end.

How to Randomize

The randomization of the order of the treatments assigned to a subject is straightforward. For each subject, a random permutation is used to define the treatment order, following the procedure described in Section 2.4. Independent permutations are selected for the different subjects.

Note

One should not confuse designs with repeated measures, discussed here, and designs with repeated observations, discussed in Section 26.7. In repeated measures designs, several or all of the treatments are applied to the same subject. Designs with repeated observations, on the other hand, are designs where several observations on the dependent variable are made for a given treatment applied to an experimental unit. It is possible to develop a repeated measures design with repeated observations, as when a given subject is exposed to each of the treatments under study and a number of observations are made at the end of each treatment application.

28.2 SINGLE-FACTOR EXPERIMENTS WITH REPEATED MEASURES ON ALL TREATMENTS

We first consider repeated measures designs where the treatments are based on a single factor, such as in the examples in Section 28.1. Almost always, the subjects in repeated measures designs (persons, stores, test markets, experimental animals) are viewed as a random sample from a population. Hence, *in all of the models for repeated measures designs to be presented in this chapter, the effects of subjects will be viewed as random.*

Table 28.1 contains the layout for a single-factor experiment with repeated measures on all treatments. Here, there are five subjects and four treatments, with the order of treatments independently randomized for each subject. Notice that this layout corresponds to the one in Table 24.1 for a randomized block design. Indeed, as we shall see next, the models for a single-factor repeated measures design are formally the same as the ones for randomized block designs, with blocks now considered to be subjects.

Model

When treatment effects are fixed, a model often appropriate for a single-factor repeated measures design is the following additive model:

(28.1) $$Y_{ij} = \mu_{..} + \rho_i + \tau_j + \varepsilon_{(ij)}$$

where:

$\mu_{..}$ is a constant
ρ_i are independent $N(0, \sigma_\rho^2)$

TABLE 28.1   Layout for Single-Factor Repeated Measures Design ($n = 5, r = 4$)

	Treatment Order			
	1	2	3	4
Subject 1	T_4	T_3	T_2	T_1
2	T_3	T_4	T_1	T_2
3	T_4	T_3	T_1	T_2
4	T_2	T_1	T_4	T_3
5	T_1	T_2	T_4	T_3

τ_j are constants subject to $\Sigma \tau_j = 0$
$\varepsilon_{(ij)}$ are independent $N(0, \sigma^2)$
ρ_i and $\varepsilon_{(ij)}$ are independent
$i = 1, \ldots, n; j = 1, \ldots, r$

Note that repeated measures model (28.1) is identical to the randomized block model (25.12) with random block effects, except that we now use the notation $\varepsilon_{(ij)}$ to show that no replications are present in this design.

Hence, we know from Section 25.4 that repeated measures model (28.1) assumes the following about the observations Y_{ij}:

(28.2a) $$E\{Y_{ij}\} = \mu_{..} + \tau_j$$

(28.2b) $$\sigma^2\{Y_{ij}\} = \sigma_Y^2 = \sigma_\rho^2 + \sigma^2$$

(28.2c) $$\sigma\{Y_{ij}, Y_{ij'}\} = \sigma_\rho^2 = \omega\sigma_Y^2 \quad j \neq j'$$

where ω is the coefficient of correlation between any two observations for the same subject:

(28.2d) $$\omega = \frac{\sigma_\rho^2}{\sigma_Y^2}$$

Thus, repeated measures model (28.1) assumes that in advance of the random trials, any two Y_{ij} treatment observations for a given subject are correlated in the same fashion for all subjects. This key assumption implies, as we saw in (25.17), that the variance-covariance matrix of the observations Y_{ij} for any given subject has compound symmetry. Any two observations from different subjects in advance of the random trials are independent according to model (28.1).

Equally important, we know from Chapter 25 that repeated measures model (28.1) assumes that, once the subjects have been selected, any two observations for a given subject are independent. Thus, model (28.1) assumes that there are no interference effects in the repeated measures study, such as order effects or carry-over effects from one treatment to the next.

Note

If interaction effects between subjects and treatments are present, interaction model (25.20) can be employed. As we noted in Chapter 25, both the interaction and no-interaction models lead to the same procedures for making inferences about the treatment effects.

Analysis of Variance and Tests

Since repeated measures model (28.1) is the same as randomized block model (25.12), the analysis of variance and the test for treatment effects will be the same as before.

Analysis of Variance.   The ANOVA sums of squares for repeated measures model (28.1) are given in (24.6), but the names of two of the sums of squares are usually changed for repeated measures applications. The sum of squares for blocks in (24.6a) will now be called the *sum of squares for subjects,* and the interaction sum of squares between blocks and treatments in (24.6c) will now be called the *interaction sum of squares between treatments and subjects.* These two sums of squares will be denoted, respectively, by SSS and $SSTR.S$. Thus, the analysis of variance decomposition for single-factor repeated measures model (28.1) is:

$$(28.3) \qquad SSTO = SSS + SSTR + SSTR.S$$

where:

$$(28.3a) \qquad SSTO = \sum_i \sum_j (Y_{ij} - \bar{Y}_{..})^2$$

$$(28.3b) \qquad SSS = r \sum_i (\bar{Y}_{i.} - \bar{Y}_{..})^2$$

$$(28.3c) \qquad SSTR = n \sum_j (\bar{Y}_{.j} - \bar{Y}_{..})^2$$

$$(28.3d) \qquad SSTR.S = \sum_i \sum_j (Y_{ij} - \bar{Y}_{i.} - \bar{Y}_{.j} + \bar{Y}_{..})^2$$

Note that no error sum of squares is present because there are no replications here.

Table 28.2 contains the analysis of variance table for repeated measures model (28.1). It is the same as the ANOVA table in Table 25.6 for the additive randomized block model (25.12), except for the change in notation. Note again that in the absence of interactions between treatments and subjects, the interaction mean square $MSTR.S$ is an unbiased estimator of the error variance σ^2.

TABLE 28.2   ANOVA Table for Single-Factor Repeated Measures Design—ANOVA Model (28.1) with Subject Effects Random and Treatment Effects Fixed

Source of Variation	SS	df	MS	E{MS}
Subjects	SSS	$n - 1$	MSS	$\sigma^2 + r\sigma_\rho^2$
Treatments	SSTR	$r - 1$	MSTR	$\sigma^2 + \dfrac{n}{r - 1} \sum \tau_j^2$
Error	SSTR.S	$(r - 1)(n - 1)$	MSTR.S	σ^2
Total	SSTO	$nr - 1$		

Note

In repeated measures studies, $SSTR$ and $SSTR.S$ are sometimes combined into a *within-subjects sum of squares* SSW:

$$(28.4) \qquad SSW = SSTR + SSTR.S$$

which can be shown to equal:

$$(28.4a) \qquad SSW = \sum_i \sum_j (Y_{ij} - \bar{Y}_{i.})^2$$

Hence, the ANOVA decomposition in (28.3) can also be expressed as follows:

$$(28.5) \qquad SSTO = \underbrace{SSS}_{\substack{\text{between-}\\\text{subjects}\\\text{variability}}} + \underbrace{SSW}_{\substack{\text{within-}\\\text{subjects}\\\text{variability}}}$$

Test for Treatment Effects.   As the $E\{MS\}$ column in Table 28.2 indicates, the appropriate statistic for the test on treatment effects:

$$(28.6a) \qquad \begin{array}{l} H_0: \text{all } \tau_j = 0 \\ H_a: \text{not all } \tau_j \text{ equal zero} \end{array}$$

is:

$$(28.6b) \qquad F^* = \frac{MSTR}{MSTR.S}$$

When H_0 holds, F^* follows the F distribution, as before, and the decision rule for controlling the Type I error at α is:

$$(28.6c) \qquad \begin{array}{l} \text{If } F^* \le F[1 - \alpha; r - 1, (r - 1)(n - 1)], \text{ conclude } H_0 \\ \text{If } F^* > F[1 - \alpha; r - 1, (r - 1)(n - 1)], \text{ conclude } H_a \end{array}$$

Example.   In a wine-judging competition, four Chardonnay wines of the same

TABLE 28.3   Data for Wine-Judging Example (ratings on a scale of 0 to 40)

Judge i	Wine (j) 1	2	3	4	$\bar{Y}_{i.}$
1	20	24	28	28	25
2	15	18	23	24	20
3	18	19	24	23	21
4	26	26	30	30	28
5	22	24	28	26	25
6	19	21	27	25	23
$\bar{Y}_{.j}$	20.00	22.00	26.67	26.00	$23.67 = \bar{Y}_{..}$

vintage were judged by six experienced judges. Each judge tasted the wines in a blind fashion, i.e., without knowing their identities. The order of the wine presentation was randomized independently for each judge. To reduce carry-over and other interference effects, the judges did not drink the wines and rinsed their mouths thoroughly between tastings. Each wine was scored on a 40-point scale; the larger the score, the greater is the excellence of the wine. The data for this competition are presented in Table 28.3. A plot of the wine scores for each judge is shown in Figure 28.1.

FIGURE 28.1   Plot of Wine Scores for Each Judge—Wine-Judging Example

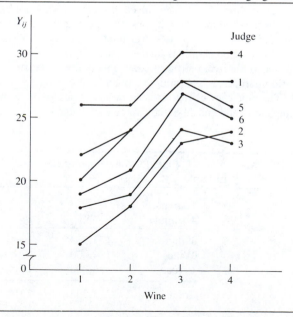

The six judges are considered to be a random sample from the population of possible judges, while the four wines tasted are of interest in themselves. Hence, the single-factor repeated measures model (28.1) is relevant, with the effects of subjects (judges) considered random and the effects of treatments (wines) considered fixed. As we shall see later, diagnostic analysis indicated that ANOVA model (28.1) is appropriate for the data.

Table 28.4 contains the analysis of variance table for the wine judging data in Table 28.3. The calculations are straightforward and were carried out by a computer package. To test for treatment effects:

$$H_0: \tau_1 = \tau_2 = \tau_3 = \tau_4 = 0$$

$$H_a: \text{not all } \tau_j \text{ equal zero}$$

we use the results of Table 28.4:

$$F^* = \frac{MSTR}{MSTR.S} = \frac{61.333}{1.067} = 57.5$$

For a level of significance of $\alpha = .01$, we require $F(.99; 3, 15) = 5.42$. Since $F^* = 57.5 > 5.42$, we conclude H_a, that the mean wine ratings for the four wines differ. The P-value for this test is $0+$.

Comments

1. As we noted in Chapter 25 (in Comment 2 on p. 959), a conservative test for treatment effects should be used if the assumption of compound symmetry in repeated measures model (28.1) is not met (i.e., if either the variances of the observations for different treatments for a given subject are not the same for all subjects or if the correlations between any two treatment observations for a given subject are not the same for all treatment pairs and for all subjects). In repeated measures studies, the compound symmetry assumption will be violated, for instance, if repeated responses over time are more highly correlated for observations close together than for observations further apart in time.

2. When the treatment effects are random, test statistic (28.6b) and decision rule (28.6c) are still appropriate for testing treatment effects.

3. The efficiency of the repeated measures design in the wine-judging example, relative to a completely randomized design where each judge is used to assess a single wine, can be mea-

TABLE 28.4   ANOVA Table for Single-Factor Repeated Measures Design—Wine-Judging Example

Source of Variation	SS	df	MS
Judges	173.333	5	34.667
Wines	184.000	3	61.333
Error	16.000	15	1.067
Total	373.333	23	

sured by means of (24.14). Using the results in Table 28.4, we obtain:

$$\hat{E} = \frac{(n - 1)MSS + n(r - 1)MSTR.S}{(nr - 1)MSTR.S} = \frac{5(34.667) + 6(3)(1.067)}{23(1.067)} = 7.85$$

Thus, almost eight times as many replications per treatment would have been required with a completely randomized design in which each judge rates a single wine as in the repeated measures design to achieve the same precision for any estimated contrast.

4. When a single-factor repeated measures design involves $r = 2$ treatments, the F^* statistic in (28.6b) is equivalent to the two-sided t test for paired observations based on test statistic (1.66).

5. Occasionally, a formal test for subject effects is desired:

$$H_0: \sigma_\rho^2 = 0$$

$$H_a: \sigma_\rho^2 > 0$$

Table 28.2 indicates that the appropriate test statistic for repeated measures model (28.1) is $F^* = MSS/MSTR.S$.

Evaluation of Aptness of Repeated Measures Model

Since repeated measures model (28.1) is equivalent to randomized block model (25.12), the earlier discussion on diagnostics for randomized block models is entirely applicable here. In particular, a plot of the responses Y_{ij} by subject, as in Figure 28.1, can be examined for indications of serious lack of parallelism, which would suggest that the additive model (28.1) may not be appropriate.

Residual sequence plots by subject can be helpful for studying constancy of the error variance and presence of interference effects. The residuals for repeated measures models (28.1) are given in (24.5):

$$(28.7) \qquad e_{ij} = Y_{ij} - \bar{Y}_{i.} - \bar{Y}_{.j} + \bar{Y}_{..}$$

A normal probability plot of the estimated subject main effects $\bar{Y}_{i.} - \bar{Y}_{..}$ can be helpful for evaluating whether the subject main effects ρ_i are normally distributed with constant variance.

In addition to these graphic diagnostics, the estimated within-subjects variance-covariance and correlation matrices for the treatment observations Y_{ij} can be examined for aptness of the repeated measures model. A typical entry in the variance-covariance matrix is the estimated within-subjects covariance between observations for treatments j and j':

$$(28.8) \qquad \frac{\sum_{i=1}^{n}(Y_{ij} - \bar{Y}_{.j})(Y_{ij'} - \bar{Y}_{.j'})}{n - 1}$$

The estimated within-subjects variance-covariance matrix should show variances of the same order of magnitude, and all of the covariances should be of similar magnitude. Of course, estimated variances and covariances tend to be subject to large sampling errors unless the sample sizes are very large. Hence, moderate differences in variances and covariances should be viewed as likely to be the result of sampling errors.

The estimated correlation matrix should show approximately similar coefficients of correlation between pairs of treatment observations within a subject.

Finally, the Tukey test described in Section 21.2 can be conducted to examine the appropriateness of the additive model. This test will need to be interpreted here as conditional on the subjects actually used in the repeated measures study.

Example. For the wine-judging example, the residuals were obtained from (28.7), and are presented in Figure 28.2a in aligned dot plots by wine. These plots support the assumption of constant error variance. Figure 28.2b presents residual sequence plots for each judge, where the residuals are plotted in the order in which the wines were tasted by the judge. These plots do not indicate any correlations of the error terms within a judge, and thus suggest that no interference effects were present. Finally, a normal probability plot of the residuals is presented in Figure 28.2c. This plot shows evidence of the effects of the rounded nature of the data, but does not suggest any major departure from normality. The correlation between the ordered residuals and their expected values under normality is .993, which also suggests that lack of normality is not a problem here.

Table 28.5 presents the estimated within-subjects variance-covariance and correlation matrices for the treatment observations. The differences found there could easily arise from sampling errors.

TABLE 28.5 Estimated Within-Subjects Variance-Covariance and Correlation Matrices between Treatment Observations—Wine-Judging Example

(a) Variance-Covariance Matrix

	j'			
j	1	2	3	4
1	14.000	11.000	9.200	8.200
2		10.000	8.200	7.600
3			7.067	6.200
4				6.800

(b) Correlation Matrix

	j'			
j	1	2	3	4
1	1	.930	.925	.840
2		1	.975	.922
3			1	.894
4				1

FIGURE 28.2   Diagnostic Residual Plots—Wine-Judging Example (SAS/GRAPH, Ref. 28.1)

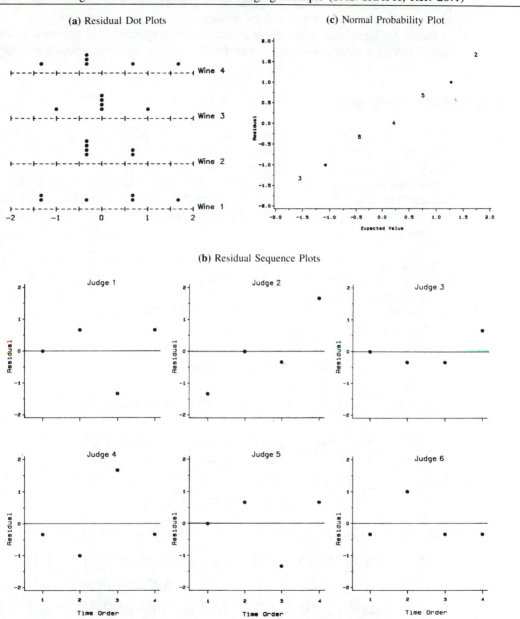

(a) Residual Dot Plots

(c) Normal Probability Plot

(b) Residual Sequence Plots

The plot of the responses by subject in Figure 28.1 also supports the appropriateness of model (28.1), since the plots for the judges are reasonably parallel. Thus, there is no indication of interactions between subjects and treatments.

Based on these and other diagnostics, it was concluded that repeated measures model (28.1) is reasonably appropriate for the data in the wine-judging example.

Analysis of Treatment Effects

The analysis of treatment effects for single-factor repeated measures model (28.1) proceeds in exactly the same fashion as described in Section 24.7 for randomized block designs with fixed treatment effects. The multiples in (24.9) for setting up confidence intervals are applicable here as they stand. The mean square used in estimating the variance of the estimated contrast is still the interaction mean square, which is now denoted by $MSTR.S$. We shall illustrate the estimation procedures by an example.

Example.   In the wine-judging example, it was desired to compare all treatment means $\mu_{.j}$ pairwise, with a 95 percent family confidence coefficient. Here $\mu_{.j}$ is the mean rating of wine j averaged over judges. The Tukey procedure was utilized for this purpose. Using (15.25) with MSE replaced by $MSTR.S$ and the results in Table 28.4, we obtain:

$$s^2\{\hat{D}\} = MSTR.S\left(\frac{1}{n} + \frac{1}{n}\right) = 1.067\left(\frac{2}{6}\right) = .3557$$

Using (24.9b), we find for a 95 percent family confidence coefficient:

$$T = \frac{1}{\sqrt{2}}q(.95; 4, 15) = \frac{1}{\sqrt{2}}(4.08) = 2.885$$

Hence:

$$Ts\{\hat{D}\} = 2.885\sqrt{.3557} = 1.72$$

Thus we obtain for the pairwise comparisons (see Table 28.3 for the $\bar{Y}_{.j}$):

$$-2.39 = (26.00-26.67) - 1.72 \leq \mu_{.4} - \mu_{.3} \leq (26.00-26.67) + 1.72 = 1.05$$

$$2.28 = (26.00-22.00) - 1.72 \leq \mu_{.4} - \mu_{.2} \leq (26.00-22.00) + 1.72 = 5.72$$

$$4.28 = (26.00-20.00) - 1.72 \leq \mu_{.4} - \mu_{.1} \leq (26.00-20.00) + 1.72 = 7.72$$

$$2.95 = (26.67-22.00) - 1.72 \leq \mu_{.3} - \mu_{.2} \leq (26.67-22.00) + 1.72 = 6.39$$

$$4.95 = (26.67-20.00) - 1.72 \leq \mu_{.3} - \mu_{.1} \leq (26.67-20.00) + 1.72 = 8.39$$

$$.28 = (22.00-20.00) - 1.72 \leq \mu_{.2} - \mu_{.1} \leq (22.00-20.00) + 1.72 = 3.72$$

We display these results graphically as follows:

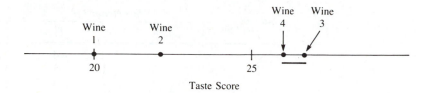

Taste Score

We conclude from these pairwise comparisons that wines 3 and 4 are judged best, and do not differ significantly from each other. Wines 1 and 2 are judged to be inferior to wines 3 and 4, with wine 1 receiving a mean rating significantly lower than that for wine 2. The family confidence coefficient of .95 applies to the entire set of comparisons.

Ranked Data

In repeated measures studies, the observations are frequently ranks, as when a number of tasters are each asked to rank recipes or when several university admissions officers are each asked to rank applicants for admission. When the data in a repeated measures study are ranks, the Friedman test described in Section 25.2 may be used for testing whether the treatment means are equal. No new principles are involved, so we shall proceed directly to an example.

Example. Six subjects were each asked to rank five coffee sweeteners according to their taste preferences, with rank 5 assigned to the most preferred sweetener. The data and basic calculational results are presented in Table 28.6. Test statistic (25.4) here is:

$$X_F^2 = \left[\frac{12}{6(5)(6)} (1,836) \right] - 3(6)(6) = 14.4$$

For a level of significance of $\alpha = .05$, we need $\chi^2(.95; 4) = 9.49$. Since $X_F^2 = 14.4 > 9.49$, we conclude that the five sweeteners are not equally liked. The P-value of the test is .006.

We now wish to make all pairwise tests by means of (25.5) with family level of significance $\alpha = .20$. For $r = 5$, we have $g = 5(4)/2 = 10$. Hence, for $n = 6$ we obtain:

$$B = z[1 - .20/2(10)] = z(.99) = 2.326$$

Thus, the right term in (25.5) is:

$$B \left[\frac{r(r + 1)}{6n} \right]^{1/2} = 2.326 \left[\frac{5(6)}{6(6)} \right]^{1/2} = 2.12$$

TABLE 28.6   Ranked Data for Coffee Sweeteners in a Repeated Measures Design

Subject	Sweetener (j)				
i	A	B	C	D	E
1	5	1	2	4	3
2	4	2	1	5	3
3	3	2	1	4	5
4	5	2	3	4	1
5	4	1	2	3	5
6	4	1	3	5	2
$R_{.j}$	25	9	12	25	19
$\bar{R}_{.j}$	4.17	1.50	2.00	4.17	3.17

$$\Sigma R_{.j}^2 = (25)^2 + (9)^2 + (12)^2 + (25)^2 + (19)^2 = 1,836$$

We note from Table 28.6 that the pairs of mean ranks whose difference does not exceed 2.12 are (B, C), (B, E), (C, E), (A, E), (D, E), and (A, D). Hence, we can set up two groups, within which the treatment means do not differ:

Group 1		Group 2	
Sweetener B	$\bar{R}_{.2} = 1.50$	Sweetener E	$\bar{R}_{.5} = 3.17$
Sweetener C	$\bar{R}_{.3} = 2.00$	Sweetener A	$\bar{R}_{.1} = 4.17$
Sweetener E	$\bar{R}_{.5} = 3.17$	Sweetener D	$\bar{R}_{.4} = 4.17$

Thus, we conclude with family level of significance of .20 that sweeteners A and D are preferred to sweeteners B and C, and that it is not clear whether sweetener E belongs in the preferred group or in the other group.

Comments

1. The Friedman test can also be used for repeated measures designs where the observations are not ranked, in case the distribution of the error terms departs far from normality. Ranks of the observations Y_{ij} are then assigned within each subject, and the Friedman test is carried out in the usual manner.

2. X_F^2 is related to Kendall's coefficient of concordance W in the following way:

$$(28.9) \qquad W = \frac{X_F^2}{n(r - 1)}$$

The coefficient of concordance W is a measure of the agreement of the rankings of the n subjects. It equals 1 if there is perfect agreement, and equals 0 if there is no agreement, that is, if all treatments receive the same mean ranking. For the coffee sweeteners example in Table 28.6, W is:

$$W = \frac{14.4}{6(4)} = .60$$

indicating a fair amount of agreement between the subjects.

28.3 TWO-FACTOR EXPERIMENTS WITH REPEATED MEASURES ON BOTH FACTORS

In the preceding section we considered single-factor repeated measures studies. The model for these designs can easily be extended when the treatments actually follow a factorial structure. For example, consider a study where four treatments are employed that represent two levels of each of two factors. Table 28.7 depicts the layout for such a design when four subjects are utilized in the study. Note that the order of the treatments is randomized within each subject. When the treatments represent a factorial structure, we can explore as usual interaction effects as well as the main effects for the two factors. The design in Table 28.7 is said to represent *repeated measures on both factors* because each subject receives all treatments defined by the factorial structure.

Model

When the factor effects are fixed and the subjects constitute a random sample, a model frequently appropriate for the case where there are repeated measures on both factors is the following one which assumes, like model (28.1), that treatments and subjects do not interact:

TABLE 28.7   Layout for Two-Factor Repeated Measures Design with Repeated Measures on Both Factors ($n = 4$, $a = 2$, $b = 2$)

	Treatment Order			
	1	2	3	4
Subject 1	$A_1 B_2$	$A_2 B_2$	$A_1 B_1$	$A_2 B_1$
2	$A_2 B_1$	$A_1 B_2$	$A_2 B_2$	$A_1 B_1$
3	$A_2 B_2$	$A_1 B_1$	$A_2 B_1$	$A_1 B_2$
4	$A_1 B_1$	$A_2 B_1$	$A_1 B_2$	$A_2 B_2$

(28.10) $$Y_{ijk} = \mu_{...} + \rho_i + \alpha_j + \beta_k + (\alpha\beta)_{jk} + \varepsilon_{(ijk)}$$

where:

$\mu_{...}$ is a constant
ρ_i are independent $N(0, \sigma_\rho^2)$
α_j are constants subject to $\Sigma\alpha_j = 0$
β_k are constants subject to $\Sigma\beta_k = 0$
$(\alpha\beta)_{jk}$ are constants subject to $\sum_j (\alpha\beta)_{jk} = 0$ for all k and $\sum_k (\alpha\beta)_{jk} = 0$ for all j
$\varepsilon_{(ijk)}$ are independent $N(0, \sigma^2)$
ρ_i and $\varepsilon_{(ijk)}$ are independent
$i = 1, \ldots, n; j = 1, \ldots, a; k = 1, \ldots, b$

The observations Y_{ijk} for repeated measures model (28.10) have the following properties:

(28.11a) $$E\{Y_{ijk}\} = \mu_{...} + \alpha_j + \beta_k + (\alpha\beta)_{jk}$$

(28.11b) $$\sigma^2\{Y_{ijk}\} = \sigma_Y^2 = \sigma_\rho^2 + \sigma^2$$

(28.11c) $$\sigma\{Y_{ijk}, Y_{ij'k'}\} = \sigma_\rho^2 \quad \text{not both } j = j' \text{ and } k = k'$$

Thus, repeated measures model (28.10) assumes that the observations Y_{ijk} have constant variance, and that any two treatment observations for the same subject in advance of the random trials have constant covariance. Any two observations from different subjects in advance of the random trials are independent according to model (28.10). Finally, all observations are assumed to be normally distributed.

Model (28.10) is a direct extension of the single-factor repeated measures model (28.1), where the treatment effect τ_j is now decomposed into factor A and factor B main effects and an AB interaction effect. Note again that the notation $\varepsilon_{(ijk)}$ is used because there are no replications present in this design.

Once the subjects have been selected, repeated measures model (28.10), like the earlier repeated measures model (28.1), assumes that all of the treatment observations for a given subject are independent—that is, that there are no interference effects.

Analysis of Variance and Tests

Analysis of Variance. The ANOVA sums of squares for model (28.10) can be readily obtained by following rule (27.3) as modified by (27.16). The sum of squares for estimating the error variance must be obtained as a remainder since there are no replications present here. This sum of squares turns out to be the interaction sum of squares $SSTR.S$, reflecting interactions between treatments and subjects. Table 28.8 presents the ANOVA decomposition, degrees of freedom, and expected mean squares for the two-factor repeated measures model (28.10). The expected mean squares can be obtained easily by following rule (27.4).

TABLE 28.8   ANOVA Table and Sums of Squares for Two-Factor Repeated Measures Design with Repeated Measures on Both Factors—Subjects Random, Factors A and B Fixed

(a) ANOVA Table

Source of Variation	SS	df	MS	E{MS}
Subjects	SSS	$n - 1$	MSS	$\sigma^2 + ab\sigma_\rho^2$
Factor A	SSA	$a - 1$	MSA	$\sigma^2 + \dfrac{nb}{a-1}\sum \alpha_j^2$
Factor B	SSB	$b - 1$	MSB	$\sigma^2 + \dfrac{na}{b-1}\sum \beta_k^2$
AB interactions	SSAB	$(a-1)(b-1)$	MSAB	$\sigma^2 + \dfrac{n}{(a-1)(b-1)}\sum\sum (\alpha\beta)_{jk}^2$
Error	SSTR.S	$(n-1)(ab-1)$	MSTR.S	σ^2
Total	SSTO	$abn - 1$		

(b) Sums of Squares

$$SSS = ab \sum_i (\bar{Y}_{i..} - \bar{Y}_{...})^2$$

$$SSA = nb \sum_j (\bar{Y}_{.j.} - \bar{Y}_{...})^2$$

$$SSB = na \sum_k (\bar{Y}_{..k} - \bar{Y}_{...})^2$$

$$SSAB = n \sum_j \sum_k (\bar{Y}_{.jk} - \bar{Y}_{.j.} - \bar{Y}_{..k} + \bar{Y}_{...})^2$$

$$SSRem = SSTR.S = \sum_i \sum_j \sum_k (Y_{ijk} - \bar{Y}_{i..} - \bar{Y}_{.jk} + \bar{Y}_{...})^2$$

Tests for Factor Effects.   It is clear from the expected mean squares column in Table 28.8 that the test for AB interaction effects:

(28.12a)
$$H_0: \text{ all } (\alpha\beta)_{jk} = 0$$
$$H_a: \text{ not all } (\alpha\beta)_{jk} \text{ equal zero}$$

uses the test statistic:

(28.12b)
$$F^* = \frac{MSAB}{MSTR.S}$$

and the decision rule for controlling the Type I error at α is:

(28.12c)   If $F^* \leq F[1 - \alpha; (a-1)(b-1), (n-1)(ab-1)]$, conclude H_0
If $F^* > F[1 - \alpha; (a-1)(b-1), (n-1)(ab-1)]$, conclude H_a

The test for factor A main effects:

(28.13a)
$$H_0: \text{all } \alpha_j = 0$$
$$H_a: \text{not all } \alpha_j \text{ equal zero}$$

uses the test statistic:

(28.13b)
$$F^* = \frac{MSA}{MSTR.S}$$

and the decision rule for controlling the Type I error at α is:

(28.13c)
$$\text{If } F^* \leq F[1 - \alpha; a - 1, (n - 1)(ab - 1)], \text{ conclude } H_0$$
$$\text{If } F^* > F[1 - \alpha; a - 1, (n - 1)(ab - 1)], \text{ conclude } H_a$$

Similarly, the test for factor B main effects:

(28.14a)
$$H_0: \text{all } \beta_k = 0$$
$$H_a: \text{not all } \beta_k \text{ equal zero}$$

uses the test statistic:

(28.14b)
$$F^* = \frac{MSB}{MSTR.S}$$

and the decision rule for controlling the Type I error at α is:

(28.14c)
$$\text{If } F^* \leq F[1 - \alpha; b - 1, (n - 1)(ab - 1)], \text{ conclude } H_0$$
$$\text{If } F^* > F[1 - \alpha; b - 1, (n - 1)(ab - 1)], \text{ conclude } H_a$$

Comments

1. When the effects of either factor A or factor B are random, the expected mean squares can be found by employing rule (27.4). In turn, these expected mean squares will identify the appropriate test statistics.

2. The conservative F test described in Chapter 25 should be used when the assumption of compound symmetry in repeated measures model (28.10) is not met.

3. Repeated measures model (28.10) assumes that treatments and subjects do not interact. This assumption can be relaxed because the treatment-subject interaction sum of squares here is made up of three components:

$$SSTR.S = SSAS + SSBS + SSABS$$

where $SSAS$ is the interaction sum of squares between factor A and subjects and the other terms are defined similarly. Thus, it is possible to allow for the first-order factor A–subject interactions and the factor B–subject interactions to be present and only assume that the second-order factor A–factor B–subject interactions equal zero. When the repeated measures model allows for some interactions between treatments and subjects, the analysis of factor effects becomes somewhat more complex.

Evaluation of Aptness of Repeated Measures Model

Our earlier discussion on the evaluation of the aptness of repeated measures model (28.1) applies here as well. In particular, residual sequence plots by subject should be constructed to examine whether interference effects are present and whether the error variance is constant. Plots of the observations by subject should be utilized to see whether the assumption of no treatment-subject interactions is appropriate.

Analysis of Factor Effects

If strong interactions between factors A and B exist that cannot be made unimportant by some simple transformation, the analysis of the factor effects should be performed in terms of the treatment means $\mu_{.jk}$, which are averaged over subjects. This analysis proceeds similar to that in Section 28.2 for a single-factor study, with $\mu_{.jk}$ corresponding there to a treatment mean $\mu_{.j}$. The mean square $MSTR.S$ will again be used in estimating the variance of any estimated contrast of the treatment means.

If factors A and B do not interact or interact only in unimportant fashion, the analysis of factor A and factor B main effects proceeds as usual. For the analysis of either factor A or factor B main effects, $MSTR.S$ will be used in the estimated variance of the estimated contrast since this mean square is the denominator of the F^* test statistic for testing factor A or factor B main effects.

The multiples for the estimated standard deviation of an estimated contrast of factor A or factor B level means are as follows:

	Main A Effect	*Main B Effect*
	Single comparison	
(28.15a)	$t[1 - \alpha/2; (n - 1)(ab - 1)]$	$t[1 - \alpha/2; (n - 1)(ab - 1)]$
	Tukey procedure (for pairwise comparisons)	
(28.15b)	$T = \dfrac{1}{\sqrt{2}}q[1 - \alpha; a, (n - 1)(ab - 1)]$	$T = \dfrac{1}{\sqrt{2}}q[1 - \alpha; b, (n - 1)(ab - 1)]$
	Scheffé procedure	
(28.15c)	$S^2 = (a - 1)F[1 - \alpha; a - 1, (n - 1)(ab - 1)]$	$S^2 = (b - 1)F[1 - \alpha; b - 1, (n - 1)(ab - 1)]$
	Bonferroni procedure	
(28.15d)	$B = t[1 - \alpha/2g; (n - 1)(ab - 1)]$	$B = t[1 - \alpha/2g; (n - 1)(ab - 1)]$

Example

A clinician studied the effects of two drugs used either alone or together on the blood flow in human subjects. Twelve healthy middle-aged males participated in the study and they are viewed as a random sample from a relevant population of middle-aged males. The four treatments used in the study are defined as follows:

$A_1 B_1$	placebo (neither drug)
$A_1 B_2$	drug B alone
$A_2 B_1$	drug A alone
$A_2 B_2$	both drugs A and B

The 12 subjects received each of the four treatments in independently randomized orders. The dependent variable is the increase in blood flow from before to shortly after the administration of the treatment. The treatments were administered on successive days. This prevented any carry-over effects because the effect of the drugs is short-lived.

The experiment was conducted in a double-blind fashion so that neither the physician nor the subject knew which treatment was administered when the change in blood flow was measured.

Table 28.9 contains the data for this study. A negative entry denotes a decrease in blood flow. The Minitab computer package (Ref. 28.2) was utilized to fit repeated measures model (28.10). Figure 28.3 contains the printout. Included in the printout are the expected mean squares for the specified ANOVA model. Each term in an expected mean square is represented by the code, in parentheses, for the variance of the model term and the preceding number which is the numerical multiple. When the model term is fixed, the letter Q is used in the printout to show that the variance is replaced by the sum of squared effects divided by degrees of freedom. For exam-

---

TABLE 28.9 Data for Blood Flow Example

Subject i	Treatment			
	$A_1 B_1$	$A_1 B_2$	$A_2 B_1$	$A_2 B_2$
1	2	10	9	25
2	−1	8	6	21
3	0	11	8	24
4	3	15	11	31
5	1	5	6	20
6	2	12	9	27
7	−2	10	8	22
8	4	16	12	30
9	−2	7	7	24
10	−2	10	10	28
11	2	8	10	25
12	−1	8	6	23

FIGURE 28.3   Computer Output for ANOVA on Blood Flow Example Data (Minitab, Ref. 28.2)

Factor	Type	Levels				Values			
S	random	12	1	2	3	4	5	6	7	8
							9	10	11	12
A	fixed	2	1	2						
B	fixed	2	1	2						

Analysis of Variance for Y

Source	DF	SS	MS	F	P
S	11	258.50	23.50	10.01	0.000
A	1	1587.00	1587.00	675.75	0.000
B	1	2028.00	2028.00	863.54	0.000
A*B	1	147.00	147.00	62.59	0.000
Error	33	77.50	2.35		
Total	47	4098.00	87.19		

Source	Variance component	Error term	Expected Mean Square (using restricted model)
1 S	5.288	5	(5) + 4(1)
2 A		5	(5) + 24Q[2]
3 B		5	(5) + 24Q[3]
4 A*B		5	(5) + 12Q[4]
5 Error	2.348		(5)

MEANS

A	B	N	Y
1	1	12	0.500
1	2	12	10.000
2	1	12	8.500
2	2	12	25.000

ple, the expected value of MSA as shown in Figure 28.3 is:

$$(5) + 24Q[2] = \sigma^2 + 24\frac{\Sigma\alpha_j^2}{a - 1}$$

which corresponds, of course, to the expected mean square shown in Table 28.8.

Various diagnostics were utilized to see if repeated measures model (28.10) is appropriate for the data in Table 28.9. The results (not shown here) supported the appropriateness of this model. The clinician expected the two drugs to interact in increasing the blood flow. To test for interaction effects:

$$H_0: \text{ all } (\alpha\beta)_{jk} = 0$$

$$H_a: \text{ not all } (\alpha\beta)_{jk} \text{ equal zero}$$

we use test statistic (28.12b) and the results from Figure 28.3:

$$F^* = \frac{MSAB}{MSTR.S} = \frac{147.000}{2.35} = 62.6$$

For a level of significance of $\alpha = .01$, we require $F(.99; 1, 33) = 7.47$. Since $F^* = 62.6 > 7.47$, we conclude H_a, that interaction effects exist. The P-value for this test is $0+$.

Figure 28.4 contains a plot of the estimated treatment means $\bar{Y}_{.jk}$, which are given in Figure 28.3. Strong interaction effects are evident. To study the nature of the interaction effects, the clinician wished to compare the joint use of the two drugs with the use of each drug alone, drug A with drug B, and each drug with no drug. Thus, the following pairwise comparisons are to be made:

$$D_1 = \mu_{.22} - \mu_{.21} \qquad D_4 = \mu_{.21} - \mu_{.11}$$

$$D_2 = \mu_{.22} - \mu_{.12} \qquad D_5 = \mu_{.12} - \mu_{.11}$$

$$D_3 = \mu_{.21} - \mu_{.12}$$

Point estimates of these pairwise comparisons are ($\bar{Y}_{.jk}$ values are in Figure 28.3):

$$\hat{D}_1 = 25.0 - 8.5 = 16.5 \qquad \hat{D}_4 = 8.5 - .5 = 8.0$$

$$\hat{D}_2 = 25.0 - 10.0 = 15.0 \qquad \hat{D}_5 = 10.0 - .5 = 9.5$$

$$\hat{D}_3 = 8.5 - 10.0 = -1.5$$

The estimated variance of each $\hat{D}$ is given in (15.13), with the relevant mean square here being $MSTR.S$. Hence, we have:

$$s^2\{\hat{D}\} = MSTR.S\left(\frac{1}{n} + \frac{1}{n}\right) = 2.348\left(\frac{2}{12}\right) = .3913$$

and $s\{\hat{D}\} = .626$. Using the Bonferroni procedure with a 95 percent family confidence coefficient, we require $B = t[1 - (.05)/2(5); 33] = t(.995; 33) =$

FIGURE 28.4   Plot of Estimated Treatment Means—Blood Flow Example

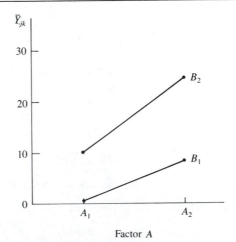

Factor A

2.733. Hence, $t(.995)s\{\hat{D}\} = 2.733(.626) = 1.71$ and the desired confidence intervals with a 95 percent family confidence coefficient are:

$$14.8 \leq \mu_{.22} - \mu_{.21} \leq 18.2 \qquad 6.3 \leq \mu_{.21} - \mu_{.11} \leq 9.7$$

$$13.3 \leq \mu_{.22} - \mu_{.12} \leq 16.7 \qquad 7.8 \leq \mu_{.12} - \mu_{.11} \leq 11.2$$

$$-3.2 \leq \mu_{.21} - \mu_{.12} \leq .2$$

It is clear from these results that either drug A alone or drug B alone tends to lead to an increase in blood flow, and that the combination of the two drugs leads to a substantial additional increase in blood flow as compared to when either drug is used alone. Finally, no significant difference exists in the mean effects of the two drugs used alone.

28.4 TWO-FACTOR EXPERIMENTS WITH REPEATED MEASURES ON ONE FACTOR

Description of Design

In many two-factor studies, repeated measures can only be made on one of the two factors. Consider, for instance, an experimenter who wished to study the effects of two types of incentives (factor A) on a person's ability to solve problems. He also wanted to study two types of problems (factor B)—abstract and concrete problems. He could ask each experimental subject to do each type of problem, but could not use more than one type of incentive stimulus on a subject because of interference effects. Thus, the design the experimenter utilized may be represented schematically as shown in Table 28.10.

In a two-factor experiment with repeated measures on one factor, two randomizations generally need to be employed. First, the level of the nonrepeated factor (A, in Table 28.10) needs to be randomly assigned to the subjects. Second, the order of the levels of the repeated factor (B, in Table 28.10) needs to be randomized independently for all subjects.

Since n subjects are randomly assigned incentive stimulus A_1 and n subjects are randomly assigned incentive stimulus A_2, as far as factor A is concerned the experiment is a completely randomized one. On the other hand, as far as factor B (type of problem) is concerned, each subject is a block. Thus, for factor B, the experiment is a randomized block design, with block effects random. We call this experimental design a *two-factor experiment with repeated measures on factor B*.

In the experiment depicted in Table 28.10, comparisons between factor A level means involve differences between groups of subjects as well as differences associated with the two factor A levels. However, comparisons between factor B level means at the same level of factor A are based on the same subject, and hence only involve differences associated with the two factor B levels. Thus, for these comparisons, each subject serves as its own control. The main effects of factor A are therefore said to be confounded with differences between groups of subjects, whereas the main effects of factor B are free of such confounding. It is for this reason that tests

TABLE 28.10   Layout for Two-Factor Design with Random Assignments of Factor A Level to Subjects and Repeated Measures on Factor B

Incentive Stimulus	Subject	Treatment Order	
		1	2
A_1	1	A_1B_1	A_1B_2
	.		.
	.		.
	.		.
	n	A_1B_1	A_1B_2
A_2	$n+1$	A_2B_2	A_2B_1
	.		.
	.		.
	.		.
	$2n$	A_2B_1	A_2B_2

on factor B main effects will generally be more sensitive than tests on the main effects for factor A.

Comments

1. A two-factor experiment with repeated measures on one factor may be viewed as an incomplete block design. With reference to the repeated measures design in Table 28.10, there are four treatments $(A_1B_1, A_1B_2, A_2B_1, \text{ and } A_2B_2)$ and one-half of the blocks (subjects) contain treatments A_1B_1 and A_1B_2 while the other half of the blocks contain treatments A_2B_1 and A_2B_2.

2. When the factor on which repeated measures are taken is time, no randomization of the levels of the repeated factor is required. Consider, for instance, a study of two different advertising campaigns in which the effect on sales is to be measured in 10 test markets during four consecutive months. Here, the only randomization required is for assigning the advertising campaigns to the test markets. Similarly, when the nonrepeated factor is a characteristic of the subjects, such as age of subject, no randomization is involved for that factor.

Model

The development of a model for a two-factor experiment with repeated measures on one factor is only a little more complex than for earlier cases. As before, we shall develop the model for random subject effects and fixed factor A and factor B effects.

Let, as usual, α_j and β_k denote the factor A and factor B main effects, respectively, and let ρ denote the subject (block) main effect. We do need to recognize, however, that the subject effect in this design is nested within factor A. Therefore, we will denote this effect by $\rho_{i(j)}$. As before, we shall assume that there are no interactions between treatments and subjects, although this condition is not essential here. A model that incorporates the above specifications is as follows:

(28.16)
$$Y_{ijk} = \mu_{...} + \rho_{i(j)} + \alpha_j + \beta_k + (\alpha\beta)_{jk} + \varepsilon_{(ijk)}$$

where:

$\mu_{...}$ is a constant
$\rho_{i(j)}$ are independent $N(0, \sigma_\rho^2)$
α_j are constants subject to $\Sigma\alpha_j = 0$
β_k are constants subject to $\Sigma\beta_k = 0$
$(\alpha\beta)_{jk}$ are constants subject to $\sum_j (\alpha\beta)_{jk} = 0$ for all k and $\sum_k (\alpha\beta)_{jk} = 0$ for all j
$\varepsilon_{(ijk)}$ are independent $N(0, \sigma^2)$
$\rho_{i(j)}$ and $\varepsilon_{(ijk)}$ are independent
$i = 1, \ldots, n; j = 1, \ldots, a; k = 1, \ldots, b$

The observations Y_{ijk} for repeated measures model (28.16) have the following properties:

(28.17a)
$$E\{Y_{ijk}\} = \mu_{...} + \alpha_j + \beta_k + (\alpha\beta)_{jk}$$

(28.17b)
$$\sigma^2\{Y_{ijk}\} = \sigma_Y^2 = \sigma_\rho^2 + \sigma^2$$

(28.17c)
$$\sigma\{Y_{ijk}, Y_{ijk'}\} = \sigma_\rho^2 \quad k \neq k'$$

Thus, the observations Y_{ijk} have constant variance. In addition, in advance of the random trials any two observations for different levels of factor B for the same subject have constant covariance, for all subjects, while observations for different subjects are independent. Also, all observations are assumed to be normally distributed.

Once the subjects have been selected, repeated measures model (28.16) assumes that any two observations for the same subject are independent, that is, that there are no interference effects.

Analysis of Variance and Tests

Analysis of Variance.   The ANOVA sums of squares for repeated measures model (28.16) can be obtained as usual by means of rule (27.3) with modification (27.16) because there are no replications for this design. The remainder sum of squares that is used for estimating the error variance turns out to be the interaction sum of squares $SSB.S(A)$. The ANOVA sums of squares are shown in Table 28.11. Also shown in Table 28.11 are the degrees of freedom for each sum of squares.

Tests for Factor Effects.   The expected mean squares for the analysis of variance in Table 28.11 are given in Table 28.12. These expected mean squares can be obtained by means of rule (27.4).

TABLE 28.11 Analysis of Variance for Two-Factor Experiment with Repeated Measures on Factor B—Model (28.16)

Source of Variation	SS	df
Factor A	$SSA = bn \sum_j (\bar{Y}_{.j.} - \bar{Y}_{...})^2$	$a - 1$
Factor B	$SSB = an \sum_k (\bar{Y}_{..k} - \bar{Y}_{...})^2$	$b - 1$
AB interactions	$SSAB = n \sum_j \sum_k (\bar{Y}_{.jk} - \bar{Y}_{.j.} - \bar{Y}_{..k} + \bar{Y}_{...})^2$	$(a - 1)(b - 1)$
Subjects (within factor A)	$SSS(A) = b \sum_i \sum_j (\bar{Y}_{ij.} - \bar{Y}_{.j.})^2$	$a(n - 1)$
Error	$SSRem =$ $SSB.S(A) = \sum_i \sum_j \sum_k (Y_{ijk} - \bar{Y}_{.jk} - \bar{Y}_{ij.} + \bar{Y}_{.j.})^2$	Remainder $=$ $a(n - 1)(b - 1)$
Total	$SSTO = \sum_i \sum_j \sum_k (Y_{ijk} - \bar{Y}_{...})^2$	$abn - 1$

It is clear from the expected mean squares in Table 28.12 that the test for AB interaction effects:

$$(28.18a) \qquad \begin{array}{l} H_0: \text{ all } (\alpha\beta)_{jk} = 0 \\ H_a: \text{ not all } (\alpha\beta)_{jk} \text{ equal zero} \end{array}$$

uses the test statistic:

$$(28.18b) \qquad F^* = \frac{MSAB}{MSB.S(A)}$$

and the decision rule for controlling the Type I error at α is:

$$(28.18c) \qquad \begin{array}{l} \text{If } F^* \le F[1 - \alpha; (a - 1)(b - 1), a(n - 1)(b - 1)], \text{ conclude } H_0 \\ \text{If } F^* > F[1 - \alpha; (a - 1)(b - 1), a(n - 1)(b - 1)], \text{ conclude } H_a \end{array}$$

The test for factor A main effects:

$$(28.19a) \qquad \begin{array}{l} H_0: \text{ all } \alpha_j = 0 \\ H_a: \text{ not all } \alpha_j \text{ equal zero} \end{array}$$

uses the test statistic:

$$(28.19b) \qquad F^* = \frac{MSA}{MSS(A)}$$

TABLE 28.12   Expected Mean Squares for Two-Factor Experiment with Repeated Measures on Factor B—Model (28.16) (A, B fixed, subjects random)

Source of Variation	MS	$E\{MS\}$
Factor A	MSA	$\sigma^2 + b\sigma_\rho^2 + bn\dfrac{\Sigma\alpha_j^2}{(a-1)}$
Factor B	MSB	$\sigma^2 + an\dfrac{\Sigma\beta_k^2}{(b-1)}$
AB interactions	$MSAB$	$\sigma^2 + n\dfrac{\Sigma\Sigma(\alpha\beta)_{jk}^2}{(a-1)(b-1)}$
Subjects (within factor A)	$MSS(A)$	$\sigma^2 + b\sigma_\rho^2$
Error	$MSB.S(A)$	σ^2

and the decision rule for controlling the Type I error at α is:

(28.19c)
$$\text{If } F^* \le F[1 - \alpha; a - 1, a(n - 1)], \text{ conclude } H_0$$
$$\text{If } F^* > F[1 - \alpha; a - 1, a(n - 1)], \text{ conclude } H_a$$

Finally, the test for factor B main effects:

(28.20a)
$$H_0: \text{ all } \beta_k = 0$$
$$H_a: \text{ not all } \beta_k \text{ equal zero}$$

uses the test statistic:

(28.20b)
$$F^* = \frac{MSB}{MSB.S(A)}$$

and the decision rule for controlling the Type I error at α is:

(28.20c)
$$\text{If } F^* \le F[1 - \alpha; b - 1, a(n - 1)(b - 1)], \text{ conclude } H_0$$
$$\text{If } F^* > F[1 - \alpha; b - 1, a(n - 1)(b - 1)], \text{ conclude } H_a$$

Comments

1. If the assumption of compound symmetry in repeated measures model (28.16) is not met, the conservative test discussed in Chapter 25 should be employed.

2. If the number of subjects within each level of factor A is not the same, problems analogous to those for unbalanced two-factor studies in a completely randomized design arise. Unless the sample sizes reflect the importance of the treatments, interest usually should center on analyses giving the cell means equal weights. While many statistical software packages handle unbalanced data, it is incumbent on the user to make sure that the package tests the hypotheses of interest to the user.

Evaluation of Aptness of Repeated Measures Model

Our earlier discussion on evaluating the aptness of a repeated measures model applies here also. The residuals for repeated measures model (28.16) are:

$$(28.21) \qquad e_{ijk} = Y_{ijk} - \bar{Y}_{.jk} - \bar{Y}_{ij.} + \bar{Y}_{.j.}$$

A special feature of repeated measures model (28.16) also warrants attention. This model requires that the variance between subjects, σ_ρ^2, be constant for all levels of factor A. This assumption can be examined by dot plots of the estimated subject effects $\bar{Y}_{ij.} - \bar{Y}_{.j.}$ for each level of factor A.

We can also conduct a formal test of the equality of the between-subjects variances by noting that the variation between subjects within factor A, $SSS(A)$, can be decomposed into components for each factor A level:

$$(28.22) \qquad SSS(A) = SSS(A_1) + SSS(A_2) + \cdots + SSS(A_a)$$

where:

$$(28.22a) \qquad SSS(A_j) = b \sum_i (\bar{Y}_{ij.} - \bar{Y}_{.j.})^2$$

Each component sum of squares has $n - 1$ degrees of freedom associated with it. We can therefore make a test of the equality of the between-subjects variances by means of the Hartley test statistic (16.13).

Similarly, the remainder variation, $SSB.S(A)$, can be decomposed into components for each factor A level:

$$(28.23) \qquad SSB.S(A) = SSB.S(A_1) + SSB.S(A_2) + \cdots + SSB.S(A_a)$$

where:

$$(28.23a) \qquad SSB.S(A_j) = \sum_i \sum_k (Y_{ijk} - \bar{Y}_{.jk} - \bar{Y}_{ij.} + \bar{Y}_{.j.})^2$$

Each component has $(n - 1)(b - 1)$ degrees of freedom associated with it. A Hartley test can be conducted here also, this time to test for the equality of the error variance σ^2 for the different factor A levels.

Both of the Hartley tests assume normality and are sensitive to this assumption. Hence, the appropriateness of the normality assumption should be established first before the Hartley tests are undertaken. If the normality assumption is not satisfied, a transformation of the data may be helpful.

Analysis of Factor Effects

When the two factors do not interact or the interactions are not important, the main effects may be analyzed in a straightforward fashion. The relevant mean square to be used in the estimated variance of a contrast of factor A level means for repeated measures model (28.16) is $MSS(A)$ because it is the denominator of the appropriate F^*

statistic for testing factor A main effects. Similarly, the mean square for estimating contrasts of factor B level means is $MSB.S(A)$.

The multiples in (28.15) need to be modified only for the degrees of freedom associated with the mean square used: $a(n - 1)$ for the analysis of factor A effects and $a(n - 1)(b - 1)$ for the analysis of factor B effects.

Note from Table 28.12 that the analysis of factor B effects can be carried out more precisely than that for factor A effects. The reason is that comparisons among factor A levels involve the variability among the subjects as well as the experimental error, while comparisons among factor B levels involve only experimental error.

When interactions exist between the two factors, the analysis of factor effects becomes considerably more complex; see, for example, Reference 28.3 for a discussion of this case.

Example

A national retail chain wanted to study the effects of two advertising campaigns (factor A) on the volume of sales of athletic shoes over time (factor B). Ten similar test markets (subjects, S) were randomly chosen to participate in this study. The two advertising campaigns (A_1 and A_2) were similar in all respects except that a different national sports personality was used in each. Sales data were collected for three two-week periods (B_1: two weeks prior to campaign; B_2: two weeks during which campaign occurred; B_3: two weeks after campaign was concluded). The experiment was conducted during a six-week period when sales of athletic shoes are usually quite stable.

The data on sales (coded) are presented in Table 28.13, and are plotted in Figure 28.5 by test market for each advertising campaign. There is no evidence in Figure 28.5 of any interactions between the test markets and the treatments. In general,

TABLE 28.13 Data for Athletic Shoes Sales Example

Advertising Campaign	Test Market	Time Period $k = 1$	$k = 2$	$k = 3$
	$i = 1$	958	1,047	933
	$i = 2$	1,005	1,122	986
$j = 1$	$i = 3$	351	436	339
	$i = 4$	549	632	512
	$i = 5$	730	784	707
	$i = 1$	780	897	718
	$i = 2$	229	275	202
$j = 2$	$i = 3$	883	964	817
	$i = 4$	624	695	599
	$i = 5$	375	436	351

FIGURE 28.5   Plots of Sales Data by Test Market and Campaign—Athletic Shoes Sales
Example

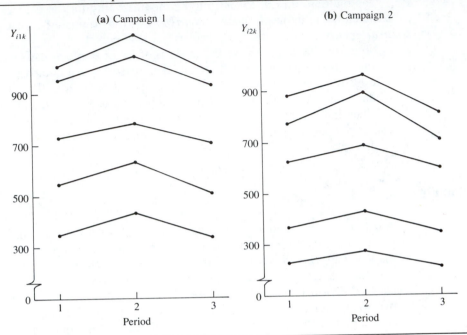

sales tended to increase during each advertising campaign, and then tended to decline to previous or lower levels than just before the campaign.

Based on Figure 28.5 and other diagnostic analyses (not shown), it was concluded
that repeated measures model (28.16) is appropriate here. A computer package for
this model was run and the printout is shown in Figure 28.6.

First we wish to test for the campaign-time interaction effects:

$$H_0: \text{all } (\alpha\beta)_{jk} = 0$$

$$H_a: \text{not all } (\alpha\beta)_{jk} \text{ equal zero}$$

We use the results from Figure 28.6 in test statistic (28.18b):

$$F^* = \frac{MSAB}{MSB.S(A)} = \frac{196}{358} = .55$$

For a level of significance of $\alpha = .05$, we require $F(.95; 2, 16) = 3.63$. Since
$F^* = .55 \leq 3.63$, we conclude H_0, that no significant interaction effects are present. The P-value for the test is .59.

Next we wish to test for advertising campaign main effects:

$$H_0: \text{all } \alpha_j = 0$$

$$H_a: \text{not all } \alpha_j \text{ equal zero}$$

FIGURE 28.6   Computer Output for ANOVA on Athletic Shoes Sales Example Data
(Minitab, Ref. 28.2)

```
Factor    Type     Levels          Values
A         fixed    2        1   2
S(A)      random   5        1   2   3   4   5
B         fixed    3        1   2   3


Analysis of Variance for Y

Source    DF        SS          MS        F       P
A          1      168151      168151    0.73   0.417
S(A)       8     1833681      229210  640.31   0.000
B          2       67073       33537   93.69   0.000
A*B        2         391         196    0.55   0.589
Error     16        5727         358
Total     29     2075023       71553


Source     Variance    Error    Expected Mean Square
           component   term     (using restricted model)
1 A                      2      (5) + 3(2) + 15Q[1]
2 S(A)      76284.0      5      (5) + 3(2)
3 B                      5      (5) + 10Q[3]
4 A*B                    5      (5) + 5Q[4]
5 Error       358.0             (5)


        MEANS

A    N         Y
1    15     739.40
2    15     589.67

B    N         Y
1    10     648.40
2    10     728.80
3    10     616.40
```

We use the results from Figure 28.6 in test statistic (28.19b):

$$F^* = \frac{MSA}{MSS(A)} = \frac{168,151}{229,210} = .73$$

For a level of significance of $\alpha = .05$, we require $F(.95; 1, 8) = 5.32$. Since $F^* = .73 \le 5.32$, we conclude H_0, that no advertising campaign main effects exist. The P-value for the test is .42. Thus, either of the two national sports personalities is equally effective in the advertising campaign.

Finally, we wish to test for time period effects:

$$H_0: \text{all } \beta_k = 0$$

$$H_a: \text{not all } \beta_k \text{ equal zero}$$

Using the results from Figure 28.6 in test statistic (28.20b), we obtain:

$$F^* = \frac{MSB}{MSB.S(A)} = \frac{33,537}{358} = 93.7$$

For a level of significance of $\alpha = .05$, we require $F(.95; 2, 16) = 3.63$. Since $F^* = 93.7 > 3.63$, we conclude H_a, that period main effects exist. The P-value for the test is $0+$.

To examine the nature of the time period effects, we shall conduct pairwise comparisons of mean sales for the three time periods:

$$D = \mu_{..k} - \mu_{..k'}$$

The Tukey procedure will be employed, with a 99 percent family confidence coefficient. We require:

$$T = \frac{1}{\sqrt{2}}q(.99; 3, 16) = \frac{1}{\sqrt{2}}(4.78) = 3.38$$

$$s^2\{\hat{D}\} = \frac{2MSB.S(A)}{an} = \frac{2(358)}{2(5)} = 71.60$$

Hence, $Ts\{\hat{D}\} = 3.38\sqrt{71.60} = 28.6$.

The point estimates of the changes in mean sales, based on the estimated factor B level means $\bar{Y}_{..k}$ in Figure 28.6, are:

$$\hat{D}_1 = \bar{Y}_{..2} - \bar{Y}_{..1} = 728.8 - 648.4 = 80.4$$
$$\hat{D}_2 = \bar{Y}_{..3} - \bar{Y}_{..1} = 616.4 - 648.4 = -32.0$$
$$\hat{D}_3 = \bar{Y}_{..3} - \bar{Y}_{..2} = 616.4 - 728.8 = -112.4$$

and the desired confidence intervals therefore are:

$$52 \leq \mu_{..2} - \mu_{..1} \leq 109$$
$$-61 \leq \mu_{..3} - \mu_{..1} \leq -3$$
$$-141 \leq \mu_{..3} - \mu_{..2} \leq -84$$

We can conclude, with family confidence coefficient .99, that the two advertising campaigns lead to an immediate increase in mean sales of between 52 and 109 (8 to 17 percent), but that mean sales in the following period fall below those for the period preceding the campaign by somewhere between 3 and 61 (.5 to 9 percent).

28.5 SPLIT-PLOT DESIGNS FOR TWO-FACTOR STUDIES

Description of Designs

Split-plot designs are frequently used in field, laboratory, industrial, and social science experiments. We shall discuss split-plot designs only for two-factor studies, but these designs can be extended to apply when three or more factors are under investigation.

Split-plot designs can be viewed as special repeated measures designs for factorial treatments when each subject can only be assigned some of the treatments. We already discussed such designs in Section 28.4, where we took up two-factor experiments with repeated measurements on one factor. Split-plot designs are an improvement over these designs by incorporating blocking of subjects. Aside from the blocking of subjects, our earlier discussion of two-factor experiments with repeated measures on one factor is applicable. We shall give three examples to illustrate split-plot designs.

Example 1.   Consider again the study of the effects of two types of incentive (factor A) and two types of problems (factor B) on a person's ability to solve a problem. The repeated measures design for this study, with repeated measurements for type of problem (factor B), was illustrated in Table 28.10. We noted when discussing this design that the analysis of the effects of type of incentive (factor A) will not be as precise usually as the analysis of the effects of type of problem because generally there is much more variability between subjects than within a subject. However, a repeated measure could not be taken on type of incentive here because of interference effects.

To improve the precision when analyzing factor A effects, we can block the subjects by some appropriate characteristic(s) so as to reduce the variability between subjects within a block. Table 28.14 illustrates the split-plot design for this example. There are n blocks, each consisting of two similar subjects. One subject in each block is assigned at random to factor level A_1, the other is assigned to factor level A_2. In the second stage of randomization, each subject is randomly assigned the order of the two problems. Thus, the only difference between the split-plot design in Table

TABLE 28.14   Layout for Split-Plot Design with Random Assignments of Factor A Level to Subjects and Repeated Measures on Factor B

		Treatment Order	
		1	2
Block 1	Subject 1	$A_2 B_1$	$A_2 B_2$
	Subject 2	$A_1 B_2$	$A_1 B_1$
Block 2	Subject 3	$A_1 B_2$	$A_1 B_1$
	Subject 4	$A_2 B_2$	$A_2 B_1$
⋮	⋮	⋮	⋮
Block n	Subject $2n - 1$	$A_1 B_1$	$A_1 B_2$
	Subject $2n$	$A_2 B_2$	$A_2 B_1$

28.14 and the design in Table 28.10 with repeated measures on one factor is the blocking of the subjects for purposes of studying factor *A* effects more precisely.

When there is a choice between which of the two factors to take repeated measures on (factor *B*), it should be the one for which more precise estimates are required. The reason is that even with blocking, the variability between subjects within a block will usually be greater than the variability within a subject.

Example 2. Split-plot designs are also used frequently when the repeated measurements are made over time. An experiment was conducted to study two methods of assembling a component (factor *A*) and the rate of learning the assembly operation (factor *B*). One-half of the employees used in the study were assigned at random to each of the two assembly methods (A_1 and A_2), and the productivity of each employee in performing the assembly operation was measured after five days (B_1) and after ten days (B_2). Table 28.14 illustrates the split-plot design for this example, as well as that for Example 1. Blocking of employees here was done by amount of experience.

In this experiment, only one randomization needs to be made (assigning employees to the assembly operation) since the second factor involves repeated observations on the subjects at different points in time. Essentially, there was no choice in this experiment as to which factor the repeated measurements were to be made on, since the employees could not be expected to handle both assembly operations effectively during the study period.

Example 3. Split-plot designs were originally developed for agricultural experiments. Consider an experiment to study the effects of two varieties of wheat (factor *A*) and two fertilizers (factor *B*) on yield, using different fields as blocks. In a randomized complete block design, the fields would be subdivided into four subunits each, and the four treatments (A_1B_1, A_1B_2, A_2B_1, A_2B_2) would then be assigned at random to the subunits.

A split-plot design is more practical here because it is difficult to plant the different varieties of wheat in small areas. With a split-plot design, each field is divided into only two parts instead of four (usually called *plots*), and the two varieties are then randomly assigned to the two plots in each field. In turn, the two plots in each field are then subdivided into two smaller areas (usually called *subplots*), and the two fertilizers are then randomly assigned to the subplots in each plot.

Table 28.14 shows the layout for this split-plot design as well. Note that randomization is here required for assigning the varieties of wheat to the plots and assigning the fertilizers to the subplots within a plot. Note also that the plots in each field correspond to the subjects, and the subplots correspond to the repeated measures units within a subject.

Comments

1. Whenever subjects can receive all treatments in a two-factor study without interference effects, a repeated measures design with repeated measures on both factors is preferable, because the factor effects for both factors can then usually be estimated more precisely than in a split-plot design.

2. Split-plot designs are useful in industrial experiments when one factor requires larger experimental units than another. Consider, for instance, a study of the effects of two additives (factor A) and two different containers (factor B) for prolonging the shelf life of a milk product. Here, it is easier to make larger batches of the milk product with a given additive, whereas the different containers can be used with smaller batches.

3. Split-plot designs may be viewed as a type of incomplete block design where the subjects are considered to be the blocks, with each subject being given only some of the full set of treatments.

Models

We shall consider two models for two-factor split-plot designs—when the block effects are fixed and when these effects are random. Throughout, we shall assume that the repeated measures are taken on factor B.

Block Effects Fixed. When the blocks are based on characteristics of the subjects, such as age of person or size of store, the block effects are usually viewed as fixed. The model we shall present also assumes that both factor A and factor B effects are fixed. In addition, the model assumes that there are no interactions between blocks and treatments except for block-factor A interactions. This latter interaction is often allowed for so as to obtain, in the case of random block effects, a reasonable correlation structure for observations from the same block.

The model to be presented contains no main effects for subjects because subjects serve as the unit of replication within each block for evaluating factor A main effects. Equivalently, it may be said that the subject effects are confounded with the factor A main effects here.

Finally, because there are no complete replications in the split-plot design, the error term is denoted as usual by $\varepsilon_{(ijk)}$.

The split-plot model for fixed block effects ρ_i is, therefore, as follows:

$$(28.24) \qquad Y_{ijk} = \mu_{...} + \rho_i + \alpha_j + \beta_k + (\rho\alpha)_{ij} + (\alpha\beta)_{jk} + \varepsilon_{(ijk)}$$

where:

$\mu_{...}$ is a constant

ρ_i are constants subject to $\Sigma\rho_i = 0$

α_j are constants subject to $\Sigma\alpha_j = 0$

β_k are constants subject to $\Sigma\beta_k = 0$

$(\rho\alpha)_{ij}$ are constants subject to $\sum\limits_{i} (\rho\alpha)_{ij} = 0$ for all j and $\sum\limits_{j} (\rho\alpha)_{ij} = 0$ for all i

$(\alpha\beta)_{jk}$ are constants subject to $\sum\limits_{j} (\alpha\beta)_{jk} = 0$ for all k and $\sum\limits_{k} (\alpha\beta)_{jk} = 0$ for all j

$\varepsilon_{(ijk)}$ are independent $N(0, \sigma^2)$

$i = 1, \ldots, n; j = 1, \ldots, a; k = 1, \ldots, b$

The observations Y_{ijk} for split-plot design model (28.24) have the following properties:

$$(28.25a) \qquad E\{Y_{ijk}\} = \mu_{...} + \rho_i + \alpha_j + \beta_k + (\rho\alpha)_{ij} + (\alpha\beta)_{jk}$$

$$(28.25b) \qquad \sigma^2\{Y_{ijk}\} = \sigma^2$$

Further, all observations are normally distributed, and any two different observations are independent. Thus, all observations have constant variance and are independent.

Block Effects Random. When the blocks are units such as schools, batches, or agricultural fields, they are usually considered to be a random sample from a relevant population. In that case, the split-plot design model is modified as follows:

$$(28.26) \qquad Y_{ijk} = \mu_{...} + \rho_i + \alpha_j + \beta_k + (\rho\alpha)_{ij} + (\alpha\beta)_{jk} + \varepsilon_{(ijk)}$$

where:

$\mu_{...}$ is a constant
ρ_i are independent $N(0, \sigma_\rho^2)$
α_j are constants subject to $\Sigma\alpha_j = 0$
β_k are constants subject to $\Sigma\beta_k = 0$
$(\rho\alpha)_{ij}$ are $N\left(0, \dfrac{a-1}{a}\sigma_{\rho\alpha}^2\right)$ subject to:

$$\sum_j (\rho\alpha)_{ij} = 0 \qquad \text{for all } i$$

$$\sigma\{(\rho\alpha)_{ij}, (\rho\alpha)_{ij'}\} = -\frac{1}{a}\sigma_{\rho\alpha}^2 \qquad \text{for } j \neq j'$$

$(\alpha\beta)_{jk}$ are constants subject to $\sum_j (\alpha\beta)_{jk} = 0$ for all k and $\sum_k (\alpha\beta)_{jk} = 0$ for all j

$\varepsilon_{(ijk)}$ are independent $N(0, \sigma^2)$
ρ_i, $(\rho\alpha)_{ij}$, and $\varepsilon_{(ijk)}$ are pairwise independent
$i = 1, \ldots, n; j = 1, \ldots, a; k = 1, \ldots, b$

The observations Y_{ijk} for split-plot design model (28.26) have the following properties:

$$(28.27a) \qquad E\{Y_{ijk}\} = \mu_{...} + \alpha_j + \beta_k + (\alpha\beta)_{jk}$$

$$(28.27b) \qquad \sigma^2\{Y_{ijk}\} = \sigma_Y^2 = \sigma_\rho^2 + \frac{a-1}{a}\sigma_{\rho\alpha}^2 + \sigma^2$$

$$(28.27c) \qquad \sigma\{Y_{ijk}, Y_{ijk'}\} = \sigma_\rho^2 + \frac{a-1}{a}\sigma_{\rho\alpha}^2 \qquad k \neq k'$$

$$\sigma\{Y_{ijk}, Y_{ij'k'}\} = \sigma_\rho^2 - \frac{1}{a}\sigma_{\rho\alpha}^2 \qquad j \neq j'$$

Thus, all observations Y_{ijk} still have constant variance with split-plot model (28.26). However, in advance of the random trials any two observations from a given block are now correlated. If the two observations are for the same subject, they are more correlated than if they are for different subjects in the same block, which is usually a reasonable property. Observations from different blocks are assumed to be independent by split-plot model (28.26).

Once the blocks have been selected, split-plot model (28.26) assumes that all observations are independent. Thus, it assumes no interference effects for the repeated measures, given the selection of the blocks.

Analysis of Variance and Tests

Table 28.15 contains the analysis of variance for split-plot models (28.24) and (28.26). The sums of squares and degrees of freedom can be obtained directly from rule (27.3) as modified by (27.16) since there are no replications here. The remainder sum of squares associated with the error term turns out to be:

$$(28.28) \qquad SSRem = SSBL.B + SSBL.AB$$

where $SSBL.B$ and $SSBL.AB$ are given by the familiar formulas for three-factor studies with one observation per cell.

The expected mean squares and appropriate test statistics for both split-plot models are given in Table 28.16. The expected mean squares can be readily obtained by means of rule (27.4).

Analysis of Factor Effects

The analysis of factor effects proceeds in an analogous fashion to that for two-factor repeated measures studies with repeated measures on one factor. When AB interactions are not present or are not important, the analysis of factor effects involves factor A level means $\mu_{.j.}$ and factor B level means $\mu_{..k}$. The appropriate mean square to be used in estimating the variance of any estimated contrast is the mean square in the denominator of the corresponding F^* test statistic. The degrees of freedom for the confidence interval multiples in (28.15) need to be modified accordingly.

TABLE 28.15 ANOVA Table for Split-Plot Design Models (28.24) and (28.26)

Source of Variation	SS	df
Blocks	$SSBL = ab \sum_i (\bar{Y}_{i..} - \bar{Y}_{...})^2$	$n - 1$
Factor A	$SSA = bn \sum_j (\bar{Y}_{.j.} - \bar{Y}_{...})^2$	$a - 1$
$BL.A$ interactions	$SSBL.A = b \sum_i \sum_j (\bar{Y}_{ij.} - \bar{Y}_{i..} - \bar{Y}_{.j.} + \bar{Y}_{...})^2$	$(a - 1)(n - 1)$
Factor B	$SSB = an \sum_k (\bar{Y}_{..k} - \bar{Y}_{...})^2$	$(b - 1)$
AB interactions	$SSAB = n \sum_j \sum_k (\bar{Y}_{.jk} - \bar{Y}_{.j.} - \bar{Y}_{..k} + \bar{Y}_{...})^2$	$(a - 1)(b - 1)$
Error	$SSRem = \sum_i \sum_j \sum_k (Y_{ijk} - \bar{Y}_{.jk} - \bar{Y}_{ij.} + \bar{Y}_{.j.})^2$	$a(b - 1)(n - 1)$
Total	$SSTO = \sum_i \sum_j \sum_k (Y_{ijk} - \bar{Y}_{...})^2$	$abn - 1$

TABLE 28.16 Expected Mean Squares and Test Statistics for Split-Plot Models (28.24) and (28.26)

Source of Variation	Mean Square	Expected Mean Square	
		Model (28.24)	Model (28.26)
Blocks	MSBL	$\sigma^2 + \dfrac{ab}{n-1}\sum \rho_i^2$	$\sigma^2 + ab\sigma_\rho^2$
Factor A	MSA	$\sigma^2 + \dfrac{nb}{a-1}\sum \alpha_j^2$	$\sigma^2 + b\sigma_{\rho\alpha}^2 + \dfrac{nb}{a-1}\sum \alpha_j^2$
BL.A interactions	MSBL.A	$\sigma^2 + \dfrac{b}{(n-1)(a-1)}\sum\sum (\rho\alpha)_{ij}^2$	$\sigma^2 + b\sigma_{\rho\alpha}^2$
Factor B	MSB	$\sigma^2 + \dfrac{na}{b-1}\sum \beta_k^2$	$\sigma^2 + \dfrac{na}{b-1}\sum \beta_k^2$
AB interactions	MSAB	$\sigma^2 + \dfrac{n}{(a-1)(b-1)}\sum\sum (\alpha\beta)_{jk}^2$	$\sigma^2 + \dfrac{n}{(a-1)(b-1)}\sum\sum (\alpha\beta)_{jk}^2$
Error	MSRem	σ^2	σ^2

Test for	F^*	
	Model (28.24)	Model (28.26)
Blocks	MSBL/MSRem	MSBL/MSRem
Factor A	MSA/MSRem	MSA/MSBL.A
BL.A interactions	MSBL.A/MSRem	MSBL.A/MSRem
Factor B	MSB/MSRem	MSB/MSRem
AB interactions	MSAB/MSRem	MSAB/MSRem

When interaction effects are present, the analysis of the treatment means $\mu_{\cdot jk}$ will be based on the estimated treatment means $\bar{Y}_{\cdot jk}$. For fixed block effects, the appropriate mean square is *MSRem*. When the block effects are random, the analysis becomes more complex; see, for example, Reference 28.3.

Some Final Comments

1. Split-plot design models containing also a block-factor B interaction term can be readily developed when these interactions are present.

2. A wide variety of split-plot designs has been developed. References 28.4 and 28.5 provide further information about these designs.

CITED REFERENCES

28.1. SAS Institute Inc. *SAS/GRAPH User's Guide*. Version 6 ed. Cary, N.C.: SAS Institute, 1988.

28.2. *MINITAB Reference Manual, Release 7*. State College, Pa.: Minitab, Inc., 1989.

28.3. Winer, B. J. *Statistical Principles in Experimental Design.* 2nd ed. New York: Mc-Graw-Hill Book Co., 1971.

28.4. Steel, R. G. D., and J. H. Torrie. *Principles and Procedures of Statistics.* 2nd ed. New York: McGraw-Hill Book Co., 1980.

28.5. Koch, G. G.; J. D. Elashoff; and I. A. Amara. "Repeated Measurements—Design and Analysis." In *Encyclopedia of Statistical Sciences,* vol. 8, ed. S. Kotz and N. L. Johnson. New York: John Wiley & Sons, 1988, pp. 46–73.

PROBLEMS

28.1. A serious potential problem with repeated measures designs is associated with carry-over effects. Describe some steps that can be taken to minimize this problem.

28.2. In designing a two-factor repeated measures study with repeated measures on one factor, does it matter which of the two factors is included as the repeated measures factor? Explain fully.

28.3. Blood pressure. The relationship between the dose of a drug that increases blood pressure and the actual amount of increase in mean diastolic blood pressure was investigated in a laboratory experiment. Twelve rabbits received in random order six different dose levels of the drug, with a suitable interval between each drug administration. The increase in blood pressure was used as the dependent variable. The data on blood pressure increase follow.

Rabbit i	Dose (j) .1	.3	.5	1.0	1.5	3.0	Rabbit i	Dose (j) .1	.3	.5	1.0	1.5	3.0
1	21	21	23	35	36	48	7	9	12	17	22	33	40
2	19	24	27	36	36	46	8	20	20	30	30	38	41
3	12	25	27	26	33	40	9	18	18	27	31	42	49
4	9	17	18	27	34	39	10	8	12	11	24	26	31
5	7	10	19	25	31	38	11	18	22	25	32	38	38
6	18	26	26	29	39	44	12	17	23	26	28	34	35

a. Obtain the residuals for repeated measures model (28.1) and plot them against the fitted values. Also prepare a normal probability plot of the residuals. What do you conclude about the appropriateness of model (28.1)?

b. Prepare aligned residual dot plots by dose level. Do these plots support the assumption of constancy of the error variance? Discuss.

c. Plot the observations Y_{ij} for each rabbit in the format of Figure 28.1. Does the assumption of no interactions between subjects (rabbits) and treatments appear to be reasonable here?

d. Conduct the Tukey test for additivity, conditional on the rabbits actually selected; use $\alpha = .005$. State the alternatives, decision rule, and conclusion. What is the P-value of the test?

28.4. Refer to **Blood pressure** Problem 28.3. Assume that repeated measures model (28.1) is appropriate.

a. Obtain the analysis of variance table.

b. Test whether or not the mean increase in blood pressure differs for the various dose levels; use significance level $\alpha = .01$. State the alternatives, decision rule, and conclusion. What is the P-value of the test?

c. Analyze the effects of the six dose levels by comparing the means for successive dose levels using the Bonferroni procedure with a 95 percent family confidence coefficient. State your findings and summarize them by a suitable line plot.

d. Based on the estimated efficiency measure (24.14), how effective was the repeated measures design here as compared to a completely randomized design?

28.5. Refer to **Blood pressure** Problems 28.3 and 28.4.

a. Develop a regression model in which the subject effects are represented by $1, -1,$ 0 indicator variables and the dose effect is represented by linear, quadratic, and cubic terms in $x = X - \bar{X}$, where X is the dose level. For instance, the x value for the first dose level ($X = .1$) is $x = .1 - 1.07 = -.97$.

b. Fit the regression model to the data.

c. Obtain the residuals and plot them against the fitted values. Does the model utilized appear to provide a reasonable fit?

d. Test whether or not the cubic effect is required in the model; use $\alpha = .05$. State the alternatives, decision rule, and conclusion. What is the P-value of the test?

28.6. **Grapefruit sales.** A supermarket chain studied the relationship between grapefruit sales and the price at which grapefruits are offered. Three price levels were studied: (1) the chief competitor's price, (2) a price slightly higher than the chief competitor's price, and (3) a price moderately higher than the chief competitor's price. Eight stores of comparable size were randomly selected for the study. Sales data were collected for three one-week periods, with the order of the three price levels randomly assigned for each store. The experiment was conducted during a time period when sales of grapefruits are usually quite stable, and no carry-over effects were anticipated for this product. Data on store sales of grapefruits during the study period follow (data coded).

Store	Price level (j)		
i	1	2	3
1	62.1	61.3	60.8
2	58.2	57.9	55.1
3	51.6	49.2	46.2
4	53.7	51.5	48.3
5	61.4	58.7	56.6
6	58.5	57.2	54.3
7	46.8	43.2	41.5
8	51.2	49.8	47.9

a. Obtain the residuals for repeated measures model (28.1) and plot them against the fitted values. Also prepare a normal probability plot of the residuals. What do you conclude about the appropriateness of model (28.1)?

b. Prepare aligned residual dot plots by price level. Do these plots support the assumption of constancy of the error variance? Discuss.

c. Plot the observations Y_{ij} for each store in the format of Figure 28.1. Does the assumption of no interactions between subjects (stores) and treatments appear to be reasonable here?

d. Conduct the Tukey test for additivity, conditional on the stores actually selected; use $\alpha = .01$. State the alternatives, decision rule, and conclusion. What is the P-value of the test?

28.7. Refer to **Grapefruit sales** Problem 28.6. Assume that repeated measures model (28.1) is appropriate.

a. Obtain the analysis of variance table.

b. Test whether or not the mean sales of grapefruits differ for the three price levels; use significance level $\alpha = .05$. State the alternatives, decision rule, and conclusion. What is the P-value of the test?

c. Analyze the effects of the three price levels by estimating all pairwise comparisons of the price level means. Use the most efficient multiple comparison procedure with a 95 percent family confidence coefficient. State your findings and summarize them by a suitable line plot.

d. Based on the estimated efficiency measure (24.14), how effective was the repeated measures design compared to a completely randomized design?

28.8. Refer to **Blood pressure** Problem 28.3. A consultant is concerned about the validity of the model assumptions and suggests that the study should be analyzed by means of the Friedman test. Rank the data within each rabbit and perform the Friedman test; use $\alpha = .01$. State the alternatives, decision rule, and conclusion. Comment on the consultant's concern here.

28.9. Refer to **Grapefruit sales** Problem 28.6. It has been suggested that the nonparametric Friedman test should be used here. Rank the data within each store and perform the Friedman test; use $\alpha = .05$. State the alternatives, decision rule, and conclusion. Is your conclusion the same as that obtained in Problem 28.7b?

28.10. Truth in advertising. A consumer research organization showed five different advertisements to 10 subjects and asked each to rank them in order of truthfulness. A rank of 1 denotes the most truthful. The results were:

Subject i	Advertisement (j)					Subject i	Advertisement (j)				
	A	B	C	D	E		A	B	C	D	E
1	3	1	2	5	4	6	4	2	1	3	5
2	4	2	1	3	5	7	4	1	2	3	5
3	4	2	3	1	5	8	5	1	3	2	4
4	3	1	2	5	4	9	4	2	3	1	5
5	4	1	2	5	3	10	5	1	2	3	4

a. Do the subjects perceive the five advertisements as having equal truthfulness? Conduct the Friedman test using a level of significance of $\alpha = .05$. State the alternatives, decision rule, and conclusion. What is the P-value of the test?

b. Use the multiple pairwise testing procedure (25.5) to group the five different advertisements according to mean perceived truthfulness; employ family significance level $\alpha = .10$. Summarize your findings.

c. Obtain the coefficient of concordance and interpret this measure.

28.11. Calculator efficiency. A computer company, to test the efficiency of its new programmable calculator, selected at random six engineers who were proficient in the use of both this calculator and an earlier model and asked them to work out two problems on both calculators. One of the problems was statistical in nature, the other was an engineering problem. The order of the four calculations was randomized independently for each engineer. The length of time (in minutes) required to solve each problem was observed. The results follow (type of problem is factor A and type of calculator model is factor B):

Engineer i	$j = 1$ Statistical Problem		$j = 2$ Engineering Problem	
	$k = 1$ New Model	$k = 2$ Earlier Model	$k = 1$ New Model	$k = 2$ Earlier Model
1 Jones	3.1	7.5	2.5	5.1
2 Williams	3.8	8.1	2.8	5.3
3 Adams	3.0	7.6	2.0	4.9
4 Dixon	3.4	7.8	2.7	5.5
5 Erickson	3.3	6.9	2.5	5.4
6 Maynes	3.6	7.8	2.4	4.8

a. Obtain the residuals for repeated measures model (28.10) and plot them against the fitted values. Also prepare a normal probability plot of the residuals. What do you conclude about the appropriateness of model (28.10)?

b. Prepare aligned residual dot plots by treatment. Do these plots support the assumption of constancy of the error variance? Discuss.

c. Plot the observations Y_{ijk} for each engineer in the format of Figure 28.1, ignoring the factorial nature of the treatments. Does the assumption of no interactions between subjects (engineers) and treatments appear to be reasonable here?

d. Conduct the Tukey test for additivity, conditional on the engineers actually selected; use $\alpha = .01$. State the alternatives, decision rule, and conclusion. What is the P-value of the test?

28.12. Refer to **Calculator efficiency** Problem 28.11. Assume that repeated measures model (28.10) is appropriate.

a. Obtain the analysis of variance table.

b. Plot the estimated treatment means in the format of Figure 28.4. Does it appear that interaction effects are present?

c. Test whether or not the two factors interact; use $\alpha = .01$. State the alternatives, decision rule, and conclusion. What is the P-value of the test?

d. It is desired to study the nature of the interaction effects by considering the three comparisons:

$$D_1 = \mu_{.12} - \mu_{.11} \qquad L_1 = D_2 - D_1$$

$$D_2 = \mu_{.22} - \mu_{.21}$$

Obtain confidence intervals for these comparisons; use the Bonferroni procedure with a 95 percent family confidence coefficient. State your findings.

e. Test whether or not subject effects are present; use $\alpha = .01$. State the alternatives, decision rule, and conclusion. What is the P-value of the test?

28.13. Migraine headaches. Two experimental pain killer drugs for relief of migraine headaches were studied at a major medical center. Ten persistent migraine sufferers were randomly selected for a pilot study and received in random order each of the four treatment combinations, with a suitable interval between drug administrations. The decrease in pain intensity was used as the dependent variable. The four treatments used in the study are defined as follows: $A_1 B_1$ = low dose of both drugs; $A_1 B_2$ = low dose of drug A, high dose of drug B; $A_2 B_1$ = high dose of drug A, low dose of drug B; $A_2 B_2$ = high dose of both drugs. The data on reduction in pain intensity follow (the higher the score, the greater the reduction in pain).

Person	A_1 ($j = 1$)		A_2 ($j = 2$)	
i	B_1 ($k = 1$)	B_2 ($k = 2$)	B_1 ($k = 1$)	B_2 ($k = 2$)
1	1.6	3.4	2.7	4.3
2	2.3	5.1	4.2	6.5
3	4.2	5.3	4.6	6.0
4	7.1	8.9	7.8	9.4
5	3.5	3.7	3.4	3.9
6	5.8	6.5	6.2	7.1
7	4.9	5.6	5.4	6.2
8	6.0	7.2	6.3	7.3
9	1.2	1.4	1.3	1.7
10	2.7	3.0	3.0	3.1

a. Obtain the residuals for repeated measures model (28.10) and plot them against the fitted values. Also prepare a normal probability plot of the residuals. What do you conclude about the appropriateness of model (28.10)?

b. Prepare aligned residual dot plots by treatment. Do these plots support the assumption of constancy of the error variance? Discuss.

c. Plot the observations Y_{ijk} for each person in the format of Figure 28.1, ignoring the factorial nature of the treatments. Does the assumption of no interactions between subjects (persons) and treatments appear to be reasonable here?

d. Conduct the Tukey test for additivity, conditional on the subjects actually selected; use $\alpha = .005$. State the alternatives, decision rule, and conclusion. What is the P-value of the test?

28.14. Refer to **Migraine headaches** Problem 28.13. Assume that repeated measures model (28.10) is appropriate.

a. Obtain the analysis of variance table.

b. Plot the estimated treatment means in the format of Figure 28.4. Does it appear that interaction effects are present? That main effects are present?

c. Test whether or not the two factors interact; use $\alpha = .005$. State the alternatives, decision rule, and conclusion. What is the P-value of the test?

d. Test separately whether or not factor A and factor B main effects are present; use $\alpha = .005$ for each test. State the alternatives, decision rule, and conclusion for each test. What is the P-value for each test?

e. Estimate the following comparisons by means of confidence intervals:

$$D_1 = \mu_{.21} - \mu_{.11} \qquad D_3 = \mu_{.21} - \mu_{.12}$$

$$D_2 = \mu_{.12} - \mu_{.11} \qquad D_4 = \mu_{.22} - \mu_{.11}$$

Use the Bonferroni procedure and family confidence coefficient of .95. Summarize your findings.

28.15. **Incentive stimulus.** Refer to the example in Section 28.4 about the effects of two types of incentives (factor A) on a person's ability to solve two types of problems (factor B); the repeated measures design is illustrated in Table 28.10. Twelve persons were randomly selected and assigned in equal numbers to the two incentive groups. The order of the two types of problems was then randomized independently for each person. The problem-solving ability scores follow (the higher the score, the greater the ability to solve problems).

| Incentive Stimulus | Subject | Problem Type | |
		Abstract $(k = 1)$	Concrete $(k = 2)$
	$i = 1$	10	18
	$i = 2$	14	19
	$i = 3$	17	18
$j = 1$	$i = 4$	8	12
	$i = 5$	12	14
	$i = 6$	15	20
	$i = 1$	16	25
	$i = 2$	19	22
	$i = 3$	22	27
$j = 2$	$i = 4$	20	23
	$i = 5$	24	29
	$i = 6$	21	22

a. Obtain the residuals for repeated measures model (28.16) and plot them against the fitted values. Also prepare a normal probability plot of the residuals. What do you conclude about the appropriateness of model (28.16)?

b. Plot the problem-solving ability scores by incentive stimulus and problem type, in the format of Figure 28.5. What do you conclude about the appropriateness of model (28.16)? Discuss.

28.16. Refer to **Incentive stimulus** Problem 28.15. Assume that repeated measures model (28.16) is appropriate.

a. Obtain the analysis of variance table.

b. Plot the estimated treatment means in the format of Figure 28.4. Does it appear that interaction effects are present? That main effects are present?

c. Test whether or not the two factors interact; use $\alpha = .05$. State the alternatives, decision rule, and conclusion. What is the P-value of the test?

d. Test separately whether or not factor A and factor B main effects are present; use $\alpha = .05$ for each test. State the alternatives, decision rule, and conclusion for each test. What is the P-value for each test?

e. The following comparisons are of interest:

$$D_1 = \mu_{.2.} - \mu_{.1.} \qquad D_2 = \mu_{..2} - \mu_{..1}$$

Estimate these comparisons by means of confidence intervals. Use the Bonferroni procedure with a 90 percent family confidence coefficient. State your findings.

28.17. Store displays. An experimental study was conducted to examine the effects of two different store displays for a household product (factor A) on sales in four successive time periods (factor B). Eight stores were randomly selected, and four were assigned at random to each display. The sales data (coded) follow.

Type of Display	Store	Time Period			
		$k = 1$	$k = 2$	$k = 3$	$k = 4$
	$i = 1$	956	953	938	1,049
$j = 1$	$i = 2$	1,008	1,032	1,025	1,123
	$i = 3$	350	352	338	438
	$i = 4$	412	449	385	532
	$i = 1$	769	766	739	859
$j = 2$	$i = 2$	880	875	860	915
	$i = 3$	176	185	168	280
	$i = 4$	209	223	217	301

a. Obtain the residuals for repeated measures model (28.16) and plot them against the fitted values. Also prepare a normal probability plot of the residuals. What do you conclude about the appropriateness of model (28.16)?

b. Plot the sales data by type of display and time period, in the format of Figure 28.5. What do you conclude about the appropriateness of model (28.16)? Discuss.

28.18. Refer to **Store displays** Problem 28.17. The experimenter wished to explore further the appropriateness of repeated measures model (28.16).

a. Conduct a formal test of the constancy of the between-subjects variance σ_ρ^2. Use (28.22) and perform the Hartley test, with $\alpha = .01$. State the alternatives, decision rule, and conclusion.

b. Decompose the remainder sum of squares $SSB.S(A)$ into components using (28.23), and perform the Hartley test for the constancy of the error variance σ^2 for the different factor A levels; use $\alpha = .01$. State the alternatives, decision rule, and conclusion.

28.19. Refer to **Store displays** Problem 28.17. Assume that repeated measures model (28.16) is appropriate.

a. Obtain the analysis of variance table.

b. Plot the estimated treatment means in the format of Figure 28.4. Does it appear that interaction effects are present? That main effects are present?

c. Test whether or not the two factors interact; use $\alpha = .025$. State the alternatives, decision rule, and conclusion. What is the P-value for the test?

d. Test separately whether or not display and time main effects are present; use $\alpha = .025$ for each test. State the alternatives, decision rule, and conclusion for each test. What is the P-value for each test?

e. To study the nature of the factor A and factor B main effects, estimate the following pairwise comparisons:

$$D_1 = \mu_{.1.} - \mu_{.2.} \qquad D_3 = \mu_{..2} - \mu_{..3}$$

$$D_2 = \mu_{..1} - \mu_{..2} \qquad D_4 = \mu_{..3} - \mu_{..4}$$

Use the Bonferroni procedure with a 90 percent family confidence coefficient. State your findings.

28.20. **Wheat yield.** Refer to Example 3 on page 1068 about the effects of two varieties of wheat (factor A) and two fertilizers (factor B) on yield. The yield data for five fields (blocks) that were randomly selected follow.

		Fertilizer	
Field	Wheat Variety	$k = 1$	$k = 2$
$i = 1$	$j = 1$	43	48
	$j = 2$	63	70
$i = 2$	$j = 1$	40	43
	$j = 2$	52	53
$i = 3$	$j = 1$	31	36
	$j = 2$	45	48
$i = 4$	$j = 1$	27	30
	$j = 2$	47	51
$i = 5$	$j = 1$	36	39
	$j = 2$	54	57

a. Obtain the residuals for split-plot model (28.26) and plot them against the fitted values. Also prepare a normal probability plot of the residuals. What do you conclude about the appropriateness of model (28.26)?

b. Plot the wheat yields by variety and fertilizer type in the format of Figure 28.5. What do you conclude about the appropriateness of model (28.26)? Discuss.

28.21. Refer to **Wheat yield** Problem 28.20. Assume that split-plot model (28.26) is appropriate.
a. Obtain the analysis of variance table.
b. Plot the estimated treatment means in the format of Figure 28.4. Does it appear that interaction effects are present? That main effects are present?
c. Test whether or not the two factors interact; use $\alpha = .05$. State the alternatives, decision rule, and conclusion. What is the P-value for the test?
d. Test separately whether or not factor A and factor B main effects are present; use $\alpha = .05$. State the alternatives, decision rule, and conclusion for each test. What is the P-value for each test?
e. To study the nature of the factor A and factor B main effects, estimate the following pairwise comparisons:

$$D_1 = \mu_{.1.} - \mu_{.2.} \qquad D_2 = \mu_{..1} - \mu_{..2}$$

Use the Bonferroni procedure with a 90 percent family confidence coefficient. State your findings.

EXERCISES

28.22. Derive the total sum of squares breakdown in (28.5).

28.23. Refer to repeated measures model (28.10).
  a. Use rule (27.3) with modification (27.16) to obtain the definitional sums of squares formulas in Table 28.8b and the associated degrees of freedom in Table 28.8a.
  b. Use rule (27.4) to obtain the expected mean squares in Table 28.8a.

28.24. Refer to repeated measures model (28.16).
  a. Use rule (27.3) with modification (27.16) to obtain the definitional sums of squares formulas in Table 28.11 and the associated degrees of freedom.
  b. Use rule (27.4) to obtain the expected mean squares in Table 28.12.

28.25. Refer to split-plot model (28.26).
  a. Use rule (27.3) with modification (27.16) to obtain the definitional sums of squares formulas in Table 28.15 and the associated degrees of freedom.
  b. Use rule (27.4) to obtain the expected mean squares in Table 28.16.

PROJECTS

28.26. Refer to **Blood pressure** Problem 28.3. Obtain the estimated within-subjects variance-covariance matrix using (28.8). Are the estimated variances and covariances of the same orders of magnitude? Is the compound symmetry assumption reasonable here?

28.27. Refer to **Grapefruit sales** Problem 28.6. Obtain the estimated within-subjects variance-covariance matrix using (28.8). Are the variances and covariances roughly of the same orders of magnitude? Is the compound symmetry assumption reasonably satisfied here?

28.28. Refer to the **Drug effect experiment** data set. Consider only Part I of the study and observation unit 1 for each drug dosage level; i.e., include only observations for which variable 2 equals 1 and variable 6 equals 1. Treat the 12 rats as subjects and ignore the classification of the rats into the three initial lever press rate groups. Assume that the subjects (rats) have random effects and the treatments (dosage levels) have fixed effects.
  a. State the additive repeated measures model for this study.
  b. Obtain the residuals and plot them against the fitted values. Also prepare a normal probability plot of the residuals. What do you conclude about the appropriateness of the model employed?
  c. Plot the responses for each rat in the format of Figure 28.1. Does the assumption of no interactions between subjects and treatments appear to be appropriate?

28.29. Refer to the **Drug effect experiment** data set and Project 28.28.
  a. Obtain the analysis of variance table.
  b. Test whether or not the drug dosage level affects the mean lever press rate; use $\alpha = .05$. State the alternatives, decision rule, and conclusion. What is the P-value of the test?
  c. Analyze the effects of the four dosage levels by comparing the mean responses for

each pair of successive dosage levels; use the Bonferroni procedure with a 90 percent family confidence coefficient. State your findings.

d. Fit a regression model in which the subject effects are represented by $1, -1, 0$ indicator variables and the dosage effect is represented by linear and quadratic terms in $x = X - \bar{X}$, where X is the dosage level. Assume that there are no interactions between subjects and treatments.

e. Obtain the residuals and plot them against the fitted values. Does the regression model appear to provide a good fit? Discuss.

f. Test whether or not the quadratic term can be dropped from the regression model; use $\alpha = .01$. State the alternatives, decision rule, and conclusion.

28.30. Refer to the **Drug effect experiment** data set. Consider the combined study. Assume that subjects (rats) and observation units have random effects, and that factor A (initial lever press rate), factor B (dosage level), and factor C (reinforcement schedule) have fixed effects. Also assume that there are no interactions between subjects and treatments.

a. Use rule (27.1) with modification (27.15) to develop the model for this experiment.

b. Use rule (27.3) with modification (27.16) to obtain the definitional sums of squares formulas and the associated degrees of freedom.

c. Use rule (27.4) to obtain the expected mean squares.

28.31. Refer to the **Drug effect experiment** data set and Project 28.30. Obtain the residuals and plot them against the fitted values. Also prepare a normal probability plot of the residuals. What do you conclude about the appropriateness of the model?

28.32. Refer to the **Drug effect experiment** data set and Projects 28.30 and 28.31. Assume that the model in Project 28.30a is appropriate.

a. Obtain the analysis of variance table.

b. Test whether or not ABC interactions are present; use $\alpha = .01$. State the alternatives, decision rule, and conclusion. What is the P-value of the test?

c. For each reinforcement schedule, plot the estimated treatment means against dosage level with different curves for the three initial lever press rate groups, in the format of Figure 22.6. Examine your plots for the nature of the interaction effects and report your findings.

28.33. Consider a repeated measures design study with $n = 3$ and $r = 3$, where each subject ranks all treatments (with no ties allowed).

a. Develop the exact sampling distribution of X_F^2 when H_0 holds. (*Hint:* All ranking permutations for a subject are equally likely under H_0 and all subjects are assumed to act independently.)

b. How does the 90th percentile of the exact sampling distribution obtained in part (a) compare with $\chi^2(.90; 2)$? What is the implication of this?

Chapter 29

Latin Square and Related Designs

In this chapter we consider latin square designs, which employ two blocking variables to reduce experimental errors, and several related designs.

29.1 BASIC ELEMENTS

Complete and Incomplete Block Designs

We saw in Section 25.6 that two blocking variables can be used simultaneously in randomized complete block designs to eliminate from the experimental error the variation associated with each of the blocking variables. Thus, the blocking variables might be age of subject and income of subject, with a block containing subjects in a given age and income group.

The full use of two blocking variables in a complete block design may at times have the disadvantage of requiring too many experimental units. For instance, if the age and income variables in our earlier illustration have six classes each, 36 blocks would be required. If six treatments were to be studied, 216 subjects would be needed for the experiment. Cost considerations may not permit the use of this many experimental units, yet precision and range of validity considerations may require the simultaneous use of two blocking variables, each with six classes, in order to reduce the experimental error variance sufficiently and to have a reasonable variety of experimental subjects. In this type of situation, an *incomplete block design* may be helpful. In such a design, all 36 blocks in our example would still be used, but now each block would not contain all six treatments.

Latin Square Designs

Taking incomplete block designs to the extreme in our example, given the employment of 36 blocks, the number of experimental units is minimized if only one treatment is run in each block. This extreme case, where each block contains only one treatment, is the type of situation for which a latin square design is appropriate. Table 29.1 provides an illustration of the difference between complete and incomplete block designs for the example considered. Column 1 shows the complete block design for this case, while columns 2 and 3 illustrate incomplete block designs, with three treatments and one treatment in each block, respectively.

There is another reason besides economy, why a latin square design with only one treatment per block is used, namely, that blocks sometimes cannot contain more than one treatment. Consider the repeated measures design discussed in Section 28.2 where each subject receives every treatment. It was stressed there that the order of treatments be randomized in case interference effects between the different treatments were to be present. If indeed interference effects due to the order position of the treatments are anticipated, it may be desirable to use the order position as another blocking variable. Thus, "subject" would be one blocking variable and "order position of treatment" a second blocking variable. Blocks would then be defined as follows for a study involving six treatments:

Block 1: Subject 1, position 1

Block 2: Subject 1, position 2

.   .
.   .
.   .

Block 6: Subject 1, position 6

Block 7: Subject 2, position 1

etc.   etc.

Notice that the blocks so defined can contain only one treatment, since the order position refers to the place of a single treatment in the sequence of treatments for a subject.

Description of Latin Square Designs

Let A, B, C represent three treatments; it is conventional with latin square designs to use Latin letters for the treatments. Suppose that day of week (Monday, Tuesday, Wednesday) and operator (1, 2, 3) are to be used as blocking variables. A latin square design might then be shown as follows:

	Operator		
Day	1	2	3
Monday	B	A	C
Tuesday	A	C	B
Wednesday	C	B	A

Operator 1 would run treatment B on Monday, treatment A on Tuesday, and treatment C on Wednesday, and so on for the other operators. Note that each operator runs each treatment, and that all treatments are run on each day.

TABLE 29.1 Complete and Incomplete Block Designs

Block Description	(1) Complete Block Design	(2) Incomplete Block Design (three treatments per block)	(3) Incomplete Block Design (one treatment per block)
Age under 25, income under $10,000	$T_1, T_2, T_3, T_4, T_5, T_6$	T_1, T_3, T_5	T_2
Age under 25, income $10,000–$19,999	$T_1, T_2, T_3, T_4, T_5, T_6$	T_2, T_4, T_6	T_5
. . .	. . .	. . .	. . .
Age 25–34, income under $10,000	$T_1, T_2, T_3, T_4, T_5, T_6$	T_2, T_4, T_5	T_3
. . .	. . .	. . .	. . .
Age 35–44, income under $10,000	$T_1, T_2, T_3, T_4, T_5, T_6$	T_3, T_4, T_6	T_2
etc.	etc.	etc.	etc.

A latin square design thus has the following features:

1. There are r treatments.
2. There are two blocking variables, each containing r classes.
3. Each row and each column in the design square contains all treatments; that is, each class of each blocking variable constitutes a replication.

Advantages and Disadvantages of Latin Square Designs

Advantages of a latin square design include:

1. The use of two blocking variables often permits greater reductions in the variability of experimental errors than can be obtained with either blocking variable alone.
2. Treatment effects can be studied from a small-scale experiment. This is particularly helpful in preliminary or pilot studies.
3. It is often helpful in repeated measures experiments to take into account the order effect of treatments by means of a latin square design.

Disadvantages of a latin square design are:

1. The number of classes of each blocking variable must equal the number of treatments. This leads to a very small number of degrees of freedom for experimental error when only a few treatments are studied. On the other hand, when many treatments are studied, the degrees of freedom for experimental error may be larger than necessary.
2. The assumptions of the model are restrictive (e.g., that there are no interactions between either blocking variable and treatments, and also none between the two blocking variables).
3. The two blocking variables cannot have different numbers of classes.
4. The randomization required is somewhat more complex than that for earlier designs considered.

Because of the limitations on the degrees of freedom for experimental error just described, latin squares are rarely used when more than eight treatments are being investigated. For the same reason, when there are only a few treatments, say, four or less, additional replications are usually required when a latin square design is employed.

Randomization of Latin Square Design

There exist many latin squares for a given number of treatments. Suppose that the number of treatments is $r = 3$; four possible latin square designs are (we omit the row and column blocking variable labels):

1			2			3			4		
A	B	C	A	C	B	B	A	C	C	B	A
B	C	A	B	A	C	C	B	A	A	C	B
C	A	B	C	B	A	A	C	B	B	A	C

For $r = 3$, there are altogether 12 different possible arrangements. This number increases rapidly as the number of treatments gets larger; for $r = 5$, there are 161,280 possible arrangements.

The objective of randomization is to select one of all possible latin squares for the given number of treatments r, such that each square has an equal probability of being selected. Clearly, it is not generally feasible to list all possible latin squares so that one can be selected at random.

Instead, we utilize *standard latin squares,* which are latin squares in which the elements of the first row and the first column are arranged alphabetically. The earlier latin square 1 is a standard latin square. Table A.13 contains all the standard squares for $r = 3$ and 4, and a single selected standard square for $r = 5, 6, 7, 8,$ and 9. The randomization procedure usually employed is as follows:

1. For $r = 3$, independently arrange the rows and columns at random.
2. For $r = 4$, select one of the standard squares at random. Then, independently arrange its rows and columns at random.
3. For $r = 5$ and higher, independently arrange the rows, columns, and treatments of the given standard square at random.

It can be shown that this procedure selects one latin square at random from all possible squares for $r = 3$ and 4. For $r = 5$ or higher, the randomization procedure is not based on all possible latin squares, but rather on very large and suitable subsets thereof.

Example 1. In our earlier illustration, the blocking variables were day (Monday, Tuesday, Wednesday) and operator (1, 2, 3). Thus, we can show the borders of the square as follows:

	Operator		
Day	1	2	3
Monday			
Tuesday			
Wednesday			

The standard latin square for $r = 3$ is:

Step 1	A	B	C
	B	C	A
	C	A	B

We obtain now a random permutation of 3 to rearrange the rows. As we explained in Chapter 2, this can be done by obtaining three 2-digit random numbers from a random number generator or from a table of random numbers. Two digits are employed to reduce the chance of ties. Suppose that the three 2-digit numbers are as follows:

Order number:	1	2	3
Random number:	87	34	63

We now rearrange the random numbers in ascending order, carrying along the original order numbers:

Random number:	34	63	87
Order number:	2	3	1

Thus, we have obtained the random permutation 2, 3, 1.

In accordance with this permutation, row 2 in the standard square comes first, row 3 second, and row 1 third. We obtain:

	Original Row Number			
	2	*B*	*C*	*A*
Step 2	3	*C*	*A*	*B*
	1	*A*	*B*	*C*

We now obtain independently another random permutation of 3 to rearrange the columns. Suppose it is: 2, 1, 3. Column 2 in the square of step 2 now becomes the first column, and so on. We obtain:

	Step 2 *column number:*	2	1	3
		C	*B*	*A*
Step 3		*A*	*C*	*B*
		B	*A*	*C*

The selected design therefore is:

Day	*Operator*		
	1	2	3
Monday	*C*	*B*	*A*
Tuesday	*A*	*C*	*B*
Wednesday	*B*	*A*	*C*

Example 2.   Consider an experiment about the effects of five different types of background music (A, B, C, D, E) on the productivity of bank tellers. A given type of music is played for one day and the productivity is observed. Day of the week and week of the experimental period are to be the two blocking variables. The borders

of the latin square design therefore are:

	Day				
Week	*M*	*T*	*W*	*Th*	*F*
1					
2					
3					
4					
5					

We start with the standard latin square of Table A.13 for $r = 5$:

Step 1

A	B	C	D	E
B	A	E	C	D
C	D	A	E	B
D	E	B	A	C
E	C	D	B	A

Next we permute the rows randomly, using a random permutation of 5, say: 4, 2, 3, 1, 5. We obtain then:

Step 2

D	E	B	A	C
B	A	E	C	D
C	D	A	E	B
A	B	C	D	E
E	C	D	B	A

We now permute the columns randomly, using an independent random permutation of 5, say: 2, 4, 1, 5, 3. We obtain:

Step 3

E	A	D	C	B
A	C	B	D	E
D	E	C	B	A
B	D	A	E	C
C	B	E	A	D

Finally, we need to permute the treatment labels randomly. Note that the treatment designations A, B, C, D, and E remain fixed. We simply wish to randomize the cells in which the specific treatments occur, given the pattern obtained in step 3. Assume the correspondence:

1	2	3	4	5
A	B	C	D	E

and that an independent random selection of a permutation of 5 yields: 3, 5, 2, 1, 4. We obtain then:

Present cell designation:	A	B	C	D	E
New cell designation:	C	E	B	A	D

Thus, every cell designated A in step 3 now is designated C, and so on. Hence, the latin square design to be used is:

	Day				
Week	*M*	*T*	*W*	*Th*	*F*
1	D	C	A	B	E
2	C	B	E	A	D
3	A	D	B	E	C
4	E	A	C	D	B
5	B	E	D	C	A

Note

For $r = 3$ and 4, it would be sufficient to randomize all r rows and the last $r - 1$ columns. It is equally good to randomize all rows and columns, the procedure suggested here.

Example

We mentioned earlier an experiment about the effects of different types of background music on the productivity of bank tellers. The treatments were defined as various combinations of tempo of music (slow, medium, fast) and style of music (instrumental and vocal, instrumental only). The treatments and Latin letter designations were as follows:

Treatment	*Latin Letter Designation*	*Tempo and Style of Music*
1	A	Slow, instrumental and vocal
2	B	Medium, instrumental and vocal
3	C	Fast, instrumental and vocal
4	D	Medium, instrumental only
5	E	Fast, instrumental only

Table 29.2 contains the results of this experiment. The treatment in each cell is shown in parentheses. Note that in this study, the experimental unit is a working day for the crew of bank tellers, and that the productivity data pertain to the performance of the entire crew. Let Y_{ijk} denote the observation in the cell defined by the ith class of the row blocking variable and the jth class of the column blocking variable. The subscript k indicates the treatment assigned to this cell by the particular latin square design employed. Thus, $Y_{123} = 17$ is the productivity on Tuesday of the first week, and Table 29.2 indicates that the type of music on that day was C.

TABLE 29.2 Latin Square Design Background Music Study—Experiment Results (productivity of crew—data coded)

Week	M	T	W	Th	F	Total
			Day			
1	18 (D)	17 (C)	14 (A)	21 (B)	17 (E)	$Y_{1..} = 87$
2	13 (C)	34 (B)	21 (E)	16 (A)	15 (D)	$Y_{2..} = 99$
3	7 (A)	29 (D)	32 (B)	27 (E)	13 (C)	$Y_{3..} = 108$
4	17 (E)	13 (A)	24 (C)	31 (D)	25 (B)	$Y_{4..} = 110$
5	21 (B)	26 (E)	26 (D)	31 (C)	7 (A)	$Y_{5..} = 111$
Total	$Y_{.1.} = 76$	$Y_{.2.} = 119$	$Y_{.3.} = 117$	$Y_{.4.} = 126$	$Y_{.5.} = 77$	$Y_{...} = 515$

$$Y_{..1} = 7 + 13 + 14 + 16 + 7 = 57 \qquad \bar{Y}_{..1} = 11.4$$
$$Y_{..2} = 21 + 34 + 32 + 21 + 25 = 133 \qquad \bar{Y}_{..2} = 26.6$$
$$Y_{..3} = 13 + 17 + 24 + 31 + 13 = 98 \qquad \bar{Y}_{..3} = 19.6$$
$$Y_{..4} = 18 + 29 + 26 + 31 + 15 = 119 \qquad \bar{Y}_{..4} = 23.8$$
$$Y_{..5} = 17 + 26 + 21 + 27 + 17 = 108 \qquad \bar{Y}_{..5} = 21.6$$

The subscript k in Y_{ijk} is actually redundant for a latin square design because the row and cell designation (i, j) determines the treatment for the particular latin square employed. However, we continue to use all three subscripts for ease of identification.

We shall analyze the results of this study in subsequent sections.

29.2 LATIN SQUARE MODEL

A latin square design model involves the main effect of the row blocking variable, denoted by ρ_i, the main effect of the column blocking variable, denoted by κ_j, and the treatment main effect, denoted by τ_k. It is assumed that no interactions exist between these three variables. Thus, the model employed is an additive one. For the case of fixed treatment and block effects, the model is:

(29.1)
$$Y_{ijk} = \mu_{...} + \rho_i + \kappa_j + \tau_k + \varepsilon_{(ijk)}$$

where:

$\mu_{...}$ is a constant

ρ_i, κ_j, τ_k are constants subject to the restrictions $\Sigma\rho_i = \Sigma\kappa_j = \Sigma\tau_k = 0$

$\varepsilon_{(ijk)}$ are independent $N(0, \sigma^2)$

$i = 1, \ldots, r; j = 1, \ldots, r; k = 1, \ldots, r$

Note again that the number of classes for each of the two blocking variables is the same as the number of treatments, and that the total number of observations is r^2. Also note the notation for the error term, $\varepsilon_{(ijk)}$, because there are no cell replications in the latin square design.

Comments

1. Sometimes, one or both of the blocking variable effects are viewed as random, as when the blocking variables refer to subjects, observers, machines, etc. We shall consider the case of random blocking variable effects in Section 29.11.

2. If the treatment effects are random, the only change in model (29.1) would be that the τ_k now are independent $N(0, \sigma_\tau^2)$ and are independent of the $\varepsilon_{(ijk)}$.

29.3 ANALYSIS OF VARIANCE AND TESTS

Notation

We shall employ the usual notation for row, column, and treatment totals and means:

$$(29.2a) \qquad Y_{i..} = \sum_j Y_{ijk} \qquad \bar{Y}_{i..} = \frac{Y_{i..}}{r}$$

$$(29.2b) \qquad Y_{.j.} = \sum_i Y_{ijk} \qquad \bar{Y}_{.j.} = \frac{Y_{.j.}}{r}$$

$$(29.2c) \qquad Y_{..k} = \sum_{i,j} Y_{ijk} \qquad \bar{Y}_{..k} = \frac{Y_{..k}}{r}$$

The overall total and mean are denoted as usual by:

$$(29.2d) \qquad Y_{...} = \sum_i \sum_j Y_{ijk} \qquad \bar{Y}_{...} = \frac{Y_{...}}{r^2}$$

Note the redundancy of any one of the three subscripts, arising from the fact that the treatment is uniquely determined by the row and column specifications for the latin square utilized. The various totals for the background music example are shown in Table 29.2.

Fitting of Model

The least squares estimators of the parameters in latin square model (29.1) are:

	Parameter	Estimator
(29.3a)	$\mu_{...}$	$\hat{\mu}_{...} = \bar{Y}_{...}$
(29.3b)	ρ_i	$\hat{\rho}_i = \bar{Y}_{i..} - \bar{Y}_{...}$
(29.3c)	κ_j	$\hat{\kappa}_j = \bar{Y}_{.j.} - \bar{Y}_{...}$
(29.3d)	τ_k	$\hat{\tau}_k = \bar{Y}_{..k} - \bar{Y}_{...}$

The fitted values therefore are:

(29.4)
$$\hat{Y}_{ijk} = \bar{Y}_{i..} + \bar{Y}_{.j.} + \bar{Y}_{..k} - 2\bar{Y}_{...}$$

and the residuals are:

(29.5)
$$e_{ijk} = Y_{ijk} - \hat{Y}_{ijk} = Y_{ijk} - \bar{Y}_{i..} - \bar{Y}_{.j.} - \bar{Y}_{..k} + 2\bar{Y}_{...}$$

Analysis of Variance

Table 29.3 presents the ANOVA table for latin square model (29.1). The sums of squares can be obtained by rule (27.3) and modification (27.16), remembering in step 3 that one subscript is redundant. The sum of squares associated with the model error term must be obtained as a remainder because there are no replications in a cell in a latin square design. The definitional forms of the sums of squares are as follows:

(29.6a)
$$SSTO = \sum_i \sum_j (Y_{ijk} - \bar{Y}_{...})^2$$

(29.6b)
$$SSROW = r \sum_i (\bar{Y}_{i..} - \bar{Y}_{...})^2$$

(29.6c)
$$SSCOL = r \sum_j (\bar{Y}_{.j.} - \bar{Y}_{...})^2$$

(29.6d)
$$SSTR = r \sum_k (\bar{Y}_{..k} - \bar{Y}_{...})^2$$

(29.6e)
$$SSRem = \sum_i \sum_j (Y_{ijk} - \bar{Y}_{i..} - \bar{Y}_{.j.} - \bar{Y}_{..k} + 2\bar{Y}_{...})^2$$

SSROW is the *row sum of squares*. The more the row means $\bar{Y}_{i..}$ differ, the larger is *SSROW*. Similarly, *SSCOL* is the *column sum of squares* and measures the variability of the column means $\bar{Y}_{.j.}$. *SSTR* and *SSRem* denote, as usual, the treatment and remainder sums of squares, respectively.

The degrees of freedom in Table 29.3 can be understood as follows. There are r^2 observations, and hence *SSTO* has $r^2 - 1$ degrees of freedom associated with it. Since there are r classes for the row and column blocking variables each, and also r treatments, each of the corresponding sums of squares has $r - 1$ degrees of freedom associated with it. The number of degrees of freedom associated with *SSRem* is the remainder, namely, $(r^2 - 1) - 3(r - 1) = (r - 1)(r - 2)$. Note that the addition of a second blocking variable has reduced the number of degrees of freedom for *SSRem* from $(r - 1)^2$ for a randomized complete block design based on the same number of experimental units to $(r - 1)(r - 2)$, a reduction of $(r - 1)$ degrees of freedom.

The $E\{MS\}$ column in Table 29.3 for latin square model (29.1) can be obtained by using rule (27.4). Again, it must be remembered in obtaining the coefficients in step 10 that one of the i, j, k subscripts is redundant.

TABLE 29.3 ANOVA Table for Latin Square Model (29.1) with Fixed Effects

Source of Variation	SS	df	MS	$E\{MS\}$
Row blocking variable	SSROW	$r-1$	$MSROW = \dfrac{SSROW}{r-1}$	$\sigma^2 + r\dfrac{\sum\rho_i^2}{r-1}$
Column blocking variable	SSCOL	$r-1$	$MSCOL = \dfrac{SSCOL}{r-1}$	$\sigma^2 + r\dfrac{\sum\kappa_j^2}{r-1}$
Treatments	SSTR	$r-1$	$MSTR = \dfrac{SSTR}{r-1}$	$\sigma^2 + r\dfrac{\sum\tau_k^2}{r-1}$
Error	SSRem	$(r-1)(r-2)$	$MSRem = \dfrac{SSRem}{(r-1)(r-2)}$	σ^2
Total	SSTO	r^2-1		

Test for Treatment Effects

To test for treatment effects in the latin square model (29.1) with fixed effects:

(29.7a)
$$H_0: \text{all } \tau_k = 0$$
$$H_a: \text{not all } \tau_k \text{ equal zero}$$

we see from the $E\{MS\}$ column in Table 29.3 that the appropriate test statistic is:

(29.7b)
$$F^* = \frac{MSTR}{MSRem}$$

The appropriate decision rule to control the risk of a Type I error at α is:

(29.7c)
$$\text{If } F^* \le F[1 - \alpha; r - 1, (r - 1)(r - 2)], \text{ conclude } H_0$$
$$\text{If } F^* > F[1 - \alpha; r - 1, (r - 1)(r - 2)], \text{ conclude } H_a$$

Example

The analysis of variance calculations for the background music data in Table 29.2 were made by using a computer package and the results are shown in Table 29.4.
To test for treatment effects:

$$H_0: \tau_1 = \tau_2 = \tau_3 = \tau_4 = \tau_5 = 0$$
$$H_a: \text{not all } \tau_k \text{ equal zero}$$

we find from Table 29.4:

$$F^* = \frac{MSTR}{MSRem} = \frac{166.1}{15.7} = 10.6$$

Assuming we are to control the risk of making a Type I error at $\alpha = .01$, we require $F(.99; 4, 12) = 5.41$. Since $F^* = 10.6 > 5.41$, we conclude H_a, that the various types of background music have differential effects on the productivity of the bank tellers. The P-value of this test is .0007.

TABLE 29.4 ANOVA Table for Background Music Example

Source of Variation	SS	df	MS
Weeks	82.0	4	20.5
Days within week	477.2	4	119.3
Type of music	664.4	4	166.1
Error	188.4	12	15.7
Total	1,412.0	24	

Comments

1. At times, it may be desired to test for the presence of blocking variable effects. The $E\{MS\}$ column in Table 29.3 indicates this is done in the usual way for fixed latin square model (29.1). The statistic for testing row blocking variable effects is:

$$(29.8a) \qquad F^* = \frac{MSROW}{MSRem}$$

and that for testing column blocking variable effects is:

$$(29.8b) \qquad F^* = \frac{MSCOL}{MSRem}$$

For instance, to test in the background music example whether productivity varies according to the day of the week, we use the test statistic $F^* = 119.3/15.7 = 7.6$. For a level of significance of $\alpha = .01$, we need $F(.99; 4, 12) = 5.41$. Hence, we would conclude that there is variation in productivity (averaged over all treatments and weeks) within a week. Since blocking variables correspond to a classification factor, the interpretation of blocking variable effects must be done with care.

2. The power of the F test for treatment effects in latin square model (29.1) involves the noncentrality parameter:

$$(29.9) \qquad \phi = \frac{1}{\sigma}\sqrt{\sum \tau_k^2}$$

with degrees of freedom $r - 1$ for numerator and $(r - 1)(r - 2)$ for denominator. Other than these modifications, no new problems are encountered in obtaining the power of the test for treatment effects in a latin square design.

3. If the treatment effects are random, the alternatives to be considered are:

$$(29.10) \qquad \begin{aligned} H_0&: \sigma_\tau^2 = 0 \\ H_a&: \sigma_\tau^2 > 0 \end{aligned}$$

but the test statistic and decision rule are the same as in (29.7) for the fixed treatment effects case.

29.4 EVALUATION OF APTNESS OF LATIN SQUARE MODEL

Residual Analysis

The use of the residuals (29.5) for examining the aptness of a latin square model presents no new issues; the basic points made earlier for other designs apply also to latin square designs.

Tukey Test for Additivity

A key question concerning the aptness of latin square model (29.1) is whether the effects of blocking variables and treatments are indeed additive. If nonadditivity is present in the data, transformations of the data should be studied to see if it can thereby be eliminated or made unimportant. The use of a model assuming additivity when in fact the effects are nonadditive will lower the level of significance and power of the test for treatment effects, or widen the confidence limits, thus making the experiment less sensitive.

The Tukey test for additivity in a randomized complete block design, discussed in Section 24.6, can be extended to latin square designs. For completeness, we outline the steps required for the Tukey test for additivity in a latin square experiment:

1. For each cell, obtain the fitted value (29.4):

$$(29.11a) \qquad \hat{Y}_{ijk} = \bar{Y}_{i..} + \bar{Y}_{.j.} + \bar{Y}_{..k} - 2\bar{Y}_{...}$$

2. Find the residual (29.5) for each cell:

$$(29.11b) \qquad e_{ijk} = Y_{ijk} - \hat{Y}_{ijk}$$

Check that the residuals sum to zero over every row, column, and treatment.

3. Calculate *SSRem* for the latin square. This can be done through (29.6e) which as always is equivalent to:

$$(29.11c) \qquad SSRem = \sum_i \sum_j e_{ijk}^2$$

4. Calculate for each cell:

$$(29.11d) \qquad U_{ijk} = (\hat{Y}_{ijk} - \bar{Y}_{...})^2$$

5. Calculate:

$$(29.11e) \qquad N = \sum_i \sum_j e_{ijk} U_{ijk}$$

6. Treating the U_{ijk} as the observations in a latin square, obtain the remainder sum of squares, denoted by *SSRem(U)* using (29.6e), with the *Y*s replaced by the *U*s.

7. The lack of fit sum of squares associated with nonadditivity is:

$$(29.11f) \qquad SSLF = \frac{N^2}{SSRem(U)}$$

This sum of squares has one degree of freedom associated with it.

8. The remainder sum of squares for the latin square model with the special interaction effects, denoted as before by *SSRem**, is:

$$(29.11g) \qquad SSRem^* = SSRem - SSLF$$

It has $(r - 1)(r - 2) - 1$ degrees of freedom associated with it.

9. The test statistic is:

$$(29.11h) \qquad F^* = \frac{SSLF}{1} \div \frac{SSRem^*}{(r - 1)(r - 2) - 1}$$

This test statistic follows the $F[1, (r - 1)(r - 2) - 1]$ distribution if no interactions exist. Large values of F^* lead to the conclusion that an additive model is not appropriate.

TABLE 29.5 Key Intermediate Calculational Results for Tukey Test for Additivity—Background Music Example of Table 29.2

Step 2—e_{ijk}

2.8	−2.6	3.0	−7.0	3.8
− .4	5.0	−2.6	.8	−2.8
0.0	1.0	1.6	− .2	−2.4
− .6	−3.0	.2	1.2	2.2
−1.8	− .4	−2.2	5.2	− .8

Step 3—SSRem
  SSRem = 188.4

Step 4—U_{ijk}

29.16	1.00	92.16	54.76	54.76
51.84	70.56	9.00	29.16	7.84
184.96	54.76	96.04	43.56	27.04
9.00	21.16	10.24	84.64	4.84
4.84	33.64	57.76	27.04	163.84

Step 5—N
  N = 530.832

Step 6—SSRem(U)
  SSRem(U) = 25,189.916

Step 7—SSLF
  $$SSLF = \frac{(530.832)^2}{25,189.916} = 11.19$$

Step 8—SSRem*
  SSRem* = 188.4 − 11.19 = 177.21

Example. For the background music example of Table 29.2, a check of the appropriateness of the additivity assumption in latin square model (29.1) was desired. The Tukey test for additivity was performed; key intermediate results are summarized in Table 29.5. The test statistic (29.11h) is:

$$F^* = \frac{11.19}{1} \div \frac{177.21}{11} = .69$$

Using a significance level of $\alpha = .05$, we need $F(.95; 1, 11) = 4.84$. Since $F^* = .69 \le 4.84$, we conclude that the effects of the blocking variables and treatments are additive. The P-value of this test is .42.

29.5 ANALYSIS OF TREATMENT EFFECTS

If differential treatment effects are found by the analysis of variance when the treatments have fixed effects, one will usually wish to estimate some contrasts involving the treatment effects, often utilizing multiple comparison procedures. The appropriate mean square to be used in the estimated variance of the contrast is $MSRem$ obtained from (29.6e), and the multiples for the estimated standard deviation of the contrast are as follows:

(29.12a)　　　Single comparison　　　　$t[1 - \alpha/2; (r - 1)(r - 2)]$

(29.12b)　　　Tukey procedure (for　　　$T = \dfrac{1}{\sqrt{2}} q[1 - \alpha; r, (r - 1)(r - 2)]$
　　　　　　　pairwise comparisons)

(29.12c)　　　Scheffé procedure　　　　$S^2 = (r - 1)F[1 - \alpha; r - 1,$
　　　　　　　　　　　　　　　　　　　　　　　　$(r - 1)(r - 2)]$

(29.12d)　　　Bonferroni procedure　　　$B = t[1 - \alpha/2g; (r - 1)(r - 2)]$
　　　　　　　(g comparisons)

Example

In the background music example, pairwise comparisons between the different kinds of music were desired with a family confidence coefficient of .90. The analyst used the Tukey procedure. Substituting into (15.25b) with $n_i = n_{i'} = r$ and using the results from Table 29.4, she obtained:

$$s^2\{\hat{D}\} = \frac{2MSRem}{r} = \frac{2(15.7)}{5} = 6.28 \qquad s\{\hat{D}\} = 2.51$$

Remember that each estimated treatment mean $\bar{Y}_{..k}$ is based on five observations here. Next, the analyst found the T multiple in (29.12b):

$$T = \frac{1}{\sqrt{2}} q(.90; 5, 12) = \frac{1}{\sqrt{2}}(3.92) = 2.77$$

so that:

$$Ts\{\hat{D}\} = 2.77(2.51) = 7.0$$

Using the estimated treatment means in Table 29.2, the pairwise comparisons were then obtained. For instance, we have for $\mu_{..2} - \mu_{..1} = \tau_2 - \tau_1$:

$$8.2 = (26.6 - 11.4) - 7.0 \leq \mu_{..2} - \mu_{..1} \leq (26.6 - 11.4) + 7.0 = 22.2$$

Here $\mu_{..1}$ is the mean productivity for treatment 1 averaged over all weeks and days of the week, and $\mu_{..2}$ has the corresponding meaning for treatment 2. The entire set of pairwise comparisons is as follows:

$$+8.2 \leq \mu_{..2} - \mu_{..1} \leq +22.2 \qquad +5.4 \leq \mu_{..4} - \mu_{..1} \leq +19.4$$

$$-14.0 \leq \mu_{..3} - \mu_{..2} \leq -.05 \qquad -9.2 \leq \mu_{..5} - \mu_{..4} \leq +4.8$$

$$+1.2 \leq \mu_{..3} - \mu_{..1} \leq +15.2 \qquad -5.0 \leq \mu_{..5} - \mu_{..3} \leq +9.0$$

$$-2.8 \leq \mu_{..4} - \mu_{..3} \leq +11.2 \qquad -12.0 \leq \mu_{..5} - \mu_{..2} \leq +2.0$$

$$-9.8 \leq \mu_{..4} - \mu_{..2} \leq +4.2 \qquad +3.2 \leq \mu_{..5} - \mu_{..1} \leq +17.2$$

These pairwise differences led to the following conclusions by the analyst with family confidence coefficient of 90 percent:

1. Treatment 2 yields greater mean productivity than treatments 1 or 3.
2. Treatments 3, 4, and 5 all yield greater mean productivity than treatment 1.
3. No pairwise differences in mean productivity for treatments 2, 4, and 5 are evident.
4. No pairwise differences in mean productivity for treatments 3, 4, and 5 are evident.

This is to say, the most promising treatment appears to be mixed instrumental-vocal music in medium tempo ($k = 2$). There is clear evidence that it is better than instrumental-vocal music in slow tempo ($k = 1$) or instrumental-vocal music in fast tempo ($k = 3$). The point estimates suggest it also is better than solely instrumental music in medium ($k = 4$) or fast ($k = 5$) tempo, but the experimental evidence on these latter two comparisons is indecisive.

29.6 FACTORIAL TREATMENTS

If the treatments in a latin square design are factorial in nature, the treatment sum of squares $SSTR$ is decomposed in the usual manner. For a two-factor experiment involving factors A and B, we would have:

$$(29.13) \qquad SSTR = SSA + SSB + SSAB$$

Estimates of fixed factor effects can be made readily since they are simply contrasts of the treatment means.

Example

In a sequel to the background music study mentioned earlier, four treatments were employed to investigate the effects of loudness of music (soft, loud) and type of music (semipopular, popular) on productivity of bank tellers. The treatments were defined as follows:

$$T_1 — \text{soft, semipopular}$$
$$T_2 — \text{loud, semipopular}$$
$$T_3 — \text{soft, popular}$$
$$T_4 — \text{loud, popular}$$

The blocking variables for the latin square design were day of week (Monday, Tuesday, Wednesday, Thursday) and week (1, 2, 3, 4). All treatment and block effects were considered to be fixed.

The analysis of variance for this latin square design experiment is shown in Table 29.6. If factors A and B do not interact, the effect of loudness may be studied from the contrast:

TABLE 29.6 ANOVA Table for Latin Square Design with Factorial Treatments—Background Music Sequel Study ($r = 4$, $a = 2$, $b = 2$)

Source of Variation	SS	df	MS
Weeks	SSROW	3	MSROW
Days within week	SSCOL	3	MSCOL
Treatments	SSTR	3	MSTR
Loudness of music (A)	SSA	1	MSA
Type of music (B)	SSB	1	MSB
AB interactions	SSAB	1	MSAB
Error	SSRem	6	MSRem
Total	SSTO	15	

$$(29.14) \qquad L_1 = \frac{\mu_{..1} + \mu_{..3}}{2} - \frac{\mu_{..2} + \mu_{..4}}{2} = \frac{\tau_1 + \tau_3}{2} - \frac{\tau_2 + \tau_4}{2}$$

where $\mu_{..k}$ is the mean productivity for the kth treatment averaged over all weeks and days of the week. The point estimator of this contrast is:

$$(29.14a) \qquad \hat{L}_1 = \frac{\overline{Y}_{..1} + \overline{Y}_{..3}}{2} - \frac{\overline{Y}_{..2} + \overline{Y}_{..4}}{2}$$

Similarly, the effect of type of music may be studied from the contrast:

$$(29.15) \qquad L_2 = \frac{\mu_{..1} + \mu_{..2}}{2} - \frac{\mu_{..3} + \mu_{..4}}{2}$$

A point estimator of this contrast is:

$$(29.15a) \qquad \hat{L}_2 = \frac{\overline{Y}_{..1} + \overline{Y}_{..2}}{2} - \frac{\overline{Y}_{..3} + \overline{Y}_{..4}}{2}$$

Confidence limits for these contrasts are obtained in the usual fashion, employing the multiples in (29.12).

29.7 PLANNING LATIN SQUARE EXPERIMENTS

Necessary Number of Replications

A latin square design provides r replications for each treatment. Power and/or estimation considerations similar to those for the randomized complete block design may indicate that r replications are too few, particularly when r is small, say, 3, 4, or 5. Two methods of increasing the number of replications with a latin square design are discussed in Section 29.10. With either method, it is necessary to assess in advance the magnitude of the experimental error variance σ^2 in order to plan the necessary number of replications.

Efficiency of Blocking Variables

The efficiency of a latin square design can be assessed relative to a completely randomized design or relative to a randomized complete block design. The efficiency relative to a completely randomized design is defined by:

$$(29.16a) \qquad E_1 = \frac{\sigma_r^2}{\sigma_L^2}$$

where σ_r^2 and σ_L^2 are the experimental error variances with a completely randomized design and a latin square design, respectively. The efficiency relative to a randomized complete block design can be measured in two ways, depending on whether the row or the column blocking variable is used in the randomized block design:

$$(29.16b) \qquad E_2 = \frac{\sigma_{br}^2}{\sigma_L^2}$$

$$(29.16c) \qquad E_3 = \frac{\sigma_{bc}^2}{\sigma_L^2}$$

where σ_{br}^2 and σ_{bc}^2 are the experimental error variances with a randomized block design if the row blocking variable or the column blocking variable is utilized, respectively.

We can estimate σ_r^2, σ_{br}^2, and σ_{bc}^2 from the results for a latin square design as follows:

$$(29.17a) \qquad s_r^2 = \frac{MSROW + MSCOL + (r - 1)MSRem}{r + 1}$$

$$(29.17b) \qquad s_{br}^2 = \frac{MSCOL + (r - 1)MSRem}{r}$$

$$(29.17c) \qquad s_{bc}^2 = \frac{MSROW + (r - 1)MSRem}{r}$$

Thus, our estimated measures of efficiency are:

$$(29.18a) \qquad \hat{E}_1 = \frac{MSROW + MSCOL + (r - 1)MSRem}{(r + 1)MSRem}$$

$$(29.18b) \qquad \hat{E}_2 = \frac{MSCOL + (r - 1)MSRem}{rMSRem}$$

$$(29.18c) \qquad \hat{E}_3 = \frac{MSROW + (r - 1)MSRem}{rMSRem}$$

If r is small, the efficiency measures may be modified by means of (24.15) to account for differences in the number of degrees of freedom associated with the mean squares used for estimating the experimental error variances for the two designs being compared.

Example. For the background music example, we obtain the following efficiency measures from the results in Table 29.4:

$$\hat{E}_1 = \frac{20.5 + 119.3 + 4(15.7)}{6(15.7)} = 2.2$$

$$\hat{E}_2 = \frac{119.3 + 4(15.7)}{5(15.7)} = 2.3$$

$$\hat{E}_3 = \frac{20.5 + 4(15.7)}{5(15.7)} = 1.1$$

We could modify the efficiency measures by (24.15) to take account of differences in the degrees of freedom associated with the mean squares used for estimating the experimental error variances for the designs being compared, but these modifications would have little effect here.

Analysis of the efficiency estimates indicates that the latin square design for the background music study was efficient relative to a completely randomized design. The latter would have required over twice as many observations as the latin square design so that the variance for any specified estimated treatment contrast would be the same with both designs. Most of this efficiency was gained by the column blocking variable (days within week), because the efficiency of the latin square design relative to a complete block design with the column blocking variable is poor, being close to 1. Hence, little was achieved by also blocking by the row blocking variable (week).

29.8 REGRESSION APPROACH TO LATIN SQUARE DESIGNS

Model (29.1) for a latin square design with fixed blocking and treatment effects:

(29.19) $$Y_{ijk} = \mu_{...} + \rho_i + \kappa_j + \tau_k + \varepsilon_{(ijk)}$$

$$i = 1, \ldots, r; j = 1, \ldots, r; k = 1, \ldots, r$$

can be expressed easily in the form of a regression model with indicator variables. As before, we shall use 1, -1, 0 indicator variables. The regression model for the background music example of Table 29.2, with $r = 5$, can be expressed as follows:

(29.20) $$Y_{ijk} = \mu_{...} + \underbrace{\rho_1 X_{ijk1} + \rho_2 X_{ijk2} + \rho_3 X_{ijk3} + \rho_4 X_{ijk4}}_{\text{Row blocking effect}}$$

$$+ \underbrace{\kappa_1 X_{ijk5} + \kappa_2 X_{ijk6} + \kappa_3 X_{ijk7} + \kappa_4 X_{ijk8}}_{\text{Column blocking effect}}$$

$$+ \underbrace{\tau_1 X_{ijk9} + \tau_2 X_{ijk10} + \tau_3 X_{ijk11} + \tau_4 X_{ijk12}}_{\text{Treatment effect}}$$

$$+ \varepsilon_{(ijk)}$$

where:

$$X_{ijk1} = \begin{array}{l} 1 \text{ if experimental unit from row blocking class 1} \\ -1 \text{ if experimental unit from row blocking class 5} \\ 0 \text{ otherwise} \end{array}$$

$X_{ijk2}, X_{ijk3}, X_{ijk4}$ are defined similarly

$$X_{ijk5} = \begin{array}{l} 1 \text{ if experimental unit from column blocking class 1} \\ -1 \text{ if experimental unit from column blocking class 5} \\ 0 \text{ otherwise} \end{array}$$

$X_{ijk6}, X_{ijk7}, X_{ijk8}$ are defined similarly

$$X_{ijk9} = \begin{array}{l} 1 \text{ if experimental unit received treatment 1} \\ -1 \text{ if experimental unit received treatment 5} \\ 0 \text{ otherwise} \end{array}$$

$X_{ijk10}, X_{ijk11}, X_{ijk12}$ are defined similarly

The **Y** observations vector and the **X** matrix for the background music example of Table 29.2 are shown in Table 29.7. Note, for instance, that for observation Y_{235}, $X_2 = 1$, $X_7 = 1$, $X_9 = X_{10} = X_{11} = X_{12} = -1$, and all other Xs equal zero. Hence, we have:

$$Y_{235} = \mu_{...} + \rho_2 + \kappa_3 - \tau_1 - \tau_2 - \tau_3 - \tau_4 + \varepsilon_{(235)}$$

$$= \mu_{...} + \rho_2 + \kappa_3 + \tau_5 + \varepsilon_{(235)}$$

which is the appropriate expression for Y_{235} according to ANOVA model (29.19) for the particular latin square used in the example.

Tests and estimation of treatment effects with the regression approach are conducted in the usual fashion.

29.9 MISSING OBSERVATIONS

Missing observations destroy the symmetry (orthogonality) of the latin square design and make the usual ANOVA calculations inappropriate. The regression approach, however, ordinarily remains appropriate with missing observations in a latin square design. We just set up the regression model for the available observations, and then fit the model to the data. The procedure is analogous to that discussed in Section 25.3 for the complete block design. Tests are conducted by also fitting appropriate reduced models. Estimation of fixed treatment effects is done in terms of the regression coefficients for the full model in the usual fashion.

TABLE 29.7 Data Matrices for Regression Model (29.20)—Background Music Example of Table 29.2

$$
\mathbf{Y} = \begin{bmatrix}
Y_{114} = 18 \\
Y_{123} = 17 \\
Y_{131} = 14 \\
Y_{142} = 21 \\
Y_{155} = 17 \\
Y_{213} = 13 \\
Y_{222} = 34 \\
Y_{235} = 21 \\
Y_{241} = 16 \\
Y_{254} = 15 \\
Y_{311} = 7 \\
Y_{324} = 29 \\
Y_{332} = 32 \\
Y_{345} = 27 \\
Y_{353} = 13 \\
Y_{415} = 17 \\
Y_{421} = 13 \\
Y_{433} = 24 \\
Y_{444} = 31 \\
Y_{452} = 25 \\
Y_{512} = 21 \\
Y_{525} = 26 \\
Y_{534} = 26 \\
Y_{543} = 31 \\
Y_{551} = 7
\end{bmatrix}
$$

	X_1	X_2	X_3	X_4	X_5	X_6	X_7	X_8	X_9	X_{10}	X_{11}	X_{12}
1	1	0	0	0	1	0	0	0	0	0	0	1
1	1	0	0	0	0	1	0	0	0	0	1	0
1	1	0	0	0	0	0	1	0	1	0	0	0
1	1	0	0	0	0	0	0	1	0	1	0	0
1	1	0	0	0	−1	−1	−1	−1	−1	−1	−1	−1
1	0	1	0	0	1	0	0	0	0	0	1	0
1	0	1	0	0	0	1	0	0	0	1	0	0
1	0	1	0	0	0	0	1	0	−1	−1	−1	−1
1	0	1	0	0	0	0	0	1	1	0	0	0
1	0	1	0	0	−1	−1	−1	−1	0	0	0	1
1	0	0	1	0	1	0	0	0	1	0	0	0
1	0	0	1	0	0	1	0	0	0	0	0	1
1	0	0	1	0	0	0	1	0	0	1	0	0
1	0	0	1	0	0	0	0	1	−1	−1	−1	−1
1	0	0	1	0	−1	−1	−1	−1	0	0	1	0
1	0	0	0	1	1	0	0	0	−1	−1	−1	−1
1	0	0	0	1	0	1	0	0	1	0	0	0
1	0	0	0	1	0	0	1	0	0	0	1	0
1	0	0	0	1	0	0	0	1	0	0	0	1
1	0	0	0	1	−1	−1	−1	−1	0	1	0	0
1	−1	−1	−1	−1	1	0	0	0	0	1	0	0
1	−1	−1	−1	−1	0	1	0	0	−1	−1	−1	−1
1	−1	−1	−1	−1	0	0	1	0	0	0	0	1
1	−1	−1	−1	−1	0	0	0	1	0	0	1	0
1	−1	−1	−1	−1	−1	−1	−1	−1	1	0	0	0

$\mathbf{X} =$ (matrix shown above)

29.10 ADDITIONAL REPLICATIONS WITH LATIN SQUARE DESIGNS

Need for Additional Replications

A latin square design, as noted earlier, provides r replications for each treatment. If power and/or estimation considerations indicate that these are too few replications, two basic methods are available for increasing the number of replications—replications within cells and additional latin squares. We consider each in turn.

Replications within Cells

This method of increasing the replications per treatment is feasible when two or more experimental units can be obtained for each cell defined by the row and column blocking variables. Consider, for instance, an experiment in which IQ (low, normal, high) and age (young, middle, old) are the blocking variables. In this type of situation, it is possible to obtain two or more experimental subjects for each cell, and each of the subjects in a cell will then receive the treatment assigned to that cell by the latin square employed.

Suppose that n experimental units are available for each cell, and let Y_{ijkm} denote the observation for the mth unit ($m = 1, \ldots, n$) in the (i, j) cell for which the assigned treatment is k. The additive fixed effects model (29.1) is modified for the n replications in each cell as follows:

$$(29.21) \qquad Y_{ijkm} = \mu_{...} + \rho_i + \kappa_j + \tau_k + \varepsilon_{m(ijk)}$$

where:

$\mu_{...}$ is a constant

ρ_i, κ_j, τ_k are constants subject to the restrictions $\Sigma\rho_i = \Sigma\kappa_j = \Sigma\tau_k = 0$

$\varepsilon_{m(ijk)}$ are independent $N(0, \sigma^2)$

$i = 1, \ldots, r; j = 1, \ldots, r; k = 1, \ldots, r; m = 1, \ldots, n$

The ANOVA sums of squares and degrees of freedom for model (29.21) can be obtained by rule (27.3) with modification (27.16) in a straightforward fashion, remembering that one of the i, j, k subscripts is redundant. The reason why the sum of squares associated with the model error term must still be obtained as a remainder here even though replications are present is because several types of interaction terms are assumed to equal zero. The treatment, row, and column sums of squares are, respectively:

$$(29.22a) \qquad SSTR = rn \sum_k (\bar{Y}_{..k.} - \bar{Y}_{....})^2$$

$$(29.22b) \qquad SSROW = rn \sum_i (\bar{Y}_{i...} - \bar{Y}_{....})^2$$

$$(29.22c) \qquad SSCOL = rn \sum_j (\bar{Y}_{.j..} - \bar{Y}_{....})^2$$

The total sum of squares as usual is:

$$(29.22d) \qquad SSTO = \sum_i \sum_j \sum_m (Y_{ijkm} - \bar{Y}_{....})^2$$

while $SSRem$ is obtained as a remainder:

$$(29.22e) \qquad SSRem = SSTO - SSROW - SSCOL - SSTR$$

The degrees of freedom for row, column, and treatment sums of squares are unchanged, while those associated with $SSRem$ are increased from $(r - 1)(r - 2)$ to $nr^2 - 3r + 2$, an increase of $(n - 1)r^2$ degrees of freedom.

The analysis of variance is shown in Table 29.8. The expected mean squares can be obtained by rule (27.4), remembering that one of the i, j, k subscripts is redundant. The test statistic for testing treatment effects is again $F^* = MSTR/MSRem$.

TABLE 29.8 ANOVA Table for Latin Square Design with n Replications per Cell

Source of Variation	SS	df	MS
Row blocking variable	SSROW	$r - 1$	MSROW
Column blocking variable	SSCOL	$r - 1$	MSCOL
Treatments	SSTR	$r - 1$	MSTR
Error	SSRem	$nr^2 - 3r + 2$	MSRem
Total	SSTO	$nr^2 - 1$	

Test for Additivity. When there are n replications within a cell for a latin square, it is possible to obtain a pure error measure irrespective of the correctness of the additive model (29.21). The pure error sum of squares is obtained in the usual manner:

$$(29.23) \qquad SSPE = \sum_i \sum_j \sum_m (Y_{ijkm} - \bar{Y}_{ijk.})^2$$

It has associated with it $(n - 1)r^2$ degrees of freedom, representing the increase in the degrees of freedom for $SSRem$ from having n observations in each cell of the latin square. The difference between $SSRem$ and $SSPE$ is a reflection of the lack of fit of the additive model:

$$(29.24) \qquad SSLF = SSRem - SSPE$$

and has associated with it $(r - 1)(r - 2)$ degrees of freedom. Let:

$$(29.25a) \qquad MSPE = \frac{SSPE}{(n - 1)r^2}$$

$$(29.25b) \qquad MSLF = \frac{SSLF}{(r - 1)(r - 2)}$$

Then:

$$(29.26) \qquad F^* = \frac{MSLF}{MSPE}$$

can be used to test whether or not the additive model is appropriate.

It can be shown that if the additive model is appropriate, F^* follows the $F[(r - 1)(r - 2), (n - 1)r^2]$ distribution, and that large values of F^* lead to the conclusion that the additive model is not appropriate. Table 29.9 contains the analysis of variance breaking down $SSRem$ into the $SSLF$ and $SSPE$ components.

Example. A state university is undertaking a prototype retraining program designed to teach general computer repair skills to persons who have been displaced from their previous occupations. Table 29.10 shows the results of an experiment to evaluate the

TABLE 29.9   ANOVA Table to Test Additivity when n Replications per Latin Square Cell

Source of Variation	SS	df	MS
Row blocking variable	SSROW	$r - 1$	
Column blocking variable	SSCOL	$r - 1$	
Treatments	SSTR	$r - 1$	
Error	SSRem	$nr^2 - 3r + 2$	
Lack of fit	SSLF	$(r - 1)(r - 2)$	MSLF
Pure error	SSPE	$(n - 1)r^2$	MSPE
Total	SSTO	$nr^2 - 1$	

TABLE 29.10   Example of Latin Square Design with Two Replications per Cell—Retraining Program Experiment

(a) Data

IQ i	Age (j)		
	Young	Middle	Old
High	(B) 19 16	(A) 20 24	(C) 25 21
Normal	(C) 24 22	(B) 14 15	(A) 14 14
Low	(A) 10 14	(C) 12 13	(B) 7 4

(b) Analysis of Variance

Source of Variation	SS	df	MS
IQ	364.3	2	182.2
Age	34.3	2	17.2
Treatments	147.0	2	73.5
Error	52.4	11	4.76
Lack of fit	16.4	2	8.2
Pure error	36.0	9	4.0
Total	598.0	17	

effects of three different incentive methods on achievement scores made by participants in the program. The blocking variables are IQ and age of subject. Two replications per cell were utilized. Table 29.10a contains the achievement scores, while Table 29.10b contains the analysis of variable table obtained from a computer package.

To test the appropriateness of the additive model, we find that test statistic (29.26) here is:

$$F^* = \frac{MSLF}{MSPE} = \frac{8.2}{4.0} = 2.05$$

For a level of significance of $\alpha = .05$, we need $F(.95; 2, 9) = 4.26$. Since $F^* = 2.05 \leq 4.26$, we conclude that the additive model (29.21) is appropriate. The P-value of the test is .18. Based on this test and various diagnostic plots, it was concluded that latin square model (29.21) is appropriate here.

Additional Latin Squares

At times, it is not possible to obtain additional experimental units within a cell. This is the case, for instance, in the background music example of Table 29.2, where only one type of music can be played in one day. When it is not possible to replicate within cells, additional replications for each treatment frequently can be obtained by adding one or more latin squares to one of the blocking variables. In the background music example of Table 29.2, the experiment could be run for another five weeks. In an experiment using plant crews as experimental units and employing as blocking variables plant shift (morning, afternoon, evening) and production department (1, 2, 3), additional replications can be obtained by running the experiment in other production departments.

The layout for the background music example of Table 29.2, when run over another five weeks, is shown in Table 29.11. The second latin square, and additional ones when required, is selected independently of the first.

Frequently, the additional squares may be viewed as classes of a third blocking variable. For instance in the background music example of Table 29.11, the two latin squares may be considered to refer to the blocking variable "time period." The first five weeks may be viewed as time period 1, and the second five weeks as time period 2. As another example, in the experiment with plant crews mentioned previously, the production departments for the first latin square may be on an hourly rate, while the departments for the second latin square may be on incentive pay, as shown in Table 29.12. Thus, with additional latin squares, one can in effect introduce a third blocking variable. As a consequence, the variation associated with the third blocking variable can be removed from the experimental error variability. In addition, one can study the interactions between the third blocking variable and the other variables and need not assume that these do not exist.

When n independent latin squares are used, representing a third blocking variable, with the same rows and columns in each latin square, a model that permits in-

TABLE 29.11   Two Latin Squares Design—Background Music Example of Table 29.2

				Day		
Square	Week	M	T	W	Th	F
	1	D	C	A	B	E
	2	C	B	E	A	D
1	3	A	D	B	E	C
	4	E	A	C	D	B
	5	B	E	D	C	A
	6	E	D	C	A	B
	7	B	A	E	D	C
2	8	D	C	A	B	E
	9	A	E	B	C	D
	10	C	B	D	E	A

teractions between the third blocking variable and the row and column blocking variables and the treatments is as follows (all factor effects are assumed to be fixed):

$$(29.27) \qquad Y_{ijkm} = \mu_{...} + \rho_i + \kappa_j + \tau_k + \delta_m + (\rho\delta)_{im} + (\kappa\delta)_{jm} + (\tau\delta)_{km} + \varepsilon_{(ijkm)}$$

where:

$\mu_{...}$ is a constant

$\rho_i, \kappa_j, \tau_k, \delta_m$ are constants subject to the restrictions:

$$\sum \rho_i = \sum \kappa_j = \sum \tau_k = \sum \delta_m = 0$$

$(\rho\delta)_{im}, (\kappa\delta)_{jm}, (\tau\delta)_{km}$ are constants subject to the restrictions:

$$\sum_i (\rho\delta)_{im} = \sum_m (\rho\delta)_{im} = \sum_j (\kappa\delta)_{jm} = 0$$

$$\sum_m (\kappa\delta)_{jm} = \sum_k (\tau\delta)_{km} = \sum_m (\tau\delta)_{km} = 0$$

$\varepsilon_{(ijkm)}$ are independent $N(0, \sigma^2)$
$i = 1, \ldots, r; j = 1, \ldots, r; k = 1, \ldots, r; m = 1, \ldots, n$

Note that δ_m denotes the third blocking variable effect. The other main effects are defined as before.

The conditions of model (29.27) might be appropriate for the background music example of Table 29.11. There it might at times be appropriate to consider the rows to refer to the position of a week within a five-week period, the columns to refer to the position of a day within a week, and the third blocking variable to refer to the time of year.

Table 29.13 shows the analysis of variance for n independent squares model (29.27). The sums of squares and degrees of freedom can be obtained by rule (27.3)

TABLE 29.12   Two Latin Squares Design—Experiment with Plant Crews

| Square | Production Department | Shift | | |
		Morning	Afternoon	Evening
1 (hourly pay)	1	C	B	A
	2	B	A	C
	3	A	C	B
2 (incentive pay)	4	A	B	C
	5	C	A	B
	6	B	C	A

with modification (27.16); remember in using this rule that one of the i, j, k subscripts is redundant. The sum of squares associated with the model error term must be obtained as a remainder because there are no within-cell replications here and some interactions are assumed to equal zero. The expected mean squares can be obtained by rule (27.4), again remembering that one of the i, j, k subscripts is redundant.

Replications in Repeated Measures Studies

We noted earlier that a latin square design is highly suitable for a repeated measures study when there are r treatments and r subjects. If additional replications are needed, however, replications within cells cannot be used since a cell pertains to an individual subject. Instead, latin square cross-over designs or independent latin squares may be used.

TABLE 29.13   ANOVA Table when n Latin Squares with Same Rows and Columns Are Employed and Interactions Involving Third Blocking Variable Are Permitted

Source of Variation	SS	df
Row blocking variable (*ROW*)	SSROW	$r - 1$
Column blocking variable (*COL*)	SSCOL	$r - 1$
Treatments (*TR*)	SSTR	$r - 1$
Third blocking variable (3)	SS3	$n - 1$
Third-row interactions (3.*ROW*)	SS3.ROW	$(n - 1)(r - 1)$
Third-column interactions (3.*COL*)	SS3.COL	$(n - 1)(r - 1)$
Third-treatment interactions (3.*TR*)	SS3.TR	$(n - 1)(r - 1)$
Error	SSRem	$n(r - 1)(r - 2)$
Total	SSTO	$nr^2 - 1$

Latin Square Cross-over Designs. These designs, also called *latin square change-over designs,* are often useful when a latin square is to be used in a repeated measures study to balance the order positions of treatments, yet more subjects are required than called for by a single latin square. With this type of design, the subjects are randomly assigned to the different treatment order patterns given by a latin square (several latin squares may be used at times). Consider an experiment in which treatments A, B, and C are to be administered to each subject, and the three treatment order patterns are given by the latin square:

	Order Position		
Pattern	1	2	3
1	A	B	C
2	B	C	A
3	C	A	B

Suppose that $3n$ subjects are available for the study. Then n subjects will be assigned at random to each of the three order patterns in a latin square cross-over design. Note that this design is a mixture of repeated measures (within subjects) and latin square (order patterns form a latin square).

Assuming that all effects are additive and fixed except that the effects for subjects are random, a relatively simple model for latin square cross-over designs can be developed for r treatments and n subjects per order pattern. In the following model, ρ_i denotes the effect of the ith treatment order pattern, κ_j denotes the effect of the jth order position, τ_k denotes the effect of the kth treatment, and $\eta_{m(i)}$ denotes the effect of subject m which is nested within the ith treatment order pattern:

(29.28) $$Y_{ijkm} = \mu_{...} + \rho_i + \kappa_j + \tau_k + \eta_{m(i)} + \varepsilon_{(ijkm)}$$

where:

$\mu_{...}$ is a constant
ρ_i, κ_j, τ_k are constants subject to the restrictions $\Sigma\rho_i = \Sigma\kappa_j = \Sigma\tau_k = 0$
$\eta_{m(i)}$ are independent $N(0, \sigma_\eta^2)$
$\varepsilon_{(ijkm)}$ are independent $N(0, \sigma^2)$ and independent of the $\eta_{m(i)}$
$i = 1, \ldots, r; j = 1, \ldots, r; k = 1, \ldots, r; m = 1, \ldots, n$

The analysis of variance sums of squares and degrees of freedom for this model can be obtained by rule (27.3) with modification (27.16); remember in using this rule that one of the i, j, k subscripts is redundant. The sum of squares associated with the model error term must again be obtained as a remainder here because there are no within-cell replications in this design. The formulas for the sums of squares follow the usual pattern. The definitional sums of squares are as follows:

(29.29a) $$SSTO = \sum_i \sum_j \sum_m (Y_{ijkm} - \bar{Y}_{....})^2$$

(29.29b) $$SSP = nr \sum_i (\bar{Y}_{i...} - \bar{Y}_{....})^2$$

(29.29c)
$$SSO = nr \sum_j (\bar{Y}_{.j..} - \bar{Y}_{....})^2$$

(29.29d)
$$SSTR = nr \sum_k (\bar{Y}_{..k.} - \bar{Y}_{....})^2$$

(29.29e)
$$SSS = r \sum_i \sum_m (\bar{Y}_{i..m} - \bar{Y}_{i...})^2$$

(29.29f)
$$SSRem = SSTO - SSP - SSO - SSTR - SSS$$

Here, SSP is the (treatment) *pattern sum of squares, SSO* is the *order position sum of squares, SSS* is the *subject sum of squares,* and the other sums of squares have their usual meanings.

Table 29.14 contains the ANOVA table. The expected mean squares can be obtained by rule (27.4), recognizing that one of the i, j, k subscripts is redundant.

Example. Table 29.15a contains data for a study of the effects of three different displays on the sale of apples, using the latin square cross-over design. Six stores were used, and two were assigned at random to each of the three treatment order patterns shown. Each display was kept for two weeks, and the observed variable was sales per 100 customers. Table 29.15b contains the analysis of variance. The sums of squares were obtained from a computer run.

To test for treatment effects, we use:

$$F^* = \frac{MSTR}{MSRem} = \frac{94.5}{2.54} = 37.2$$

For $\alpha = .05$, we require $F(.95; 2, 8) = 4.46$. Since $F^* = 37.2 > 4.46$, we conclude that there are differential sales effects for the three displays. The P-value of the test is $0+$. Tests for pattern effects, order position effects, and store effects were also

---

TABLE 29.14 ANOVA Table for Latin Square Cross-over Design

Source of Variation	SS	df	MS	E{MS}
Patterns (P)	SSP	$r - 1$	MSP	$\sigma^2 + r\sigma_\eta^2 + nr\dfrac{\Sigma\rho_i^2}{r - 1}$
Order positions (O)	SSO	$r - 1$	MSO	$\sigma^2 + nr\dfrac{\Sigma\kappa_j^2}{r - 1}$
Treatments (TR)	SSTR	$r - 1$	MSTR	$\sigma^2 + nr\dfrac{\Sigma\tau_k^2}{r - 1}$
Subjects (S) (within patterns)	SSS	$r(n - 1)$	MSS	$\sigma^2 + r\sigma_\eta^2$
Error	SSRem	$(r - 1)(nr - 2)$	MSRem	σ^2
Total	SSTO	$nr^2 - 1$		

TABLE 29.15 Latin Square Cross-over Design—Apple Sales Example

(a) Data (coded)

Pattern i	Store	Two-Week Period (j)		
		1	2	3
1	m = 1	9 (B)	12 (C)	15 (A)
	m = 2	4 (B)	12 (C)	9 (A)
2	m = 1	12 (A)	14 (B)	3 (C)
	m = 2	13 (A)	14 (B)	3 (C)
3	m = 1	7 (C)	18 (A)	6 (B)
	m = 2	5 (C)	20 (A)	4 (B)

(b) Analysis of Variance

Source of Variation	SS	df	MS
Patterns	.33	2	.17
Order positions	233.33	2	116.67
Displays	189.00	2	94.50
Stores (within patterns)	21.00	3	7.00
Error	20.33	8	2.54
Total	464.0	17	

carried out. They indicated that order position effects were present, but no pattern or store effects. Order position effects here are associated with the three time periods in which the displays were studied, and may reflect seasonal effects as well as the results of special events, such as unusually hot weather in one period.

Use of Independent Latin Squares. If the order position effects are not approximately constant for all subjects (stores, etc.), a cross-over design is not fully effective. It may then be preferable to place the subjects into homogeneous groups with respect to the order position effects and use independent latin squares for each group. Suppose that four treatments are to be administered to eight subjects each, four males and four females, and that the experimenter expects the fatigue effect to be strong for females but only mild for males. The use of two independent latin squares, one for male subjects and the other for female subjects, may then be advisable.

Carry-over Effects. If carry-over effects from one treatment to another are anticipated, that is, if not only the order position but also the preceding treatment has an effect, one may balance out these carry-over effects by choosing a latin square in

TABLE 29.16   Illustration of a Latin Square Double Cross-over Design

Square	Store	Two-Week Period		
		1	2	3
1	1	A	B	C
	2	B	C	A
	3	C	A	B
2	4	A	C	B
	5	B	A	C
	6	C	B	A

which every treatment follows every other treatment an equal number of times. For $r = 4$, an example of such a latin square is:

Subject	Period			
	1	2	3	4
1	A	B	D	C
2	B	C	A	D
3	C	D	B	A
4	D	A	C	B

Note that treatment A follows each of the other treatments once, and similarly for the other treatments. This design is suited when the carry-over effects do not persist for more than one period.

When r is odd, the sequence balance can be obtained by using a pair of latin squares with the property that the treatment sequences in one are reversed in the other square. Indeed, even when r is even, it is usually desirable to use a pair of such squares so that the degrees of freedom associated with $MSRem$ are reasonably large. Such a design is sometimes called a *latin square double cross-over design*. This type of design retains the advantages of employing two blocking variables in a latin square, while enabling the experimenter also to balance and measure the carry-over effects.

For the earlier apple display illustration in which three displays were studied in six stores, the two latin squares might be as shown in Table 29.16. The stores should first be placed into two homogeneous groups and these should then be assigned to the two latin squares.

29.11   RANDOM BLOCKING VARIABLE EFFECTS

If the row and/or column blocking variable(s) in a latin square design have classes that should be viewed as random selections from a population, the fixed effects latin square model (29.1) no longer is applicable.

Both Blocking Variables Random

Consider the case where the row blocking variable is subject and the column blocking variable observer, and both the subjects and the observers included in the study are viewed as random samples from relevant populations. In that case, assuming treatment effects are fixed, the additive model for a latin square design is:

$$(29.30) \qquad\qquad Y_{ijk} = \mu_{...} + \rho_i + \kappa_j + \tau_k + \varepsilon_{(ijk)}$$

where:

$\mu_{...}$ is a constant
ρ_i are independent $N(0, \sigma_\rho^2)$
κ_j are independent $N(0, \sigma_\kappa^2)$
τ_k are constants subject to the restriction $\Sigma \tau_k = 0$
$\varepsilon_{(ijk)}$ are independent $N(0, \sigma^2)$
ρ_i, κ_j, and $\varepsilon_{(ijk)}$ are pairwise independent
$i = 1, \ldots, r; j = 1, \ldots, r; k = 1, \ldots, r$

The analysis of variance is the same as for the fixed blocking variable effects model. The expected mean squares for the two blocking variables are obtained by replacing the sums of squared effects divided by degrees of freedom in Table 29.3 by variance terms. Consequently, all tests and estimates of treatment effects are conducted as for fixed blocking variable effects.

One Blocking Variable Random, Other Fixed

When one of the blocking variables has random effects while the other has fixed effects, the additive model with fixed treatment effects becomes a mixture of models (29.1) and (29.30). Again, there will be no change in the analysis of variance or in tests or estimates of treatment effects. An instance where this mixed model would be appropriate is a repeated measures study where the row blocking variable is subject and the column blocking variable is order position of treatment.

29.12 YOUDEN AND GRAECO-LATIN SQUARES

When it is not possible to use a latin square design because the number of column classes is less than the number of row classes, a *Youden square design* may be helpful. Consider a repeated measures study involving four treatments and four subjects (row blocking variable). Suppose now that a subject can be given only three treatments because of serious fatigue effects. Hence, the column blocking variable (order position of treatments) can have only three classes. A Youden square design suitable for this occasion is shown in Table 29.17. Note that this layout would become a latin square with the addition of the column (D, C, B, A). Also note that each treatment

TABLE 29.17   Illustration of a Youden Square Design

Subject	Order Position of Treatment		
	1	2	3
1	A	B	C
2	D	A	B
3	C	D	A
4	B	C	D

TABLE 29.18   Illustration of a Graeco-Latin Square Design ($r = 4$)

Row Blocking Variable	Column Blocking Variable			
	1	2	3	4
1	$\alpha : A$	$\beta : B$	$\gamma : C$	$\delta : D$
2	$\beta : C$	$\alpha : D$	$\delta : A$	$\gamma : B$
3	$\delta : B$	$\gamma : A$	$\beta : D$	$\alpha : C$
4	$\gamma : D$	$\delta : C$	$\alpha : B$	$\beta : A$

occurs once in each order position, and that every pair of treatments appears together an equal number of times within subjects. These are characteristics present in all Youden squares. The analysis of Youden square designs is more complex than that of latin squares because not all treatments are run in each class of the row blocking variable. A reference, such as Reference 29.1, should be consulted if a Youden square design is to be used.

A *graeco-latin square design* is an extension of a latin square design when three blocking variables are to be used simultaneously. Table 29.18 illustrates a graeco-latin square design for $r = 4$. The symbols α, β, γ, and δ represent the four levels of the third blocking variable. Thus, the cell corresponding to the first class of each of the three blocking variables is to receive treatment A, and so on. Note that the levels of the third blocking variable appear once in each row and once in each column, and they appear once only with each treatment. Graeco-latin square designs are used much less frequently in practice than the other designs we have discussed. Reference 29.1 discusses the analysis of graeco-latin square designs.

CITED REFERENCE

29.1. Cochran, W. G., and G. M. Cox. *Experimental Designs*. 2nd ed. New York: John Wiley & Sons, 1957.

PROBLEMS

29.1. A behavioral scientist explained why latin square designs are used so frequently: "Many times in behavioral science, we require the use of repeated measures designs because variability between human subjects is so great. Since an order effect may be present in this situation, we employ latin square designs to eliminate any bias due to order effects." Comment.

29.2. a. Using random permutations, select randomly a 3 by 3 latin square. Show all steps.
   b. Using random permutations, select randomly a 6 by 6 latin square. Show all steps.

29.3. Hardware sales. A manufacturer conducted a small pilot study of the effect of the price of one of its products on sales of this product in hardware stores. Since it might be confusing to customers if prices were switched repeatedly within a store, only one price was used for any one store during the six-month study period. Sixteen stores were employed in the study. To reduce experimental error variability, they were chosen so that there would be one store for each sales volume-geographic location class. The four price levels (A: \$1.79; B: \$1.69; C: \$1.59; D: \$1.49) were assigned to the stores according to the latin square design shown below. Data on sales during the six-month period (in thousand dollars) follow.

Sales Volume Class i	Geographic Location Class (j)			
	Northeast	Northwest	Southeast	Southwest
1 (smallest)	1.2 (B)	1.5 (C)	1.0 (A)	1.7 (D)
2	1.4 (A)	1.9 (D)	1.6 (B)	1.5 (C)
3	2.8 (C)	2.1 (B)	2.7 (D)	2.0 (A)
4 (largest)	3.4 (D)	2.5 (A)	2.9 (C)	2.7 (B)

   a. Obtain the residuals for latin square model (29.1) and plot them against the fitted values. Also prepare a normal probability plot of the residuals and calculate the coefficient of correlation between the ordered residuals and their expected values under normality. Summarize your findings about the appropriateness of model (29.1) here.
   b. Conduct the Tukey test for additivity; use $\alpha = .01$. State the alternatives, decision rule, and conclusion. What is the P-value of the test?

29.4. Refer to **Hardware sales** Problem 29.3. Assume that latin square model (29.1) is appropriate.
   a. Prepare a normal probability plot of the estimated treatment means. What does the plot suggest about the effects of the four price levels on sales?
   b. Test whether or not price level affects mean sales. Use significance level $\alpha = .05$. State the alternatives, decision rule, and conclusion. What is the P-value of the test?
   c. Analyze the nature of the price effect on sales by making all pairwise comparisons among the treatment means. Use the Tukey procedure and a 90 percent family confidence coefficient. Summarize your findings.
   d. Does there appear to be a linear relationship between price level and mean sales? Could you formally test for linearity? Explain.

29.5. Refer to **Hardware sales** Problems 29.3 and 29.4.
   a. Calculate the three estimated efficiency measures in (29.18).

b. Would a randomized block design have been adequate here? If so, which blocking variable would have been best?

29.6. Summary reports. A management information systems consultant conducted a small-scale study of five different daily summary reports (*A*: greatest amount of detail; *B*; *C*; *D*; *E*: least amount of detail). She used five sales executives in the study. Each was given one type of daily report for a month and then was asked to rate its helpfulness on a 25-point scale (0: no help; 25: extremely helpful). Over a five-month period, each executive received each type of report for one month according to the latin square design shown below. The helpfulness ratings follow.

Executive	Month (*j*)				
i	*March*	*April*	*May*	*June*	*July*
Harrison	21 (*D*)	8 (*A*)	17 (*C*)	9 (*B*)	16 (*E*)
Smith	5 (*A*)	10 (*E*)	3 (*B*)	12 (*C*)	15 (*D*)
Carmichael	20 (*C*)	10 (*B*)	15 (*E*)	22 (*D*)	12 (*A*)
Loeb	4 (*B*)	17 (*D*)	3 (*A*)	9 (*E*)	10 (*C*)
Munch	17 (*E*)	16 (*C*)	20 (*D*)	7 (*A*)	11 (*B*)

a. Obtain the residuals for latin square model (29.1) and plot them against the fitted values. Also prepare a normal probability plot of the residuals and calculate the coefficient of correlation between the ordered residuals and their expected values under normality. Summarize your findings about the appropriateness of model (29.1) here.

b. Conduct the Tukey test for additivity; use $\alpha = .05$. State the alternatives, decision rule, and conclusion. What is the *P*-value of the test?

29.7. Refer to **Summary reports** Problem 29.6. Assume that latin square model (29.1) is appropriate.

a. Prepare a normal probability plot of the estimated treatment means. What does the plot suggest about the effects of the five types of reports?

b. Test whether or not the five types of reports differ in mean helpfulness; use significance level $\alpha = .01$. State the alternatives, decision rule, and conclusion. What is the *P*-value of the test?

c. Analyze the effectiveness of the five types of reports by making all pairwise comparisons among the treatment means. Use the Tukey procedure and a 95 percent family confidence coefficient. Summarize your findings.

29.8. Refer to **Summary reports** Problems 29.6 and 29.7.

a. Calculate the three estimated efficiency measures in (29.18).

b. How effective was the use of the latin square design here?

29.9. Refer to **Hardware sales** Problems 29.3 and 29.4. Assume that $\sigma = .15$. What is the power of the test for treatment effects in Problem 29.4b if $\tau_1 = -.4$, $\tau_2 = 0$, $\tau_3 = .1$, and $\tau_4 = .3$?

29.10. Refer to **Summary reports** Problems 29.6 and 29.7. Assume that $\sigma = 1.4$. What is the power of the test for treatment effects in Problem 29.7b if $\tau_1 = -2$, $\tau_2 = -1$, $\tau_3 = 0$, $\tau_4 = 1.5$, $\tau_5 = 1.5$?

29.11. Drugs interaction. A pilot study was undertaken on the interaction effects of two drugs to stimulate growth in girls who are of short stature because of a particular syndrome. Each drug was known to be modestly effective singly, but the combination of the two drugs had never been investigated. Blocking by both subject and time period was desired whereby repeated measures for different treatments applied to the same subject are obtained. A 4 by 4 latin square design, shown below, was utilized for four subjects, four time periods, and four treatments. The four time periods consisted of one month each, separated by an intervening month during which no treatment was given. The four treatments were A: no treatment (placebo); B: drug X alone; C: drug Y alone; D: both drugs X and Y. The dependent variable was the difference in the growth rates (in centimeters per month) during the treatment period and the base period before the experiment began. The results of the study follow.

Subject	Period (j)			
i	1	2	3	4
1	.02 (A)	.15 (B)	.45 (D)	.18 (C)
2	.27 (B)	.24 (C)	−.01 (A)	.58 (D)
3	.11 (C)	.35 (D)	.14 (B)	−.03 (A)
4	.48 (D)	.04 (A)	.18 (C)	.22 (B)

a. Obtain the residuals in (29.5) and plot them against the fitted values. Also prepare a normal probability plot of the residuals and calculate the coefficient of correlation between the ordered residuals and their expected values under normality. Summarize your findings.

b. Conduct the Tukey test for additivity, assuming that all effects are fixed and ignoring the factorial structure of the treatments; use $\alpha = .05$. State the alternatives, decision rule, and conclusion. What is the P-value of the test?

29.12. Refer to **Drugs interaction** Problem 29.11. Assume that an appropriate model is latin square model (29.1), modified so that subjects have random effects and a factorial structure for the treatments is incorporated (factor A: drug X; factor B: drug Y).

a. State the model to be employed.

b. Test for interaction effects between the two drugs; use significance level $\alpha = .10$. State the alternatives, decision rule, and conclusion. What is the P-value of the test?

c. Estimate the interaction contrast:

$$L = \left(\frac{\mu_{..2} + \mu_{..3}}{2} - \mu_{..1}\right) - \left(\mu_{..4} - \frac{\mu_{..2} + \mu_{..3}}{2}\right) = \mu_{..2} - \mu_{..1} - \mu_{..4} + \mu_{..3}$$

using a 90 percent confidence interval. Interpret your result.

29.13. Refer to **Hardware sales** Problem 29.3.

a. Set up the regression model equivalent to latin square model (29.1) using $1, -1, 0$ indicator variables.

b. Test by means of the regression approach whether or not price level affects mean sales. Use significance level $\alpha = .05$. State the alternatives, decision rule, and conclusion.

c. Obtain a 95 percent confidence interval by the regression approach for $D = \tau_3 - \tau_4$. Interpret your interval estimate.

d. Suppose that observation $Y_{232} = 1.6$ were missing.
  (i) Use the regression approach to test whether price level affects mean sales; control the α risk at .05. State the alternatives, decision rule, and conclusion.

(ii) Use the regression approach to estimate $D = \tau_1 - \tau_2$ by means of a 95 percent confidence interval.

29.14. Refer to **Summary reports** Problem 29.6. Suppose that observations $Y_{114} = 21$ and $Y_{453} = 10$ were missing.

    a. Use the regression approach to test whether the five types of reports differ in mean effectiveness; employ significance level $\alpha = .01$. State the alternatives, decision rule, and conclusion.

    b. Use the regression approach to estimate $D = \tau_4 - \tau_1$ by means of a 99 percent confidence interval.

29.15. TV commercials. A study was undertaken to determine whether the volume of sound of a television commercial affects recall and whether this effect varies by product. Thirty-two subjects were chosen, two each for 16 groups defined according to age (class 1: youngest; 2; 3; 4: oldest) and amount of education (class 1: lowest education level; 2; 3; 4: highest education level). Each subject was exposed to one of four television commercial showings (A: high volume, product X; B: low volume, product X; C: high volume, product Y; D: low volume, product Y) according to the latin square design shown below. Two different commercials were involved, one for each product. During the following week, the subjects were asked to mention everything they could remember about the advertisement. Scores were based on the number of learning points mentioned, suitably standardized. The results follow.

Age Class	Education Level (j)			
i	1	2	3	4
1	(D) 83 86	(A) 64 69	(C) 78 75	(B) 76 74
2	(B) 70 76	(C) 81 75	(A) 64 60	(D) 87 81
3	(C) 67 74	(B) 67 61	(D) 76 81	(A) 64 57
4	(A) 56 60	(D) 72 67	(B) 63 67	(C) 64 66

    a. Obtain the residuals for latin square model (29.21) and plot them against the fitted values. Also prepare a normal probability plot of the residuals and calculate the coefficient of correlation between the ordered residuals and their expected values under normality. Summarize your findings about the appropriateness of the model utilized here.

    b. Conduct a formal test of whether or not the effects of blocking variables and treatments are additive; ignore the factorial structure of the treatments. Use a level of significance of $\alpha = .01$. State the alternatives, decision rule, and conclusion. What is the P-value of the test?

29.16. Refer to **TV commercials** Problem 29.15. Assume that an appropriate model is latin square model (29.21), modified to allow for factorial treatments (factor A: volume; factor B: product).

a. State the model to be employed.
b. Test for volume-product interaction effects; use $\alpha = .01$. State the alternatives, decision rule, and conclusion. What is the P-value of the test?
c. Test for volume main effects and product main effects. For each test, use $\alpha = .01$ and state the alternatives, decision rule, and conclusion. What is the P-value of each test?
d. To study the nature of the volume and product main effects, estimate the difference between the two factor level means for each factor. Use the Bonferroni procedure and a 95 percent family confidence coefficient. State your findings.

29.17. **Recall decay.** In an experiment to study recall decay with three different questionnaires (A, B, C), nine subjects were questioned at three different times three months apart about the number of trips to a shopping center during the preceding three months. Each time a different questionnaire was used. The latin square design shown below was used to determine the questionnaire order for each subject, with three subjects assigned randomly to each of the three treatment order patterns. The data on number of shopping trips reported follow.

Pattern i	Subject	Time Period (j) 1	2	3
	$m = 1$	40 (C)	18 (A)	30 (B)
1	$m = 2$	35 (C)	25 (A)	37 (B)
	$m = 3$	31 (C)	22 (A)	28 (B)
	$m = 1$	10 (B)	43 (C)	33 (A)
2	$m = 2$	18 (B)	49 (C)	37 (A)
	$m = 3$	15 (B)	48 (C)	29 (A)
	$m = 1$	7 (A)	19 (B)	59 (C)
3	$m = 2$	11 (A)	24 (B)	51 (C)
	$m = 3$	19 (A)	21 (B)	62 (C)

Obtain the residuals for latin square cross-over model (29.28) and plot them against the fitted values. Also prepare a normal probability plot of the residuals and calculate the coefficient of correlation between the ordered residuals and their expected values under normality. Summarize your findings about the appropriateness of model (29.28) here.

29.18. Refer to **Recall decay** Problem 29.17. Assume that latin square cross-over model (29.28) is appropriate.
a. Test for the presence of treatment order pattern, time period, and questionnaire effects. For each test, use a level of significance of $\alpha = .05$ and state the alternatives, decision rule, and conclusion. What is the P-value of each test?
b. Analyze the questionnaire main effects by estimating all pairwise comparisons of treatment means. Use the Tukey procedure and a 90 percent family confidence coefficient. Summarize your findings.

EXERCISES

29.19. Derive the expected mean squares in Table 29.3 for latin square model (29.1) by using rule (27.4).

29.20. Derive the expected mean squares for latin square model (29.21) with n replications by using rule (27.4).

29.21. Derive the expected mean squares for latin square model (29.27) with n independent squares by using rule (27.4).

29.22. Derive the expected mean squares in Table 29.14 for latin square cross-over model (29.28) with n subjects for each treatment order pattern by using rule (27.4).

Appendix A

Tables

TABLE A.1   Cumulative Probabilities of the Standard Normal Distribution

Entry is area A under the standard normal curve from $-\infty$ to $z(A)$

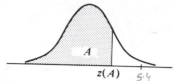

z	.00	.01	.02	.03	.04	.05	.06	.07	.08	.09
.0	.5000	.5040	.5080	.5120	.5160	.5199	.5239	.5279	.5319	.5359
.1	.5398	.5438	.5478	.5517	.5557	.5596	.5636	.5675	.5714	.5753
.2	.5793	.5832	.5871	.5910	.5948	.5987	.6026	.6064	.6103	.6141
.3	.6179	.6217	.6255	.6293	.6331	.6368	.6406	.6443	.6480	.6517
.4	.6554	.6591	.6628	.6664	.6700	.6736	.6772	.6808	.6844	.6879
.5	.6915	.6950	.6985	.7019	.7054	.7088	.7123	.7157	.7190	.7224
.6	.7257	.7291	.7324	.7357	.7389	.7422	.7454	.7486	.7517	.7549
.7	.7580	.7611	.7642	.7673	.7704	.7734	.7764	.7794	.7823	.7852
.8	.7881	.7910	.7939	.7967	.7995	.8023	.8051	.8078	.8106	.8133
.9	.8159	.8186	.8212	.8238	.8264	.8289	.8315	.8340	.8365	.8389
1.0	.8413	.8438	.8461	.8485	.8508	.8531	.8554	.8577	.8599	.8621
1.1	.8643	.8665	.8686	.8708	.8729	.8749	.8770	.8790	.8810	.8830
1.2	.8849	.8869	.8888	.8907	.8925	.8944	.8962	.8980	.8997	.9015
1.3	.9032	.9049	.9066	.9082	.9099	.9115	.9131	.9147	.9162	.9177
1.4	.9192	.9207	.9222	.9236	.9251	.9265	.9279	.9292	.9306	.9319
1.5	.9332	.9345	.9357	.9370	.9382	.9394	.9406	.9418	.9429	.9441
1.6	.9452	.9463	.9474	.9484	.9495	.9505	.9515	.9525	.9535	.9545
1.7	.9554	.9564	.9573	.9582	.9591	.9599	.9608	.9616	.9625	.9633
1.8	.9641	.9649	.9656	.9664	.9671	.9678	.9686	.9693	.9699	.9706
1.9	.9713	.9719	.9726	.9732	.9738	.9744	.9750	.9756	.9761	.9767
2.0	.9772	.9778	.9783	.9788	.9793	.9798	.9803	.9808	.9812	.9817
2.1	.9821	.9826	.9830	.9834	.9838	.9842	.9846	.9850	.9854	.9857
2.2	.9861	.9864	.9868	.9871	.9875	.9878	.9881	.9884	.9887	.9890
2.3	.9893	.9896	.9898	.9901	.9904	.9906	.9909	.9911	.9913	.9916
2.4	.9918	.9920	.9922	.9925	.9927	.9929	.9931	.9932	.9934	.9936
2.5	.9938	.9940	.9941	.9943	.9945	.9946	.9948	.9949	.9951	.9952
2.6	.9953	.9955	.9956	.9957	.9959	.9960	.9961	.9962	.9963	.9964
2.7	.9965	.9966	.9967	.9968	.9969	.9970	.9971	.9972	.9973	.9974
2.8	.9974	.9975	.9976	.9977	.9977	.9978	.9979	.9979	.9980	.9981
2.9	.9981	.9982	.9982	.9983	.9984	.9984	.9985	.9985	.9986	.9986
3.0	.9987	.9987	.9987	.9988	.9988	.9989	.9989	.9989	.9990	.9990
3.1	.9990	.9991	.9991	.9991	.9992	.9992	.9992	.9992	.9993	.9993
3.2	.9993	.9993	.9994	.9994	.9994	.9994	.9992	.9992	.9993	.9993
3.3	.9995	.9995	.9995	.9996	.9996	.9994	.9994	.9995	.9995	.9995
3.4	.9997	.9997	.9997	.9997	.9997	.9996	.9996	.9996	.9996	.9997
						.9997	.9997	.9997	.9997	.9998

Selected Percentiles

Cumulative probability A:	.90	.95	.975	.98	.99	.995	.999
$z(A)$:	1.282	1.645	1.960	2.054	2.326	2.576	3.090

TABLE A.2   Percentiles of the t Distribution

Entry is $t(A; \nu)$ where $P\{t(\nu) \leq t(A; \nu)\} = A$

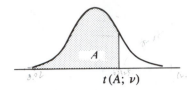

$t(A; \nu)$

				A			
ν	.60	.70	.80	.85	.90	.95	.975
1	0.325	0.727	1.376	1.963	3.078	6.314	12.706
2	0.289	0.617	1.061	1.386	1.886	2.920	4.303
3	0.277	0.584	0.978	1.250	1.638	2.353	3.182
4	0.271	0.569	0.941	1.190	1.533	2.132	2.776
5	0.267	0.559	0.920	1.156	1.476	2.015	2.571
6	0.265	0.553	0.906	1.134	1.440	1.943	2.447
7	0.263	0.549	0.896	1.119	1.415	1.895	2.365
8	0.262	0.546	0.889	1.108	1.397	1.860	2.306
9	0.261	0.543	0.883	1.100	1.383	1.833	2.262
10	0.260	0.542	0.879	1.093	1.372	1.812	2.228
11	0.260	0.540	0.876	1.088	1.363	1.796	2.201
12	0.259	0.539	0.873	1.083	1.356	1.782	2.179
13	0.259	0.537	0.870	1.079	1.350	1.771	2.160
14	0.258	0.537	0.868	1.076	1.345	1.761	2.145
15	0.258	0.536	0.866	1.074	1.341	1.753	2.131
16	0.258	0.535	0.865	1.071	1.337	1.746	2.120
17	0.257	0.534	0.863	1.069	1.333	1.740	2.110
18	0.257	0.534	0.862	1.067	1.330	1.734	2.101
19	0.257	0.533	0.861	1.066	1.328	1.729	2.093
20	0.257	0.533	0.860	1.064	1.325	1.725	2.086
21	0.257	0.532	0.859	1.063	1.323	1.721	2.080
22	0.256	0.532	0.858	1.061	1.321	1.717	2.074
23	0.256	0.532	0.858	1.060	1.319	1.714	2.069
24	0.256	0.531	0.857	1.059	1.318	1.711	2.064
25	0.256	0.531	0.856	1.058	1.316	1.708	2.060
26	0.256	0.531	0.856	1.058	1.315	1.706	2.056
27	0.256	0.531	0.855	1.057	1.314	1.703	2.052
28	0.256	0.530	0.855	1.056	1.313	1.701	2.048
29	0.256	0.530	0.854	1.055	1.311	1.699	2.045
30	0.256	0.530	0.854	1.055	1.310	1.697	2.042
40	0.255	0.529	0.851	1.050	1.303	1.684	2.021
60	0.254	0.527	0.848	1.045	1.296	1.671	2.000
120	0.254	0.526	0.845	1.041	1.289	1.658	1.980
∞	0.253	0.524	0.842	1.036	1.282	1.645	1.960

TABLE A.2 (*concluded*)   Percentiles of the *t* Distribution

				A			
ν	.98	.985	.99	.9925	.995	.9975	.9995
1	15.895	21.205	31.821	42.434	63.657	127.322	636.590
2	4.849	5.643	6.965	8.073	9.925	14.089	31.598
3	3.482	3.896	4.541	5.047	5.841	7.453	12.924
4	2.999	3.298	3.747	4.088	4.604	5.598	8.610
5	2.757	3.003	3.365	3.634	4.032	4.773	6.869
6	2.612	2.829	3.143	3.372	3.707	4.317	5.959
7	2.517	2.715	2.998	3.203	3.499	4.029	5.408
8	2.449	2.634	2.896	3.085	3.355	3.833	5.041
9	2.398	2.574	2.821	2.998	3.250	3.690	4.781
10	2.359	2.527	2.764	2.932	3.169	3.581	4.587
11	2.328	2.491	2.718	2.879	3.106	3.497	4.437
12	2.303	2.461	2.681	2.836	3.055	3.428	4.318
13	2.282	2.436	2.650	2.801	3.012	3.372	4.221
14	2.264	2.415	2.624	2.771	2.977	3.326	4.140
15	2.249	2.397	2.602	2.746	2.947	3.286	4.073
16	2.235	2.382	2.583	2.724	2.921	3.252	4.015
17	2.224	2.368	2.567	2.706	2.898	3.222	3.965
18	2.214	2.356	2.552	2.689	2.878	3.197	3.922
19	2.205	2.346	2.539	2.674	2.861	3.174	3.883
20	2.197	2.336	2.528	2.661	2.845	3.153	3.849
21	2.189	2.328	2.518	2.649	2.831	3.135	3.819
22	2.183	2.320	2.508	2.639	2.819	3.119	3.792
23	2.177	2.313	2.500	2.629	2.807	3.104	3.768
24	2.172	2.307	2.492	2.620	2.797	3.091	3.745
25	2.167	2.301	2.485	2.612	2.787	3.078	3.725
26	2.162	2.296	2.479	2.605	2.779	3.067	3.707
27	2.158	2.291	2.473	2.598	2.771	3.057	3.690
28	2.154	2.286	2.467	2.592	2.763	3.047	3.674
29	2.150	2.282	2.462	2.586	2.756	3.038	3.659
30	2.147	2.278	2.457	2.581	2.750	3.030	3.646
40	2.123	2.250	2.423	2.542	2.704	2.971	3.551
60	2.099	2.223	2.390	2.504	2.660	2.915	3.460
120	2.076	2.196	2.358	2.468	2.617	2.860	3.373
∞	2.054	2.170	2.326	2.432	2.576	2.807	3.291

TABLE A.3 Percentiles of the χ^2 Distribution

Entry is $\chi^2(A; \nu)$ where $P\{\chi^2(\nu) \le \chi^2(A; \nu)\} = A$

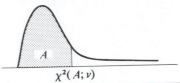

$\chi^2(A; \nu)$

ν	.005	.010	.025	.050	.100	.900	.950	.975	.990	.995
1	0.0^4393	0.0^3157	0.0^3982	0.0^2393	0.0158	2.71	3.84	5.02	6.63	7.88
2	0.0100	0.0201	0.0506	0.103	0.211	4.61	5.99	7.38	9.21	10.60
3	0.072	0.115	0.216	0.352	0.584	6.25	7.81	9.35	11.34	12.84
4	0.207	0.297	0.484	0.711	1.064	7.78	9.49	11.14	13.28	14.86
5	0.412	0.554	0.831	1.145	1.61	9.24	11.07	12.83	15.09	16.75
6	0.676	0.872	1.24	1.64	2.20	10.64	12.59	14.45	16.81	18.55
7	0.989	1.24	1.69	2.17	2.83	12.02	14.07	16.01	18.48	20.28
8	1.34	1.65	2.18	2.73	3.49	13.36	15.51	17.53	20.09	21.96
9	1.73	2.09	2.70	3.33	4.17	14.68	16.92	19.02	21.67	23.59
10	2.16	2.56	3.25	3.94	4.87	15.99	18.31	20.48	23.21	25.19
11	2.60	3.05	3.82	4.57	5.58	17.28	19.68	21.92	24.73	26.76
12	3.07	3.57	4.40	5.23	6.30	18.55	21.03	23.34	26.22	28.30
13	3.57	4.11	5.01	5.89	7.04	19.81	22.36	24.74	27.69	29.82
14	4.07	4.66	5.63	6.57	7.79	21.06	23.68	26.12	29.14	31.32
15	4.60	5.23	6.26	7.26	8.55	22.31	25.00	27.49	30.58	32.80
16	5.14	5.81	6.91	7.96	9.31	23.54	26.30	28.85	32.00	34.27
17	5.70	6.41	7.56	8.67	10.09	24.77	27.59	30.19	33.41	35.72
18	6.26	7.01	8.23	9.39	10.86	25.99	28.87	31.53	34.81	37.16
19	6.84	7.63	8.91	10.12	11.65	27.20	30.14	32.85	36.19	38.58
20	7.43	8.26	9.59	10.85	12.44	28.41	31.41	34.17	37.57	40.00
21	8.03	8.90	10.28	11.59	13.24	29.62	32.67	35.48	38.93	41.40
22	8.64	9.54	10.98	12.34	14.04	30.81	33.92	36.78	40.29	42.80
23	9.26	10.20	11.69	13.09	14.85	32.01	35.17	38.08	41.64	44.18
24	9.89	10.86	12.40	13.85	15.66	33.20	36.42	39.36	42.98	45.56
25	10.52	11.52	13.12	14.61	16.47	34.38	37.65	40.65	44.31	46.93
26	11.16	12.20	13.84	15.38	17.29	35.56	38.89	41.92	45.64	48.29
27	11.81	12.88	14.57	16.15	18.11	36.74	40.11	43.19	46.96	49.64
28	12.46	13.56	15.31	16.93	18.94	37.92	41.34	44.46	48.28	50.99
29	13.12	14.26	16.05	17.71	19.77	39.09	42.56	45.72	49.59	52.34
30	13.79	14.95	16.79	18.49	20.60	40.26	43.77	46.98	50.89	53.67
40	20.71	22.16	24.43	26.51	29.05	51.81	55.76	59.34	63.69	66.77
50	27.99	29.71	32.36	34.76	37.69	63.17	67.50	71.42	76.15	79.49
60	35.53	37.48	40.48	43.19	46.46	74.40	79.08	83.30	88.38	91.95
70	43.28	45.44	48.76	51.74	55.33	85.53	90.53	95.02	100.4	104.2
80	51.17	53.54	57.15	60.39	64.28	96.58	101.9	106.6	112.3	116.3
90	59.20	61.75	65.65	69.13	73.29	107.6	113.1	118.1	124.1	128.3
100	67.33	70.06	74.22	77.93	82.36	118.5	124.3	129.6	135.8	140.2

Source: Reprinted, with permission, from C. M. Thompson, "Table of Percentage Points of the Chi-Square Distribution," *Biometrika* 32 (1941), pp. 188–89.

TABLE A.4 Percentiles of the F Distribution

Entry is $F(A; \nu_1, \nu_2)$ where $P\{F(\nu_1, \nu_2) \le F(A; \nu_1, \nu_2)\} = A$

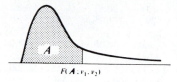

$$F(A; \nu_1, \nu_2) = \frac{1}{F(1 - A; \nu_2, \nu_1)}$$

Den. df	α	Numerator df SSR								
		1	2	3	4	5	6	7	8	9
1	.50	1.00	1.50	1.71	1.82	1.89	1.94	1.98	2.00	2.03
	.90	39.9	49.5	53.6	55.8	57.2	58.2	58.9	59.4	59.9
	.95	161	200	216	225	230	234	237	239	241
	.975	648	800	864	900	922	937	948	957	963
	.99	4,052	5,000	5,403	5,625	5,764	5,859	5,928	5,981	6,022
	.995	16,211	20,000	21,615	22,500	23,056	23,437	23,715	23,925	24,091
	.999	405,280	500,000	540,380	562,500	576,400	585,940	592,870	598,140	602,280
2	.50	0.667	1.00	1.13	1.21	1.25	1.28	1.30	1.32	1.33
	.90	8.53	9.00	9.16	9.24	9.29	9.33	9.35	9.37	9.38
	.95	18.5	19.0	19.2	19.2	19.3	19.3	19.4	19.4	19.4
	.975	38.5	39.0	39.2	39.2	39.3	39.3	39.4	39.4	39.4
	.99	98.5	99.0	99.2	99.2	99.3	99.3	99.4	99.4	99.4
	.995	199	199	199	199	199	199	199	199	199
	.999	998.5	999.0	999.2	999.2	999.3	999.3	999.4	999.4	999.4
3	.50	0.585	0.881	1.00	1.06	1.10	1.13	1.15	1.16	1.17
	.90	5.54	5.46	5.39	5.34	5.31	5.28	5.27	5.25	5.24
	.95	10.1	9.55	9.28	9.12	9.01	8.94	8.89	8.85	8.81
	.975	17.4	16.0	15.4	15.1	14.9	14.7	14.6	14.5	14.5
	.99	34.1	30.8	29.5	28.7	28.2	27.9	27.7	27.5	27.3
	.995	55.6	49.8	47.5	46.2	45.4	44.8	44.4	44.1	43.9
	.999	167.0	148.5	141.1	137.1	134.6	132.8	131.6	130.6	129.9
4	.50	0.549	0.828	0.941	1.00	1.04	1.06	1.08	1.09	1.10
	.90	4.54	4.32	4.19	4.11	4.05	4.01	3.98	3.95	3.94
	.95	7.71	6.94	6.59	6.39	6.26	6.16	6.09	6.04	6.00
	.975	12.2	10.6	9.98	9.60	9.36	9.20	9.07	8.98	8.90
	.99	21.2	18.0	16.7	16.0	15.5	15.2	15.0	14.8	14.7
	.995	31.3	26.3	24.3	23.2	22.5	22.0	21.6	21.4	21.1
	.999	74.1	61.2	56.2	53.4	51.7	50.5	49.7	49.0	48.5
5	.50	0.528	0.799	0.907	0.965	1.00	1.02	1.04	1.05	1.06
	.90	4.06	3.78	3.62	3.52	3.45	3.40	3.37	3.34	3.32
	.95	6.61	5.79	5.41	5.19	5.05	4.95	4.88	4.82	4.77
	.975	10.0	8.43	7.76	7.39	7.15	6.98	6.85	6.76	6.68
	.99	16.3	13.3	12.1	11.4	11.0	10.7	10.5	10.3	10.2
	.995	22.8	18.3	16.5	15.6	14.9	14.5	14.2	14.0	13.8
	.999	47.2	37.1	33.2	31.1	29.8	28.8	28.2	27.6	27.2
6	.50	0.515	0.780	0.886	0.942	0.977	1.00	1.02	1.03	1.04
	.90	3.78	3.46	3.29	3.18	3.11	3.05	3.01	2.98	2.96
	.95	5.99	5.14	4.76	4.53	4.39	4.28	4.21	4.15	4.10
	.975	8.81	7.26	6.60	6.23	5.99	5.82	5.70	5.60	5.52
	.99	13.7	10.9	9.78	9.15	8.75	8.47	8.26	8.10	7.98
	.995	18.6	14.5	12.9	12.0	11.5	11.1	10.8	10.6	10.4
	.999	35.5	27.0	23.7	21.9	20.8	20.0	19.5	19.0	18.7
7	.50	0.506	0.767	0.871	0.926	0.960	0.983	1.00	1.01	1.02
	.90	3.59	3.26	3.07	2.96	2.88	2.83	2.78	2.75	2.72
	.95	5.59	4.74	4.35	4.12	3.97	3.87	3.79	3.73	3.68
	.975	8.07	6.54	5.89	5.52	5.29	5.12	4.99	4.90	4.82
	.99	12.2	9.55	8.45	7.85	7.46	7.19	6.99	6.84	6.72
	.995	16.2	12.4	10.9	10.1	9.52	9.16	8.89	8.68	8.51
	.999	29.2	21.7	18.8	17.2	16.2	15.5	15.0	14.6	14.3

Den. df	A	Numerator df								
		10	12	15	20	24	30	60	120	∞
1	.50	2.04	2.07	2.09	2.12	2.13	2.15	2.17	2.18	2.20
	.90	60.2	60.7	61.2	61.7	62.0	62.3	62.8	63.1	63.3
	.95	242	244	246	248	249	250	252	253	254
	.975	969	977	985	993	997	1,001	1,010	1,014	1,018
	.99	6,056	6,106	6,157	6,209	6,235	6,261	6,313	6,339	6,366
	.995	24,224	24,426	24,630	24,836	24,940	25,044	25,253	25,359	25,464
	.999	605,620	610,670	615,760	620,910	623,500	626,100	631,340	633,970	636,620
2	.50	1.34	1.36	1.38	1.39	1.40	1.41	1.43	1.43	1.44
	.90	9.39	9.41	9.42	9.44	9.45	9.46	9.47	9.48	9.49
	.95	19.4	19.4	19.4	19.4	19.5	19.5	19.5	19.5	19.5
	.975	39.4	39.4	39.4	39.4	39.5	39.5	39.5	39.5	39.5
	.99	99.4	99.4	99.4	99.4	99.5	99.5	99.5	99.5	99.5
	.995	199	199	199	199	199	199	199	199	200
	.999	999.4	999.4	999.4	999.4	999.5	999.5	999.5	999.5	999.5
3	.50	1.18	1.20	1.21	1.23	1.23	1.24	1.25	1.26	1.27
	.90	5.23	5.22	5.20	5.18	5.18	5.17	5.15	5.14	5.13
	.95	8.79	8.74	8.70	8.66	8.64	8.62	8.57	8.55	8.53
	.975	14.4	14.3	14.3	14.2	14.1	14.1	14.0	13.9	13.9
	.99	27.2	27.1	26.9	26.7	26.6	26.5	26.3	26.2	26.1
	.995	43.7	43.4	43.1	42.8	42.6	42.5	42.1	42.0	41.8
	.999	129.2	128.3	127.4	126.4	125.9	125.4	124.5	124.0	123.5
4	.50	1.11	1.13	1.14	1.15	1.16	1.16	1.18	1.18	1.19
	.90	3.92	3.90	3.87	3.84	3.83	3.82	3.79	3.78	3.76
	.95	5.96	5.91	5.86	5.80	5.77	5.75	5.69	5.66	5.63
	.975	8.84	8.75	8.66	8.56	8.51	8.46	8.36	8.31	8.26
	.99	14.5	14.4	14.2	14.0	13.9	13.8	13.7	13.6	13.5
	.995	21.0	20.7	20.4	20.2	20.0	19.9	19.6	19.5	19.3
	.999	48.1	47.4	46.8	46.1	45.8	45.4	44.7	44.4	44.1
5	.50	1.07	1.09	1.10	1.11	1.12	1.12	1.14	1.14	1.15
	.90	3.30	3.27	3.24	3.21	3.19	3.17	3.14	3.12	3.11
	.95	4.74	4.68	4.62	4.56	4.53	4.50	4.43	4.40	4.37
	.975	6.62	6.52	6.43	6.33	6.28	6.23	6.12	6.07	6.02
	.99	10.1	9.89	9.72	9.55	9.47	9.38	9.20	9.11	9.02
	.995	13.6	13.4	13.1	12.9	12.8	12.7	12.4	12.3	12.1
	.999	26.9	26.4	25.9	25.4	25.1	24.9	24.3	24.1	23.8
6	.50	1.05	1.06	1.07	1.08	1.09	1.10	1.11	1.12	1.12
	.90	2.94	2.90	2.87	2.84	2.82	2.80	2.76	2.74	2.72
	.95	4.06	4.00	3.94	3.87	3.84	3.81	3.74	3.70	3.67
	.975	5.46	5.37	5.27	5.17	5.12	5.07	4.96	4.90	4.85
	.99	7.87	7.72	7.56	7.40	7.31	7.23	7.06	6.97	6.88
	.995	10.2	10.0	9.81	9.59	9.47	9.36	9.12	9.00	8.88
	.999	18.4	18.0	17.6	17.1	16.9	16.7	16.2	16.0	15.7
7	.50	1.03	1.04	1.05	1.07	1.07	1.08	1.09	1.10	1.10
	.90	2.70	2.67	2.63	2.59	2.58	2.56	2.51	2.49	2.47
	.95	3.64	3.57	3.51	3.44	3.41	3.38	3.30	3.27	3.23
	.975	4.76	4.67	4.57	4.47	4.42	4.36	4.25	4.20	4.14
	.99	6.62	6.47	6.31	6.16	6.07	5.99	5.82	5.74	5.65
	.995	8.38	8.18	7.97	7.75	7.65	7.53	7.31	7.19	7.08
	.999	14.1	13.7	13.3	12.9	12.7	12.5	12.1	11.9	11.7

Den. df	*A*	\multicolumn{9}{c}{Numerator df}								
		1	2	3	4	5	6	7	8	9
8	.50	0.499	0.757	0.860	0.915	0.948	0.971	0.988	1.00	1.01
	.90	3.46	3.11	2.92	2.81	2.73	2.67	2.62	2.59	2.56
	.95	5.32	4.46	4.07	3.84	3.69	3.58	3.50	3.44	3.39
	.975	7.57	6.06	5.42	5.05	4.82	4.65	4.53	4.43	4.36
	.99	11.3	8.65	7.59	7.01	6.63	6.37	6.18	6.03	5.91
	.995	14.7	11.0	9.60	8.81	8.30	7.95	7.69	7.50	7.34
	.999	25.4	18.5	15.8	14.4	13.5	12.9	12.4	12.0	11.8
9	.50	0.494	0.749	0.852	0.906	0.939	0.962	0.978	0.990	1.00
	.90	3.36	3.01	2.81	2.69	2.61	2.55	2.51	2.47	2.44
	.95	5.12	4.26	3.86	3.63	3.48	3.37	3.29	3.23	3.18
	.975	7.21	5.71	5.08	4.72	4.48	4.32	4.20	4.10	4.03
	.99	10.6	8.02	6.99	6.42	6.06	5.80	5.61	5.47	5.35
	.995	13.6	10.1	8.72	7.96	7.47	7.13	6.88	6.69	6.54
	.999	22.9	16.4	13.9	12.6	11.7	11.1	10.7	10.4	10.1
10	.50	0.490	0.743	0.845	0.899	0.932	0.954	0.971	0.983	0.992
	.90	3.29	2.92	2.73	2.61	2.52	2.46	2.41	2.38	2.35
	.95	4.96	4.10	3.71	3.48	3.33	3.22	3.14	3.07	3.02
	.975	6.94	5.46	4.83	4.47	4.24	4.07	3.95	3.85	3.78
	.99	10.0	7.56	6.55	5.99	5.64	5.39	5.20	5.06	4.94
	.995	12.8	9.43	8.08	7.34	6.87	6.54	6.30	6.12	5.97
	.999	21.0	14.9	12.6	11.3	10.5	9.93	9.52	9.20	8.96
12	.50	0.484	0.735	0.835	0.888	0.921	0.943	0.959	0.972	0.981
	.90	3.18	2.81	2.61	2.48	2.39	2.33	2.28	2.24	2.21
	.95	4.75	3.89	3.49	3.26	3.11	3.00	2.91	2.85	2.80
	.975	6.55	5.10	4.47	4.12	3.89	3.73	3.61	3.51	3.44
	.99	9.33	6.93	5.95	5.41	5.06	4.82	4.64	4.50	4.39
	.995	11.8	8.51	7.23	6.52	6.07	5.76	5.52	5.35	5.20
	.999	18.6	13.0	10.8	9.63	8.89	8.38	8.00	7.71	7.48
15	.50	0.478	0.726	0.826	0.878	0.911	0.933	0.949	0.960	0.970
	.90	3.07	2.70	2.49	2.36	2.27	2.21	2.16	2.12	2.09
	.95	4.54	3.68	3.29	3.06	2.90	2.79	2.71	2.64	2.59
	.975	6.20	4.77	4.15	3.80	3.58	3.41	3.29	3.20	3.12
	.99	8.68	6.36	5.42	4.89	4.56	4.32	4.14	4.00	3.89
	.995	10.8	7.70	6.48	5.80	5.37	5.07	4.85	4.67	4.54
	.999	16.6	11.3	9.34	8.25	7.57	7.09	6.74	6.47	6.26
20	.50	0.472	0.718	0.816	0.868	0.900	0.922	0.938	0.950	0.959
	.90	2.97	2.59	2.38	2.25	2.16	2.09	2.04	2.00	1.96
	.95	4.35	3.49	3.10	2.87	2.71	2.60	2.51	2.45	2.39
	.975	5.87	4.46	3.86	3.51	3.29	3.13	3.01	2.91	2.84
	.99	8.10	5.85	4.94	4.43	4.10	3.87	3.70	3.56	3.46
	.995	9.94	6.99	5.82	5.17	4.76	4.47	4.26	4.09	3.96
	.999	14.8	9.95	8.10	7.10	6.46	6.02	5.69	5.44	5.24
24	.50	0.469	0.714	0.812	0.863	0.895	0.917	0.932	0.944	0.953
	.90	2.93	2.54	2.33	2.19	2.10	2.04	1.98	1.94	1.91
	.95	4.26	3.40	3.01	2.78	2.62	2.51	2.42	2.36	2.30
	.975	5.72	4.32	3.72	3.38	3.15	2.99	2.87	2.78	2.70
	.99	7.82	5.61	4.72	4.22	3.90	3.67	3.50	3.36	3.26
	.995	9.55	6.66	5.52	4.89	4.49	4.20	3.99	3.83	3.69
	.999	14.0	9.34	7.55	6.59	5.98	5.55	5.23	4.99	4.80

Den. df	A	Numerator df								
		10	12	15	20	24	30	60	120	∞
8	.50	1.02	1.03	1.04	1.05	1.06	1.07	1.08	1.08	1.09
	.90	2.54	2.50	2.46	2.42	2.40	2.38	2.34	2.32	2.29
	.95	3.35	3.28	3.22	3.15	3.12	3.08	3.01	2.97	2.93
	.975	4.30	4.20	4.10	4.00	3.95	3.89	3.78	3.73	3.67
	.99	5.81	5.67	5.52	5.36	5.28	5.20	5.03	4.95	4.86
	.995	7.21	7.01	6.81	6.61	6.50	6.40	6.18	6.06	5.95
	.999	11.5	11.2	10.8	10.5	10.3	10.1	9.73	9.53	9.33
9	.50	1.01	1.02	1.03	1.04	1.05	1.05	1.07	1.07	1.08
	.90	2.42	2.38	2.34	2.30	2.28	2.25	2.21	2.18	2.16
	.95	3.14	3.07	3.01	2.94	2.90	2.86	2.79	2.75	2.71
	.975	3.96	3.87	3.77	3.67	3.61	3.56	3.45	3.39	3.33
	.99	5.26	5.11	4.96	4.81	4.73	4.65	4.48	4.40	4.31
	.995	6.42	6.23	6.03	5.83	5.73	5.62	5.41	5.30	5.19
	.999	9.89	9.57	9.24	8.90	8.72	8.55	8.19	8.00	7.81
10	.50	1.00	1.01	1.02	1.03	1.04	1.05	1.06	1.06	1.07
	.90	2.32	2.28	2.24	2.20	2.18	2.16	2.11	2.08	2.06
	.95	2.98	2.91	2.84	2.77	2.74	2.70	2.62	2.58	2.54
	.975	3.72	3.62	3.52	3.42	3.37	3.31	3.20	3.14	3.08
	.99	4.85	4.71	4.56	4.41	4.33	4.25	4.08	4.00	3.91
	.995	5.85	5.66	5.47	5.27	5.17	5.07	4.86	4.75	4.64
	.999	8.75	8.45	8.13	7.80	7.64	7.47	7.12	6.94	6.76
12	.50	0.989	1.00	1.01	1.02	1.03	1.03	1.05	1.05	1.06
	.90	2.19	2.15	2.10	2.06	2.04	2.01	1.96	1.93	1.90
	.95	2.75	2.69	2.62	2.54	2.51	2.47	2.38	2.34	2.30
	.975	3.37	3.28	3.18	3.07	3.02	2.96	2.85	2.79	2.72
	.99	4.30	4.16	4.01	3.86	3.78	3.70	3.54	3.45	3.36
	.995	5.09	4.91	4.72	4.53	4.43	4.33	4.12	4.01	3.90
	.999	7.29	7.00	6.71	6.40	6.25	6.09	5.76	5.59	5.42
15	.50	0.977	0.989	1.00	1.01	1.02	1.02	1.03	1.04	1.05
	.90	2.06	2.02	1.97	1.92	1.90	1.87	1.82	1.79	1.76
	.95	2.54	2.48	2.40	2.33	2.29	2.25	2.16	2.11	2.07
	.975	3.06	2.96	2.86	2.76	2.70	2.64	2.52	2.46	2.40
	.99	3.80	3.67	3.52	3.37	3.29	3.21	3.05	2.96	2.87
	.995	4.42	4.25	4.07	3.88	3.79	3.69	3.48	3.37	3.26
	.999	6.08	5.81	5.54	5.25	5.10	4.95	4.64	4.48	4.31
20	.50	0.966	0.977	0.989	1.00	1.01	1.01	1.02	1.03	1.03
	.90	1.94	1.89	1.84	1.79	1.77	1.74	1.68	1.64	1.61
	.95	2.35	2.28	2.20	2.12	2.08	2.04	1.95	1.90	1.84
	.975	2.77	2.68	2.57	2.46	2.41	2.35	2.22	2.16	2.09
	.99	3.37	3.23	3.09	2.94	2.86	2.78	2.61	2.52	2.42
	.995	3.85	3.68	3.50	3.32	3.22	3.12	2.92	2.81	2.69
	.999	5.08	4.82	4.56	4.29	4.15	4.00	3.70	3.54	3.38
24	.50	0.961	0.972	0.983	0.994	1.00	1.01	1.02	1.02	1.03
	.90	1.88	1.83	1.78	1.73	1.70	1.67	1.61	1.57	1.53
	.95	2.25	2.18	2.11	2.03	1.98	1.94	1.84	1.79	1.73
	.975	2.64	2.54	2.44	2.33	2.27	2.21	2.08	2.01	1.94
	.99	3.17	3.03	2.89	2.74	2.66	2.58	2.40	2.31	2.21
	.995	3.59	3.42	3.25	3.06	2.97	2.87	2.66	2.55	2.43
	.999	4.64	4.39	4.14	3.87	3.74	3.59	3.29	3.14	2.97

TABLE A.4 *(continued)* Percentiles of the *F* Distribution

Den. df	*A*	Numerator df								
		1	2	3	4	5	6	7	8	9
30	.50	0.466	0.709	0.807	0.858	0.890	0.912	0.927	0.939	0.948
	.90	2.88	2.49	2.28	2.14	2.05	1.98	1.93	1.88	1.85
	.95	4.17	3.32	2.92	2.69	2.53	2.42	2.33	2.27	2.21
	.975	5.57	4.18	3.59	3.25	3.03	2.87	2.75	2.65	2.57
	.99	7.56	5.39	4.51	4.02	3.70	3.47	3.30	3.17	3.07
	.995	9.18	6.35	5.24	4.62	4.23	3.95	3.74	3.58	3.45
	.999	13.3	8.77	7.05	6.12	5.53	5.12	4.82	4.58	4.39
60	.50	0.461	0.701	0.798	0.849	0.880	0.901	0.917	0.928	0.937
	.90	2.79	2.39	2.18	2.04	1.95	1.87	1.82	1.77	1.74
	.95	4.00	3.15	2.76	2.53	2.37	2.25	2.17	2.10	2.04
	.975	5.29	3.93	3.34	3.01	2.79	2.63	2.51	2.41	2.33
	.99	7.08	4.98	4.13	3.65	3.34	3.12	2.95	2.82	2.72
	.995	8.49	5.80	4.73	4.14	3.76	3.49	3.29	3.13	3.01
	.999	12.0	7.77	6.17	5.31	4.76	4.37	4.09	3.86	3.69
120	.50	0.458	0.697	0.793	0.844	0.875	0.896	0.912	0.923	0.932
	.90	2.75	2.35	2.13	1.99	1.90	1.82	1.77	1.72	1.68
	.95	3.92	3.07	2.68	2.45	2.29	2.18	2.09	2.02	1.96
	.975	5.15	3.80	3.23	2.89	2.67	2.52	2.39	2.30	2.22
	.99	6.85	4.79	3.95	3.48	3.17	2.96	2.79	2.66	2.56
	.995	8.18	5.54	4.50	3.92	3.55	3.28	3.09	2.93	2.81
	.999	11.4	7.32	5.78	4.95	4.42	4.04	3.77	3.55	3.38
∞	.50	0.455	0.693	0.789	0.839	0.870	0.891	0.907	0.918	0.927
	.90	2.71	2.30	2.08	1.94	1.85	1.77	1.72	1.67	1.63
	.95	3.84	3.00	2.60	2.37	2.21	2.10	2.01	1.94	1.88
	.975	5.02	3.69	3.12	2.79	2.57	2.41	2.29	2.19	2.11
	.99	6.63	4.61	3.78	3.32	3.02	2.80	2.64	2.51	2.41
	.995	7.88	5.30	4.28	3.72	3.35	3.09	2.90	2.74	2.62
	.999	10.8	6.91	5.42	4.62	4.10	3.74	3.47	3.27	3.10

TABLE A.4 (concluded)   Percentiles of the *F* Distribution

Den. df	*A*	Numerator df								
		10	12	15	20	24	30	60	120	∞
30	.50	0.955	0.966	0.978	0.989	0.994	1.00	1.01	1.02	1.02
	.90	1.82	1.77	1.72	1.67	1.64	1.61	1.54	1.50	1.46
	.95	2.16	2.09	2.01	1.93	1.89	1.84	1.74	1.68	1.62
	.975	2.51	2.41	2.31	2.20	2.14	2.07	1.94	1.87	1.79
	.99	2.98	2.84	2.70	2.55	2.47	2.39	2.21	2.11	2.01
	.995	3.34	3.18	3.01	2.82	2.73	2.63	2.42	2.30	2.18
	.999	4.24	4.00	3.75	3.49	3.36	3.22	2.92	2.76	2.59
60	.50	0.945	0.956	0.967	0.978	0.983	0.989	1.00	1.01	1.01
	.90	1.71	1.66	1.60	1.54	1.51	1.48	1.40	1.35	1.29
	.95	1.99	1.92	1.84	1.75	1.70	1.65	1.53	1.47	1.39
	.975	2.27	2.17	2.06	1.94	1.88	1.82	1.67	1.58	1.48
	.99	2.63	2.50	2.35	2.20	2.12	2.03	1.84	1.73	1.60
	.995	2.90	2.74	2.57	2.39	2.29	2.19	1.96	1.83	1.69
	.999	3.54	3.32	3.08	2.83	2.69	2.55	2.25	2.08	1.89
120	.50	0.939	0.950	0.961	0.972	0.978	0.983	0.994	1.00	1.01
	.90	1.65	1.60	1.55	1.48	1.45	1.41	1.32	1.26	1.19
	.95	1.91	1.83	1.75	1.66	1.61	1.55	1.43	1.35	1.25
	.975	2.16	2.05	1.95	1.82	1.76	1.69	1.53	1.43	1.31
	.99	2.47	2.34	2.19	2.03	1.95	1.86	1.66	1.53	1.38
	.995	2.71	2.54	2.37	2.19	2.09	1.98	1.75	1.61	1.43
	.999	3.24	3.02	2.78	2.53	2.40	2.26	1.95	1.77	1.54
∞	.50	0.934	0.945	0.956	0.967	0.972	0.978	0.989	0.994	1.00
	.90	1.60	1.55	1.49	1.42	1.38	1.34	1.24	1.17	1.00
	.95	1.83	1.75	1.67	1.57	1.52	1.46	1.32	1.22	1.00
	.975	2.05	1.94	1.83	1.71	1.64	1.57	1.39	1.27	1.00
	.99	2.32	2.18	2.04	1.88	1.79	1.70	1.47	1.32	1.00
	.995	2.52	2.36	2.19	2.00	1.90	1.79	1.53	1.36	1.00
	.999	2.96	2.74	2.51	2.27	2.13	1.99	1.66	1.45	1.00

Source: Reprinted from Table 5 of Pearson and Hartley, *Biometrika Tables for Statisticians,* Volume 2, 1972, published by the Cambridge University Press, on behalf of The Biometrika Society, by permission of the authors and publishers.

TABLE A.5   Power Function for Two-Sided t Test

$$\alpha = .05$$

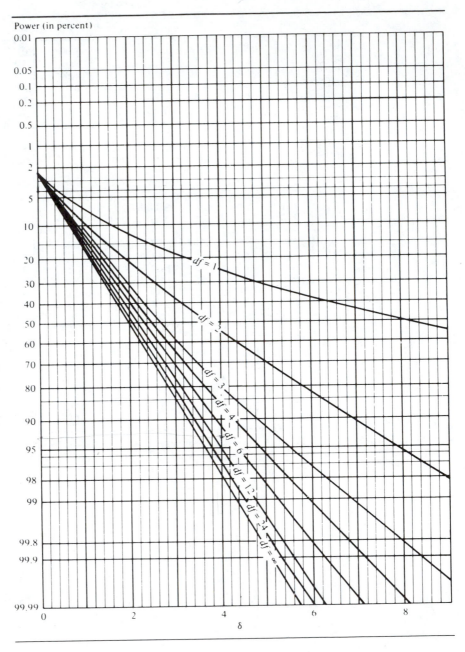

TABLE A.5 (concluded)   Power Function for Two-Sided t Test

$\alpha = .01$

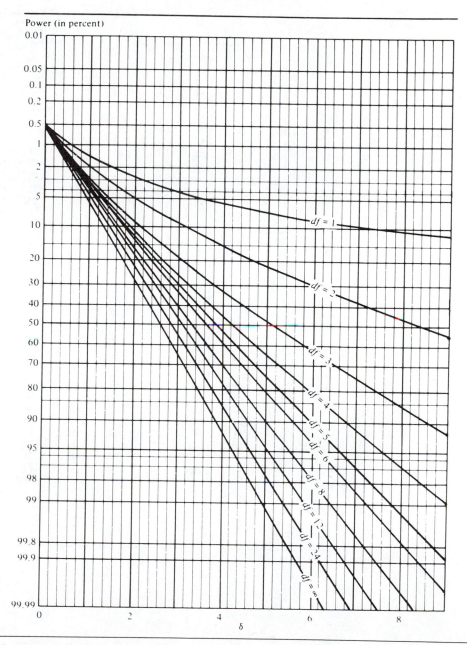

Power (in percent)

Source: Reprinted, with permission, from D. B. Owen, *Handbook of Statistical Tables* (Reading, Mass.: Addison-Wesley Publishing, 1962), pp. 32, 34. Courtesy of U.S. Atomic Energy Commission.

TABLE A.6   Durbin-Watson Test Bounds

Level of Significance $\alpha = .05$

n	$p-1=1$ d_L	d_U	$p-1=2$ d_L	d_U	$p-1=3$ d_L	d_U	$p-1=4$ d_L	d_U	$p-1=5$ d_L	d_U
15	1.08	1.36	0.95	1.54	0.82	1.75	0.69	1.97	0.56	2.21
16	1.10	1.37	0.98	1.54	0.86	1.73	0.74	1.93	0.62	2.15
17	1.13	1.38	1.02	1.54	0.90	1.71	0.78	1.90	0.67	2.10
18	1.16	1.39	1.05	1.53	0.93	1.69	0.82	1.87	0.71	2.06
19	1.18	1.40	1.08	1.53	0.97	1.68	0.86	1.85	0.75	2.02
20	1.20	1.41	1.10	1.54	1.00	1.68	0.90	1.83	0.79	1.99
21	1.22	1.42	1.13	1.54	1.03	1.67	0.93	1.81	0.83	1.96
22	1.24	1.43	1.15	1.54	1.05	1.66	0.96	1.80	0.86	1.94
23	1.26	1.44	1.17	1.54	1.08	1.66	0.99	1.79	0.90	1.92
24	1.27	1.45	1.19	1.55	1.10	1.66	1.01	1.78	0.93	1.90
25	1.29	1.45	1.21	1.55	1.12	1.66	1.04	1.77	0.95	1.89
26	1.30	1.46	1.22	1.55	1.14	1.65	1.06	1.76	0.98	1.88
27	1.32	1.47	1.24	1.56	1.16	1.65	1.08	1.76	1.01	1.86
28	1.33	1.48	1.26	1.56	1.18	1.65	1.10	1.75	1.03	1.85
29	1.34	1.48	1.27	1.56	1.20	1.65	1.12	1.74	1.05	1.84
30	1.35	1.49	1.28	1.57	1.21	1.65	1.14	1.74	1.07	1.83
31	1.36	1.50	1.30	1.57	1.23	1.65	1.16	1.74	1.09	1.83
32	1.37	1.50	1.31	1.57	1.24	1.65	1.18	1.73	1.11	1.82
33	1.38	1.51	1.32	1.58	1.26	1.65	1.19	1.73	1.13	1.81
34	1.39	1.51	1.33	1.58	1.27	1.65	1.21	1.73	1.15	1.81
35	1.40	1.52	1.34	1.58	1.28	1.65	1.22	1.73	1.16	1.80
36	1.41	1.52	1.35	1.59	1.29	1.65	1.24	1.73	1.18	1.80
37	1.42	1.53	1.36	1.59	1.31	1.66	1.25	1.72	1.19	1.80
38	1.43	1.54	1.37	1.59	1.32	1.66	1.26	1.72	1.21	1.79
39	1.43	1.54	1.38	1.60	1.33	1.66	1.27	1.72	1.22	1.79
40	1.44	1.54	1.39	1.60	1.34	1.66	1.29	1.72	1.23	1.79
45	1.48	1.57	1.43	1.62	1.38	1.67	1.34	1.72	1.29	1.78
50	1.50	1.59	1.46	1.63	1.42	1.67	1.38	1.72	1.34	1.77
55	1.53	1.60	1.49	1.64	1.45	1.68	1.41	1.72	1.38	1.77
60	1.55	1.62	1.51	1.65	1.48	1.69	1.44	1.73	1.41	1.77
65	1.57	1.63	1.54	1.66	1.50	1.70	1.47	1.73	1.44	1.77
70	1.58	1.64	1.55	1.67	1.52	1.70	1.49	1.74	1.46	1.77
75	1.60	1.65	1.57	1.68	1.54	1.71	1.51	1.74	1.49	1.77
80	1.61	1.66	1.59	1.69	1.56	1.72	1.53	1.74	1.51	1.77
85	1.62	1.67	1.60	1.70	1.57	1.72	1.55	1.75	1.52	1.77
90	1.63	1.68	1.61	1.70	1.59	1.73	1.57	1.75	1.54	1.78
95	1.64	1.69	1.62	1.71	1.60	1.73	1.58	1.75	1.56	1.78
100	1.65	1.69	1.63	1.72	1.61	1.74	1.59	1.76	1.57	1.78

TABLE A.6 (*concluded*)   Durbin-Watson Test Bounds

Level of Significance $\alpha = .01$

n	$p - 1 = 1$		$p - 1 = 2$		$p - 1 = 3$		$p - 1 = 4$		$p - 1 = 5$	
	d_L	d_U	d_L	d_U	d_L	d_U	d_L	d_U	d_L	d_U
15	0.81	1.07	0.70	1.25	0.59	1.46	0.49	1.70	0.39	1.96
16	0.84	1.09	0.74	1.25	0.63	1.44	0.53	1.66	0.44	1.90
17	0.87	1.10	0.77	1.25	0.67	1.43	0.57	1.63	0.48	1.85
18	0.90	1.12	0.80	1.26	0.71	1.42	0.61	1.60	0.52	1.80
19	0.93	1.13	0.83	1.26	0.74	1.41	0.65	1.58	0.56	1.77
20	0.95	1.15	0.86	1.27	0.77	1.41	0.68	1.57	0.60	1.74
21	0.97	1.16	0.89	1.27	0.80	1.41	0.72	1.55	0.63	1.71
22	1.00	1.17	0.91	1.28	0.83	1.40	0.75	1.54	0.66	1.69
23	1.02	1.19	0.94	1.29	0.86	1.40	0.77	1.53	0.70	1.67
24	1.04	1.20	0.96	1.30	0.88	1.41	0.80	1.53	0.72	1.66
25	1.05	1.21	0.98	1.30	0.90	1.41	0.83	1.52	0.75	1.65
26	1.07	1.22	1.00	1.31	0.93	1.41	0.85	1.52	0.78	1.64
27	1.09	1.23	1.02	1.32	0.95	1.41	0.88	1.51	0.81	1.63
28	1.10	1.24	1.04	1.32	0.97	1.41	0.90	1.51	0.83	1.62
29	1.12	1.25	1.05	1.33	0.99	1.42	0.92	1.51	0.85	1.61
30	1.13	1.26	1.07	1.34	1.01	1.42	0.94	1.51	0.88	1.61
31	1.15	1.27	1.08	1.34	1.02	1.42	0.96	1.51	0.90	1.60
32	1.16	1.28	1.10	1.35	1.04	1.43	0.98	1.51	0.92	1.60
33	1.17	1.29	1.11	1.36	1.05	1.43	1.00	1.51	0.94	1.59
34	1.18	1.30	1.13	1.36	1.07	1.43	1.01	1.51	0.95	1.59
35	1.19	1.31	1.14	1.37	1.08	1.44	1.03	1.51	0.97	1.59
36	1.21	1.32	1.15	1.38	1.10	1.44	1.04	1.51	0.99	1.59
37	1.22	1.32	1.16	1.38	1.11	1.45	1.06	1.51	1.00	1.59
38	1.23	1.33	1.18	1.39	1.12	1.45	1.07	1.52	1.02	1.58
39	1.24	1.34	1.19	1.39	1.14	1.45	1.09	1.52	1.03	1.58
40	1.25	1.34	1.20	1.40	1.15	1.46	1.10	1.52	1.05	1.58
45	1.29	1.38	1.24	1.42	1.20	1.48	1.16	1.53	1.11	1.58
50	1.32	1.40	1.28	1.45	1.24	1.49	1.20	1.54	1.16	1.59
55	1.36	1.43	1.32	1.47	1.28	1.51	1.25	1.55	1.21	1.59
60	1.38	1.45	1.35	1.48	1.32	1.52	1.28	1.56	1.25	1.60
65	1.41	1.47	1.38	1.50	1.35	1.53	1.31	1.57	1.28	1.61
70	1.43	1.49	1.40	1.52	1.37	1.55	1.34	1.58	1.31	1.61
75	1.45	1.50	1.42	1.53	1.39	1.56	1.37	1.59	1.34	1.62
80	1.47	1.52	1.44	1.54	1.42	1.57	1.39	1.60	1.36	1.62
85	1.48	1.53	1.46	1.55	1.43	1.58	1.41	1.60	1.39	1.63
90	1.50	1.54	1.47	1.56	1.45	1.59	1.43	1.61	1.41	1.64
95	1.51	1.55	1.49	1.57	1.47	1.60	1.45	1.62	1.42	1.64
100	1.52	1.56	1.50	1.58	1.48	1.60	1.46	1.63	1.44	1.65

Source: Reprinted, with permission, from J. Durbin and G. S. Watson, "Testing for Serial Correlation in Least Squares Regression. II," *Biometrika* 38 (1951), pp. 159–78.

TABLE A.7   Table of z' Transformation of Correlation Coefficient

r ρ	z' ζ	r ρ	z' ζ	r ρ	z' ζ	r ρ	z' ζ
.00	.0000	.25	.2554	.50	.5493	.75	.973
.01	.0100	.26	.2661	.51	.5627	.76	.996
.02	.0200	.27	.2769	.52	.5763	.77	1.020
.03	.0300	.28	.2877	.53	.5901	.78	1.045
.04	.0400	.29	.2986	.54	.6042	.79	1.071
.05	.0500	.30	.3095	.55	.6184	.80	1.099
.06	.0601	.31	.3205	.56	.6328	.81	1.127
.07	.0701	.32	.3316	.57	.6475	.82	1.157
.08	.0802	.33	.3428	.58	.6625	.83	1.188
.09	.0902	.34	.3541	.59	.6777	.84	1.221
.10	.1003	.35	.3654	.60	.6931	.85	1.256
.11	.1104	.36	.3769	.61	.7089	.86	1.293
.12	.1206	.37	.3884	.62	.7250	.87	1.333
.13	.1307	.38	.4001	.63	.7414	.88	1.376
.14	.1409	.39	.4118	.64	.7582	.89	1.422
.15	.1511	.40	.4236	.65	.7753	.90	1.472
.16	.1614	.41	.4356	.66	.7928	.91	1.528
.17	.1717	.42	.4477	.67	.8107	.92	1.589
.18	.1820	.43	.4599	.68	.8291	.93	1.658
.19	.1923	.44	.4722	.69	.8480	.94	1.738
.20	.2027	.45	.4847	.70	.8673	.95	1.832
.21	.2132	.46	.4973	.71	.8872	.96	1.946
.22	.2237	.47	.5101	.72	.9076	.97	2.092
.23	.2342	.48	.5230	.73	.9287	.98	2.298
.24	.2448	.49	.5361	.74	.9505	.99	2.647

Source: Abridged from Table 14 of Pearson and Hartley, *Biometrika Tables for Statisticians*, Volume 1, 1966, published by the Cambridge University Press, on behalf of The Biometrika Society, by permission of the authors and publishers.

TABLE A.8 Power Function for Analysis of Variance (fixed factor levels model)

$\nu_1 = 2$

Power $= 1 - \beta$

ϕ (for $\alpha = 0.01$)

ϕ (for $\alpha = 0.05$)

$\nu_2 = \infty$

$\alpha = 0.05$

$\alpha = 0.01$

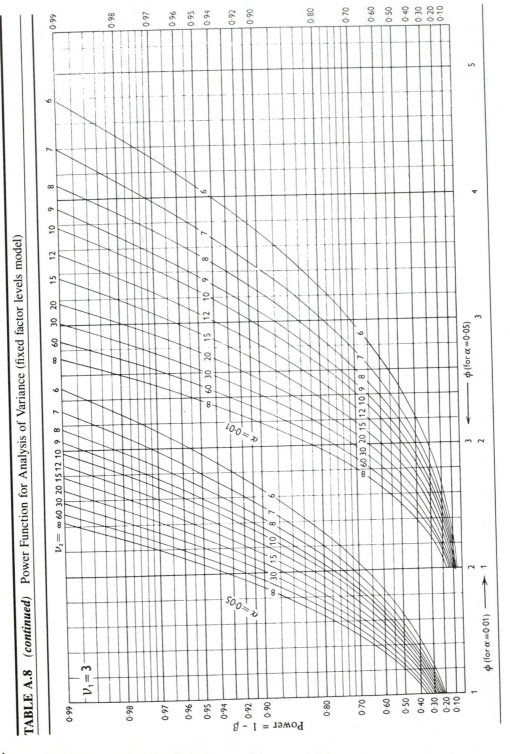

$\nu_1 = 4$

$\nu_2 = \infty$ 60 30 20 15 12 10 9 8

$\alpha = 0.05$

$\alpha = 0.01$

Power $= 1 - \beta$

ϕ (for $\alpha = 0.01$) ——→

ϕ (for $\alpha = 0.05$) ——→

TABLE A.8 *(continued)* Power Function for Analysis of Variance (fixed factor levels model)

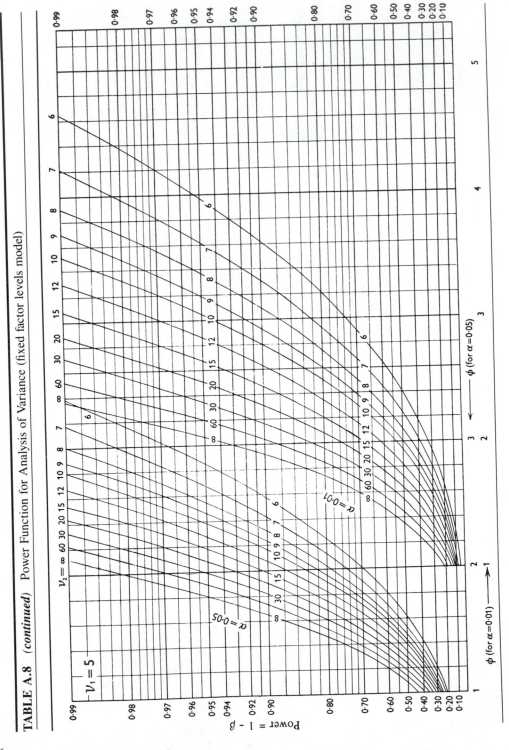

TABLE A.8 (*concluded*) Power Function for Analysis of Variance (fixed factor levels model)

Source: Reprinted, with permission, from E. S. Pearson and H. O. Hartley, "Charts of the Power Function for Analysis of Variance Tests, Derived from the Non-Central F-Distribution," *Biometrika* 38 (1951), pp. 112–30.

1147

TABLE A.9 Percentiles of the Studentized Range Distribution

Entry is $q(1 - \alpha; r, \nu)$ where $P\{q(r, \nu) \leq q(1 - \alpha; r, \nu)\} = 1 - \alpha$

$$1 - \alpha = .90$$

ν	2	3	4	5	6	7	8	9	10	11	12	13	14	15	16	17	18	19	20
1	8.93	13.4	16.4	18.5	20.2	21.5	22.6	23.6	24.5	25.2	25.9	26.5	27.1	27.6	28.1	28.5	29.0	29.3	29.7
2	4.13	5.73	6.77	7.54	8.14	8.63	9.05	9.41	9.72	10.0	10.3	10.5	10.7	10.9	11.1	11.2	11.4	11.5	11.7
3	3.33	4.47	5.20	5.74	6.16	6.51	6.81	7.06	7.29	7.49	7.67	7.83	7.98	8.12	8.25	8.37	8.48	8.58	8.68
4	3.01	3.98	4.59	5.03	5.39	5.68	5.93	6.14	6.33	6.49	6.65	6.78	6.91	7.02	7.13	7.23	7.33	7.41	7.50
5	2.85	3.72	4.26	4.66	4.98	5.24	5.46	5.65	5.82	5.97	6.10	6.22	6.34	6.44	6.54	6.63	6.71	6.79	6.86
6	2.75	3.56	4.07	4.44	4.73	4.97	5.17	5.34	5.50	5.64	5.76	5.87	5.98	6.07	6.16	6.25	6.32	6.40	6.47
7	2.68	3.45	3.93	4.28	4.55	4.78	4.97	5.14	5.28	5.41	5.53	5.64	5.74	5.83	5.91	5.99	6.06	6.13	6.19
8	2.63	3.37	3.83	4.17	4.43	4.65	4.83	4.99	5.13	5.25	5.36	5.46	5.56	5.64	5.72	5.80	5.87	5.93	6.00
9	2.59	3.32	3.76	4.08	4.34	4.54	4.72	4.87	5.01	5.13	5.23	5.33	5.42	5.51	5.58	5.66	5.72	5.79	5.85
10	2.56	3.27	3.70	4.02	4.26	4.47	4.64	4.78	4.91	5.03	5.13	5.23	5.32	5.40	5.47	5.54	5.61	5.67	5.73
11	2.54	3.23	3.66	3.96	4.20	4.40	4.57	4.71	4.84	4.95	5.05	5.15	5.23	5.31	5.38	5.45	5.51	5.57	5.63
12	2.52	3.20	3.62	3.92	4.16	4.35	4.51	4.65	4.78	4.89	4.99	5.08	5.16	5.24	5.31	5.37	5.44	5.49	5.55
13	2.50	3.18	3.59	3.88	4.12	4.30	4.46	4.60	4.72	4.83	4.93	5.02	5.10	5.18	5.25	5.31	5.37	5.43	5.48
14	2.49	3.16	3.56	3.85	4.08	4.27	4.42	4.56	4.68	4.79	4.88	4.97	5.05	5.12	5.19	5.26	5.32	5.37	5.43
15	2.48	3.14	3.54	3.83	4.05	4.23	4.39	4.52	4.64	4.75	4.84	4.93	5.01	5.08	5.15	5.21	5.27	5.32	5.38
16	2.47	3.12	3.52	3.80	4.03	4.21	4.36	4.49	4.61	4.71	4.81	4.89	4.97	5.04	5.11	5.17	5.23	5.28	5.33
17	2.46	3.11	3.50	3.78	4.00	4.18	4.33	4.46	4.58	4.68	4.77	4.86	4.93	5.01	5.07	5.13	5.19	5.24	5.30
18	2.45	3.10	3.49	3.77	3.98	4.16	4.31	4.44	4.55	4.65	4.75	4.83	4.90	4.98	5.04	5.10	5.16	5.21	5.26
19	2.45	3.09	3.47	3.75	3.97	4.14	4.29	4.42	4.53	4.63	4.72	4.80	4.88	4.95	5.01	5.07	5.13	5.18	5.23
20	2.44	3.08	3.46	3.74	3.95	4.12	4.27	4.40	4.51	4.61	4.70	4.78	4.85	4.92	4.99	5.05	5.10	5.16	5.20
24	2.42	3.05	3.42	3.69	3.90	4.07	4.21	4.34	4.44	4.54	4.63	4.71	4.78	4.85	4.91	4.97	5.02	5.07	5.12
30	2.40	3.02	3.39	3.65	3.85	4.02	4.16	4.28	4.38	4.47	4.56	4.64	4.71	4.77	4.83	4.89	4.94	4.99	5.03
40	2.38	2.99	3.35	3.60	3.80	3.96	4.10	4.21	4.32	4.41	4.49	4.56	4.63	4.69	4.75	4.81	4.86	4.90	4.95
60	2.36	2.96	3.31	3.56	3.75	3.91	4.04	4.16	4.25	4.34	4.42	4.49	4.56	4.62	4.67	4.73	4.78	4.82	4.86
120	2.34	2.93	3.28	3.52	3.71	3.86	3.99	4.10	4.19	4.28	4.35	4.42	4.48	4.54	4.60	4.65	4.69	4.74	4.78
∞	2.33	2.90	3.24	3.48	3.66	3.81	3.93	4.04	4.13	4.21	4.28	4.35	4.41	4.47	4.52	4.57	4.61	4.65	4.69

TABLE A.9 (continued)   Percentiles of the Studentized Range Distribution

$1 - \alpha = .95$

ν									r										
	2	3	4	5	6	7	8	9	10	11	12	13	14	15	16	17	18	19	20
1	18.0	27.0	32.8	37.1	40.4	43.1	45.4	47.4	49.1	50.6	52.0	53.2	54.3	55.4	56.3	57.2	58.0	58.8	59.6
2	6.08	8.33	9.80	10.9	11.7	12.4	13.0	13.5	14.0	14.4	14.7	15.1	15.4	15.7	15.9	16.1	16.4	16.6	16.8
3	4.50	5.91	6.82	7.50	8.04	8.48	8.85	9.18	9.46	9.72	9.95	10.2	10.3	10.5	10.7	10.8	11.0	11.1	11.2
4	3.93	5.04	5.76	6.29	6.71	7.05	7.35	7.60	7.83	8.03	8.21	8.37	8.52	8.66	8.79	8.91	9.03	9.13	9.23
5	3.64	4.60	5.22	5.67	6.03	6.33	6.58	6.80	6.99	7.17	7.32	7.47	7.60	7.72	7.83	7.93	8.03	8.12	8.21
6	3.46	4.34	4.90	5.30	5.63	5.90	6.12	6.32	6.49	6.65	6.79	6.92	7.03	7.14	7.24	7.34	7.43	7.51	7.59
7	3.34	4.16	4.68	5.06	5.36	5.61	5.82	6.00	6.16	6.30	6.43	6.55	6.66	6.76	6.85	6.94	7.02	7.10	7.17
8	3.26	4.04	4.53	4.89	5.17	5.40	5.60	5.77	5.92	6.05	6.18	6.29	6.39	6.48	6.57	6.65	6.73	6.80	6.87
9	3.20	3.95	4.41	4.76	5.02	5.24	5.43	5.59	5.74	5.87	5.98	6.09	6.19	6.28	6.36	6.44	6.51	6.58	6.64
10	3.15	3.88	4.33	4.65	4.91	5.12	5.30	5.46	5.60	5.72	5.83	5.93	6.03	6.11	6.19	6.27	6.34	6.40	6.47
11	3.11	3.82	4.26	4.57	4.82	5.03	5.20	5.35	5.49	5.61	5.71	5.81	5.90	5.98	6.06	6.13	6.20	6.27	6.33
12	3.08	3.77	4.20	4.51	4.75	4.95	5.12	5.27	5.39	5.51	5.61	5.71	5.80	5.88	5.95	6.02	6.09	6.15	6.21
13	3.06	3.73	4.15	4.45	4.69	4.88	5.05	5.19	5.32	5.43	5.53	5.63	5.71	5.79	5.86	5.93	5.99	6.05	6.11
14	3.03	3.70	4.11	4.41	4.64	4.83	4.99	5.13	5.25	5.36	5.46	5.55	5.64	5.71	5.79	5.85	5.91	5.97	6.03
15	3.01	3.67	4.08	4.37	4.59	4.78	4.94	5.08	5.20	5.31	5.40	5.49	5.57	5.65	5.72	5.78	5.85	5.90	5.96
16	3.00	3.65	4.05	4.33	4.56	4.74	4.90	5.03	5.15	5.26	5.35	5.44	5.52	5.59	5.66	5.73	5.79	5.84	5.90
17	2.98	3.63	4.02	4.30	4.52	4.70	4.86	4.99	5.11	5.21	5.31	5.39	5.47	5.54	5.61	5.67	5.73	5.79	5.84
18	2.97	3.61	4.00	4.28	4.49	4.67	4.82	4.96	5.07	5.17	5.27	5.35	5.43	5.50	5.57	5.63	5.69	5.74	5.79
19	2.96	3.59	3.98	4.25	4.47	4.65	4.79	4.92	5.04	5.14	5.23	5.31	5.39	5.46	5.53	5.59	5.65	5.70	5.75
20	2.95	3.58	3.96	4.23	4.45	4.62	4.77	4.90	5.01	5.11	5.20	5.28	5.36	5.43	5.49	5.55	5.61	5.66	5.71
24	2.92	3.53	3.90	4.17	4.37	4.54	4.68	4.81	4.92	5.01	5.10	5.18	5.25	5.32	5.38	5.44	5.49	5.55	5.59
30	2.89	3.49	3.85	4.10	4.30	4.46	4.60	4.72	4.82	4.92	5.00	5.08	5.15	5.21	5.27	5.33	5.38	5.43	5.47
40	2.86	3.44	3.79	4.04	4.23	4.39	4.52	4.63	4.73	4.82	4.90	4.98	5.04	5.11	5.16	5.22	5.27	5.31	5.36
60	2.83	3.40	3.74	3.98	4.16	4.31	4.44	4.55	4.65	4.73	4.81	4.88	4.94	5.00	5.06	5.11	5.15	5.20	5.24
120	2.80	3.36	3.68	3.92	4.10	4.24	4.36	4.47	4.56	4.64	4.71	4.78	4.84	4.90	4.95	5.00	5.04	5.09	5.13
∞	2.77	3.31	3.63	3.86	4.03	4.17	4.29	4.39	4.47	4.55	4.62	4.68	4.74	4.80	4.85	4.89	4.93	4.97	5.01

TABLE A.9 *(concluded)*   Percentiles of the Studentized Range Distribution

$1 - \alpha = .99$

r

v	2	3	4	5	6	7	8	9	10	11	12	13	14	15	16	17	18	19	20
1	90.0	135	164	186	202	216	227	237	246	253	260	266	272	277	282	286	290	294	298
2	14.0	19.0	22.3	24.7	26.6	28.2	29.5	30.7	31.7	32.6	33.4	34.1	34.8	35.4	36.0	36.5	37.0	37.5	37.9
3	8.26	10.6	12.2	13.3	14.2	15.0	15.6	16.2	16.7	17.1	17.5	17.9	18.2	18.5	18.8	19.1	19.3	19.5	19.8
4	6.51	8.12	9.17	9.96	10.6	11.1	11.5	11.9	12.3	12.6	12.8	13.1	13.3	13.5	13.7	13.9	14.1	14.2	14.4
5	5.70	6.97	7.80	8.42	8.91	9.32	9.67	9.97	10.2	10.5	10.7	10.9	11.1	11.2	11.4	11.6	11.7	11.8	11.9
6	5.24	6.33	7.03	7.56	7.97	8.32	8.61	8.87	9.10	9.30	9.49	9.65	9.81	9.95	10.1	10.2	10.3	10.4	10.5
7	4.95	5.92	6.54	7.01	7.37	7.68	7.94	8.17	8.37	8.55	8.71	8.86	9.00	9.12	9.24	9.35	9.46	9.55	9.65
8	4.74	5.63	6.20	6.63	6.96	7.24	7.47	7.68	7.87	8.03	8.18	8.31	8.44	8.55	8.66	8.76	8.85	8.94	9.03
9	4.60	5.43	5.96	6.35	6.66	6.91	7.13	7.32	7.49	7.65	7.78	7.91	8.03	8.13	8.23	8.32	8.41	8.49	8.57
10	4.48	5.27	5.77	6.14	6.43	6.67	6.87	7.05	7.21	7.36	7.48	7.60	7.71	7.81	7.91	7.99	8.07	8.15	8.22
11	4.39	5.14	5.62	5.97	6.25	6.48	6.67	6.84	6.99	7.13	7.25	7.36	7.46	7.56	7.65	7.73	7.81	7.88	7.95
12	4.32	5.04	5.50	5.84	6.10	6.32	6.51	6.67	6.81	6.94	7.06	7.17	7.26	7.36	7.44	7.52	7.59	7.66	7.73
13	4.26	4.96	5.40	5.73	5.98	6.19	6.37	6.53	6.67	6.79	6.90	7.01	7.10	7.19	7.27	7.34	7.42	7.48	7.55
14	4.21	4.89	5.32	5.63	5.88	6.08	6.26	6.41	6.54	6.66	6.77	6.87	6.96	7.05	7.12	7.20	7.27	7.33	7.39
15	4.17	4.83	5.25	5.56	5.80	5.99	6.16	6.31	6.44	6.55	6.66	6.76	6.84	6.93	7.00	7.07	7.14	7.20	7.26
16	4.13	4.78	5.19	5.49	5.72	5.92	6.08	6.22	6.35	6.46	6.56	6.66	6.74	6.82	6.90	6.97	7.03	7.09	7.15
17	4.10	4.74	5.14	5.43	5.66	5.85	6.01	6.15	6.27	6.38	6.48	6.57	6.66	6.73	6.80	6.87	6.94	7.00	7.05
18	4.07	4.70	5.09	5.38	5.60	5.79	5.94	6.08	6.20	6.31	6.41	6.50	6.58	6.65	6.72	6.79	6.85	6.91	6.96
19	4.05	4.67	5.05	5.33	5.55	5.73	5.89	6.02	6.14	6.25	6.34	6.43	6.51	6.58	6.65	6.72	6.78	6.84	6.89
20	4.02	4.64	5.02	5.29	5.51	5.69	5.84	5.97	6.09	6.19	6.29	6.37	6.45	6.52	6.59	6.65	6.71	6.76	6.82
24	3.96	4.54	4.91	5.17	5.37	5.54	5.69	5.81	5.92	6.02	6.11	6.19	6.26	6.33	6.39	6.45	6.51	6.56	6.61
30	3.89	4.45	4.80	5.05	5.24	5.40	5.54	5.65	5.76	5.85	5.93	6.01	6.08	6.14	6.20	6.26	6.31	6.36	6.41
40	3.82	4.37	4.70	4.93	5.11	5.27	5.39	5.50	5.60	5.69	5.77	5.84	5.90	5.96	6.02	6.07	6.12	6.17	6.21
60	3.76	4.28	4.60	4.82	4.99	5.13	5.25	5.36	5.45	5.53	5.60	5.67	5.73	5.79	5.84	5.89	5.93	5.98	6.02
120	3.70	4.20	4.50	4.71	4.87	5.01	5.12	5.21	5.30	5.38	5.44	5.51	5.56	5.61	5.66	5.71	5.75	5.79	5.83
∞	3.64	4.12	4.40	4.60	4.76	4.88	4.99	5.08	5.16	5.23	5.29	5.35	5.40	5.45	5.49	5.54	5.57	5.61	5.65

Source: Reprinted, with permission, from Henry Scheffé, *The Analysis of Variance* (New York: John Wiley & Sons, 1959), pp. 434–36.

TABLE A.10 Table for Determining Sample Size for Analysis of Variance (fixed factor levels model)

Power $1 - \beta = .70$

| | $\Delta/\sigma = 1.0$ | | | | $\Delta/\sigma = 1.25$ | | | | $\Delta/\sigma = 1.50$ | | | | $\Delta/\sigma = 1.75$ | | | | $\Delta/\sigma = 2.0$ | | | | $\Delta/\sigma = 2.5$ | | | | $\Delta/\sigma = 3.0$ | | | |
| | α | | | | α | | | | α | | | | α | | | | α | | | | α | | | | α | | | |
r	.2	.1	.05	.01	.2	.1	.05	.01	.2	.1	.05	.01	.2	.1	.05	.01	.2	.1	.05	.01	.2	.1	.05	.01	.2	.1	.05	.01
2	7	11	14	21	5	7	9	15	4	6	7	11	3	4	6	9	3	4	5	7	2	3	4	5	2	3	3	5
3	9	13	17	25	6	9	11	17	5	7	8	12	4	5	7	10	3	4	5	8	3	3	4	6	2	3	3	5
4	11	15	19	28	7	10	13	19	5	7	9	13	4	6	7	10	4	5	6	8	3	3	4	6	2	3	4	5
5	12	17	21	30	8	11	14	20	6	8	10	14	5	6	8	11	4	5	6	9	3	4	5	6	3	3	4	5
6	13	18	22	32	9	12	15	21	6	9	11	15	5	7	8	12	4	5	7	9	3	4	5	7	3	3	4	5
7	14	19	24	34	9	13	16	22	7	9	11	16	5	7	9	12	4	6	7	10	3	4	5	7	3	3	4	5
8	15	20	25	35	10	13	16	23	7	10	12	17	6	7	9	13	5	6	7	10	3	4	5	7	3	3	4	5
9	15	21	26	37	10	14	17	24	7	10	12	17	6	8	9	13	5	6	8	10	3	4	5	7	3	4	4	6
10	16	22	27	38	11	14	18	25	8	10	13	18	6	8	10	14	5	6	8	11	4	5	6	7	3	4	4	6

Power $1 - \beta = .80$

| | $\Delta/\sigma = 1.0$ | | | | $\Delta/\sigma = 1.25$ | | | | $\Delta/\sigma = 1.50$ | | | | $\Delta/\sigma = 1.75$ | | | | $\Delta/\sigma = 2.0$ | | | | $\Delta/\sigma = 2.5$ | | | | $\Delta/\sigma = 3.0$ | | | |
| | α | | | | α | | | | α | | | | α | | | | α | | | | α | | | | α | | | |
r	.2	.1	.05	.01	.2	.1	.05	.01	.2	.1	.05	.01	.2	.1	.05	.01	.2	.1	.05	.01	.2	.1	.05	.01	.2	.1	.05	.01
2	10	14	17	26	7	9	12	17	5	7	9	13	4	5	7	10	3	4	6	8	3	3	4	6	2	3	4	5
3	12	17	21	30	8	11	14	20	6	8	10	14	5	6	8	11	4	5	6	9	3	4	5	7	3	3	4	5
4	14	19	23	33	9	13	15	22	7	9	11	16	5	7	9	12	4	6	7	10	3	4	5	7	3	3	4	5
5	16	21	25	35	10	14	17	23	8	10	12	17	6	8	9	13	5	6	7	10	4	4	5	7	3	4	4	6
6	17	22	27	38	11	15	18	25	8	11	13	18	6	8	10	13	5	7	8	11	4	5	6	8	3	4	4	6
7	18	24	29	39	12	16	19	26	9	11	14	18	7	9	10	14	5	7	8	11	4	5	6	8	3	4	5	6
8	19	25	30	41	12	16	20	27	9	12	14	19	7	9	11	15	6	7	9	12	4	5	6	8	3	4	5	6
9	20	26	31	43	13	17	21	28	9	12	15	20	7	9	11	15	6	7	9	12	4	5	6	8	3	4	5	6
10	21	27	33	44	14	18	21	29	10	13	15	21	8	10	12	16	6	8	9	12	4	5	6	8	3	4	5	6

TABLE A.10 (concluded) Table for Determining Sample Size for Analysis of Variance (fixed factor levels model)

Power $1 - \beta = .90$

r	Δ/σ = 1.0				Δ/σ = 1.25				Δ/σ = 1.50				Δ/σ = 1.75				Δ/σ = 2.0				Δ/σ = 2.5				Δ/σ = 3.0			
	.2	.1	.05	.01	.2	.1	.05	.01	.2	.1	.05	.01	.2	.1	.05	.01	.2	.1	.05	.01	.2	.1	.05	.01	.2	.1	.05	.01
2	14	18	23	32	9	12	15	21	7	9	11	15	5	7	8	12	4	6	7	10	3	4	5	7	3	3	4	6
3	17	22	27	37	11	15	18	24	8	11	13	18	6	8	10	13	5	7	8	11	4	5	6	8	3	3	5	6
4	20	25	30	40	13	16	20	27	9	12	14	19	7	9	11	15	6	7	9	12	4	5	6	8	3	4	5	6
5	21	27	32	43	14	18	21	28	10	13	15	20	8	10	12	15	6	8	9	12	4	5	6	9	4	4	5	7
6	22	29	34	46	15	19	23	30	11	14	16	21	8	11	12	16	7	8	10	13	5	6	7	9	4	4	5	7
7	24	31	36	48	16	20	25	31	11	14	17	22	9	11	13	17	7	9	10	13	5	6	7	9	4	5	5	7
8	26	32	38	50	17	21	25	33	12	15	18	23	9	11	13	17	7	9	11	14	5	6	7	9	4	5	6	7
9	27	33	40	52	17	22	26	34	13	16	18	24	9	12	14	18	8	9	11	14	5	6	8	10	4	5	6	7
10	28	35	41	54	18	23	27	35	13	16	19	25	10	12	14	19	8	10	11	15	5	7	8	10	4	5	6	7

Power $1 - \beta = .95$

r	Δ/σ = 1.0				Δ/σ = 1.25				Δ/σ = 1.50				Δ/σ = 1.75				Δ/σ = 2.0				Δ/σ = 2.5				Δ/σ = 3.0			
	.2	.1	.05	.01	.2	.1	.05	.01	.2	.1	.05	.01	.2	.1	.05	.01	.2	.1	.05	.01	.2	.1	.05	.01	.2	.1	.05	.01
2	18	23	27	38	12	15	18	25	9	11	13	18	7	8	10	14	5	7	8	11	4	5	6	8	3	4	5	6
3	22	27	32	43	14	18	21	29	10	13	15	20	8	10	12	16	6	8	9	12	5	6	7	9	4	4	5	7
4	25	30	36	47	16	20	23	31	12	14	17	22	9	11	13	17	7	9	10	13	5	6	7	9	4	5	5	7
5	27	33	39	51	18	22	25	33	13	15	18	23	10	12	14	18	8	10	11	14	5	7	7	10	4	5	5	7
6	29	35	41	53	19	23	27	35	13	16	19	25	10	12	14	19	8	10	11	15	6	7	8	10	5	5	6	8
7	30	37	43	56	20	24	28	36	14	17	20	26	11	13	15	19	8	11	12	15	6	7	8	10	5	5	6	8
8	32	39	45	58	21	25	29	38	15	18	21	27	11	14	16	20	9	11	12	16	6	8	8	11	5	6	6	8
9	33	40	47	60	22	26	30	39	15	19	22	28	12	14	16	21	9	11	13	16	6	8	9	11	5	6	6	8
10	34	42	48	62	22	27	31	40	16	19	22	29	12	15	17	21	9	11	13	17	6	8	9	11	5	6	7	8

Source: Reprinted, with permission, from T. L. Bratcher, M. A. Moran, and W. J. Zimmer, "Tables of Sample Sizes in the Analysis of Variance," *Journal of Quality Technology* 2 (1970), pp. 156–64. Copyright American Society for Quality Control, Inc.

TABLE A.11   Table of $\lambda \sqrt{n}/\sigma$ for Determining Sample Size to Find "Best" of r
Population Means

Number of Populations (r)	Probability of Correct Identification $(1 - \alpha)$		
	.90	.95	.99
2	1.8124	2.3262	3.2900
3	2.2302	2.7101	3.6173
4	2.4516	2.9162	3.7970
5	2.5997	3.0552	3.9196
6	2.7100	3.1591	4.0121
7	2.7972	3.2417	4.0861
8	2.8691	3.3099	4.1475
9	2.9301	3.3679	4.1999
10	2.9829	3.4182	4.2456

Source: Reprinted, with permission, from R. E. Bechhofer, "A Single-Sample Multiple Decision Procedure for Ranking Means of Normal Populations with Known Variances," *The Annals of Mathematical Statistics* 25 (1954), pp. 16–39.

TABLE A.12 Percentiles of H Statistic Distribution

Entry is $H(1 - \alpha; r, df)$ where $P\{H \leq H(1 - \alpha; r, df)\} = 1 - \alpha$

$$1 - \alpha = .95$$

					r						
df	2	3	4	5	6	7	8	9	10	11	12
2	39.0	87.5	142	202	266	333	403	475	550	626	704
3	15.4	27.8	39.2	50.7	62.0	72.9	83.5	93.9	104	114	124
4	9.60	15.5	20.6	25.2	29.5	33.6	37.5	41.1	44.6	48.0	51.4
5	7.15	10.8	13.7	16.3	18.7	20.8	22.9	24.7	26.5	28.2	29.9
6	5.82	8.38	10.4	12.1	13.7	15.0	16.3	17.5	18.6	19.7	20.7
7	4.99	6.94	8.44	9.70	10.8	11.8	12.7	13.5	14.3	15.1	15.8
8	4.43	6.00	7.18	8.12	9.03	9.78	10.5	11.1	11.7	12.2	12.7
9	4.03	5.34	6.31	7.11	7.80	8.41	8.95	9.45	9.91	10.3	10.7
10	3.72	4.85	5.67	6.34	6.92	7.42	7.87	8.28	8.66	9.01	9.34
12	3.28	4.16	4.79	5.30	5.72	6.09	6.42	6.72	7.00	7.25	7.48
15	2.86	3.54	4.01	4.37	4.68	4.95	5.19	5.40	5.59	5.77	5.93
20	2.46	2.95	3.29	3.54	3.76	3.94	4.10	4.24	4.37	4.49	4.59
30	2.07	2.40	2.61	2.78	2.91	3.02	3.12	3.21	3.29	3.36	3.39
60	1.67	1.85	1.96	2.04	2.11	2.17	2.22	2.26	2.30	2.33	2.36
∞	1.00	1.00	1.00	1.00	1.00	1.00	1.00	1.00	1.00	1.00	1.00

$$1 - \alpha = .99$$

					r						
df	2	3	4	5	6	7	8	9	10	11	12
2	199	448	729	1,036	1,362	1,705	2,063	2,432	2,813	3,204	3,605
3	47.5	85	120	151	184	216	249	281	310	337	361
4	23.2	37	49	59	69	79	89	97	106	113	120
5	14.9	22	28	33	38	42	46	50	54	57	60
6	11.1	15.5	19.1	22	25	27	30	32	34	36	37
7	8.89	12.1	14.5	16.5	18.4	20	22	23	24	26	27
8	7.50	9.9	11.7	13.2	14.5	15.8	16.9	17.9	18.9	19.8	21
9	6.54	8.5	9.9	11.1	12.1	13.1	13.9	14.7	15.3	16.0	16.6
10	5.85	7.4	8.6	9.6	10.4	11.1	11.8	12.4	12.9	13.4	13.9
12	4.91	6.1	6.9	7.6	8.2	8.7	9.1	9.5	9.9	10.2	10.6
15	4.07	4.9	5.5	6.0	6.4	6.7	7.1	7.3	7.5	7.8	8.0
20	3.32	3.8	4.3	4.6	4.9	5.1	5.3	5.5	5.6	5.8	5.9
30	2.63	3.0	3.3	3.4	3.6	3.7	3.8	3.9	4.0	4.1	4.2
60	1.96	2.2	2.3	2.4	2.4	2.5	2.5	2.6	2.6	2.7	2.7
∞	1.00	1.0	1.0	1.0	1.0	1.0	1.0	1.0	1.0	1.0	1.0

Source: Reprinted, with permission, from H. A. David, "Upper 5 and 1% Points of the Maximum F-Ratio," *Biometrika* 39 (1952), pp. 422–24.

TABLE A.13 Selected Standard Latin Squares

3 × 3

```
A  B  C
B  C  A
C  A  B
```

4 × 4

1	2	3	4

```
    1              2              3              4
A  B  C  D    A  B  C  D    A  B  C  D    A  B  C  D
B  A  D  C    B  C  D  A    B  D  A  C    B  A  D  C
C  D  B  A    C  D  A  B    C  A  D  B    C  D  A  B
D  C  A  B    D  A  B  C    D  C  B  A    D  C  B  A
```

5 × 5

```
A  B  C  D  E
B  A  E  C  D
C  D  A  E  B
D  E  B  A  C
E  C  D  B  A
```

6 × 6

```
A  B  C  D  E  F
B  F  D  C  A  E
C  D  E  F  B  A
D  A  F  E  C  B
E  C  A  B  F  D
F  E  B  A  D  C
```

7 × 7

```
A  B  C  D  E  F  G
B  C  D  E  F  G  A
C  D  E  F  G  A  B
D  E  F  G  A  B  C
E  F  G  A  B  C  D
F  G  A  B  C  D  E
G  A  B  C  D  E  F
```

8 × 8

```
A  B  C  D  E  F  G  H
B  C  D  E  F  G  H  A
C  D  E  F  G  H  A  B
D  E  F  G  H  A  B  C
E  F  G  H  A  B  C  D
F  G  H  A  B  C  D  E
G  H  A  B  C  D  E  F
H  A  B  C  D  E  F  G
```

9 × 9

```
A  B  C  D  E  F  G  H  I
B  C  D  E  F  G  H  I  A
C  D  E  F  G  H  I  A  B
D  E  F  G  H  I  A  B  C
E  F  G  H  I  A  B  C  D
F  G  H  I  A  B  C  D  E
G  H  I  A  B  C  D  E  F
H  I  A  B  C  D  E  F  G
I  A  B  C  D  E  F  G  H
```

Appendix B

Data Sets

DATA SET B.1 SENIC

The primary objective of the Study on the Efficacy of Nosocomial Infection Control (**SENIC** Project) was to determine whether infection surveillance and control programs have reduced the rates of nosocomial (hospital-acquired) infection in United States hospitals. This data set consists of a random sample of 113 hospitals selected from the original 338 hospitals surveyed.

Each line of the data set has an identification number and provides information on 11 other variables for a single hospital. The data presented here are for the 1975–76 study period. The 12 variables are:

Variable Number	Variable Name	Description
1	Identification number	1-113
2	Length of stay	Average length of stay of all patients in hospital (in days)
3	Age	Average age of patients (in years)
4	Infection risk	Average estimated probability of acquiring infection in hospital (in percent)
5	Routine culturing ratio	Ratio of number of cultures performed to number of patients without signs or symptoms of hospital-acquired infection, times 100
6	Routine chest X-ray ratio	Ratio of number of X-rays performed to number of patients without signs or symptoms of pneumonia, times 100
7	Number of beds	Average number of beds in hospital during study period
8	Medical school affiliation	1 = Yes, 2 = No

Variable Number	Variable Name	Description
9	Region	Geographic region, where: 1 = NE, 2 = NC, 3 = S, 4 = W
10	Average daily census	Average number of patients in hospital per day during study period
11	Number of nurses	Average number of full-time equivalent registered and licensed practical nurses during study period (number full time plus one half the number part time)
12	Available facilities and services	Percent of 35 potential facilities and services that are provided by the hospital

Reference: Special Issue, "The SENIC Project," *American Journal of Epidemiology* 111 (1980), pp. 465–653.
Data obtained from: Robert W. Haley, M.D., Hospital Infections Program, Center for Infectious Diseases, Centers for Disease Control, Atlanta, Georgia 30333.

1	2	3	4	5	6	7	8	9	10	11	12
1	7.13	55.7	4.1	9.0	39.6	279	2	4	207	241	60.0
2	8.82	58.2	1.6	3.8	51.7	80	2	2	51	52	40.0
3	8.34	56.9	2.7	8.1	74.0	107	2	3	82	54	20.0
4	8.95	53.7	5.6	18.9	122.8	147	2	4	53	148	40.0
5	11.20	56.5	5.7	34.5	88.9	180	2	1	134	151	40.0
6	9.76	50.9	5.1	21.9	97.0	150	2	2	147	106	40.0
7	9.68	57.8	4.6	16.7	79.0	186	2	3	151	129	40.0
8	11.18	45.7	5.4	60.5	85.8	640	1	2	399	360	60.0
9	8.67	48.2	4.3	24.4	90.8	182	2	3	130	118	40.0
10	8.84	56.3	6.3	29.6	82.6	85	2	1	59	66	40.0
11	11.07	53.2	4.9	28.5	122.0	768	1	1	591	656	80.0
12	8.30	57.2	4.3	6.8	83.8	167	2	3	105	59	40.0
13	12.78	56.8	7.7	46.0	116.9	322	1	1	252	349	57.1
14	7.58	56.7	3.7	20.8	88.0	97	2	2	59	79	37.1
15	9.00	56.3	4.2	14.6	76.4	72	2	3	61	38	17.1
16	11.08	50.2	5.5	18.6	63.6	387	2	3	326	405	57.1
17	8.28	48.1	4.5	26.0	101.8	108	2	4	84	73	37.1
18	11.62	53.9	6.4	25.5	99.2	133	2	1	113	101	37.1
19	9.06	52.8	4.2	6.9	75.9	134	2	2	103	125	37.1
20	9.35	53.8	4.1	15.9	80.9	833	2	3	547	519	77.1
21	7.53	42.0	4.2	23.1	98.9	95	2	4	47	49	17.1
22	10.24	49.0	4.8	36.3	112.6	195	2	2	163	170	37.1
23	9.78	52.3	5.0	17.6	95.9	270	1	1	240	198	57.1
24	9.84	62.2	4.8	12.0	82.3	600	2	3	468	497	57.1
25	9.20	52.2	4.0	17.5	71.1	298	1	4	244	236	57.1
26	8.28	49.5	3.9	12.0	113.1	546	1	2	413	436	57.1
27	9.31	47.2	4.5	30.2	101.3	170	2	1	124	173	37.1
28	8.19	52.1	3.2	10.8	59.2	176	2	1	156	88	37.1
29	11.65	54.5	4.4	18.6	96.1	248	2	1	217	189	37.1
30	9.89	50.5	4.9	17.7	103.6	167	2	2	113	106	37.1
31	11.03	49.9	5.0	19.7	102.1	318	2	1	270	335	57.1
32	9.84	53.0	5.2	17.7	72.6	210	2	2	200	239	54.3
33	11.77	54.1	5.3	17.3	56.0	196	2	1	164	165	34.3
34	13.59	54.0	6.1	24.2	111.7	312	2	1	258	169	54.3
35	9.74	54.4	6.3	11.4	76.1	221	2	2	170	172	54.3
36	10.33	55.8	5.0	21.2	104.3	266	2	1	181	149	54.3
37	9.97	58.2	2.8	16.5	76.5	90	2	2	69	42	34.3
38	7.84	49.1	4.6	7.1	87.9	60	2	3	50	45	34.3
39	10.47	53.2	4.1	5.7	69.1	196	2	2	168	153	54.3
40	8.16	60.9	1.3	1.9	58.0	73	2	3	49	21	14.3
41	8.48	51.1	3.7	12.1	92.8	166	2	3	145	118	34.3
42	10.72	53.8	4.7	23.2	94.1	113	2	3	90	107	34.3
43	11.20	45.0	3.0	7.0	78.9	130	2	3	95	56	34.3
44	10.12	51.7	5.6	14.9	79.1	362	1	3	313	264	54.3
45	8.37	50.7	5.5	15.1	84.8	115	2	2	96	88	34.3
46	10.16	54.2	4.6	8.4	51.5	831	1	4	581	629	74.3
47	19.56	59.9	6.5	17.2	113.7	306	2	1	273	172	51.4
48	10.90	57.2	5.5	10.6	71.9	593	2	2	446	211	51.4
49	7.67	51.7	1.8	2.5	40.4	106	2	3	93	35	11.4
50	8.88	51.5	4.2	10.1	86.9	305	2	3	238	197	51.4
51	11.48	57.6	5.6	20.3	82.0	252	2	1	207	251	51.4
52	9.23	51.6	4.3	11.6	42.6	620	2	2	413	420	71.4
53	11.41	61.1	7.6	16.6	97.9	535	2	3	330	273	51.4
54	12.07	43.7	7.8	52.4	105.3	157	2	2	115	76	31.4
55	8.63	54.0	3.1	8.4	56.2	76	2	1	39	44	31.4
56	11.15	56.5	3.9	7.7	73.9	281	2	1	217	199	51.4

1	2	3	4	5	6	7	8	9	10	11	12
57	7.14	59.0	3.7	2.6	75.8	70	2	4	37	35	31.4
58	7.65	47.1	4.3	16.4	65.7	318	2	4	265	314	51.4
59	10.73	50.6	3.9	19.3	101.0	445	1	2	374	345	51.4
60	11.46	56.9	4.5	15.6	97.7	191	2	3	153	132	51.4
61	10.42	58.0	3.4	8.0	59.0	119	2	1	67	64	31.4
62	11.18	51.0	5.7	18.8	55.9	595	1	2	546	392	68.6
63	7.93	64.1	5.4	7.5	98.1	68	2	4	42	49	28.6
64	9.66	52.1	4.4	9.9	98.3	83	2	2	66	95	28.6
65	7.78	45.5	5.0	20.9	71.6	489	2	3	391	329	48.6
66	9.42	50.6	4.3	24.8	62.8	508	2	1	421	528	48.6
67	10.02	49.5	4.4	8.3	93.0	265	2	2	191	202	48.6
68	8.58	55.0	3.7	7.4	95.9	304	2	3	248	218	48.6
69	9.61	52.4	4.5	6.9	87.2	487	2	3	404	220	48.6
70	8.03	54.2	3.5	24.3	87.3	97	2	1	65	55	28.6
71	7.39	51.0	4.2	14.6	88.4	72	2	2	38	67	28.6
72	7.08	52.0	2.0	12.3	56.4	87	2	3	52	57	28.6
73	9.53	51.5	5.2	15.0	65.7	298	2	3	241	193	48.6
74	10.05	52.0	4.5	36.7	87.5	184	1	1	144	151	68.6
75	8.45	38.8	3.4	12.9	85.0	235	2	2	143	124	48.6
76	6.70	48.6	4.5	13.0	80.8	76	2	4	51	79	28.6
77	8.90	49.7	2.9	12.7	86.9	52	2	1	37	35	28.6
78	10.23	53.2	4.9	9.9	77.9	752	1	2	595	446	68.6
79	8.88	55.8	4.4	14.1	76.8	237	2	2	165	182	48.6
80	10.30	59.6	5.1	27.8	88.9	175	2	2	113	73	45.7
81	10.79	44.2	2.9	2.6	56.6	461	1	2	320	196	65.7
82	7.94	49.5	3.5	6.2	92.3	195	2	4	109	110	45.7
83	7.63	52.1	5.5	11.6	61.1	197	2	4	109	110	45.7
84	8.77	54.5	4.7	5.2	47.0	143	2	4	85	87	25.7
85	8.09	56.9	1.7	7.6	56.9	92	2	3	61	61	45.7
86	9.05	51.2	4.1	20.5	79.8	195	2	3	127	112	45.7
87	7.91	52.8	2.9	11.9	79.5	477	2	3	349	188	65.7
88	10.39	54.6	4.3	14.0	88.3	353	2	2	223	200	65.7
89	9.36	54.1	4.8	18.3	90.6	165	2	1	127	158	45.7
90	11.41	50.4	5.8	23.8	73.0	424	1	3	359	335	45.7
91	8.86	51.3	2.9	9.5	87.5	100	2	3	65	53	25.7
92	8.93	56.0	2.0	6.2	72.5	95	2	3	59	56	25.7
93	8.92	53.9	1.3	2.2	79.5	56	2	2	40	14	5.7
94	8.15	54.9	5.3	12.3	79.8	99	2	4	55	71	25.7
95	9.77	50.2	5.3	15.7	89.7	154	2	2	123	148	25.7
96	8.54	56.1	2.5	27.0	82.5	98	2	1	57	75	45.7
97	8.66	52.8	3.8	6.8	69.5	246	2	3	178	177	45.7
98	12.01	52.8	4.8	10.8	96.9	298	2	1	237	115	45.7
99	7.95	51.8	2.3	4.6	54.9	163	2	3	128	93	42.9
100	10.15	51.9	6.2	16.4	59.2	568	1	3	452	371	62.9
101	9.76	53.2	2.6	6.9	80.1	64	2	4	47	55	22.9
102	9.89	45.2	4.3	11.8	108.7	190	2	1	141	112	42.9
103	7.14	57.6	2.7	13.1	92.6	92	2	4	40	50	22.9
104	13.95	65.9	6.6	15.6	133.5	356	2	1	308	182	62.9
105	9.44	52.5	4.5	10.9	58.5	297	2	3	230	263	42.9
106	10.80	63.9	2.9	1.6	57.4	130	2	3	69	62	22.9
107	7.14	51.7	1.4	4.1	45.7	115	2	3	90	1>	22.9
108	8.02	55.0	2.1	3.8	46.5	91	2	2	44	32	22.9
109	11.80	53.8	5.7	9.1	116.9	571	1	2	441	469	62.9
110	9.50	49.3	5.8	42.0	70.9	98	2	3	68	46	22.9
111	7.70	56.9	4.4	12.2	67.9	129	2	4	85	136	62.9
112	17.94	56.2	5.9	26.4	91.8	835	1	1	791	407	62.9
113	9.41	59.5	3.1	20.6	91.7	29	2	3	20	22	22.9

DATA SET B.2   SMSA

This data set provides information for 141 large Standard Metropolitan Statistical Areas (**SMSA**s) in the United States. A standard metropolitan statistical area includes a city (or cities) of specified population size which constitutes the central city and the county (or counties) in which it is located, as well as contiguous counties when the economic and social relationships between the central and contiguous counties meet specified criteria of metropolitan character and integration. An SMSA may have up to three central cities and may cross state lines.

Each line of the data set has an identification number and provides information on 11 other variables for a single SMSA. The information generally pertains to the years 1976 and 1977. The 12 variables are:

Variable Number	Variable Name	Description
1	Identification number	1-141
2	Land area	In square miles
3	Total population	Estimated 1977 population (in thousands)
4	Percent of population in central cities	Percent of 1976 SMSA population in central city or cities
5	Percent of population 65 or older	Percent of 1976 SMSA population 65 years old or older
6	Number of active physicians	Number of professionally active nonfederal physicians as of December 31, 1977
7	Number of hospital beds	Total number of beds, cribs, and bassinettes during 1977
8	Percent high school graduates	Percent of adult population (persons 25 years old or older) who completed 12 or more years of school, according to the 1970 Census of the Population
9	Civilian labor force	Total number of persons in civilian labor force (persons 16 years old or older classified as employed or unemployed) in 1977 (in thousands)
10	Total personal income	Total current income received in 1976 by residents of the SMSA from all sources, before deduction of income and other personal taxes but after deduction of personal contributions to social security and other social insurance programs (in millions of dollars)

Variable Number	Variable Name	Description
11	Total serious crimes	Total number of serious crimes in 1977, including murder, rape, robbery, aggravated assault, burglary, larceny-theft, and motor vehicle theft, as reported by law enforcement agencies
12	Geographic region	Geographic region classification is that used by the U.S. Bureau of the Census, where: 1 = NE, 2 = NC, 3 = S, 4 = W

Data obtained from: U.S. Bureau of the Census, *State and Metropolitan Area Data Book, 1979* (a Statistical Abstract Supplement).

1	2	3	4	5	6	7	8	9	10	11	12
1	1384	9387	78.1	12.3	25627	69678	50.1	4083.9	72100	709234	1
2	4069	7031	44.0	10.0	15389	39699	62.0	3353.6	52737	499813	4
3	3719	7017	43.9	9.4	13326	43292	53.9	3305.9	54542	393162	2
4	3553	4794	37.4	10.7	9724	33731	50.6	2066.3	33216	198102	1
5	3916	4370	29.9	8.8	6402	24167	52.2	1966.7	32906	294466	2
6	2480	3182	31.5	10.5	8502	16751	66.1	1514.5	26573	255162	4
7	2815	3033	23.1	6.7	7340	16941	68.3	1541.9	25663	177355	3
8	1218	2688	0.0	8.8	5255	22137	62.9	1213.3	21524	127567	1
9	8360	2673	46.3	8.2	4047	14347	53.6	1321.2	18350	193125	3
10	6794	2512	60.1	6.3	4562	14333	51.7	1272.7	18221	162976	3
11	4935	2380	21.8	11.0	4071	17752	47.8	1061.2	16120	137479	2
12	3049	2294	19.5	12.1	4005	21149	53.4	967.5	15826	69989	1
13	2259	2147	38.6	9.3	5141	16485	44.6	966.8	14246	138214	3
14	4647	2037	31.5	9.2	3916	12815	65.1	1032.2	14542	112642	2
15	1008	1969	16.6	10.3	4006	16704	55.9	935.5	15953	106646	1
16	1519	1950	31.8	10.5	4094	12545	54.6	906.0	14684	102816	2
17	4326	1832	23.6	7.3	3064	9976	50.4	867.2	12107	106482	3
18	782	1801	28.4	7.8	3119	8656	70.5	915.2	12591	113821	4
19	4261	1683	48.6	9.7	3396	7552	65.3	644.3	10392	112359	4
20	4651	1464	38.8	7.7	3380	8517	67.4	729.2	10375	116861	4
21	2042	1441	24.5	16.5	4071	10039	51.9	681.7	10166	116304	3
22	4226	1427	38.1	9.8	3285	5392	67.8	699.8	10918	91399	4
23	1456	1427	46.7	10.4	2484	8555	56.8	710.4	10104	63695	2
24	2045	1380	37.2	21.4	1949	8863	50.7	543.2	7989	89257	3
25	2149	1375	29.8	10.6	2530	8354	48.4	617.6	9037	68319	2
26	1590	1313	30.1	10.9	2296	9988	50.4	565.7	8411	67965	1
27	27293	1306	25.3	12.3	2018	6323	57.4	510.6	7399	99293	4
28	3341	1293	35.8	10.1	2289	7593	59.9	656.3	9106	81510	2
29	9155	1254	53.8	11.1	2280	6450	60.1	575.2	7766	107370	4
30	1300	1217	47.6	6.8	2794	4989	69.0	610.8	9215	76570	4
31	3072	1144	68.0	9.3	2181	7497	56.0	549.6	7736	61381	2
32	1967	1133	51.1	8.8	2520	8467	45.8	460.5	7038	69285	3
33	3650	1121	34.6	11.1	2358	6224	62.9	539.3	7792	77316	4
34	2460	1087	49.6	8.4	1874	7706	59.9	510.7	6658	62603	2
35	2527	1025	78.7	8.4	1760	7664	46.5	391.1	5582	62694	3
36	2966	970	26.9	10.3	2053	6604	56.3	450.4	6966	54854	1
37	3434	929	28.9	8.3	1844	3215	65.1	422.6	5909	72410	4
38	1392	883	37.2	9.8	1579	6087	46.5	396.8	5705	45642	3
39	2298	886	76.2	9.0	1644	7673	48.2	394.6	5185	52094	3
40	1219	864	31.7	20.6	1396	6158	55.4	352.8	5879	68109	3
41	1708	833	24.0	8.8	1062	5315	56.2	367.5	5489	52606	2
42	8565	822	29.7	7.3	1604	3485	67.6	349.3	4655	49111	4
43	3358	805	35.1	11.3	1649	5512	44.9	359.1	4941	42786	3
44	2624	794	30.4	12.2	1532	4730	55.2	356.5	5094	30771	1
45	2187	777	47.0	10.2	1098	4342	51.9	355.4	5142	46213	2
46	3214	774	47.7	9.4	1285	3459	40.3	401.7	4924	34941	3
47	3491	769	48.5	9.7	1496	5620	59.6	362.3	4798	44513	3
48	4080	773	59.6	9.9	1597	7496	47.3	380.9	4600	33936	3
49	596	723	100.0	6.0	1260	2819	66.0	319.9	5181	46984	4
50	3199	694	80.6	8.7	983	4749	50.8	292.4	4127	43010	3
51	903	661	37.3	9.6	948	4064	55.6	293.3	4102	34725	2
52	2419	647	27.8	9.9	1250	2870	57.8	286.8	3860	30829	1
53	938	644	48.1	7.4	614	3016	50.0	280.9	4177	35106	2
54	1951	629	28.4	14.5	696	4843	47.9	271.5	3667	14868	1
55	1490	624	33.1	11.9	827	3818	47.4	300.2	4144	19090	1
56	5677	610	55.8	10.5	760	3883	56.2	292.0	4035	32146	3
57	1525	597	55.7	8.3	751	3234	44.9	318.5	3777	37070	3
58	2528	593	19.2	10.2	798	3135	55.4	274.1	3489	44442	3
59	312	594	19.5	7.5	769	2463	55.0	298.7	4352	29100	1
60	1537	581	63.8	8.7	1234	5160	62.7	272.6	3725	32271	2
61	1420	576	32.6	9.5	833	2950	54.0	280.8	3553	26645	2
62	47	564	41.9	11.9	745	3352	36.3	258.9	3915	29157	1
63	1023	541	35.1	10.0	639	3144	52.1	234.1	3437	22111	2
64	2115	526	19.9	9.1	676	2296	38.8	253.3	2962	30684	3
65	1182	514	32.4	7.4	518	2515	52.4	216.8	3627	35201	2
66	1165	516	14.5	8.6	746	4277	54.4	237.1	3724	31358	3
67	476	492	8.9	10.9	787	2778	60.1	218.4	3603	24787	1
68	1553	487	50.0	8.0	2207	4931	52.0	257.2	2991	24269	3
69	2023	477	22.1	21.8	752	2317	55.7	194.2	3283	36418	3
70	2766	474	67.9	7.7	679	3873	56.3	224.0	2598	29967	3

1	2	3	4	5	6	7	8	9	10	11	12
71	5966	472	39.5	9.6	737	1907	52.7	246.6	3007	38205	4
72	1863	468	50.4	7.7	674	2989	63.8	194.8	2747	25159	4
73	192	462	60.5	10.8	617	1789	44.1	212.6	3158	27161	1
74	9240	455	67.0	10.3	1123	2347	63.1	183.6	2598	41649	4
75	2277	455	39.5	7.5	512	1788	61.9	221.1	2853	20053	2
76	1630	449	41.9	10.7	724	4395	50.0	198.0	2445	17596	3
77	1617	435	71.0	6.9	518	2031	54.1	197.9	2617	31539	3
78	1057	435	90.7	6.1	479	2551	51.1	163.4	2012	25650	3
79	1624	429	13.4	11.0	832	2938	55.4	207.8	2885	16985	1
80	1676	423	36.6	9.2	505	3297	60.7	156.3	2689	24266	4
81	2818	425	48.5	9.3	540	2694	42.3	172.8	2162	22374	3
82	2866	408	24.9	10.7	427	2864	39.1	169.1	1987	10425	3
83	4883	402	72.4	7.3	873	2236	64.9	185.2	2353	28171	4
84	966	401	24.9	10.6	427	3192	52.2	174.7	2446	15981	2
85	2109	403	41.2	10.3	520	2539	45.2	183.1	2308	16240	3
86	2449	395	68.4	9.6	681	2864	63.2	207.4	2651	25149	2
87	2618	385	31.7	6.1	836	2159	48.0	145.6	1992	25046	3
88	1465	374	30.3	6.8	598	6456	50.6	164.7	2201	26428	3
89	1704	375	52.1	10.5	379	2491	55.6	173.2	2662	18599	2
90	1750	370	49.3	9.7	446	3472	58.2	176.5	2439	16529	2
91	1489	369	58.8	9.5	911	5720	56.5	175.1	2264	26032	3
92	8152	363	22.3	9.1	405	1254	51.7	165.6	2257	28351	4
93	2207	364	57.3	9.7	356	2167	45.5	165.9	2331	19138	3
94	7874	360	44.4	6.9	398	1365	65.2	174.2	2410	33687	4
95	655	364	75.2	6.6	425	3879	51.6	163.0	2088	15623	3
96	1803	362	35.3	10.4	483	2137	53.7	168.9	2666	16405	2
97	2363	356	53.1	10.6	565	2717	49.3	146.4	1996	19212	3
98	1435	352	13.4	11.7	342	1076	44.7	156.8	2165	11273	1
99	946	348	16.4	11.1	366	1455	43.9	163.8	2178	8116	1
100	1136	333	58.6	9.7	448	2630	68.1	171.4	2396	20465	2
101	2658	327	39.0	12.2	365	5430	49.9	136.9	1862	9325	1
102	228	317	31.1	10.2	667	3179	52.8	156.5	2264	19410	1
103	1758	310	56.8	11.5	565	2081	65.3	131.2	1939	17379	4
104	1198	313	55.1	8.0	1171	3877	71.2	172.3	2038	18676	2
105	1412	311	39.2	11.3	436	1837	49.4	154.2	2098	25714	4
106	2071	306	19.9	11.3	470	2531	58.9	133.1	1782	11161	1
107	862	302	26.3	13.4	423	1929	43.3	145.5	2010	7699	1
108	1526	303	71.7	7.7	413	1636	47.1	125.8	1692	20038	3
109	1758	297	33.2	11.6	296	2652	45.3	114.4	1641	12467	3
110	1651	296	64.6	8.9	774	5431	56.1	136.9	1724	14468	3
111	1493	294	64.8	8.9	863	3289	53.7	154.7	1787	15871	3
112	1610	294	59.8	9.5	471	4633	62.9	116.1	1851	18651	4
113	2710	288	63.7	6.2	357	1277	72.8	110.9	1639	18173	4
114	1975	291	46.5	12.6	405	2896	51.5	133.8	1853	12787	2
115	1920	291	49.8	7.8	283	1306	53.2	126.9	1553	12315	3
116	1404	289	38.5	10.0	299	1766	56.2	138.6	1776	11715	2
117	2737	287	45.0	10.5	602	1462	71.3	131.4	1980	18208	4
118	1700	287	18.8	8.0	739	3381	45.9	120.4	1616	14534	3
119	909	277	41.2	11.5	307	1309	54.2	131.9	1762	13722	2
120	1858	277	24.3	13.7	354	1562	46.3	116.9	1507	19133	3
121	3324	275	49.7	8.4	373	929	62.5	120.5	1918	14776	4
122	1697	274	23.8	7.2	338	1610	51.0	105.9	1354	19317	3
123	813	272	46.0	9.8	293	1693	58.4	119.9	1688	10402	1
124	7397	267	47.3	12.5	355	2042	56.2	113.7	1654	12273	2
125	1165	268	43.7	9.4	450	2070	57.5	129.4	1719	16226	2
126	802	268	52.6	9.8	392	1425	52.2	129.6	1816	13230	2
127	1770	268	14.8	12.2	285	2804	44.1	106.7	1537	4205	1
128	495	264	50.7	7.8	220	1177	52.6	119.5	1661	8398	2
129	1255	261	26.0	10.7	458	1646	51.6	113.0	1725	10208	3
130	1148	589	45.3	11.1	891	5790	54.0	277.0	3510	29237	1
131	1509	643	37.6	12.0	1087	4900	51.4	319.6	3982	29058	1
132	2013	254	61.7	9.7	273	1484	50.9	106.7	1412	14446	3
133	711	250	42.4	6.1	1411	3659	67.5	131.0	1790	16228	2
134	471	251	46.3	8.6	219	1128	47.8	105.3	1458	13474	2
135	4552	249	54.4	9.1	329	719	61.9	118.0	1386	15596	4
136	1400	242	50.8	8.0	290	1271	45.7	104.4	1351	10391	3
137	1511	236	38.7	10.7	348	1093	50.4	127.2	1452	16676	4
138	1543	232	39.6	8.1	159	481	30.3	80.6	769	8436	3
139	1011	233	37.8	10.5	264	964	70.7	93.2	1337	14018	3
140	813	232	13.4	10.9	371	4355	58.0	97.0	1589	8428	1
141	654	231	28.8	3.9	140	1296	55.1	66.9	1148	15884	3

SMSA Identifications

1	NEW YORK, NY	48	NASHVILLE, TN	95	NEWPORT NEWS, VA
2	LOS ANGELES, CA	49	HONOLULU, HI	96	PEORIA, IL
3	CHICAGO, IL	50	JACKSONVILLE, FL	97	SHREVEPORT, LA
4	PHILADELPHIA, PA	51	AKRON, OH	98	YORK, PA
5	DETROIT, MI	52	SYRACUSE, NY	99	LANCASTER, PA
6	SAN FRANCISCO, CA	53	GARY, IN	100	DES MOINES, IA
7	WASHINGTON, DC	54	NORTHEAST, PA	101	UTICA, NY
8	NASSAU, NY	55	ALLENTOWN, PA	102	TRENTON, NJ
9	DALLAS, TX	56	TULSA, OK	103	SPOKANE, WA
10	HOUSTON, TX	57	CHARLOTTE, NC	104	MADISON, WI
11	ST. LOUIS, MO	58	ORLANDO, FL	105	STOCKTON, CA
12	PITTSBURG, PA	59	NEW BRUNSWICK, NJ	106	BINGHAMTON, NY
13	BALTIMORE, MD	60	OMAHA, NE	107	READING, PA
14	MINNEAPOLIS, MN	61	GRAND RAPIDS, MI	108	CORPUS CHRISTI, TX
15	NEWARK, NJ	62	JERSEY CITY, NJ	109	HUNTINGTON, WV
16	CLEVELAND, OH	63	YOUNGSTOWN, OH	110	JACKSON, MS
17	ATLANTA, GA	64	GREENVILLE, SC	111	LEXINGTON, KY
18	ANAHEIM, CA	65	FLINT, MI	112	VALLEJO, CA
19	SAN DIEGO, CA	66	WILMINGTON, DE	113	COLORADO SPRINGS, CO
20	DENVER, CO	67	LONG BRANCH, NJ	114	EVANSVILLE, IN
21	MIAMI, FL	68	RALEIGH, NC	115	HUNTSVILLE, AL
22	SEATTLE, WA	69	W. PALM BEACH, FL	116	APPLETON, WI
23	MILWAUKEE, WI	70	AUSTIN, TX	117	SANTA BARBARA, CA
24	TAMPA, FL	71	FRESNO, CA	118	AUGUSTA, GA
25	CINCINNATI, OH	72	OXNARD, CA	119	SOUTH BEND, IN
26	BUFFALO, NY	73	PATERSON, NJ	120	LAKELAND, FL
27	RIVERSIDE, CA	74	TUCSON, AZ	121	SALINAS, CA
28	KANSAS CITY, MO	75	LANSING, MI	122	PENSACOLA, FL
29	PHOENIX, AZ	76	KNOXVILLE, TN	123	ERIE, PA
30	SAN JOSE, CA	77	BATON ROUGE, LA	124	DULUTH, MN
31	INDIANAPOLIS, IN	78	EL PASO, TX	125	KALAMAZOO, MI
32	NEW ORLEANS, LA	79	HARRISBURG, PA	126	ROCKFORD, IL
33	PORTLAND, OR	80	TACOMA, WA	127	JOHNSTOWN, PA
34	COLUMBUS, OH	81	MOBILE, AL	128	LORAIN, OH
35	SAN ANTONIO, TX	82	JOHNSON CITY, TN	129	CHARLESTON, WV
36	ROCHESTER, NY	83	ALBUQUERQUE, NM	130	SPRINGFIELD, MA
37	SACRAMENTO, CA	84	CANTON, OH	131	WORCESTER, MA
38	LOUISVILLE, KY	85	CHATANOOGA, TN	132	MONTGOMERY, AL
39	MEMPHIS, TN	86	WICHITA, KS	133	ANN ARBOR, MI
40	FT. LAUDERDALE, FL	87	CHARLESTON, SC	134	HAMILTON, OH
41	DAYTON, OH	88	COLUMBIA, SC	135	EUGENE, OR
42	SALT LAKE CITY, UT	89	DAVENPORT, IA	136	MACON, GA
43	BIRMINGHAM, AL	90	FORT WAYNE, IN	137	MODESTO, CA
44	ALBANY, NY	91	LITTLE ROCK, AR	138	MCALLEN, TX
45	TOLEDO, OH	92	BAKERSFIELD, CA	139	MELBOURNE, FL
46	GREENSBORO, NC	93	BEAUMONT, TX	140	POUGHKEEPSIE, NY
47	OKLAHOMA CITY, OK	94	LAS VEGAS, NV	141	FAYETTEVILLE, NC

DATA SET B.3   DRUG EFFECT EXPERIMENT

This data set provides results adapted from an experiment in which the effects of a drug on the behavior of rats were studied. The behavior under consideration was the rate at which a rat deprived of water presses a lever to obtain water. The experiment was carried out in two parts. Variable 2 identifies the two parts of the study (1, 2).

In Part I of the study, 12 male albino rats of the same strain and approximately the same weight were utilized. Variable 3 identifies each rat (1, . . . , 12). Prior to the experiment, each rat was trained to press a lever for water until a stable rate of pressing was reached. Two factors were studied in this experiment—initial lever press rate (factor A) and dosage of the drug (factor B). The 12 rats were classified into one of three groups according to their initial lever press rate. Variable 4 identifies the level of the initial lever press rate (1, 2, 3). Level 1 is a slow rate, level 2 a moderate

rate, and level 3 a fast rate. The levels were defined such that one third of the rats were classified into each of the three levels.

Four dosage levels of the drug were studied, including a zero level consisting of a saline solution. Variable 5 identifies the drug dosage (1, . . . , 4). All dosage levels were specified in terms of milligrams of drug per kilogram of weight of the rat.

One hour after a drug dosage injection was administered, an experimental session began during which the rat received water each time after the second lever press. This reinforcement schedule will be denoted by FR-2. Each rat received all four drug dosage levels in a random order. Each of the four drug dosages was administered twice, thus providing two observation units for each treatment. Variable 6 identifies the observation unit (1, 2).

The response variable was defined as the total number of lever presses divided by the elapsed time (in seconds) during a session for the given treatment. Variable 7 is the response variable.

In Part II of the study, another 12 albino male rats of the same strain and approximately the same weight as the rats used in Part I were used. Variable 2 identifies this part of the study, and variable 3 identifies the 12 additional rats (13, . . . , 24). The experimental design for Part II of the study was exactly the same as for Part I, except that each rat received water each time after the fifth lever press. This reinforcement schedule will be denoted by FR-5. Variable 2 identifies the reinforcement schedule since Part I of the study used schedule FR-2 while Part II of the study used schedule FR-5. The reinforcement schedule thus is another factor (factor C) that was studied in the combined experiment.

To summarize, the variables for this experimental design are:

Variable Number	Description	Levels
1	Identification number	1–192
2	Part of study (factor C: reinforcement schedule)	1: Part I (FR-2) 2: Part II (FR-5)
3	Rat identification	1–24
4	Initial lever press rate (factor A)	1: Slow 2: Moderate 3: Fast
5	Dosage level (mg/kg) (factor B)	1: 0 (saline solution) 2: .5 3: 1.0 4: 1.8
6	Observation unit	1, 2
7	Response variable—lever press rate (total number of lever presses divided by elapsed time in seconds)	

Reference: T. G. Heffner, R. B. Drawbaugh, and M. J. Zigmond, "Amphetamine and Operant Behavior in Rats: Relationship between Drug Effect and Control Response Rate," *Journal of Comparative and Physiological Psychology* 86 (1974), pp. 1031–43.

1	2	3	4	5	6	7	1	2	3	4	5	6	7
1	1	1	1	1	1	.81	49	1	1	1	1	2	.84
2	1	1	1	2	1	.80	50	1	1	1	2	2	.85
3	1	1	1	3	1	.82	51	1	1	1	3	2	.88
4	1	1	1	4	1	.50	52	1	1	1	4	2	.58
5	1	2	1	1	1	.77	53	1	2	1	1	2	.72
6	1	2	1	2	1	.78	54	1	2	1	2	2	.73
7	1	2	1	3	1	.79	55	1	2	1	3	2	.74
8	1	2	1	4	1	.51	56	1	2	1	4	2	.42
9	1	3	1	1	1	.80	57	1	3	1	1	2	.73
10	1	3	1	2	1	.82	58	1	3	1	2	2	.76
11	1	3	1	3	1	.83	59	1	3	1	3	2	.75
12	1	3	1	4	1	.52	60	1	3	1	4	2	.48
13	1	4	1	1	1	.95	61	1	4	1	1	2	.89
14	1	4	1	2	1	.95	62	1	4	1	2	2	.90
15	1	4	1	3	1	.91	63	1	4	1	3	2	.97
16	1	4	1	4	1	.60	64	1	4	1	4	2	.67
17	1	5	2	1	1	1.03	65	1	5	2	1	2	1.11
18	1	5	2	2	1	1.13	66	1	5	2	2	2	1.02
19	1	5	2	3	1	1.04	67	1	5	2	3	2	1.12
20	1	5	2	4	1	.82	68	1	5	2	4	2	.75
21	1	6	2	1	1	.96	69	1	6	2	1	2	1.01
22	1	6	2	2	1	.93	70	1	6	2	2	2	1.05
23	1	6	2	3	1	1.02	71	1	6	2	3	2	.95
24	1	6	2	4	1	.63	72	1	6	2	4	2	.72
25	1	7	2	1	1	.98	73	1	7	2	1	2	1.05
26	1	7	2	2	1	1.00	74	1	7	2	2	2	1.07
27	1	7	2	3	1	.98	75	1	7	2	3	2	1.05
28	1	7	2	4	1	.74	76	1	7	2	4	2	.79
29	1	8	2	1	1	1.17	77	1	8	2	1	2	1.12
30	1	8	2	2	1	1.20	78	1	8	2	2	2	1.13
31	1	8	2	3	1	1.18	79	1	8	2	3	2	1.11
32	1	8	2	4	1	.91	80	1	8	2	4	2	.83
33	1	9	3	1	1	1.20	81	1	9	3	1	2	1.28
34	1	9	3	2	1	1.24	82	1	9	3	2	2	1.17
35	1	9	3	3	1	1.27	83	1	9	3	3	2	1.21
36	1	9	3	4	1	.96	84	1	9	3	4	2	.91
37	1	10	3	1	1	1.25	85	1	10	3	1	2	1.21
38	1	10	3	2	1	1.23	86	1	10	3	2	2	1.31
39	1	10	3	3	1	1.30	87	1	10	3	3	2	1.22
40	1	10	3	4	1	1.01	88	1	10	3	4	2	.93
41	1	11	3	1	1	1.23	89	1	11	3	1	2	1.16
42	1	11	3	2	1	1.20	90	1	11	3	2	2	1.15
43	1	11	3	3	1	1.18	91	1	11	3	3	2	1.23
44	1	11	3	4	1	.95	92	1	11	3	4	2	1.02
45	1	12	3	1	1	1.31	93	1	12	3	1	2	1.40
46	1	12	3	2	1	1.42	94	1	12	3	2	2	1.33
47	1	12	3	3	1	1.41	95	1	12	3	3	2	1.35
48	1	12	3	4	1	1.08	96	1	12	3	4	2	1.20

1	2	3	4	5	6	7	1	2	3	4	5	6	7
97	2	13	1	1	1	2.18	145	2	13	1	1	2	2.26
98	2	13	1	2	1	2.44	146	2	13	1	2	2	2.40
99	2	13	1	3	1	1.92	147	2	13	1	3	2	1.99
100	2	13	1	4	1	.92	148	2	13	1	4	2	.99
101	2	14	1	1	1	2.02	149	2	14	1	1	2	1.96
102	2	14	1	2	1	2.20	150	2	14	1	2	2	2.18
103	2	14	1	3	1	1.75	151	2	14	1	3	2	1.81
104	2	14	1	4	1	.82	152	2	14	1	4	2	.78
105	2	15	1	1	1	2.06	153	2	15	1	1	2	2.10
106	2	15	1	2	1	2.28	154	2	15	1	2	2	2.24
107	2	15	1	3	1	1.86	155	2	15	1	3	2	1.92
108	2	15	1	4	1	.80	156	2	15	1	4	2	.88
109	2	16	1	1	1	2.28	157	2	16	1	1	2	2.35
110	2	16	1	2	1	2.46	158	2	16	1	2	2	2.49
111	2	16	1	3	1	1.90	159	2	16	1	3	2	1.95
112	2	16	1	4	1	.90	160	2	16	1	4	2	.96
113	2	17	2	1	1	2.62	161	2	17	2	1	2	2.68
114	2	17	2	2	1	2.58	162	2	17	2	2	2	2.64
115	2	17	2	3	1	2.21	163	2	17	2	3	2	2.17
116	2	17	2	4	1	1.03	164	2	17	2	4	2	.96
117	2	18	2	1	1	2.60	165	2	18	2	1	2	2.66
118	2	18	2	2	1	2.60	166	2	18	2	2	2	2.62
119	2	18	2	3	1	2.34	167	2	18	2	3	2	2.28
120	2	18	2	4	1	1.14	168	2	18	2	4	2	1.23
121	2	19	2	1	1	2.39	169	2	19	2	1	2	2.43
122	2	19	2	2	1	2.41	170	2	19	2	2	2	2.48
123	2	19	2	3	1	2.09	171	2	19	2	3	2	2.16
124	2	19	2	4	1	.90	172	2	19	2	4	2	.84
125	2	20	2	1	1	2.70	173	2	20	2	1	2	2.66
126	2	20	2	2	1	2.64	174	2	20	2	2	2	2.70
127	2	20	2	3	1	2.23	175	2	20	2	3	2	2.27
128	2	20	2	4	1	1.02	176	2	20	2	4	2	.98
129	2	21	3	1	1	2.98	177	2	21	3	1	2	2.94
130	2	21	3	2	1	2.64	178	2	21	3	2	2	2.70
131	2	21	3	3	1	2.34	179	2	21	3	3	2	2.44
132	2	21	3	4	1	1.28	180	2	21	3	4	2	1.33
133	2	22	3	1	1	3.10	181	2	22	3	1	2	3.20
134	2	22	3	2	1	2.85	182	2	22	3	2	2	2.91
135	2	22	3	3	1	2.40	183	2	22	3	3	2	2.45
136	2	22	3	4	1	1.35	184	2	22	3	4	2	1.39
137	2	23	3	1	1	2.80	185	2	23	3	1	2	2.84
138	2	23	3	2	1	2.48	186	2	23	3	2	2	2.53
139	2	23	3	3	1	2.16	187	2	23	3	3	2	2.23
140	2	23	3	4	1	1.01	188	2	23	3	4	2	1.07
141	2	24	3	1	1	3.21	189	2	24	3	1	2	3.31
142	2	24	3	2	1	2.92	190	2	24	3	2	2	2.98
143	2	24	3	3	1	2.56	191	2	24	3	3	2	2.47
144	2	24	3	4	1	1.40	192	2	24	3	4	2	1.51

Appendix C

Selected Bibliography

The selected references are grouped into the following categories:

1. General regression books
2. Diagnostics and model building
3. Statistical computing
4. General experimental design and analysis of variance books
5. Miscellaneous topics

1. GENERAL REGRESSION BOOKS

Allen, D. M., and F. B. Cady. *Analyzing Experimental Data by Regression*. New York: Van Nostrand Reinhold, 1982.

Bowerman, B. L.; R. T. O'Connell; and D. A. Dickey. *Linear Statistical Models: An Applied Approach*. Boston: Duxbury Press, 1986.

Brook, R. J., and G. C. Arnold. *Applied Regression Analysis and Experimental Design*. New York: Marcel Dekker, 1985.

Chatterjee, S., and B. Price. *Regression Analysis by Example*. New York: John Wiley & Sons, 1977.

Cohen, J., and P. Cohen. *Applied Multiple Regression/Correlation Analysis for the Behavioral Sciences*. 2nd ed. Hillsdale, N.J.: Lawrence Erlbaum Associates, 1983.

Daniel, C., and F. S. Wood. *Fitting Equations to Data*. 2nd ed. New York: Wiley-Interscience, 1980.

Draper, N.R., and H. Smith. *Applied Regression Analysis*. 2nd ed. New York: John Wiley & Sons, 1981.

Dunn, O. J., and V. A. Clark. *Applied Statistics: Analysis of Variance and Regression*. 2nd ed. New York: John Wiley & Sons, 1987.

Edwards, A. L. *An Introduction to Linear Regression and Correlation*. 2nd ed. New York: W. H. Freeman & Co., 1984.

Edwards, A. L. *Multiple Regression and the Analysis of Variance and Covariance*. 2nd ed. New York: W. H. Freeman & Co., 1985.

Gunst, R. F., and R. L. Mason. *Regression Analysis and Its Application*. New York: Marcel Dekker, 1980.

Kleinbaum, D. G.; L. L. Kupper; and K. E. Muller. *Applied Regression Analysis and Other Multivariate Methods*. 2nd ed. Boston: PWS-Kent Publishing Co., 1988.

Mendenhall, W., and T. Sincich. *A Second Course in Business Statistics: Regression Analysis*. 2nd ed. San Francisco: Dellen Publishing Co., 1986.

Montgomery, D. C., and E. A. Peck. *Introduction to Linear Regression Analysis*. New York: John Wiley & Sons, 1982.

Mosteller, F., and J. W. Tukey. *Data Analysis and Regression*. Reading, Pa.: Addison-Wesley Publishing, 1977.

Myers, R. H. *Classical and Modern Regression with Applications*. Boston: Duxbury Press, 1986.

Pedhazur, E. J. *Multiple Regression in Behavioral Research*. 2nd ed. New York: Holt, Rinehart & Winston, 1982.

Seber, G. A. F. *Linear Regression Analysis*. New York: John Wiley & Sons, 1977.

Weisberg, S. *Applied Linear Regression*. 2nd ed. New York: John Wiley & Sons, 1985.

Younger, M. S. *A First Course in Linear Regression*. 2nd ed. Boston: Duxbury Press, 1985.

2. DIAGNOSTICS AND MODEL BUILDING

Allen, D. M. "Mean Square Error of Prediction as a Criterion for Selecting Variables." *Technometrics* 13 (1971), pp. 469–75.

Anscombe, F. J., and J. W. Tukey. "The Examination and Analysis of Residuals." *Technometrics* 5 (1963), pp. 141–60.

Atkinson, A. C. *Plots, Transformations and Regression*. Oxford: Clarendon Press, 1985.

Barnett, V., and T. Lewis. *Outliers in Statistical Data*. 2nd ed. New York: John Wiley & Sons, 1984.

Belsley, D. A.; E. Kuh; and R. E. Welsch. *Regression Diagnostics: Identifying Influential Data and Sources of Collinearity*. New York: John Wiley & Sons, 1980.

Box, G. E. P., and D. R. Cox. "An Analysis of Transformations." *Journal of the Royal Statistical Society B* 26 (1964), pp. 211–43.

Box, G. E. P., and N. R. Draper. *Empirical Model-Building and Response Surfaces*. New York: John Wiley & Sons, 1987.

Box, G. E. P., and P. W. Tidwell. "Transformations of the Independent Variables." *Technometrics* 4 (1962), pp. 531–50.

Chatterjee, S., and A. S. Hadi. *Sensitivity Analysis in Linear Regression*. New York: John Wiley & Sons, 1988.

Cook, R. D., and S. Weisberg. *Residuals and Influence in Regression*. London: Chapman and Hall, 1982.

Durbin, J., and G. S. Watson. "Testing for Serial Correlation in Least Squares Regression. II." *Biometrika* 38 (1951), pp. 159–78.

Flack, V. F., and P. C. Chang. "Frequency of Selecting Noise Variables in Subset-Regression Analysis: A Simulation Study." *The American Statistician* 41 (1987), pp. 84–86.

Freedman, D. A. "A Note on Screening Regression Equations." *The American Statistician* 37 (1983), pp. 152–55.

Glaser, R. E. "Bartlett's Test of Homogeneity of Variances." In *Encyclopedia of Statistical Sciences,* vol. 1, ed. S. Kotz and N. L. Johnson. New York: John Wiley & Sons, 1982, pp. 189–91.

Hoaglin, D. C., and R. Welsch. "The Hat Matrix in Regression and ANOVA." *The American Statistician* 32 (1978), pp. 17–22.

Hocking, R. R. "The Analysis and Selection of Variables in Linear Regression." *Biometrics* 32 (1976), pp. 1–49.

Hoerl, A. E., and R. W. Kennard. "Ridge Regression: Applications to Nonorthogonal Problems." *Technometrics* 12 (1970), pp. 69–82.

Mallows, C. L. "Some Comments on C_p." *Technometrics* 15 (1973), pp. 661–75.

Mansfield, E. R., and M. D. Conerly. "Diagnostic Value of Residual and Partial Residual Plots." *The American Statistician* 41 (1987), pp. 107–16.

Mantel, N. "Why Stepdown Procedures in Variable Selection." *Technometrics* 12 (1970), pp. 621–25.

Pope, P. T., and J. T. Webster. "The Use of an *F*-Statistic in Stepwise Regression Procedures." *Technometrics* 14 (1972), pp. 327–40.

Rousseeuw, P. J., and A. M. Leroy. *Robust Regression and Outlier Detection.* New York: John Wiley & Sons, 1987.

Snee, R. D. "Validation of Regression Models: Methods and Examples." *Technometrics* 19 (1977), pp. 415–28.

Stone, M. "Cross-Validatory Choice and Assessment of Statistical Prediction." *Journal of the Royal Statistical Society B* 36 (1974), pp. 111–47.

Theil, H., and A. L. Nagar. "Testing the Independence of Regression Disturbances." *Journal of the American Statistical Association* 56 (1961), pp. 793–806.

3. STATISTICAL COMPUTING

Dixon, W. J., chief editor. *BMDP Statistical Software Manual,* vols. 1 and 2. Berkeley, Calif.: University of California Press, 1988.

IMSL, Inc. *STAT/LIBRARY User's Manual, Version 1.1.* Houston: IMSL, 1989.

Kennedy, W. J., Jr., and J. E. Gentle. *Statistical Computing.* New York: Marcel Dekker, 1980.

MINITAB Reference Manual, Release 7. State College, Pa.: Minitab, Inc., 1989.

NAG, *The Generalized Linear Interactive Modelling (GLIM) System, Release 3.77.* Downers Grove, Ill.: Numerical Algorithms Group, Inc., 1986.

SAS User's Guide: Statistics. Version 6 ed. Cary, N.C.: SAS Institute, 1987.

SPSS^X User's Guide. 2nd ed. Chicago: SPSS, 1986.

4. GENERAL EXPERIMENTAL DESIGN AND ANALYSIS OF VARIANCE BOOKS

Anderson, V. L., and R. A. McLean. *Design of Experiments.* New York: Marcel Dekker, Inc., 1974.

Box, G. E. P.; W. G. Hunter; and J. S. Hunter. *Statistics for Experimenters.* New York: John Wiley & Sons, 1978.

Cochran, W. G., and G. M. Cox. *Experimental Designs*. 2nd ed. New York: John Wiley & Sons, 1957.

Cox, D. R. *Planning of Experiments*. New York: John Wiley & Sons, 1958.

Fisher, R. A. *The Design of Experiments*. 8th ed. New York: Hafner Publishing Co., 1966.

Gill, J. L. *Design and Analysis of Experiments*, vols. I and II. Ames, Iowa: Iowa State University Press, 1978.

Graybill, F. A. *Theory and Application of the Linear Model*. Boston: Duxbury Press, 1976.

Hicks, C. R. *Fundamental Concepts in the Design of Experiments*. 3rd ed. New York: Holt, Rinehart and Winston, 1982.

Hocking, R. R. *The Analysis of Linear Models*. Monterey, Calif.: Brooks/Cole Publishing Co., 1985.

John, P. W. M. *Statistical Design and Analysis of Experiments*. New York: Macmillan Co., 1971.

Johnson, N. L., and F. C. Leone. *Statistics and Experimental Design in Engineering and the Physical Sciences*, vols. I and II. 2nd ed. New York: John Wiley & Sons, 1966.

Kempthorne, O. *The Design and Analysis of Experiments*. New York: John Wiley & Sons, 1952.

Keppel, G. *Design and Analysis: A Researcher's Handbook*. 2nd ed. Englewood Cliffs, N.J.: Prentice-Hall, 1982.

Kirk, R. E. *Experimental Design: Procedures for the Behavioral Sciences*. 2nd ed. Monterey, Calif.: Brooks/Cole Publishing Co., 1982.

Mendenhall, W. *Introduction to Linear Models and the Design and Analysis of Experiments*. Boston: Duxbury Press, 1968.

Montgomery, D. C. *Design and Analysis of Experiments*. 2nd ed. New York: John Wiley & Sons, 1983.

Myers, J. L. *Fundamentals of Experimental Design*. 3rd ed. Boston: Allyn and Bacon, Inc., 1979.

Peterson, R. G. *Design and Analysis of Experiments*. New York: Marcel Dekker, Inc., 1985.

Scheffé, H. *The Analysis of Variance*. New York: John Wiley & Sons, 1959.

Searle, S. R. *Linear Models for Unbalanced Data*. New York: John Wiley & Sons, 1987.

Seber, G. A. F. *The Linear Hypothesis*. 2nd ed. London: Charles Griffin, 1980.

Steel, R. G. D., and J. H. Torrie. *Principles and Procedures of Statistics*. 2nd ed. New York: McGraw-Hill Book Co., 1980.

Winer, B. J. *Statistical Principles in Experimental Design*. 2nd ed. New York: McGraw-Hill Book Co., 1971.

5. MISCELLANEOUS TOPICS

Berkson, J. "Are There Two Regressions?" *Journal of the American Statistical Association* 45 (1950), pp. 164–80.

Bishop, Y. M. M.; S. E. Fienberg; and P. W. Holland. *Discrete Multivariate Analysis: Theory and Practice*. Cambridge, Mass.: MIT Press, 1975.

Box, G. E. P. "Use and Abuse of Regression." *Technometrics* 8 (1966), pp. 625–29.

Box, G. E. P., and G. M. Jenkins. *Times Series Analysis: Forecasting and Control*. Rev. ed. San Francisco: Holden-Day, 1976.

Cox, D. R. "Notes on Some Aspects of Regression Analysis." *Journal of the Royal Statistical Society A* 131 (1968), pp. 265–79.

Federer, W. T., and M. Zelen. "Analysis of Multifactor Classifications with Unequal Numbers of Observations." *Biometrics* 22 (1966), pp. 525–52.

Fuller, W. A. *Measurement Error Models*. New York: John Wiley & Sons, 1987.

Gibbons, J. D. *Nonparametric Methods for Quantitative Analysis*. 2nd ed. Columbus, Ohio: American Sciences Press, 1985.

Graybill, F. A. *Matrices with Applications in Statistics*. 2nd ed. Belmont, Calif.: Wadsworth, 1983.

Greenhouse, S. W., and S. Geisser. "On Methods in the Analysis of Profile Data." *Psychometrika* 24 (1959), pp. 95–112.

Hocking, R. R. "A Discussion of the Two-Way Mixed Model." *The American Statistician* 27 (1973), pp. 148–52.

Hogg, R. V. "Statistical Robustness: One View of Its Use in Applications Today." *The American Statistician* 33 (1979), pp. 108–15.

Huynh, H., and L. Feldt. "Estimation of the Box Correction for Degrees of Freedom from Sample Data in the Randomized Block and Split-plot Designs." *Journal of Educational Statistics* 1 (1976), pp. 69–82.

Johnson, D. E., and F. A. Graybill. "Estimation of σ^2 in a Two-Way Classification Model with Interaction." *Journal of the American Statistical Association* 67 (1972), pp. 388–94.

Johnson, R. A., and D. W. Wichern. *Applied Multivariate Statistical Analysis*. 2nd ed. Englewood Cliffs, N.J.: Prentice-Hall, 1988.

Koch, G. G.; J. D. Elashoff; and I. A. Amara. "Repeated Measurements—Design and Analysis." In *Encyclopedia of Statistical Sciences*, vol. 8, ed. S. Kotz and N. L. Johnson. New York: John Wiley & Sons, 1988, pp. 46–73.

Miller, R. G., Jr. *Simultaneous Statistical Inference*. 2nd ed. New York: Springer-Verlag, 1981.

Owen, D. B. *Handbook of Statistical Tables*. Reading, Mass.: Addison-Wesley Publishing, 1962.

Pindyck, R. S., and D. L. Rubinfeld. *Econometric Models and Economic Forecasts*. 2nd ed. New York: McGraw-Hill, 1981.

Satterthwaite, F. E. "An Approximate Distribution of Estimates of Variance Components." *Biometrics Bulletin* 2 (1946), pp. 110–14.

Searle, S. R. *Matrix Algebra Useful for Statistics*. New York: John Wiley & Sons, 1982.

Snedecor, G. W., and W. G. Cochran. *Statistical Methods*. 7th ed. Ames, Iowa: Iowa State University Press, 1980.

Index